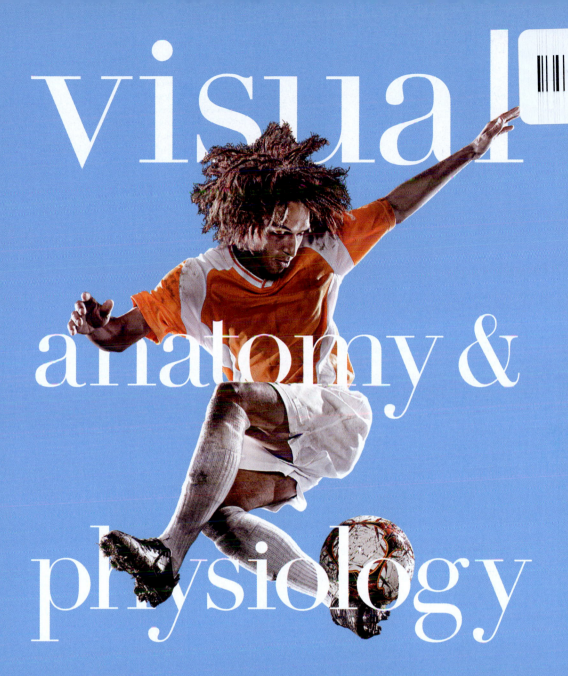

visual
anatomy &
physiology

Frederic H. Martini, Ph.D.
University of Hawaii at Manoa

William C. Ober, M.D.
Washington and Lee University

Judi L. Nath, Ph.D.
Lourdes College

Edwin F. Bartholomew, M.S.
Contributing Author

Claire W. Garrison, R.N.
Medical & Scientific Illustration
Contributing Illustrator

Kathleen Welch, M.D.
Fellow, American Academy
of Family Practice
Clinical Consultant

Benjamin Cummings

Boston Columbus Indianapolis New York San Francisco Upper Saddle River
Amsterdam Cape Town Dubai London Madrid Milan Munich Paris Montréal Toronto
Delhi Mexico City São Paulo Sydney Hong Kong Seoul Singapore Taipei Tokyo

Executive Editor: *Leslie Berriman*
Project Editor: *Robin Pille*
Editorial Development Manager: *Barbara Yien*
Development Editor: *Alan Titche*
Associate Editors: *Katie Seibel and Kelly Reed*
Editorial Assistant: *Nicole McFadden*
Senior Managing Editor: *Deborah Cogan*
Senior Production Project Manager: *Nancy Tabor*
Production Management and
 Composition: *S4Carlisle Publishing Services, Inc.*
Copyeditor: *Michael Rossa*

Director of Media Development: *Lauren Fogel*
Media Producer: *Lucinda Bingham*
Design Manager: *Mark Ong*
Interior Designer: *Gibson Design Associates*
Cover Designer: *Jana Anderson*
Senior Photo Editor: *Donna Kalal*
Photo Researcher: *Maureen Spuhler*
Senior Manufacturing Buyer: *Stacey Weinberger*
Market Development Manager: *Brooke Suchomel*
Marketing Manager: *Derek Perrigo*

Cover Photo Credit: © Tim Tadder/Corbis All Rights Reserved.

Library of Congress Cataloging-in-Publication Data
Martini, Frederic.
 Visual anatomy & physiology / Frederic H. Martini, William C. Ober ;
with Judi L. Nath . . . [et al.].
  p. ; cm.
 Other title: Visual anatomy and physiology
 Includes bibliographical references and index.
 ISBN 978-0-321-78667-X (student ed.) -- ISBN 978-0-321-56018-6
(professional copy) 1. Human anatomy--Programmed instruction. 2.
Human physiology -- Programmed instruction. I. Ober, William C. II.
Nath, Judi L. III. Title. IV. Title: Visual anatomy and
physiology.
 [DNLM: 1. Anatomy. 2. Physiological Phenomena. QS 4 M3855v 2011]
 QP31.2.M373 2011
 612 -- dc22
                                    2010025028

ISBN 10: 0-321-78667-X (Student edition)
ISBN 13: 978-0-321-78667-8 (Student edition)
ISBN 10: 0-321-56018-3 (Professional copy)
ISBN 13: 978-0-321-56018-6 (Professional copy)

Benjamin Cummings
is an imprint of

www.pearsonhighered.com

4 5 6 7 8 9 10—CKV—14 13 12

To my son, PK, for convincing me it was time to look at teaching and learning in new ways, and to the A&P students and instructors who helped shape the resulting text.

— *Ric Martini*

To my sons Todd and Carl, whose warmth and humor have enriched my life in countless ways.

— *Bill Ober*

To my nieces Kelsey, Kyley, Katrina, and Marissa, who represent the next generation of eager students.

— *Judi Nath*

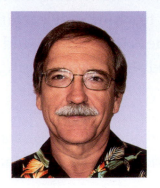

Frederic (Ric) H. Martini, Ph.D.
Author

Dr. Martini received his Ph.D. from Cornell University in comparative and functional anatomy for work on the pathophysiology of stress. In addition to professional publications that include journal articles and contributed chapters, technical reports, and magazine articles, he is the lead author of nine undergraduate texts on anatomy and physiology or anatomy. Dr. Martini is currently affiliated with the University of Hawaii at Manoa and has a long-standing bond with the Shoals Marine Laboratory, a joint venture between Cornell University and the University of New Hampshire. He has been active in the Human Anatomy and Physiology Society (HAPS) for 18 years and was a member of the committee that established the course curriculum guidelines for A&P. He is now a President Emeritus of HAPS after serving as President-Elect, President, and Past-President over 2005–2007. Dr. Martini is also a member of the American Physiological Society, the American Association of Anatomists, the Society for Integrative and Comparative Biology, the Australia/New Zealand Association of Clinical Anatomists, the Hawaii Academy of Science, the American Association for the Advancement of Science, and the International Society of Vertebrate Morphologists.

Judi L. Nath, Ph.D.
Author

Dr. Nath is a biology professor at Lourdes College, where she teaches anatomy and physiology, pathophysiology, medical terminology, and pharmacology. She received her Bachelor's and Master's degrees from Bowling Green State University and her Ph.D from the University of Toledo. Dr. Nath is devoted to her students and strives to convey the intricacies of science in a captivating way that students find meaningful, interactive, and exciting. She is a multiple recipient of the Faculty Excellence Award, granted by the college to recognize her effective teaching, scholarship, and community service. She is active in many professional organizations, notably the Human Anatomy and Physiology Society (HAPS), where she has served several terms on the board of directors. On a personal note, Dr. Nath enjoys family life with her husband, Mike, and their three dogs. Piano playing and cycling are welcome diversions from authoring, and her favorite charities include the local Humane Society, the Cystic Fibrosis Foundation, and Real Partners Uganda.

Edwin F. Bartholomew, M.S.
Contributing Author

Edwin F. Bartholomew received his undergraduate degree from Bowling Green State University in Ohio and his M.S. from the University of Hawaii. Mr. Bartholomew has taught human anatomy and physiology at both the secondary and undergraduate levels and a wide variety of other science courses (from botany to zoology) at Maui Community College and at historic Lahainaluna High School, the oldest high school west of the Rockies. Working with Dr. Martini, he coauthored *Essentials of Anatomy & Physiology, Structure and Function of the Human Body,* and *The Human Body in Health and Disease* (all published by Pearson Benjamin Cummings). Mr. Bartholomew is a member of the Human Anatomy and Physiology Society (HAPS), the National Association of Biology Teachers, the National Science Teachers Association, the Hawaii Science Teachers Association, and the American Association for the Advancement of Science.

William C. Ober, M.D.
Author

Dr. Ober received his undergraduate degree from Washington and Lee University and his M.D. from the University of Virginia. He also studied in the Department of Art as Applied to Medicine at John Hopkins University. After graduation, Dr. Ober completed a residency in Family Practice and later was on the faculty at the University of Virginia in the Department of Family Medicine and in the Department of Sports Medicine. He also served as Chief of Medicine of Martha Jefferson Hospital in Charlottesville, VA. He is currently a Visiting Professor of Biology at Washington and Lee University, where he has taught several courses and led student trips to the Galapagos Islands. He is on the Core Faculty at Shoals Marine Laboratory, where he teaches Biological Illustration every summer. Dr. Ober has collaborated with Dr. Martini on all of his textbooks.

Claire W. Garrison, R.N.
Contributing Illustrator

Claire W. Garrison, R.N., B.A., practiced pediatric and obstetric nursing before turning to medical illustration as a full-time career. She returned to school at Mary Baldwin College, where she received her degree with distinction in studio art. Following a five-year apprenticeship, she has worked as Dr. Ober's partner in Medical & Scientific Illustration since 1986. She is on the Core Faculty at Shoals Marine Laboratory and co-teaches the Biological Illustration course with Dr. Ober every summer. The textbooks illustrated by Medical & Scientific Illustration have won numerous design and illustration awards.

Kathleen Welch, M.D.
Clinical Consultant

Dr. Welch received her M.D. from the University of Washington in Seattle and did her residency in Family Practice at the University of North Carolina in Chapel Hill. For two years, she served as Director of Maternal and Child Health at the LBJ Tropical Medical Center in American Samoa and subsequently was a member of the Department of Family Practice at the Kaiser Permanente Clinic in Lahaina, Hawaii. She has been in private practice since 1987 and is licensed to practice in Hawaii, Washington, and New Zealand. Dr. Welch is a Fellow of the American Academy of Family Practice and a member of the Hawaii Medical Association and the Human Anatomy and Physiology Society (HAPS). With Dr. Martini, she has coauthored both a textbook on anatomy and physiology and the *A&P Applications Manual*. She and Dr. Martini were married in 1979, and they have one son, PK, to whom this book is dedicated.

Anatomy & Physiology for the 21st Century

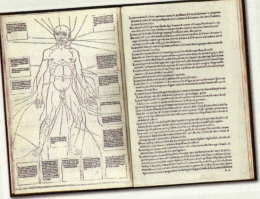

There are already so many anatomy & physiology textbooks, why create another one? The simple answer is that none of them seem to work for the entire range of students taking the A&P course. *Visual Anatomy & Physiology* is the product of a complete re-analysis of the concept of "textbook." While you will find the scope and content familiar, the organization and approach are unique.

A&P texts are still following a format that was established 500 years ago.

The anatomy text on the left, printed in 1491, conveys information in lengthy blocks of narrative text occasionally supplemented by illustrations. The same format is used in college textbooks today (although the narrative is no longer in Latin). It is certainly a "tried and true" method of instruction. But is it the best method for reaching the minds of today's students?

23.6

Filtration occurs at the renal corpuscle

The renal corpuscle, the start of the nephron, is responsible for the filtration of blood. This is the vital first step in the formation of urine.

1 At the renal corpuscle, the capillary knot of the glomerulus projects into the capsular space like the heart projects into the pericardial cavity. Like the pericardium, the glomerular capsule has an outer parietal layer and an inner visceral layer.

The **efferent arteriole** delivers blood to peritubular capillaries. It has a smaller diameter than the afferent arteriole; this elevates the blood pressure within the glomerulus.

The **juxtaglomerular complex** consists of specialized cells that secrete renin when glomerular blood pressure falls.

The **afferent arteriole** delivers blood from a cortical radiate artery.

The glomerular capsule forms the outer wall of the renal corpuscle and covers the glomerular capillaries.

The **capsular space** separates the parietal and visceral layers of the glomerular capsule.

Initial segment of renal tubule

DCT

Parietal layer

Visceral layer

2 The visceral layer consists of large cells with complex processes, or "feet," that wrap around the specialized dense layer of the glomerular capillaries. These unusual cells are called **podocytes** (PŌ-dō-sīts; *podos*, foot + *-cyte*, cell), and their feet are known as **pedicels**. Materials passing out of the blood at the glomerulus must be small enough to pass between the narrow gaps, or **filtration slits**, between adjacent pedicels.

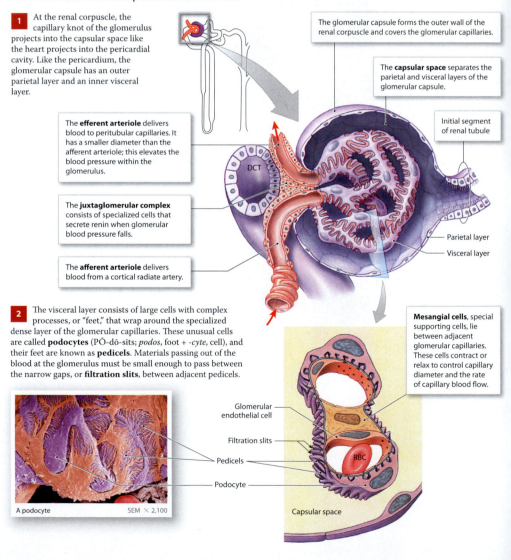

Mesangial cells, special supporting cells, lie between adjacent glomerular capillaries. These cells contract or relax to control capillary diameter and the rate of capillary blood flow.

Glomerular endothelial cell

Filtration slits

Pedicels

Podocyte

RBC

Capsular space

A podocyte   SEM × 2,100

Students of the 21st century have been shaped by an integrated media environment.

Everywhere you look today, information is being presented through a fusion of narrative and images. Owner's manuals for cars and electronics, emergency instruction cards on airplanes, social networking Websites—all integrate text and visuals in effective ways. We have applied this contemporary presentation style to this new A&P text. By eliminating redundancies between the narrative and the illustrations, the book can be shorter, clearer, and easier to follow.

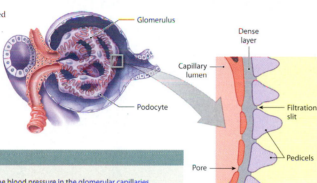

3 The glomerular capillaries are fenestrated capillaries containing large-diameter pores. The endothelium covers a **dense layer**, a specialized type of basal lamina. Together, the fenestrated endothelium, the dense layer, and the filtration slits form the **filtration membrane**. Under normal circumstances only a few plasma proteins— such as albumin molecules, with an average diameter of 7 nm—can cross the filtration membrane and enter the capsular space.

Glomerulus
Podocyte
Dense layer
Capillary lumen
Filtration slit
Pedicels
Pore
Capsular space
Filtration membrane

Factors Controlling Glomerular Filtration

The **glomerular hydrostatic pressure (GHP)** is the blood pressure in the glomerular capillaries. This pressure tends to push water and solute molecules out of the plasma and into the filtrate. The GHP, which averages 50 mm Hg, is significantly higher than capillary pressures elsewhere in the systemic circuit, due to the different diameters of the afferent and efferent capillaries.

Filtrate in capsular space
Plasma proteins
Solutes

50
25
15
10 mm Hg

The **blood colloid osmotic pressure (BCOP)** tends to draw water out of the filtrate and into the plasma; it thus opposes filtration. Over the entire length of the glomerular capillary bed, the BCOP averages about 25 mm Hg.

The **net filtration pressure (NFP)** is the pressure acting across the glomerular capillaries. It represents the sum of the hydrostatic pressures and the colloid osmotic pressures. Under normal circumstances, the net filtration pressure is approximately...

The **capsular collo...** is usually 0 because ... proteins enter the c...

4 The primary fa... basically the sa... movement across cap... balance between **hyd...** **colloid osmotic pres...** solution) on either si...

.6 Review

...ular space separates which layers of the ...ar capsule?

...hy blood pressure is higher in glomerular ...than in other systemic capillaries.

...oidal osmotic pressure tends to draw ...of the filtrate and into the plasma. Why ...occur?

Section 2: Overview of Renal Physiology • 861

This has been an exciting, innovative, and challenging project, and the many talented and dedicated people involved are listed in the Acknowledgments section. If you have questions, comments, or suggestions for improvement, please contact us at the e-mail address below right.

Sincerely,

Frederic H. Martini, Ph.D.   William C. Ober, MD
Judi L. Nath, Ph.D.

Visual Anatomy & Physiology abandons the 15th-century format and adopts a modern, integrated approach.

The complete integration of text and art within an efficient organization makes it easier for students to access information.

Organizational overview:

- The concepts are presented in 381 modules—361 in two-page spreads and 20 as single pages, each with a focused topic statement.

- Each module ends with a series of review questions to provide reinforcement and emphasize key concepts and relationships.

- Groups of related modules are clustered in sections, and each section opens with a one-page introduction and closes with a one-page section review.

- The end-of-chapter material is organized into a highly visual review exercise capped by an applied scenario or case study that integrates the material from the entire chapter.

These structural and organizational innovations make *Visual Anatomy & Physiology* a completely new kind of textbook—one that students find both exciting and easier to understand. The pages that follow will walk you through the specific features of this text in more detail.

Send questions or comments to:
martini@maui.net

How Does the Modular Organization Work?

The time-saving modular organization presents topics in two-page spreads. These two-page spreads give students an efficient organization for managing their time. Students can study each module during the limited time they have in their busy schedules—ten minutes for one module now, ten minutes for another module later—checking off each module as they complete it.

First, the top left page begins with a full-sentence topic heading that teaches the major point of the module. (These topic headings are correlated by number to the learning outcomes on the chapter-opening page. The learning outcomes are derived from the learning outcomes recommended by the Human Anatomy & Physiology Society.)

4.2

Epithelial cells are extensively interconnected, both structurally and functionally

To be effective as a barrier, an epithelium must form a complete cover or lining, and have the ability to replace lost or damaged cells through the divisions of stem cells. The physical integrity of an epithelium depends on intercellular connections and attachment to adjacent tissues.

1 The detailed structure of each form of intercellular attachment demonstrates the linkage between structure and function at all levels.

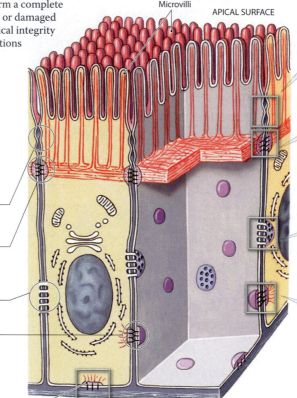

Microvilli

APICAL SURFACE

Intercellular Attachments

Occluding junctions form a barrier that isolates the basolateral surfaces and deeper tissues from the contents of the lumen.

An **adhesion belt** locks together the terminal webs of neighboring cells, strengthening the apical region and preventing distortion and leakage at the occluding junctions.

Gap junctions permit chemical communication that coordinates the activities of adjacent cells.

Desmosomes (DEZ-mō-sōms; *desmos*, ligament + *soma*, body) provide firm attachment between neighboring cells by interlocking their cytoskeletons.

BASE

Next, the red-boxed numbers guide students through the presentation of the topic.

Intermediate filaments of the cytoskeleton

Hemidesmosome

2 **Hemidesmosomes** attach the deepest epithelial cells to the basal lamina. At a hemidesmosome, the basal cytoskeleton is locked to peripheral proteins and to transmembrane proteins that are firmly attached to a layer of extracellular protein filaments and fibers.

Basal Lamina

The **basal lamina**, or basement membrane, is a complex structure produced by the basal surface of the epithelium and the underlying connective tissue.

The **clear layer**, or lamina lucida (LAM-i-nah LOO-si-dah; *lamina*, thin layer + *lucida*, clear) contains glycoproteins and a network of fine protein filaments.

The **dense layer**, or lamina densa, containing bundles of coarse protein fibers, gives the basal lamina its strength and acts as a filter that restricts diffusion between the adjacent tissues and the epithelium.

3 At an occluding junction, the attachment is so tight that it prevents the passage of water and solutes between cells.

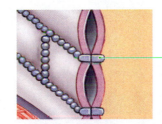

At an occluding junction, the lipid portions of the two plasma membranes are tightly bound together by interlocking membrane proteins.

4 A continuous adhesion belt forms a band that encircles cells and binds them to their neighbors. The bands are dense proteins that are attached to the microfilaments of the terminal web.

5 At a gap junction, two cells are held together by interlocking junctional proteins called **connexons**. Gap junctions between epithelial cells are common where the movement of ions helps coordinate functions such as secretion or the beating of cilia. Gap junctions occur in other tissues as well; in cardiac muscle tissue, for example, they help coordinate contractions of the heart muscle.

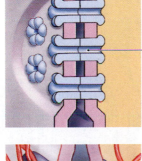

Connexons are channel proteins that form a narrow passageway and let small molecules and ions pass from cell to cell.

6 At a desmosome, the opposing plasma membranes are locked together. Desmosomes are very strong and resist stretching and twisting.

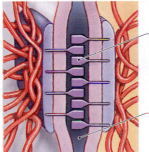

Cell adhesion molecules (CAMs) are transmembrane proteins that bind to each other and to extracellular materials.

The membranes of adjacent cells may also be bonded by **intercellular cement**, a thin layer of proteoglycans that contain polysaccharide derivatives, most notably **hyaluronan**.

Epithelia lack blood vessels, and for this reason they are said to be **avascular** (*a*, without). The cells forming the deepest layer of an epithelium must remain firmly attached to underlying connective tissues, because the blood vessels in those tissues nourish the entire epithelium.

Module 4.2 Review

a. Identify the various types of epithelial intercellular connections.

b. How do epithelial tissues, which are avascular, obtain needed nutrients?

c. What is the functional significance of gap junctions?

Then, instead of long columns of narrative text that refer to visuals, brief text is built right into the visuals. Students read while looking at the corresponding visual, which means:

- No long paragraphs
- No flipping of pages
- Everything in one place

Wh t Is th Visu l Appro ch?

The unique visual approach allows the illustrations to be the central teaching and learning element, with the text built directly around them—creating true text-art integration. This approach matches how students naturally want to use their A&P textbook. Our extensive research with A&P students—via student reviews, student focus groups, and student class tests—reveals that A&P students go first to the visuals and then to the corresponding text.

Descriptions and key terminology are embedded in the art.

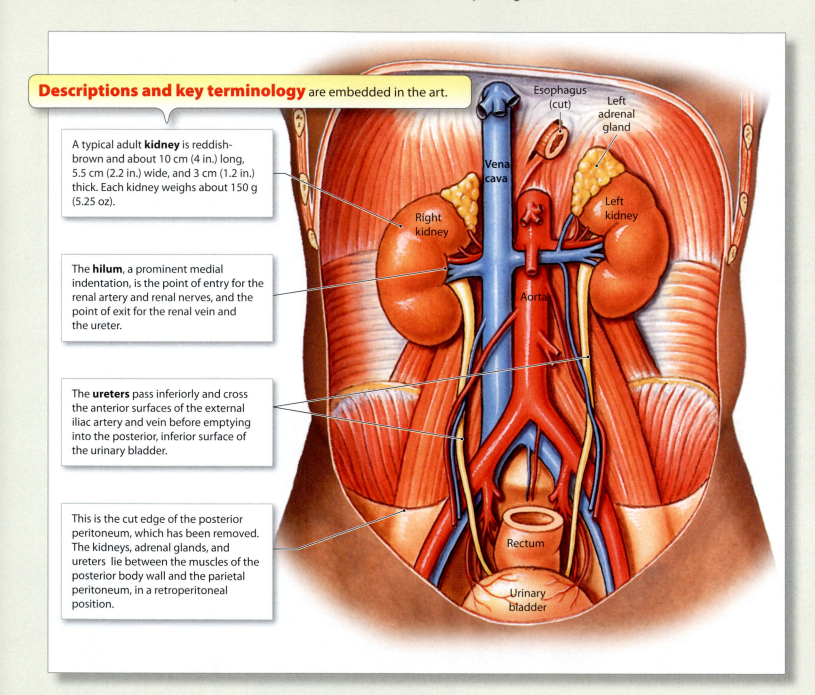

A typical adult **kidney** is reddish-brown and about 10 cm (4 in.) long, 5.5 cm (2.2 in.) wide, and 3 cm (1.2 in.) thick. Each kidney weighs about 150 g (5.25 oz).

The **hilum**, a prominent medial indentation, is the point of entry for the renal artery and renal nerves, and the point of exit for the renal vein and the ureter.

The **ureters** pass inferiorly and cross the anterior surfaces of the external iliac artery and vein before emptying into the posterior, inferior surface of the urinary bladder.

This is the cut edge of the posterior peritoneum, which has been removed. The kidneys, adrenal glands, and ureters lie between the muscles of the posterior body wall and the parietal peritoneum, in a retroperitoneal position.

Esophagus (cut)

Left adrenal gland

Vena cava

Right kidney

Left kidney

Aorta

Rectum

Urinary bladder

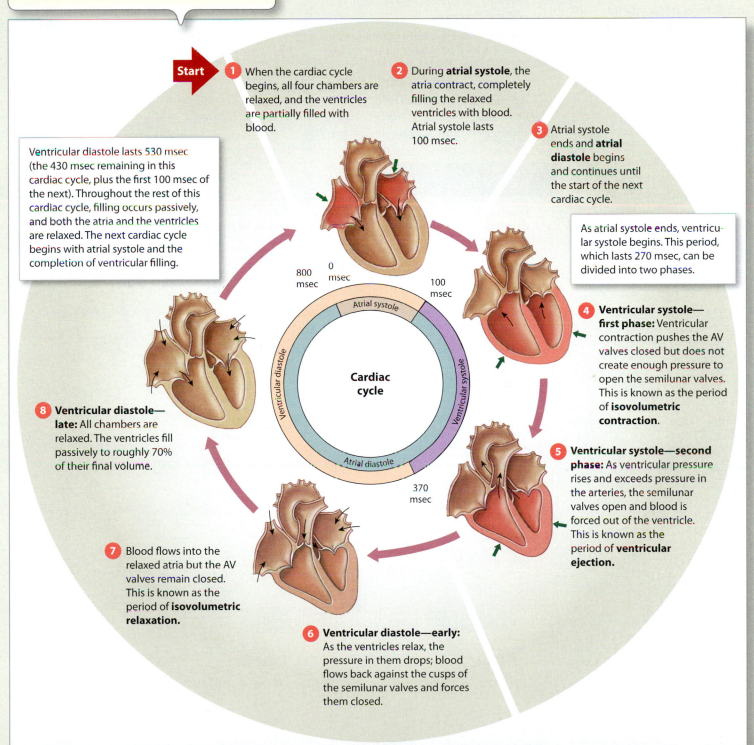

Start

1. When the cardiac cycle begins, all four chambers are relaxed, and the ventricles are partially filled with blood.

2. During **atrial systole**, the atria contract, completely filling the relaxed ventricles with blood. Atrial systole lasts 100 msec.

3. Atrial systole ends and **atrial diastole** begins and continues until the start of the next cardiac cycle.

As atrial systole ends, ventricular systole begins. This period, which lasts 270 msec, can be divided into two phases.

4. **Ventricular systole—first phase:** Ventricular contraction pushes the AV valves closed but does not create enough pressure to open the semilunar valves. This is known as the period of **isovolumetric contraction**.

5. **Ventricular systole—second phase:** As ventricular pressure rises and exceeds pressure in the arteries, the semilunar valves open and blood is forced out of the ventricle. This is known as the period of **ventricular ejection**.

6. **Ventricular diastole—early:** As the ventricles relax, the pressure in them drops; blood flows back against the cusps of the semilunar valves and forces them closed.

7. Blood flows into the relaxed atria but the AV valves remain closed. This is known as the period of **isovolumetric relaxation**.

8. **Ventricular diastole—late:** All chambers are relaxed. The ventricles fill passively to roughly 70% of their final volume.

Ventricular diastole lasts 530 msec (the 430 msec remaining in this cardiac cycle, plus the first 100 msec of the next). Throughout the rest of this cardiac cycle, filling occurs passively, and both the atria and the ventricles are relaxed. The next cardiac cycle begins with atrial systole and the completion of ventricular filling.

800 msec   0 msec   100 msec   370 msec

Atrial systole
Ventricular systole
Atrial diastole
Ventricular diastole

Cardiac cycle

Where Is the Practice?

Three predictable places to stop and check understanding help students pace their learning throughout the chapter.

Module Reviews appear at the end of every module for frequent and consistent self-assessment.

Module 23.9 Review

a. Define countercurrent multiplication as it occurs in the kidneys.

b. The thick ascending limb of the nephron loop actively pumps what substances into the peritubular fluid?

c. An increase in sodium and chloride ions in the peritubular fluid affects the descending limb in what way?

Section Reviews appear after groups of related modules and include "workbook-style" review activities, such as labeling and concept mapping.

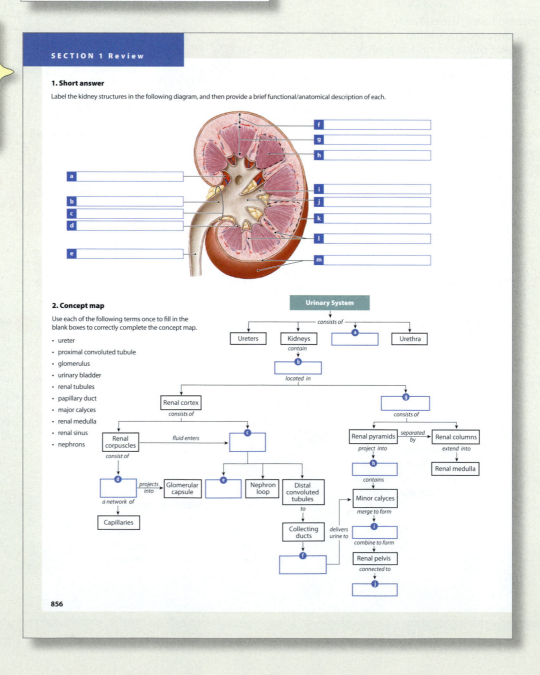

SECTION 1 Review

1. Short answer

Label the kidney structures in the following diagram, and then provide a brief functional/anatomical description of each.

a
b
c
d
e
f
g
h
i
j
k
l
m

2. Concept map

Use each of the following terms once to fill in the blank boxes to correctly complete the concept map.

- ureter
- proximal convoluted tubule
- glomerulus
- urinary bladder
- renal tubules
- papillary duct
- major calyces
- renal medulla
- renal sinus
- nephrons

Urinary System

consists of

Ureters   Kidneys   a   Urethra

contain

b

located in

Renal cortex   g

consists of   consists of

Renal corpuscles — fluid enters → c   Renal pyramids — separated by → Renal columns

consist of   project into   extend into

d   h   Renal medulla

projects into → Glomerular capsule   e   Nephron loop   Distal convoluted tubules   contains

a network of   to   Minor calyces

Capillaries   Collecting ducts   delivers urine to   merge to form

                                                    f   i

                                                                         combine to form

                                                                         Renal pelvis

                                                                         connected to

                                                                         j

856

Visual Outline with Key Terms

Summarize the content of each module using the terms in the order provided.

SECTION 1

Anatomy of the Urinary System

- urinary system
- urinary tract
- kidneys
- urine
- ureters
- urinary bladder
- urethra
- urination

23.1

The kidneys are paired retroperitoneal organs

- kidney
- hilum
- ureters
- retroperitoneal
- fibrous capsule
- perinephric fat capsule
- renal fascia

23.2

The kidneys are complex at the gross and microscopic levels

- hilum
- fibrous capsule
- renal sinus
- renal cortex
- renal medulla
- renal pyramid
- renal papilla
- renal column
- kidney lobe
- minor calyc...
- renal ...
- nephr...
- cortic...
- juxtar...
- nephr...
- nephr...

23.3

A nephron can be divided into r... each region has specific functio...

- nephron
- renal corpuscle
- renal tubule
- capsular space
- filtrate
- glomerular capsule
- glomerulus
- proximal convoluted tubule (PCT)
- nephron loop
- distal tubule...
- collec...
- collec...
- papilla...

* = Term boldfaced in this module

23.4

The kidneys are highly vascular, and the circulation patterns are complex

- renal artery and renal vein
- segmental arteries
- interlobar arteries and veins
- arcuate arteries and veins
- cortical radiate arteries and veins
- afferent and efferent arterioles
- glomerulus
- peritubular capillaries
- vasa recta

SECTION 2

Overview of Renal Physiology

- urea
- creatinine
- uric acid
- filtration
- reabsorption

Chapter Integration: Applying what you've learned

For consumers, some recent developments have turned the purity of food and food products, for humans and animals alike, from a matter of trust into a matter of concern.

In 2007, the unintentional adulteration of pet food made in China with the nitrogen-containing industrial chemical melamine caused pet illnesses and deaths in the United States and resulted in massive recalls of contaminated products. In 2008, milk intentionally contaminated with melamine—put there to skew laboratory tests that measure nitrogen as an index of the protein content in a food or drink powder—sickened at least 64,000 children in China, several of whom died. Clearly, vigilance concerning our food supply is crucial, even while standards and regulations are in place.

In the body, melamine contamination has two major effects: (1) the formation of crystalline masses in the filtrate and/or urine, and (2) acidification of the tubular fluid. Resulting clinical problems range from blood in the urine, to acid–base and electrolyte disorders, to urinary obstruction and (in severe cases) kidney failure.

1. Propose a linkage between the stated effects of melamine poisoning and the clinical problems observed.

2. Primary therapeutic options for melamine poisoning include infusion of fluids, dialysis, and medication. Explain how these options address specific problems caused by melamine poisoning and suggest possible follow-up tests.

MasteringA&P™

Access more review material online in the Study Area at **www.masteringaandp.com**.

There, you'll find:
- Chapter guides
- Chapter quizzes
- Practice tests
- Labeling activities
- MP3 Tutor Sessions
- Animations
- Flashcards
- A glossary with pronunciations

IP Use *Interactive Physiology®* (IP) to help you understand difficult physiological concepts in this chapter. Go to **Urinary System** and find the following topics:
- Anatomy Review
- Glomerular Filtration
- Early Filtrate Processing
- Late Filtrate Processing

Chapter Reviews

include a Visual Outline with Key Terms and a Chapter Integration section. In the Visual Outline, each entry begins with the module title and an image to remind students of the general topic, followed by a list of the key terms needed to summarize the module content. In the Chapter Integration section, a clinical scenario is followed by critical thinking questions that help students tie important concepts together.

Additional practice is available online in the MasteringA&P Study Area. (See pages xvi–xvii.)

MasteringA&P™

An Assignment and Assessment System

Get your students ready for the A&P course.

MasteringA&P allows you to assign tutorials and assessments on *Get Ready* topics:

- Study Skills
- Basic Math Review
- Terminology
- Body Basics
- Chemistry
- Cell Biology

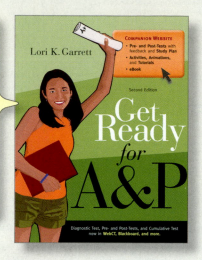

Get your students to come to class prepared.

Assignable Reading Quizzes motivate your students to read the textbook before coming to class.

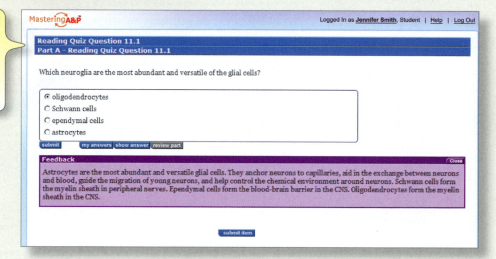

Assign art from the textbook.

Assign and assess figures from the textbook.

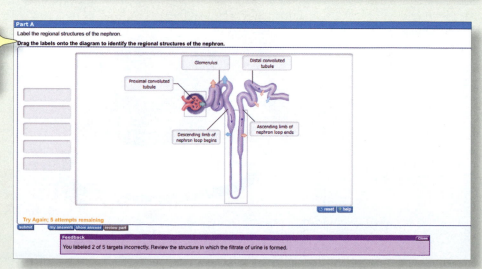

Give your students extra coaching.

Assign tutorials from your favorite media—such as Interactive Physiology® (IP) and A&P Flix™—to help students understand and visualize tough topics. MasteringA&P provides coaching through helpful wrong-answer feedback and hints.

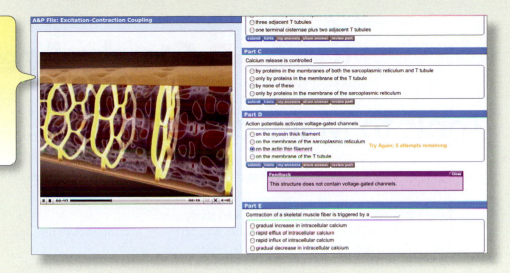

Give your students 24/7 lab practice.

Practice Anatomy Lab™ (PAL™) 3.0 is a tool that helps students study for their lab practicals outside of the lab.

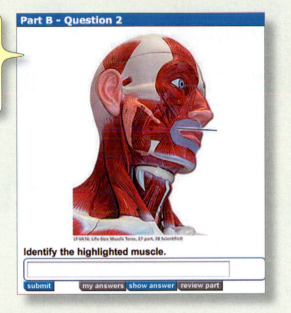

Identify struggling students before it's too late.

MasteringA&P has a color-coded gradebook that helps you identify vulnerable students at a glance. Assignments in MasteringA&P are automatically graded, and grades can be easily exported to course management systems or spreadsheets.

Go to

www.masteringaandp.com to watch the demo movie.

Mastering A&P™ Study Area

Tools to Make the Grade

MasteringA&P includes a Study Area that will help students get ready for tests with its simple three-step approach. Students can:

1. **Take a pre-test** and obtain a personalized study plan.
2. **Learn and practice** with animations, labeling activities, and interactive tutorials.
3. **Self-test** with quizzes and a chapter post-test.

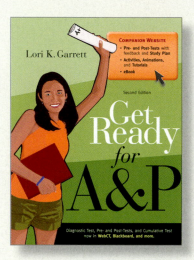

Get Ready for A&P

Students can access the **Get Ready for A&P** e-book, activities, and diagnostic tests for these important topics:

- Study Skills
- Basic Math Review
- Terminology
- Body Basics
- Chemistry
- Cell Biology

MP3 Tutor Sessions

Study on the go!

Students can download the MP3 Tutor Sessions for specific chapters of the textbook and study wherever, whenever. They can listen to mini-lectures about the toughest topics and take audio quizzes to check their understanding.

Practice Anatomy Lab™ (PAL™) 3.0

Practice Anatomy Lab (PAL) 3.0 is a virtual anatomy study and practice tool that gives students 24/7 access to the most widely used lab specimens, including the human cadaver, anatomical models, histology, cat, and fetal pig. PAL 3.0 retains all of the key advantages of version 2.0, including ease-of-use, built-in audio pronunciations, rotatable bones, and simulated fill-in-the-blank lab practical exams.

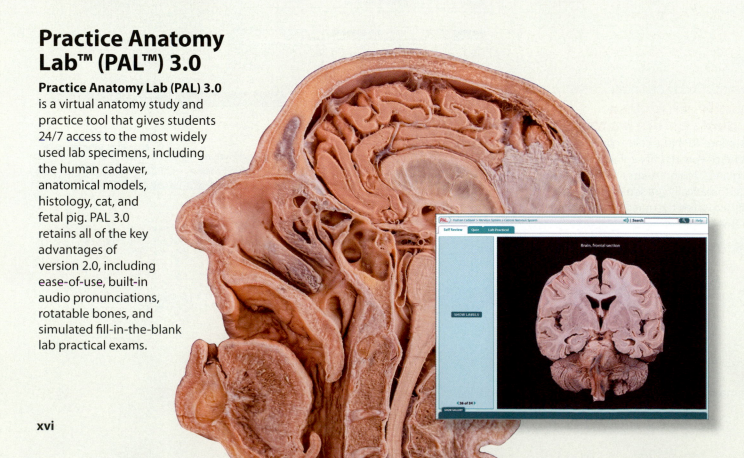

A&P Flix™

A&P Flix are 3-D movie-quality animations with self-paced tutorials and gradable quizzes that help students master the toughest topics in A&P:

Cell Physiology
Membrane Transport
DNA Replication
Protein Synthesis
Mitosis

Muscle Physiology
Events at the Neuromuscular Junction
Excitation-Contraction Coupling
Cross-Bridge Cycle

Neurophysiology
Resting Membrane Potential
Generation of an Action Potential
Propagation of an Action Potential

Origins, Insertions, Actions, Innervations
Over 50 animations on this topic

Group Muscle Actions & Joints
Over 60 animations on this topic

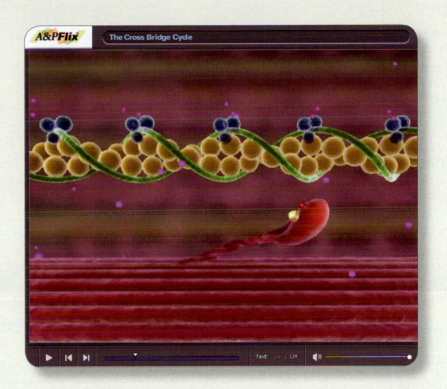

Interactive Physiology® (IP)

IP helps students understand the hardest part of A&P: physiology. Fun, interactive tutorials, games, and quizzes give students additional explanations to help them grasp difficult concepts.

Modules:
- Muscular System
- Nervous System I
- Nervous System II
- Cardiovascular System
- Respiratory System
- Urinary System
- Fluids & Electrolytes
- Endocrine System
- Digestive System
- Immune System

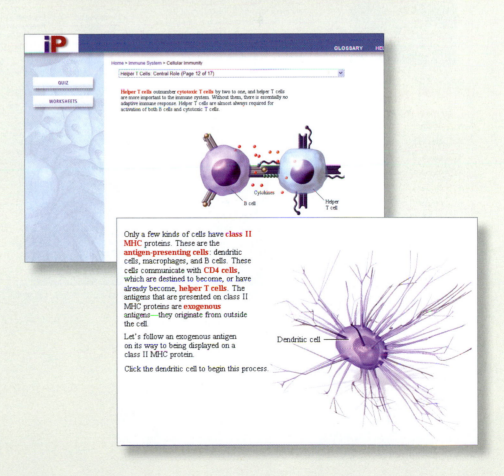

Support for Students

eText

MasteringA&P (www.masteringaandp.com) includes an eText. Students can access their textbook wherever and whenever they are online. eText pages look exactly like the printed text yet offer additional functionality. Students can:

- Create notes.
- Highlight text in different colors.
- Create bookmarks.
- Zoom in and out.
- View in single-page or two-page view.
- Click hyperlinked words and phrases to view definitions.
- Link directly to relevant animations.
- Search quickly and easily for specific content.

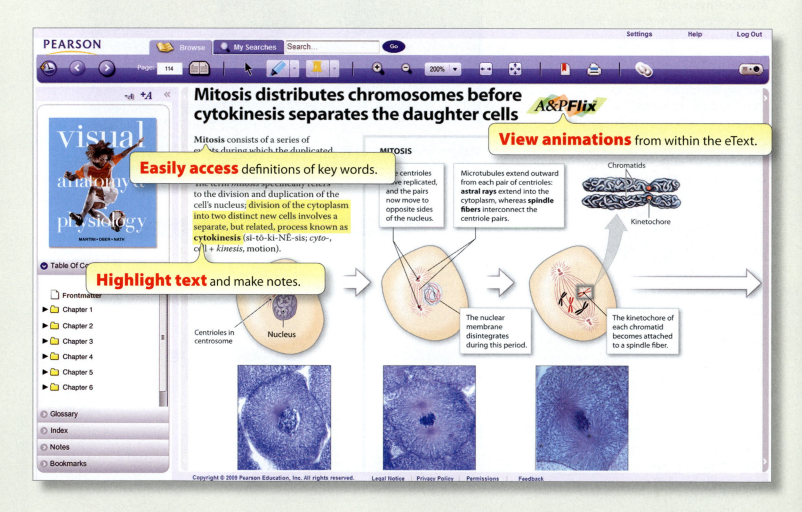

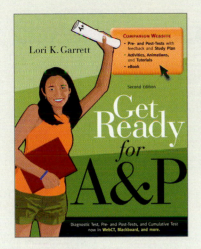

Get Ready for A&P
by Lori K. Garrett

This book and online component were created to help students be better prepared for their A&P course. Features include pre-tests, guided explanations followed by interactive quizzes and exercises, and end-of-chapter cumulative tests. Also available in the Study Area of www.masteringaandp.com.

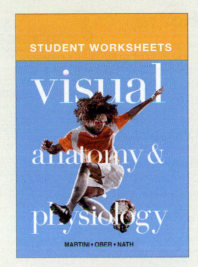

Student Worksheets for Visual Anatomy & Physiology

This booklet contains all of the Section Review pages from the book for students who would prefer to mark their answers on separate pages rather than in the book itself. In addition, the Visual Outline with Key Terms from the end of each chapter is reprinted with space for students to summarize the content of each module using the key terms in the order provided.

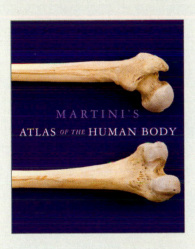

Martini's Atlas of the Human Body
by Frederic H. Martini

The Atlas offers an abundant collection of anatomy photographs, radiology scans, and embryology summaries, helping students visualize structures and become familiar with the types of images seen in a clinical setting.

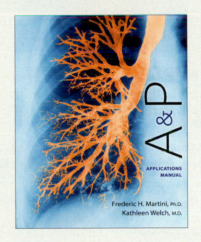

A&P Applications Manual
by Frederic H. Martini and Kathleen Welch

This manual contains extensive discussions on clinical topics and disorders to help students apply the concepts of anatomy and physiology to daily life and their future health professions.

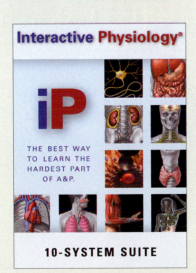

Interactive Physiology® 10-System Suite (IP-10) CD-ROM

IP-10 helps students understand the hardest part of A&P: physiology. Fun, interactive tutorials, games, and quizzes give students additional explanations to help them grasp difficult physiological concepts.

Practice Anatomy Lab™ (PAL™) 3.0 DVD

PAL 3.0 is an indispensable virtual anatomy study and practice tool that gives students 24/7 access to the most widely used lab specimens, including the human cadaver, anatomical models, histology, cat, and fetal pig. Standalone DVD available for purchase: 978-0-321-68211-6/ 0-321-68211-4

See pages xvi–xvii for the MasteringA&P Study Area

Support for Instructors

eText with Whiteboard Mode

The *Visual Anatomy & Physiology* eText within MasteringA&P comes with Whiteboard Mode, allowing instructors to use the eText for dynamic classroom presentations. Instructors can show one-page or two-page views from the book, zoom in or out to focus on select topics, and use the Whiteboard Mode to point to structures, circle parts of a process, trace pathways, and customize their presentations.

Instructors can also add notes to guide students, upload documents, and share their custom-enhanced eText with the whole class.

Instructors can find the eText with Whiteboard Mode on MasteringA&P.

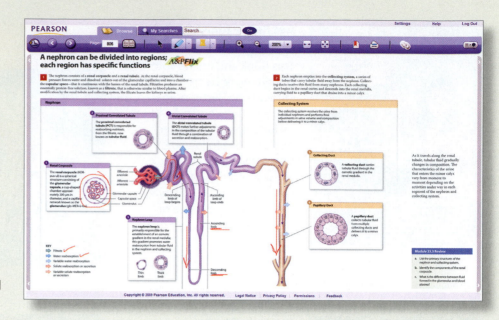

Instructor Resource DVD (IRDVD)

with Lecture Outlines by Alexander G. Cheroske and Clicker Questions and Quiz Shows by Margaret (Betsy) Ott
978-0-321-56017-9 / 0-321-56017-5

The IRDVD offers a wealth of instructor media resources, including presentation art, lecture outlines, test items, and answer keys—all in one convenient location. The IRDVD includes:

- Textbook images in JPEG format (in two versions—one with labels and one without)
- Customizable textbook images embedded in PowerPoint slides (in three versions—one with editable labels, one without labels, and one as step-edit art)
- Customizable PowerPoint lecture outlines, including figures and tables from the book and links to the A&P Flix
- A&P Flix™ 3-D movie-quality animations on tough topics
- PRS-enabled Active Lecture Clicker Questions
- PRS-enabled Quiz Show Clicker Questions
- *Interactive Physiology*® 10-System Suite (IP-10) Exercise Sheets and Answer Key
- *Martini's Atlas of the Human Body* images
- MRI/CT scans
- Histology slides
- Muscle Origins and Insertions images
- The Test Bank in TestGen® format and Microsoft® Word format
- The Instructor's Manual in Microsoft® Word format
- PDF files of Transparency Acetate masters

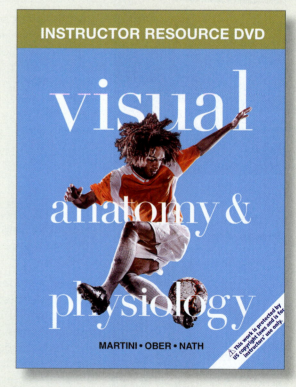

Instructor's Manual
by Jeff Schinske

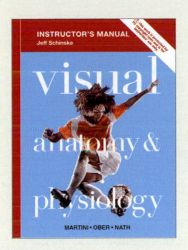

978-0-321-58177-8 / 0-321-58177-6

This useful resource includes a wealth of materials to help instructors organize their lectures, such as lecture ideas, visual analogies, suggested classroom demonstrations, vocabulary aids, applications, and common student misconceptions/problems.

Printed Test Bank
by Jason LaPres

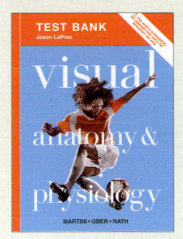

978-0-321-58179-2 / 0-321-58179-2

The test bank of more than 3,000 questions tied to the Learning Outcomes in each chapter helps instructors design a variety of tests and quizzes. The test bank includes text-based and art-based questions. This supplement is the print version of TestGen that is in the IRDVD package.

Instructor's Visual Guide

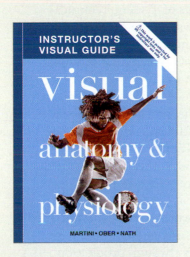

978-0-321-58182-2 / 0-321-58182-2

This guide is a printed and bound collection of thumbnails of the images and media on the IRDVD. (See previous page.) With this take-anywhere supplement, instructors can plan lectures when away from their computers.

Practice Anatomy Lab™ (PAL™) IRDVD

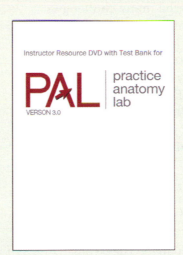

978-0-321-74963-5/ 0-321-74963-4

This DVD includes everything instructors need to present and assess PAL 3.0 in lecture and lab. It includes images in PowerPoint® and JPEG formats, links to animations, and a test bank with more than 4,000 lab practical questions.

Transparency Acetates
978-0-321-58181-5 / 0-321-58181-4

All figures and tables from the text are included in the printed Transparency Acetates. Complex figures are broken out for readable projected display. A full set of Transparency Acetate masters of all figures and tables is also available on the IRDVD. (See previous page.)

CourseCompass™/ WebCT / Blackboard

Pre-loaded book-specific content and test item files accompanying the text are available in several course management formats.

See pages xiv–xv for MasteringA&P

Acknowledgments

We feel very fortunate that we've been given the opportunity to do something new and different. Our visual approach involves creating the text and art as a unit, and this means that the traditional sequential publishing process—manuscript generation to art development to editing to design to paging—could not apply. As a result, not only the authors but also the publisher needed to develop an all-new process to create this book. So at the outset we would like to thank Pearson Science for their willingness to think "outside the box." Paul Corey, President of Pearson Science, Frank Ruggirello, VP/Editorial Director of Applied Sciences, and Leslie Berriman, Executive Editor of Applied Sciences, showed unwavering support for this effort despite the many challenges it posed to established publishing systems.

Once the project was officially under way, a team of talented people was drafted to assist us in turning our vision of this text into a physical reality. Ric Martini worked closely with Judi Nath, Ed Bartholomew, and Kathleen Welch as the pedagogical framework and clinical coverage evolved. Their talents, patience, and good humor helped immeasurably. This book required a completely different approach to illustration and design. Once we drafted a two-page module, Bill Ober worked with Claire Garrison and Anita Impagliazzo to produce the illustrations and a preliminary layout. Jim Gibson and Gibson Design Associates, who shaped the design, then adjusted and finalized the layout for optimal flow. This process was repeated for each module and every iteration as chapters were created, reviewed, and revised. Without the creative talents of this team, this project would not have been possible.

We would also like to thank our development editor, Alan Titche, who assisted us by organizing and summarizing the review comments and steering us toward the necessary improvements to the modules. Throughout this process, Robin Pille, our project editor at Pearson Benjamin Cummings, kept the wheels turning in-house by coordinating reviews, manuscript trafficking, photo research, and a million other details. Barbara Yien oversaw the development and helped to keep the book on track. Deborah Cogan and Nancy Tabor invented new processes to shepherd this unique book through production and into print. Thanks also to Susan Malloy, Cassandra Cummings, Katie Seibel, Kelly Reed, Steven

Le, and Beth Collins for investing their time and effort to help with the reviewing and production of the book. Michael Rossa's careful attention to detail in his copyedit resulted in many important changes for improved consistency and accuracy. Tiffany Timmerman led the skilled team at S4Carlisle in moving the book smoothly through composition. Maureen Spuhler and Donna Kalal did an excellent job researching the photos for this text.

Thanks are due to Jason LaPres of Lone Star College—North Harris for the Test Bank, to Alexander Cheroske of Mesa Community College for the Lecture Presentations, to Jeff Schinske of De Anza College for the Instructor's Manual, to Margaret (Betsy) Ott of Tyler Junior College for the Clicker Questions and Quiz Shows, to Angela Edwards of Trident Technical College for her accuracy reviews, and to Mike Yard of Indiana University—Purdue University Indianapolis and Agnes Yard of the University of Indianapolis for the assessments in MasteringA&P. We are grateful to Lucinda Bingham, media producer at Pearson Benjamin Cummings, for her expert management of the media resources for instructors and students, especially the new MasteringA&P Website. Thanks also to Lorin Hawley, Bradford Threlkeld, and Katherine Foley for their work on MasteringA&P. We are also grateful to Nicole McFadden, Editorial Assistant at Pearson Benjamin Cummings, for her dedicated work on the print and media supplements and to Shannon Kong and Dorothy Cox for shepherding them smoothly through production.

Finally, we would like to thank Brooke Suchomel, Market Development Manager, Derek Perrigo, Marketing Manager, and all of the enthusiastic Pearson Science sales representatives who worked tirelessly to present this book to instructors and students over the course of its development to garner invaluable feedback and relay it to us so that we could integrate it into every page.

Because this is the first text to use this visual/modular format, we needed confirmation that our vision matched the needs and expectations of instructors and students. This confirmation was provided through instructor reviews, student reviews, and student focus groups of draft chapters followed by class testing of more final chapters. We are indebted to the following instructors and students for their comments and suggestions throughout the development of the program:

Instructor Reviewers

Benja Allen
El Centro College

Kara Battle
Durham Technical Community College

Nina Beaman
Bryant & Stratton College

Felicia Brenoe
Glendale Community College

Janet Brodsky
Ivy Tech Community College – Lafayette

Steve Byrne
St. Petersburg College

Maura Cavanagh-Dick
Salem Community College

Alexander Cheroske
Mesa Community College at Red Mountain

Robert Clark
Ozarks Technical Community College

Gerard Cronin
Salem Community College

Lynnette Danzl-Tauer
Rock Valley College

Martha Dixon
Diablo Valley College

Sharon Ellerton
Queensborough Community College

Seema Endley
Blinn College – Bryan

Linda Falkow
Mercer County Community College

Carl Frailey
Johnson County Community College

Patrick Galliart
North Iowa Area Community College

William Gressett
Holmes Community College

Michael Harman
Lone Star College – North Harris

Lisa Hawthorne
Brown Mackie College

Chris Hazzi
State College of Florida Manatee-Sarasota

Stuart Hill
Blinn College – Bryan

Dale Horeth
Tidewater Community College

Julie Huggins
Arkansas State University – Jonesboro

Jason Hunt
Brigham Young University – Idaho

Alexander Ibe
Weatherford College

Jeba Inbarasu
Metropolitan Community College

Jason Jennings
Southwest Tennessee Community College – Union Avenue

Thomas Jordan
Pima Community College – Northwest Campus

Leslie King
University of San Francisco

William Kleinelp
Middlesex County College

Chad Knights
Northern Virginia Community College – Alexandria

Michael LaPointe
Indiana University –Northwest

Thomas McDonald
Pima Community College – East Campus & West Campus

Abraham Miller
University of Tampa

Claire Miller
Community College of Denver

Michele N. Moore
Ivy Tech Community College – East Central

David Moyer
Piedmont Virginia Community College & University of Virginia School of Medicine

Hong Nguyen
Northern Virginia Community College – Alexandria

Phillip Nicotera
St. Petersburg College

Margaret (Betsy) Ott
Tyler Junior College

Thomas Pilat
Illinois Central College – East Peoria Campus

Jacqueline Quiros
Suffolk County Community College – Grant Campus

Susan Rohde
Triton College

Hiranya Roychowdhury
New Mexico State University

Karla Rues
Ozarks Technical Community College

Hope Sasway
Suffolk County Community College – Grant Campus

Jeff Schinske
De Anza College

Donald Shaw
The University of Tennessee at Martin

Marilyn Shopper
Johnson County Community College

Mark Slivkoff
Collin College

Scott Smidt
Laramie County Community College – Albany County Campus

Phillip Snider
Gadsden State Community College

Asha Stephens
College of the Mainland

Shelia Taylor
Ozarks Technical Community College

Keti Venovski
Lake-Sumter Community College – South Lake Campus

Patricia Visser
Jackson Community College

Delon Washo-Krupps
Arizona State University

Alan Wasmoen
Metropolitan Community College

Jen Wortham
University of Tampa

Patricia Wu
Chabot College

Janice Yoder Smith
Tarrant County College – Northwest Campus

Instructor Class Testers

Melissa Bailey
Emporia State University

Verona Barr
Heartland Community College

Claudia Barreto
University of New Mexico

Janet Brodsky
Ivy Tech Community College – Lafayette

Robert Brozanski
Community College of Allegheny County – North

Peter Bushnell
Indiana University – South Bend

Nickolas Butkevich
Schoolcraft College

Zinnia Callueng
Central Florida Community College

Alexander Cheroske
Mesa Community College at Red Mountain

Ron Clark
Mercer County Community College

Debra Claypool
Mid Michigan Community College

Jan Clifton
Ivy Tech Community College – Muncie

Vicki Clouse
Montana State University – Northern

Judy Cunningham
Montgomery County Community College

Lynette Danzl-Tauer
Rock Valley College

Mario DeLaHaye
Kennedy King College

Robin Dodson
Parkland College

Patricia Dolan
Pacific Lutheran University

Sondra Dubowsky
McLennan Community College

(Instructor Class Testers, continued)

Sharon Ellerton
Queensborough Community College

Theresia Elrod
State College of Florida – Venice

Seema Endley
Blinn College – Bryan

Jeff Engel
Western Illinois University

Gregory Erianne
Naugatuck Valley Community College

Linda Falkow
Mercer County Community College

Alyssa Farnsworth
Ivy Tech Community College – Muncie

Carol Flora
Indiana University – Purdue University Indianapolis

Lori Garrett
Parkland College

Jennifer Gibbs
Hinds Community College – Raymond

John Gillen
Hostos Community College

Theresa Gillian
Virginia Polytechnic Institute and State University

Evan Goldman
Philadelphia University

Suzanne Gould
Ivy Tech Community College – Muncie

Joanna Greene
Ivy Tech Community College – Muncie

David Griffith
Ferris State University

Rebecca Harris
Hinds Community College – Raymond

Amy Harwell
Oregon State University

Susan Holland
Wilson Community College

Kevin Holt
Northeast Alabama Community College

Alexander Ibe
Weatherford College

Anthony Jones
Tallahassee Community College

Philip Jones
Manchester Community College

Warren Jones
Loyola University

Leslie King
University of San Francisco

David Klarberg
Queensborough Community College

William Kleinelp
Middlesex County College

Chad Knights
Northern Virginia Community College - Alexandria

Jeff Laborda
State College of Florida – Manatee-Sarasota

Michael LaPointe
Indiana University – Northwest

Ray Larsen
Bowling Green State University

Andrey Lebed
Lee College

Curtis Lee
Dallas Baptist University

Carlos Liachovitzky
Bronx Community College

Mitch Lockhart
Valdosta State University

Jodi Long
Santa Fe College

James Ludden
College of DuPage

John Moore
Parkland College

Judi Nath
Lourdes College

Tammy Oliver
Eastfield College

Margaret (Betsy) Ott
Tyler Junior College

Keith Overbaugh
Northwestern Michigan College

Paul Passalacqua
Owens Community College

Krya Perry
Mesa Community College at Red Mountain

Harry Pierre
Keiser University

Thomas Pilat
Illinois Central College – East Peoria Campus

Brandon Poe
Springfield Technical Community College

Eugenie Pool
Lee College

Julie Porterfield
Tulsa Community College – Southeast

Mark Robbins
Ivy Tech Community College – Marion

Dan Roberts
Owens Community College

Susan Rohde
Triton College

Nick Roster
Northwestern Michigan College

Thomas Ruehlmann
College of DuPage

Chrisanna Saums
Hinds Community College – Raymond

Steve Schenk
Truckee Meadows Community College

Ralph Schwartz
Hostos Community College

Joanne Settel
Baltimore City Community College

Donald Shaw
The University of Tennessee at Martin

Marilyn Shopper
Johnson County Community College

Mark Slivkoff
Collin College

Jane Slone
Cedar Valley College

Scott Smidt
Laramie County Community College – Albany County Campus

Janice Smith
Tarrant County College – Northwest

Joy Smoots
Cape Fear Community College

Phillip Snider
Gadsden State Community College

Julian Stark
Queensborough Community College

Olga Steinberg
Hostos Community College

Yung Su
Florida State University

James Timbilla
Queensborough Community College

Corinne Ulbright
Indiana University – Purdue University Indianapolis

Patricia Visser
Jackson Community College

Jane Walden
Chattahoochee Valley Community College

Delon Washo-Krupps
Arizona State University

Patricia Wu
Chabot College

Janice Yoder Smith
Tarrant County College – Northwest Campus

Robert Zdor
Andrews University

Student Reviewers

Arizona State University
Elija Armstead
Bret Bierman
Mohammed Brini
Jennifer Busenkell
Reese Dare
Jonathon DeJeu
Sally Deng
Erin Donahue
Julian Dudley
Cynthia Fernandez
Ramito Garcia
Ajanii Gibson
Anessa Haws
Jessalyn Italiano
Rachel Johnson
Anne Keller

Tori Kottwitz
Bonnie Leman
Kevin Marshall
Staci Mitchell
Renee Morales
Marielle Mori
Kyle Popelka
Sara Ramirez
Eric Ramos
Mustafa Salih
Zaher Sbai
Rind Shai
Tobin Short
Christina Sugwitan
Blake Testa
Jessica Valdez
Krupa Venkatesh
Ben Waggener

California State University – Sacramento
Ripal Shah

Collin College
Anna Berry
Tonya Higgins
Kayla Hovey
Erin McGranahan

De Anza College
Evelyn Nguyen

Jackson Community College
King Baltimore
Amelia Damon
Amy Higgins
Andrea Ortell
Jamie Ulch

Lake-Sumter Community College
Tammy Gibson
Juliana Green
Jason Robinson
Kristen Thomas

Lone Star College – North Harris
Erica Adlakha
Jessica Basa
Gary Williams

Lourdes College
Melissa Bailey
Susan Behrens
Carrie Bittner
Kayleah Burchett
Kimberly Conger
Sarah Cooper
Ashley Filsinger
Lindsay Holcomb
Lori Jones
Sarah Jones
Vanessa Link
Kenneth Lowery
Antoinette Miller
Carmalita Neiding
Jessica Niese
Justin Ohm
Taryn Paule
Rebecca Powell
Jackie Ruttino
Alaina Schirg
Laura Swander
Brianne Troike
Sarah Waschpusch

Troy Wetzel
Nichole White
Karen Wonderly
Megan Wroblewski

Mesa Community College at Red Mountain
Peter Buehler
Simon Cordeiro
Leslie Hermes
Kayla Hutchings
Reyna Lopez
Suzanne Lorentzen
Satya Merrill
Christy Muntz
Julie Rogers
Beatrice Saavedra
Neda Shahidinejad
Mary Shockley
Rondel Simmons
Heather Sullivan
Candace Robertson Wheeler
Lena Yeary

Middlesex County College
Sandra Adamska
Katrina Baptista
Kelsey Ford
Carlisa Jimenez
Tara Petersen
Nancy Protasenia
William Wekesa
Ashley Wilson
Dyan Yu

Northern Virginia Community College
Katherine Dinterman
Minh Do
Sarah Gregory

Pima Community College
Daniela Lopez
Alanna Moldenhauer
Rachel Rinckey

Queensborough Community College
Lou Ann Cettina
Donna Hunt
Grace Lee
Marie Lissa Metayer
Kaye-Diane Stewart

Tyler Junior College
Nykole Kafka
Rachel Loughmiller
Ria Prakashbhai

University of Arizona
Emily Wolfarth

University of San Francisco
Carmela Dulanas
Nadine Huang
Laura Thornton

Student Focus Group Participants

Instructor: Delon Washo-Krupps
Arizona State University

Jeffrey R. Baker
Samuel Becker
Katherina Carballo
Jerret L. Carnes

Ali Chamseddine
Samantha K. Coe
André M. Colbert
Sierra Cook
Danielle Davis
Zachary Doyle
Erika Finn
Gillian Claire Gentry
Braden L. Haldeman
Tara Herbert
Nicholas Hodgeman
Alexandra Johnson
Taylor Jones
Erica Kokemueller
Dana Lee
Edgar Medrano
Lauren McKinley
Ashley Nicole Phelps
Alexandra Pierre-Bez
Traci Pochardt
Kade Rapier
Desiré M. Richard
Michelle Robles
Karissa Rogers
Jaime Sublett
Breawna Swenson
Hillary Vance
Alyssa VanderMolen
Michael Voight
Karoline Wang
Alex Worley
Callan Yarkosky

Instructor: Jeff Schinske
De Anza College

Lindsay K. Contreras
Mary Fabriquer
Ademir Gallardo
Lori S. Joe
Tramy Quach
Ripal M. Shah
Michele Shareef

Instructor: Thomas Pilat
Illinois Central College

Jenn Bailey
Samantha M. Carroll
Heather N. Groeper
Niki Huette
Simeon S. Johnson
Jeff Jones
Amanda Loudermilk
Matthew Morris
Dianna Pruett
Amber Rasbury
Jennifer A. Redmond
Carmen Y. Reyna
Emily Shular
Michael Smith
Chelsey Steiger
Amy Jo Williamson
William C. Young

Instructor: Janet Brodsky
Ivy Tech Community College – Lafayette

Vanessa C. Bores
Rebekah Joy Bremer
Phillip Brown
Chassity H. Burchfield

Hailey Campbell
Nickeytha Evans
Rissa Halsema
Marianne Harmas
Lucas G. Keay
Gina M. Martin
Keri Mills
Becky Mitchell
Jamie S. Johnson
Matt Johnson
Jennifer Birmingham Reitz
Dorian Rodriguez
Adam J. Smith
Michele Spottswood
Carolyn Vazquez
Denise Verhey

Instructor: Moges Bizuneh
Ivy Tech Community College – Lawrence

Mark Cathcart
Jayme Clark
Sira Conde
Emily Croke
Christy A. Durbin
Marla Fowler
Lorraine Harmon
Emily Hearn-Hortin
Stacy Hopson
Brenda K. Perry
Amanda Peterman
Jamie Williams

Instructor: Alexander Cheroske
Mesa Community College at Red Mountain

Maria Baglio
Anthony Champagne
Simon Cordeiro
Kiley A. Fountain
Krista Gillette
Kelliann Glowacka
Bonnie Holman-Arnow
Caitlin M. Kosecki
Britni Manderfield

Sophia Miguel
Wendy Mulhern-Garcia
Chad William Pioske
Mari Rubens
Beatrice Saavedra
Mary B. Shockley
Rondel Simmons
Jeanette Smith
Terri Werner
Candace Robertson Wheeler

Instructor: Peter Porter
Moraine Valley Community College

Agnes Abramowicz
Yazmin Alayan
Haneya M. Allan
Mariam M. Allan
Ayman Atieh
Niveen Atieh
Kelly Binkowski
Ashley Capone
Paula Cichowicz-Kida
Cindy Cruz
Rana Deis
Jonathan Eiber
Dominika Fafrowicz
Agata Golasc
Julia Gonzalez
Stephanie Hamby
Linsey Heimann
Katrina L. Hopman
Ashley Kats
Marta Kurnat
Magdalena Lach
Anna Lenart
Amana Mohammad
Lindsey Mulrenin
Rocio Nunez
Leonel Ortega
Julie Pickar
Ashley E. Sapinski
Roksana Skupiewska
Dora Svensson

Carrie A. Ternand
Shawna A. Underwood
Cindy Zhang

Instructor: Gregory Erianne
Naugatuck Valley Community College

Sandra Centurelli
Marilyn Clemon
Ilfet Klenja
Dawn Nierenburg-Reynolds
Kevin Outlaw-Miller

Instructor: Patrick Galliart
North Iowa Area Community College

Miranda Y. Balster
Cynthia Cmelik-Reams
Angela Dockum
Carl Galliart
Kaitlyn Goplerud
Kristy Heimer
Justine Hejlik
Carrin Landau
Juston Meier
Melody Molln
Megan Simon
Mark Luhan Sinniah
Staci Smith
Joe Speth
Holly Stone

Instructor: Leslie King
University of San Francisco

Brianna Barnett
Karen Chen
Laura Feinberg
Daniela H. Mak
Emma McDermott
Jessica Navarro
Chris Panaligan
Lauren Quibodeaux
Nicole F. Spitaleri
Laura Thornton
Stephanie S. Willis
Lindsey Wreden

Contents

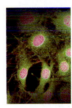

4 Tissue Level of Organization 122

5 The Integumentary System 158

6 Osseous Tissue and Bone Structure 182

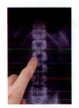

10 The Muscular System 312

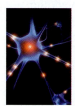

11 Neural Tissue 362

17 Blood and Blood Vessels 574

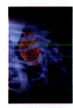

18 The Heart and Cardiovascular Function 624

19 The Lymphatic System and Immunity 676

20 The Respiratory System 720

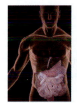

21 The Digestive System 760

22 Metabolism and Energetics 810

26 Development and Inheritance   944

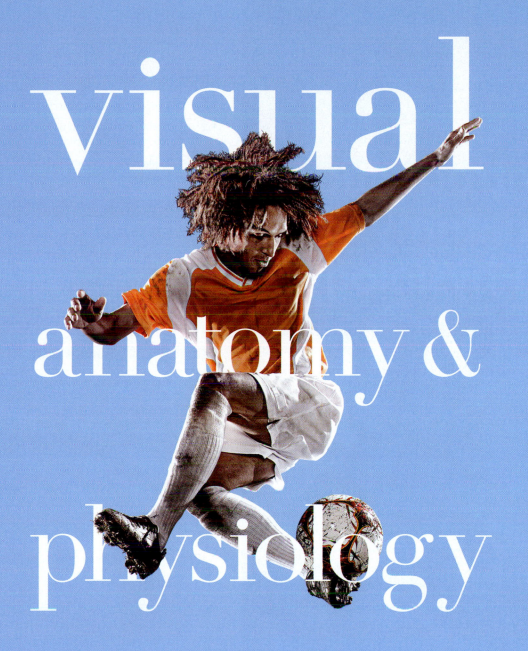

visual

anatomy &

physiology

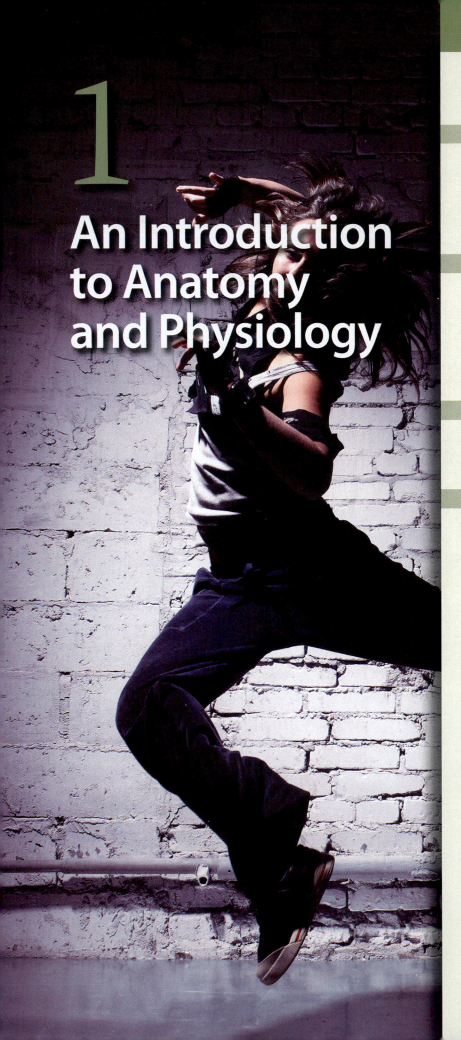

1

An Introduction to Anatomy and Physiology

A&P in Perspective

Human anatomy and physiology considers how the human body performs the functions that keep you alive and alert. You will learn many interesting and important facts about the human body as we proceed. However, the approach you learn and the attitude you develop will be at least as important as the things you memorize. The basic approach in A&P can be summed up as "What is that structure, and how does it work?" The complexity of the answer depends on the level of detail you need. In science, if we know that something works but we don't know how, it's usually called a "Black Box." The more you learn, the smaller (and more numerous) those Black Boxes become; the more you learn, the more you realize how much you don't know.

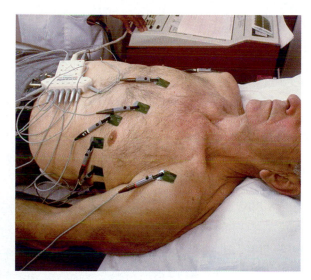

We will devote considerable time to explaining how the body responds to normal and abnormal conditions and maintains **homeostasis**, a relatively constant internal environment. As we proceed, you will see how your body copes with injury, disease, or anything that threatens homeostasis.

Tips on How to Succeed in Your A&P Course

- **Approach the information in different ways**. For example, you might visualize the information, talk it over with or "teach" a fellow student, or spend additional time in lab asking questions of your lab instructor.

- **Set up a study schedule** and stick to it.

- **Devote a block of time each day** to your A&P course.

- **Practice memorization.** Memorization is an important skill, and an integral part of the course. You are going to have to memorize all sorts of things—among them muscle names, directional terms, and the names of bones and brain parts. Realize that this is an important study skill, and that the more you practice, the better you will be at remembering terms and definitions. We will try to give you handles and tricks along the way, to help you keep the information in mind.

- **Avoid shortcuts.** Actually there are no shortcuts. (Sorry.) You won't get the grade you want if you don't put in the time and do the work. This requires preparation throughout the term.

- **Attend all lectures, labs, and study sessions**. Ask questions and participate in discussions.

- **Read your lecture and lab assignments** before coming to class.

- **Do not procrastinate**! Do not do all your studying the night before the exam! Actually STUDY the material several times throughout the week. Marathon study sessions are often counterproductive. There is no easy button; you must push yourself.

- **Seek assistance immediately if you have a problem understanding the material.** Do not wait until the end of the term when it is too late to salvage your grade.

1.1

Biology is the study of life

The world around us contains a variety of living organisms with different appearances and lifestyles. Despite this diversity, all living things perform the same basic functions:

1. Living things respond to changes in their immediate environment—Plants orient to the sun, you move your hand away from a hot stove, and your dog barks at passing strangers.

2. Organisms show adaptability—Their internal operations and responses to stimulation can vary from moment to moment.

3. Over time, organisms grow and reproduce—This creates subsequent generations of similar, but not identical, organisms.

4. Many organisms are capable of some degree of movement. If that movement takes them from one place to another, we call the process locomotion.

Responsiveness, adaptability, growth, reproduction, and locomotion are active processes that require energy. This energy must continually be replaced as it is used. For animals, energy capture typically involves the absorption of oxygen from the atmosphere through respiration and the absorption of various chemicals from the surrounding environment. Each living organism also generates waste products that are discharged into the environment in the process of excretion. These are the basic characteristics of living things, both plant and animal.

For very small organisms, absorption, respiration, and excretion simply involve transferring materials across exposed surfaces. But for larger creatures like dogs, cats, or human beings, this is not possible. For example, human beings cannot absorb steaks or ice cream without processing them first. That processing, called digestion, occurs in specialized areas where complex foods are broken down into simpler components that can be easily absorbed. Finally, because absorption, respiration, and excretion are performed in different portions of the body, most animals have an internal distribution system, or circulation, that transports materials from one place to another.

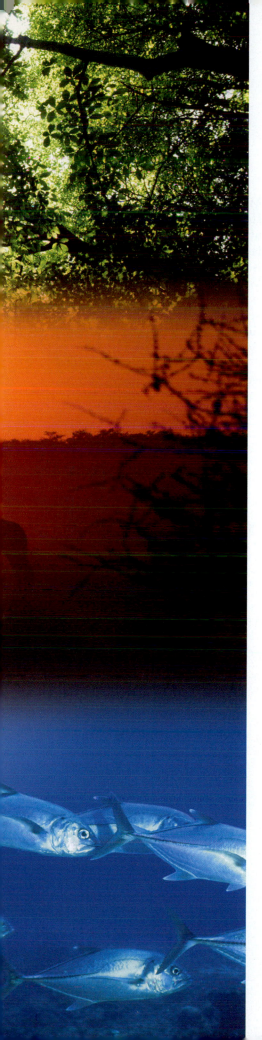

Characteristics of Living Organisms

Characteristic	Importance	Notes
Responsiveness	Indicates that the organism recognizes changes in its internal or external environment	Required for adaptability
Adaptability	Changes the organism's behavior, capabilities, or structure	Required for survival in a constantly changing world
Growth and reproduction	Indicate that the organism is successful; growth must occur before reproduction	Organisms that do not respond, or do not adapt, will not grow and cannot reproduce
Movement	Distributes materials throughout large organisms; changes orientation or position of a plant or immobile animal; moves mobile animals around the environment (locomotion)	Animals show locomotion at some point in their lives
Respiration*	Usually refers to the absorption and utilization of oxygen, and the generation and release of carbon dioxide	Oxygen is required for chemical processes that release energy in a usable form; carbon dioxide is released as a waste product
Circulation*	Movement of fluid within the organism; may involve a pump and a network of special vessels	The circulation provides an internal distribution network
Digestion*	The chemical breakdown of complex materials for absorption and use by the organism	The chemicals released can be used to generate energy or to support growth
Excretion*	The elimination of chemical waste products generated by the organism	The waste products are often toxic, so their removal is essential

* The mechanics of the process depends on the size and complexity of the organism.

In the next 26 chapters we will be considering the mechanics of each of these vital processes. Although we will be examining the functions of the human body, the basic concepts have broad application.

Module 1.1 Review

a. Define biology.

b. List the basic functions shared by all living things.

c. Explain why most animals have an internal circulation system that transports materials from place to place.

Anatomy is the study of form . . .

Anatomy, which means "a cutting open," is the study of internal and external structures of the body and the physical relationships among body parts. Here is an overview of the anatomy of the heart, with the walls opened so that you can see the complexity of its internal structure.

1 **Gross anatomy**, or **macroscopic anatomy**, involves the examination of relatively large structures and features usually visible with the unaided eye. This illustration of a dissected heart is an example of gross anatomy.

2 **Microscopic anatomy** deals with structures that cannot be seen without magnification, and thus the boundaries of microscopic anatomy are established by the limits of the equipment used. With a dissecting microscope, you can see tissue structure; with a light microscope, you can see basic details of cell structure; with an electron microscope, you can see individual molecules that are only a few nanometers (billionths of a meter) across.

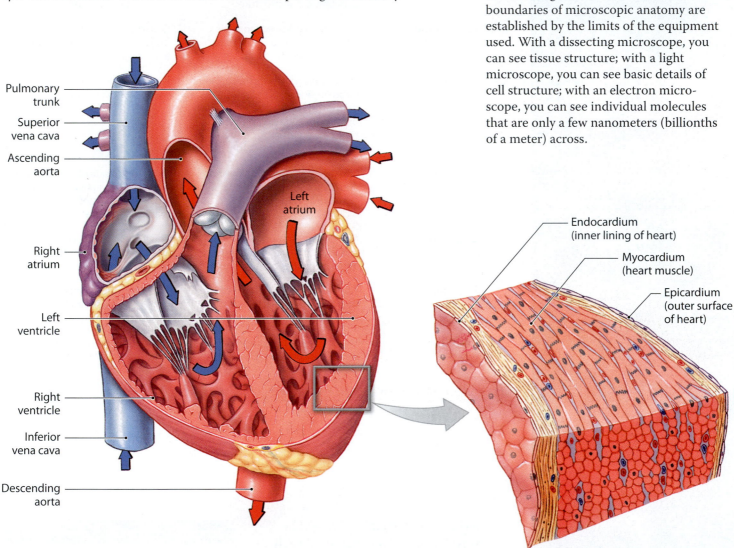

Pulmonary trunk
Superior vena cava
Ascending aorta
Left atrium
Right atrium
Left ventricle
Right ventricle
Inferior vena cava
Descending aorta

Endocardium (inner lining of heart)
Myocardium (heart muscle)
Epicardium (outer surface of heart)

All specific functions are performed by specific structures. The link between structure and function is always present, but not always understood. For example, although the anatomy of the heart was clearly described in the 15th century, almost 200 years passed before the heart's pumping action was demonstrated.

... physiology is the study of function

Physiology is the study of function, and considers the functions of the human body. These functions are complex and much more difficult to examine than most anatomical structures. A physiologist looking at the heart focuses on its functional properties, such as the timing and sequence of the heartbeat, and its effects on blood pressure in the major arteries.

3 The heartbeat is coordinated by electrical events within the heart muscle. Those electrical events can be detected by monitoring electrodes placed on the body surface. A record of these electrical events is called an electrocardiogram, or ECG.

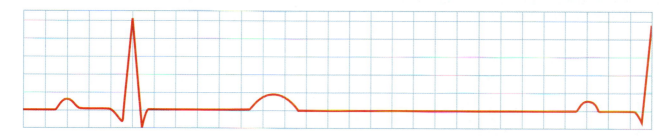

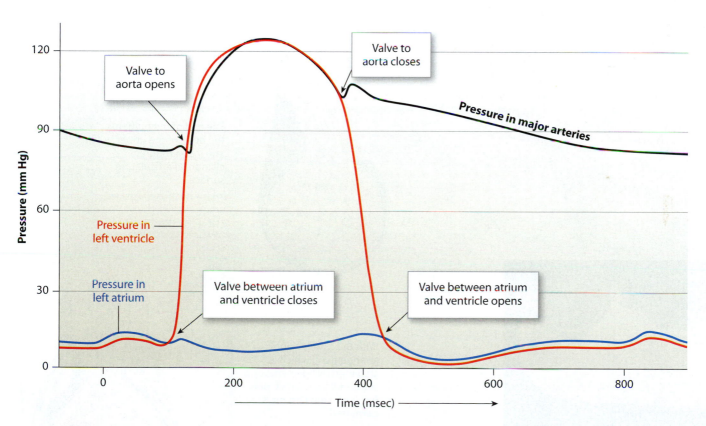

Valve to aorta opens

Valve to aorta closes

Pressure in major arteries

Pressure in left ventricle

Pressure in left atrium

Valve between atrium and ventricle closes

Valve between atrium and ventricle opens

Pressure (mm Hg)

Time (msec)

4 As the heart beats, pressure rises and falls within the major arteries and the chambers of the heart. Blood pressure in the major arteries must be maintained within normal limits to prevent vessel damage (from high pressures) or vessel collapse (from low pressures).

Module 1.2 Review

a. Define anatomy and physiology.

b. What are the differences between gross anatomy and microscopic anatomy?

c. Explain the link between anatomy and physiology.

Form and function are interrelated

Physiology and anatomy are closely interrelated both theoretically and practically, for anatomical details are significant only because each has an effect on function, and physiological mechanisms can be fully understood only in terms of the underlying structural relationships.

1 This relationship is easily understood at the gross anatomical level. You are well aware that your elbow joint functions like a hinge. It lets your forearm move toward or away from your shoulder, but it does not allow twisting at the joint. These functional limits are imposed by the internal structure of the joint.

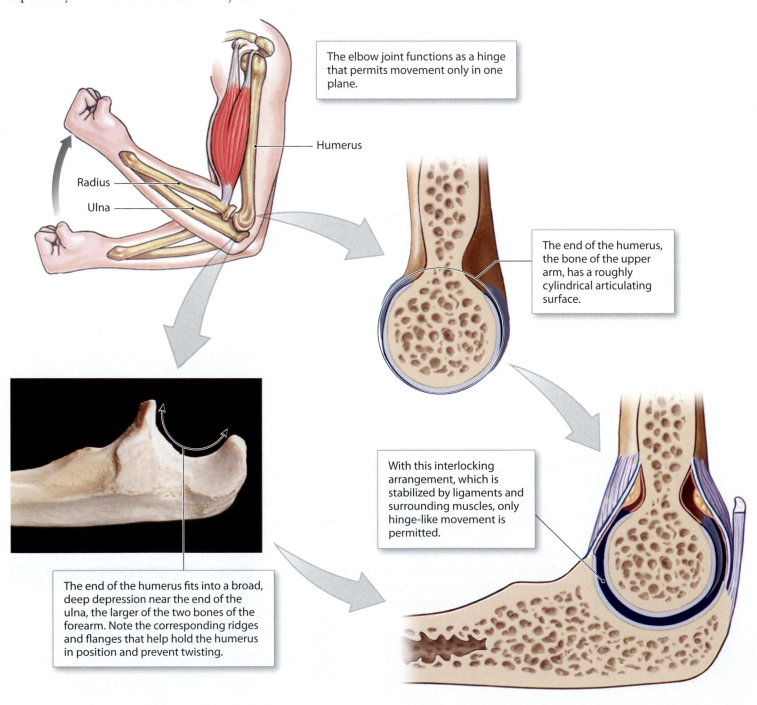

The elbow joint functions as a hinge that permits movement only in one plane.

Humerus

Radius

Ulna

The end of the humerus, the bone of the upper arm, has a roughly cylindrical articulating surface.

With this interlocking arrangement, which is stabilized by ligaments and surrounding muscles, only hinge-like movement is permitted.

The end of the humerus fits into a broad, deep depression near the end of the ulna, the larger of the two bones of the forearm. Note the corresponding ridges and flanges that help hold the humerus in position and prevent twisting.

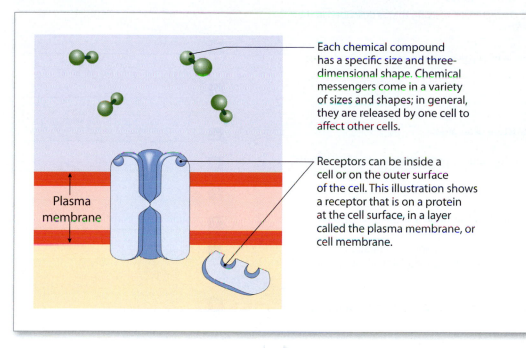

Each chemical compound has a specific size and three-dimensional shape. Chemical messengers come in a variety of sizes and shapes; in general, they are released by one cell to affect other cells.

Receptors can be inside a cell or on the outer surface of the cell. This illustration shows a receptor that is on a protein at the cell surface, in a layer called the plasma membrane, or cell membrane.

Plasma membrane

2 The relationship between form and function also applies at the microscopic level of detail. Cells throughout your body communicate with one another through the use of chemical messengers that you will learn more about in later chapters. The detection of and response to these messengers usually involves the attachment of the chemical messenger released by one cell to a receptor protein at another cell. That attachment depends in large part on the three-dimensional shapes of the messenger and the receptor, and how well they fit together.

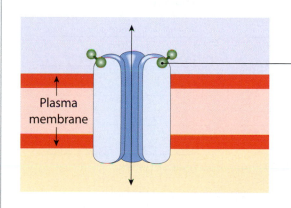

Plasma membrane

Chemical messengers are detected when they attach, or bind, to a receptor that has the proper shape. Binding creates a new structure—messenger and receptor—and the entire complex often changes shape as a result. This can change the function of the receptor. In this case, the binding of the messenger leads to the opening of a passageway through the plasma membrane.

It's important to realize that no mysterious forces are involved in the workings of the body. Although our knowledge is incomplete, it is quite clear that living systems are subject to the same laws of physics and chemistry as buildings, oceans, and mountain ranges. In fact, many advances in our understanding of the human body came only after advances in one of the physical or applied sciences. For example, the action and purpose of the heart valves remained a mystery until the 1600s, when pumps containing valves were developed to remove the water from flooded coal mines. An English physician, William Harvey, was then astute enough to demonstrate that those design principles explained the function of the heart and the circulation of the blood.

Module 1.3 Review

a. Describe how structure and function are interrelated.

b. Compare the functioning of the elbow joint with a door on a hinge.

c. Predict what would happen to the function of a structure if its anatomy were altered.

1. Vocabulary

For each of the following descriptions, write the characteristic of living things described in the corresponding blank.

a   Usually refers to the absorption and utilization of oxygen and the generation and release of carbon dioxide

b   Indications that an organism is successful

c   Changes in the behavior, capabilities, or structure of an organism

d   Movement of fluid within the body; may involve a pump and a network of special vessels

e   The elimination of chemical waste products generated by the body

f   The chemical breakdown of complex structures for absorption and use by the body

g   Transports materials around the body of a large organism; changes orientation or position of a plant or immobile animal; moves mobile animals around the environment (locomotion)

h   Indicates that the organism recognizes changes in the internal or external environment

a   _____

b   _____

c   _____

d   _____

e   _____

f   _____

g   _____

h   _____

2. Matching

Write each of the following terms under the proper heading.

- Right atrium
- Myocardium
- Valve to aorta opens
- Left ventricle
- Valve between left atrium and left ventricle closes
- Pressure in left atrium
- Electrocardiogram
- Endocardium
- Superior vena cava

Anatomy	Physiology
_____	_____
_____	_____
_____	_____
_____	_____
_____	_____

3. Short answer

Briefly describe how the relationship of form and function of a house key and its front door lock are both similar to and different from a chemical messenger and its receptor protein.

_____

_____

_____

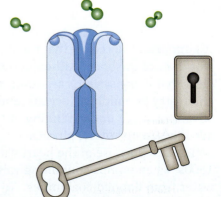

4. Section integration

How might a large organism's survival be affected by an inadequate internal circulation network?

_____

_____

Levels of Organization

You have probably either played with Legos or helped a child build something from Legos at one time or another. It is quite astonishing what you can create using a Lego set that consists of only a half-dozen different types of blocks—buildings, airplanes, pirate ships, and spacecraft can be built following reasonably simple directions. The individual blocks snap together to create larger structures that in turn combine to form complex shapes. As you work from creating simple structures to combined structures to larger forms, you are working your way through successive levels of organization. Each level is more complex than the previous one, but all can be broken down into similar components. The human body is far more complex than a Lego ship, but the same basic principle applies to its structure: The apparent complexity represents multiple levels of organization. Each level is more complex than the previous one, but all can be broken down into similar components.

The Organism Level. An **organism**—in this case, a human—is the highest level of organization. All organ systems of the body must work together to maintain life and health.

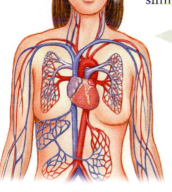

The Organ System Level. (Chapters 5–26) Organs interact in organ systems. Each time it contracts, the heart pushes blood into a network of blood vessels. Together, the heart, blood, and blood vessels form the cardiovascular system, one of 11 **organ systems** in the body.

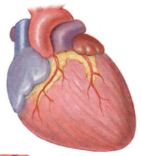

The Organ Level. An **organ** consists of two or more tissues working in combination to perform several functions. Layers of cardiac muscle tissue, in combination with connective tissue, another type of tissue, form the bulk of the wall of the heart, a hollow, three-dimensional organ.

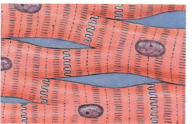

The Tissue Level. (Chapter 4) A **tissue** is a group of cells working together to perform one or more specific functions. Heart muscle cells, or cardiac muscle cells (*cardium*, heart), interact with other types of cells and with extracellular materials to form cardiac muscle tissue.

The Cellular Level. (Chapter 3) **Cells** are the smallest living units in the body. Their functions depend on organelles composed of interacting molecules. Interactions among protein filaments, for example, produce the contractions of muscle cells in the heart.

  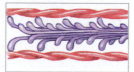

Atoms in combination   Complex protein molecule   Protein filaments

The Chemical (or Molecular) Level. (Chapter 2) **Atoms**, the smallest stable units of matter, can combine to form molecules with complex shapes. The functional properties of a particular molecule are determined by its unique three-dimensional shape and atomic components.

Cells are the smallest units of life

Free-living **cells** are the smallest independent organisms, with all of the characteristics described earlier in the chapter. Most of the plants and animals you are familiar with are multicellular—they consist of thousands to billions of cells. These cells do not exist as independent entities, but instead they work together, each with its own characteristics and functions. Cells are the living building blocks of our bodies. There are literally trillions of cells in your body, but there are only an estimated 200 different types of cells. Nevertheless, those 200 types show remarkable diversity in appearance and function.

1 The human body contains roughly 200 different cell types. The dimensions of cells are usually given in terms of micrometers (μm). One micrometer is one-millionth of a meter, or approximately 1/25,000th of an inch. All the cells illustrated here are shown with the dimensions they would have if they were magnified about 1500 times.

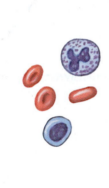

Smooth muscle cells

Smooth muscle cells, found in many organs, are long and slender. The skeletal muscle cells that give you the ability to move around are too large to illustrate here; at this magnification a large skeletal muscle cell would have the diameter of a small dinner plate and be over 300 m long.

Blood cells

Blood cells are either flattened discs (red blood cells) or roughly spherical (white blood cells). Red blood cells—the most abundant cells in the body—transport oxygen and carbon dioxide in the bloodstream. White blood cells are responsible for fighting off infection and combating disease.

Bone cells

Bone cells reside within small cavities inside the mass of a bone. These cells are responsible for the mainte-nance of the bone and for recycling the calcium and phosphate stored there.

Fat cells

Fat cells are roughly spherical storage containers. Whenever we take in more energy than we expend, the excess energy obtained from the food gets stored as fat, and these cells get larger and more numerous.

2 The importance of cells is apparent in the **cell theory**, one of the foundations of modern biology.

Basic Principles of the Cell Theory

- Cells are the structural building blocks of all plants and animals.
- Cells produced by the divisions of pre-existing cells.
- Cells are the smallest structural units that perform all vital functions.

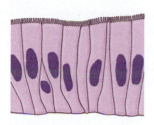

Cells lining the digestive tract

Cells lining the digestive tract are relatively
delicate. The nutrients, vitamins, minerals,
and water we need are absorbed by these
cells.

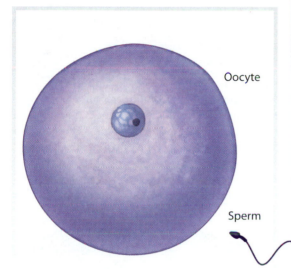

Oocyte

Sperm

Reproductive cells (sex cells)

Cells involved in sexual reproduction are
called sex cells. Women produce relatively
large oocytes in very small numbers, usually
at monthly intervals. Males continuously
produce relatively tiny sperm in enormous
numbers.

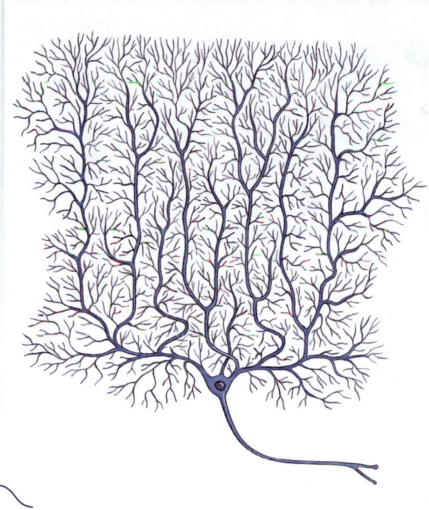

Nerve cells (neurons)

Neurons, or nerve cells, are the equivalent of computer chips—they process
information. Thought, memory, consciousness, and muscle control are all
based on the actions of, and interactions among, neurons. There are many
different types and shapes of neurons. This is a neuron from a part of the
brain involved with the control of balance and movement. The extensive
branching provides a huge surface area for communicating with other
neurons.

The cells of the body work together, and our lives ultimately depend on
their actions. If they don't do the right thing at the right time, we're in
trouble; if our cells can't survive, we're doomed. Yet each individual cell
remains unaware of its role in the "big picture"—it simply responds
and adapts to changes in its local environment. How the responses of
cells in different parts of the body are coordinated and controlled is
obviously a key question, and we will spend considerable time in later
chapters considering the answers.

Module 1.4 Review

a. List the three basic principles of the cell
theory.

b. Name and define the unit used to
measure cell size.

c. Relate the functions of a fat cell and a
neuron to their shapes.

Tissues are specialized groups of cells and cell products

The roughly 200 different cell types in the body combine to form tissues, collections of cells and cell products that perform specific functions. **Histology** (*histos*, tissue) is the study of tissues. This module will introduce the four **primary tissue types** that, in various combinations, form the tissues of the body: epithelial tissue, connective tissue, muscle tissue, and neural tissue.

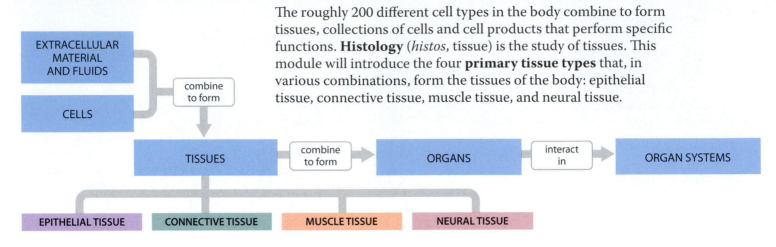

1 The most common type of **epithelial** (ep-i-THĒ-lē-ul) **tissue** is a layer of cells that forms a barrier with specific properties. Epithelia cover every exposed body surface; line the digestive, respiratory, reproductive, and urinary tracts; surround internal cavities such as the chest cavity or the fluid-filled chambers in the brain, eye, and inner ear; and line the inner surfaces of the blood vessels and heart.

2 **Connective tissue** is quite diverse in appearance. All forms of connective tissue contain cells and an extracellular matrix that consists of protein fibers and a liquid known as the ground substance. The amount and consistency of the matrix depends on the particular type of connective tissue. In blood, the cells are suspended in a watery matrix called plasma. Bone has a more durable matrix, with crystals of calcium salts organized around a fibrous framework, and very little ground substance.

EPITHELIAL TISSUE

- Covers and protects exposed surfaces
- Lines internal passageways and chambers
- Produces glandular secretions

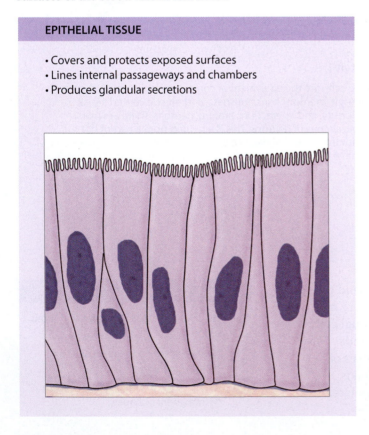

CONNECTIVE TISSUE

- Fills internal spaces
- Provides structural support
- Stores energy

Matrix

Fibers   Ground substance

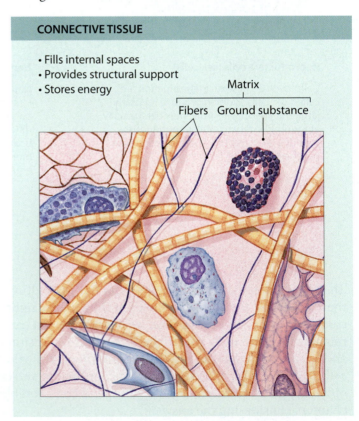

3 **Muscle tissue** is unique because individual muscle cells have the ability to contract forcefully. Major functions of muscle tissue include skeletal movement, soft tissue support, maintenance of blood flow, movement of materials along internal passageways, and the stabilization of normal body temperature. There are three different types of muscle tissue.

MUSCLE TISSUE

• Contracts to produce active movement

Nuclei

Skeletal muscle tissue is usually attached to the skeleton, directly or indirectly, and its contractions move or stabilize the position of bones or internal organs.

Nucleus   Muscle cell

Cardiac muscle tissue is found only in the heart, where its coordinated contractions propel blood through the blood vessels.

Smooth muscle tissue can be found in the walls of blood vessels, within glands, and along the respiratory, circulatory, digestive, and reproductive tracts.

4 **Neural tissue** is specialized to carry information or instructions from one place in the body to another. Two basic types of cells are present: nerve cells, or **neurons** (NOO-rons; *neuro*, nerve), and supporting cells, or **neuroglia** (noo-RŌG-lē-a; *glia*, glue). Neurons transmit information in the form of electrical impulses. Neuroglia isolate and protect neurons while forming a supporting framework. The neural tissue in the body can be divided on anatomical grounds into the **central nervous system**, or brain and spinal cord, and the **peripheral nervous system**, which includes the nerves connecting the central nervous system with other tissues and organs.

NEURAL TISSUE

• Conducts electrical impulses
• Carries information

Neurons   Neuroglia

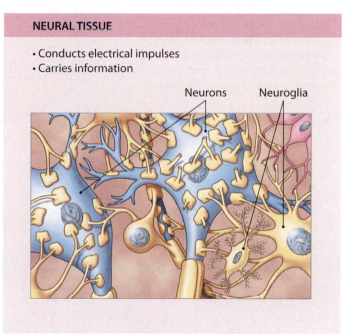

Module 1.5 Review

a. Define histology.

b. Identify the four primary tissue types.

c. Explain the functions of each of the primary tissue types.

Organs and organ systems perform vital functions

1 An **organ** is a functional unit composed of more than one tissue type. The particular combination and organization of tissues within an organ both determines and limits the organ's functions. The three-dimensional physical relationships are significant; a flat piece of cardboard can protect a table from scratching, but if you fold it into a box you can pack things inside it. An **organ system** consists of organs that interact to perform a specific range of functions, often in a coordinated fashion.

Organism level

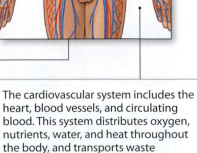

The heart, for example, is an organ that contains cardiac muscle tissue, epithelia, connective tissues, and neural tissue. The interconnections between cardiac muscle cells ensure that the contractions are coordinated, producing a heartbeat; the neural tissue adjusts the heart rate. When the heart beats, the internal anatomy of the heart, largely composed of cardiac muscle tissue and connective tissue, enables it to function as a pump.

Organ level

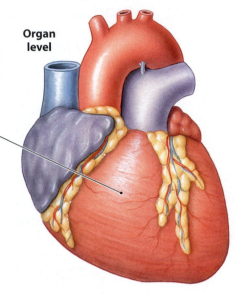

Cardiovascular System

Endocrine

Nervous

Muscular

Organ system level

Skeletal

Integumentary

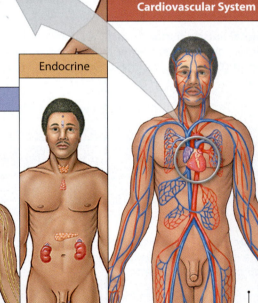

The cardiovascular system includes the heart, blood vessels, and circulating blood. This system distributes oxygen, nutrients, water, and heat throughout the body, and transports waste products to sites where they can be excreted.

2 The table at right lists the 11 organ systems in the human body. Although this categorization is a convenient way to organize information, the concept of separate "organ systems" is artificial and somewhat misleading. Nothing in the body functions in isolation—not cells, not tissues, not organs, and certainly not organ systems. Organs and organ systems are interdependent, and something that affects one organ will affect the functioning of the body as a whole. For example, the heart cannot pump blood effectively after massive blood loss. If the heart cannot pump and blood cannot flow, oxygen and nutrients cannot be distributed. Very soon, cardiac muscle tissue begins to break down as individual muscle cells die from oxygen and nutrient starvation. These changes will not be restricted to the cardiovascular system; all cells, tissues, and organs in the body will be damaged, with potentially fatal results.

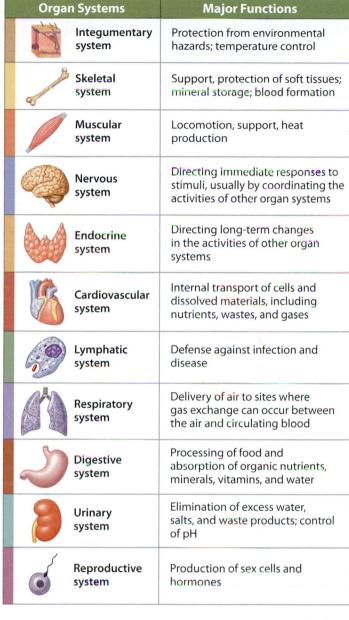

Organ Systems		Major Functions
	Integumentary system	Protection from environmental hazards; temperature control
	Skeletal system	Support, protection of soft tissues; mineral storage; blood formation
	Muscular system	Locomotion, support, heat production
	Nervous system	Directing immediate responses to stimuli, usually by coordinating the activities of other organ systems
	Endocrine system	Directing long-term changes in the activities of other organ systems
	Cardiovascular system	Internal transport of cells and dissolved materials, including nutrients, wastes, and gases
	Lymphatic system	Defense against infection and disease
	Respiratory system	Delivery of air to sites where gas exchange can occur between the air and circulating blood
	Digestive system	Processing of food and absorption of organic nutrients, minerals, vitamins, and water
	Urinary system	Elimination of excess water, salts, and waste products; control of pH
	Reproductive system	Production of sex cells and hormones

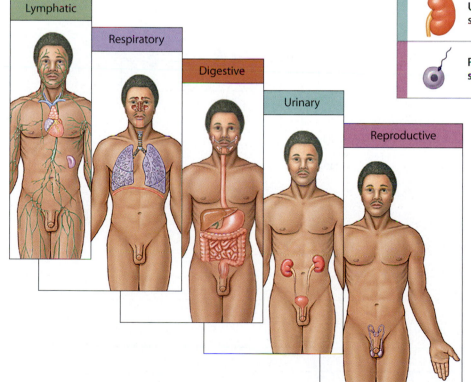

Lymphatic

Respiratory

Digestive

Urinary

Reproductive

Module 1.6 Review

a. List the 11 major organ systems of the body.

b. Explain the relationship between the skeletal system and the digestive system.

c. Using the table as a reference, describe how falling down a flight of stairs could affect at least six of the organ systems.

1. Short answer

For five different organ systems in the human body, identify a specialized cell type found in that system.

_____

_____

_____

_____

_____

2. Concept map

Use each of the following terms once to fill in the blank boxes to correctly complete the extracellular materials and fluids concept map.

- organs
- epithelial tissue
- cells
- connective tissue
- muscle tissue
- nervous tissue
- organ systems
- external and internal surfaces
- matrix
- glandular secretions
- bones of the skeleton
- neuroglia
- blood
- materials within digestive tract
- protein fibers
- ground substance
- movement

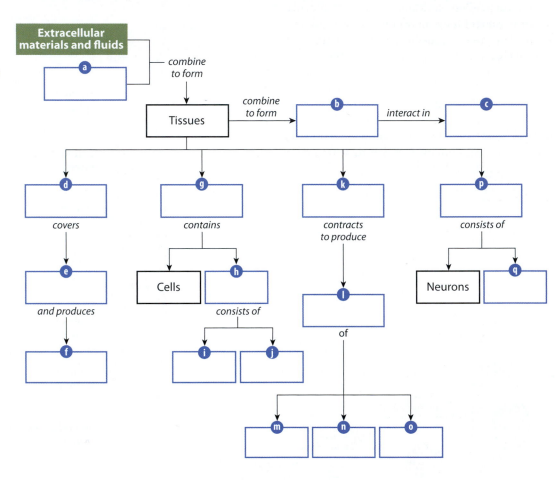

3. Matching

Order these six levels of organization of the human body from smallest (a) to largest (f).

_____ tissue   _____ cell   _____ organ   _____ molecule   _____ organism   _____ organ system

4. Short answer

Summarize the major functions of each of the following organ systems.

a skeletal system   _____

b digestive system   _____

c integumentary system   _____

d urinary system   _____

e nervous system   _____

Homeostasis

Homeostasis (*homeo*, unchanging + *stasis*, standing) is the presence of a stable internal environment. Maintaining homeostasis is absolutely vital to an organism's survival; failure to maintain homeostasis soon leads to illness or even death. The principle of homeostasis is the central theme of this text and the foundation of all modern physiology. **Homeostatic regulation** is the adjustment of physiological systems to preserve homeostasis in environments that are often inconsistent, unpredictable, and potentially dangerous. An understanding of homeostatic regulation is crucial to making accurate predictions about the body's responses to both normal and abnormal conditions.

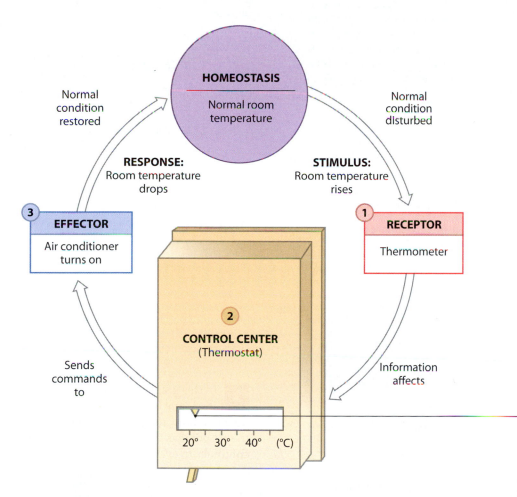

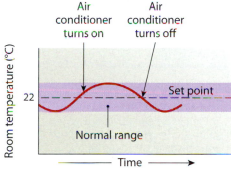

2 Homeostatic control is not precise—it maintains a normal range rather than an absolute value. The same is true for controlling room temperature—a house may have the thermostat on one wall of one room, and the air conditioning outlets at multiple locations. Over time, the temperature in the house will oscillate around the set point.

The setting on a thermostat establishes the **set point**, or desired value, which in this case is the temperature you select. (In our example, the set point is 22°C, or about 72°F.) The function of the thermostat is to keep room temperature within acceptable limits, usually within a degree or so of the set point.

1 The maintenance of a relatively constant temperature in your living space is a familiar example of homeostasis. Like all homeostatic mechanisms, it consists of (1) a **receptor** or sensor—in this case, a thermometer—that is sensitive to a particular environmental change, or stimulus; (2) a **control center** or integration center—in this case, a thermostat—which receives and processes the information supplied by the receptor, and which sends out commands; and (3) an **effector**—in this case, an air conditioner—which responds to these commands by opposing the stimulus. The net effect is that any variation outside normal limits triggers a response that restores normal conditions.

Negative feedback provides stability . . .

Feedback occurs when receptor stimulation triggers a response that changes the environment at the receptor. In the case of temperature control by a thermostat, temperature variation outside the desired range triggers an automatic response that corrects the situation. This method of homeostatic regulation is called **negative feedback**, because an effector activated by the control center opposes, or negates, the original stimulus. Negative feedback thus tends to minimize change, keeping variation in key body systems within limits compatible with our long-term survival.

Start

HOMEOSTASIS

At normal body temperature (set point: 37°C or 98.6°F), the temperature control center is relatively inactive; superficial blood flow and sweat gland activity are at normal levels.

Homeostasis restored

Homeostasis disturbed

Homeostasis and body temperature

3

EFFECTORS

Increased activity in the control center targets two effectors: (1) smooth muscle in the walls of blood vessels supplying the skin and (2) sweat glands. The smooth muscle relaxes and the blood vessels dilate, increasing blood flow through vessels near the body surface; the sweat glands accelerate their secretion. The skin then acts like a radiator by losing heat to the environment, and the evaporation of sweat speeds the process.

1

RECEPTORS

If body temperature rises above 37.2° C (99° F), two sets of temperature receptors, one in the skin and the other within the brain, send signals to the homeostatic control center.

2

CONTROL CENTER

The temperature control center receives information from the two sets of temperature receptors and sends commands to the effectors.

2 In this graph of body temperature over time in a warm environment, note that body temperature declines past the set point as the sweat already secreted continues to evaporate.

1 Negative feedback is the primary mechanism of homeostatic regulation, and it provides long-term control over the body's internal conditions and systems. Homeostatic mechanisms using negative feedback normally ignore minor variations, and they maintain a normal range rather than a fixed value. The regulatory process itself is dynamic, because the set point may vary with changing environments or differing activity levels. For example, when you are asleep, your thermoregulatory set point is lower, whereas when you work outside on a hot day (or when you have a fever), it is higher. Thus, body temperature can vary from moment to moment or from day to day for any individual, due to either (1) small oscillations around the set point or (2) changes in the set point. Comparable variations occur in all other aspects of physiology.

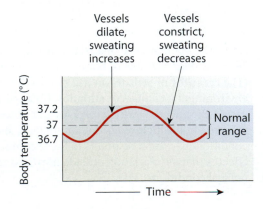

Vessels dilate, sweating increases

Vessels constrict, sweating decreases

Body temperature (°C)

37.2
37
36.7

Normal range

Time

...whereas positive feedback accelerates a process to completion

In **positive feedback**, an initial stimulus produces a response that exaggerates or enhances the change in the original conditions, rather than opposing it. You seldom encounter positive feedback in your daily life, simply because it tends to produce extreme responses. For example, suppose that the thermostat in your house was accidentally connected to a heater rather than to an air conditioner. Now, when room temperature exceeds the set point, the thermostat turns on the heater, causing a further rise in room temperature. Room temperature will continue to increase until someone switches off the thermostat, turns off the heater, or intervenes in some other way. This kind of escalating cycle is often called a **positive feedback loop**.

A break in a blood vessel wall causes bleeding

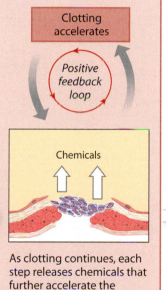

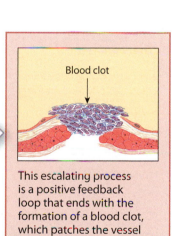

Clotting accelerates

Positive feedback loop

Chemicals

Blood clot

Chemicals

Damage to cells in the blood vessel wall releases chemicals that begin the process of blood clotting.	The chemicals start chain reactions in which cells, cell fragments, and soluble proteins in the blood begin to form a clot.	As clotting continues, each step releases chemicals that further accelerate the process.	This escalating process is a positive feedback loop that ends with the formation of a blood clot, which patches the vessel wall and stops the bleeding.

3 In the body, positive feedback loops are typically found when a potentially dangerous or stressful process must be completed quickly before homeostasis can be restored. For example, the immediate danger from a severe cut is loss of blood, which can lower blood pressure and reduce the efficiency of the heart.

Module 1.7 Review

a. Identify the components of homeostatic regulation.

b. Explain the function of negative feedback systems.

c. Why is positive feedback helpful in blood clotting but unsuitable for the regulation of body temperature?

1. Vocabulary

Write the term for each of the following descriptions in the space provided.

a _____ Mechanism that increases a deviation from normal limits after an initial stimulus

b _____ Adjustment of physiological systems to preserve homeostasis

c _____ The maintenance of a relatively constant internal environment

d _____ Concentration of hormones circulating in the blood

e _____ Corrective mechanism that opposes or cancels a variation from normal limits

2. Matching

Indicate whether each of the following processes matches with the process of negative feedback or positive feedback.

a A rise in the level of calcium dissolved in the blood stimulates the release of a hormone that causes bone cells to deposit more of the calcium in bone. _____

b Labor contractions become increasingly forceful during childbirth. _____

c An increase in blood pressure triggers a nervous system response that results in lowering the blood pressure. _____

d Blood vessel cells damaged by a break in the vessel release chemicals that accelerate the blood clotting process. _____

3. Short answer

Assuming a normal body temperature range of 36.7°–37.2° C (98°–99° F), identify from the graph below what would happen if there were an increase or decrease in body temperature beyond the normal limits. Use the following descriptive terms to explain what would happen at (a) and (b) on the graph.

- body surface cools
- shivering occurs
- sweating increases
- temperature declines
- body heat is conserved
- blood flow to skin increases
- blood flow to skin decreases
- temperature rises

37.8° C /100° F
36.7°–37.2° C /98°–99° F] Normal range
36.1° C /97° F

a _____

_____

_____

_____

b _____

_____

_____

_____

4. Section integration

It is a warm day and you feel a little chilled. On checking your temperature, you find that your body temperature is 1.5 degrees below normal. Suggest some possible reasons for this situation.

_____

_____

Anatomical Terms

Early anatomists created maps of the human body, and we still rely on maps for orientation. The landmarks are prominent anatomical structures; distances are measured in centimeters or inches; and specialized directional terms are used. In effect, anatomy uses a special language that must be learned almost at the start. Many terms are based on Latin or Greek words used by ancient anatomists. However, Latin and Greek terms are not the only ones that have been imported into the anatomical vocabulary over the centuries, and the vocabulary continues to expand. Many anatomical structures and clinical conditions were initially named after either the discoverer or, in the case of diseases, the most famous victim. Most of these commemorative names, or eponyms, have been replaced by more precise terms, but a few persist. The table below summarizes key word roots and derivatives you will want to learn before proceeding in this course.

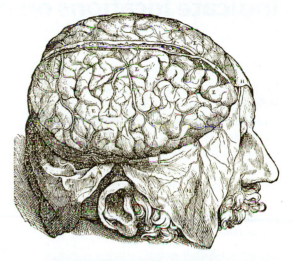

Important Word Roots, Prefixes, Suffixes, and Combining Forms in Anatomy

a-, *a-,* without: avascular	**ex-,** *ex,* out of, away from: exocytosis	**ost-, oste-, osteo-,** *osteon,* bone: osteal, ostealgia, osteocyte
aer-, *aeros,* air: aerobic metabolism	**hemo-,** *haima,* blood: hemopoiesis	**oto-,** *otikos,* ear: otolith
-algia, *algos,* pain: neuralgia	**hemi-,** *hemi,* one-half: hemisphere	**path-, -pathy, patho-,** *pathos,* disease: pathergy, idiopathy, pathogenesis
arter-, *arteria,* artery: arterial	**histo-,** *histos,* tissue: histology	**peri-,** *peri,* around: perineurium
arthro-, *arthros,* joint: arthroscopy	**homo-,** *homos,* same: homozygous	**phago-,** *phago,* to eat: phagocyte
auto-, *auto,* self: autonomic	**hyper-,** *hyper,* above: hyperpolarization	**-phil, -philia,** *philo,* love: neutrophil, hemophilia
bio-, *bios,* life: biology	**hypo-,** *hypo,* under: hypothyroid	**-phot, -photo,** *phos,* light: photalgia, photoreceptor
-blast, *blastos,* germ: osteoblast	**inter-,** *inter,* between: interventricular	**physio-,** *physis,* nature: physiology
bronch-, *bronchus,* windpipe, airway: bronchial	**iso-,** *isos,* equal: isotonic	**pre-,** *prae,* before: precapillary sphincter
cardi-, cardio-, -cardia, *kardia,* heart: cardiac, cardiopulmonary	**leuk-, leuko-,** *leukos,* white: leukemia, leukocyte	**pulmo-,** *pulmo,* lung: pulmonary
cerebr-, *cerebrum,* brain: cerebral hemispheres	**lyso-, -lysis, -lyze,** *lysis,* a loosening: hydrolysis	**retro-,** *retro,* backward: retroperitoneal
cervic-, *cervicis,* neck: cervical vertebrae	**meso-,** *mesos,* middle: mesoderm	**sarco-,** *sarkos,* flesh: sarcomere
chondro-, *chondros,* cartilage: chondrocyte	**micr-,** *mikros,* small: microscope	**scler-, sclero-,** *skleros,* hard: sclera, sclerosis
cranio-, *cranium,* skull: craniosacral	**morph-, morpho-,** *morphe,* form: morphology, morphotype	**-scope,** *skopeo,* to view: colonoscope
cyt-, cyto-, *kytos,* a hollow cell: cytology, cytokine	**myo-,** *mys,* muscle: myofilament	**sub-,** *sub,* below: subcutaneous
derm-, *derma,* skin: dermatome	**nephr-,** *nephros,* kidney: nephron	**super-,** *super,* above or beyond: superficial
-ectomy, *ektome,* excision: appendectomy	**neur-, neuri-, neuro-,** *neuron,* nerve: neural, neurilemma, neuromuscular	**-trophy,** *trophe,* nourishment: atrophy
end-, endo-, *endon,* within: endergonic, endometrium	**-ology,** *logos,* the study of: physiology	**vas -,** *vas,* vessel: vascular
epi-, *epi,* on: epimysium	**-osis, -osis,** state, condition: neurosis	

Superficial anatomy and regional anatomy indicate locations on or in the body

1 This illustration shows the body in the **anatomical position**. In this position, the hands are at the sides with the palms facing forward, and the feet are together. Unless otherwise noted, all descriptions in this text refer to the body in the anatomical position. A person lying down in the anatomical position is said to be **supine** (soo-PĪN) when face up, and **prone** when face down.

Regions of the Human Body

Structure	Region
Cephalon (head)	Cephalic region
Cervicis (neck)	Cervical region
Thoracis (thorax or chest)	Thoracic region
Brachium (arm)	Brachial region
Antebrachium (forearm)	Antebrachial region
Carpus (wrist)	Carpal region
Manus (hand)	Manual region
Abdomen	Abdominal region
Lumbus (loin)	Lumbar region
Gluteus (buttock)	Gluteal region
Pelvis	Pelvic region
Pubis (anterior pelvis)	Pubic region
Inguen (groin)	Inguinal region
Femur (thigh)	Femoral region
Crus (anterior leg)	Crural region
Sura (calf)	Sural region
Tarsus (ankle)	Tarsal region
Pes (foot)	Pedal region
Planta (sole)	Plantar region

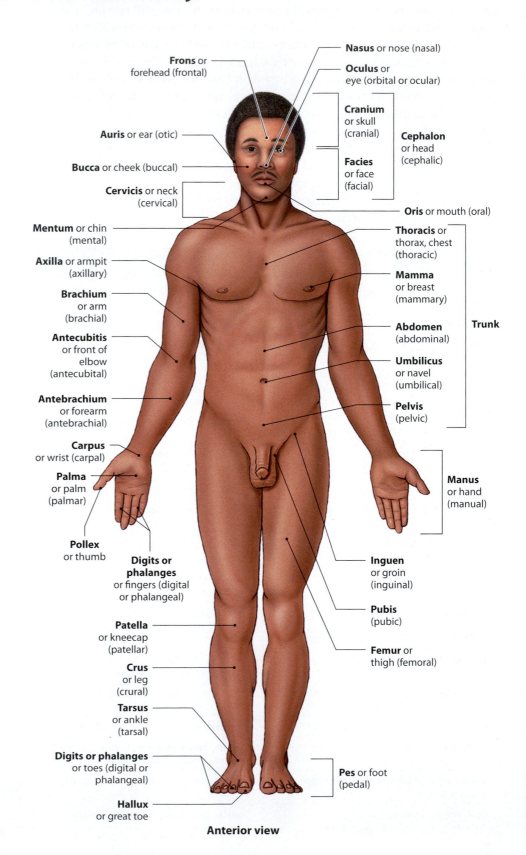

Frons or forehead (frontal)

Nasus or nose (nasal)

Oculus or eye (orbital or ocular)

Auris or ear (otic)

Cranium or skull (cranial)

Cephalon or head (cephalic)

Bucca or cheek (buccal)

Facies or face (facial)

Cervicis or neck (cervical)

Oris or mouth (oral)

Mentum or chin (mental)

Thoracis or thorax, chest (thoracic)

Axilla or armpit (axillary)

Mamma or breast (mammary)

Brachium or arm (brachial)

Abdomen (abdominal)

Antecubitis or front of elbow (antecubital)

Umbilicus or navel (umbilical)

Trunk

Antebrachium or forearm (antebrachial)

Pelvis (pelvic)

Carpus or wrist (carpal)

Palma or palm (palmar)

Manus or hand (manual)

Pollex or thumb

Digits or phalanges or fingers (digital or phalangeal)

Inguen or groin (inguinal)

Pubis (pubic)

Patella or kneecap (patellar)

Femur or thigh (femoral)

Crus or leg (crural)

Tarsus or ankle (tarsal)

Digits or phalanges or toes (digital or phalangeal)

Pes or foot (pedal)

Hallux or great toe

Anterior view

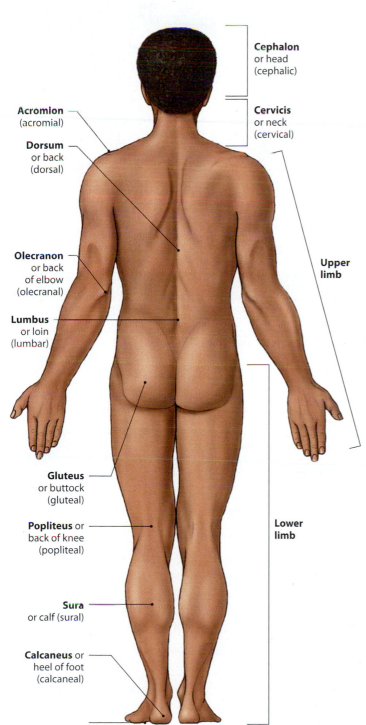

Cephalon or head (cephalic)

Cervicis or neck (cervical)

Acromion (acromial)

Dorsum or back (dorsal)

Olecranon or back of elbow (olecranal)

Lumbus or loin (lumbar)

Gluteus or buttock (gluteal)

Popliteus or back of knee (popliteal)

Upper limb

Lower limb

Sura or calf (sural)

Calcaneus or heel of foot (calcaneal)

Posterior view

2 Clinicians refer to four **abdominopelvic quadrants** formed by a pair of imaginary perpendicular lines that intersect at the umbilicus (navel). This simple method provides useful references for the description of aches, pains, and injuries. The location can help physicians determine the possible cause.

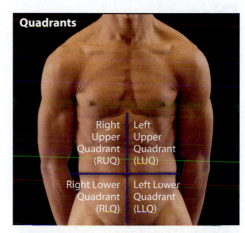

Quadrants

Right Upper Quadrant (RUQ)

Left Upper Quadrant (LUQ)

Right Lower Quadrant (RLQ)

Left Lower Quadrant (LLQ)

3 Anatomists prefer more precise terms to describe the location and orientation of internal organs. Nine **abdominopelvic regions** are recognized by anatomists.

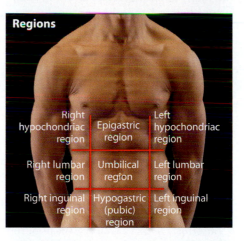

Regions

Right hypochondriac region

Epigastric region

Left hypochondriac region

Right lumbar region

Umbilical region

Left lumbar region

Right inguinal region

Hypogastric (pubic) region

Left inguinal region

4 The image at the lower right shows the relationships among quadrants, regions, and internal organs.

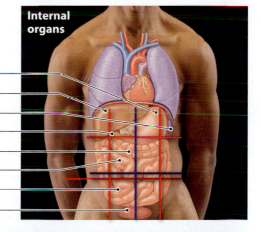

Internal organs

Stomach

Liver

Spleen

Gallbladder

Large intestine

Small intestine

Appendix

Urinary bladder

Module 1.8 Review

a. Describe a person in the anatomical position.

b. Contrast the descriptions used by clinicians and anatomists when referring to the positions of injuries or internal organs of the abdomen and pelvis.

c. A masseuse often begins a massage by asking patrons to lie face down with their arms at their sides. What anatomical term describes that position?

Directional and sectional terms describe specific points of reference

1 The figure and table on this page introduce the principal directional terms and examples of their use. There are many different terms, and some can be used interchangeably. As you learn these directional terms, it is important to remember that all anatomical directions utilize the anatomical position as the standard point of reference.

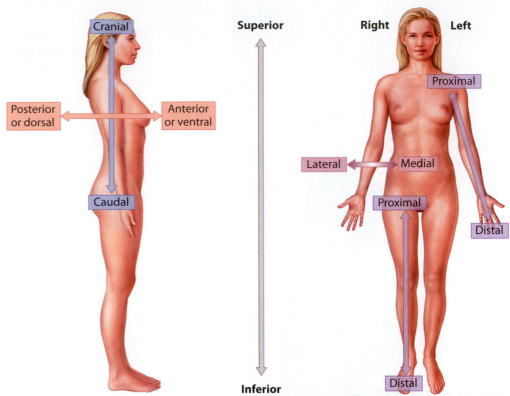

Directional Terms

Term	Region or Reference	Example
Anterior	The front surface	The navel is on the *anterior* surface of the trunk.
Ventral	The belly side (equivalent to anterior when referring to the human body)	The navel is on the *ventral* surface of the trunk.
Posterior or **dorsal**	The back surface	The shoulder blade is located *posterior* to the rib cage.
Cranial or **cephalic**	The head	The *cranial*, or *cephalic*, border of the pelvis is on the side toward the head rather than toward the thigh.
Superior	Above; at a higher level (in the human body, toward the head)	In humans, the cranial border of the pelvis is *superior* to the thigh.
Caudal	The tail (coccyx in humans)	The hips are *caudal* to the waist.
Inferior	Below; at a lower level	The knees are *inferior* to the hips.
Medial	Toward the body's longitudinal axis; toward the midsagittal plane	The *medial* surfaces of the thighs may be in contact; moving medially from the arm across the chest surface brings you to the sternum.
Lateral	Away from the body's longitudinal axis; away from the midsagittal plane	The thigh articulates with the *lateral* surface of the pelvis; moving laterally from the nose brings you to the cheeks.
Proximal	Toward an attached base	The thigh is *proximal* to the foot; moving proximally from the wrist brings you to the elbow.
Distal	Away from an attached base	The fingers are *distal* to the wrist; moving distally from the elbow brings you to the wrist.
Superficial	At, near, or relatively close to the body surface	The skin is *superficial* to underlying structures.
Deep	Farther from the body surface	The bone of the thigh is *deep* to the surrounding skeletal muscles.

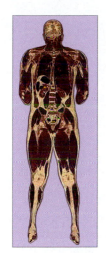

Frontal plane

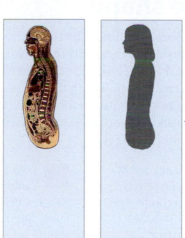

Sagittal plane

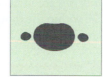

Transverse plane

Terms That Indicate Sectional Planes

Orientation of Plane	Plane	Directional Reference	Description
Perpendicular to long axis	Transverse or horizontal	Transversely or horizontally	A **transverse**, or **horizontal**, **section** separates superior and inferior portions of the body. A cut in this plane is called a **cross section**.
Parallel to long axis	Sagittal	Sagittally	A **sagittal section** separates right and left portions. You examine a sagittal section, but you section sagittally.
Parallel to long axis	Midsagittal	Sagittally	In a **midsagittal section** or **median section**, the plane passes through the midline, dividing the body into right and left halves.
Parallel to long axis	Parasagittal	Sagittally	A **parasagittal section**, which is a cut parallel to the midsagittal plane, separates the body into right and left portions of unequal size.
Parallel to long axis	Frontal or coronal	Frontally or coronally	A **frontal**, or **coronal**, **section** separates anterior and posterior portions of the body; *coronal* usually refers to sections passing through the skull.

2 A presentation in sectional view is sometimes the only way to illustrate the relationships between the parts of a three-dimensional object. An understanding of sectional views and the terminology presented here has become increasingly important since the development of medical imaging techniques.

Module 1.9 Review

a. What is the purpose of directional and sectional terms?

b. In the anatomical position, describe an anterior view and a posterior view.

c. What type of section would separate the two eyes?

Body cavities protect internal organs and allow them to change shape

The interior of the body is often subdivided into regions established by the body wall. For example, everything deep to the chest wall is considered to be within the **thoracic cavity**, and all of the structures deep to the abdominal and pelvic walls are said to lie within the **abdominopelvic cavity**. Many vital internal organs within these regions are suspended within fluid-filled chambers that are true **body cavities** with two essential functions: (1) They protect delicate organs from shocks and impacts; and (2) they permit significant changes in the size and shape of internal organs.

1 The internal organs that are partially or completely enclosed by body cavities are called **viscera** (VIS-e-ruh) or visceral organs. Viscera do not float within the body cavities—they remain connected to the rest of the body. To understand the physical relationships, we will examine the smallest subdivision of the ventral body cavity, the **pericardial cavity** that surrounds the heart.

2 During embryological development, a single **ventral body cavity,** or **coelom** (SĒ-lōm; *koila*, cavity), forms, and it contains organs of the respiratory, cardiovascular, digestive, urinary, and reproductive systems. The ventral body cavity is later subdivided into separate body cavities whose boundaries are indicated in red. Three of these subdivisions lie within the thoracic cavity and one lies in the abdominopelvic cavity.

The relationship between the heart and the pericardial cavity resembles that of a fist pushing into a balloon. The wrist corresponds to the base (attached portion) of the heart, and the balloon corresponds to the lining of the pericardial cavity.

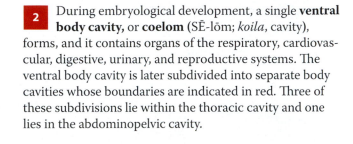

The **pericardium** (*peri-*, around + *cardium*, heart) is a delicate membrane, called a **serous membrane**, lining the pericardial cavity.

During each beat, the heart changes size and shape. The pericardial cavity permits these changes, and the slippery pericardial lining prevents friction between the heart and adjacent structures.

Cardiac muscle of the heart wall

A serous membrane covers the viscera and lines the subdivisions of the ventral body cavity.

A watery fluid coats the walls of these internal cavities and covers the surfaces of the enclosed viscera. It keeps the surfaces moist and reduces friction.

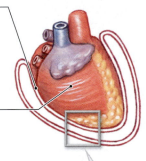

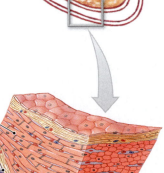

BODY CAVITIES

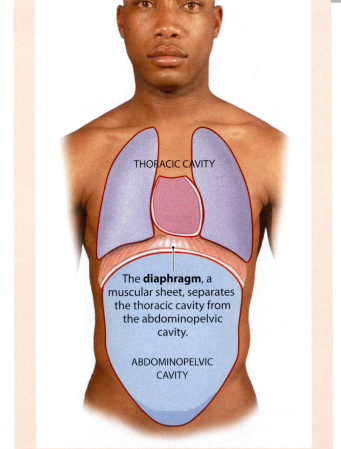

THORACIC CAVITY

The **diaphragm,** a muscular sheet, separates the thoracic cavity from the abdominopelvic cavity.

ABDOMINOPELVIC CAVITY

3 The thoracic cavity contains the lungs, heart, and other structures. Its boundaries are established by the chest wall and diaphragm.

THORACIC CAVITY

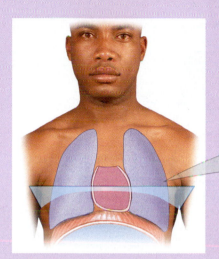

A horizontal section through the thoracic cavity shows the relationship between the subdivisions of the ventral body cavity in this region.

Each lung is enclosed within a **pleural cavity**, lined by a shiny, slippery serous membrane called the **pleura** (PLOO-ra).

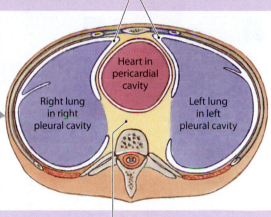

Heart in pericardial cavity

Right lung in right pleural cavity

Left lung in left pleural cavity

Note the orientation of the section. Unless otherwise noted, all cross sections are shown as if the viewer were standing at the feet of a supine person and looking toward the head.

The pericardial cavity is embedded within the **mediastinum**, a mass of connective tissue that separates the two pleural cavities and stabilizes the positions of embedded organs and blood vessels.

ABDOMINOPELVIC CAVITY

During development, the portion of the original ventral body cavity extending into the abdominopelvic cavity remains intact as the **peritoneal** (per-i-tō-NĒ-al) **cavity**, a chamber lined by a serous membrane known as the **peritoneum** (per-i-tō-NĒ-um). A few organs, such as the kidneys and pancreas, lie between the peritoneal lining and the muscular wall of the abdominal cavity. Those organs are said to be **retroperitoneal** (re-trō-per-i-tō-NĒ-al; *retro*, behind).

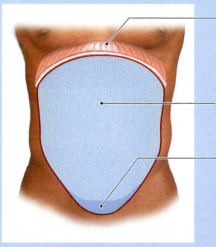

Diaphragm

Peritoneum (red) showing the boundaries of the peritoneal cavity

The abdominal cavity contains many digestive glands and organs

Retroperitoneal area

The pelvic cavity contains the urinary bladder, reproductive organs, and the last portion of the digestive tract; many of these structures lie posterior to, or inferior to, the peritoneal cavity.

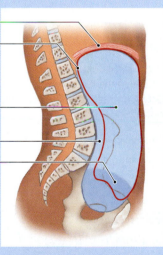

4 The boundaries of the abdominopelvic cavity are established by the diaphragm, the muscles of the abdominal wall, the trunk muscles and inferior portions of the vertebral column, and the bones and muscles of the pelvis. It may be subdivided into the **abdominal cavity** and the **pelvic cavity**.

Module 1.10 Review

a. Describe two essential functions of body cavities.

b. Identify the subdivisions of the ventral body cavity.

c. If a surgeon makes an incision just inferior to the diaphragm, what body cavity will be opened?

1. Labeling

Label the directional terms in the figures at right.

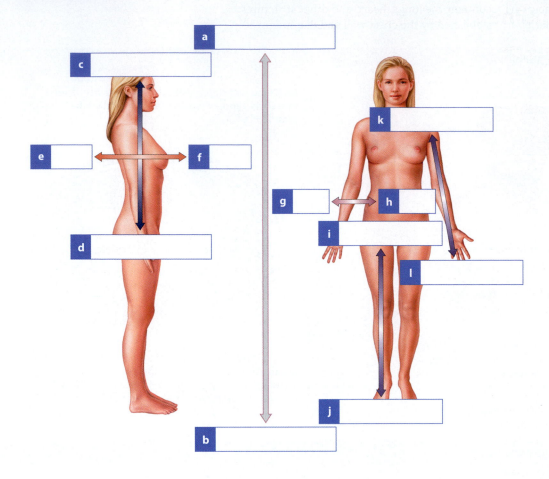

a

c

e   f

d

k

g   h

i

l

j

b

2. Concept map

Use each of the following terms once to fill in the blank boxes to correctly complete the body cavities concept map.

- digestive glands and organs
- abdominopelvic cavity
- thoracic cavity
- heart
- mediastinum
- diaphragm
- pelvic cavity
- trachea, esophagus
- reproductive organs
- left lung
- peritoneal cavity

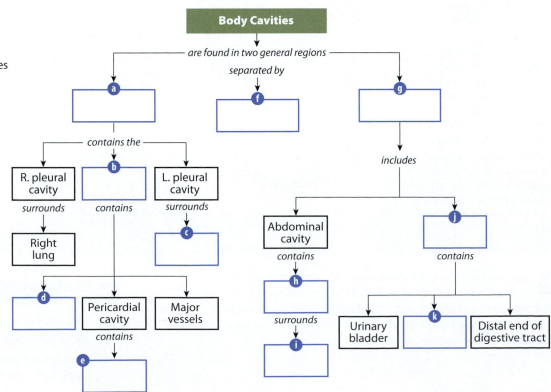

Body Cavities

are found in two general regions

separated by

a

f

g

contains the

b

R. pleural cavity

L. pleural cavity

surrounds

contains

surrounds

Right lung

c

includes

Abdominal cavity

j

contains

contains

h

d

Pericardial cavity

Major vessels

Urinary bladder

k

Distal end of digestive tract

contains

surrounds

e

i

Visual Outline with Key Terms

Summarize the content of each module using the terms in the order provided.

SECTION 1

A&P in Perspective

- homeostasis

1.1

Biology is the study of life

- responsiveness
- adaptability
- growth
- reproduction
- movement
- respiration
- circulation
- digestion
- excretion

1.2

Anatomy is the study of form; physiology is the study of function

- anatomy
- gross anatomy
- macroscopic anatomy
- microscopic anatomy
- physiology

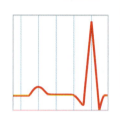

1.3

Form and function are interrelated

- elbow joint
- humerus
- radius
- ulna
- chemical messengers
- receptors

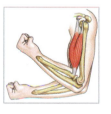

SECTION 2

Levels of Organization

- organism
- organ systems
- organ
- tissue
- cells
- atoms

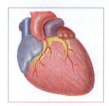

1.4

Cells are the smallest units of life

- cells
- cell theory

1.5

Tissues are specialized groups of cells and cell products

- histology
- primary tissue types
- epithelial tissue
- connective tissue
- muscle tissue
- skeletal muscle tissue
- cardiac muscle tissue
- smooth muscle tissue
- neural tissue
- neurons
- neuroglia
- central nervous system
- peripheral nervous system

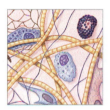

1.6

Organs and organ systems perform vital functions

- organ
- organ system
- integumentary system
- skeletal system
- muscular system
- nervous system
- endocrine system
- cardiovascular system
- lymphatic system
- respiratory system
- digestive system
- urinary system
- reproductive system

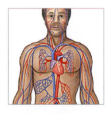

● = *Term boldfaced in this module*

SECTION 3

Homeostasis

- homeostasis
- homeostatic regulation
- receptor
- control center
- effector
- set point

HOMEOSTASIS

Normal room temperature

1.7

Negative feedback provides stability, whereas positive feedback accelerates a process to completion

- negative feedback
- positive feedback
- positive feedback loop

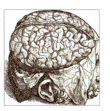

SECTION 4

Anatomical Terms

○ eponyms
○ word roots

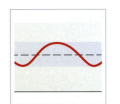

1.8

Superficial anatomy and regional anatomy indicate locations on or in the body

- anatomical position
- supine
- prone
- abdominopelvic quadrants
- abdominopelvic regions

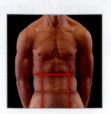

1.9

Directional and sectional terms describe specific points of reference

- anterior
- ventral
- posterior
- dorsal
- cranial
- cephalic
- superior
- caudal
- inferior
- medial

- lateral
- proximal
- distal
- superficial
- deep
- transverse section
- horizontal section
- sagittal section
- midsagittal section

- median section
- parasagittal section
- frontal section
- coronal section

1.10

Body cavities protect internal organs and allow them to change shape

- thoracic cavity
- abdominopelvic cavity
- body cavities
- viscera
- pericardial cavity
- pericardium
- serous membrane
- ventral body cavity

- coelom
- diaphragm
- pleural cavity
- pleura
- mediastinum
- abdominal cavity
- pelvic cavity
- peritoneal cavity
- peritoneum
- retroperitoneal

• = *Term boldfaced in this module*

Chapter Integration: Applying what you've learned

The school year and your A&P studies are just beginning. The reading workload is massive, the terms seem like a foreign language, and the instructor seems to think that your entire life should revolve around her class. But you really do love anatomy and are lucky enough to be able to experience the study of the human body firsthand. At your last class you were introduced to the human anatomy lab, which you now approach with a little trepidation, as you have never had to study a deceased human body. Body donation for scientific purposes is a noble gesture on the part of the deceased, and you fully understand your instructor's expectations for showing complete respect to the cadaver.

Your instructor has you glove up, put your apron and eye protection on, and presents the cadaver to you. Your fears prove groundless as this amazing experience begins. Your instructor explains that this cadaver was the victim of a gunshot wound to the lower right abdomen with no exit wound. There was no autopsy, for some unknown reason, so the anatomical/physiological cause of death has yet to be determined. The information on the body donor card provides the following information: male, 45 years old, no known medical history, and his occupation is unspecified. As a beginning anatomy student, use your knowledge to answer the following questions.

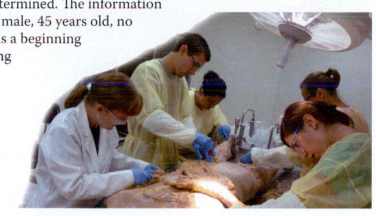

1. Upon superficial examination, you discover that the bullet entered about 2 centimeters inferior to the navel. Using anatomical terminology, describe the location.

2. Before opening the abdominal cavity, predict what organs may have been affected.

3. Propose a plausible cause of death.

Access more review material online in the Study Area at **www.masteringaandp.com.**

There, you'll find:
- **Chapter guides**
- **Chapter quizzes**
- **Practice tests**
- **Animations**
- **Labeling activities**
- **MP3 Tutor Sessions**
- **Flashcards**
- **Quizzes**
- **A glossary with pronunciations**

For more information about all the extra practice available to you in the MasteringA&P Study Area, turn to page xvi at the front of the book.

2

Chemical Level of Organization

LEARNING OUTCOMES

These Learning Outcomes correspond by number to this chapter's modules and indicate what you should be able to do after completing the chapter.

SECTION 1 • Atoms and Molecules

2.1 Describe an atom and how atomic structure affects the mass number and atomic weight of the various chemical elements.

2.2 Explain the relationship between electrons and energy levels.

2.3 Compare the ways in which atoms combine to form molecules and compounds.

2.4 Describe the three states of matter and the importance of hydrogen bonds in liquid water.

SECTION 2 • Chemical Reactions

2.5 Use chemical notation to symbolize chemical reactions.

2.6 Distinguish among the major types of chemical reactions that are important for studying physiology.

2.7 Describe the crucial role of enzymes in metabolism.

SECTION 3 • The Importance of Water in the Body

2.8 Explain how the chemical properties of water affect the solubility of inorganic and organic molecules.

2.9 Discuss the importance of pH and the role of buffers in body fluids.

SECTION 4 • Organic Compounds

2.10 Discuss the structures and functions of carbohydrates.

2.11 Discuss the structures and functions of lipids.

2.12 Discuss the structures and diverse functions of eicosanoids, steroids, phospholipids, and glycolipids.

2.13 Discuss protein structure and the essential functions of proteins within the body.

2.14 Explain how enzymes function within the body.

2.15 Discuss the structure and function of high-energy compounds.

2.16 Compare and contrast the structures and functions of DNA and RNA.

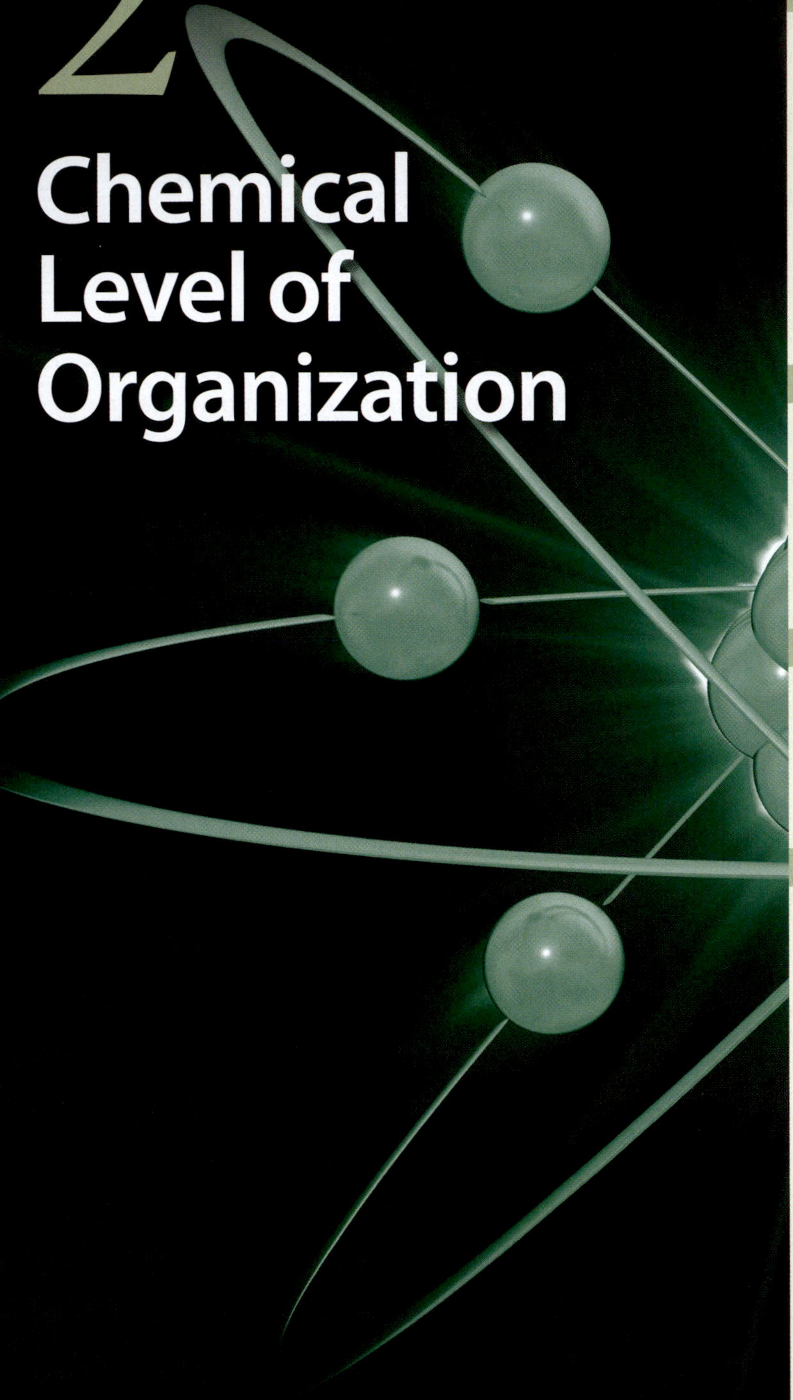

Atoms and Molecules

Our study of the human body begins at the chemical level of organization. Chemistry is the science that studies the structure of matter, which is defined as anything that takes up space and has mass. **Mass**, the amount of material in matter, is a physical property that determines the weight of an object in Earth's gravitational field. For our purposes, the mass of an object is the same as its weight. However, the two are not always equivalent: In orbit you would be weightless, but your mass would remain unchanged.

1 **Atoms** are the smallest stable units of matter. They are composed of **subatomic particles**, only three of which are important for understanding the basic chemical properties of matter.

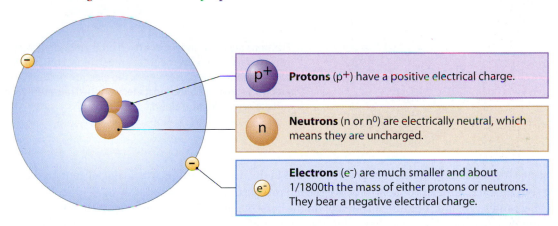

p⁺ **Protons** (p⁺) have a positive electrical charge.

n **Neutrons** (n or n⁰) are electrically neutral, which means they are uncharged.

e⁻ **Electrons** (e⁻) are much smaller and about 1/1800th the mass of either protons or neutrons. They bear a negative electrical charge.

2 Atoms can be subdivided into the nucleus and the electron cloud.

The **nucleus** of an atom lies at its center. The nucleus contains one or more protons and it may contain neutrons as well. The mass of the atom is primarily determined by the numbers of protons and neutrons in the nucleus.

The electrons in the atom whirl around the nucleus, creating an **electron cloud**.

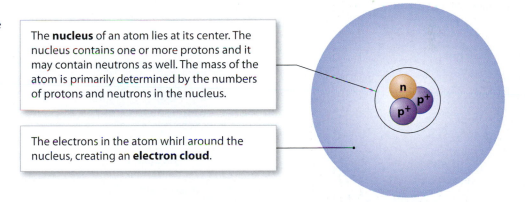

3 A **molecule** forms when atoms interact and produce larger, more complex structures.

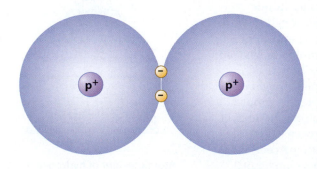

Everything around us is composed of atoms in varying combinations. The unique characteristics of each object, living or nonliving, result from the types of atoms involved and the ways those atoms combine and interact. The mass of any object represents the sum of the masses of its component atoms.

Typical atoms contain protons, neutrons, and electrons

Atoms normally contain equal numbers of protons and electrons. The number of protons in an atom is known as the **atomic number**; the total number of protons and neutrons is its **mass number**. An **element** is a pure substance consisting only of atoms with the same atomic number.

1 Hydrogen (H) is the simplest atom, with an atomic number of 1. Thus, an atom of hydrogen contains one proton and one electron. Hydrogen's proton is located in the center of the atom and forms the nucleus.

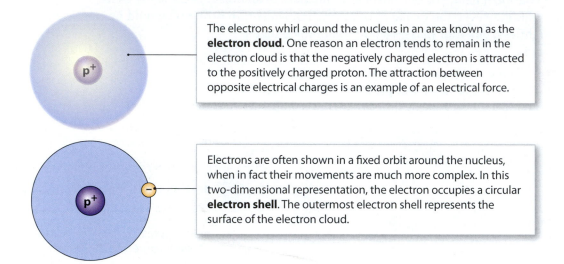

The electrons whirl around the nucleus in an area known as the **electron cloud**. One reason an electron tends to remain in the electron cloud is that the negatively charged electron is attracted to the positively charged proton. The attraction between opposite electrical charges is an example of an electrical force.

Electrons are often shown in a fixed orbit around the nucleus, when in fact their movements are much more complex. In this two-dimensional representation, the electron occupies a circular **electron shell**. The outermost electron shell represents the surface of the electron cloud.

2 The atoms of a single element can differ in the number of neutrons in the nucleus. Atoms whose nuclei contain the same number of protons, but different numbers of neutrons, are called **isotopes**. Different isotopes of an element have essentially identical chemical properties, and so are indistinguishable except on the basis of mass. The mass number is therefore used to designate isotopes. Mass numbers are useful because they tell us the number of subatomic particles in the nuclei of different atoms.

Electron-shell model

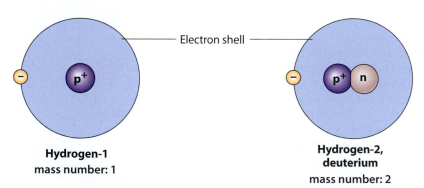

Electron shell

Hydrogen-1
mass number: 1

Hydrogen-2, deuterium
mass number: 2

Hydrogen-3, tritium
mass number: 3

3 The actual mass of an atom is known as its **atomic weight.** The unit used to express atomic weight is the **dalton** (also known as the atomic mass unit, or amu). One dalton is very close to the weight of a single proton or neutron. Thus, the atomic weight of the most common isotope of hydrogen is very close to 1. However, the atomic weight of an element is an average mass number that reflects the proportions of different isotopes. For example, the atomic number of hydrogen is 1, but the atomic weight of hydrogen is 1.0079, primarily because some hydrogen atoms (0.015 percent) have a mass number of 2, and even fewer have a mass number of 3.

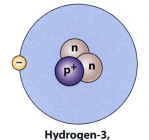

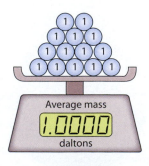

Average mass
1.0000
daltons

Atomic weight of hydrogen-1 = 1

Average mass
1.0079
daltons

Atomic weight of a mixture of hydrogen isotopes = 1.0079

Principal Elements of the Human Body

Element (% of total body weight)	Significance
Oxygen, O (65)	A component of water and other compounds; gaseous form is essential for respiration
Carbon, C (18.6)	Found in all organic molecules
Hydrogen, H (9.7)	A component of water and most other compounds in the body
Nitrogen, N (3.2)	Found in proteins, nucleic acids, and other organic compounds
Calcium, Ca (1.8)	Found in bones and teeth; important for membrane function, nerve impulses, muscle contraction, and blood clotting
Phosphorus, P (1.0)	Found in bones and teeth, nucleic acids, and high-energy compounds
Potassium, K (0.4)	Important for proper membrane function, nerve impulses, and muscle contraction
Sodium, Na (0.2)	Important for blood volume, membrane function, nerve impulses, and muscle contraction
Chlorine, Cl (0.2)	Important for blood volume, membrane function, and water absorption
Magnesium, Mg (0.06)	A cofactor for many enzymes
Sulfur, S (0.04)	Found in many proteins
Iron, Fe (0.007)	Essential for oxygen transport and energy capture
Iodine, I (0.0002)	A component of hormones of the thyroid gland
Trace elements: silicon (Si), fluorine (F), copper (Cu), manganese (Mn), zinc (Zn), selenium (Se), cobalt (Co), molybdenum (Mo), cadmium (Cd), chromium (Cr), tin (Sn), aluminum (Al), boron (B), and vanadium (V)	Some function as cofactors; the functions of many trace elements are poorly understood

4 This table shows the relative contributions of the 13 most abundant elements in the human body to total body weight. The human body also contains atoms of another 14 elements—called **trace elements**—that are present in very small amounts. Only 92 elements exist in nature, although about two dozen additional elements have been created through nuclear reactions in research laboratories. Every element has a **chemical symbol**, an abbreviation recognized by scientists everywhere. Most of the symbols are easily connected with the English names of the elements (O for oxygen, N for nitrogen, C for carbon, and so on), but a few are abbreviations of their names in other languages. For example, the symbol for sodium, Na, comes from the Latin word *natrium*.

Module 2.1 Review

a. Define atom.

b. Describe trace elements.

c. How is it possible for two samples of hydrogen to contain the same number of atoms yet have different weights?

Electrons occupy various energy levels

Atoms are electrically neutral; every positively charged proton is balanced by a negatively charged electron. Thus, each increase in atomic number is accompanied by a comparable increase in the number of electrons traveling around the nucleus. Within the electron cloud, electrons occupy an orderly series of energy levels. Although the electrons in an energy level may travel in complex patterns around the nucleus, for our purposes the patterns can be diagrammed as a series of concentric electron shells. The first electron shell (the one closest to the nucleus) corresponds to the lowest energy level.

Reactive elements

1 The outermost energy level forms the "surface" of the atom. Atoms with unfilled energy levels, such as hydrogen and lithium, will react with other atoms, usually in ways that give them full outer energy levels. An atom with a filled outermost energy level is stable and does not readily react with other atoms.

Inert elements

2 Elements that do not readily participate in chemical processes are said to be **inert**. Helium and neon, which have filled outermost energy levels, are called **inert gases** because their atoms neither react with one another nor combine with atoms of other elements.

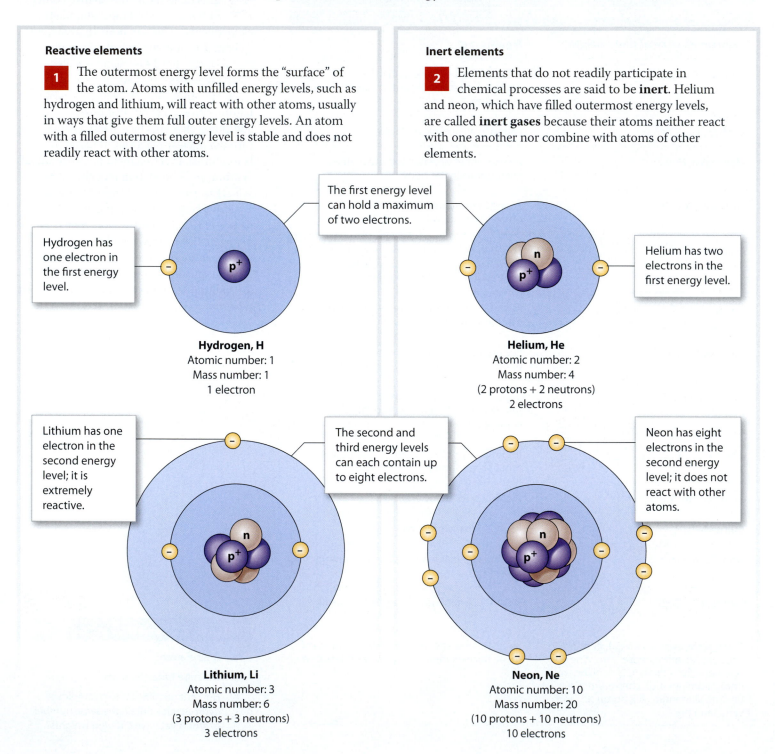

The first energy level can hold a maximum of two electrons.

Hydrogen has one electron in the first energy level.

Helium has two electrons in the first energy level.

Hydrogen, H
Atomic number: 1
Mass number: 1
1 electron

Helium, He
Atomic number: 2
Mass number: 4
(2 protons + 2 neutrons)
2 electrons

Lithium has one electron in the second energy level; it is extremely reactive.

The second and third energy levels can each contain up to eight electrons.

Neon has eight electrons in the second energy level; it does not react with other atoms.

Lithium, Li
Atomic number: 3
Mass number: 6
(3 protons + 3 neutrons)
3 electrons

Neon, Ne
Atomic number: 10
Mass number: 20
(10 protons + 10 neutrons)
10 electrons

3   Elements with unfilled outermost energy levels, such as hydrogen, lithium, or sodium, are called **reactive**, because they readily interact or combine with other atoms. In doing so, these atoms achieve stability by gaining, losing, or sharing electrons to fill their outermost energy level. When this involves the loss of electrons from the outer energy level, the result is an atom that is no longer electrically neutral—it now has more protons than electrons. The atom has a net positive charge, and it is called a positive ion or **cation**. A single missing electron gives the ion a charge of +1; some ions carry charges of +2, +3, or +4, depending on how many electrons are lost to achieve stability.

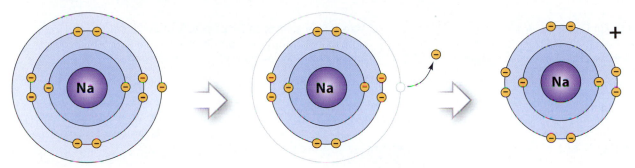

Sodium atom, Na (reactive)   **Sodium ion, Na$^+$** (stable)

4   At other times atoms achieve stability by filling their outer energy level with electrons obtained from other atoms. This also creates an atom that is no longer electrically neutral—it has more electrons than protons. The atom now has a net negative charge, and it is called a negative ion or **anion**. A single extra electron gives the ion a charge of −1; some ions carry charges of −2, −3, or −4, depending on how many electrons are needed to achieve stability.

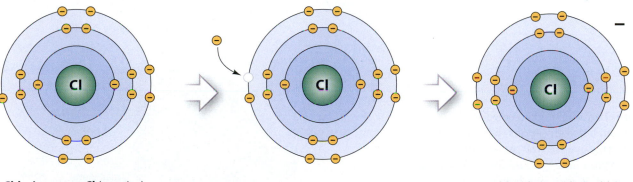

Chlorine atom, Cl (reactive)   **Chloride ion, Cl$^-$** (stable)

The interactions that stabilize the outer energy levels of atoms often result in the formation of **chemical bonds**. These bonds hold the participating atoms together once the reaction has ended.

Module 2.2 Review

a. Indicate the maximum number of electrons that can occupy each of the first three electron shells (energy levels) of an atom.

b. Explain why the atoms of inert elements do not react with one another or combine with atoms of other elements.

c. Explain how cations and anions form.

The most common chemical bonds are ionic bonds and covalent bonds

When chemical bonding occurs, the result is the creation of new chemical entities called compounds and molecules. A **compound** is a chemical substance made up of atoms of two or more different elements, regardless of the type of bond joining them.

Step 1: Formation of sodium and chloride ions. The sodium atom loses an electron to the chlorine atom. This produces two stable ions with filled outer energy levels.

Step 2: Formation of an ionic bond. Because these ions form close together, and have opposite charges, they are attracted to one another. This creates NaCl, an ionic compound.

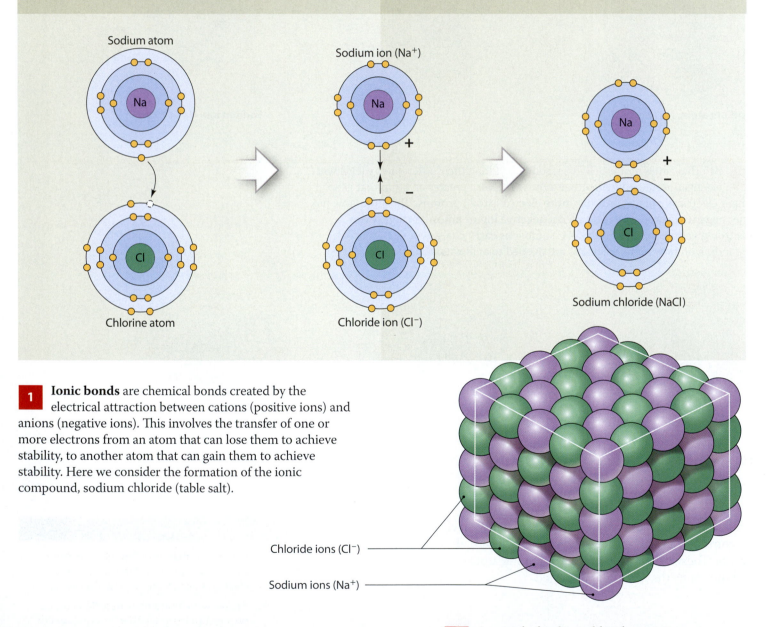

Sodium atom

Na

Chlorine atom

Cl

Sodium ion (Na⁺)

Na

+

Chloride ion (Cl⁻)

Cl

−

Na

+

Cl

−

Sodium chloride (NaCl)

1 **Ionic bonds** are chemical bonds created by the electrical attraction between cations (positive ions) and anions (negative ions). This involves the transfer of one or more electrons from an atom that can lose them to achieve stability, to another atom that can gain them to achieve stability. Here we consider the formation of the ionic compound, sodium chloride (table salt).

Chloride ions (Cl⁻)

Sodium ions (Na⁺)

2 A crystal of sodium chloride contains a large number of sodium and chloride ions packed closely together.

3 Some atoms can complete their outer electron shells not by gaining or losing electrons, but by sharing electrons with other atoms. Such sharing creates **covalent** (kō-VA-lent) **bonds** between the atoms involved. A **molecule** is a chemical structure consisting of atoms of one or more elements held together by covalent bonds.

Molecule	Description
Hydrogen (H_2)	Hydrogen atoms aren't found as individuals—they exist as molecules, each containing a pair of hydrogen atoms. The two atoms share their electrons to fill their outer energy levels, and the electron pair orbits both nuclei. One electron is contributed by each atom, so this is called a **single covalent bond**.
Oxygen (O_2) 	An oxygen atom has 6 electrons in its outer energy level. By forming a **double covalent bond** with another oxygen atom, an oxygen molecule is created with a stable outer energy level.
Carbon dioxide (CO_2)	A carbon atom has 4 electrons in its outer energy level, so it needs to gain 4 from other atoms to achieve stability. In a molecule of carbon dioxide, a carbon atom shares a pair of electrons with each of two oxygen atoms and forms two double covalent bonds.

4 These are space-filling models of oxygen and carbon dioxide molecules. In a typical covalent bond, the participating atoms share the electrons equally, and there is no electrical charge on the molecule. Due to this lack of electrical charge, such molecules are called **nonpolar molecules**.

Oxygen (O_2)

Carbon dioxide (CO_2)

5 Some molecules, however, are formed by covalent bonds that involve an unequal sharing of electrons. This energy level model shows the formation of a water molecule.

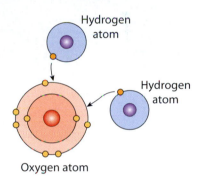

Hydrogen atom

Hydrogen atom

Oxygen atom

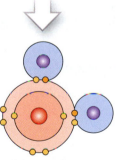

Water molecule

6 In a water molecule, the electron clouds of the hydrogen atoms are distorted because the 8 protons in the oxygen atom exert a much stronger attraction for the electrons than do the single protons of the hydrogen atoms. As a result, each hydrogen atom carries a slightly positive charge (δ^+), and the oxygen atom carries a slightly negative charge (δ^-). This creates an asymmetrical **polar molecule**. Covalent bonds that produce polar molecules are called **polar covalent bonds**.

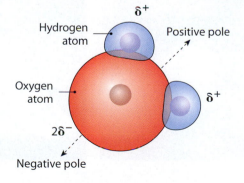

δ^+

Hydrogen atom — Positive pole

Oxygen atom

δ^+

$2\delta^-$

Negative pole

Module 2.3 Review

a. Name and distinguish between the two most common types of chemical bonds.

b. Describe the kind of bonds that hold the atoms in a water molecule together.

c. Relate why we can apply the term *molecule* to the smallest particle of water but not to that of table salt.

Matter may exist as a solid, a liquid, or a gas

Most matter in our environment exists in one of three states: solid, liquid, or gas. Whether a particular substance is a solid, a liquid, or a gas depends on the degree of interaction among its atoms or molecules. The particles of a solid are held tightly together, while those of a liquid less so, and the particles of a gas are independent of each other.

1 Solids maintain their volume and their shape at ordinary temperatures and pressures. A lump of granite, a brick, and a textbook are solid objects.

2 Liquids have a constant volume but no fixed shape. The shape of a liquid is determined by the shape of its container. Water, brewed coffee, and soda are liquids.

3 A gas has neither a constant volume nor a fixed shape. Gases can be compressed or expanded; unlike liquids they will fill a container of any size. The air of our atmosphere is the gas with which we are most familiar.

4 Water is the only substance that occurs as a solid (ice), a liquid (water), and a gas (water vapor) at temperatures compatible with life. Water exists in the liquid state over a broad range of temperatures primarily because of significant interactions among the polar water molecules.

5 The small positive charges on the hydrogen atoms of one polar molecule can be attracted to the negative charges on another polar molecule, and this can change the shapes of the molecules or pull adjacent molecules together. This weak attractive force is called a **hydrogen bond**.

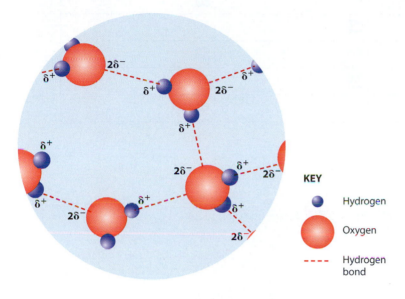

KEY

● Hydrogen

● Oxygen

- - - Hydrogen bond

6 At the water surface, the hydrogen bonds between water molecules slow the rate of evaporation and create the phenomenon known as **surface tension**. Surface tension acts as a barrier that keeps small objects from entering the water. For example, it allows insects to walk across the surface of a pond or puddle. Similarly, small objects such as dust particles are prevented from touching the surface of the eye by the surface tension produced by a layer of tears.

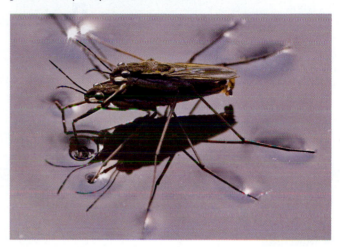

7 The polar charges on water molecules give water the ability to disrupt the ionic bonds of a variety of inorganic compounds and cause them to dissolve.

Almost all naturally occurring elements are found in seawater; at least 29 elements are dissolved in our body fluids.

Module 2.4 Review

a. Describe the different states of matter in terms of shape and volume.

b. By what means are water molecules attracted to each other?

c. Explain why small insects can walk on the surface of a pond, and tears protect the surface of the eye from dust particles.

1. Short answer

Fill in the missing information in the following table.

Element	Number of protons	Number of electrons	Number of neutrons	Mass number
Helium	a	2	2	b
Hydrogen	1	c	d	1
Carbon	6	e	6	f
Nitrogen	g	7	h	14
Calcium	i	j	20	40

2. Short answer

Indicate which of the following molecules are compounds and which are elements.

H_2 (hydrogen) H_2O (water) O_2 (oxygen)   CO (carbon monoxide)

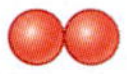

a _____   b _____   c _____   d _____

3. Matching

Match the following terms with the most closely related description.

- atomic number
- electrons
- protons
- neutrons
- isotopes
- ions
- ionic bond
- covalent bond
- mass number
- element
- compound
- hydrogen bond

a _____ Atoms that have gained or lost electrons

b _____ Located in the nucleus, have no charge

c _____ Atoms of two or more different elements bonded together

d _____ The number of protons in an atom

e _____ Attractive force between water molecules

f _____ Type of chemical bond within a water molecule

g _____ The number of subatomic particles in the nucleus

h _____ Substance composed only of atoms with same atomic number

i _____ Subatomic particles in the nucleus, have charge

j _____ Atoms of the same element with different masses

k _____ Type of chemical bond in table salt

l _____ Subatomic particles outside the nucleus, have charge

4. Section integration

Describe how the following pairs of terms concerning atomic interactions are similar and how they are different.

a   inert element/reactive element   _____

b   polar molecules/nonpolar molecules   _____

c   covalent bond/ionic bond   _____

Chemical Reactions

Cells remain alive and functional by controlling chemical reactions. In a chemical reaction, new chemical bonds form between atoms, or existing bonds between atoms are broken. These changes occur as atoms in the reacting substances, or **reactants**, are rearranged to form different substances, or **products.** All of the reactions under way in the cells and tissues of the body at any given moment constitute its **metabolism** (me-TAB-ō-lizm).

1 In effect, each cell is a chemical factory. Growth, maintenance and repair, secretion, and contraction all involve complex chemical reactions. Cells use chemical reactions to provide the energy needed to maintain homeostasis and to perform essential functions.

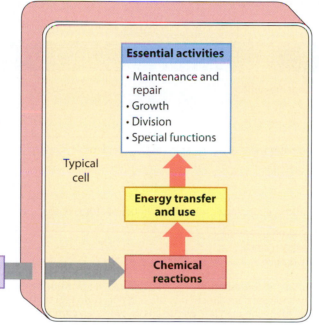

Essential activities
- Maintenance and repair
- Growth
- Division
- Special functions

Typical cell

Energy transfer and use

Substances absorbed

Chemical reactions

2 **Work** is the movement of an object or a change in the physical structure of matter. In your body, work includes movements like walking or running, and also the synthesis of molecules and the conversion of liquid water to water vapor (evaporation).

Energy is the capacity to perform work; movement or physical change cannot occur unless energy is provided. **Kinetic energy** is the energy of motion—energy that can be transferred to another object and perform work. **Potential energy** is stored energy—energy that has the potential to do work. It may derive from an object's position (you standing on a ladder) or from its physical or chemical structure (a stretched spring or a charged battery).

Cells perform work as they synthesize complex molecules and move materials into, out of, and within the cell. The cells of a skeletal muscle at rest, for example, contain potential energy in the form of the positions of protein filaments and the covalent bonds between molecules within the cells. When a muscle contracts, it performs work, and potential energy is converted into kinetic energy. Such a conversion is never 100 percent efficient. Each time an energy exchange or transfer occurs, some of the energy is released as heat. That is why your body temperature rises when you exercise.

Chemical notation is a concise method of describing chemical reactions

Before we can consider the specific compounds that occur in the human body, we must be able to describe chemical compounds and reactions effectively. The use of sentences to describe chemical structures and events often leads to confusion. A simple form of "chemical shorthand" makes communication much more efficient. The chemical shorthand we use is known as **chemical notation**.

1 Chemical notation enables us to describe complex events briefly and precisely. Many important rules of chemical notation are summarized in this table. It is relatively easy to use chemical notation to calculate the weights of the reactants involved in a particular reaction.

Rules of Chemical Notation

The symbol of an element indicates one atom of that element:

> H = one atom of hydrogen
> O = one atom of oxygen

A number preceding the symbol of an element indicates more than one atom of that element:

> 2 H = two atoms of hydrogen
> 2 O = two atoms of oxygen

A subscript following the symbol of an element indicates a molecule with that number of atoms of that element:

> H_2 = hydrogen molecule, composed of two hydrogen atoms
> O_2 = oxygen molecule, composed of two oxygen atoms
> H_2O = water molecule, composed of two hydrogen atoms and one oxygen atom

In a description of a chemical reaction, the participants at the start of the reaction are called reactants, and the reaction generates one or more products. An arrow indicates the direction of the reaction, from reactants (usually on the left) to products (usually on the right). In the following reaction, two atoms of hydrogen combine with one atom of oxygen to produce a single molecule of water:

> $2 H + O \longrightarrow H_2O$

A superscript plus or minus sign following the symbol of an element indicates an ion. A single plus sign indicates a cation with a charge of +1. (The original atom has lost one electron.) A single minus sign indicates an anion with a charge of –1. (The original atom has gained one electron.) If more than one electron has been lost or gained, the charge on the ion is indicated by a number preceding the plus or minus sign:

> Na^+ = sodium ion (the sodium atom has lost one electron)
> Cl^- = chloride ion (the chlorine atom has gained one electron)
> Ca^{2+} = calcium ion (the calcium atom has lost two electrons)

Chemical reactions neither create nor destroy atoms; they merely rearrange them into new combinations. Therefore, the numbers of atoms of each element must always be the same on both sides of the equation for a chemical reaction. When this is the case, the equation is **balanced**:

> Unbalanced: $H_2 + O_2 \longrightarrow H_2O$
> Balanced: $2 H_2 + O_2 \longrightarrow 2 H_2O$

2 A **mole** (abbreviated mol) is a quantity with a weight in grams equal to an element's atomic weight. One mole of a given element always contains the same number of atoms as one mole of any other element. The atomic weight of oxygen is 16 and the atomic weight of hydrogen is 1. So a mole of oxygen will weigh 16 grams and contain the same number of atoms as a mole of hydrogen, which weighs 1 gram.

3 The **molecular weight** of a molecule is the sum of the atomic weights of its component atoms. Molecular weights are important because you can neither handle individual molecules nor easily count the billions of molecules involved in chemical reactions in the body.

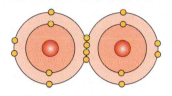

Molecular weight of O_2 = 32   Molecular weight of H_2 = 2

grams

1 mole of oxygen

1.00
grams

1 mole of hydrogen

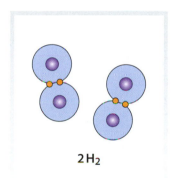 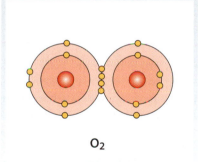 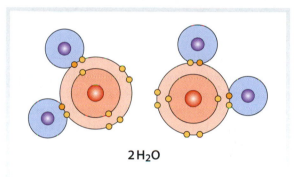

2 H₂   +   **O₂**   =   **2 H₂O**

Atomic weight of H = 1
Molecular weight of H_2 = 2
Two moles of H_2 weigh 4 g

Atomic weight of O = 16
Molecular weight of O_2 = 32
One mole of O_2 weighs 32 g

Molecular weight of 2 H_2O:
2 x (2 + 16)
2 x 18 = 36
Two moles of H_2O weigh 36 g

4 Using molecular weights, you can calculate the quantities of reactants needed for a specific reaction and determine the amount of product generated. For example, suppose you want to form water from hydrogen and oxygen according to this equation: $2 H_2 + O_2 = 2 H_2O$. The first step would be to calculate the molecular weights involved. As detailed above, combining 4 g of hydrogen with 32 g of oxygen yields 36 g of water. Although by convention we use grams, you could also work with ounces, pounds, or tons, as long as the proportions remained the same. Notice that when you do the calculation correctly, the molecular weights of reactants and products are balanced.

Module 2.5 Review

a. The chemical shorthand used to describe chemical compounds and reactions effectively is known as

_____.

b. Using the rules for chemical notation, write the molecular formula for glucose, a compound composed of 6 carbon atoms, 12 hydrogen atoms, and 6 oxygen atoms.

c. Calculate the weight of one mole of glucose. (The atomic weight of carbon = 12.)

There are three basic types of chemical reactions

Decomposition Reactions

Decomposition is a reaction that breaks a molecule into smaller fragments. You could represent a simple decomposition reaction as

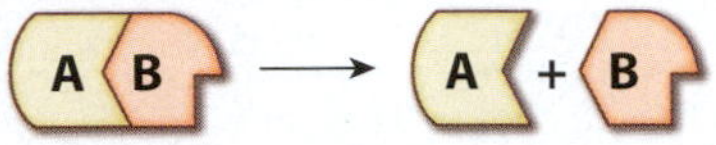

AB ⟶ A + B

Decomposition reactions occur outside cells as well as inside them. For example, a typical meal contains molecules of fats, sugars, and proteins that are too large and too complex to be absorbed and used by your body. Decomposition reactions in the digestive tract break these molecules down into smaller fragments before absorption begins.

Decomposition reactions involving water are important in the breakdown of complex molecules in the body. In **hydrolysis** (hī-DROL-i-sis; *hydro-*, water + *lysis*, a loosening), one of the bonds in a complex molecule is broken, and the components of a water molecule (H and OH) are added to the resulting fragments:

A-B + H_2O ⟶ A-H + OH-B

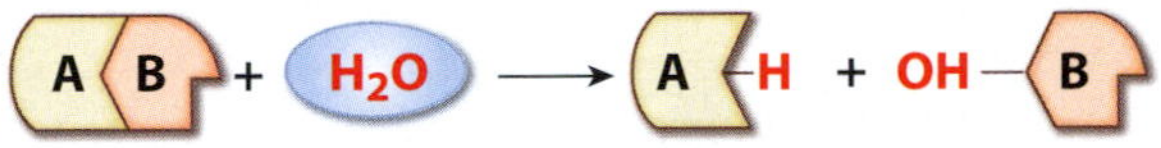

Collectively, the decomposition reactions of complex molecules within the body's cells and tissues are referred to as **catabolism** (ka-TAB-ō-lizm; *katabole*, a throwing down). When a covalent bond—a form of potential energy—is broken, it releases kinetic energy that can perform work. By harnessing the energy released in this way, cells perform vital functions such as growth, movement, and reproduction.

CD ⟶ C + D + **ENERGY**

Synthesis Reactions

Synthesis (SIN-the-sis) is the opposite of decomposition. A synthesis reaction assembles smaller molecules into larger molecules. A simple synthetic reaction could be diagrammed as:

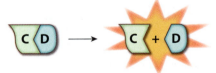

A + B ⟶ AB

Synthesis reactions may involve the combining of atoms or molecules to form even larger products. The formation of water from hydrogen and oxygen molecules is a synthesis reaction. Synthesis always involves the formation of new chemical bonds, whether the reactants are atoms or molecules.

Dehydration synthesis, or condensation, is the formation of a complex molecule by the removal of a water molecule:

$$\text{A-H} + \text{OH-B} \longrightarrow \text{A-B} + \text{H}_2\text{O}$$

Dehydration synthesis is therefore the opposite of hydrolysis. We will encounter examples of both reactions in later sections.

Collectively, the synthesis of new molecules within the body's cells and tissues is known as **anabolism** (a-NAB-ō-lizm; *anabole*, a throwing upward). Because it takes energy to create a chemical bond, anabolism is usually considered an "uphill" process. Cells must balance their energy budgets, with catabolism providing the energy to support anabolism and other vital functions.

Chemical reactions are reversible (at least theoretically), so if

$$\text{A} + \text{B} \longrightarrow \text{AB}, \text{ then } \text{AB} \longrightarrow \text{A} + \text{B}$$

Many important biological reactions are freely reversible. Such reactions can be represented as an equation:

$$\text{A} + \text{B} \rightleftharpoons \text{AB}$$

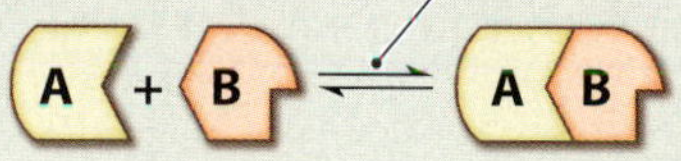

This equation indicates that, in a sense, two reactions are occurring simultaneously, one a synthesis and the other a decomposition. At **equilibrium**, the rates at which the two reactions proceed are in balance: As fast as one molecule of AB forms, another degrades into A + B.

Exchange Reactions

In an **exchange reaction**, parts of the reacting molecules are shuffled around to produce new products:

$$\text{AB} + \text{CD} \longrightarrow \text{AD} + \text{CB}$$

Although the reactants and products contain the same components (A, B, C, and D), those components are present in different combinations. In an exchange reaction, the reactant molecules AB and CD must break apart (a decomposition) before they can interact with each other to form AD and CB (a synthesis).

Module 2.6 Review

a. Identify and describe three types of chemical reactions important in human physiology.

b. Distinguish the roles of water in hydrolysis and dehydration synthesis reactions.

c. In cells, glucose, a six-carbon molecule, is converted into two three-carbon molecules by a reaction that releases energy. What is the source of the energy?

Enzymes lower the activation energy requirements of chemical reactions

Most chemical reactions do not occur spontaneously, or occur so slowly that they would be of little value to cells. Before a reaction can proceed, enough energy must be provided to activate the reactants. The amount of energy required to start a reaction is called the **activation energy**. Although many reactions can be activated by changes in temperature or acidity, such changes are deadly to cells. Instead, your cells use special proteins called **enzymes** to perform most of the complex synthesis and decomposition reactions in your body.

1 Enzymes promote chemical reactions by lowering the activation energy requirements. In doing so, they make it possible for chemical reactions, such as the breakdown of sugars, to proceed under conditions compatible with life. Enzymes belong to a class of substances called **catalysts** (KAT-uh-lists; *katalysis*, dissolution), compounds that accelerate chemical reactions without themselves being permanently changed or consumed. Enzymatic reactions, which are generally reversible, proceed until an equilibrium becomes established.

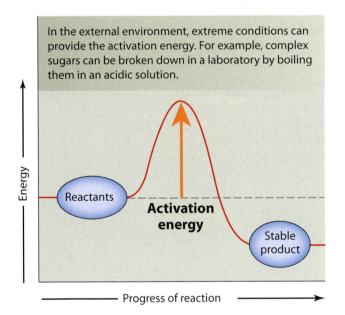

In the external environment, extreme conditions can provide the activation energy. For example, complex sugars can be broken down in a laboratory by boiling them in an acidic solution.

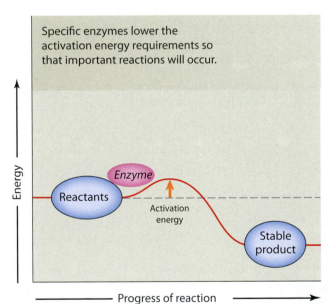

Specific enzymes lower the activation energy requirements so that important reactions will occur.

2 The complex reactions that support life proceed in a series of interlocking steps, each controlled by a specific enzyme. Such a reaction sequence is called a **metabolic pathway**. A synthetic pathway can be diagrammed as:

It takes activation energy to start a chemical reaction, but once it has begun, the reaction as a whole may absorb or release energy as it proceeds to completion. Reactions that release energy are said to be **exergonic** (*exo-*, outside). If more energy is required to begin the reaction than is released as it proceeds, the reaction is called **endergonic** (*endo-*, inside). Exergonic reactions are relatively common in the body; they are responsible for generating the heat that maintains your body temperature.

3 Enzymatic reactions are essential to the processing of **metabolites** (me-TAB-ō-līts; *metabole*, change), which include all the molecules that can be synthesized or broken down by chemical reactions inside our bodies. **Nutrients** are essential metabolites that are normally obtained from the diet. Nutrients and metabolites can be broadly categorized as either *organic* or *inorganic*.

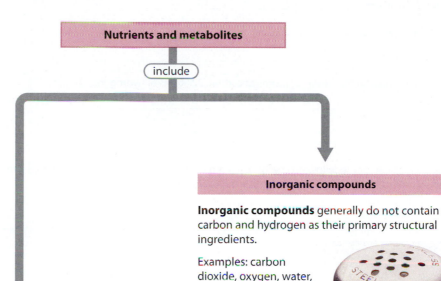

Nutrients and metabolites

include

Inorganic compounds

Inorganic compounds generally do not contain carbon and hydrogen as their primary structural ingredients.

Examples: carbon dioxide, oxygen, water, acids, bases, and salts

Organic compounds

Organic compounds always contain carbon and hydrogen as their primary structural ingredients.

Examples: sugars, fats, proteins, and nucleic acids (RNA, DNA), which are produced by living organisms.

Module 2.7 Review

a. What is an enzyme?

b. Explain the differences between metabolites and nutrients.

c. Why are enzymes needed in our cells?

1. Short answer

Using chemical notation, write the formula of each of the following.

a One molecule of hydrogen _____

b Two atoms of hydrogen _____

c Six molecules of water _____

d One molecule of sucrose (in this order: 12 atoms of carbon, 22 atoms of hydrogen,

and 11 atoms of oxygen) _____

2. Short answer

Write the chemical equation for the following chemical reaction: one molecule of glucose combined with six molecules of oxygen produce six molecules of carbon dioxide and six molecules of water.

_____

3. Short answer

Indicate which of the following reactions is a hydrolysis reaction, and which is a dehydration synthesis reaction.

a  A B + H_2O ⟶ A H + HO B _____

b A H + HO B ⟶ A B + H_2O _____

4. Matching

Match the following terms with the most closely related description.

- exergonic
- activation energy
- organic compounds
- exchange reaction
- hydrolysis
- endergonic
- reactants
- enzyme

a _____ Catalyst

b _____ Starting substances in a chemical reaction

c _____ Chemical reaction involving water

d _____ Reactions that absorb energy

e _____ Shuffles parts of reactants

f _____ Primary components are carbon and hydrogen

g _____ Reactions that release energy

h _____ Requirement for starting a chemical reaction

5. Section integration

In a metabolic pathway that consists of four steps, how would decreasing the amount of enzyme that catalyzes the second step affect the amount of product produced at the end of the pathway?

_____

_____

The Importance of Water in the Body

Water, H_2O, is the most important constituent of the body, accounting for up to two-thirds of total body weight. A change in the body's water content can have fatal consequences because virtually all physiological systems will be affected. Although water is familiar to everyone, it has some highly unusual properties.

Important Properties of Water

Lubrication

Water is an effective lubricant because there is little friction between water molecules. Thus even a thin layer of water between two opposing surfaces will greatly reduce friction between them; water reduces friction within joints and in body cavities.

Reactivity

In our bodies, chemical reactions occur in water, and water molecules are also participants in some reactions, including hydrolysis and dehydration synthesis.

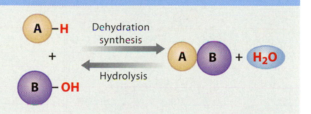

High heat capacity

Heat capacity is the ability to absorb and retain heat. Water has an unusually high heat capacity, because water molecules in the liquid state are attracted to one another through hydrogen bonding.

- The temperature of water must be high before individual molecules have enough energy to break free to become water vapor, a gas.

- Water carries a great deal of heat away with it when it finally does change from a liquid to a gas. This feature accounts for the cooling effect of perspiration on the skin.

- A large mass of water changes temperature very slowly. This property is called **thermal inertia**.

Solubility

A remarkable number of inorganic and organic molecules will dissolve in water. The individual particles become dispersed within the water, and the result is a **solution**—a uniform mixture of two or more substances. The medium in which other atoms, ions, or molecules are dispersed is called the **solvent**; the dispersed substances are the **solutes**. In **aqueous solutions**, water is the solvent.

Physiological systems depend on water

Many inorganic compounds are held together partially or completely by ionic bonds. In water, these compounds undergo **ionization** (ī-on-ī-ZĀ-shun), or **dissociation** (di-sō-sē-Ā-shun). In this process, ionic bonds are broken as the individual ions interact with the positive or negative poles of polar water molecules.

1 A water molecule is said to be polar because it has positive and negative poles. This polarity is due to the asymmetrical positions of the hydrogen atoms that are attached by polar covalent bonds.

2 In solution, an ionic compound dissociates as water molecules break them apart. The anions are surrounded by the positive poles of water molecules, and the cations are surrounded by the negative poles of water molecules. The sheath of water molecules around an ion in solution is called a **hydration sphere.**

3 Hydration spheres also form around an organic molecule containing polar covalent bonds. If the molecule binds water strongly, as does glucose, it will be carried into solution—in other words, it will dissolve. Molecules that interact readily with water molecules in this way are called **hydrophilic** (hī-drō-FIL-ik; *hydro-*, water + *philos*, loving).

Negative pole
2δ⁻
O
δ⁺
H
Positive pole
δ⁺

Sodium chloride crystal

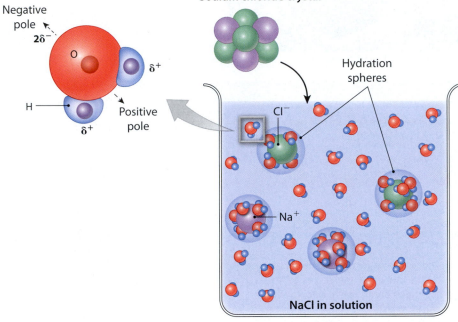

Hydration spheres

Cl⁻

Na⁺

NaCl in solution

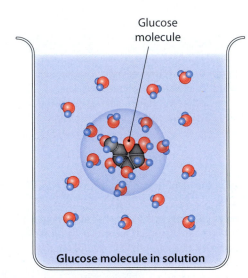

Glucose molecule

Glucose molecule in solution

4 An aqueous solution containing anions and cations will conduct an electrical current. Soluble inorganic molecules whose ions will conduct an electrical current in solution are called **electrolytes** (e-LEK-trō-līts). Sodium chloride is an important electrolyte in body fluids. In an electrical field, cations in solution will move toward the negative side, or negative terminal, and anions will move toward the positive terminal. Electrical forces across plasma membranes affect the functioning of all cells. Small electrical currents carried by ions are essential to muscle contraction and nerve function, two topics that will be investigated in later chapters.

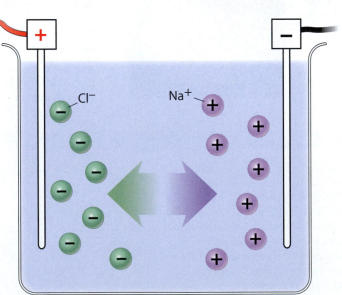

Cl⁻   Na⁺

5 This table lists the most important electrolytes. The dissociation of electrolytes in blood and other body fluids releases a variety of ions. Changes in the concentrations of electrolytes in body fluids will disturb almost every vital function. For example, declining potassium levels will lead to a general muscular paralysis, and rising concentrations will cause weak and irregular heartbeats. The concentrations of ions in body fluids are carefully regulated, primarily by the coordination of activities at the kidneys (ion excretion), the digestive tract (ion absorption), and the skeletal system (ion storage or release).

Important Electrolytes That Dissociate in Body Fluids

Electrolyte		Ions Released
NaCl (sodium chloride)	$\longrightarrow$	$Na^+ + Cl^-$
KCl (potassium chloride)	$\longrightarrow$	$K^+ + Cl^-$
CaPO₄ (calcium phosphate)	$\longrightarrow$	$Ca^{2+} + PO_4^{2-}$
NaHCO₃ (sodium bicarbonate)	$\longrightarrow$	$Na^+ + HCO_3^-$
MgCl₂ (magnesium chloride)	$\longrightarrow$	$Mg^{2+} + 2\,Cl^-$
Na₂HPO₄ (sodium hydrogen phosphate)	$\longrightarrow$	$2\,Na^+ + HPO_4^{2-}$
Na₂SO₄ (sodium sulfate)	$\longrightarrow$	$2\,Na^+ + SO_4^{2-}$

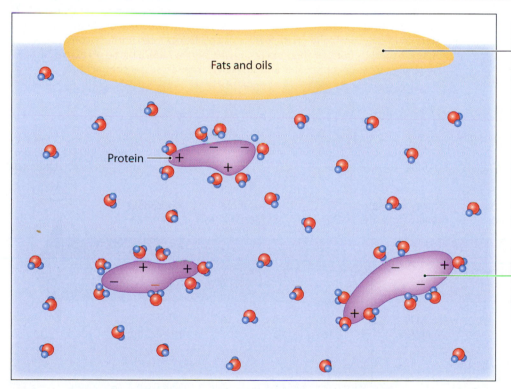

Fats and oils

Protein

Many organic molecules either lack polar covalent bonds or have very few. Such molecules do not have positive and negative poles and are said to be nonpolar. When nonpolar molecules are exposed to water, hydration spheres do not form and the molecules do not dissolve. Molecules that do not readily interact with water are called **hydrophobic** (hī-drō-FŌ-bik; *hydro-*, water + *phobos*, fear). Among the most familiar hydrophobic molecules are fats and oils of all kinds.

Body fluids typically contain large and complex organic molecules, such as proteins, that are held in solution by their association with water molecules.

6 A solution containing dispersed proteins or other large molecules is called a **colloid**. The particles or molecules in a colloid will remain in solution indefinitely. Liquid Jell-O is a familiar, viscous (thick) colloid. A **suspension** contains even larger particles that will, if undisturbed, settle out of solution due to the force of gravity. Whole blood is a temporary suspension, because the blood cells are suspended in the blood plasma. If clotting is prevented, the cells in a blood sample will gradually settle to the bottom of the container.

Module 2.8 Review

a. Define electrolytes.

b. Distinguish between hydrophilic and hydrophobic molecules.

c. Explain how the ionic compound sodium chloride dissolves in water.

Regulation of body fluid pH is vital for homeostasis

A hydrogen atom involved in a chemical bond or participating in a chemical reaction can easily lose its electron to become a **hydrogen ion**, **H⁺**. Hydrogen ions are extremely reactive in solution. In excessive numbers, they will break chemical bonds, change the shapes of complex molecules, and generally disrupt cell and tissue functions. As a result, the concentration of hydrogen ions in body fluids must be regulated precisely.

1 A few hydrogen ions are normally present even in a sample of pure water, because some of the water molecules dissociate spontaneously, releasing a hydrogen ion, H⁺, and a **hydroxide** (hī-DROK-sīd) **ion, OH⁻**.

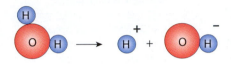

2 The hydrogen ion concentration in body fluids is so important to physiological processes that a special shorthand is used to express it. The **pH** of a solution is defined as the negative logarithm of the hydrogen ion concentration in moles per liter (mol/L). For common liquids, the pH scale ranges from 0 to 14.

Blood

The pH of blood normally ranges from 7.35 to 7.45. Abnormal fluctuations in pH can damage cells and tissues by breaking chemical bonds, changing the shapes of proteins, and altering cellular functions. **Acidosis** is an abnormal physiological state caused by low blood pH (below 7.35); a pH below 7 can produce coma. **Alkalosis** results from an abnormally high pH (above 7.45); a blood pH above 7.8 generally causes uncontrollable and sustained skeletal muscle contractions.

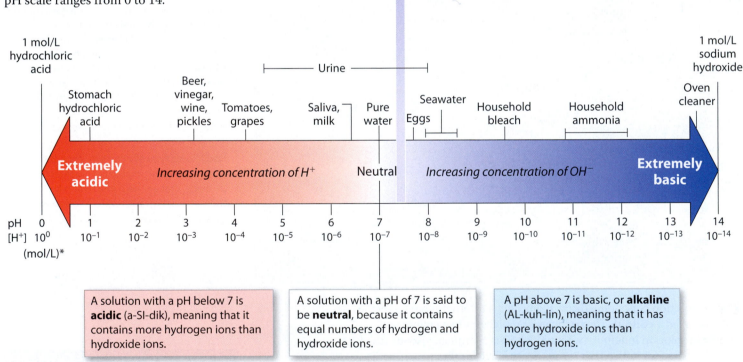

A solution with a pH below 7 is **acidic** (a-SI-dik), meaning that it contains more hydrogen ions than hydroxide ions.

A solution with a pH of 7 is said to be **neutral**, because it contains equal numbers of hydrogen and hydroxide ions.

A pH above 7 is basic, or **alkaline** (AL-kuh-lin), meaning that it has more hydroxide ions than hydrogen ions.

* One liter of pure water contains about 0.0000001 mol of hydrogen ions and an equal number of hydroxide ions. In other words, the concentration of hydrogen ions in a solution of pure water is 0.0000001 mol per liter. This can be written as [H⁺] = 10^{-7} mol/L. The brackets around the H⁺ signify "the concentration of," another example of chemical notation.

Acids and Bases

3 An **acid** is any solute that dissociates in solution and releases hydrogen ions, thereby lowering the pH. Because a hydrogen atom that loses its electron consists solely of a proton, hydrogen ions are often referred to simply as protons, and acids as proton donors. A strong acid dissociates completely in solution, and the reaction occurs essentially in one direction only. **Hydrochloric acid** (HCl) is a representative strong acid; in water, it ionizes as follows:

$$HCl \longrightarrow H^+ + Cl^-$$

4 A **base** is a solute that removes hydrogen ions from a solution and thereby acts as a proton acceptor. In solution, many bases release a hydroxide ion. Hydroxide ions have a strong affinity for hydrogen ions and react quickly with them to form water molecules. A strong base dissociates completely in solution. **Sodium hydroxide** (NaOH) is a strong base; in solution, it releases sodium ions and hydroxide ions:

$$NaOH \longrightarrow Na^+ + OH^-$$

5 Weak acids and weak bases fail to dissociate completely. At equilibrium, a significant number of molecules remain intact in the solution. For a given number of molecules in solution, weak acids and weak bases therefore have less of an impact on pH than do strong acids and strong bases. **Carbonic acid** (H_2CO_3) is a weak acid found in body fluids. In solution, carbonic acid reversibly dissociates into a hydrogen ion and a **bicarbonate ion** (HCO_3^-).

$$H_2CO_3 \rightleftharpoons H^+ + HCO_3^-$$

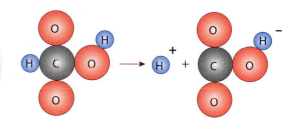

Salts

6 A **salt** is an ionic compound consisting of any cation except a hydrogen ion and any anion except a hydroxide ion. Because they are held together by ionic bonds, many salts dissociate completely in water, releasing cations and anions. For example, sodium chloride (table salt) dissociates immediately in water, releasing Na^+ and Cl^-, the most abundant ions in body fluids. The ionization of sodium chloride does not affect the local concentrations of hydrogen ions or hydroxide ions, so NaCl, like many salts, is a "neutral" solute. Other salts may indirectly affect the concentrations of H^+ and OH^-, making a solution slightly acidic or slightly basic.

$$NaCl \longrightarrow Na^+ + Cl^-$$

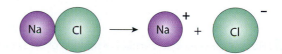

Buffers

Buffers are compounds that stabilize the pH of a solution by removing or replacing hydrogen ions. **Buffer systems** typically involve a weak acid and its related salt, which functions as a weak base. For example, the body's carbonic acid–bicarbonate buffer system consists of carbonic acid (H_2CO_3) and sodium bicarbonate, (NaHCO$_3$), otherwise known as baking soda. Buffers and buffer systems in body fluids help maintain pH within normal limits.

Module 2.9 Review

a. Define pH.

b. Explain the differences among an acid, a base, and a salt.

c. What is the significance of pH in physiological systems?

1. Short answer

List four properties of water important to the functioning of the human body.

a _____

b _____

c _____

d _____

2. Matching

Match the following terms with the most closely related description.

- solvent
- water
- buffers
- hydrophilic
- inorganic compounds
- hydrophobic
- acid
- solute
- alkaline
- salt

a _____ HCl, NaOH, and NaCl

b _____ A dissolved substance

c _____ A solution with a pH greater than 7

d _____ Molecules that readily interact with water

e _____ Fluid medium of a solution

f _____ Ionic compound not containing hydrogen ions or hydroxide ions

g _____ Compounds that stabilize pH in body fluids

h _____ Solution with a pH of 6.5

i _____ Molecules that do not interact with water

j _____ Makes up two-thirds of human body weight

3. Short answer

Identify the regions a–c on the pH scale below.

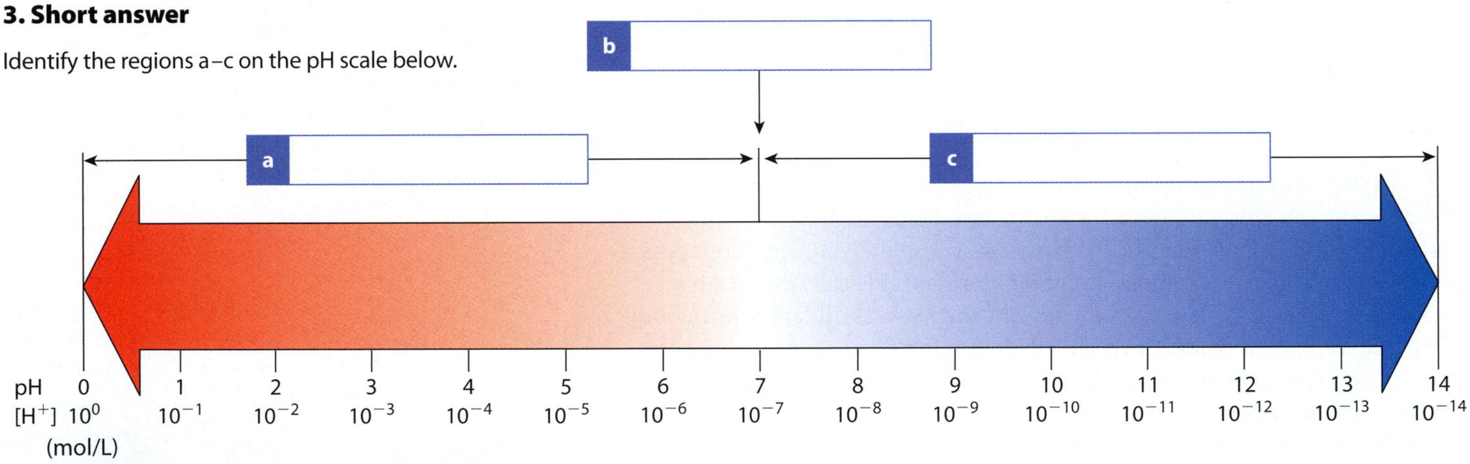

b [_____]

a [_____]   c [_____]

pH	0	1	2	3	4	5	6	7	8	9	10	11	12	13	14
[H$^+$] (mol/L)	10^0	10^{-1}	10^{-2}	10^{-3}	10^{-4}	10^{-5}	10^{-6}	10^{-7}	10^{-8}	10^{-9}	10^{-10}	10^{-11}	10^{-12}	10^{-13}	10^{-14}

d How much more or less acidic is a solution of pH 3 compared to one with a pH of 6? _____

e Describe three negative effects of abnormal pH fluctuations in the human body.

_____

4. Section integration

The addition of table salt to pure water does not result in a change in its pH. Why?

_____

_____

Organic Compounds

Organic compounds always contain the elements carbon and hydrogen, and generally oxygen as well. Many organic molecules are made up of long chains of carbon atoms linked by covalent bonds. The carbon atoms typically form additional covalent bonds with hydrogen or oxygen atoms and, less commonly, with nitrogen, phosphorus, sulfur, iron, or other elements. Many organic molecules are soluble in water. Although organic compounds are diverse, certain groupings of atoms occur again and again, even in very different types of molecules. These **functional groups** greatly influence the properties of any molecule in which they occur.

Important Functional Groups of Organic Compounds

Functional Group	Structural Formula*		Importance	Examples
Carboxylic acid group, – COOH	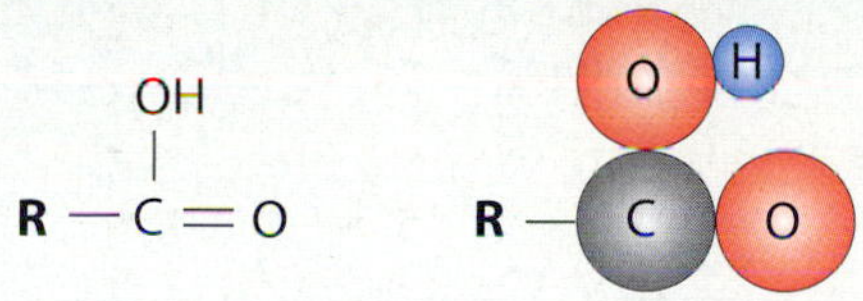 R — C = O (with OH above)	(O, H, C, O ball diagram) R — C ⟨O, O—H⟩	Acts as an acid, releasing H^+ to become R–COO⁻	• Fatty acids • Amino acids
Amino group, – NH₂	R — N ⟨H, H⟩	R — N (with H, H ball diagram)	Can accept or release H^+, depending on pH; can form bonds with other molecules	• Amino acids
Hydroxyl group, – OH	R — O — H	R — O — H (ball diagram)	May link molecules through dehydration synthesis; hydrogen bonding between hydroxyl groups and water molecules affect solubility	• Carbohydrates • Fatty acids • Amino acids
Phosphate group, – PO₄	R — O — P — O (with O above and O below)	R — O — P — O (ball diagram with O above and below)	May link other molecules to form larger structures; may store energy	• Phospholipids • Nucleic acids • High-energy compounds

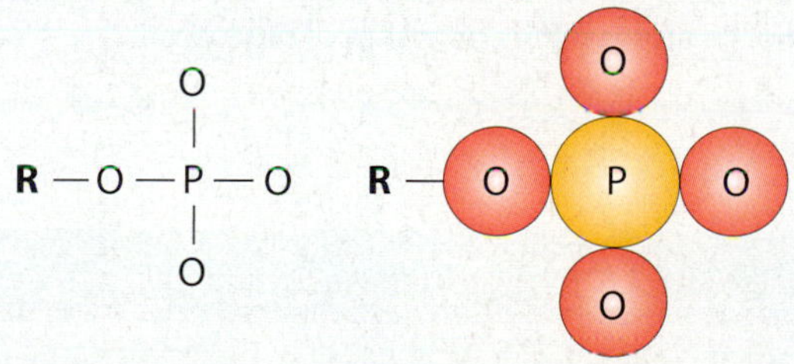

* The term R group is used to denote the rest of the molecule, whatever that might be. The R group is also known as a side chain.

In this section we will introduce the major classes of organic compounds. We will also consider how enzymes facilitate essential reactions within living cells, and how cells capture and transfer energy with special **high-energy compounds**.

Carbohydrates contain carbon, hydrogen, and oxygen, usually in a 1:2:1 ratio

A **carbohydrate** is an organic molecule that contains carbon, hydrogen, and oxygen in a ratio near 1:2:1. Familiar carbohydrates include the sugars and starches that make up roughly half of the typical U.S. diet. Carbohydrates typically account for less than 1.5 percent of total body weight. Although they may have other functions, carbohydrates are most important as energy sources that are catabolized rather than stored.

Monosaccharides

1 A simple sugar, or **monosaccharide** (mon-ō-SAK-uh-rīd; *mono-*, single + *sakcharon*, sugar), is a carbohydrate containing from three to seven carbon atoms. A monosaccharide can be called a triose (three-carbon), tetrose (four-carbon), pentose (five-carbon), hexose (six-carbon), or heptose (seven-carbon). The hexose **glucose** (GLOO-kōs) is the most important metabolic "fuel" in the body.

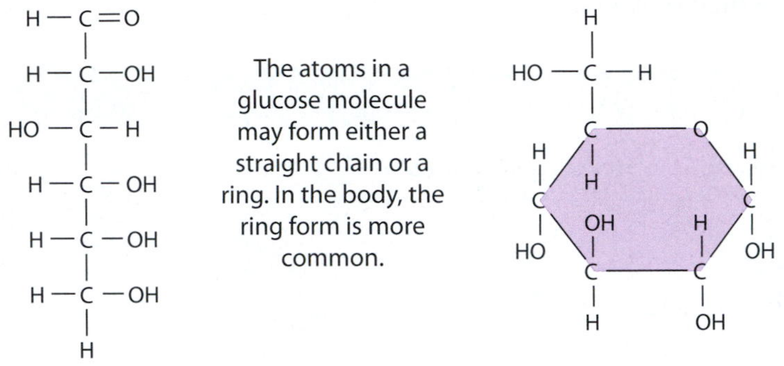

The atoms in a glucose molecule may form either a straight chain or a ring. In the body, the ring form is more common.

2 The three-dimensional structure of an organic molecule is an important characteristic, because it usually determines the molecule's fate or function. Some molecules have the same molecular formula—in other words, the same types and numbers of atoms—but different structures. Such molecules are called **isomers**. The body usually treats different isomers as distinct molecules.

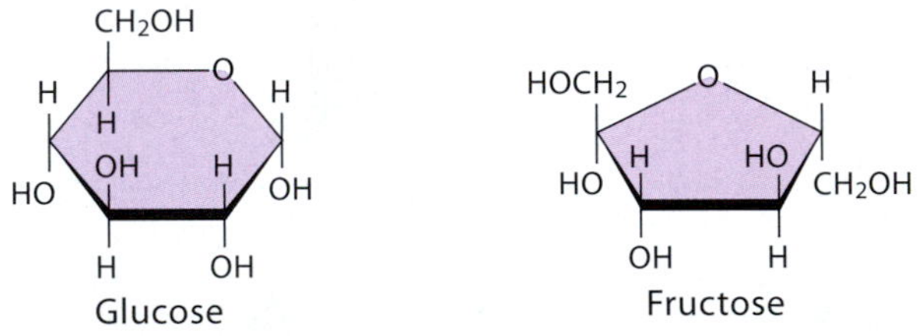

Glucose

Fructose

The monosaccharides glucose and **fructose** are isomers. Fructose is a hexose found in many fruits. Although its chemical formula, $C_6H_{12}O_6$, is the same as that of glucose, the arrangement of its atoms differs from that of glucose.

Carbohydrates in the Body

Structural Class	Examples	Primary Function	Remarks
Monosaccharides (simple sugars)	Glucose, fructose	Energy source	Manufactured in the body and obtained from food; distributed in body fluids
Disaccharides	Sucrose, lactose, maltose	Energy source	Sucrose is table sugar, lactose is in milk, and maltose is malt sugar; all must be broken down to monosaccharides before absorption
Polysaccharides	Glycogen	Storage of glucose	Glycogen is in animal cells; other starches and cellulose are within or around plant cells

Disaccharides

Two monosaccharides joined together form a **disaccharide** (dī-SAK-uh-rīd; *di-*, two). Disaccharides such as **sucrose** (table sugar) have a sweet taste and, like monosaccharides, are quite soluble in water. The formation of sucrose involves dehydration synthesis.

3 During dehydration synthesis, two molecules are joined by the removal of a water molecule.

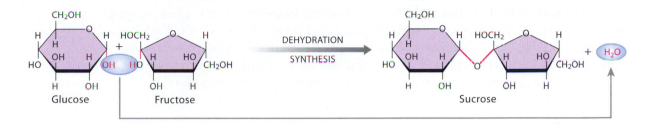

4 Hydrolysis reverses the steps of dehydration synthesis; a complex molecule is broken down by the addition of a water molecule.

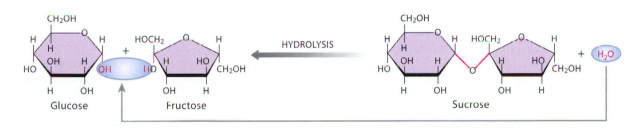

Polysaccharides

5 More-complex carbohydrates result when repeated dehydration synthesis reactions add additional monosaccharides or disaccharides. These large molecules are called **polysaccharides** (pol-ē-SAK-uh-rīdz; *poly-*, many). **Starches** are large polysaccharides formed from glucose molecules. Your digestive tract can break these molecules into monosaccharides. Starches such as those in potatoes and grains are a major dietary energy source.

Glucose molecules

The polysaccharide **glycogen** (GLĪ-kō-jen), or animal starch, has many side branches consisting of chains of glucose molecules. Muscle cells make and store glycogen. When these cells have a high demand for glucose, glycogen molecules are broken down; when the demand is low, they absorb glucose from the bloodstream and rebuild glycogen reserves.

Module 2.10 Review

a. List the three structural classes of carbohydrates, and give an example of each.

b. A food contains organic molecules with the elements C, H, and O in a ratio of 1:2:1. What class of compounds do these molecules belong to, and what are their major functions in the body?

c. Predict the reactants and the type of chemical reaction involved when muscle cells make and store glycogen.

Lipids often contain a carbon-to-hydrogen ratio of 1:2

Like carbohydrates, **lipids** (*lipos*, fat) contain carbon, hydrogen, and oxygen, and the carbon-to-hydrogen ratio is typically near 1:2. However, lipids contain much less oxygen than do carbohydrates with the same number of carbon atoms. The hydrogen-to-oxygen ratio is therefore very large; a representative lipid, such as lauric acid, has a formula of $C_{12}H_{24}O_2$. Lipids may also contain small quantities of phosphorus, nitrogen, or sulfur. Familiar lipids include fats, oils, and waxes. Most lipids are insoluble in water, but special transport mechanisms carry them in the circulating blood.

Fatty Acids

1 **Fatty acids** are long carbon chains with hydrogen atoms attached. One end of the carbon chain, called the head, always bears a **carboxylic** (kar-bok-SIL-ik) **acid group**: —COOH.

2 In a **saturated fatty acid**, each carbon atom in the tail has four single covalent bonds.

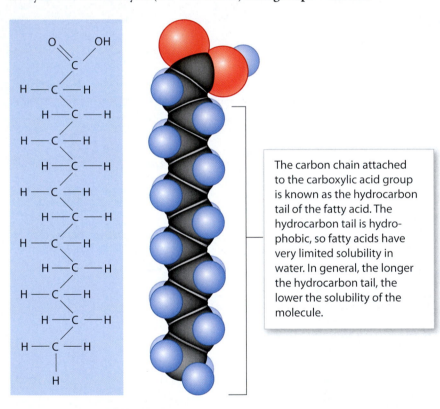

The carbon chain attached to the carboxylic acid group is known as the hydrocarbon tail of the fatty acid. The hydrocarbon tail is hydrophobic, so fatty acids have very limited solubility in water. In general, the longer the hydrocarbon tail, the lower the solubility of the molecule.

Lauric acid ($C_{12}H_{24}O_2$)

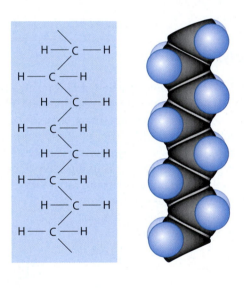

3 In an **unsaturated fatty acid**, one or more of the single covalent bonds between carbon atoms has been replaced by a double covalent bond. As a result, each carbon atom involved will bind only one hydrogen atom rather than two. This changes both the shape of the hydrocarbon tail and the way the fatty acid is metabolized. A monounsaturated fatty acid has a single double bond in the hydrocarbon tail. A polyunsaturated fatty acid contains multiple double bonds.

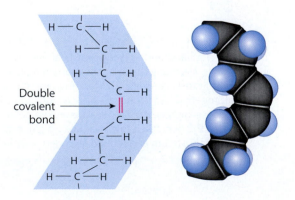

Double covalent bond

Representative Lipids in the Body

Lipid Type	Example(s)	Primary Function(s)	Remarks
Fatty acids	Lauric acid	Energy sources	Absorbed from food or synthesized in cells; transported in the blood
Glycerides	Monoglycerides, diglycerides, triglycerides	Energy sources, energy storage, insulation, and physical protection	Stored in fat deposits; must be broken down to fatty acids and glycerol before they can be used as an energy source
Eicosanoids	Prostaglandins, leukotrienes	Chemical messengers coordinating local cellular activities	Prostaglandins are produced in most body tissues
Steroids	Cholesterol	Structural components of cell membranes, hormones, digestive secretions in bile	All steroids have the same carbon ring framework
Phospholipids, glycolipids	Lecithin (a phospholipid)	Structural components of cell membranes	Derived from fatty acids and nonlipid components

Glycerides

4 Individual fatty acids cannot be strung together in a chain by dehydration synthesis. But they can be attached to another compound, **glycerol** (GLIS-er-ol), through a similar reaction. The result is a lipid known as a **glyceride** (GLIS-er-īd).

Glycerol   Fatty acids

Fatty Acid 1 — Saturated

Fatty Acid 2 — Saturated

Fatty Acid 3 — Unsaturated

HYDROLYSIS   DEHYDRATION SYNTHESIS

Triglyceride

Lipids form essential structural components of all cells. In addition, lipid deposits are important as energy reserves. On average, lipids provide roughly twice as much energy as carbohydrates do, gram for gram, when broken down in the body. Lipids normally account for 12–18 percent of total body weight of adult men, and 18–24 percent in adult women. You are unable to synthesize all of the lipids needed by the body, and several fatty acids must be obtained from the diet.

Dehydration synthesis can produce a **monoglyceride** (mon-ō-GLI-ser-īd), consisting of glycerol + one fatty acid. Subsequent reactions can yield a **diglyceride** (glycerol + two fatty acids) and then a **triglyceride** (glycerol + three fatty acids). Hydrolysis breaks the glycerides into fatty acids and glycerol.

Module 2.11 Review

a. Describe lipids.

b. Describe the structures of saturated and unsaturated fatty acids.

c. In the hydrolysis of a triglyceride, what are the reactants and the products?

Eicosanoids, steroids, phospholipids, and glycolipids have diverse functions

Lipids are important as chemical messengers and as components of cellular structures. **Structural lipids** help form and maintain intracellular and extracellular membranes. At the cellular level, membranes are sheets or layers composed primarily of hydrophobic lipids. Functionally, a membrane is an effective barrier that can separate two aqueous solutions of differing composition.

Eicosanoids

1 **Eicosanoids** (ī-KŌ-sa-noydz) are lipids derived from arachidonic (ah-rak-i-DON-ik) acid, a fatty acid that must be absorbed in the diet because it cannot be synthesized by the body. **Leukotrienes** are produced primarily by cells involved with coordinating the responses to injury or disease, and they will be considered in later chapters.

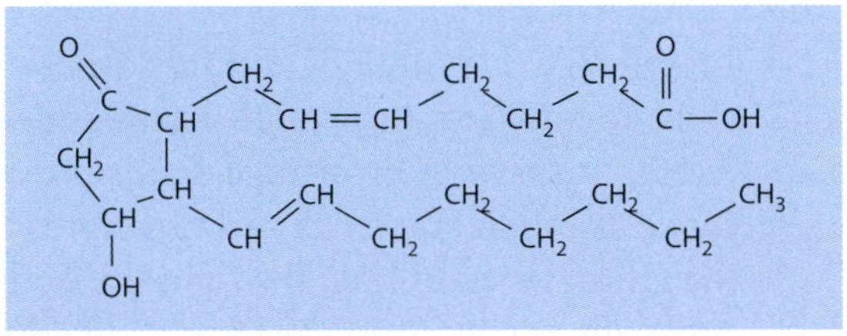

Prostaglandins (pros-tuh-GLAN-dinz) are short-chain fatty acids in which five of the carbon atoms are joined in a ring. These compounds are released by cells to coordinate or direct local cellular activities, and they are extremely powerful, even in minute quantities. Small amounts of prostaglandins released by damaged tissues, for example, stimulate nerve endings and produce the sensation of pain.

Steroids

2 **Steroids** are large lipid molecules that share a distinctive carbon-ring framework. They differ in the functional groups that are attached to the basic ring structure.

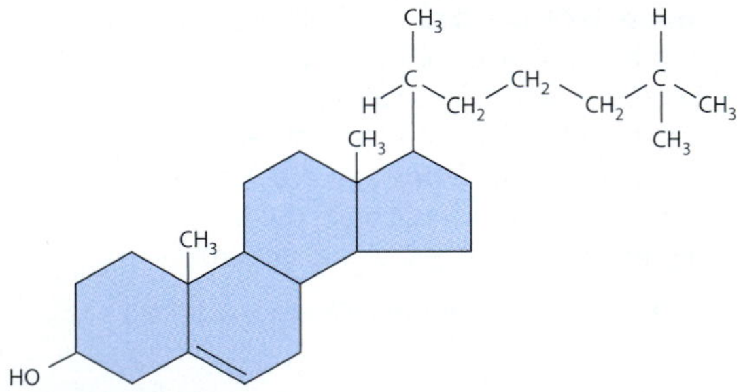

All animal plasma membranes contain **cholesterol** (kōh-LES-ter-ol; *chole-*, bile + *stereos*, solid). Cells need cholesterol to maintain their plasma membranes, as well as for cell growth and division.

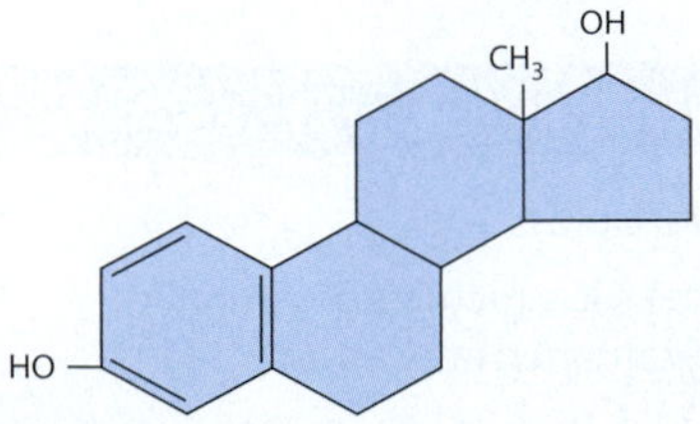

Cholesterol

Steroid hormones are involved in the regulation of sexual function. Examples include the sex hormones estrogen and testosterone.

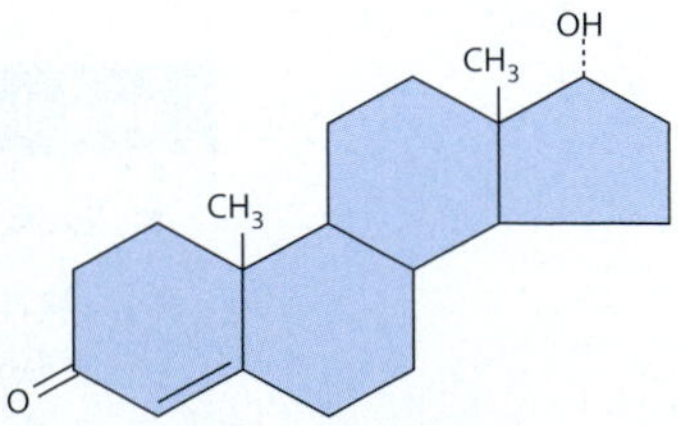

Estrogen   **Testosterone**

Phospholipids and Glycolipids

3 **Phospholipids** (FOS-fō-lip-idz) and **glycolipids** (GLĪ-kō-lip-idz) are structurally related, and our cells can synthesize both types of lipids, primarily from fatty acids.

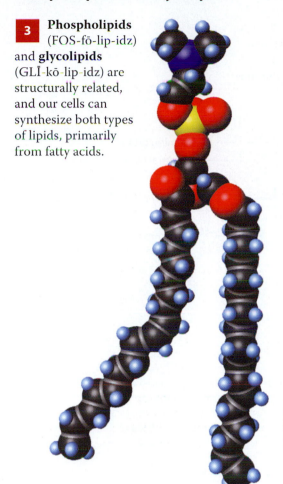

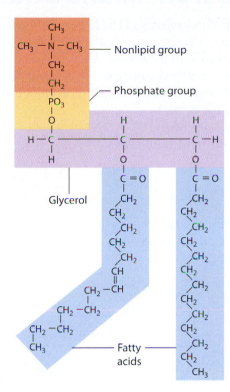

Nonlipid group

Phosphate group

Glycerol

Fatty acids

In a phospholipid, a phosphate group links a diglyceride to a nonlipid group.

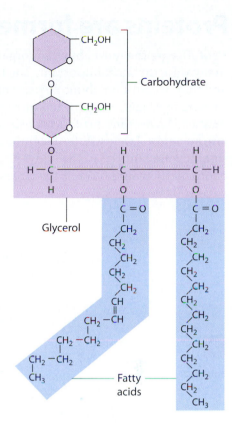

Carbohydrate

Glycerol

Fatty acids

In a glycolipid, a carbohydrate is attached to a diglyceride.

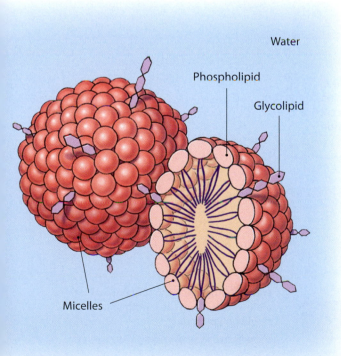

Water

Phospholipid

Glycolipid

Micelles

4 The long hydrocarbon tails of phospholipids and glycolipids are hydrophobic, but the opposite ends, the nonlipid heads, are hydrophilic. In water, large numbers of these molecules tend to form droplets, or **micelles** (mī-SELZ), with the hydrophilic portions on the outside. Most meals contain a mixture of lipids and other organic molecules, and micelles form as the digestive tract breaks the food down.

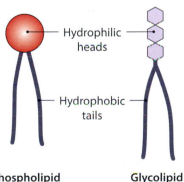

Hydrophilic heads

Hydrophobic tails

Phospholipid

Glycolipid

Module 2.12 Review

a. Why is cholesterol necessary in the body?

b. Describe the basic functions of eicosanoids, steroids, phospholipids, and glycolipids.

c. Describe the orientations of phospholipids and glycolipids when they form a micelle.

Proteins are formed from amino acids

Proteins are the most abundant organic components of the human body, and in many ways the most important. The human body contains many different proteins, and they account for about 20 percent of total body weight. All proteins contain carbon, hydrogen, oxygen, and nitrogen; smaller quantities of sulfur and phosphorus may also be present. Proteins consist of long chains of organic molecules called **amino acids**. Twenty different amino acids occur in significant quantities in the body. A typical protein contains 1000 amino acids; the largest protein complexes have 100,000 or more.

1 Every amino acid has the same basic structural elements.

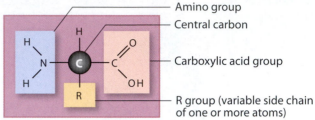

- Amino group
- Central carbon
- Carboxylic acid group
- R group (variable side chain of one or more atoms)

2 Two amino acids can be linked together by dehydration synthesis.

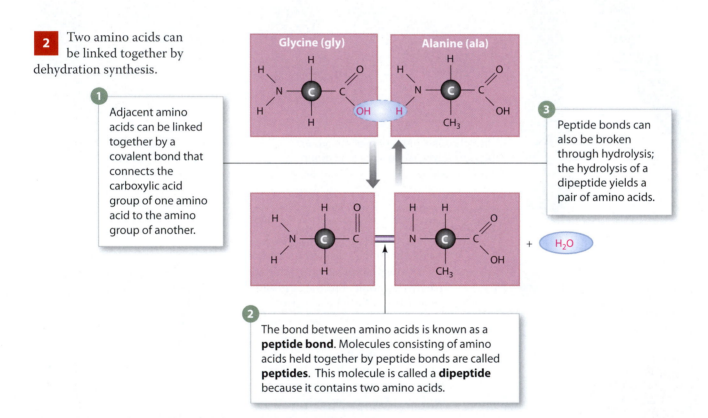

1 Adjacent amino acids can be linked together by a covalent bond that connects the carboxylic acid group of one amino acid to the amino group of another.

3 Peptide bonds can also be broken through hydrolysis; the hydrolysis of a dipeptide yields a pair of amino acids.

2 The bond between amino acids is known as a **peptide bond**. Molecules consisting of amino acids held together by peptide bonds are called **peptides**. This molecule is called a **dipeptide** because it contains two amino acids.

3 The chain can be lengthened by the addition of more amino acids. Attaching a third amino acid produces a tripeptide. Tripeptides and larger peptide chains are called **polypeptides**. Polypeptides containing more than 100 amino acids are usually called proteins. Proteins can have four levels of structural complexity. The first level of complexity is called the **primary structure**.

Primary structure is the sequence of amino acids along the length of a single polypeptide.

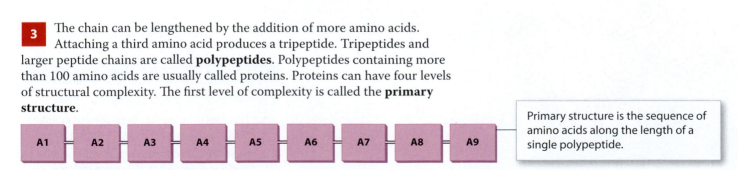

Primary Structure

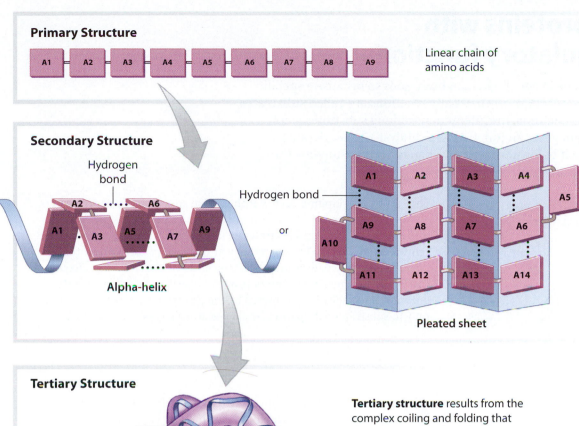

A1 | A2 | A3 | A4 | A5 | A6 | A7 | A8 | A9

Linear chain of amino acids

Secondary Structure

Hydrogen bond

Hydrogen bond

A2 | A6
A1 | A3 | A5 | A7 | A9

Alpha-helix

or

A1 | A2 | A3 | A4
A5
A9 | A8 | A7 | A6
A10
A11 | A12 | A13 | A14

Pleated sheet

Secondary structure results from bonds between atoms at different parts of the polypeptide chain. Hydrogen bonding, for example, may create either a simple spiral, known as an alpha-helix, or a flat pleated sheet.

Tertiary Structure

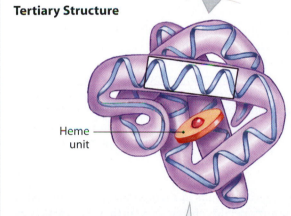

Heme unit

Tertiary structure results from the complex coiling and folding that gives a protein its final three-dimensional shape. Tertiary structure results primarily from interactions between the polypeptide chain and the surrounding water molecules, and to a lesser extent from interactions between the R groups of amino acids in different parts of the molecule.

As temperatures rise, protein shape changes and enzyme function deteriorates. Eventually the protein undergoes **denaturation**, a change in tertiary or quaternary structure that makes it nonfunctional. Death occurs at very high body temperatures (above 43°C, or 110°F) because the denaturation of structural proteins and enzymes soon causes irreparable damage to organs and organ systems.

Quaternary Structure

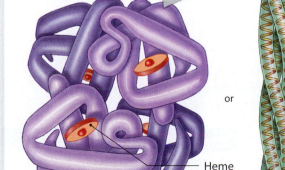

Heme unit

Hemoglobin
(globular protein)

or

Keratin
(fibrous protein)

Quaternary structure results from the interaction between individual polypeptide chains to form a protein complex. The protein **hemoglobin** contains four polypeptide subunits. Hemoglobin is found within red blood cells, where it binds and transports oxygen. It is an example of a globular protein. In **keratin** and **collagen**, three linear subunits intertwine, forming a fibrous protein. The three-dimensional shape of a protein plays an essential role in determining its functional properties.

Module 2.13 Review

a. Describe proteins.

b. What kind of bond forms during the dehydration synthesis of two amino acids?

c. Why does boiling a protein affect its structural and functional properties?

Enzymes are proteins with important regulatory functions

Almost everything that happens inside the human body does so because a specific enzyme makes it possible. The reactants in enzymatic reactions are called **substrates**. As in other types of chemical reactions, the interactions among substrates yield specific products. Before an enzyme can function as a catalyst—to accelerate a chemical reaction without itself being permanently changed or consumed—the substrates must bind to a special region of the enzyme.

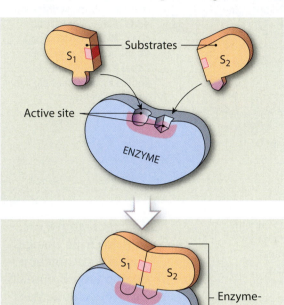

1 Substrate binding occurs at the **active site**, typically a groove or pocket into which one or more substrates nestle, like a key fitting into a lock. Weak electrical attractive forces, such as hydrogen bonding, reinforce the physical fit. The tertiary or quaternary structure of the enzyme molecule determines the shape of the active site. Each enzyme catalyzes only one type of reaction, a characteristic called **specificity**. An enzyme's specificity is determined by the ability of its active sites to bind only to substrates with particular shapes and charges.

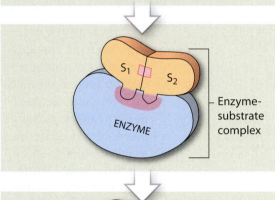

2 Substrate binding produces an **enzyme-substrate complex**. Each cell contains an assortment of enzymes, and any particular enzyme may be active under one set of conditions and inactive under another. Virtually anything that changes the tertiary or quaternary shape of an enzyme can turn it "on" or "off" by changing the properties of the active site and preventing formation of an enzyme-substrate complex. Because the change is immediate, enzyme activation or inactivation is an important method of short-term control over reaction rates and pathways.

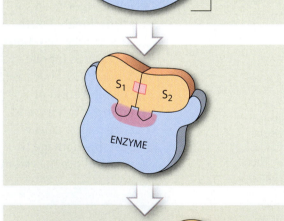

3 Substrate binding typically results in a temporary, reversible change in the shape of the enzyme; this change may further the reaction by placing physical stresses on the substrate molecules. The enzyme then promotes product formation. In some cases, the change in enzyme shape that accompanies substrate binding is sufficient to catalyze the reaction. In other cases, an external source must provide the activation energy required.

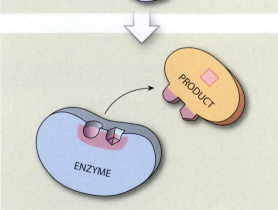

4 The completed product then detaches from the active site, and the enzyme is free to repeat the process. When every enzyme molecule is cycling through its reaction sequence at top speed, further increases in substrate concentration will not affect the rate of reaction. The substrate concentration required to have the maximum rate of reaction is called the **saturation limit**.

Module 2.14 Review

a. Define active site.

b. What are the reactants in an enzymatic reaction called?

c. Relate an enzyme's structure to its reaction specificity.

2.15

High-energy compounds may store and transfer a portion of energy released during enzymatic reactions

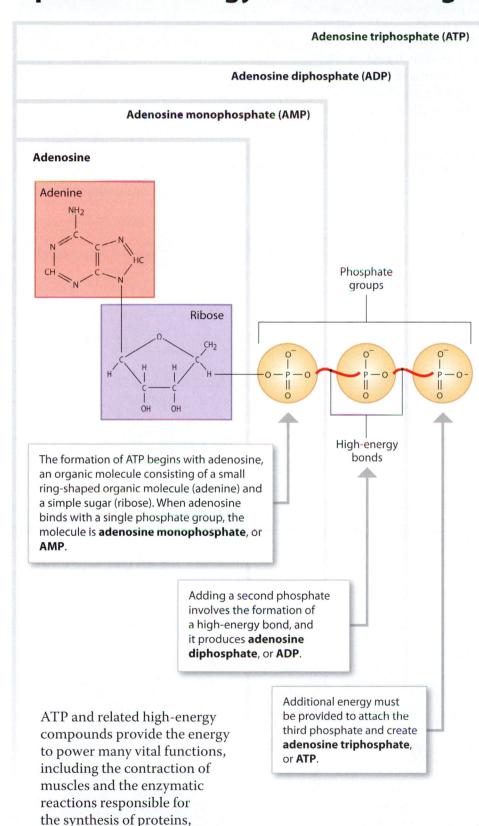

Adenosine triphosphate (ATP)

Adenosine diphosphate (ADP)

Adenosine monophosphate (AMP)

Adenosine

Adenine

Ribose

Phosphate groups

High-energy bonds

The formation of ATP begins with adenosine, an organic molecule consisting of a small ring-shaped organic molecule (adenine) and a simple sugar (ribose). When adenosine binds with a single phosphate group, the molecule is **adenosine monophosphate**, or **AMP**.

Adding a second phosphate involves the formation of a high-energy bond, and it produces **adenosine diphosphate**, or **ADP**.

Additional energy must be provided to attach the third phosphate and create **adenosine triphosphate**, or **ATP**.

ATP and related high-energy compounds provide the energy to power many vital functions, including the contraction of muscles and the enzymatic reactions responsible for the synthesis of proteins, carbohydrates, and lipids.

1 Enzymes may catalyze synthesis, decomposition, or exchange reactions. When product formation requires an energy donor, that donor is typically a **high-energy compound**. High-energy compounds contain **high-energy bonds**, covalent bonds whose breakdown releases energy under controlled conditions. The most common high-energy compound is **adenosine triphosphate**, or **ATP**.

2 The formation of ATP from ADP is a reversible reaction; the energy stored when ATP forms is released when it breaks down to ADP. Cells can synthesize ATP in one location and then break it down in another, harnessing the energy released to power essential activities.

ADP + ENERGY + P ⇌ ATP

Module 2.15 Review

a. Where do cells obtain the energy needed for their vital functions?

b. Describe ATP.

c. Compare AMP with ADP.

DNA and RNA are nucleic acids

Nucleic (noo-KLĀ-ik) **acids** are large organic molecules composed of carbon, hydrogen, oxygen, nitrogen, and phosphorus. The two classes of nucleic acid molecules are **deoxyribonucleic** (dē-oks-ē-rī-bō-noo-KLĀ-ik) **acid**, or **DNA**, and **ribonucleic** (rī-bō-noo-KLĀ-ik) **acid**, or **RNA**. The primary role of nucleic acids is the storage and transfer of information—specifically, information essential to the synthesis of proteins by cells. A nucleic acid consists of one or two long chains that are formed by dehydration synthesis. The individual subunits of a nucleic acid are called **nucleotides**.

1 A typical nucleotide consists of a phosphate group, a sugar, and an organic molecule known as a **nitrogenous base** that may be either a **purine** or a **pyrimidine**. Adenosine monophosphate (AMP) is an example of a nucleotide that you have already encountered.

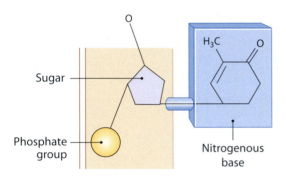

2 The phosphate and sugar of adjacent nucleotides can be strung together by dehydration synthesis, creating the long chains that comprise functional nucleic acids. The "backbone" of this molecule is a linear sugar-to-phosphate-to-sugar sequence, with the nitrogenous bases projecting to one side. In both DNA and RNA, it is the sequence of nitrogenous bases that carries the information.

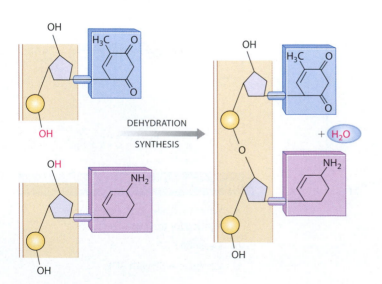

Nitrogenous bases

The purines **adenine** and **guanine** are found in both DNA and RNA.

There are three important pyrimidines. DNA and RNA both contain **cytosine**. **Thymine** is found only in DNA, and **uracil** is found only in RNA.

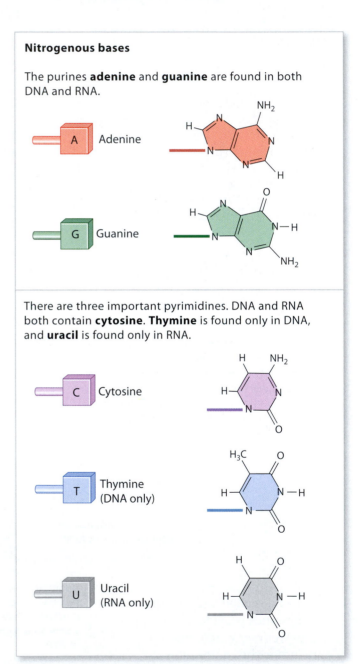

A Comparison of DNA with RNA

Characteristic	DNA	RNA
Sugar	Deoxyribose	Ribose
Nitrogenous bases	Adenine (A), guanine (G), cytosine (C), thymine (T)	Adenine, guanine, cytosine, uracil (U)
Number of nucleotides in typical molecule	Always more than 45 million	Varies from fewer than 100 to about 50,000
Shape of molecule	Paired strands coiled in a double helix	Varies with hydrogen bonding along the length of the strand of each of the three main types (mRNA, tRNA, rRNA)
Function	Stores genetic information that controls protein synthesis	Performs protein synthesis as directed by DNA

3 A DNA molecule consists of a pair of nucleotide chains. Hydrogen bonding between opposing nitrogenous bases holds the two strands together. The shapes of the nitrogenous bases allow adenine to bond only to thymine, and cytosine to bond only to guanine. As a result, the combinations adenine–thymine (A-T) and cytosine–guanine (C-G) are known as **complementary base pairs**, and the two nucleotide chains of the DNA molecule are known as **complementary strands**.

4 A molecule of RNA consists of a single chain of nucleotides. Its shape, and thus its function, depends on the order of the nucleotides and the interactions among them. Our cells have three types of RNA: (1) **messenger RNA** (**mRNA**), (2) **transfer RNA** (**tRNA**), and (3) **ribosomal RNA** (**rRNA**). Their specific functions will be considered in Chapter 3.

Deoxyribose

Phosphate group

Hydrogen bond

Adenine Thymine

The two strands of DNA twist around one another in a double helix that resembles a spiral staircase.

DNA strand 1

DNA strand 2

Hydrogen bond

Cytosine Guanine

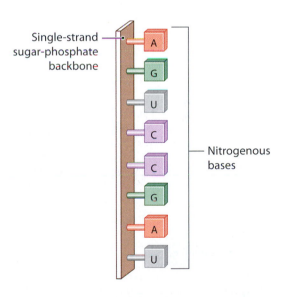

Single-strand sugar-phosphate backbone

Nitrogenous bases

Module 2.16 Review

a. Describe nucleic acids.

b. Explain how the complementary strands of DNA are held together.

c. A large organic molecule composed of ribose, nitrogenous bases, and phosphate groups is which kind of nucleic acid?

1. Matching

Match the following terms with the most closely related description.

- monosaccharide
- ATP
- polyunsaturated
- glycerol
- cholesterol
- isomers
- glycogen
- active site
- nucleotide
- RNA
- peptide

a _____ Polysaccharide with an energy-storage role in animal tissues

b _____ Molecules with same chemical formula but different structure

c _____ A fatty acid with more than one C-to-C double covalent bond

d _____ The region of an enzyme that binds the substrate

e _____ Three-carbon molecule that combines with fatty acids

f _____ A steroid essential to plasma membranes

g _____ A high-energy compound consisting of adenosine and three phosphate groups

h _____ A nucleic acid that contains the sugar ribose

i _____ The covalent bond between the carboxylic acid and amino groups of adjacent amino acids

j _____ Organic molecule consisting of a sugar, a phosphate group, and a nitrogenous base

k _____ A simple sugar

2. Vocabulary

In the space provided, write the boldfaced terms introduced in this section that contain the indicated word part.

Word Part	Meaning	Terms
a poly-	many	_____
b tri-	three	_____
c di-	two	_____
d glyco-	sugar	_____

3. Concept map

Use each of the following terms once to fill in the blank boxes to correctly complete the organic compounds concept map.

- lipids
- carbohydrates
- nucleic acids
- disaccharides
- RNA
- fatty acids
- phosphate groups
- glycerol
- polysaccharides
- proteins
- monosaccharides
- ATP
- amino acids
- DNA
- nucleotides

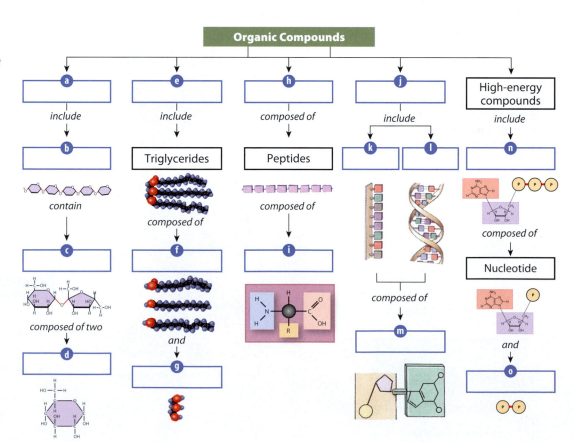

Visual Outline with Key Terms

Summarize the content of each module using the terms in the order provided.

SECTION 1

Atoms and Molecules

- mass
- atoms
- subatomic particles
- protons
- neutrons
- electrons
- nucleus
- electron cloud
- molecule

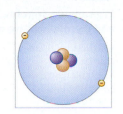

SECTION 2

Chemical Reactions

- reactants
- products
- metabolism
- work
- energy
- kinetic energy
- potential energy

2.1

Typical atoms contain protons, neutrons, and electrons

- atomic number
- mass number
- element
- electron cloud
- electron shell
- isotopes
- atomic weight
- dalton
- trace elements
- chemical symbol

2.5

Chemical notation is a concise method of describing chemical reactions

- chemical notation
- balanced
- mole (mol)
- molecular weight

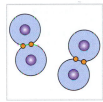

2.2

Electrons occupy various energy levels

- inert
- inert gases
- reactive
- cation
- anion
- chemical bonds

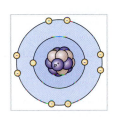

2.6

There are three basic types of chemical reactions

- decomposition
- hydrolysis
- catabolism
- synthesis
- dehydration synthesis
- anabolism
- equilibrium
- exchange reaction

2.3

The most common chemical bonds are ionic bonds and covalent bonds

- compound
- ionic bonds
- covalent bonds
- molecule
- single covalent bond
- double covalent bond
- nonpolar molecules
- polar molecule
- polar covalent bonds

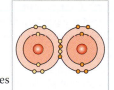

2.7

Enzymes lower the activation energy requirements of chemical reactions

- activation energy
- enzymes
- catalysts
- metabolic pathway
- exergonic
- endergonic
- metabolites
- nutrients
- organic compounds
- inorganic compounds

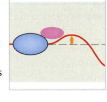

2.4

Matter may exist as a solid, a liquid, or a gas

- hydrogen bond
- surface tension

SECTION 3

The Importance of Water in the Body

- lubrication
- reactivity
- heat capacity
- thermal inertia
- solubility
- solution
- solvent
- solutes
- aqueous solutions

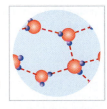

● = Term boldfaced in this module

2.8

Physiological systems depend on water

- ionization
- dissociation
- hydration sphere
- hydrophilic
- electrolytes
- colloid
- suspension
- hydrophobic

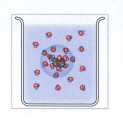

2.9

Regulation of body fluid pH is vital for homeostasis

- hydrogen ion (H⁺)
- hydroxide ion (OH⁻)
- pH
- acidosis
- alkalosis
- acidic
- neutral
- alkaline
- acid
- hydrochloric acid
- base
- sodium hydroxide
- carbonic acid
- bicarbonate ion
- salt
- buffers
- buffer systems

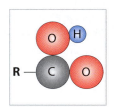

SECTION 4

Organic Compounds

- organic compounds
- functional groups
- carboxylic acid group
- amino group
- hydroxyl group
- phosphate group
- high-energy compounds

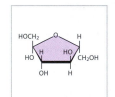

2.10

Carbohydrates contain carbon, hydrogen, and oxygen, usually in a 1:2:1 ratio

- carbohydrate
- monosaccharide
- glucose
- isomers
- fructose
- disaccharide
- sucrose
- polysaccharides
- starches
- glycogen

2.11

Lipids often contain a carbon-to-hydrogen ratio of 1:2

- lipids
- fatty acids
- carboxylic acid group
- saturated fatty acid
- unsaturated fatty acid
- glycerol
- glyceride
- monoglyceride
- diglyceride
- triglyceride

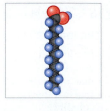

2.12

Eicosanoids, steroids, phospholipids, and glycolipids have diverse functions

- structural lipids
- eicosanoids
- leukotrienes
- prostaglandins
- steroids
- cholesterol
- phospholipids
- glycolipids
- micelles

2.13

Proteins are formed from amino acids

- proteins
- amino acids
- peptide bond
- peptides
- dipeptide
- polypeptides
- primary structure
- secondary structure
- tertiary structure
- quaternary structure
- hemoglobin
- keratin
- collagen
- denaturation

2.14

Enzymes are proteins with important regulatory functions

- substrates
- active site
- enzyme-substrate complex
- saturation limit

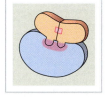

2.15

High-energy compounds may store and transfer a portion of energy released during enzymatic reactions

- high-energy compound
- high-energy bonds
- adenosine triphosphate (ATP)
- adenosine monophosphate (AMP)
- adenosine diphosphate (ADP)

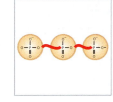

2.16

DNA and RNA are nucleic acids

- nucleic acids
- deoxyribonucleic acid (DNA)
- ribonucleic acid (RNA)
- nucleotides
- nitrogenous base
- purine
- pyrimidine
- adenine
- guanine
- cytosine
- thymine
- uracil
- complementary base pairs
- complementary strands
- messenger RNA (mRNA)
- transfer RNA (tRNA)
- ribosomal RNA (rRNA)

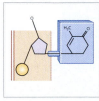

● = *Term boldfaced in this module*

Chapter Integration: Applying what you've learned

Susan, age 19, is a student-athlete in her first year of college. In high school she finished second in her state cross-country meet and was recruited by a number of colleges. She needs her athletic scholarship to continue attending college, so she feels a lot of pressure to do well in her event. During some free time, Susan read in a sports magazine that elite athletes sometimes consume sodium bicarbonate (baking soda) prior to a race to diminish the buildup of lactate in muscle tissue.

As a young woman concerned about weight management, Susan is conscientious about her diet. Thus she consumes no more than 1500 calories per day and keeps her intake of carbohydrates to approximately 55–60 percent of her daily caloric intake. Upon returning to school from a recent vacation, she noticed a weight gain of about 10 pounds due to taking a break from her daily running regime and indulging in many foods she normally avoids. As a quick weight-loss measure, she's considering eliminating all fats from her diet for a week and replacing them with a fat substitute, *Olestra,* a compound the body cannot absorb. While talking with Susan, you learn of her plan. Based on your understanding of basic chemistry, provide answers to the following questions.

1. In principle, what effect would consuming baking soda have on Susan?

2. If Susan consumes roughly 55–60 percent of her calories as carbohydrates, how can carbohydrates only account for less than 1.5 percent of normal body weight?

3. If Susan swallowed 2 tablespoons of sodium bicarbonate dissolved in water on an empty stomach and vomited 5 minutes later, would you expect the vomitus to be slightly acidic or alkaline? Why? (Assume that the natural pH of the stomach is 3.0 because it contains hydrochloric acid, HCl.)

4. Is it prudent for Susan to eliminate all fats from her diet and instead consume a fat substitute such as *Olestra*?

MasteringA&P™

Access more review material online in the Study Area at **www.masteringaandp.com.**

There, you'll find:
- **Chapter guides**
- **Chapter quizzes**
- **Practice tests**
- **Animations**
- **Labeling activities**

- **MP3 Tutor Sessions**
- **Flashcards**
- **Quizzes**
- **A glossary with pronunciations**

3

Cellular Level of Organization

LEARNING OUTCOMES

These Learning Outcomes correspond by number to this chapter's modules and indicate what you should be able to do after completing the chapter.

SECTION 1 • An Introduction to Cells

3.1 Describe the cell and its organelles, including the composition and function of each.

3.2 Describe the chief structural features of the plasma membrane.

3.3 Differentiate among the structures and functions of the cytoskeleton.

3.4 Describe the ribosome and indicate its specific functions.

3.5 Describe the Golgi apparatus and indicate its specific functions.

3.6 Describe mitochondria, indicate their functions, and explain their significance to cellular function.

SECTION 2 • Structure and Function of the Nucleus

3.7 Explain the functions of the cell nucleus, and discuss the nature and importance of the genetic code.

3.8 Summarize the process of protein synthesis.

3.9 Summarize the process of transcription.

3.10 Summarize the process of translation.

SECTION 3 • How Things Enter and Leave the Cell

3.11 Explain the process of diffusion, and identify its significance in the body.

3.12 Explain the process of osmosis, and identify its significance in the body.

3.13 Describe carrier-mediated transport and its role in the absorption and removal of specific substances.

3.14 Describe vesicular transport as a mechanism for facilitating the absorption or removal of specific substances from cells.

SECTION 4 • The Cell Life Cycle

3.15 Describe interphase, and explain its significance.

3.16 Describe the process of mitosis, and the cell life cycle.

3.17 Discuss the relationship between cell division and cancer.

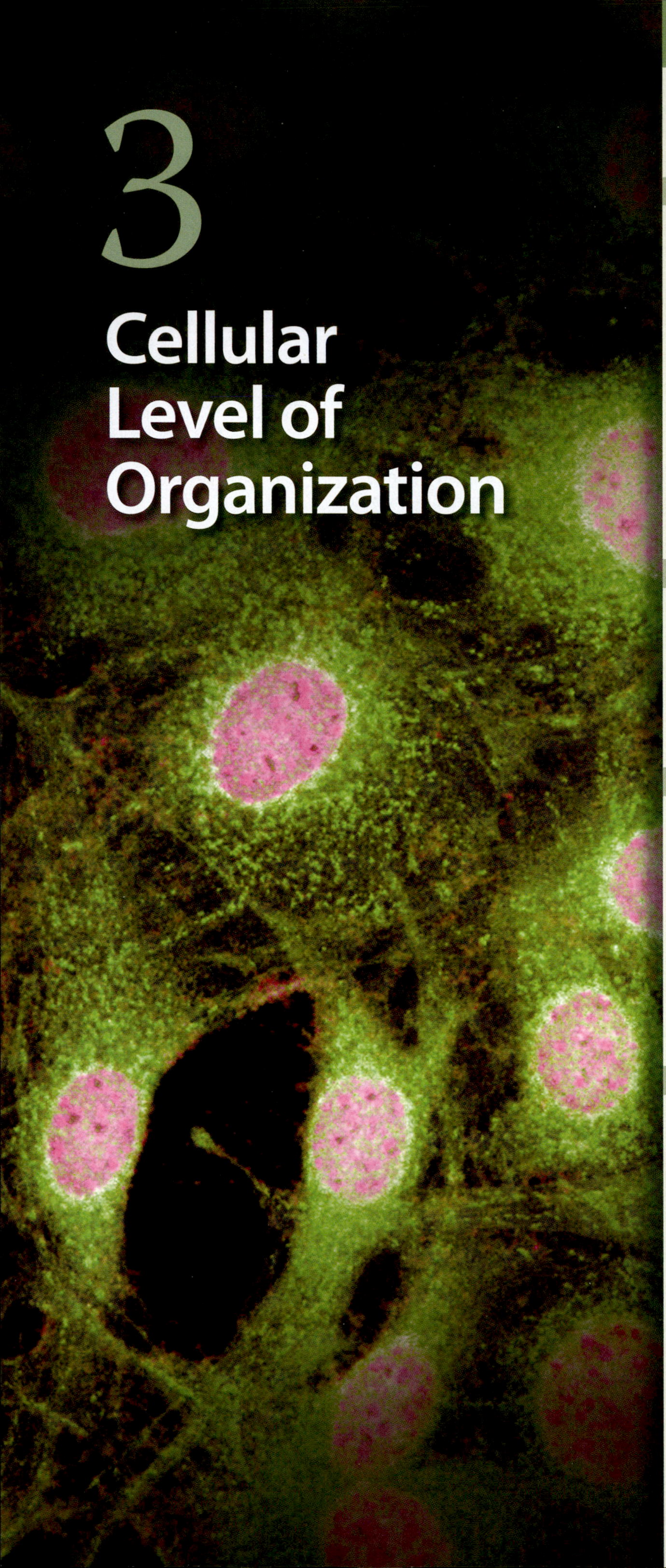

An Introduction to Cells

A typical cell—the smallest living unit in the human body—is only about 0.1 mm in diameter. Thus no one could examine the structure of a cell until relatively effective microscopes were invented in the 17th century. Research in the ensuing years has produced the cell theory, which can be summarized as follows:

Cells are the building blocks of all plants and animals.

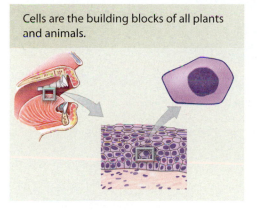

All new cells come from the division of pre-existing cells.

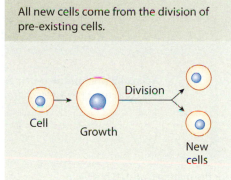

Cells are the smallest units that perform all vital physiological functions.

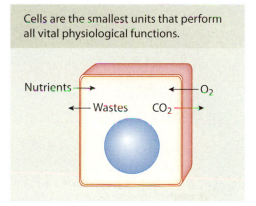

Each cell maintains homeostasis at the cellular level, but it requires the combined and coordinated actions of many cells to achieve homeostasis at higher levels of organization. Although cells vary widely in size, shape, and function, all of them are the descendants of a single cell: the fertilized ovum.

1 At fertilization, the fertilized ovum—which is very large—contains the genetic potential to become any cell in the body.

2 As divisions occur, the cytoplasm is subdivided into smaller parcels. These parcels differ from one another because there were regional differences in the composition of the cytoplasm at fertilization.

3 The cytoplasmic differences affect the DNA of the cells, turning specific genes on or off. The daughter cells begin to develop specialized structural and functional characteristics. This process of gradual specialization is called **differentiation**.

4 Differentiation produces the specialized cells that form the tissues responsible for all body functions.

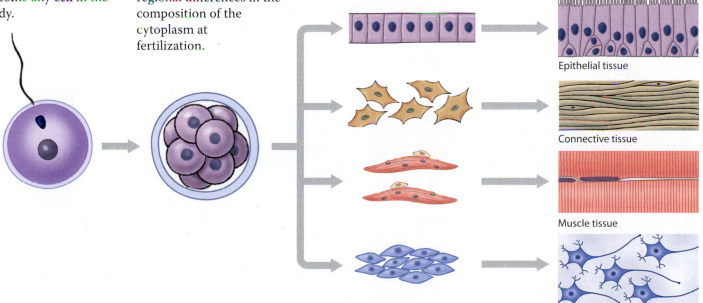

Epithelial tissue

Connective tissue

Muscle tissue

Neural tissue

Cells are the smallest living units of life

Our body cells are surrounded by a watery medium known as the **extracellular fluid**. The extracellular fluid in most tissues is called **interstitial** (in-ter-STISH-ul) **fluid** (*interstitium*, something standing between). This module introduces the major components of our cells, although not all of these components are found in every cell of the body.

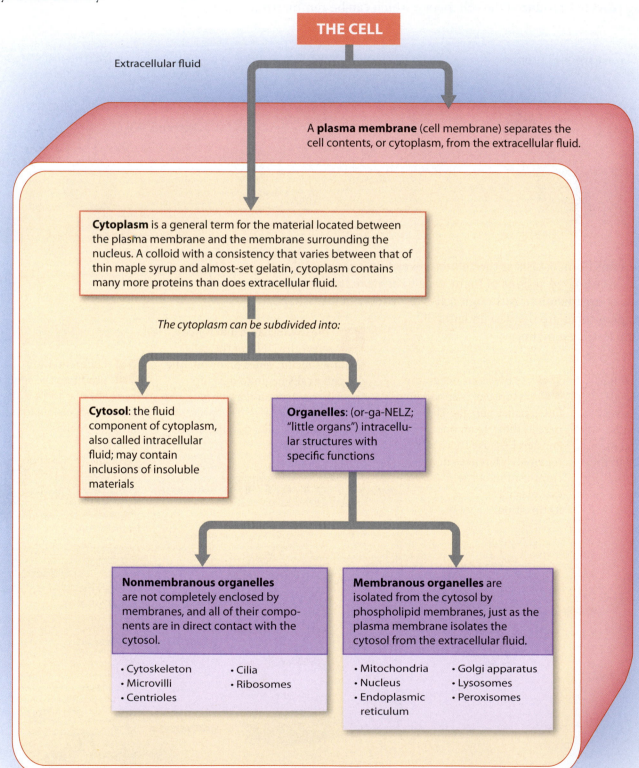

THE CELL

Extracellular fluid

A **plasma membrane** (cell membrane) separates the cell contents, or cytoplasm, from the extracellular fluid.

Cytoplasm is a general term for the material located between the plasma membrane and the membrane surrounding the nucleus. A colloid with a consistency that varies between that of thin maple syrup and almost-set gelatin, cytoplasm contains many more proteins than does extracellular fluid.

The cytoplasm can be subdivided into:

Cytosol: the fluid component of cytoplasm, also called intracellular fluid; may contain inclusions of insoluble materials

Organelles: (or-ga-NELZ; "little organs") intracellular structures with specific functions

Nonmembranous organelles are not completely enclosed by membranes, and all of their components are in direct contact with the cytosol.

- Cytoskeleton
- Microvilli
- Centrioles
- Cilia
- Ribosomes

Membranous organelles are isolated from the cytosol by phospholipid membranes, just as the plasma membrane isolates the cytosol from the extracellular fluid.

- Mitochondria
- Nucleus
- Endoplasmic reticulum
- Golgi apparatus
- Lysosomes
- Peroxisomes

Peroxisome
STRUCTURE: Vesicles containing degradative enzymes. FUNCTION: Breakdown of organic compounds; neutralization of toxic compounds generated in the process

Lysosome
STRUCTURE: Vesicles containing digestive enzymes. FUNCTION: Breakdown of organic compounds and damaged organelles or pathogens

Microvilli
STRUCTURE: Membrane extensions containing microfilaments. FUNCTION: Increase surface area to facilitate absorption of extracellular materials

Golgi apparatus
STRUCTURE: Stacks of flattened membranes (cisternae) containing chambers. FUNCTION: Storage, alteration, and packaging of synthesized products

Nucleus
STRUCTURE: A fluid nucleoplasm containing enzymes, proteins, DNA, and nucleotides; surrounded by a double membrane, the **nuclear envelope.** FUNCTION: Control of metabolism; storage and processing of genetic information; control of protein synthesis

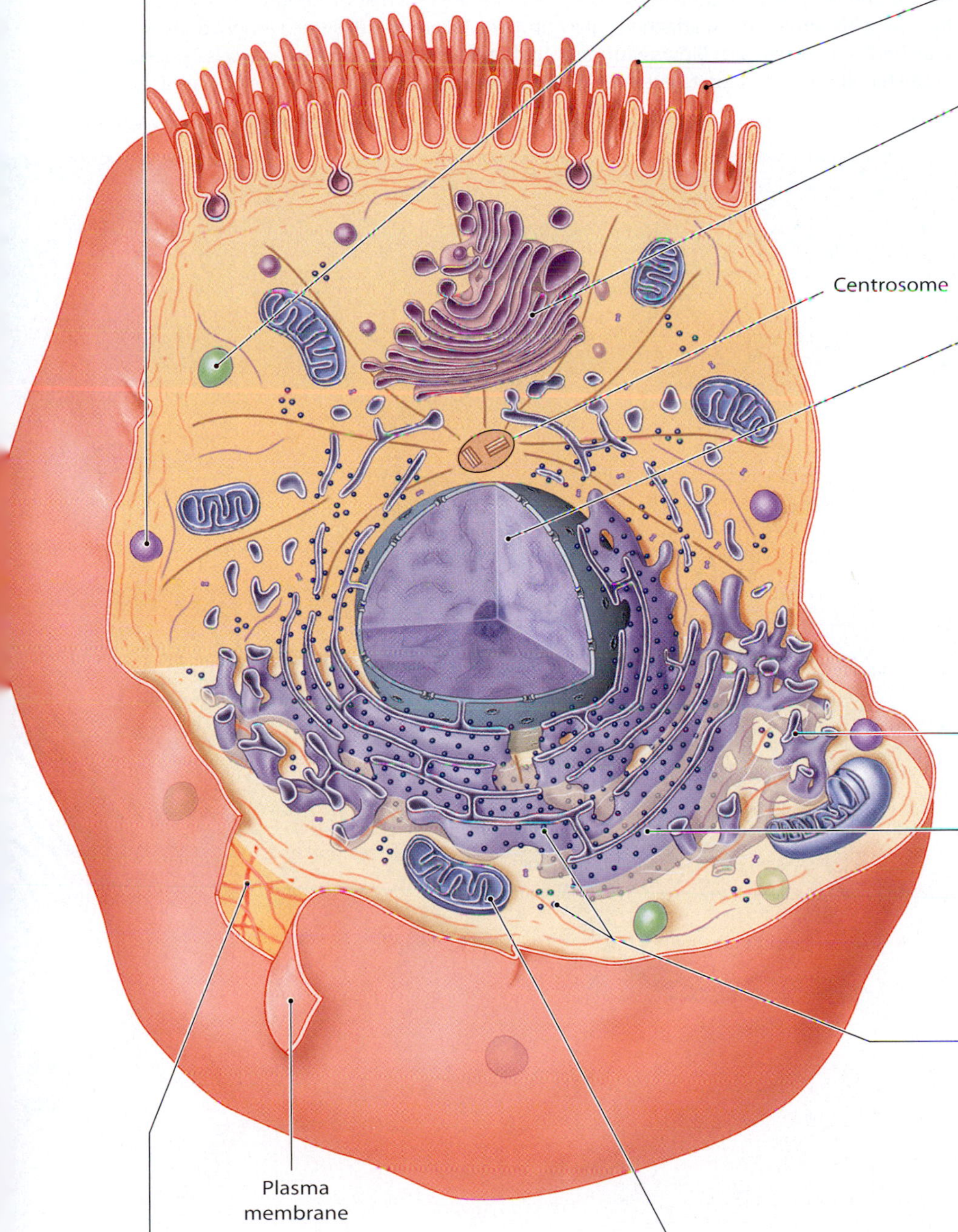

Endoplasmic reticulum (ER)
STRUCTURE: Network of membranous sheets and channels extending throughout the cytoplasm. FUNCTION: Synthesis of secretory products; intracellular storage and transport

- **Smooth ER**, which has no attached ribosomes, synthesizes lipids and carbohydrates.

- **Rough ER**, which has ribosomes bound to the membranes, modifies and packages newly synthesized proteins.

Ribosomes
STRUCTURE: RNA and proteins; fixed ribosomes bound to rough ER, free ribosomes scattered in cytoplasm. FUNCTION: Protein synthesis

Centrosome

Plasma membrane

Cytoskeleton
STRUCTURE: Proteins organized in fine filaments or slender tubes; organizing center located at the **centrosome**, a region that contains a pair of centrioles. FUNCTION: Strength and support; movement of cellular structures and materials

Mitochondria
STRUCTURE: Double membrane, with inner membrane folds enclosing important metabolic enzymes. FUNCTION: Production of 95 percent of the ATP required by the cell

Module 3.1 Review

a. Distinguish between the cytoplasm and cytosol.

b. Describe the functions of the cytoskeleton.

c. Identify the membranous organelles and describe their functions.

The plasma membrane isolates the cell from its environment and performs varied functions

1 The **plasma membrane** is a physical barrier that separates the inside of the cell from the surrounding extracellular fluid. It is a selectively permeable barrier that controls the entry of ions and nutrients, such as glucose; the elimination of wastes; and the release of secretions.

Superficial membrane carbohydrates form a layer known as the **glycocalyx** (glī -kō-KĀ-liks; *calyx*, cup). Carbohydrates account for roughly 3 percent of the weight of a plasma membrane; they are components of complex molecules such as proteoglycans (carbohydrate with some protein attached), glycoproteins (protein with some carbohydrate attached), and glycolipids (lipids with carbohydrates attached). The glycocalyx is important in cell recognition, binding to extracellular structures, and lubrication of the cell surface.

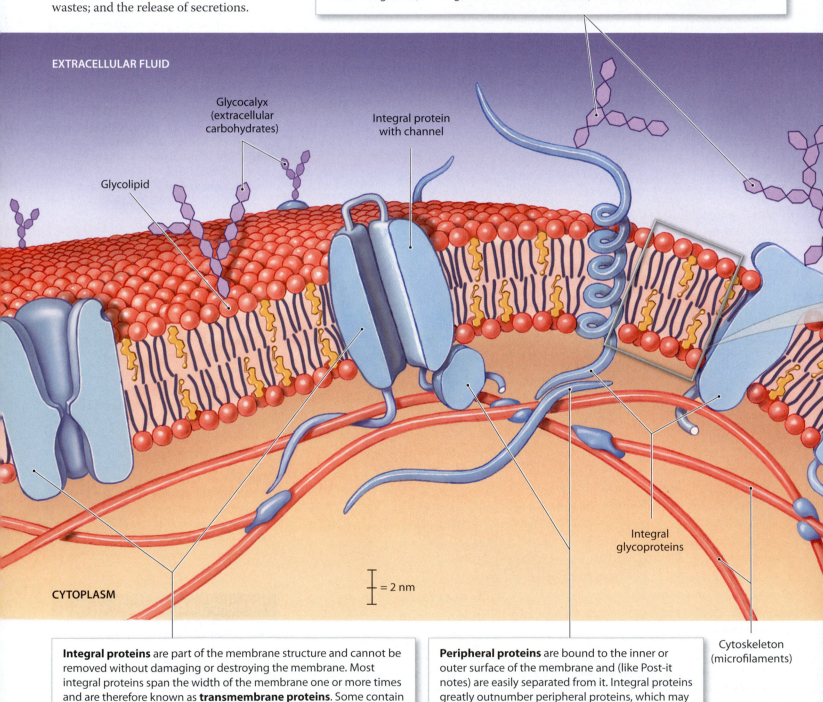

EXTRACELLULAR FLUID

Glycocalyx (extracellular carbohydrates)

Integral protein with channel

Glycolipid

Glycocalyx

Integral glycoproteins

Cytoskeleton (microfilaments)

= 2 nm

CYTOPLASM

Integral proteins are part of the membrane structure and cannot be removed without damaging or destroying the membrane. Most integral proteins span the width of the membrane one or more times and are therefore known as **transmembrane proteins**. Some contain pores or channels through which water and solutes may pass.

Peripheral proteins are bound to the inner or outer surface of the membrane and (like Post-it notes) are easily separated from it. Integral proteins greatly outnumber peripheral proteins, which may have regulatory or enzymatic functions.

2 The plasma membrane is extremely thin (6–10 nm) and very delicate. It is called a **phospholipid bilayer**, because the phospholipid molecules in it form two layers. In each half of the bilayer, the phospholipids lie with their hydrophilic heads at the membrane surface and their hydrophobic tails on the inside, just as in a micelle. The hydrophobic layer in the center of the membrane isolates the cytoplasm from the extracellular fluid. Such isolation is important because the composition of cytoplasm is very different from that of extracellular fluid.

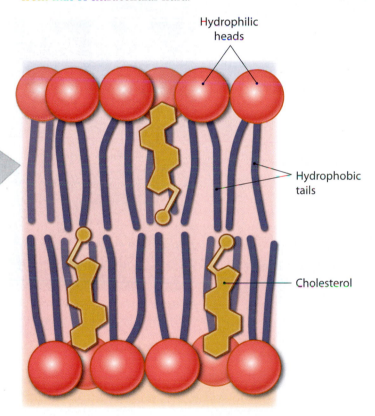

Hydrophilic heads

Hydrophobic tails

Cholesterol

3 The general functions of the plasma membrane include physical isolation, regulation of exchange with the environment, sensitivity to the environment, and structural support. The lipid bilayer provides isolation, and the membrane proteins perform most of the other functions. This table describes the five major functional classes of plasma membrane proteins.

Functional Classes of Membrane Proteins

Anchoring proteins attach the plasma membrane to other structures and stabiliize its position. Inside the cell, membrane proteins are bound to the cytoskeleton.

Recognition proteins are detected by cells of the immune system. Enzymes in plasma membranes may be integral or peripheral proteins.

Receptor proteins bind to specific extracellular molecules called **ligands** (LI-gandz). A ligand can be anything from a small ion like calcium, to a relatively large and complex hormone.

Carrier proteins bind solutes and transport them across the plasma membrane.

Channels are integral proteins containing a central pore (channel) that forms a passageway completely across the plasma membrane. The channel permits the passage of water and small solutes that cannot otherwise cross the lipid bilayer of the plasma membrane.

Module 3.2 Review

a. List the general functions of the plasma membrane.

b. Which structural component of the plasma membrane is mostly responsible for its ability to isolate a cell from its external environment?

c. Which type of integral protein allows water and small ions to pass through the plasma membrane?

The cytoskeleton plays both a structural and a functional role

The **cytoskeleton** functions as the cell's skeleton. It provides an internal protein framework that gives the cytoplasm strength and flexibility.

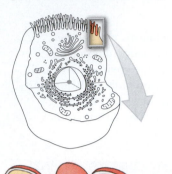

1 The cytoskeleton of all cells includes microfilaments, intermediate filaments, and microtubules.

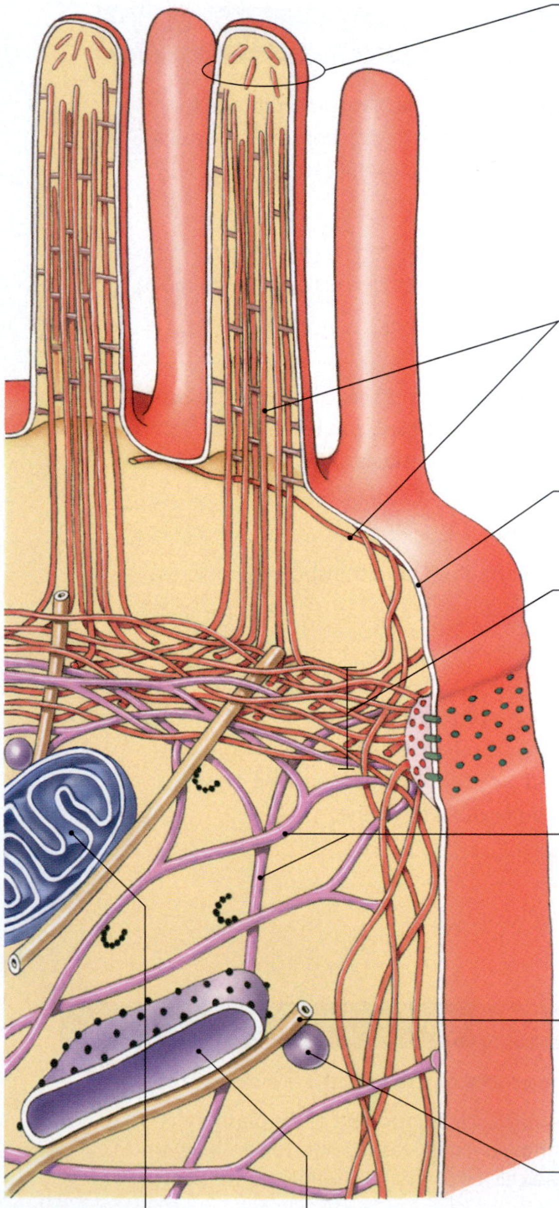

Microvilli (singular, *microvillus*) are finger-shaped extensions of the plasma membrane. A core of microfilaments stiffens each microvillus and anchors it to the cytoskeleton at the terminal web. Microvilli greatly increase the surface area of the cell and enhance its ability to absorb materials from the extracellular fluid.

The smallest of the cytoskeletal elements are the **microfilaments**. These protein strands are generally less than 6 nm in diameter. Typical microfilaments are composed of the protein **actin**. They are common in the periphery of the cell, but relatively rare in the region immediately surrounding the nucleus.

Plasma membrane

The **terminal web** is a layer of microfilaments just inside the plasma membrane at the exposed surface of a cell that forms a layer or lining, such as that of the intestinal tract.

Intermediate filaments, which range from 7 to 11 nm in diameter, are the strongest and most durable cytoskeletal elements.

Microtubules are the largest components of the cytoskeleton, with diameters of about 25 nm. Microtubules extend outward into the periphery of the cell from a region near the nucleus called the centrosome.

Secretory vesicle

Mitochondrion   Endoplasmic reticulum

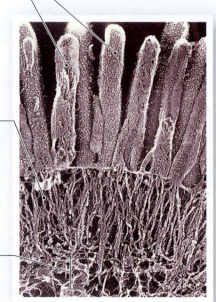

Microvilli   SEM × 30,000

2 This table summarizes the general components of the cytoskeleton.

The Cytoskeleton

Structure	Remarks	Location	Functions
Microfilaments	Present in most cells; best organized in skeletal and cardiac muscle cells	In bundles beneath the cell membrane and throughout the cytoplasm	Provide strength, alter cell shape, bind the cytoskeleton to the plasma membrane, tie cells together, involved in muscle contraction
Intermediate filaments	Present in most cells; at least five types known	In cytoplasm	Provide strength, move materials through cytoplasm
Thick filaments	Found in skeletal and cardiac muscle cells	In cytoplasm	Interact with actin microfilaments to produce muscle contraction
Microtubules	Present in most cells	In cytoplasm radiating away from centrosome	Provide strength, move organelles
Centrioles	Nine groups of microtubule triplets form a short cylinder	In centrosome near nucleus	Organize microtubules in the spindle to move chromosomes during cell division
Cilia	Nine groups of long microtubule doublets form a cylinder around a central pair	At cell surface	Propel fluids or solids across cell surface

3 **Centrioles** are cylindrical structures composed of short microtubules. The microtubules form nine groups, three in each group. Each of these nine "triplets" is connected to its nearest neighbors on either side. Two centrioles are located in the **centrosome**, and the microtubules of the cytoskeleton generally begin at the centrosome and radiate through the cytoplasm. During cell division, the centrioles are associated with the movement of DNA strands. Cells that lack centrioles, such as red blood cells and skeletal muscle cells, cannot divide.

4 **Cilia** (singular, *cilium*) are relatively long, slender extensions of the plasma membrane. They are especially common on cells lining portions of the respiratory tract and the reproductive tract. Cilia resemble centrioles but have nine pairs of microtubules (rather than triplets) surrounding a central pair. The microtubules are anchored to a compact **basal body** situated just beneath the cell surface.

5 Cilia beat rhythmically to move fluids or secretions across the cell surface. The ciliated cells lining the trachea beat their cilia in synchronized waves to move sticky mucus and trapped dust particles toward the throat and away from delicate respiratory surfaces.

Power stroke

Return stroke

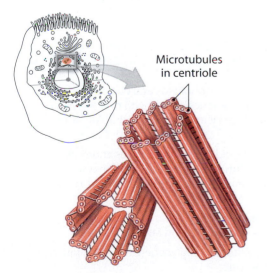

Microtubules in centriole

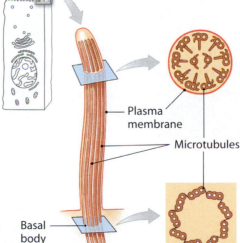

Plasma membrane

Microtubules

Basal body

Module 3.3 Review

a. List the three basic components of the cytoskeleton.

b. Which cytoskeletal component is common to both centrioles and cilia?

c. What is the function of cilia?

Ribosomes are responsible for protein synthesis and are often associated with the endoplasmic reticulum

Ribosomes are the organelles responsible for protein synthesis. The more proteins a cell synthesizes, the more ribosomes it has. Liver cells have many; fat cells have relatively few. Free ribosomes are scattered throughout the cytoplasm. The proteins they manufacture enter the cytosol.

1 A functional ribosome consists of two subunits that are normally separate and distinct. These subunits contain special proteins and ribosomal RNA (rRNA), one of the RNA types introduced in Module 2.16. Before protein synthesis can begin, large and small ribosomal subunits must join together, as shown here.

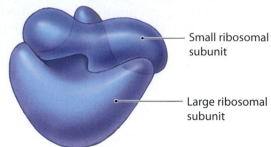

Small ribosomal subunit

Large ribosomal subunit

2 The **endoplasmic reticulum** (en-dō-PLAZ-mik re-TIK-ū-lum), or **ER**, is a network of intracellular membranes connected to the nuclear envelope, which surrounds the nucleus.

3 The **smooth endoplasmic reticulum**, or **SER**, lacks ribosomes, and the cisternae are often tubular.

The ER forms hollow tubes, flattened sheets, and chambers called **cisternae** (sis-TUR-nē; singular, *cisterna*, a reservoir for water).

Nuclear envelope

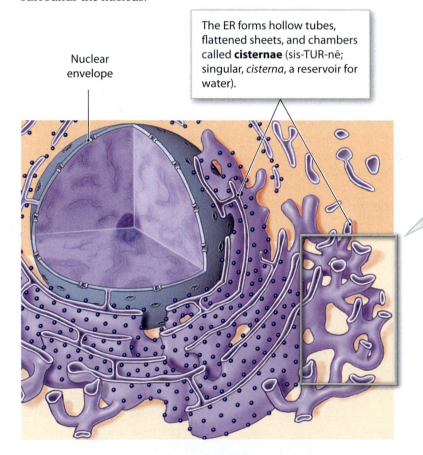

Tubular cisternae

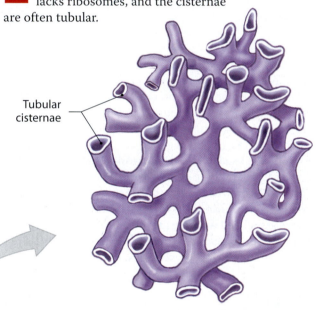

The Functions of the SER

- Synthesis of the phospholipids and cholesterol needed for maintenance and growth of the plasma membrane, ER, nuclear envelope, and Golgi apparatus
- Synthesis of steroid hormones, such as androgens and estrogens (the dominant sex hormones in males and in females, respectively) in the reproductive organs
- Synthesis and storage of glycerides, especially triglycerides, in liver cells and fat cells
- Synthesis and storage of glycogen in skeletal muscle cells and liver cells

4 **The rough endoplasmic reticulum (RER)** functions as a combination workshop and shipping depot. It is where many newly synthesized proteins are chemically modified and packaged for export to the Golgi apparatus, described in Module 3.5.

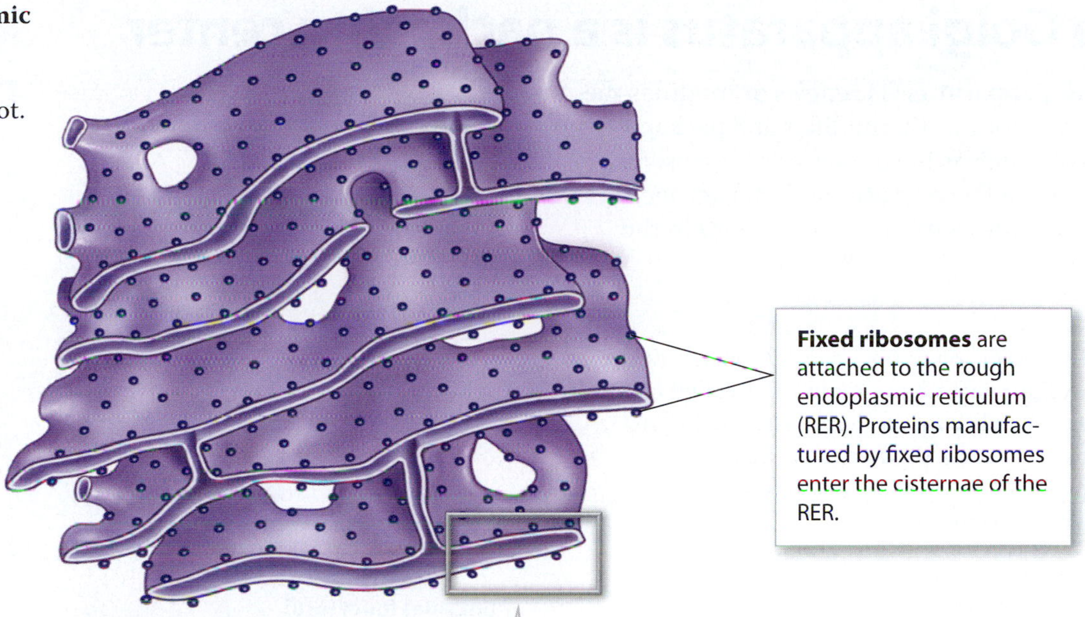

Fixed ribosomes are attached to the rough endoplasmic reticulum (RER). Proteins manufactured by fixed ribosomes enter the cisternae of the RER.

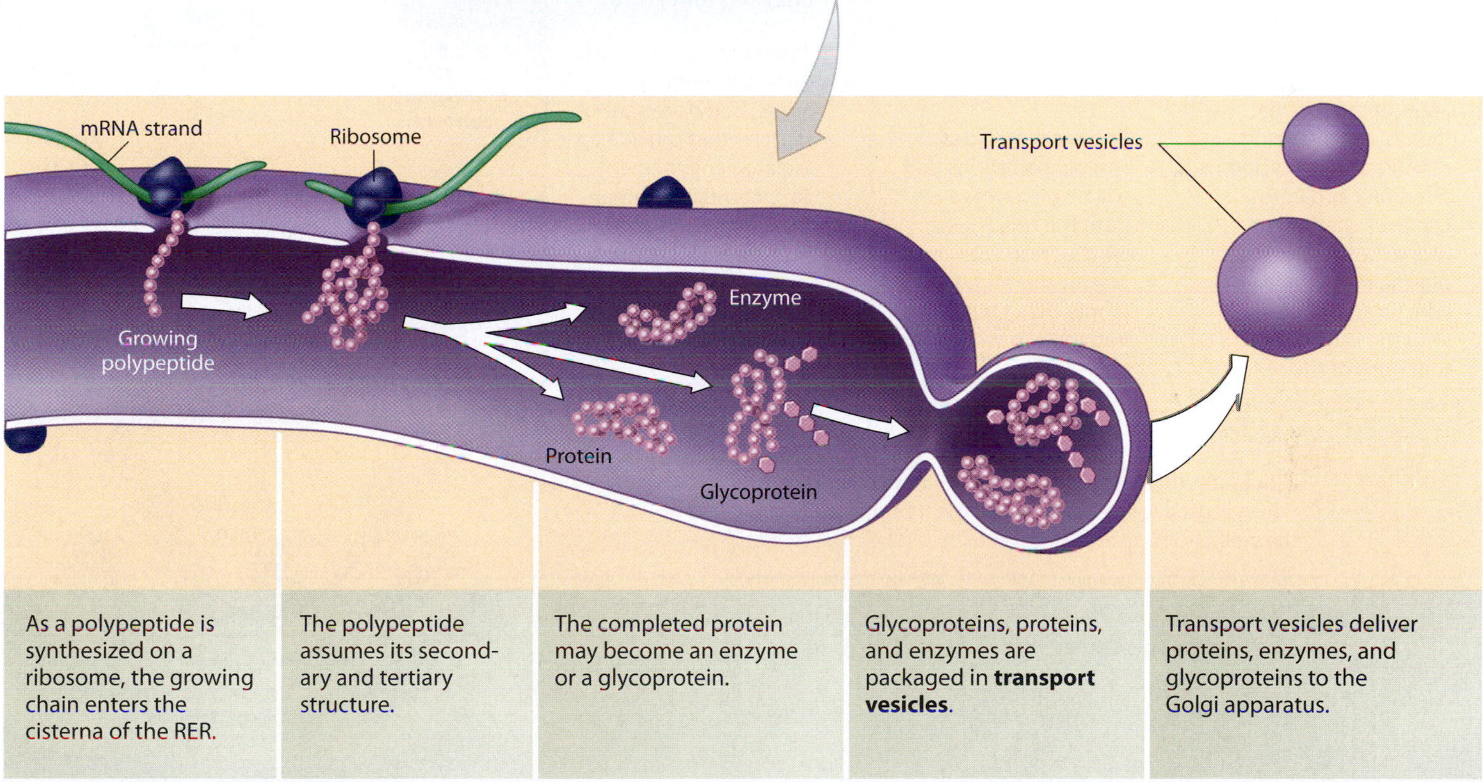

mRNA strand

Ribosome

Transport vesicles

Growing polypeptide

Enzyme

Protein

Glycoprotein

| As a polypeptide is synthesized on a ribosome, the growing chain enters the cisterna of the RER. | The polypeptide assumes its secondary and tertiary structure. | The completed protein may become an enzyme or a glycoprotein. | Glycoproteins, proteins, and enzymes are packaged in **transport vesicles**. | Transport vesicles deliver proteins, enzymes, and glycoproteins to the Golgi apparatus. |

The amount of endoplasmic reticulum and the proportion of RER to SER vary with the type of cell and its ongoing activities. For example, pancreatic cells that manufacture digestive enzymes contain an extensive RER, but the SER is relatively small. The situation is just the reverse in reproductive system cells that synthesize steroid hormones.

Module 3.4 Review

a. Describe the immediate cellular destinations of newly synthesized proteins from free ribosomes and fixed ribosomes.

b. Describe the structure of smooth endoplasmic reticulum.

c. Why do certain cells in the ovaries and testes contain large amounts of smooth endoplasmic reticulum (SER)?

The Golgi apparatus is a packaging center

The **Golgi apparatus** (1) renews or modifies the plasma membrane; (2) modifies and packages secretions, such as hormones or enzymes, for release through exocytosis; and (3) packages special enzymes within vesicles for use in the cytosol.

1 The Golgi apparatus typically consists of five or six flattened membranous discs called **cisternae**. A single cell may contain more than one Golgi apparatus, typically situated near the nucleus.

Extracellular Fluid

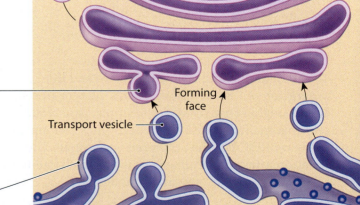

Membrane renewal

Membrane renewal vesicles add to the surface area of the plasma membrane. At the same time, other areas of the plasma membrane are being removed and recycled. The Golgi apparatus can thus change the properties of the plasma membrane, which can profoundly alter the sensitivity and functions of the cell.

Secretion

Secretory vesicles contain secretions that will be discharged from the cell. These vesicles fuse with the plasma membrane and empty their contents into the extracellular environment.

Enzymes for cytosol

Lysosomes (LĪ-sō-sōmz; *lyso-*, a loosening + *soma*, body) are special vesicles that provide an isolated environment for potentially dangerous chemical reactions. These vesicles contain digestive enzymes whose varied functions are described on the facing page.

Cytosol

Membrane renewal

Secretion

Secretory vesicle

Enzymes for cytosol

Lysosome

Cisternae

Maturing face

4 Ultimately, the product arrives at the **maturing face**, which is usually oriented toward the free surface of the cell.

3 Small vesicles move material from one cisterna to the next.

2 The transport vesicles then fuse with the Golgi membrane, emptying their contents into the cisternae. Inside the Golgi apparatus, enzymes modify the arriving proteins and glycoproteins. For example, the enzymes may change the carbohydrate structure of a glycoprotein, or they may attach a phosphate group, sugar, or fatty acid to a protein.

Forming face

Transport vesicle

Start **1** Some proteins and glycoproteins synthesized in the RER are delivered to the Golgi apparatus by **transport vesicles**. The vesicles generally arrive at a cisterna known as the **forming face**.

2 Cells often need to break down and recycle large organic molecules, and even complex structures like organelles. The breakdown process requires the use of powerful enzymes, and it often generates toxic chemicals that could damage or kill the cell. Lysosomes allow those chemical reactions to be isolated from the rest of the cytoplasm. The three basic functions of the lysosomes are shown below.

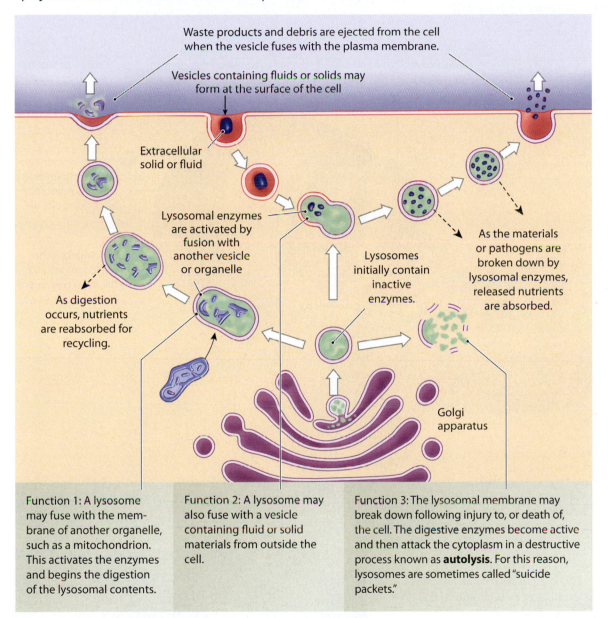

Waste products and debris are ejected from the cell when the vesicle fuses with the plasma membrane.

Vesicles containing fluids or solids may form at the surface of the cell

Extracellular solid or fluid

Lysosomal enzymes are activated by fusion with another vesicle or organelle

Lysosomes initially contain inactive enzymes.

As the materials or pathogens are broken down by lysosomal enzymes, released nutrients are absorbed.

As digestion occurs, nutrients are reabsorbed for recycling.

Golgi apparatus

Function 1: A lysosome may fuse with the membrane of another organelle, such as a mitochondrion. This activates the enzymes and begins the digestion of the lysosomal contents.

Function 2: A lysosome may also fuse with a vesicle containing fluid or solid materials from outside the cell.

Function 3: The lysosomal membrane may break down following injury to, or death of, the cell. The digestive enzymes become active and then attack the cytoplasm in a destructive process known as **autolysis**. For this reason, lysosomes are sometimes called "suicide packets."

With the exception of mitochondria, all membranous organelles in the cell are either interconnected or in communication through the movement of vesicles. This continuous movement and exchange is called **membrane flow.** In an actively secreting cell, an area equal to the entire membrane surface may be replaced *each hour.* Membrane flow is an example of the dynamic nature of cells. It gives cells a way to change the characteristics of their plasma membranes—the lipids, receptors, channels, anchors, and enzymes—as they grow, mature, or respond to a specific environmental stimulus.

Module 3.5 Review

a. List the three major functions of the Golgi apparatus.

b. The Golgi apparatus produces lysosomes. What do these lysosomes contain?

c. Describe three functions of lysosomes.

Mitochondria are the powerhouses of the cell

All living cells require energy to carry out the functions of life. The organelles responsible for energy production are the **mitochondria** (mī-tō-KON-drē-uh; singular, mitochondrion; *mitos*, thread + *chondrion*, granule).

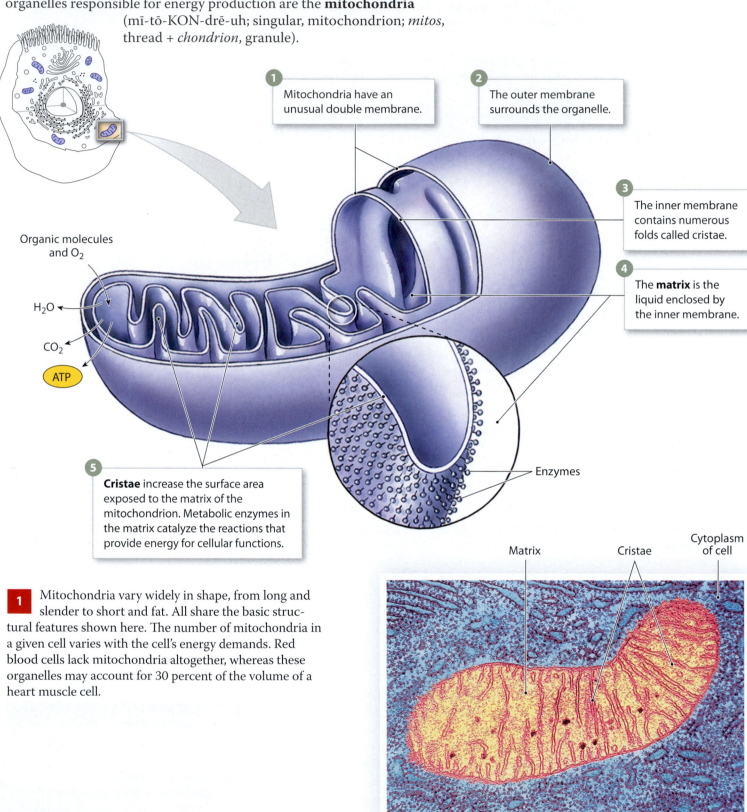

1 Mitochondria have an unusual double membrane.

2 The outer membrane surrounds the organelle.

3 The inner membrane contains numerous folds called cristae.

4 The **matrix** is the liquid enclosed by the inner membrane.

Organic molecules and O_2

H_2O

CO_2

ATP

Enzymes

5 **Cristae** increase the surface area exposed to the matrix of the mitochondrion. Metabolic enzymes in the matrix catalyze the reactions that provide energy for cellular functions.

Matrix Cristae Cytoplasm of cell

1 Mitochondria vary widely in shape, from long and slender to short and fat. All share the basic structural features shown here. The number of mitochondria in a given cell varies with the cell's energy demands. Red blood cells lack mitochondria altogether, whereas these organelles may account for 30 percent of the volume of a heart muscle cell.

Mitochondrion TEM × 50,000

2 Typical cells generate ATP and other high-energy compounds through the breakdown of carbohydrates, especially glucose. The major steps in the process are shown here.

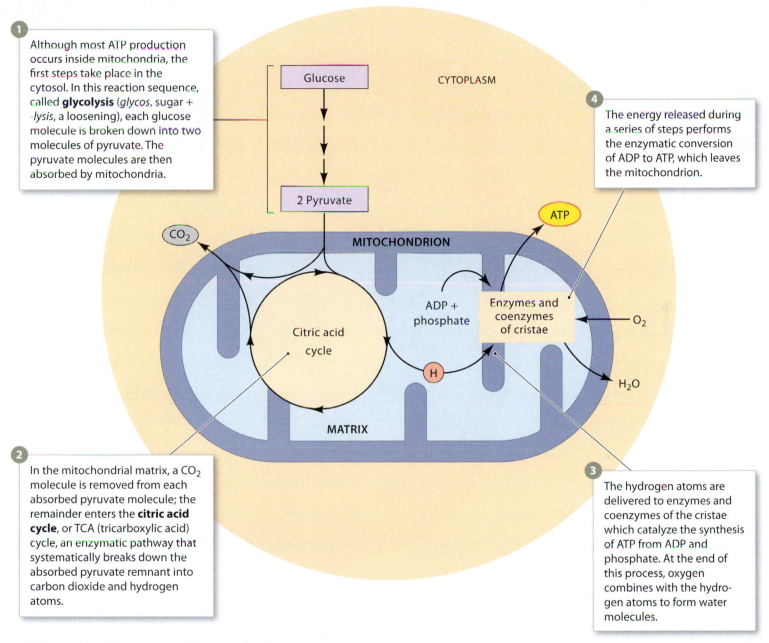

1 Although most ATP production occurs inside mitochondria, the first steps take place in the cytosol. In this reaction sequence, called **glycolysis** (*glycos*, sugar + *-lysis*, a loosening), each glucose molecule is broken down into two molecules of pyruvate. The pyruvate molecules are then absorbed by mitochondria.

Glucose

CYTOPLASM

2 Pyruvate

CO_2

ATP

MITOCHONDRION

Citric acid cycle

ADP + phosphate

Enzymes and coenzymes of cristae

O_2

H

H_2O

MATRIX

4 The energy released during a series of steps performs the enzymatic conversion of ADP to ATP, which leaves the mitochondrion.

2 In the mitochondrial matrix, a CO_2 molecule is removed from each absorbed pyruvate molecule; the remainder enters the **citric acid cycle**, or TCA (tricarboxylic acid) cycle, an enzymatic pathway that systematically breaks down the absorbed pyruvate remnant into carbon dioxide and hydrogen atoms.

3 The hydrogen atoms are delivered to enzymes and coenzymes of the cristae which catalyze the synthesis of ATP from ADP and phosphate. At the end of this process, oxygen combines with the hydrogen atoms to form water molecules.

Although ATP is generated during glycolysis, the amount is very small in comparison to the amount produced by the breakdown of pyruvate within mitochondria. Because mitochondrial activity requires oxygen, this method of ATP production is known as **aerobic metabolism** (*aer*, air + *bios*, life), or *cellular respiration*. Aerobic metabolism in mitochondria produces about 95 percent of the ATP needed to keep a cell alive. (Enzymatic reactions in the cytosol produce the rest.) Note that although the chemical reactions that release energy occur in the mitochondria, most of the cellular activities that require energy occur in the surrounding cytoplasm. So mitochondria are like little cellular batteries or fuel cells that provide the energy needed to power cellular functions.

Module 3.6 Review

a. Describe the structure of a mitochondrion.

b. Most of a cell's ATP is produced within its mitochondria. What gas do mitochondria require to produce ATP?

c. What does the presence of many mitochondria imply about a cell's energy requirements?

1. Short answer

Correctly label the indicated structures on the accompanying diagram of a cell and then describe the functions of each.

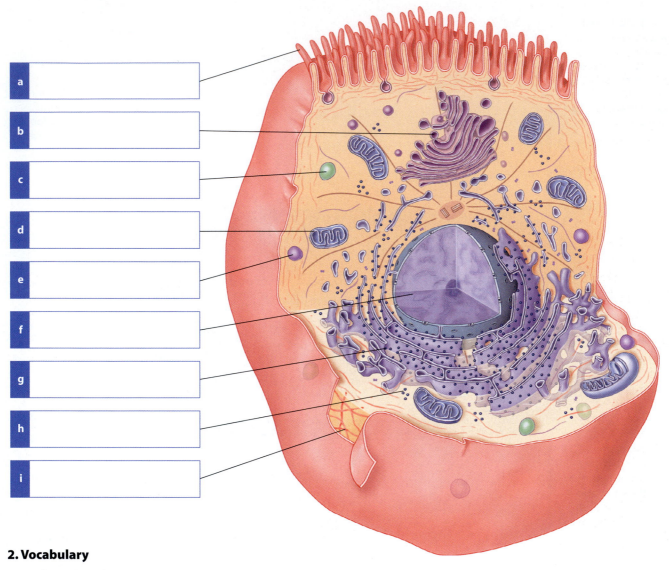

a

b

c

d

e

f

g

h

i

2. Vocabulary

For each word part listed below, write the boldfaced term(s) introduced in this section.

Word Part	Meaning	Terms
a glycos-	sugar	
b aero-	air	
c micro-	small	
d lyso-	a loosening	

3. Section integration

What is the advantage of having some of the cellular organelles enclosed by a membrane similar to the plasma membrane? _____

_____

Structure and Function of the Nucleus

The nucleus is usually the largest and most conspicuous structure in a cell; under a light microscope, it is often the only organelle visible. The nucleus serves as the control center for cellular operations. A single nucleus stores all the information needed to direct the synthesis of the more than 100,000 different proteins in the human body. This genetic information is coded in the sequence of nucleotides in DNA. By controlling what proteins are synthesized and when, the nucleus determines the structure of the cell and what functions it can perform. Most cells contain a single nucleus, but exceptions exist. For example, skeletal muscle cells have many nuclei, whereas mature red blood cells have none. A cell without a nucleus cannot repair itself, and it will disintegrate within three or four months.

1 This is an overview of the role the nucleus plays in preserving homeostasis at the cellular level. When the extracellular environment changes, the cell has mechanisms that can provide rapid short-term adjustments. If the ECF quickly returns to normal, those short-term adjustments may be sufficient to save the cell. But if the environmental stresses persist, the cell must adapt, and this involves long-term changes in its biochemical and physical characteristics. Dramatic, long-term changes in cell structure and function also occur as part of growth, development, and aging.

In this section we will consider topics of extreme importance to the rest of this text: the structure of the nucleus, the nature of the genetic material, how protein synthesis is controlled and directed, and how genetic information is partitioned during cell division.

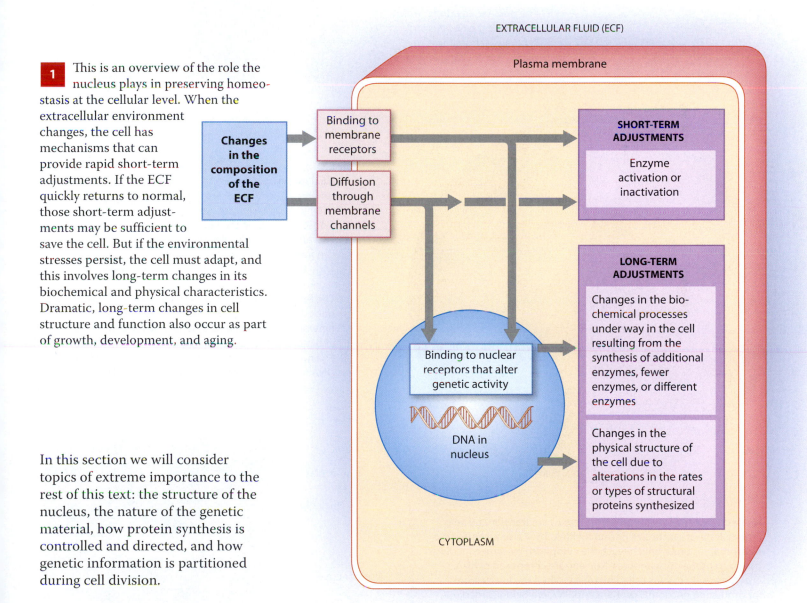

EXTRACELLULAR FLUID (ECF)

Plasma membrane

Changes in the composition of the ECF

Binding to membrane receptors

Diffusion through membrane channels

SHORT-TERM ADJUSTMENTS

Enzyme activation or inactivation

Binding to nuclear receptors that alter genetic activity

DNA in nucleus

LONG-TERM ADJUSTMENTS

Changes in the biochemical processes under way in the cell resulting from the synthesis of additional enzymes, fewer enzymes, or different enzymes

Changes in the physical structure of the cell due to alterations in the rates or types of structural proteins synthesized

CYTOPLASM

The nucleus contains DNA, RNA, organizing proteins, and enzymes

The **nucleus** is often the most prominent and visible organelle. In standard histological sections and micrographs, the nucleus stains darkly and is easily identified.

1 The nucleus serves as the control center for cellular processes. The surrounding nuclear envelope, with its nuclear pores, controls chemical communication between the cytoplasm and the nucleus.

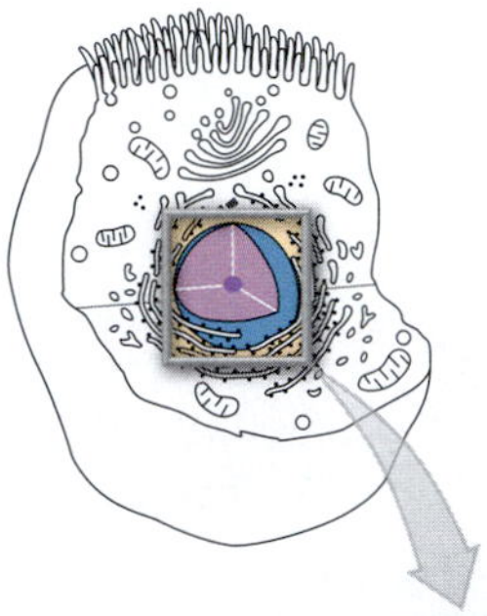

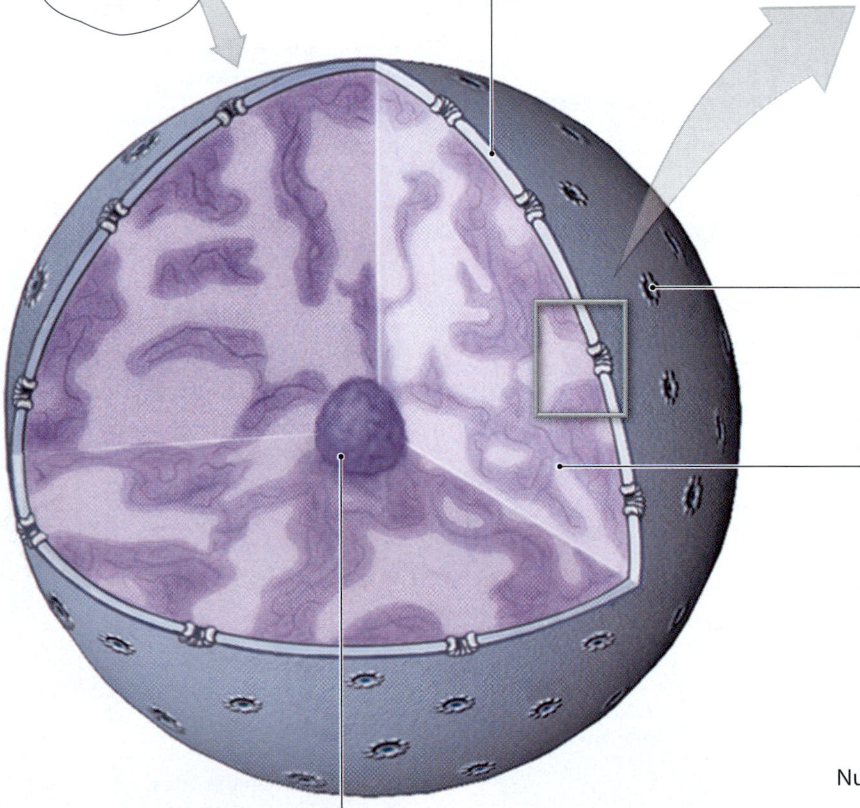

The **nuclear envelope**, which surrounds the nucleus and separates it from the cytoplasm, is a double membrane.

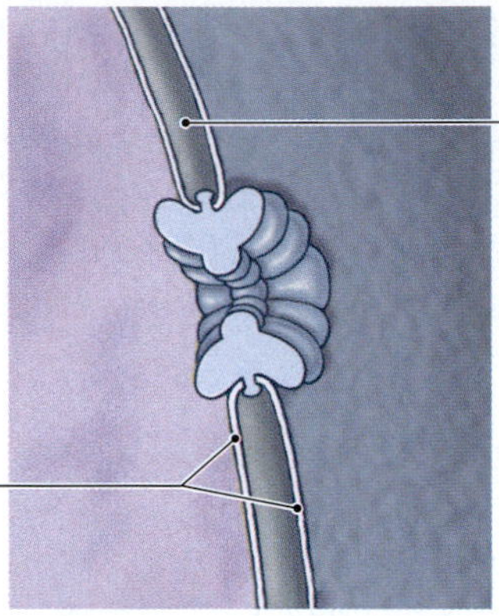

The narrow **perinuclear space** (*peri-*, around) separates the two layers of the nuclear membrane.

Nuclear pores, which account for about 10 percent of the surface of the nucleus, are passageways that permit chemical communication between the nucleus and the cytosol. Proteins at the pores regulate the movement of ions and small molecules, and neither proteins nor DNA can freely cross the nuclear envelope.

The term **nucleoplasm** refers to the fluid contents of the nucleus. The nucleoplasm contains the nuclear matrix, a network of fine filaments that provides structural support and may be involved in the regulation of genetic activity. The nucleoplasm also contains ions, enzymes, RNA and DNA nucleotides, small amounts of RNA, and DNA.

Nucleoli (noo-KLĒ-ō-lī; singular, *nucleolus*) are transient nuclear organelles that synthesize ribosomal RNA. They also assemble the ribosomal subunits, which reach the cytoplasm by carrier-mediated transport at the nuclear pores. Nucleoli are composed of RNA, enzymes, and proteins called **histones**. When the instructions for producing ribosomal proteins and RNA are being carried out, nucleoli form around the portions of DNA containing those instructions. Nucleoli are most prominent in cells that manufacture large amounts of proteins, such as liver, nerve, and muscle cells.

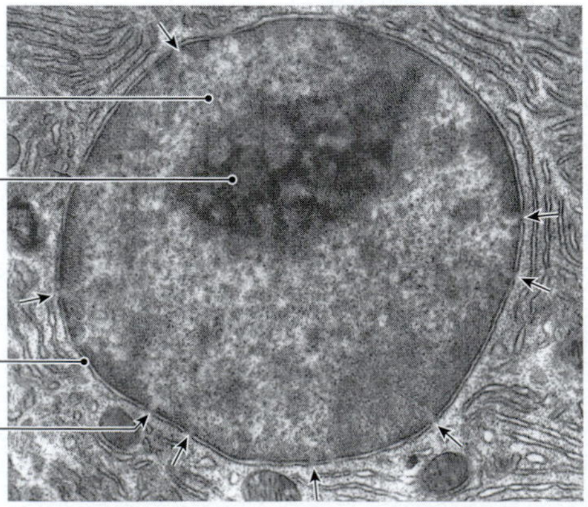

Nucleoplasm

Nucleolus

Nuclear envelope

Nuclear pore

Nucleus   TEM × 34,800

2 It is the DNA in the nucleus that stores the instructions for protein synthesis. In the nucleus, the DNA strands are coiled, rather than straight, allowing a great deal of DNA to be packaged in a relatively small space. The coils wrap around histone molecules, forming complexes known as **nucleosomes**. The coiling can be tight or loose, and the entire chain of nucleosomes may coil around other proteins.

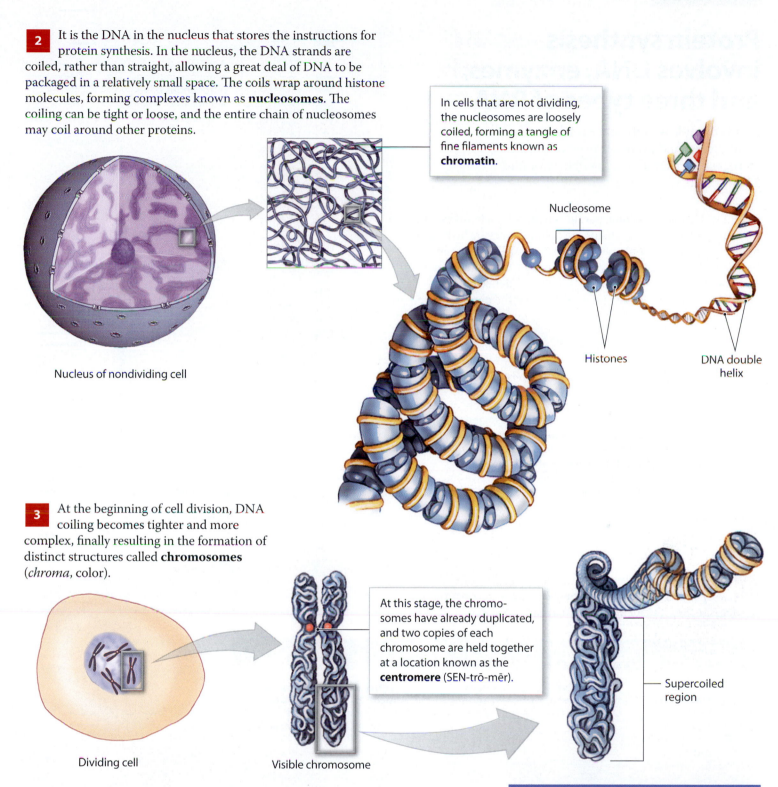

In cells that are not dividing, the nucleosomes are loosely coiled, forming a tangle of fine filaments known as **chromatin**.

Nucleosome

Nucleus of nondividing cell

Histones

DNA double helix

3 At the beginning of cell division, DNA coiling becomes tighter and more complex, finally resulting in the formation of distinct structures called **chromosomes** (*chroma*, color).

At this stage, the chromosomes have already duplicated, and two copies of each chromosome are held together at a location known as the **centromere** (SEN-trō-mēr).

Supercoiled region

Dividing cell

Visible chromosome

In humans, the nuclei of somatic cells (general body cells, as opposed to sperm and oocytes, which are called sex cells) contain 23 pairs of chromosomes. One member of each pair is derived from the mother, and one from the father. The DNA in these chromosomes not only carries the instructions for synthesizing proteins, but for synthesizing RNA. In addition, some DNA segments have a regulatory function, and others have as yet undetermined functions.

Module 3.7 Review

a. What molecule in the nucleus contains instructions for making proteins?

b. Describe the contents and the structure of the nucleus.

c. How many chromosomes are contained within a typical somatic cell?

Protein synthesis involves DNA, enzymes, and three types of RNA

In this module we introduce the key components and events of protein synthesis, a process we examine in detail in Modules 3.9 and 3.10.

1 DNA consists of long, parallel chains of nucleotides. The two adjacent DNA chains are held together by hydrogen bonding between complementary base pairs. The nitrogenous bases involved are adenine (A), thymine (T), cytosine (C), and guanine (G). Information is stored in the sequence of base pairs; the chemical "language" used is called the **genetic code**.

> A **gene** is the functional unit of heredity; it contains all the DNA nucleotides needed to produce specific proteins. The number of nucleotides in a gene depends on the size of the polypeptide represented. A relatively short polypeptide chain might require fewer than 300 nucleotides, whereas the instructions for building a large protein might involve 3000 or more nucleotides.

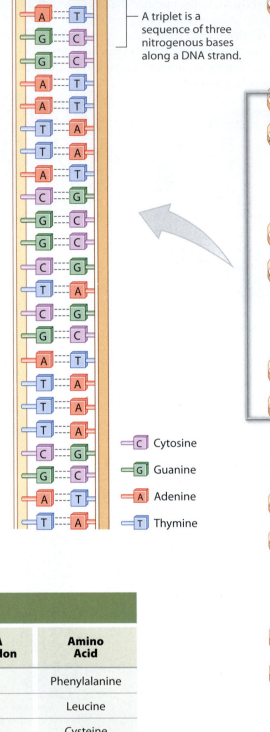

A triplet is a sequence of three nitrogenous bases along a DNA strand.

C Cytosine
G Guanine
A Adenine
T Thymine

2 The genetic code is called a **triplet code**, because a sequence of three nitrogenous bases, which constitutes a **triplet**, specifies the identity of a single amino acid.

Examples of the Triplet Code

DNA Triplet Template Strand	DNA Triplet Coding Strand	mRNA Codon	tRNA Anticodon	Amino Acid
AAA	TTT	UUU	AAA	Phenylalanine
AAT	TTA	UUA	AAU	Leucine
ACA	TGT	UGU	ACA	Cysteine
CAA	GTT	GUU	CAA	Valine
TAC	ATG	AUG	UAC	Methionine
TCG	AGC	AGC	UCG	Serine
GGC	CCG	CCG	GGC	Proline
CGG	GCC	GCC	CGG	Alanine

3 Before a gene can affect a cell, the portion of the DNA molecule containing that gene must be uncoiled, and the histones temporarily removed. The factors controlling this process, called **gene activation**, are only partially understood. We know, however, that every gene contains segments responsible for regulating its own activity.

The two DNA strands separate in the region containing the activated gene, exposing the triplets to the nucleoplasm.

Paired DNA strands

4 Enzymes then assemble a strand of **messenger RNA (mRNA)** by interconnecting complementary RNA nucleotides (A, G, C, U). This produces an RNA strand that contains information in triplets of RNA nucleotides. Those triplets are called **codons**.

Enzyme

The mRNA strand containing the complementary codons passes through a nuclear pore and enters the cytoplasm.

Codon on mRNA

5 At a ribosome in the cytoplasm, nucleotide triplets called **anticodons** on **transfer RNA (tRNA)** bind to the mRNA codons. There are many different types of tRNA, and each type carries a specific amino acid.

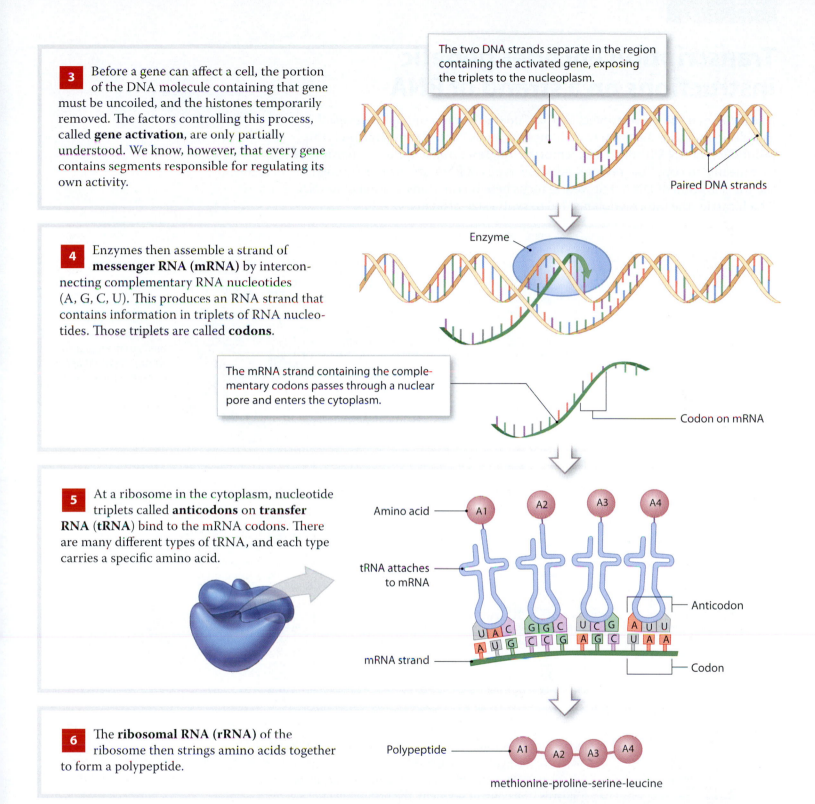

Amino acid — A1   A2   A3   A4

tRNA attaches to mRNA

Anticodon

| U A C | G G C | U C G | A U U |
| A U G | C C G | A G C | U A A |

mRNA strand

Codon

6 The **ribosomal RNA (rRNA)** of the ribosome then strings amino acids together to form a polypeptide.

Polypeptide — A1 — A2 — A3 — A4

methionine-proline-serine-leucine

So to summarize:

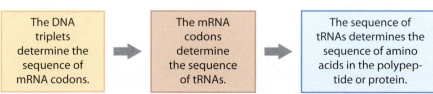

| The DNA triplets determine the sequence of mRNA codons. | → | The mRNA codons determine the sequence of tRNAs. | → | The sequence of tRNAs determines the sequence of amino acids in the polypeptide or protein. |

Transcription encodes genetic instructions on a strand of RNA

In effect, a gene contains triplets that say "do (or do not) read this message," as well as "message starts here," and "message ends here." **Transcription** is the production of RNA from a DNA template. The term *transcription* is appropriate, as it means "to copy" or "rewrite." All three types of RNA are formed through the transcription of DNA, but we will focus here on the transcription of mRNA, which carries the information needed to synthesize proteins.

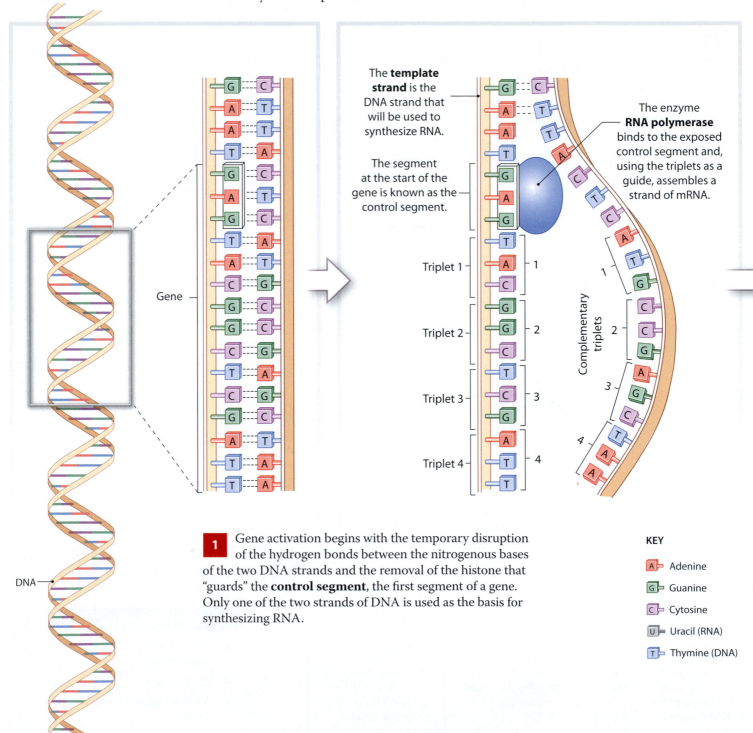

The **template strand** is the DNA strand that will be used to synthesize RNA.

The segment at the start of the gene is known as the control segment.

The enzyme **RNA polymerase** binds to the exposed control segment and, using the triplets as a guide, assembles a strand of mRNA.

Gene

DNA

Triplet 1

Triplet 2

Triplet 3

Triplet 4

Complementary triplets

1 Gene activation begins with the temporary disruption of the hydrogen bonds between the nitrogenous bases of the two DNA strands and the removal of the histone that "guards" the **control segment**, the first segment of a gene. Only one of the two strands of DNA is used as the basis for synthesizing RNA.

KEY

A⊢ Adenine

G⊢ Guanine

C⊢ Cytosine

U⊢ Uracil (RNA)

T⊢ Thymine (DNA)

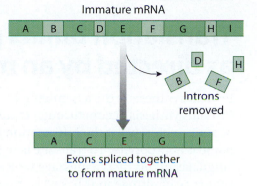

Immature mRNA

Introns removed

Exons spliced together to form mature mRNA

RNA polymerase works only on RNA nucleotides—it can attach adenine, guanine, cytosine, or uracil, but never thymine. If the DNA triplet is TAC, the corresponding mRNA codon will be AUG.

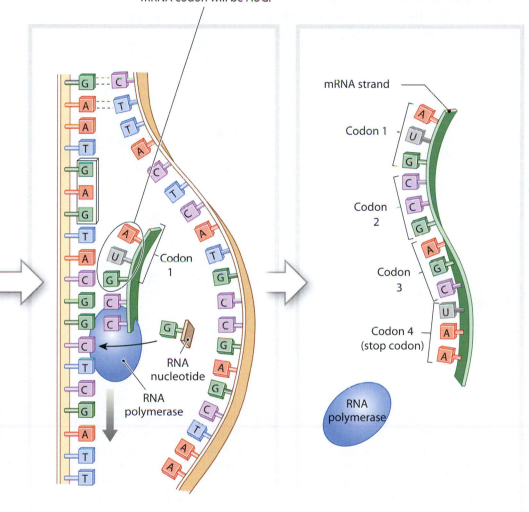

mRNA strand

Codon 1

Codon 2

Codon 3

Codon 4 (stop codon)

RNA polymerase

Codon 1

RNA nucleotide

RNA polymerase

4 Each gene includes a number of triplets that are not needed to build a functional protein. As a result, the mRNA strand assembled during transcription, sometimes called **immature mRNA** or pre-mRNA, must be "edited" before it leaves the nucleus to direct protein synthesis. In this RNA processing, nonsense regions, called **introns**, are snipped out, and the remaining coding segments, or **exons**, are spliced together.

The process creates a shorter, functional strand of mRNA that then enters the cytoplasm through a nuclear pore. Intron removal is extremely important and tightly regulated. By changing the editing instructions and removing different introns, a single gene can produce mRNAs that code for several different proteins.

2 RNA polymerase promotes hydrogen bonding between the nitrogenous bases of the template strand of DNA and complementary RNA nucleotides in the nucleoplasm. The complementary nucleotides are then strung together by covalent bonding.

3 At the "stop" signal, the enzyme and the mRNA strand detach from the DNA strand, and transcription ends. The complementary DNA strands now reassociate as hydrogen bonding occurs between complementary base pairs.

Module 3.9 Review

a. Define DNA template strand.

b. What is transcription?

c. What process would be affected if a cell could not synthesize the enzyme RNA polymerase?

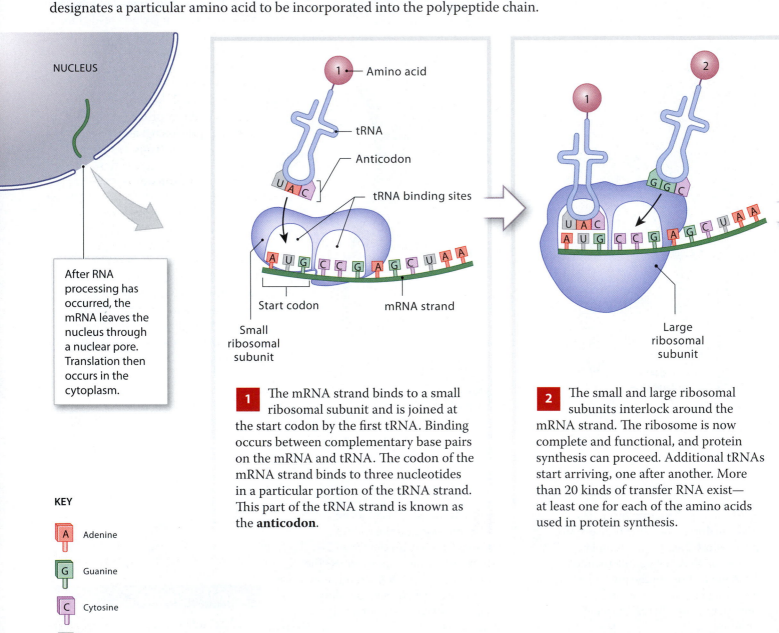

Translation builds polypeptides as directed by an mRNA strand

Protein synthesis is the assembling of functional polypeptides in the cytoplasm. Protein synthesis occurs through **translation**, the formation of a linear chain of amino acids, using the information provided by an mRNA strand. Again, the name is appropriate: To translate is to present the same information in a different language; in this case, a message written in the "language" of nucleic acids (the sequence of nitrogenous bases) is translated by ribosomes into the "language" of proteins (the sequence of amino acids in a polypeptide chain). Each mRNA codon designates a particular amino acid to be incorporated into the polypeptide chain.

NUCLEUS

After RNA processing has occurred, the mRNA leaves the nucleus through a nuclear pore. Translation then occurs in the cytoplasm.

1 — Amino acid

tRNA

Anticodon

tRNA binding sites

Start codon   mRNA strand

Small ribosomal subunit

Large ribosomal subunit

1 The mRNA strand binds to a small ribosomal subunit and is joined at the start codon by the first tRNA. Binding occurs between complementary base pairs on the mRNA and tRNA. The codon of the mRNA strand binds to three nucleotides in a particular portion of the tRNA strand. This part of the tRNA strand is known as the **anticodon**.

2 The small and large ribosomal subunits interlock around the mRNA strand. The ribosome is now complete and functional, and protein synthesis can proceed. Additional tRNAs start arriving, one after another. More than 20 kinds of transfer RNA exist— at least one for each of the amino acids used in protein synthesis.

KEY

A — Adenine

G — Guanine

C — Cytosine

U — Uracil

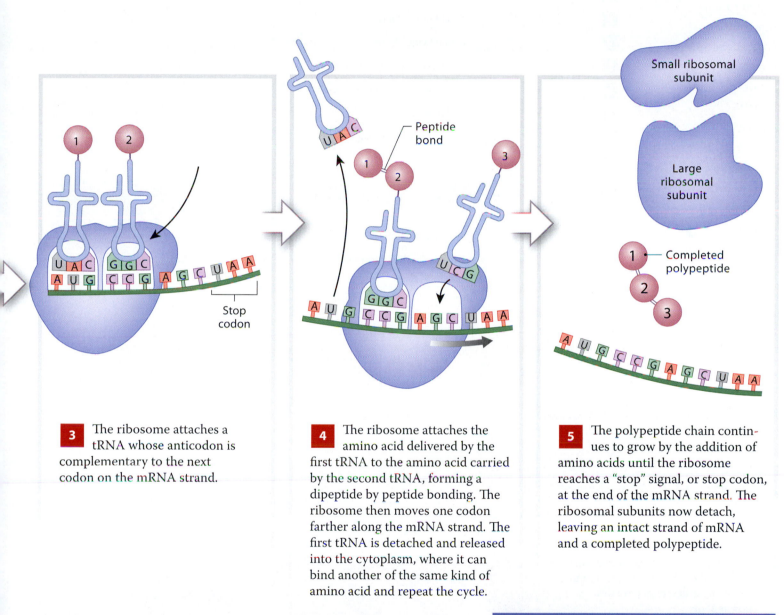

3 The ribosome attaches a tRNA whose anticodon is complementary to the next codon on the mRNA strand.

Stop codon

Peptide bond

4 The ribosome attaches the amino acid delivered by the first tRNA to the amino acid carried by the second tRNA, forming a dipeptide by peptide bonding. The ribosome then moves one codon farther along the mRNA strand. The first tRNA is detached and released into the cytoplasm, where it can bind another of the same kind of amino acid and repeat the cycle.

Small ribosomal subunit

Large ribosomal subunit

Completed polypeptide

5 The polypeptide chain continues to grow by the addition of amino acids until the ribosome reaches a "stop" signal, or stop codon, at the end of the mRNA strand. The ribosomal subunits now detach, leaving an intact strand of mRNA and a completed polypeptide.

Translation proceeds swiftly, producing a typical protein in about 20 seconds. The mRNA strand remains intact, and it can interact with other ribosomes to create additional copies of the same polypeptide chain. The process does not continue indefinitely, however, because after a few minutes to a few hours, mRNA strands are broken down, and the nucleotides recycled. However, large numbers of protein chains can be produced during that time. Although only two mRNA codons are "read" by a ribosome at any one time, many ribosomes can bind to a single mRNA strand, and a multitude of identical proteins may be quickly and efficiently produced.

Module 3.10 Review

a. What is translation?

b. The nucleotide sequence of three mRNA codons is AUU-GCA-CUA. What is the complementary anticodon sequence for the second codon?

c. During the process of transcription, a nucleotide was deleted from an mRNA sequence that coded for a protein. What effect would this deletion have on the amino acid sequence of the protein?

1. Matching

Match the following terms with the most closely related description.

- introns
- transcription
- tRNA
- chromosomes
- exons
- genetic information
- nucleus
- thymine
- mRNA
- gene
- uracil
- nuclear envelope
- nuclear pore
- nucleoli

a _____	DNA strands and histones
b _____	DNA nitrogen base
c _____	double membrane
d _____	mRNA noncoding regions
e _____	RNA nitrogen base
f _____	assemble ribosomal subunits
g _____	passageway for functional mRNA
h _____	mRNA formation
i _____	functional unit of heredity
j _____	mRNA coding regions
k _____	codon
l _____	anticodon
m _____	cell control center
n _____	DNA nucleotide sequence

2. Short answer

This is a sequence of DNA bases in a protein-coding gene.

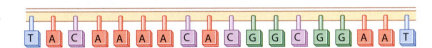

T A C A A A A C A C G G C G G A A T

a Provide the corresponding mRNA base sequence and insert a slash mark (/) between the codons:

_____

b Convert the mRNA codons above to tRNA anticodons:

_____

c Using the triplet code table in Module 3.8, translate the anticodon sequence into the amino acid sequence of this polypeptide:

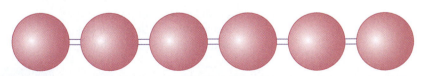

3. Section integration

The nucleus is often described as the control center of the cell. Explain the role of the nucleus in maintaining homeostasis.

_____

_____

_____

How Things Enter and Leave the Cell

Because the plasma membrane is an effective barrier, conditions inside the cell can be considerably different from conditions outside the cell. However, the barrier cannot be absolute, because cells are not self-sufficient, and their activities must be coordinated. In this section we will consider how the plasma membrane selectively regulates the movement of materials into and out of the cell.

1 **Permeability** is the property of the plasma membrane that determines precisely which substances can enter or leave the cytoplasm.

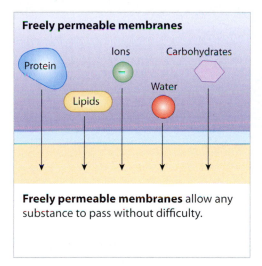

Freely permeable membranes

Freely permeable membranes allow any substance to pass without difficulty.

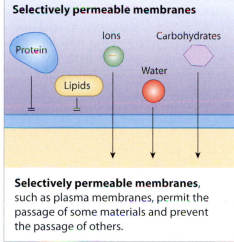

Selectively permeable membranes

Selectively permeable membranes, such as plasma membranes, permit the passage of some materials and prevent the passage of others.

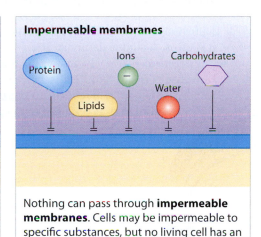

Impermeable membranes

Nothing can pass through **impermeable membranes**. Cells may be impermeable to specific substances, but no living cell has an impermeable membrane.

2 **Selectively permeable membranes** permit the free passage of some materials and restrict the passage of others. The distinction may be based on size, electrical charge, molecular shape, lipid solubility, or other factors. Cells differ in their permeabilities, depending on what lipids and proteins are present in the plasma membrane and how these components are arranged.

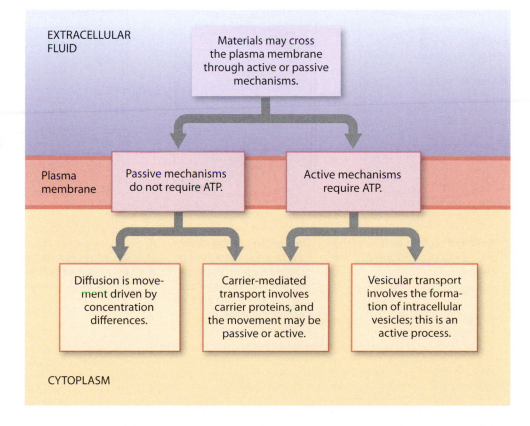

EXTRACELLULAR FLUID

Materials may cross the plasma membrane through active or passive mechanisms.

Plasma membrane

Passive mechanisms do not require ATP.

Active mechanisms require ATP.

Diffusion is movement driven by concentration differences.

Carrier-mediated transport involves carrier proteins, and the movement may be passive or active.

Vesicular transport involves the formation of intracellular vesicles; this is an active process.

CYTOPLASM

Diffusion is movement driven by concentration differences

Ions and molecules in liquids and gases are constantly in motion, colliding and bouncing off one another and off obstacles in their paths. The movement is random: A molecule can bounce in any direction. One result of this continuous random motion is that, over time, the molecules in any given space will tend to become evenly distributed. This distribution process is called **diffusion**. When molecules are not evenly distributed, a concentration difference, or **gradient** exists. Unless prevented from doing so, given enough time diffusion will eliminate a concentration gradient. After the gradient has been eliminated, the molecular motion continues, but net movement no longer occurs in any particular direction.

1 Diffusion in air and water is slow, and it is most important over very small distances. A simple demonstration—observing a colored sugar cube placed in a beaker of water—can give you a good mental picture of what happens at the cellular level. However, compared to a cell, a beaker of water is enormous, and additional factors (which we will ignore) account for dye distribution over distances of centimeters as opposed to micrometers.

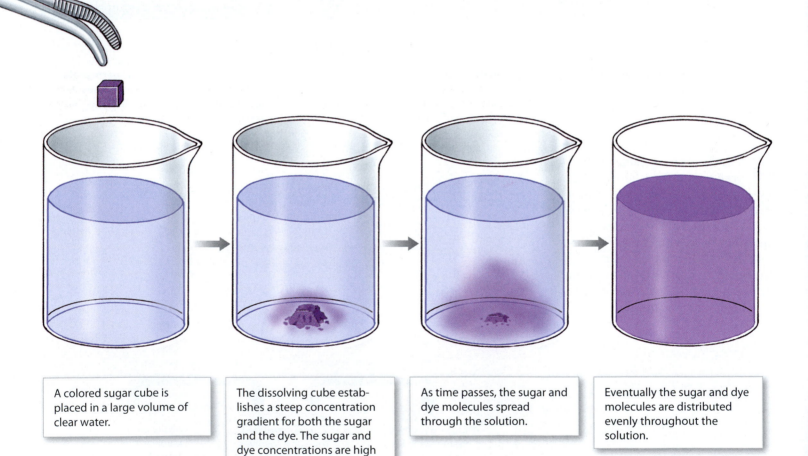

A colored sugar cube is placed in a large volume of clear water.

The dissolving cube establishes a steep concentration gradient for both the sugar and the dye. The sugar and dye concentrations are high near the cube and negligible elsewhere.

As time passes, the sugar and dye molecules spread through the solution.

Eventually the sugar and dye molecules are distributed evenly throughout the solution.

2 In extracellular fluids, water and dissolved solutes diffuse freely. A plasma membrane, however, acts as a barrier that selectively restricts diffusion: Some substances pass through easily, while others cannot penetrate the membrane. An ion or a molecule can diffuse across a plasma membrane only by (1) crossing the lipid portion of the membrane or (2) passing through a membrane channel.

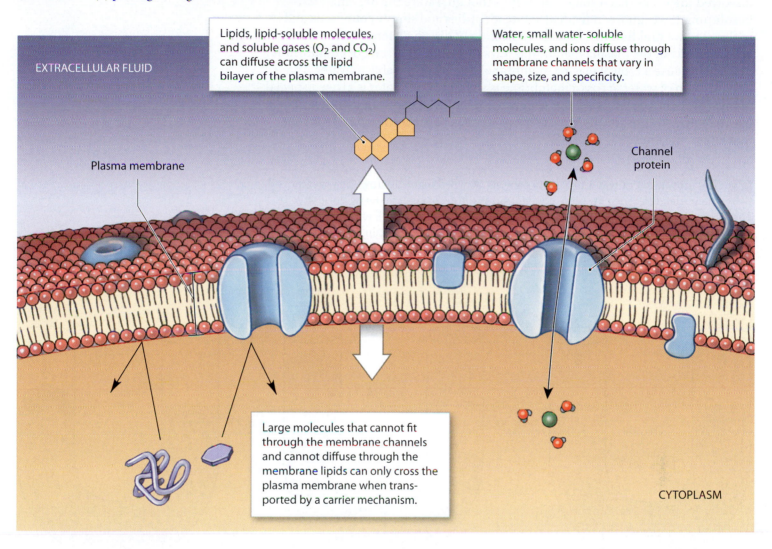

EXTRACELLULAR FLUID

Lipids, lipid-soluble molecules, and soluble gases (O_2 and CO_2) can diffuse across the lipid bilayer of the plasma membrane.

Water, small water-soluble molecules, and ions diffuse through membrane channels that vary in shape, size, and specificity.

Plasma membrane

Channel protein

Large molecules that cannot fit through the membrane channels and cannot diffuse through the membrane lipids can only cross the plasma membrane when transported by a carrier mechanism.

CYTOPLASM

The diffusion of nutrients, waste products, and dissolved gases must keep pace with the demands of active cells. Important factors that influence diffusion rates include:

- Distance: The shorter the distance, the more quickly concentration gradients are eliminated. In the human body, few cells are farther than 25 μm from a blood vessel.

- Molecule size: Ions and small organic molecules, such as glucose, diffuse more rapidly than do large proteins.

- Temperature: The higher the temperature, the faster the diffusion rate.

- Gradient size: The larger the concentration gradient, the faster diffusion proceeds.

- Electrical forces: Opposite electrical charges (+ and −) attract each other; like charges (+ and + or − and −) repel each other. Electrical attraction or repulsion can accelerate or reduce the rate of ion diffusion.

Module 3.11 Review

a. Define diffusion.

b. Identify factors that influence diffusion rates.

c. How would a decrease in the oxygen concentration in the lungs affect the diffusion of oxygen into the blood?

Osmosis is the passive movement of water

Intracellular and extracellular fluids are solutions that contain a variety of dissolved materials. Each solute diffuses as though it were the only material in solution. If we ignore individual solute identities and simply count ions and molecules, we find that the total concentration of dissolved ions and molecules on either side of the plasma membrane stays the same. This state of equilibrium persists because a typical plasma membrane is freely permeable to water. The net diffusion of water across a membrane is so important that it is given a special name: **osmosis** (oz-MŌ-sis; *osmos*, a push). For convenience, we will always use the term *osmosis* for the movement of water, and the term *diffusion* for the movement of solutes.

1 The movement of water driven by osmosis is called **osmotic flow**. The greater the initial difference in solute concentrations, the stronger is the osmotic flow. The **osmotic pressure** of a solution is an indication of the force with which pure water moves into that solution as a result of its solute concentration. We can measure a solution's osmotic pressure in several ways. For example, an opposing pressure can prevent the osmotic flow of water into the solution.

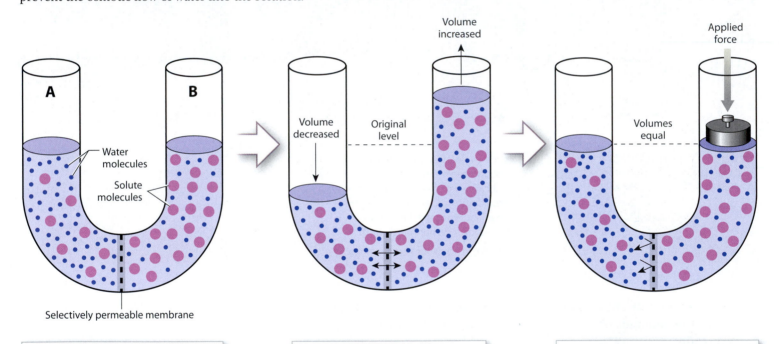

A selectively permeable membrane separates these two solutions, which have different solute concentrations. Water molecules (small blue dots) begin to cross the membrane toward solution B, the solution with the higher concentration of solutes (larger pink circles).

At equilibrium, the solute concentrations on the two sides of the membrane are equal. Note that the volume of solution B has increased at the expense of that of solution A.

Pushing against a fluid generates **hydrostatic pressure**. The osmotic pressure of solution B is equal to the amount of hydrostatic pressure, indicated by the weight, required to stop the osmotic flow.

2 The total solute concentration in an aqueous solution is the solution's **osmolarity**, or **osmotic concentration**. The nature of the solutes, however, is often as important as the total osmolarity. When we describe the effects of various osmotic solutions on cells, we usually use the term **tonicity** instead of osmolarity. Although often used interchangeably, the terms *osmolarity* and *tonicity* do not always mean the same thing. Osmolarity refers to the solute concentration of the solution, whereas tonicity is a description of how the solution affects a cell. Tonicity may have one of three effects, as illustrated here.

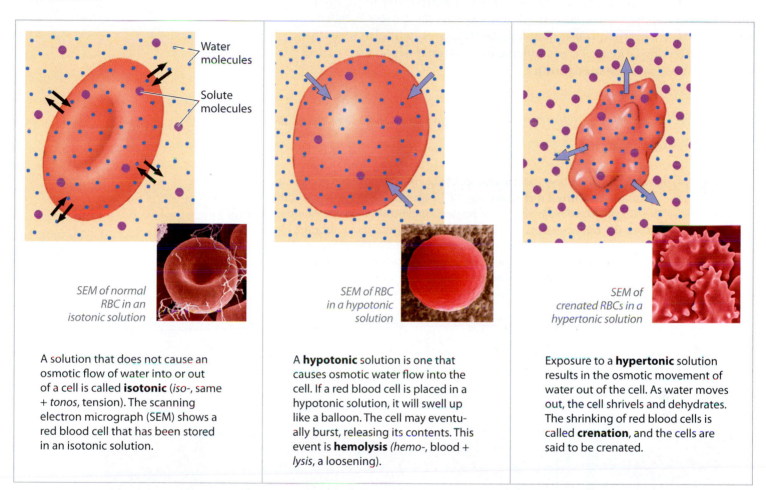

Water molecules

Solute molecules

SEM of normal RBC in an isotonic solution

SEM of RBC in a hypotonic solution

SEM of crenated RBCs in a hypertonic solution

A solution that does not cause an osmotic flow of water into or out of a cell is called **isotonic** (*iso-*, same + *tonos*, tension). The scanning electron micrograph (SEM) shows a red blood cell that has been stored in an isotonic solution.

A **hypotonic** solution is one that causes osmotic water flow into the cell. If a red blood cell is placed in a hypotonic solution, it will swell up like a balloon. The cell may eventually burst, releasing its contents. This event is **hemolysis** (*hemo-*, blood + *lysis*, a loosening).

Exposure to a **hypertonic** solution results in the osmotic movement of water out of the cell. As water moves out, the cell shrivels and dehydrates. The shrinking of red blood cells is called **crenation**, and the cells are said to be crenated.

It is often necessary to give patients large volumes of fluid to combat severe blood loss or dehydration. At such times the difference between osmolarity and tonicity is critically important. Consider a solution that has the same osmolarity as the intracellular fluid, but a higher concentration of one or more individual ions. If any of those ions can cross the plasma membrane and diffuse into the cell, the osmolarity of the intracellular fluid will gradually increase, and that of the extracellular solution will gradually decrease. Osmosis will then occur, moving water into the cell. If the process continues, the cell will gradually inflate like a water balloon. In this case, the extracellular solution and the intracellular fluid were initially equal in osmolarity, but the extracellular fluid was hypotonic rather than isotonic. In clinical emergencies, the fluid often administered is a 0.9 percent (0.9 g/dL) solution of sodium chloride (NaCl). This isotonic solution is called **normal saline**.

Module 3.12 Review

a. Describe osmosis.

b. Contrast the effects of a hypotonic solution and a hypertonic solution on a red blood cell.

c. Some pediatricians recommend using a 10 percent salt solution to relieve nasal congestion in infants. Explain the effects this treatment would have on the cells lining the nasal cavity. Would it be effective?

In carrier-mediated transport, integral proteins facilitate membrane passage

Nutrients that are insoluble in lipids and too large to fit through membrane channels may be transported across the plasma membrane by **carrier proteins**. Many carrier proteins move a specific substance in one direction only, either into or out of the cell. Some carrier proteins move more than one substance in the same direction simultaneously; this proces is called **cotransport**. If a carrier protein simultaneously moves two substances in opposite directions, the process is called **countertransport** and the carrier protein is called an **exchange pump**.

Facilitated Diffusion

1 Many substances can be passively transported across the plasma membrane in a process called **facilitated diffusion.** No ATP is expended in facilitated diffusion. The molecules simply move from an area of higher concentration to one of lower concentration. However, the molecules involved cannot cross the plasma membrane by diffusion through the membrane lipids or through membrane channels. The number of suitable carrier proteins limits the rate of transport into the cell at any given moment. Once all the carrier proteins are saturated, the rate of transport cannot increase, regardless of further increases in the concentration of that substance in the ECF.

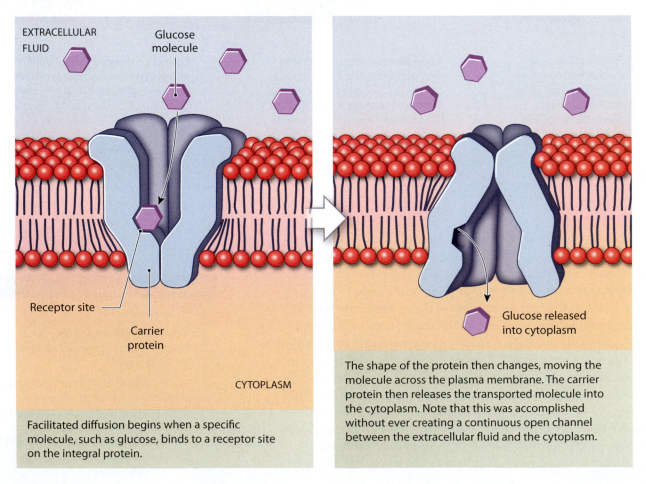

EXTRACELLULAR FLUID

Glucose molecule

Receptor site

Carrier protein

CYTOPLASM

Glucose released into cytoplasm

Facilitated diffusion begins when a specific molecule, such as glucose, binds to a receptor site on the integral protein.

The shape of the protein then changes, moving the molecule across the plasma membrane. The carrier protein then releases the transported molecule into the cytoplasm. Note that this was accomplished without ever creating a continuous open channel between the extracellular fluid and the cytoplasm.

Active Transport

2 In **active transport**, ATP (or another high-energy compound) provides the energy needed to move ions or molecules across the plasma membrane. Active transport offers one great advantage: It is not dependent on a concentration gradient. As a result, the cell can import or export specific substances, regardless of their intracellular or extracellular concentrations. All cells contain carrier proteins called **ion pumps**, which actively transport the cations sodium (Na^+), potassium (K^+), calcium (Ca^{2+}), and magnesium (Mg^{2+}) across their plasma membranes, and specialized cells transport other ions as well.

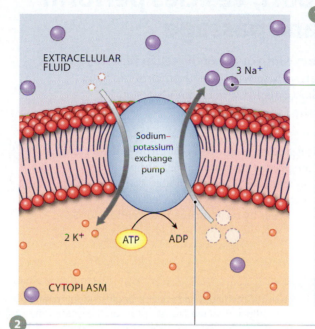

1 Sodium ion concentrations are high in the extracellular fluids, but low in the cytoplasm. The distribution of potassium in the body is just the opposite: low in the extracellular fluids and high in the cytoplasm. As a result, sodium ions slowly diffuse into the cell, and potassium ions diffuse out through leak channels. Homeostasis within the cell depends on the ejection of sodium ions and the recapture of lost potassium ions. The sodium–potassium exchange pump is a carrier protein called **sodium–potassium ATPase**. It exchanges intracellular sodium for extracellular potassium.

2 On average, for each ATP molecule consumed, three sodium ions are ejected and the cell reclaims two potassium ions. The energy demands are impressive: Sodium–potassium ATPase may use up to 40 percent of the ATP produced by a resting cell!

Secondary Active Transport

3 In **secondary active transport**, the transport mechanism itself does not require energy from ATP, but the cell often needs to expend ATP at a later time to preserve homeostasis. Movement across the plasma membrane follows an existing concentration gradient for one of two substances transported, and transport does not require an energy input.

To preserve homeostasis, the cell must then expend ATP to pump the arriving sodium ions out of the cell by using the sodium–potassium exchange pump. It thus "costs" the cell one ATP for every three glucose molecules it transports into the cell.

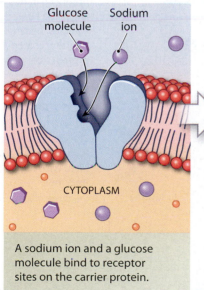

A sodium ion and a glucose molecule bind to receptor sites on the carrier protein.

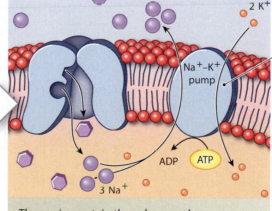

The carrier protein then changes shape, opening a path to the cytoplasm and releasing the transported materials. It then reassumes its original shape and is ready to repeat the process.

Module 3.13 Review

a. Describe the process of carrier-mediated transport.

b. What do the transport processes of facilitated diffusion and active transport have in common?

c. During digestion, the concentration of hydrogen ions (H^+) in the stomach contents increases to many times that in cells lining the stomach. Which transport process could be responsible?

In vesicular transport, vesicles perform selective membrane passage

In **vesicular transport**, materials move into or out of the cell in vesicles, small membranous sacs that form at, or fuse with, the plasma membrane. The two major categories of vesicular transport are **endocytosis** and **exocytosis**; both require energy in the form of ATP.

1 **Receptor-mediated endocytosis** is a selective process that involves the formation of small vesicles at the surface of the plasma membrane. Many important substances, including cholesterol and iron ions (Fe^{2+}), are distributed throughout the body attached to special transport proteins. These proteins are too large to pass through membrane pores, but they can and do enter cells by receptor-mediated endocytosis.

Endocytosis

Endocytosis is the formation of vesicles at the cell surface; these vesicles are generally known as **endosomes**.

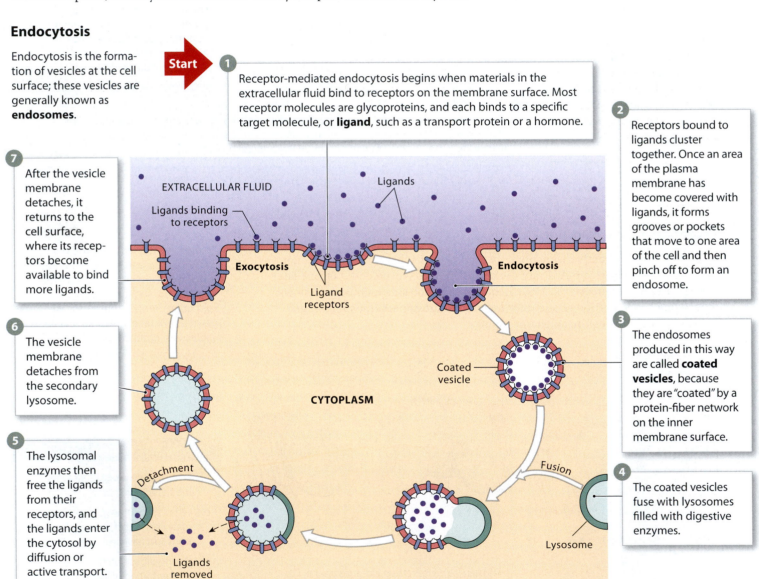

Start

1 Receptor-mediated endocytosis begins when materials in the extracellular fluid bind to receptors on the membrane surface. Most receptor molecules are glycoproteins, and each binds to a specific target molecule, or **ligand**, such as a transport protein or a hormone.

2 Receptors bound to ligands cluster together. Once an area of the plasma membrane has become covered with ligands, it forms grooves or pockets that move to one area of the cell and then pinch off to form an endosome.

3 The endosomes produced in this way are called **coated vesicles**, because they are "coated" by a protein-fiber network on the inner membrane surface.

4 The coated vesicles fuse with lysosomes filled with digestive enzymes.

5 The lysosomal enzymes then free the ligands from their receptors, and the ligands enter the cytosol by diffusion or active transport.

6 The vesicle membrane detaches from the secondary lysosome.

7 After the vesicle membrane detaches, it returns to the cell surface, where its receptors become available to bind more ligands.

EXTRACELLULAR FLUID

Ligands

Ligands binding to receptors

Exocytosis

Endocytosis

Ligand receptors

Coated vesicle

CYTOPLASM

Fusion

Lysosome

Detachment

Ligands removed

2 "Cell drinking," or **pinocytosis** (pi-nō-sī-TŌ-sis), is the formation of endosomes filled with extracellular fluid. This process is not as selective as receptor-mediated endocytosis, because no receptor proteins are involved. The target appears to be the fluid contents in general, rather than specific ligands. In a few specialized cells, pinocytosis produces vesicles on one side of the cell that are discharged on the other. This method of bulk transport is common in cells lining small blood vessels.

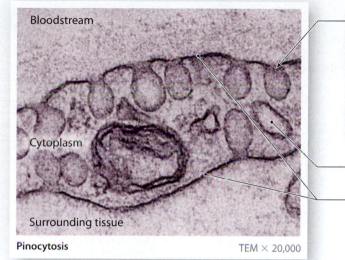

Bloodstream

Cytoplasm

Surrounding tissue

Pinocytosis

TEM × 20,000

Pinocytosis begins with the formation of deep grooves or pockets that then pinch off and enter the cytoplasm. The steps are similar to those of receptor-mediated endocytosis, but they occur in the absence of ligand binding.

Endosome

Plasma membrane

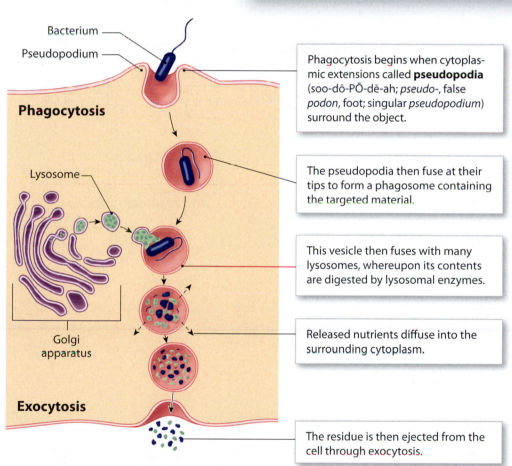

Bacterium

Pseudopodium

Phagocytosis

Lysosome

Golgi apparatus

Exocytosis

Phagocytosis begins when cytoplasmic extensions called **pseudopodia** (soo-dō-PŌ-dē-ah; *pseudo-*, false *podon*, foot; singular *pseudopodium*) surround the object.

The pseudopodia then fuse at their tips to form a phagosome containing the targeted material.

This vesicle then fuses with many lysosomes, whereupon its contents are digested by lysosomal enzymes.

Released nutrients diffuse into the surrounding cytoplasm.

The residue is then ejected from the cell through exocytosis.

3 "Cell eating," or **phagocytosis** (fag-ō-sī-TŌ-sis), produces **phagosomes** containing solid objects that may be as large as the cell itself. Although most cells display pinocytosis, only specialized cells perform phagocytosis; these cells are called **phagocytes** or **macrophages**.

4 **Exocytosis** (ek-sō-sī-TŌ-sis) is the functional opposite of endocytosis. In exocytosis, a vesicle created inside the cell fuses with the plasma membrane and discharges its contents into the extracellular environment. The ejected material may be waste products (such as those accumulating in lysosomes) or secretory products (such as mucins or hormones).

Module 3.14 Review

a. Describe endocytosis.

b. Describe exocytosis.

c. When they encounter bacteria, certain types of white blood cells engulf the bacteria and bring them into the cell. What is this process called?

1. Concept map

Use each of the following terms once to fill in the blank boxes to correctly complete the membrane permeability concept map.

- exocytosis
- diffusion
- "cell eating"
- molecular size
- pinocytosis
- facilitated diffusion
- vesicular transport
- net diffusion of water
- active transport
- specificity

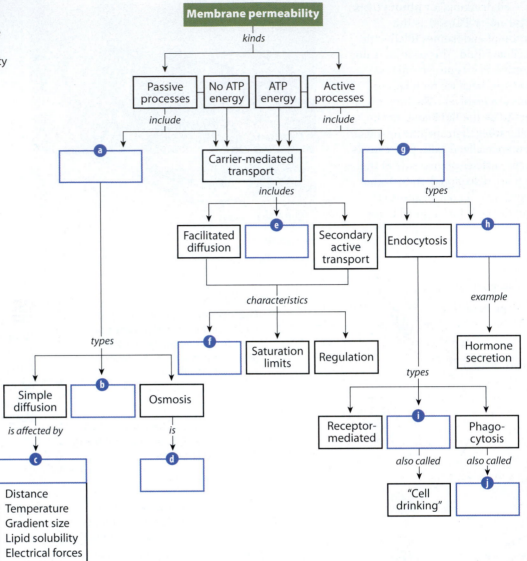

2. Short answer

Classify each of the following situations as an example of diffusion, osmosis, or neither.

- You walk into a room and smell a balsam-scented candle.

- A drop of food coloring disperses within a liquid medium.

- Water flows through a garden hose.

- A sugar cube placed in a cup of hot tea dissolves.

- Grass in the yard wilts after being exposed to excess chemical fertilizer.

- After soaking several hours in water containing sodium chloride, a stalk of celery weighs less than before it was placed in the salty water.

a _____

b _____

c _____

d _____

e _____

f _____

The Cell Life Cycle

The period between fertilization and physical maturity involves tremendous changes in organization and complexity. At fertilization, a single cell is all there is; at maturity, your body has roughly 75 trillion cells. **Cell division** makes this transformation possible.

Even when development is complete, cell division continues to be essential to survival. Physical wear and tear, toxic chemicals, temperature changes, and other environmental stresses damage cells. And, like individuals, cells age. The life span of a cell varies from hours to decades, depending on the type of cell and the stresses involved. Many cells apparently self-destruct after a certain period of time as a result of the activation of specific "suicide genes" in the nucleus. The genetically controlled death of cells is called **apoptosis** (ap-op-TŌ-sis or ap-ō-TŌ-sis; *apo-*, separated from + *ptosis*, a falling).

There are actually two different forms of cell division. **Mitosis** (mī-TŌ-sis), the focus of this section, produces two daughter cells, each containing a complete set of 46 chromosomes. **Meiosis** (mī-Ō-sis), which we will examine in Chapter 25, produces sex cells (sperm or oocytes) containing only 23 chromosomes.

1 The division of a single cell produces a pair of **daughter cells**, each half the size of the original. Before dividing, each of the daughter cells will grow to the size of the original cell.

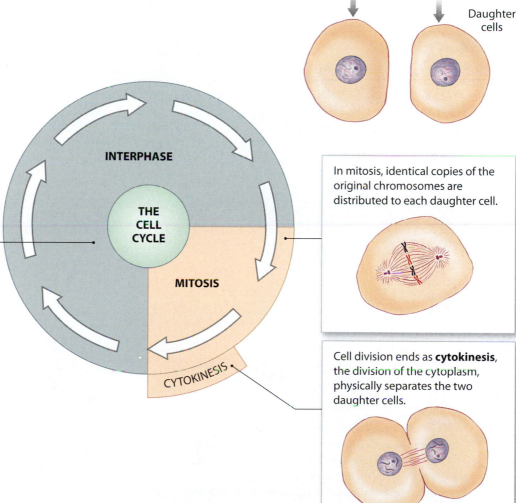

Original cell

Cell division

Daughter cells

2 This is a diagram detailing the life cycle of a typical cell. The cycle ends only upon the death of the cell.

INTERPHASE

THE CELL CYCLE

MITOSIS

CYTOKINESIS

Interphase is the period in which the cell is performing normal functions and not actively engaged in cell division. Some cells are in interphase indefinitely; others are always either dividing or preparing to divide.

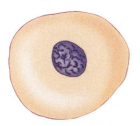

During interphase in a cell preparing to divide, the chromosomes of the cell are duplicated and associated proteins are synthesized.

In mitosis, identical copies of the original chromosomes are distributed to each daughter cell.

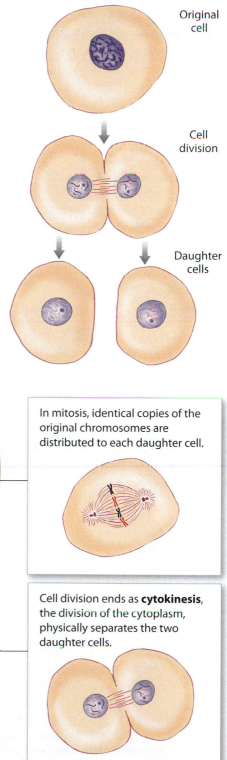

Cell division ends as **cytokinesis**, the division of the cytoplasm, physically separates the two daughter cells.

During interphase, the cell prepares for cell division

Most cells spend only a small part of their time actively engaged in cell division. **Somatic** (*soma*, body) **cells** spend the majority of their functional lives in a state known as **interphase**. During interphase, a cell performs all its normal functions and, if necessary, prepares for cell division.

1 In a cell preparing to divide, interphase can be divided into the **G_1**, **S**, and **G_2 phases.**

When the activities of the G_1 phase have been completed, the cell enters the S phase. Over the next 6–8 hours, the cell duplicates its chromosomes. This involves DNA replication and the synthesis of histones and other proteins in the nucleus.

A cell that is ready to divide first enters the G_1 phase. In this phase, the cell makes enough mitochondria, cytoskeletal elements, endoplasmic reticula, ribosomes, Golgi membranes, and cytosol for two functional cells. Centriole replication begins in G_1 and commonly continues until G_2. In cells dividing at top speed, G_1 may last just 8–12 hours. Such cells pour all their energy into mitosis, and all other activities cease. If G_1 lasts for days, weeks, or months, preparation for mitosis occurs as the cells perform their normal functions.

Once DNA replication has ended, there is a brief (2–5-hour) G_2 phase devoted to last-minute protein synthesis and to the completion of centriole replication.

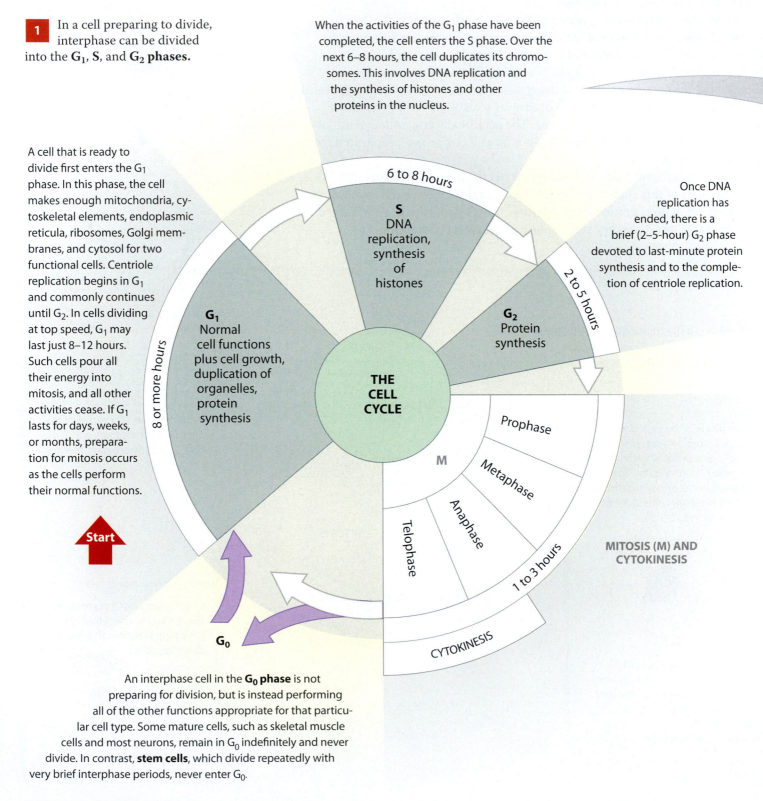

6 to 8 hours

S DNA replication, synthesis of histones

2 to 5 hours

G_1 Normal cell functions plus cell growth, duplication of organelles, protein synthesis

8 or more hours

THE CELL CYCLE

G_2 Protein synthesis

M

Prophase

Metaphase

Anaphase

Telophase

1 to 3 hours

CYTOKINESIS

MITOSIS (M) AND CYTOKINESIS

Start

G_0

An interphase cell in the **G_0 phase** is not preparing for division, but is instead performing all of the other functions appropriate for that particular cell type. Some mature cells, such as skeletal muscle cells and most neurons, remain in G_0 indefinitely and never divide. In contrast, **stem cells**, which divide repeatedly with very brief interphase periods, never enter G_0.

2 The goal of DNA replication, which occurs in cells preparing to undergo either mitosis or meiosis, is to copy the genetic information in the nucleus.

1 **DNA replication** begins when enzymes unwind the strands and disrupt the hydrogen bonds between the bases. As the strands unwind, molecules of **DNA polymerase** bind to the exposed nitrogenous bases. This enzyme (1) promotes bonding between the nitrogenous bases of the DNA strand and complementary DNA nucleotides dissolved in the nucleoplasm and (2) links the nucleotides by covalent bonds.

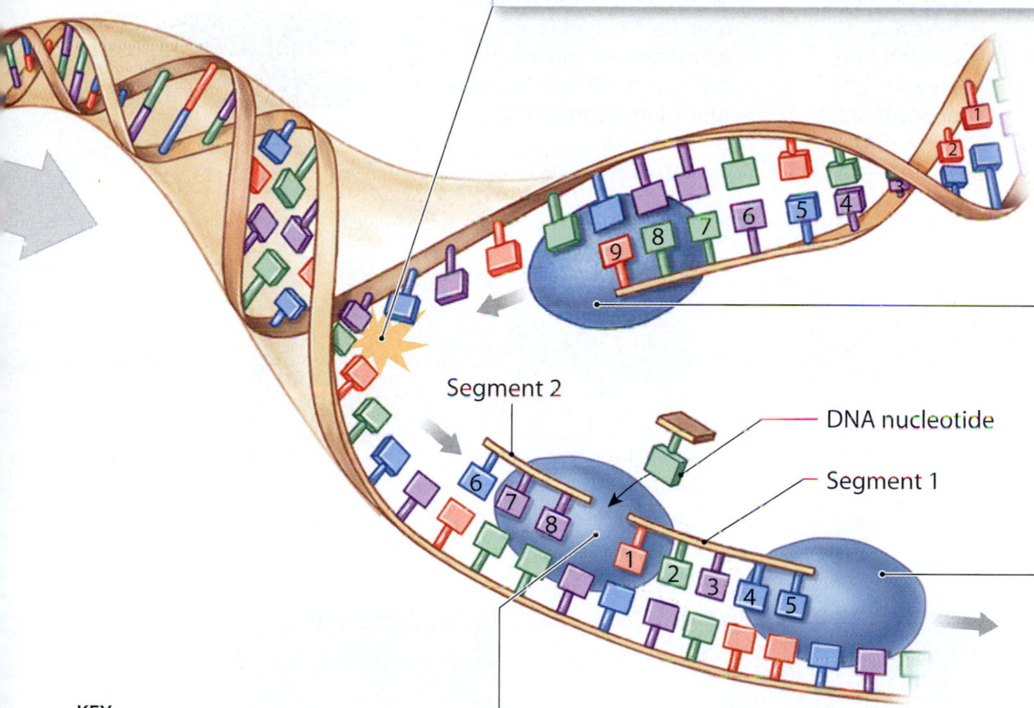

2 As the two original strands gradually separate, DNA polymerase binds to the strands. DNA polymerase can work in only one direction along a strand of DNA, but the two strands in a DNA molecule are oriented in opposite directions. The DNA polymerase bound to the upper strand shown here adds nucleotides to make a single, continuous complementary copy that grows toward the "zipper."

Segment 2

DNA nucleotide

Segment 1

3 DNA polymerase on the lower strand can work only away from the zipper. So the first DNA polymerase to bind to this strand must add nucleotides and build a complementary DNA strand moving from left to right. As the two original strands continue to unzip, additional nucleotides are continuously being exposed to the nucleoplasm. The first DNA polymerase on this strand cannot go into reverse; it can only continue to elongate the strand it already started.

KEY

🟥 Adenine

🟩 Guanine

🟪 Cytosine

🟦 Thymine

4 Thus, a second DNA polymerase must bind closer to the point of unzipping and assemble a complementary copy (segment 2) that grows until it "bumps into" segment 1 created by the first DNA polymerase. The two segments are then spliced together by enzymes called **ligases** (LĪ-gās-ez; *liga*, to tie).

3 Eventually, the unzipping completely separates the original strands. The copying ends, the last splicing is done, and two identical DNA molecules have formed. Once the DNA has been replicated, the centrioles duplicated, and the necessary enzymes and proteins have been synthesized, the cell leaves interphase and is ready to proceed to mitosis (M).

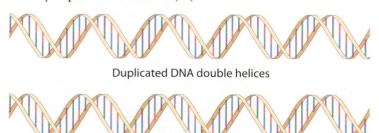

Duplicated DNA double helices

Module 3.15 Review

a. Describe interphase, and identify its stages.

b. What enzymes must be present for DNA replication to proceed normally?

c. A cell is actively manufacturing enough organelles to serve two functional cells. This cell is probably in what phase of interphase?

Mitosis distributes chromosomes before cytokinesis separates the daughter cells

Mitosis consists of a series of events during which the duplicated chromosomes of a cell are separated and distributed into two identical nuclei. The term *mitosis* specifically refers to the division and duplication of the cell's nucleus; division of the cytoplasm into two distinct new cells involves a separate, but related, process known as **cytokinesis** (sī-tō-ki-NĒ-sis; *cyto-*, cell + *kinesis*, motion).

MITOSIS

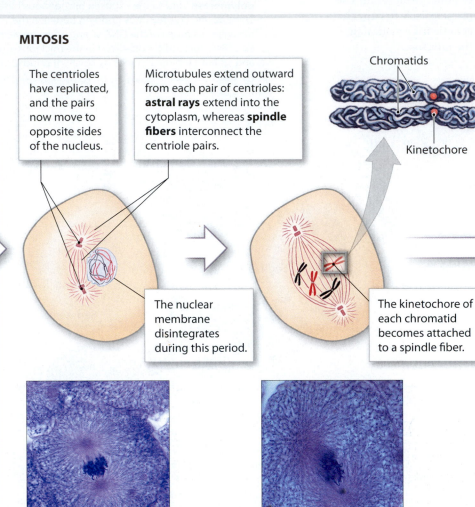

The centrioles have replicated, and the pairs now move to opposite sides of the nucleus.

Microtubules extend outward from each pair of centrioles: **astral rays** extend into the cytoplasm, whereas **spindle fibers** interconnect the centriole pairs.

Chromatids

Kinetochore

The nuclear membrane disintegrates during this period.

The kinetochore of each chromatid becomes attached to a spindle fiber.

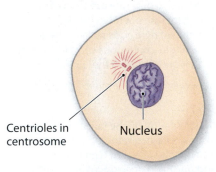

Centrioles in centrosome

Nucleus

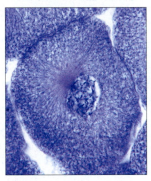

1 During interphase, the DNA strands are loosely coiled and chromosomes cannot be seen.

2 **Prophase** (PRŌ-fāz; *pro*, before) begins when the chromosomes coil so tightly that they become visible as individual structures under a light microscope. As a result of DNA replication during the S phase, two copies of each chromosome now exist. Each copy is called a **chromatid**, and the paired chromatids are connected at a region known as the centromere. A raised region on the centromere, called the **kinetochore** (ki-NE-tō-kor), is where the spindle fibers attach to the chromatids.

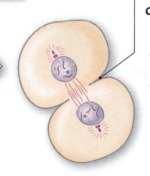

The two chromatids are now pulled apart and drawn to opposite ends of the cell along the **spindle apparatus** (the complex of spindle fibers). Anaphase ends when the chromatids arrive near the centrioles at opposite ends of the cell.

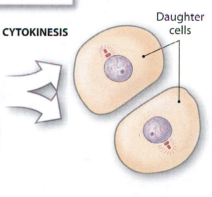

As the chromatids approach the ends of the spindle apparatus, the cytoplasm constricts along the plane of the metaphase plate, forming a **cleavage furrow**.

CYTOKINESIS

Daughter cells

Metaphase plate

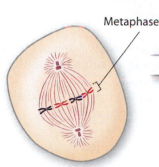
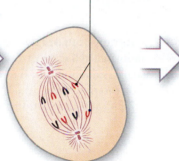
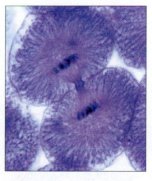
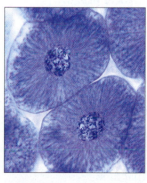

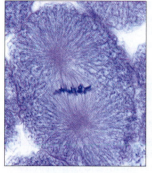
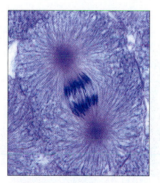

3 **Metaphase**
(MET-a-fãz; *meta*, after) begins as the chromatids move to a narrow central zone called the **metaphase plate**. Metaphase ends when all the chromatids are aligned in the plane of the metaphase plate.

4 **Anaphase**
(AN-a-fãz; *ana-*, apart) begins when the centromere of each chromatid pair splits and the chromatids separate. The chromatids are now pulled along the spindle fibers toward opposite sides of the dividing cell.

5 During **telophase** (TĒL-ō-fãz; *telo-*, end), each new cell prepares to return to the interphase state. The nuclear membranes re-form, the nuclei enlarge, and the chromosomes gradually uncoil to the chromatin state. This stage marks the end of mitosis.

6 **Cytokinesis** usually begins with the formation of a cleavage furrow during anaphase and continues throughout telophase. The completion of cytokinesis marks the end of cell division.

Module 3.16 Review

a. Define mitosis, and list its four stages.

b. What is a chromatid, and how many would be present during normal mitosis in a human cell?

c. What would happen if spindle fibers failed to form in a cell during mitosis?

Tumors and cancer are characterized by abnormal cell growth and division

When the rates of cell division and growth exceed the rate of cell death, a tissue begins to enlarge. The enlargement often results from the divisions of a single abnormal cell that no longer responds to the factors that regulate the rate of cell division. **Cancer** is an illness characterized by **mutations**—permanent changes in DNA nucleotide sequence and function—that disrupt normal control mechanisms that regulate the rates of cell division.

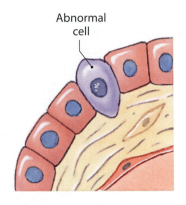

Abnormal cell

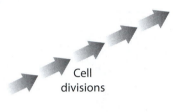

Cell divisions

1 Cancer usually begins with a single abnormal cell. Every time a cell divides, there is a chance that something will go wrong with the control mechanism. As a result, cancers are most common in tissues where cells are dividing rapidly and continuously, such as the epithelium of the skin or the intestinal lining.

2 A **tumor**, or **neoplasm**, is a mass or swelling produced by abnormal cell growth and division. In a **benign tumor**, the cells usually remain within the tissue of origin. Such a tumor seldom threatens an individual's life and can usually be surgically removed if its size or position disturbs tissue function.

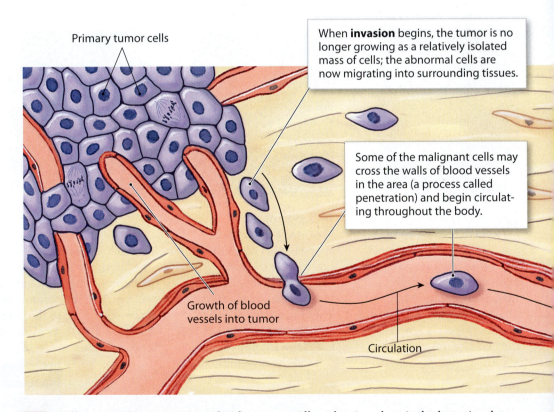

Primary tumor cells

When **invasion** begins, the tumor is no longer growing as a relatively isolated mass of cells; the abnormal cells are now migrating into surrounding tissues.

Some of the malignant cells may cross the walls of blood vessels in the area (a process called penetration) and begin circulating throughout the body.

Growth of blood vessels into tumor

Circulation

3 Cells in a **malignant tumor** divide very rapidly, releasing chemicals that stimulate the growth of blood vessels into the area. The availability of additional nutrients accelerates tumor growth, and malignant cells then begin migrating into surrounding tissues and nearby blood vessels. This process—**metastasis**—can produce secondary tumors in tissues remote from the site of the primary tumor.

4 Malignant cells may no longer perform their original functions, or they may perform normal functions in an abnormal way. For example, endocrine cancer cells may produce normal hormones, but in excessively large amounts. This photo shows a patient with a malignant tumor of the thyroid gland. The large amounts of thyroid hormone produced dramatic changes in metabolic activity throughout the body.

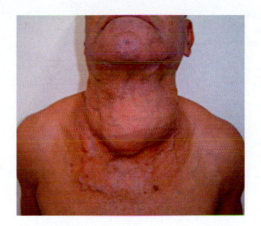

Responding to cues that are as yet unknown, cancer cells in the bloodstream ultimately escape out of blood vessels to establish secondary tumors at other sites. These tumors are extremely active metabolically, and their presence stimulates the growth of blood vessels into the area. The increased vascular supply provides additional nutrients to the cancer cells and further accelerates tumor growth and metastasis.

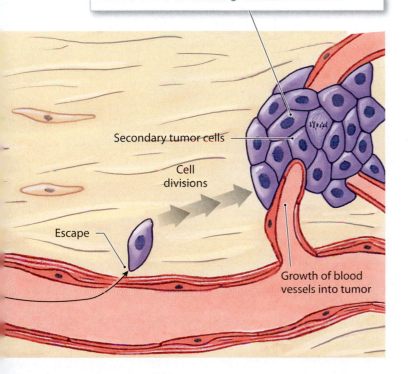

Secondary tumor cells

Cell divisions

Escape

Growth of blood vessels into tumor

Cancer cells do not use energy very efficiently. They grow and multiply at the expense of healthy tissues, competing for space and nutrients with normal cells. This competition contributes to the starved appearance of many patients in the late stages of cancer. Death may occur as a result of the compression of vital organs when nonfunctional cancer cells have killed or replaced the healthy cells in those organs, or when the cancer cells have starved normal tissues of essential nutrients.

Module 3.17 Review

a. Define metastasis.

b. What is a benign tumor?

c. Define cancer.

1. Concept map

Use each of the following terms once to fill in the blank boxes to correctly complete the cell cycle concept map.

- metaphase
- DNA replication
- somatic cells
- G_2 phase
- telophase
- mitosis
- cytokinesis
- G_1 phase

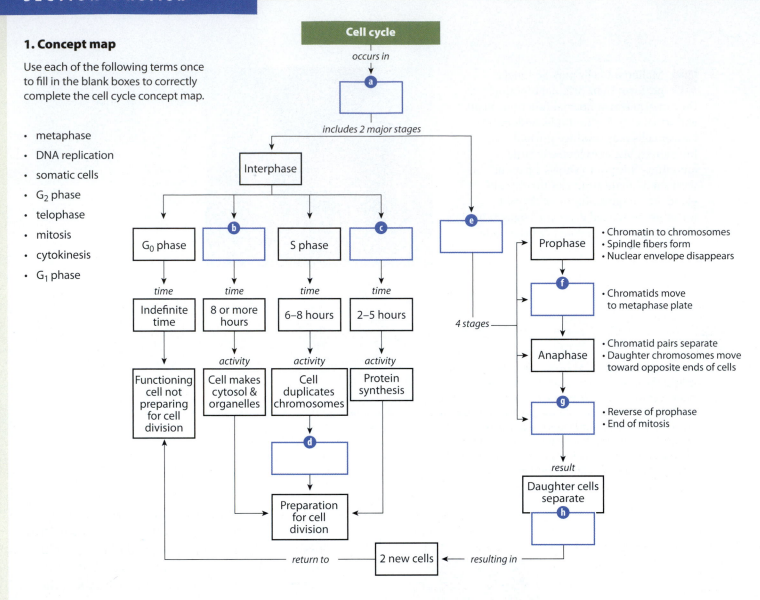

2. Vocabulary

In the space provided, write the boldfaced terms introduced in this section that contain the indicated word part.

Word Part	Meaning	Terms
a telo-	end	
b pro-	before	
c centro-	in the middle	

3. Section integration

The muscle cells that that make up skeletal muscle tissue are large and multinucleated. They form early in development as groups of embryonic cells fuse together, contributing their nuclei and losing their individual plasma membranes. Describe an alternate mechanism that would also result in a large, multinucleated cell.

Visual Outline with Key Terms

Summarize the content of each module using the terms in the order provided.

SECTION 1

An Introduction to Cells

- cell theory
- differentiation

3.1

Cells are the smallest living units of life

- extracellular fluid
- interstitial fluid
- plasma membrane
- cytoplasm
- cytosol
- organelles
- nonmembranous organelles
- membranous organelles
- peroxisome
- lysosome
- microvilli
- Golgi apparatus
- nucleus
- nuclear envelope
- endoplasmic reticulum (ER)
- smooth ER
- rough ER
- ribosomes
- cytoskeleton
- centrosome
- mitochondria

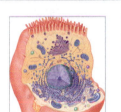

3.2

The plasma membrane isolates the cell from its environment and performs varied functions

- plasma membrane
- glycocalyx
- integral proteins
- transmembrane proteins
- peripheral proteins
- phospholipid bilayer
- anchoring proteins
- recognition proteins
- receptor proteins
- ligands
- carrier proteins
- channels

3.3

The cytoskeleton plays both a structural and a functional role

- cytoskeleton
- microvilli
- microfilaments
- actin
- terminal web
- intermediate filaments
- microtubules
- thick filaments
- centrioles
- centrosome
- cilia
- basal body

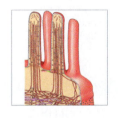

3.4

Ribosomes are responsible for protein synthesis and are often associated with the endoplasmic reticulum

- ribosomes
- ribosomal RNA (rRNA)
- endoplasmic reticulum (ER)
- cisternae
- smooth endoplasmic reticulum (SER)
- rough endoplasmic reticulum (RER)
- fixed ribosomes
- transport vesicles

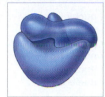

3.5

The Golgi apparatus is a packaging center

- Golgi apparatus
- cisternae
- transport vesicles
- forming face
- maturing face
- membrane renewal vesicles
- secretory vesicles
- lysosomes
- autolysis
- membrane flow

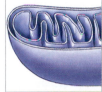

3.6

Mitochondria are the powerhouses of the cell

- mitochondria
- matrix
- cristae
- glycolysis
- citric acid cycle
- aerobic metabolism

SECTION 2

Structure and Function of the Nucleus

- nucleus
- cellular-level homeostasis
- short-term adjustments
- long-term adjustments

● = Term boldfaced in this module

3.7

The nucleus contains DNA, RNA, organizing proteins, and enzymes

- nucleus
- nuclear envelope
- perinuclear space
- nuclear pores
- nucleoplasm
- nucleoli
- histones
- nucleosomes
- chromatin
- chromosomes
- centromere

3.8

Protein synthesis involves DNA, enzymes, and three types of RNA

- genetic code
- gene
- triplet code
- triplet
- gene activation
- messenger RNA (mRNA)
- codons
- anticodons
- transfer RNA (tRNA)
- ribosomal RNA (rRNA)

3.9

Transcription encodes genetic instructions on a strand of RNA

- transcription
- control segment
- template strand
- RNA polymerase
- immature mRNA
- introns
- exons

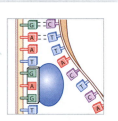

3.10

Translation builds polypeptides as directed by an mRNA strand

- translation
- anticodon

SECTION 3

How Things Enter and Leave the Cell

- permeability
- freely permeable membranes
- selectively permeable membranes
- impermeable membranes

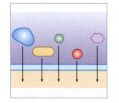

3.11

Diffusion is movement driven by concentration differences

- diffusion
- gradient

3.12

Osmosis is the passive movement of water

- osmosis
- osmotic flow
- osmotic pressure
- hydrostatic pressure
- osmolarity (osmotic concentration)
- tonicity
- isotonic
- hypotonic
- hemolysis
- hypertonic
- crenation
- normal saline

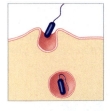

3.13

In carrier-mediated transport, integral proteins facilitate membrane passage

- carrier-mediated transport
- carrier proteins
- cotransport
- countertransport
- exchange pump
- facilitated diffusion
- active transport
- ion pumps
- sodium–potassium ATPase
- secondary active transport

3.14

In vesicular transport, vesicles perform selective membrane passage

- vesicular transport
- endocytosis
- exocytosis
- receptor-mediated endocytosis
- endosomes
- ligand
- coated vesicles
- pinocytosis
- phagocytosis
- phagosomes
- phagocytes
- macrophages
- pseudopodia

SECTION 4

The Cell Life Cycle

- cell division
- apoptosis
- mitosis
- meiosis
- daughter cells
- interphase
- cytokinesis

• = *Term boldfaced in this module*

3.15

During interphase, the cell prepares for cell division

- somatic cells
- interphase
- G_1 phase
- S phase
- G_2 phase
- G_0 phase
- stem cells
- DNA replication
- DNA polymerase
- ligases

3.17

Tumors and cancer are characterized by abnormal cell growth and division

- cancer
- mutations
- tumor
- neoplasm
- benign tumor
- malignant tumor
- metastasis
- invasion

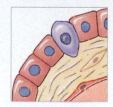

3.16

Mitosis distributes chromosomes before cytokinesis separates the daughter cells

- mitosis
- cytokinesis
- prophase
- chromatid
- kinetochore
- astral rays
- spindle fibers
- metaphase
- metaphase plate
- anaphase
- spindle apparatus
- telophase
- cleavage furrow

● = *Term boldfaced in this module*

Chapter Integration: Applying what you've learned

Corrine is a world-class bodybuilder who has competed quite successfully in the Ms. Fitness USA contest. She works as a certified personal trainer and is the owner of a fitness center in her hometown. Five days per week she works out in her gym, often lifting free weights three hours per day. When she is not working on sculpting her own body, she assists others in changing their body image working as a personal trainer. Corrine also provides advice on exercise, health behavior modification, and nutritional supplements. She often touts the benefits of electrolyte-laden sports drinks, claiming that they are good for your cells.

1. Corrine tells her clients that the muscle cells of people training for endurance sports (cycling, marathon running, swimming, tennis, etc.) have the potential for storing more energy and improving muscle performance, because those cells contain increased numbers of mitochondria. Is this true? Why or why not?

2. Use the principles of tonicity discussed in this chapter to explain the benefits of drinking an energy drink after strenuous exercise.

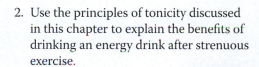

4

Tissue Level of Organization

Epithelial Tissue

Our perspective has gradually changed over the preceding chapters. Atoms and molecules can only be examined through special imaging techniques and experimental procedures. Cellular details often escape detection unless an electron microscope is used. Tissue structure, however, can be examined with a light microscope, and based on your experiences in the laboratory you may already be able to identify some tissues with the unaided eye.

Although the human body contains trillions of cells, there are only about 200 different cell types. For the human body to work efficiently, cells must coordinate their efforts. Cells working together form **tissues**—collections of cells and cell products that perform a relatively limited number of specialized functions. The study of tissues is called **histology**. Histologists recognize four basic types of tissue, epithelial tissue, connective tissue, muscle tissue, and neural tissue. This chapter examines each of these tissue types, and in doing so sets the stage for our study of organs and organ systems.

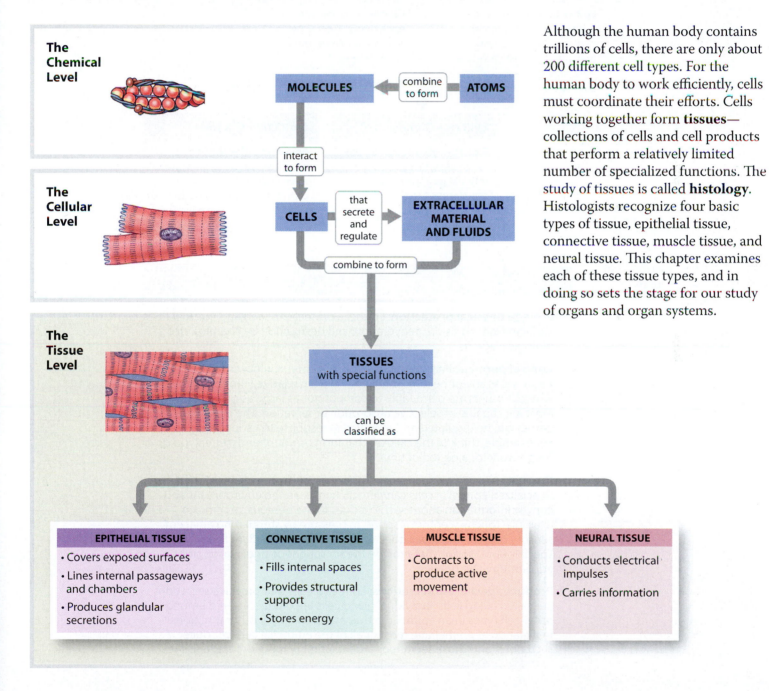

The Chemical Level

MOLECULES ← combine to form ← ATOMS

interact to form ↓

The Cellular Level

CELLS → that secrete and regulate → EXTRACELLULAR MATERIAL AND FLUIDS

combine to form ↓

The Tissue Level

TISSUES with special functions

can be classified as ↓

EPITHELIAL TISSUE	CONNECTIVE TISSUE	MUSCLE TISSUE	NEURAL TISSUE
• Covers exposed surfaces • Lines internal passageways and chambers • Produces glandular secretions	• Fills internal spaces • Provides structural support • Stores energy	• Contracts to produce active movement	• Conducts electrical impulses • Carries information

Epithelial tissue covers surfaces, lines structures, and forms secretory glands

1 It is convenient to begin our discussion with epithelial tissue because it includes the surface of your skin, a relatively familiar feature. Epithelial tissue includes epithelia and glands.

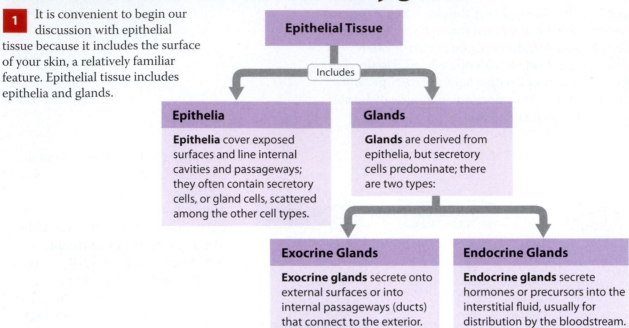

Epithelial Tissue

Includes

Epithelia

Epithelia cover exposed surfaces and line internal cavities and passageways; they often contain secretory cells, or gland cells, scattered among the other cell types.

Glands

Glands are derived from epithelia, but secretory cells predominate; there are two types:

Exocrine Glands

Exocrine glands secrete onto external surfaces or into internal passageways (ducts) that connect to the exterior.

Endocrine Glands

Endocrine glands secrete hormones or precursors into the interstitial fluid, usually for distribution by the bloodstream.

2 This table describes four essential functions of epithelial tissue.

Functions of Epithelial Tissue

- **Provide physical protection**: Epithelia protect exposed and internal surfaces from abrasion, dehydration, and destruction by chemical or biological agents.

- **Control permeability**: Any substance that enters or leaves the body has to cross an epithelium. Some epithelia are relatively impermeable, whereas others are permeable to compounds as large as proteins. Most are capable of selective absorption or secretion. The epithelial barrier can be regulated and modified in response to various stimuli. For example, think of the calluses that form on your hands when you do rough work for a period of time.

- **Provide sensation**: Sensory nerves extensively innervate most epithelia. Specialized epithelial cells can detect changes in the environment and convey information about such changes to the nervous system. For example, touch receptors respond to pressure by stimulating adjacent sensory nerves. A neuroepithelium is a sensory epithelium found in special sense organs that provide the sensations of smell, taste, sight, equilibrium, and hearing.

- **Produce specialized secretions**: Epithelial cells that produce secretions are called gland cells. Individual gland cells are often scattered among other cell types in an epithelium that may have many other functions. In a glandular epithelium, most or all of the epithelial cells produce secretions, and those secretions are the primary function of the tissue.

3 The cells of any epithelium share a number of basic features. An epithelium has an **apical** (Ā-pi-kal) **surface**, which faces the exterior of the body or some internal space, and a **base**, which is attached to adjacent tissues. The term **polarity** refers to the presence of structural and functional differences between the exposed and attached surfaces.

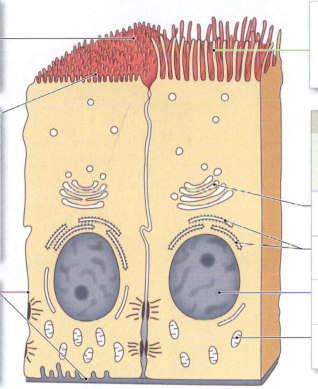

Microvilli are often found on the apical surfaces of epithelial cells that line internal passageways of the digestive, urinary, and reproductive tracts.

The apical surface is the region of the cell exposed to an internal or external environment. When the epithelium lines a tube, such as the intestinal tract, the apical surfaces of the epithelial cells are exposed to the space inside the tube, a passageway called the **lumen** (LOO-men).

The **basolateral surfaces** include both the basal surface, where the cell attaches to underlying epithelial cells or deeper tissues, and the lateral surfaces, where the cell contacts its neighbors.

Cilia cover the apical surfaces in portions of the respiratory and reproductive tracts. A typical ciliated cell contains about 250 cilia that beat in a coordinated fashion.

Membranous Organelles

Most epithelial cells have membranous organelles comparable to those of other cell types.

Golgi apparatus (facing apical surface)

Endoplasmic reticulum (often extensive around the nucleus)

Nucleus (in a tall cell, located closer to the base than the apical surface)

Mitochondria (may be apical or basal, depending on cell functions)

Simple epithelia

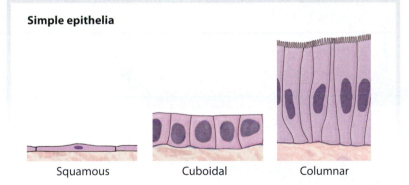

Squamous   Cuboidal   Columnar

Stratified epithelia

Squamous   Cuboidal   Columnar

4 Three epithelial cell shapes are identified: **squamous**, **cuboidal**, and **columnar**. For classification purposes, one looks at the superficial cells in a section perpendicular to both the exposed surface and the basal surface. In sectional view, squamous cells appear thin and flat, cuboidal cells look like little boxes, and columnar cells look like tall, relatively slender rectangles. If only one layer of cells is present, that layer is a **simple epithelium**. In contrast, a **stratified epithelium** contains several layers of cells. Stratified epithelia are generally located in areas that need protection from mechanical or chemical stresses, such as the surface of the skin and the lining of the mouth.

Module 4.1 Review

a. List four essential functions of epithelial tissue.

b. Summarize the classification of an epithelium based on cell shape and number of cell layers.

c. What is the probable function of an epithelial surface whose cells bear many cilia?

Epithelial cells are extensively interconnected, both structurally and functionally

To be effective as a barrier, an epithelium must form a complete cover or lining, and have the ability to replace lost or damaged cells through the divisions of stem cells. The physical integrity of an epithelium depends on intercellular connections and attachment to adjacent tissues.

1 The detailed structure of each form of intercellular attachment demonstrates the linkage between structure and function at all levels.

Microvilli

APICAL SURFACE

Intercellular Attachments

Occluding junctions form a barrier that isolates the basolateral surfaces and deeper tissues from the contents of the lumen.

An **adhesion belt** locks together the terminal webs of neighboring cells, strengthening the apical region and preventing distortion and leakage at the occluding junctions.

Gap junctions permit chemical communication that coordinates the activities of adjacent cells.

Desmosomes (DEZ-mō-sōms; *desmos*, ligament + *soma*, body) provide firm attachment between neighboring cells by interlocking their cytoskeletons.

BASE

Intermediate filaments of the cytoskeleton

Hemidesmosome

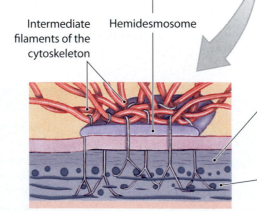

2 **Hemidesmosomes** attach the deepest epithelial cells to the basal lamina. At a hemidesmosome, the basal cytoskeleton is locked to peripheral proteins and to transmembrane proteins that are firmly attached to a layer of extracellular protein filaments and fibers.

Basal Lamina

The **basal lamina**, or basement membrane, is a complex structure produced by the basal surface of the epithelium and the underlying connective tissue.

The **clear layer**, or lamina lucida (LAM-i-nah LOO-si-dah; *lamina*, thin layer + *lucida*, clear) contains glycoproteins and a network of fine protein filaments.

The **dense layer**, or lamina densa, containing bundles of coarse protein fibers, gives the basal lamina its strength and acts as a filter that restricts diffusion between the adjacent tissues and the epithelium.

3 At an occluding junction, the attachment is so tight that it prevents the passage of water and solutes between cells.

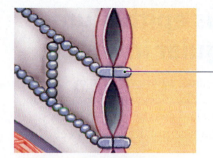

At an occluding junction, the lipid portions of the two plasma membranes are tightly bound together by interlocking membrane proteins.

4 A continuous adhesion belt forms a band that encircles cells and binds them to their neighbors. The bands are dense proteins that are attached to the microfilaments of the terminal web.

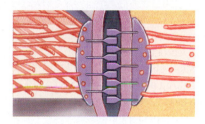

5 At a gap junction, two cells are held together by interlocking junctional proteins called **connexons**. Gap junctions between epithelial cells are common where the movement of ions helps coordinate functions such as secretion or the beating of cilia. Gap junctions occur in other tissues as well; in cardiac muscle tissue, for example, they help coordinate contractions of the heart muscle.

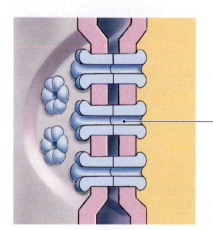

Connexons are channel proteins that form a narrow passageway and let small molecules and ions pass from cell to cell.

6 At a desmosome, the opposing plasma membranes are locked together. Desmosomes are very strong and resist stretching and twisting.

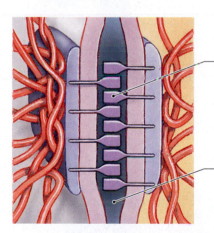

Cell adhesion molecules (CAMs) are transmembrane proteins that bind to each other and to extracellular materials.

The membranes of adjacent cells may also be bonded by **intercellular cement**, a thin layer of proteoglycans that contain polysaccharide derivatives, most notably **hyaluronan**.

Epithelia lack blood vessels, and for this reason they are said to be **avascular** (*a*, without). The cells forming the deepest layer of an epithelium must remain firmly attached to underlying connective tissues, because the blood vessels in those tissues nourish the entire epithelium.

Module 4.2 Review

a. Identify the various types of epithelial intercellular connections.

b. How do epithelial tissues, which are avascular, obtain needed nutrients?

c. What is the functional significance of gap junctions?

The cells in a squamous epithelium are flattened and irregular in shape

The cells in a **squamous epithelium** (SKWĀ-mus; *squama*, plate or scale) are thin, flat, and somewhat irregular in shape, like pieces of a jigsaw puzzle. From the surface, the cells resemble fried eggs laid side by side. In sectional view, the disc-shaped nucleus occupies the thickest portion of each cell.

1 A **simple squamous epithelium** is the body's most delicate type of epithelium. This type of epithelium is located in protected regions where absorption or diffusion takes place, or where a slick, slippery surface reduces friction. Simple squamous epithelia are found along passageways in the kidneys, inside the eye, and at the gas exchange surfaces (alveoli) of the lungs. Simple squamous epithelia in certain locations have special names. A **mesothelium** lines each ventral body cavity, whereas an **endothelium** lines the heart and blood vessels.

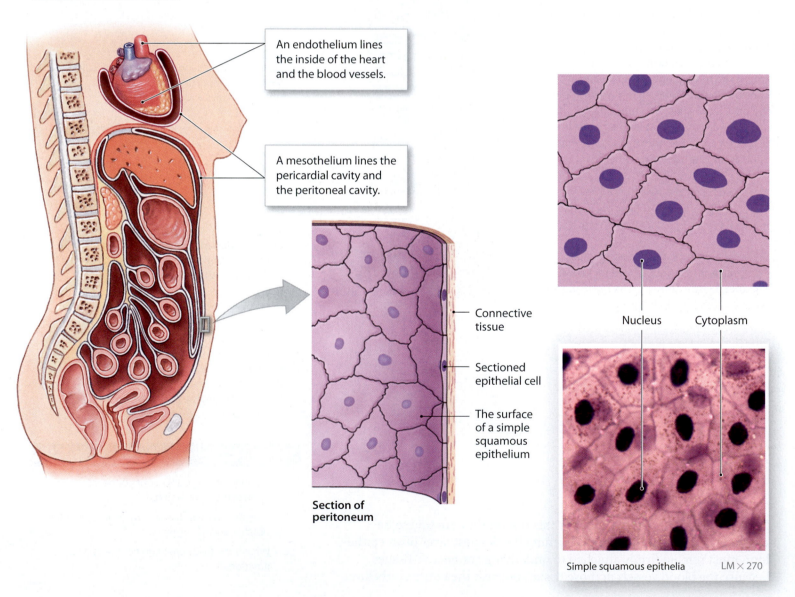

An endothelium lines the inside of the heart and the blood vessels.

A mesothelium lines the pericardial cavity and the peritoneal cavity.

Connective tissue

Sectioned epithelial cell

The surface of a simple squamous epithelium

Section of peritoneum

Nucleus   Cytoplasm

Simple squamous epithelia   LM × 270

2 A **stratified squamous epithelium** is generally located where mechanical or chemical stresses are severe. The cells form a series of layers, like the layers in a sheet of plywood. Stratified squamous epithelia form the surface of the skin and line the mouth, throat, esophagus, rectum, anus, and vagina.

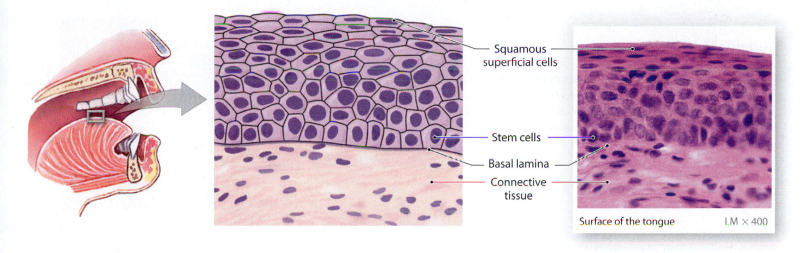

Squamous superficial cells

Stem cells

Basal lamina

Connective tissue

Surface of the tongue   LM × 400

3 On exposed body surfaces, where mechanical stress and dehydration are potential problems, apical layers of epithelial cells are packed with filaments of the protein keratin. As a result, superficial layers are both tough and water resistant, and the epithelium is said to be **keratinized**. A **nonkeratinized** stratified squamous epithelium resists abrasion but will dry out and deteriorate unless kept moist. Nonkeratinized stratified squamous epithelia are found in the oral cavity, pharynx, esophagus, anus, and vagina.

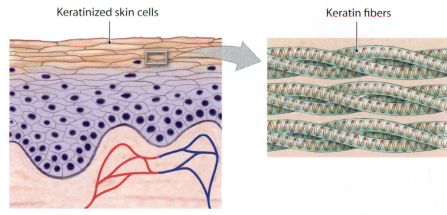

Keratinized skin cells

Keratin fibers

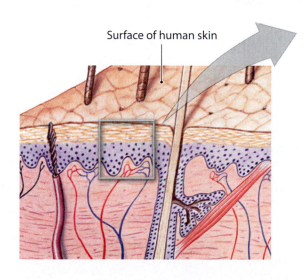

Surface of human skin

Module 4.3 Review

a. What properties are common to keratinized epithelia?

b. Why do the pharynx, esophagus, anus, and vagina have a similar epithelial organization?

c. Under a light microscope, simple squamous epithelium is seen on the outer surface. Could this be a skin surface sample? Why or why not?

Cuboidal and transitional epithelia are found along several passageways and chambers connected to the exterior

Cuboidal Epithelium

1 The cells of a **cuboidal epithelium** resemble hexagonal boxes. (In typical sectional views they appear square.) The spherical nuclei are near the center of each cell, and the distance between adjacent nuclei is roughly equal to the height of the epithelium.

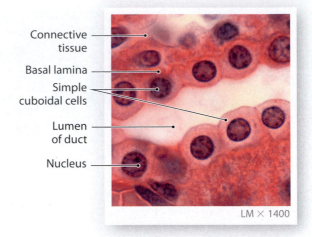

Connective tissue

Basal lamina

Simple cuboidal cells

Lumen of duct

Nucleus

LM × 1400

2 **Simple cuboidal epithelia** are found lining exocrine glands and ducts (passageways that carry secretions). They are also found in portions of the kidneys and lining secretory chambers in the thyroid gland.

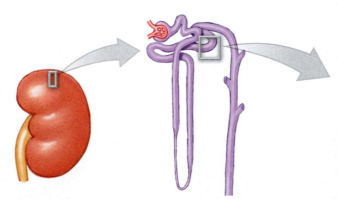

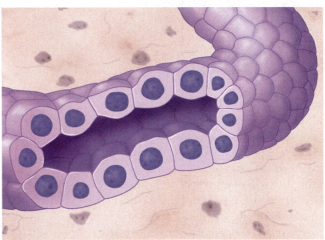

Sectioned kidney tubule

3 **Stratified cuboidal epithelia** are relatively rare. They are most common along the ducts of sweat glands, mammary glands, and other exocrine glands.

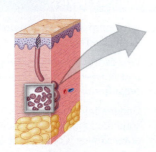

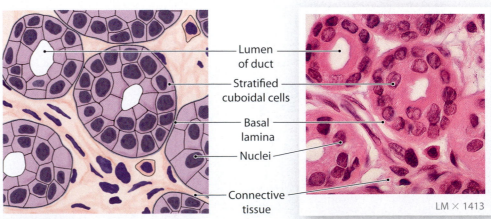

Lumen of duct

Stratified cuboidal cells

Basal lamina

Nuclei

Connective tissue

LM × 1413

Sweat gland duct

Transitional Epithelium

4 A **transitional epithelium** is an unusual stratified epithelium because, unlike most epithelia, it tolerates repeated cycles of stretching and recoiling (returning to its previous shape) without damage. It is called *transitional* because the appearance of the epithelium changes as stretching occurs. Transitional epithelia line the urinary bladder, the ureters, and the urine-collecting chambers within the kidneys.

Epithelium in a Relaxed Bladder

In an empty urinary bladder, the plump superficial cells are cuboidal with a dome-shaped surface.

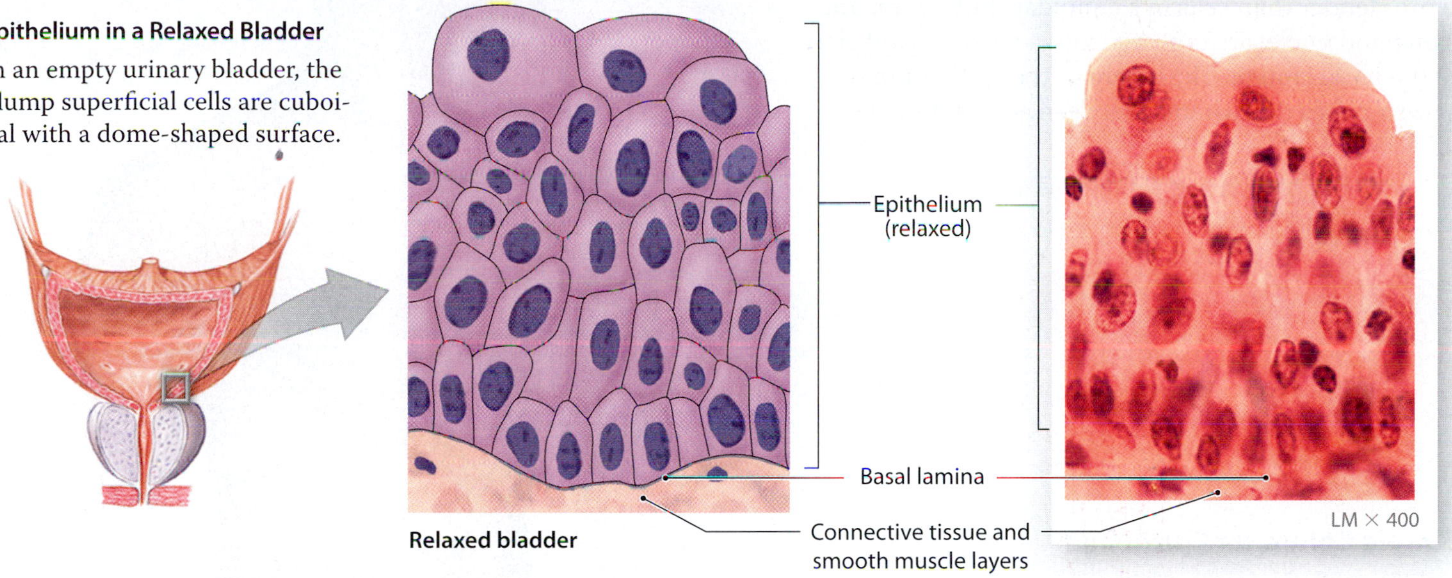

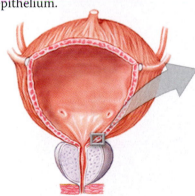

Relaxed bladder

Epithelium (relaxed)

Basal lamina

Connective tissue and smooth muscle layers

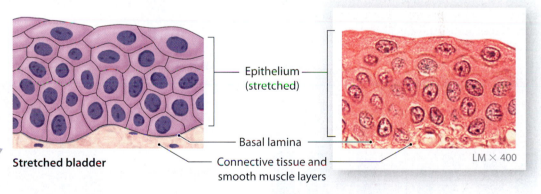

LM × 400

Epithelium in a Stretched Bladder

When the urinary bladder is full, the volume of urine has stretched the lining to such a degree that the epithelium appears flattened, and more like a stratified squamous epithelium.

Stretched bladder

Epithelium (stretched)

Basal lamina

Connective tissue and smooth muscle layers

LM × 400

Module 4.4 Review

a. Identify the epithelium that lines the urinary bladder and changes in appearance as stretching occurs.

b. Describe the appearance of simple cuboidal epithelial cells in sectional view.

c. Stratified cuboidal epithelia are associated with what epithelial structures?

Columnar epithelia typically perform absorption or provide protection from chemical or environmental stresses

In a typical sectional view, the cells of a **columnar epithelium** appear rectangular. In reality, the densely packed cells are hexagonal, but they are taller and more slender than cells in a cuboidal epithelium. The elongated nuclei are crowded into a narrow band close to the basal lamina. The height of the epithelium is several times the distance between adjacent nuclei.

Simple Columnar Epithelium

1 **Simple columnar epithelia** are found lining the stomach, intestine, gallbladder, uterine tubes, and ducts within the kidneys. These cells may have microvilli, which increase surface area for absorption, or cilia that move substances across the apical surface.

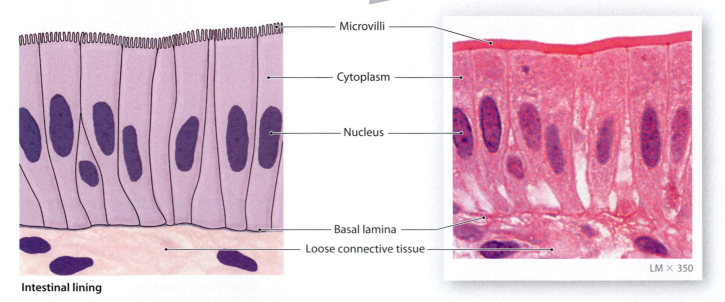

Microvilli

Cytoplasm

Nucleus

Basal lamina

Loose connective tissue

LM × 350

Intestinal lining

Pseudostratified Columnar Epithelium

Pseudostratified epithelia line the nasal cavities, the trachea, and larger airways of the lungs. They are also found along portions of the male reproductive tract.

2 A **pseudostratified columnar epithelium** includes several types of cells with varying shapes and functions. The distances between the cell nuclei and the exposed surface vary, so the epithelium appears to be layered, or stratified. It is not truly stratified, though, because every epithelial cell contacts the basal lamina. Pseudostratified columnar epithelial cells typically possess cilia.

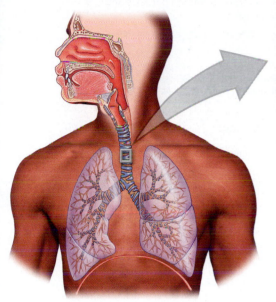

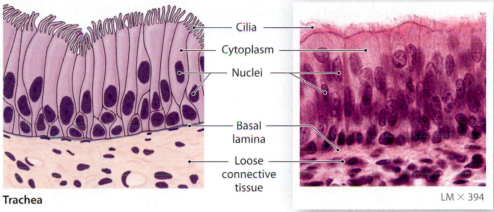

Cilia

Cytoplasm

Nuclei

Basal lamina

Loose connective tissue

Trachea

LM × 394

Stratified Columnar Epithelium

3 **Stratified columnar epithelia** are relatively rare. These epithelia may have either two layers or multiple layers. In the latter case, only the superficial cells are columnar in shape. Stratified columnar epithelia are most often found lining large ducts such as those of the salivary glands or pancreas.

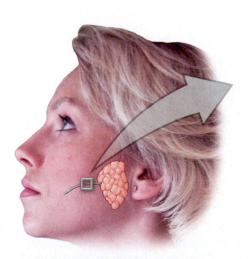

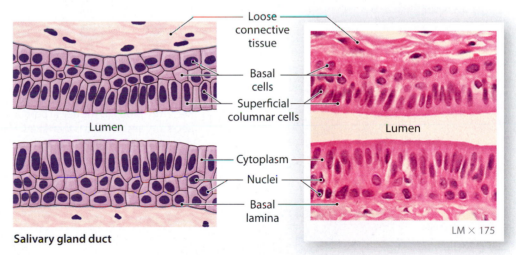

Loose connective tissue

Basal cells

Superficial columnar cells

Lumen

Cytoplasm

Nuclei

Basal lamina

Salivary gland duct

Lumen

LM × 175

Module 4.5 Review

a. Describe the appearance of simple columnar epithelial cells in a sectional view.

b. Explain why a pseudostratified columnar epithelium is not truly stratified.

c. The columnar epithelium lining the intestine typically has _____ on its apical surface.

Glandular epithelia are specialized for secretion

Methods of Secretion

Many epithelia contain gland cells that are specialized for secretion. Collections of epithelial cells (or structures derived from epithelial cells) that produce secretions are called **glands**. They range from scattered cells to complex glandular organs. Some of these glands, called **endocrine glands**, release their secretions into the interstitial fluid. Others, known as **exocrine glands**, release their secretions into passageways called ducts that open onto an epithelial surface. We will consider exocrine glands here, and endocrine glands in a later chapter.

1 Glandular epithelial cells may release their secretions in one of three ways; the methods are illustrated here.

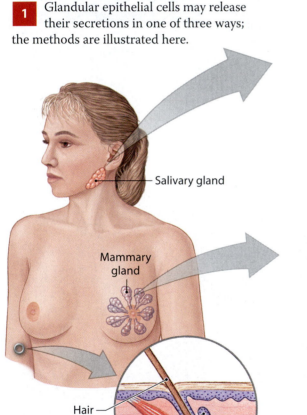

- Salivary gland
- Mammary gland
- Hair
- Sebaceous gland
- Hair follicle

Mucin is a merocrine secretion that mixes with water to form mucus. **Mucus** is an effective lubricant, a protective barrier, and a sticky trap for foreign particles and microorganisms.

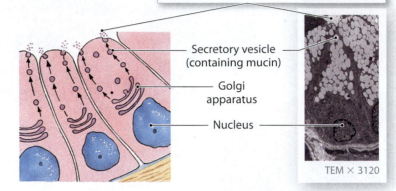

- Secretory vesicle (containing mucin)
- Golgi apparatus
- Nucleus

TEM × 3120

In **merocrine secretion** (MER-u-krin; *meros*, part), the product is released from secretory vesicles by exocytosis. This is the most common mode of secretion.

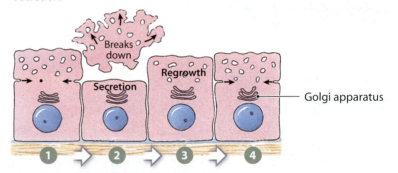

Breaks down

Regrowth

Secretion

Golgi apparatus

1 → 2 → 3 → 4

Apocrine secretion (AP-ō-krin; *apo-*, off) involves the loss of cytoplasm as well as the secretory product. The apical portion of the cytoplasm becomes packed with secretory vesicles and is then shed. Milk production in the mammary glands involves a combination of merocrine and apocrine secretions.

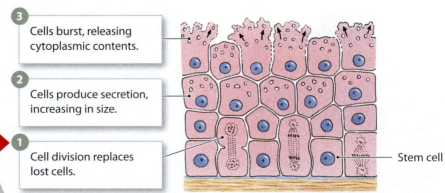

3 Cells burst, releasing cytoplasmic contents.

2 Cells produce secretion, increasing in size.

Start

1 Cell division replaces lost cells.

Stem cell

Holocrine secretion (HOL-ō-krin; *holos*, entire), by contrast, destroys the gland cell. During holocrine secretion, the entire cell becomes packed with secretory products and then bursts, releasing the secretion and killing the cell. Further secretion depends on the replacement of destroyed gland cells by the division of stem cells.

Exocrine Gland Structure

Three characteristics are used to describe the structure of multicellular exocrine glands: the structure of the duct, the shape of the secretory area of the gland, and the relationship between the duct and the secretory areas.

2 A gland is **simple** if it has a single duct that does not divide on its way to the gland cells.

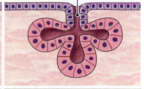

A gland is **branched** if several secretory areas (tubular or acinar) share a duct. Note that "branched" refers to the glandular areas and not to the duct.

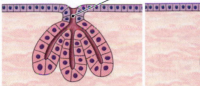

Duct

Gland cells

SIMPLE TUBULAR	SIMPLE COILED TUBULAR	SIMPLE BRANCHED TUBULAR	SIMPLE ALVEOLAR (ACINAR)	SIMPLE BRANCHED ALVEOLAR
Examples: • Intestinal glands	Examples: • Merocrine sweat glands	Examples: • Gastric glands • Mucous glands of esophagus, tongue, duodenum	Examples: • A stage in the embryonic development of simple branched glands	Examples: • Sebaceous (oil) glands

Glands whose glandular cells form tubes are **tubular**; the tubes may be straight or coiled.

Glands whose glandular cells form sac-like pockets are **alveolar** (al-VĒ-ō-lar; *alveolus*, sac) or **acinar** (AS-i-nar; *acinus*, chamber).

3 A gland is **compound** if the duct divides one or more times on its way to the gland cells.

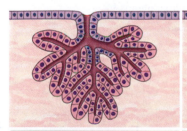
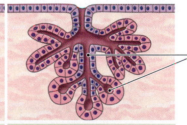

Glands whose secretory cells form both tubes and sacs are called **tubuloalveolar**.

COMPOUND TUBULAR	COMPOUND ALVEOLAR (ACINAR)	COMPOUND TUBULOALVEOLAR
Examples: • Mucous glands (in mouth) • Bulbo-urethral glands (in male reproductive system) • Seminiferous tubules of testes	Examples: • Mammary glands	Examples: • Salivary glands • Glands of respiratory passages • Pancreas

4 In epithelia that have independent, scattered gland cells, the individual secretory cells are called **mucous cells**, and they secrete mucin. The apical cytoplasm is filled with large secretory vesicles that look clear or foamy in a light micrograph.

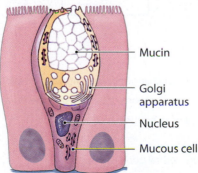

Mucin

Golgi apparatus

Nucleus

Mucous cell

Module 4.6 Review

a. Name the two primary types of glands.

b. What mode of secretion occurs in the secretory cells of sebaceous glands, which fill with secretions and then rupture, releasing their contents?

c. Which type of gland has no ducts to carry the glandular secretions, and the gland's secretions are released directly into the interstitial fluid?

1. Labeling

Label the types of epithelial tissues shown in the drawing to the right.

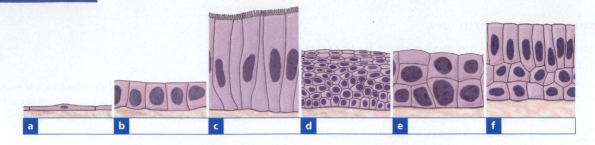

a   b   c   d   e   f

2. Short answer

Complete the following table by writing the missing epithelium type or structure.

Type of Epithelium	Structure (or Organ)
a	Lining of the trachea
Transitional epithelium	b
c	Surface of the skin
d	Lining of the small intestine
Simple squamous epithelium	e
Simple cuboidal epithelium	f
g	Ducts of sweat glands and mammary glands

3. Concept map

Use the following terms to fill in the blank spaces to complete the glandular epithelia concept map.

- endocrine glands
- ducts
- mucous cells
- apocrine secretion
- interstitial fluid
- mucus
- merocrine secretion
- epithelial surfaces
- mucin
- exocrine glands
- holocrine secretion

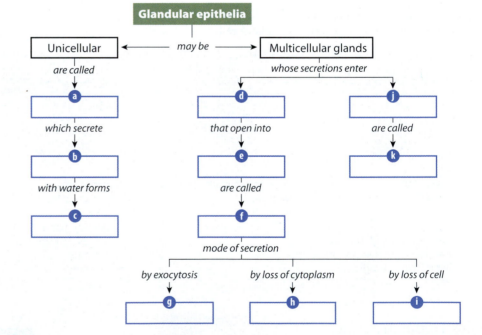

Glandular epithelia

Unicellular ← *may be* → Multicellular glands

are called a

whose secretions enter d   j

which secrete b

that open into e   *are called* k

with water forms c

are called f

mode of secretion

by exocytosis g   *by loss of cytoplasm* h   *by loss of cell* i

4. Vocabulary

Fill in the blanks with the appropriate term.

a _____ A term meaning no blood vessels

b _____ A gland whose glandular cells form blind pockets

c _____ A type of epithelium that withstands stretching and that changes in appearance as stretching occurs

d _____ The cell junction formed by the partial fusion of the lipid portions of two plasma membranes

e _____ The complex structure attached to the basal surface of an epithelium

f _____ A gland that has a single duct

g _____ The type of epithelium lining the subdivisions of the ventral body cavity

Connective Tissue

Connective tissue varies widely in appearance and function, but all forms share three basic components: (1) specialized cells, (2) extracellular protein fibers, and (3) a fluid known as **ground substance**. Together, the extracellular fibers and ground substance constitute the **matrix** that surrounds the cells. Whereas cells make up the bulk of epithelial tissue, the matrix typically accounts for most of the volume of connective tissue.

Functions of Connective Tissue

- Establishing a structural framework for the body

- Transporting fluids and dissolved materials

- Protecting delicate organs

- Supporting, surrounding, and interconnecting other types of tissue

- Storing energy reserves, especially in the form of triglycerides

- Defending the body from invading microorganisms

Connective Tissue

The various types of connective tissue are situated throughout the body but they are never exposed to the outside environment. Many types of connective tissue are highly vascular (that is, they have many blood vessels) and contain sensory receptors that detect pain, pressure, temperature, and other stimuli.

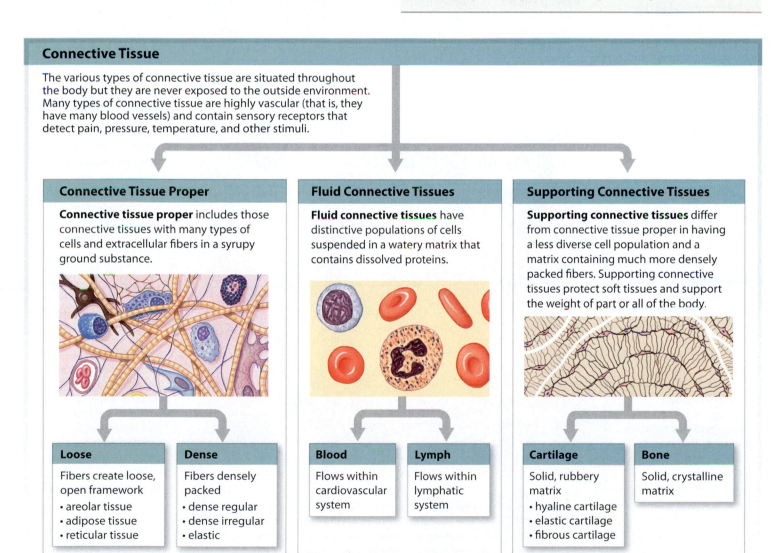

Connective Tissue Proper

Connective tissue proper includes those connective tissues with many types of cells and extracellular fibers in a syrupy ground substance.

Fluid Connective Tissues

Fluid connective tissues have distinctive populations of cells suspended in a watery matrix that contains dissolved proteins.

Supporting Connective Tissues

Supporting connective tissues differ from connective tissue proper in having a less diverse cell population and a matrix containing much more densely packed fibers. Supporting connective tissues protect soft tissues and support the weight of part or all of the body.

Loose	**Dense**
Fibers create loose, open framework	Fibers densely packed
• areolar tissue	• dense regular
• adipose tissue	• dense irregular
• reticular tissue	• elastic

Blood	**Lymph**
Flows within cardiovascular system	Flows within lymphatic system

Cartilage	**Bone**
Solid, rubbery matrix	Solid, crystalline matrix
• hyaline cartilage	
• elastic cartilage	
• fibrous cartilage	

Loose connective tissues provide padding and a supporting framework for other tissue types

Connective tissue proper contains extracellular protein fibers, a viscous ground substance, and two classes of cells. Fixed cells are stationary and are involved primarily with local maintenance, repair, and energy storage. Wandering cells are concerned primarily with the defense and repair of damaged tissues. The number of cells at any given moment varies depending on local conditions.

Loose Connective Tissues

The body contains three types of **loose connective tissues:** areolar tissue, adipose tissue, and reticular tissue.

1 **Areolar tissue** is the most common form of connective tissue proper in adults. It is the general packing material in the body. All of the cell types found in other forms of connective tissue proper can be found in areolar tissue.

Fibers

Reticular fibers are strong and form a branching network.

Collagen fibers are thick, straight or wavy, and often form bundles. They are very strong and resist stretching.

Elastic fibers are slender, unbranching, and very stretchy. They recoil to their original length after stretching or distortion.

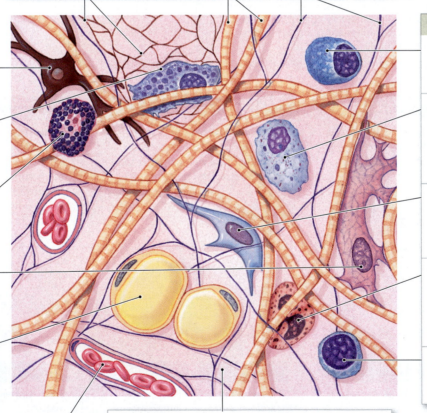

Fixed Cells

A **melanocyte** is a fixed pigment cell that synthesizes melanin, a brownish-yellow pigment.

A **fixed macrophage** is a stationary phagocytic cell that engulfs cell debris and pathogens.

Mast cells are fixed cells that stimulate local inflammation and mobilize tissue defenses.

Fibroblasts are fixed cells that synthesize the extracellular fibers of the connective tissue.

Adipocytes (fat cells) are fixed cells that store lipid reserves in large intracellular vesicles.

Wandering Cells

A **plasma cell** is an active, mobile immune cell that produces antibodies.

Free macrophages are wandering phagocytic cells that patrol the tissue, engulfing debris or pathogens.

Mesenchymal cells are mobile stem cells that participate in the repair of damaged tissues.

Neutrophils and **eosinophils** are small, mobile, phagocytic blood cells that enter tissues during infection or injury.

Lymphocytes are mobile cells of the immune system.

Red blood cell in vessel

Ground substance fills the spaces between cells and surrounds connective tissue fibers. In all forms of connective tissue proper, ground substance is clear, colorless, and viscous (syrupy) due to the presence of proteoglycans and glycoproteins.

2 **Adipose tissue** is found deep to the skin, especially at the flanks, buttocks, and breasts. It also forms a layer that provides padding within the orbit of the eyes, in the abdominopelvic cavity, and around the kidneys. The distinction between areolar tissue and adipose tissue is somewhat arbitrary. Adipocytes account for most of the volume of adipose tissue, but only a fraction of the volume of areolar tissue.

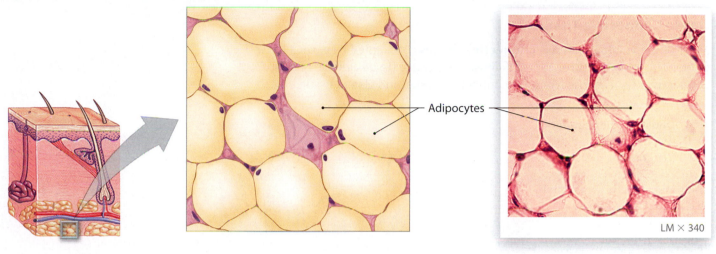

Adipocytes

LM × 340

3 **Reticular tissue** is found in the liver, kidney, spleen, lymph nodes, and bone marrow, where it forms a tough, flexible network that provides support and resists distortion. In reticular tissue, reticular fibers create a complex three-dimensional supporting network known as a **stroma**. Fixed macrophages and fibroblasts are present in reticular tissues, but these cells are seldom visible because the stroma is dominated by cells that are unique to the organ under consideration and have specialized functions.

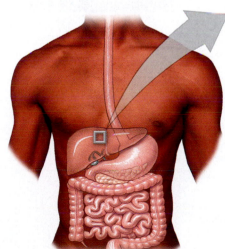

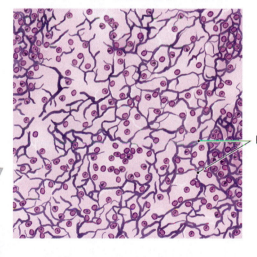

Reticular fibers

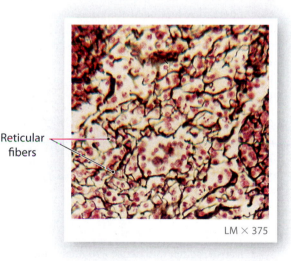

LM × 375

Module 4.7 Review

a. Identify the types of cells found in connective tissue proper.

b. Which type of connective tissue contains primarily lipids?

c. Describe the role of fibroblasts in connective tissue.

Dense connective tissues are dominated by extracellular fibers, whereas fluid connective tissues have a liquid matrix

Dense Connective Tissues

Most of the volume of **dense connective tissues** is occupied by extracellular fibers. The body has three types of dense connective tissues: dense regular connective tissue, dense irregular connective tissue, and elastic tissue.

1 **Dense regular connective tissue** is found in cords (tendons) or sheets connecting skeletal muscles to bone, and in cords (ligaments) that interconnect bones or stabilize the positions of internal organs. The forces applied to these cords and sheets arrive from a consistent direction: parallel to the long axis of the collagen fibers.

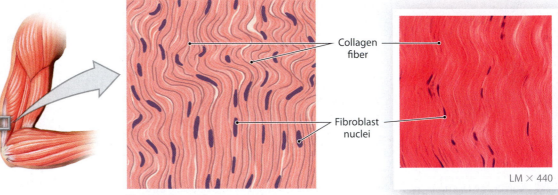

Collagen fiber

Fibroblast nuclei

LM × 440

Tendon from triceps muscle

2 In **dense irregular connective tissue**, the fibers form an interwoven meshwork in no consistent pattern. These tissues strengthen and support areas subjected to stresses from many directions. Dense irregular connective tissue forms (1) a covering, or capsule, that sheathes visceral organs; (2) a superficial layer covering bones, cartilages, and peripheral nerves; and (3) a thick supporting layer in the skin (the dermis).

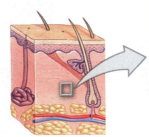

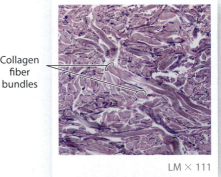

Collagen fiber bundles

LM × 111

Dermis

3 Dense regular and dense irregular connective tissues contain variable amounts of elastic fibers. When elastic fibers outnumber collagen fibers, the tissue has a springy, resilient nature that allows it to tolerate cycles of extension and recoil. This **elastic tissue** is found between vertebrae of the spinal column, in the walls of large blood vessels, and in ligaments supporting transitional epithelia and the erectile tissues of the penis.

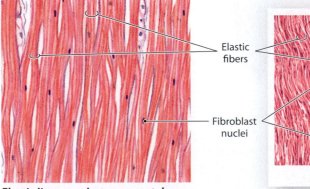

Elastic fibers

Fibroblast nuclei

LM × 887

Elastic ligament between vertebrae

Fluid Connective Tissues

4 **Fluid connective tissues** have a fluid matrix that includes many types of suspended proteins that under normal conditions do not form insoluble fibers. In blood, the watery matrix is called **plasma**. Suspended in the plasma are blood cells and fragments of cells, collectively known as **formed elements**.

Red Blood Cells

Red blood cells are formed elements responsible for the transport of oxygen (and, to a lesser degree, of carbon dioxide) in the blood.

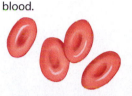

Red blood cells account for roughly half the volume of whole blood and give blood its color.

White Blood Cells

White blood cells are formed elements that help defend the body from infection and disease.

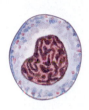

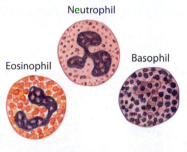

Eosinophil

Neutrophil

Basophil

Monocytes are phagocytes similar to the free macrophages in other tissues.

Lymphocytes are uncommon in the blood but they are the dominant cell type in lymph, the second type of fluid connective tissue.

Eosinophils and **neutrophils** are phagocytes. **Basophils** promote inflammation much like mast cells in other connective tissues.

Platelets

Platelets are formed elements consisting of membrane-enclosed packets of cytoplasm.

These cell fragments are involved in the clotting response that seals leaks in damaged or broken blood vessels.

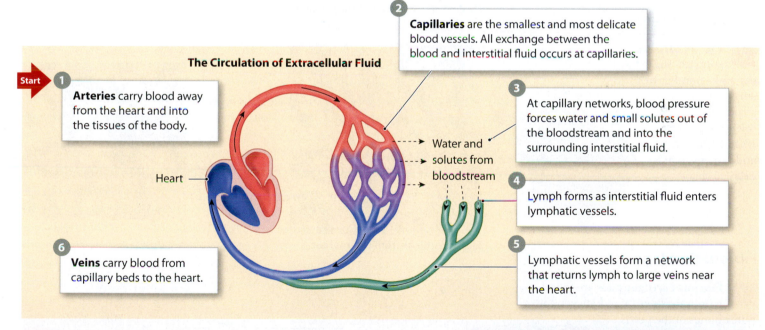

The Circulation of Extracellular Fluid

2 **Capillaries** are the smallest and most delicate blood vessels. All exchange between the blood and interstitial fluid occurs at capillaries.

Start **1** **Arteries** carry blood away from the heart and into the tissues of the body.

3 At capillary networks, blood pressure forces water and small solutes out of the bloodstream and into the surrounding interstitial fluid.

Water and solutes from bloodstream

Heart

4 Lymph forms as interstitial fluid enters lymphatic vessels.

6 **Veins** carry blood from capillary beds to the heart.

5 Lymphatic vessels form a network that returns lymph to large veins near the heart.

5 **Blood** is normally confined to the vessels of the cardiovascular system, and contractions of the heart keep it in motion. As blood flows through body tissues, water and solutes move from the plasma into the surrounding interstitial fluid. **Lymph**, the second type of fluid connective tissue, is formed as interstitial fluid drains into lymphatic vessels that begin in peripheral tissues and empty into the venous system. The continuous recirculation of extracellular fluid is essential to homeostasis. It helps eliminate local differences in the levels of nutrients, wastes, or toxins; maintains blood volume; and alerts the immune system to infections that may be under way in peripheral tissues.

Module 4.8 Review

a. Which two types of connective tissue have a liquid matrix?

b. Lack of vitamin C in the diet interferes with the ability of fibroblasts to produce collagen. How might this affect connective tissue function?

c. Summarize the role of extracellular fluid in maintaining homeostasis.

Cartilage provides a flexible supporting framework

In **cartilage**, the matrix is a firm gel that contains polysaccharide derivatives called **chondroitin sulfates** (kon-DROY-tin; *chondros*, cartilage). Chondroitin sulfates form complexes with proteins in the ground substance, producing proteoglycans. Cartilage cells, or **chondrocytes** (KON-drō-sīts), are the only cells in the avascular cartilage matrix. They occupy small chambers known as **lacunae** (la-KOO-nē; *lacus*, lake). The physical properties of cartilage depend on the proteoglycans in the matrix, and on the type and abundance of extracellular fibers.

Hyaline Cartilage

1 **Hyaline cartilage** is found between the tips of the ribs and the bones of the sternum, covering bone surfaces at mobile joints, supporting the respiratory passageways, and forming part of the nasal septum. It provides stiff but somewhat flexible support and reduces friction between bony surfaces.

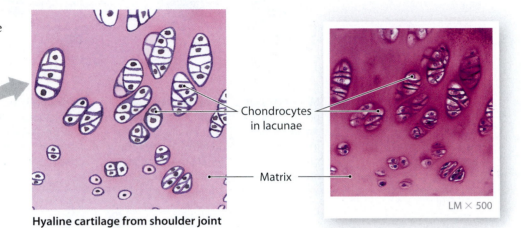

Chondrocytes in lacunae

Matrix

LM × 500

Hyaline cartilage from shoulder joint

Elastic Cartilage

2 **Elastic cartilage** supports the external ear and a number of smaller internal structures. It tolerates distortion without damage and returns to its original shape.

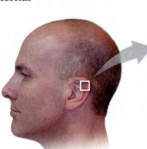

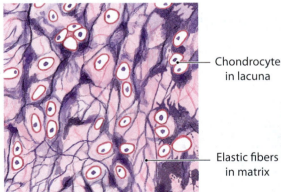

Chondrocyte in lacuna

Elastic fibers in matrix

LM × 358

Elastic cartilage from external ear

Fibrous Cartilage

3 **Fibrous cartilage** pads are found within the knee joint, between the pubic bones of the pelvis, and in the intervertebral discs of the vertebral column. It resists compression, prevents bone-to-bone contact, and limits relative movement.

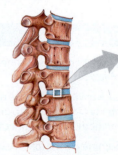

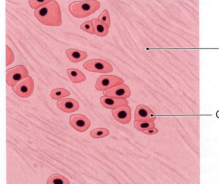

Collagen fibers in matrix

Chondrocytes

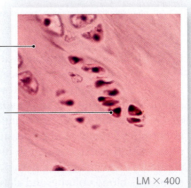

LM × 400

Fibrous cartilage from intervertebral disc

4 A cartilage is generally set apart from surrounding tissues by a **perichondrium** (per-i-KON-drē-um; *peri-*, around). The perichondrium contains two distinct layers: an outer, fibrous layer of dense irregular connective tissue, and an inner, cellular layer. The fibrous layer provides mechanical support and protection and attaches the cartilage to other structures. The cellular layer is important to the growth and maintenance of the cartilage.

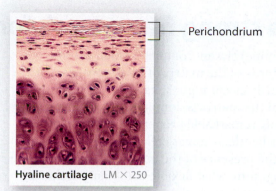

Perichondrium

Hyaline cartilage   LM × 250

Appositional Growth

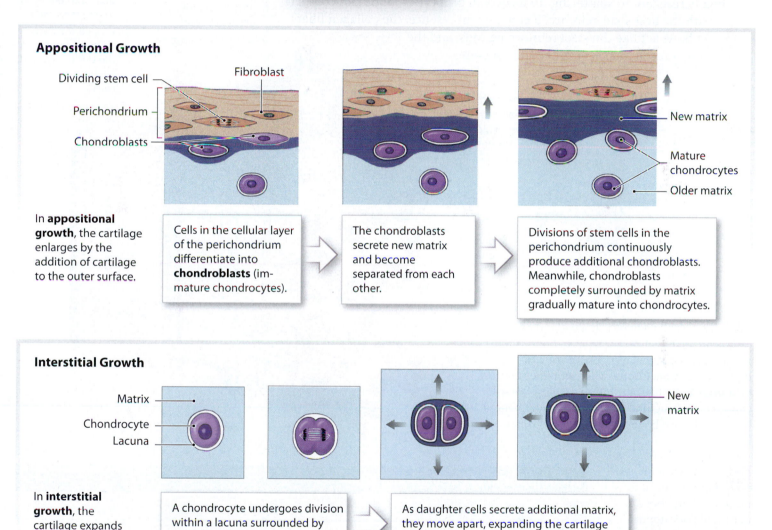

Dividing stem cell

Fibroblast

Perichondrium

Chondroblasts

New matrix

Mature chondrocytes

Older matrix

In **appositional growth**, the cartilage enlarges by the addition of cartilage to the outer surface.

Cells in the cellular layer of the perichondrium differentiate into **chondroblasts** (immature chondrocytes).

The chondroblasts secrete new matrix and become separated from each other.

Divisions of stem cells in the perichondrium continuously produce additional chondroblasts. Meanwhile, chondroblasts completely surrounded by matrix gradually mature into chondrocytes.

Interstitial Growth

Matrix

Chondrocyte

Lacuna

New matrix

In **interstitial growth**, the cartilage expands from within.

A chondrocyte undergoes division within a lacuna surrounded by cartilage matrix.

As daughter cells secrete additional matrix, they move apart, expanding the cartilage from within.

5 Both interstitial and appositional growth occur during development, but interstitial growth predominates. In normal adults, cartilage growth no longer occurs. However, in unusual circumstances—such as after cartilage has been slightly damaged or subjected to excessive hormonal stimulation—some repair may occur by appositional growth. After severe damage, cartilage is not repaired, and a fibrous patch replaces the damaged area.

Module 4.9 Review

a. Mature cartilage cells are called _____.

b. Which connective tissue fiber is characteristic of the cartilage supporting the ear?

c. If a person has a herniated intervertebral disc, which type of cartilage has been damaged?

Bone provides a strong framework for the body

Bone, or **osseous** (OS-ē-us; *os*, bone) **tissue**, contains a small volume of ground substance. About two-thirds of the matrix of bone consists of a mixture of calcium salts—primarily calcium phosphate, with lesser amounts of calcium carbonate. The rest of the matrix is dominated by collagen fibers. This combination gives bone truly remarkable properties. By themselves, calcium salts are hard but rather brittle, whereas collagen fibers are strong and relatively flexible. In bone, the presence of the minerals surrounding the collagen fibers produces a strong, somewhat flexible combination that is highly resistant to shattering. In its overall properties, bone can compete with the best steel-reinforced concrete. In essence, the collagen fibers in bone act like the steel reinforcing rods, and the mineralized matrix acts like the concrete.

Compact bone also has a superficial layer of bone that was deposited during appositional growth of the bone. Because the matrix is solid and calcified, interstitial growth cannot occur in bone.

Unlike cartilage, bone is highly vascular. Large vessels outside the bone are connected to smaller vessels that supply areas of compact bone and the soft tissues that fill the interior cavity.

In compact bone the matrix is organized in concentric layers around branches of blood vessels within the bone.

Spongy bone

Compact bone

1 A typical long bone is hollow, and its walls contain two different types of bone. The weight-bearing outer layer consists of well-organized **compact bone**, whereas a finer network of **spongy bone** lines the internal cavity. We will consider the structural and functional differences between the two bone types in a later chapter.

2 Cartilage and bone share a number of functions, but these two supporting connective tissues are organized very differently.

A Comparison of Cartilage and Bone		
Characteristic	**Cartilage**	**Bone**
Cells	Chondrocytes in lacunae	Osteocytes in lacunae
Ground substance	Chondroitin sulfate (in proteoglycan) and water	A small volume of liquid surrounding insoluble crystals of calcium salts
Fibers	Collagen, elastic, and reticular fibers (proportions vary)	Collagen fibers predominate
Vascularity	None	Extensive
Covering	Perichondrium (two layers)	Periosteum (two layers)
Strength	Limited: bends easily, but hard to break	Strong: resists distortion until breaking point

3 Although the arrangement of the layers of bone may vary, the superficial and deeper layers share common structural features.

The fibrous layer assists in the attachment of a bone to surrounding tissues and to associated tendons and ligaments.

The cellular layer functions in appositional bone growth and participates in repairs after an injury.

Except in joint cavities, where they are covered by a layer of hyaline cartilage, bone surfaces are sheathed by a **periosteum** (per-ē-OS-tē-um) composed of fibrous (outer) and cellular (inner) layers.

Lacunae in the matrix contain **osteocytes** (OS-tē-ō-sīts), or bone cells. The lacunae are typically organized around blood vessels that branch through the bony matrix.

Layers of matrix separate the lacunae. These layers are oriented along the main axis of the bone.

Canaliculi (kan-a-LIK-ū-lē; little canals) are fine passageways that form a branching network for the exchange of materials between blood vessels and osteocytes. This is important because diffusion cannot occur through the calcified matrix of bone.

A **central canal** at the center of an osteon contains the blood vessels that provide oxygen and nutrients to the osteocytes.

4 The functional unit of compact bone is called an **osteon**.

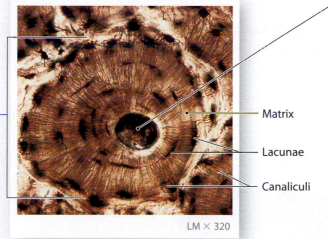

Osteon

Matrix

Lacunae

Canaliculi

LM × 320

Unlike cartilage, bone undergoes extensive remodeling throughout life, and complete repairs can be made even after severe damage has occurred. Bones also respond to the stresses placed on them, growing thicker and stronger with exercise and becoming thin and brittle with inactivity.

Module 4.10 Review

a. Mature bone cells in lacunae are called _____ .

b. Distinguish between the two types of supporting connective tissues with respect to their characteristic fibers.

c. Explain why bone does not undergo interstitial growth.

Membranes are physical barriers, and fasciae create internal compartments and divisions

Membranes

A membrane is a physical barrier. There are many different types of anatomical membranes—you encountered plasma membranes in Chapter 3, and you will find many other kinds of membranes in later chapters. Here we consider membranes that line or cover body surfaces. They typically consist of an epithelium supported by connective tissue. Four of these membranes occur in the body: (1) mucous membranes, (2) serous membranes, (3) the cutaneous membrane, and (4) synovial membranes.

Mucous Membranes

Mucous membranes, or *mucosae* (mū-KŌ-sē), line passageways and chambers that communicate with the exterior, including those in the digestive, respiratory, reproductive, and urinary tracts. These epithelial surfaces must be kept moist to reduce friction and, in many cases, to facilitate absorption or secretion. The epithelial surfaces are lubricated either by mucus (produced by mucous cells or multicellular glands) or by exposure to fluids such as urine or semen.

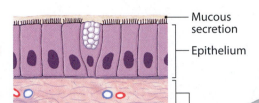

Mucous secretion

Epithelium

The **lamina propria** (PRŌ-prē-uh) is a layer of areolar tissue that supports the mucous epithelium.

Serous Membranes

Serous membranes consist of a mesothelium supported by areolar tissue. They are extremely delicate, and never directly connected to the exterior. Three serous membranes line the subdivisions of the ventral body cavity: (1) the **pleura**, which lines the pleural cavities and covers the lungs; (2) the **peritoneum**, which lines the peritoneal cavity and covers the surfaces of visceral organs; and (3) the **pericardium**, which lines the pericardial cavity and covers the heart.

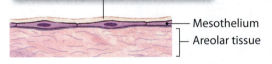

A **transudate** is a liquid layer that coats the surfaces of a serous membrane and prevents friction.

Mesothelium

Areolar tissue

The Cutaneous Membrane

The **cutaneous membrane** covers the surface of the body. It consists of a stratified squamous epithelium and a layer of areolar tissue reinforced by underlying dense irregular connective tissue. In contrast to serous and mucous membranes, the cutaneous membrane is thick, relatively waterproof, and usually dry.

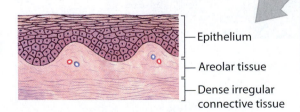

Epithelium

Areolar tissue

Dense irregular connective tissue

Synovial Membrane

A **synovial membrane** lines mobile joint cavities but does not cover the opposing joint surfaces. Although the covering of the synovial membrane is often called an epithelium, it differs from true epithelia in four respects: (1) It develops within a connective tissue, (2) no basal lamina is present, (3) gaps of up to 1 mm may separate adjacent cells, and (4) the synovial fluid and capillaries in the underlying connective tissue are continuously exchanging fluid and solutes.

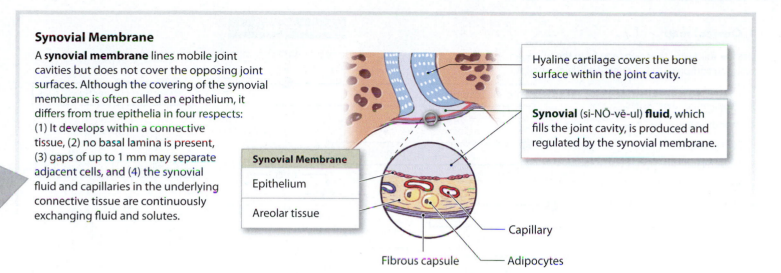

Hyaline cartilage covers the bone surface within the joint cavity.

Synovial (si-NŌ-vē-ul) **fluid**, which fills the joint cavity, is produced and regulated by the synovial membrane.

Synovial Membrane
Epithelium
Areolar tissue

Capillary

Fibrous capsule — Adipocytes

Fasciae

Fasciae (FASH-ē-ē; singular, fascia) are connective tissue layers and wrappings that support and surround organs. The fasciae consist of three types of layers: the **superficial fascia**, the **deep fascia**, and the **subserous fascia**.

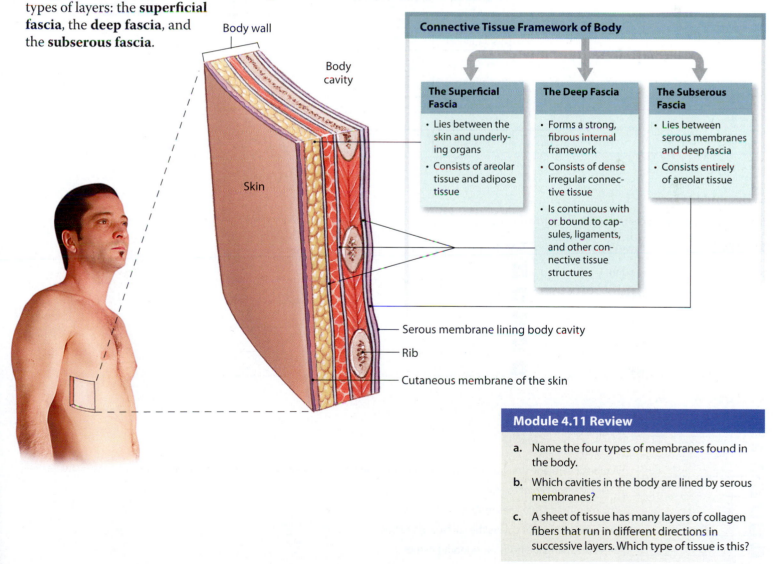

Body wall

Body cavity

Skin

Connective Tissue Framework of Body

The Superficial Fascia	The Deep Fascia	The Subserous Fascia
• Lies between the skin and underlying organs • Consists of areolar tissue and adipose tissue	• Forms a strong, fibrous internal framework • Consists of dense irregular connective tissue • Is continuous with or bound to capsules, ligaments, and other connective tissue structures	• Lies between serous membranes and deep fascia • Consists entirely of areolar tissue

Serous membrane lining body cavity

Rib

Cutaneous membrane of the skin

Module 4.11 Review

a. Name the four types of membranes found in the body.

b. Which cavities in the body are lined by serous membranes?

c. A sheet of tissue has many layers of collagen fibers that run in different directions in successive layers. Which type of tissue is this?

1. Concept map

Use the following terms once to fill in the blank boxes to complete the connective tissue concept map.

- loose connective tissues
- chondrocytes in lacunae
- fluid connective tissue
- tendons
- ligaments
- regular
- hyaline
- blood
- adipose
- bone

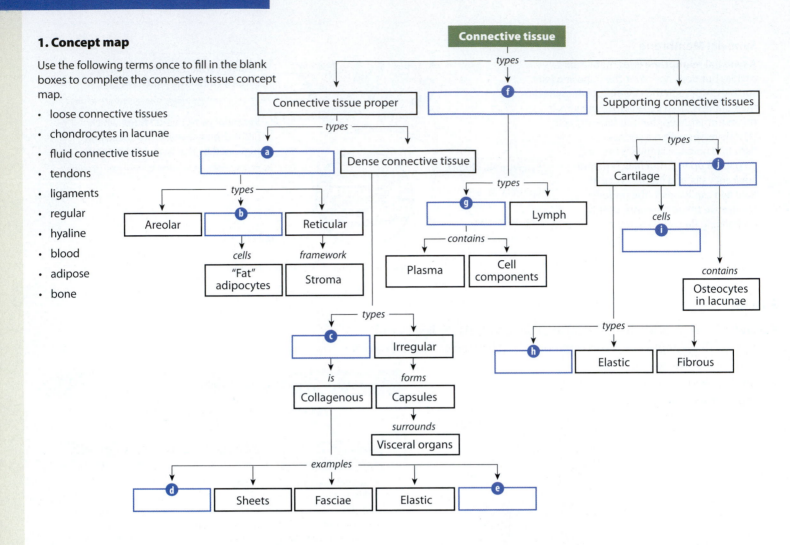

2. Vocabulary

In the space provided, write the boldfaced terms introduced in this section that contain the indicated word part.

Word Part	Meaning	Terms
peri-	around	**a** _____
os-	bone	**b** _____
chondro-	cartilage	**c** _____
inter-	between	**d** _____
lacus-	lake	**e** _____

Fill in the blanks with the appropriate term.

f _____ A cartilage cell

g _____ Bone tissue

h _____ A type of cartilage that has a matrix with little ground substance and large amounts of collagen fibers

i _____ Cells that store lipid reserves

j _____ The membrane that lines mobile joint cavities

k _____ The membrane that covers the surface of the body

l _____ Separates cartilage from surrounding tissues

Muscle Tissue and Neural Tissue

Epithelia cover surfaces and line passageways; connective tissues support weight and interconnect parts of the body. Together, these tissues provide a strong, interwoven framework within which the organs of the body can function. However, they account for less than half of the weight of the body.

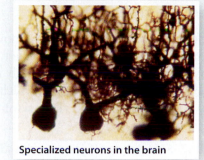

Specialized neurons in the brain

Epithelial tissue
3%

Connective tissue 45%

Neural tissue 2%

Muscle tissue
50%

1 This pie chart shows the relative contributions of the four tissue types to the body as a whole.

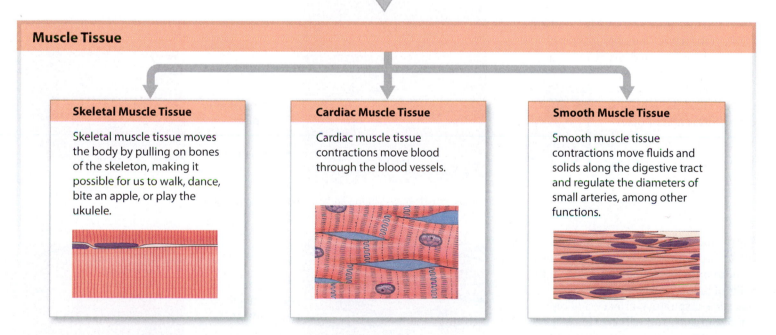

Muscle Tissue

Skeletal Muscle Tissue

Skeletal muscle tissue moves the body by pulling on bones of the skeleton, making it possible for us to walk, dance, bite an apple, or play the ukulele.

Cardiac Muscle Tissue

Cardiac muscle tissue contractions move blood through the blood vessels.

Smooth Muscle Tissue

Smooth muscle tissue contractions move fluids and solids along the digestive tract and regulate the diameters of small arteries, among other functions.

In this section we will consider the histology of muscle tissue and neural tissue. Of the four primary tissue types, muscle tissue contributes the most, and neural tissue the least, to body weight. This section provides only an overview of these tissue types because additional details will be provided in later chapters dealing with the muscular system, nervous system, cardiovascular system, and digestive system.

Muscle tissue is specialized for contraction; neural tissue is specialized for communication

Muscle Tissue

Several vital functions involve movement of one kind or another—movement of materials along the digestive tract, movement of blood around the cardiovascular system, or movement of the body from one place to another. Movement is produced by **muscle tissue**, which is specialized for contraction. There are three types of muscle tissue: skeletal muscle, cardiac muscle, and smooth muscle.

1 **Skeletal muscle tissue** is found in skeletal muscles, organs that also contain connective tissues and neural tissue. The cells are long, cylindrical, banded (**striated**), and have multiple nuclei (**multinucleate**).

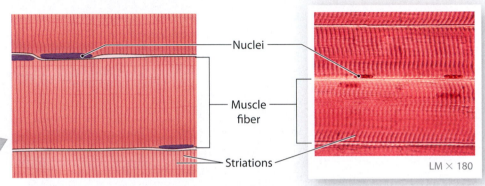

Nuclei

Muscle fiber

Striations

LM × 180

Skeletal muscles move or stabilize the position of the skeleton; guard entrances and exits to the digestive, respiratory, and urinary tracts; generate heat; and protect internal organs.

2 **Cardiac muscle tissue** is found only in the heart. The cells are short, branched, and striated, usually with a single nucleus; cells are interconnected at specialized intercellular junctions called **intercalated discs**.

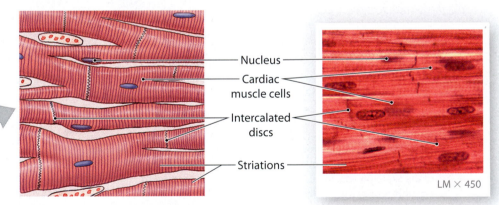

Nucleus

Cardiac muscle cells

Intercalated discs

Striations

LM × 450

Cardiac muscle moves blood and maintains blood pressure.

3 **Smooth muscle tissue** is found throughout the body. For example, smooth muscle is found in the skin, in the walls of blood vessels, and in many digestive, respiratory, urinary, and reproductive organs. The cells are short, spindle-shaped, and nonstriated, with a single, central nucleus.

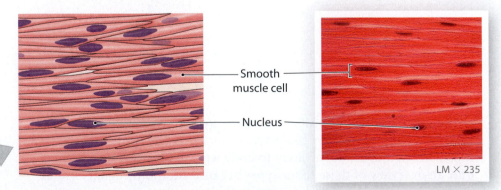

Smooth muscle cell

Nucleus

LM × 235

Smooth muscle moves food, urine, and reproductive tract secretions; controls diameter of respiratory passageways and regulates diameter of blood vessels.

Neural Tissue

Neural tissue, which is also known as nervous tissue, is specialized for the conduction of electrical impulses from one region of the body to another. Ninety-eight percent of the neural tissue in the body is concentrated in the brain and spinal cord, which are the control centers of the nervous system. Neural tissue contains two basic types of cells: (1) **neurons** (NOOR-onz; *neuro*, nerve) and (2) several kinds of supporting cells, collectively called **neuroglia** (noo-ROG-lē-uh or noo-rō-GLĒ-uh), or **glial cells** (*glia*, glue).

4 Neurons transfer information from place to place and perform information processing like living computer chips. Their sizes and shapes vary widely. The longest cells in your body are neurons; they can reach a meter (39 in.) in length.

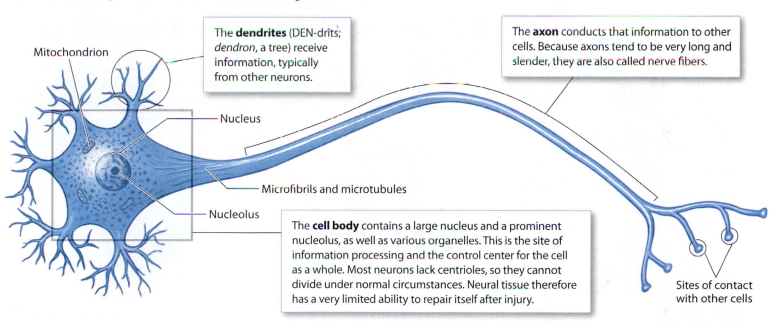

Mitochondrion

The **dendrites** (DEN-drīts; *dendron*, a tree) receive information, typically from other neurons.

The **axon** conducts that information to other cells. Because axons tend to be very long and slender, they are also called nerve fibers.

Nucleus

Microfibrils and microtubules

Nucleolus

The **cell body** contains a large nucleus and a prominent nucleolus, as well as various organelles. This is the site of information processing and the control center for the cell as a whole. Most neurons lack centrioles, so they cannot divide under normal circumstances. Neural tissue therefore has a very limited ability to repair itself after injury.

Sites of contact with other cells

5 There are several different types of neuroglia, each with specific functions. Each type has a distinctive appearance related to its primary function. In general, neuroglia protect, support, and repair neural tissue and maintain the nutrient supply to neurons.

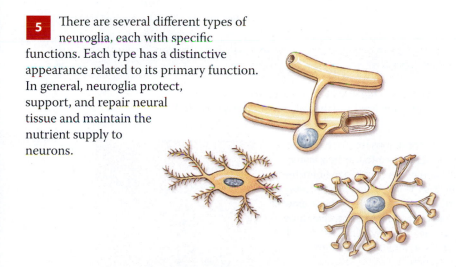

Functions of Neuroglia

- Maintain physical structure of neural tissue
- Repair neural tissue framework after injury
- Perform phagocytosis
- Provide nutrients to neurons
- Regulate the composition of the interstitial fluid surrounding neurons

Module 4.12 Review

a. Identify the three types of muscle tissue in the body.

b. Which type of muscle tissue has small, tapering cells with single nuclei and no obvious striations?

c. Irregularly shaped cells with many fibrous projections, some several centimeters long, are probably which type of cell?

Our conscious and unconscious thought processes reflect the communication among neurons in the brain. Such communication involves the propagation of electrical impulses that are generated at the cell body and travel along the axon to reach other cells.

The response to tissue injury involves inflammation and regeneration

Tissues are not isolated; they combine to form organs with diverse functions. Therefore, any injury affects several types of tissue simultaneously. These tissues must respond in a coordinated way to preserve homeostasis. The restoration of homeostasis after an injury involves two related processes: inflammation and regeneration.

Injury

When a tissue is injured, a general defense mechanism is activated.

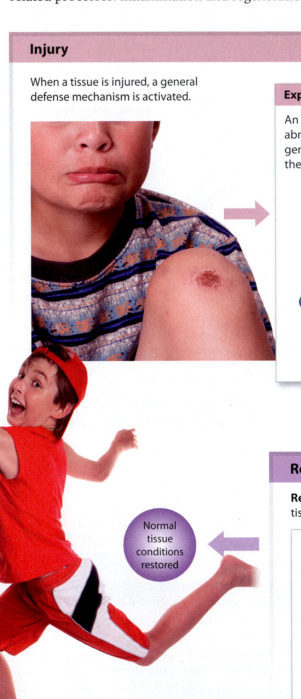

Exposure to Pathogens and Toxins

An injured tissue contains an abnormal concentration of pathogens, toxins, waste products, and the chemicals from injured cells.

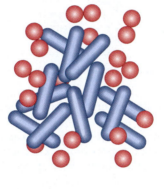

stimulates

Mast Cell Activation

When an injury damages connective tissue, mast cells release a variety of chemicals. This process, called mast cell activation, stimulates inflammation.

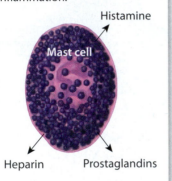

Histamine

Mast cell

Heparin   Prostaglandins

Regeneration inhibits mast cell activation

Regeneration

Regeneration is the repair that occurs after the damaged tissue has been stabilized and the inflammation has subsided.

Normal tissue conditions restored

As tissue conditions return to normal, fibroblasts move into the area, laying down a network of collagen fibers that stabilizes the injury site. This process produces a dense, collagenous framework known as **scar tissue**. Over time, scar tissue is usually "remodeled" and gradually assumes a more normal appearance.

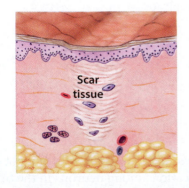

Scar tissue

Inflammation

Inflammation produces several familiar indications of injury, including swelling, redness, warmth, and pain. Inflammation may also result from the presence of pathogens, such as harmful bacteria, within the tissues; the presence of these pathogens constitutes an **infection**. Because all organs have connective tissues, inflammation can occur anywhere in the body.

Increased Blood Flow

In response to the released chemicals, the smooth muscle tissue that surrounds local blood vessels relaxes, and the vessels **dilate**, or enlarge in diameter. This dilation increases blood flow through the damaged tissue.

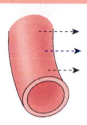

Increased Vessel Permeability

The dilation is accompanied by an increase in the permeability of the capillary walls. Plasma, including blood proteins, now diffuses into the injured tissue, so the area becomes swollen.

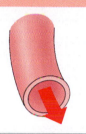

Pain

The combination of abnormal conditions within the tissue and the chemicals released by mast cells stimulates nerve endings that produce the sensation of pain.

PAIN

Increased local temperature
The increased blood flow and permeability causes the tissue to become warm and red.

Increased oxygen and nutrients
Vessel dilation, increased blood flow, and increased vessel permeability result in enhanced delivery of oxygen and nutrients.

O_2

Increased phagocytosis
Phagocytes in the tissue are activated, and they begin engulfing tissue debris and pathogens. Additional phagocytes migrate into the tissue from the bloodstream, drawn to the site by the abnormal local conditions.

Removal of toxins and wastes
The enhanced circulation that increases the volume of interstitial fluid also carries away toxins and waste products, distributing them to the kidneys for excretion, or to the liver for inactivation.

Toxins and wastes

Over a period of hours to days, the cleanup process generally succeeds in eliminating the inflammatory stimuli.

Each organ has a different ability to regenerate after injury—an ability that can be directly linked to the pattern of tissue organization in the injured organ. Epithelia, connective tissues (except cartilage), and smooth muscle tissue usually regenerate well, whereas other muscle tissues and neural tissue regenerate relatively poorly if at all, and damaged areas are often replaced by scar tissue. The permanent replacement of normal tissue by scar tissue is called **fibrosis** (fī-BRŌ-sis). Fibrosis in muscle and other tissues may occur in response to injury, disease, or aging.

Module 4.13 Review

a. Identify the two processes in the response to tissue injury.

b. What are the four indications of inflammation that occur following an injury?

c. Why can inflammation occur in any organ in the body?

1. Concept map

Fill in the blank boxes with boldfaced terms and concepts introduced in this section to complete the muscle tissue concept map.

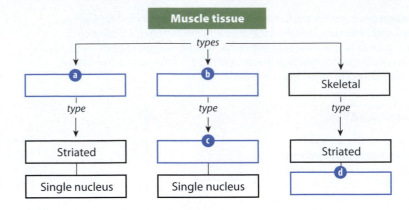

2. Concept map

Fill in the blank boxes with the boldfaced terms and concepts introduced in this section to complete the neural tissue concept map.

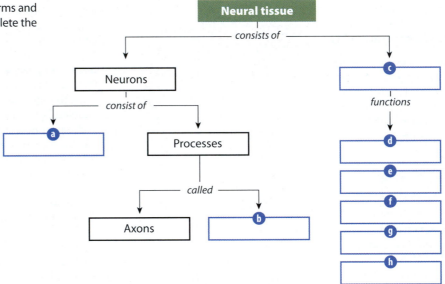

3. Vocabulary

Fill in the blanks with the appropriate term.

a	_____	A single process that extends from the cell body of a neuron and carries information to other cells
b	_____	A specialized intercellular junction between cardiac muscle cells
c	_____	The supporting cells found in nervous tissue
d	_____	Muscle tissue that contains large, multinucleate, striated cells
e	_____	Muscle tissue that regulates the diameter of blood vessels and respiratory passageways
f	_____	The repair process that occurs after inflammation has subsided
g	_____	The first process in a tissue's response to injury

4. Section integration

During inflammation, both blood flow and blood vessel permeability increase in the injured area.
Describe how these responses aid the cleanup process and eliminate the inflammatory stimuli in the injured area.

_____

_____

_____

Visual Outline with Key Terms

Summarize the content of each module using the terms in the order provided.

SECTION 1

Epithelial Tissue

- tissues
- histology

EPITHELIAL TISSUE
- Covers exposed surfaces
- Lines internal passageways and chambers
- Produces glandular secretions

4.4

Cuboidal and transitional epithelia are found along several passageways and chambers connected to the exterior

- cuboidal epithelium
- simple cuboidal epithelium
- stratified cuboidal epithelium
- transitional epithelium

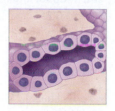

4.1

Epithelial tissue covers surfaces, lines structures, and forms secretory glands

- epithelia
- glands
- exocrine glands
- endocrine glands
- apical surface
- base
- polarity
- lumen
- basolateral surfaces
- squamous
- cuboidal
- columnar
- simple epithelium
- stratified epithelium

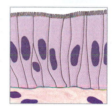

4.5

Columnar epithelia typically perform absorption or provide protection from chemical or environmental stresses

- columnar epithelium
- simple columnar epithelium
- pseudostratified columnar epithelium
- stratified columnar epithelium

4.2

Epithelial cells are extensively interconnected, both structurally and functionally

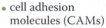

- occluding junction
- adhesion belt
- gap junction
- desmosomes
- hemidesmosomes
- basal lamina
- clear layer
- dense layer
- connexons
- cell adhesion molecules (CAMs)
- intercellular cement
- hyaluronan
- avascular

4.6

Glandular epithelia are specialized for secretion

- glands
- endocrine glands
- exocrine glands
- merocrine secretion
- mucin
- mucus
- apocrine secretion
- holocrine secretion
- simple gland
- branched gland
- tubular gland
- alveolar (acinar) gland
- compound gland
- tubuloalveolar gland
- mucous cells

4.3

The cells in a squamous epithelium are flattened and irregular in shape

- squamous epithelium
- simple squamous epithelium
- mesothelium
- endothelium
- stratified squamous epithelium
- keratininzed
- nonkeratinized

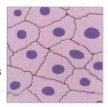

SECTION 2

Connective Tissue

- ground substance
- matrix
- connective tissue proper
- fluid connective tissues
- supporting connective tissues

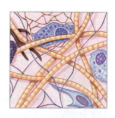

● = *Term boldfaced in this module*

4.7

Loose connective tissues provide padding and a supporting framework for other tissue types

- loose connective tissues
- areolar tissue
- reticular fibers
- collagen fibers
- elastic fibers
- melanocyte
- fixed macrophage
- mast cells
- fibroblasts
- adipocytes
- plasma cell
- free macrophages
- mesenchymal cells
- neutrophils
- eosinophils
- lymphocytes
- adipose tissue
- reticular tissue
- stroma

4.8

Dense connective tissues are dominated by extracellular fibers, whereas fluid connective tissues have a liquid matrix

- dense connective tissues
- dense regular connective tissue
- dense irregular connective tissue
- elastic tissue
- fluid connective tissues
- plasma
- formed elements
- red blood cells
- white blood cells
- monocytes
- lymphocytes
- eosinophils
- neutrophils
- basophils
- platelets
- blood
- lymph
- arteries
- capillaries
- veins

4.9

Cartilage provides a flexible supporting framework

- cartilage
- chondroitin sulfates
- chondrocytes
- lacunae
- hyaline cartilage
- elastic cartilage
- fibrous cartilage
- perichondrium
- appositional growth
- chondroblasts
- interstitial growth

4.10

Bone provides a strong framework for the body

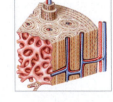

- osseous tissue
- compact bone
- spongy bone
- periosteum
- osteocytes
- canaliculi
- central canal
- osteon

4.11

Membranes are physical barriers, and fasciae create internal compartments and divisions

- mucous membranes
- lamina propria
- serous membranes
- pleura
- peritoneum
- pericardium
- transudate
- cutaneous membrane
- synovial membrane
- synovial fluid
- fasciae
- superficial fascia
- deep fascia
- subserous fascia

SECTION 3

Muscle Tissue and Neural Tissue

- muscle tissue
- neural tissue

4.12

Muscle tissue is specialized for contraction; neural tissue is specialized for communication

- muscle tissue
- skeletal muscle tissue
- striated
- multinucleate
- cardiac muscle tissue
- intercalated discs
- smooth muscle tissue
- neural tissue
- neurons
- neuroglia (glial cells)
- dendrites
- axon
- cell body

4.13

The response to tissue injury involves inflammation and regeneration

- mast cell activation
- inflammation
- infection
- dilate
- regeneration
- scar tissue
- fibrosis

 = *Term boldfaced in this module*

Chapter Integration: Applying what you've learned

As we have seen, tissues are made up of specialized cells and their products. Athletes work very hard to ensure that their tissues—especially their skeletal muscle, cardiac muscle, and connective tissues, including bones, ligaments, and tendons—are able to withstand the stresses involved in athletic competition. Many of these tissues function together near and within joints: The joints themselves contain considerable amounts of cartilage and fibrous connective tissue, and tendons often connect muscle tissue with osseous tissue at or near joints. Muscle tissue and the many kinds of connective tissue are inherently strong, and even more so in trained athletes. But even the strongest tissues have their limits.

During a college football game, Brandon injured the anterior cruciate ligament (ACL) in his knee while trying to recover his own fumble. During the wild scramble for the ball, several linemen drove their 300-pound bodies into him, and while his cleated football shoe was firmly planted in the turf, his knee was twisted in an unnatural way. The muscles of his calf were also injured when they were stepped on by someone else's cleated shoe. The team's physician and athletic trainer quickly attended to Brandon. Once his injuries were stabilized, he was taken to the university hospital's emergency room for a complete physical evaluation.

1. Based on what you have learned in this chapter, describe the expected generalized response to damaged tissues at the injury sites, including how that response produces the signs and symptoms the examining physician will evaluate.

2. The orthopedic surgeon on call at the hospital informs Brandon that the injury to the cartilage in his knee joint is going to take more time to heal than the injury to his calf muscles. Explain why.

3. The physician also advises Brandon to increase his intake of protein and vitamin C. Why?

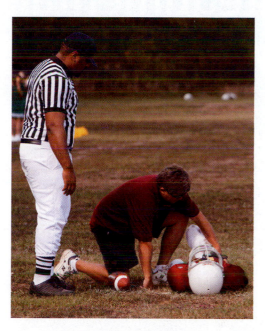

Access more review material online in the Study Area at www.masteringaandp.com.

There, you'll find:
- Chapter guides
- Chapter quizzes
- Practice tests
- Labeling activities
- MP3 Tutor Sessions
- Animations
- Flashcards
- A glossary with pronunciations

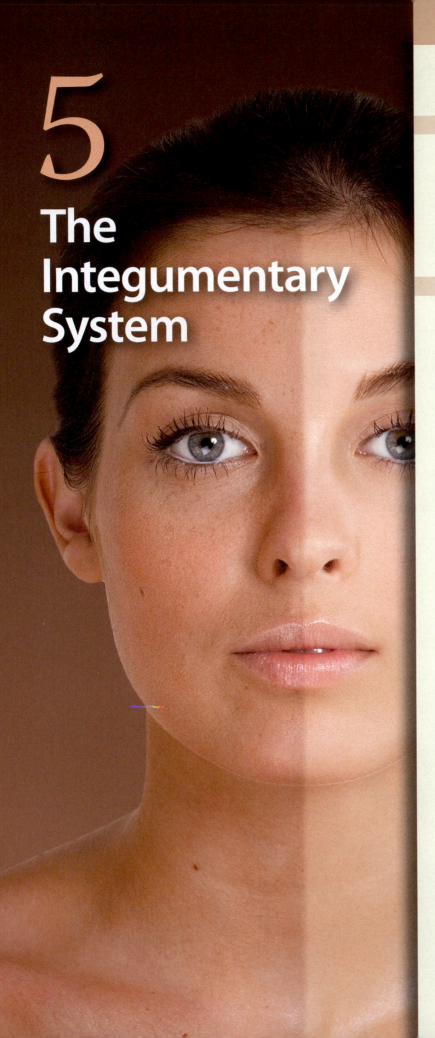

5

The Integumentary System

These Learning Outcomes correspond by number to this chapter's modules and indicate what you should be able to do after completing the chapter.

SECTION 1 · Functional Anatomy of the Skin

5.1 Describe the main structural features of the epidermis, and explain the functional significance of each.

5.2 Explain what accounts for individual differences in skin color and compare basal cell carcinoma with malignant melanoma.

5.3 Describe the structure and functions of the dermis and hypodermis.

SECTION 2 · Accessory Organs of the Skin

5.4 Describe the mechanisms of hair production, and explain the structural basis for hair texture and color.

5.5 Describe the various kinds of exocrine glands in the skin, and discuss the secretions of each.

5.6 Explain the anatomy of a typical nail.

5.7 ➕ CLINICAL MODULE Summarize the effects of aging on the skin.

5.8 Describe the interaction between sunlight and endocrine functioning as they relate to the skin.

5.9 ➕ CLINICAL MODULE Explain how the skin responds to injury and is able to repair itself.

Functional Anatomy of the Skin

The **integumentary system** is the most accessible and often the least appreciated organ system. Often referred to simply as the skin or **integument** (in-TEG-ū-ment), this system accounts for about 16 percent of your total body weight. Its surface, 1.5–2 m² in area, is constantly worn away, attacked by microorganisms, irradiated by sunlight, and exposed to environmental chemicals. The integumentary system is the place where you and the outside world meet—and your body's first line of defense against an often hostile environment.

Cutaneous Membrane

The **epidermis** (*epi*, above) consists of a stratified squamous epithelium.

Dermis —
  Papillary layer
  Reticular layer

The **dermis** consists of a papillary layer of areolar tissue and a reticular layer of dense irregular connective tissue.

The **hypodermis** (subcutaneous layer or superficial fascia) separates the integument from the fascia around deeper organs. Note that this tissue layer is not part of the integument.

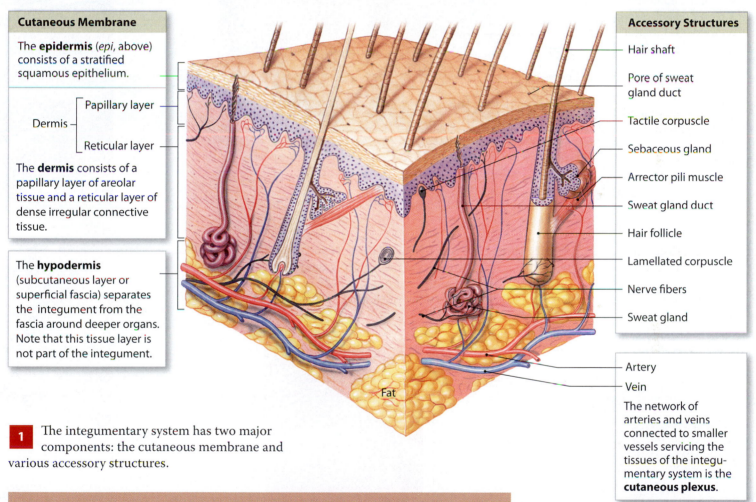

Fat

Accessory Structures

Hair shaft

Pore of sweat gland duct

Tactile corpuscle

Sebaceous gland

Arrector pili muscle

Sweat gland duct

Hair follicle

Lamellated corpuscle

Nerve fibers

Sweat gland

Artery

Vein

The network of arteries and veins connected to smaller vessels servicing the tissues of the integumentary system is the **cutaneous plexus**.

1 The integumentary system has two major components: the cutaneous membrane and various accessory structures.

Functions of the Integumentary System

- Protection of underlying tissues and organs against impact, abrasion, fluid loss, and chemical attack
- Excretion of salts, water, and organic wastes by integumentary glands
- Maintenance of normal body temperature through either insulation or evaporative cooling, as needed
- Production of melanin, which protects underlying tissue from ultraviolet radiation
- Production of keratin, which protects against abrasion and serves as a water repellent
- Synthesis of vitamin D_3, a steroid that is subsequently converted to calcitriol, a hormone important to normal calcium metabolism
- Storage of lipids in adipocytes in the dermis and in adipose tissue in the hypodermis
- Detection of touch, pressure, pain, and temperature stimuli, and the relaying of that information to the nervous system

The epidermis is composed of layers with various functions

The epidermis is dominated by **keratinocytes** (ke-RAT-i-nō-sīts), the body's most abundant epithelial cells. These cells form several layers, or strata. Keratinocytes are continuously produced by stem cell divisions in the deepest layers and are shed at the exposed surface.

1 The deeper layers of the epidermis form **epidermal ridges**, which extend into the dermis and are adjacent to dermal projections called **dermal papillae** (singular, *papilla*; a nipple-shaped mound) that project into the epidermis. These ridges and papillae are significant because they greatly increase the surface area for attachment, firmly binding the epidermis to the dermis.

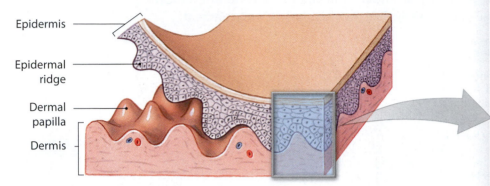

Epidermis

Epidermal ridge

Dermal papilla

Dermis

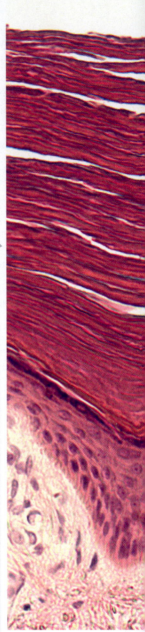

2 **Thin skin**, which covers most of the body surface, contains four strata, and is about as thick as the wall of a plastic sandwich bag (roughly 0.08 mm).

3 **Thick skin**, which occurs on the palms of the hands and the soles of the feet, contains a fifth stratum. It is about as thick as a standard paper towel (roughly 0.5 mm). Note that the terms "thick" and "thin" refer to the relative thickness of the epidermis, not to the integument as a whole.

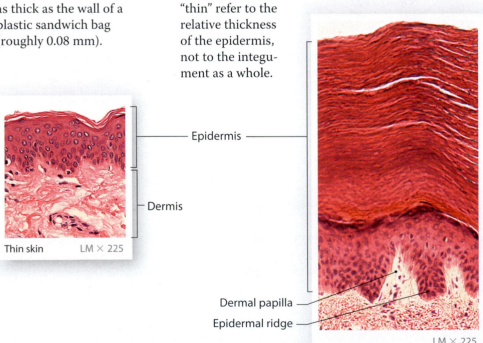

Epidermis

Dermis

Thin skin   LM × 225

Dermal papilla

Epidermal ridge

LM × 225

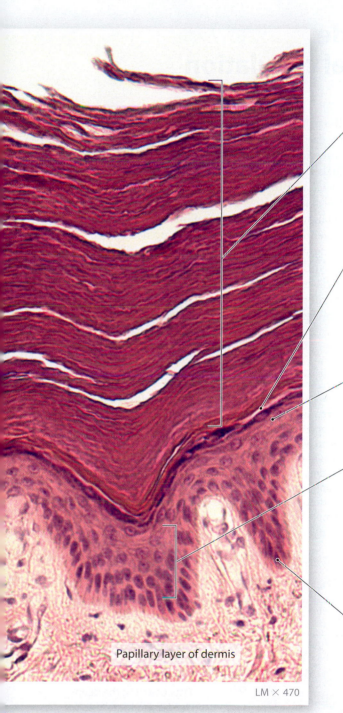

Papillary layer of dermis

LM × 470

4 Like all other epithelia, the epidermis lacks local blood vessels. Epidermal cells rely on the diffusion of nutrients and oxygen from capillaries within the dermis. As a result, the cells with the highest metabolic demand are closest to the underlying dermis.

Layers of the Epidermis

At the exposed surface of both thick skin and thin skin is the **stratum corneum** (STRA-tum KOR-nē-um; *cornu*, horn), which normally contains 15 to 30 layers of keratinized cells. Keratinization is the formation of protective, superficial layers of cells filled with **keratin** (KER-a-tin; *keros*, horn). The dead cells in each layer of the stratum corneum remain tightly interconnected by desmosomes. It takes 7 to 10 days for a cell to move from the stratum basale to the stratum corneum. The dead cells generally remain in the exposed stratum corneum for an additional two weeks before they are shed or washed away.

In the thick skin of the palms and soles, a **stratum lucidum** ("clear layer") separates the stratum corneum from deeper layers. The cells in the stratum lucidum are flattened, densely packed, largely devoid of organelles, and filled with the proteins keratin and **keratohyalin** (ker-a-tō-HĪ-a-lin). These proteins were synthesized while the cells were in the underlying stratum granulosum. By the time they reach the stratum lucidum, they are dead and undergoing dehydration.

The **stratum granulosum**, or "grainy layer," consists of three to five layers of keratinocytes. By the time cells reach this layer, most have stopped dividing and have started making large amounts of keratin and keratohyalin. As these fibers develop, the cells grow thinner and flatter, and their membranes thicken and become less permeable.

The **stratum spinosum** consists of 8 to 10 layers of keratinocytes bound together by desmosomes. The name *stratum spinosum*, which means "spiny layer," refers to the fact that the cells look like miniature pincushions in standard histological sections. They look that way because the keratinocytes were processed with chemicals that shrank the cytoplasm but left the cytoskeletal elements and desmosomes intact. The stratum spinosum also contains **dendritic cells**, which participate in the immune response by stimulating a defense against (1) microorganisms that manage to penetrate the superficial layers of the epidermis and (2) superficial skin cancers.

The **stratum basale** (STRA-tum bā-SA-lā) is the basal layer of the epidermis. Hemidesmosomes attach the cells of this layer to the basal lamina that separates the epidermis from the areolar tissue of the adjacent papillary layer of the dermis. **Basal cells**, or germinative cells, dominate the stratum basale. Basal cells are stem cells whose divisions replace the more superficial keratinocytes that are lost or shed at the epithelial surface. Skin surfaces that lack hair also contain specialized epithelial cells known as **Merkel cells** scattered among the basal cells. Merkel cells are sensitive to touch; when compressed, they release chemicals that stimulate sensory nerve endings.

SEM × 25

5 Ridge patterns in the thick skin of the surface of fingertips produce fingerprints, which have been used to identify individuals in criminal investigations for more than a century. This is a scanning electron micrograph of the tip of an index finger.

Module 5.1 Review

a. Identify the layers of the epidermis (from deep to superficial).

b. Dandruff is caused by excessive shedding of cells from the outer layers of the skin of the scalp. Thus, dandruff is composed of cells from which epidermal layer?

c. A splinter that penetrates to the third layer of the epidermis of the palm is lodged in which layer?

Factors influencing skin color include epidermal pigmentation and dermal circulation

Skin color is influenced by the presence of pigments in the skin, the degree of dermal circulation, and the thickness and degree of keratinization in the epidermis.

1 The primary pigments involved in skin coloration are carotene and melanin. The micrograph at left shows a section through the thin skin of a light-skinned individual. The ratio of melanocytes to basal cells ranges between 1:4 and 1:20, depending on the region of the body. The skin covering most areas of the body has about 1000 melanocytes per square millimeter. Differences in skin pigmentation among individuals do not reflect different numbers of melanocytes, but instead different levels of synthetic activity. Even the melanocytes of albino individuals are distributed normally, although the cells are incapable of producing melanin. The figure below depicts a single melanocyte in the stratum basale.

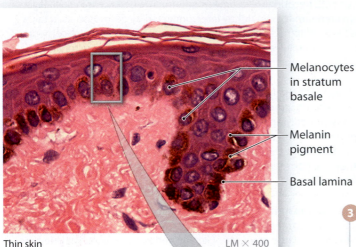

Melanocytes in stratum basale

Melanin pigment

Basal lamina

Thin skin   LM × 400

3 Melanosomes travel within the processes of melanocytes and are transferred intact to keratinocytes. The transfer of pigmentation colors the keratinocyte temporarily, until the melanosomes are destroyed by fusion with lysosomes. In individuals with pale skin, this transfer occurs in the stratum basale and stratum spinosum, and the cells of more superfical layers lose their pigmentation. In dark-skinned people, the melanosomes are larger, and the transfer may occur in the stratum granulosum as well; skin pigmentation is thus darker and more persistent.

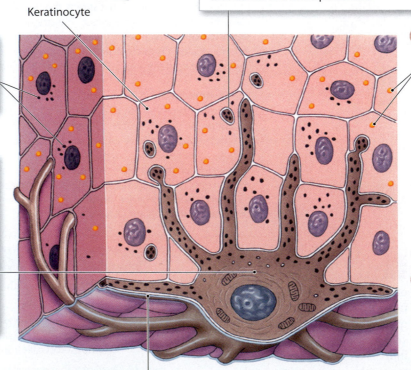

Keratinocyte

1 **Melanin** is a brown, yellow-brown, or black pigment produced by **melanocytes**.

2 The melanocytes are located in the stratum basale, squeezed between or deep to the epithelial cells. Melanocytes manufacture melanin from the amino acid tyrosine, and package it in intracellular vesicles called **melanosomes**.

4 **Carotene** (KAR-uh-tēn) is an orange-yellow pigment that normally accumulates in epidermal cells. It is most apparent in cells of the stratum corneum of light-skinned individuals, but it also accumulates in fatty tissues in the deep dermis and hypodermis. Carotene is found in a variety of orange vegetables.

5 The color of one's skin is genetically programmed. However, increased pigmentation, or tanning, can result in response to ultraviolet (UV) radiation.

Basal lamina

2 The blood supply affects skin color because blood contains red blood cells filled with the red pigment hemoglobin. When bound to oxygen, hemoglobin is bright red, giving capillaries in the dermis a reddish tint that is most apparent in lightly pigmented individuals. If those vessels are dilated, the red tones become much more pronounced. For example, your skin becomes flushed and red when your body temperature rises because the superficial blood vessels dilate so that the skin can act like a radiator and lose heat. When the blood flow decreases, oxygen levels in the tissues decline, and under these conditions hemoglobin releases oxygen and turns a much darker red. Seen from the surface, the skin then takes on a bluish color. This coloration is called **cyanosis** (sī-uh-NŌ-sis; *kyanos*, blue). In individuals of any skin color, cyanosis is most apparent in areas of very thin skin, such as the lips or beneath the nails.

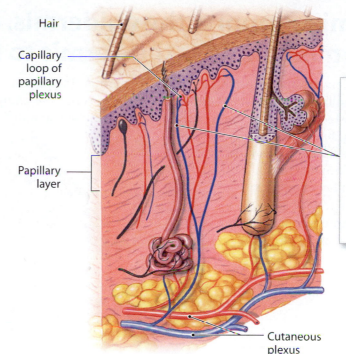

Hair

Capillary loop of papillary plexus

Papillary layer

Cutaneous plexus

The papillary layer of the skin contains the **papillary plexus**, a network of capillaries that provides oxygen and nutrients to the dermis and epidermis.

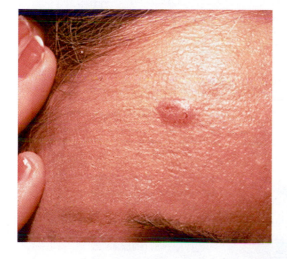

3 Skin cancers are the most common types of cancers. The most common form of skin cancer is a **basal cell carcinoma**. This is a cancer that originates in the stratum basale, due to mutations caused by overexposure to the ultraviolet (UV) radiation in sunlight. Metastasis virtually never occurs in basal cell carcinomas, and most people survive these cancers. The melanin in keratinocytes provides some protection against the effects of UV radiation because the melanosomes are concentrated around the nucleus, where they act like sunshades for the enclosed DNA.

4 In contrast, a **malignant melanoma** (mel-a-NŌ-muh) is extremely dangerous. In this condition, cancerous melanocytes grow rapidly and metastasize through the lymphatic system. The outlook for long-term survival is in many cases determined by how early the condition is diagnosed. If the cancer is detected early, while it is still localized, the affected area can be surgically removed, and the 5-year survival rate is 99 percent. If the condition is not detected until extensive metastasis has occurred, the 5-year survival rate drops to 14 percent.

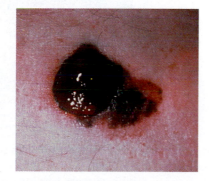

Module 5.2 Review

a. Name the two pigments contained in the epidermis.

b. Why does exposure to sunlight or sunlamps darken skin?

c. Why does the skin of a fair-skinned person appear red during exercise in hot weather?

The dermis supports the epidermis, and the hypodermis connects the dermis to the rest of the body

The **dermis** lies between the epidermis and the hypodermis. The presence of two types of fibers—collagen fibers, which are very strong and resist stretching but are easily bent or twisted, and elastic fibers, which permit stretching and then recoil to their original length—enables the dermis to tolerate limited stretching. The elastic fibers provide flexibility, and the collagen fibers limit that flexibility to prevent damage to the tissue. Aging, hormonal changes, and the destructive effects of ultraviolet radiation permanently reduce the amount of elastin in the dermis; the results are wrinkles and sagging skin.

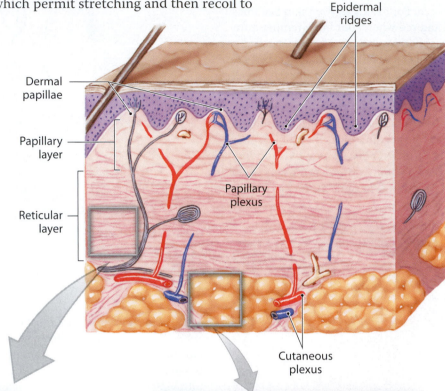

Epidermal ridges

Dermal papillae

Papillary layer

Reticular layer

Papillary plexus

Cutaneous plexus

1 The **papillary layer** consists of a highly vascularized areolar tissue with all of the typical cell types within it. This layer also contains the capillaries, lymphatic vessels, and sensory neurons that supply the surface of the skin. The papillary layer gets its name from the dermal papillae that project between the epidermal ridges.

Reticular layer of dermis   SEM × 1500

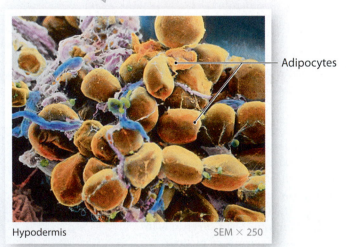

Adipocytes

Hypodermis   SEM × 250

2 The **reticular layer** consists of an interwoven meshwork of dense irregular connective tissue containing both collagen and elastic fibers. Bundles of collagen fibers extend superficially beyond the reticular layer to blend into those of the papillary layer, so the boundary between the two layers is indistinct. Collagen fibers of the reticular layer also extend into the deeper hypodermis. In addition, the reticular and papillary layers of the dermis contain networks of blood vessels, lymphatic vessels, nerve fibers, and accessory organs such as hair follicles and sweat glands.

3 The **hypodermis** separates the skin from deeper structures. It stabilizes the position of the skin in relation to underlying tissues (such as skeletal muscles or other organs) while permitting independent movement. Because it is often dominated by adipose tissue, this layer also represents an important site for the storage of energy reserves.

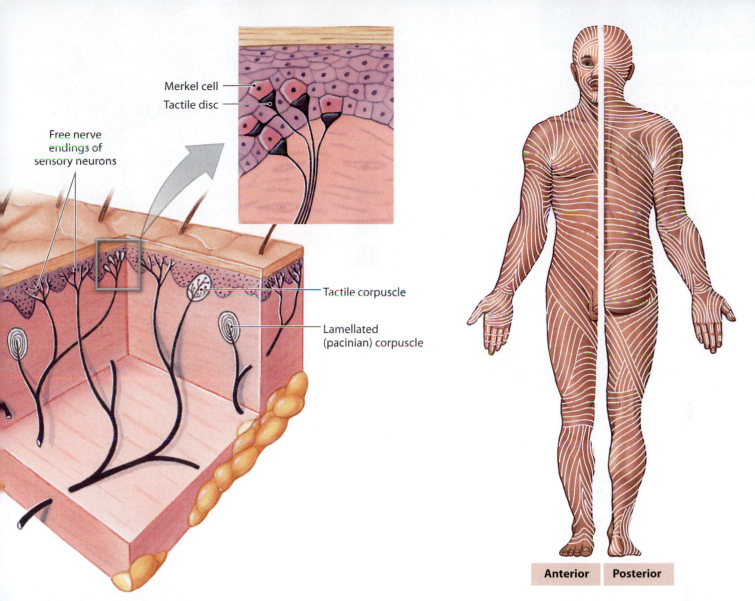

Merkel cell

Tactile disc

Free nerve endings of sensory neurons

Tactile corpuscle

Lamellated (pacinian) corpuscle

Anterior   Posterior

4 The integument contains many sensory receptors, and anything that comes in contact with the skin—from the lightest touch of a mosquito to the weight of a loaded backpack—initiates a nerve impulse that can reach our conscious awareness. We have already noted that the deeper layers of the epidermis contain Merkel cells; sensory terminals known as **tactile discs** monitor these cells. The epidermis also contains the extensions of sensory neurons that provide sensations of pain and temperature. The dermis contains similar receptors, as well as other more specialized receptors. Examples include receptors sensitive to light touch—**tactile corpuscles**, located in dermal papillae—and receptors sensitive to deep pressure and vibration—**lamellated corpuscles**, also called *pacinian corpuscles*, in the reticular layer.

Beginning at puberty, men accumulate subcutaneous fat at the neck, on the arms, along the lower back, and over the buttocks. In contrast, women accumulate subcutaneous fat at the breasts, buttocks, hips, and thighs. In adults of either gender, the hypodermis of the backs of the hands and the upper surfaces of the feet contains few fat cells, whereas distressing amounts of adipose tissue can accumulate in the abdominal region, producing a prominent "potbelly."

5 The banding shown here indicates the **lines of cleavage** in the skin. Most of the collagen and elastic fibers at any location are arranged in parallel bundles oriented to resist the forces applied to the skin during normal movement. The resulting pattern of fiber bundles establishes lines of cleavage. These lines are clinically significant: a cut parallel to a cleavage line will usually remain closed and heal with little scarring, whereas a cut at right angles to a cleavage line will be pulled open as severed elastic fibers recoil and will result in greater scarring. For these reasons, surgeons choose to make neat incisions parallel to the lines of cleavage.

Module 5.3 Review

a. Describe the location of the dermis.

b. Where are the capillaries and sensory neurons that supply the epidermis located?

c. What accounts for the ability of the dermis to undergo repeated stretching?

1. Short answer

Identify and describe the layers of the cutaneous membrane and the underlying layer of loose connective tissue in the diagram at right.

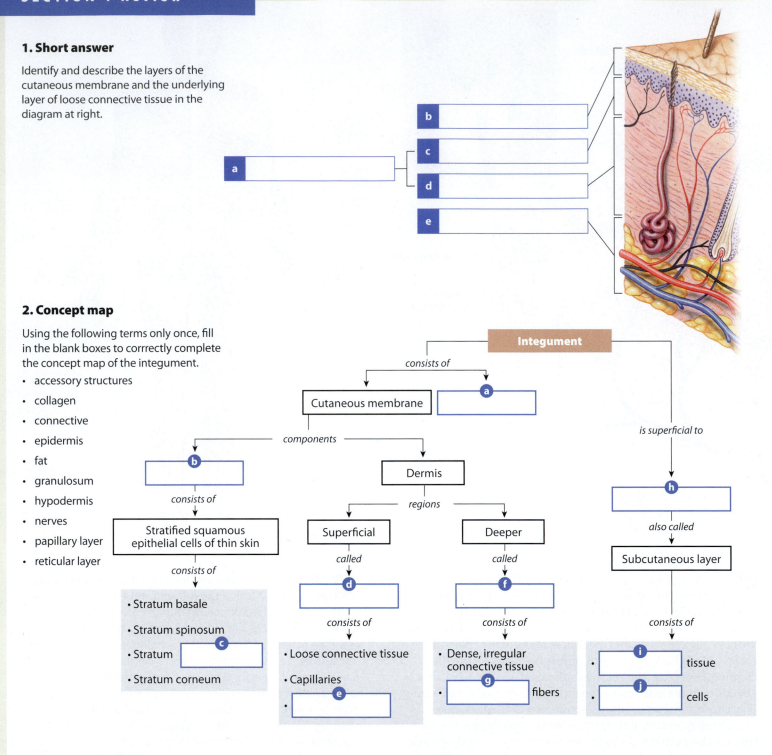

a _____

b _____

c _____

d _____

e _____

2. Concept map

Using the following terms only once, fill in the blank boxes to corrrectly complete the concept map of the integument.

- accessory structures
- collagen
- connective
- epidermis
- fat
- granulosum
- hypodermis
- nerves
- papillary layer
- reticular layer

Integument

consists of

Cutaneous membrane — **a** _____

is superficial to

h _____

also called

Subcutaneous layer

components

b _____

consists of

Stratified squamous epithelial cells of thin skin

consists of

- Stratum basale
- Stratum spinosum
- Stratum **c** _____
- Stratum corneum

Dermis

regions

Superficial

called

d _____

consists of

- Loose connective tissue
- Capillaries
- **e** _____

Deeper

called

f _____

consists of

- Dense, irregular connective tissue
- **g** _____ fibers

consists of

- **i** _____ tissue
- **j** _____ cells

3. System integration

a. Describe why malignancies of melanocytes are so often fatal.

_____

_____

b. After taking a long bubble bath or doing dishes by hand, your fingers often appear wrinkled, or "pruny." Explain why.

_____

_____

Accessory Organs of the Skin

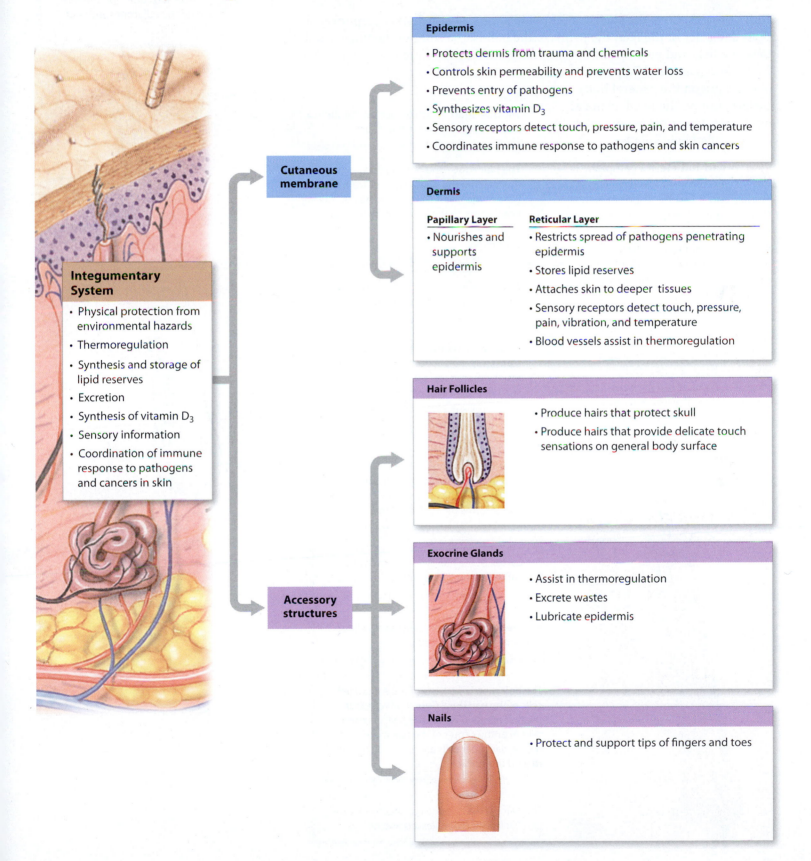

Integumentary System

- Physical protection from environmental hazards
- Thermoregulation
- Synthesis and storage of lipid reserves
- Excretion
- Synthesis of vitamin D_3
- Sensory information
- Coordination of immune response to pathogens and cancers in skin

Cutaneous membrane

Epidermis

- Protects dermis from trauma and chemicals
- Controls skin permeability and prevents water loss
- Prevents entry of pathogens
- Synthesizes vitamin D_3
- Sensory receptors detect touch, pressure, pain, and temperature
- Coordinates immune response to pathogens and skin cancers

Dermis

Papillary Layer	**Reticular Layer**
• Nourishes and supports epidermis	• Restricts spread of pathogens penetrating epidermis • Stores lipid reserves • Attaches skin to deeper tissues • Sensory receptors detect touch, pressure, pain, vibration, and temperature • Blood vessels assist in thermoregulation

Accessory structures

Hair Follicles

- Produce hairs that protect skull
- Produce hairs that provide delicate touch sensations on general body surface

Exocrine Glands

- Assist in thermoregulation
- Excrete wastes
- Lubricate epidermis

Nails

- Protect and support tips of fingers and toes

Hair is composed of dead, keratinized cells produced in a specialized hair follicle

Hairs project above the surface of the skin almost everywhere, except over the sides and soles of the feet, the palms of the hands, the sides of the fingers and toes, the lips, and portions of the external genitalia. The human body has about 2.5 million hairs, and 75 percent of them are on the general body surface, not on the head. Hairs are nonliving structures produced in organs called hair follicles.

1 A **hair follicle** is a complex structure composed of epithelial and connective tissues that is responsible for the formation of a single hair. **Terminal hairs** are large, coarse, often darkly pigmented hairs such as those found on the scalp or the armpits. **Vellus hairs** are smaller, shorter, and more delicate hairs found on the general body surface.

2 Hair production begins at the base of a hair follicle. Here a mass of epithelial cells forms a cap, called the **hair bulb**, that surrounds a small **hair papilla**, a peg of connective tissue containing capillaries and nerves.

A **sebaceous gland** produces secretions that coat the hair and the adjacent surface of the skin.

The **hair shaft**, which we see on the surface, begins deep within the hair follicle.

The **hair root**—the portion that anchors the hair into the skin—extends from the base of the hair follicle, where hair production begins, to the point where the hair shaft loses its connection with the walls of the follicle.

A connective tissue sheath surrounds the epithelial cells of the hair follicle.

A **root hair plexus** of sensory nerves surrounds the base of each hair follicle. As a result, you can feel the movement of the shaft of even a single hair.

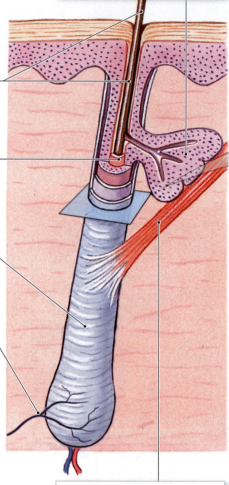

The **arrector pili** (a-REK-tor PĪ-lē; plural, arrectores pilorum) is a smooth muscle whose contraction pulls on the follicle, forcing the hair to stand erect.

The **cuticle** of the hair consists of daughter cells produced at the edges of the hair matrix. The cuticle forms the surface of the hair.

The **cortex** of the hair is an intermediate layer of cells deep to the cuticle.

The **medulla** of the hair consists of cells at the center of the hair matrix.

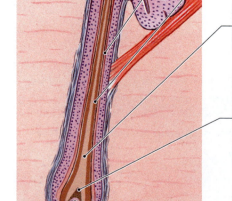

The **hair matrix** consists of the superficial cells of the hair bulb. Germinative cells in the hair matrix produce the hair. As stem cells near the center of the hair matrix divide, daughter cells are gradually pushed toward the surface.

The hair papilla is a small connective tissue peg filled with blood vessels and nerves.

Hair Structure

The medulla, or core, of the hair contains a flexible **soft keratin**.

The cortex contains thick layers of **hard keratin**, which give the hair its stiffness.

The cuticle, although thin, is very tough and contains hard keratin.

Hair Follicle Structure

The **internal root sheath** surrounds the hair root and the deeper portion of the shaft. The cells of this sheath disintegrate quickly, and this layer does not extend the entire length of the hair follicle.

The **external root sheath** consists of epithelial cells and extends from the skin surface to the hair matrix.

The **glassy membrane** is a thickened, clear basal lamina.

Connective tissue sheath

3 A cross-section through a hair follicle reveals the internal structure of both the hair and the follicle.

4 Hairs grow and are shed according to a **hair growth cycle**. A hair in the scalp grows for 2 to 5 years, at a rate of about 0.33 mm per day. Variations in the growth rate and in the duration of the hair growth cycle account for individual differences in the length of uncut hair.

2 The follicle then begins to undergo regression and transitions to the resting phase.

1 The **active phase** lasts 2–5 years. During the active phase, the hair grows continuously at a rate of approximately 0.33 mm/day.

3 During the **resting phase** the hair loses its attachment to the follicle and becomes a **club hair**.

4 When follicle reactivation occurs, the club hair is shed and the hair matrix begins producing a replacement hair.

Variations in hair color reflect differences in structure and variations in the pigment produced by melanocytes at the hair papilla. Different forms of melanin give a dark brown, yellow-brown, or red color to the hair. As pigment production decreases with age, hair color lightens. White hair results from the combination of a lack of pigment and the presence of air bubbles in the medulla of the hair shaft.

Module 5.4 Review

a. Describe a typical strand of hair.

b. What happens when an arrector pili muscle contracts?

c. Why is pulling a hair painful, yet cutting a hair is not?

Sebaceous glands and sweat glands are exocrine glands found in the skin

In this module we examine the structure and function of the accessory structures that produce exocrine secretions, with particular attention to sebaceous glands and sweat glands. These exocrine glands assist in thermoregulation, excrete wastes, and lubricate the epidermis.

1 **Sebaceous** (se-BĀ-shus) **glands**, or oil glands, are holocrine glands that discharge an oily lipid secretion. Sebaceous glands that communicate with a single follicle share a duct and thus are classified as simple branched alveolar glands. The lipids released from gland cells enter the lumen (open passageway) of the gland. Contractions of the arrector pili muscle squeeze the sebaceous gland and force the sebum, a mixture of triglycerides, cholesterol, proteins, and electrolytes, into the hair follicle and onto the surface of the skin. The sebum moisturizes and conditions the hair and surrounding skin.

Sebaceous Glands

Sebaceous glands secrete **sebum** (SĒ-bum), which coats the hair shaft and surrounding epidermal surfaces. Sebum provides lubrication and keeps the hair shaft from becoming dry and brittle. It also has antibacterial properties.

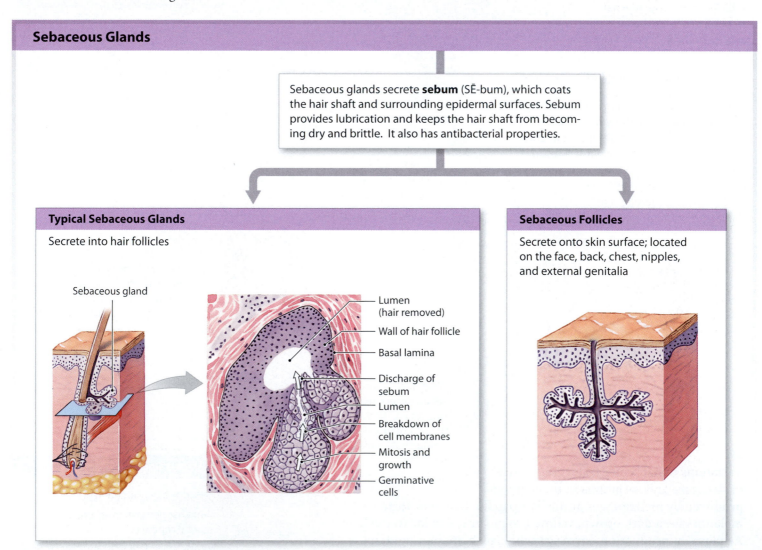

Typical Sebaceous Glands

Secrete into hair follicles

Sebaceous gland

- Lumen (hair removed)
- Wall of hair follicle
- Basal lamina
- Discharge of sebum
- Lumen
- Breakdown of cell membranes
- Mitosis and growth
- Germinative cells

Sebaceous Follicles

Secrete onto skin surface; located on the face, back, chest, nipples, and external genitalia

Sweat Glands

Sweat glands produce a watery solution by merocrine secretion, flush the epidermal surface, and perform other special functions.

Apocrine Sweat Glands

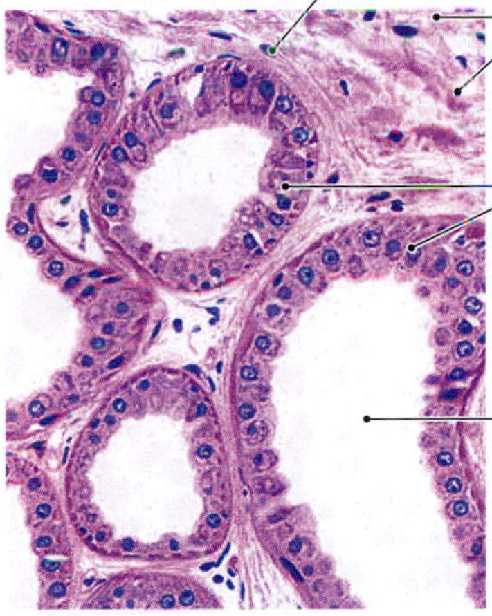

Connective tissue of dermis

Apocrine gland cells

Lumen

Section of apocrine sweat gland

LM × 375

Myoepithelial cells (*myo*-, muscle) are contractile cells that squeeze the gland and discharge the accumulated secretion.

Merocrine Sweat Glands

Merocrine gland cells

Lumen

Section of merocrine sweat gland

LM × 210

- Limited distribution (axillae, groin, nipples)
- Produce a viscous secretion of complex composition
- Possible function in olfactory communication
- Strongly influenced by hormones
- Include ceruminous glands of the external ear and mammary glands that produce milk

- Found in most areas of the skin
- Produce watery secretions containing electrolytes
- Merocrine secretion mechanism
- Controlled primarily by nervous system
- Important in thermoregulation and excretion
- Some antibacterial action

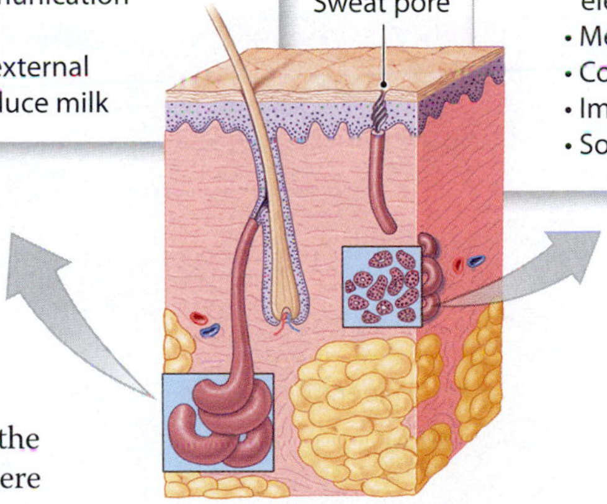

Sweat pore

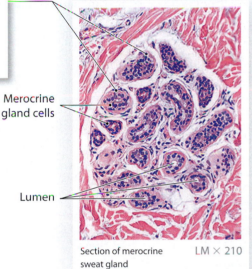

2   **Apocrine sweat glands** are found in the armpits, around the nipples, and in the pubic region, where they secrete into hair follicles. These glands produce a sticky, cloudy, and potentially odorous secretion. Despite their name, these glands rely on merocrine secretion. The adult integument also contains 2–5 million **merocrine sweat glands** that discharge their secretions directly onto the surface of the skin. The palms and soles have the highest numbers, with the palm possessing an estimated 500 merocrine sweat glands per square centimeter (3000 per square inch).

Module 5.5 Review

a. Identify two types of exocrine glands found in the skin.

b. What are the functions of sebaceous secretions?

c. Deodorants are used to mask the effects of secretions from which type of skin gland?

Nails are thick sheets of keratinized epidermal cells that protect the tips of fingers and toes

Nails protect the exposed dorsal surfaces of the tips of the fingers and toes. They also help limit distortion of the digits when they are subjected to mechanical stress—for example, when you run or grasp objects.

1 These superficial and sectional views illustrate common landmarks of nail structure. The **nail body** consists of dead, tightly compressed cells packed with keratin. The nail body is recessed deep to the level of the surrounding epithelium and is bounded on either side by **lateral nail grooves** (depressions) and **lateral nail folds** (ridges).

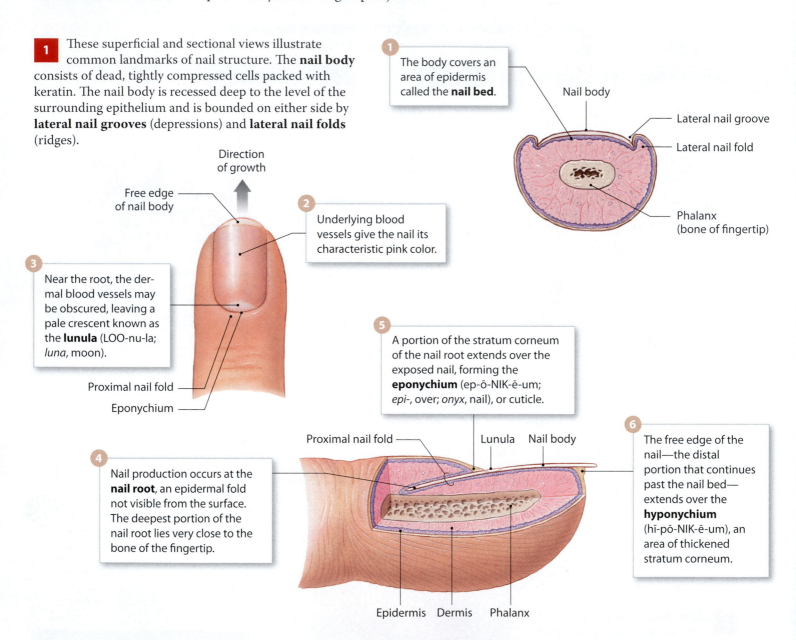

1 The body covers an area of epidermis called the **nail bed**.

Nail body

Lateral nail groove

Lateral nail fold

Phalanx (bone of fingertip)

Direction of growth

Free edge of nail body

2 Underlying blood vessels give the nail its characteristic pink color.

3 Near the root, the dermal blood vessels may be obscured, leaving a pale crescent known as the **lunula** (LOO-nu-la; *luna*, moon).

Proximal nail fold

Eponychium

5 A portion of the stratum corneum of the nail root extends over the exposed nail, forming the **eponychium** (ep-ō-NIK-ē-um; *epi-*, over; *onyx*, nail), or cuticle.

Proximal nail fold   Lunula   Nail body

6 The free edge of the nail—the distal portion that continues past the nail bed—extends over the **hyponychium** (hī-pō-NIK-ē-um), an area of thickened stratum corneum.

4 Nail production occurs at the **nail root**, an epidermal fold not visible from the surface. The deepest portion of the nail root lies very close to the bone of the fingertip.

Epidermis   Dermis   Phalanx

2 The cells producing the nails can be affected by conditions that alter body metabolism, so changes in the shape, structure, or appearance of the nails can provide useful diagnostic information. For example, the nails may become pitted and distorted as a result of psoriasis (a condition marked by rapid stem cell division in the stratum basale), and concave as a result of some blood disorders.

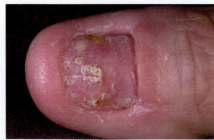

Module 5.6 Review

a. Define hyponychium.

b. Describe a typical fingernail.

c. Where does nail production occur?

Age-related changes alter the integument

Fewer Melanocytes

Melanocyte activity declines, and in light-skinned individuals the skin becomes very pale. With less melanin in the skin, people become more sensitive to sun exposure and more likely to experience sunburn.

Drier Epidermis

Sebaceous gland secretion decreases, and the skin becomes dry and often scaly.

Thinning Epidermis

The epidermis thins as germinative cell activity declines, and the connections between the epidermis and dermis weaken, making older people more prone to injury, skin tears, and skin infections. The metabolic activity of the skin decreases as well. The epidermis normally produces vitamin D3, and decreased production leads to muscle weakness and brittle bones.

Diminished Immune Response

The number of dendritic cells decreases to about half the levels seen at maturity (roughly age 21). This reduction in cells may decrease the sensitivity of the immune response and further encourage skin damage and infection.

Thinning Dermis

The dermis becomes thinner and has fewer elastic fibers, making the integument weaker and less resilient. The results—sagging and wrinkling—are most pronounced in body regions with the most exposure to the sun.

Decreased Perspiration

Merocrine sweat glands become less active, and with impaired perspiration, older people cannot lose heat as fast as younger people. Thus, the elderly are at greater risk of overheating in warm environments.

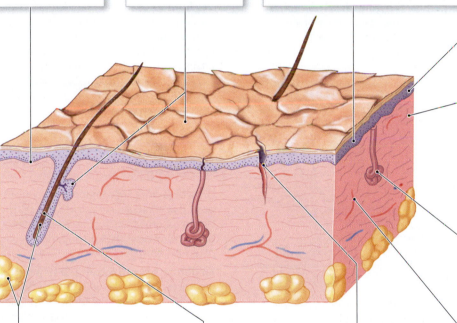

Altered Hair and Fat Distribution

With declining levels of sex hormones, differences in secondary sexual characteristics with respect to hair distribution and body-fat distribution begin to fade. As a consequence, people age 90–100 of both sexes tend to look alike.

Fewer Active Follicles

Hair follicles stop functioning or produce thinner, finer hairs. With decreased melanocyte activity, these hairs are gray or white.

Slower Skin Repair

Skin repairs proceed more slowly. Thus, whereas repairs to an uninfected blister might take three to four weeks in a young adult, the same repairs could take six to eight weeks at age 65–75.

Reduced Blood Supply

A reduction in dermal blood supply cools the skin, which can stimulate thermoreceptors and make a person feel cold even in a warm room. Reduced circulation and sweat gland function in the elderly lessens their ability to lose body heat, which can allow body temperatures to soar dangerously high with overexertion.

Aging affects all of the structures in the integument. The major structural and functional changes are summarized in this illustration. The gradual changes in the superficial appearance of the skin provide visual cues that we use unconsciously to estimate a person's age. This accounts for the popularity of cosmetic facial surgery.

Module 5.7 Review

a. Identify some common effects of the aging process on skin.

b. Why does hair turn white or gray with age?

c. Why do individuals tolerate summer heat less well and become more susceptible to heat-related illness when they become older?

The integument both responds to circulating hormones and has endocrine functions that are dependent on ultraviolet radiation

1 The integumentary system is physically and functionally tied to all other body systems. As a result, the state of the skin can be an indication of the health of the individual. Much of the communication between the skin and the rest of the body is chemical communication, as the specialized cells of the integument respond to levels of circulating hormones.

Steroid Hormones

Steroid hormones called **glucocorticoids** are released during times of stress. These hormones loosen the connections between keratinocytes and reduce the effectiveness of the epidermis as a barrier to infection.

Thyroid Hormones

Thyroid gland hormones maintain normal blood flow to the papillary plexus.

Sex Hormones

Sex hormones stimulate epidermal cell divisions, increasing epidermal thickness and accelerating wound repair. These hormones also increase the number of dendritic cells that provide defense against cancer cells and pathogens.

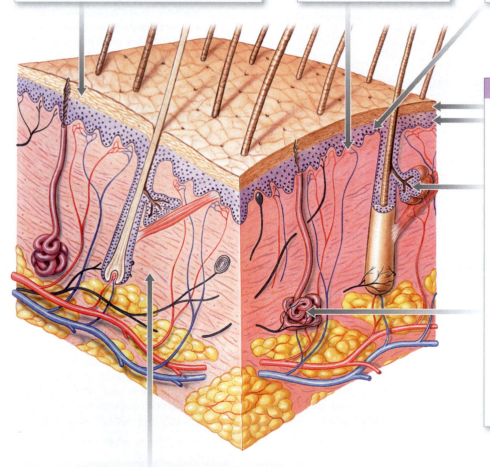

Growth Factors

Growth factors are compounds produced in the body that stimulate cell growth and cell division. Dozens of growth factors have been identified, but in many cases their origins and mechanisms of regulation are unknown.

Epidermal growth factor (EGF) is a peptide growth factor that has widespread effects on epithelia throughout the body. This peptide is produced by the salivary glands and glands of the duodenum, the initial segment of the small intestine.

Among the roles of EGF are the following:
- Promoting the divisions of germinative cells in the stratum basale and stratum spinosum
- Accelerating the production of keratin in differentiating keratinocytes
- Stimulating epidermal development and epidermal repair after injury
- Stimulating synthetic activity and secretion by epithelial glands

Growth Hormone

Growth hormone (**GH**) stimulates fibroblast activity and collagen synthesis. Acting through intermediary compounds, GH also stimulates germinative cell divisions, thickens the epidermis, and promotes wound repair.

2 Although too much sunlight can damage epithelial cells and deeper tissues, limited exposure to sunlight is beneficial because UV radiation plays a vital role in the synthesis of an important vitamin: **cholecalciferol** (kō-le-kal-SIF-er-ol), or **vitamin D₃**.

Sources of Vitamin D₃

Sunlight: When exposed to ultraviolet radiation, epidermal cells in the stratum spinosum and stratum basale convert a cholesterol-related steroid into cholecalciferol. This vitamin then diffuses across the basal lamina and enters capillaries of the papillary plexus.

Diet: Cholecalciferol can be obtained from the diet, but few foods contain it other than fish, fish oils, and shellfish, and even then the presence and quantities are variable. Today, many food products are "fortified with vitamin D"—most notably milk, soy milk, and orange juice.

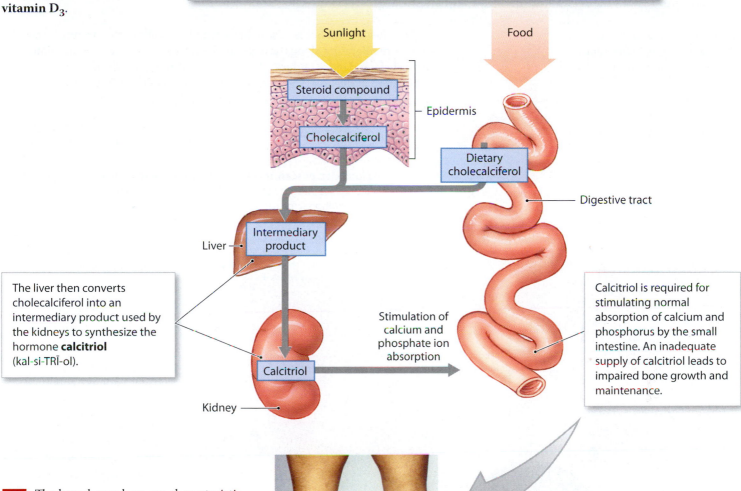

The liver then converts cholecalciferol into an intermediary product used by the kidneys to synthesize the hormone **calcitriol** (kal-si-TRĪ-ol).

Calcitriol is required for stimulating normal absorption of calcium and phosphorus by the small intestine. An inadequate supply of calcitriol leads to impaired bone growth and maintenance.

3 The legs shown here are characteristic of **rickets**, a condition that results in flexible, poorly mineralized bones. Rickets develops in a growing child whose skin is not exposed to sunlight, and whose diet does not include a source of cholecalciferol (vitamin D₃). The bones have the proper shape, but they lack rigidity because the bone matrix contains inadequate amounts of calcium and phosphate. Rickets has largely been eliminated in the United States because cholecalciferol has been added to many different foods. However, cholecalciferol is required throughout life. Adults are less likely to develop rickets due to a cholecalciferol deficiency, but bone density decreases, and this leads to a greater risk of fractures and slows the healing process. The risks are especially acute for the elderly because skin production of cholecalciferol decreases by about 75 percent, even when exposed to sunlight.

Module 5.8 Review

a. List some hormones that are necessary for maintaining a healthy integument.

b. Explain the relationship between sunlight exposure and vitamin D₃.

c. In some cultures, females must be covered from head to toe when they go outdoors. Explain why these women are at increased risk of developing bone problems later in life.

The integument can often repair itself, even after extensive damage

The integumentary system displays a significant degree of functional independence—it often responds directly and automatically to local influences without the involvement of the nervous or endocrine systems. A dramatic display of local regulation can be seen after an injury to the skin. These four panels illustrate the steps in the regeneration of the skin after injury.

1 An initial injury to the skin results in bleeding and mast cell activation.

2 After several hours, a scab has formed and cells of the stratum basale are migrating along the edges of the wound. Macrophages are removing debris, and more of these phagocytes are arriving via the enhanced circulation in the area. Clotting around the edges of the affected area partially isolates the region from adjacent undamaged tissues.

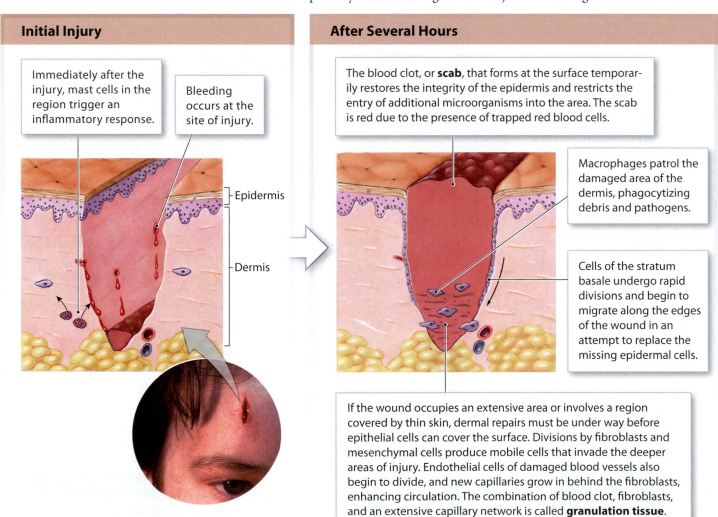

Initial Injury

Immediately after the injury, mast cells in the region trigger an inflammatory response.

Bleeding occurs at the site of injury.

Epidermis

Dermis

After Several Hours

The blood clot, or **scab**, that forms at the surface temporarily restores the integrity of the epidermis and restricts the entry of additional microorganisms into the area. The scab is red due to the presence of trapped red blood cells.

Macrophages patrol the damaged area of the dermis, phagocytizing debris and pathogens.

Cells of the stratum basale undergo rapid divisions and begin to migrate along the edges of the wound in an attempt to replace the missing epidermal cells.

If the wound occupies an extensive area or involves a region covered by thin skin, dermal repairs must be under way before epithelial cells can cover the surface. Divisions by fibroblasts and mesenchymal cells produce mobile cells that invade the deeper areas of injury. Endothelial cells of damaged blood vessels also begin to divide, and new capillaries grow in behind the fibroblasts, enhancing circulation. The combination of blood clot, fibroblasts, and an extensive capillary network is called **granulation tissue**.

3 One week after the injury, the scab has been undermined by epidermal cells migrating over a meshwork produced by fibroblast activity. Phagocytic activity around the site has almost ended, and the blood clot is disintegrating.

4 After several weeks, the scab has been shed, and the epidermis is complete. A shallow depression marks the injury site, but fibroblasts in the dermis continue to create scar tissue that will gradually elevate the overlying epidermis.

After One Week

Over time, deeper portions of the clot dissolve, and the number of capillaries declines. Fibroblast activity leads to the appearance of collagen fibers and typical ground substance. The repairs do not restore the integument to its original condition, however, because the dermis will contain an abnormally large number of collagen fibers and relatively few blood vessels.

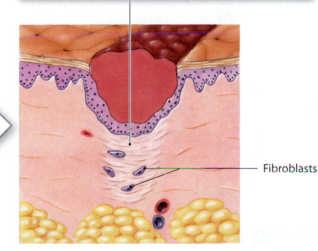

Fibroblasts

After Several Weeks

Severely damaged hair follicles, sebaceous or sweat glands, muscle cells, and nerves are seldom repaired, and they too are replaced by fibrous tissue. The formation of this rather inflexible, fibrous, noncellular **scar tissue** completes the repair process but fails to restore the tissue to its original condition.

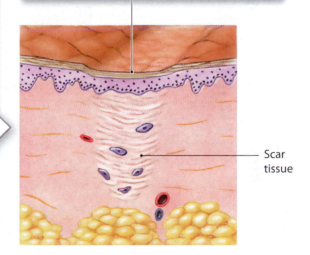

Scar tissue

5 In some adults, most often those with dark skin, scar tissue formation may continue beyond the requirements of tissue repair. The result is a **keloid** (KĒ-loyd), a raised, thickened mass of scar tissue that begins at the site of injury and grows into the surrounding dermis. Keloids are covered by a shiny, smooth epidermal surface and most commonly develop on the upper back, shoulders, anterior chest, or earlobes. They are harmless; some aboriginal cultures intentionally produce keloids as a form of body decoration.

Module 5.9 Review

a. Identify the first step in the repair of tissue.

b. Describe granulation tissue.

c. Why can skin regenerate effectively even after considerable damage?

1. Matching

Match the following terms with the most closely related description.

- malignant melanoma
- wrinkled skin
- nail root
- sebum
- apocrine sweat glands
- eponychium
- EGF
- vitamin D$_3$
- reticular layer of dermis
- merocrine sweat glands

a _____ Produced by epidermal cells stimulated by UV radiation

b _____ Epithelial fold not visible from the surface

c _____ Found in the armpit

d _____ Peptide produced by salivary glands

e _____ Site of hair production

f _____ Decrease in elastic fibers

g _____ Oily lipid secretion

h _____ Melanocytes metastasize through the lymphatic system

i _____ Abundant in the palms and soles

j _____ Cuticle

2. Labeling

Label the structures of a typical nail in the accompanying figures.

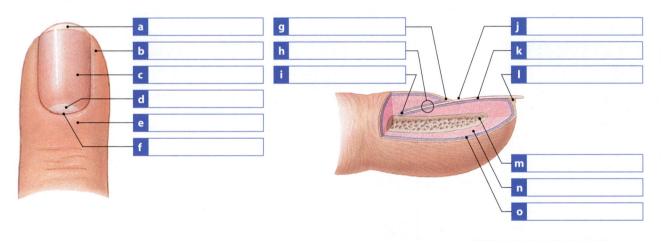

a _____

b _____

c _____

d _____

e _____

f _____

g _____

h _____

i _____

j _____

k _____

l _____

m _____

n _____

o _____

3. Labeling

Label the structures of a hair follicle in the accompanying figure.

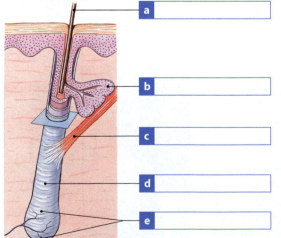

a _____

b _____

c _____

d _____

e _____

4. Section integration

Many people change the natural appearance of their hair, either by coloring it or by altering the degree of curl in it. Which layers of the hair do you suppose are affected by the chemicals added during these procedures? Why are the effects of the procedures not permanent?

Visual Outline with Key Terms   Summarize the content of each module using the terms in the order provided.

SECTION 1

Functional Anatomy of the Skin

- integumentary system
- integument
- cutaneous membrane
- epidermis
- dermis
- hypodermis
- accessory structures
- cutaneous plexus

SECTION 2

Accessory Organs of the Skin

- hair follicles
- exocrine glands
- nails

5.1

The epidermis is composed of layers with various functions

- keratinocytes
- epidermal ridges
- dermal papillae
- thin skin
- thick skin
- stratum corneum
- keratin
- stratum lucidum
- keratohyalin
- stratum granulosum
- stratum spinosum
- dendritic cells
- stratum basale
- basal cells
- Merkel cells

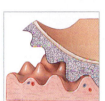

5.4

Hair is composed of dead, keratinized cells produced in a specialized hair follicle

- hairs
- hair follicle
- terminal hairs
- vellus hairs
- hair shaft
- hair root
- root hair plexus
- sebaceous gland
- arrector pili
- hair bulb
- hair papilla
- cuticle
- cortex
- medulla
- hair matrix
- soft keratin
- hard keratin
- internal root sheath
- external root sheath
- glassy membrane
- hair growth cycle
- active phase
- resting phase
- club hair

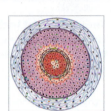

5.2

Factors influencing skin color include epidermal pigmentation and dermal circulation

- melanin
- melanocyte
- melanosomes
- carotene
- cyanosis
- papillary plexus
- basal cell carcinoma
- malignant melanoma

5.5

Sebaceous glands and sweat glands are exocrine glands found in the skin

- sebaceous glands
- sebum
- sebaceous follicles
- sweat glands
- apocrine sweat glands
- merocrine sweat glands
- myoepithelial cells

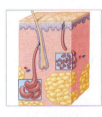

5.3

The dermis supports the epidermis, and the subcutaneous layer connects the dermis to the rest of the body

- dermis
- papillary layer
- reticular layer
- hypodermis
- tactile discs
- tactile corpuscles
- lamellated corpuscles
- lines of cleavage

5.6

Nails are thick sheets of keratinized epidermal cells that protect the tips of fingers and toes

- nails
- nail body
- lateral nail grooves
- lateral nail folds
- nail bed
- lunula
- nail root
- eponychium
- hyponychium

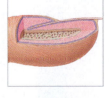

• = *Term boldfaced in this module*

5.7

Age-related changes alter the integument

- fewer melanocytes
- drier epidermis
- thinning epidermis
- diminished immune response
- thinning dermis
- decreased perspiration
- altered hair and fat distribution
- fewer active hair follicles
- slower skin repair
- reduced blood supply

5.9

The integument can often repair itself, even after extensive damage

- scab
- granulation tissue
- scar tissue
- keloid

5.8

The integument both responds to circulating hormones and has endocrine functions that are dependent on ultraviolet radiation

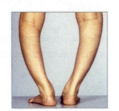

- glucocorticoids
- thyroid gland hormones
- sex hormones
- growth factors
- epidermal growth factor (EGF)
- growth hormone (GH)
- cholecalciferol (vitamin D$_3$)
- sunlight
- diet
- calcitriol
- rickets

• = *Term boldfaced in this module*

Chapter Integration:
Applying what you've learned

The skin is the body's first line of defense, so when scratches, cuts, scrapes, or burns compromise its integrity, the risk of bacterial infection increases. Consider for a moment the various skin injuries you have experienced. All of us have had minor breaks in the skin that we washed off and forgot about, and that healed in just a few days. Treatment of more serious skin disorders is targeted at reducing pain, preventing a worsening of the condition, minimizing infections, and promoting healing.

Sunburn is a skin injury that many people have experienced to some degree. Severe sunburn can lead to acute inflammation and the formation of blisters that disrupt epidermal barriers. Sunburn is also a risk factor for developing the two most common types of skin cancer: basal cell carcinoma and melanoma. Fair-skinned individuals who live in the tropics are most susceptible to all forms of skin cancer, because their melanocytes are unable to shield them from the ultraviolet radiation. Sun damage can be prevented by avoiding exposure to the sun during the hours of 10 a.m. to 4 p.m., wearing a wide-brimmed hat, covering up with light clothing, and applying

a sunblock (not tanning lotion or sunscreen). Sunblocks reflect and scatter UV light and act as a barrier between the sun and skin, whereas sunscreens act as a filter and allow a certain range of UV light to be absorbed into the skin.

Many of us have experienced more serious skin injuries, such as deep cuts—perhaps even one that required a trip to the emergency room and some stitches. These wounds need greater care and take longer to heal than even a severe sunburn. But very few of us have had life-threatening deep-tissue injuries or burns that required extensive treatment and an extended hospital stay. Maintaining skin health is key for decreasing mortality rates in both burn victims and immobile hospital patients, many of whom must lie still in one position for days or even weeks. Many such patients develop bedsores—also called pressure ulcers or decubitus ulcers—which can be provide a conduit for bacterial invasion into deeper tissues and lead to lethal systemic infections.

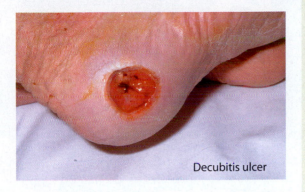

Decubitis ulcer

1. Explain why sun protection is important all year round, but why avoiding sunlight completely can be dangerous.

2. With common minor skin injuries, such as a paper cut or scratch, bacteria rarely invade the wound beyond the superficial skin layers. What features of the skin help protect the body from deep-tissue invasion?

3. To alleviate pain, a nurse might give a patient an injection into the loose connective tissues of the hypodermis. Beginning at the superficial layer of the skin of the buttocks, identify the tissue layers, in order, that a hypodermic needle would penetrate to reach the hypodermis.

4. Briefly describe the factors that contribute to the development of bedsores on the skin of bedridden patients.

Access more review material online in the Study Area at **www.masteringaandp.com.**

There, you'll find:

- **Chapter guides**   • **Tutorials**
- **Chapter quizzes**   • **Animations**
- **Practice tests**   • **Flashcards**
- **Labeling activities**   • **A glossary with**
- **MP3 Tutor Sessions**   **pronunciations**

6

Osseous Tissue and Bone Structure

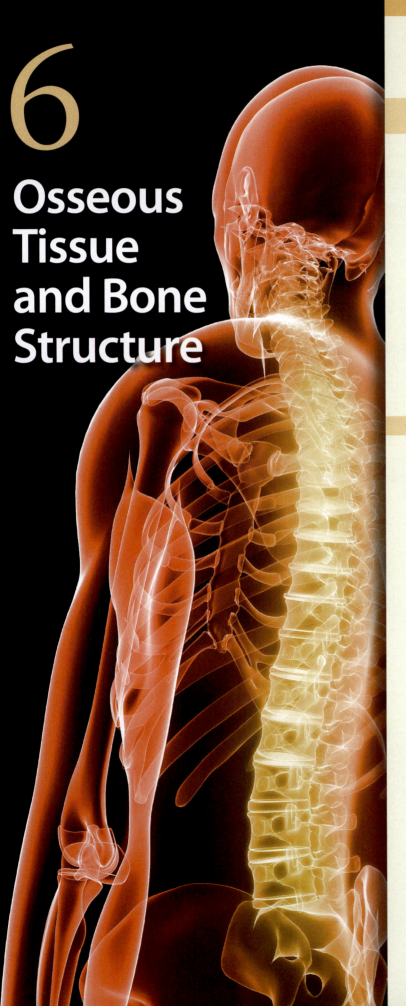

An Introduction to the Bones of the Skeletal System

The skeletal system includes the varied bones of the skeleton and the cartilages, ligaments, and other connective tissues that stabilize or interconnect them. This chapter expands upon the introduction to bone tissue, presented in Chapter 4, by considering the gross anatomy of bones and examining the mechanisms involved in bone growth, remodeling, and repair.

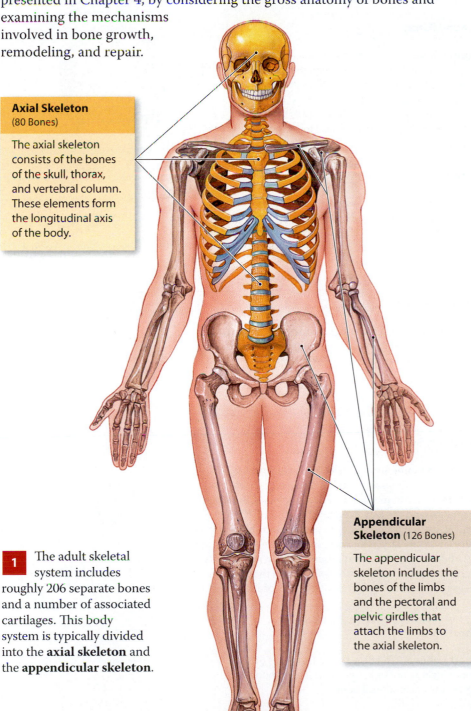

Axial Skeleton
(80 Bones)

The axial skeleton consists of the bones of the skull, thorax, and vertebral column. These elements form the longitudinal axis of the body.

1 The adult skeletal system includes roughly 206 separate bones and a number of associated cartilages. This body system is typically divided into the **axial skeleton** and the **appendicular skeleton**.

Appendicular Skeleton (126 Bones)

The appendicular skeleton includes the bones of the limbs and the pectoral and pelvic girdles that attach the limbs to the axial skeleton.

Functions of the Skeletal System

- **Support:** The skeletal system provides structural support for the entire body. Individual bones or groups of bones provide a framework for the attachment of soft tissues and organs.

- **Storage of Minerals:** The calcium salts of bone represent a valuable mineral reserve that maintains normal concentrations of calcium and phosphate ions in body fluids. Calcium is the most abundant mineral in the human body. A typical human body contains 1–2 kg (2.2–4.4 lb) of calcium, with more than 98 percent of it deposited in the bones of the skeleton.

- **Blood Cell Production:** Red blood cells, white blood cells, and platelets are produced in the red marrow, which fills the internal cavities of many bones.

- **Protection:** Delicate tissues and organs are often surrounded by skeletal elements. The ribs protect the heart and lungs, the skull encloses the brain, the vertebrae shield the spinal cord, and the pelvis cradles delicate digestive and reproductive organs.

- **Leverage:** Many bones of the skeleton function as levers that can change the magnitude and direction of the forces generated by skeletal muscles. The movements produced range from the delicate motions of a fingertip to powerful changes in the position of the entire body.

2 The skeleton has many vital functions that are summarized in the table above. All of these functions ultimately depend on the unique and dynamic properties of bone tissue. The bone specimens that you study in lab or that you are familiar with from skeletons of dead animals are only the dry remains of this living tissue. They bear the same relationship to the bone in a living organism as kiln-dried lumber does to a living oak.

Bones are classified according to shape and structure, and display surface features

Flat Bones

Flat bones have thin, roughly parallel surfaces. Flat bones form the roof of the skull, the sternum, the ribs, and the scapulae. They provide protection for underlying soft tissues and offer an extensive surface area for the attachment of skeletal muscles.

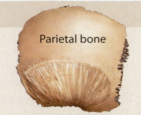

Parietal bone

Sutural Bones

Sutural bones, or Wormian bones, are small, flat, irregularly shaped bones between the flat bones of the skull. There are individual variations in the number, shape, and position of sutural bones. Their borders are like pieces of a jigsaw puzzle, and they range in size from a grain of sand to a quarter.

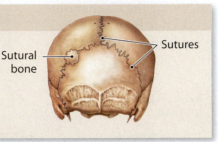

Sutural bone

Sutures

Long Bones

Long bones are relatively long and slender. They are located in the arm and forearm, thigh and leg, palms, soles, fingers, and toes. The femur, the long bone of the thigh, is the largest and heaviest bone in the body.

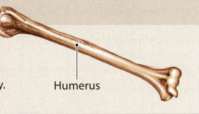

Humerus

Irregular Bones

Irregular bones have complex shapes with short, flat, notched, or ridged surfaces. The spinal vertebrae, the bones of the pelvis, and several skull bones are irregular bones.

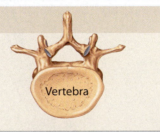

Vertebra

Sesamoid Bones

Sesamoid bones are generally small, flat, and shaped somewhat like a sesame seed. They develop inside tendons and are most commonly located near joints at the knees, the hands, and the feet. Everyone has sesamoid patellae (pa-TEL-ē; singular, *patella*, a small shallow dish), or kneecaps, but individuals vary in the location and abundance of other sesamoid bones. This variation, among others, accounts for disparities in the total number of bones in the skeleton. (Sesamoid bones may form in at least 26 locations.)

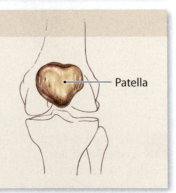

Patella

Short Bones

Short bones are small and boxy. Examples of short bones include bones in the wrists (carpal bones) and in the ankles (tarsal bones).

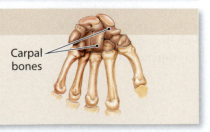

Carpal bones

1 The 206 major bones in a typical adult human skeleton can be divided into six broad categories based on their individual shapes. The skeleton also contains a variable number of minor bones (typically sutural bones and sesamoid bones) whose numbers vary from individual to individual.

2 Illustrated here are the major types of **surface features**, also known as bone markings. Each bone in the body has characteristic external and internal features that are related to its particular functions. Elevations or projections form where tendons and ligaments attach, and where adjacent bones articulate (that is, at joints). Depressions, grooves, and tunnels in bone indicate sites where blood vessels or nerves lie alongside or penetrate the bone.

1 **Surface Features of the Skull**

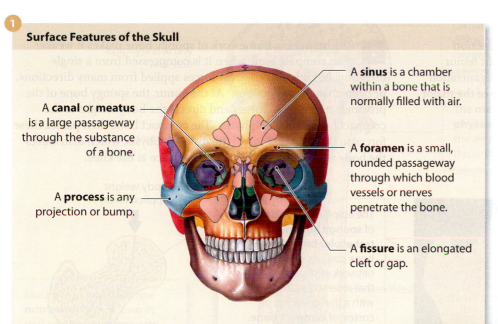

A **canal** or **meatus** is a large passageway through the substance of a bone.

A **process** is any projection or bump.

A **sinus** is a chamber within a bone that is normally filled with air.

A **foramen** is a small, rounded passageway through which blood vessels or nerves penetrate the bone.

A **fissure** is an elongated cleft or gap.

2 **Surface Features of the Humerus**

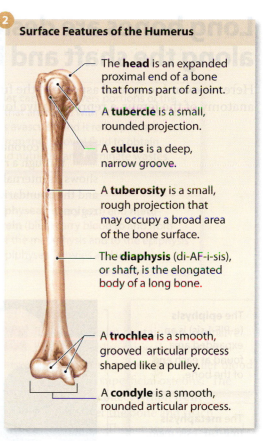

The **head** is an expanded proximal end of a bone that forms part of a joint.

A **tubercle** is a small, rounded projection.

A **sulcus** is a deep, narrow groove.

A **tuberosity** is a small, rough projection that may occupy a broad area of the bone surface.

The **diaphysis** (di-AF-i-sis), or shaft, is the elongated body of a long bone.

A **trochlea** is a smooth, grooved articular process shaped like a pulley.

A **condyle** is a smooth, rounded articular process.

3 **Surface Features of the Femur**

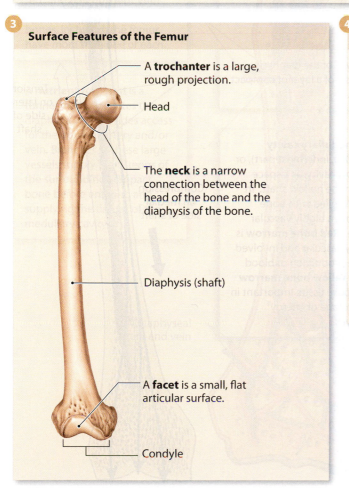

A **trochanter** is a large, rough projection.

Head

The **neck** is a narrow connection between the head of the bone and the diaphysis of the bone.

Diaphysis (shaft)

A **facet** is a small, flat articular surface.

Condyle

4 **Surface Features of the Pelvis**

A **crest** is a prominent ridge.

A **fossa** is a shallow depression or recess in the surface of the bone.

A **line** is a low ridge, more delicate than a crest.

A **spine** is a pointed or narrow process.

A **ramus** is an extension of a bone that makes an angle with the rest of the structure.

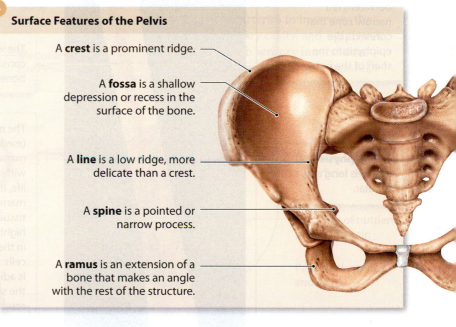

Module 6.1 Review

a. Define surface feature.

b. Identify the six broad categories for classifying a bone according to shape.

c. Compare a tubercle with a tuberosity.

Bone has a calcified matrix associated with osteocytes, osteoblasts, osteoprogenitor cells, and osteoclasts

Both compact bone and spongy bone contain the same four cell types.

1 Mature bone cells called **osteocytes** (OS-tē-ō-sīts; (*osteo-*, bone + *cyte*, cell) maintain the protein and mineral content of the surrounding matrix through the turnover of matrix components. Osteocytes secrete chemicals that dissolve the adjacent matrix, and the released minerals enter the circulation. The osteocytes then rebuild the matrix, stimulating the deposition of mineral crystals. Osteocytes also participate in the repair of damaged bone.

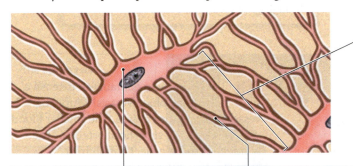

The layers of matrix are called **lamellae** (lah-MEL-lē; singular, *lamella*, a thin plate).

Osteocytes account for most of the cell population in bone. Each osteocyte occupies a **lacuna**, a pocket sandwiched between layers of matrix. Osteocytes cannot divide, and a lacuna never contains more than one osteocyte.

Processes of the osteocytes extend into narrow passageways called **canaliculi** that penetrate the lamellae. The canaliculi interconnect the lacunae and reach vascular passageways, providing a route for nutrient diffusion.

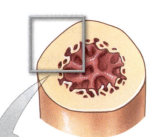

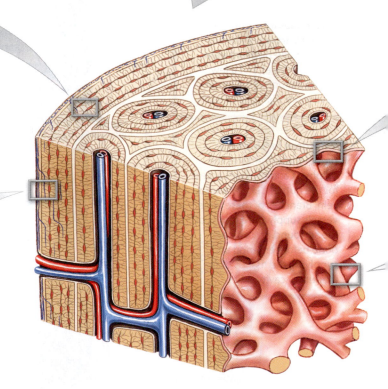

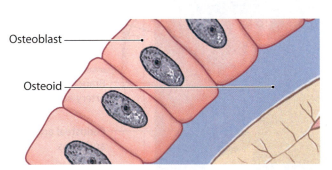

Osteoblast

Osteoid

2 **Osteoblasts** (OS-tē-ō-blasts; *blast*, precursor) produce new bone matrix in a process called **osteogenesis** (os-tē-ō-JEN-e-sis; *gennan*, to produce), or **ossification**. Osteoblasts make and release the proteins and other organic components of the matrix. Before calcium salts are deposited, this organic matrix is called **osteoid** (OS-tē-oyd). Osteoblasts also assist in elevating local concentrations of calcium phosphate to the point where calcium salts are deposited in the organic matrix. This process converts osteoid to bone. Osteocytes develop from osteoblasts that have become completely surrounded by bone matrix.

3 **Osteoprogenitor** (os-tē-ō-prō-JEN-i-tor, *progenitor*, ancestor) **cells** are mesenchymal cells present in small numbers in the inner, cellular layer of the periosteum; in an inner layer, or endosteum, that lines medullary cavities; and in the lining of passageways, containing blood vessels, that penetrate the matrix of compact bone. These stem cells divide to produce daughter cells that differentiate into osteoblasts, and they are thus important in the repair of a fracture (a break or a crack in a bone).

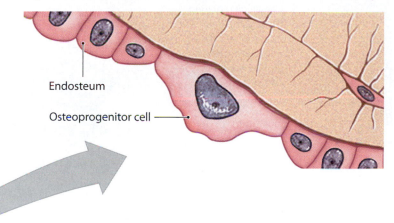

Endosteum

Osteoprogenitor cell

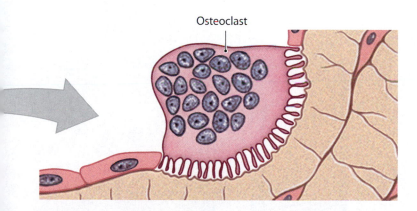

Osteoclast

4 **Osteoclasts** (OS-tē-ō-clasts; *clast*, to break) are cells that remove and recycle bone matrix. These are giant cells with 50 or more nuclei. Osteoclasts are not related to osteoprogenitor cells or their descendants. Instead, they are derived from the same stem cells that produce monocytes and macrophages. Acids and proteolytic (protein-digesting) enzymes secreted by osteoclasts dissolve the matrix and release the stored minerals. This process, called **osteolysis** (os-tē-OL-i-sis; *lysis*, a loosening) or resorption, is important in the regulation of calcium and phosphate ion concentrations in body fluids.

5 A bone without a calcified matrix looks normal, but is very flexible. Roughly one-third of the weight of bone is contributed by collagen fibers, and cells account for only 2 percent of the weight of a typical bone. Calcium phosphate, $Ca_3(PO_4)_2$, accounts for almost

two-thirds of the weight of bone. Calcium phosphate interacts with calcium hydroxide, $Ca(OH)_2$, to form crystals of **hydroxyapatite**, $Ca_{10}(PO_4)_6(OH)_2$. As they form, these crystals incorporate other calcium salts, such as calcium carbonate ($CaCO_3$), and ions such as sodium, magnesium, and fluoride. Hydroxyapatite is very hard but inflexible and brittle. Collagen fibers are strong and flexible, but if they are compressed they bend. The protein–crystal combination in bone is strong, somewhat flexible, and highly resistant to shattering. In its overall properties, bone is on a par with the best steel-reinforced concrete. In fact, bone is far superior to concrete, because it can undergo remodeling (cycles of bone formation and resorption) as needed and can repair itself after injury.

Module 6.3 Review

a. Define osteocyte, osteoblast, osteoprogenitor cell, and osteoclast.

b. How would the compressive strength of a bone be affected if the ratio of collagen to hydroxyapatite increased?

c. If osteoclast activity exceeds osteoblast activity in a bone, what would be the effect on the bone?

Compact bone consists of parallel osteons, and spongy bone consists of a network of trabeculae

The basic functional unit of mature compact bone is the **osteon** (OS-tē-on), or Haversian system. Many of the important features of osteons were introduced in Module 4.10, which you may wish to revisit before proceeding.

1 The lamellae of each osteon form a series of nested cylinders around the central canal. In transverse section, these **concentric lamellae** resemble a target, with the **central canal** as the bull's-eye. You might think of a single osteon as a drinking straw with very thick walls: When you attempt to push the ends of the straw together or to pull them apart, the straw is quite strong. But if you hold the ends and push from the side, the straw will bend sharply with relative ease. The osteons in the diaphysis of a long bone are parallel to the long axis of the shaft. Thus, the shaft does not bend, even when extreme forces are applied to either end. The femur can withstand 10–15 times the body's weight without breaking. Yet a much smaller force applied to the side of the shaft can break any long bone, even the femur.

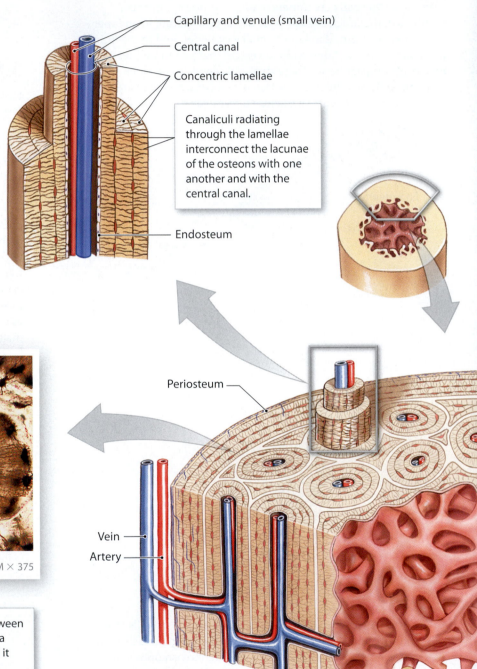

Capillary and venule (small vein)

Central canal

Concentric lamellae

Canaliculi radiating through the lamellae interconnect the lacunae of the osteons with one another and with the central canal.

Endosteum

Periosteum

Vein

Artery

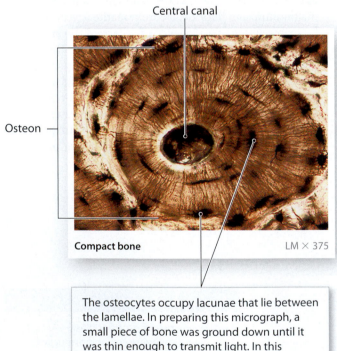

Central canal

Osteon

Compact bone LM × 375

The osteocytes occupy lacunae that lie between the lamellae. In preparing this micrograph, a small piece of bone was ground down until it was thin enough to transmit light. In this process, the lacunae and canaliculi are filled with bone dust, and thus appear black.

2 A section cut from the shaft of a long bone shows the organization of osteons and blood vessels, and allows a comparison between compact and spongy bone.

3 Spongy bone is located where bones are not heavily stressed or where stresses arrive from many directions. Below is the head of the femur, which transmits body weight to the shaft of that bone. The trabeculae are oriented along stress lines and are cross-braced extensively. In addition to being able to withstand stresses applied from many directions, spongy bone is much lighter than compact bone. Spongy bone thus reduces the weight of the skeleton, making it easier for muscles to move the bones.

4 In spongy bone, lamellae are not arranged in osteons. The matrix in spongy bone forms struts and plates called **trabeculae**. The thin trabeculae branch, creating an open network. There are no capillaries or venules in the matrix of spongy bone. Nutrients reach the osteocytes by diffusion along canaliculi that open onto the surfaces of trabeculae.

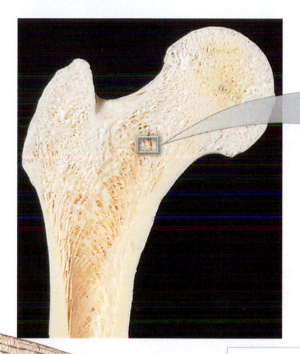

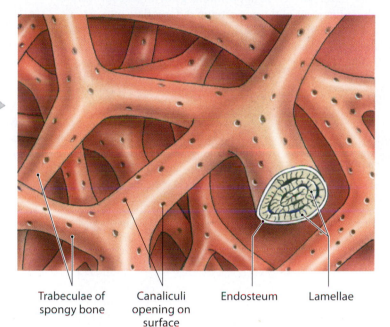

Trabeculae of spongy bone   Canaliculi opening on surface   Endosteum   Lamellae

Circumferential lamellae (*circum-*, around + *ferre*, to bear) are found at the outer and inner surfaces of the bone, where they are covered by the periosteum and endosteum, respectively. These lamellae are produced during the growth and maintenance of the bone.

Interstitial lamellae fill in the spaces between the osteons in compact bone. These lamellae are remnants of osteons whose matrix components have been almost completely recycled by osteoclasts.

Central canal

Perforating canal

Central canals generally run parallel to the surface of the bone. Other passageways, known as **perforating canals**, extend roughly perpendicular to the surface. Blood vessels in these canals supply blood to osteons deeper in the bone and to tissues of the medullary cavity.

Module 6.4 Review

a. Define osteon.

b. Compare the structures and functions of compact bone and spongy bone.

c. A sample of bone has lamellae that are not arranged in osteons. Is the sample more likely from the epiphysis or from the diaphysis?

In appositional bone growth, layers of compact bone are added to the bone's outer surface

The appositional growth of cartilage was detailed in Module 4.9. As you will see in this module, appositional bone growth has many similarities to appositional cartilage growth.

1 The diameter of a bone enlarges through **appositional growth** at the outer surface. In appositional growth, osteoprogenitor cells in the inner layer of the periosteum differentiate into osteoblasts and add bone matrix to the surface. This adds successive layers of circumferential lamellae to the outer surface of the bone. Osteoblasts trapped between these lamellae differentiate into osteocytes. Over time, the deeper lamellae are recycled and replaced with the osteons typical of compact bone.

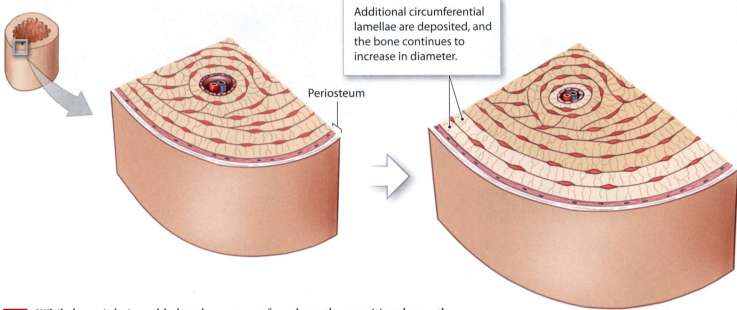

Additional circumferential lamellae are deposited, and the bone continues to increase in diameter.

Periosteum

2 While bone is being added to the outer surface through appositional growth, osteoclasts are removing and recycling lamellae at the inner surface. As a result, the medullary cavity gradually enlarges as the bone increases in diameter.

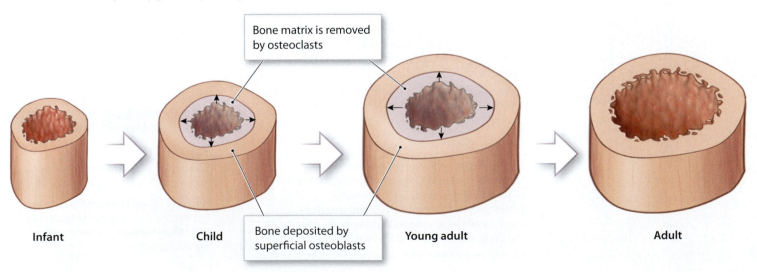

Bone matrix is removed by osteoclasts

Bone deposited by superficial osteoblasts

Infant   Child   Young adult   Adult

3 Except within joint cavities, the superficial layer of compact bone that covers all bones is wrapped by a **periosteum**, which has a fibrous outer layer and a cellular inner layer. The periosteum (1) isolates the bone from surrounding tissues, (2) provides a route for the blood and nervous supply, and (3) actively participates in bone growth and repair. As the bone grows, the collagen fibers from tendons, ligaments, and joint capsules are cemented into the circumferential lamellae by osteoblasts from the cellular layer of the periosteum. These are called **perforating fibers**, and this method of attachment is extremely strong. An excessive pull on a tendon or ligament will usually break a bone rather than snap the collagen fibers at the bone surface.

4 The **endosteum** is an incomplete cellular layer that lines the medullary cavity. This layer, which is active during bone growth, repair, and remodeling, covers the trabeculae of spongy bone and lines the inner surfaces of the central canals. It consists of a simple flattened layer of osteoprogenitor cells that covers the bone matrix, generally without any intervening connective tissue fibers. Where the cellular layer is incomplete, the matrix is exposed. At these exposed sites, osteoclasts and osteoblasts can remove or deposit matrix components. The osteoclasts generally occur in shallow depressions (Howship lacunae) that they have eroded into the matrix. As a result of their activities, the inner circumferential lamellae are often incomplete or interrupted.

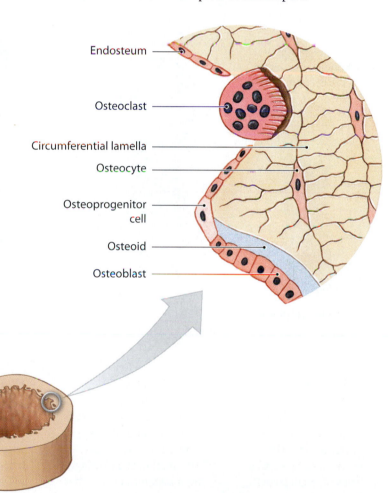

Circumferential lamellae

Fibrous layer of periosteum

Cellular layer of periosteum

Canaliculi

Osteocyte in lacuna

Perforating fibers

Endosteum

Osteoclast

Circumferential lamella

Osteocyte

Osteoprogenitor cell

Osteoid

Osteoblast

Appositional bone growth is important in increasing the diameters of existing bones, but it does not form the original bones. Bone formation begins roughly six weeks after fertilization, when the embryo is 12 mm (0.5 in.) long. Two major processes are involved: endochondral ossification (Module 6.6) and intramembranous ossification (Module 6.7).

Module 6.5 Review

a. Define appositional growth.

b. Distinguish between the periosteum and the endosteum.

c. As a bone increases in diameter, what happens to the medullary cavity?

Endochondral ossification is the replacement of a cartilaginous model with bone

When bone formation begins in the embryo, all existing skeletal elements are cartilaginous. These cartilages are gradually replaced by bone through the process of **endochondral** (en-dō- KON-drul, *endo-*, inside + *chondros*, cartilage) **ossification**. This sequence illustrates the key steps in endochondral ossification. This process begins with a small cartilage that is basically a miniature model of the corresponding bone of the adult skeleton. As it forms, the bone grows in length and in diameter. The increase in diameter involves appositional bone deposition.

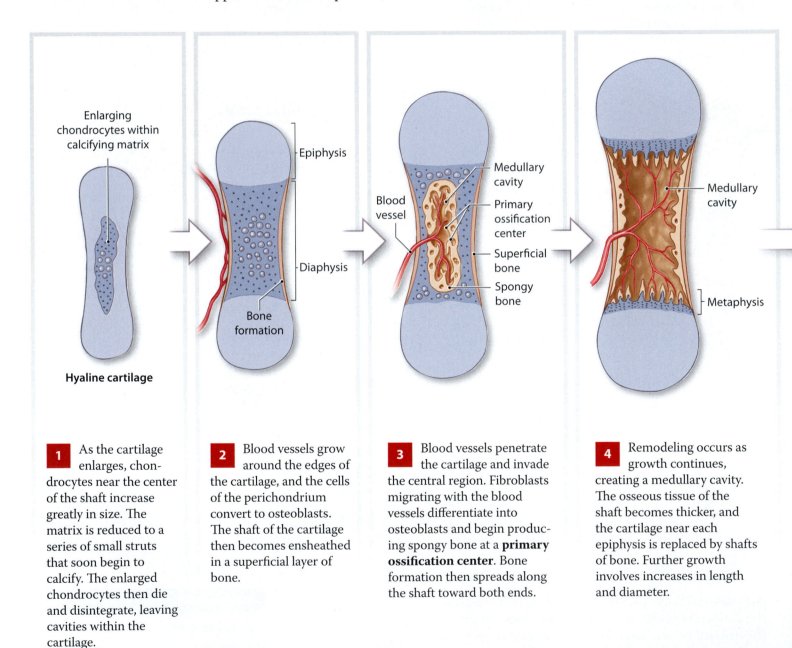

Enlarging chondrocytes within calcifying matrix

Hyaline cartilage

Epiphysis

Diaphysis

Bone formation

Blood vessel

Medullary cavity

Primary ossification center

Superficial bone

Spongy bone

Medullary cavity

Metaphysis

1 As the cartilage enlarges, chondrocytes near the center of the shaft increase greatly in size. The matrix is reduced to a series of small struts that soon begin to calcify. The enlarged chondrocytes then die and disintegrate, leaving cavities within the cartilage.

2 Blood vessels grow around the edges of the cartilage, and the cells of the perichondrium convert to osteoblasts. The shaft of the cartilage then becomes ensheathed in a superficial layer of bone.

3 Blood vessels penetrate the cartilage and invade the central region. Fibroblasts migrating with the blood vessels differentiate into osteoblasts and begin producing spongy bone at a **primary ossification center**. Bone formation then spreads along the shaft toward both ends.

4 Remodeling occurs as growth continues, creating a medullary cavity. The osseous tissue of the shaft becomes thicker, and the cartilage near each epiphysis is replaced by shafts of bone. Further growth involves increases in length and diameter.

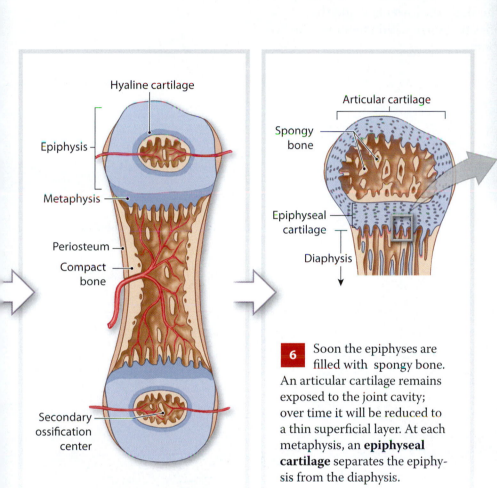

Epiphyseal cartilage matrix

Cartilage cells undergoing division and secreting additional cartilage matrix

LM × 250

Medullary cavity Osteoblasts Osteoid

Hyaline cartilage

Epiphysis

Metaphysis

Periosteum

Compact bone

Secondary ossification center

Articular cartilage

Spongy bone

Epiphyseal cartilage

Diaphysis

5 Capillaries and osteoblasts migrate into the epiphyses, creating **secondary ossification centers**.

6 Soon the epiphyses are filled with spongy bone. An articular cartilage remains exposed to the joint cavity; over time it will be reduced to a thin superficial layer. At each metaphysis, an **epiphyseal cartilage** separates the epiphysis from the diaphysis.

7 The light micrograph above shows the ossifying surface of an epiphyseal cartilage. The pink material is osteoid, deposited by osteoblasts in the medullary cavity. On the shaft side of the epiphyseal cartilage, osteoblasts are continuously invading the cartilage and replacing it with bone. On the epiphyseal side, new cartilage is continuously being added. The osteoblasts are therefore moving toward the epiphysis, which is being pushed away by the expansion of the epiphyseal cartilage. The osteoblasts won't catch up to the epiphysis, as long as both the osteoblasts and the epiphysis "run away" from the primary ossification center at the same rate. Meanwhile, the bone grows longer and longer.

At puberty, the combination of rising levels of sex hormones, growth hormone, and thyroid hormones stimulates bone growth dramatically. Osteoblasts now begin producing bone faster than chondrocytes are producing new epiphyseal cartilage. As a result, the osteoblasts "catch up," and the epiphyseal cartilage gets narrower and narrower until it ultimately disappears. The completion of epiphyseal growth is called epiphyseal closure. In adults, the former location of this cartilage is often detectable in x-rays as a distinct **epiphyseal line**, which remains after epiphyseal growth has ended.

Module 6.6 Review

a. Define endochondral ossification.

b. In endochondral ossification, what is the original source of osteoblasts?

c. How could x-rays of the femur be used to determine whether a person has reached full height?

Intramembranous ossification is the formation of bone without a prior cartilaginous model

Intramembranous (in-tra-MEM-bra-nus) **ossification** begins when mesenchymal cells differentiate into osteoblasts within embryonic or fibrous connective tissue. This type of ossification normally occurs in the deeper layers of the dermis, and the bones that result are often called **dermal bones** or membrane bones. Examples of dermal bones include the roofing bones of the skull, the lower jaw, and the collarbone. Sesamoid bones are membrane bones that form within tendons; the patella (kneecap) is an example of a sesamoid bone.

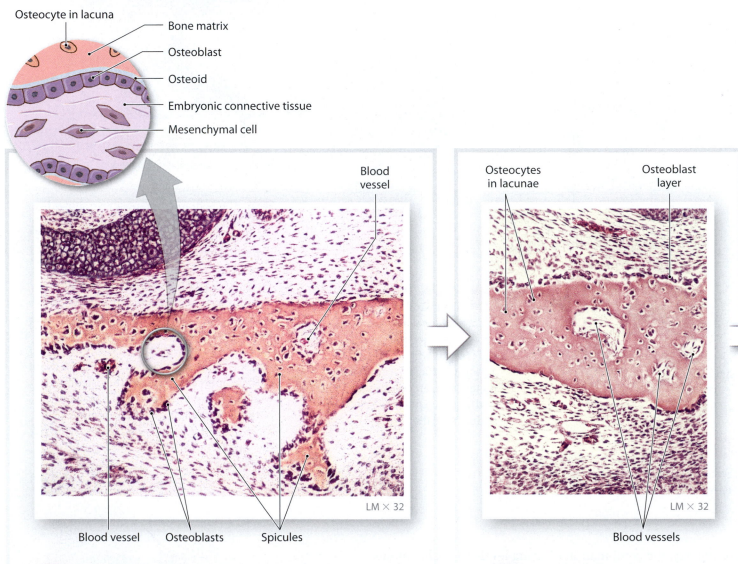

Osteocyte in lacuna
Bone matrix
Osteoblast
Osteoid
Embryonic connective tissue
Mesenchymal cell

Blood vessel

Osteocytes in lacunae
Osteoblast layer

LM × 32

Blood vessel   Osteoblasts   Spicules

Blood vessels

LM × 32

1 Mesenchymal cells first cluster together and start to secrete the organic components of the matrix. The resulting osteoid then becomes mineralized through the crystallization of calcium salts, and the mesenchymal cells differentiate into osteoblasts. The location in a tissue where ossification begins is called an **ossification center**. The developing bone grows outward from the ossification center in small struts called **spicules**. As ossification proceeds, it traps some osteoblasts inside bony pockets; these cells differentiate into osteocytes. Meanwhile, mesenchymal cell divisions continue to produce additional osteoblasts.

2 Bone growth is an active process, and osteoblasts require oxygen and a reliable supply of nutrients. Blood vessels begin to grow into the area. As spicules meet and fuse together, some of these blood vessels become trapped within the developing bone.

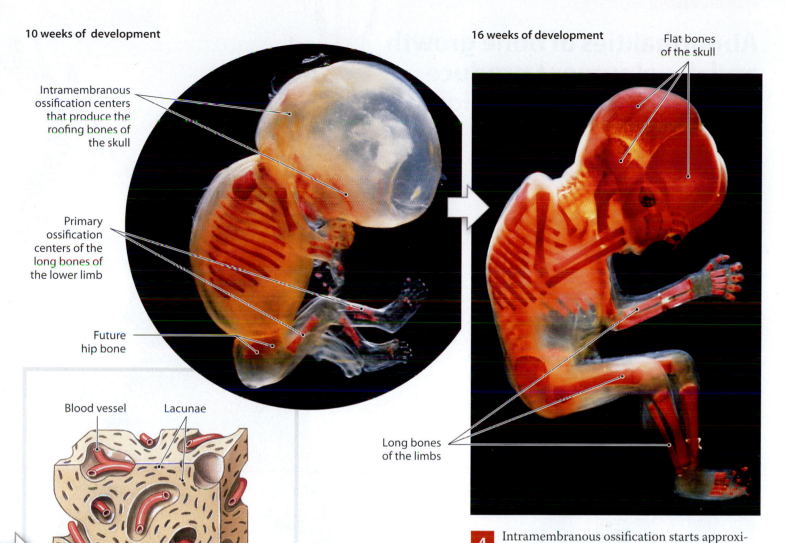

10 weeks of development

Intramembranous ossification centers that produce the roofing bones of the skull

Primary ossification centers of the long bones of the lower limb

Future hip bone

Blood vessel   Lacunae

16 weeks of development

Flat bones of the skull

Long bones of the limbs

4 Intramembranous ossification starts approximately during the eighth week of embryonic development. These photos show the extent of intramembranous and endochondral ossification that occurs between 10 and 16 weeks of development. At 10 weeks, bone formation is under way but the skeleton is incomplete. At 16 weeks, most of the bones of the adult skeleton can be identified.

3 Initially, the intramembranous bone consists of spongy bone only. Subsequent remodeling around trapped blood vessels can produce osteons typical of compact bone. As the rate of growth slows, the connective tissue around the bone becomes organized into the fibrous layer of the periosteum. The osteoblasts closest to the bone surface become less active but remain as the inner, cellular layer of the periosteum.

Module 6.7 Review

a. Define intramembranous ossification.

b. During intramembranous ossification, which type(s) of tissue is (are) replaced by bone?

c. Explain the primary difference between endochondral ossification and intramembranous ossification.

Abnormalities of bone growth and development produce recognizable physical signs

A variety of endocrine or metabolic problems can result in atypical skeletal growth. Here we consider several conditions that affect the skeleton as a whole.

1 In **pituitary growth failure**, inadequate production of growth hormone leads to reduced epiphyseal cartilage activity and abnormally short bones. This condition is becoming increasingly rare in the United States, because children can be treated with synthetic human growth hormone.

2 **Achondroplasia** (a-kon-drō-PLĀ-sē-uh) results from abnormal epiphyseal activity. In this case the epiphyseal cartilages of the long bones grow unusually slowly and are replaced by bone early in life. As a result, the individual develops short, stocky limbs. Although other skeletal abnormalities occur, the trunk is normal in size, and sexual and mental development remain unaffected.

3 Several inherited metabolic conditions that affect many systems influence the growth and development of the skeletal system. These conditions produce characteristic variations in body proportions. For example, many individuals with **Marfan syndrome** are very tall and have long, slender limbs, due to excessive cartilage formation at the epiphyseal cartilages. Although this is an obvious physical distinction, the characteristic body proportions are not in themselves dangerous. However, the underlying mutation, which affects the structure of connective tissue throughout the body, commonly causes life-threatening cardiovascular problems.

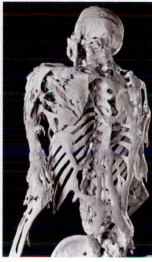

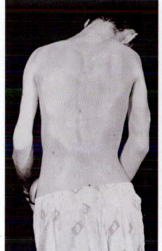

5 Under abnormal conditions, mesenchymal stem cells in any connective tissue can develop into osteoblasts that begin producing bone. The individual on the left has **fibrodysplasia ossificans progressiva** (FOP), a rare single gene mutation disorder that involves the deposition of bone around skeletal muscles. The skeleton on the right shows the extent of abnormal ossification that can occur. Bones that develop in unusual places are called **heterotopic** (*hetero*, place), or **ectopic** (*ektos*, outside), **bones**. There is no effective treatment for this painful and debilitating condition, and patients seldom survive into their 40s.

6 If growth hormone levels rise abnormally after epiphyseal cartilages close, the skeleton does not grow longer, but bones get thicker, especially those in the face, jaw, and hands. Cartilage growth and alterations in soft-tissue structure lead to changes in physical features, such as the contours of the face. These physical changes occur in the disorder called **acromegaly**. Early diagnosis and treatment through pituitary surgery and/or drugs that reduce growth hormone levels may prevent irreversible physical changes and extend life expectancy in both acromegaly and gigantism.

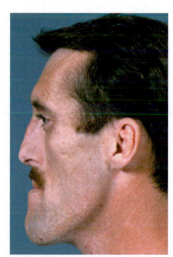

4 **Gigantism** results from an overproduction of growth hormone before puberty. Individuals can reach heights of over 2.7 m (8 ft 11 in.) and weights of over 200 kg (440 lb). Puberty is often delayed, and the facial features in adults resemble those of acromegaly (**6**). The most common cause is a pituitary tumor, which may be treated by surgery, radiation, or drugs that suppress GH release.

Module 6.8 Review

a. Describe Marfan syndrome.

b. Compare gigantism with acromegaly.

c. Why is pituitary dwarfism less common today in the United States?

1. Vocabulary

In the space provided, write the term for each of the following definitions.

a _____ Bones with complex shapes

b _____ The expanded ends of a long bone

c _____ A shallow depression in the surface of a bone

d _____ The marrow-filled space within a bone

e _____ The strut- and plate-shaped matrix of spongy bone

f _____ Cells that remove and recycle bone matrix

g _____ Bones that develop in tendons

h _____ The process that forms new bone matrix

i _____ The basic functional unit of compact bone

j _____ Type of bone growth that increases bone diameter

k _____ Process by which cartilage is replaced by bone

2. Concept map

Use each of the following terms once to fill in the blank boxes to correctly complete the bone formation concept map.

- lacunae
- osteocytes
- collagen
- intramembranous ossification
- compact bone
- periosteum
- hyaline cartilage

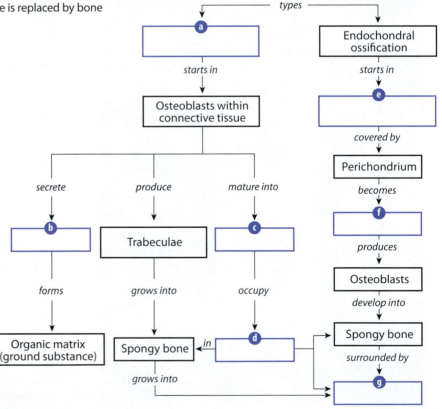

3. Section integration

While playing on her swing set, 10-year-old Rebecca falls and breaks her right leg. At the emergency room, the doctor tells her parents that the proximal end of the tibia where the epiphysis meets the diaphysis is fractured. The fracture is properly set and eventually heals. During a routine physical when she is 18, Rebecca learns that her right leg is 2 cm shorter than her left. What might account for this difference? _____

_____

_____

_____

The Physiology of Bones

In this section we will consider the dynamic relationship between calcium concentrations in body fluids and calcium reserves in the skeletal system. We will also examine the role of hormones in regulating calcium balance in the body.

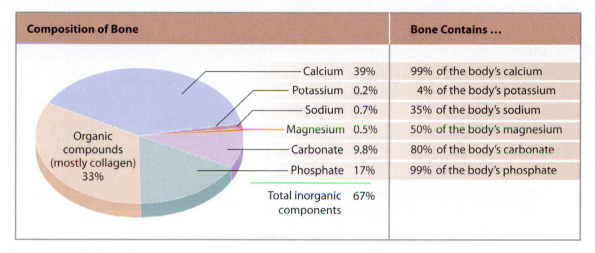

Composition of Bone		Bone Contains ...
Calcium	39%	99% of the body's calcium
Potassium	0.2%	4% of the body's potassium
Sodium	0.7%	35% of the body's sodium
Magnesium	0.5%	50% of the body's magnesium
Carbonate	9.8%	80% of the body's carbonate
Phosphate	17%	99% of the body's phosphate
Total inorganic components	67%	

Organic compounds (mostly collagen) 33%

1 This chemical analysis reveals the importance of bones as mineral reservoirs. In this section we will focus on the homeostatic regulation of calcium ion concentrations in body fluids; we will consider other minerals in later chapters.

At the intestines, calcium and phosphate ions are absorbed from the diet. The rate of absorption is hormonally regulated.

Normal Ca^{2+} levels in plasma

At the kidneys, the levels of calcium and phosphate ions lost in the urine is hormonally regulated.

Bone

Within the skeleton, osteoblasts are continuously depositing new bone matrix. At the same time, osteoclasts are eroding existing matrix and releasing calcium and phosphate ions that enter the circulation. The balance between osteoblast and osteoclast activity is hormonally regulated.

2 **Calcium** is the most abundant mineral in the human body. A typical human body contains 1–2 kg (2.2–4.4 lb) of calcium, with roughly 99 percent of it deposited in the skeleton. Calcium ions play a role in a variety of physiological processes, and even small variations from the normal concentration affect cellular operations. The homeostatic regulation of calcium ion levels is a juggling act that balances activities under way in the intestines, in the bones of the skeleton, and in the kidneys.

If the calcium concentration of body fluids increases or decreases by more than 30–35 percent, neuron and muscle cell function is disrupted, with potentially lethal results. Calcium ion concentration is so closely regulated, however, that daily fluctuations of more than 10 percent are highly unusual.

The primary hormones regulating calcium ion metabolism are parathyroid hormone and calcitonin; calcitriol is also involved

Calcium ion homeostasis is maintained by hormones that target the skeletal system, the digestive tract, and the kidneys.

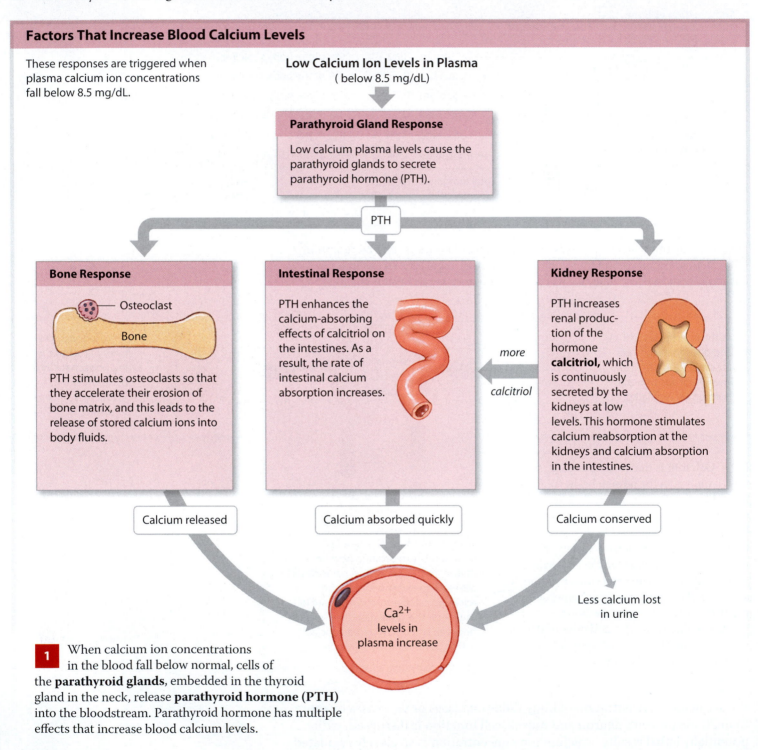

Factors That Increase Blood Calcium Levels

These responses are triggered when plasma calcium ion concentrations fall below 8.5 mg/dL.

Low Calcium Ion Levels in Plasma
(below 8.5 mg/dL)

Parathyroid Gland Response

Low calcium plasma levels cause the parathyroid glands to secrete parathyroid hormone (PTH).

PTH

Bone Response

Osteoclast

Bone

PTH stimulates osteoclasts so that they accelerate their erosion of bone matrix, and this leads to the release of stored calcium ions into body fluids.

Intestinal Response

PTH enhances the calcium-absorbing effects of calcitriol on the intestines. As a result, the rate of intestinal calcium absorption increases.

more

calcitriol

Kidney Response

PTH increases renal production of the hormone **calcitriol,** which is continuously secreted by the kidneys at low levels. This hormone stimulates calcium reabsorption at the kidneys and calcium absorption in the intestines.

Calcium released

Calcium absorbed quickly

Calcium conserved

Less calcium lost in urine

Ca^{2+} levels in plasma increase

1 When calcium ion concentrations in the blood fall below normal, cells of the **parathyroid glands**, embedded in the thyroid gland in the neck, release **parathyroid hormone (PTH)** into the bloodstream. Parathyroid hormone has multiple effects that increase blood calcium levels.

Factors That Decrease Blood Calcium Levels

These responses are triggered when plasma calcium ion concentrations rise above 11 mg/dL.

High Calcium Ion Levels in Plasma
(above 11 mg/dL)

Thyroid Gland Response

C cells in the thyroid gland secrete calcitonin.

Calcitonin

Bone Response

Bone

Calcitonin inhibits osteoclasts but does not affect osteoblasts, which continue to deposit calcium ions within the matrix of bones.

Intestinal Response

Decreasing PTH or calcitriol levels result in a decrease in the rate of calcium ion absorption from the intestinal contents.

Kidney Response

less

calcitriol

Increased calcitonin levels have an inhibitory effect on the kidneys and suppresses calcium ion reabsorption.

Calcium stored

Calcium absorbed slowly

Calcium excreted

Ca^{2+} levels in plasma decrease

More calcium lost in urine

2 If the calcium ion concentrations of the blood rise above normal, **C cells** of the **thyroid gland** secrete **calcitonin**. This hormone decreases calcium ion concentrations in body fluids through two different mechanisms. Under these conditions, less calcium enters body fluids because osteoclasts leave the mineral matrix alone. More calcium is removed from body fluids because osteoblasts continue to produce new bone matrix while calcium ion excretion at the kidneys accelerates. The net result is a decline in the calcium ion concentration of body fluids, restoring homeostasis.

By providing a calcium reserve, the skeleton plays the primary role in the homeostatic maintenance of normal calcium ion concentrations of body fluids. This function can have a direct effect on the shape and strength of the bones in the skeleton. When large numbers of calcium ions are mobilized in body fluids, the bones become weaker; when calcium salts are deposited, the bones become denser and stronger.

Module 6.9 Review

a. Identify the hormones involved in stimulating and inhibiting the release of calcium ions from bone matrix.

b. What effect would increased PTH secretion have on blood calcium levels?

c. How does calcitonin lower the calcium ion concentration of blood?

A fracture is a crack or a break in a bone

Despite its mineral strength, bone can crack or even break if subjected to extreme loads, sudden impacts, or stresses from unusual directions. The damage produced constitutes a **fracture**. Most fractures heal even after severe damage, provided that the blood supply and the cellular components of the endosteum and periosteum survive.

Repair of a Fracture

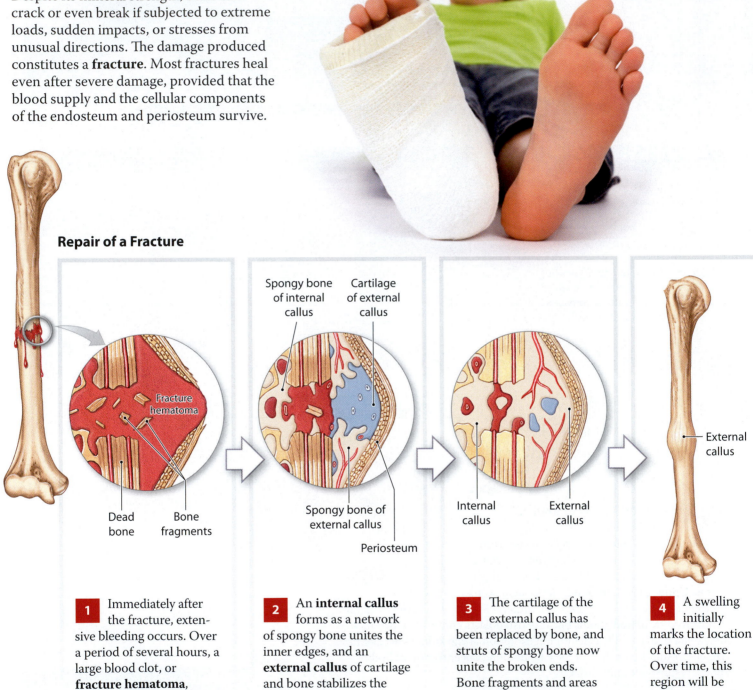

Spongy bone of internal callus

Cartilage of external callus

Fracture hematoma

Dead bone

Bone fragments

Spongy bone of external callus

Periosteum

Internal callus

External callus

External callus

1 Immediately after the fracture, extensive bleeding occurs. Over a period of several hours, a large blood clot, or **fracture hematoma**, develops.

2 An **internal callus** forms as a network of spongy bone unites the inner edges, and an **external callus** of cartilage and bone stabilizes the outer edges.

3 The cartilage of the external callus has been replaced by bone, and struts of spongy bone now unite the broken ends. Bone fragments and areas of dead bone closest to the break have been removed and replaced.

4 A swelling initially marks the location of the fracture. Over time, this region will be remodeled, and little evidence of the fracture will remain.

Types of Fractures

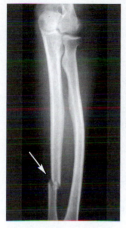

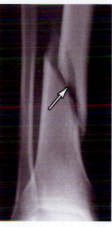

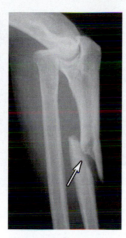

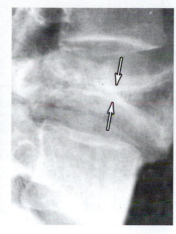

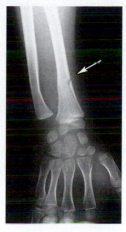

Transverse fractures, such as this fracture of the ulna, break a bone shaft across its long axis.

Spiral fractures, such as this fracture of the tibia, are produced by twisting stresses that spread along the length of the bone.

Displaced fractures produce new and abnormal bone alignments; **nondisplaced fractures** retain the normal alignment of the bones or fragments.

Compression fractures occur in vertebrae subjected to extreme stresses, such as those produced by the forces that arise when you land on your seat in a fall.

In a **greenstick fracture**, such as this fracture of the radius, only one side of the shaft is broken, and the other is bent. This type of fracture generally occurs in children, whose long bones have yet to ossify fully.

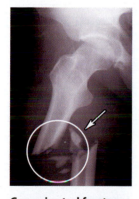

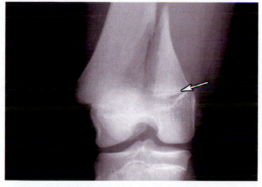

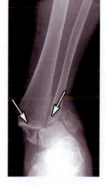

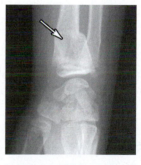

Comminuted fractures, such as this fracture of the femur, shatter the affected area into a multitude of bony fragments.

Epiphyseal fractures, such as this fracture of the femur, tend to occur where the bone matrix is undergoing calcification and chondrocytes are dying. A clean transverse fracture along this line generally heals well. Unless carefully treated, fractures between the epiphysis and the epiphyseal cartilage can permanently stop growth at this site.

A **Pott fracture** occurs at the ankle and affects both bones of the leg.

A **Colles fracture**, a break in the distal portion of the radius, is typically the result of reaching out to cushion a fall.

5 Fractures are named using various criteria, including their external appearance, their location, and the nature of the crack or break in the bone. Important types of fractures are illustrated here by representative x-rays. The broadest general categories are closed fractures and open fractures. **Closed**, or **simple**, **fractures** are completely internal. They can be seen only on x-rays, because they do not involve a break in the skin. **Open**, or **compound**, **fractures** project through the skin. These fractures, which are obvious on inspection, are more dangerous than closed fractures, due to the possibility of infection or uncontrolled bleeding. Many fractures fall into more than one category, because the terms overlap.

Module 6.10 Review

a. Define open fracture and closed fracture.

b. List the steps involved in fracture repair, beginning just after the fracture occurs.

c. When during fracture repair does an external callus form?

1. Concept map

Use each of the following terms once to fill in the blank boxes to correctly complete the regulation of calcium ion concentration concept map.

- ↓ Ca^{2+} level
- homeostasis
- release of stored Ca^{2+} from bone
- ↓ Ca^{2+} concentration in body fluids
- parathyroid glands
- calcitonin
- ↑ Ca^{2+} concentration in body fluids

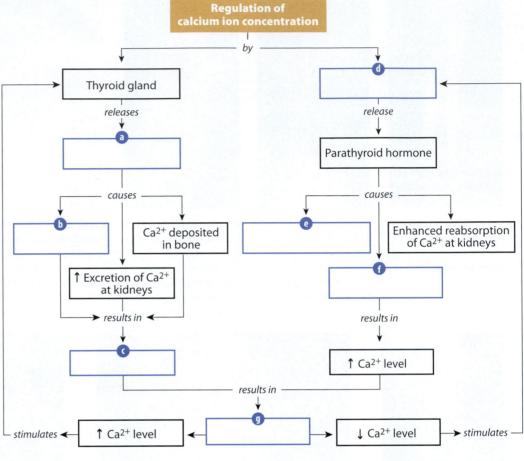

Regulation of calcium ion concentration

by

Thyroid gland

releases

a

causes

b

Ca^{2+} deposited in bone

↑ Excretion of Ca^{2+} at kidneys

results in

c

d

release

Parathyroid hormone

causes

e

Enhanced reabsorption of Ca^{2+} at kidneys

f

results in

↑ Ca^{2+} level

results in

g

stimulates ← ↑ Ca^{2+} level ← → ↓ Ca^{2+} level → *stimulates*

2. Short answer

Identify the type of fracture and the bones involved in each of the following x-ray images. Circle the fracture that typically results from cushioning a fall.

a

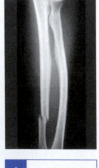

b

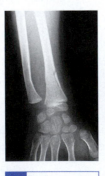

c

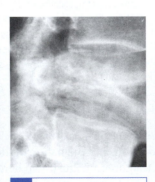

d

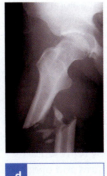

e

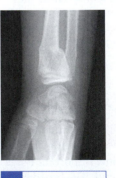

f

Visual Outline with Key Terms

Summarize the content of each module using the terms in the order provided.

SECTION 1

An Introduction to the Bones of the Skeletal System

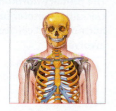

- axial skeleton
- appendicular skeleton

6.1

Bones are classified according to shape and structure, and display surface features

- flat bones
- sutural bones
- long bones
- irregular bones
- sesamoid bones
- short bones
- surface features
- canal (meatus)
- process
- sinus
- foramen
- fissure
- head
- tubercle
- sulcus
- tuberosity
- diaphysis
- trochlea
- condyle
- trochanter
- neck
- facet
- crest
- fossa
- line
- spine
- ramus

6.2

Long bones are designed to transmit forces along the shaft and have a rich blood supply

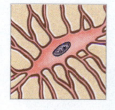

- epiphysis
- metaphysis
- diaphysis
- spongy (trabecular) bone
- compact bone
- medullary cavity
- red bone marrow
- yellow bone marrow
- nutrient artery
- nutrient vein
- nutrient foramen
- articular cartilage
- metaphyseal artery
- metaphyseal vein

6.3

Bone has a calcified matrix associated with osteocytes, osteoblasts, osteoprogenitor cells, and osteoclasts

- osteocytes
- lacuna
- canaliculi
- lamellae
- osteoblasts
- osteogenesis
- ossification
- osteoid
- osteoprogenitor cells
- osteoclasts
- osteolysis
- hydroxyapatite

6.4

Compact bone consists of parallel osteons, and spongy bone consists of a network of trabeculae

- osteon
- concentric lamellae
- central canal
- circumferential lamellae
- interstitial lamellae
- perforating canals
- trabeculae

6.5

In appositional bone growth, layers of compact bone are added to the bone's outer surface

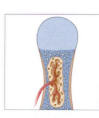

- appositional growth
- periosteum
- perforating fibers
- endosteum

6.6

Endochondral ossification is the replacement of a cartilaginous model with bone

- endochondral ossification
- primary ossification center
- secondary ossification center
- epiphyseal cartilage
- epiphyseal line

6.7

Intramembranous ossification is the formation of bone without a prior cartilaginous model

- intramembranous ossification
- dermal bones
- ossification center
- spicules

- = Term boldfaced in this module

6.8

Abnormalities of bone growth and development produce recognizable physical signs

- pituitary growth failure
- achondroplasia
- Marfan syndrome
- gigantism
- fibrodysplasia ossificans progressiva (FOP)
- heterotopic (ectopic) bones
- acromegaly

6.9

The primary hormones regulating calcium ion metabolism are parathyroid hormone and calcitonin; calcitriol is also involved

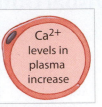

- parathyroid glands
- parathyroid hormone (PTH)
- kidneys
- calcitriol
- thyroid gland
- C cells
- calcitonin

SECTION 2

The Physiology of Bones

- composition of bone
- homeostatic regulation of calcium
- osteoblasts
- osteoclasts
- calcium

6.10

A fracture is a crack or a break in a bone

- fracture
- fracture hematoma
- internal callus
- external callus
- closed (simple) fractures
- open (compound) fractures
- transverse fractures
- spiral fractures
- displaced fractures
- nondisplaced fractures
- compression fractures
- greenstick fracture
- comminuted fractures
- epiphyseal fractures
- Pott fracture
- Colles fracture

• = *Term boldfaced in this module*

Chapter Integration: Applying what you've learned

Ensuring normal bone growth and development, and maintaining healthy bones, are contingent on a variety of factors. Sufficient amounts of vitamin D_3 (from sunlight exposure and from dietary intake) and calcium (also from the diet) are essential if osteoblasts are to be able to produce new bone matrix. Engaging in weight-bearing exercise is essential for establishing and maintaining normal bone mass. The potential loss of bone mass is a problem for astronauts spending extended time in the weightlessness of space—the standard stay at the international space station is 6 months.

The factors affecting bone growth and maintenance can also provide clues about the lives of people dead for centuries. Physical anthropologists often draw conclusions about ancient peoples from examining their long-buried skeletal remains. For example, bone thickness and density indicate aspects of nutrition and lifestyle; surface features can reveal information on the individual's muscular strength; and bone analysis can provide information regarding ancient environmental and living conditions.

1. Astronauts on extended space flights are not subjected to gravitational forces, and thus performing weight-bearing exercise is not very effective. What dietary supplements and types of exercise would be beneficial to astronauts in ensuring that their skeletal system maintains its integrity while enduring long periods of weightlessness?

2. What dietary supplement might be necessary to ensure adequate bone development for a child who consumes a diet rich in calcium but rarely receives exposure to sunlight?

3. What sorts of clues might bones provide about the lives and lifestyles of ancient peoples?

Access more review material online in the Study Area at **www.masteringaandp.com**.

There you'll find:
- **Chapter guides**
- **Chapter quizzes**
- **Practice tests**
- **Labeling activities**
- **MP3 Tutor Sessions**
- **Animations**
- **Flashcards**
- **A glossary with pronunciations**

7

The Skeleton

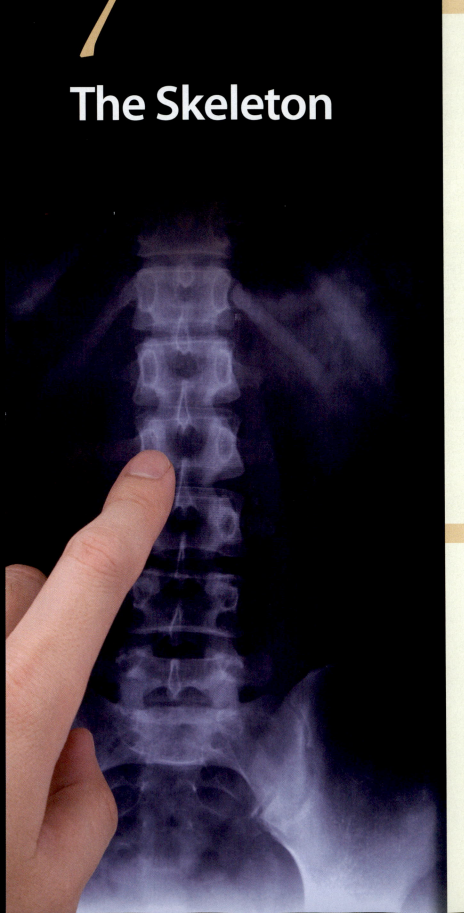

The Axial Skeleton

The **axial skeleton** forms the longitudinal axis of the body. This division of the skeletal system includes the skull and associated bones, the thoracic cage, the vertebral column, and various supplemental cartilages. There are typically 80 bones in the axial skeleton—roughly 40 percent of the bones in the human body.

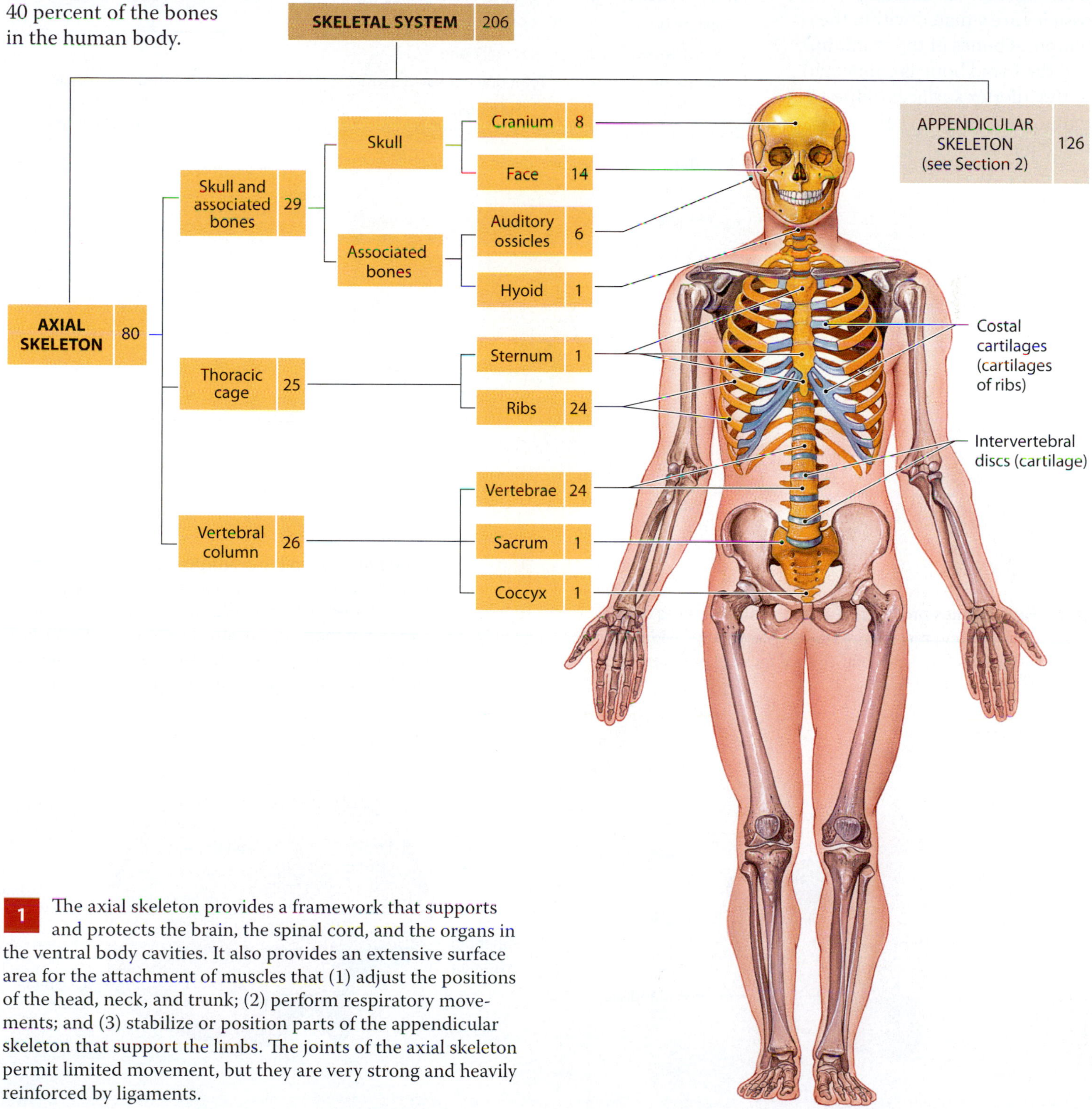

| SKELETAL SYSTEM | 206 |

| AXIAL SKELETON | 80 |

Skull and associated bones	29
Thoracic cage	25
Vertebral column	26

| Skull |
| Associated bones |

Cranium	8
Face	14
Auditory ossicles	6
Hyoid	1
Sternum	1
Ribs	24
Vertebrae	24
Sacrum	1
Coccyx	1

| APPENDICULAR SKELETON (see Section 2) | 126 |

Costal cartilages (cartilages of ribs)

Intervertebral discs (cartilage)

1 The axial skeleton provides a framework that supports and protects the brain, the spinal cord, and the organs in the ventral body cavities. It also provides an extensive surface area for the attachment of muscles that (1) adjust the positions of the head, neck, and trunk; (2) perform respiratory movements; and (3) stabilize or position parts of the appendicular skeleton that support the limbs. The joints of the axial skeleton permit limited movement, but they are very strong and heavily reinforced by ligaments.

The skull has cranial and facial components that are usually bound together by sutures

The skull contains 22 bones: 8 form the cranium, or braincase, and 14 form the face. Seven additional bones are associated with the skull: six auditory ossicles are situated within the temporal bones of the cranium, and the hyoid bone is connected to the inferior surfaces of the temporal bones by a pair of ligaments.

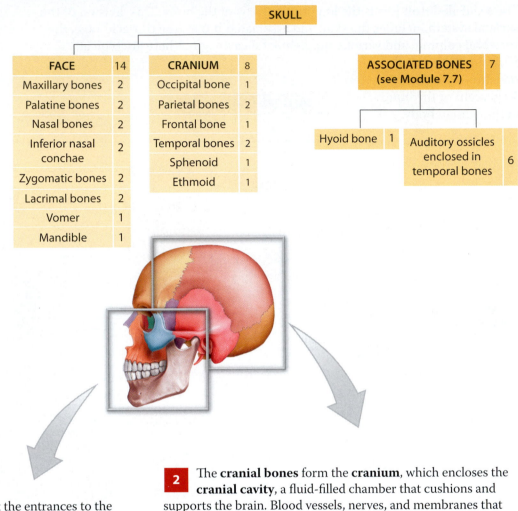

SKULL		

FACE	14
Maxillary bones	2
Palatine bones	2
Nasal bones	2
Inferior nasal conchae	2
Zygomatic bones	2
Lacrimal bones	2
Vomer	1
Mandible	1

CRANIUM	8
Occipital bone	1
Parietal bones	2
Frontal bone	1
Temporal bones	2
Sphenoid	1
Ethmoid	1

ASSOCIATED BONES (see Module 7.7)	7
Hyoid bone	1
Auditory ossicles enclosed in temporal bones	6

1 **Facial bones** protect and support the entrances to the digestive and respiratory tracts. They also provide areas for the attachment of muscles that control facial expressions and assist in manipulating food.

2 The **cranial bones** form the **cranium**, which encloses the **cranial cavity**, a fluid-filled chamber that cushions and supports the brain. Blood vessels, nerves, and membranes that stabilize the position of the brain are attached to the inner surface of the cranium. Its outer surface provides an extensive area for the attachment of muscles that move the eyes, jaws, and head.

Facial Bones

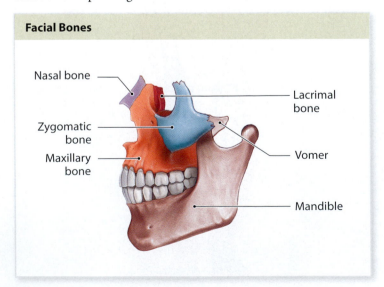

Nasal bone
Lacrimal bone
Zygomatic bone
Maxillary bone
Vomer
Mandible

Cranial Bones

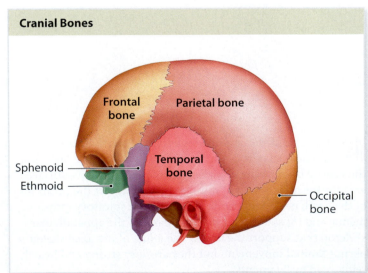

Frontal bone
Parietal bone
Sphenoid
Ethmoid
Temporal bone
Occipital bone

3 Joints, or articulations, form where two bones interconnect. Except where the mandible contacts the cranium, the connections between the skull bones of adults are immovable joints called **sutures**. At a suture, bones are tied firmly together with dense fibrous connective tissue. Each suture of the skull has a name, but at this point you need to know only four major sutures.

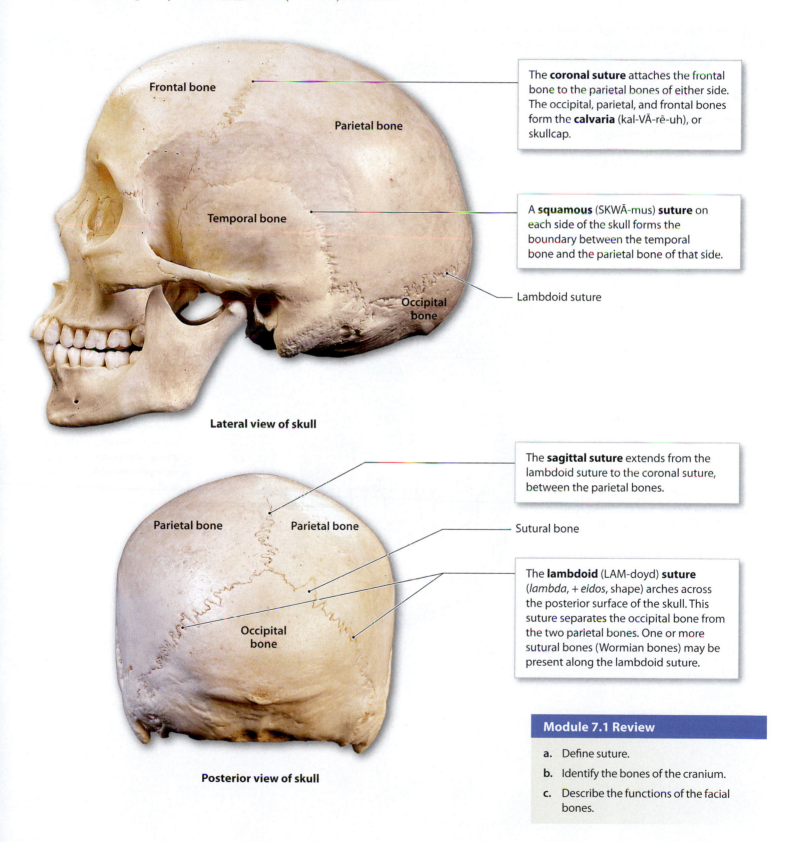

Frontal bone

Parietal bone

The **coronal suture** attaches the frontal bone to the parietal bones of either side. The occipital, parietal, and frontal bones form the **calvaria** (kal-VĀ-rē-uh), or skullcap.

Temporal bone

A **squamous** (SKWĀ-mus) **suture** on each side of the skull forms the boundary between the temporal bone and the parietal bone of that side.

Occipital bone

Lambdoid suture

Lateral view of skull

The **sagittal suture** extends from the lambdoid suture to the coronal suture, between the parietal bones.

Parietal bone   Parietal bone

Sutural bone

The **lambdoid** (LAM-doyd) **suture** (*lambda*, + *eidos*, shape) arches across the posterior surface of the skull. This suture separates the occipital bone from the two parietal bones. One or more sutural bones (Wormian bones) may be present along the lambdoid suture.

Occipital bone

Posterior view of skull

Module 7.1 Review

a. Define suture.

b. Identify the bones of the cranium.

c. Describe the functions of the facial bones.

Facial bones dominate the anterior aspect of the skull, and cranial bones dominate the posterior surface

1 We begin by examining the skull in anterior view. If you consider the cranium as the home of the brain, the facial bones form the front porch.

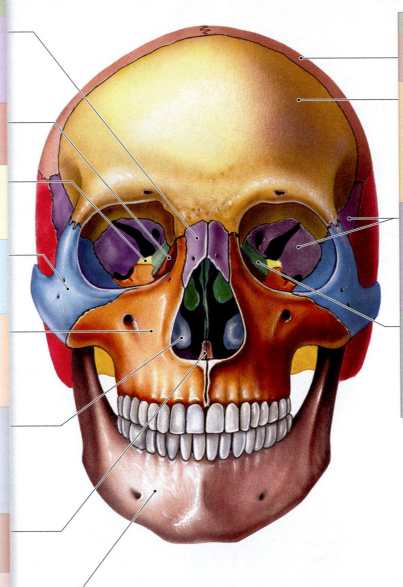

Facial Bones

The **nasal bones** support the superior portion of the bridge of the nose. They are connected to cartilages that support the distal portions of the nose.

The **lacrimal bones** form part of the medial wall of the orbit (eye socket).

The **palatine bones** form the posterior portion of the hard palate and contribute to the floor of each orbit.

The **zygomatic bones** contribute to the rim and lateral wall of the orbit and form part of the cheekbone.

The **maxillae** support the upper teeth and form the inferior orbital rim, the lateral margins of the external nares, the upper jaw, and most of the hard palate.

The **inferior nasal conchae** create turbulence in air passing through the nasal cavity, and increase the epithelial surface area to promote warming and humidification of inhaled air.

The **vomer** forms the inferior portion of the bony nasal septum.

The **mandible** forms the lower jaw.

Cranial Bones

Parietal bone

The **frontal bone** forms the anterior portion of the cranium and the roof of the orbits. Mucous secretions of the frontal sinuses within this bone help flush the surfaces of the nasal cavities.

The **sphenoid** forms part of the floor of the cranium, unites the cranial and facial bones, and acts as a cross-brace that strengthens the sides of the skull.

The **ethmoid** forms the anteromedial floor of the cranium, the roof of the nasal cavity, and part of the nasal septum and medial orbital wall.

2 Whereas the anterior view is dominated by facial bones, the posterior view is dominated by bones of the cranium. Several prominent landmarks on the occipital and temporal bones are identified here.

Cranial Bones

The **parietal bone** on each side forms part of the superior and lateral surfaces of the cranium.

The **occipital bone** contributes to the posterior, lateral, and inferior surfaces of the cranium.

The **temporal bone** on either side (1) forms part of the lateral wall of the cranium and articulates with facial bones, (2) forms an articulation with the mandible, (3) surrounds and protects the sense organs of the inner ear, and (4) is an attachment site for muscles that close the jaws and move the head.

The **external occipital crest**, which begins at the external occipital protuberance, extends inferiorly and marks the attachment of a ligament that helps stabilize the vertebrae of the neck.

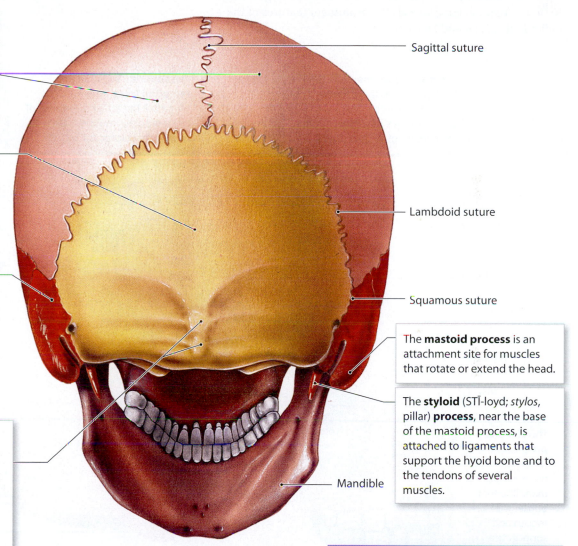

Sagittal suture

Lambdoid suture

Squamous suture

The **mastoid process** is an attachment site for muscles that rotate or extend the head.

The **styloid** (STĪ-loyd; *stylos*, pillar) **process**, near the base of the mastoid process, is attached to ligaments that support the hyoid bone and to the tendons of several muscles.

Mandible

Module 7.2 Review

a. Identify the facial bones.

b. Quincy suffers a hit to the skull that fractures the right superior lateral surface of his cranium. Which bone is fractured?

c. Identify the following bones as either a facial bone or a cranial bone: vomer, ethmoid, sphenoid, temporal, and inferior nasal conchae.

The lateral and medial aspects of the skull share many surface features

1 This lateral view of the skull shows how the large bones interconnect and reveals the prominent features of the individual cranial and facial bones.

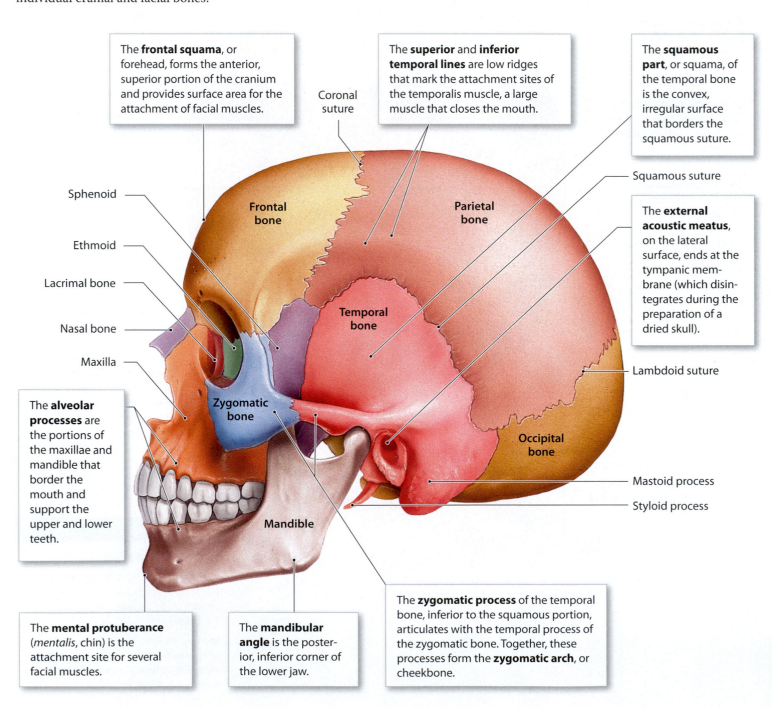

The **frontal squama**, or forehead, forms the anterior, superior portion of the cranium and provides surface area for the attachment of facial muscles.

The **superior** and **inferior temporal lines** are low ridges that mark the attachment sites of the temporalis muscle, a large muscle that closes the mouth.

The **squamous part**, or squama, of the temporal bone is the convex, irregular surface that borders the squamous suture.

Coronal suture

Squamous suture

Sphenoid

Frontal bone

Parietal bone

The **external acoustic meatus**, on the lateral surface, ends at the tympanic membrane (which disintegrates during the preparation of a dried skull).

Ethmoid

Lacrimal bone

Nasal bone

Temporal bone

Maxilla

Lambdoid suture

The **alveolar processes** are the portions of the maxillae and mandible that border the mouth and support the upper and lower teeth.

Zygomatic bone

Occipital bone

Mandible

Mastoid process

Styloid process

The **mental protuberance** (*mentalis*, chin) is the attachment site for several facial muscles.

The **mandibular angle** is the posterior, inferior corner of the lower jaw.

The **zygomatic process** of the temporal bone, inferior to the squamous portion, articulates with the temporal process of the zygomatic bone. Together, these processes form the **zygomatic arch**, or cheekbone.

2 This sagittal section passes slightly to the left of the midline, leaving the vomer and the perpendicular plate of the ethmoid intact.

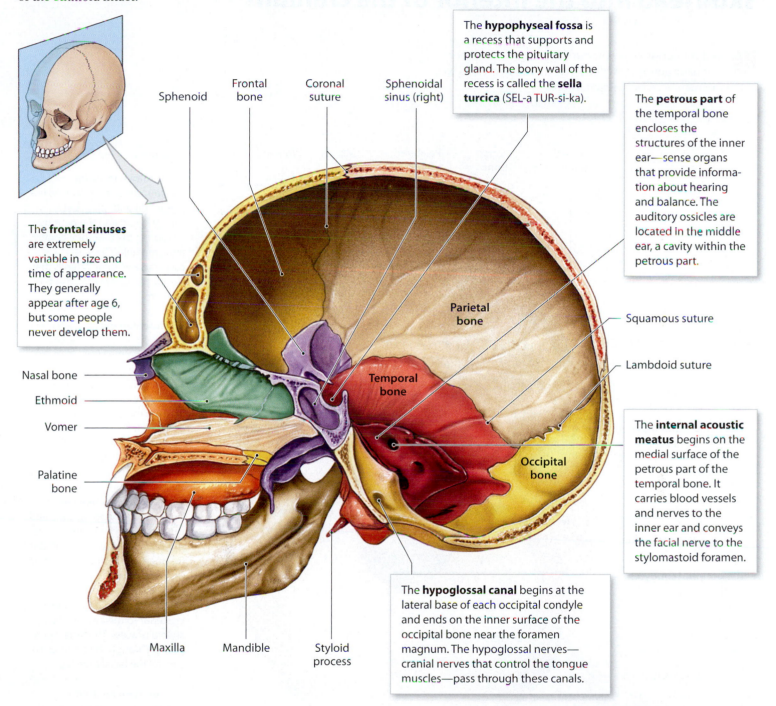

The **frontal sinuses** are extremely variable in size and time of appearance. They generally appear after age 6, but some people never develop them.

The **hypophyseal fossa** is a recess that supports and protects the pituitary gland. The bony wall of the recess is called the **sella turcica** (SEL-a TUR-si-ka).

The **petrous part** of the temporal bone encloses the structures of the inner ear—sense organs that provide information about hearing and balance. The auditory ossicles are located in the middle ear, a cavity within the petrous part.

The **internal acoustic meatus** begins on the medial surface of the petrous part of the temporal bone. It carries blood vessels and nerves to the inner ear and conveys the facial nerve to the stylomastoid foramen.

The **hypoglossal canal** begins at the lateral base of each occipital condyle and ends on the inner surface of the occipital bone near the foramen magnum. The hypoglossal nerves—cranial nerves that control the tongue muscles—pass through these canals.

Sphenoid

Frontal bone

Coronal suture

Sphenoidal sinus (right)

Parietal bone

Squamous suture

Lambdoid suture

Temporal bone

Occipital bone

Nasal bone

Ethmoid

Vomer

Palatine bone

Maxilla

Mandible

Styloid process

Module 7.3 Review

a. Name the meatuses found in the temporal bone.

b. What is the function of the internal acoustic meatus?

c. The alveolar processes perform what functions in which bones?

The foramina on the inferior surface of the skull lead into the interior of the cranium

1 In this inferior view you can see many of the important passageways for blood vessels and nerves, as well as the foramen used by the spinal cord.

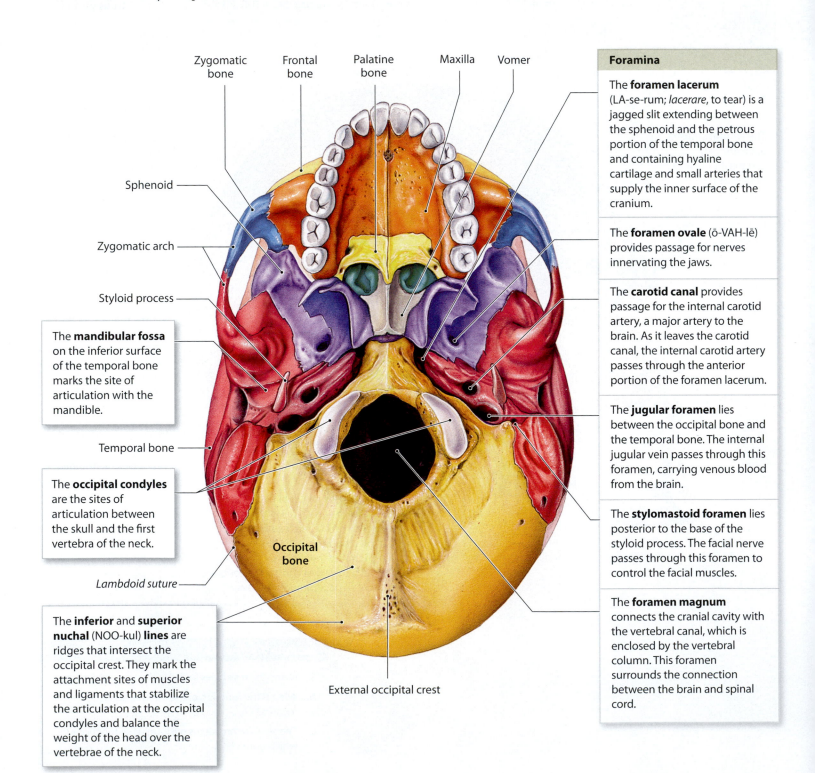

Zygomatic bone

Frontal bone

Palatine bone

Maxilla

Vomer

Sphenoid

Zygomatic arch

Styloid process

The **mandibular fossa** on the inferior surface of the temporal bone marks the site of articulation with the mandible.

Temporal bone

The **occipital condyles** are the sites of articulation between the skull and the first vertebra of the neck.

Occipital bone

Lambdoid suture

The **inferior** and **superior nuchal** (NOO-kul) **lines** are ridges that intersect the occipital crest. They mark the attachment sites of muscles and ligaments that stabilize the articulation at the occipital condyles and balance the weight of the head over the vertebrae of the neck.

External occipital crest

Foramina

The **foramen lacerum** (LA-se-rum; *lacerare*, to tear) is a jagged slit extending between the sphenoid and the petrous portion of the temporal bone and containing hyaline cartilage and small arteries that supply the inner surface of the cranium.

The **foramen ovale** (ō-VAH-lē) provides passage for nerves innervating the jaws.

The **carotid canal** provides passage for the internal carotid artery, a major artery to the brain. As it leaves the carotid canal, the internal carotid artery passes through the anterior portion of the foramen lacerum.

The **jugular foramen** lies between the occipital bone and the temporal bone. The internal jugular vein passes through this foramen, carrying venous blood from the brain.

The **stylomastoid foramen** lies posterior to the base of the styloid process. The facial nerve passes through this foramen to control the facial muscles.

The **foramen magnum** connects the cranial cavity with the vertebral canal, which is enclosed by the vertebral column. This foramen surrounds the connection between the brain and spinal cord.

2 This horizontal section should be compared to the inferior view on the previous page.

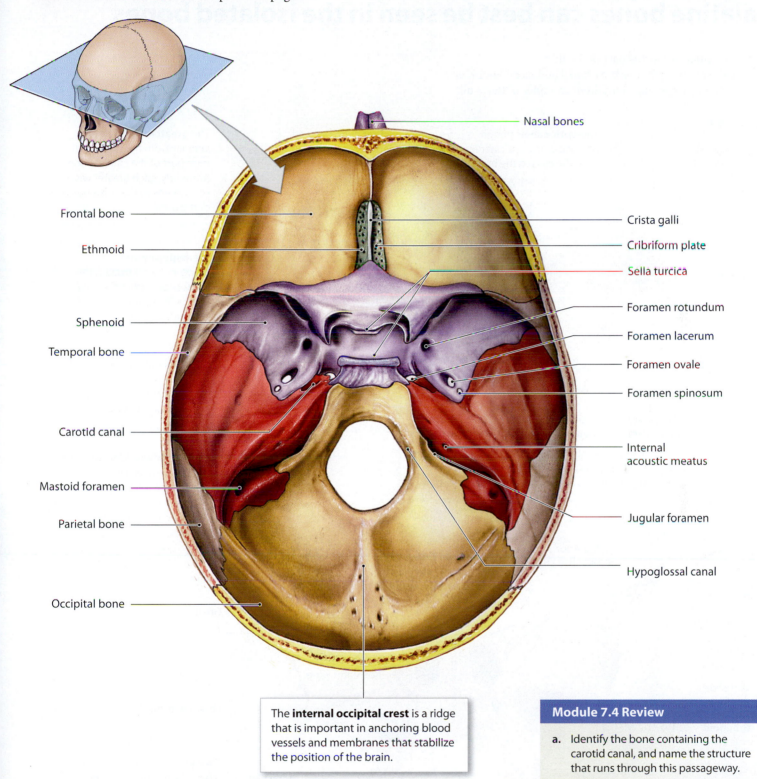

Nasal bones

Frontal bone

Ethmoid

Sphenoid

Temporal bone

Carotid canal

Mastoid foramen

Parietal bone

Occipital bone

Crista galli

Cribriform plate

Sella turcica

Foramen rotundum

Foramen lacerum

Foramen ovale

Foramen spinosum

Internal acoustic meatus

Jugular foramen

Hypoglossal canal

The **internal occipital crest** is a ridge that is important in anchoring blood vessels and membranes that stabilize the position of the brain.

Module 7.4 Review

a. Identify the bone containing the carotid canal, and name the structure that runs through this passageway.

b. Which foramen provides a passageway for nerves innervating the jaw?

c. In which bone is the foramen magnum located, and what is significant about this opening?

The shapes and landmarks of the sphenoid, ethmoid, and palatine bones can best be seen in the isolated bones

1 The **sphenoid** forms part of the floor of the cranium, unites the cranial and facial bones, and acts as a cross-bridge that stengthens the sides of the skull.

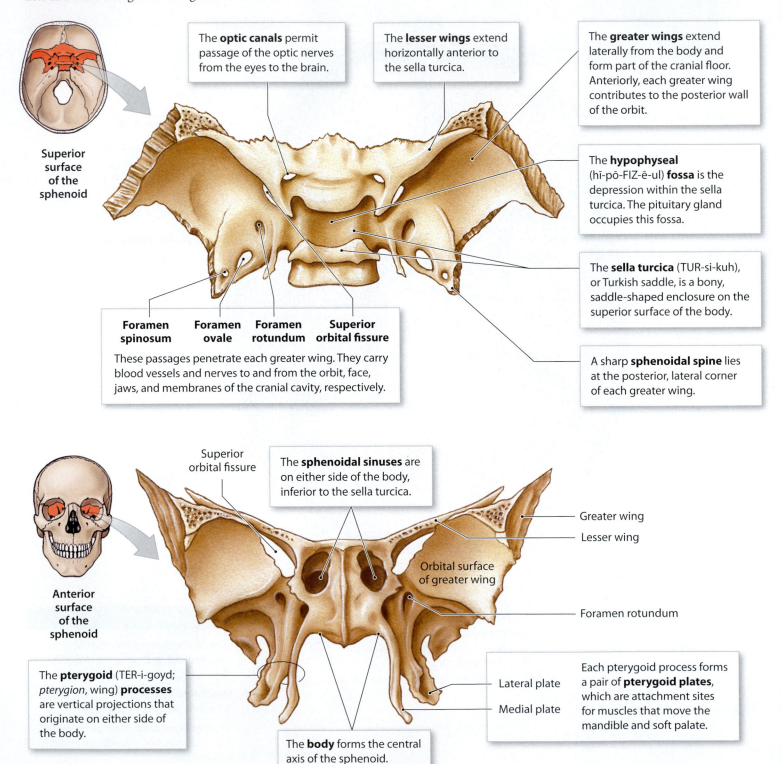

Superior surface of the sphenoid

The **optic canals** permit passage of the optic nerves from the eyes to the brain.

The **lesser wings** extend horizontally anterior to the sella turcica.

The **greater wings** extend laterally from the body and form part of the cranial floor. Anteriorly, each greater wing contributes to the posterior wall of the orbit.

The **hypophyseal** (hī-pō-FIZ-ē-ul) **fossa** is the depression within the sella turcica. The pituitary gland occupies this fossa.

The **sella turcica** (TUR-si-kuh), or Turkish saddle, is a bony, saddle-shaped enclosure on the superior surface of the body.

A sharp **sphenoidal spine** lies at the posterior, lateral corner of each greater wing.

Foramen spinosum   Foramen ovale   Foramen rotundum   Superior orbital fissure

These passages penetrate each greater wing. They carry blood vessels and nerves to and from the orbit, face, jaws, and membranes of the cranial cavity, respectively.

Anterior surface of the sphenoid

Superior orbital fissure

The **sphenoidal sinuses** are on either side of the body, inferior to the sella turcica.

Greater wing

Lesser wing

Orbital surface of greater wing

Foramen rotundum

The **pterygoid** (TER-i-goyd; *pterygion*, wing) **processes** are vertical projections that originate on either side of the body.

Lateral plate

Medial plate

Each pterygoid process forms a pair of **pterygoid plates**, which are attachment sites for muscles that move the mandible and soft palate.

The **body** forms the central axis of the sphenoid.

2 The **ethmoid** forms part of the floor of the anterior cranium, the roof of the nasal cavity, and part of the nasal septum and medial orbital wall. The ethmoid has three parts: (1) the cribriform plate, (2) the paired lateral masses, and (3) the perpendicular plate.

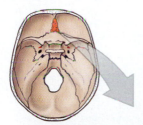

The **cribriform plate** (*cribrum*, sieve) forms the anteromedial floor of the cranium and the roof of the nasal cavity. The olfactory foramina in the cribriform plate permit passage of the olfactory nerves, which provide the sense of smell.

The **crista galli** (*crista*, crest + *gallus*, chicken; cock's comb) is a bony ridge that projects superior to the cribriform plate. The falx cerebri, a membrane that stabilizes the position of the brain, attaches to this ridge.

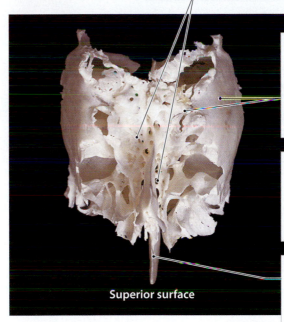

The **lateral masses** contain the **ethmoidal labyrinth**, which consists of the interconnected ethmoidal air cells, small chambers that open into the nasal cavity on each side.

The **superior nasal conchae** (KONG-kē; singular, *concha*, a snail shell) and the **middle nasal conchae** are delicate projections of the lateral masses.

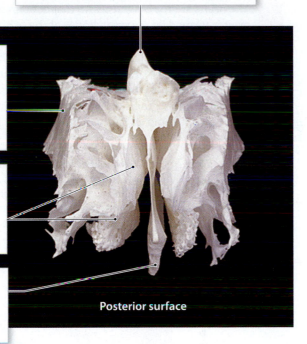

The **perpendicular plate** forms part of the nasal septum, along with the vomer and a piece of hyaline cartilage.

Superior surface

Posterior surface

3 The **palatine bones** form the posterior portion of the hard palate and contribute to the floor of each orbit.

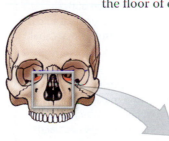

The **orbital process**, which forms part of the floor of the orbit. This process contains a small sinus that normally opens into the sphenoidal sinus.

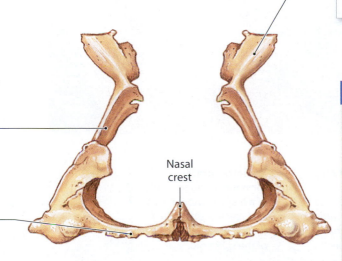

Nasal crest

The **perpendicular plate** of the palatine bone extends from the horizontal plate to the orbital process.

The **horizontal plate** forms the posterior part of the hard palate.

Module 7.5 Review

a. Identify the bone containing the optic canal, and cite the structures using this passageway.

b. Which bone contains the sella turcica? What structure is in the depression (fossa) within the sella turcica?

c. Identify the bone containing the cribriform plate. What is significant about this structure?

Each orbital complex contains one eye, and the nasal complex encloses the nasal cavities

The facial bones not only protect and support the openings of the digestive and respiratory systems, but also protect the delicate sense organs responsible for vision and smell. Together, certain cranial bones and facial bones form an orbital complex containing one eye, and the nasal complex which surrounds the nasal cavities.

1 The orbits are the bony recesses that contain the eyes. Each orbit is formed by the seven bones of the **orbital complex**. The frontal bone forms the roof, the zygomatic bone forms the lateral wall, and the maxilla provides most of the orbital floor. The orbital rim and the first portion of the medial wall are formed by the maxilla, the lacrimal bone, and the lateral mass of the ethmoid. The lateral mass articulates with the sphenoid and a small process of the palatine bone.

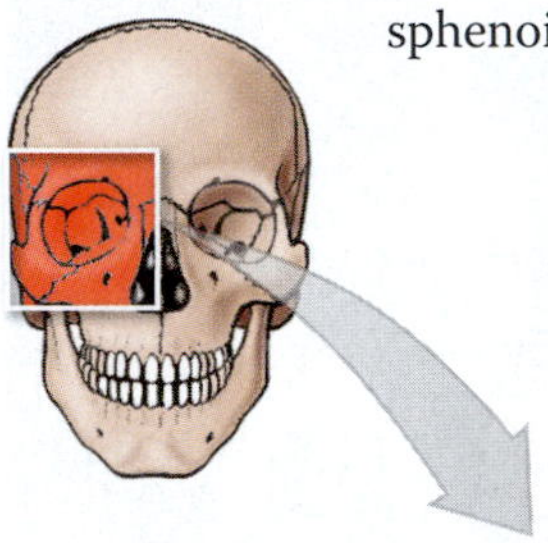

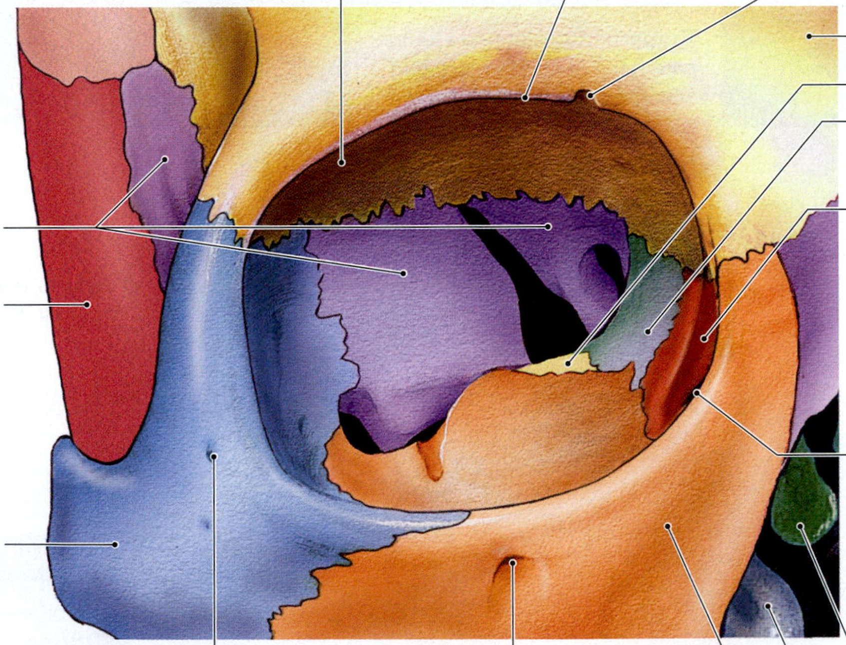

The **lacrimal fossa** on the superior and lateral surface of the orbit is a shallow depression in the frontal bone that marks the location of the lacrimal (tear) gland, which lubricates the surface of the eye.

The **supra-orbital margin** is a thickening of the frontal bone that helps protect the eye.

Either a **supra-orbital notch** or a fully enclosed supra-orbital foramen provides passage for blood vessels that supply the eyebrow, eyelids, and frontal sinuses.

Frontal bone

Palatine bone

Ethmoid

The **lacrimal sulcus**, a groove along the anterior, lateral surface of the lacrimal bone, marks the location of the lacrimal sac. The lacrimal sulcus leads to the nasolacrimal canal.

The **nasolacrimal canal**, formed by a maxilla and lacrimal bone, protects the lacrimal sac and the nasolacrimal duct, which carries tears from the orbit to the nasal cavity.

Sphenoid

Temporal bone

Zygomatic bone

Middle nasal concha

Inferior nasal concha

Maxilla

The **zygomaticofacial foramen** on the anterior surface of each zygomatic bone carries a sensory nerve that innervates the cheek.

The **infra-orbital foramen** marks the path of a major sensory nerve that reaches the brain via the foramen rotundum of the sphenoid.

2 The **nasal complex** includes the bones that enclose the nasal cavities and the paranasal sinuses, air-filled chambers connected to the nasal cavities. The sphenoid, ethmoid, frontal bone, palatine bone, and maxillae contain the **paranasal sinuses**. (The tiny palatine sinuses, not shown, generally open into the sphenoidal sinuses.) The paranasal sinuses lighten the skull bones and provide an extensive area of mucous epithelium.

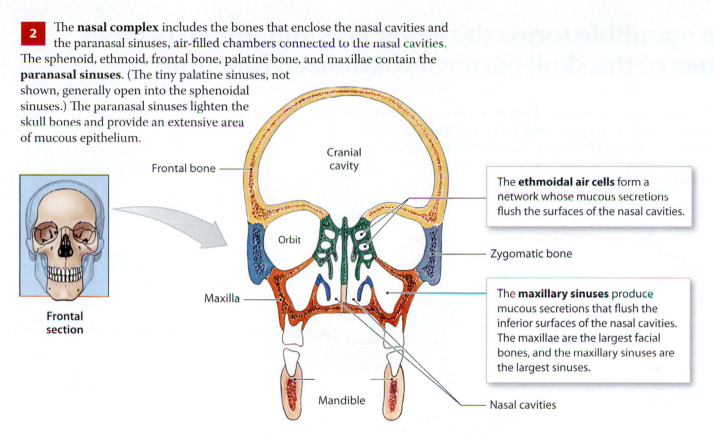

Frontal section

Frontal bone

Cranial cavity

Orbit

Maxilla

Mandible

The **ethmoidal air cells** form a network whose mucous secretions flush the surfaces of the nasal cavities.

Zygomatic bone

The **maxillary sinuses** produce mucous secretions that flush the inferior surfaces of the nasal cavities. The maxillae are the largest facial bones, and the maxillary sinuses are the largest sinuses.

Nasal cavities

3 This is a sagittal section with the nasal septum removed. The frontal bone, sphenoid, and ethmoid form the superior wall of the nasal cavities. The lateral walls are formed by the maxillae and the lacrimal bones, the ethmoid (the superior and middle nasal conchae), and the inferior nasal conchae. Much of the anterior margin of the nasal cavity is formed by the soft tissues of the nose, but the bridge of the nose is supported by the maxillae and nasal bones.

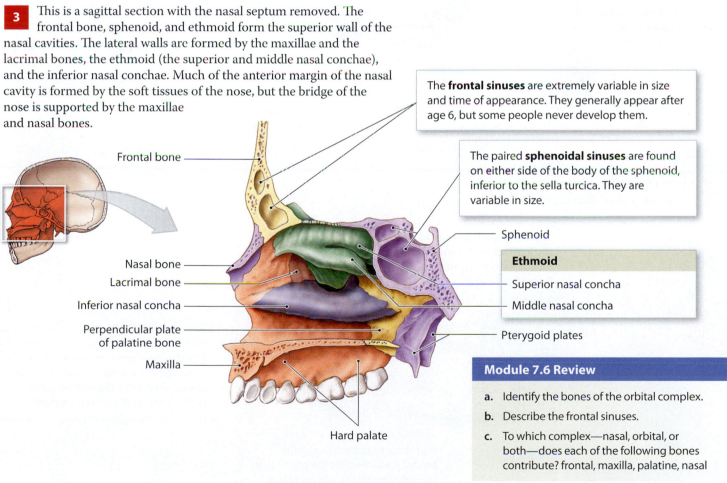

Frontal bone

Nasal bone

Lacrimal bone

Inferior nasal concha

Perpendicular plate of palatine bone

Maxilla

Hard palate

The **frontal sinuses** are extremely variable in size and time of appearance. They generally appear after age 6, but some people never develop them.

The paired **sphenoidal sinuses** are found on either side of the body of the sphenoid, inferior to the sella turcica. They are variable in size.

Sphenoid

Ethmoid

Superior nasal concha

Middle nasal concha

Pterygoid plates

Module 7.6 Review

a. Identify the bones of the orbital complex.

b. Describe the frontal sinuses.

c. To which complex—nasal, orbital, or both—does each of the following bones contribute? frontal, maxilla, palatine, nasal

The mandible forms the lower jaw, and the associated bones of the skull perform specialized functions

1 The mandible can be subdivided into the horizontal body and the ascending rami (singular *ramus*) of the mandible. The lower teeth are supported by the mandibular body. The mandible forms the entire lower jaw and articulates with the mandibular fossae of the temporal bones.

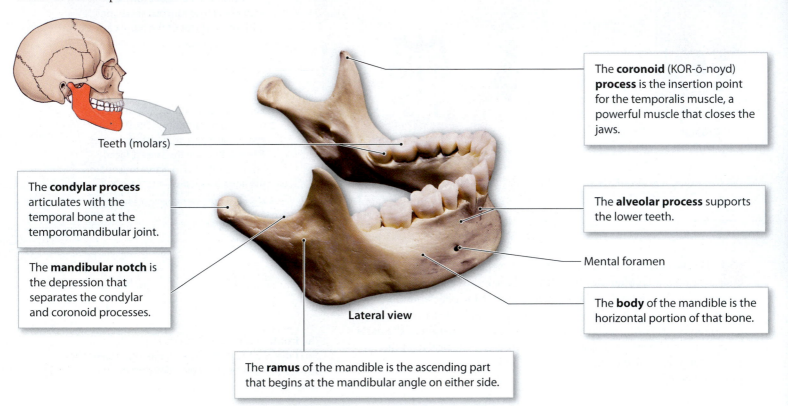

Teeth (molars)

The **coronoid** (KOR-ō-noyd) **process** is the insertion point for the temporalis muscle, a powerful muscle that closes the jaws.

The **condylar process** articulates with the temporal bone at the temporomandibular joint.

The **mandibular notch** is the depression that separates the condylar and coronoid processes.

The **alveolar process** supports the lower teeth.

Mental foramen

The **body** of the mandible is the horizontal portion of that bone.

Lateral view

The **ramus** of the mandible is the ascending part that begins at the mandibular angle on either side.

2 The medial surface of the mandible has prominent surface features.

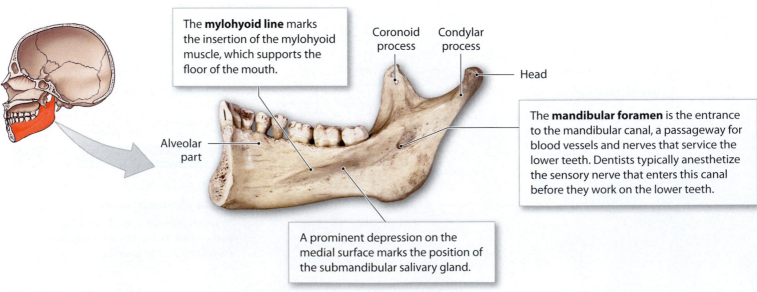

The **mylohyoid line** marks the insertion of the mylohyoid muscle, which supports the floor of the mouth.

Coronoid process

Condylar process

Head

Alveolar part

The **mandibular foramen** is the entrance to the mandibular canal, a passageway for blood vessels and nerves that service the lower teeth. Dentists typically anesthetize the sensory nerve that enters this canal before they work on the lower teeth.

A prominent depression on the medial surface marks the position of the submandibular salivary gland.

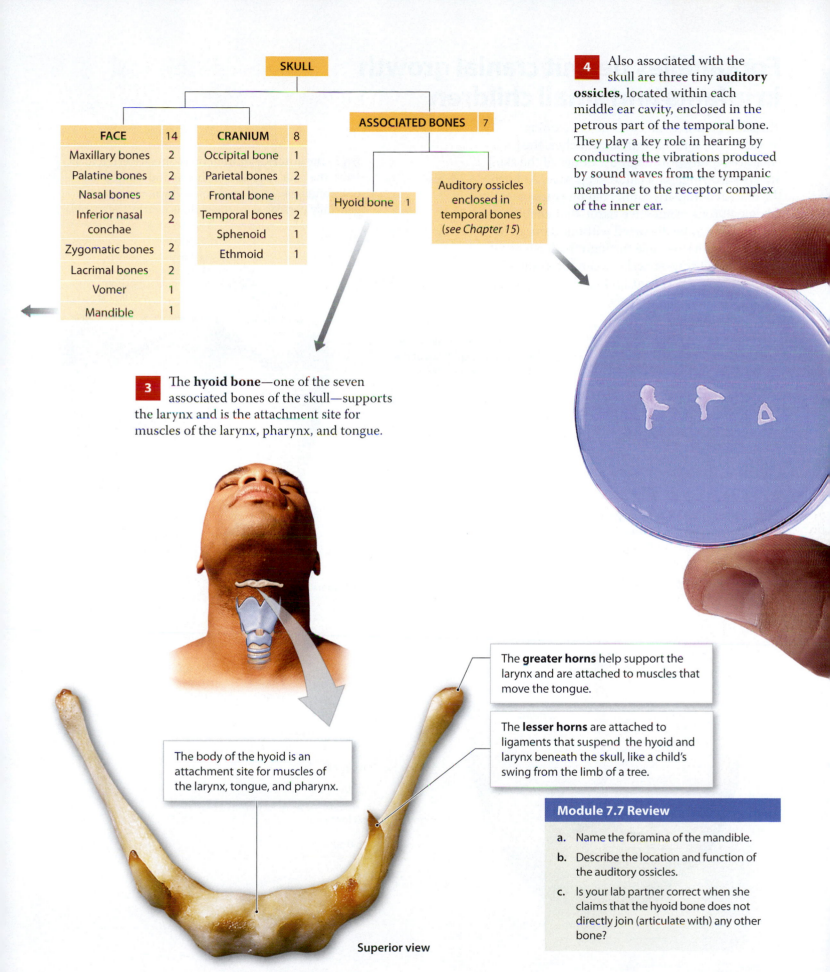

```
                    SKULL
        ┌─────────────┴──────────────────────┐
```

FACE	14
Maxillary bones	2
Palatine bones	2
Nasal bones	2
Inferior nasal conchae	2
Zygomatic bones	2
Lacrimal bones	2
Vomer	1
Mandible	1

CRANIUM	8
Occipital bone	1
Parietal bones	2
Frontal bone	1
Temporal bones	2
Sphenoid	1
Ethmoid	1

ASSOCIATED BONES	7

Hyoid bone	1

Auditory ossicles enclosed in temporal bones (*see Chapter 15*)	6

4 Also associated with the skull are three tiny **auditory ossicles**, located within each middle ear cavity, enclosed in the petrous part of the temporal bone. They play a key role in hearing by conducting the vibrations produced by sound waves from the tympanic membrane to the receptor complex of the inner ear.

3 The **hyoid bone**—one of the seven associated bones of the skull—supports the larynx and is the attachment site for muscles of the larynx, pharynx, and tongue.

The **greater horns** help support the larynx and are attached to muscles that move the tongue.

The **lesser horns** are attached to ligaments that suspend the hyoid and larynx beneath the skull, like a child's swing from the limb of a tree.

The body of the hyoid is an attachment site for muscles of the larynx, tongue, and pharynx.

Superior view

Module 7.7 Review

a. Name the foramina of the mandible.

b. Describe the location and function of the auditory ossicles.

c. Is your lab partner correct when she claims that the hyoid bone does not directly join (articulate with) any other bone?

Fontanelles permit cranial growth in infants and small children

The skull organizes around the developing brain. As the time of birth approaches, the brain enlarges rapidly. Although the bones of the skull are also growing, they fail to keep pace. At birth, the cranial bones are connected by areas of flexible fibrous connective tissue that enable the skull's shape to be distorted without damage to ease the infant's passage through the birth canal. The largest fibrous areas between the cranial bones are known as **fontanelles** (fon-tuh-NELZ; sometimes spelled *fontanels*).

1 The **anterior fontanelle** is often referred to as the "soft spot" on newborns and is often the only fontanelle easily recognized by new parents. It generally persists until the child is nearly 2 years old.

The anterior fontanelle is the largest fontanelle. It lies at the intersection of the frontal, sagittal, and coronal sutures in the anterior portion of the skull.

2 Over time, the sutures gradually narrow and the fontanelles become smaller and smaller. In this posterior view of the skull of an infant, the occipital fontanelle has almost disappeared.

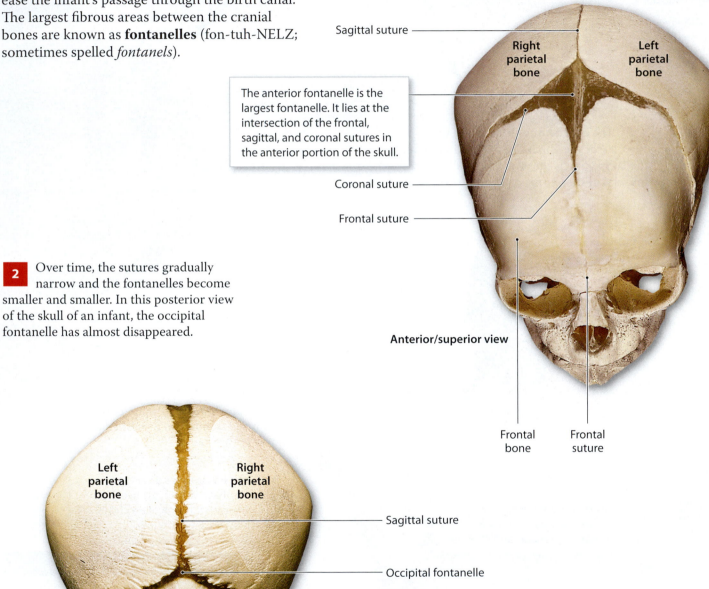

Sagittal suture

Right parietal bone

Left parietal bone

Coronal suture

Frontal suture

Anterior/superior view

Frontal bone

Frontal suture

Left parietal bone

Right parietal bone

Sagittal suture

Occipital fontanelle

Lambdoid suture

Occipital bone

Posterior view

3 This lateral view shows how small an infant's facial bones are compared to the cranium.

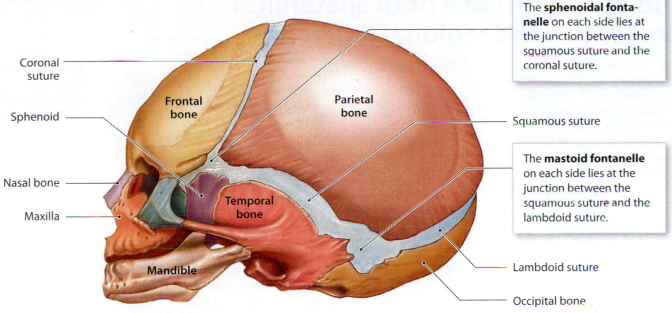

Coronal suture

Sphenoid

Nasal bone

Maxilla

Frontal bone

Parietal bone

Temporal bone

Mandible

The **sphenoidal fontanelle** on each side lies at the junction between the squamous suture and the coronal suture.

Squamous suture

The **mastoid fontanelle** on each side lies at the junction between the squamous suture and the lambdoid suture.

Lambdoid suture

Occipital bone

4 This superior view shows the size of the anterior fontanelle relative to the width of the fibrous areas destined to become the sutures of the adult skull.

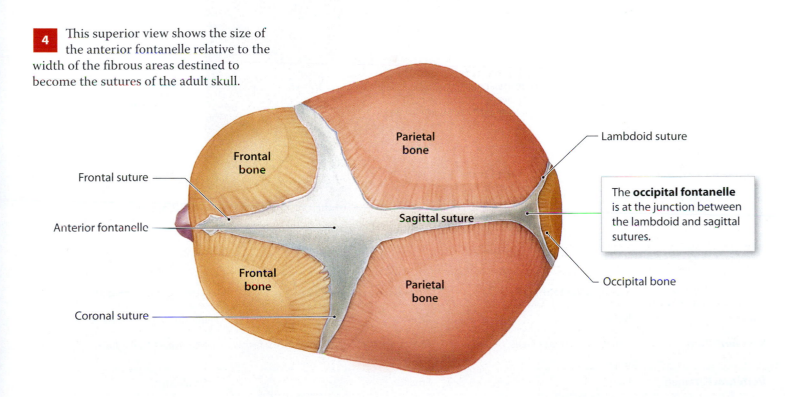

Frontal suture

Anterior fontanelle

Coronal suture

Frontal bone

Parietal bone

Frontal bone

Sagittal suture

Parietal bone

Lambdoid suture

The **occipital fontanelle** is at the junction between the lambdoid and sagittal sutures.

Occipital bone

The occipital, sphenoidal, and mastoid fontanelles disappear within a month or two after birth. Even after the fontanelles disappear, the bones of the skull remain separated by fibrous connections. The most significant growth in the skull occurs before age 5, because at that time the brain stops growing and the cranial sutures ossify.

Module 7.8 Review

a. Define fontanelle.

b. Identify the major fontanelles.

c. What purposes do fontanelles serve?

The vertebral column has four spinal curves, and vertebrae have both anatomical similarities and regional differences

1 The adult vertebral column (or spine) consists of 26 bones: the 24 vertebrae, the sacrum, and the coccyx (KOK-siks), or tailbone. The vertebrae provide a column of support that bears the weight of the head, neck, and trunk and ultimately transfers the body's weight to the appendicular skeleton of the lower limbs. The vertebrae also protect the spinal cord and help maintain an upright body position, as in sitting or standing. The total length of the vertebral column of an adult averages 71 cm (28 in.).

Spinal Curves

Primary curves develop before birth, and secondary curves after birth.

The **cervical curve**, a secondary curve, develops as the infant learns to balance the weight of the head on the vertebrae of the neck.

The **thoracic curve**, a primary curve, accommodates the thoracic organs.

The **lumbar curve**, a secondary curve, balances the weight of the trunk over the lower limbs; it develops with the ability to stand.

The **sacral curve,** a primary curve, accommodates the abdominopelvic organs.

Vertebral Regions

Regions are defined by anatomical characteristics of individual vertebrae.

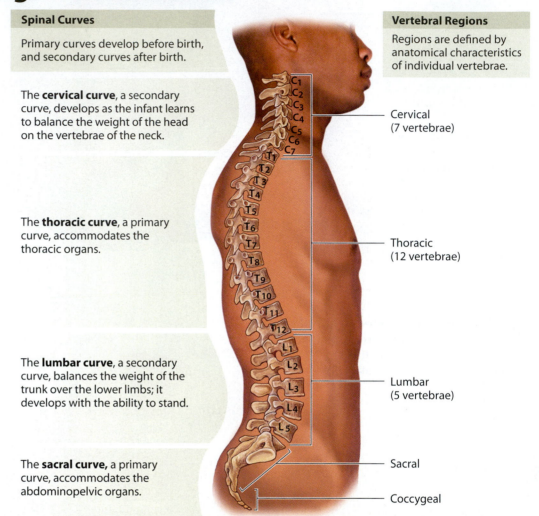

Cervical (7 vertebrae)

Thoracic (12 vertebrae)

Lumbar (5 vertebrae)

Sacral

Coccygeal

Regional Differences in Vertebral Structure and Function

	Cervical Vertebrae (7)	**Thoracic Vertebrae (12)**	**Lumbar Vertebrae (5)**
Location	Neck	Chest	Inferior portion of back
Vertebral Body	Small, oval, curved faces	Medium, heart-shaped, flat faces; facets for rib articulations	Massive, oval, flat faces
Vertebral Foramen	Large	Smaller	Smallest
Spinous Process	Long; split tip; points inferiorly	Long, slender; not split; points inferiorly	Blunt, broad; points posteriorly
Transverse Processes	Have transverse foramina	All but two have facets for rib articulations	Short; no articular facets or transverse foramina
Functions	Support skull, stabilize relative positions of brain and spinal cord, and allow controlled head movements	Support weight of head, neck, upper limbs, and chest; articulate with ribs to allow changes in volume of thoracic cage	Support weight of head, neck, upper limbs, and trunk

2 Each vertebra consists of three basic parts: (1) articular processes, (2) a vertebral arch, and (3) a vertebral body.

Parts of a Vertebra

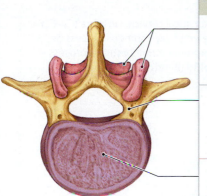

Superior view

The **articular processes** extend superiorly and inferiorly to articulate with adjacent vertebrae.

The **vertebral arch** forms the posterior and lateral margins of the vertebral foramen.

The **vertebral body** is the part of a vertebra that transfers weight along the axis of the vertebral column.

3 The vertebral arch can be divided into four different parts: the spinous process, laminae, transverse processes, and pedicles.

The **vertebral foramen** is framed by the vertebral body and the vertebral arch.

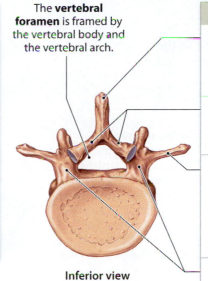

Inferior view

The Vertebral Arch

A **spinous process** projects posteriorly from the point where the vertebral laminae fuse to complete the vertebral arch.

The **laminae** (LAM-i-nē; singular, *lamina*, a thin plate) form the "roof" of the vertebral foramen.

Transverse processes project laterally on both sides from the point where the laminae join the pedicles. These processes are sites of muscle attachment, and they may also articulate with the ribs.

The **pedicles** (PED-i-kulz) form the sides of the vertebral arch.

4 This lateral view of three vertebrae shows how they combine to form the vertebral canal.

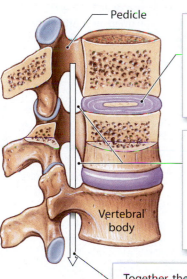

Pedicle

Vertebral body

The bodies of adjacent vertebrae are interconnected by ligaments but are separated by pads of fibrous cartilage, the **intervertebral discs**.

The spaces between successive pedicles are called **intervertebral foramina**. Nerves and blood vessels running to or from the spinal cord pass through these foramina.

Together, the vertebral foramina of successive vertebrae form the **vertebral canal**, which encloses the spinal cord.

5 Each articular process has a smooth, concave surface called an **articular facet**. This posterior view shows that the vertebral arches form the roof of the vertebral canal.

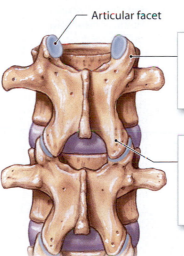

Articular facet

The **superior articular processes** articulate with the inferior articular processes of a more superior vertebra.

The **inferior articular processes** articulate with the superior articular processes of a more inferior vertebra (or the sacrum, in the case of the last lumbar vertebra).

Note that when referring to a specific vertebra, we use the capital letters C, T, L, S, and C_o to indicate the cervical, thoracic, lumbar, sacral, and coccygeal regions, respectively. In addition, we use a subscript number to indicate the relative position of the vertebra within that region, with 1 indicating the vertebra closest to the skull. For example, C_3 is the third cervical vertebra; C_1 is in contact with the skull. Similarly, L_4 is the fourth lumbar vertebra; L_1 is in contact with T_{12}.

Module 7.9 Review

a. Name the major components of a typical vertebra.

b. What is the importance of the secondary curves of the spine?

c. To which part of the vertebra do the intervertebral discs attach?

There are seven cervical vertebrae and twelve thoracic vertebrae

Cervical Vertebrae

The cervical vertebrae, the smallest in the vertebral column, extend from the occipital bone of the skull to the thorax.

1 The vertebral foramen of a cervical vertebra is very large. At this level, the spinal cord still contains most of the axons that connect the brain to the rest of the body. However, cervical vertebrae support only the weight of the head, so the vertebral body can be relatively small and light. As you continue toward the sacrum, the loading increases and the vertebral bodies gradually enlarge.

2 The first two cervical vertebrae are specialized to support and stabilize the cranium while permitting head movement. The **atlas**, C_1, has no vertebral body and no spinous process, but it has a large, round vertebral foramen bounded by anterior and posterior arches. The **axis**, C_2, resembles more inferior cervical vertebrae except for the presence of the prominent dens on the superior surface of the body.

3 The last cervical vertebra has a very robust spinous process that ends in a solid tubercle that can easily be felt through the skin. For this reason C_7 is known as the **vertebra prominens**. The **ligamentum nuchae** (lig-uh-MEN-tum NOO-kē; *nucha*, nape), a stout elastic ligament, begins at the vertebra prominens and extends to an insertion along the external occipital crest of the skull. When your head is upright, this ligament acts like the string on a bow to maintain the cervical curvature without muscular effort.

In a typical cervical vertebra (C_2–C_6), the tip of each spinous process bears a prominent notch; such a process is a **bifid** (BĪ-fid) spinous process.

The **transverse foramen,** formed by a connection of the costal and transverse processes, protects the vertebral arteries and vertebral veins that service the brain.

The transverse process is relatively short and stumpy.

A slender **costal process** extends anterolaterally from either side of the vertebral body.

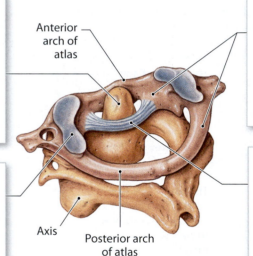

Vertebral foramen

Vertebral body

Cervical vertebra (superior view)

During development, the body of the atlas fuses to the body of the axis, where it forms the prominent **dens** (DENZ; *dens*, tooth), or **odontoid** (ō-DON-toyd; *odontos*, tooth) **process**.

The articulation between the skull's occipital condyles and the atlas is a joint that permits nodding (such as when you indicate "yes").

Anterior arch of atlas

The atlas holds up the skull, articulating with the occipital condyles of the skull. This vertebra is named after Atlas, who, according to Greek myth, holds the world on his shoulders.

A transverse ligament binds the dens to the inner surface of the anterior arch of the atlas. This articulation permits rotation (as when you shake your head to indicate "no").

Axis

Posterior arch of atlas

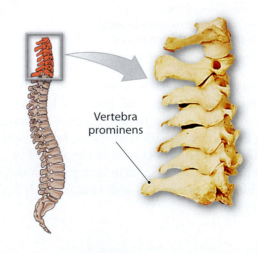

Vertebra prominens

Thoracic Vertebrae

4 There are 12 thoracic vertebrae. Between T_1 toward T_{12}, each vertebral body is slightly larger and more massive than the one superior to it, because the weight being transmitted along the vertebral column is steadily increasing.

5 A typical thoracic vertebra has a distinctive heart-shaped body that is much larger and more massive than that of a cervical vertebra, and the vertebral foramen is considerably smaller.

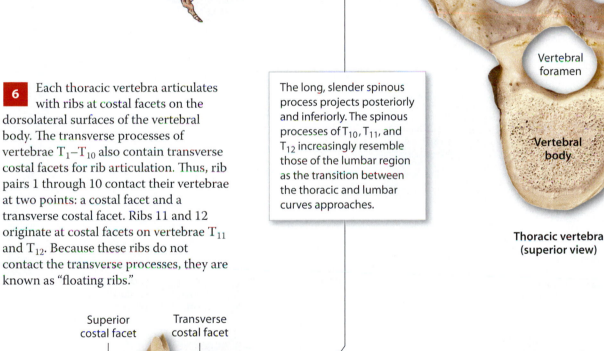

Transverse process

Superior articular facet

Vertebral foramen

Superior costal facet

Vertebral body

Thoracic vertebra (superior view)

6 Each thoracic vertebra articulates with ribs at costal facets on the dorsolateral surfaces of the vertebral body. The transverse processes of vertebrae T_1–T_{10} also contain transverse costal facets for rib articulation. Thus, rib pairs 1 through 10 contact their vertebrae at two points: a costal facet and a transverse costal facet. Ribs 11 and 12 originate at costal facets on vertebrae T_{11} and T_{12}. Because these ribs do not contact the transverse processes, they are known as "floating ribs."

The long, slender spinous process projects posteriorly and inferiorly. The spinous processes of T_{10}, T_{11}, and T_{12} increasingly resemble those of the lumbar region as the transition between the thoracic and lumbar curves approaches.

Superior costal facet

Transverse costal facet

Vertebral body

Inferior costal facet

Transverse process

Thoracic vertebra (lateral view)

Module 7.10 Review

a. Joe suffered a hairline fracture at the base of the dens. Which bone is fractured and where is the fractured bone located?

b. Examining a human vertebra, you notice that, in addition to the large foramen for the spinal cord, two smaller foramina are on either side of the bone in the region of the transverse processes. From which region of the vertebral column is this vertebra?

c. When you run your finger down the middle of a person's spine, what part of each vertebra are you feeling just beneath the skin?

There are five lumbar vertebrae; the sacrum and coccyx consist of fused vertebrae

Lumbar Vertebrae

1 The five lumbar vertebrae are the largest vertebrae, and they transmit the most weight. Compression fractures of these vertebrae may occur at any age, but they are most often seen after aging and osteoporosis have reduced the strength of bones throughout the body.

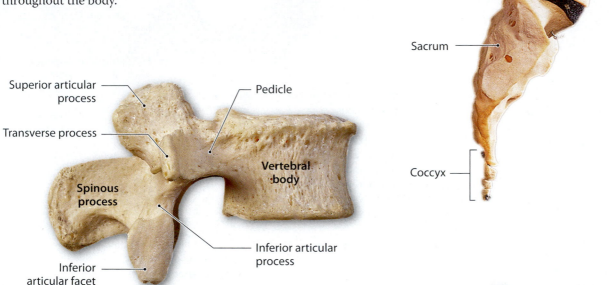

L₁
L₂
L₃
L₄
L₅

Sacrum

Coccyx

Superior articular process

Transverse process

Pedicle

Spinous process

Vertebral body

Inferior articular process

Inferior articular facet

Lateral view

Spinous process

Lamina

Transverse process

Superior articular facet

Superior articular process

Vertebral foramen

Pedicle

Vertebral body

Superior view

2 The body of a typical lumbar vertebra is thicker than that of a thoracic vertebra, and the superior and inferior surfaces are oval rather than heart shaped. Other noteworthy features are that (1) lumbar vertebrae do not have costal facets; (2) the slender transverse processes, which lack transverse costal facets, project dorsolaterally; (3) the vertebral foramen is triangular; (4) the stumpy spinous processes project posteriorly; (5) the superior articular processes face medially ("up and in"); and (6) the inferior articular processes face laterally ("down and out").

Sacrum

The sacrum consists of the fused components of five sacral vertebrae. These vertebrae begin fusing shortly after puberty and, in general, are completely fused at age 25–30. The sacrum protects the reproductive, digestive, and urinary organs and, via paired articulations, attaches the axial skeleton to the pelvic girdle of the appendicular skeleton.

3 The anterior surface of the sacrum is concave, and the posterior surface is convex. The degree of curvature is more pronounced in males than in females.

At the base of the sacrum, a broad sacral **ala**, or wing, extends on either side. The anterior and superior surfaces of each ala provide an extensive area for muscle attachment.

Four pairs of **sacral foramina** extend between the posterior and anterior surfaces. The intervertebral foramina of the fused sacral vertebrae open into these passageways.

The **base** of the sacrum is the broad superior surface.

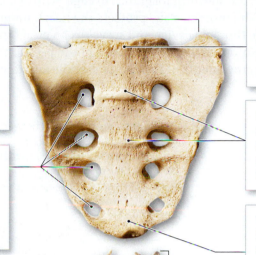

The **sacral promontory** is an important landmark in females during pelvic examinations and during labor and delivery.

Prominent transverse lines mark the former boundaries of individual vertebrae that fuse during the formation of the sacrum.

The **apex** is the narrow, inferior portion of the sacrum.

Coccyx

4 A posterior view of the sacrum shows the broad surface that provides an extensive area for the attachment of muscles.

The **sacral canal** is a passageway that extends the length of the sacrum. Nerves and membranes that line the vertebral canal in the spinal cord continue into the sacral canal.

The **superior articular process** of the sacrum articulates with the last lumbar vertebra.

The **median sacral crest** is a ridge formed by the fused spinous processes of the sacral vertebrae.

The **sacral hiatus** (hī-Ā-tus) is the opening at the inferior end of the sacral canal.

The small **coccyx** consists of three to five coccygeal vertebrae that have generally begun fusing by age 26. The **coccygeal cornua** curve to meet the sacral cornua.

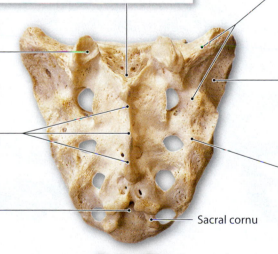

Sacral cornu

Coccygeal cornu

The **sacral tuberosity** is a roughened area dorsal to the auricular surface. It marks the attachment site of ligaments that stabilize the sacro-iliac joint.

The **auricular surface** is a thickened, flattened area lateral and anterior to the superior portion of the lateral sacral crest. The auricular surface is the site of articulation with the pelvic girdle (the sacro-iliac joint).

The **lateral sacral crest** is a ridge that represents the fused transverse processes of the sacral vertebrae.

Module 7.11 Review

a. How many vertebrae are present in the lumbar region? In the sacrum?

b. What structure forms the posterior wall of the pelvic girdle?

c. Why are the bodies of the lumbar vertebrae so large?

The thoracic cage protects organs in the chest and provides sites for muscle attachment

The skeleton of the chest, or **thoracic cage**, provides bony support to the walls of the thoracic cavity. It consists of the thoracic vertebrae, the ribs, and the sternum (breastbone). The ribs and the sternum form the rib cage, whose movements are important in respiration. The thoracic cage protects the heart, lungs, thymus, and other structures in the thoracic cavity, and serves as an attachment point for muscles involved in (1) respiration, (2) maintenance of the position of the vertebral column, and (3) movements of the pectoral girdle and upper limbs.

1 An anterior view of the thoracic cage shows the sternum, the costal cartilages, and the major classes of ribs.

The **jugular notch**, located between the clavicular articulations, is a shallow indentation on the superior surface of the manubrium.

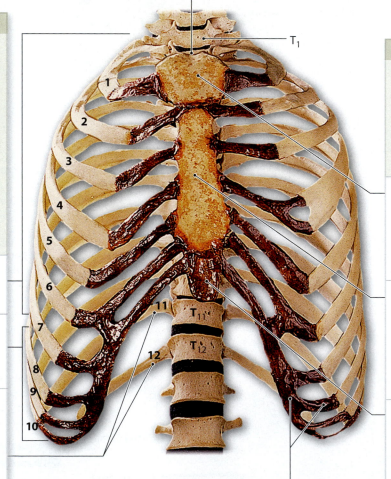

Ribs

The **ribs** reinforce the posterior and lateral walls of the thoracic cavity. Ribs 1–7 gradually increase in length and radius of curvature. Although the ribs are quite mobile and are among the most flexible of bones, they can be broken by a sharp blow or crushing impact. Fortunately, they are so stabilized by connective tissues and surrounding muscles that open fractures are rare, and splinting is unnecessary.

Vertebrosternal ribs (ribs 1–7) are connected to the sternum by individual **costal cartilages**.

The **vertebrochondral ribs** (ribs 8–10) are connected to the sternum by shared costal cartilages.

The last two pairs of ribs (11 and 12) are called **floating ribs**, because they have no connection with the sternum, or **vertebral ribs**, because they are attached only to the vertebrae and muscles of the body wall.

Sternum

The adult **sternum**, or breastbone, is a flat bone that forms in the anterior midline of the thoracic wall. It has three distinct regions that usually fuse together during adulthood.

Manubrium: The broad, trapezoid-shaped manubrium (ma-NOO-brē-um) articulates with the clavicles (collarbones) and the cartilages of the first pair of ribs.

Body: The body attaches to the inferior surface of the manubrium and extends inferiorly along the midline. Individual costal cartilages from rib pairs 2–7 are attached to this portion of the sternum.

Xiphoid process: The xiphoid (ZĪ-foyd) process, the smallest part of the sternum, is attached to the inferior surface of the body of the sternum.

Costal cartilages connect ribs to the sternum, either individually or in groups.

2 Ribs are elongated, curved, flattened bones that originate on or between the thoracic vertebrae and end in the wall of the thoracic cavity. Each of us, regardless of sex, has 12 pairs of ribs. Major landmarks on a representative rib (ribs 2–9) are shown in this posterior view.

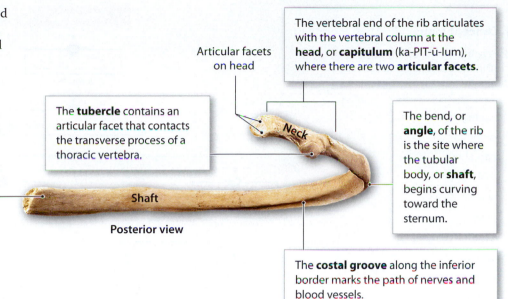

The vertebral end of the rib articulates with the vertebral column at the **head**, or **capitulum** (ka-PIT-ū-lum), where there are two **articular facets**.

Articular facets on head

The **tubercle** contains an articular facet that contacts the transverse process of a thoracic vertebra.

The bend, or **angle**, of the rib is the site where the tubular body, or **shaft**, begins curving toward the sternum.

The superficial surface is convex and provides an attachment site for muscles of the pectoral girdle and trunk. The intercostal muscles, which move the ribs, are attached to the superior and inferior surfaces.

Shaft

Neck

Posterior view

The **costal groove** along the inferior border marks the path of nerves and blood vessels.

3 This is a superior view of a representative rib showing sites of articulations with a thoracic vertebra. The heads of ribs 2–9 articulate with **costal facets** on two adjacent vertebrae, and their tubercular facets articulate with the transverse costal facets of the inferior vertebra. The heads of ribs 1, 10, 11, and 12 articulate with individual vertebrae at single costal facets. The tubercular facets of ribs 1 and 10 articulate with the costal facets of vertebrae T_1 and T_{10}, respectively. Ribs 11 and 12 articulate only at their heads, and there are no tubercular facets.

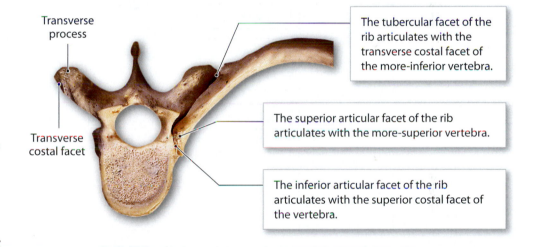

Transverse process

Transverse costal facet

The tubercular facet of the rib articulates with the transverse costal facet of the more-inferior vertebra.

The superior articular facet of the rib articulates with the more-superior vertebra.

The inferior articular facet of the rib articulates with the superior costal facet of the vertebra.

4 A typical rib acts as if it were a bucket's handle at a position just below horizontal. Pushing the handle down forces it inward; pulling it up swings it outward. Because of the curvature of the ribs, the same movements change the position of the sternum. Depression of the ribs pulls the sternum inward, whereas elevation moves it outward. As a result, movements of the ribs affect both the width and the depth of the thoracic cage, increasing or decreasing its volume accordingly.

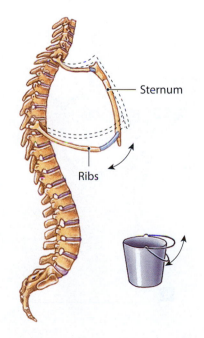

Sternum

Ribs

Module 7.12 Review

a. How are vertebrosternal ribs distinguished from vertebrochondral ribs?

b. Improper chest compressions during cardiopulmonary resuscitation (CPR) can result in fractures of which bones?

c. In addition to the ribs and sternum, what other bones make up the thoracic cage?

1. Concept map

Use each of the following terms once to fill in the blank boxes to correctly complete the skeleton concept map.

- floating
- temporal
- mandible
- axial
- hyoid
- sacral
- vertebral column
- lacrimal
- xiphoid process
- occipital
- sternum
- skull
- thoracic
- longitudinal
- lumbar

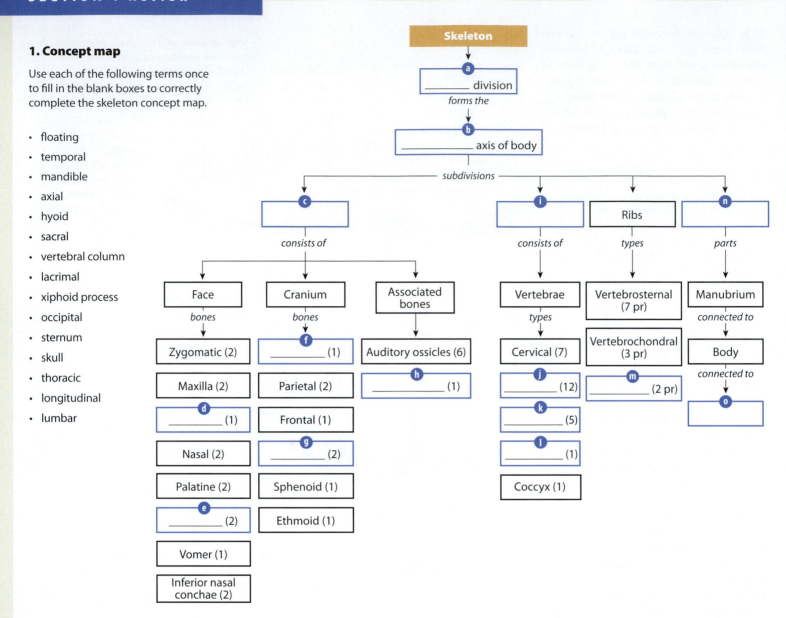

Skeleton

a _____ division

forms the

b _____ axis of body

subdivisions

c _____ **i** _____ Ribs **n** _____

consists of *consists of* *types* *parts*

| Face | Cranium | Associated bones | Vertebrae | Vertebrosternal (7 pr) | Manubrium |

bones *bones* *types* *connected to*

| Zygomatic (2) | **f** _____ (1) | Auditory ossicles (6) | Cervical (7) | Vertebrochondral (3 pr) | Body |

connected to

| Maxilla (2) | Parietal (2) | **h** _____ (1) | **j** _____ (12) | **m** _____ (2 pr) | **o** _____ |

| **d** _____ (1) | Frontal (1) | | **k** _____ (5) | | |

| Nasal (2) | **g** _____ (2) | | **l** _____ (1) | | |

| Palatine (2) | Sphenoid (1) | | Coccyx (1) | | |

| **e** _____ (2) | Ethmoid (1) | | | | |

| Vomer (1) | | | | | |

| Inferior nasal conchae (2) | | | | | |

2. Short answer

For each of the following vertebrae, identify its vertebral region.

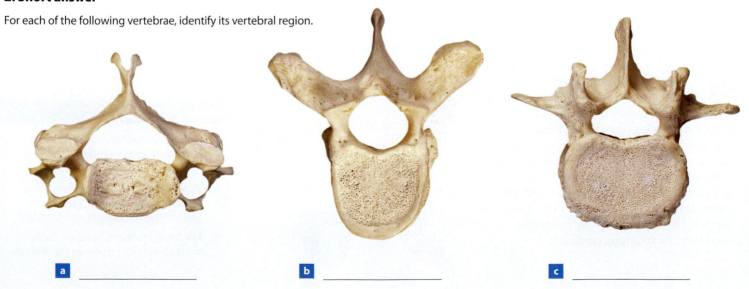

a _____

b _____

c _____

The Appendicular Skeleton

The **appendicular skeleton** includes the bones of the limbs and the supporting elements, or girdles, that connect them to the trunk. The descriptions in this section emphasize surface features that either have functional importance (such as the attachment sites for skeletal muscles and the paths of major nerves and blood vessels) or provide landmarks that define areas and locate structures of the body.

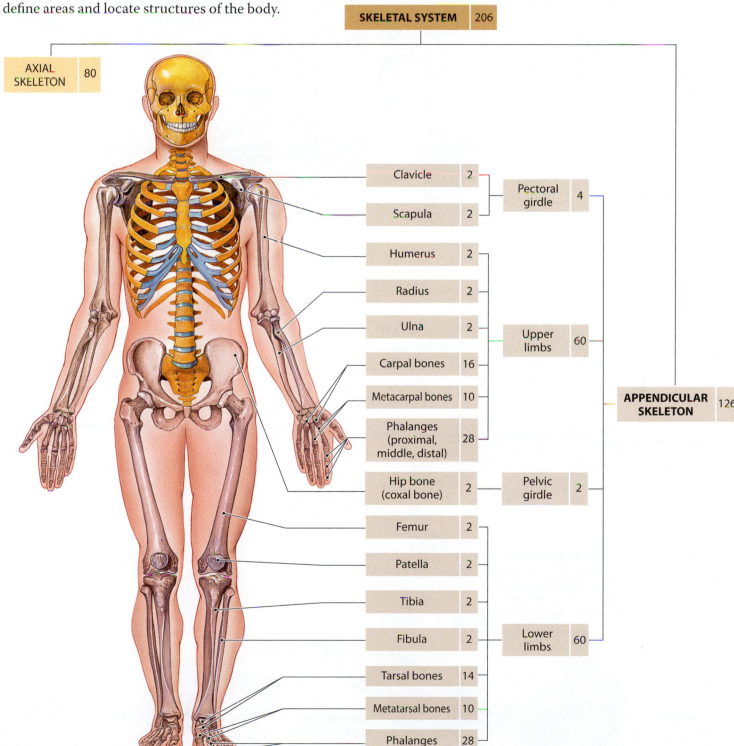

| SKELETAL SYSTEM | 206 |

| AXIAL SKELETON | 80 |

Clavicle	2
Scapula	2
Pectoral girdle	4

Humerus	2
Radius	2
Ulna	2
Carpal bones	16
Metacarpal bones	10
Phalanges (proximal, middle, distal)	28
Upper limbs	60

| Hip bone (coxal bone) | 2 |
| Pelvic girdle | 2 |

Femur	2
Patella	2
Tibia	2
Fibula	2
Tarsal bones	14
Metatarsal bones	10
Phalanges	28
Lower limbs	60

| APPENDICULAR SKELETON | 126 |

The pectoral girdles—the clavicles and scapulae—connect the upper limbs to the axial skeleton

Each arm articulates (that is, forms a joint) with the trunk at the **pectoral girdle**, or shoulder girdle. The pectoral girdle consists of two S-shaped **clavicles** (KLAV-i-kulz; collarbones) and two broad, flat **scapulae** (SKAP-ū-lē; singular, *scapula*, SKAP-ū-luh; shoulder blades).

1 The clavicles originate at the superior, lateral border of the manubrium of the sternum, lateral to the jugular notch.

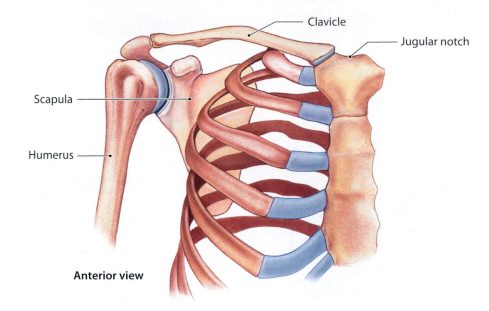

Clavicle

Jugular notch

Scapula

Humerus

Anterior view

2 From the sternal articulation at the pyramid-shaped **sternal end**, each clavicle curves laterally and posteriorly for roughly half its length. It then forms a smooth posterior curve to articulate with a process of the scapula, the **acromion** (a-KRŌ-mē-on).

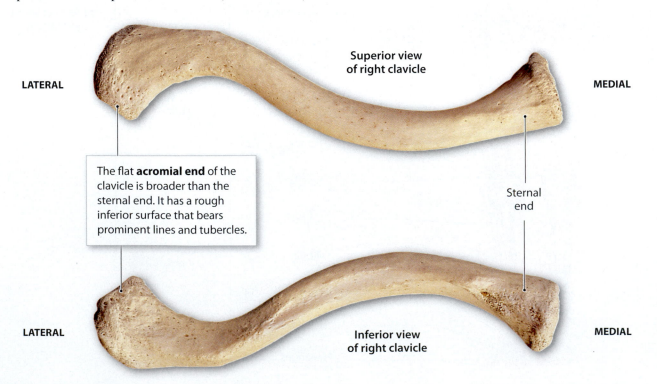

Superior view of right clavicle

LATERAL

MEDIAL

The flat **acromial end** of the clavicle is broader than the sternal end. It has a rough inferior surface that bears prominent lines and tubercles.

Sternal end

LATERAL

Inferior view of right clavicle

MEDIAL

3 The anterior surface of the **body** of each scapula forms a broad, smooth triangle. The three sides of the triangle are the **superior border**; the **medial border**, or vertebral border; and the **lateral border**, or axillary (*axilla*, armpit) border. Muscles that position the scapula attach along these edges. The corners of the triangle are called the **superior angle**, the **inferior angle**, and the **lateral angle**. The depression in the anterior surface is called the **subscapular fossa**.

4 The posterior surface of the scapula is convex and has prominent ridges and processes for muscle attachment.

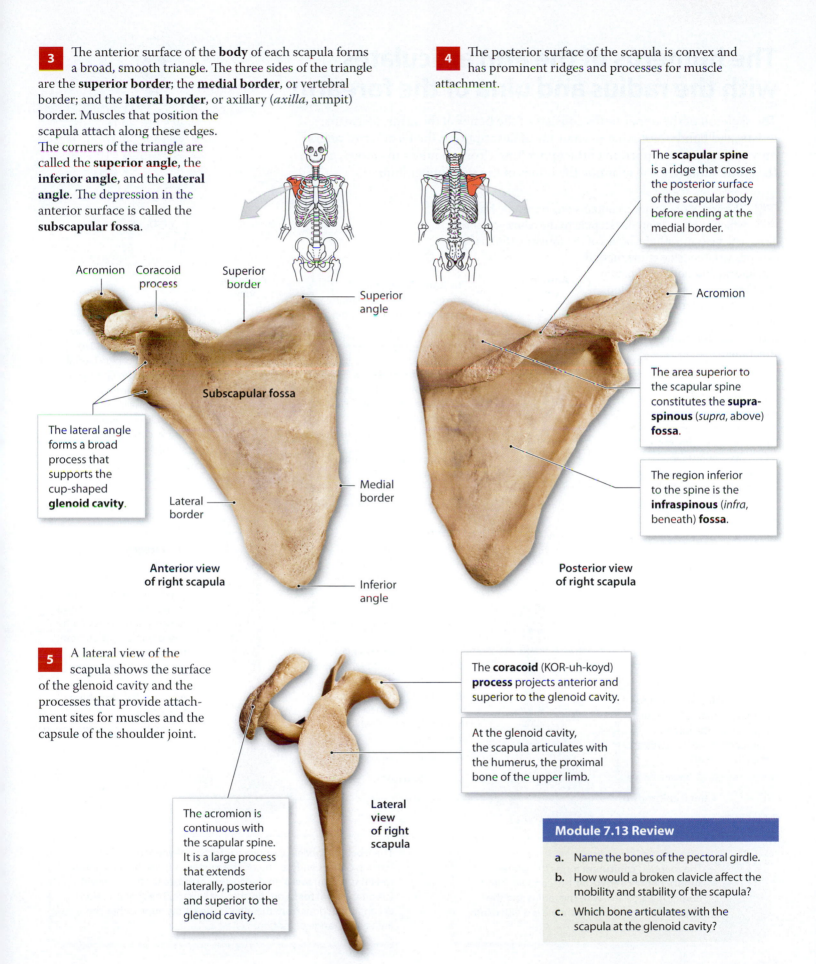

The **scapular spine** is a ridge that crosses the posterior surface of the scapular body before ending at the medial border.

Acromion

Coracoid process

Superior border

Superior angle

Acromion

Subscapular fossa

The area superior to the scapular spine constitutes the **supraspinous** (*supra*, above) **fossa**.

The lateral angle forms a broad process that supports the cup-shaped **glenoid cavity**.

Lateral border

Medial border

The region inferior to the spine is the **infraspinous** (*infra*, beneath) **fossa**.

Anterior view of right scapula

Posterior view of right scapula

Inferior angle

5 A lateral view of the scapula shows the surface of the glenoid cavity and the processes that provide attachment sites for muscles and the capsule of the shoulder joint.

The **coracoid** (KOR-uh-koyd) **process** projects anterior and superior to the glenoid cavity.

At the glenoid cavity, the scapula articulates with the humerus, the proximal bone of the upper limb.

The acromion is continuous with the scapular spine. It is a large process that extends laterally, posterior and superior to the glenoid cavity.

Lateral view of right scapula

Module 7.13 Review

a. Name the bones of the pectoral girdle.

b. How would a broken clavicle affect the mobility and stability of the scapula?

c. Which bone articulates with the scapula at the glenoid cavity?

The humerus of the arm articulates with the radius and ulna of the forearm

The skeleton of the upper limbs consists of the bones of the arms, forearms, wrists, and hands. Note that in anatomical descriptions, the term "arm" refers only to the proximal portion of the upper limb (from shoulder to elbow), not to the entire limb. We will examine the bones of the right upper limb.

1 The arm, or brachium, contains one bone, the **humerus**, which extends from the scapula to the elbow. Near the distal articulation with the bones of the forearm, the shaft expands to either side at the **medial** and **lateral epicondyles**. Epicondyles are processes that develop proximal to an articulation and provide additional surface area for muscle attachment.

Anterior view

Posterior view

Greater tubercle

The round **head** at the proximal end of the humerus articulates with the glenoid cavity of the scapula.

The prominent **greater tubercle** is a rounded projection on the lateral surface of the epiphysis, near the margin of the humeral head. The greater tubercle establishes the lateral contour of the shoulder.

The **lesser tubercle** is a smaller projection that lies on the anterior, medial surface of the epiphysis.

The **intertubercular groove** lies between the greater and lesser tubercles. Both tubercles are important sites for muscle attachment; a large tendon runs along the groove.

The **anatomical neck** marks the extent of the joint capsule.

Shaft

The **surgical neck** corresponds to the metaphysis of the growing bone. The name reflects the fact that fractures typically occur at this site.

The **radial groove** crosses the inferior end of the deltoid tuberosity. This depression marks the path of the radial nerve, a large nerve that provides both sensory information from the posterior surface of the limb and motor control over the large muscles that straighten the elbow.

The **radial fossa** accommodates a portion of the radial head when the forearm approaches the humerus as the elbow bends.

The **deltoid tuberosity** is a large, rough elevation on the lateral surface of the shaft, approximately halfway along its length. It is named after the deltoid muscle, which attaches to it.

Coronoid fossa

Olecranon fossa

Lateral epicondyle

Medial epicondyle

Trochlea

The rounded **capitulum** forms the lateral surface of the **condyle**. At the condyle, the humerus articulates with the radius and the ulna of the forearm. The condyle is divided into two articular regions.

The **trochlea** (*trochlea*, a pulley) is the spool-shaped medial portion of the condyle. The trochlea extends from the **olecranon** (ō-LEK-ruh-non) **fossa** on the posterior surface to the **coronoid** (*corona*, crown) **fossa** on the anterior surface. These depressions accept projections from the ulna as the elbow approaches the limits of its range of motion.

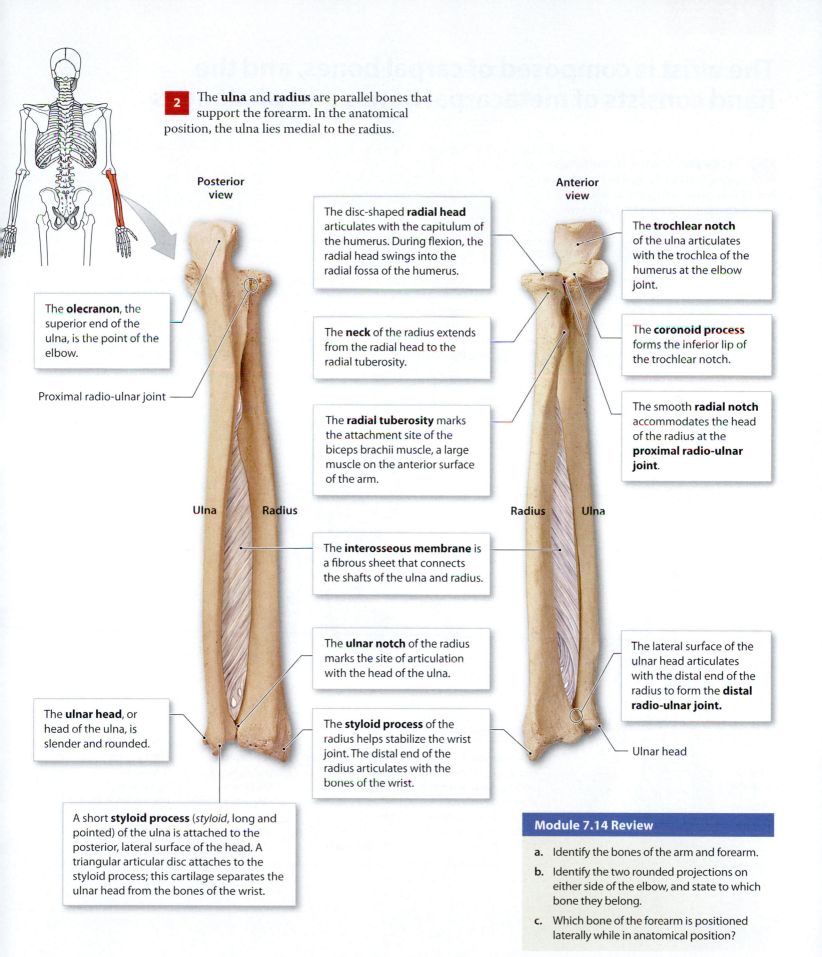

2 The **ulna** and **radius** are parallel bones that support the forearm. In the anatomical position, the ulna lies medial to the radius.

Posterior view

Anterior view

The disc-shaped **radial head** articulates with the capitulum of the humerus. During flexion, the radial head swings into the radial fossa of the humerus.

The **trochlear notch** of the ulna articulates with the trochlea of the humerus at the elbow joint.

The **olecranon**, the superior end of the ulna, is the point of the elbow.

Proximal radio-ulnar joint

The **neck** of the radius extends from the radial head to the radial tuberosity.

The **coronoid process** forms the inferior lip of the trochlear notch.

The smooth **radial notch** accommodates the head of the radius at the **proximal radio-ulnar joint**.

The **radial tuberosity** marks the attachment site of the biceps brachii muscle, a large muscle on the anterior surface of the arm.

Ulna   Radius   Radius   Ulna

The **interosseous membrane** is a fibrous sheet that connects the shafts of the ulna and radius.

The **ulnar notch** of the radius marks the site of articulation with the head of the ulna.

The lateral surface of the ulnar head articulates with the distal end of the radius to form the **distal radio-ulnar joint.**

The **ulnar head**, or head of the ulna, is slender and rounded.

The **styloid process** of the radius helps stabilize the wrist joint. The distal end of the radius articulates with the bones of the wrist.

Ulnar head

A short **styloid process** (*styloid*, long and pointed) of the ulna is attached to the posterior, lateral surface of the head. A triangular articular disc attaches to the styloid process; this cartilage separates the ulnar head from the bones of the wrist.

Module 7.14 Review

a. Identify the bones of the arm and forearm.

b. Identify the two rounded projections on either side of the elbow, and state to which bone they belong.

c. Which bone of the forearm is positioned laterally while in anatomical position?

The wrist is composed of carpal bones, and the hand consists of metacarpal bones and phalanges

1 The **carpus**, or wrist, contains eight carpal bones arranged in two rows: a row of four proximal carpal bones, and a row containing four distal carpal bones.

Proximal Carpal Bones

The **scaphoid** (*skaphe*, boat) is the proximal carpal bone on the lateral border of the wrist; it is the carpal bone closest to the styloid process of the radius.

The comma-shaped **lunate** (*luna*, moon) lies medial to the scaphoid and, like the scaphoid, articulates with the radius.

The small, pea-shaped **pisiform** (PIS-i-form; *pisum*, pea) sits anterior to the triquetrum.

The **triquetrum** (*triquetrus*, three-cornered) is a small pyramid-shaped bone medial to the lunate. The triquetrum articulates with the articular disc that separates the ulnar head from the wrist.

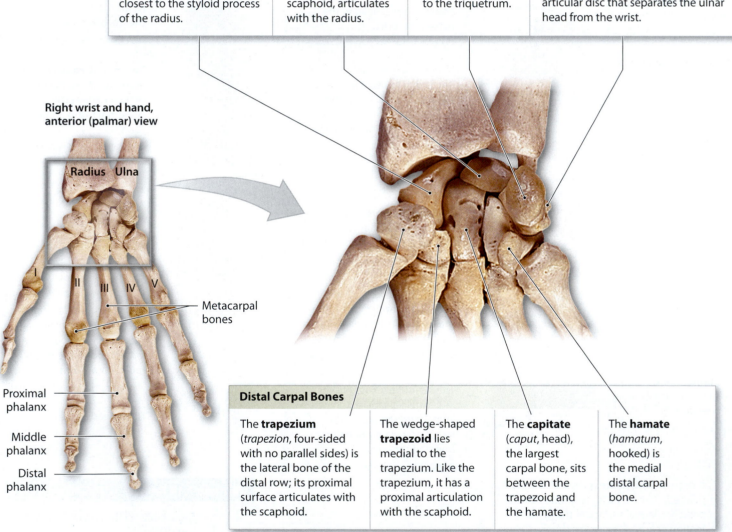

Right wrist and hand, anterior (palmar) view

Radius Ulna

I II III IV V

Metacarpal bones

Proximal phalanx

Middle phalanx

Distal phalanx

Distal Carpal Bones

The **trapezium** (*trapezion*, four-sided with no parallel sides) is the lateral bone of the distal row; its proximal surface articulates with the scaphoid.

The wedge-shaped **trapezoid** lies medial to the trapezium. Like the trapezium, it has a proximal articulation with the scaphoid.

The **capitate** (*caput*, head), the largest carpal bone, sits between the trapezoid and the hamate.

The **hamate** (*hamatum*, hooked) is the medial distal carpal bone.

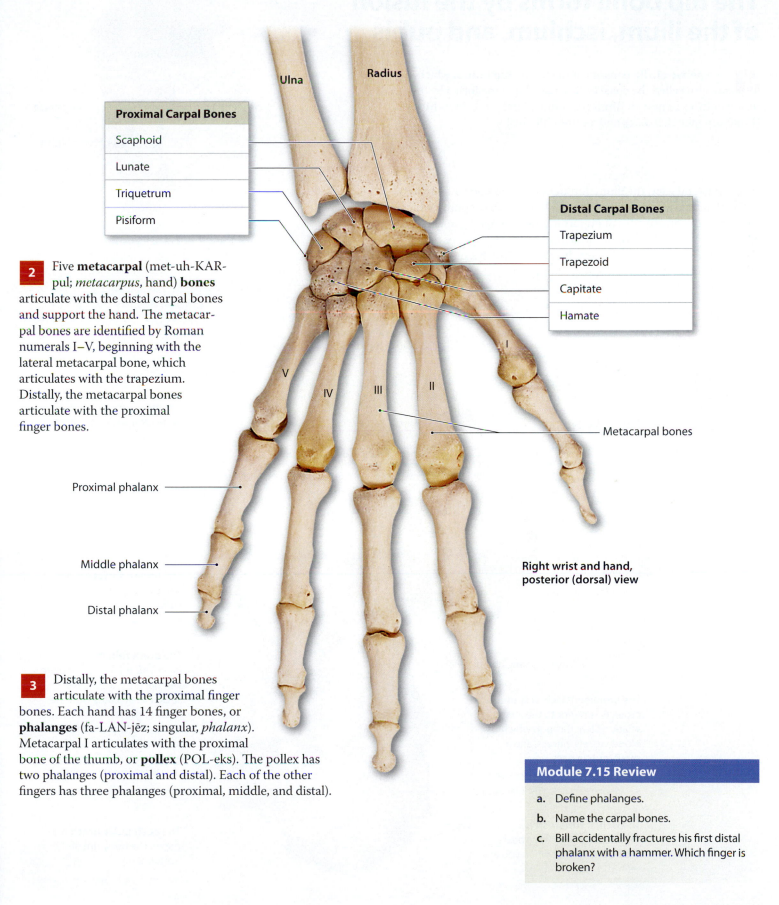

Proximal Carpal Bones

Proximal Carpal Bones
Scaphoid
Lunate
Triquetrum
Pisiform

Ulna

Radius

Distal Carpal Bones

Distal Carpal Bones
Trapezium
Trapezoid
Capitate
Hamate

2 Five **metacarpal** (met-uh-KAR-pul; *metacarpus*, hand) **bones** articulate with the distal carpal bones and support the hand. The metacarpal bones are identified by Roman numerals I–V, beginning with the lateral metacarpal bone, which articulates with the trapezium. Distally, the metacarpal bones articulate with the proximal finger bones.

Metacarpal bones

Proximal phalanx

Middle phalanx

Right wrist and hand, posterior (dorsal) view

Distal phalanx

3 Distally, the metacarpal bones articulate with the proximal finger bones. Each hand has 14 finger bones, or **phalanges** (fa-LAN-jēz; singular, *phalanx*). Metacarpal I articulates with the proximal bone of the thumb, or **pollex** (POL-eks). The pollex has two phalanges (proximal and distal). Each of the other fingers has three phalanges (proximal, middle, and distal).

Module 7.15 Review

a. Define phalanges.

b. Name the carpal bones.

c. Bill accidentally fractures his first distal phalanx with a hammer. Which finger is broken?

The hip bone forms by the fusion of the ilium, ischium, and pubis

1 The **pelvic girdle** consists of the paired **hip bones**, which are also called the coxal bones. Each hip bone forms by the fusion of three bones: an **ilium** (IL-ē-um; plural, *ilia*), an **ischium** (IS-kē-um; plural, *ischia*), and a **pubis** (PŪ-bis).

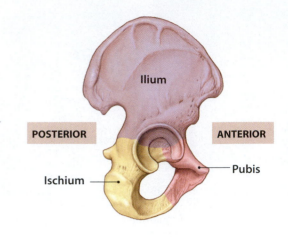

POSTERIOR

ANTERIOR

Ilium

Ischium

Pubis

2 In lateral view, the ilium dominates the hip bone. Landmarks along the margin of the ilium include the **iliac spines**, which mark the attachment sites of important muscles and ligaments; the **gluteal lines**, which mark the attachment of large hip muscles; and the **greater sciatic** (sī-AT-ik) **notch**, through which a major nerve (the sciatic nerve) reaches the lower limb.

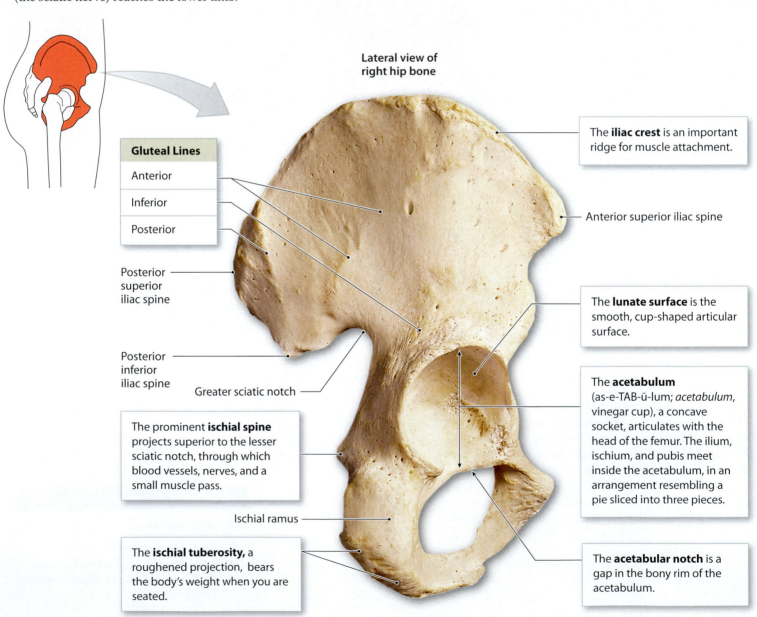

Lateral view of right hip bone

Gluteal Lines
Anterior
Inferior
Posterior

Posterior superior iliac spine

Posterior inferior iliac spine

Greater sciatic notch

The prominent **ischial spine** projects superior to the lesser sciatic notch, through which blood vessels, nerves, and a small muscle pass.

Ischial ramus

The **ischial tuberosity,** a roughened projection, bears the body's weight when you are seated.

The **iliac crest** is an important ridge for muscle attachment.

Anterior superior iliac spine

The **lunate surface** is the smooth, cup-shaped articular surface.

The **acetabulum** (as-e-TAB-ū-lum; *acetabulum*, vinegar cup), a concave socket, articulates with the head of the femur. The ilium, ischium, and pubis meet inside the acetabulum, in an arrangement resembling a pie sliced into three pieces.

The **acetabular notch** is a gap in the bony rim of the acetabulum.

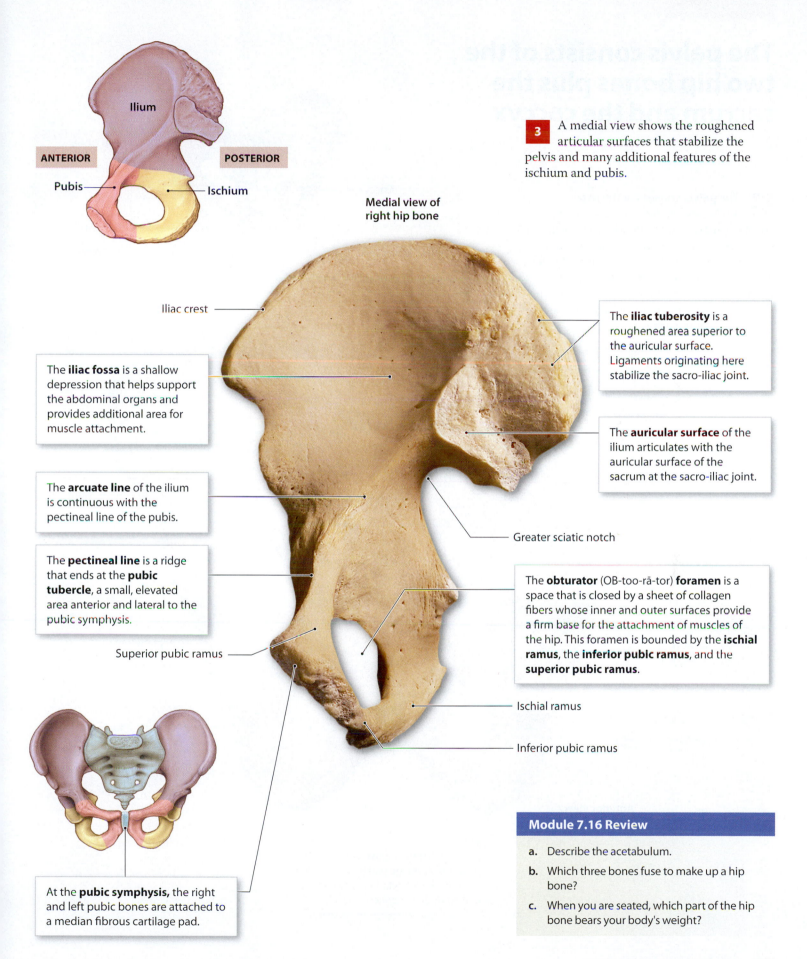

Ilium

ANTERIOR   POSTERIOR

Pubis   Ischium

3 A medial view shows the roughened articular surfaces that stabilize the pelvis and many additional features of the ischium and pubis.

Medial view of right hip bone

Iliac crest

The **iliac tuberosity** is a roughened area superior to the auricular surface. Ligaments originating here stabilize the sacro-iliac joint.

The **iliac fossa** is a shallow depression that helps support the abdominal organs and provides additional area for muscle attachment.

The **auricular surface** of the ilium articulates with the auricular surface of the sacrum at the sacro-iliac joint.

The **arcuate line** of the ilium is continuous with the pectineal line of the pubis.

Greater sciatic notch

The **pectineal line** is a ridge that ends at the **pubic tubercle**, a small, elevated area anterior and lateral to the pubic symphysis.

The **obturator** (OB-too-rā-tor) **foramen** is a space that is closed by a sheet of collagen fibers whose inner and outer surfaces provide a firm base for the attachment of muscles of the hip. This foramen is bounded by the **ischial ramus**, the **inferior pubic ramus**, and the **superior pubic ramus**.

Superior pubic ramus

Ischial ramus

Inferior pubic ramus

At the **pubic symphysis,** the right and left pubic bones are attached to a median fibrous cartilage pad.

Module 7.16 Review

a. Describe the acetabulum.

b. Which three bones fuse to make up a hip bone?

c. When you are seated, which part of the hip bone bears your body's weight?

The pelvis consists of the two hip bones plus the sacrum and the coccyx

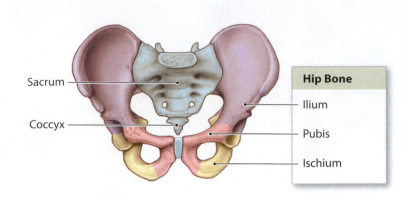

Sacrum

Coccyx

Hip Bone

Ilium

Pubis

Ischium

1 The **pelvis** consists of the two hip bones, the sacrum, and the coccyx. An extensive network of ligaments connects the lateral borders of the sacrum with the iliac crest, the ischial tuberosity, the ischial spine, and the arcuate line. Other ligaments tie the ilia to the posterior lumbar vertebrae. These interconnections increase the stability of the pelvis.

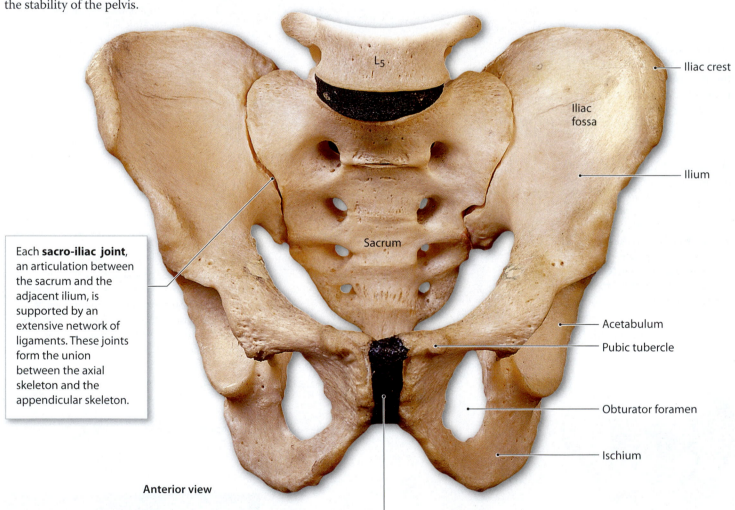

L₅

Iliac crest

Iliac fossa

Ilium

Sacrum

Each **sacro-iliac joint**, an articulation between the sacrum and the adjacent ilium, is supported by an extensive network of ligaments. These joints form the union between the axial skeleton and the appendicular skeleton.

Acetabulum

Pubic tubercle

Obturator foramen

Ischium

Anterior view

The **pubic symphysis** is a fibrous cartilage pad that forms the articulation between the two pubic bones.

2 The pelvis may be divided into the **true** (lesser) **pelvis** (shown in purple) and the **false** (greater) **pelvis.** The true pelvis encloses the pelvic cavity, a subdivision of the abdominopelvic cavity. The superior limit of the true pelvis is a line that extends from either side of the base of the sacrum, along the arcuate line and pectineal line to the pubic symphysis.

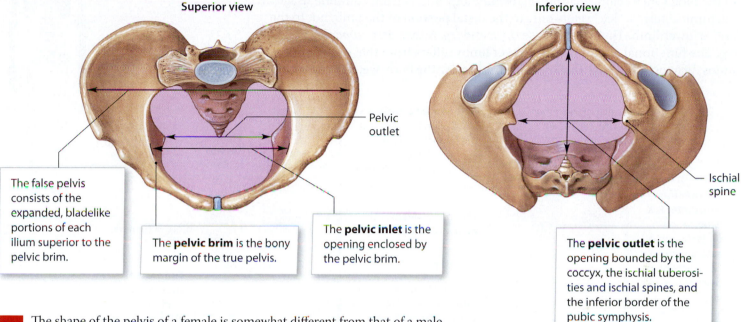

Superior view

Inferior view

Pelvic outlet

The false pelvis consists of the expanded, bladelike portions of each ilium superior to the pelvic brim.

The **pelvic brim** is the bony margin of the true pelvis.

The **pelvic inlet** is the opening enclosed by the pelvic brim.

Ischial spine

The **pelvic outlet** is the opening bounded by the coccyx, the ischial tuberosities and ischial spines, and the inferior border of the pubic symphysis.

3 The shape of the pelvis of a female is somewhat different from that of a male. Some of the differences are the result of variations in body size and muscle mass. For example, in females, the pelvis is generally smoother and lighter and has less prominent markings. Females have other skeletal adaptations for childbearing, including:

- An enlarged pelvic outlet.
- A broader pubic angle (the inferior angle between the pubic bones), greater than 100°.
- Less curvature on the sacrum and coccyx, which, in males, arcs into the pelvic outlet.
- A wider, more circular pelvic inlet.
- A relatively broad pelvis that does not extend as far superiorly (a "low pelvis").
- Ilia that project farther laterally but do not extend as far superior to the sacrum.

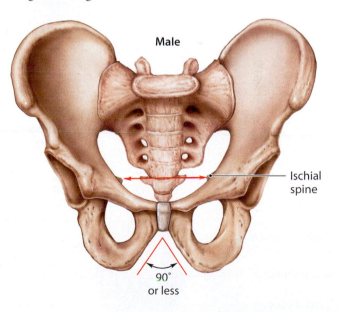

Male

Ischial spine

90° or less

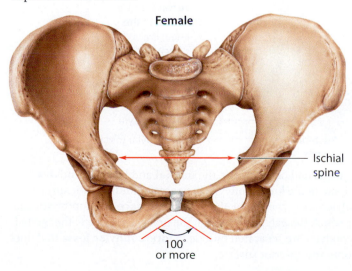

Female

Ischial spine

100° or more

Module 7.17 Review

a. Name the bones of the pelvis.

b. The pubic bones are joined anteriorly by what structure?

c. How is the pelvis of females adapted for childbearing?

The femur articulates with the patella and tibia

The skeleton of each lower limb consists of a **femur** (thigh), a **patella** (kneecap), a **tibia** and a **fibula** (leg), and the **tarsal bones**, **metatarsal bones**, and **phalanges** of the foot. Once again, anatomical terminology differs from common usage. In anatomical terms, "leg" refers only to the distal portion of the limb, not to the entire lower limb. Thus, we will use *thigh* and *leg,* rather than *upper leg* and *lower leg.* The functional anatomy of the lower limbs differs from that of the upper limbs, primarily because the lower limbs transfer the body weight to the ground.

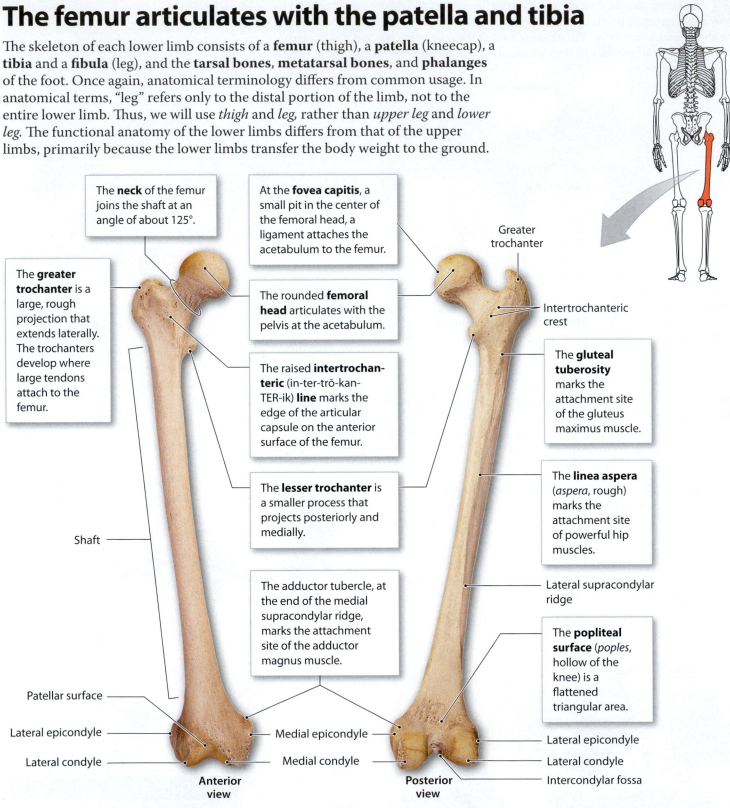

The **neck** of the femur joins the shaft at an angle of about 125°.

At the **fovea capitis,** a small pit in the center of the femoral head, a ligament attaches the acetabulum to the femur.

The **greater trochanter** is a large, rough projection that extends laterally. The trochanters develop where large tendons attach to the femur.

The rounded **femoral head** articulates with the pelvis at the acetabulum.

The raised **intertrochanteric** (in-ter-trō-kan-TER-ik) **line** marks the edge of the articular capsule on the anterior surface of the femur.

The **lesser trochanter** is a smaller process that projects posteriorly and medially.

Shaft

The adductor tubercle, at the end of the medial supracondylar ridge, marks the attachment site of the adductor magnus muscle.

Patellar surface

Lateral epicondyle

Lateral condyle

Medial epicondyle

Medial condyle

Anterior view

Greater trochanter

Intertrochanteric crest

The **gluteal tuberosity** marks the attachment site of the gluteus maximus muscle.

The **linea aspera** (*aspera*, rough) marks the attachment site of powerful hip muscles.

Lateral supracondylar ridge

The **popliteal surface** (*poples*, hollow of the knee) is a flattened triangular area.

Lateral epicondyle

Lateral condyle

Intercondylar fossa

Posterior view

1 The femur is the longest and heaviest bone in the body. It articulates with the hip bone at the hip joint and with the tibia of the leg at the knee joint. Major landmarks are shown here on the anterior and posterior surfaces of the right femur.

2 At the distal end of the femur, the **medial** and **lateral condyles** participate in the knee joint. On the anterior and inferior surfaces, the two condyles are separated by the **patellar surface**, a smooth articular surface over which the patella glides. On the posterior surface, the medial and lateral condyles are separated by a deep **intercondylar fossa** that does not extend onto the anterior surface.

3 The patella is a large sesamoid bone that forms within the tendon of the quadriceps femoris, a group of muscles that extend (straighten) the knee.

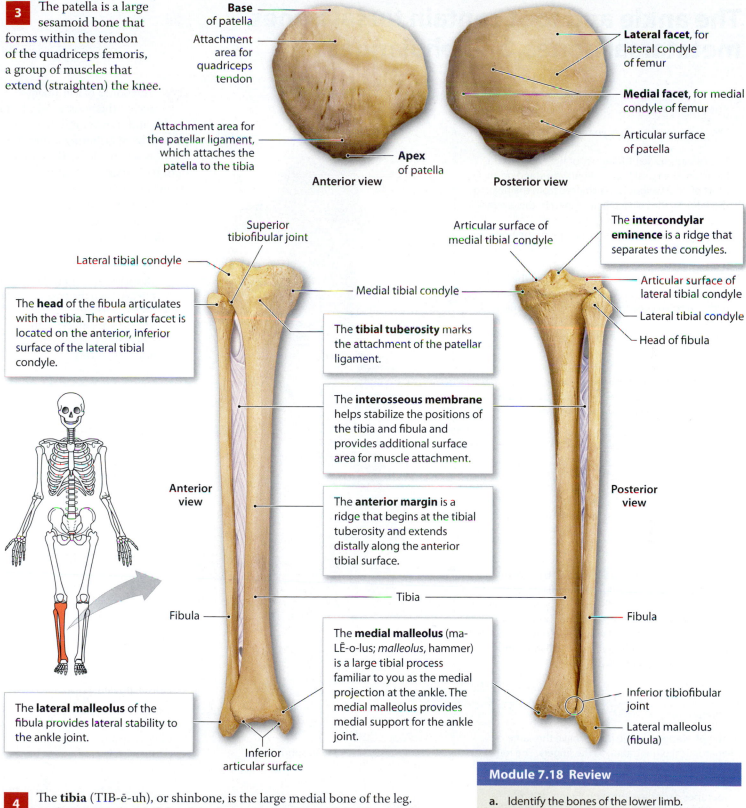

Base of patella

Attachment area for quadriceps tendon

Attachment area for the patellar ligament, which attaches the patella to the tibia

Apex of patella

Anterior view

Lateral facet, for lateral condyle of femur

Medial facet, for medial condyle of femur

Articular surface of patella

Posterior view

Superior tibiofibular joint

Lateral tibial condyle

The **head** of the fibula articulates with the tibia. The articular facet is located on the anterior, inferior surface of the lateral tibial condyle.

Medial tibial condyle

The **tibial tuberosity** marks the attachment of the patellar ligament.

Anterior view

Fibula

The **interosseous membrane** helps stabilize the positions of the tibia and fibula and provides additional surface area for muscle attachment.

The **anterior margin** is a ridge that begins at the tibial tuberosity and extends distally along the anterior tibial surface.

Tibia

The **medial malleolus** (ma-LĒ-o-lus; *malleolus*, hammer) is a large tibial process familiar to you as the medial projection at the ankle. The medial malleolus provides medial support for the ankle joint.

The **lateral malleolus** of the fibula provides lateral stability to the ankle joint.

Inferior articular surface

Articular surface of medial tibial condyle

The **intercondylar eminence** is a ridge that separates the condyles.

Articular surface of lateral tibial condyle

Lateral tibial condyle

Head of fibula

Posterior view

Fibula

Inferior tibiofibular joint

Lateral malleolus (fibula)

4 The **tibia** (TIB-ē-uh), or shinbone, is the large medial bone of the leg. At the proximal end of the tibia, the **medial** and **lateral tibial condyles** articulate with the medial and lateral condyles of the femur. The slender **fibula** (FIB-ū-luh) parallels the lateral border of the tibia but does not participate in the knee joint and bears no weight. However, the fibula is important as a site for the attachment of muscles that move the foot and toes. In addition, the distal tip of the fibula extends lateral to the ankle, providing important stability to that joint.

Module 7.18 Review

a. Identify the bones of the lower limb.

b. Which structure articulates with the acetabulum?

c. The fibula neither participates in the knee joint nor bears weight. Yet, when it is fractured, walking becomes difficult. Why?

The ankle and foot contain tarsal bones, metatarsal bones, and phalanges

The Ankle (Tarsus)

The ankle consists of seven **tarsal bones**.

The **calcaneus** (kal-KĀ-nē-us), or heel bone, is the largest of the tarsal bones. When you stand normally, most of your weight is transmitted from the tibia, to the talus, to the calcaneus, and then to the ground.

The large **talus** transmits the weight of the body from the tibia toward the toes.

The **navicular** is anterior to the talus, on the medial side of the ankle. It articulates with the talus and with the three cuneiform (kū-NĒ-i-form) bones.

The **cuboid** articulates with the anterior surface of the calcaneus.

The three **cuneiform bones** are arranged in a row, with articulations between them. They are named according to their relative positions: medial, intermediate, and lateral.

Metatarsals

The distal surfaces of the cuboid and the cuneiform bones articulate with the **metatarsal bones** of the foot. The five long metatarsal bones form the distal portion of the foot.

The metatarsal bones are identified by Roman numerals I–V, proceeding from medial to lateral across the sole. Proximally, metatarsal bones I–III articulate with the three cuneiform bones, and metatarsal bones IV and V articulate with the cuboid. Distally, each metatarsal bone articulates with a different proximal phalanx.

Phalanges

The **phalanges**, or toe bones, have the same anatomical organization as the fingers. The toes contain 14 phalanges. The **hallux**, or great toe, has two phalanges (proximal and distal), and the other four toes each have three phalanges (proximal, middle, and distal).

1 The bones of the ankle accept the body weight from the leg and transfer it to the ground by distributing it through the bones of the foot. The combination of ankle and foot must be strong enough yet flexible enough to deal with the changes in loading that occur during walking, running, and jumping.

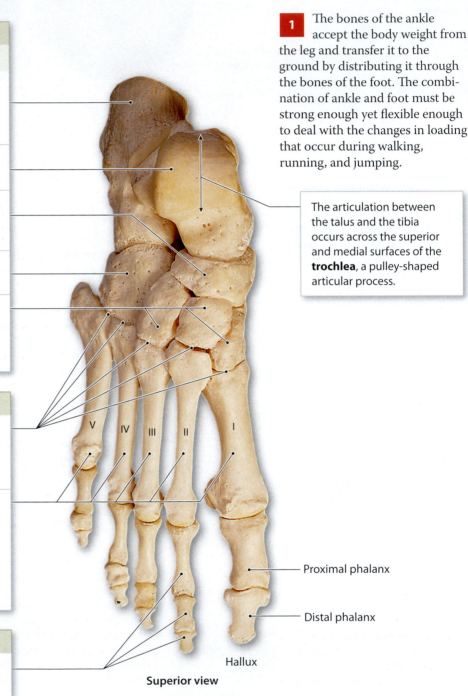

The articulation between the talus and the tibia occurs across the superior and medial surfaces of the **trochlea**, a pulley-shaped articular process.

V IV III II I

Proximal phalanx

Distal phalanx

Hallux

Superior view

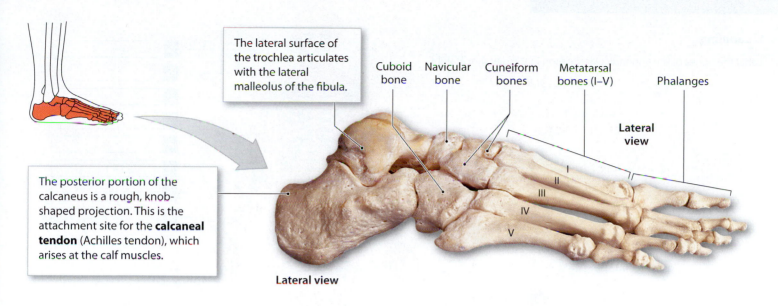

The lateral surface of the trochlea articulates with the lateral malleolus of the fibula.

Cuboid bone

Navicular bone

Cuneiform bones

Metatarsal bones (I–V)

Phalanges

Lateral view

I
II
III
IV
V

Lateral view

The posterior portion of the calcaneus is a rough, knob-shaped projection. This is the attachment site for the **calcaneal tendon** (Achilles tendon), which arises at the calf muscles.

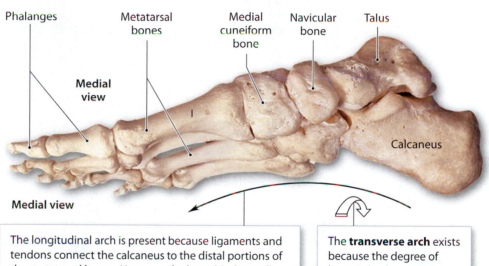

Phalanges

Metatarsal bones

Medial cuneiform bone

Navicular bone

Talus

Medial view

I

Calcaneus

Medial view

2 Weight transfer occurs along the **longitudinal arch** of the foot. The amount of weight transferred forward depends on the position of the foot and the placement of one's body weight. When you "dig in your heels" all your body weight rests on the calcaneus, but when you are on tiptoes all of your weight is transferred to the metatarsal bones and phalanges. When you stand normally, your body weight is distributed evenly between the calcaneus and the distal ends of the metatarsal bones.

The longitudinal arch is present because ligaments and tendons connect the calcaneus to the distal portions of the metatarsal bones. However, the lateral (calcaneal) portion of the longitudinal arch has much less curvature than the medial (talar) portion, in part because the talar portion has considerably more elasticity. As a result, the medial plantar surface of the foot remains elevated, so that the muscles, nerves, and blood vessels that supply the inferior surface are not squeezed between the metatarsal bones and the ground.

The **transverse arch** exists because the degree of longitudinal curvature changes from the medial border to the lateral border of the foot.

The arches of the foot are usually present at birth. Sometimes, however, they fail to develop properly. In **congenital talipes equinovarus** (clubfoot), abnormal muscle development distorts growing bones and joints. One or both feet may be involved, and the condition can be mild, moderate, or severe. In most cases, the tibia, ankle, and foot are affected; the longitudinal arch is exaggerated, and the feet are turned medially and inverted. If both feet are involved, the soles face one another. This condition, which affects 1 in 1000 births, is roughly twice as common in boys as girls. Prompt treatment with casts or other supports in infancy helps alleviate the problem, and fewer than half the cases require surgery.

Module 7.19 Review

a. Identify the tarsal bones.

b. Which foot bone transmits the weight of the body from the tibia toward the toes?

c. While jumping off the back steps at his house, 10-year-old Joey lands on his right heel and breaks his foot. Which foot bone is most likely broken?

1. Labeling

Label the bones of the appendicular skeleton in the diagram at right.

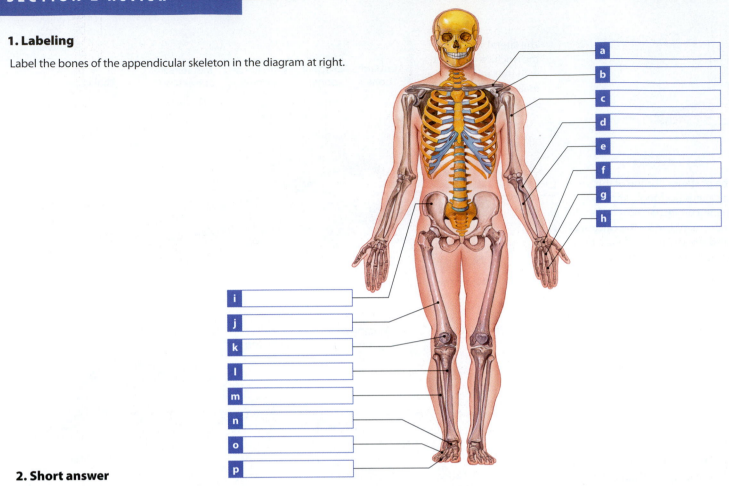

a
b
c
d
e
f
g
h

i
j
k
l
m
n
o
p

2. Short answer

In the following photographs of a scapula, identify the three views (a–c) and the indicated bone markings (d–j).

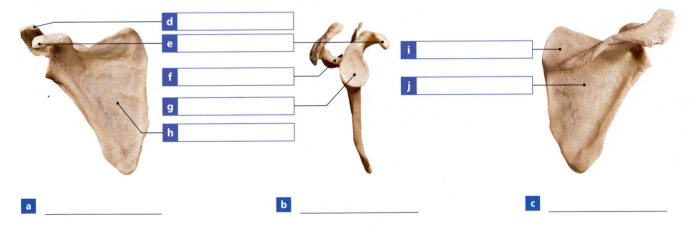

d
e
f
g
h

i
j

a _____

b _____

c _____

3. Labeling

In the following drawing of the pelvis, label the indicated structures.

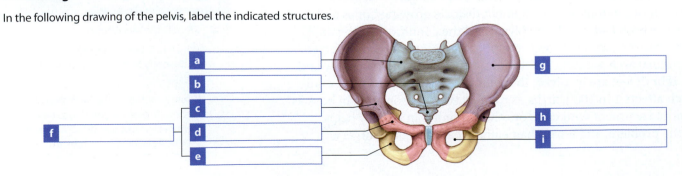

a
b
c
f
d
e

g
h
i

Visual Outline with Key Terms

Summarize the content of each module using the terms in the order provided.

SECTION 1

The Axial Skeleton

- axial skeleton
  - skull and associated bones
  - thoracic cage
  - vertebral column

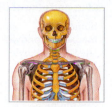

7.1

The skull has cranial and facial components that are usually bound together by sutures

- facial bones
- cranial bones
- cranium
- cranial cavity
- sutures
- coronal suture
- calvaria
- squamous suture
- sagittal suture
- lambdoid suture

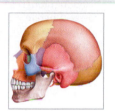

7.2

Facial bones dominate the anterior aspect of the skull, and cranial bones dominate the posterior surface

- nasal bones
- lacrimal bones
- palatine bones
- zygomatic bones
- maxillae
- inferior nasal conchae
- vomer
- mandible
- frontal bone
- sphenoid
- ethmoid
- parietal bones
- occipital bone
- temporal bones
- external occipital crest
- mastoid process
- styloid process

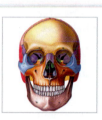

7.3

The lateral and medial aspects of the skull share many surface features

- frontal squama
- superior temporal lines
- inferior temporal lines
- squamous part
- external acoustic meatus
- zygomatic process
- zygomatic arch
- mandibular angle
- mental protuberance
- alveolar processes
- frontal sinuses
- hypophyseal fossa
- sella turcica
- petrous part
- internal acoustic meatus
- hypoglossal canal

7.4

The foramina on the inferior surface of the skull lead into the interior of the cranium

- mandibular fossa
- occipital condyles
- inferior and superior nuchal lines
- foramen lacerum
- foramen ovale
- carotid canal
- jugular foramen
- stylomastoid foramen
- foramen magnum
- internal occipital crest

• = *Term boldfaced in this module*

7.5

The shapes and landmarks of the sphenoid, ethmoid, and palatine bones can best be seen in the isolated bones

- sphenoid
- optic canals
- lesser wings
- greater wings
- hypophyseal fossa
- sella turcica
- sphenoidal spine
- foramen spinosum
- foramen ovale
- foramen rotundum
- superior orbital fissure
- sphenoidal sinuses
- pterygoid plates
- body
- pterygoid processes
- ethmoid
- cribriform plate
- crista galli
- lateral masses
- ethmoidal labyrinth
- superior nasal conchae
- middle nasal conchae
- perpendicular plate (ethmoid)
- palatine bones
- perpendicular plate (palatine)
- horizontal plate
- orbital process

7.6

Each orbital complex contains one eye, and the nasal complex encloses the nasal cavities

- orbital complex
- lacrimal fossa
- supra-orbital margin
- supra-orbital notch
- lacrimal sulcus
- nasolacrimal canal
- infra-orbital foramen
- zygomaticofacial foramen
- nasal complex
- paranasal sinuses
- ethmoidal air cells
- maxillary sinuses
- frontal sinuses
- sphenoidal sinuses

7.7

The mandible forms the lower jaw, and the associated bones of the skull perform specialized functions

- condylar process
- mandibular notch
- coronoid process
- alveolar process
- body
- ramus
- mylohyoid line
- mandibular foramen
- hyoid bone
- greater horns
- lesser horns
- auditory ossicles

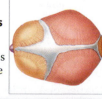

7.8

Fontanelles permit cranial growth in infants and small children

- fontanelles
- anterior fontanelle
- sphenoidal fontanelles
- mastoid fontanelles
- occipital fontanelle

7.9

The vertebral column has four spinal curves, and vertebrae have both anatomical similarities and regional differences

- cervical curve
- thoracic curve
- lumbar curve
- sacral curve
- articular processes
- vertebral arch
- vertebral body
- vertebral foramen
- spinous process
- laminae
- transverse processes
- pedicles
- intervertebral discs
- intervertebral foramina
- vertebral canal
- articular facet
- superior articular processes
- inferior articular processes

7.10

There are seven cervical vertebrae and twelve thoracic vertebrae

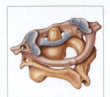

- cervical vertebrae
- transverse foramen
- bifid
- costal process
- atlas
- axis
- dens (odontoid process)
- vertebra prominens
- ligamentum nuchae
- thoracic vertebrae

7.11

There are five lumbar vertebrae; the sacrum and coccyx consist of fused vertebrae

- lumbar vertebrae
- sacrum
- ala
- sacral foramina
- base of sacrum
- sacral promontory
- apex of sacrum
- sacral canal
- superior articular processes
- median sacral crest
- sacral hiatus
- coccyx
- coccygeal cornua
- sacral tuberosity
- auricular surface
- lateral sacral crest

7.12

The thoracic cage protects organs in the chest and provides sites for muscle attachment

- thoracic cage
- ribs
- vertebrosternal ribs
- costal cartilages
- vertebrochondral ribs
- floating ribs
- vertebral ribs
- jugular notch
- sternum
- manubrium
- body (of sternum)
- xiphoid process
- tubercle
- head (capitulum)
- articular facets
- angle of rib
- shaft of rib
- costal groove
- costal facets

SECTION 2

The Appendicular Skeleton

- appendicular skeleton
- pectoral girdle
- upper limbs
- pelvic girdles
- lower limbs

7.13

The pectoral girdles—the clavicles and scapulae—connect the upper limbs to the axial skeleton

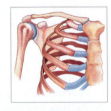

- pectoral girdle
- clavicles
- scapulae
- sternal end
- acromion
- acromial end (of clavicle)
- body (of scapula)
- superior border
- medial border
- lateral border
- superior angle
- inferior angle
- lateral angle
- subscapular fossa
- glenoid cavity
- scapular spine
- supraspinous fossa
- infraspinous fossa
- coracoid process

7.14

The humerus of the arm articulates with the radius and ulna of the forearm

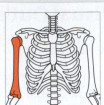

- humerus
- medial epicondyle
- lateral epicondyle
- intertubercular groove
- radial fossa
- capitulum
- condyle
- head (of humerus)
- lesser tubercle
- anatomical neck
- surgical neck
- deltoid tuberosity
- greater tubercle
- radial groove
- trochlea
- olecranon fossa
- coronoid fossa
- ulna
- radius
- olecranon
- ulnar head
- styloid process (ulna)
- radial head
- neck (of radius)
- radial tuberosity
- interosseous membrane
- ulnar notch
- styloid process (radius)
- trochlear notch
- coronoid process
- radial notch
- proximal radio-ulnar joint
- distal radio-ulnar joint

7.15

The wrist is composed of carpal bones, and the hand consists of metacarpal bones and phalanges

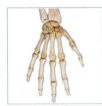

- carpus
- scaphoid
- lunate
- pisiform
- triquetrum
- trapezium
- trapezoid
- capitate
- hamate
- metacarpal bones
- phalanges
- pollex

7.16

The hip bone forms by the fusion of the ilium, ischium, and pubis

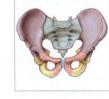

- pelvic girdle
- hip bones
- ilium
- ischium
- pubis
- iliac spines
- gluteal lines
- greater sciatic notch
- ischial spine
- ischial tuberosity
- iliac crest
- lunate surface
- acetabulum
- acetabular notch
- iliac fossa
- arcuate line
- pectineal line
- pubic tubercle
- pubic symphysis
- iliac tuberosity
- auricular surface (ilium)
- obturator foramen
- ischial ramus
- inferior pubic ramus
- superior pubic ramus

7.17

The pelvis consists of the two hip bones plus the sacrum and the coccyx

- pelvis
- sacro-iliac joint
- pubic symphysis
- true pelvis
- false pelvis
- pelvic brim
- pelvic inlet
- pelvic outlet

• = *Term boldfaced in this module*

7.18

The femur articulates with the patella and tibia

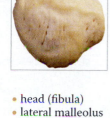

- femur
- patella
- tibia
- fibula
- tarsal bones
- metatarsal bones
- phalanges
- greater trochanter
- neck (femur)
- fovea capitis
- femoral head
- intertrochanteric line
- lesser trochanter
- gluteal tuberosity
- linea aspera

- popliteal surface
- medial condyle
- lateral condyle
- patellar surface
- intercondylar fossa
- base (patella)
- apex (patella)
- lateral facet (patella)
- medial facet (patella)
- tibia
- medial tibial condyle
- lateral tibial condyle
- fibula

- head (fibula)
- lateral malleolus
- tibial tuberosity
- interosseous membrane
- anterior margin (tibia)
- medial malleolus
- intercondylar eminence

7.19

The ankle and foot contain tarsal bones, metatarsal bones, and phalanges

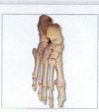

- ankle (tarsus)
- tarsal bones
- calcaneus
- talus
- trochlea
- navicular
- cuboid
- cuneiform bones

- metatarsal bones
- phalanges
- hallux
- longitudinal arch
- calcaneal tendon
- transverse arch
- congenital talipes equinovarus

• = *Term boldfaced in this module*

Chapter Integration: Applying what you've learned

Wayne plays right wing on his college hockey team. He grew up in a hockey family in Peterborough, Canada. His dad played Juniors and had hopes of an NHL career, until marriage and children caused him to give up his dream. But he put skates on Wayne as soon as Wayne could walk, and taught him the game. After high school, Wayne followed in his father's footsteps by choosing to play college hockey. He too hopes to have a career in professional hockey.

During a collegiate hockey game against his archrivals, Wayne was handling the puck well and out-skating the defense, when he was checked from behind. Checking is a maneuver in which an opponent is neutralized through the proper use of one's body or stick; checking from behind is illegal. While lying on the ice being examined by the team's athletic trainer, Wayne was experiencing chest and shoulder pain. His teammates helped him to the locker room, where a thorough physical exam revealed a badly bruised chest; dyspnea (difficulty breathing); and immobility of his left arm, especially in the anterior direction. X-rays subsequently taken at a nearby hospital revealed two fractured ribs.

1. Why might Wayne be experiencing difficulty breathing?

2. What is the probable reason for Wayne's inability to move his left arm?

3. State whether Wayne's injuries are related to the axial skeleton or to the appendicular skeleton.

Access more review material online in the Study Area at **www.masteringaandp.com**.

There, you'll find:
- **Chapter guides**
- **Chapter quizzes**
- **Practice tests**
- **Bone reviews**
- **Animations**
- **Flashcards**
- **A glossary with pronunciations**

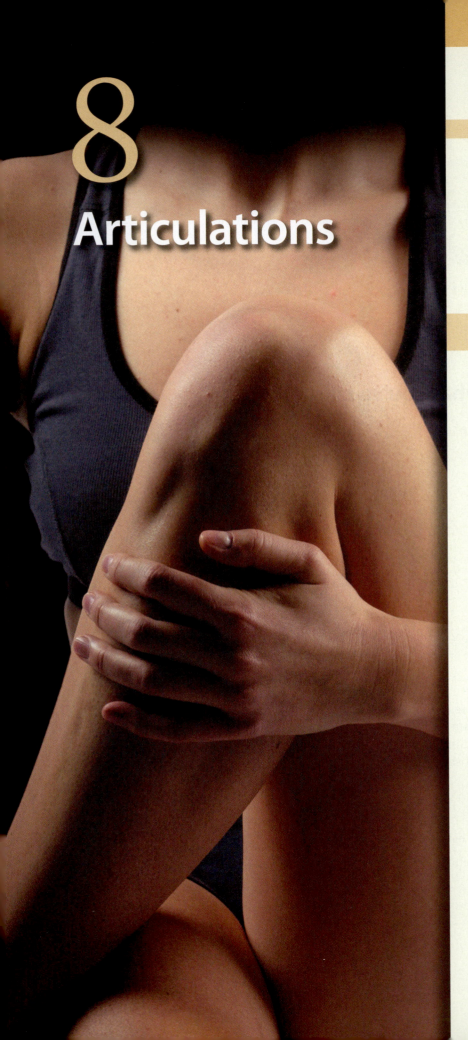

8

Articulations

Joint Design and Movement

Because the bones of the skeleton are relatively inflexible, movements can occur only at **articulations**, or joints, where two bones interconnect. The anatomical structure of a joint determines the type and amount of movement that may occur. Each joint reflects a compromise between the need for strength and the need for mobility. As a result, articulations differ in the amount of movement permitted, and this property is known as **range of motion (ROM)**. Articulations are often categorized by range of motion, with subgroups based on anatomical structure.

Functional and Structural Classifications of Articulations

Functional Category	Structural Category and Type		Description
Synarthrosis (no movement) At a synarthrosis, the bony edges are quite close together and may even interlock. These extremely strong joints are located where movement between the bones must be prevented.	**Fibrous** Suture		A **suture** (*sutura*, a sewing together) is a synarthrotic joint located only between the bones of the skull. The edges of the bones are interlocked and bound together at the suture by dense fibrous connective tissue.
	Gomphosis		A **gomphosis** (gom-FŌ-sis; *gomphosis*, a bolting together) is a synarthrosis that binds the teeth to bony sockets in the maxillae and mandible. The fibrous connection between a tooth and its socket is a periodontal (per-ē-ō-DON-tal) ligament (*peri*, around + *odontos*, tooth).
	Cartilaginous Synchondrosis		A **synchondrosis** (sin-kon-DRŌ-sis; *syn*, together + *chondros*, cartilage) is a rigid, cartilaginous bridge between two articulating bones. The cartilaginous connection between the ends of the first pair of vertebrosternal ribs and the sternum is a synchondrosis.
	Bony fusion Synostosis		A **synostosis** (sin-os-TŌ-sis) is a totally rigid, immovable joint created when two bones fuse and the boundary between them disappears. The frontal suture of the frontal bone and the epiphyseal lines of mature long bones are synostoses.
Amphiarthrosis (little movement) An amphiarthrosis permits more movement than a synarthrosis, but is much stronger than a freely movable joint. The articulating bones are connected by collagen fibers or cartilage.	**Fibrous** Syndesmosis		At a **syndesmosis** (sin-dez-MŌ-sis; *desmos*, a band or ligament), bones are connected by a ligament. One example is the distal articulation between the tibia and fibula.
	Cartilaginous Symphysis		At a **symphysis**, the articulating bones are separated by a wedge or pad of fibrous cartilage. The articulation between the two pubic bones (the pubic symphysis) is an example of a symphysis.
Diarthrosis (free movement)	**Synovial**		**Diarthroses**, or **synovial** (si-NŌ-ve-ul) **joints**, permit a wider range of motion than do other types of joints. They are typically located at the ends of long bones, such as those of the upper and lower limbs.

Synarthrotic and amphiarthrotic joints are relatively simple in structure, with direct connections between the articulating bones. Diarthrotic joints are quite complex in structure, and they permit the greatest range of motion. In this section, we will consider the structure and function of synovial joints.

Synovial joints are freely movable diarthroses containing synovial fluid

In this module we review the general structure of a synovial joint, which was introduced in our earlier discussion of synovial membranes. We will focus attention on the key components of synovial joints: the articular cartilages and accessory structures.

1 Under normal conditions, the opposing bony surfaces within a synovial joint cannot contact one another, because these surfaces are covered by special articular cartilages, which are slick and smooth. This feature alone can reduce friction during movement at the joint. However, even when pressure is applied across a joint, the smooth articular cartilages do not touch one another, because they are separated by a thin film of synovial fluid within the joint cavity.

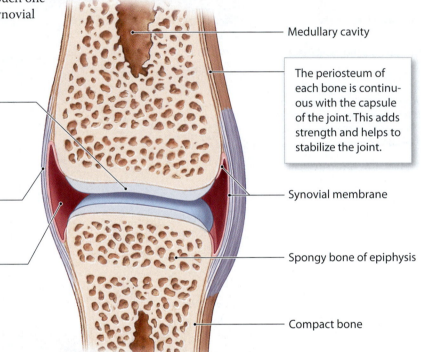

Medullary cavity

The periosteum of each bone is continuous with the capsule of the joint. This adds strength and helps to stabilize the joint.

Synovial membrane

Spongy bone of epiphysis

Compact bone

Components of Synovial Joints

Articular cartilages resemble hyaline cartilages elsewhere in the body. However, articular cartilages have no perichondrium, and the matrix contains more water than that of other cartilages.

The **joint capsule**, or articular capsule, is dense and fibrous, and it may be reinforced with various accessory structures such as tendons or ligaments.

Synovial fluid within the joint cavity provides lubrication, cushions shocks, prevents abrasion, and supports the chondrocytes of the articular cartilages. Even in a large joint such as the knee, the total quantity of synovial fluid in a joint is normally less than 3 mL.

2 This table details the major functions of synovial fluid, which is similar in composition to the ground substance in loose connective tissues. Synovial fluid is produced at the synovial membrane that lines the joint cavity. During normal movement, synovial fluid circulates from the areolar tissue into the joint cavity and percolates through the articular cartilages, providing oxygen and nutrients to the chondrocytes and carrying away their metabolic wastes.

Functions of Synovial Fluid

- **Lubrication**. When part of an articular cartilage is compressed during movement, some of the synovial fluid is squeezed out of the cartilage and into the space between the opposing surfaces. This thin layer of fluid markedly reduces friction between moving surfaces, just as a thin film of water reduces friction between a car's tires and a highway.

- **Nutrient Distribution**. The synovial fluid in a joint must circulate continuously to provide nutrients and waste disposal for the chondrocytes of the articular cartilages. It circulates whenever the joint moves, and the compression and reexpansion of the articular cartilages pump synovial fluid into and out of the cartilage matrix.

- **Shock Absorption**. Synovial fluid cushions shocks in joints that are subjected to compression. For example, when you jog your knees are severely compressed and the synovial fluid distributes that force evenly across the articular surfaces and outward to the joint capsule.

3 In complex synovial joints, such as the knee, a variety of accessory structures provide support and additional stability. Many of these accessory structures can be seen in this diagrammatic view of a sagittal section of the knee.

The tendon of the quadriceps muscles attaches to the base of the patella. Although not part of the articulation itself, tendons passing across or around a joint can limit the joint's range of motion and provide mechanical support for it.

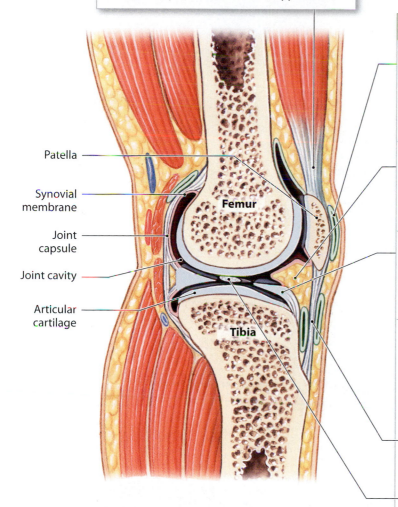

Patella

Synovial membrane

Joint capsule

Joint cavity

Articular cartilage

Femur

Tibia

Accessory Structures Supporting the Knee

A **bursa** (BUR-sa; a pouch; plural, *bursae*) is a small, fluid-filled pocket that forms in connective tissue. It contains synovial fluid and is lined by a synovial membrane. Bursae often form where a tendon or ligament rubs against other tissues. Located around most synovial joints, including the knee joint, bursae reduce friction and act as shock absorbers.

Fat pads are localized masses of adipose tissue covered by a layer of synovial membrane. They are commonly superficial to the joint capsule. Fat pads protect the articular cartilages and act as packing material for the joint. When the bones move, the fat pads fill in the spaces created as the joint cavity changes shape.

A **meniscus** (me-NIS-kus; a crescent; plural, *menisci*) is a pad of fibrous cartilage situated between opposing bones within a synovial joint. Menisci, or articular discs, may subdivide a synovial cavity, channel the flow of synovial fluid, or allow for variations in the shapes of the articular surfaces.

Accessory ligaments support, strengthen, and reinforce synovial joints. **Intrinsic ligaments**, or capsular ligaments, are localized thickenings of the joint capsule. **Extrinsic ligaments** are separate from the joint capsule. Extrinsic ligaments may pass outside or inside the joint capsule, and are called extracapsular or intracapsular ligaments, respectively.

The patellar ligament extends from the apex of the patella to the tibial tuberosity. This is an example of an **extracapsular ligament**.

The cruciate ligaments that run through the interior of the knee joint are examples of **intracapsular ligaments**.

A joint cannot be both highly mobile and very strong. The greater the range of motion at a joint, the weaker it becomes. A synarthrosis, the strongest type of joint, permits no movement, whereas a diarthrosis, such as the shoulder, is far weaker but permits a broad range of motion. Any mobile diarthrosis will be damaged by movement beyond its normal range of motion. When reinforcing structures cannot protect a joint from extreme stresses, a **dislocation,** or **luxation** (luk-SĀ-shun), results. In a dislocation, the articulating surfaces are forced out of position. The displacement can damage the articular cartilages, tear ligaments, or distort the joint capsule. Although the inside of a joint has no pain receptors, nerves that monitor the capsule, ligaments, and tendons are quite sensitive, so dislocations are very painful.

Module 8.1 Review

a. Define a joint dislocation (luxation).

b. Describe the components of a synovial joint, and identify the functions of each.

c. Why would improper circulation of synovial fluid lead to the degeneration of articular cartilages in the affected joint?

Anatomical organization determines the functional properties of synovial joints

An accurate description of the functions of a joint includes terms that indicate the types of motion permitted. To demonstrate the general types of movement, we will use a simple model that you can try for yourself.

1 Take a pencil and stand it upright on the surface of a desk or table. The pencil represents a bone, and the desktop represents an articular surface. A little imagination and a lot of twisting, pushing, and pulling will demonstrate that there are only three ways to move the pencil. Considering them one at a time will provide a frame of reference for us to analyze complex movements.

2 If you hold the pencil upright, without securing the point, you can push the pencil point across the surface. This kind of motion, **gliding**, is an example of linear motion. You could slide the point forward or backward, from side to side, or diagonally. However you move the pencil, the motion can be described by using two lines of reference (axes). One line represents forward–backward motion, the other left–right movement. For example, a simple movement along one axis could be described as "forward 1 cm" or "left 2 cm." A diagonal movement could be described with both axes, as in "backward 1 cm and to the right 2.5 cm."

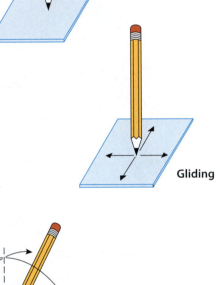

Gliding

3 With the tip held in position, you can move the free (eraser) end of the pencil forward and backward, from side to side, or at some intermediate angle. These movements, which change the angle between the shaft and the desktop, are examples of **angular motion**. We can describe such motion by the angle the pencil shaft makes with the surface.

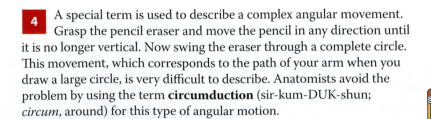

Angular motion

4 A special term is used to describe a complex angular movement. Grasp the pencil eraser and move the pencil in any direction until it is no longer vertical. Now swing the eraser through a complete circle. This movement, which corresponds to the path of your arm when you draw a large circle, is very difficult to describe. Anatomists avoid the problem by using the term **circumduction** (sir-kum-DUK-shun; *circum*, around) for this type of angular motion.

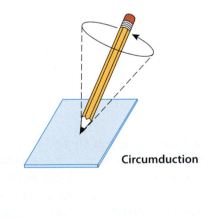

Circumduction

5 If you keep the shaft vertical and the point at one location, you can still spin the pencil around its longitudinal axis. This movement is called **rotation**. Several articulations permit partial rotation, but none can rotate freely without limit because such a movement would hopelessly tangle the blood vessels, nerves, and muscles that cross the joint.

Rotation

Types of Synovial Joints	Models of Joint Motion	Examples
Gliding joint 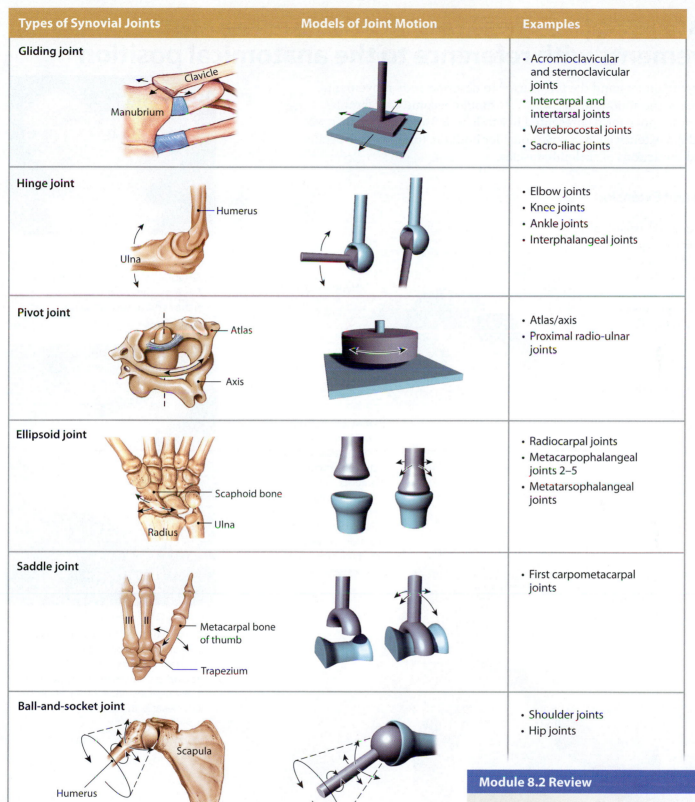 Clavicle, Manubrium		• Acromioclavicular and sternoclavicular joints • Intercarpal and intertarsal joints • Vertebrocostal joints • Sacro-iliac joints
Hinge joint Humerus, Ulna		• Elbow joints • Knee joints • Ankle joints • Interphalangeal joints
Pivot joint Atlas, Axis		• Atlas/axis • Proximal radio-ulnar joints
Ellipsoid joint Scaphoid bone, Ulna, Radius		• Radiocarpal joints • Metacarpophalangeal joints 2–5 • Metatarsophalangeal joints
Saddle joint III, II, Metacarpal bone of thumb, Trapezium		• First carpometacarpal joints
Ball-and-socket joint Scapula, Humerus		• Shoulder joints • Hip joints

6 This visual summary gives representative examples of the various anatomical classes of synovial joints based on the shapes of the articulating surfaces. It relates articular structure to simplified joint models and lists examples of each group of joints.

Module 8.2 Review

a. Identify the types of synovial joints based on the shapes of the articulating surfaces.

b. What type of synovial joint permits the widest range of motion?

c. Indicate the type of synovial joint for each of the following: shoulder, elbow, ankle, and thumb.

Broad descriptive terms are used to describe movements with reference to the anatomical position

It is easiest to understand the terms used to describe body movements when you see the actions under way. You should become very familiar with the descriptive terms presented in Modules 8.3 and 8.4, because we will use them extensively when we consider both the functions of specific joints and the actions of skeletal muscles.

Flexion and Extension

1 Flexion and extension are usually applied to the movements of the long bones of the limbs, but they are also used to describe movements of the axial skeleton. For example, when you bring your head toward your chest, you flex the inter-vertebral joints of the neck.

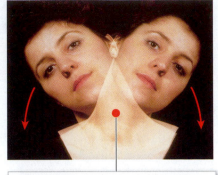

Lateral flexion occurs when your vertebral column bends to the side. This movement is most pronounced in the cervical and thoracic regions.

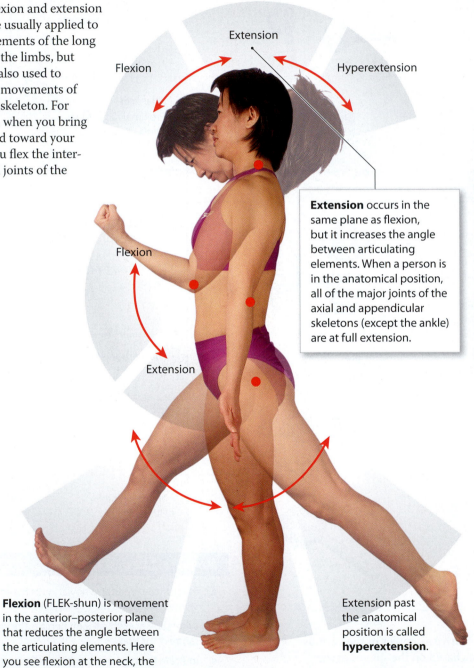

Extension

Flexion

Hyperextension

Flexion

Extension occurs in the same plane as flexion, but it increases the angle between articulating elements. When a person is in the anatomical position, all of the major joints of the axial and appendicular skeletons (except the ankle) are at full extension.

Extension

Flexion (FLEK-shun) is movement in the anterior–posterior plane that reduces the angle between the articulating elements. Here you see flexion at the neck, the elbow, and the hip.

Extension past the anatomical position is called **hyperextension**.

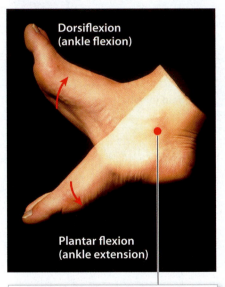

Dorsiflexion (ankle flexion)

Plantar flexion (ankle extension)

Dorsiflexion is flexion at the ankle joint and elevation of the sole, as when you dig in your heel. **Plantar flexion** (*planta*, sole), the opposite movement, extends the ankle joint and elevates the heel, as when you stand on tiptoe.

Abduction and Adduction

Spreading the fingers or toes apart abducts them, because they move away from a central digit. Bringing them together constitutes adduction. (Fingers move toward or away from the middle finger; toes move toward or away from the second toe.)

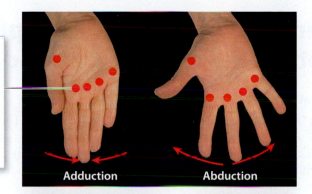

Adduction   **Abduction**

2 Abduction and adduction always refer to the movements of the appendicular skeleton, not to those of the axial skeleton.

Abduction   **Adduction**

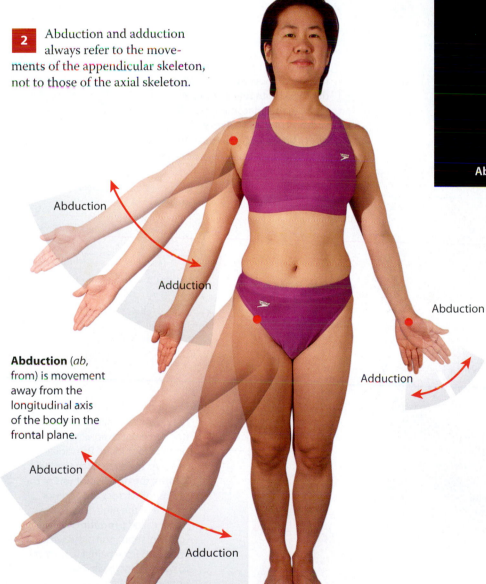

Abduction

Adduction

Abduction (*ab*, from) is movement away from the longitudinal axis of the body in the frontal plane.

Abduction

Adduction

Abduction

Adduction

Abduction

Adduction

Adduction (*ad*, to) is movement toward the longitudinal axis of the body in the frontal plane.

Circumduction

3 Moving your arm as if to draw a big circle on the wall is **circumduction**. In this movement your hand moves in a circle, but your arm does not rotate.

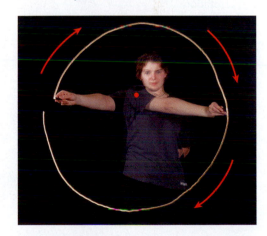

Module 8.3 Review

a. When doing jumping jacks, which lower limb movements are necessary?

b. Which movements are associated with hinge joints?

c. Compare dorsiflexion to plantar flexion.

Terms of more limited application describe rotational movements and special movements

Rotation

1 Rotational movements of the trunk are described as left or right rotation. Rotation of the limbs can be described as medial or lateral rotation; special terms are used to describe the rotational movements of the forearm.

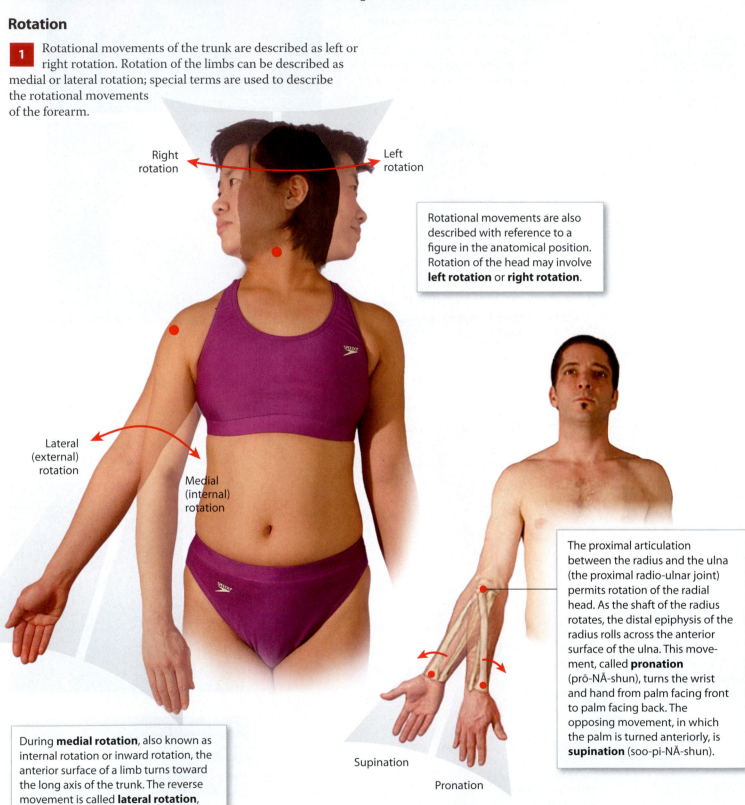

Right rotation

Left rotation

Rotational movements are also described with reference to a figure in the anatomical position. Rotation of the head may involve **left rotation** or **right rotation**.

Lateral (external) rotation

Medial (internal) rotation

The proximal articulation between the radius and the ulna (the proximal radio-ulnar joint) permits rotation of the radial head. As the shaft of the radius rotates, the distal epiphysis of the radius rolls across the anterior surface of the ulna. This movement, called **pronation** (prō-NĀ-shun), turns the wrist and hand from palm facing front to palm facing back. The opposing movement, in which the palm is turned anteriorly, is **supination** (soo-pi-NĀ-shun).

During **medial rotation**, also known as internal rotation or inward rotation, the anterior surface of a limb turns toward the long axis of the trunk. The reverse movement is called **lateral rotation**, external rotation, or outward rotation.

Supination

Pronation

Special Movements

2 There are several special terms that apply to specific articulations or unusual types of movement.

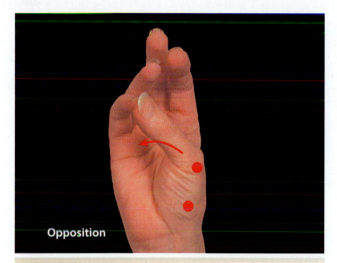

Opposition

Opposition is movement of the thumb toward the surface of the palm or the pads of other fingers. Opposition enables you to grasp and hold objects between your thumb and palm. It involves movement at the first carpometacarpal and metacarpophalangeal joints. Flexion at the fifth metacarpophalangeal joint can assist this movement.

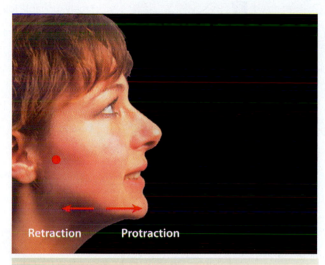

Retraction   Protraction

Protraction entails moving a part of the body anteriorly in the horizontal plane. **Retraction** is the reverse movement. You protract your jaw when you grasp your upper lip with your lower teeth, and you protract your clavicles when you cross your arms.

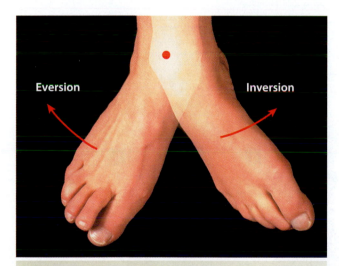

Eversion   Inversion

Inversion (*in*, into + *vertere*, to turn) is a twisting motion of the foot that turns the sole inward, elevating the medial edge of the sole. The opposite movement is called **eversion** (ē-VER-zhun; *e*, out).

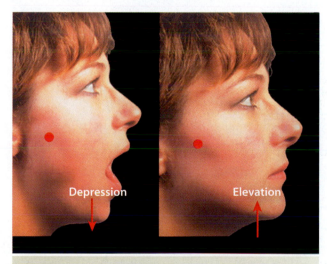

Depression   Elevation

Depression and **elevation** occur when a structure moves in an inferior or a superior direction, respectively.

Module 8.4 Review

a. Snapping your fingers involves what movement with the thumb and third metacarpophalangeal joint?

b. What movements are made possible by the rotation of the radius head?

c. What hand movements occur when wriggling into tight-fitting gloves?

1. Matching

Match the following terms with the most closely related description.

- amphiarthrosis
- synarthrosis
- dislocation
- pronation-supination
- diarthrosis
- shoulder
- articular discs
- fluid-filled pouch

a	_____	Freely movable joint
b	_____	Movements of forearm bones
c	_____	Ball and socket
d	_____	Menisci
e	_____	Immovable joint
f	_____	Luxation
g	_____	Bursa
h	_____	Slightly movable joint

2. Labeling

Label the structures in the synovial joint figure at right.

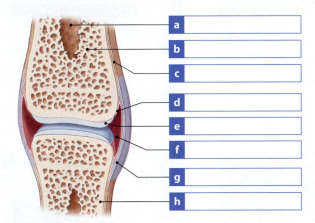

a _____
b _____
c _____
d _____
e _____
f _____
g _____
h _____

3. Labeling

Identify each of the following movements.

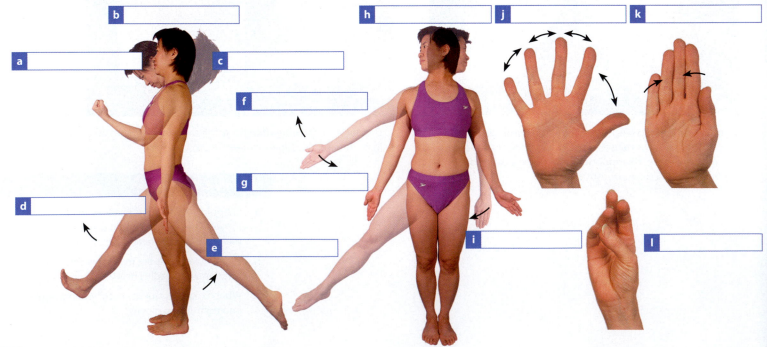

a _____
b _____
c _____
d _____
e _____
f _____
g _____
h _____
i _____
j _____
k _____
l _____

Articulations of the Axial and Appendicular Skeletons

1 The majority of the joints of the axial skeleton—several of which have been noted in earlier chapters—are strong joints that permit very little movement. We begin this section by examining the structure of the intervertebral joints, which permit limited but important movements of the vertebral column.

2 A typical articulation in the appendicular skeleton has an extensive range of motion. Because an articulation cannot be both strong and highly mobile, these joints are often weaker than those of the axial skeleton. We will examine two joints in each limb—the shoulder and elbow in the upper limb, and the hip and knee in the lower limb. This will be sufficient to demonstrate the structural and functional patterns that determine mobility at a synovial joint.

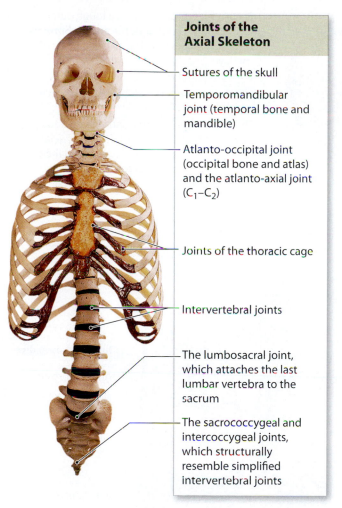

Joints of the Axial Skeleton

Sutures of the skull

Temporomandibular joint (temporal bone and mandible)

Atlanto-occipital joint (occipital bone and atlas) and the atlanto-axial joint (C_1–C_2)

Joints of the thoracic cage

Intervertebral joints

The lumbosacral joint, which attaches the last lumbar vertebra to the sacrum

The sacrococcygeal and intercoccygeal joints, which structurally resemble simplified intervertebral joints

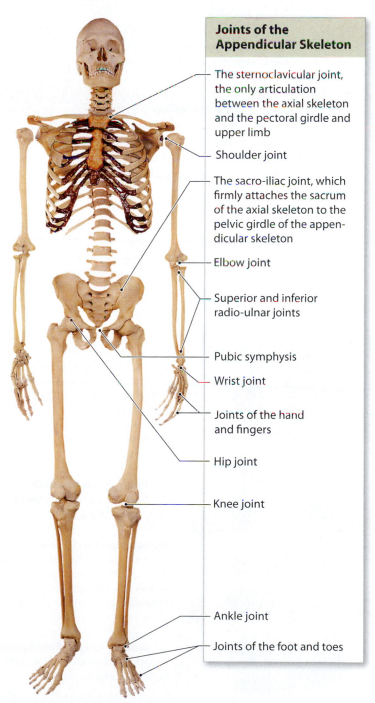

Joints of the Appendicular Skeleton

The sternoclavicular joint, the only articulation between the axial skeleton and the pectoral girdle and upper limb

Shoulder joint

The sacro-iliac joint, which firmly attaches the sacrum of the axial skeleton to the pelvic girdle of the appendicular skeleton

Elbow joint

Superior and inferior radio-ulnar joints

Pubic symphysis

Wrist joint

Joints of the hand and fingers

Hip joint

Knee joint

Ankle joint

Joints of the foot and toes

Adjacent vertebrae have gliding diarthroses between their articular processes, and symphyseal joints between their vertebral bodies

The articulations between the superior and inferior articular processes of adjacent vertebrae are gliding joints that permit flexion and rotation movements of the vertebral column. Little gliding occurs between adjacent vertebral bodies. From axis to sacrum, the bodies of adjacent vertebrae form symphyseal joints. At these joints the vertebrae are separated and cushioned by pads of fibrous cartilage called **intervertebral discs**.

1 The intervertebral discs make a significant contribution to an individual's height: they account for roughly one-quarter the length of the vertebral column superior to the sacrum. As we grow older, the water content of the nucleus pulposus decreases and it becomes less effective as a cushion. This increases the chances for vertebral injury. Water loss from the intervertebral discs also causes shortening of the vertebral column, accounting for the characteristic decrease in height with advancing age.

Each intervertebral disc has a tough outer layer of fibrous cartilage, the **anulus fibrosus** (AN-ū-lus fī-BRŌ-sus). The collagen fibers of this layer attach the disc to the bodies of adjacent vertebrae.

The anulus fibrosus surrounds the **nucleus pulposus** (pul- PŌ-sus), a soft, elastic, gelatinous core. The nucleus pulposus gives the disc resiliency and enables it to absorb shocks.

Superior view

2 Numerous ligaments are attached to the bodies and processes of all vertebrae, binding them together and stabilizing the vertebral column. The primary ligaments have been identified in the anterior view (left) and lateral and sectional views (right).

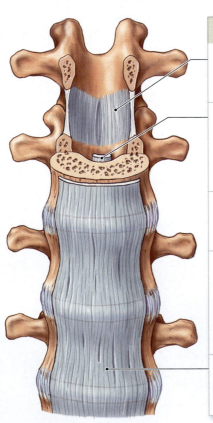

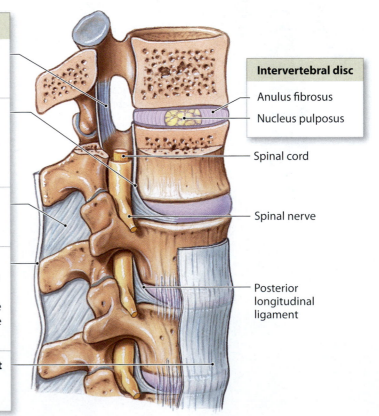

Primary Vertebral Ligaments

The **ligamentum flavum** (plural, *ligamenta flava*) connects the laminae of adjacent vertebrae.

The **posterior longitudinal ligament** parallels the anterior longitudinal ligament and connects the posterior surfaces of adjacent vertebral bodies.

The **interspinous ligament** connects the spinous processes of adjacent vertebrae.

The **supraspinous ligament** interconnects the tips of the spinous processes from the sacrum to vertebra C₇. The ligamentum nuchae extends from vertebra C₇ to the base of the skull.

The **anterior longitudinal ligament** connects the anterior surfaces of adjacent vertebral bodies.

Intervertebral disc

Anulus fibrosus

Nucleus pulposus

Spinal cord

Spinal nerve

Posterior longitudinal ligament

3 If the posterior longitudinal ligaments are weakened, as often occurs with advancing age, the compressed nucleus pulposus may distort the anulus fibrosus, forcing it partway into the vertebral canal. This condition, seen here in lateral view, is called a **slipped disc**, although the disc does not actually slip.

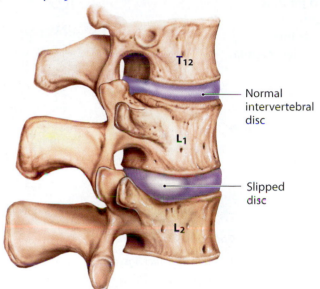

T₁₂

Normal intervertebral disc

L₁

Slipped disc

L₂

4 If the nucleus pulposus breaks through the anulus fibrosus, it too may protrude into the vertebral canal. This condition, shown here in superior view, is called a **herniated disc**. When a disc herniates, sensory nerves are distorted, and the protruding mass can also compress the nerves passing through the adjacent intervertebral foramen.

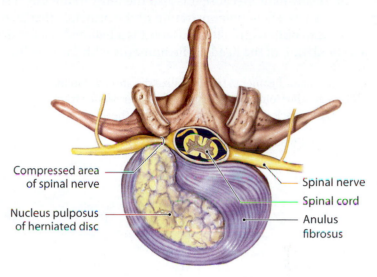

Compressed area of spinal nerve

Nucleus pulposus of herniated disc

Spinal nerve

Spinal cord

Anulus fibrosus

5 The bones of the skeleton become thinner and weaker as a normal part of the aging process. Inadequate ossification is called **osteopenia** (os-tē-ō-PĒ-nē-uh; *penia*, lacking). This reduction in bone mass begins between the ages of 30 and 40 as osteoblast activity begins to decline, while osteoclast activity continues at previous levels. Thereafter, women lose roughly 8 percent of their skeletal mass every decade; men lose roughly 3 percent per decade. When the reduction in bone mass is sufficient to compromise normal function, the condition is known as **osteoporosis** (os-tē-ō-po-RŌ-sis; *porosus*, porous). The combination of osteopenia or osteoporosis and the reduction in the cushioning properties of the intervertebral discs makes vertebral fractures a common problem among the elderly.

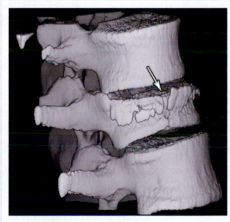

Clinical scan of a compression fracture in a lumbar vertebra

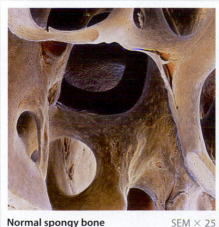

Normal spongy bone   SEM × 25

Spongy bone with osteoporosis   SEM × 21

Module 8.5 Review

a. Identify the primary vertebral ligaments.

b. Describe the nucleus pulposus and anulus fibrosus of an intervertebral disc.

c. Compare a slipped disc with a herniated disc.

The shoulder and hip are ball-and-socket joints

Shoulder Joint

The shoulder joint, or **glenohumeral joint**, permits the greatest range of motion of any joint. Because it is also the most frequently dislocated joint, it provides an excellent demonstration of the principle that stability must be sacrificed to obtain mobility. This joint is a ball-and-socket diarthrosis formed by the articulation of the head of the humerus with the glenoid cavity of the scapula.

1 Five major ligaments help stabilize the shoulder joint. As at other joints, bursae at the shoulder reduce friction where large muscles and tendons pass across the joint capsule.

Coracoid process

Clavicle

Acromion

Bursae

A tendon of the biceps brachii muscle runs through the shoulder joint. As it passes through the articular capsule, it is surrounded by a tubular bursa that is continuous with the joint cavity.

Articular capsule

Scapula

Ligaments Stabilizing the Shoulder

Coracoclavicular ligaments

Acromioclavicular ligament

Coraco-acromial ligament

Coracohumeral ligament

Glenohumeral ligaments

Humerus

2 This frontal section through the joint shows that the relatively loose articular capsule extends from the scapula to the anatomical neck of the humerus. The somewhat oversized articular capsule permits an extensive range of motion.

Subdeltoid bursa | Articular capsule | Coraco-acromial ligament | Coracoclavicular ligaments

Clavicle

Articular cartilages

Humerus

Scapula

Synovial membrane

The extent of the glenoid cavity is increased by a fibrous-cartilaginous **glenoid labrum** (*labrum*, lip or edge), which continues beyond the bony rim and deepens the socket.

3 In lateral view, you can see how small the glenoid cavity is compared to the size of the articular capsule. As a result, most of the stability at this joint is provided by the surrounding skeletal muscles, associated tendons, and various ligaments.

Acromioclavicular ligament

Clavicle

Acromion

Articular capsule

Glenoid cavity

Tendon of supraspinatus muscle

Tendon of infraspinatus muscle

Tendon of biceps brachii muscle

Coracohumeral ligament (cut)

Glenohumeral ligaments

Subscapularis muscle

Teres minor muscle

Scapula

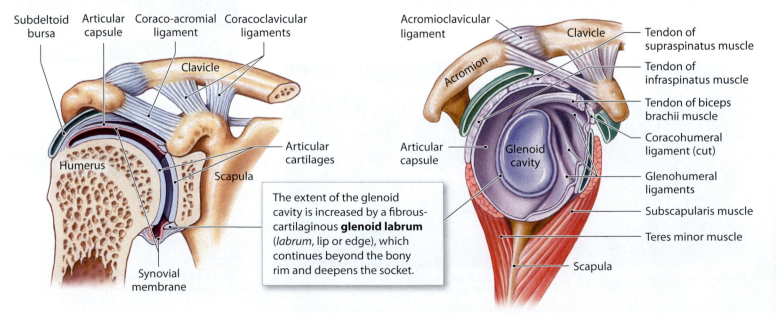

Hip Joint

The hip joint is a sturdy ball-and-socket diarthrosis that permits flexion and extension, adduction and abduction, circumduction, and rotation.

4 This lateral view shows the hip joint with the femur removed. Within the acetabulum, a fibrous cartilage pad extends like a horseshoe to either side of the acetabular notch. Two of the five ligaments that reinforce the articular capsule are seen here: the ligament of the femoral head and the transverse acetabular ligament.

Iliofemoral ligament   Fibrous cartilage pad

The **acetabular labrum**, a projecting rim of fibrous cartilage, increases the depth of the joint cavity.

The acetabulum, a deep fossa, accommodates the head of the femur.

Fat pad

The **ligamentum teres** (*teres*, long and round), or the **ligament of the femoral head**, originates along the transverse acetabular ligament and attaches to the fovea capitis, a small pit at the center of the femoral head.

The **transverse acetabular ligament** crosses the acetabular notch, filling in the gap in the inferior border of the acetabulum.

5 The articular capsule of the hip joint extends from the lateral and inferior surfaces of the pelvic girdle to the intertrochanteric line and intertrochanteric crest of the femur, enclosing both the head and neck of the femur. The three remaining broad ligaments that reinforce the articular capsule can be seen in these views.

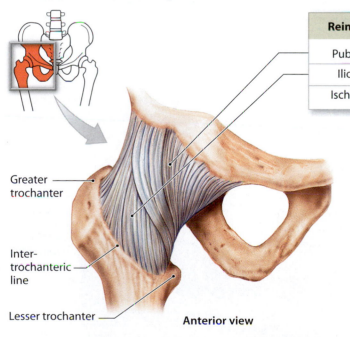

Reinforcing Ligaments
Pubofemoral ligament
Iliofemoral ligament
Ischiofemoral ligament

Greater trochanter

Intertrochanteric line

Lesser trochanter

Anterior view

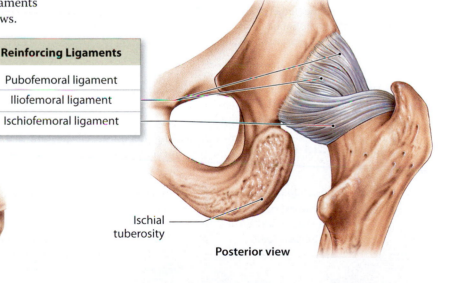

Ischial tuberosity

Posterior view

Although the head of the femur is well supported, the ball-and-socket joint is not directly aligned with the weight distribution along the shaft. As a result, fractures of the femoral neck or between the greater and lesser trochanters of the femur are much more common than hip dislocations.

The elbow and the knee are hinge joints

The Elbow

1 The elbow joint is a complex hinge joint that involves the humerus, radius, and ulna. It is extremely stable because (1) the bony surfaces of the humerus and ulna interlock, (2) a single, thick articular capsule surrounds both the humero-ulnar and proximal radio-ulnar joints, and (3) the articular capsule is reinforced by strong ligaments.

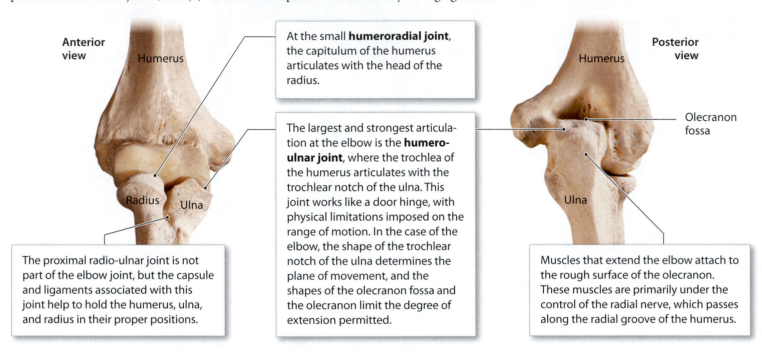

Anterior view

Humerus

At the small **humeroradial joint**, the capitulum of the humerus articulates with the head of the radius.

Radius   Ulna

The largest and strongest articulation at the elbow is the **humero-ulnar joint**, where the trochlea of the humerus articulates with the trochlear notch of the ulna. This joint works like a door hinge, with physical limitations imposed on the range of motion. In the case of the elbow, the shape of the trochlear notch of the ulna determines the plane of movement, and the shapes of the olecranon fossa and the olecranon limit the degree of extension permitted.

The proximal radio-ulnar joint is not part of the elbow joint, but the capsule and ligaments associated with this joint help to hold the humerus, ulna, and radius in their proper positions.

Posterior view

Humerus

Olecranon fossa

Ulna

Muscles that extend the elbow attach to the rough surface of the olecranon. These muscles are primarily under the control of the radial nerve, which passes along the radial groove of the humerus.

2 Although the elbow is extremely strong and stable, severe stresses can produce dislocations or other injuries, especially if epiphyseal growth has not been completed. For example, parents in a rush may drag a toddler along behind them, exerting an upward, twisting pull on the elbow joint that can result in a partial dislocation known as **nursemaid's elbow**.

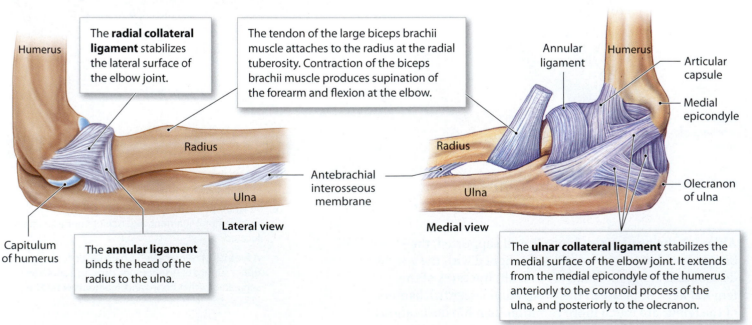

Humerus

The **radial collateral ligament** stabilizes the lateral surface of the elbow joint.

The tendon of the large biceps brachii muscle attaches to the radius at the radial tuberosity. Contraction of the biceps brachii muscle produces supination of the forearm and flexion at the elbow.

Radius

Ulna

Antebrachial interosseous membrane

Lateral view

Capitulum of humerus

The **annular ligament** binds the head of the radius to the ulna.

Annular ligament   Humerus

Articular capsule

Medial epicondyle

Radius

Ulna

Olecranon of ulna

Medial view

The **ulnar collateral ligament** stabilizes the medial surface of the elbow joint. It extends from the medial epicondyle of the humerus anteriorly to the coronoid process of the ulna, and posteriorly to the olecranon.

The Knee

The knee joint contains three separate articulations: two between the femur and tibia (medial condyle to medial condyle, and lateral condyle to lateral condyle) and one between the patella and the patellar surface of the femur. Together, these articulations permit flexion, extension, and very limited rotation.

3 The tendon from the quadriceps muscle passes over the anterior surface of the joint, embedding the patella, and the patellar ligament then continues to its attachment on the anterior surface of the tibia. The patellar ligament and adjacent ligamentous bands support the anterior surface of the knee joint.

4 The posterior surface of the joint is reinforced by the popliteal ligaments and several muscles that originate or insert on the femoral or tibial epiphyses.

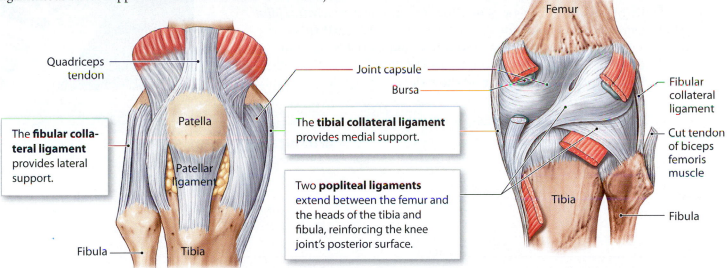

Quadriceps tendon

Patella

Patellar ligament

The **fibular collateral ligament** provides lateral support.

Fibula

Tibia

Joint capsule

Bursa

The **tibial collateral ligament** provides medial support.

Two **popliteal ligaments** extend between the femur and the heads of the tibia and fibula, reinforcing the knee joint's posterior surface.

Femur

Fibular collateral ligament

Cut tendon of biceps femoris muscle

Tibia

Fibula

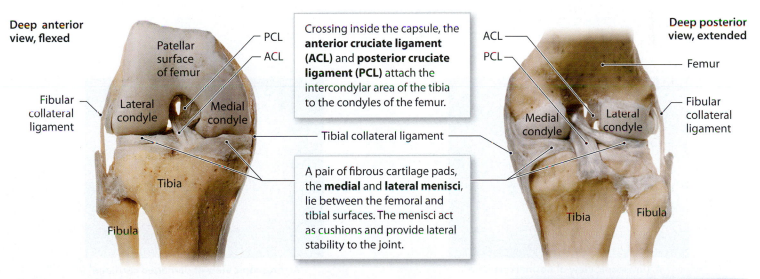

Deep anterior view, flexed

Patellar surface of femur

PCL

ACL

Fibular collateral ligament

Lateral condyle

Medial condyle

Tibia

Fibula

Crossing inside the capsule, the **anterior cruciate ligament (ACL)** and **posterior cruciate ligament (PCL)** attach the intercondylar area of the tibia to the condyles of the femur.

Tibial collateral ligament

A pair of fibrous cartilage pads, the **medial** and **lateral menisci**, lie between the femoral and tibial surfaces. The menisci act as cushions and provide lateral stability to the joint.

ACL

PCL

Deep posterior view, extended

Femur

Fibular collateral ligament

Medial condyle

Lateral condyle

Tibia

Fibula

Within the joint, the anterior cruciate ligament (ACL) and posterior cruciate ligament (PCL) limit the anterior and posterior movement of the femur and maintain the alignment of the femoral and tibial condyles. At full extension, a slight lateral rotation of the tibia tightens the ACL and forces the lateral meniscus between the tibia and femur. The knee joint is then locked in the extended position; unlocking requires medial rotation of the tibia or lateral rotation of the femur.

Module 8.7 Review

a. Between the elbow and knee joints, which have menisci?

b. What signs and symptoms would you expect in an individual who has damaged the menisci of the knee joint?

Arthritis can disrupt normal joint structure and function

Joints are subjected to heavy wear and tear throughout our lifetimes, and problems with joint function are relatively common, especially in older individuals. **Rheumatism** (ROO-muh-tiz-um) is a general term that indicates pain and stiffness affecting the skeletal system, the muscular system, or both. Several major forms of rheumatism exist. **Arthritis** (ar-THRĪ-tis) encompasses all the rheumatic diseases that affect synovial joints. Arthritis always involves damage to the articular cartilages, but the specific cause can vary. **Osteoarthritis** (os-tē-ō-ar-THRĪ-tis), also known as *degenerative arthritis* or *degenerative joint disease (DJD)*, generally affects individuals age 60 or older. Osteoarthritis can result from the cumulative effects of wear and tear at joint surfaces or from genetic factors affecting collagen formation. In the U.S. population, 25 percent of women and 15 percent of men over age 60 show signs of this disease.

Normal Joint

Articular cartilage

LM × 180

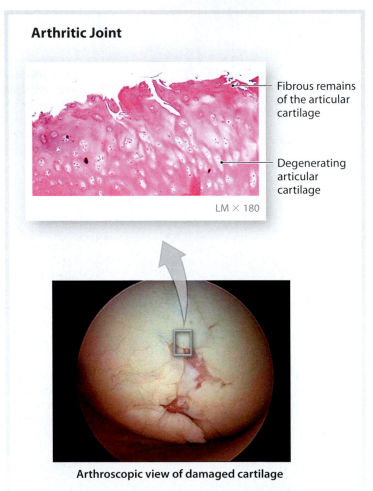

Arthritic Joint

Fibrous remains of the articular cartilage

Degenerating articular cartilage

LM × 180

Arthroscopic view of normal cartilage

Arthroscopic view of damaged cartilage

1 This is a normal articular cartilage. The cartilage is thick, the matrix homogeneous, and the surface smooth and slick.

2 This is an articular cartilage damaged by osteoarthritis. The exposed surfaces change from a slick, smooth-gliding surface to a surface composed of a rough feltwork of bristly collagen fibers. Such a change drastically increases friction at the joint, which then promotes further degeneration of the articular cartilage.

3 There are several options available that enable clinicians to examine the structure of problematic joints. This is an arthroscopic view of the interior of the left knee, showing injuries to the anterior and posterior cruciate ligaments. An **arthroscope** uses fiber optics within a narrow tube to permit exploration of a joint without major surgery. Optical fibers are thin threads of glass or plastic that conduct light. The fibers can be bent around corners, so they can be introduced into a knee or other joint and moved around, enabling the physician to see what is going on inside the joint. If necessary, a second small incision can be made to insert flexible instruments that permit surgery inside the joint, within view of the arthroscope. This procedure, called **arthroscopic surgery**, has greatly improved the treatment of knee and other joint injuries.

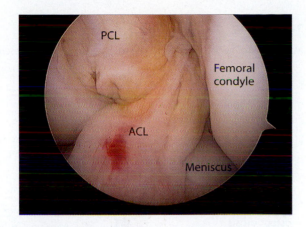

PCL

Femoral condyle

ACL

Meniscus

4 An arthroscope cannot show soft tissue details outside the joint cavity, and repeated arthroscopy eventually leads to the formation of scar tissue and other joint problems. Magnetic resonance imaging (MRI) is a cost-effective and noninvasive method of viewing, without injury, and examining soft tissues around the joint. These are MRI views of the knee joint in horizontal and frontal section. Note the clarity of the soft tissue detail.

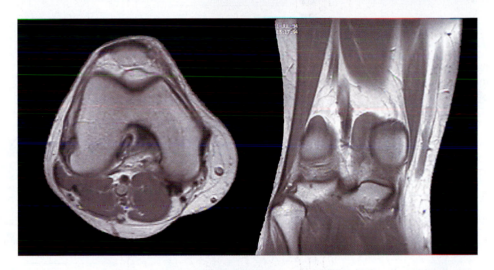

5 Artificial joints such as these may be the method of last resort when regular exercise, physical therapy, and anti-inflammatory drugs (such as aspirin) fail to slow the progress of arthritis. Artificial joints can restore mobility and relieve pain, but they are not as strong as natural joints, so they are most suitable for the elderly, who are less active and place less stress on their joints. Unfortunately, artificial joints typically have a service life of only about 10 years, and replacing an artificial joint in a patient 80-90 years old is a very stressful procedure.

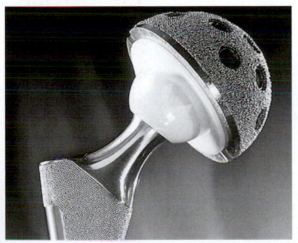

Artificial hip

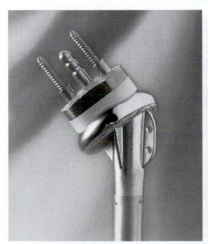

Artificial shoulder

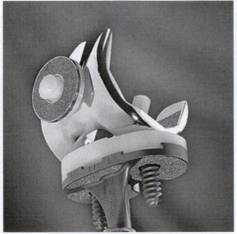

Artificial knee

Module 8.8 Review

a. Compare rheumatism to osteoarthritis.

b. Explain the use of an arthroscope.

c. What can a person do to slow the progression of arthritis?

1. Matching

Match the following terms with the most closely related description.

- acetabulum
- popliteal ligament
- disc outer layer
- dislocation
- arthritis
- disc inner layer
- reinforce knee joint
- osteoporosis

a	_____	Knee joint posterior
b	_____	Articular cartilage damage
c	_____	Reduced bone mass
d	_____	Cruciate ligaments
e	_____	Anulus fibrosus
f	_____	Deep fossa
g	_____	Nucleus pulposus
h	_____	Nursemaid's elbow

2. Labeling

Label each of the structures in the accompanying diagram of the shoulder joint.

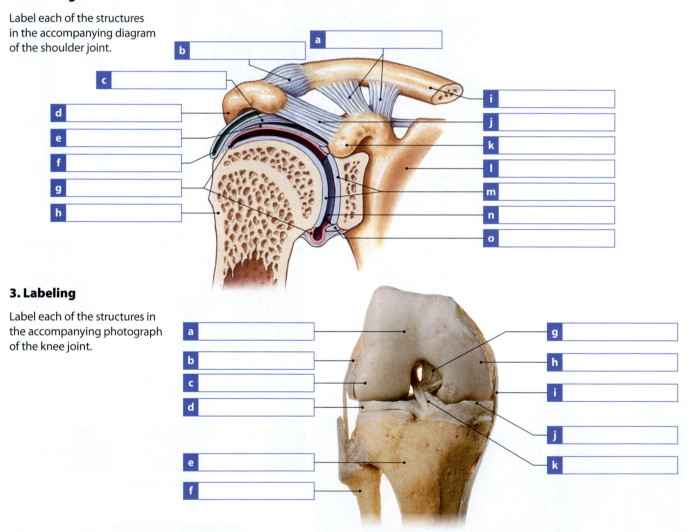

3. Labeling

Label each of the structures in the accompanying photograph of the knee joint.

4. Short answer

Identify the basic articulations between the pectoral and pelvic girdles (of the appendicular skeleton) and the axial skeleton.

_____

_____

Visual Outline with Key Terms

Summarize the content of each module using the terms in the order provided.

SECTION 1

Joint Design and Movement

- articulations
- range of motion (ROM)
- synarthrosis
- suture
- gomphosis
- synchondrosis
- synostosis
- amphiarthrosis
- syndesmosis
- symphysis
- diarthrosis
- diarthroses (synovial joints)

8.1

Synovial joints are freely movable diarthroses containing synovial fluid

- articular cartilages
- joint capsule
- synovial fluid
- bursa
- fat pads
- meniscus
- accessory ligaments
- intrinsic ligaments
- extrinsic ligaments
- extracapsular ligaments
- intracapsular ligaments
- dislocation
- luxation

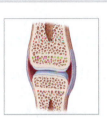

8.2

Anatomical organization determines the functional properties of synovial joints

- gliding
- angular motion
- circumduction
- rotation
- gliding joint
- hinge joint
- pivot joint
- ellipsoid joint
- saddle joint
- ball-and-socket joint

8.3

Broad descriptive terms are used to describe movements with reference to the anatomical position

- flexion
- extension
- hyperextension
- lateral flexion
- dorsiflexion
- plantar flexion
- abduction
- adduction
- circumduction

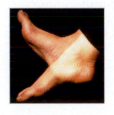

8.4

Terms of more limited application describe rotational movements and special movements

- left rotation
- right rotation
- medial rotation
- lateral rotation
- pronation
- supination
- opposition
- protraction
- retraction
- inversion
- eversion
- depression
- elevation

SECTION 2

Articulations of the Axial and Appendicular Skeletons

- joints of the axial skeleton
- joints of the appendicular skeleton

8.5

Adjacent vertebrae have gliding diarthroses between their articular processes, and symphyseal joints between their vertebral bodies

- intervertebral discs
- anulus fibrosus
- nucleus pulposus
- ligamentum flavum
- posterior longitudinal ligament
- interspinous ligament
- supraspinous ligament
- anterior longitudinal ligament
- slipped disc
- herniated disc
- osteopenia
- osteoporosis

8.6

The shoulder and hip are ball-and-socket joints

- glenohumeral joint
- glenoid labrum
- ligament of the femoral head
- transverse acetabular ligament
- acetabular labrum
- ligamentum teres

 = *Term boldfaced in this module*

8.7

The elbow and knee are hinge joints

- humeroradial joint
- humero-ulnar joint
- nursemaid's elbow
- radial collateral ligament
- annular ligament
- ulnar collateral ligament
- fibular collateral ligament
- tibial collateral ligament
- popliteal ligaments
- anterior cruciate ligament (ACL)
- posterior cruciate ligament (PCL)
- medial menisci
- lateral menisci

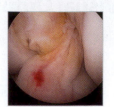

8.8

Arthritis can disrupt normal joint structure and function

- rheumatism
- arthritis
- osteoarthritis
- arthroscope
- arthroscopic surgery

● = *Term boldfaced in this module*

Chapter Integration: Applying what you've learned

Jon loved his golf. He loved the feel of just being out on a golf course. Even into his early 60s he stilled played; his only concession to age—and to a chronically aching knee—was pushing his bag on a wheeled cart instead of carrying his bag over his shoulder. He chalked the bum knee up to aging and managed the pain with wraps, ibuprofen, ice, and hot baths. For years his family physician had been urging him to consider a knee replacement, because advances in the procedure made it almost miraculous in relieving pain and allowing near-normal activity. Finally realizing that the pain would only get worse and that he might be prevented from enjoying his golf, Jon did some research on knee replacement surgery, or knee arthroplasty.

Arthroplasty is used as a last resort when other treatments, such as physical therapy and anti-inflammatory drugs, do not relieve the signs and symptoms, and when the patient's discomfort becomes intense. In knee arthroplasty, a total knee dissection is performed, and damaged cartilage and bone, including the patella, are removed. The removed natural structures are replaced by synthetic structures made of metal and plastic—typically, a plastic patella rubs against the metal surfaces over the distal femur and proximal tibia. In the United States, the average age for a patient undergoing joint replacement is about 65–70 years.

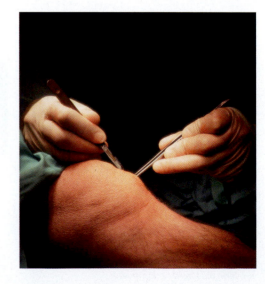

After reading about the surgery and speaking with several friends who had had the procedure, Jon decided to undergo knee arthroplasty.

Here is an x-ray of Jon's knees taken 6 months after surgery. He can now walk without pain, and has returned to golf—although his game remains as bad as ever.

1. Which bones comprise the functioning knee joint?

2. What purpose would physical therapy serve for a person suffering from joint pain?

3. Classify the knee joint in terms of function and structure, and indicate the types of movements possible at this joint.

4. Predict which activities may be restricted following a total knee arthroplasty.

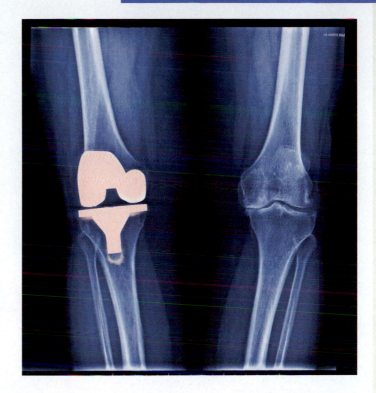

Access more review material online in the Study Area at **www.masteringaandp.com.**

There, you'll find:
- **Chapter guides**
- **Chapter quizzes**
- **Practice tests**
- **Labeling activities**
- **A&P Flix: Group Muscle Actions and Joints**
- **MP3 Tutor Sessions**
- **Animations**
- **Flashcards**
- **A glossary with pronunciations**

9

Skeletal Muscle Tissue

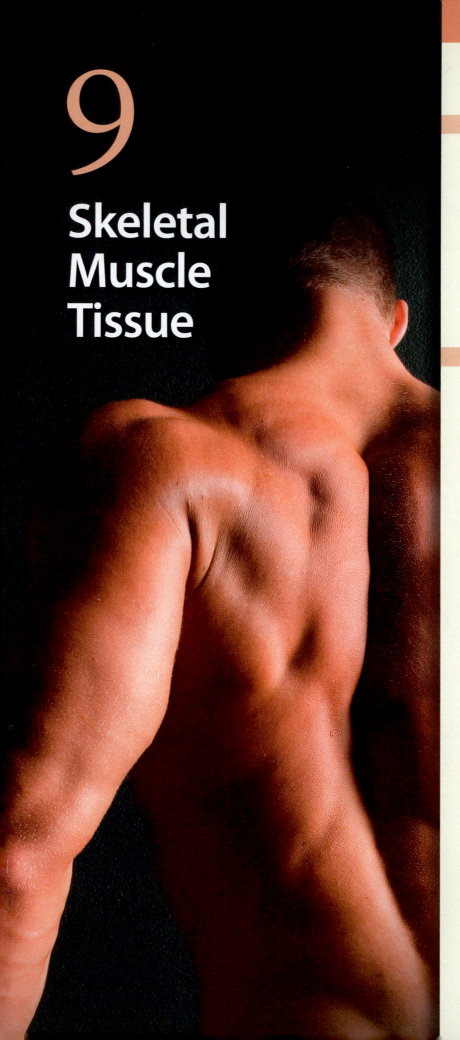

Functional Anatomy of Skeletal Muscle Tissue

Muscle tissue, one of the four primary tissue types, consists chiefly of muscle cells that are highly specialized for contraction. Without the three types of muscle tissue—skeletal muscle, cardiac muscle, and smooth muscle—nothing in the body would move, and no body movement could occur. In this chapter we consider the structure and function of skeletal muscle tissue. Each cell in skeletal muscle tissue is a single muscle fiber. **Skeletal muscles** are organs composed primarily of skeletal muscle tissue plus connective tissues, nerves, and blood vessels. Skeletal muscles are directly or indirectly attached to the bones of the skeleton and have several functions, as detailed in the table at right.

Muscle Tissue

Skeletal Muscle Tissue

Skeletal muscle tissue contractions move the body by pulling on bones of the skeleton, making it possible for us to walk, dance, bite an apple, or play the ukulele.

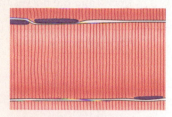

Cardiac Muscle Tissue

Cardiac muscle tissue contractions in the heart propel blood through the blood vessels.

See Chapter 18.

Smooth Muscle Tissue

Smooth muscle tissue contractions move fluids and solids along the digestive tract and regulate the diameters of small arteries, among other functions.

See Chapter 21.

In the first module we examine the functional anatomy of a typical skeletal muscle. In this section we will place particular emphasis on the microscopic structural features that make contractions possible.

Functions of Skeletal Muscle Tissue

- **Produce Skeletal Movement**. Skeletal muscle contractions pull on tendons and move the bones of the skeleton. The effects range from simple motions such as extending the arm or breathing, to the highly coordinated movements of swimming, skiing, or typing.

- **Maintain Posture and Body Position**. Tension in skeletal muscles maintains body posture—for example, holding your head still when you read a book, or balancing your body weight above your feet when you walk. Without constant muscular activity, you could neither sit upright nor stand.

- **Support Soft Tissues**. The abdominal wall and the floor of the pelvic cavity consist of layers of skeletal muscle. These muscles support the weight of visceral organs and shield internal tissues from injury.

- **Guard Entrances and Exits**. The openings of the digestive and urinary tracts are encircled by skeletal muscles. These muscles (sphincters) provide voluntary control over swallowing, defecation, and urination.

- **Maintain Body Temperature**. Muscle contractions require energy; whenever energy is used in the body, some of it is converted to heat. The heat released by working muscles keeps body temperature in the range required for normal functioning.

- **Provide Nutrient Reserves**. When the diet contains inadequate proteins or calories, the contractile proteins in skeletal muscles are broken down into amino acids, which are released into the circulation. Some of these amino acids can be used by the liver to synthesize glucose; others can be broken down to provide energy.

A skeletal muscle contains skeletal muscle tissue, connective tissues, blood vessels, and nerves

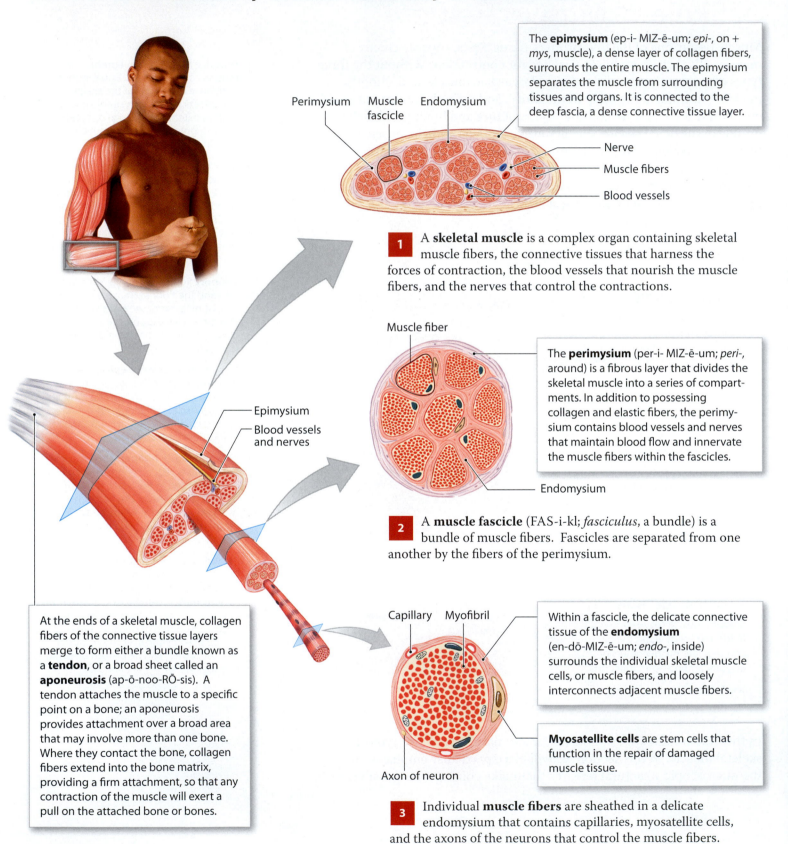

The **epimysium** (ep-i- MIZ-ē-um; *epi-*, on + *mys*, muscle), a dense layer of collagen fibers, surrounds the entire muscle. The epimysium separates the muscle from surrounding tissues and organs. It is connected to the deep fascia, a dense connective tissue layer.

Perimysium Muscle fascicle Endomysium

Nerve
Muscle fibers
Blood vessels

1 A **skeletal muscle** is a complex organ containing skeletal muscle fibers, the connective tissues that harness the forces of contraction, the blood vessels that nourish the muscle fibers, and the nerves that control the contractions.

Muscle fiber

The **perimysium** (per-i- MIZ-ē-um; *peri-*, around) is a fibrous layer that divides the skeletal muscle into a series of compartments. In addition to possessing collagen and elastic fibers, the perimysium contains blood vessels and nerves that maintain blood flow and innervate the muscle fibers within the fascicles.

Endomysium

2 A **muscle fascicle** (FAS-i-kl; *fasciculus*, a bundle) is a bundle of muscle fibers. Fascicles are separated from one another by the fibers of the perimysium.

Epimysium
Blood vessels and nerves

At the ends of a skeletal muscle, collagen fibers of the connective tissue layers merge to form either a bundle known as a **tendon**, or a broad sheet called an **aponeurosis** (ap-ō-noo-RŌ-sis). A tendon attaches the muscle to a specific point on a bone; an aponeurosis provides attachment over a broad area that may involve more than one bone. Where they contact the bone, collagen fibers extend into the bone matrix, providing a firm attachment, so that any contraction of the muscle will exert a pull on the attached bone or bones.

Capillary Myofibril

Within a fascicle, the delicate connective tissue of the **endomysium** (en-dō-MIZ-ē-um; *endo-*, inside) surrounds the individual skeletal muscle cells, or muscle fibers, and loosely interconnects adjacent muscle fibers.

Myosatellite cells are stem cells that function in the repair of damaged muscle tissue.

Axon of neuron

3 Individual **muscle fibers** are sheathed in a delicate endomysium that contains capillaries, myosatellite cells, and the axons of the neurons that control the muscle fibers.

4 During development, groups of embryonic cells called **myoblasts** (*myo-*, muscle + *blastos*, formative cell) fuse, forming multinucleate cells. These large cells then develop into distinctive skeletal muscle fibers. Each nucleus in a skeletal muscle fiber represents the contribution of a single myoblast. Some myoblasts, however, do not fuse with developing muscle fibers. These unfused cells remain in the endomysium of adult skeletal muscle tissue as myosatellite cells. After an injury, myosatellite cells may enlarge, divide, and fuse with damaged muscle fibers, thereby assisting in the repair of the tissue.

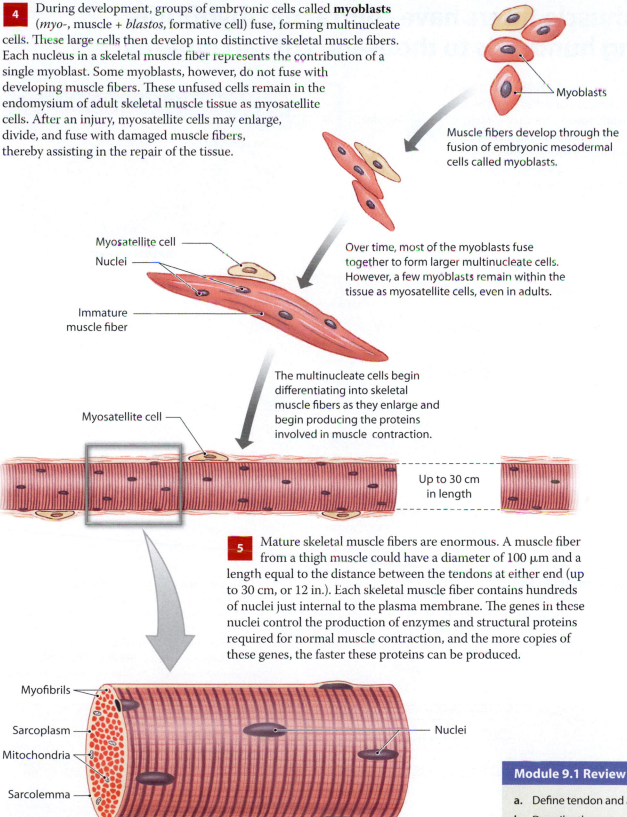

Myoblasts

Muscle fibers develop through the fusion of embryonic mesodermal cells called myoblasts.

Myosatellite cell
Nuclei

Immature muscle fiber

Over time, most of the myoblasts fuse together to form larger multinucleate cells. However, a few myoblasts remain within the tissue as myosatellite cells, even in adults.

Myosatellite cell

The multinucleate cells begin differentiating into skeletal muscle fibers as they enlarge and begin producing the proteins involved in muscle contraction.

Up to 30 cm in length

5 Mature skeletal muscle fibers are enormous. A muscle fiber from a thigh muscle could have a diameter of 100 μm and a length equal to the distance between the tendons at either end (up to 30 cm, or 12 in.). Each skeletal muscle fiber contains hundreds of nuclei just internal to the plasma membrane. The genes in these nuclei control the production of enzymes and structural proteins required for normal muscle contraction, and the more copies of these genes, the faster these proteins can be produced.

Myofibrils

Sarcoplasm

Mitochondria

Sarcolemma

Nuclei

6 Because skeletal muscle fibers are so unusual in size and appearance, special terms are used to describe them. The plasma membrane is called the **sarcolemma** (sar-kō-LEM-uh; *sarkos*, flesh + *lemma*, husk), and the cytoplasm surrounding the myofibrils is called **sarcoplasm** (SAR-kō-plazm).

Module 9.1 Review

a. Define tendon and aponeurosis.

b. Describe the connective tissue layers associated with skeletal muscle tissue.

c. How would severing the tendon attached to a muscle affect the muscle's ability to move a body part?

Skeletal muscle fibers have contractile myofibrils containing hundreds to thousands of sarcomeres

1 A **myofibril** is a cylindrical structure 1–2 μm in diameter and as long as the entire cell. The sarcoplasm of a single skeletal muscle fiber may contain hundreds to thousands of myofibrils. Because each myofibril has a banded appearance and the cell is jam-packed with myofibrils lying side-by-side, the entire muscle fiber appears to have bands, or striations.

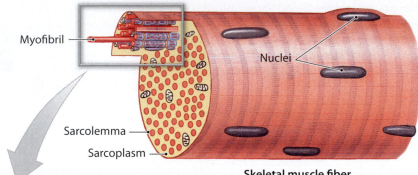

Myofibril

Nuclei

Sarcolemma

Sarcoplasm

Skeletal muscle fiber

2 Myofibrils consist of bundles of protein filaments called **myofilaments**. The most abundant myofilaments are **thin filaments** composed primarily of actin, and **thick filaments** composed primarily of myosin.

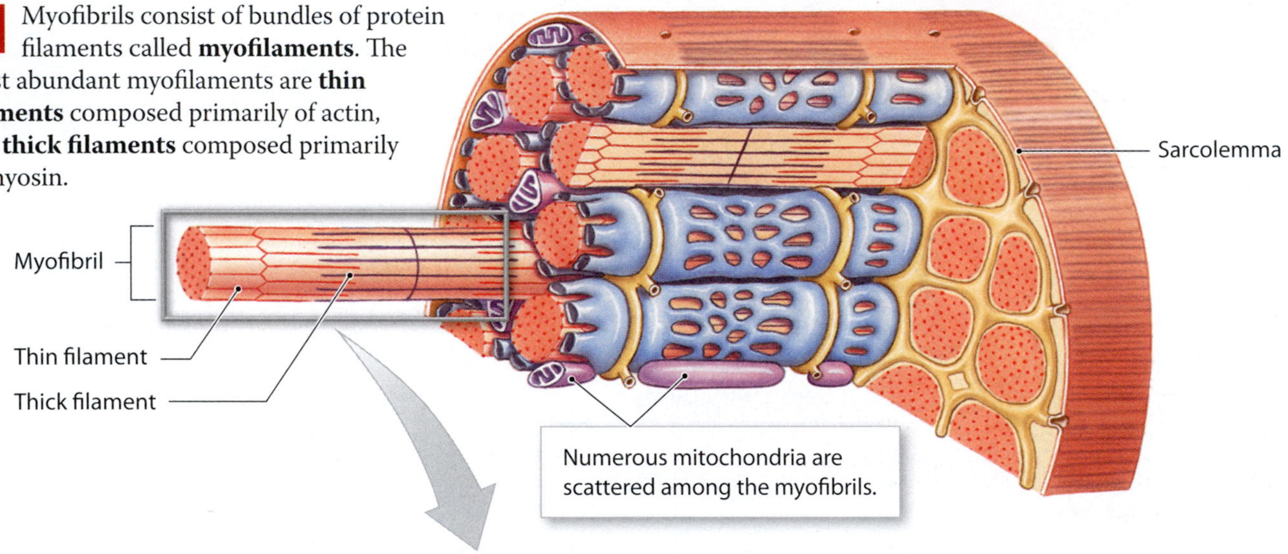

Myofibril

Thin filament

Thick filament

Sarcolemma

Numerous mitochondria are scattered among the myofibrils.

3 Myofilaments have repeating functional units called **sarcomeres** (SAR-kō-mērz; *sarkos*, flesh + *meros*, part). Each myofibril consists of approximately 10,000 sarcomeres, each with a resting length of about 2 μm.

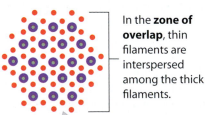

In the **zone of overlap**, thin filaments are interspersed among the thick filaments.

The **A band** is the dense region of the sarcomere that contains thick filaments.

Each **I band**, which contains thin filaments but no thick filaments, extends from the A band of one sarcomere to the A band of the next.

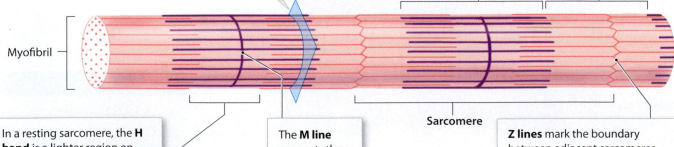

Myofibril

Sarcomere

In a resting sarcomere, the **H band** is a lighter region on either side of the M line. The H band contains thick filaments, but no thin filaments.

The **M line** connects the central portion of each thick filament.

Z lines mark the boundary between adjacent sarcomeres. Z lines consist of proteins called **actinins**, which interconnect thin filaments of adjacent sarcomeres.

4 The sarcolemma separates the sarcoplasm from the surrounding interstitial fluid, and there are differences in the distribution of positive and negative charges on either side of the sarcolemma. This uneven distribution of charges is called the **transmembrane potential**. All living cells have a characteristic transmembrane potential whose presence depends on the selective permeability of the plasma membrane. In skeletal muscle fibers, the transmembrane potential can change markedly, and a sudden change in the transmembrane potential is the first step that leads to a muscle contraction. These changes, triggered by the activity of a nerve cell at one location on the sarcolemma, are swiftly propagated across the entire surface of the muscle fiber.

Transverse tubules, or **T tubules**, are narrow tubes that are continuous with the sarcolemma and extend into the sarcoplasm at right angles to the cell surface. T tubules form passageways through the muscle fiber, like a network of tunnels through a mountain. Changes in the transmembrane potential of the sarcolemma travel along the T tubules into the cell interior.

Inside the sarcoplasm, T tubules encircle each sarcomere at the zones of overlap. Wherever a transverse tubule encircles a myofibril, the tubule is tightly bound to the membranes of the sarcoplasmic reticulum.

The **sarcoplasmic reticulum (SR)**, similar to the smooth endoplasmic reticulum of other cells, forms a tubular network around each individual myofibril. On either side of a T tubule, the tubules of the SR enlarge, fuse, and form expanded chambers called **terminal cisternae** (sis-TUR-nē).

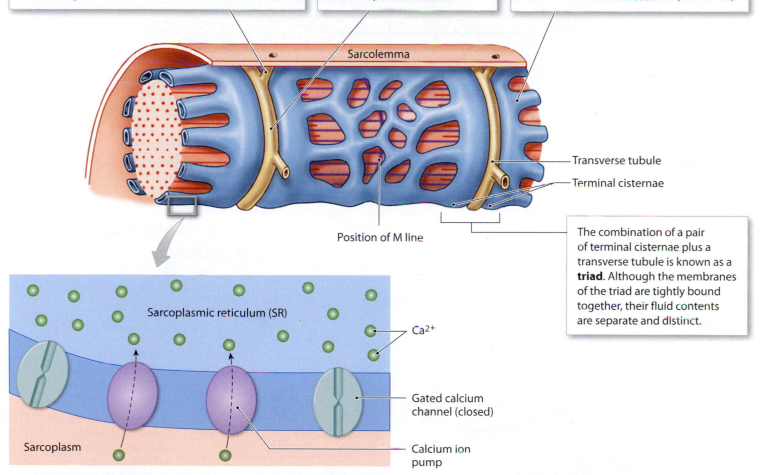

Sarcolemma

Transverse tubule

Terminal cisternae

Position of M line

The combination of a pair of terminal cisternae plus a transverse tubule is known as a **triad**. Although the membranes of the triad are tightly bound together, their fluid contents are separate and distinct.

Sarcoplasmic reticulum (SR)

Ca^{2+}

Gated calcium channel (closed)

Sarcoplasm

Calcium ion pump

5 The membrane of the SR contains ion pumps that pump calcium ions from the sarcoplasm and into the SR. In the SR, some of the calcium ions remain free in solution; the rest become reversibly bound to storage proteins. Including both the free calcium and the bound calcium, the total concentration of Ca^{2+} within cisternae in a resting (inactive) skeletal muscle fiber can be 40,000 times that of the surrounding sarcoplasm. A muscle contraction begins when stored calcium ions are released into the sarcoplasm through gated calcium channels.

Module 9.2 Review

a. Define transverse tubules.

b. Describe the structural components of a sarcomere.

c. Where would you expect the greatest concentration of Ca^{2+} to be in a resting skeletal muscle?

The sliding filament theory of muscle contraction involves thin and thick filaments

A longitudinal section of a sarcomere enables us to examine the structure and organization of the thin and thick filaments in greater detail.

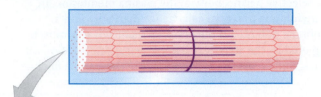

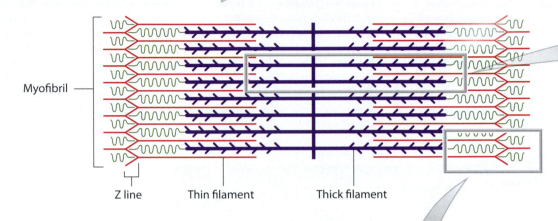

Myofibril

Z line   Thin filament   Thick filament

Structure of Thin Filaments

1 At either end of the sarcomere, the thin filaments are attached to the Z line. **Actinin** interconnects the thin filaments at the Z line, creating an open meshwork at the end of the sarcomere. A typical thin filament is 5–6 nm in diameter and 1 μm in length. As indicated in the magnified view below, each filament is primarily composed of actin, which is associated with other interacting proteins.

Actinin   Z line

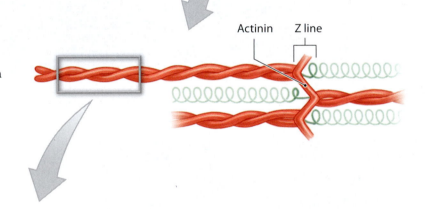

Active site

F-actin (filamentous actin) is a twisted strand composed of two rows of 300–400 individual molecules of G-actin (globular actin).

A **troponin** (TRŌ-pō-nin) molecule consists of three globular subunits. One subunit binds to tropomyosin, locking them together as a troponin–tropomyosin complex; a second subunit binds to one G-actin, holding the troponin–tropomyosin complex in position; and the third subunit has a receptor that binds two calcium ions.

A long strand of **nebulin** extends along the F-actin strand in the cleft between the rows of G-actin molecules. Nebulin holds the F-actin strand together.

Each **G-actin** molecule contains an **active site** that can bind to myosin, much as a substrate molecule binds to the active site of an enzyme.

Strands of **tropomyosin** (trō-pō-MĪ-ō-sin; *tropos*, turning) cover the active sites on G-actin and prevent actin–myosin interaction. A tropomyosin molecule is a double-stranded protein that is bound to one molecule of troponin midway along its length.

Structure of Thick Filaments

2 Thick filaments are 10–12 nm in diameter and 1.6 μm long. A thick filament contains roughly 300 **myosin molecules**, each made up of a pair of myosin subunits twisted around one another. All the myosin molecules are arranged with their tails pointing toward the M line; the myosin heads are arranged in a spiral, each facing one of the surrounding thin filaments. There are no myosin heads in a small area on either side of the M line.

Each thick filament has a core of **titin**. From either end of the thick filament, a strand of titin continues across the I band to the Z line on that side. The portion of the titin strand exposed within the I band is elastic and recoils after stretching. In the normal resting sarcomere, the titin strands are completely relaxed; they become tense only when some external force stretches the sarcomere.

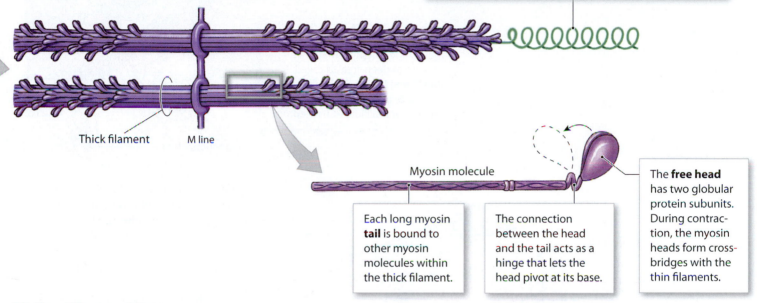

Thick filament   M line

Myosin molecule

Each long myosin **tail** is bound to other myosin molecules within the thick filament.

The connection between the head and the tail acts as a hinge that lets the head pivot at its base.

The **free head** has two globular protein subunits. During contraction, the myosin heads form cross-bridges with the thin filaments.

Sliding Filament Theory

3 When a skeletal muscle fiber contracts, thin filaments slide past the thick filaments. In this process, (1) the H bands and I bands get smaller, (2) the zones of overlap get larger, (3) the Z lines move closer together, and (4) the width of the A band remains constant. This explanation is known as the **sliding filament theory**.

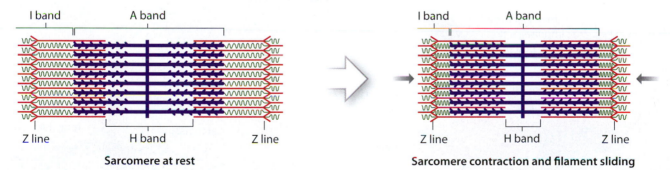

I band	A band

Z line   H band   Z line

Sarcomere at rest

I band	A band

Z line   H band   Z line

Sarcomere contraction and filament sliding

4 During a contraction, sliding occurs in every sarcomere along a myofibril, so the myofibril gets shorter. Because myofibrils are attached to the sarcolemma at each Z line and at either end of the muscle fiber, when myofibrils shorten, so does the muscle fiber.

Module 9.3 Review

a. Describe the components of thin filaments and thick filaments.

b. What gives skeletal muscle its striated appearance?

c. Briefly describe the sliding filament theory.

A skeletal muscle fiber contracts when stimulated by a motor neuron

Skeletal muscle fibers contract only under the control of the nervous system. Each skeletal muscle fiber is controlled by the nervous system at a specialized site known as a **neuromuscular junction (NMJ)**, located midway along the muscle fiber's length.

The **synaptic cleft**, a narrow space, separates the synaptic terminal of the neuron from the opposing motor end plate.

Vesicles containing ACh (red)

The motor end plate contains membrane receptors that bind ACh. It has deep creases called **junctional folds**, which increase its surface area and thus the number of available ACh receptors.

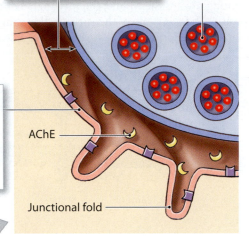

AChE

Junctional fold

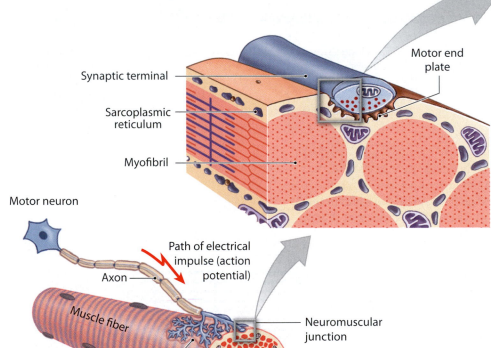

Synaptic terminal

Sarcoplasmic reticulum

Myofibril

Motor end plate

2 The cytoplasm of the synaptic terminal contains vesicles filled with molecules of **acetylcholine** (as-e-til-KŌ-lēn), or **ACh**. Acetylcholine is a neurotransmitter, a chemical released by a neuron to change the permeability or other properties of another cell's plasma membrane. The synaptic cleft and the sarcolemma contain molecules of the enzyme **acetylcholinesterase** (**AChE**), which breaks down ACh.

Motor neuron

Path of electrical impulse (action potential)

Axon

Muscle fiber

Neuromuscular junction

Myofibril

Motor end plate

1 A single axon may branch to control more than one skeletal muscle fiber, but each muscle fiber has only one NMJ. At the NMJ, the **synaptic terminal** of the neuron lies near the **motor end plate** of the muscle fiber.

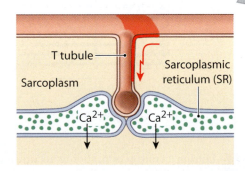

T tubule

Sarcoplasm

Sarcoplasmic reticulum (SR)

Ca^{2+} Ca^{2+}

8 As the action potential sweeps down each T tubule and passes between the terminal cisternae, the SR's permeability changes, and calcium ions are dumped onto the sarcomeres at the zones of overlap. This event, called **excitation–contraction coupling**, triggers the contraction of the muscle fiber (see Module 9.5).

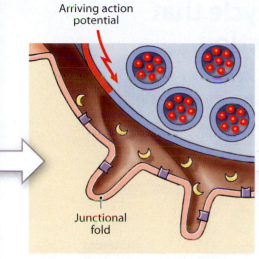

Arriving action potential

Junctional fold

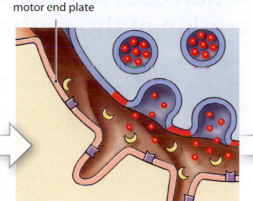

Sarcolemma of motor end plate

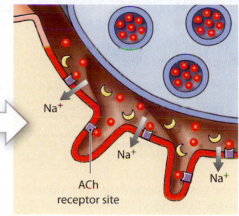

Na⁺

Na⁺

Na⁺

ACh receptor site

3   The stimulus for ACh release is the arrival of an electrical impulse, or **action potential**, at the synaptic terminal. An action potential is a sudden change in the transmembrane potential that travels along the length of the axon.

4   When the action potential reaches the neuron's synaptic terminal, permeability changes in the membrane trigger the exocytosis of ACh into the synaptic cleft. Exocytosis occurs as vesicles fuse with the neuron's plasma membrane.

5   ACh molecules diffuse across the synaptic cleft and bind to ACh receptors on the surface of the motor end plate. ACh binding alters the membrane's permeability to sodium ions. Because the extracellular fluid contains a high concentration of sodium ions, and sodium ion concentration inside the cell is very low, sodium ions rush into the sarcoplasm.

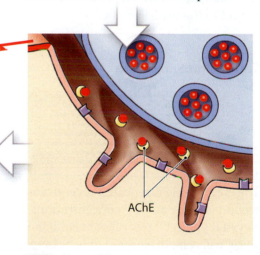

Action potential

AChE

6   The sudden inrush of sodium ions results in the generation of an action potential in the sarcolemma. AChE quickly breaks down the ACh in the synaptic cleft, thus inactivating the ACh receptor sites.

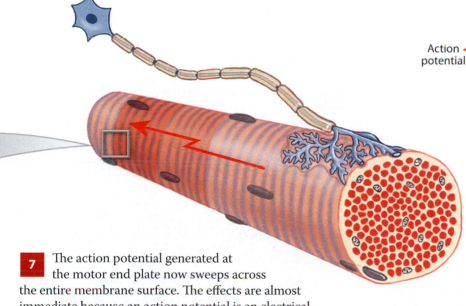

7   The action potential generated at the motor end plate now sweeps across the entire membrane surface. The effects are almost immediate because an action potential is an electrical event that flashes like a spark across the sarcolemmal surface. The effects are brief because the ACh has been removed, and no further stimulus acts upon the motor end plate until another action potential arrives at the synaptic terminal.

Module 9.4 Review

a. Describe the neuromuscular junction.

b. How would a drug that blocks acetylcholine release affect muscle contraction?

c. Predict what would happen if there were no AChE in the synaptic cleft.

A muscle fiber contraction uses ATP in a cycle that is repeated for the duration of the contraction

Resting Sarcomere

In the resting sarcomere, each myosin head is already "energized"—charged with the energy that will be used to power a contraction. Each myosin head points away from the M line. In this position, the myosin head is "cocked" like the spring in a mousetrap. Cocking the myosin head requires energy, which is obtained by breaking down ATP; in doing so, the myosin head functions as ATPase, an enzyme that breaks down ATP. At the start of the contraction cycle, the breakdown products, ADP and phosphate (often represented as P), remain bound to the myosin head.

Contracted Sarcomere

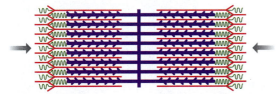

The entire cycle is repeated several times each second, as long as Ca^{2+} concentrations remain elevated and ATP reserves are sufficient. Calcium ion levels will remain elevated only as long as action potentials continue to pass along the T tubules and stimulate the terminal cisternae. Once that stimulus is removed, the calcium channels in the SR close and calcium ion pumps pull Ca^{2+} from the sarcoplasm and store it within the terminal cisternae. Troponin molecules then shift position, swinging the tropomyosin strands over the active sites and preventing further cross-bridge formation.

Contraction Cycle Begins

1 The **contraction cycle**, which involves a series of interrelated steps, begins with the arrival of calcium ions within the zone of overlap.

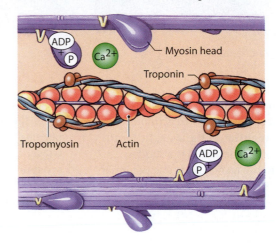

Myosin Reactivation

6 Myosin reactivation occurs when the free myosin head splits ATP into ADP and P. The energy released is used to "recock" the myosin head.

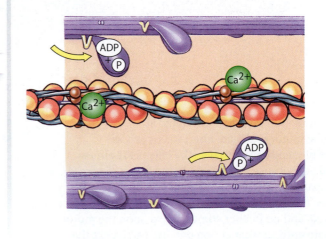

Active-Site Exposure

2 Calcium ions bind to troponin, weakening the bond between actin and the troponin–tropomyosin complex. The troponin molecule then changes position, rolling the tropomyosin molecule away from the active sites on actin and allowing interaction with the energized myosin heads.

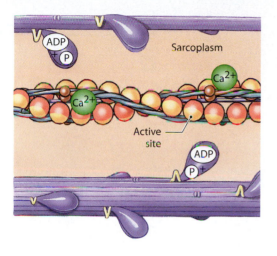

Cross-Bridge Formation

3 Once the active sites are exposed, the energized myosin heads bind to them, forming cross-bridges.

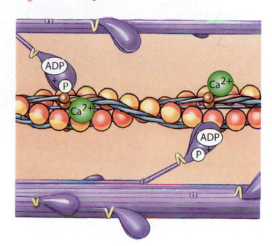

Cross-Bridge Detachment

5 When another ATP binds to the myosin head, the link between the myosin head and the active site on the actin molecule is broken. The active site is now exposed and able to form another cross-bridge.

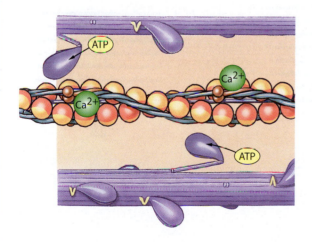

Myosin Head Pivoting

4 After cross-bridge formation, the energy that was stored in the resting state is released as the myosin head pivots toward the M line. This action is called the power stroke; when it occurs, the bound ADP and phosphate group are released.

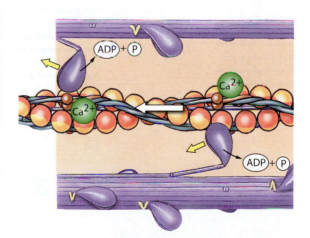

Module 9.5 Review

a. What molecule supplies the energy for a muscle contraction?

b. List the interrelated steps that occur once the contraction cycle has begun.

c. What triggers myosin reactivation?

1. Labeling

Label the structures in the following figure of a skeletal muscle fiber.

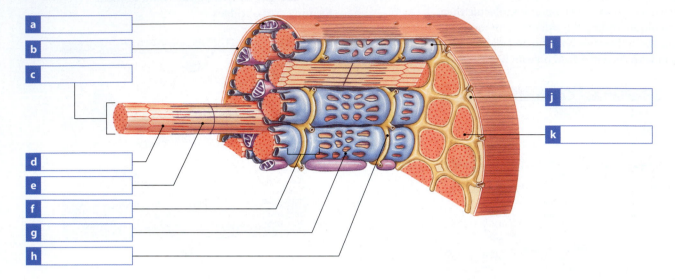

a
b
c
d
e
f
g
h
i
j
k

2. Labeling

Label the structures in the following diagram of adjacent sarcomeres.

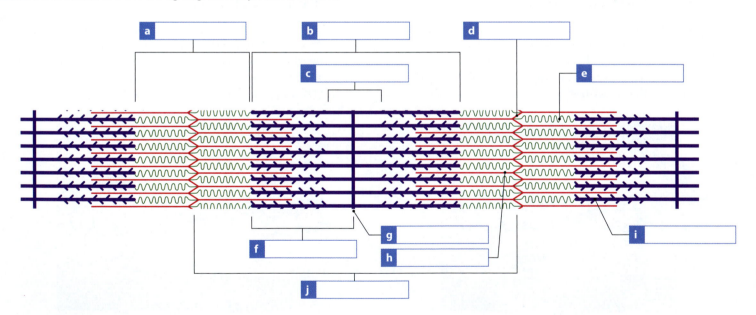

a
b
c
d
e
f
g
h
i
j

3. Vocabulary

In the space provided, write the boldfaced terms introduced in this section that contain the indicated word part.

Word Part	Meaning	Terms
myo-	muscle	a _____
sarko-	flesh	b _____

Functional Properties of Skeletal Muscle Tissue

Before we examine how skeletal muscle tissue works, recall the sequence of events we have considered thus far.

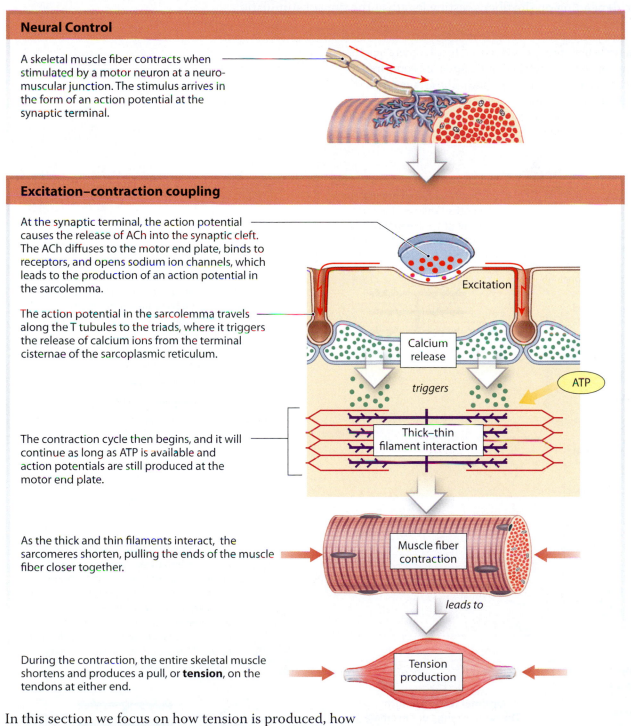

Neural Control

A skeletal muscle fiber contracts when stimulated by a motor neuron at a neuro-muscular junction. The stimulus arrives in the form of an action potential at the synaptic terminal.

Excitation–contraction coupling

At the synaptic terminal, the action potential causes the release of ACh into the synaptic cleft. The ACh diffuses to the motor end plate, binds to receptors, and opens sodium ion channels, which leads to the production of an action potential in the sarcolemma.

The action potential in the sarcolemma travels along the T tubules to the triads, where it triggers the release of calcium ions from the terminal cisternae of the sarcoplasmic reticulum.

The contraction cycle then begins, and it will continue as long as ATP is available and action potentials are still produced at the motor end plate.

As the thick and thin filaments interact, the sarcomeres shorten, pulling the ends of the muscle fiber closer together.

During the contraction, the entire skeletal muscle shortens and produces a pull, or **tension**, on the tendons at either end.

Excitation

Calcium release

triggers

ATP

Thick–thin filament interaction

Muscle fiber contraction

leads to

Tension production

In this section we focus on how tension is produced, how muscle contractions are classified and fueled, and how different muscle fibers respond from one moment to the next.

Tension production is greatest when a muscle is stimulated at its optimal length

There is no mechanism to regulate the amount of tension produced in a contraction by changing the number of contracting sarcomeres. When calcium ions are released, they are released from all triads in the muscle fiber. Thus, a muscle fiber is either "on" (producing tension) or "off" (relaxed). Tension production at the level of the individual muscle fiber does vary, however, depending on the fiber's resting length at the time of stimulation.

1 The tension a muscle fiber produces is related to sarcomere length. When sarcomeres are either stretched or compressed compared to optimal resting length, tension production declines. The arrangement of skeletal muscles, connective tissues, and bones normally prevents too much compression or stretching. During walking, for example, leg muscle fibers are stretched very close to "ideal length" before contractions occur.

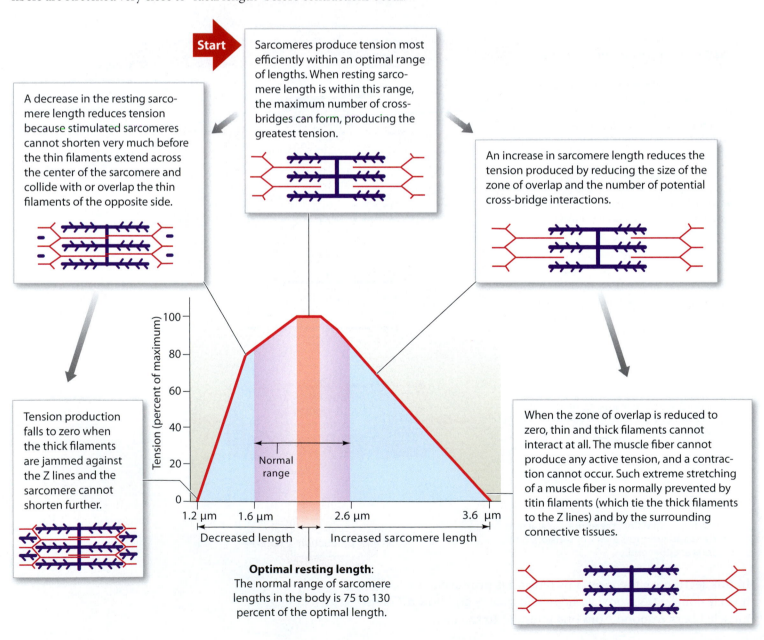

Start Sarcomeres produce tension most efficiently within an optimal range of lengths. When resting sarcomere length is within this range, the maximum number of cross-bridges can form, producing the greatest tension.

A decrease in the resting sarcomere length reduces tension because stimulated sarcomeres cannot shorten very much before the thin filaments extend across the center of the sarcomere and collide with or overlap the thin filaments of the opposite side.

An increase in sarcomere length reduces the tension produced by reducing the size of the zone of overlap and the number of potential cross-bridge interactions.

Tension production falls to zero when the thick filaments are jammed against the Z lines and the sarcomere cannot shorten further.

When the zone of overlap is reduced to zero, thin and thick filaments cannot interact at all. The muscle fiber cannot produce any active tension, and a contraction cannot occur. Such extreme stretching of a muscle fiber is normally prevented by titin filaments (which tie the thick filaments to the Z lines) and by the surrounding connective tissues.

Tension (percent of maximum)

100
80
60
40
20
0

Normal range

1.2 μm   1.6 μm   2.6 μm   3.6 μm

Decreased length   Increased sarcomere length

Optimal resting length: The normal range of sarcomere lengths in the body is 75 to 130 percent of the optimal length.

2 This **myogram**—a graph of tension development in muscle fibers—shows two examples of a **twitch**, a single stimulus-contraction-relaxation sequence in a muscle fiber. Twitches vary in duration, depending on muscle type and location, internal and external environmental conditions, and other factors. Although muscle twitches can be produced by electrical stimulation in a laboratory, they are too brief to be part of any normal activity. However, they reveal some interesting differences in the functional properties of various skeletal muscles.

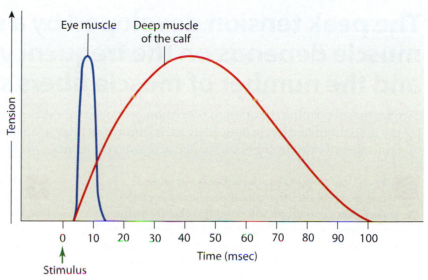

3 This myogram shows the phases of a 40-msec twitch in a muscle fiber from the gastrocnemius muscle, a prominent superficial muscle of the calf.

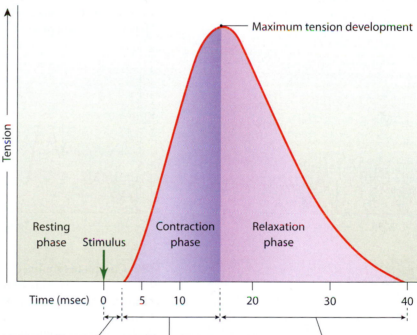

The **latent period** begins at stimulation and typically lasts about 2 msec. During this period, an action potential sweeps across the sarcolemma, and the sarcoplasmic reticulum releases calcium ions. The muscle fiber does not produce tension during the latent period, because the contraction cycle has yet to begin.

In the **contraction phase**, tension rises to a peak. As the tension rises, calcium ions are binding to troponin, active sites on thin filaments are being exposed, and cross-bridge interactions are occurring.

The **relaxation phase** lasts about 25 msec. During this period, calcium levels are falling, active sites are being covered by tropomyosin, and the number of active cross-bridges is declining as they detach. As a result, tension returns to resting levels.

Module 9.6 Review

a. Name a factor that affects the amount of tension produced when a skeletal muscle contracts.

b. Explain two key concepts of the length–tension relationship.

c. For each portion of a myogram tracing a twitch in a stimulated gastrocnemius (calf) muscle fiber, describe the events that occur within the muscle.

The peak tension developed by a skeletal muscle depends on the frequency of stimulation and the number of muscle fibers stimulated

Two factors determine the amount of tension produced by a skeletal muscle: (1) the amount of tension produced by each stimulated muscle fiber, and (2) the total number of muscle fibers stimulated at a given moment.

1 Each time a skeletal muscle fiber is stimulated immediately after the relaxation phase has ended, the subsequent contraction will develop a slightly higher maximum tension than did the previous contraction. The increase in peak tension will continue over the first 30–50 stimulations. Because the tension rises like the steps in a staircase, this phenomenon is called **treppe** (TREP-eh, German for *staircase*). Most skeletal muscles do not demonstrate treppe.

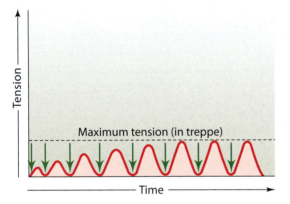

2 If a second stimulus arrives before the relaxation phase has ended, a second, more powerful contraction occurs. The addition of one twitch to another in this way constitutes **wave summation**. The duration of a single twitch determines the maximum time available to produce wave summation. For example, if a twitch lasts 20 msec (1/50 sec), a stimulus frequency of greater than 50 per second produces wave summation, whereas a stimulus frequency of less than 50 per second will produce individual twitches and treppe.

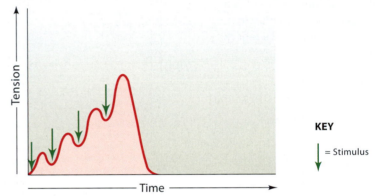

KEY

↓ = Stimulus

4 **Complete tetanus** occurs when a higher stimulation frequency eliminates the relaxation phase. During complete tetanus, action potentials arrive so rapidly that the sarcoplasmic reticulum cannot reclaim calcium ions. The high Ca^{2+} concentration in the cytoplasm prolongs the contraction, making it continuous. Although muscles can be forced into tetanus in the laboratory, they seldom if ever develop peak tension in the course of normal activities, in part because normal movements require precise control over, and continuous variation in, the amount of tension produced.

3 A muscle producing almost peak tension during rapid cycles of contraction and relaxation is said to be in **incomplete tetanus** (*tetanos*, convulsive tension).

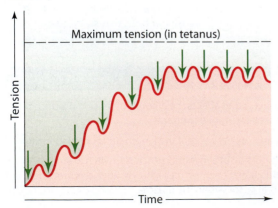

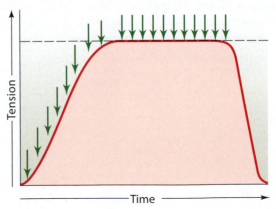

5 A typical skeletal muscle contains thousands of muscle fibers. Although some motor neurons control just a few muscle fibers, most control hundreds of them. The amount of tension produced is controlled at the subconscious level through variations in the number of muscle fibers stimulated.

All the muscle fibers controlled by a single motor neuron constitute a **motor unit**. The size of a motor unit is an indication of how fine the control of movement can be. In the muscles of the eye, where precise control is extremely important, a motor neuron may control 4–6 muscle fibers. We have much less precise control over our leg muscles, where a single motor neuron may control 1000–2000 muscle fibers.

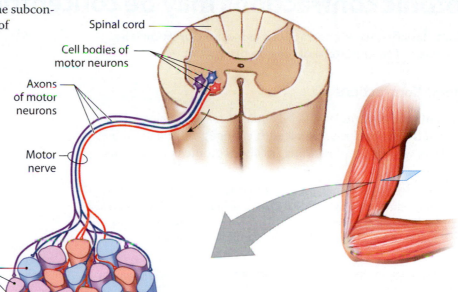

Spinal cord

Cell bodies of motor neurons

Axons of motor neurons

Motor nerve

The muscle fibers of each motor unit are intermingled with those of other motor units. Because of this intermingling, the direction of pull exerted on the tendon does not change when the number of activated motor units changes. When you decide to perform a specific movement, the contraction begins with the activation of the smallest motor units in the stimulated muscle. As the movement continues, larger motor units containing faster and more powerful muscle fibers are activated, and tension production rises steeply. The smooth but steady increase in muscular tension produced by increasing the number of active motor units is called **recruitment**.

KEY

▪ Motor unit 1

▪ Motor unit 2

▪ Motor unit 3

6 During a sustained contraction, motor units are activated on a rotating basis, so some of them are resting and recovering while others are actively contracting. In this "relay team" approach, called **asynchronous motor unit summation**, each motor unit can recover somewhat before it is stimulated again. As a result, when your muscles contract for sustained periods, they produce slightly less than maximal tension.

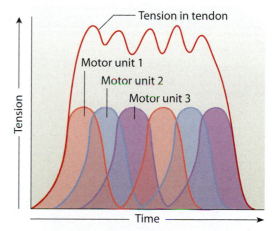

Tension in tendon

Motor unit 1
Motor unit 2
Motor unit 3

Tension

Time

A variable number of motor units is always active, even when the entire muscle is not contracting. Their contractions do not produce enough tension to cause movement, but they do tense and firm the muscle. This resting tension in a skeletal muscle is called **muscle tone**, and it is regulated at the subconscious level. The activity level of each motor neuron changes constantly, and individual muscle fibers can relax while a constant tension is maintained in the attached tendon. Activated muscle fibers use energy, so the greater the muscle tone, the higher the "resting" rate of metabolism. Elevated muscle tone increases resting energy consumption by a small amount, but the effects are cumulative, and they continue 24 hours per day.

Module 9.7 Review

a. Define motor unit.

b. Describe the relationship between the number of fibers in a motor unit and the precision of body movements.

c. Compare incomplete tetanus with wave summation.

Muscle contractions may be isotonic or isometric; isotonic contractions may be concentric or eccentric

We can classify muscle contractions as isotonic or isometric on the basis of their pattern of tension production.

Isotonic Contractions

In an **isotonic contraction** (*iso-*, equal + *tonos*, tension), tension rises and the skeletal muscle's length changes. Lifting an object off a desk, walking, and running involve isotonic contractions.

1 Consider a skeletal muscle that is 1 cm² in cross-sectional area and can produce roughly 4 kg (8.8 lb) of tension in complete tetanus. If we hang a load of 2 kg (4.4 lb) from that muscle and stimulate it, the muscle will shorten. This is called a **concentric contraction**.

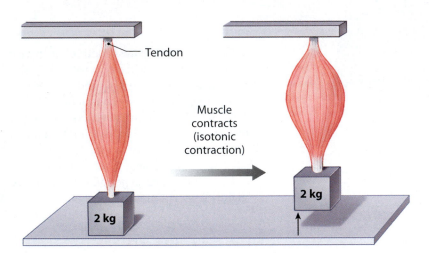

2 Before the muscle can shorten, the cross-bridges must produce enough tension to overcome the load—in this case, the 2-kg weight. During this initial period, tension in the muscle fibers rises until the tension in the tendon exceeds the load. As the muscle shortens, the tension in the skeletal muscle remains constant.

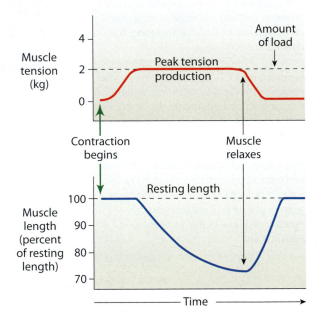

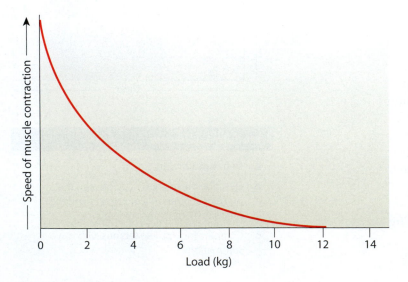

3 The speed, or rate, of muscle contraction varies inversely with the load on the muscle. If the load is relatively light, the muscle will contract quickly; if the load is heavy, the muscle will contract slowly or not at all. So, the speed of muscle contraction is fastest when the load equals zero.

4 If the load is greater than the peak tension the muscle can develop, the muscle will elongate. This is similar to a tug-of-war team trying to stop a moving car and having the rope slip through their fingers. This elongation is called an **eccentric contraction**; an example is tensing the biceps brachii while lowering a barbell.

5 The rate of elongation during an eccentric contraction depends on the difference between the tension developed by the active muscle fibers and the size of the load. By varying the tension in an eccentric contraction, you can control the rate of elongation.

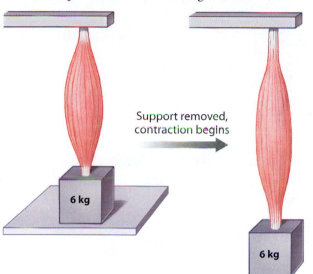

Support removed, contraction begins

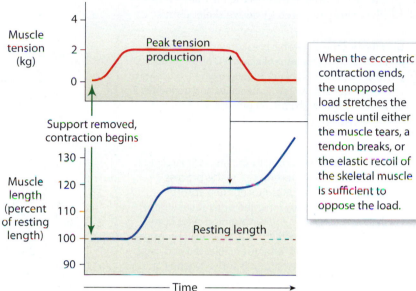

When the eccentric contraction ends, the unopposed load stretches the muscle until either the muscle tears, a tendon breaks, or the elastic recoil of the skeletal muscle is sufficient to oppose the load.

Isometric Contractions

In contrast to an isotonic contraction, in an **isometric contraction** (*metric*, measure), the muscle as a whole does not change length, and the tension produced never exceeds the load. Many of the reflexive muscle contractions that keep your body upright when you stand or sit involve isometric contractions of muscles that oppose the force of gravity.

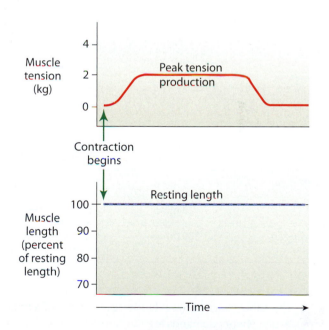

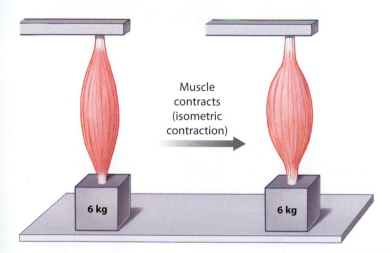

Muscle contracts (isometric contraction)

6 If the load equals the peak tension, the load won't move when the muscle contracts. The contracting muscle bulges, but not as much as it does during an isotonic contraction. In an isometric contraction, although the muscle as a whole does not shorten, the individual muscle fibers shorten as connective tissues stretch. The muscle fibers cannot shorten further, because the tension does not exceed the load.

Module 9.8 Review

a. Define isotonic contraction and isometric contraction.

b. Can a skeletal muscle contract without shortening? Why or why not?

c. Explain the relationship between load and speed of muscle contraction.

Muscle contraction requires large amounts of ATP that may be produced aerobically or anaerobically

1 Mitochondrial activity is the ultimate source of the energy required by active skeletal muscles.

Glycolysis is the anaerobic breakdown of glucose to pyruvate in the cytoplasm of a cell. It is an anaerobic process, because it does not require oxygen. Glycolysis provides a net gain of 2 ATP molecules and generates 2 pyruvate molecules from each glucose molecule.

Aerobic metabolism normally provides 95 percent of the ATP demands of a resting cell. In this process, introduced in Module 3.6, mitochondria absorb oxygen, ADP, phosphate ions, and organic substrates (such as pyruvate) from the surrounding cytoplasm. Primarily through the activity of the electron transport chain, large amounts of energy are released and used to make ATP. The entire process is very efficient: For each molecule of pyruvate "fed" into the citric acid cycle, the cell gains 17 ATP molecules.

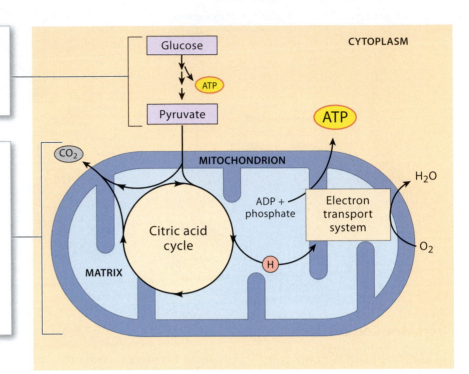

Sources of Energy Stored in a Typical Muscle Fiber

Energy Source	Initial Quantity	Utilization Process	Number of Twitches Supported by Each Energy Source Alone	Duration of Isometric Tetanic Contraction Supported by Each Energy Source Alone
ATP	3 mmol	ATP $\longrightarrow$ ADP + P	10	2 sec
CP	20 mmol	ADP + CP $\longrightarrow$ ATP + C	70	15 sec
Glycogen	100 mmol	Glycolysis (anaerobic) Aerobic metabolism	670 12,000	130 sec 2400 sec (40 min)

2 This table characterizes the energy reserves of a typical skeletal muscle fiber. When energy demands are low and oxygen is abundant, a muscle fiber absorbs nutrients from the interstitial fluid and builds up its energy reserves. In addition to ATP, muscle fibers store an additional high-energy compound called **creatine phosphate** (**CP**). (Creatine is a compound assembled within skeletal muscles from catabolized amino acids.) However, muscle fibers store only a relatively small amount of energy in high-energy compounds; most energy is stored as glycogen, which may account for 1.5 percent of total muscle weight and enables muscle contractions to continue for extended periods.

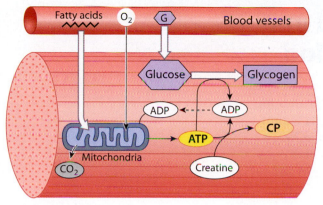

Muscle at rest

3 In a resting skeletal muscle, the demand for ATP is low. More than enough oxygen is available for the mitochondria to meet that demand, and they produce a surplus of ATP. The extra ATP is used to build up reserves of CP and glycogen. Resting muscle fibers absorb fatty acids and glucose delivered by the bloodstream. The fatty acids are broken down in the mitochondria, and the ATP that is generated is used to convert creatine to creatine phosphate and glucose to glycogen.

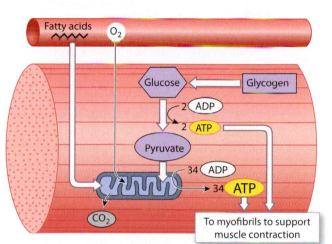

Muscle at moderate activity levels

4 At moderate levels of activity, the demand for ATP increases. This demand is met by the mitochondria, and the rate of oxygen consumption increases as a result. The skeletal muscle now relies primarily on the aerobic metabolism of pyruvate to generate ATP, and the pyruvate is provided by glycolysis, using glucose obtained from stored glycogen reserves. As long as the demand for ATP can be met by mitochondrial activity, the muscle will not become fatigued until glycogen, lipid, and amino acid reserves are exhausted. This type of fatigue affects the muscles of endurance athletes, such as marathon runners, after hours of exertion.

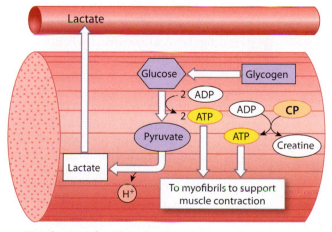

Muscle at peak activity levels

5 At peak levels of activity, ATP demands are enormous and mitochondrial ATP production rises to a maximum rate determined by the availability of oxygen. At peak levels of exertion, mitochondrial activity can provide only about one-third of the ATP needed. The remainder is produced through glycolysis. When glycolysis produces pyruvate faster than it can be utilized by the mitochondria, pyruvate levels rise in the sarcoplasm. Under these conditions, pyruvate is converted to **lactic acid**, a related three-carbon molecule that dissociates into a three-carbon **lactate** molecule and a hydrogen ion (H^+). The production of lactic acid during peak activity lowers the intracellular and extracellular pH. After only a few seconds of peak activity, changes in pH will alter the functional characteristics of key enzymes so that the muscle fiber cannot continue to contract. Sprinters usually experience this type of muscle fatigue.

Module 9.9 Review

a. Identify three sources of energy utilized by muscle fibers.

b. How do muscle cells continuously synthesize ATP?

c. Under what conditions do muscle fibers produce lactic acid?

Muscles are subject to fatigue and may require an extended recovery period

An active skeletal muscle is said to be **fatigued** when it can no longer continue to perform at the required level of activity. Many factors are involved in promoting muscle fatigue, but after peak activity a major factor is the decline in pH within the muscle fibers and the muscle as a whole, decreasing calcium ion binding to troponin and altering enzyme activities. In the **recovery period**, the conditions in muscle fibers are returned to normal, pre-exertion levels. After a period of moderate activity, it may take several hours for muscle fibers to recover. After sustained activity at higher levels, complete recovery can take a week.

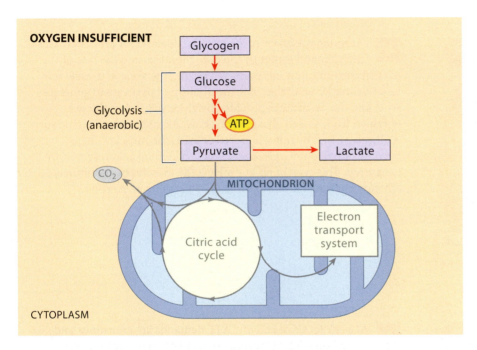

1 Glycolysis enables a skeletal muscle to continue contracting even when mitochondrial activity is limited by the availability of oxygen. As we have seen, glycolysis is not the most efficient way to generate ATP. It squanders the glucose reserves of the muscle fibers, and it is potentially dangerous because the dissociation of lactic acid can lower the pH of the blood and tissues. Glycolysis can produce ATP faster than aerobic metabolism, but only until glycogen reserves are depleted (1–2 minutes). In terms of energy efficiency, glycolysis captures some 4–6 percent of the energy as ATP in the conversion of glucose to pyruvate. The reactions of glycolysis also elevate body temperature, triggering increased sweat gland activity.

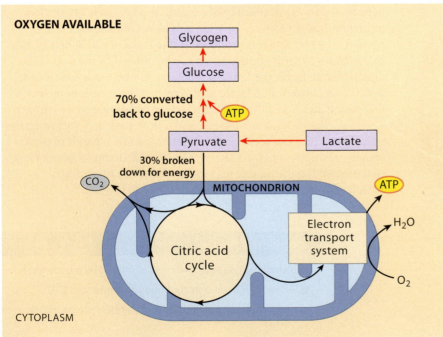

2 During the recovery period, when oxygen is available in abundance, lactate can be recycled by conversion back to pyruvate. The pyruvate can then be used either by mitochondria to generate ATP or as a substrate for enzyme pathways that synthesize glucose and rebuild glycogen reserves. Aerobic metabolism is much more efficient than glycolysis—the muscle fiber can capture about 42 percent of the energy released. However, whenever they are using ATP, muscles are continuously releasing heat. Roughly 85 percent of the heat needed to maintain normal body temperature is provided by resting skeletal muscles.

Much of the large amounts of lactate produced during peak exertion diffuses out of the muscle fibers and into the bloodstream. The liver absorbs this lactate and begins converting it into pyruvate.

This process continues after exertion has ended, because lactate levels within muscle fibers remain relatively high, and lactate continues to diffuse into the bloodstream. After the absorbed lactate is converted to pyruvate in the liver, roughly 30 percent of the new pyruvate molecules are broken down in the mitochondria, providing the ATP needed to convert the remaining 70 percent of pyruvate molecules into glucose. The glucose molecules are then released into the circulation, where they are absorbed by skeletal muscle fibers and used to rebuild their glycogen reserves.

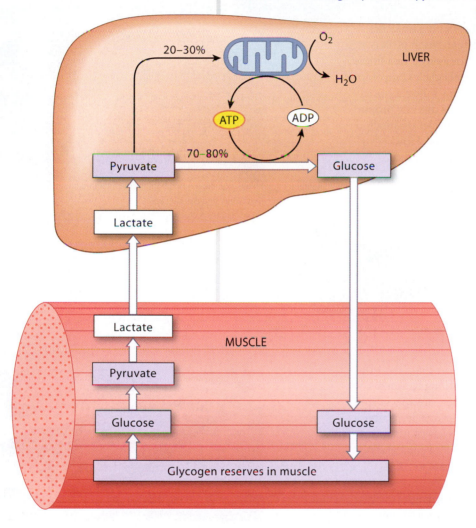

3 Much of the lactate released by muscle fibers during strenuous activity diffuses from the muscle tissue and into the bloodstream. The liver absorbs and recycles it, and releases glucose into the circulation. This shuffling of lactate to the liver and of glucose back to muscle cells is called the **Cori cycle**. Throughout the recovery period, the body's oxygen demand remains elevated above normal resting levels. The more ATP required, the more oxygen will be needed. The amount of oxygen required to restore normal, pre-exertion conditions is called the **oxygen debt**, or **excess postexercise oxygen consumption (EPOC)**. Most of the additional oxygen consumption occurs in skeletal muscle fibers, which must restore ATP, creatine phosphate, and glycogen concentrations to their former levels, and in liver cells, which generate the ATP needed to convert excess lactate to glucose.

Module 9.10 Review

a. Define oxygen debt (excess postexercise oxygen consumption).

b. What two processes are crucial in repaying a muscle's oxygen debt during the recovery period?

c. After strenuous exercise, what causes the "burning" sensation in skeletal muscles?

Fast, slow, and intermediate skeletal muscle fibers differ in size, internal structure, metabolism, and resistance to fatigue

The human body has three major types of skeletal muscle fibers: fast fibers, slow fibers, and intermediate fibers.

1 Most of the skeletal muscle fibers in the body are called **fast fibers**, because they can reach peak twitch tension in 0.01 second or less after stimulation. Fast fibers have large diameters and contain densely packed myofibrils, large glycogen reserves, and relatively few mitochondria. The tension produced by a muscle fiber is directly proportional to the number of myofibrils, so muscles dominated by fast fibers produce powerful contractions. However, fast fibers fatigue rapidly because their contractions use ATP in massive amounts, and they have relatively few mitochondria to generate ATP. As a result, prolonged activity is supported primarily by anaerobic metabolism.

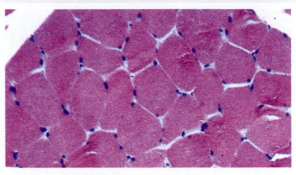

Fast fibers in cross section   LM × 171

2 **Slow fibers** have only about half the diameter of fast fibers and take three times as long to reach peak tension after stimulation. Slow fibers are specialized to continue contracting for extended periods, long after a fast fiber would become fatigued. Slow muscle fibers are surrounded by a more extensive network of capillaries than fast muscle tissue, so they have a dramatically higher oxygen supply to support mitochondrial activity. Slow fibers also contain the red pigment **myoglobin** (MĪ-ō-glō-bin), a globular protein that is structurally related to hemoglobin, the red oxygen-carrying pigment in blood. Both myoglobin and hemoglobin reversibly bind oxygen molecules. Myoglobin is most abundant in slow fibers, so resting slow fibers contain substantial oxygen reserves that can be mobilized during a contraction. Because slow fibers have both an extensive capillary supply and a high concentration of myoglobin, skeletal muscles dominated by slow fibers are dark red.

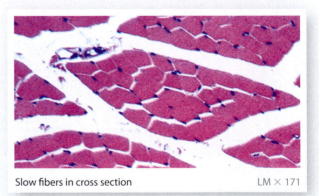

Slow fibers in cross section   LM × 171

3 This light micrograph (a longitudinal section) shows that a fast muscle fiber (W, for white) and a slow muscle fiber (R, for red) have very different sizes and densities, and that slow fibers have more mitochondria (M) and a more extensive capillary supply (cap). Our bodies also contain a third group of muscle fibers—**intermediate fibers**—that more closely resemble fast fibers, for they contain little myoglobin and are relatively pale. Intermediate fibers have a more extensive capillary network around them and are more resistant to fatigue than are fast fibers.

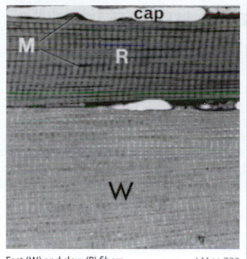

Fast (W) and slow (R) fibers in longitudinal section   LM × 783

Properties of Skeletal Muscle Fiber Types

Property	Slow Fibers	Intermediate Fibers	Fast Fibers
Cross-sectional diameter	Small	Intermediate	Large
Time to peak tension	Prolonged	Medium	Rapid
Contraction speed	Slow	Fast	Fast
Fatigue resistance	High	Intermediate	Low
Color	Red	Pink	White
Myoglobin content	High	Low	Low
Capillary supply	Dense	Intermediate	Scarce
Mitochondria	Many	Intermediate	Few
Glycolytic enzyme concentration in sarcoplasm	Low	High	High
Substrates used for ATP generation during contraction	Lipids, carbohydrates, amino acids (aerobic)	Primarily carbohydrates (anaerobic)	Carbohydrates (anaerobic)
Alternative names	Type I, S (slow), red, SO (slow oxidative), slow-twitch oxidative	Type II-A, FR (fast resistant), fast-twitch oxidative	Type II-B, FF (fast fatigue), white, fast-twitch glycolytic

4 This table compares the properties of fast, intermediate, and slow muscle fibers. The percentages of fast, intermediate, and slow fibers in a skeletal muscle can be quite variable. Most human muscles contain a mixture of fiber types and so appear pink. However, there are no slow fibers in muscles of the eye or hand, where swift but brief contractions are required. Many back and calf muscles are dominated by slow fibers; these muscles contract almost continuously to maintain an upright posture. The percentage of fast versus slow fibers in each muscle is genetically determined, but the ratio of intermediate fibers to fast fibers can increase as a result of athletic training. For example, if a muscle is used repeatedly for endurance events, some of the fast fibers will develop the appearance and functional capabilities of intermediate fibers.

Module 9.11 Review

a. Identify the three types of skeletal muscle fibers.

b. Why would a sprinter experience muscle fatigue before a marathon runner would?

c. Which type of muscle fiber would you expect to predominate in the large leg muscles of someone who excels at endurance activities, such as cycling or long-distance running?

Many factors can result in muscle hypertrophy, atrophy, or paralysis

1 As a result of repeated, exhaustive stimulation, muscle fibers develop more mitochondria, a higher concentration of glycolytic enzymes, and larger glycogen reserves. Such muscle fibers have more myofibrils than do fibers that are less used, and each myofibril contains more thick and thin filaments. The net effect is **hypertrophy**, or an enlargement of the stimulated muscle. The number of muscle fibers does not change significantly, but the muscle as a whole enlarges because each muscle fiber increases in diameter. The muscle also becomes stronger, since tension production is proportional to the cross-sectional area of the muscle. The muscles of a bodybuilder are excellent examples of muscular hypertrophy. Although hypertrophy can be promoted by administering steroid hormones such as androgens (male sex hormones), exhaustive training is still required, and the side effects of steroid use can be very dangerous.

2 A skeletal muscle that is not regularly stimulated by a motor neuron loses muscle tone and mass. The muscle becomes flaccid, and the muscle fibers become smaller and weaker. This reduction in muscle size, tone, and power is called **atrophy**. A variable degree of muscle atrophy is a normal consequence of aging. Individuals of any age who are paralyzed by spinal injuries or other damage to the nervous system will gradually lose muscle tone and size in the areas affected. Even a temporary reduction in muscle use can lead to muscular atrophy; you can easily observe this effect by comparing "before and after" limb muscles in someone who has worn a cast. Muscle atrophy is initially reversible, but dying muscle fibers are not replaced. In extreme atrophy, the functional losses are permanent. That is why physical therapy is crucial for people who are temporarily unable to move normally.

3 This figure considers a few of the many clinical conditions that can affect skeletal muscles. Because skeletal muscles depend on motor neurons for stimulation, disorders that affect the nervous system can indirectly affect the muscular system, and are thus of particular interest here.

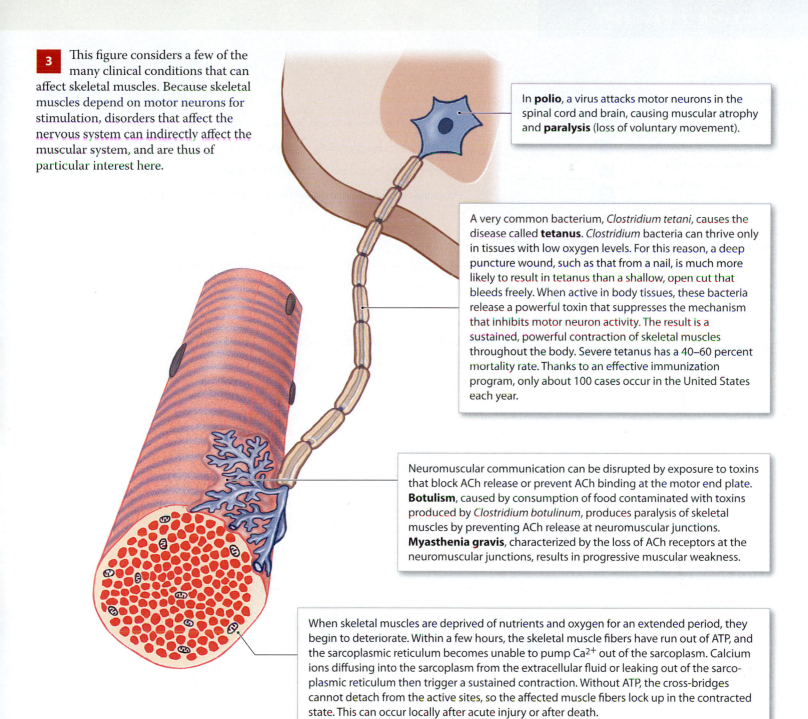

In **polio**, a virus attacks motor neurons in the spinal cord and brain, causing muscular atrophy and **paralysis** (loss of voluntary movement).

A very common bacterium, *Clostridium tetani*, causes the disease called **tetanus**. *Clostridium* bacteria can thrive only in tissues with low oxygen levels. For this reason, a deep puncture wound, such as that from a nail, is much more likely to result in tetanus than a shallow, open cut that bleeds freely. When active in body tissues, these bacteria release a powerful toxin that suppresses the mechanism that inhibits motor neuron activity. The result is a sustained, powerful contraction of skeletal muscles throughout the body. Severe tetanus has a 40–60 percent mortality rate. Thanks to an effective immunization program, only about 100 cases occur in the United States each year.

Neuromuscular communication can be disrupted by exposure to toxins that block ACh release or prevent ACh binding at the motor end plate. **Botulism**, caused by consumption of food contaminated with toxins produced by *Clostridium botulinum*, produces paralysis of skeletal muscles by preventing ACh release at neuromuscular junctions. **Myasthenia gravis**, characterized by the loss of ACh receptors at the neuromuscular junctions, results in progressive muscular weakness.

When skeletal muscles are deprived of nutrients and oxygen for an extended period, they begin to deteriorate. Within a few hours, the skeletal muscle fibers have run out of ATP, and the sarcoplasmic reticulum becomes unable to pump Ca^{2+} out of the sarcoplasm. Calcium ions diffusing into the sarcoplasm from the extracellular fluid or leaking out of the sarcoplasmic reticulum then trigger a sustained contraction. Without ATP, the cross-bridges cannot detach from the active sites, so the affected muscle fibers lock up in the contracted state. This can occur locally after acute injury or after death.

Shortly after death, a generalized skeletal muscle contraction, called **rigor mortis**, occurs throughout the body, beginning with the smaller muscles of the face, neck, and arms. As the SR deteriorates, calcium ions are released and a sustained contraction begins. As ATP reserves are exhausted, the muscles become locked in the contracted state. Because all skeletal muscles are involved during rigor mortis, the individual becomes "stiff as a board." It typically begins 2–7 hours after death and disappears after 1–6 days or when decomposition begins; but the timing depends on environmental factors such as temperature. Forensic pathologists base time of death estimates on the degree of rigor mortis and environmental conditions.

Module 9.12 Review

a. Define muscle hypertrophy and muscle atrophy.

b. Six weeks after Fred broke his leg the cast is removed, and as he steps down from the exam table, his leg gives way and he falls. Propose a logical explanation.

c. Explain how the flexibility or rigidity of a dead body can provide a clue about a murder victim's time of death.

1. Matching

Match the following terms with the most closely related description.

- eccentric contraction
- isometric contraction
- isotonic contraction
- concentric contraction

a	_____	Muscle does not change length during contraction
b	_____	Muscle changes length during contraction
c	_____	Peak tension less than load, and muscle elongates
d	_____	Peak tension greater than load, and muscle shortens

2. Labeling

Use the following terms to correctly label the structures in the diagram representing a blood vessel and a skeletal muscle at rest.

- creatine
- glycogen
- CP
- glucose
- O_2
- fatty acids

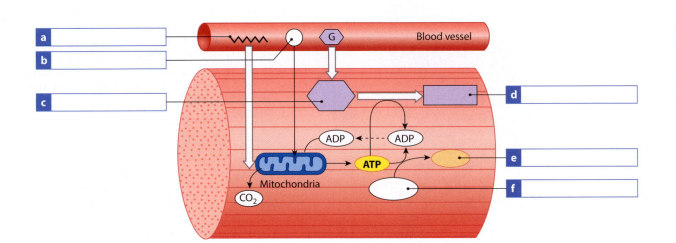

3. Short answer

Complete the following table by writing in the anatomical and physiological properties of the three types of skeletal muscle fibers.

Property	Slow Fibers	Intermediate Fibers	Fast Fibers
Cross-sectional diameter	a	b	c
Color	d	pink	e
Myoglobin content	high	f	g
Capillary supply	h	i	scarce
Mitochondria	j	k	l
Time to peak tension	m	medium	n
Contraction speed	o	p	q
Fatigue resistance	high	r	s
Glycolytic enzyme concentration in sarcoplasm	t	u	high

Visual Outline with Key Terms
Summarize the content of each module using the terms in the order provided.

SECTION 1

Functional Anatomy of Skeletal Muscle Tissue

- skeletal muscles
- skeletal muscle tissue
- cardiac muscle tissue
- smooth muscle tissue

9.1

A skeletal muscle contains skeletal muscle tissue, connective tissues, blood vessels, and nerves

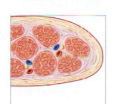

- skeletal muscle
- epimysium
- tendon
- aponeurosis
- muscle fascicle
- perimysium
- muscle fibers
- endomysium
- myosatellite cells
- myoblasts
- sarcolemma
- sarcoplasm

9.2

Skeletal muscle fibers have contractile myofibrils containing hundreds to thousands of sarcomeres

- skeletal muscle fiber
- myofibril
- myofilaments
- thin filaments
- thick filaments
- sarcomeres
- H band
- M line
- zone of overlap
- A band
- I band
- Z lines
- actinins
- transmembrane potential
- transverse tubules (T tubules)
- sarcoplasmic reticulum (SR)
- terminal cisternae
- triad

9.3

The sliding filament theory of muscle contraction involves thin and thick filaments

- thin filaments
- actinin
- F-actin
- nebulin
- G-actin
- active site
- tropomyosin
- troponin
- thick filaments
- myosin molecules
- titin
- tail
- free head
- sliding filament theory

9.4

A skeletal muscle fiber contracts when stimulated by a motor neuron

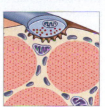

- neuromuscular junction (NMJ)
- synaptic terminal
- motor end plate
- acetylcholine (ACh)
- acetylcholinesterase (AChE)
- junctional folds
- synaptic cleft
- action potential
- excitation–contraction coupling

9.5

A muscle fiber contraction uses ATP in a cycle that is repeated for the duration of the contraction

- contraction cycle
- active-site exposure
- cross-bridge formation
- myosin head pivoting
- cross-bridge detachment
- myosin reactivation

SECTION 2

Functional Properties of Skeletal Muscle Tissue

- neural control
- excitation–contraction coupling
- tension

9.6

Tension production is greatest when a muscle is stimulated at its optimal length

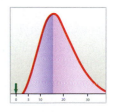

- optimal resting length
- myogram
- twitch
- latent period
- contraction phase
- relaxation phase

• = *Term boldfaced in this module*

9.7

The peak tension developed by a skeletal muscle depends on the frequency of stimulation and the number of muscle fibers stimulated

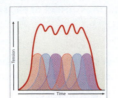

- treppe
- wave summation
- incomplete tetanus
- complete tetanus
- motor unit
- recruitment
- asynchronous motor unit summation
- muscle tone

9.8

Muscle contractions may be isotonic or isometric; isotonic contractions may be concentric or eccentric

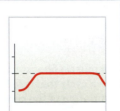

- isotonic contraction
- concentric contraction
- eccentric contraction
- isometric contraction

9.9

Muscle contraction requires large amounts of ATP that may be produced aerobically or anaerobically

- glycolysis
  ○ pyruvate
- aerobic metabolism
- creatine phosphate (CP)
- lactic acid
- lactate

9.10

Muscles are subject to fatigue and may require an extended recovery period

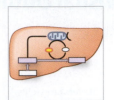

- fatigued
- recovery period
- Cori cycle
- oxygen debt
- excess postexercise oxygen consumption (EPOC)

9.11

Fast, slow, and intermediate skeletal muscle fibers differ in size, internal structure, metabolism, and resistance to fatigue

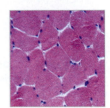

- fast fibers
- slow fibers
- myoglobin
- intermediate fibers

9.12

Many factors can result in muscle hypertrophy, atrophy, or paralysis

- hypertrophy
- atrophy
- polio
- paralysis
- tetanus
- botulism
- myasthenia gravis
- rigor mortis

● = *Term boldfaced in this module*

Chapter Integration: Applying what you've learned

Pesticides are commonly used to control pests, such as insects, in the environment. The most effective pesticides contain organophosphates, which are biological toxins that interfere with the action of the enzyme acetylcholinesterase (AChE). Although the goal is to paralyze and kill insects, the drug can potentially affect other animals, including humans. The Environmental Protection Agency (EPA) in the United States and the Pest Management Regulatory Agency (PMRA) of Canada regulate pesticides, and the American Medical Association (AMA) recommends limiting a person's exposure to organophosphate-based pesticides and using safer alternatives. Regrettably, the latest recommendations do no good for someone who has had a lifetime of exposure.

Harry has been the groundskeeper for the athletic fields of a local school system for 34 years. Prior to that he worked for a private lawn maintenance company doing very similar work. In the normal course of work, Harry used the common pesticides and weed killers. Periodically over the years, Harry has had minor skin irritations. In each case, the problem resolved within a short period, although one episode lasted about 10 days. But just two weeks ago Harry was working with a new employee when a cloud of chemicals engulfed him. He had a coughing fit and felt as if he could not catch his breath. His muscles then started twitching uncontrollably, and he soon was unable to stand. By the time an ambulance arrived, the twitching had become extremely violent, but when he arrived at the hospital Harry was unconscious and barely breathing. He was put on a respirator, given oxygen, and treated with the chemical *atropine,* which blocks ACh receptor sites. Harry was fortunate—he survived—but he is now looking for a different line of work.

1. Relate the signs and symptoms Harry experienced to the specific actions of organophosphates.

2. Explain how a pesticide could cause skin problems.

Access more review material online in the Study Area at www.masteringaandp.com.

There, you'll find:
- **Chapter guides**
- **Chapter quizzes**
- **Practice tests**
- **Labeling activities**
- **A&P Flix:**
  - **Events at the Neuromuscular Junction**
  - **Excitation–Contraction Coupling**
  - **Cross-Bridge Cycle**
- **MP3 Tutor Sessions**
- **Tutorials**
- **Animations**
- **Flashcards**
- **A glossary with pronunciations**

iP® Use *Interactive Physiology®* (IP) to help you understand difficult physiological concepts in this chapter. Go to Muscular System and find the following topics:

- **Anatomy Review: Skeletal Muscle Tissue**
- **The Neuromuscular Junction**
- **Sliding Filament Theory**
- **Muscle Metabolism**
- **Contraction of Motor Units**
- **Contraction of Whole Muscle**

The names of muscles can provide clues to their appearance and/or function

1 In most cases one end of a muscle is fixed in position, and the other end moves during a contraction. The place where the fixed end attaches is called the **origin** of the muscle. Most muscles originate at a bone, but some originate at a connective tissue sheath or band such as the **intermuscular septa** (components of the deep fascia that may separate adjacent skeletal muscles) or at the interosseous membranes of the forearm or leg. The site where the movable end attaches to another structure is called the **insertion** of the muscle. The origin is typically proximal to the insertion when the body is in the anatomical position. However, knowing which end is the origin and which is the insertion is ultimately less important than knowing where the two ends attach and what the muscle accomplishes when it contracts. When a muscle contracts, it produces a specific movement or **action**. In general, we will describe actions in terms of movement at specific joints.

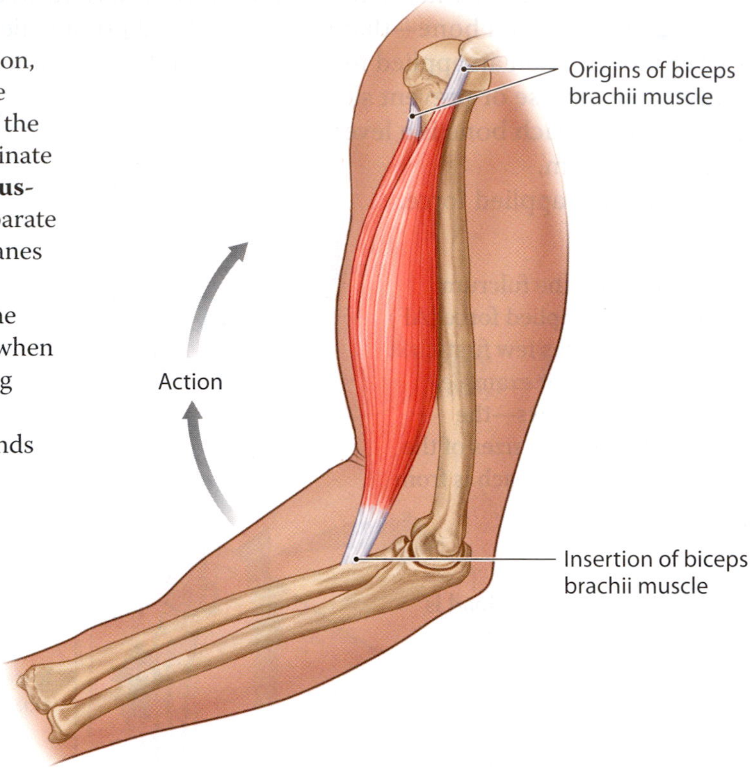

Origins of biceps brachii muscle

Action

Insertion of biceps brachii muscle

2 When complex movements occur, muscles commonly work in groups rather than individually. Their cooperation improves the efficiency of a particular movement. For example, large muscles of the limbs produce flexion or extension over an extended range of motion. Based on their functions, muscles may be described as agonists, antagonists, or synergists.

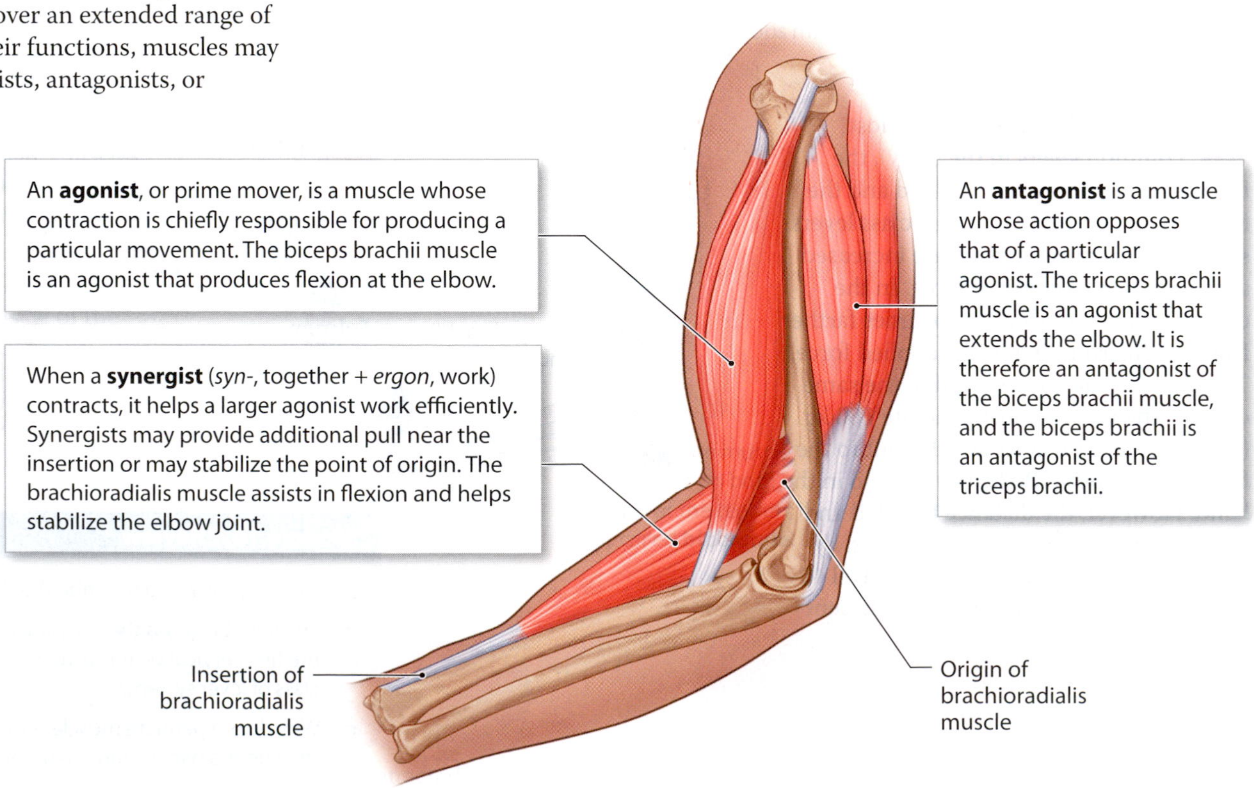

An **agonist**, or prime mover, is a muscle whose contraction is chiefly responsible for producing a particular movement. The biceps brachii muscle is an agonist that produces flexion at the elbow.

When a **synergist** (*syn-*, together + *ergon*, work) contracts, it helps a larger agonist work efficiently. Synergists may provide additional pull near the insertion or may stabilize the point of origin. The brachioradialis muscle assists in flexion and helps stabilize the elbow joint.

An **antagonist** is a muscle whose action opposes that of a particular agonist. The triceps brachii muscle is an agonist that extends the elbow. It is therefore an antagonist of the biceps brachii muscle, and the biceps brachii is an antagonist of the triceps brachii.

Insertion of brachioradialis muscle

Origin of brachioradialis muscle

Muscle Terminology

Terms Indicating Specific Regions of the Body*	Terms Indicating Position, Direction, or Fascicle Organization	Terms Indicating Structural Characteristics of the Muscle	Terms Indicating Actions
Abdominis (abdomen)	Anterior (front)	**Nature of Origin**	**General**
Anconeus (elbow)	Externus (superficial)	Biceps (two heads)	Abductor
Auricularis (auricle of ear)	Extrinsic (outside)	Triceps (three heads)	Adductor
Brachialis (brachium)	Inferioris (inferior)	Quadriceps (four heads)	Depressor
Capitis (head)	Internus (deep, internal)		Extensor
Carpi (wrist)	Intrinsic (inside)		Flexor
Cervicis (neck)	Lateralis (lateral)	**Shape**	Levator
Cleido-/-clavius (clavicle)	Medialis/medius (medial, middle)	Deltoid (triangle)	Pronator
Coccygeus (coccyx)	Oblique	Orbicularis (circle)	Rotator
Costalis (ribs)	Posterior (back)	Pectinate (comblike)	Supinator
Cutaneous (skin)	Profundus (deep)	Piriformis (pear-shaped)	Tensor
Femoris (femur)	Rectus (straight, parallel)	Platy- (flat)	
Genio- (chin)	Superficialis (superficial)	Pyramidal (pyramid)	**Specific**
Glosso-/-glossal (tongue)	Superioris (superior)	Rhomboid	Buccinator (trumpeter)
Hallucis (great toe)	Transversus (transverse)	Serratus (serrated)	Risorius (laugher)
Ilio- (ilium)		Splenius (bandage)	Sartorius (like a tailor)
Inguinal (groin)		Teres (long and round)	
Lumborum (lumbar region)		Trapezius (trapezoid)	
Nasalis (nose)			
Nuchal (back of neck)		**Other Striking Features**	
Oculo- (eye)		Alba (white)	
Oris (mouth)		Brevis (short)	
Palpebrae (eyelid)		Gracilis (slender)	
Pollicis (thumb)		Lata (wide)	
Popliteus (posterior to knee)		Latissimus (widest)	
Psoas (loin)		Longissimus (longest)	
Radialis (radius)		Longus (long)	
Scapularis (scapula)		Magnus (large)	
Temporalis (temples)		Major (larger)	
Thoracis (thoracic region)		Maximus (largest)	
Tibialis (tibia)		Minimus (smallest)	
Ulnaris (ulna)		Minor (smaller)	
Uro- (urinary)		Vastus (great)	

* For other regional terms, refer to Module 1.8.

3 This table includes a useful summary of the most important terms used in naming skeletal muscles. A familiarity with these terms will help you to identify and remember specific muscles. Except for the platysma and the diaphragm, the complete names of all skeletal muscles include the term "muscle". Although the full name, such as the biceps brachii muscle, will usually appear in the text, for simplicity only the descriptive name (biceps brachii) will be used in figures and tables.

Module 10.2 Review

a. Define the term *synergist* as it relates to muscle action.

b. Muscle A abducts the humerus, and muscle B adducts the humerus. What is the relationship between these two muscles?

c. What does the name *flexor carpi radialis longus* tell you about this muscle?

The skeletal muscles can be assigned to the axial division or the appendicular division based on origins and functions

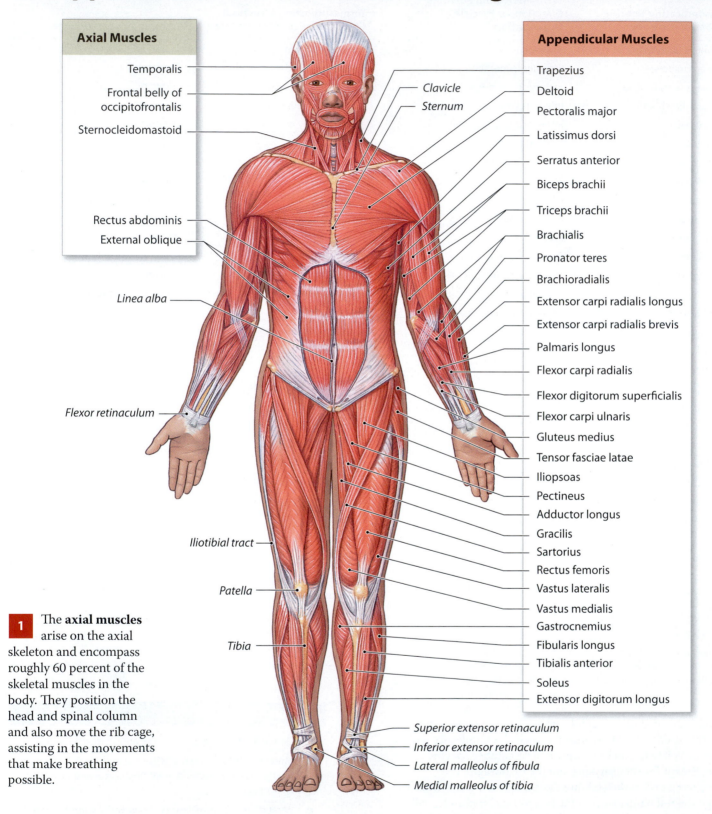

Axial Muscles

- Temporalis
- Frontal belly of occipitofrontalis
- Sternocleidomastoid
- Rectus abdominis
- External oblique
- Linea alba
- Flexor retinaculum
- Iliotibial tract
- Patella
- Tibia

Clavicle
Sternum

Appendicular Muscles

- Trapezius
- Deltoid
- Pectoralis major
- Latissimus dorsi
- Serratus anterior
- Biceps brachii
- Triceps brachii
- Brachialis
- Pronator teres
- Brachioradialis
- Extensor carpi radialis longus
- Extensor carpi radialis brevis
- Palmaris longus
- Flexor carpi radialis
- Flexor digitorum superficialis
- Flexor carpi ulnaris
- Gluteus medius
- Tensor fasciae latae
- Iliopsoas
- Pectineus
- Adductor longus
- Gracilis
- Sartorius
- Rectus femoris
- Vastus lateralis
- Vastus medialis
- Gastrocnemius
- Fibularis longus
- Tibialis anterior
- Soleus
- Extensor digitorum longus

Superior extensor retinaculum
Inferior extensor retinaculum
Lateral malleolus of fibula
Medial malleolus of tibia

1 The **axial muscles** arise on the axial skeleton and encompass roughly 60 percent of the skeletal muscles in the body. They position the head and spinal column and also move the rib cage, assisting in the movements that make breathing possible.

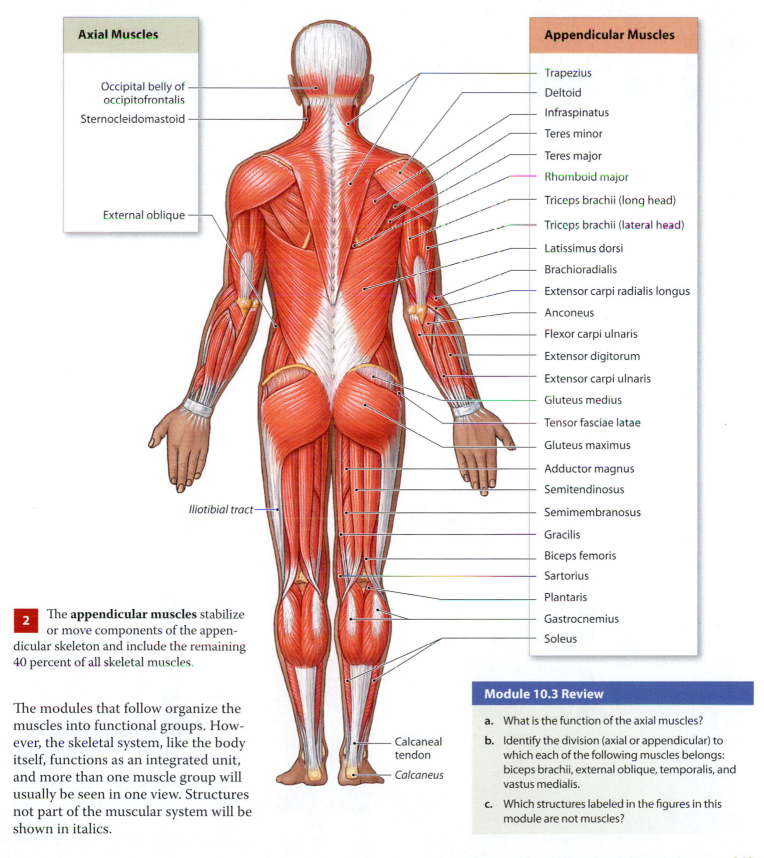

Axial Muscles

Occipital belly of occipitofrontalis

Sternocleidomastoid

External oblique

Iliotibial tract

Appendicular Muscles

Trapezius

Deltoid

Infraspinatus

Teres minor

Teres major

Rhomboid major

Triceps brachii (long head)

Triceps brachii (lateral head)

Latissimus dorsi

Brachioradialis

Extensor carpi radialis longus

Anconeus

Flexor carpi ulnaris

Extensor digitorum

Extensor carpi ulnaris

Gluteus medius

Tensor fasciae latae

Gluteus maximus

Adductor magnus

Semitendinosus

Semimembranosus

Gracilis

Biceps femoris

Sartorius

Plantaris

Gastrocnemius

Soleus

Calcaneal tendon

Calcaneus

2 The **appendicular muscles** stabilize or move components of the appendicular skeleton and include the remaining 40 percent of all skeletal muscles.

The modules that follow organize the muscles into functional groups. However, the skeletal system, like the body itself, functions as an integrated unit, and more than one muscle group will usually be seen in one view. Structures not part of the muscular system will be shown in italics.

Module 10.3 Review

a. What is the function of the axial muscles?

b. Identify the division (axial or appendicular) to which each of the following muscles belongs: biceps brachii, external oblique, temporalis, and vastus medialis.

c. Which structures labeled in the figures in this module are not muscles?

1. Short answer

Label the pennate muscles in the following diagram, and for each indicate the type of pennate muscle based on the relationship between the tendon(s) and fascicle organization.

a

b

c

2. Labeling

Label each of the indicated superficial muscles in the diagram to the right.

a

b

c

d

e

f

g

h

i

j

k

l

m

n

o

p

q

r

s

t

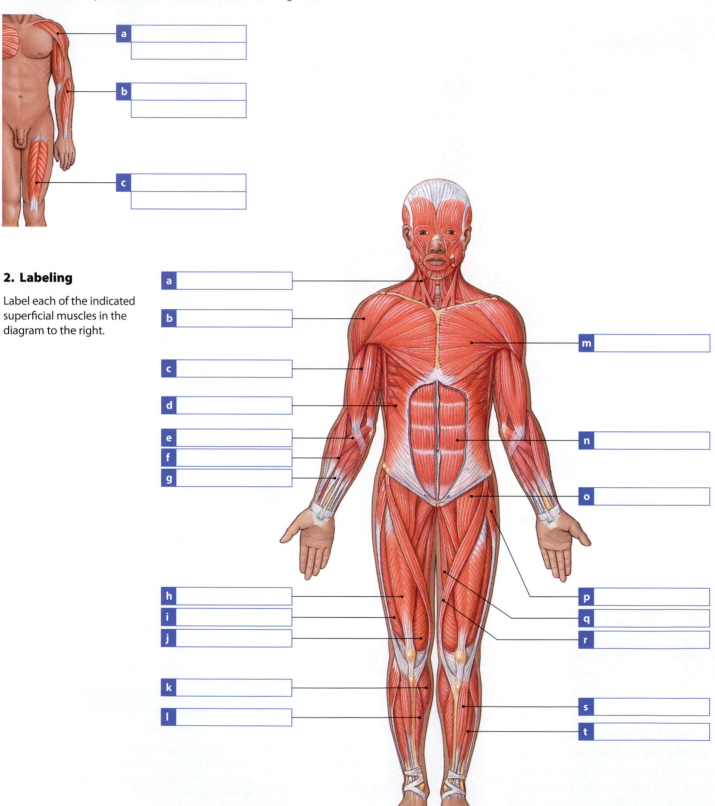

The Axial Muscles

The axial musculature is involved in stabilizing and positioning the head, neck, and trunk. Based on location and/or function, we can divide the axial muscles into the four groups shown here. The groups do not always have distinct anatomical boundaries. For example, a function such as the extension of the vertebral column involves muscles along its entire length.

1 The first group contains muscles of the head and neck that are not associated with the vertebral column. These muscles include the muscles of facial expression (Module 10.4), the extrinsic eye muscles (Module 10.5), and the muscles of the tongue, pharynx, and neck (Module 10.6).

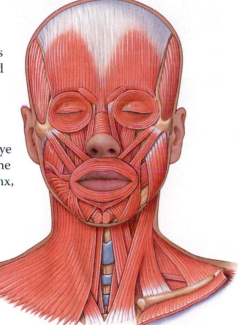

2 The second group—the muscles of the vertebral column—includes numerous muscles of varied size that stabilize, flex, extend, or rotate the vertebral column (Module 10.7).

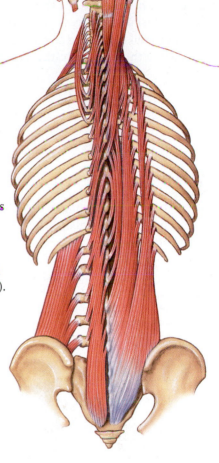

3 The third group consists of the oblique and rectus muscles of the trunk (Module 10.8). These muscles form broad sheets or bands that form the muscular walls of the thoracic and abdomino-pelvic cavities.

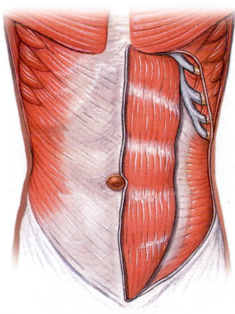

4 The fourth group—the muscles of the pelvic floor (Module 10.9)—spans the pelvic outlet and supports the organs of the pelvis.

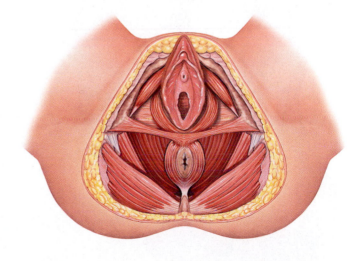

The muscles of facial expression are important in eating and useful for communication

The **muscles of facial expression** originate on the surface of the skull. At their insertions, the fibers of the epimysium are woven into those of the superficial fascia and the dermis of the skin. Thus, when they contract, the skin moves.

1 This anterior view shows superficial muscles on the right side of the face and deeper muscles on the left side of the face.

2 A corresponding lateral view shows the major facial muscles; note the abundance of muscles involved in movements of the lips.

The occipitofrontalis muscle, which forms the scalp, has two bellies that are connected by a collagenous sheet, the **epicranial aponeurosis**.

Occipital belly   Frontal belly

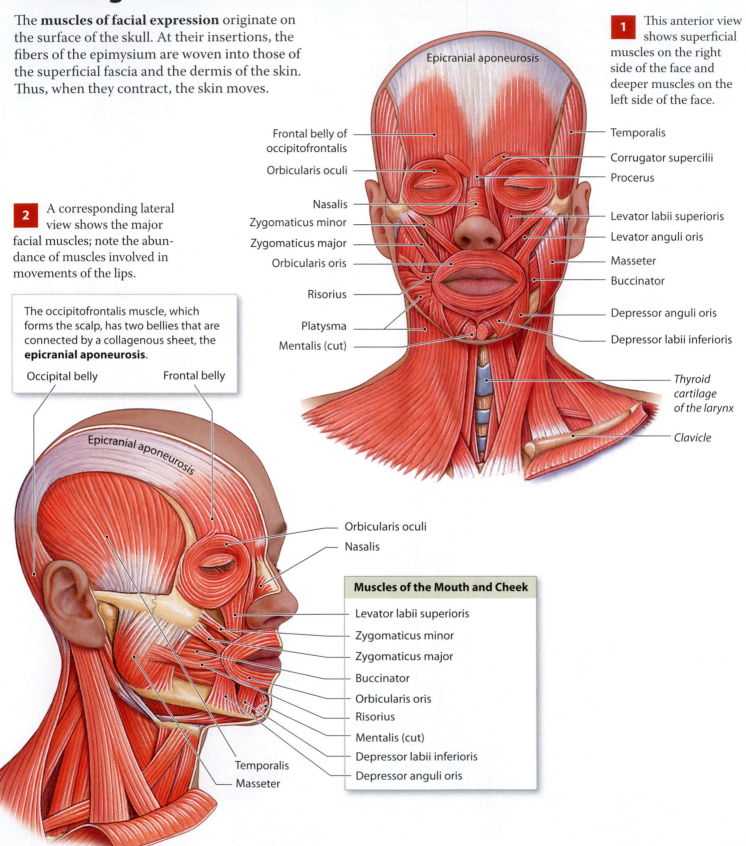

Epicranial aponeurosis

Frontal belly of occipitofrontalis
Orbicularis oculi
Nasalis
Zygomaticus minor
Zygomaticus major
Orbicularis oris
Risorius
Platysma
Mentalis (cut)

Temporalis
Corrugator supercilii
Procerus
Levator labii superioris
Levator anguli oris
Masseter
Buccinator
Depressor anguli oris
Depressor labii inferioris
Thyroid cartilage of the larynx
Clavicle

Epicranial aponeurosis

Orbicularis oculi
Nasalis

Temporalis
Masseter

Muscles of the Mouth and Cheek

Levator labii superioris
Zygomaticus minor
Zygomaticus major
Buccinator
Orbicularis oris
Risorius
Mentalis (cut)
Depressor labii inferioris
Depressor anguli oris

Muscles of Facial Expression

Group and Muscle(s)	Origin	Insertion	Action
Mouth			
Buccinator	Alveolar processes of maxillary bone and mandible	Blends into fibers of orbicularis oris	Compresses cheeks
Depressor labii inferioris	Mandible between the anterior midline and the mental foramen	Skin of lower lip	Depresses lower lip
Levator labii superioris	Inferior margin of orbit, superior to the infra-orbital foramen	Orbicularis oris	Elevates upper lip
Levator anguli oris	Maxillary bone below the infra-orbital foramen	Corner of mouth	Elevates the corner of the mouth
Mentalis	Incisive fossa of mandible	Skin of chin	Elevates and protrudes lower lip
Orbicularis oris	Maxillary bone and mandible	Lips	Compresses, purses lips
Risorius	Fascia surrounding parotid salivary gland	Angle of mouth	Draws corner of mouth to the side
Depressor anguli oris	Anterolateral surface of mandibular body	Skin at angle of mouth	Depresses corner of mouth
Zygomaticus major	Zygomatic bone near zygomaticomaxillary suture	Angle of mouth	Retracts and elevates corner of mouth
Zygomaticus minor	Zygomatic bone posterior to zygomaticotemporal suture	Upper lip	Retracts and elevates upper lip
Eye			
Corrugator supercilii	Orbital rim of frontal bone near nasal suture	Eyebrow	Pulls skin inferiorly and anteriorly; wrinkles brow
Levator palpebrae superioris (see Module 10.5)	Tendinous band around optic foramen	Upper eyelid	Elevates upper eyelid
Orbicularis oculi	Medial margin of orbit	Skin around eyelids	Closes eye
Nose			
Procerus	Nasal bones and lateral nasal cartilages	Aponeurosis at bridge of nose and skin of forehead	Moves nose, changes position and shape of nostrils
Nasalis	Maxillary bone and alar cartilage of nose	Bridge of nose	Compresses bridge, depresses tip of nose; elevates corners of nostrils
Scalp (epicranium)			
Occipitofrontalis			
Frontal belly	Epicranial aponeurosis	Skin of eyebrow and bridge of nose	Raises eyebrows, wrinkles forehead
Occipital belly	Occipital and temporal bone	Epicranial aponeurosis	Tenses and retracts scalp
Neck			
Platysma	Superior thorax between cartilage of 2nd rib and acromion of scapula	Mandible and skin of cheek	Tenses skin of neck; depresses mandible

Module 10.4 Review

a. Identify the muscles associated with the mouth.

b. State whether the following muscles involve the mouth, eye, nose, ear, scalp, or neck: buccinator, corrugator supercilii, mentalis, nasalis, platysma, procerus, and risorius.

c. Explain how an individual is able to consciously move the skin on the scalp but is not able to consciously move the skin of the thigh.

The extrinsic eye muscles position the eye, and the muscles of mastication move the lower jaw

1 Five of the six extrinsic eye muscles are visible in this lateral view of the right eye.

2 One additional muscle can be seen in a medial view of the right eye.

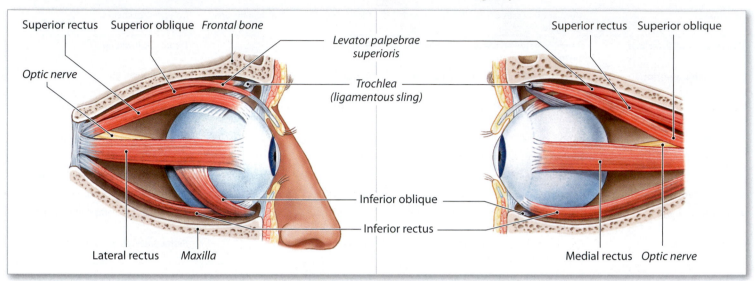

3 This anterior view of the right eye shows the direction of eye movements produced by the contraction of each extrinsic eye muscle operating independently.

4 This anterior view of the right orbit shows the origins of the extrinsic eye muscles.

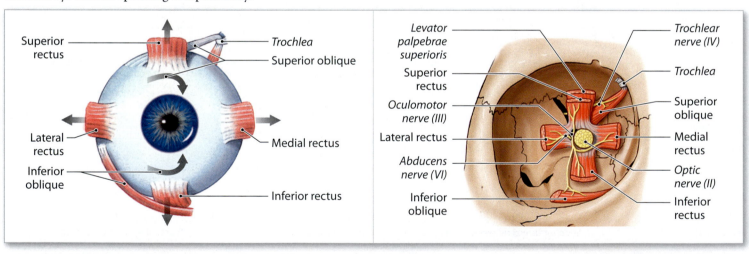

Extrinsic Eye Muscles

Muscle	Origin	Insertion	Action
Inferior rectus	Sphenoid around optic canal	Inferior, medial surface of eyeball	Eye looks down
Medial rectus	As above	Medial surface of eyeball	Eye looks medially
Superior rectus	As above	Superior surface of eyeball	Eye looks up
Lateral rectus	As above	Lateral surface of eyeball	Eye looks laterally
Inferior oblique	Maxillary bone at anterior portion of orbit	Inferior, lateral surface of eyeball	Eye rolls, looks up and laterally
Superior oblique	Sphenoid around optic canal	Superior, lateral surface of eyeball	Eye rolls, looks down and laterally

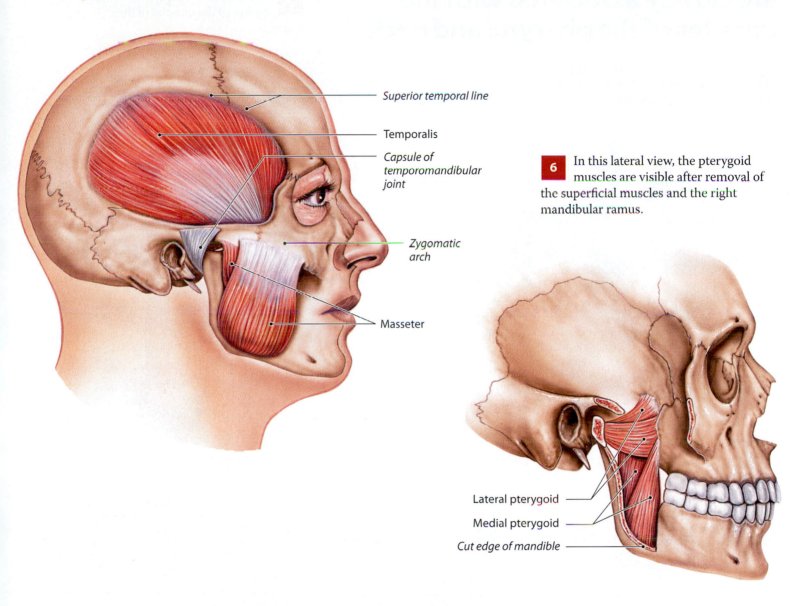

5 This superficial lateral view shows the largest muscles of mastication of the right side of the head.

- Superior temporal line
- Temporalis
- *Capsule of temporomandibular joint*
- *Zygomatic arch*
- Masseter

6 In this lateral view, the pterygoid muscles are visible after removal of the superficial muscles and the right mandibular ramus.

- Lateral pterygoid
- Medial pterygoid
- *Cut edge of mandible*

Muscles of Mastication

Muscle	Origin	Insertion	Action
Masseter	Zygomatic arch	Lateral surface of mandibular ramus	Elevates mandible and closes the jaws
Temporalis	Along temporal lines of skull	Coronoid process of mandible	Elevates mandible
Pterygoids (medial and lateral)	Lateral pterygoid plate	Medial surface of mandibular ramus	*Medial:* Elevates the mandible and closes the jaws, or slides the mandible from side to side *Lateral:* Opens jaws, protrudes the mandible, or slides the mandible from side to side

Module 10.5 Review

a. Name the extrinsic eye muscles.

b. Which muscles have their origin on the lateral pterygoid plates and their insertion on the medial surface of the mandibular ramus?

c. If you were contracting and relaxing your masseter muscle, what would you probably be doing?

The muscles of the tongue are closely associated with the muscles of the pharynx and neck

1 This lateral view shows the tongue muscles as dissected after removal of the left half of the mandible.

2 This lateral view shows the major groups of the muscles of the pharynx.

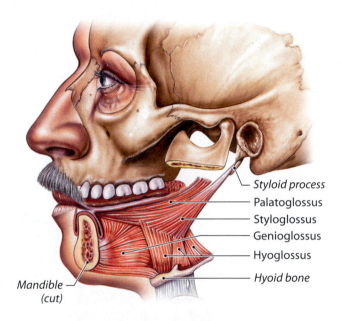

Styloid process
Palatoglossus
Styloglossus
Genioglossus
Hyoglossus
Hyoid bone

Mandible (cut)

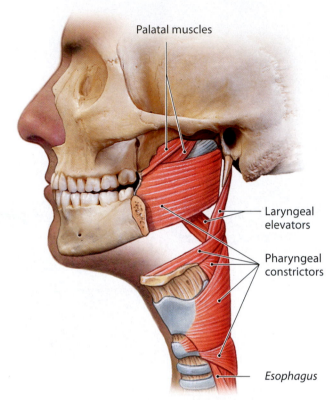

Palatal muscles

Laryngeal elevators

Pharyngeal constrictors

Esophagus

Muscles of the Tongue

Muscle	Origin	Insertion	Action
Genioglossus	Medial surface of mandible around chin	Body of tongue, hyoid bone	Depresses and protracts tongue
Hyoglossus	Body and greater horn of hyoid bone	Side of tongue	Depresses and retracts tongue
Palatoglossus	Anterior surface of soft palate	As above	Elevates tongue, depresses soft palate
Styloglossus	Styloid process of temporal bone	Along the side to tip and base of tongue	Retracts tongue, elevates side of tongue

Muscles of the Pharynx

Muscle	Origin	Insertion	Action
Pharyngeal constrictors	Pterygoid process of sphenoid, medial surfaces of mandible, horns of hyoid bone, cricoid and thyroid cartilages of larynx	Median raphe attached to occipital bone	Constrict pharynx to propel bolus into esophagus
Laryngeal elevators	Soft palate, cartilage around inferior portion of auditory tube, styloid process	Thyroid cartilage	Elevate larynx
Palatal muscles	Petrous part of temporal bone and adjacent soft tissues, sphenoidal spine and adjacent soft tissues	Soft palate	Elevate soft palate

3 The anterior muscles of the neck are primarily involved in positioning the mandible, hyoid bone, and larynx.

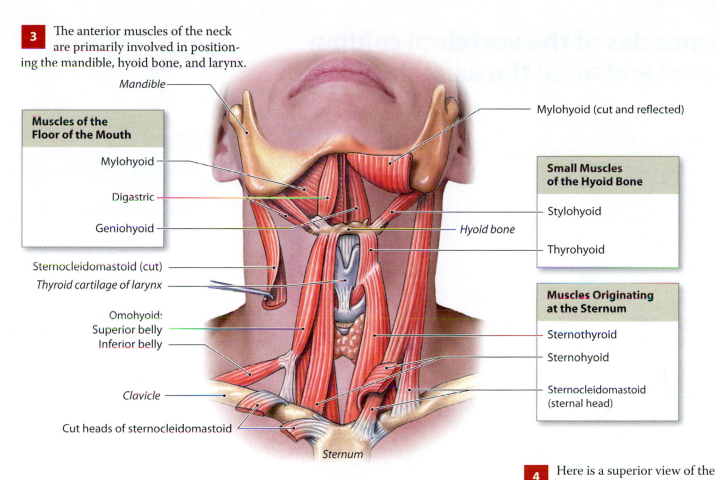

Mandible

Mylohyoid (cut and reflected)

Muscles of the Floor of the Mouth

Mylohyoid

Digastric

Geniohyoid

Small Muscles of the Hyoid Bone

Stylohyoid

Hyoid bone

Thyrohyoid

Sternocleidomastoid (cut)

Thyroid cartilage of larynx

Omohyoid:
Superior belly
Inferior belly

Muscles Originating at the Sternum

Sternothyroid

Sternohyoid

Sternocleidomastoid (sternal head)

Clavicle

Cut heads of sternocleidomastoid

Sternum

4 Here is a superior view of the isolated mandible. Several muscles extending from the hyoid bone to the mandible form the muscular floor of the mouth and support the tongue.

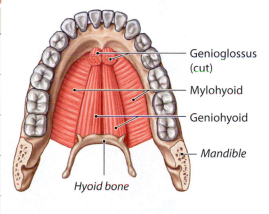

Genioglossus (cut)

Mylohyoid

Geniohyoid

Mandible

Hyoid bone

Anterior Muscles of the Neck

Muscle	Origin	Insertion	Action
Digastric	Inferior surface of mandible at chin and mastoid region	Hyoid bone	Depresses mandible or elevates larynx
Geniohyoid	Medial surface of mandible at chin	Hyoid bone	As above and pulls hyoid bone anteriorly
Mylohyoid	Mylohyoid line of mandible	Median raphe that runs to hyoid bone	Elevates hyoid bone or depresses mandible
Omohyoid	Superior border of scapula near scapular notch	Hyoid bone	Depresses hyoid bone and larynx
Sternohyoid	Clavicle and manubrium	Hyoid bone	As above
Sternothyroid	Manubrium and first costal cartilage	Thyroid cartilage of larynx	As above
Stylohyoid	Styloid process	Hyoid bone	Elevates larynx
Thyrohyoid	Thyroid cartilage of larynx	Hyoid bone	Elevates thyroid, depresses hyoid
Sternocleidomastoid	One head attaches to sternal end of clavicle; the other head attaches to manubrium	Mastoid region of skull and lateral portion of superior nuchal line	Flexes the neck; one alone bends head toward shoulder and rotates neck

Module 10.6 Review

a. List the muscles of the tongue.

b. Which muscles associated with the hyoid form the floor of the mouth?

c. Which muscles elevate the soft palate?

The muscles of the vertebral column support and align the axial skeleton

1 The **muscles of the vertebral column** are arranged in several layers. They include muscles originating or inserting on the ribs and the processes of the vertebrae. Although this mass of muscles extends from the sacrum to the skull, each muscle group is composed of numerous separate muscles of various lengths.

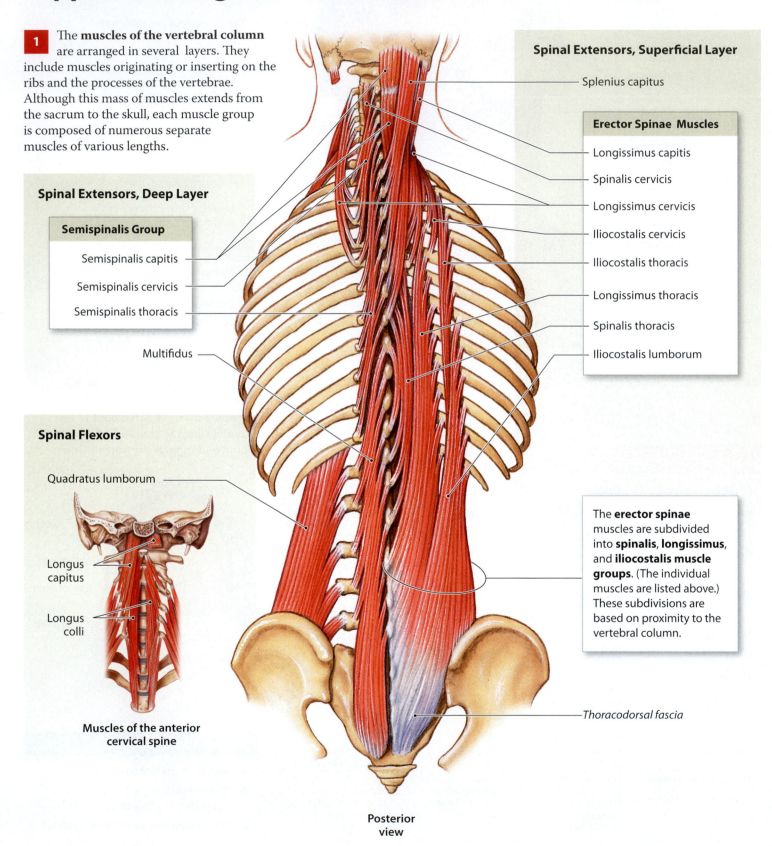

Spinal Extensors, Superficial Layer

Splenius capitus

Erector Spinae Muscles

Longissimus capitis

Spinalis cervicis

Longissimus cervicis

Iliocostalis cervicis

Iliocostalis thoracis

Longissimus thoracis

Spinalis thoracis

Iliocostalis lumborum

Spinal Extensors, Deep Layer

Semispinalis Group

Semispinalis capitis

Semispinalis cervicis

Semispinalis thoracis

Multifidus

Spinal Flexors

Quadratus lumborum

Longus capitus

Longus colli

Muscles of the anterior cervical spine

The **erector spinae** muscles are subdivided into **spinalis, longissimus,** and **iliocostalis muscle groups.** (The individual muscles are listed above.) These subdivisions are based on proximity to the vertebral column.

Thoracodorsal fascia

Posterior view

Muscles of the Vertebral Column

Group and Muscle(s)	Origin	Insertion	Action
Spinal Extensors — Superficial Layer			
Splenius (splenius capitis, splenius cervicis)	Spinous processes and ligaments connecting inferior cervical and superior thoracic vertebrae	Mastoid process, occipital bone of skull, and superior cervical vertebrae	Together, the two sides extend neck; alone, each rotates and laterally flexes neck to that side
Erector spinae			
SPINALIS GROUP — **Spinalis cervicis**	Inferior portion of ligamentum nuchae and spinous process of C_7	Spinous process of axis	Extends neck
SPINALIS GROUP — **Spinalis thoracis**	Spinous processes of inferior thoracic and superior lumbar vertebrae	Spinous processes of superior thoracic vertebrae	Extends vertebral column
LONGISSIMUS GROUP — **Longissimus capitis**	Transverse processes of inferior cervical and superior thoracic vertebrae	Mastoid process of temporal bone	Together, the two sides extend head; alone, each rotates and laterally flexes neck to that side
LONGISSIMUS GROUP — **Longissimus cervicis**	Transverse processes of superior thoracic vertebrae	Transverse processes of middle and superior cervical vertebrae	As above
LONGISSIMUS GROUP — **Longissimus thoracis**	Broad aponeurosis and transverse processes of inferior thoracic and superior lumbar vertebrae; joins iliocostalis	Transverse processes of superior vertebrae and inferior surfaces of ribs	Extends vertebral column; alone, each produces lateral flexion to that side
ILIOCOSTALIS GROUP — **Iliocostalis cervicis**	Superior borders of vertebrosternal ribs near the angles	Transverse processes of middle and inferior cervical vertebrae	Extends or laterally flexes neck, elevates ribs
ILIOCOSTALIS GROUP — **Iliocostalis thoracis**	Superior borders of inferior seven ribs medial to the angles	Upper ribs and transverse process of last cervical vertebra	Stabilizes thoracic vertebrae in extension
ILIOCOSTALIS GROUP — **Iliocostalis lumborum**	Iliac crest, sacral crests, and spinous processes	Inferior surfaces of inferior seven ribs near their angles	Extends vertebral column, depresses ribs
Spinal Extensors — Deep Layer			
SEMISPINALIS GROUP — **Semispinalis capitis**	Articular processes of inferior cervical and transverse processes of superior thoracic vertebrae	Occipital bone, between nuchal lines	Together, the two sides extend head; alone, each extends and laterally flexes neck
SEMISPINALIS GROUP — **Semispinalis cervicis**	Transverse processes of T_1–T_5 or T_6	Spinous processes of C_2–C_5	Extends vertebral column and rotates toward opposite side
SEMISPINALIS GROUP — **Semispinalis thoracis**	Transverse processes of T_6–T_{10}	Spinous processes of C_5–T_4	As above
SEMISPINALIS GROUP — **Multifidus**	Sacrum and transverse processes of each vertebra	Spinous processes of the third or fourth more superior vertebrae	As above
Spinal Flexors			
Longus capitis	Transverse processes of cervical vertebrae	Base of the occipital bone	Together, the two sides flex the neck; alone, each rotates head to that side
Longus colli	Anterior surfaces of cervical and superior thoracic vertebrae	Transverse processes of superior cervical vertebrae	Flexes or rotates neck; limits hyperextension
Quadratus lumborum	Iliac crest and iliolumbar ligament	Last rib and transverse processes of lumbar vertebrae	Together, they depress ribs; alone, each side laterally flexes vertebral column

2 Note that this table lists many extensors of the vertebral column, but few flexors. The vertebral column does not need a massive series of flexor muscles because (1) many of the large trunk muscles flex the vertebral column when they contract, and (2) most of the body weight lies anterior to the vertebral column, and gravity tends to flex the spine when unopposed by the extensor muscles.

Module 10.7 Review

a. List the spinal flexor muscles.

b. Which muscles enable you to extend the neck?

The oblique and rectus muscles form the muscular walls of the trunk

1 The scalene muscles are oblique muscles that extend into the neck.

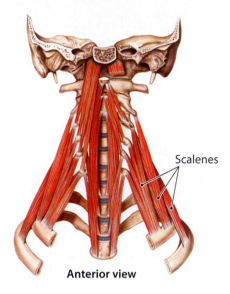

Anterior view

2 Superficial muscles are shown on the right side of the body, and deeper muscles of the oblique and rectus groups are shown on the left side.

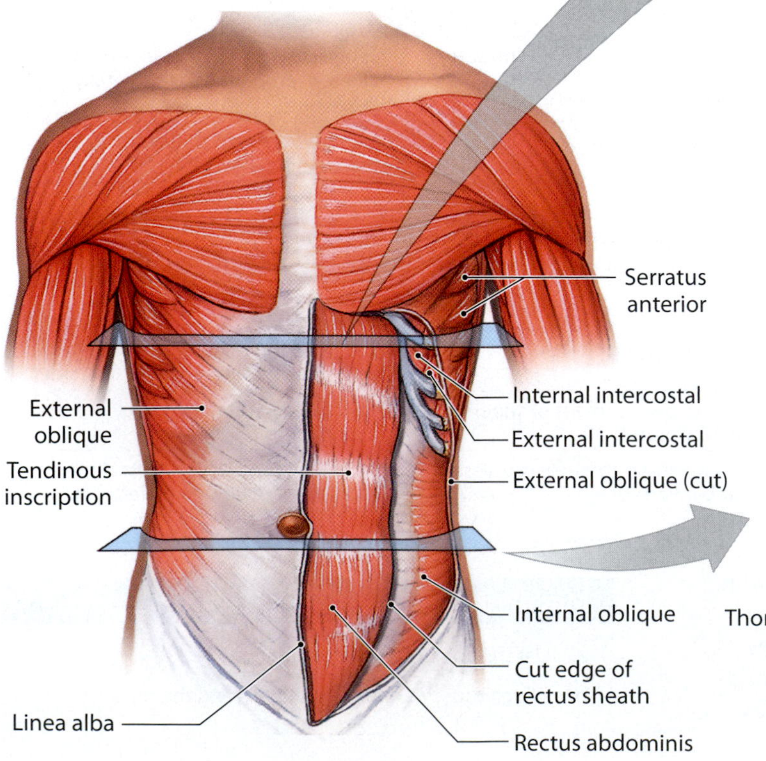

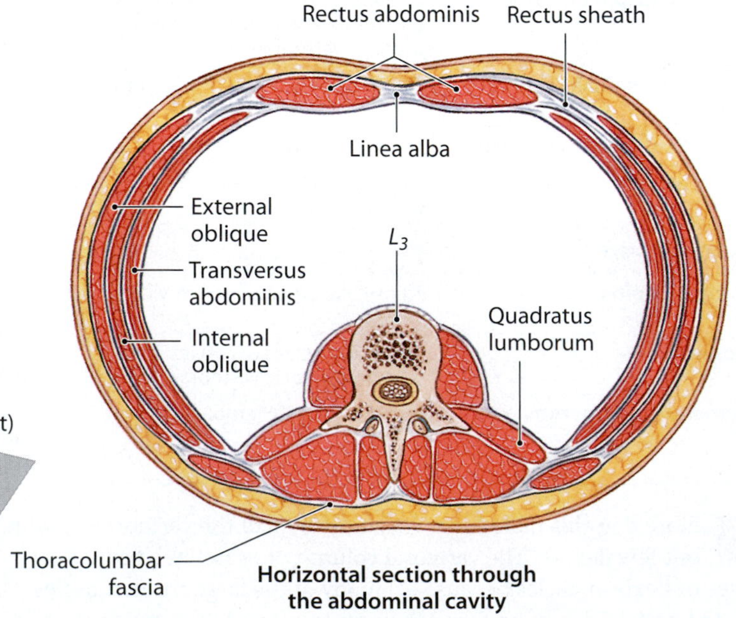

External oblique

Tendinous inscription

Linea alba

Serratus anterior

Internal intercostal

External intercostal

External oblique (cut)

Internal oblique

Cut edge of rectus sheath

Rectus abdominis

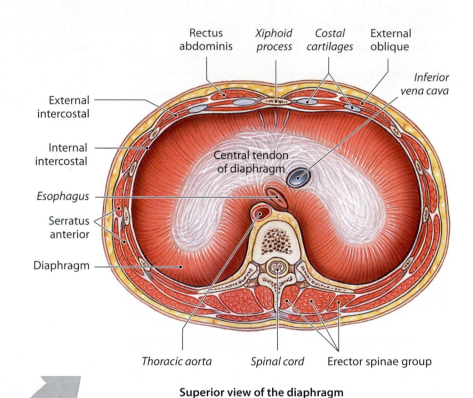

Rectus abdominis   Xiphoid process   Costal cartilages   External oblique

External intercostal

Internal intercostal

Esophagus

Serratus anterior

Diaphragm

Central tendon of diaphragm

Inferior vena cava

Thoracic aorta   Spinal cord   Erector spinae group

Superior view of the diaphragm

Rectus abdominis   Rectus sheath

Linea alba

External oblique

Transversus abdominis

Internal oblique

L_3

Quadratus lumborum

Thoracolumbar fascia

Horizontal section through the abdominal cavity

Oblique and Rectus Muscles

Group and Muscle(s)	Origin	Insertion	Action
Oblique Group			
CERVICAL REGION — **Scalenes**	Transverse and costal processes of cervical vertebrae	Superior surfaces of first two ribs	Elevate ribs or flex neck
THORACIC REGION — **External intercostals**	Inferior border of each rib	Superior border of more inferior rib	Elevate ribs
THORACIC REGION — **Internal intercostals**	Superior border of each rib	Inferior border of the preceding rib	Depress ribs
THORACIC REGION — **Transversus thoracis**	Posterior surface of sternum	Cartilages of ribs	As above
ABDOMINAL REGION — **External oblique**	External and inferior borders of ribs 5–12	Linea alba and iliac crest	Compresses abdomen, depresses ribs, flexes or bends spine
ABDOMINAL REGION — **Internal oblique**	Lumbodorsal fascia and iliac crest	Inferior ribs, xiphoid process, and linea alba	As above
ABDOMINAL REGION — **Transversus abdominis**	Cartilages of ribs 6–12, iliac crest, and lumbodorsal fascia	Linea alba and pubis	Compresses abdomen
Rectus Group			
THORACIC REGION — **Diaphragm**	Xiphoid process, cartilages of ribs 4–10, and anterior surfaces of lumbar vertebrae	Central tendinous sheet	Contraction expands thoracic cavity, compresses abdominopelvic cavity
ABDOMINAL REGION — **Rectus abdominis**	Superior surface of pubis around symphysis	Inferior surfaces of costal cartilages (ribs 5–7) and xiphoid process	Depresses ribs, flexes vertebral column, compresses abdomen

3 This is an inferior view of the diaphragm, showing its major landmarks.

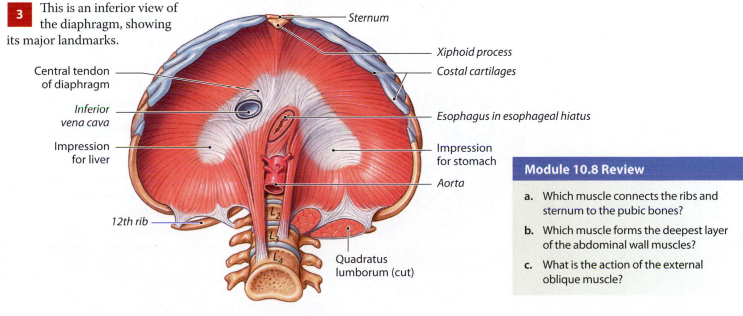

Central tendon of diaphragm
Inferior vena cava
Impression for liver
12th rib

Sternum
Xiphoid process
Costal cartilages
Esophagus in esophageal hiatus
Impression for stomach
Aorta
Quadratus lumborum (cut)

L₂ L₃ L₄

Module 10.8 Review

a. Which muscle connects the ribs and sternum to the pubic bones?

b. Which muscle forms the deepest layer of the abdominal wall muscles?

c. What is the action of the external oblique muscle?

The muscles of the pelvic floor support the organs of the abdominopelvic cavity

Superficial Dissections

Deep Dissections

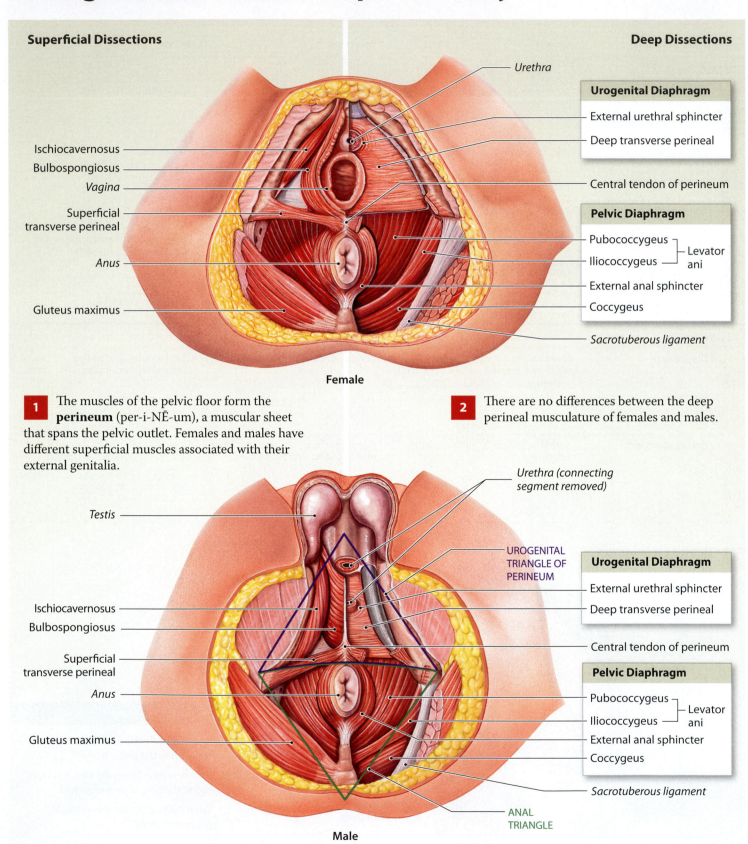

Ischiocavernosus
Bulbospongiosus
Vagina
Superficial transverse perineal
Anus
Gluteus maximus

Urethra

Urogenital Diaphragm

External urethral sphincter

Deep transverse perineal

Central tendon of perineum

Pelvic Diaphragm

Pubococcygeus — Levator ani
Iliococcygeus —

External anal sphincter

Coccygeus

Sacrotuberous ligament

Female

1 The muscles of the pelvic floor form the **perineum** (per-i-NĒ-um), a muscular sheet that spans the pelvic outlet. Females and males have different superficial muscles associated with their external genitalia.

2 There are no differences between the deep perineal musculature of females and males.

Testis

Urethra (connecting segment removed)

Ischiocavernosus
Bulbospongiosus
Superficial transverse perineal
Anus
Gluteus maximus

UROGENITAL TRIANGLE OF PERINEUM

Urogenital Diaphragm

External urethral sphincter

Deep transverse perineal

Central tendon of perineum

Pelvic Diaphragm

Pubococcygeus — Levator ani
Iliococcygeus —

External anal sphincter

Coccygeus

Sacrotuberous ligament

ANAL TRIANGLE

Male

Muscles of the Pelvic Floor

Group and Muscle(s)	Origin	Insertion	Action
Urogenital Triangle			
SUPERFICIAL MUSCLES — **Bulbospongiosus**			
Females	Collagen sheath at base of clitoris; fibers run on either side of urethral and vaginal opening	Central tendon of perineum	Compresses and stiffens clitoris; narrows vaginal opening
Males	Collagen sheath at base of penis; fibers cross over urethra	Median raphe and central tendon of perineum	Compresses base and stiffens penis; ejects urine or semen
Ischiocavernosus	Ischial ramus and tuberosity	Pubic symphysis anterior to base of penis or clitoris	Compresses and stiffens penis or clitoris
Superficial transverse perineal	Ischial ramus	Central tendon of perineum	Stabilizes central tendon of perineum
DEEP MUSCLES — **Urogenital diaphragm**			
Deep transverse perineal	Ischial ramus	Median raphe of urogenital diaphragm	As above
External urethral sphincter			
Females	Ischial and pubic rami	To median raphe; inner fibers encircle urethra	Closes urethra; compresses vagina and greater vestibular glands
Males	Ischial and pubic rami	To median raphe at base of penis; inner fibers encircle urethra	Closes urethra; compresses prostate and bulbourethral glands
Anal Triangle			
Pelvic diaphragm			
Coccygeus	Ischial spine	Lateral, inferior borders of sacrum and coccyx	Flexes coccygeal joints; tenses and supports pelvic floor
Levator ani			
Iliococcygeus	Ischial spine, pubis	Coccyx and median raphe	Tenses floor of pelvis; flexes coccygeal joints; elevates and retracts anus
Pubococcygeus	Inner margins of pubis	As above	As above
External anal sphincter	Via tendon from coccyx	Encircles anal opening	Closes anal opening

The urogenital and pelvic diaphragms do not completely close the pelvic outlet, because the urethra, vagina, and anus pass through them to open at the external surface. Muscular sphincters surround their openings and permit voluntary control of urination and defecation. Muscles, nerves, and blood vessels also pass through the pelvic outlet as they travel to or from the lower limbs.

Module 10.9 Review

a. Which muscles make up the urogenital diaphragm?

b. In females, what is the action of the bulbospongiosus muscle?

c. The coccygeus muscle extends from the sacrum and coccyx to which structure?

1. Labeling

Label each of the indicated muscles of the face in the following diagram.

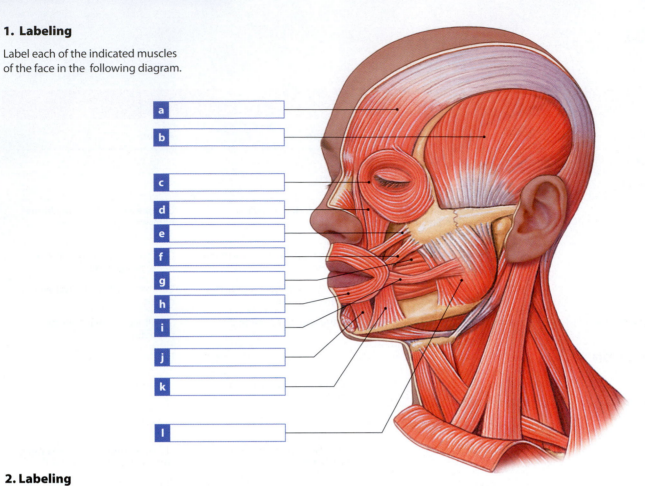

a
b
c
d
e
f
g
h
i
j
k
l

2. Labeling

Label each of the indicated muscles of the neck in the following diagram.

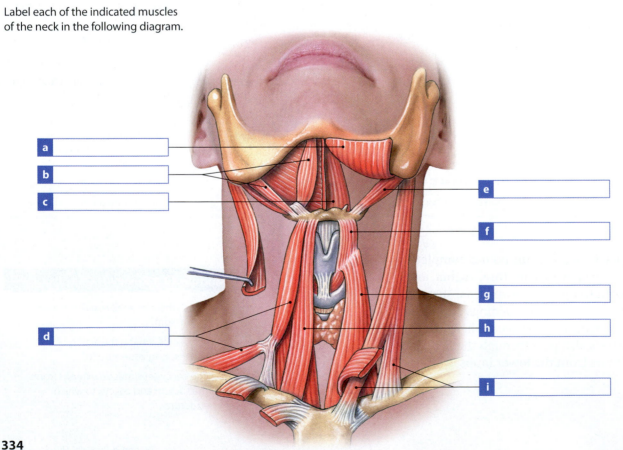

a
b
c
d
e
f
g
h
i

The Appendicular Muscles

The appendicular muscles stabilize, position, and/or support the limbs. In the following modules, we will group these muscles by their actions and origins.

Upper Limb

Muscles That Position the Pectoral Girdle

These muscles originate on the axial skeleton and insert on the clavicle and scapula.

Muscles That Move the Arm

These muscles originate on the pectoral girdle and the thoracic cage and insert on the humerus.

Muscles That Move the Forearm and Hand

These muscles primarily originate on the pectoral girdle and arm, and insert on the radius, ulna, and/or carpals.

Extrinsic Muscles of the Hand and Fingers

These muscles primarily originate on the humerus, radius, and ulna, and insert on the metacarpals and phalanges.

Intrinsic Muscles of the Hand

These are the muscles that perform fine movements. They originate primarily on the carpal and metacarpal bones, and insert on the phalanges.

Lower Limb

Muscles That Move the Thigh

These muscles originate in the pelvic region, and typically insert on the femur.

Muscles That Move the Leg

These muscles originate on the pelvis and femur, and insert on the tibia and/or fibula.

Extrinsic Muscles That Move the Foot and Toes

These muscles originate on the tibia and fibula, and insert on the tarsals, metatarsals, and/or phalanges.

Intrinsic Muscles of the Foot

These muscles originate primarily on the tarsal and metatarsal bones, and insert on the phalanges.

The largest appendicular muscles originate on the trunk

1 In general, muscles originating on the trunk control gross movements of the limbs. These muscles are often large and powerful; distally, the limb muscles get smaller and more numerous, and their movements become more precise.

Superficial Dissection

Deep Dissection

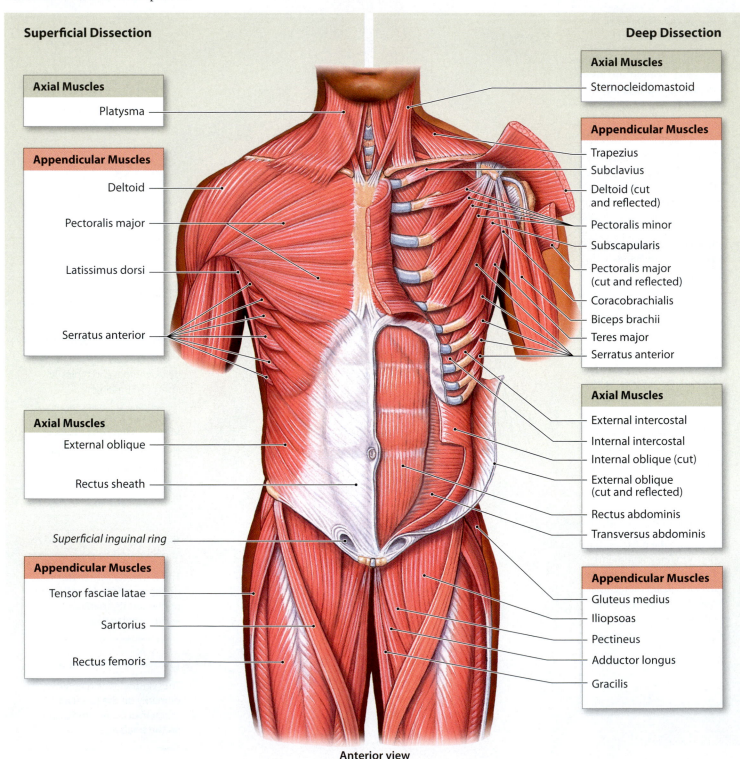

Axial Muscles

Platysma

Appendicular Muscles

Deltoid

Pectoralis major

Latissimus dorsi

Serratus anterior

Axial Muscles

External oblique

Rectus sheath

Superficial inguinal ring

Appendicular Muscles

Tensor fasciae latae

Sartorius

Rectus femoris

Axial Muscles

Sternocleidomastoid

Appendicular Muscles

Trapezius

Subclavius

Deltoid (cut and reflected)

Pectoralis minor

Subscapularis

Pectoralis major (cut and reflected)

Coracobrachialis

Biceps brachii

Teres major

Serratus anterior

Axial Muscles

External intercostal

Internal intercostal

Internal oblique (cut)

External oblique (cut and reflected)

Rectus abdominis

Transversus abdominis

Appendicular Muscles

Gluteus medius

Iliopsoas

Pectineus

Adductor longus

Gracilis

Anterior view

2 The posterior surface of the trunk is dominated by appendicular muscles that originate on the large bones of the limb girdles and the proximal bones of the limbs.

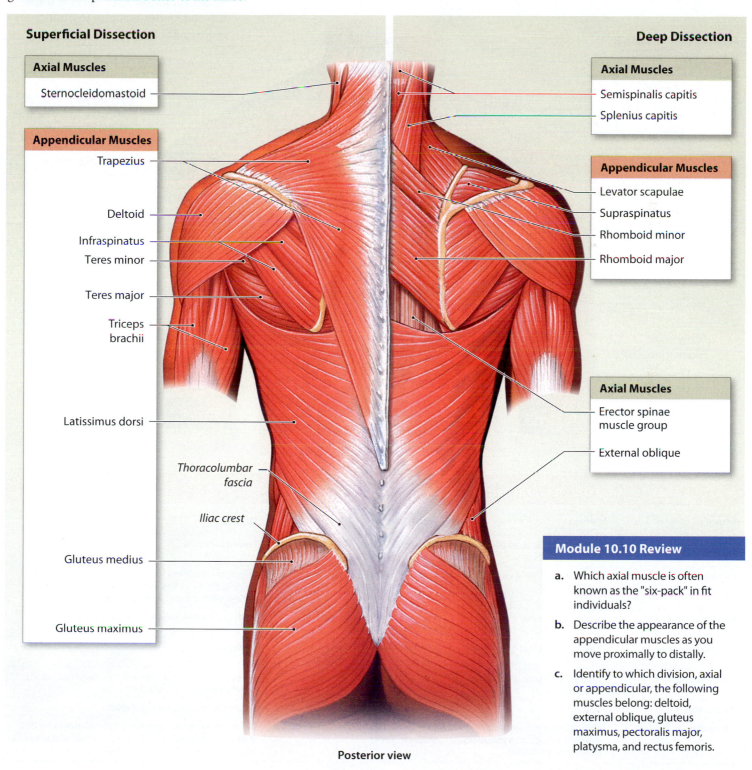

Superficial Dissection

Axial Muscles

Sternocleidomastoid

Appendicular Muscles

Trapezius

Deltoid

Infraspinatus

Teres minor

Teres major

Triceps brachii

Latissimus dorsi

Thoracolumbar fascia

Iliac crest

Gluteus medius

Gluteus maximus

Deep Dissection

Axial Muscles

Semispinalis capitis

Splenius capitis

Appendicular Muscles

Levator scapulae

Supraspinatus

Rhomboid minor

Rhomboid major

Axial Muscles

Erector spinae muscle group

External oblique

Posterior view

Module 10.10 Review

a. Which axial muscle is often known as the "six-pack" in fit individuals?

b. Describe the appearance of the appendicular muscles as you move proximally to distally.

c. Identify to which division, axial or appendicular, the following muscles belong: deltoid, external oblique, gluteus maximus, pectoralis major, platysma, and rectus femoris.

Muscles that position the pectoral girdle originate on the occipital bone, superior vertebrae, and ribs

The muscles that position the pectoral girdle also serve to anchor the pectoral girdle to the axial skeleton. Although these muscles have a smaller range of motion compared with other appendicular muscles, they help to increase upper limb mobility.

1 Several deep muscles of the chest, notably the pectoralis minor and serratus anterior, work with the trapezius and levator scapulae to position and stabilize the pectoral girdle.

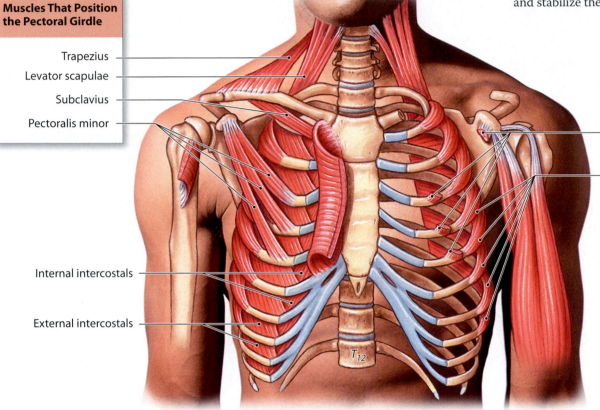

Muscles That Position the Pectoral Girdle

- Trapezius
- Levator scapulae
- Subclavius
- Pectoralis minor
- Internal intercostals
- External intercostals

T₁₂

Muscles That Position the Pectoral Girdle

- Pectoralis minor (cut)
- Serratus anterior

Anterior view

Muscles That Position the Pectoral Girdle

Muscle	Origin	Insertion	Action
Levator scapulae	Transverse processes of first four cervical vertebrae	Vertebral border of scapula near superior angle	Elevates scapula
Pectoralis minor	Anterior-superior surfaces of ribs 3–5	Coracoid process of scapula	Depresses and protracts shoulder; rotates scapula so glenoid cavity moves inferiorly (downward rotation); elevates ribs if scapula is stationary
Rhomboid major	Spinous processes of superior thoracic vertebrae	Vertebral border of scapula from spine to inferior angle	Adducts scapula and performs downward rotation
Rhomboid minor	Spinous processes of vertebrae C_7–T_1	Vertebral border of scapula near spine	As above
Serratus anterior	Anterior and superior margins of ribs 1–8 or 1–9	Anterior surface of vertebral border of scapula	Protracts shoulder; rotates scapula so glenoid cavity moves superiorly (upward rotation)
Subclavius	First rib	Clavicle (inferior border)	Depresses and protracts shoulder
Trapezius	Occipital bone, ligamentum nuchae, and spinous processes of thoracic vertebrae	Clavicle and scapula (acromion and scapular spine)	Depends on active region and state of other muscles; may (1) elevate, retract, depress, or rotate scapula upward, (2) elevate clavicle, or (3) extend neck

2 The broad trapezius is the largest muscle in this group. The rhomboid major and rhomboid minor muscles lie deep to the trapezius.

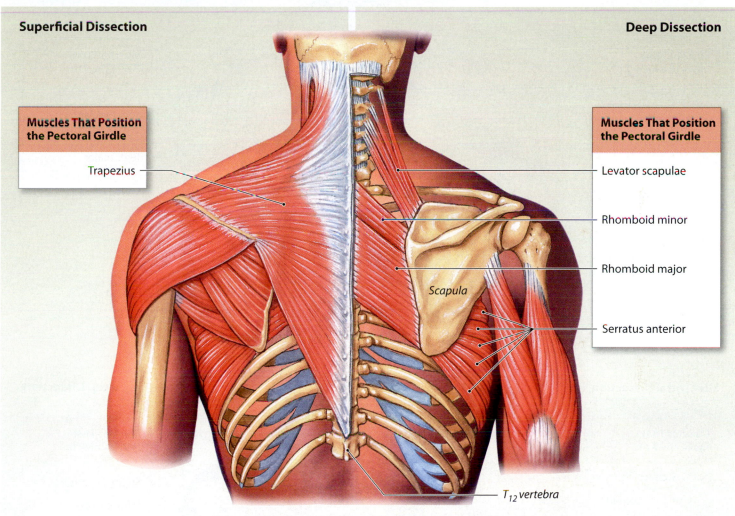

Superficial Dissection

Deep Dissection

Muscles That Position the Pectoral Girdle

Trapezius

Muscles That Position the Pectoral Girdle

Levator scapulae

Rhomboid minor

Rhomboid major

Scapula

Serratus anterior

T_{12} vertebra

Posterior view

Module 10.11 Review

a. Identify the largest of the superficial muscles that position the pectoral girdle.

b. Which muscles enable you to shrug your shoulders?

c. Which muscle originates on the first rib and inserts on the inferior border of the clavicle?

Muscles that move the arm originate on the clavicle, scapula, thoracic cage, and vertebral column

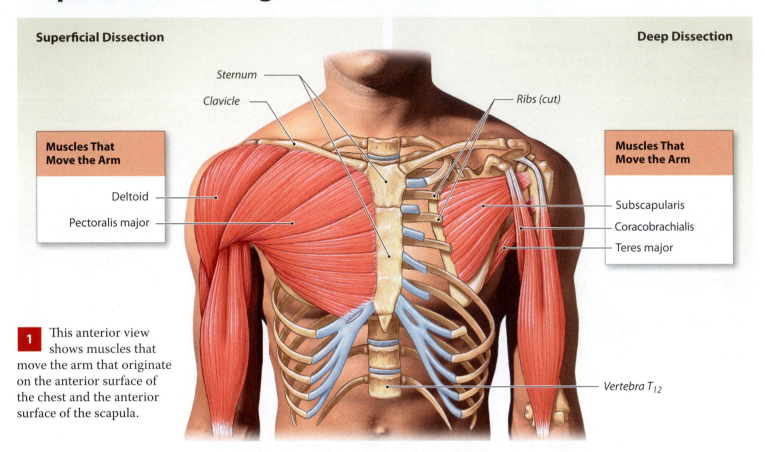

Superficial Dissection

Deep Dissection

Sternum

Clavicle

Ribs (cut)

Muscles That Move the Arm
Deltoid
Pectoralis major

Muscles That Move the Arm
Subscapularis
Coracobrachialis
Teres major

1 This anterior view shows muscles that move the arm that originate on the anterior surface of the chest and the anterior surface of the scapula.

Vertebra T_{12}

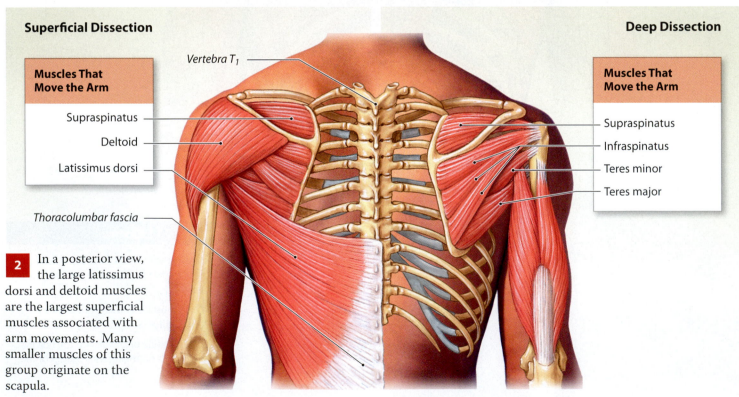

Superficial Dissection

Deep Dissection

Vertebra T_1

Muscles That Move the Arm
Supraspinatus
Deltoid
Latissimus dorsi

Muscles That Move the Arm
Supraspinatus
Infraspinatus
Teres minor
Teres major

Thoracolumbar fascia

2 In a posterior view, the large latissimus dorsi and deltoid muscles are the largest superficial muscles associated with arm movements. Many smaller muscles of this group originate on the scapula.

Muscles That Move the Arm

Muscle	Origin	Insertion	Action
Deltoid	Clavicle and scapula (acromion and adjacent scapular spine)	Deltoid tuberosity of humerus	Whole muscle: abduction at shoulder; anterior part: flexion and medial rotation; posterior part: extension and lateral rotation
Supraspinatus	Supraspinous fossa of scapula	Greater tubercle of humerus	Abduction at the shoulder
Subscapularis	Subscapular fossa of scapula	Lesser tubercle of humerus	Medial rotation at shoulder
Teres major	Inferior angle of scapula	Passes medially to reach the medial lip of intertubercular groove of humerus	Extension, adduction, and medial rotation at shoulder
Infraspinatus	Infraspinous fossa of scapula	Greater tubercle of humerus	Lateral rotation at shoulder
Teres minor	Lateral border of scapula	Passes laterally to reach the greater tubercle of humerus	Lateral rotation at shoulder
Coracobrachialis	Coracoid process	Medial margin of shaft of humerus	Adduction and flexion at shoulder
Pectoralis major	Cartilages of ribs 2–6, body of sternum, and inferior, medial portion of clavicle	Crest of greater tubercle and lateral lip of intertubercular groove of humerus	Flexion, adduction, and medial rotation at shoulder
Latissimus dorsi	Spinous processes of inferior thoracic and all lumbar vertebrae, ribs 8–12, and lumbodorsal fascia	Floor of intertubercular groove of the humerus	Extension, adduction, and medial rotation at shoulder

3 The action of muscles positioning the arm can be understood by considering their direction of pull relative to the center of the glenoid cavity. The arrows indicate the line of force, or **action line**, produced when a muscle contracts.

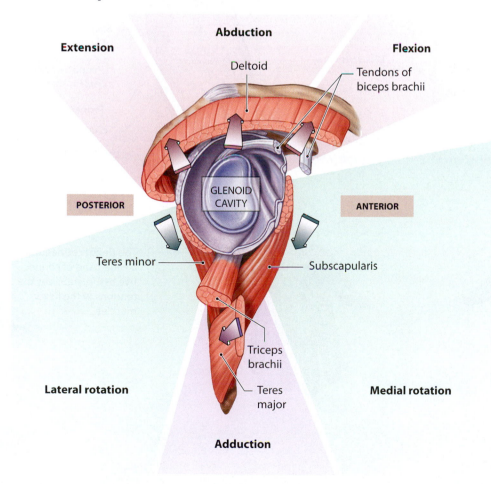

Collectively, the supraspinatus, infraspinatus, teres minor, and subscapularis muscles and their associated tendons form the **rotator cuff**. The acronym SITS assists in remembering these four muscles. Sports that involve throwing a ball, such as baseball or football, place considerable strain on the rotator cuff, and rotator cuff injuries are relatively common.

Module 10.12 Review

a. Define action line.

b. Name the muscle that abducts the upper arm.

c. Which muscle originates on the anterior surface of the scapula and inserts on the lesser tubercle of the humerus?

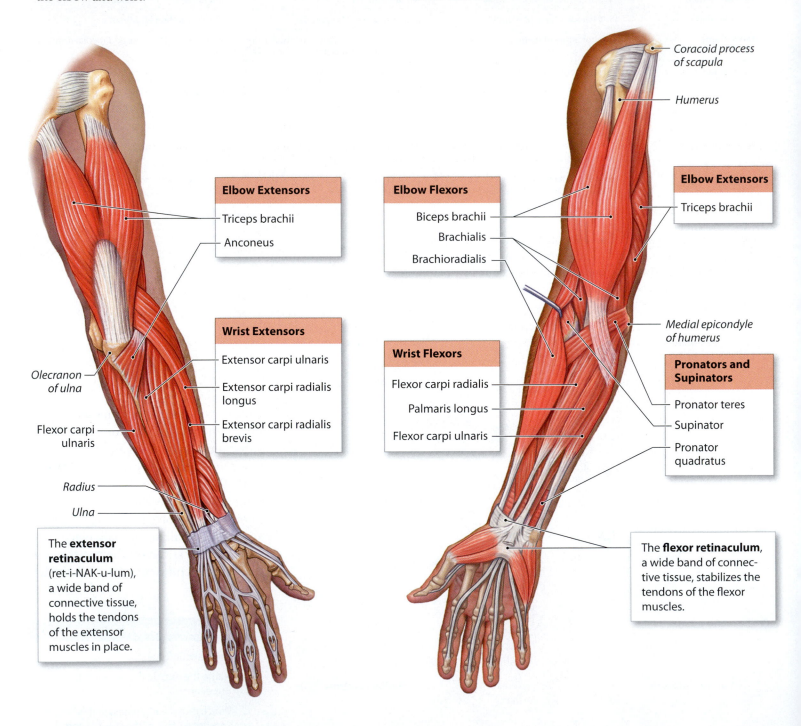

Muscles that move the forearm and hand originate on the scapula, humerus, radius, or ulna

1 This posterior view shows the superficial layer of muscles involved in extension at the elbow and wrist.

2 This anterior view shows the superficial muscles involved in flexion at the elbow and wrist.

Coracoid process of scapula

Humerus

Elbow Extensors

Triceps brachii

Anconeus

Elbow Flexors

Biceps brachii

Brachialis

Brachioradialis

Elbow Extensors

Triceps brachii

Medial epicondyle of humerus

Wrist Extensors

Extensor carpi ulnaris

Extensor carpi radialis longus

Extensor carpi radialis brevis

Wrist Flexors

Flexor carpi radialis

Palmaris longus

Flexor carpi ulnaris

Pronators and Supinators

Pronator teres

Supinator

Pronator quadratus

Olecranon of ulna

Flexor carpi ulnaris

Radius

Ulna

The **extensor retinaculum** (ret-i-NAK-u-lum), a wide band of connective tissue, holds the tendons of the extensor muscles in place.

The **flexor retinaculum**, a wide band of connective tissue, stabilizes the tendons of the flexor muscles.

Muscles That Move the Forearm and Hand

Group and Muscle(s)		Origin	Insertion	Action
Action at the Elbow				
FLEXORS	Biceps brachii	One head from the coracoid process; the other from the supraglenoid tubercle (both on the scapula)	Tuberosity of radius	Flexion at elbow and shoulder; supination
	Brachialis	Anterior, distal surface of humerus	Tuberosity of ulna	Flexion at elbow
	Brachioradialis	Ridge superior to the lateral epicondyle of humerus	Lateral aspect of styloid process of radius	As above
EXTENSORS	Anconeus	Posterior, inferior surface of lateral epicondyle of humerus	Lateral margin of olecranon on ulna	Extension at elbow
	Triceps brachii	One head from the superior, lateral margin of humerus, one from infraglenoid tubercle of scapula, and one from posterior surface of humerus inferior to radial groove	Olecranon of ulna	Extension at elbow, plus extension and adduction at the shoulder
PRONATORS / SUPINATORS	Pronator quadratus	Anterior and medial surfaces of distal portion of ulna	Anterolateral surface of distal portion of radius	Pronation
	Pronator teres	Medial epicondyle of humerus and coronoid process of ulna	Midlateral surface of radius	As above
	Supinator	Lateral epicondyle of humerus, annular ligament, and ridge near radial notch of ulna	Anterolateral surface of radius distal to the radial tuberosity	Supination
Action at the Hand				
FLEXORS	Flexor carpi radialis	Medial epicondyle of humerus	Bases of second and third metacarpal bones	Flexion and abduction at wrist
	Flexor carpi ulnaris	Medial epicondyle of humerus; adjacent medial surface of olecranon and anteromedial portion of ulna	Pisiform bone, hamate bone, and base of fifth metacarpal bone	Flexion and adduction at wrist
	Palmaris longus	Medial epicondyle of humerus	Palmar aponeurosis and flexor retinaculum	Flexion at wrist
EXTENSORS	Extensor carpi radialis longus	Lateral supracondylar ridge of humerus	Base of second metacarpal bone	Extension and abduction at wrist
	Extensor carpi radialis brevis	Lateral epicondyle of humerus	Base of third metacarpal bone	As above
	Extensor carpi ulnaris	Lateral epicondyle of humerus; adjacent dorsal surface of ulna	Base of fifth metacarpal bone	Extension and adduction at wrist

3 **Synovial tendon sheaths** are tubular bursae that surround tendons where they cross bony surfaces. The tendons of the flexor muscles pass through such sheaths as they pass deep to the flexor retinaculum. Inflammation of the flexor retinaculum and synovial tendon sheaths can restrict movement and irritate the distal portions of the median nerve, a mixed (sensory and motor) nerve that innervates the hand. This condition, known as **carpal tunnel syndrome**, causes chronic pain in the wrist and hand.

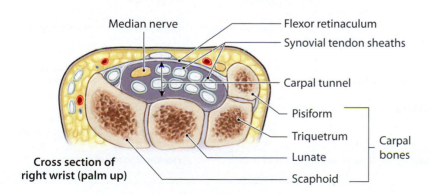

Median nerve — Flexor retinaculum
— Synovial tendon sheaths
— Carpal tunnel
— Pisiform
— Triquetrum
— Lunate
— Scaphoid
Carpal bones

Cross section of right wrist (palm up)

Module 10.13 Review

a. Define retinaculum.

b. On which surface, anterior or posterior, are the wrist extensors located?

c. Which muscles are involved in turning a doorknob?

Muscles that move the hand and fingers originate on the humerus, radius, ulna, and interosseous membrane

1 In anterior view, the large flexor digitorum superficialis covers the smaller digital flexors.

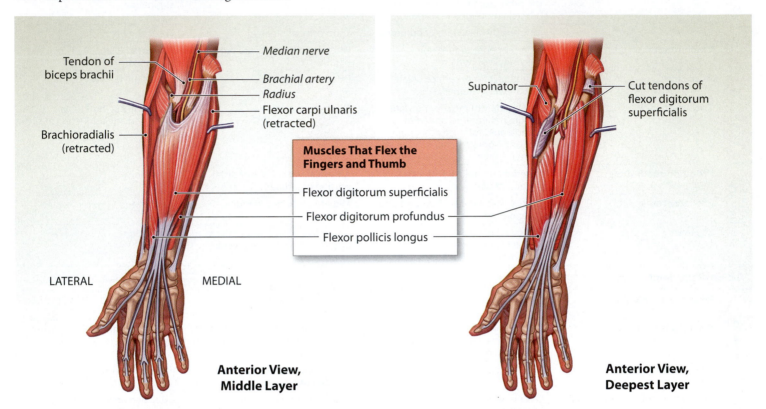

Tendon of biceps brachii

Median nerve

Brachial artery

Radius

Flexor carpi ulnaris (retracted)

Brachioradialis (retracted)

Supinator

Cut tendons of flexor digitorum superficialis

Muscles That Flex the Fingers and Thumb

Flexor digitorum superficialis

Flexor digitorum profundus

Flexor pollicis longus

LATERAL   MEDIAL

Anterior View, Middle Layer

Anterior View, Deepest Layer

Muscles That Move the Hand and Fingers

	Muscle	Origin	Insertion	Action
	Abductor pollicis longus	Proximal dorsal surfaces of ulna and radius	Lateral margin of first metacarpal bone	Abduction at joints of thumb and wrist
EXTENSORS	**Extensor digitorum**	Lateral epicondyle of humerus	Posterior surfaces of the phalanges, fingers 2–5	Extension at finger joints and wrist
	Extensor pollicis brevis	Shaft of radius distal to origin of adductor pollicis longus	Base of proximal phalanx of thumb	Extension at joints of thumb; abduction at wrist
	Extensor pollicis longus	Posterior and lateral surfaces of ulna and interosseous membrane	Base of distal phalanx of thumb	As above
	Extensor indicis	Posterior surface of ulna and interosseous membrane	Posterior surface of phalanges of index finger (2), with tendon of extensor digitorum	Extension and adduction at joints of index finger
	Extensor digiti minimi	Via extensor tendon to lateral epicondyle of humerus and from intermuscular septa	Posterior surface of proximal phalanx of little finger (5)	Extension at joints of little finger
FLEXORS	**Flexor digitorum superficialis**	Medial epicondyle of humerus; adjacent anterior surfaces of ulna and radius	Midlateral surfaces of middle phalanges of fingers 2–5	Flexion at proximal interphalangeal, metacarpophalangeal, and wrist joints
	Flexor digitorum profundus	Medial and posterior surfaces of ulna, medial surface of coronoid process, and interosseus membrane	Bases of distal phalanges of fingers 2–5	Flexion at distal interphalangeal joints and, to a lesser degree, proximal interphalangeal joints and wrist
	Flexor pollicis longus	Anterior shaft of radius, interosseous membrane	Base of distal phalanx of thumb	Flexion at joints of thumb

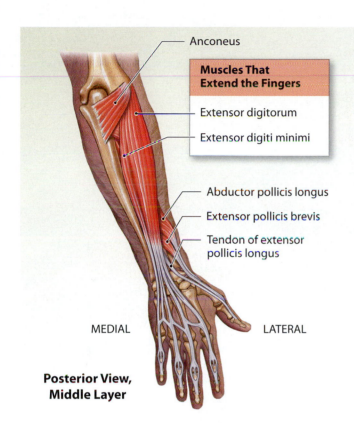

Anconeus

Muscles That Extend the Fingers

Extensor digitorum

Extensor digiti minimi

Abductor pollicis longus

Extensor pollicis brevis

Tendon of extensor pollicis longus

MEDIAL   LATERAL

Posterior View, Middle Layer

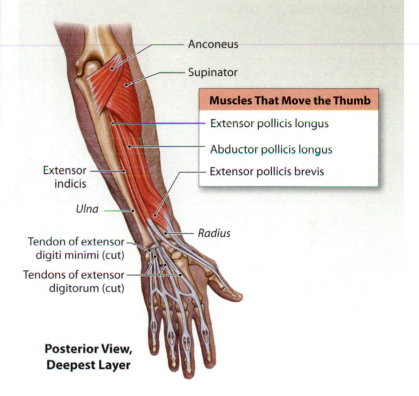

Anconeus

Supinator

Muscles That Move the Thumb

Extensor pollicis longus

Abductor pollicis longus

Extensor pollicis brevis

Extensor indicis

Ulna

Radius

Tendon of extensor digiti minimi (cut)

Tendons of extensor digitorum (cut)

Posterior View, Deepest Layer

2 In posterior view, the muscles that extend the fingers can only be seen after removal of the muscles involved in wrist movements. The deepest digital extensor muscles are those associated with movements of the thumb.

As you study these muscles you will notice that extensor muscles usually lie along the posterior and lateral surfaces of the forearm, whereas flexors are typically found on the anterior and medial surfaces. This information can be quite useful when you are trying to identify a particular muscle on a quiz or a lab practical.

Module 10.14 Review

a. List the muscles that extend the fingers.

b. Name the muscles that abduct the wrist.

c. The names of muscles associated with the thumb frequently include what term?

The intrinsic muscles of the hand originate on the carpal and metacarpal bones and associated tendons and ligaments

1 Dissection of the palmar surface of the hand reveals numerous small muscles involved with delicate positioning of the thumb and fingers. More powerful movements are controlled by the many tendons originating at the muscles of the forearm considered in Module 10.14.

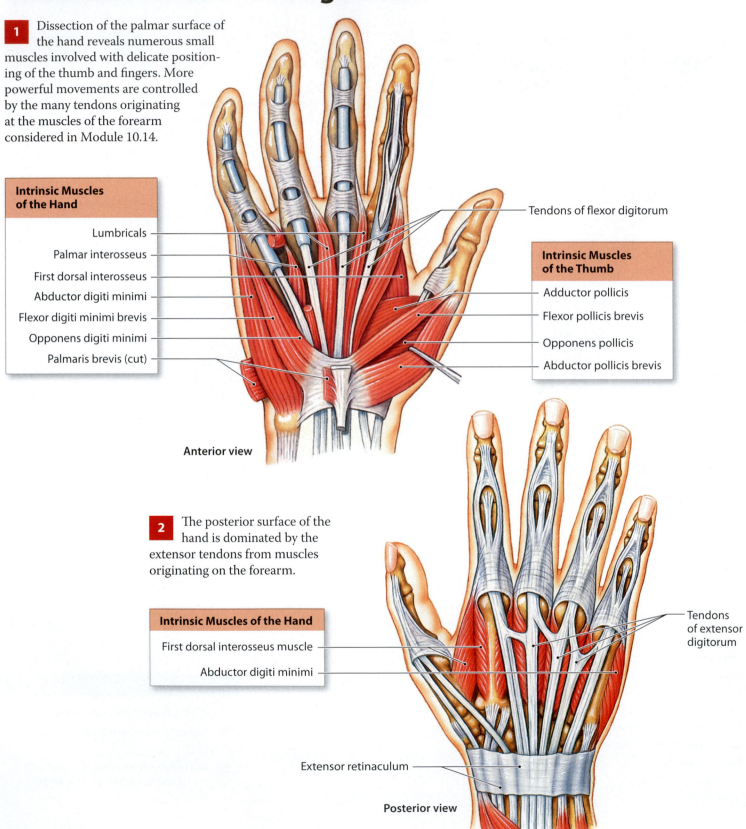

Intrinsic Muscles of the Hand

Lumbricals

Palmar interosseus

First dorsal interosseus

Abductor digiti minimi

Flexor digiti minimi brevis

Opponens digiti minimi

Palmaris brevis (cut)

Tendons of flexor digitorum

Intrinsic Muscles of the Thumb

Adductor pollicis

Flexor pollicis brevis

Opponens pollicis

Abductor pollicis brevis

Anterior view

2 The posterior surface of the hand is dominated by the extensor tendons from muscles originating on the forearm.

Intrinsic Muscles of the Hand

First dorsal interosseus muscle

Abductor digiti minimi

Tendons of extensor digitorum

Extensor retinaculum

Posterior view

Intrinsic Muscles of the Hand

Muscle	Origin	Insertion	Action
Palmaris brevis	Palmar aponeurosis	Skin of medial border of hand	Moves skin on medial border toward midline of palm
Adductor pollicis	Metacarpal and carpal bones	Proximal phalanx of thumb	Adduction of thumb
Palmar interosseus (3–4)	Sides of metacarpal bones II, IV, and V	Bases of proximal phalanges of fingers 2, 4, and 5	Adduction at metacarpophalangeal joints of fingers 2, 4, and 5; flexion at metacarpophalangeal joints; extension at interphalangeal joints
Abductor pollicis brevis	Transverse carpal ligament, scaphoid bone, and trapezium	Radial side of base of proximal phalanx of thumb	Abduction of thumb
Dorsal interosseus (4)	Each originates from opposing faces of two metacarpal bones (I and II, II and III, III and IV, IV and V)	Bases of proximal phalanges of fingers 2–4	Abduction at metacarpophalangeal joints of fingers 2 and 4; flexion at metacarpophalangeal joints; extension at interphalangeal joints
Abductor digiti minimi	Pisiform bone	Proximal phalanx of little finger	Abduction of little finger and flexion at its metacarpophalangeal joint
Flexor pollicis brevis	Flexor retinaculum, trapezium, capitate bone, and ulnar side of first metacarpal bone	Radial and ulnar sides of proximal phalanx of thumb	Flexion and adduction of thumb
Lumbrical (4)	Tendons of flexor digitorum profundus	Tendons of extensor digitorum to digits 2–5	Flexion at metacarpophalangeal joints 2–5; extension at proximal and distal interphalangeal joints, digits 2–5
Flexor digiti minimi brevis	Hamate bone	Proximal phalanx of little finger	Flexion at joints of little finger
Opponens pollicis	Trapezium and flexor retinaculum	First metacarpal bone	Opposition of thumb
Opponens digiti minimi	As above	Fifth metacarpal bone	Opposition of fifth metacarpal bone

ADDUCTION / ABDUCTION applies to rows: Adductor pollicis, Palmar interosseus, Abductor pollicis brevis, Dorsal interosseus, Abductor digiti minimi.

FLEXION applies to rows: Flexor pollicis brevis, Lumbrical, Flexor digiti minimi brevis.

Fine control of the hand involves small intrinsic muscles of the hand that originate on the carpal and metacarpal bones. These intrinsic muscles are responsible for (1) flexion and extension of the fingers at the metacarpophalangeal joints, (2) abduction and adduction of the fingers at the metacarpophalangeal joints, and (3) opposition and reposition (relaxed position) of the thumb. No muscles originate on the phalanges, and only tendons extend across the distal joints of the fingers.

Module 10.15 Review

a. Name the intrinsic muscles of the thumb.

b. Which muscles originate on the phalanges?

c. If there are no muscles in the fingers, how are we able to move them?

The muscles that move the thigh originate on the pelvis and associated ligaments and fasciae

1 The **gluteal group** covers the posterior and lateral surfaces of the pelvis.

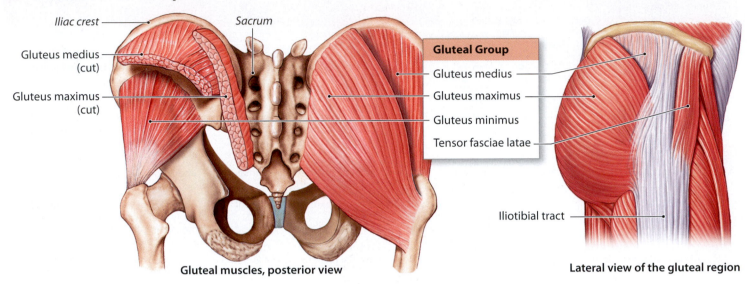

Iliac crest

Gluteus medius (cut)

Gluteus maximus (cut)

Sacrum

Gluteal Group

Gluteus medius

Gluteus maximus

Gluteus minimus

Tensor fasciae latae

Iliotibial tract

Gluteal muscles, posterior view

Lateral view of the gluteal region

2 This dissection of the gluteal region shows the orientation of the six lateral rotators.

3 This anterior view shows the isolated iliopsoas muscle group and the adductor group.

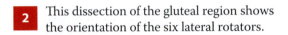

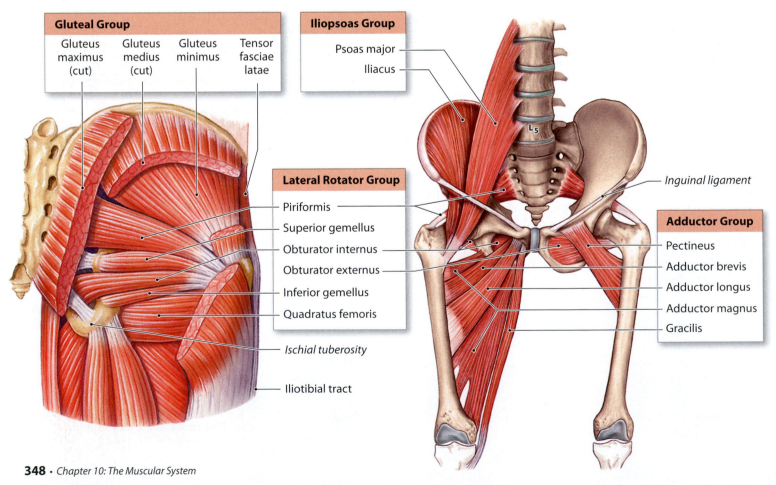

Gluteal Group			
Gluteus maximus (cut)	Gluteus medius (cut)	Gluteus minimus	Tensor fasciae latae

Lateral Rotator Group

Piriformis

Superior gemellus

Obturator internus

Obturator externus

Inferior gemellus

Quadratus femoris

Ischial tuberosity

Iliotibial tract

Iliopsoas Group

Psoas major

Iliacus

L5

Inguinal ligament

Adductor Group

Pectineus

Adductor brevis

Adductor longus

Adductor magnus

Gracilis

Muscles That Move the Thigh

Group and Muscle	Origin	Insertion	Action
Gluteal Group			
Gluteus maximus	Iliac crest, posterior gluteal line, and lateral surface of ilium; sacrum, coccyx, and lumbodorsal fascia	Iliotibial tract and gluteal tuberosity of femur	Extension and lateral rotation at hip
Gluteus medius	Anterior iliac crest of ilium, lateral surface between posterior and anterior gluteal lines	Greater trochanter of femur	Abduction and medial rotation at hip
Gluteus minimus	Lateral surface of ilium between inferior and anterior gluteal lines	As above	As above
Tensor fasciae latae	Iliac crest and lateral surface of anterior superior iliac spine	Iliotibial tract	Flexion and medial rotation at hip; tenses fascia lata, which laterally supports the knee
Lateral Rotator Group			
Obturator (externus and internus)	Lateral and medial margins of obturator foramen	Trochanteric fossa of femur (externus); medial surface of greater trochanter (internus)	Lateral rotation at hip
Piriformis	Anterolateral surface of sacrum	Greater trochanter of femur	Lateral rotation and abduction at hip
Gemellus (superior and inferior)	Ischial spine and tuberosity	Medial surface of greater trochanter with tendon of obturator internus	Lateral rotation at hip
Quadratus femoris	Lateral border of ischial tuberosity	Intertrochanteric crest of femur	As above
Adductor Group			
Adductor brevis	Inferior ramus of pubis	Linea aspera of femur	Adduction, flexion, and medial rotation at hip
Adductor longus	Inferior ramus of pubis anterior to adductor brevis	As above	As above
Adductor magnus	Inferior ramus of pubis posterior to adductor brevis and ischial tuberosity	Linea aspera and adductor tubercle of femur	Adduction at hip; superior part produces flexion and medial rotation; inferior part produces extension and lateral rotation
Pectineus	Superior ramus of pubis	Pectineal line inferior to lesser trochanter of femur	Flexion, medial rotation, and adduction at hip
Gracilis	Inferior ramus of pubis	Medial surface of tibia inferior to medial condyle	Flexion at knee; adduction and medial rotation at hip
Iliopsoas Group			
Iliacus	Iliac fossa of ilium	Femur distal to lesser trochanter; tendon fused with that of psoas major	Flexion at hip
Psoas major	Anterior surfaces and transverse processes of vertebrae (T_{12}–L_5)	Lesser trochanter in company with iliacus	Flexion at hip or lumbar intervertebral joints

One method for understanding the actions of these diverse muscles is to consider their orientation around the hip joint. Muscles originating on the surface of the pelvis and inserting on the femur will produce characteristic movements determined by their position relative to the acetabulum. Many of the muscles that act on the hip are very large, and they have insertions that extend over a broad area. As a result, these muscles often have more than one action line, and therefore produce more than one action at the hip. For example, the action of the adductor magnus varies depending on what portion of the muscle is activated; when the entire muscle contracts, it produces a combination of flexion, extension, and adduction at the hip.

Module 10.16 Review

a. Name the muscles that compose the gluteal group.

b. Identify the muscle whose origin is the lateral border of the ischial tuberosity and whose insertion is the intertrochanteric crest of the femur.

c. Which leg movement would be impaired by injury to the obturator muscles?

The muscles that move the leg originate on the pelvis and femur

1 **Flexors of the knee** originate on the pelvic girdle and extend along the posterior and medial surfaces of the thigh.

2 Most of the **extensors of the knee** originate on the femoral surface and extend along the anterior and lateral surfaces of the thigh. Collectively the knee extensors are called the **quadriceps muscles** or the **quadriceps femoris**.

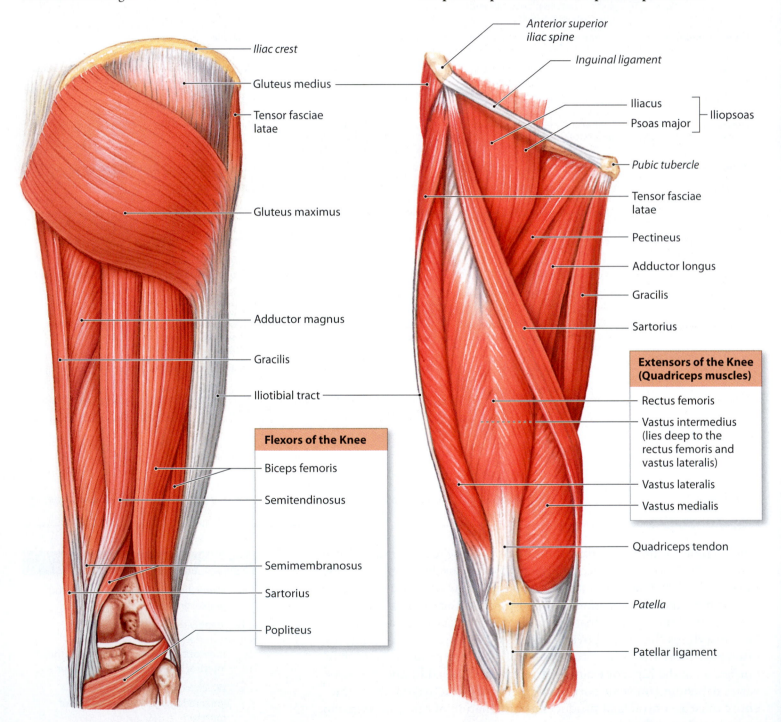

Iliac crest
Gluteus medius
Tensor fasciae latae
Gluteus maximus
Adductor magnus
Gracilis
Iliotibial tract

Flexors of the Knee
Biceps femoris
Semitendinosus
Semimembranosus
Sartorius
Popliteus

Anterior superior iliac spine
Inguinal ligament
Iliacus
Psoas major — Iliopsoas
Pubic tubercle
Tensor fasciae latae
Pectineus
Adductor longus
Gracilis
Sartorius

Extensors of the Knee (Quadriceps muscles)
Rectus femoris
Vastus intermedius (lies deep to the rectus femoris and vastus lateralis)
Vastus lateralis
Vastus medialis
Quadriceps tendon
Patella
Patellar ligament

Muscles That Move the Leg

Group and Muscle(s)	Origin	Insertion	Action
Flexors of the Knee			
Biceps femoris	Ischial tuberosity and linea aspera of femur	Head of fibula, lateral condyle of tibia	Flexion at knee; extension and lateral rotation at hip
Semimembranosus	Ischial tuberosity	Posterior surface of medial condyle of tibia	Flexion at knee; extension and medial rotation at hip
Semitendinosus	As above	Proximal, medial surface of tibia near insertion of gracilis	As above
Sartorius	Anterior superior iliac spine	Medial surface of tibia near tibial tuberosity	Flexion at knee; flexion and lateral rotation at hip
Popliteus	Lateral condyle of femur	Posterior surface of proximal tibial shaft	Medial rotation of tibia (or lateral rotation of femur); flexion at knee
Extensors of the Knee			
Rectus femoris	Anterior inferior iliac spine and superior acetabular rim of ilium	Tibial tuberosity via patellar ligament	Extension at knee; flexion at hip
Vastus intermedius	Anterolateral surface of femur and linea aspera (distal half)	As above	Extension at knee
Vastus lateralis	Anterior and inferior to greater trochanter of femur and along linea aspera (proximal half)	As above	As above
Vastus medialis	Entire length of linea aspera of femur	As above	As above

3 This cross-sectional view shows the positions of the major thigh muscles relative to the femur.

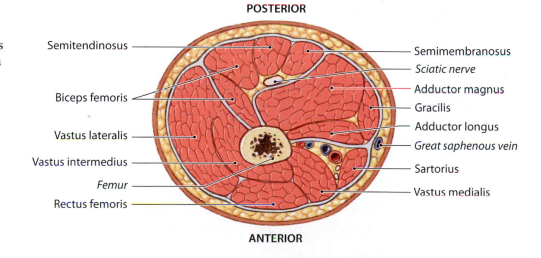

POSTERIOR

Semitendinosus — Semimembranosus — *Sciatic nerve* — Adductor magnus — Gracilis — Adductor longus — *Great saphenous vein* — Sartorius — Vastus medialis

Biceps femoris — Vastus lateralis — Vastus intermedius — *Femur* — Rectus femoris

ANTERIOR

Module 10.17 Review

a. Name the quadriceps muscles.

b. Which muscles flex the knee?

c. Identify the muscle whose origin is on the lateral condyle of the femur.

The extrinsic muscles that move the foot and toes originate on the tibia and fibula

1 These views show the multiple muscle layers in the posterior leg.

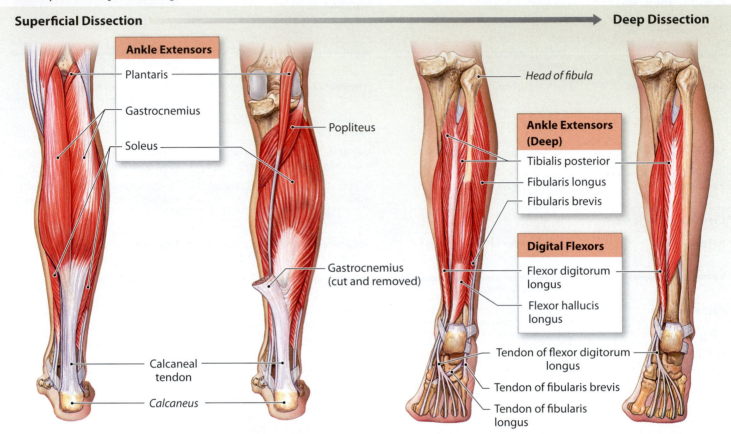

Superficial Dissection → **Deep Dissection**

Ankle Extensors
- Plantaris
- Gastrocnemius
- Soleus

Popliteus

Gastrocnemius (cut and removed)

Calcaneal tendon

Calcaneus

Head of fibula

Ankle Extensors (Deep)
- Tibialis posterior
- Fibularis longus
- Fibularis brevis

Digital Flexors
- Flexor digitorum longus
- Flexor hallucis longus

Tendon of flexor digitorum longus

Tendon of fibularis brevis

Tendon of fibularis longus

2 These lateral and medial views show the arrangement of the major superficial muscles.

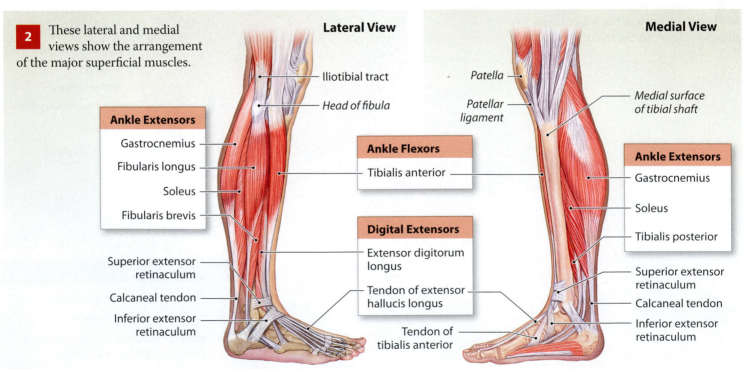

Lateral View

Iliotibial tract

Head of fibula

Ankle Extensors
- Gastrocnemius
- Fibularis longus
- Soleus
- Fibularis brevis

Superior extensor retinaculum

Calcaneal tendon

Inferior extensor retinaculum

Ankle Flexors
- Tibialis anterior

Digital Extensors
- Extensor digitorum longus

Tendon of extensor hallucis longus

Tendon of tibialis anterior

Medial View

Patella

Patellar ligament

Medial surface of tibial shaft

Ankle Extensors
- Gastrocnemius
- Soleus
- Tibialis posterior

Superior extensor retinaculum

Calcaneal tendon

Inferior extensor retinaculum

Extrinsic Muscles That Move the Foot and Toes

Group and Muscle	Origin	Insertion	Action
Action at the Ankle			
FLEXORS (Dorsiflexors)			
Tibialis anterior	Lateral condyle and proximal shaft of tibia	Base of first metatarsal bone and medial cuneiform bone	Flexion (dorsiflexion) at ankle; inversion of foot
EXTENSORS (Plantar flexors)			
Gastrocnemius	Femoral condyles	Calcaneus via calcaneal tendon	Extension (plantar flexion) at ankle; inversion of foot; flexion at knee
Fibularis brevis	Midlateral margin of fibula	Base of fifth metatarsal bone	Eversion of foot and extension (plantar flexion) at ankle
Fibularis longus	Lateral condyle of tibia, head and proximal shaft of fibula	Base of first metatarsal bone and medial cuneiform bone	Eversion of foot and extension (plantar flexion) at ankle; supports longitudinal arch
Plantaris	Lateral supracondylar ridge	Posterior portion of calcaneus	Extension (plantar flexion) at ankle; flexion at knee
Soleus	Head and proximal shaft of fibula and adjacent posteromedial shaft of tibia	Calcaneus via calcaneal tendon (with gastrocnemius)	Extension (plantar flexion) at ankle
Tibialis posterior	Interosseous membrane and adjacent shafts of tibia and fibula	Tarsal and metatarsal bones	Adduction and inversion of foot; extension (plantar flexion) at ankle
Action at the Toes			
DIGITAL FLEXORS			
Flexor digitorum longus	Posteromedial surface of tibia	Inferior surfaces of distal phalanges, toes 2–5	Flexion at joints of toes 2–5
Flexor hallucis longus	Posterior surface of fibula	Inferior surface, distal phalanx of great toe	Flexion at joints of great toe
DIGITAL EXTENSORS			
Extensor digitorum longus	Lateral condyle of tibia, anterior surface of fibula	Superior surfaces of phalanges, toes 2–5	Extension at joints of toes 2–5
Extensor hallucis longus	Anterior surface of fibula	Superior surface, distal phalanx of great toe	Extension at joints of great toe

The largest muscles associated with ankle movement are the **gastrocnemius** and **soleus muscles**. These muscles produce ankle extension (plantar flexion), a movement essential to walking and running. The muscles that move the toes are much smaller, and they originate on the surface of the tibia, the fibula, or both. Large tendon sheaths surround the tendons of the tibialis anterior, extensor digitorum longus, and extensor hallucis longus muscles where they cross the ankle joint. The positions of these sheaths are stabilized by the **superior** and **inferior retinacula**, tough supporting bands of collagen fibers.

Module 10.18 Review

a. Name the muscles involved in flexing the toes.

b. Name the muscles involved in extending the ankle.

c. How would a torn calcaneal tendon affect movement of the foot?

The intrinsic muscles of the foot originate on the tarsal and metatarsal bones and associated tendons and ligaments

1 This superior view introduces some of the **intrinsic muscles of the foot**. It also reveals the importance of the retinacula in stabilizing the positions of the tendons descending from the leg.

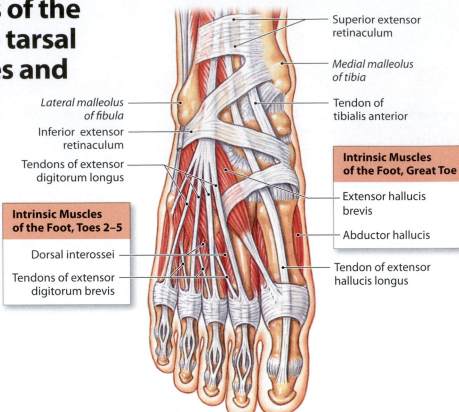

Superior extensor retinaculum

Medial malleolus of tibia

Tendon of tibialis anterior

Lateral malleolus of fibula

Inferior extensor retinaculum

Tendons of extensor digitorum longus

Intrinsic Muscles of the Foot, Great Toe

Extensor hallucis brevis

Abductor hallucis

Tendon of extensor hallucis longus

Intrinsic Muscles of the Foot, Toes 2–5

Dorsal interossei

Tendons of extensor digitorum brevis

2 Intrinsic muscles are more numerous on the inferior surface of the foot and occur in several layers.

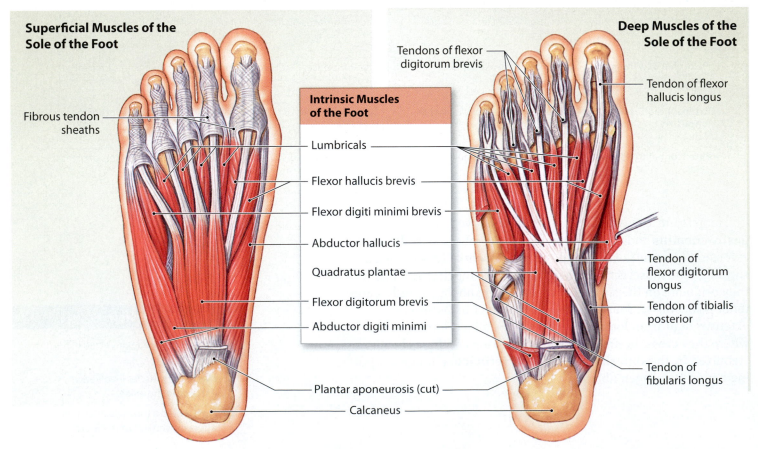

Superficial Muscles of the Sole of the Foot

Fibrous tendon sheaths

Deep Muscles of the Sole of the Foot

Tendons of flexor digitorum brevis

Tendon of flexor hallucis longus

Intrinsic Muscles of the Foot

Lumbricals

Flexor hallucis brevis

Flexor digiti minimi brevis

Abductor hallucis

Quadratus plantae

Flexor digitorum brevis

Abductor digiti minimi

Plantar aponeurosis (cut)

Calcaneus

Tendon of flexor digitorum longus

Tendon of tibialis posterior

Tendon of fibularis longus

Intrinsic Muscles of the Foot

	Muscle	Origin	Insertion	Action
FLEXION / EXTENSION	Flexor hallucis brevis	Cuboid and lateral cuneiform bones	Proximal phalanx of great toe	Flexion at metatarsophalangeal joint of great toe
	Flexor digitorum brevis	Inferior surface of calcaneus	Sides of middle phalanges, toes 2–5	Flexion at proximal interphalangeal joints of toes 2–5
	Quadratus plantae	Calcaneus (medial, inferior surfaces)	Tendon of flexor digitorum longus	Flexion at joints of toes 2–5
	Lumbrical (4)	Tendons of flexor digitorum longus	Tendons of extensor digitorum longus, toes 2 to 5	Flexion at metatarsophalangeal joints; extension at proximal interphalangeal joints of toes 2–5
	Flexor digiti minimi brevis	Base of metatarsal bone V	Lateral side of proximal phalanx of toe 5	Flexion at metatarsophalangeal joint of toe 5
	Extensor digitorum brevis	Calcaneus (superior and lateral surfaces)	Dorsal surfaces of toes 1–4	Extension at metatarsophalangeal joints of toes 1–4
	Extensor hallucis brevis	Superior surface of anterior calcaneus	Dorsal surface of the base of proximal phalanx of great toe	Extension of great toe
ADDUCTION / ABDUCTION	Adductor hallucis	Bases of metatarsal bones II–IV and plantar ligaments	Proximal phalanx of great toe	Adduction at metatarsophalangeal joint of great toe
	Abductor hallucis	Calcaneus (tuberosity on inferior surface)	Medial side of proximal phalanx of great toe	Abduction at metatarsophalangeal joint of great toe
	Plantar interosseus (3)	Bases and medial sides of metatarsal bones	Medial sides of toes 3–5	Adduction at metatarsophalangeal joints of toes 3–5
	Dorsal interosseus (4)	Sides of metatarsal bones	Medial and lateral sides of toe 2; lateral sides of toes 3 and 4	Abduction at metatarsophalangeal joints of toes 3 and 4
	Abductor digiti minimi	Inferior surface of calcaneus	Lateral side of proximal phalanx, toe 5	Abduction at metatarsophalangeal joint of toe 5

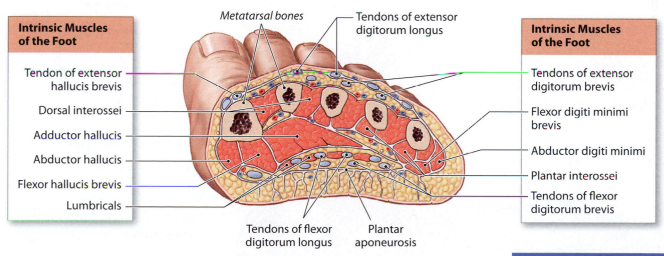

Intrinsic Muscles of the Foot

- Tendon of extensor hallucis brevis
- Dorsal interossei
- Adductor hallucis
- Abductor hallucis
- Flexor hallucis brevis
- Lumbricals

- Metatarsal bones
- Tendons of extensor digitorum longus
- Tendons of flexor digitorum longus
- Plantar aponeurosis

Intrinsic Muscles of the Foot

- Tendons of extensor digitorum brevis
- Flexor digiti minimi brevis
- Abductor digiti minimi
- Plantar interossei
- Tendons of flexor digitorum brevis

3 As you see in this cross section, most of the muscle mass in the foot lies inferior to the metatarsal bones. Many of these muscles are flexors that tense during ankle extension and help you "push off" when walking. This anatomical arrangement provides padding and assists in maintaining the arches of the foot.

Module 10.19 Review

a. Identify the intrinsic muscle that flexes the great toe.

b. What are the functions of the superior and inferior retinacula of the foot?

c. Describe the origin, insertion, and action of the lumbrical muscles.

The deep fascia separates the limb muscles into separate compartments

The muscles of the limbs are organized into **compartments** created by fibrous partitions of the deep fascia. The muscles within each compartment have roughly compatible functions. Each compartment has a characteristic blood supply and innervation.

1 Here are sectional views at intervals along the length of an upper and lower limb. In each section the compartments are labeled and representative muscles and important reference structures are indicated.

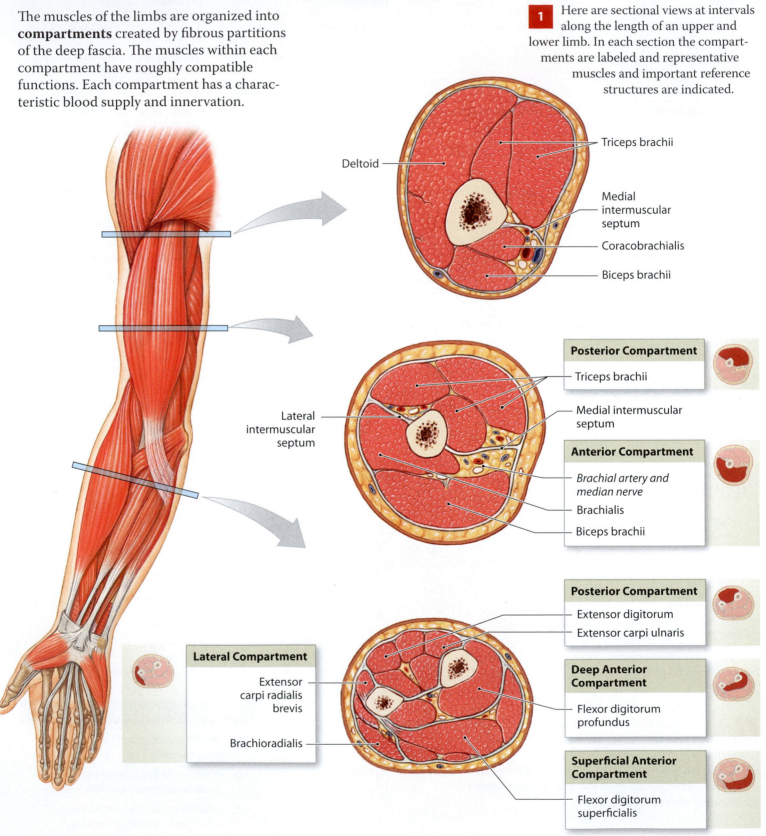

Deltoid

Triceps brachii

Medial intermuscular septum

Coracobrachialis

Biceps brachii

Lateral intermuscular septum

Posterior Compartment

Triceps brachii

Medial intermuscular septum

Anterior Compartment

Brachial artery and median nerve

Brachialis

Biceps brachii

Lateral Compartment

Extensor carpi radialis brevis

Brachioradialis

Posterior Compartment

Extensor digitorum

Extensor carpi ulnaris

Deep Anterior Compartment

Flexor digitorum profundus

Superficial Anterior Compartment

Flexor digitorum superficialis

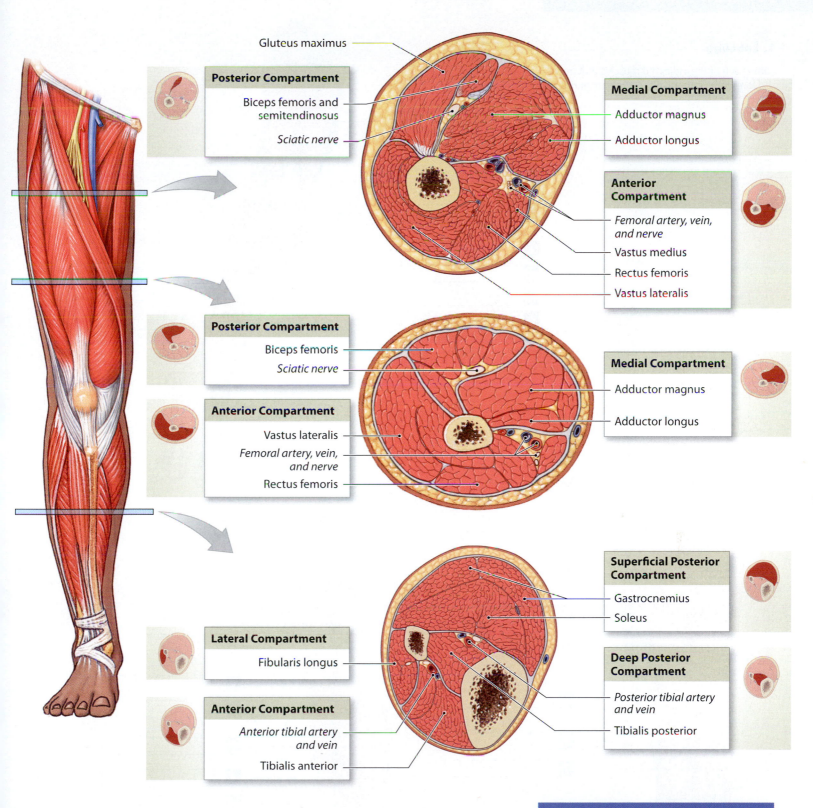

Gluteus maximus

Posterior Compartment

Biceps femoris and semitendinosus

Sciatic nerve

Medial Compartment

Adductor magnus

Adductor longus

Anterior Compartment

Femoral artery, vein, and nerve

Vastus medius

Rectus femoris

Vastus lateralis

Posterior Compartment

Biceps femoris

Sciatic nerve

Anterior Compartment

Vastus lateralis

Femoral artery, vein, and nerve

Rectus femoris

Medial Compartment

Adductor magnus

Adductor longus

Lateral Compartment

Fibularis longus

Anterior Compartment

Anterior tibial artery and vein

Tibialis anterior

Superficial Posterior Compartment

Gastrocnemius

Soleus

Deep Posterior Compartment

Posterior tibial artery and vein

Tibialis posterior

Compartments are clinically important because trauma to a limb can cause bleeding into a compartment. This elevates pressures and compresses blood vessels and nerves within the compartment. This **compartment syndrome** can lead to the paralysis or death of the muscles within the damaged compartment if the pressure is not relieved within 2–4 hours.

Module 10.20 Review

a. Name the six possible compartments of the muscles of the limbs.

b. Define compartment syndrome.

c. Propose a reason why compartment syndrome can be life threatening.

1. Labeling

Label each of the indicated muscles that move the forearm and hand in the diagram at right.

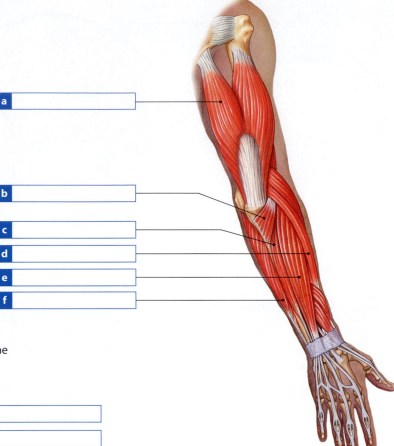

a [ ]

b [ ]

c [ ]

d [ ]

e [ ]

f [ ]

2. Labeling

Label each of the indicated muscles that move the thigh and leg in the diagram below.

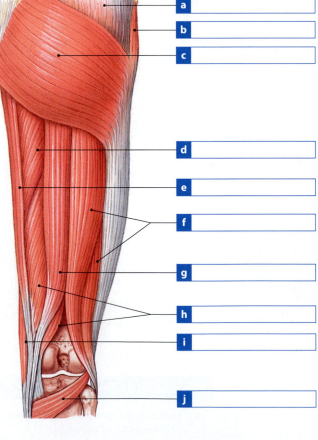

a [ ]

b [ ]

c [ ]

d [ ]

e [ ]

f [ ]

g [ ]

h [ ]

i [ ]

j [ ]

h [ ]

3. Labeling

Label each of the indicated muscles that move the foot and toe in the diagram below.

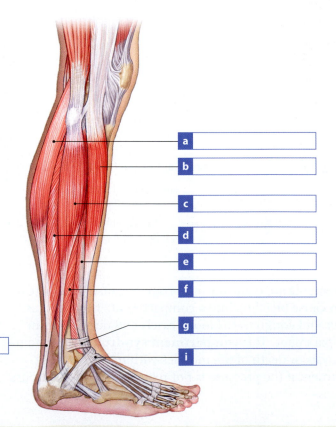

a [ ]

b [ ]

c [ ]

d [ ]

e [ ]

f [ ]

g [ ]

i [ ]

Visual Outline with Key Terms

Summarize the content of each module using the terms in the order provided.

SECTION 1

Functional Organization of the Muscular System

- muscular system
- axial muscles
- appendicular muscles

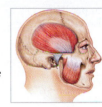
Muscular system

10.1

Muscular power and range of motion are influenced by fascicle organization and leverage

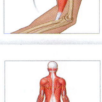

- parallel muscle
- body
- convergent muscle
- pennate muscle
- unipennate
- bipennate
- multipennate
- circular muscle (sphincter)
- first-class lever
- second-class lever
- third-class lever

10.2

The names of muscles can provide clues to their appearance and/or function

- origin
- intermuscular septa
- insertion
- action
- agonist
- synergist
- antagonist

10.3

The skeletal muscles can be assigned to the axial division or the appendicular division based on origins and functions

- axial muscles
- appendicular muscles

SECTION 2

The Axial Muscles

- muscles of the head and neck
- muscles of the vertebral column
- muscles of the trunk
- muscles of the pelvic floor

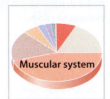

10.4

The muscles of facial expression are important in eating and useful for communication

- muscles of facial expression
- epicranial aponeurosis
- buccinator
- depressor labii inferioris
- levator labii superioris
- levator anguli oris
- mentalis
- orbicularis oris
- risorius
- depressor anguli oris
- zygomaticus major
- zygomaticus minor
- corrugator supercilii
- levator palpebrae superioris
- orbicularis oculi
- procerus
- nasalis
- occipitofrontalis
- platysma

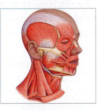

10.5

The extrinsic eye muscles position the eye, and the muscles of mastication move the lower jaw

- inferior rectus
- medial rectus
- superior rectus
- lateral rectus
- inferior oblique
- superior oblique
- masseter
- temporalis
- pterygoids

10.6

The muscles of the tongue are closely associated with the muscles of the pharynx and neck

- genioglossus
- hyoglossus
- palatoglossus
- styloglossus
- pharyngeal constrictors
- laryngeal elevators
- palatal muscles
- digastric
- geniohyoid
- mylohyoid
- omohyoid
- sternohyoid
- sternothyroid
- stylohyoid
- thyrohyoid
- sternocleidomastoid

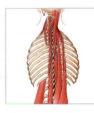

10.7

The muscles of the vertebral column support and align the axial skeleton

- muscles of the vertebral column
- spinal extensors
- splenius
- erector spinae muscles
- spinalis group
- longissimus group
- iliocostalis group
- semispinalis group
- spinal flexors

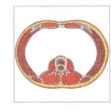

10.8

The oblique and rectus muscles form the muscular walls of the trunk

- oblique group
- scalenes
- external intercostals
- internal intercostals
- transversus thoracis
- external oblique
- internal oblique
- transversus abdominis
- rectus group
- diaphragm
- rectus abdominis

10.9

The muscles of the pelvic floor support the organs of the abdominopelvic cavity

- perineum
- urogenital triangle
- bulbospongiosus
- ischiocavernosus
- superficial transverse perineal
- urogenital diaphragm
- deep transverse perineal
- external urethral sphincter
- anal triangle
- pelvic diaphragm
- coccygeus
- levator ani
- iliococcygeus
- pubococcygeus
- external anal sphincter

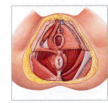

● = *Term boldfaced in this module*

SECTION 3

The Appendicular Muscles

- upper limb
- lower limb

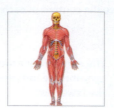

10.10

The largest appendicular muscles originate on the trunk

- ○ deltoid
- ○ pectoralis major
- ○ latissimus dorsi
- ○ serratus anterior
- ○ tensor fasciae latae
- ○ sartorius
- ○ rectus femoris
- ○ trapezius
- ○ subclavius
- ○ pectoralis minor
- ○ subscapularis

- ○ coracobrachialis
- ○ biceps brachii
- ○ teres major
- ○ gluteus medius
- ○ iliopsoas
- ○ pectineus
- ○ adductor longus
- ○ gracilis
- ○ infraspinatus
- ○ teres minor
- ○ triceps brachii

- ○ latissimus dorsi
- ○ gluteus maximus
- ○ levator scapulae
- ○ supraspinatus
- ○ rhomboid minor
- ○ rhomboid major

10.11

Muscles that position the pectoral girdle originate on the occipital bone, superior vertebrae, and ribs

- ○ pectoral girdle
- • levator scapulae
- • pectoralis minor
- • rhomboid major

- • rhomboid minor
- • serratus anterior
- • subclavius
- • trapezius

10.12

Muscles that move the arm originate on the clavicle, scapula, thoracic cage, and vertebral column

- • deltoid
- • supraspinatus
- • subscapularis
- • teres major
- • infraspinatus
- • teres minor

- • coracobrachialis
- • pectoralis major
- • latissimus dorsi
- • action line
- • rotator cuff

10.13

Muscles that move the forearm and hand originate on the scapula, humerus, radius, or ulna

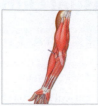

- • extensor retinaculum
- • flexor retinaculum
- • biceps brachii
- • brachialis
- • brachioradialis
- • anconeus
- • triceps brachii
- • pronator quadratus
- • pronator teres

- • supinator
- • flexor carpi radialis
- • flexor carpi ulnaris
- • palmaris longus
- • extensor carpi radialis longus
- • extensor carpi radialis brevis

- • extensor carpi ulnaris
- • synovial tendon sheaths
- • carpal tunnel syndrome

10.14

Muscles that move the hand and fingers originate on the humerus, radius, ulna, and interosseus membrane

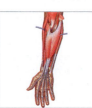

- • abductor pollicis longus
- • extensor digitorum
- • extensor pollicis brevis
- • extensor pollicis longus

- • extensor indicis
- • extensor digiti minimi
- • flexor digitorum superficialis
- • flexor digitorum profundus

- • flexor pollicis longus

10.15

The intrinsic muscles of the hand originate on the carpal and metacarpal bones and associated tendons and ligaments

- ○ intrinsic muscles of the hand
- ○ intrinsic muscles of the thumb
- • palmaris brevis
- • adductor pollicis
- • palmar interosseus

- • abductor pollicis brevis
- • dorsal interosseus
- • abductor digiti minimi
- • flexor pollicis brevis
- • lumbricals

- • flexor digiti minimi brevis
- • opponens pollicis
- • opponens digiti minimi

10.16

The muscles that move the thigh originate on the pelvis and associated ligaments and fasciae

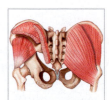

- • gluteal group
- • gluteus maximus
- • gluteus medius
- • gluteus minimus
- • tensor fasciae latae
- • lateral rotator group
- • obturator externus
- • obturator internus

- • piriformis
- • superior gemellus
- • inferior gemellus
- • quadratus femoris
- • adductor group
- • adductor brevis
- • adductor longus
- • adductor magnus

- • pectineus
- • gracilis
- • iliopsoas group
- • iliacus
- • psoas major

• = *Term boldfaced in this module*

10.17

The muscles that move the leg originate on the pelvis and femur

- flexors of the knee
- biceps femoris
- semimembranosus
- semitendinosus
- sartorius
- popliteus
- extensors of the knee
- quadriceps muscles (quadriceps femoris)
- rectus femoris
- vastus intermedius
- vastus lateralis
- vastus medialis

10.19

The intrinsic muscles of the foot originate on the tarsal and metatarsal bones and associated tendons and ligaments

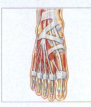

- intrinsic muscles of the foot
- flexor hallucis brevis
- flexor digitorum brevis
- quadratus plantae
- lumbrical (4)
- flexor digiti minimi brevis
- extensor digitorum brevis
- extensor hallucis brevis
- adductor hallucis
- abductor hallucis
- plantar interosseus (3)
- dorsal interosseus (4)
- abductor digiti minimi

10.18

The extrinsic muscles that move the foot and toes originate on the tibia and fibula

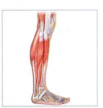

- tibialis anterior
- gastrocnemius
- fibularis brevis
- fibularis longus
- plantaris
- soleus
- tibialis posterior
- flexor digitorum longus
- flexor hallucis longus
- extensor digitorum longus
- extensor hallucis longus
- superior retinaculum
- inferior retinaculum

10.20

The deep fascia separates the limb muscles into separate compartments

- compartments
- compartment syndrome

• = *Term boldfaced in this module*

Chapter Integration: Applying what you've learned

Bodybuilders spend many hours in the gym lifting free weights to develop their muscles. Larger muscles and greater muscle definition are the goals, and looking "ripped" requires a lot of dedication. To sculpt their arms, for example, bodybuilders do a lot of biceps curls (flexion at the elbow holding weights in the anatomical position) and triceps curls (extension at the elbow).

As a 10-year-old, Jerry and his friends would go to the beach at Lion's Park, known locally as "Muscle Beach" because all the local bodybuilders would go there in the summer to work out and show off for the girls. Some female bodybuilders even started going there. Jerry was always amazed by the size, shape, and strength of these musclemen and vowed that someday he would become one of them. As he reached puberty, Jerry became a fitness fanatic who worked out many hours a day. As he learned more about bodybuilding, he eschewed the massive, heavily muscled look for one of a more athletic, lean, and well-defined musculature. Everyone came to admire his "six-pack abs," especially his girlfriend, DJ.

1. Explain why doing both biceps curls and triceps curls helps achieve larger, well-toned arms.

2. Which exercises would be best for shaping your abdominal muscles into "six-pack abs"?

Access more review material online in the Study Area at **www.masteringaandp.com**.

There, you'll find:
- **Chapter guides**
- **Chapter quizzes**
- **Practice tests**
- **Labeling activities**

- **A&P Flix:**
  - **Origins, Insertions, Actions, Innervations**
  - **Group Muscle Actions and Joints**
- **Animations**
- **Flashcards**
- **A glossary with pronunciations**

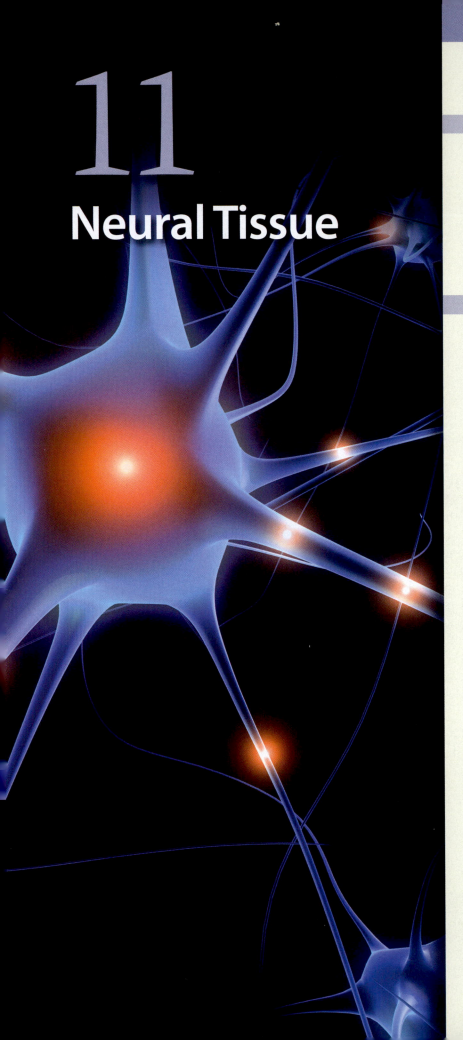

11
Neural Tissue

These Learning Outcomes correspond by number to this chapter's modules and indicate what you should be able to do after completing the chapter.

SECTION 1 • Neurons and Neuroglia

11.1 Sketch and label the structure of a typical neuron, and describe the functions of each component.

11.2 Classify and describe neurons on the basis of their structure and function.

11.3 Describe the locations and functions of neuroglia in the CNS.

11.4 Describe the locations and functions of Schwann cells and satellite cells.

SECTION 2 • Neurophysiology

11.5 Explain how the resting potential is created and maintained.

11.6 Describe the functions of gated channels with respect to the permeability of the plasma membrane.

11.7 Describe graded potentials.

11.8 Describe the events involved in the generation of an action potential.

11.9 Describe continuous propagation and saltatory propagation, and discuss the factors that affect the speed with which action potentials are propagated.

11.10 Describe the general structure of synapses in the CNS and PNS, and discuss the events that occur at a chemical synapse.

11.11 Discuss the significance of postsynaptic potentials, including the roles of excitatory postsynaptic potentials and inhibitory postsynaptic potentials.

11.12 Discuss the interactions that make the processing of information in neural tissue possible.

Neurons and Neuroglia

This chapter, the first of three on the nervous system, considers the structure of neural tissue and introduces basic principles of neurophysiology. Our discussion of neural function requires a basic understanding of the functional organization of the nervous system. Major components and functions of the nervous system are indicated in this flowchart.

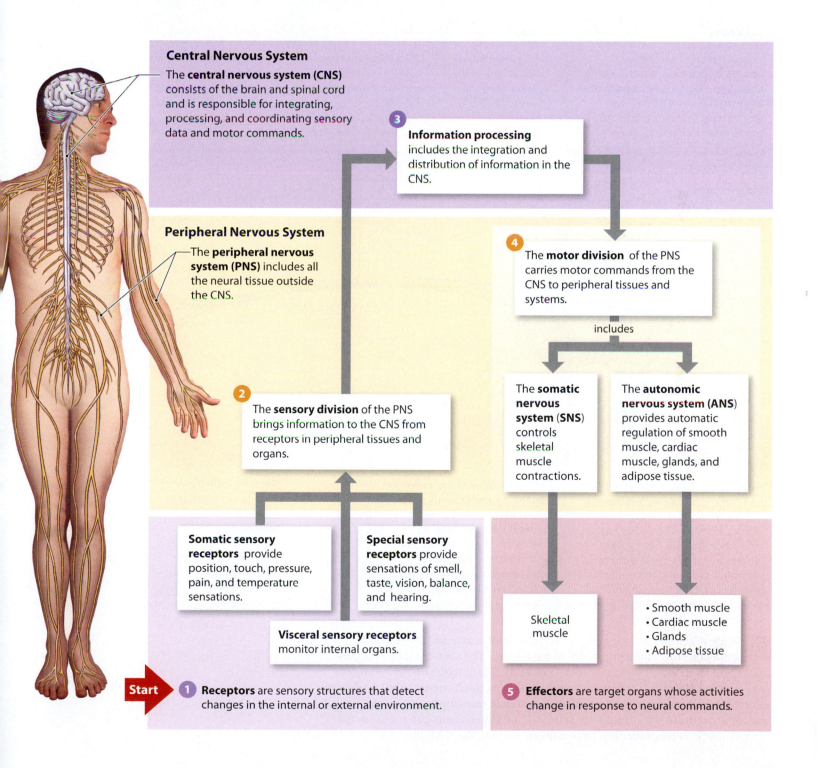

Central Nervous System

The **central nervous system (CNS)** consists of the brain and spinal cord and is responsible for integrating, processing, and coordinating sensory data and motor commands.

3 **Information processing** includes the integration and distribution of information in the CNS.

Peripheral Nervous System

The **peripheral nervous system (PNS)** includes all the neural tissue outside the CNS.

4 The **motor division** of the PNS carries motor commands from the CNS to peripheral tissues and systems.

includes

2 The **sensory division** of the PNS brings information to the CNS from receptors in peripheral tissues and organs.

The **somatic nervous system (SNS)** controls skeletal muscle contractions.

The **autonomic nervous system (ANS)** provides automatic regulation of smooth muscle, cardiac muscle, glands, and adipose tissue.

Somatic sensory receptors provide position, touch, pressure, pain, and temperature sensations.

Special sensory receptors provide sensations of smell, taste, vision, balance, and hearing.

Visceral sensory receptors monitor internal organs.

Skeletal muscle

• Smooth muscle
• Cardiac muscle
• Glands
• Adipose tissue

Start

1 **Receptors** are sensory structures that detect changes in the internal or external environment.

5 **Effectors** are target organs whose activities change in response to neural commands.

Neurons are nerve cells specialized for intercellular communication

In this module we examine the structural features of neurons. As we saw in Chapter 4, neurons have three general regions: **dendrites**, which receive stimuli from the environment or from other neurons; a **cell body**, which contains the nucleus and other organelles; and one or more **axons**, which carry information toward other cells.

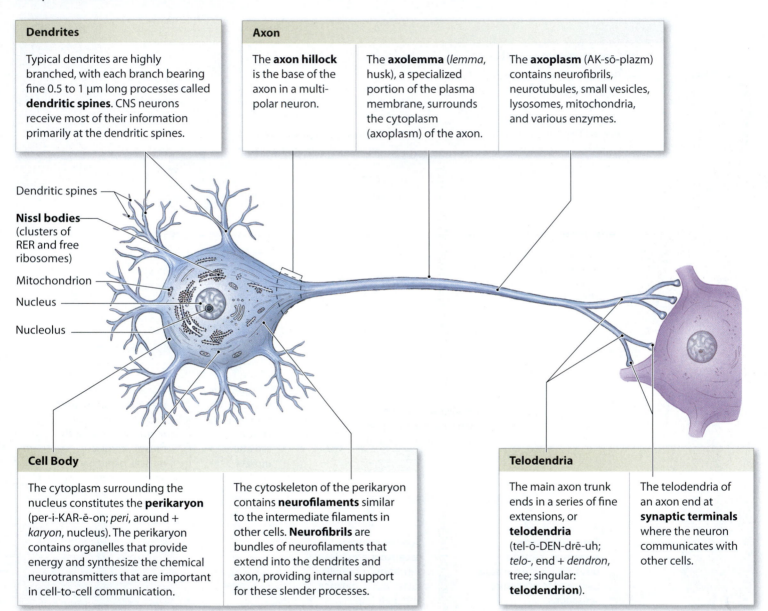

Dendrites

Typical dendrites are highly branched, with each branch bearing fine 0.5 to 1 μm long processes called **dendritic spines**. CNS neurons receive most of their information primarily at the dendritic spines.

Axon

The **axon hillock** is the base of the axon in a multi-polar neuron.

The **axolemma** (*lemma*, husk), a specialized portion of the plasma membrane, surrounds the cytoplasm (axoplasm) of the axon.

The **axoplasm** (AK-sō-plazm) contains neurofibrils, neurotubules, small vesicles, lysosomes, mitochondria, and various enzymes.

Dendritic spines —

Nissl bodies (clusters of RER and free ribosomes)

Mitochondrion —

Nucleus —

Nucleolus —

Cell Body

The cytoplasm surrounding the nucleus constitutes the **perikaryon** (per-i-KAR-ē-on; *peri*, around + *karyon*, nucleus). The perikaryon contains organelles that provide energy and synthesize the chemical neurotransmitters that are important in cell-to-cell communication.

The cytoskeleton of the perikaryon contains **neurofilaments** similar to the intermediate filaments in other cells. **Neurofibrils** are bundles of neurofilaments that extend into the dendrites and axon, providing internal support for these slender processes.

Telodendria

The main axon trunk ends in a series of fine extensions, or **telodendria** (tel-ō-DEN-drē-uh; *telo-*, end + *dendron*, tree; singular: **telodendrion**).

The telodendria of an axon end at **synaptic terminals** where the neuron communicates with other cells.

1 Here is a diagrammatic view of a representative neuron. The cell body contains most of the organelles of the neuron. The axon is a long cytoplasmic process that extends away from the cell body. Many materials, including enzymes and lysosomes, travel the length of the axon along **neurotubules** (neuron microtubules) through a process called **axoplasmic transport**. Axoplasmic transport occurs in both directions. If debris or unusual chemicals appear in the synaptic knob, **retrograde flow** soon delivers them to the cell body.

2 Each synaptic terminal is part of a **synapse**, a specialized site where the neuron communicates with another cell. Every synapse involves a **presynaptic cell** and a **postsynaptic cell**. Communication between those cells most commonly involves the release of chemicals called **neurotransmitters** into the **synaptic cleft**, a narrow space separating the two cells. Synaptic terminals receive (through axoplasmic transport) a continuous supply of neurotransmitters synthesized in the cell body. They also reabsorb and reassemble fragments of neurotransmitters broken down in the synaptic cleft.

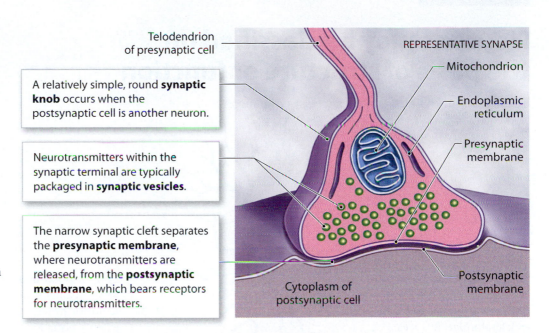

Telodendrion of presynaptic cell

REPRESENTATIVE SYNAPSE

A relatively simple, round **synaptic knob** occurs when the postsynaptic cell is another neuron.

Neurotransmitters within the synaptic terminal are typically packaged in **synaptic vesicles**.

The narrow synaptic cleft separates the **presynaptic membrane**, where neurotransmitters are released, from the **postsynaptic membrane**, which bears receptors for neurotransmitters.

Mitochondrion

Endoplasmic reticulum

Presynaptic membrane

Cytoplasm of postsynaptic cell

Postsynaptic membrane

3 Except in a few special cases involving sensory receptors, a presynaptic cell is always a neuron. The postsynaptic cell can be either a neuron or another type of cell. Notice that axons may branch along their length, producing side branches collectively known as **collateral branches**. Collateral branches enable a single neuron to communicate with several other cells.

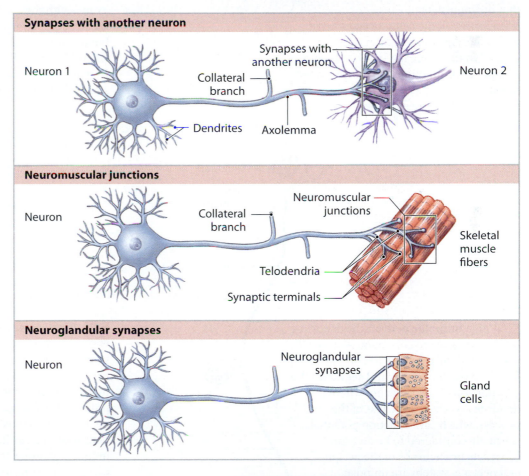

Synapses with another neuron

Neuron 1

Synapses with another neuron

Neuron 2

Collateral branch

Dendrites

Axolemma

Neuromuscular junctions

Neuron

Collateral branch

Neuromuscular junctions

Skeletal muscle fibers

Telodendria

Synaptic terminals

Neuroglandular synapses

Neuron

Neuroglandular synapses

Gland cells

Most CNS neurons lack centrioles and cannot divide. As a result, neurons lost to injury or disease are seldom replaced. Although neural stem cells persist in the adult nervous system, these cells are typically inactive (except in the epithelium responsible for our sense of smell, in the retina of the eye, and in the hippocampus, a portion of the brain involved with memory storage).

Module 11.1 Review

a. Name the structural components of a typical neuron.

b. Describe a synapse.

c. Why is a CNS neuron not usually replaced after it is injured?

Neurons may be classified on the basis of structure or function

There are four major anatomical classes of neurons.

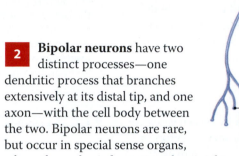

Dendrites

Dendritic process

Cell body

Axon

Synaptic terminals

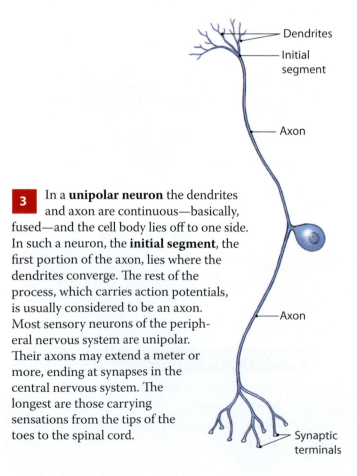

1 **Anaxonic** (an-AKS-on-ic) **neurons** are small and lack anatomical features that distinguish dendrites from axons; all the cell processes look alike. Anaxonic neurons are located in the brain and in special sense organs. Their functions are poorly understood.

2 **Bipolar neurons** have two distinct processes—one dendritic process that branches extensively at its distal tip, and one axon—with the cell body between the two. Bipolar neurons are rare, but occur in special sense organs, where they relay information about sight, smell, or hearing from receptor cells to other neurons. Bipolar neurons are small; the largest measure less than 30 μm from end to end.

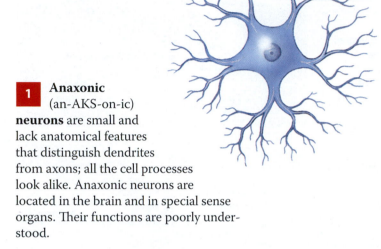

Dendrites

Initial segment

Axon

Axon

Synaptic terminals

3 In a **unipolar neuron** the dendrites and axon are continuous—basically, fused—and the cell body lies off to one side. In such a neuron, the **initial segment**, the first portion of the axon, lies where the dendrites converge. The rest of the process, which carries action potentials, is usually considered to be an axon. Most sensory neurons of the peripheral nervous system are unipolar. Their axons may extend a meter or more, ending at synapses in the central nervous system. The longest are those carrying sensations from the tips of the toes to the spinal cord.

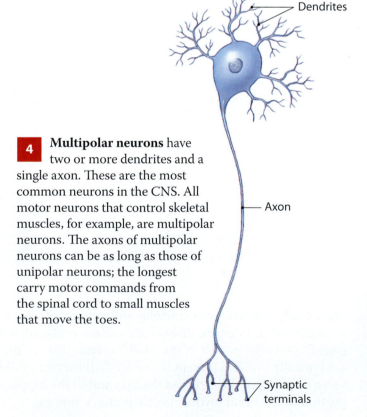

Dendrites

Axon

Synaptic terminals

4 **Multipolar neurons** have two or more dendrites and a single axon. These are the most common neurons in the CNS. All motor neurons that control skeletal muscles, for example, are multipolar neurons. The axons of multipolar neurons can be as long as those of unipolar neurons; the longest carry motor commands from the spinal cord to small muscles that move the toes.

Central Nervous System (CNS)

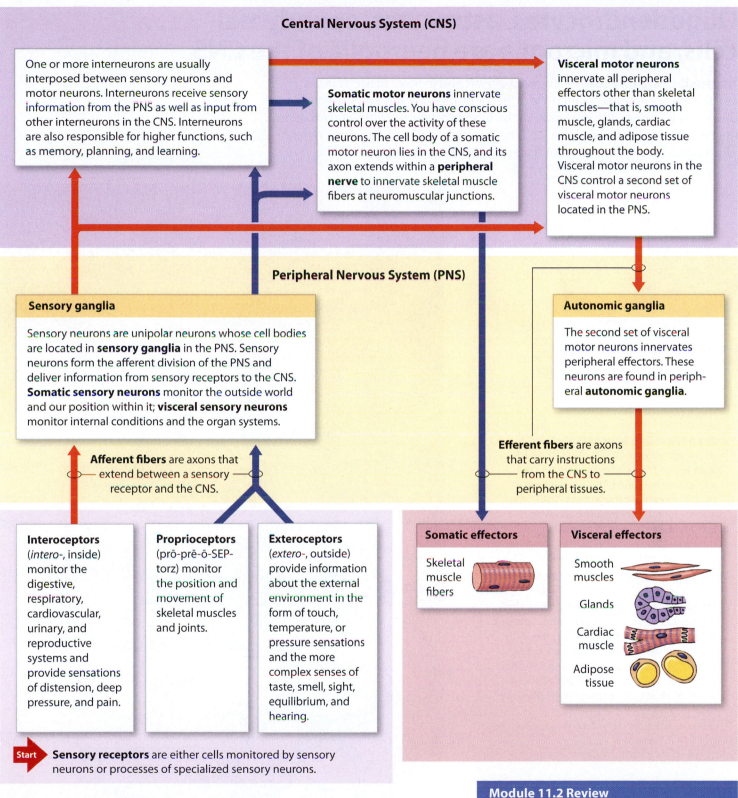

One or more interneurons are usually interposed between sensory neurons and motor neurons. Interneurons receive sensory information from the PNS as well as input from other interneurons in the CNS. Interneurons are also responsible for higher functions, such as memory, planning, and learning.

Somatic motor neurons innervate skeletal muscles. You have conscious control over the activity of these neurons. The cell body of a somatic motor neuron lies in the CNS, and its axon extends within a **peripheral nerve** to innervate skeletal muscle fibers at neuromuscular junctions.

Visceral motor neurons innervate all peripheral effectors other than skeletal muscles—that is, smooth muscle, glands, cardiac muscle, and adipose tissue throughout the body. Visceral motor neurons in the CNS control a second set of visceral motor neurons located in the PNS.

Peripheral Nervous System (PNS)

Sensory ganglia

Sensory neurons are unipolar neurons whose cell bodies are located in **sensory ganglia** in the PNS. Sensory neurons form the afferent division of the PNS and deliver information from sensory receptors to the CNS. **Somatic sensory neurons** monitor the outside world and our position within it; **visceral sensory neurons** monitor internal conditions and the organ systems.

Autonomic ganglia

The second set of visceral motor neurons innervates peripheral effectors. These neurons are found in peripheral **autonomic ganglia**.

Afferent fibers are axons that extend between a sensory receptor and the CNS.

Efferent fibers are axons that carry instructions from the CNS to peripheral tissues.

Interoceptors (*intero-*, inside) monitor the digestive, respiratory, cardiovascular, urinary, and reproductive systems and provide sensations of distension, deep pressure, and pain.

Proprioceptors (prō-prē-ō-SEP-torz) monitor the position and movement of skeletal muscles and joints.

Exteroceptors (*extero-*, outside) provide information about the external environment in the form of touch, temperature, or pressure sensations and the more complex senses of taste, smell, sight, equilibrium, and hearing.

Somatic effectors

Skeletal muscle fibers

Visceral effectors

Smooth muscles

Glands

Cardiac muscle

Adipose tissue

Start **Sensory receptors** are either cells monitored by sensory neurons or processes of specialized sensory neurons.

5 This simplified flowchart indicates the basic relationships among the three functional classes of neurons: **sensory neurons**, **interneurons**, and **motor neurons**. The human body has about 10 million sensory neurons, half a million motor neurons, and an estimated 20 billion interneurons.

KEY

→ = Somatic (sensory & motor)

→ = Visceral (sensory & motor)

Module 11.2 Review

a. Classify neurons according to their structure.

b. Classify neurons according to their function.

c. Are unipolar neurons in a tissue sample of the PNS more likely to function as sensory neurons or motor neurons?

Oligodendrocytes, astrocytes, ependymal cells, and microglia are neuroglia of the CNS

Neuroglia (or **glial cells**)—cells that support and protect neurons in the PNS and CNS—are abundant and diverse, accounting for about half the volume of the nervous system. Neural tissue in the CNS is organized differently than in the PNS, primarily because the CNS has a greater variety of glial cell types. This illustration summarizes information about neuroglia in the CNS.

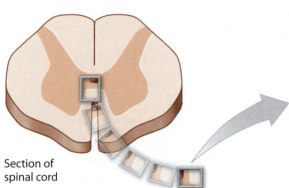

Section of spinal cord

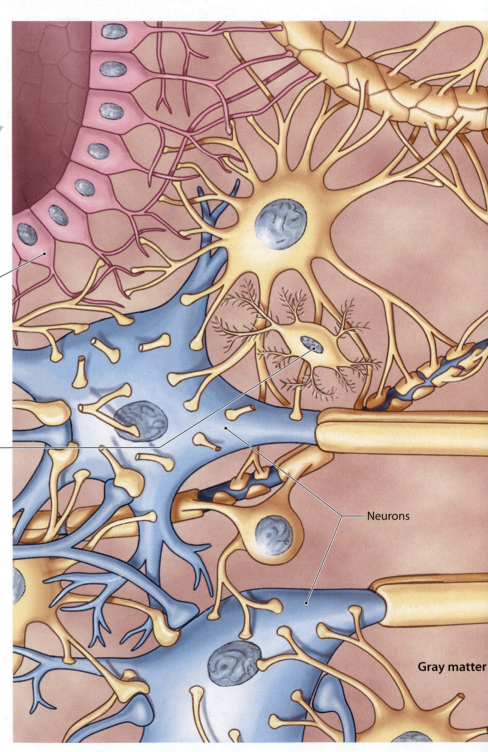

Ependymal cells form an epithelium known as the **ependyma** (ep-EN-di-muh), which lines a fluid-filled passageway within the spinal cord and brain. This passageway is filled with **cerebrospinal fluid (CSF)**, which also surrounds the brain and spinal cord. Ependymal cells assist in producing, monitoring, and circulating the CSF; some cells in the ventricles may be ciliated.

Microglia (mī-KRŌG-lē-uh) are embryologically related to monocytes and macrophages. Microglia migrate into the CNS as the nervous system forms and they persist as mobile cells, continuously moving through the neural tissue, removing cellular debris, waste products, and pathogens by phagocytosis.

Neurons

Gray matter

Astrocytes maintain the **blood–brain barrier** that isolates the CNS from the chemicals and hormones circulating in the blood. They also provide structural support within neural tissue; regulate ion, nutrient, and dissolved gas concentrations in the interstitial fluid surrounding the neurons; absorb and recycle neurotransmitters that are not broken down or reabsorbed at synapses; and form scar tissue after CNS injury.

Oligodendrocytes (ol-i-gō-DEN-drō-sīts; *oligo-*, few) provide a structural framework within the CNS by stabilizing the positions of axons. They also produce **myelin** (MĪ-e-lin), a membranous wrapping that coats axons and increases the speed of nerve impulse transmission. When myelinating an axon, the tip of an oligodendrocyte process expands to form an enormous membranous pad containing very little cytoplasm. This flattened "pancake" somehow gets wound around the axon, forming concentric layers of plasma membrane. These layers constitute a **myelin sheath**.

Many oligodendrocytes cooperate in the formation of a myelin sheath along the length of an axon. Such an axon is said to be **myelinated**. Each oligodendrocyte myelinates segments of several axons. The relatively large areas of the axon that are thus wrapped in myelin are called **internodes** (*inter*, between).

The small gaps of a few micrometers that separate adjacent internodes are called **nodes**. In dissection, myelinated axons appear glossy white, primarily because of the lipids within the myelin. As a result, regions dominated by myelinated axons constitute the **white matter** of the CNS.

Not all axons in the CNS are myelinated. **Unmyelinated axons** may not be completely covered by the processes of neuroglia. Such axons are common where relatively short axons and collaterals form synapses with densely packed neuron cell bodies. Areas containing neuron cell bodies, dendrites, and unmyelinated axons have a dusky gray color, and they constitute the **gray matter** of the CNS.

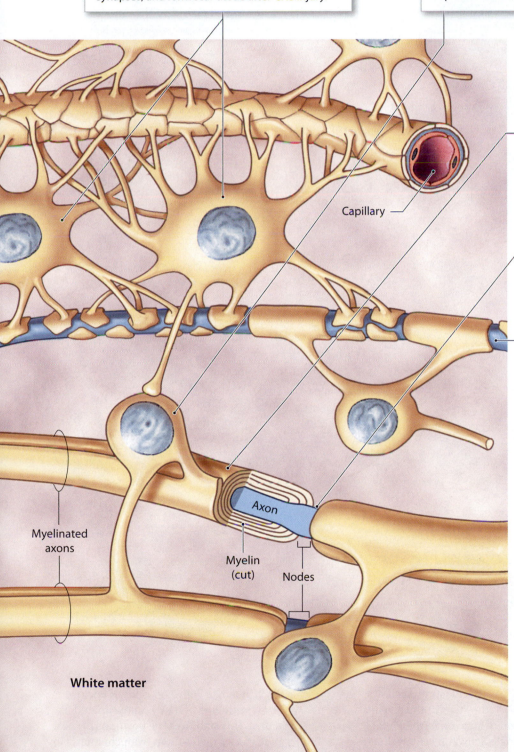

Capillary

Myelinated axons

Axon

Myelin (cut)

Nodes

White matter

Module 11.3 Review

a. Identify the neuroglia of the central nervous system.

b. Which glial cell protects the CNS from chemicals and hormones circulating in the blood?

c. Which type of neuroglia would occur in increased numbers in the brain tissue of a person with a CNS infection?

Schwann cells and satellite cells are the neuroglia of the PNS

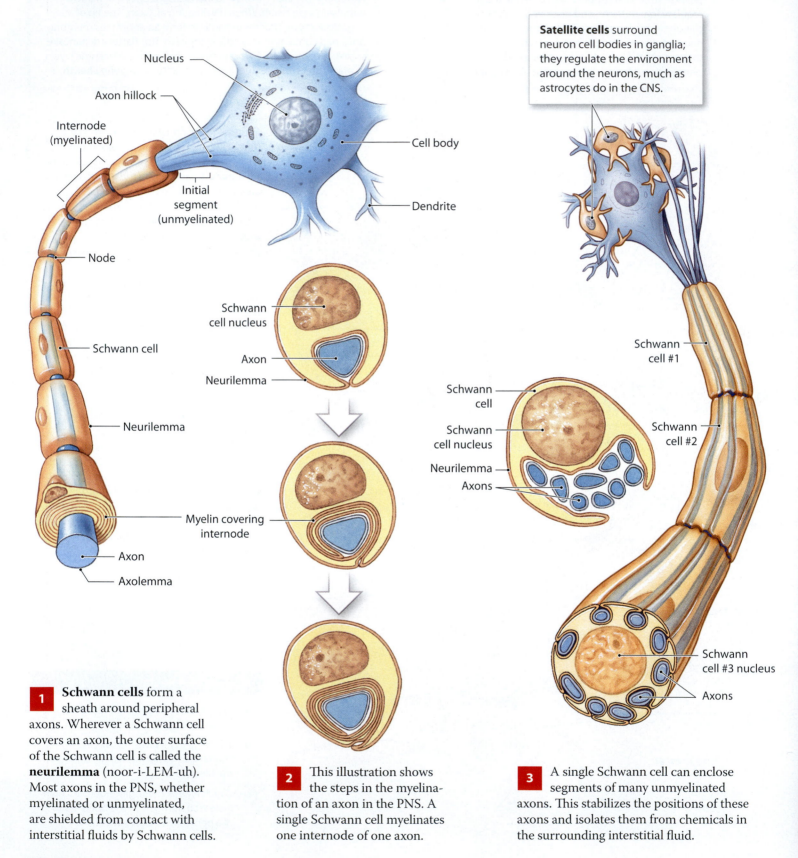

Satellite cells surround neuron cell bodies in ganglia; they regulate the environment around the neurons, much as astrocytes do in the CNS.

Nucleus

Axon hillock

Internode (myelinated)

Cell body

Initial segment (unmyelinated)

Dendrite

Node

Schwann cell nucleus

Axon

Neurilemma

Schwann cell

Schwann cell nucleus

Neurilemma

Axons

Schwann cell #1

Schwann cell #2

Neurilemma

Myelin covering internode

Axon

Axolemma

Schwann cell #3 nucleus

Axons

1 **Schwann cells** form a sheath around peripheral axons. Wherever a Schwann cell covers an axon, the outer surface of the Schwann cell is called the **neurilemma** (noor-i-LEM-uh). Most axons in the PNS, whether myelinated or unmyelinated, are shielded from contact with interstitial fluids by Schwann cells.

2 This illustration shows the steps in the myelination of an axon in the PNS. A single Schwann cell myelinates one internode of one axon.

3 A single Schwann cell can enclose segments of many unmyelinated axons. This stabilizes the positions of these axons and isolates them from chemicals in the surrounding interstitial fluid.

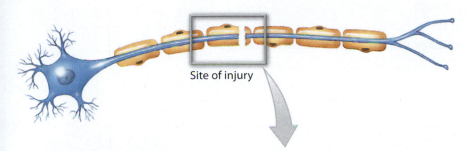

Site of injury

Step 1: Distal to the injury site, the axon and myelin degenerate and fragment.

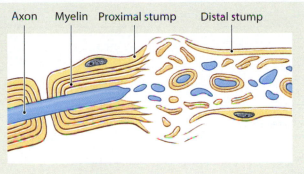

Axon   Myelin   Proximal stump   Distal stump

4 Schwann cells participate in the repair of damaged nerves in the PNS. The repair process, which often fails to restore full function, is known as **Wallerian degeneration**.

Step 2: The Schwann cells do not degenerate; instead, they proliferate along the path of the original axon. Over this period, macrophages move into the area and remove the degenerating debris distal to the injury site.

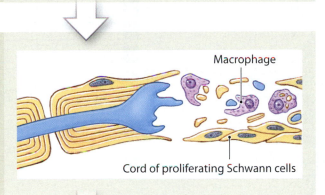

Macrophage

Cord of proliferating Schwann cells

Step 3: As the neuron recovers, its axon grows into the site of injury and then distally, along the path created by the Schwann cells.

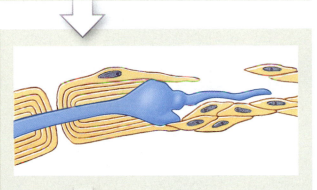

Limited regeneration can occur in the CNS, but there the situation is more complicated because (1) many more axons are likely to be involved, (2) astrocytes produce scar tissue that can prevent axon growth across the damaged area, and (3) astrocytes release chemicals that block the regrowth of axons.

Step 4: As the axon elongates, the Schwann cells wrap around it. If the axon reestablishes its normal synaptic contacts, normal function may be regained. However, if it stops growing or wanders off in some new direction, normal function will not return.

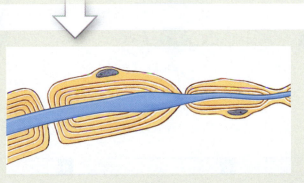

Module 11.4 Review

a. Identify the neuroglia of the peripheral nervous system.

b. Describe the neurilemma.

c. In which part of the nervous system does Wallerian degeneration occur?

1. Vocabulary

In the space provided, write the boldfaced terms introduced in this section that contain the indicated word part.

Word Part	Meaning	Term(s)
a neur-	nerve	
b dendr-	tree	
c ef-	away from	
d af-	toward	

2. Labeling

Label each of the structures in the following diagram of a neuron.

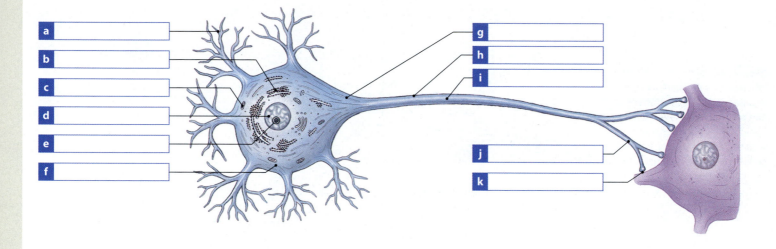

3. Labeling

Label the types of neurons depicted below.

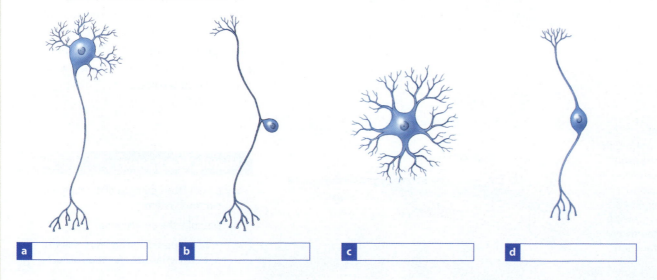

Neurophysiology

1 Plasma membranes are selectively permeable, and the cytosol and extracellular fluid differ in composition. Under normal circumstances, the inside of the plasma membrane has a slight negative charge with respect to the outside. The cause is a slight excess of positively charged ions outside the plasma membrane, and a slight excess of negatively charged ions and proteins inside the plasma membrane. This unequal charge distribution is called the **transmembrane potential**, and it is a characteristic of all living cells. It results from differences in the permeability of the membrane to various ions, as well as by active transport mechanisms.

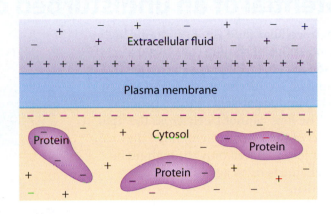

Extracellular fluid

Plasma membrane

Protein   Cytosol   Protein

Protein

---

1 The transmembrane potential of a resting cell is called the **resting potential**. All neural activities begin with a change in the resting potential of a neuron.

See Modules 11.5 and 11.6

2 A typical stimulus produces a temporary, localized change in the resting potential. The effect, which decreases with distance from the stimulus, is called a **graded potential**.

See Module 11.7

3 If the graded potential is sufficiently large, it triggers an **action potential** in the membrane of the axon. An action potential is an electrical event that involves one location on the membrane. Once an action potential develops in one location, it is propagated along the surface of an axon toward the synaptic terminals.

See Modules 11.8 and 11.9

4 **Synaptic activity** then produces graded potentials in the plasma membrane of the postsynaptic cell. The process typically involves the release of neurotransmitters, such as ACh, by the presynaptic cell. These compounds bind to receptors on the postsynaptic plasma membrane, changing its permeability.

See Module 11.10

5 The response of the postsynaptic cell ultimately depends on what the stimulated receptors do and what other stimuli are influencing the cell at the same time. The integration of stimuli at the level of the individual cell is the simplest form of **information processing** in the nervous system.

See Modules 11.11 and 11.12

Graded potential

Resting potential

stimulus produces

may produce

Action potential

triggers

Synaptic activity

Information processing

Presynaptic neuron

Postsynaptic cell

2 This figure provides an overview of the role of the transmembrane potential in neural activity. Changes in the transmembrane potential have many important functions; for example, they can trigger muscle contraction and gland secretion as well as transfer information in the nervous system.

The resting potential is the transmembrane potential of an undisturbed cell

The extracellular fluid (ECF) contains high concentrations of sodium ions (Na⁺) and chloride ions (Cl⁻), whereas the cytosol contains high concentrations of potassium ions (K⁺) and negatively charged proteins (Pr⁻). The ions cannot freely cross the lipid portions of the plasma membrane; they can enter or leave the cell only through membrane channels or by active transport mechanisms.

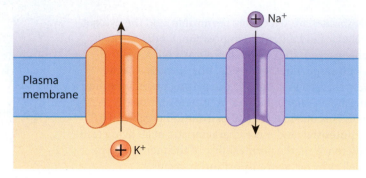

1 The transmembrane potential exists primarily because plasma membranes contain large numbers of passive **leak channels**, which are always open. The size, shape, and structure of a leak channel determines what ions can pass through it.

Potassium ions can diffuse out of the cell through potassium leak channels.

The sodium–potassium exchange pump ejects 3 Na⁺ for every 2 K⁺ recovered from the extracellular fluid. At a transmembrane potential of −70 mV, the rate of Na⁺ entry versus K⁺ loss is 3:2, and the exchange pump maintains a stable resting potential.

Sodium ions can diffuse into the cell through sodium leak channels.

The unit of measurement of potential difference is the **volt (V)**, and the transmembrane potential of a neuron is usually near 0.07 V. Such a value is usually expressed as −70 mV (or −70 millivolts—thousandths of a volt) with the minus sign indicating that the interior is negatively charged.

The cytosol contains an abundance of negatively charged proteins, whereas the extracellular fluid contains relatively few. These proteins cannot cross the plasma membrane.

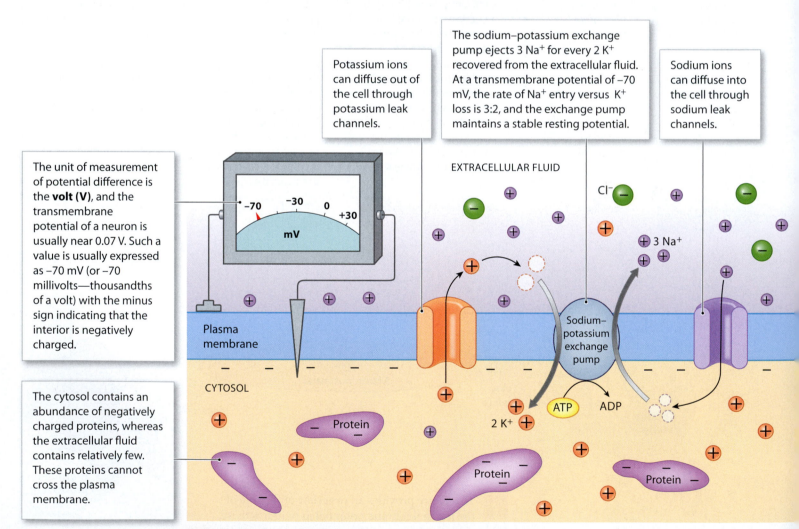

2 The **resting potential** is the transmembrane potential of an undisturbed cell. This is an overview of the events responsible for the normal resting potential of a neuron.

Potassium Ion Gradients

At normal resting potential, an electrical gradient opposes the chemical gradient for potassium ions (K⁺). The net electrochemical gradient tends to force potassium ions out of the cell.

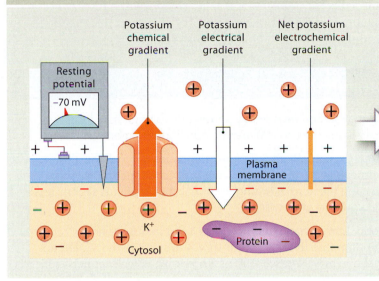

If the plasma membrane were freely permeable to potassium ions, the outflow of K⁺ would continue until the equilibrium potential (–90 mV) was reached.

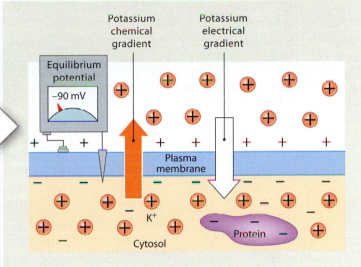

Sodium Ion Gradients

At the normal resting potential, chemical and electrical gradients combine to drive sodium ions (Na⁺) into the cell.

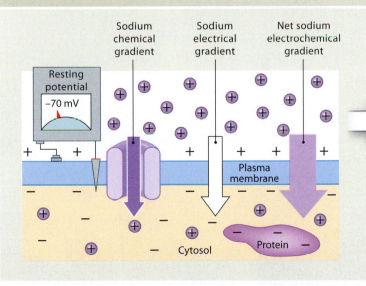

If the plasma membrane were freely permeable to sodium ions, the influx of Na⁺ would continue until the equilibrium potential (+66 mV) was reached.

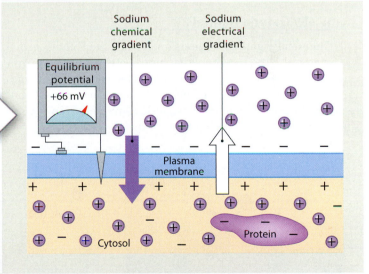

3 Any ion's **chemical gradient** is the concentration gradient for that ion across the plasma membrane. The **electrical gradient** is created by the attraction between opposite charges, or the repulsion between like charges (+/+ or –/–). The two ions we need to be concerned with are potassium and sodium. At a transmembrane potential known as the **equilibrium potential**, the electrical and chemical gradients are equal and opposite, and there is no net movement of ions across the membrane.

Module 11.5 Review

a. Define resting potential.

b. What effect would decreasing the concentration of extracellular potassium ions have on the transmembrane potential of a neuron?

c. What happens at the sodium–potassium exchange pump?

Gated channels can change the permeability of the plasma membrane

This table summarizes the most important information about the resting potential. The resting potential remains stable until the cell is disturbed or stimulated. The transmembrane potential then changes, primarily because the plasma membrane contains **gated channels** that open or close in response to specific stimuli. Three different classes of gated channels exist, and we will introduce them in this module.

The Resting Potential

- Because the plasma membrane is highly permeable to potassium ions, the resting potential is fairly close to −90 mV, the equilibrium potential for K^+.

- Although the electrochemical gradient for sodium ions is very large, the membrane's permeability to these ions is very low. Consequently, Na^+ has only a small effect on the normal resting potential, making it just slightly less negative than it would be otherwise.

- The sodium–potassium exchange pump ejects 3 Na^+ ions for every 2 K^+ ions that it brings into the cell. It thus serves to stabilize the resting potential when the ratio of Na^+ entry to K^+ loss through passive channels is 3:2.

- At the normal resting potential, these passive and active mechanisms are in balance. A typical neuron has a resting potential of approximately −70 mV.

Chemically Gated Channels

1 **Chemically gated channels** open when they bind specific chemicals. The receptors that bind acetylcholine (ACh) at the neuromuscular junction are chemically gated channels. Chemically gated channels are most abundant on the dendrites and cell body of a neuron, the areas where most synaptic communication occurs.

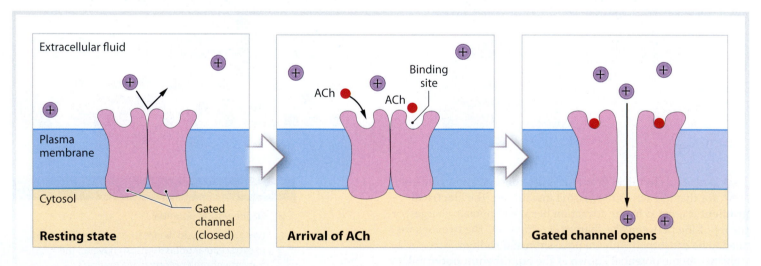

Voltage-Gated Channels

2 **Voltage-gated channels** are characteristic of areas of excitable membrane, a membrane capable of generating and conducting an action potential. Voltage-gated channels open or close in response to changes in the transmembrane potential. The most important voltage-gated channels, for our purposes, are voltage-gated sodium channels, potassium channels, and calcium channels. Sodium channels (shown here) have two gates that function independently: an activation gate that opens on stimulation, letting sodium ions into the cell, and an inactivation gate that closes to stop the entry of sodium ions.

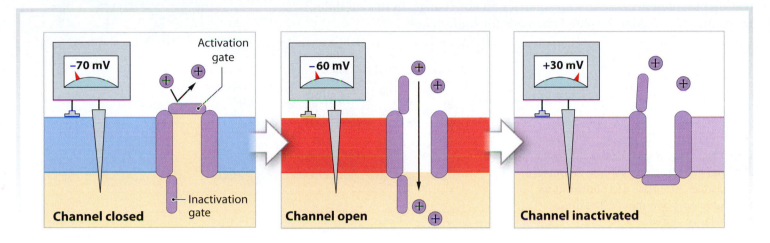

Mechanically Gated Channels

3 **Mechanically gated channels** open in response to physical distortion of the membrane surface. Such channels are important in sensory receptors that respond to touch, pressure, or vibration.

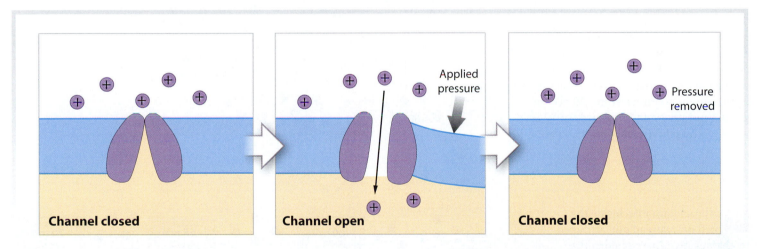

Most gated channels are closed at the resting potential. The opening of gated channels increases the rate of ion movement across the plasma membrane and this changes the transmembrane potential.

Module 11.6 Review

a. Define gated channels.

b. Identify the three types of gated channels, and state the conditions under which each operates.

c. What effect would a chemical that blocks voltage-gated sodium channels in a neuron's plasma membrane have on its transmembrane potential?

Graded potentials are localized changes in the transmembrane potential

Graded potentials, or local potentials, are changes in the transmembrane potential that cannot spread far from the site of stimulation. Any stimulus that opens a gated channel will produce a graded potential. Note that when illustrating graded potentials, we can ignore the leak channels responsible for the resting potential because their properties do not change. In this module we primarily consider the gated channels for sodium ions.

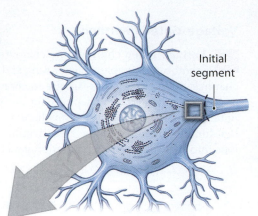

1 This diagrammatic view shows the chemically gated sodium channels in the plasma membrane of a neuron at its normal resting potential. All of these channels are closed.

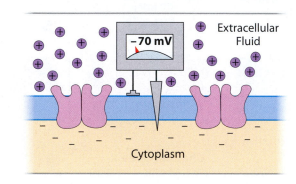

2 When the membrane is exposed to a chemical that opens the chemically gated sodium channels, sodium ions enter the cell. This inrush of positive charges reduces the transmembrane potential in this area. Any shift from the resting potential toward a more positive value is called a **depolarization**, a term that applies to changes in potential from −70 mV to smaller negative values (toward 0 mV) as well as to membrane potentials above 0 mV.

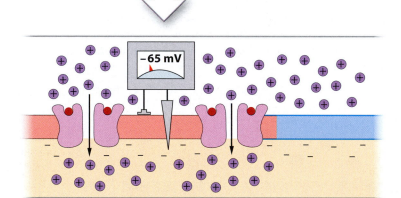

3 The sodium ions inside the cell now spread out, attracted by the negative charges along the inner surface of the membrane. The extent of depolarization spreads as well. As the plasma membrane depolarizes, extracellular sodium ions move toward the open channels, replacing ions that have entered the cell. This movement of positive charges parallel to the inner and outer surfaces of the membrane is called a **local current**.

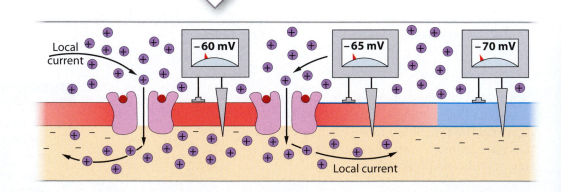

4 The degree of depolarization decreases with distance away from the stimulation site, primarily because the ions are entering only in one location but spreading in all directions. The impact on the transmembrane potential is proportional to the size of the stimulus, because the intensity of the stimulus determines the number of open sodium channels. The more open channels there are, the more sodium ions enter the cell, the greater the membrane area affected, and the greater the degree of depolarization.

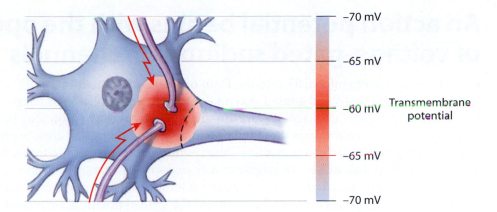

5 This graph shows the changes in the transmembrane potential that occur over time when different chemical stimuli are applied to the axon hillock.

The application of a chemical such as ACh opens chemically gated sodium ion channels. This leads to membrane depolarization.

When the chemical stimulus is removed, the membrane returns to its normal resting potential as the excess sodium ions are transported out of the cytoplasm. This process is called **repolarization**.

It is also possible to expose the neuron to a chemical that opens chemically gated potassium channels. This results in a more negative transmembrane potential because additional potassium ions leave the cytoplasm. A shift in the transmembrane potential past resting levels is called **hyperpolarization**.

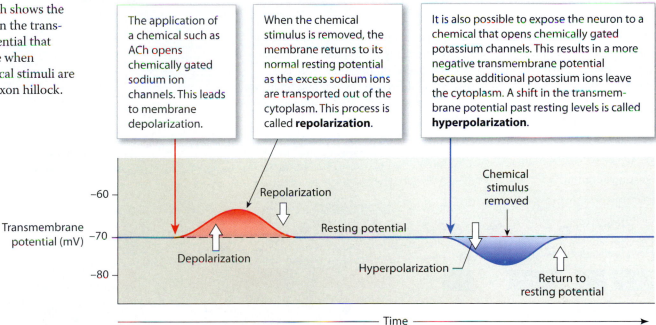

Graded Potentials

Graded potentials, whether depolarizing or hyperpolarizing, share four basic characteristics:

1. The transmembrane potential is most affected at the site of stimulation, and the effect decreases with distance.

2. The effect spreads passively through local currents.

3. The graded change in membrane potential may involve either depolarization or hyperpolarization. The nature of the change is determined by the properties of the membrane channels involved. For example, in a resting membrane, the opening of sodium channels will cause depolarization, whereas the opening of potassium channels will cause hyperpolarization. Thus, the change in membrane potential reflects whether positive charges enter or leave the cell.

4. The stronger the stimulus, the greater is the change in the transmembrane potential, and the larger is the area affected.

6 This table summarizes the most important characteristics of graded potentials.

Module 11.7 Review

a. Define graded potential.

b. Describe depolarization, repolarization, and hyperpolarization.

c. What factors account for the local currents associated with graded potentials?

An action potential begins with the opening of voltage-gated sodium ion channels

Each neuron receives information in the form of graded potentials on its dendrites and cell body, and graded potentials at the synaptic terminals trigger the release of neurotransmitters. However, the two ends of the neuron may be a meter apart, and even the largest graded potentials affect only a tiny area. Such relatively long-range communication requires a different mechanism—the action potential. **Action potentials** are propagated changes in the transmembrane potential that, once initiated, affect an entire excitable membrane. Whereas the resting potential depends on leak channels and the graded potential we considered depends on chemically gated channels, action potentials depend on voltage-gated channels.

Axon hillock

Initial segment

1 This figure illustrates steps in the formation of an action potential at the initial segment of an axon. The first step is a graded depolarization caused by the opening of chemically gated sodium ion channels, usually at the axon hillock. Note that when illustrating action potentials, we can ignore both the leak channels and the chemically gated channels, because their properties do not change.

Resting Potential

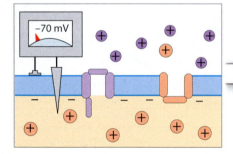

The axolemma contains both voltage-gated sodium channels and voltage-gated potassium channels that are closed when the membrane is at the resting potential.

KEY

= Sodium ion

= Potassium ion

1 **Depolarization to Threshold**

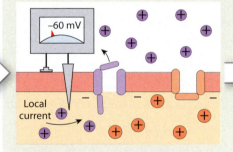

Local current

The stimulus that initiates an action potential is a graded depolarization large enough to open voltage-gated sodium channels. The opening of the channels occurs at a transmembrane potential known as the **threshold**.

2 **Activation of Sodium Channels and Rapid Depolarization**

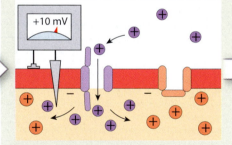

When the sodium channel activation gates open, the plasma membrane becomes much more permeable to Na^+. Driven by the large electrochemical gradient, sodium ions rush into the cytoplasm, and rapid depolarization occurs. The inner membrane surface now contains more positive ions than negative ones, and the transmembrane potential has changed from −60 mV to a positive value.

2 This graph plots the changes in the transmembrane potential at one location during the generation of an action potential. The numbers in the graph correspond to the steps illustrated below.

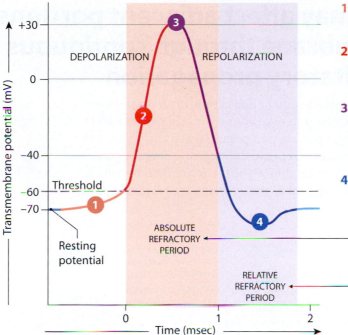

DEPOLARIZATION   REPOLARIZATION

Threshold

Resting potential

ABSOLUTE REFRACTORY PERIOD

RELATIVE REFRACTORY PERIOD

Time (msec)

1 A graded depolarization brings an area of excitable membrane to threshold (–60 mV).

2 Voltage-gated sodium channels open and sodium ions move into the cell. The transmembrane potential rises to +30 mV.

3 Sodium channels close, voltage-gated potassium channels open, and potassium ions move out of the cell. Repolarization begins.

4 Potassium channels close, and both sodium and potassium channels return to their normal states.

During the **absolute refractory period**, the membrane cannot respond to further stimulation.

During the **relative refractory period**, the membrane can respond only to a larger-than-normal stimulus.

---

3 **Inactivation of Sodium Channels and Activation of Potassium Channels**

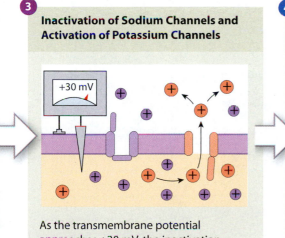

+30 mV

As the transmembrane potential approaches +30 mV, the inactivation gates of the voltage-gated sodium channels close. This step is known as **sodium channel inactivation**, and it coincides with the opening of voltage-gated potassium channels. Positively charged potassium ions move out of the cytosol, shifting the transmembrane potential back toward resting levels. Repolarization now begins.

4 **Potassium Channels Close**

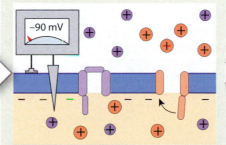

–90 mV

The voltage-gated sodium channels remain inactivated until the membrane has repolarized to near threshold levels. At this time, they regain their normal status: closed but capable of opening. The voltage-gated potassium channels begin closing as the membrane reaches the normal resting potential (about –70 mV). Until all of these potassium channels have closed, potassium ions continue to leave the cell. This produces a brief hyperpolarization.

Resting Potential

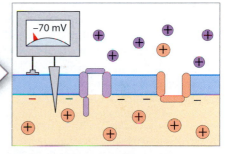

–70 mV

As the voltage-gated potassium channels close, the transmembrane potential returns to normal resting levels. The action potential is now over, and the membrane is once again at the resting potential.

In an excitable membrane, a graded depolarization is like the pressure on the trigger of a gun, and the action potential is like the firing of the gun. As long as the trigger is pulled hard enough, the gun fires, and it fires the same way every time. Similarly, all stimuli that bring the membrane to threshold generate identical action potentials. This concept is called the **all-or-none principle**, because a given stimulus either triggers a typical action potential, or it does not trigger one at all.

Module 11.8 Review

a. Define action potential.

b. List the events involved in the generation of an action potential.

c. Compare the absolute refractory period with the relative refractory period.

Action potentials may affect adjacent portions of the plasma membrane through continuous propagation or saltatory propagation

An action potential generated at the initial segment doesn't move along the axon like a car on a highway. Instead, the action potential at one site triggers an action potential at an adjacent site, which triggers an action potential at a third site, and so forth, along the entire length of the axon. Because the same events take place over and over, the term **propagation** is preferable to the term *conduction*, which is used for electrical wiring. (In fact, compared to wires, axons are relatively poor conductors of electricity.)

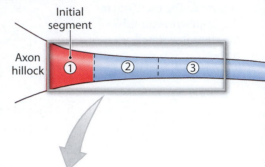

Continuous Propagation

1 In **continuous propagation**, which occurs along unmyelinated axons, the action potential appears to "move" in a series of tiny steps. This diagram shows how an action potential generated in the initial segment ① affects more distant portions of the axon ② and ③. Each step takes only a millisecond, but the steps must be repeated along the entire axon, so propagation along an unmyelinated axon only occurs at a speed of about 1 meter per second.

Step 1
As an action potential develops at the initial segment ①, the transmembrane potential at this site depolarizes to +30 mV.

Step 2
As the sodium ions entering at ① spread away from the open voltage-gated channels, a graded depolarization quickly brings the membrane in segment ② to threshold.

Step 3
An action potential now occurs in segment ② while segment ① begins repolarization.

Step 4
As the sodium ions entering at segment ② spread laterally, a graded depolarization quickly brings the membrane in segment ③ to threshold. The action potential can only move forward, not backward, because the membrane at segment ① is in the absolute refractory period of repolarization.

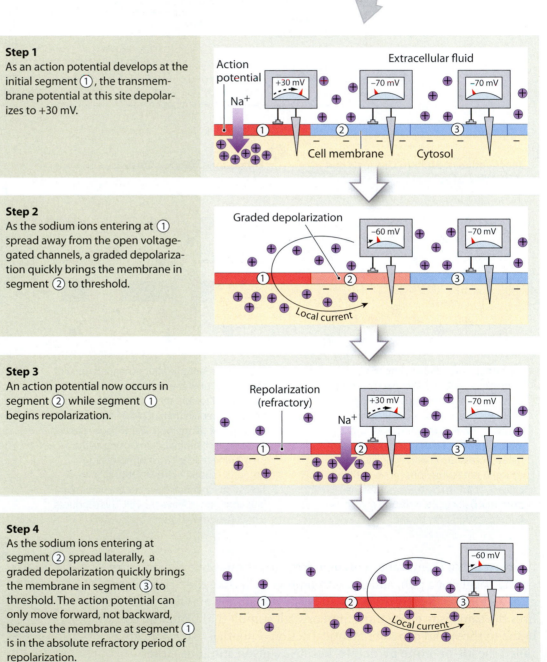

Saltatory Propagation

2 Continuous propagation cannot occur along a myelinated axon, because myelin blocks the flow of ions across the membrane. Ions can only cross the plasma membrane at the nodes. As a result, only the nodes can respond to a depolarizing stimulus. When an action potential appears at the initial segment of a myelinated axon, the local current skips the internode and depolarizes the closest node in a process called **saltatory propagation** (*saltare*, leaping).

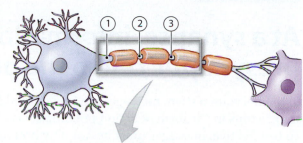

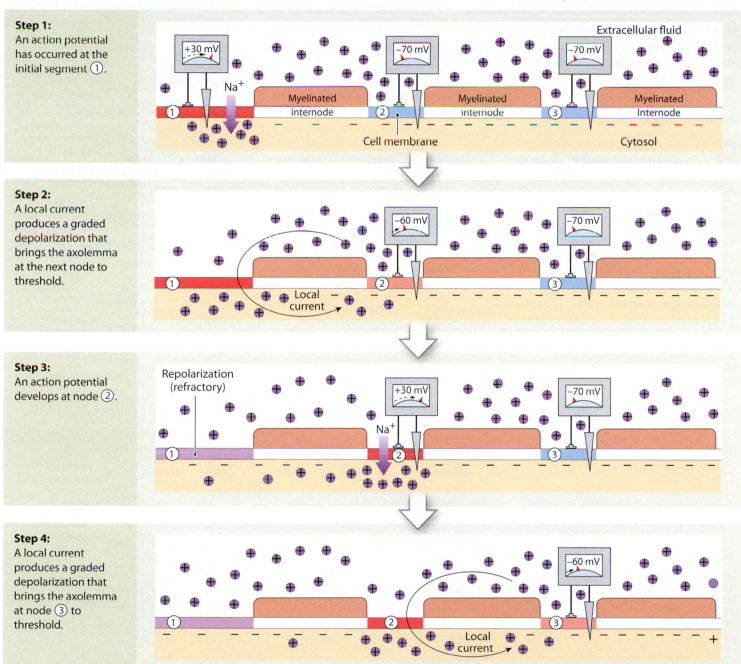

Step 1:
An action potential has occurred at the initial segment ①.

Extracellular fluid

+30 mV   −70 mV   −70 mV

Na⁺   Myelinated internode   Myelinated internode   Myelinated internode

Cell membrane   Cytosol

Step 2:
A local current produces a graded depolarization that brings the axolemma at the next node to threshold.

−60 mV   −70 mV

Local current

Step 3:
An action potential develops at node ②.

Repolarization (refractory)

+30 mV   −70 mV

Na⁺

Step 4:
A local current produces a graded depolarization that brings the axolemma at node ③ to threshold.

−60 mV

Local current

Saltatory propagation is much faster than continuous propagation, but there are differences among myelinated axons. That is because the speed varies with axon diameter. The larger the diameter, the lower the resistance to ion movement, and the faster the current travels.

Module 11.9 Review

a. Define continuous propagation and saltatory propagation.

b. What is the relationship between myelin and the propagation speed of action potentials?

At a synapse, information travels from the presynaptic cell to the postsynaptic cell

In the nervous system, messages are transmitted from one location to another along axons in the form of action potentials, also known as "nerve impulses." To be effective, messages must be not only propagated along an axon but also transferred in some way to another neuron or an effector cell. That transfer occurs at a **synapse**. At a synapse involving two neurons, information is relayed from a presynaptic neuron to a postsynaptic neuron.

Chemical Synapses

1 **Chemical synapses**, which rely on neurotransmitter release, are by far the most abundant type of synapse. Most synapses between neurons, and all synapses between neurons and other types of cells, involve chemical synapses. Synapses that release acetylcholine as a neurotransmitter are known as **cholinergic synapses**. This figure illustrates the events that occur at a cholinergic synapse when an action potential arrives at a synaptic knob. The same steps are involved in communication across a neuromuscular junction, a process described in Module 9.4.

Step 1
The normal stimulus for neurotransmitter release is the depolarization of the synaptic knob by the arrival of an action potential.

Presynaptic neuron
Synaptic vesicles
ER
Action potential
EXTRACELLULAR FLUID
Synaptic knob
AChE
POSTSYNAPTIC NEURON

Step 2
The depolarization of the synaptic knob opens voltage-gated calcium channels. Calcium ions rush into the knob and trigger the exocytosis of synaptic vesicles and the release of ACh into the synaptic cleft. The calcium ions that triggered exocytosis are rapidly removed, ending the release of ACh.

ACh
Ca^{2+} Ca^{2+}
Synaptic cleft
Chemically gated sodium ion channels

Step 3
The released ACh diffuses across the synaptic cleft and binds to the chemically gated Na^+ receptors on the postsynaptic membrane. The greater the amount of ACh released, the more receptors respond, and the larger the depolarization. If the depolarization is great enough, an action potential will appear in the postsynaptic neuron.

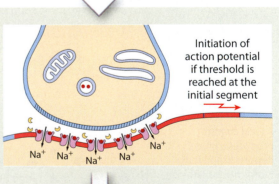

Initiation of action potential if threshold is reached at the initial segment

Na^+ Na^+ Na^+ Na^+

Step 4
The effects on the postsynaptic membrane are temporary, because of the presence of the enzyme acetylcholinesterase (AChE) in the synaptic cleft. ACh molecules that bind to receptor sites are generally broken down within 20 msec of their arrival.

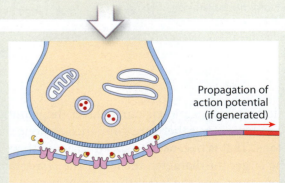

Propagation of action potential (if generated)

2 This figure summarizes the events that occur at a cholinergic synapse each time an action potential arrives at the synaptic knob. Because neurotransmitters are reabsorbed and recycled, the synaptic knob can continue to function for an extended period. However, after extended stimulation it may be unable to keep pace with the demand for neurotransmitter. **Synaptic fatigue** then occurs, and the synapse will be unable to function normally until its supply of ACh has been replenished.

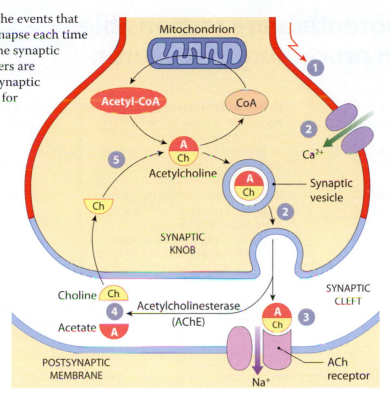

Events Occurring at Synapse

1 An arriving action potential depolarizes the synaptic knob.

2 Calcium ions enter the cytoplasm, and after a brief delay, ACh is released through the exocytosis of synaptic vesicles.

3 ACh binds to sodium channel receptors on the postsynaptic membrane, producing a graded depolarization.

4 Depolarization ends as ACh is broken down into acetate and choline by AChE.

5 The synaptic knob reabsorbs choline from the synaptic cleft and uses it to synthesize new molecules of ACh.

A **synaptic delay** lasting 0.2–0.5 msec occurs between the arrival of the action potential at the synaptic knob and the effect on the postsynaptic membrane. Although a delay of 0.5 msec is not very long, when information is being passed along a chain of interneurons the cumulative synaptic delay may be considerable. This is why reflexes are important for survival—they involve only a few synapses and thus provide rapid and automatic responses to stimuli. There is only one way to eliminate synaptic delay entirely—directly couple the presynaptic neuron to the postsynaptic cell at an electrical synapse.

Electrical Synapses

3 At an **electrical synapse**, the presynaptic and postsynaptic membranes are locked together by gap junctions. As a result, changes in the transmembrane potential of one cell will produce local currents that affect the other cell as if the two shared a common membrane. Electrical synapses are located in both the CNS and PNS, but they are extremely rare. They are present in some areas of the brain, in the eye, and in at least one pair of PNS ganglia (the ciliary ganglia). An action potential reaching an electrical synapse will always be propagated to the next cell, an arrangement that is efficient but not versatile. Much of the complexity and adaptability of the nervous system results from the fact that the responses of a postsynaptic cell can vary depending on the local chemical environment or the activities of synapses that release multiple neurotransmitters with varied effects.

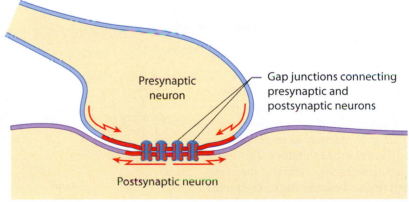

Module 11.10 Review

a. Describe the parts of a chemical synapse.

b. Contrast an electrical synapse with a chemical synapse.

c. What is synaptic fatigue, and how does the synapse recover?

Postsynaptic potentials are responsible for information processing in a neuron

An **excitatory postsynaptic potential**, or **EPSP**, is a graded depolarization caused by the arrival of a neurotransmitter at the postsynaptic membrane that shifts the transmembrane potential closer to the threshold. When that occurs, the membrane is said to be **facilitated**. The larger the degree of facilitation, the smaller is the additional stimulus needed to trigger an action potential.

An **inhibitory postsynaptic potential**, or **IPSP**, is a graded hyperpolarization of the postsynaptic membrane. An IPSP may result, for example, from the opening of chemically gated potassium channels. When membrane hyperpolarization occurs, the neuron is said to be inhibited, because a larger-than-usual depolarizing stimulus must be provided to bring the membrane potential to threshold.

Summation is the integration of the effects of graded potentials on a segment of the plasma membrane. EPSPs and IPSPs result from the activation of different types of chemically gated channels. If two different neurotransmitters arrive simultaneously, and both sets of channels open, the net effect may be no change in the membrane potential.

Time 2: Hyperpolarizing stimulus applied

Stimulus removed

Time 3: Hyperpolarizing stimulus applied

−60 mV

−70 mV EPSP Resting potential IPSP Resting potential EPSP / IPSP

−80 mV

Time 1: Depolarizing stimulus applied

Stimulus removed

Time 3: Depolarizing stimulus applied

Stimuli removed

Time

1 **Postsynaptic potentials** are graded potentials that develop in the postsynaptic membrane in response to a neurotransmitter. Two major types of postsynaptic potentials develop at neuron-to-neuron synapses: excitatory postsynaptic potentials and inhibitory postsynaptic potentials.

2 A single neuron may receive information across thousands of synapses. Some of the neurotransmitters arriving at the postsynaptic cell at any moment may be excitatory, whereas others may be inhibitory. The net effect on the transmembrane potential at the axon hillock determines how the neuron responds from moment to moment. This is because (1) the axon hillock is closest to the initial segment, and graded potentials are what trigger the action potential, and (2) the threshold at the boundary between the axon hillock and the initial segment is lower than it is elsewhere on the cell body. Thus it is the axon hillock that integrates the excitatory and inhibitory stimuli affecting the cell body and dendrites at any given moment and determines the rate of action potential generation at the initial segment. This integration process is the simplest level of information processing in the nervous system.

Axon hillock

Initial segment

Glial cell processes

Dendrite

Synaptic knobs

3   An individual EPSP or IPSP has a small effect on the transmembrane potential; a typical EPSP produces a depolarization of about 0.5 mV at the postsynaptic membrane. Before an action potential will arise in the initial segment, local currents must depolarize that region by at least 10 mV. Therefore, a single EPSP will not result in an action potential, even if the synapse is on the axon hillock. However, individual EPSPs combine through the process of summation. There are two forms of summation that can generate an action potential.

Temporal Summation

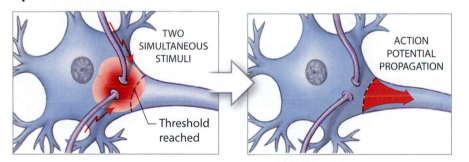

FIRST STIMULUS

Initial segment

SECOND STIMULUS

Threshold reached

ACTION POTENTIAL PROPAGATION

Temporal summation (*tempus*, time) occurs when a single synapse is active repeatedly. For example, a typical EPSP lasts about 20 msec, but under maximum stimulation an action potential can reach the synaptic knob each millisecond. When a second EPSP arrives before the effects of the first EPSP have disappeared, another group of vesicles discharges ACh into the synaptic cleft, more ACh molecules arrive at the postsynaptic membrane, and the degree of depolarization increases. In this way, a series of small steps can eventually bring the initial segment to threshold.

Spatial Summation

TWO SIMULTANEOUS STIMULI

Threshold reached

ACTION POTENTIAL PROPAGATION

Spatial summation involves multiple synapses that are active simultaneously. The effects on the transmembrane potential are cumulative, because each active synapse opens gated channels that allow ion movement in or out of the cell. In this example, the active synapses are generating EPSPs and opening gated channels that let sodium ions enter the cell. The effects are cumulative, and the degree of depolarization at the initial segment depends on (1) how many excitatory synapses are active at any given moment, and (2) how far they are from the initial segment. As in temporal summation, an action potential results when the transmembrane potential at the initial segment reaches threshold.

Each neuron integrates the information arriving across synapses, but also monitors and responds to a variety of factors in the local environment. Neurons are dynamic, active cells, and their sensitivity to stimulation changes in response to changes in body temperature, changes in the availability of oxygen or nutrients, or the presence of abnormal chemicals.

Module 11.11 Review

a. Define excitatory postsynaptic potential (EPSP) and inhibitory postsynaptic potential (IPSP).

b. Compare temporal summation with spatial summation.

c. If a single EPSP depolarizes the initial segment from a resting potential of −70 mV to −65 mV, and threshold is at −60 mV, will an action potential be generated?

Information processing involves interacting groups of neurons, and information is encoded in the frequency and pattern of action potentials

1 Cellular information processing occurs at the postsynaptic membrane, as the transmembrane potential at the axon hillock rises and falls. Higher levels of information processing involve **regulatory neurons** that facilitate or inhibit the activities of presynaptic neurons by affecting the membrane of the cell body or by altering the sensitivity of synaptic knobs.

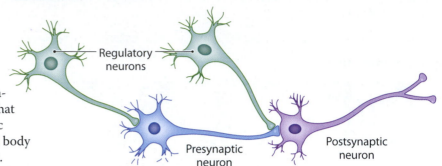

Regulatory neurons

Presynaptic neuron

Postsynaptic neuron

2 One of the reasons the nervous system is so complex and versatile is that it uses more than 100 different neurotransmitters that each work in different ways. This table introduces a few of the important neurotransmitters you will encounter in later chapters. As you see, they may have indirect effects as well as direct effects on ion channels. The indirect effects usually involve binding to **G proteins**, a diverse family of receptor proteins. Activated G proteins trigger the formation or release of **second messengers** that alter conditions in the cytoplasm and change the activity of the postsynaptic cell.

Selected Neurotransmitters

Neurotransmitter	Chemical Structure	Mechanism of Action	Location(s)	Comments
Acetylcholine (ACh)	$CH_3-N-CH_2-CH_2-O-C-CH_3$ (with CH_3 groups and O)	Primarily direct, through binding to chemically gated channels	CNS: Synapses throughout brain and spinal cord / PNS: Neuromuscular junctions, neuroglandular junctions, and synapses in autonomic ganglia	Widespread in CNS and PNS; best known and most studied of the neurotransmitters
Norepinephrine (NE)	$NH_2-CH_2-CH(OH)-$ (ring with OH, OH)	Indirect, through G proteins and second messengers	CNS: Cerebral cortex, hypothalamus, brain stem, cerebellum, and spinal cord / PNS: Most neuromuscular and neuroglandular junctions of sympathetic division of ANS	Involved in attention and consciousness, control of body temperature, and regulation of pituitary gland secretion
Epinephrine (E)	$CH_2-NH-CH_2-CH(OH)-$ (ring with OH, OH)	Indirect: G proteins and second messengers	CNS: Thalamus, hypothalamus, midbrain, and spinal cord	Generally excitatory effect along autonomic pathways
Serotonin	$NH_2-CH_2-CH_2-$ (indole ring with OH, N)	Primarily indirect: G proteins and second messengers	CNS: Hypothalamus, limbic system, cerebellum, spinal cord, and retina	Important in emotional states, moods, and body temperature; several illicit hallucinogenic drugs, such as Ecstasy, target serotonin receptors
Glutamate	$HO-C(O)-CH(NH_2)-CH_2-CH_2-C(O)-OH$	Indirect: G proteins and second messengers / Direct: opens calcium/sodium channels	CNS: Cerebral cortex and brain stem	Important in memory and learning; most important excitatory neurotransmitter in the brain
Gamma-aminobutyric acid (GABA)	$NH_2-CH_2-CH_2-CH_2-C(O)-OH$	Direct or indirect (G proteins), depending on type of receptor	CNS: Cerebral cortex, cerebellum, interneurons throughout brain and spinal cord	Direct inhibitory effects: opens Cl^- channels; indirect effects: opens K^+ channels and blocks entry of Ca^{2+}

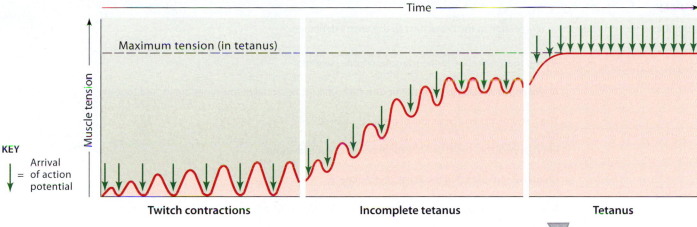

Maximum tension (in tetanus)

Muscle tension

KEY

↓ = Arrival of action potential

Twitch contractions **Incomplete tetanus** **Tetanus**

Time

3 In the nervous system, complex information is translated into action potentials that are propagated along axons. On arrival, the message is often interpreted solely on the basis of the frequency of action potentials. This figure shows how the rate of action potentials arriving at a neuro-muscular junction determines the nature of the resulting contraction.

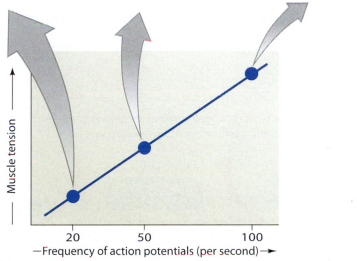

Muscle tension

20   50   100
—Frequency of action potentials (per second)→

4 This table summarizes the key concepts for understanding information processing in the nervous system.

Information Processing

- The neurotransmitters released at a synapse may have either excitatory or inhibitory effects. The effect on the axon's initial segment reflects a summation of the stimuli arriving at any moment. The frequency of action potentials generated is an indication of the degree of sustained depolarization at the axon hillock.

- Neurons may be facilitated or inhibited by extracellular chemicals other than neurotransmitters.

- The response of a postsynaptic neuron to the activation of a pre-synaptic neuron can be altered by (1) the presence of chemicals that cause facilitation or inhibition at the synapse, (2) activity under way at other synapses affecting the postsynaptic cell, and (3) modification of the rate of neurotransmitter release through facilitation or inhibition by regulatory neurons.

- Information is relayed in the form of action potentials. In general, the degree of sensory stimulation or the strength of the motor response is proportional to the frequency of action potentials.

Module 11.12 Review

a. Describe the role of regulatory neurons.

b. What determines the frequency of action potential generation?

c. The greater the degree of sustained depolarization at the axon hillock, the _____ (higher or lower) the frequency of generation of action potentials.

1. Vocabulary

Write the boldfaced term introduced in this section in the blank next to the term's definition.

a _____ A propagated change in the transmembrane potential

b _____ A synapse in which the presynaptic and postsynaptic neuronal membranes are locked together by gap junctions

c _____ The transmembrane potential of a nonstimulated cell

d _____ Ion channels that open or close in response to specific stimuli

e _____ Chemical synapses that release acetylcholine

f _____ A shift in the transmembrane potential from –70 mV to –85 mV

g _____ The movement of positive charges parallel to the inner and outer membrane surfaces

h _____ A shift in the transmembrane potential from –70 mV to +30 mV

2. Short answer

For the following diagram of a cholinergic synapse, write the names of components a–f in the boxes at left, and then fill in the table at right with descriptions of the events represented by g–l.

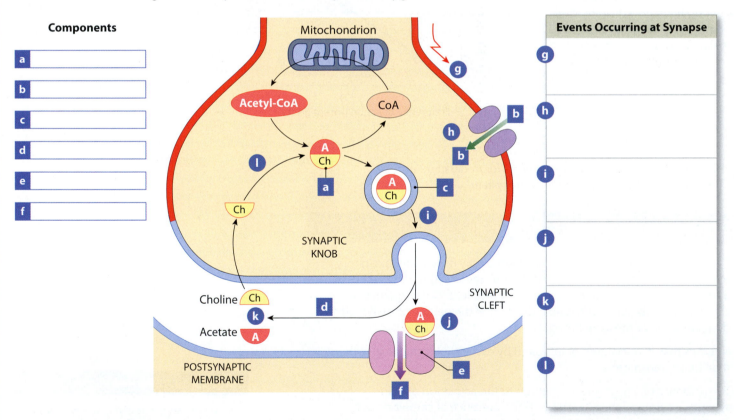

Components

a [_____]
b [_____]
c [_____]
d [_____]
e [_____]
f [_____]

Events Occurring at Synapse

g
h
i
j
k
l

3. Section integration

Guillain-Barré (ghee-yan bah-ray) syndrome is a degeneration of myelin sheaths that ultimately may result in paralysis. Propose a mechanism by which myelin sheath degeneration can cause muscular paralysis.

_____

_____

Visual Outline with Key Terms

Summarize the content of each module using the terms in the order provided.

SECTION 1

Neurons and Neuroglia

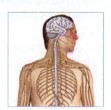

- receptors
- visceral sensory receptors
- somatic sensory receptors
- special sensory receptors
- sensory division
- peripheral nervous system (PNS)
- information processing
- central nervous system (CNS)
- motor division
- somatic nervous system (SNS)
- autonomic nervous system (ANS)
- effectors

11.1

Neurons are nerve cells specialized for intercellular communication

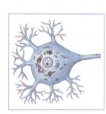

- dendrites
- cell body
- axons
- neurotubules
- axoplasmic transport
- retrograde flow
- dendritic spines
- axon hillock
- axolemma
- axoplasm
- nissl bodies
- perikaryon
- neurofilaments
- neurofibrils
- telodendria/ telodendrion
- synaptic terminals
- synapse
- presynaptic cell
- postsynaptic cell
- neurotransmitters
- synaptic cleft
- synaptic knob
- synaptic vesicles
- presynaptic membrane
- postsynaptic membrane
- collateral branches

11.2

Neurons may be classified on the basis of structure or function

- anaxonic neurons
- bipolar neurons
- unipolar neuron
- initial segment
- multipolar neurons
- sensory neurons
- interneurons
- motor neurons
- sensory receptors
- interoceptors
- proprioceptors
- exteroceptors
- afferent fibers
- sensory ganglia
- visceral sensory neurons
- somatic sensory neurons
- somatic motor neurons
- peripheral nerve
- visceral motor neurons
- efferent fibers
- autonomic ganglia
  - somatic effectors
  - visceral effectors

11.3

Oligodendrocytes, astrocytes, ependymal cells, and microglia are neuroglia of the CNS

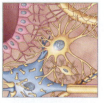

- neuroglia (glial cells)
- ependymal cells
- ependyma
- cerebrospinal fluid (CSF)
- microglia
- astrocytes
- blood–brain barrier
- oligodendrocytes
- myelin
- myelin sheath
- myelinated
- internodes
- nodes
- white matter
- unmyelinated axons
- gray matter

11.4

Schwann cells and satellite cells are the neuroglia of the PNS

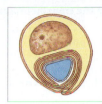

- Schwann cells
- neurilemma
- satellite cells
- Wallerian degeneration

SECTION 2

Neurophysiology

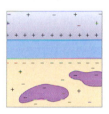

- transmembrane potential
- resting potential
- graded potential
- action potential
- synaptic activity
- information processing

11.5

The resting potential is the transmembrane potential of an undisturbed cell

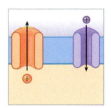

- leak channels
- resting potential
- volt (V)
  - millivolt (mV)
- chemical gradient
- electrical gradient
- equilibrium potential

11.6

Gated channels can change the permeability of the plasma membrane

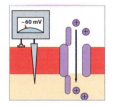

- gated channels
- chemically gated channels
- voltage-gated channels
- mechanically gated channels

• = *Term boldfaced in this module*

11.7

Graded potentials are localized changes in the transmembrane potential

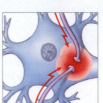

- graded potentials
- depolarization
- local current
- repolarization
- hyperpolarization

11.8

An action potential begins with the opening of voltage-gated sodium ion channels

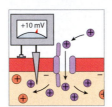

- action potentials
- threshold
- sodium channel inactivation
- absolute refractory period
- relative refractory period
- all-or-none principle

11.9

Action potentials may affect adjacent portions of the plasma membrane through continuous propagation or saltatory propagation

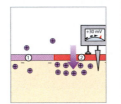

- propagation
- continuous propagation
- saltatory propagation

11.10

At a synapse, information travels from the presynaptic cell to the postsynaptic cell

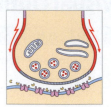

- synapse
- chemical synapses
- cholinergic synapses
- synaptic fatigue
- synaptic delay
- electrical synapse

11.11

Postsynaptic potentials are responsible for information processing in a neuron

- postsynaptic potentials
- excitatory postsynaptic potential (EPSP)
- facilitated
- inhibitory postsynaptic potential (IPSP)
- summation
- temporal summation
- spatial summation

11.12

Information processing involves interacting groups of neurons, and information is encoded in the frequency and pattern of action potentials

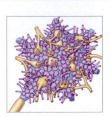

- regulatory neurons
- G proteins
- second messengers

● = *Term boldfaced in this module*

Chapter Integration: Applying what you've learned

Multiple sclerosis (MS) is a progressive, debilitating autoimmune disease in which the body's immune system attacks myelinated portions of the central nervous system, leading to demyelination of affected axons. The disease is so named because *scleroses*—also known as scars, plaques, or lesions—form in many places within myelinated regions (white matter). The age at onset is most commonly between 20 and 40 years. The cause of the disease is unknown, but may involve some combination of environmental agents, genetic factors, and viral infections. Common signs and symptoms include partial loss of vision and problems with speech, balance, and general motor coordination, including loss of bowel and urinary bladder control. The incidence among women is about twice that of men. Individuals with MS experience unpredictable, recurrent cycles of deterioration, remission,

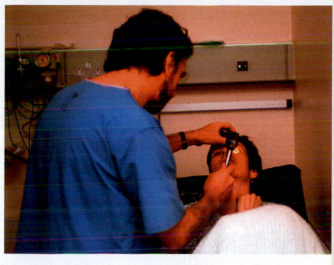

and relapse. There is no cure for MS, although drugs that alter the sensitivity or responses of the immune system can slow the progression of the disease.

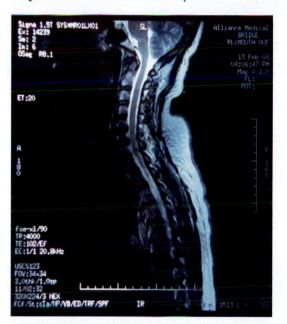

1. Define demyelination.

2. Why would individuals with MS experience generalized motor coordination dysfunction?

3. Which glial cells would be affected in MS?

Access more review material online in the Study Area at **www.masteringaandp.com.**

There, you'll find:
- **Chapter guides**
- **Chapter quizzes**
- **Practice tests**
- **Labeling activities**
- **MP3 Tutor Sessions**
- **Tutorials**
- **A&P Flix:**
  - **Resting Membrane Potential**
  - **Generation of an Action Potential**
  - **Propagation of an Action Potential**
- **Animations**
- **Flashcards**
- **A glossary with pronunciations**

iP® Use *Interactive Physiology*® (IP) to help you understand difficult physiological concepts in this chapter. Go to Nervous System and find the following topics:

- **The Membrane Potential**
- **Ion Channels**
- **The Action Potential**
- **Synaptic Transmission**
- **Synaptic Potentials and Cellular Integration**

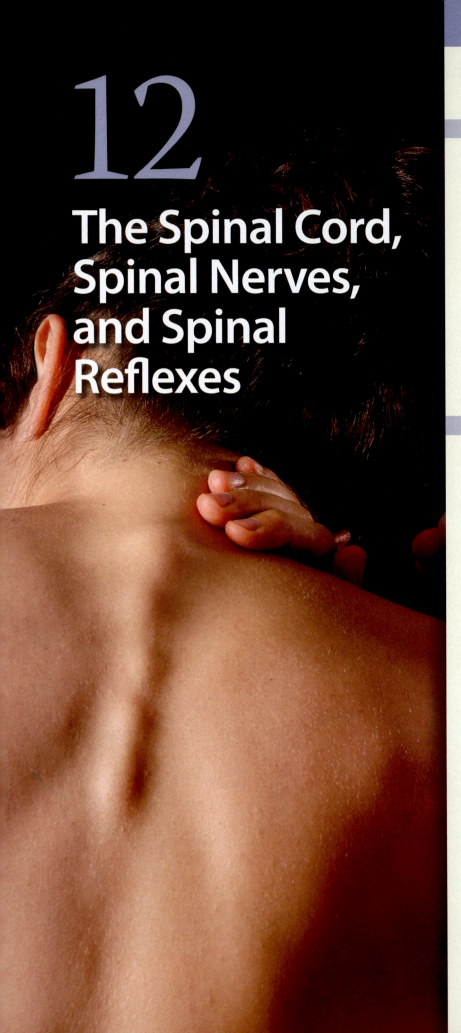

12

The Spinal Cord, Spinal Nerves, and Spinal Reflexes

The Functional Organization of the Spinal Cord

Organization is usually the key to success in any complex environment. In a large corporation, for example, only the most complex and important problems reach the desk of the president. The nervous system works in much the same way: Only the most important information reaches our conscious awareness. We are consciously aware of only a small fraction of the day-to-day activities monitored by sensory receptors and controlled by motor neurons. Throughout our lives, input pathways are routing sensations, processing centers are prioritizing and distributing information, and motor centers are directing responses to stimuli, most often outside of our awareness. This is possible only because the nervous system is so highly organized. Because our primary interest here is how the nervous system functions, we will consider the system from a functional perspective. Our approach is represented in the diagram below.

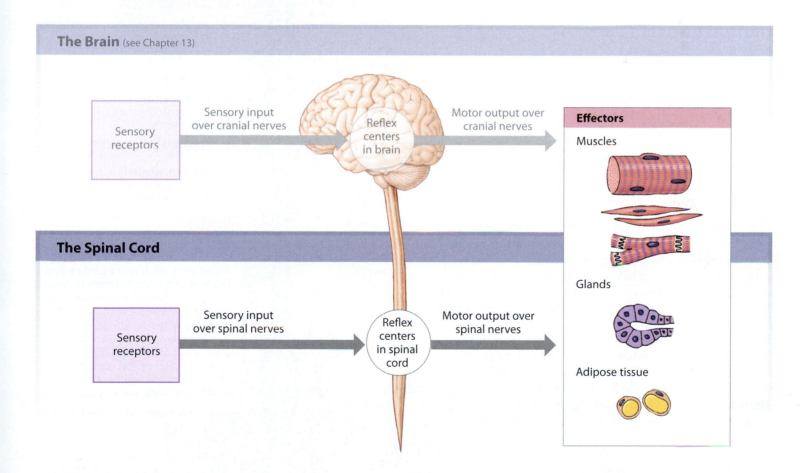

The Brain (see Chapter 13)

Sensory receptors

Sensory input over cranial nerves

Reflex centers in brain

Motor output over cranial nerves

Effectors

Muscles

Glands

Adipose tissue

The Spinal Cord

Sensory receptors

Sensory input over spinal nerves

Reflex centers in spinal cord

Motor output over spinal nerves

We will begin with the **spinal cord**, the simplest part of the CNS, because the functional and structural relationships are relatively easy to understand. Once you are familiar with the basic principles, it will be much easier to sort out the complexities of the brain. This chapter considers the spinal cord, spinal nerves, and spinal reflexes. Chapter 13 examines the brain, cranial nerves, cranial reflexes, and the major pathways that interconnect the brain and spinal cord.

The spinal cord contains gray matter and white matter

The adult spinal cord measures approximately 45 cm (18 in.) in length and has a maximum width of roughly 14 mm (0.55 in.). In sectional view, it has an outer layer of white matter and an inner layer of gray matter surrounding a small central canal.

Cervical spinal nerves

C₁
C₂
C₃
C₄
C₅
C₆
C₇
C₈

The **cervical enlargement** supplies nerves to the shoulder and upper limbs.

T₁
T₂
T₃
T₄
T₅
T₆
T₇
T₈

Thoracic spinal nerves

T₉

Posterior median sulcus

T₁₀

The **lumbar enlargement** provides innervation to structures of the pelvis and lower limbs.

T₁₁
T₁₂

L₁

The **conus medullaris** is the tapered, conical portion of the spinal cord inferior to the lumbar enlargement.

L₂

Inferior tip of spinal cord

1 There are 31 pairs of spinal nerves, each identified by its association with adjacent vertebrae. Each spinal nerve inferior to the first thoracic vertebra takes its name from the vertebra immediately superior to it. Thus, spinal nerve T₁ emerges immediately inferior to vertebra T₁, spinal nerve T₂ follows vertebra T₂, and so forth. The arrangement differs in the cervical region, because the first pair of spinal nerves, C₁, passes between the skull and the first cervical vertebra. For this reason, each cervical nerve takes its name from the vertebra immediately inferior to it. In other words, cervical nerve C₂ precedes vertebra C₂, and the same system is used for the rest of the cervical series. The transition from one numbering system to another occurs between the last cervical vertebra and the first thoracic vertebra. The spinal nerve found at this location has been designated C₈. Therefore, although there are only seven cervical vertebrae, there are eight cervical nerves.

Lumbar spinal nerves

L₃
L₄
L₅

Because the adult spinal cord ends at the level of the first or second lumbar vertebra, the dorsal and ventral roots of spinal segments L₂ to S₅ extend inferiorly. When seen in gross dissection, the filum terminale and the long ventral and dorsal roots resemble a horse's tail. Hence, this complex is called the **cauda equina** (KAW-duh ek-WĪ-nuh; *cauda*, tail + *equus*, horse).

Sacral spinal nerves

S₁
S₂
S₃
S₄
S₅

Coccygeal nerve (Co₁)

The **filum terminale** ("terminal thread") is a slender strand of fibrous tissue that extends from the inferior tip of the conus medullaris to the second sacral vertebra. It provides longitudinal support to the spinal cord as a component of the coccygeal ligament.

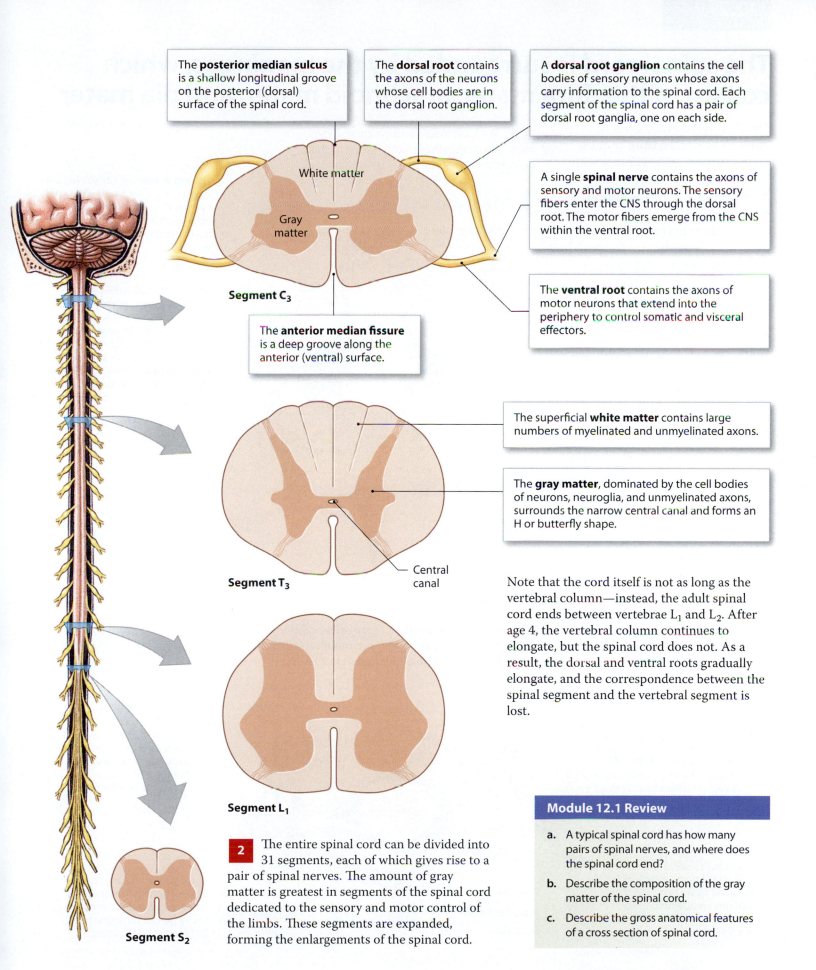

The **posterior median sulcus** is a shallow longitudinal groove on the posterior (dorsal) surface of the spinal cord.

The **dorsal root** contains the axons of the neurons whose cell bodies are in the dorsal root ganglion.

A **dorsal root ganglion** contains the cell bodies of sensory neurons whose axons carry information to the spinal cord. Each segment of the spinal cord has a pair of dorsal root ganglia, one on each side.

White matter

Gray matter

A single **spinal nerve** contains the axons of sensory and motor neurons. The sensory fibers enter the CNS through the dorsal root. The motor fibers emerge from the CNS within the ventral root.

The **ventral root** contains the axons of motor neurons that extend into the periphery to control somatic and visceral effectors.

Segment C$_3$

The **anterior median fissure** is a deep groove along the anterior (ventral) surface.

The superficial **white matter** contains large numbers of myelinated and unmyelinated axons.

The **gray matter**, dominated by the cell bodies of neurons, neuroglia, and unmyelinated axons, surrounds the narrow central canal and forms an H or butterfly shape.

Central canal

Segment T$_3$

Note that the cord itself is not as long as the vertebral column—instead, the adult spinal cord ends between vertebrae L$_1$ and L$_2$. After age 4, the vertebral column continues to elongate, but the spinal cord does not. As a result, the dorsal and ventral roots gradually elongate, and the correspondence between the spinal segment and the vertebral segment is lost.

Segment L$_1$

2 The entire spinal cord can be divided into 31 segments, each of which gives rise to a pair of spinal nerves. The amount of gray matter is greatest in segments of the spinal cord dedicated to the sensory and motor control of the limbs. These segments are expanded, forming the enlargements of the spinal cord.

Segment S$_2$

Module 12.1 Review

a. A typical spinal cord has how many pairs of spinal nerves, and where does the spinal cord end?

b. Describe the composition of the gray matter of the spinal cord.

c. Describe the gross anatomical features of a cross section of spinal cord.

The spinal cord is surrounded by the meninges, which consist of the dura mater, arachnoid mater, and pia mater

The delicate neural tissues must be protected from shocks, including damaging contact with the surrounding bony walls of the vertebral canal. The **spinal meninges** (me-NIN-jēz; singular, *meninx*, membrane), a series of specialized membranes surrounding the spinal cord, provide the necessary physical stability and shock absorption. Blood vessels branching within these layers deliver oxygen and nutrients to the spinal cord. The spinal meninges consist of three layers: (1) the dura mater, (2) the arachnoid mater, and (3) the pia mater. At the foramen magnum of the skull, the spinal meninges are continuous with the cranial meninges, which surround the brain.

1 The basic relationships among the spinal meninges can be seen in this posterior view of the dissected spinal cord.

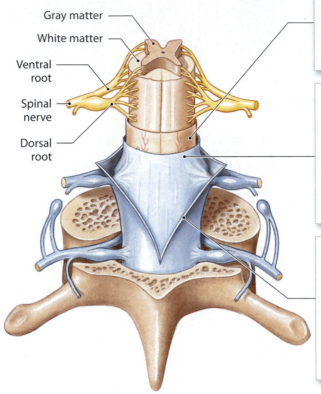

Gray matter
White matter
Ventral root
Spinal nerve
Dorsal root

The **pia mater** (PĒ-uh MĀ-ter; *pia*, delicate + *mater*, mother) consists of a meshwork of elastic and collagen fibers that is firmly bound to the underlying neural tissue.

The **arachnoid** (a-RAK-noyd; *arachne*, spider) **mater** is the middle meningeal layer. It includes a simple squamous epithelium, called the arachnoid membrane, and the **subarachnoid space** that extends between the arachnoid membrane and the outer surface of the pia mater.

The tough, fibrous **dura mater** (DOO-ruh; *dura*, hard) is the outermost covering of the spinal cord. It contains dense collagen fibers that are oriented along the longitudinal axis of the cord. A narrow subdural space separates the dura mater from the arachnoid mater.

2 A cross-sectional view provides additional information about the surrounding structures and the spaces between the meningeal layers.

The subarachnoid space contains the arachnoid trabeculae, a network of collagen and elastic fibers that attaches the arachnoid mater to the pia mater. It is filled with **cerebrospinal fluid (CSF)**, which acts as a shock absorber and a diffusion medium for dissolved gases, nutrients, chemical messengers, and waste products.

Between the dura mater and the walls of the vertebral canal lies the **epidural space**, a region that contains areolar tissue, blood vessels, and a protective padding of adipose tissue.

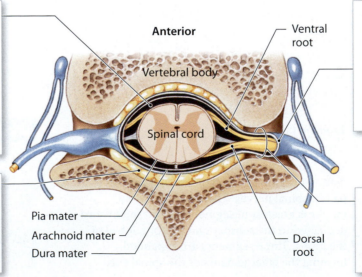

Anterior
Ventral root
Vertebral body
Spinal cord
Pia mater
Arachnoid mater
Dura mater
Dorsal root

The spinal meninges accompany the dorsal and ventral roots as they pass through the intervertebral foramina. The meningeal membranes are continuous with the connective tissues that surround the spinal nerves and their peripheral branches.

The dorsal root ganglion lies between the pedicles of the adjacent vertebrae, and the spinal nerve passes through the intervertebral foramen.

3 This anterior view of the cervical spinal cord shows the actual appearance of the meninges, supporting ligaments, and the roots of the spinal nerves.

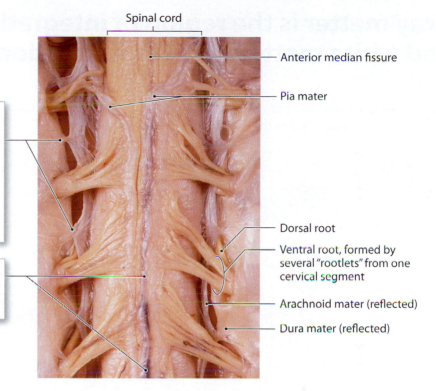

Spinal cord

Anterior median fissure

Pia mater

Along the length of the spinal cord, paired **denticulate ligaments** extend from the pia mater through the arachnoid mater to the dura mater. Denticulate ligaments prevent lateral movement of the spinal cord. Together, the dural connections at the foramen magnum and the coccygeal ligament at the sacrum prevent superior–inferior movement of the spinal cord.

The blood vessels servicing the spinal cord run along the surface of the spinal pia mater, within the subarachnoid space.

Dorsal root

Ventral root, formed by several "rootlets" from one cervical segment

Arachnoid mater (reflected)

Dura mater (reflected)

4 This is an x-ray of the inferior lumbar vertebrae. The vertebral canal inferior to vertebra L_1 contains the cauda equina bathed in the CSF of the subarachnoid space.

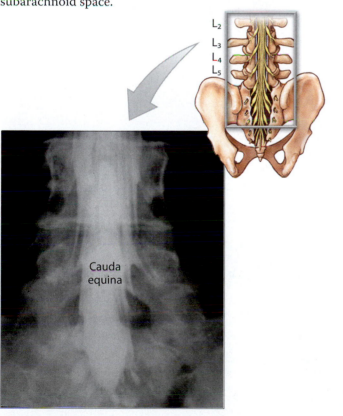

L_2
L_3
L_4
L_5

Cauda equina

5 In adults, cerebrospinal fluid can be safely withdrawn in a procedure known as a **lumbar puncture** or spinal tap. A needle is inserted into the subarachnoid space in the lumbar region inferior to the tip of the conus medullaris.

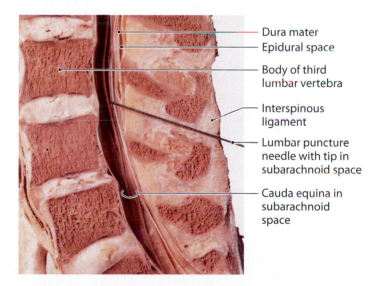

Dura mater
Epidural space

Body of third lumbar vertebra

Interspinous ligament

Lumbar puncture needle with tip in subarachnoid space

Cauda equina in subarachnoid space

Module 12.2 Review

a. Identify and describe the three spinal meninges.

b. Where is the cerebrospinal fluid that surrounds the spinal cord located?

c. Name the structures and spinal coverings that would be penetrated during a lumbar puncture procedure.

Gray matter is the region of integration, and white matter carries information

1 This cross section shows most of the anatomical landmarks of the spinal cord; compare this view to the diagrammatic views in this module.

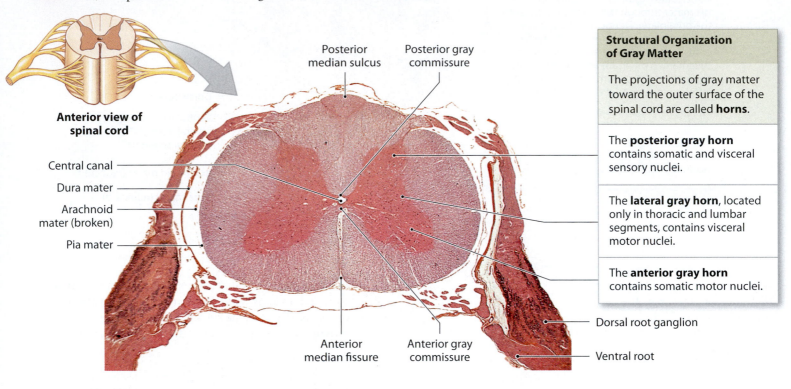

Anterior view of spinal cord

Posterior median sulcus

Posterior gray commissure

Central canal

Dura mater

Arachnoid mater (broken)

Pia mater

Anterior median fissure

Anterior gray commissure

Dorsal root ganglion

Ventral root

Structural Organization of Gray Matter

The projections of gray matter toward the outer surface of the spinal cord are called **horns**.

The **posterior gray horn** contains somatic and visceral sensory nuclei.

The **lateral gray horn**, located only in thoracic and lumbar segments, contains visceral motor nuclei.

The **anterior gray horn** contains somatic motor nuclei.

2 This diagrammatic view focuses on the organization of the gray matter of the spinal cord.

A frontal section along the length of the central canal (dashed line) of the spinal cord separates the sensory (posterior, or dorsal) nuclei from the motor (anterior, or ventral) nuclei.

The **gray commissures** (*commissura*, a joining together) are posterior to and anterior to the central canal. They contain axons that cross from one side of the cord to the other before they reach a destination in the gray matter.

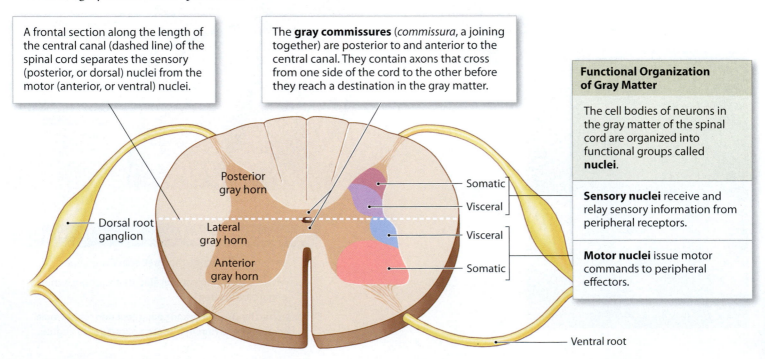

Posterior gray horn

Dorsal root ganglion

Lateral gray horn

Anterior gray horn

Somatic

Visceral

Visceral

Somatic

Ventral root

Functional Organization of Gray Matter

The cell bodies of neurons in the gray matter of the spinal cord are organized into functional groups called **nuclei**.

Sensory nuclei receive and relay sensory information from peripheral receptors.

Motor nuclei issue motor commands to peripheral effectors.

3 Like the gray horns, the white matter is organized according to the region of the body innervated. The white matter on each side of the spinal cord can be divided into three regions called **columns**. These columns contain **tracts**. A tract is a bundle of axons in the CNS that is relatively uniform with respect to diameter, myelination, and conduction speed. All the axons within a tract relay the same type of information (sensory or motor) in the same direction. **Ascending tracts** carry sensory information toward the brain, and **descending tracts** convey motor commands to the spinal cord.

Structural and Functional Organization of White Matter

The **posterior white column** lies between the posterior gray horns and the posterior median sulcus.

The **lateral white column** includes the white matter on either side of the spinal cord, between the anterior and posterior columns.

The **anterior white column** lies between the anterior gray horns and the anterior median fissure.

Organization of Tracts in the Posterior White Column

The posterior white column contains ascending tracts providing sensations from the trunk and limbs.

- Leg
- Hip
- Trunk
- Arm

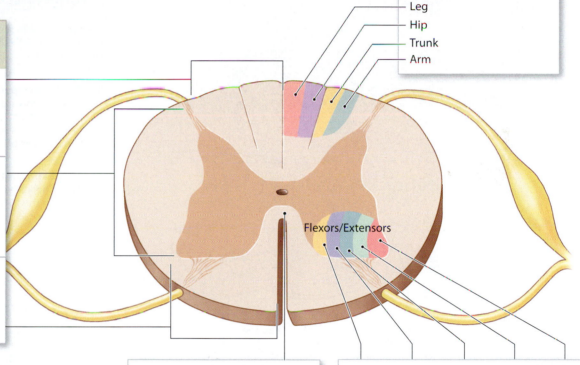

Flexors/Extensors

The **anterior white commissure** interconnects the anterior white columns. This is where axons cross from one side of the spinal cord to the other.

Trunk   Shoulder   Arm   Forearm   Hand

In the cervical enlargement, which contains neurons involved with sensations and motor control of the upper limbs, the motor nuclei of the anterior gray horn are grouped by region, with motor neurons controlling flexor muscles medial to those controlling extensor muscles.

Module 12.3 Review

a. Differentiate between sensory nuclei and motor nuclei.

b. A person with polio has lost the use of his leg muscles. In which area of his spinal cord would you expect the virus-infected motor neurons to be?

c. A disease that damages myelin sheaths would affect which portion of the spinal cord?

Spinal nerves have a relatively consistent anatomical structure and pattern of distribution

1 Every segment of the spinal cord is connected to a pair of spinal nerves. Surrounding each spinal nerve is a series of connective tissue layers continuous with those of their associated peripheral nerves. These layers, best seen in sectional view, are comparable to those associated with skeletal muscles.

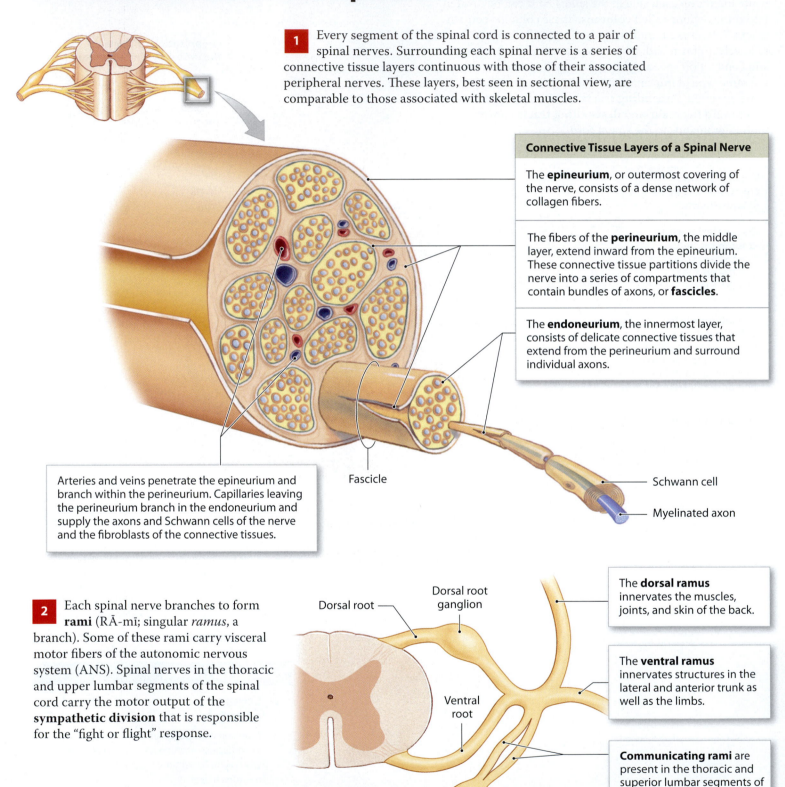

Connective Tissue Layers of a Spinal Nerve

The **epineurium**, or outermost covering of the nerve, consists of a dense network of collagen fibers.

The fibers of the **perineurium**, the middle layer, extend inward from the epineurium. These connective tissue partitions divide the nerve into a series of compartments that contain bundles of axons, or **fascicles**.

The **endoneurium**, the innermost layer, consists of delicate connective tissues that extend from the perineurium and surround individual axons.

Arteries and veins penetrate the epineurium and branch within the perineurium. Capillaries leaving the perineurium branch in the endoneurium and supply the axons and Schwann cells of the nerve and the fibroblasts of the connective tissues.

Fascicle

Schwann cell

Myelinated axon

2 Each spinal nerve branches to form **rami** (RĀ-mī; singular *ramus*, a branch). Some of these rami carry visceral motor fibers of the autonomic nervous system (ANS). Spinal nerves in the thoracic and upper lumbar segments of the spinal cord carry the motor output of the **sympathetic division** that is responsible for the "fight or flight" response.

Dorsal root

Dorsal root ganglion

The **dorsal ramus** innervates the muscles, joints, and skin of the back.

Ventral root

The **ventral ramus** innervates structures in the lateral and anterior trunk as well as the limbs.

Communicating rami are present in the thoracic and superior lumbar segments of the spinal cord. These rami contain the axons of sympathetic neurons.

Autonomic nerve

Sympathetic ganglion

3 The specific bilateral region of the skin surface monitored by a single pair of spinal nerves is known as a **dermatome**. Each pair of spinal nerves services its own dermatome, but the boundaries of adjacent dermatomes overlap to some degree. Spinal nerve C_1 typically lacks a sensory branch to the skin; when present, it innervates the scalp with C_2 and C_3. The face is monitored by a pair of cranial nerves.

4 Dermatomes are clinically important because damage or infection of a spinal nerve or dorsal root ganglion produces a characteristic loss of sensation in the corresponding region of the skin. Additionally, characteristic signs may appear on the skin supplied by that specific nerve. The skin eruptions shown here are characteristic of **shingles**, a viral infection of dorsal root ganglia. Shingles (derived from the Latin *cingulum*, girdle) is caused by the varicella-zoster virus (VZV), the same herpes virus that causes chickenpox. This herpes virus attacks neurons within the dorsal roots of spinal nerves and sensory ganglia of cranial nerves. This disorder produces a painful rash and blisters whose distribution corresponds to that of the affected sensory nerve and its associated dermatome. Any person who has had chickenpox is at risk of developing shingles, because the virus can remain dormant within the anterior gray horns of the spinal cord. It is not known what triggers the reactivation of the virus. In 2006, the U.S. Food and Drug Administration approved a VZV vaccine (Zostavax) for use in people ages 60 and above who have had chickenpox.

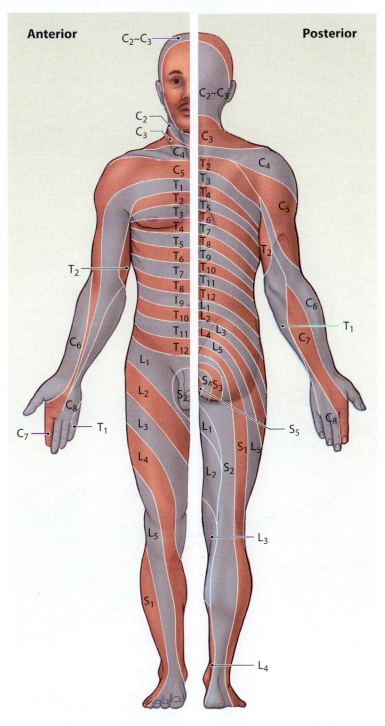

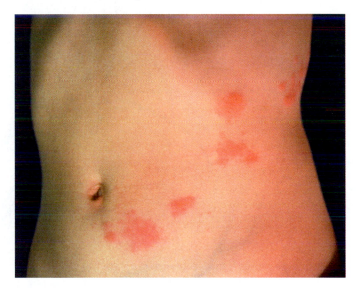

Module 12.4 Review

a. Identify the three layers of connective tissue of a spinal nerve and identify the major peripheral branches of a spinal nerve.

b. Describe a dermatome.

c. Explain the etiology (cause) of shingles.

Each ramus (branch) of a spinal nerve provides motor and sensory innervation to a specific region

1 This diagram illustrates how a spinal nerve distributes motor commands that originate in motor nuclei of the thoracic or superior lumbar segments of the spinal cord.

Motor Commands

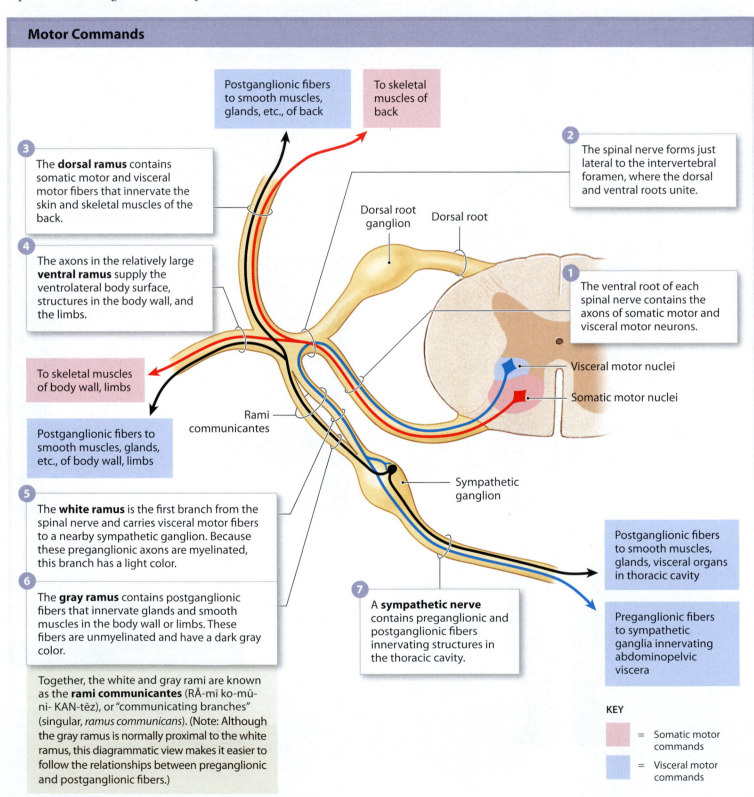

Postganglionic fibers to smooth muscles, glands, etc., of back

To skeletal muscles of back

3 The **dorsal ramus** contains somatic motor and visceral motor fibers that innervate the skin and skeletal muscles of the back.

2 The spinal nerve forms just lateral to the intervertebral foramen, where the dorsal and ventral roots unite.

Dorsal root ganglion   Dorsal root

4 The axons in the relatively large **ventral ramus** supply the ventrolateral body surface, structures in the body wall, and the limbs.

1 The ventral root of each spinal nerve contains the axons of somatic motor and visceral motor neurons.

Visceral motor nuclei

Somatic motor nuclei

To skeletal muscles of body wall, limbs

Postganglionic fibers to smooth muscles, glands, etc., of body wall, limbs

Rami communicantes

Sympathetic ganglion

5 The **white ramus** is the first branch from the spinal nerve and carries visceral motor fibers to a nearby sympathetic ganglion. Because these preganglionic axons are myelinated, this branch has a light color.

6 The **gray ramus** contains postganglionic fibers that innervate glands and smooth muscles in the body wall or limbs. These fibers are unmyelinated and have a dark gray color.

7 A **sympathetic nerve** contains preganglionic and postganglionic fibers innervating structures in the thoracic cavity.

Postganglionic fibers to smooth muscles, glands, visceral organs in thoracic cavity

Preganglionic fibers to sympathetic ganglia innervating abdominopelvic viscera

Together, the white and gray rami are known as the **rami communicantes** (RĂ-mī ko-mū-ni- KAN-tēz), or "communicating branches" (singular, *ramus communicans*). (Note: Although the gray ramus is normally proximal to the white ramus, this diagrammatic view makes it easier to follow the relationships between preganglionic and postganglionic fibers.)

KEY
= Somatic motor commands
= Visceral motor commands

2 This diagram illustrates how a spinal nerve collects sensory information from peripheral structures and delivers it to sensory nuclei in the thoracic or superior lumbar segments of the spinal cord. The dorsal, ventral, and white rami also contain sensory fibers. Somatic sensory information arrives over the dorsal and ventral rami; visceral sensory information reaches the dorsal root through the dorsal, ventral, and white rami.

Sensory Information

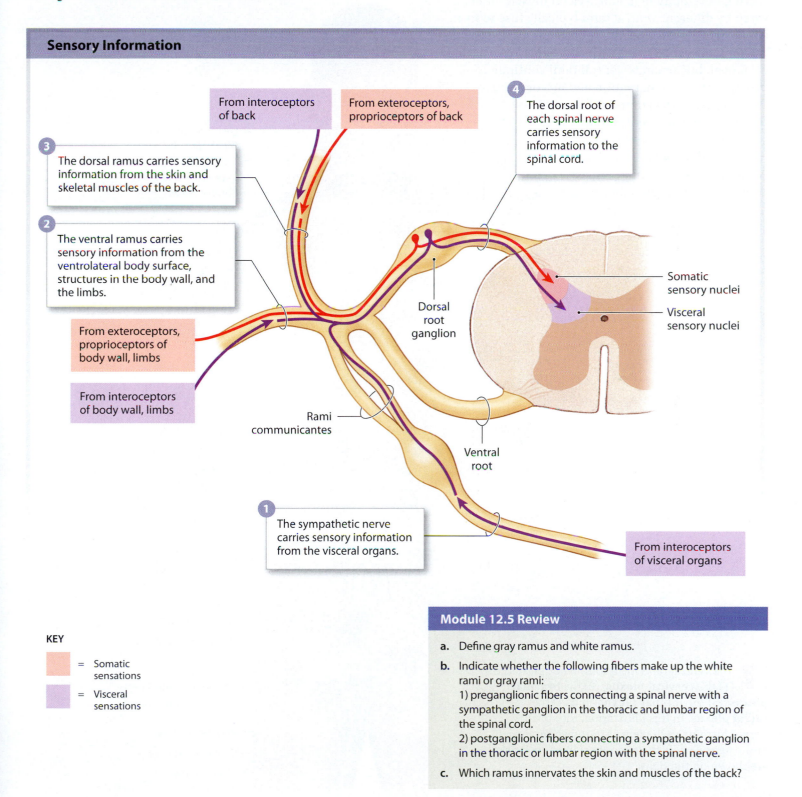

From interoceptors of back

From exteroceptors, proprioceptors of back

4 The dorsal root of each spinal nerve carries sensory information to the spinal cord.

3 The dorsal ramus carries sensory information from the skin and skeletal muscles of the back.

2 The ventral ramus carries sensory information from the ventrolateral body surface, structures in the body wall, and the limbs.

From exteroceptors, proprioceptors of body wall, limbs

From interoceptors of body wall, limbs

Dorsal root ganglion

Somatic sensory nuclei

Visceral sensory nuclei

Rami communicantes

Ventral root

1 The sympathetic nerve carries sensory information from the visceral organs.

From interoceptors of visceral organs

KEY

= Somatic sensations

= Visceral sensations

Module 12.5 Review

a. Define gray ramus and white ramus.

b. Indicate whether the following fibers make up the white rami or gray rami:
1) preganglionic fibers connecting a spinal nerve with a sympathetic ganglion in the thoracic and lumbar region of the spinal cord.
2) postganglionic fibers connecting a sympathetic ganglion in the thoracic or lumbar region with the spinal nerve.

c. Which ramus innervates the skin and muscles of the back?

Spinal nerves form nerve plexuses that innervate the skin and skeletal muscles; the cervical plexus is the smallest of these nerve plexuses

During development, small skeletal muscles innervated by different ventral rami typically fuse to form larger muscles with compound origins. The anatomical distinctions between the component muscles may disappear, but separate ventral rami continue to provide sensory innervation and motor control to each part of the compound muscle. As they converge, the ventral rami of adjacent spinal nerves blend their fibers, producing a series of compound nerve trunks. Such a complex interwoven network of nerves is called a **nerve plexus** (PLEK-sus; *plexus*, braid).

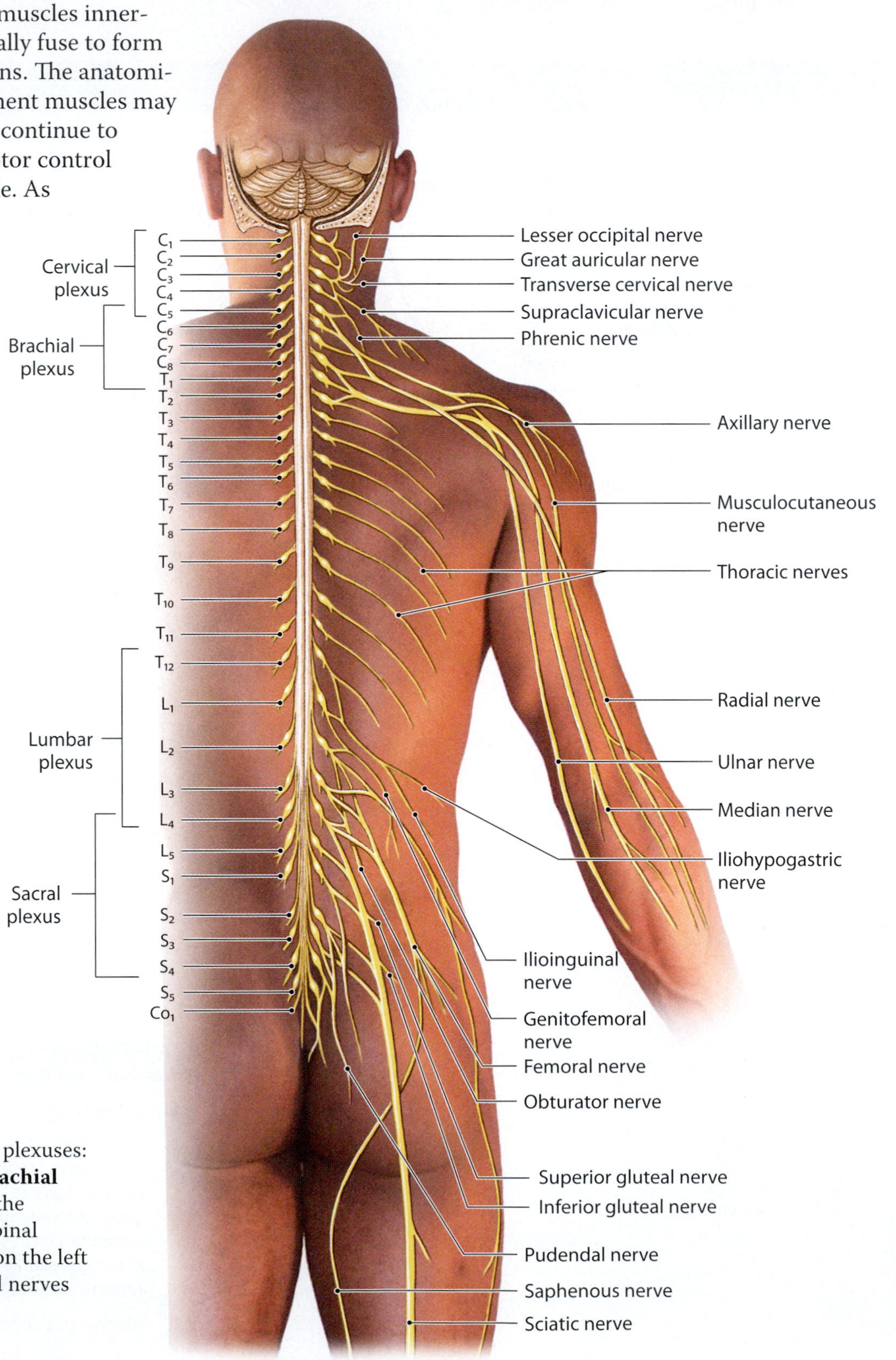

1 The ventral rami form four major plexuses: (1) the **cervical plexus**, (2) the **brachial plexus**, (3) the **lumbar plexus**, and (4) the **sacral plexus**. In this illustration, the spinal nerves forming each plexus are shown on the left of the figure, while the major peripheral nerves are shown on the right.

2 The cervical plexus consists of the ventral rami of spinal nerves C_1–C_5. The branches of the cervical plexus innervate the muscles of the neck and extend into the thoracic cavity, where they control the diaphragmatic muscles. The **phrenic nerve**, the major nerve of the cervical plexus, provides the entire nerve supply to the diaphragm, a key respiratory muscle. Other branches of this nerve plexus are distributed to the skin of the neck and the superior part of the chest.

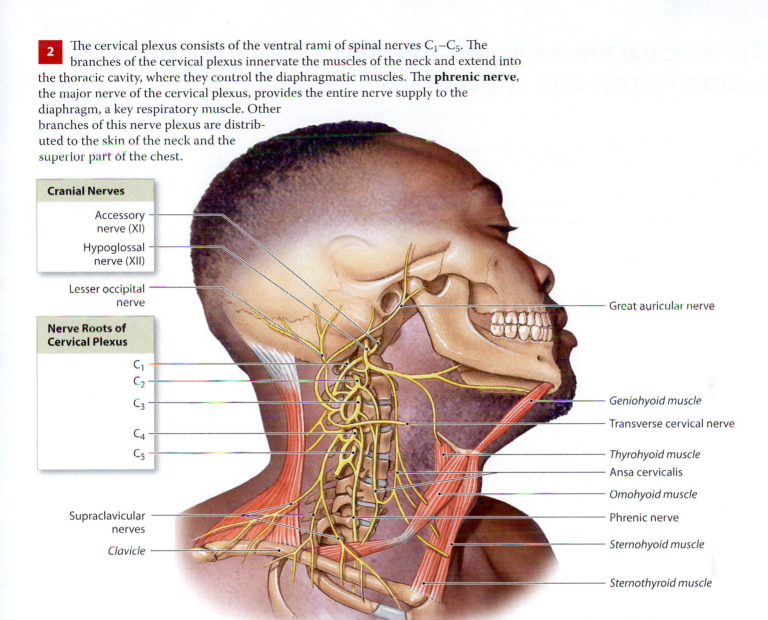

Cranial Nerves
Accessory nerve (XI)
Hypoglossal nerve (XII)

Lesser occipital nerve

Nerve Roots of Cervical Plexus
C_1
C_2
C_3
C_4
C_5

Supraclavicular nerves
Clavicle

Great auricular nerve

Geniohyoid muscle
Transverse cervical nerve

Thyrohyoid muscle
Ansa cervicalis
Omohyoid muscle
Phrenic nerve
Sternohyoid muscle

Sternothyroid muscle

The Cervical Plexus

Nerve(s)	Spinal Segments	Distribution
Ansa cervicalis (superior and inferior branches)	C_1–C_4	Five of the extrinsic laryngeal muscles: sternothyroid, sternohyoid, omohyoid, geniohyoid, and thyrohyoid muscles (via cranial nerve XII)
Lesser occipital, transverse cervical, supraclavicular, and great auricular nerves	C_2–C_3	Skin of upper chest, shoulder, neck, and ear
Phrenic nerve	C_3–C_5	Diaphragm
Cervical nerves	C_1–C_5	Levator scapulae, scalene, sternocleidomastoid, and trapezius muscles (with cranial nerve XI)

Module 12.6 Review

a. Define nerve plexus, and list the major nerve plexuses.

b. An anesthetic blocks the function of the dorsal rami of the cervical spinal nerves. Which areas of the body will be affected?

c. Injury to which of the nerve plexuses would interfere with the ability to breathe?

The brachial plexus innervates the pectoral girdle and upper limbs

1 The brachial plexus innervates the pectoral girdle and upper limbs, with contributions from the ventral rami of spinal nerves C_4–T_1.

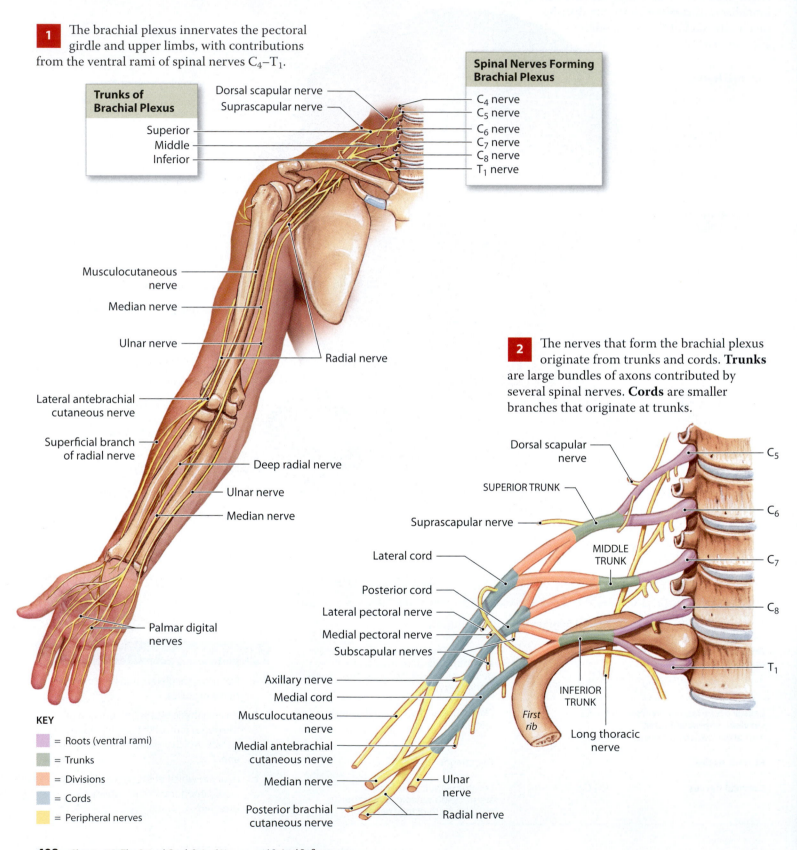

Trunks of Brachial Plexus

Dorsal scapular nerve
Suprascapular nerve

Superior
Middle
Inferior

Spinal Nerves Forming Brachial Plexus

C_4 nerve
C_5 nerve
C_6 nerve
C_7 nerve
C_8 nerve
T_1 nerve

Musculocutaneous nerve

Median nerve

Ulnar nerve

Radial nerve

Lateral antebrachial cutaneous nerve

Superficial branch of radial nerve

Deep radial nerve

Ulnar nerve

Median nerve

Palmar digital nerves

KEY

= Roots (ventral rami)

= Trunks

= Divisions

= Cords

= Peripheral nerves

2 The nerves that form the brachial plexus originate from trunks and cords. **Trunks** are large bundles of axons contributed by several spinal nerves. **Cords** are smaller branches that originate at trunks.

Dorsal scapular nerve

SUPERIOR TRUNK

Suprascapular nerve

Lateral cord

Posterior cord

Lateral pectoral nerve

Medial pectoral nerve

Subscapular nerves

Axillary nerve

Medial cord

Musculocutaneous nerve

Medial antebrachial cutaneous nerve

Median nerve

Posterior brachial cutaneous nerve

C_5

C_6

MIDDLE TRUNK

C_7

C_8

INFERIOR TRUNK

First rib

Long thoracic nerve

T_1

Ulnar nerve

Radial nerve

3 The distribution of the cutaneous nerves of the wrist and hand is very important in clinical medicine. Nerve damage or injury in this region can be precisely localized by carefully testing the sensory function of the hand.

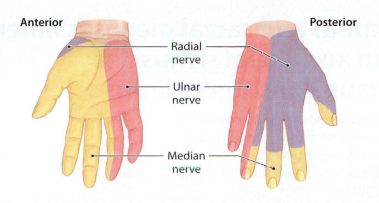

Anterior — Posterior

Radial nerve
Ulnar nerve
Median nerve

The Brachial Plexus

Nerve	Spinal Segments	Distribution
Nerve to subclavius	C_4–C_6	Subclavius muscle
Dorsal scapular nerve	C_5	Rhomboid and levator scapulae muscles
Long thoracic nerve	C_5–C_7	Serratus anterior muscle
Suprascapular nerve	C_5, C_6	Supraspinatus and infraspinatus muscles; sensory from shoulder joint and scapula
Pectoral nerves (medial and lateral)	C_5–T_1	Pectoralis muscles
Subscapular nerves	C_5, C_6	Subscapularis and teres major muscles
Thoracodorsal nerve	C_6–C_8	Latissimus dorsi muscle
Axillary nerve	C_5, C_6	Deltoid and teres minor muscles; sensory from the skin of the shoulder
Medial antebrachial cutaneous nerve	C_8, T_1	Sensory from skin over anterior, medial surface of arm and forearm
Radial nerve	C_5–T_1	Many extensor muscles on the arm and forearm (triceps brachii, anconeus, extensor carpi radialis, extensor carpi ulnaris, and brachioradialis muscles); supinator muscle, digital extensor muscles, and abductor pollicis muscle via the deep branch; sensory from skin over the posterolateral surface of the limb through the posterior brachial cutaneous nerve (arm), posterior antebrachial cutaneous nerve (forearm), and the superficial branch (radial half of hand)
Musculocutaneous nerve	C_5–T_1	Flexor muscles on the arm (biceps brachii, brachialis, and coracobrachialis muscles); sensory from skin over lateral surface of the forearm through the lateral antebrachial cutaneous nerve
Median nerve	C_6–T_1	Flexor muscles on the forearm (flexor carpi radialis and palmaris longus muscles); pronator quadratus and pronator teres muscles; digital flexors (through the anterior interosseous nerve); sensory from skin over anterolateral surface of the hand
Ulnar nerve	C_8, T_1	Flexor carpi ulnaris muscle, flexor digitorum profundus muscle, adductor pollicis muscle, and small digital muscles via the deep branch; sensory from skin over medial surface of the hand through the superficial branch

Module 12.7 Review

a. Define a nerve plexus trunk and cord.

b. Describe the brachial plexus.

c. Name the major nerves associated with the brachial plexus.

The lumbar and sacral plexuses innervate the skin and skeletal muscles of the trunk and lower limbs

The **lumbar plexus** and the **sacral plexus** arise from the lumbar and sacral segments of the spinal cord, respectively. The nerves arising at these plexuses innervate the pelvic girdle and lower limbs.

1 These illustrations show the origins of the spinal nerves of the lumbar and sacral plexuses. Notice that the lumbosacral trunk provides some nerve input from the L_4 spinal nerve to the sacral plexus.

Spinal Nerves Forming the Lumbar Plexus

- T_{12} nerve
- L_1 nerve
- L_2 nerve
- L_3 nerve
- L_4 nerve

Nerves of the Lumbar Plexus

- Iliohypogastric
- Ilioinguinal
- Genitofemoral
- Lateral femoral cutaneous
- Femoral
- Obturator

L_5

Lumbosacral trunk

Lumbar plexus, anterior view

Spinal Nerves Forming the Sacral Plexus

- L_4 nerve
- L_5 nerve
- S_1 nerve
- S_2 nerve
- S_3 nerve
- S_4 nerve

S_5

Co_1

Lumbosacral trunk

Nerves of the Sacral Plexus

- Superior gluteal
- Inferior gluteal
- Sciatic
- Posterior femoral cutaneous
- Pudendal

Sacral plexus, anterior view

2 This posterior view of the lower limb shows the distribution of the nerves of the sacral plexus. The **sciatic nerve** is the largest and longest nerve in the body.

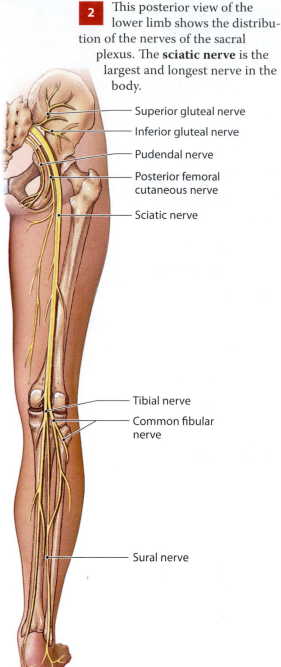

- Superior gluteal nerve
- Inferior gluteal nerve
- Pudendal nerve
- Posterior femoral cutaneous nerve
- Sciatic nerve
- Tibial nerve
- Common fibular nerve
- Sural nerve

3 These illustrations show the dermatomes of the sensory nerves innervating the ankle and foot.

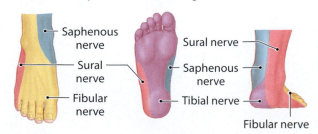

- Saphenous nerve
- Sural nerve
- Sural nerve
- Saphenous nerve
- Fibular nerve
- Tibial nerve
- Fibular nerve

4 In this anterior view of the lower trunk and lower limb, you can see the distribution of the nerves of both the lumbar and sacral plexuses.

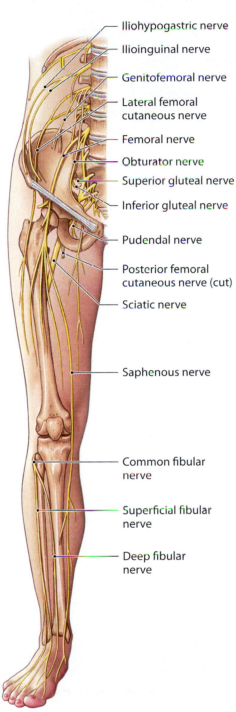

- Iliohypogastric nerve
- Ilioinguinal nerve
- Genitofemoral nerve
- Lateral femoral cutaneous nerve
- Femoral nerve
- Obturator nerve
- Superior gluteal nerve
- Inferior gluteal nerve
- Pudendal nerve
- Posterior femoral cutaneous nerve (cut)
- Sciatic nerve
- Saphenous nerve
- Common fibular nerve
- Superficial fibular nerve
- Deep fibular nerve

The Lumbar and Sacral Plexuses

Nerve	Spinal Segment(s)	Distribution
Lumbar Plexus		
Iliohypogastric nerve	T_{12}, L_1	Abdominal muscles (external and internal oblique muscles, transversus abdominis muscle); skin over inferior abdomen and buttocks
Ilioinguinal nerve	L_1	Abdominal muscles (with iliohypogastric nerve); skin over superior, medial thigh and portions of external genitalia
Genitofemoral nerve	L_1, L_2	Skin over anteromedial surface of thigh and portions of external genitalia
Lateral femoral cutaneous nerve	L_2, L_3	Skin over anterior, lateral, and posterior surfaces of thigh
Femoral nerve	L_2–L_4	Anterior muscles of thigh (sartorius muscle and quadriceps group); flexors and adductors of hip (pectineus and iliopsoas muscles); skin over anteromedial surface of thigh, medial surface of leg and foot
Obturator nerve	L_2–L_4	Adductors of hip (adductors magnus, brevis, and longus muscles); gracilis muscle; skin over medial surface of thigh
Saphenous nerve	L_2–L_4	Skin over medial surface of leg
Sacral Plexus		
Gluteal nerves:	L_4–S_2	
Superior		Abductors of hip (gluteus minimus, gluteus medius, and tensor fasciae latae muscles)
Inferior		Extensor of hip (gluteus maximus muscle)
Posterior femoral cutaneous nerve	S_1–S_3	Skin of perineum and posterior surfaces of thigh and leg
Sciatic nerve:	L_4–S_3	Two of the hamstrings (semimembranosus and semitendinosus muscles); adductor magnus muscle (with obturator nerve)
Tibial nerve		Flexors of knee and extensors (plantar flexors) of ankle (popliteus, gastrocnemius, soleus, and tibialis posterior muscles, and the long head of the biceps femoris muscle); flexors of toes; skin over posterior surface of leg, plantar surface of foot
Fibular nerve		Biceps femoris muscle (short head); fibularis muscles (brevis and longus) and tibialis anterior muscle; extensors of toes; skin over anterior surface of leg and dorsal surface of foot; skin over lateral portion of foot (through the sural nerve)
Pudendal nerve	S_2–S_4	Muscles of perineum, including urogenital diaphragm and external anal and urethral sphincter muscles; skin of external genitalia and related skeletal muscles (bulbospongiosus and ischiocavernosus muscles)

Module 12.8 Review

a. Describe the lumbar plexus and sacral plexus.

b. List the major nerves of the sacral plexus.

c. Compression of which nerve produces the sensation that your lower limb has "fallen asleep"?

1. Labeling

Label each of the structures in the following cross-sectional diagram of the spinal cord.

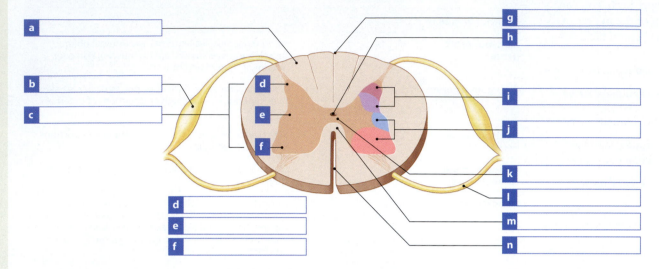

a		g
b		h
c		i
	d	j
	e	k
	f	l
d		m
e		n
f		

2. Labeling

Label the views (a, e) and the nerves that innervate the indicated regions of the hands (b–d).

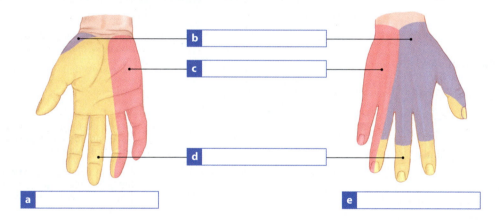

3. Vocabulary

Write the boldfaced term introduced in this section that matches each of the following definitions.

a _____ The regions of white matter in the spinal cord

b _____ The terminal portion of the spinal cord

c _____ Bundles of axons in the PNS plus their associated blood vessels and connective tissues

d _____ Specialized membranes that provide stability and support for the spinal cord and brain

e _____ The complex made up of the filum terminale and the long dorsal and ventral spinal nerve roots inferior to the spinal cord

f _____ The plexus from which the radial, median, and ulnar nerves originate

g _____ The outermost covering of the spinal cord

h _____ The connective tissue partition that separates adjacent bundles of nerve fibers in a spinal nerve or a peripheral nerve

i _____ A bundle of unmyelinated, postganglionic fibers that innervates glands and smooth muscles in the body wall or limbs

An Introduction to Reflexes

The human body has about 10 million sensory neurons, one-half million motor neurons, and 20 billion interneurons. The interneurons of the CNS are organized into a much smaller number of **neuronal pools**—functional groups of interconnected neurons. A neuronal pool may involve neurons in several regions of the brain, or the neurons may be in one specific location in the brain or spinal cord. Estimates of the number of neuronal pools range between a few hundred and a few thousand. The pattern of interaction among neurons provides clues to the functional characteristics of a neuronal pool. It is customary to refer to the "wiring diagram" as a **neural circuit**, just as we refer to electrical circuits in the wiring of a house. Here are a few common circuit patterns.

Divergence

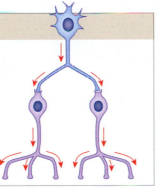

Divergence is the spread of information from one neuron to several neurons, or from one pool to multiple pools. Divergence permits the broad distribution of a specific input. Considerable divergence occurs when sensory neurons bring information into the CNS: The information is distributed to neuronal pools throughout the spinal cord and brain.

Parallel Processing

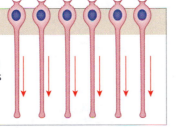

Parallel processing occurs when several neurons or neuronal pools process the same information simultaneously. Many responses can occur simultaneously.

Serial Processing

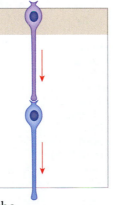

In **serial processing**, information is relayed in a stepwise fashion, from one neuron to another or from one neuronal pool to the next. This pattern occurs as sensory information is relayed from one part of the brain to another.

The most complex neural processing occurs in the brain. The simplest circuits, which occur within the PNS and the spinal cord, control **reflexes**—preprogrammed responses to specific stimuli. Reflexes are a bit like Legos: Individually, they are quite simple, but they can be combined in a great variety of ways to create very complex motor responses.

Convergence

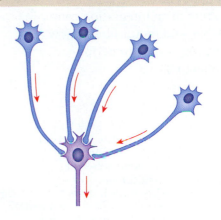

In **convergence**, several neurons synapse on a single postsynaptic neuron. Several patterns of activity in the presynaptic neurons can therefore have the same effect on the postsynaptic neuron. Through convergence, the same motor neurons can be subject to both conscious and subconscious control. For example, the movements of your diaphragm and ribs are now being controlled by your brain at the subconscious level. But the same motor neurons can also be controlled consciously, as when you take a deep breath and hold it.

Reverberation

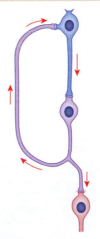

In **reverberation**, collateral branches of axons somewhere along the circuit extend back toward the source of an impulse and further stimulate the presynaptic neurons. Reverberation is like a positive feedback loop involving neurons: Once a reverberating circuit has been activated, it will continue to function until synaptic fatigue or inhibitory stimuli break the cycle.

Reflexes are rapid, automatic responses to stimuli

Reflexes are rapid, automatic responses to specific stimuli. Reflexes preserve homeostasis by making rapid adjustments in the function of organs or organ systems. The response shows little variability: Each time a particular reflex is activated, it usually produces the same motor response. In neural reflexes, sensory fibers deliver information from peripheral receptors to an integration center in the CNS, and motor fibers carry motor commands to peripheral effectors.

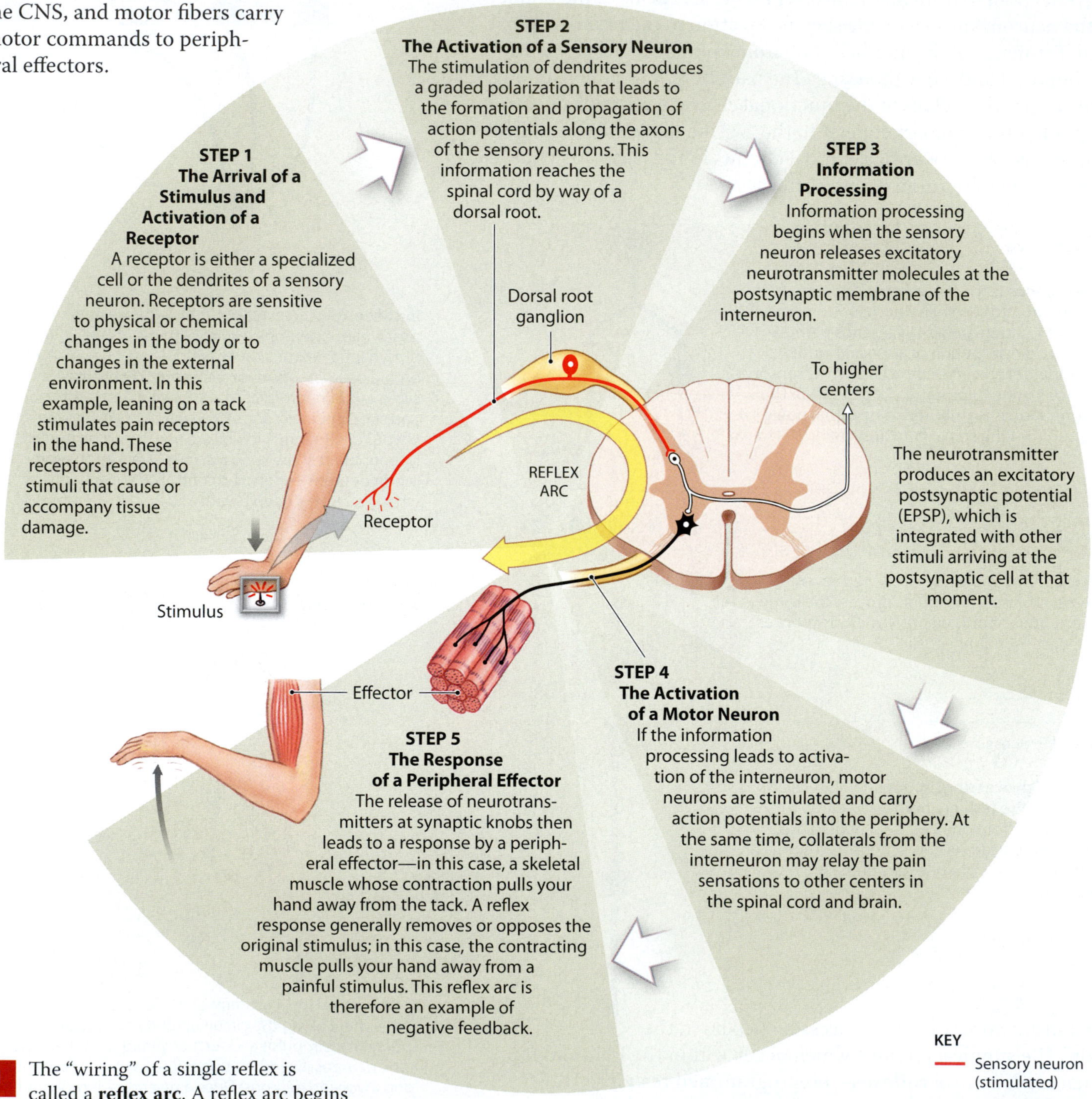

STEP 1
The Arrival of a Stimulus and Activation of a Receptor
A receptor is either a specialized cell or the dendrites of a sensory neuron. Receptors are sensitive to physical or chemical changes in the body or to changes in the external environment. In this example, leaning on a tack stimulates pain receptors in the hand. These receptors respond to stimuli that cause or accompany tissue damage.

STEP 2
The Activation of a Sensory Neuron
The stimulation of dendrites produces a graded polarization that leads to the formation and propagation of action potentials along the axons of the sensory neurons. This information reaches the spinal cord by way of a dorsal root.

STEP 3
Information Processing
Information processing begins when the sensory neuron releases excitatory neurotransmitter molecules at the postsynaptic membrane of the interneuron.

The neurotransmitter produces an excitatory postsynaptic potential (EPSP), which is integrated with other stimuli arriving at the postsynaptic cell at that moment.

STEP 4
The Activation of a Motor Neuron
If the information processing leads to activation of the interneuron, motor neurons are stimulated and carry action potentials into the periphery. At the same time, collaterals from the interneuron may relay the pain sensations to other centers in the spinal cord and brain.

STEP 5
The Response of a Peripheral Effector
The release of neurotransmitters at synaptic knobs then leads to a response by a peripheral effector—in this case, a skeletal muscle whose contraction pulls your hand away from the tack. A reflex response generally removes or opposes the original stimulus; in this case, the contracting muscle pulls your hand away from a painful stimulus. This reflex arc is therefore an example of negative feedback.

Dorsal root ganglion

To higher centers

REFLEX ARC

Receptor

Stimulus

Effector

KEY
— Sensory neuron (stimulated)
═ Excitatory interneuron
— Motor neuron (stimulated)

1 The "wiring" of a single reflex is called a **reflex arc**. A reflex arc begins at a receptor and ends at a peripheral effector, such as a muscle fiber or a gland cell. This figure shows the steps in one example: a simple withdrawal reflex.

Reflexes Can Be Classified by

Development

Innate reflexes result from the connections that form between neurons during development and are genetically or developmentally programmed. Such reflexes generally appear in a predictable sequence, from the simplest reflex responses (withdrawal from pain) to more complex motor patterns (chewing, suckling, or tracking objects with the eyes).

Acquired reflexes are rapid and automatic, but they were learned rather than preestablished. Such reflexes are enhanced by repetition.

The Nature of the Response

Somatic reflexes provide a mechanism for the involuntary control of the skeletal muscles. (The withdrawal reflex is one example.) Somatic reflexes are essential because they are immediate. Somatic reflexes provide a rapid response that can later be supplemented by voluntary motor commands.

Visceral reflexes, or autonomic reflexes, control or adjust the activities of smooth muscle, cardiac muscle, glands, and adipose tissues.

The Complexity of the Circuit

Polysynaptic reflexes involve at least one interneuron in addition to a sensory neuron and a motor neuron. These reflexes have a longer delay between stimulus and response, with the length of the delay proportional to the number of synapses. Polysynaptic reflexes can produce far more complicated responses than monosynaptic reflexes, because the interneurons can control motor neurons that activate several muscle groups simultaneously.

A **monosynaptic reflex** is the simplest reflex arc, involving only one synapse in the CNS. The sensory neuron synapses directly on a motor neuron, which serves as the processing center. Transmission across a chemical synapse always involves a synaptic delay, but with only one synapse, the delay between the stimulus and the response is minimized.

The Processing Site

In **spinal reflexes**, the important interconnections and processing events occur in nuclei of the spinal cord. In a simple segmental reflex, all of the processing occurs in a single spinal segment. **Intersegmental reflexes** involve multiple segments of the spinal cord, and they can produce very complex and coordinated responses.

In **cranial reflexes**, the important interconnections and processing events occur in nuclei of the brain.

2 Reflexes can be classified according to (1) their development, (2) the nature of the resulting motor response, (3) the complexity of the neural circuit involved, or (4) the site of information processing. These categories are not mutually exclusive—they represent different ways of describing a single reflex.

Module 12.9 Review

a. Define reflex and list the components of a reflex arc.

b. What are common characteristics of reflexes?

c. Describe the various classifications of reflexes.

The stretch reflex is a monosynaptic reflex involving muscle spindles

1 The best-known monosynaptic reflex is the **stretch reflex**, which provides automatic regulation of skeletal muscle length. The patellar reflex shown here is an example.

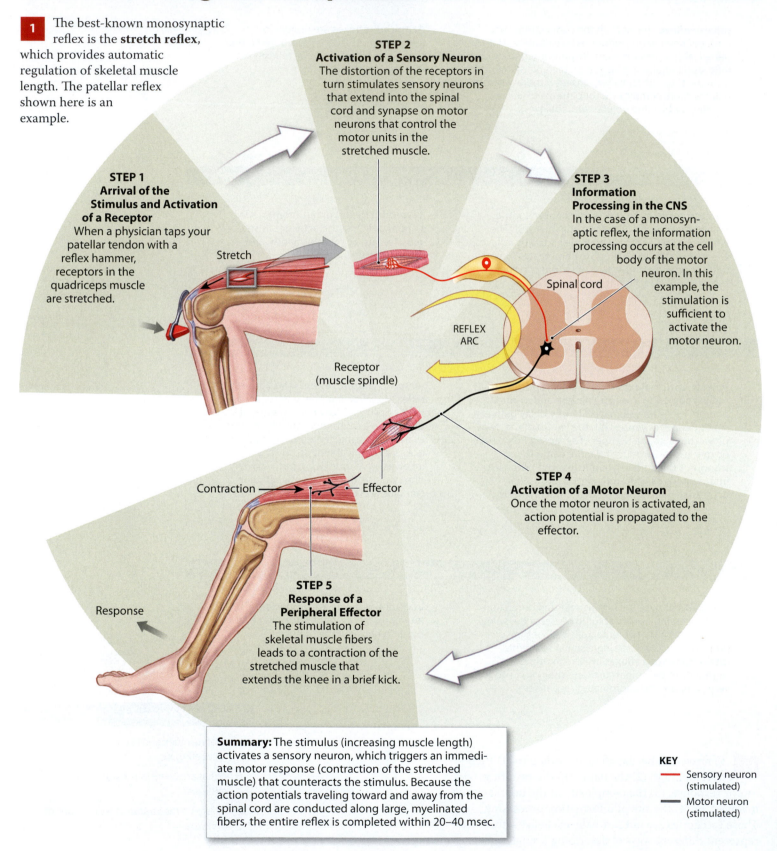

STEP 1
Arrival of the Stimulus and Activation of a Receptor
When a physician taps your patellar tendon with a reflex hammer, receptors in the quadriceps muscle are stretched.

Stretch

STEP 2
Activation of a Sensory Neuron
The distortion of the receptors in turn stimulates sensory neurons that extend into the spinal cord and synapse on motor neurons that control the motor units in the stretched muscle.

STEP 3
Information Processing in the CNS
In the case of a monosynaptic reflex, the information processing occurs at the cell body of the motor neuron. In this example, the stimulation is sufficient to activate the motor neuron.

Spinal cord

REFLEX ARC

Receptor (muscle spindle)

Contraction — Effector

STEP 4
Activation of a Motor Neuron
Once the motor neuron is activated, an action potential is propagated to the effector.

STEP 5
Response of a Peripheral Effector
The stimulation of skeletal muscle fibers leads to a contraction of the stretched muscle that extends the knee in a brief kick.

Response

Summary: The stimulus (increasing muscle length) activates a sensory neuron, which triggers an immediate motor response (contraction of the stretched muscle) that counteracts the stimulus. Because the action potentials traveling toward and away from the spinal cord are conducted along large, myelinated fibers, the entire reflex is completed within 20–40 msec.

KEY
— Sensory neuron (stimulated)
— Motor neuron (stimulated)

2 The sensory receptors involved in the stretch reflex are **muscle spindles**. Each muscle spindle consists of a bundle of small, specialized skeletal muscle fibers called **intrafusal muscle fibers**. The muscle spindle is surrounded by larger skeletal muscle fibers responsible for resting muscle tone and, at greater levels of stimulation, for the contraction of the entire muscle.

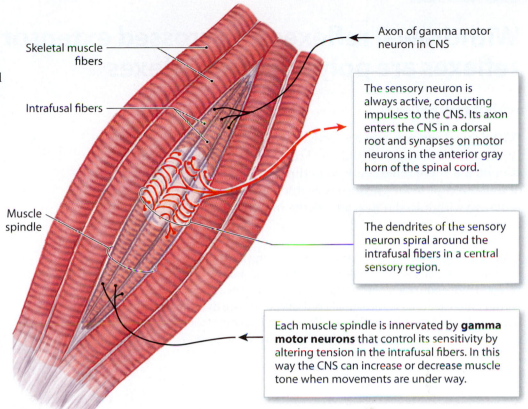

Skeletal muscle fibers

Intrafusal fibers

Muscle spindle

Axon of gamma motor neuron in CNS

The sensory neuron is always active, conducting impulses to the CNS. Its axon enters the CNS in a dorsal root and synapses on motor neurons in the anterior gray horn of the spinal cord.

The dendrites of the sensory neuron spiral around the intrafusal fibers in a central sensory region.

Each muscle spindle is innervated by **gamma motor neurons** that control its sensitivity by altering tension in the intrafusal fibers. In this way the CNS can increase or decrease muscle tone when movements are under way.

3 Stretching the central portion of the intrafusal fiber distorts the dendrites and stimulates the sensory neuron. This in turn stimulates the motor neurons, increasing muscle tone. Compressing the central portion inhibits the sensory neuron, which reduces stimulation of the motor neurons, and muscle tone decreases.

Sensory Region	Action Potential Frequency in Sensory Neuron	Effect on Skeletal Muscle
Resting length		Normal muscle tone persists
Stretched		Muscle tone increases
Compressed		Muscle tone decreases

Many stretch reflexes are **postural reflexes**—reflexes that help us maintain a normal upright posture. Standing, for example, requires cooperation among many muscle groups. Some of these muscles work in opposition to one another, exerting forces that keep the body's weight balanced over the feet. If the body leans forward, stretch receptors in the calf muscles are stimulated. Those muscles then respond by increasing muscle tone, returning the body to an upright position. Postural muscles generally have a firm muscle tone and extremely sensitive stretch receptors. As a result, very fine adjustments are continually being made, and you are not aware of the cycles of contraction and relaxation that occur.

Module 12.10 Review

a. Define stretch reflex.

b. In the patellar reflex, identify the response observed and the effectors involved.

c. In the patellar reflex, in what way does stimulation of the muscle spindle by gamma motor neurons affect the speed of the reflex?

Withdrawal reflexes and crossed extensor reflexes are polysynaptic reflexes

Withdrawal Reflexes

Withdrawal reflexes move affected parts of the body away from a stimulus. The strongest withdrawal reflexes are triggered by painful stimuli, but these reflexes are sometimes initiated by the stimulation of touch receptors or pressure receptors. The **flexor reflex**, a representative withdrawal reflex, affects the muscles of a limb.

1 If you grab an unexpectedly hot pan on the stove, a dramatic flexor reflex will occur. When the pain receptors in your hand are stimulated, the sensory neurons activate interneurons in the spinal cord that stimulate motor neurons in the anterior gray horns. The result is a contraction of flexor muscles that yanks your hand away from the stove.

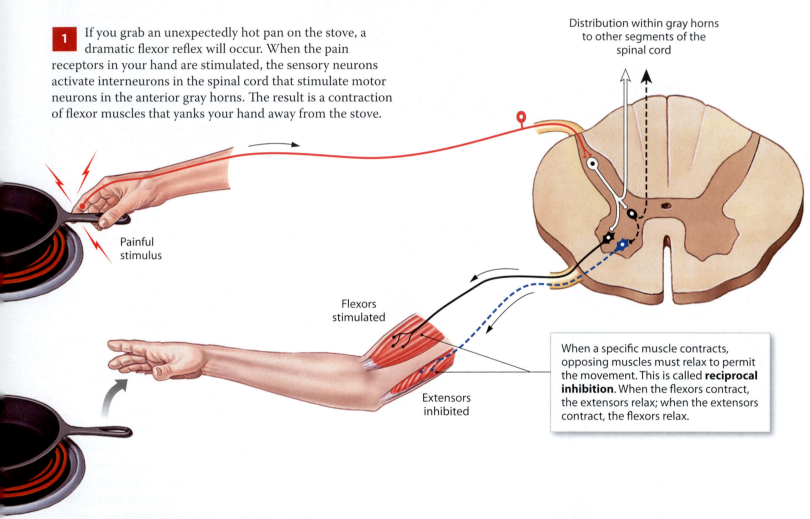

Distribution within gray horns to other segments of the spinal cord

Painful stimulus

Flexors stimulated

Extensors inhibited

When a specific muscle contracts, opposing muscles must relax to permit the movement. This is called **reciprocal inhibition**. When the flexors contract, the extensors relax; when the extensors contract, the flexors relax.

Withdrawal reflexes show tremendous versatility, because the sensory neurons activate many pools of interneurons. The distribution of the effects and the strength and character of the motor responses depend on the intensity and location of the stimulus. Mild discomfort might provoke a brief contraction in muscles of your hand and wrist. More powerful stimuli would produce coordinated muscular contractions affecting the positions of your hand, wrist, forearm, and arm. Severe pain would also stimulate contractions of your shoulder, trunk, and arm muscles. These contractions could persist for several seconds, owing to the activation of reverberating circuits.

KEY

— Sensory neuron (stimulated)

▭ Excitatory interneuron

— Motor neuron (stimulated)

--- Motor neuron (inhibited)

---- Inhibitory interneuron

Crossed Extensor Reflexes

The stretch reflexes and withdrawal reflexes involve **ipsilateral reflex arcs** (*ipsi*, same + *lateral*, side): The sensory stimulus and the motor response occur on the same side of the body. The **crossed extensor reflex** involves a **contralateral reflex arc** (*contra*, opposite), because an additional motor response occurs on the side opposite the stimulus.

2 When you step on a tack, the flexor reflex pulls the affected foot away from the ground as excitatory interneurons stimulate the flexor muscles of that limb and inhibitory interneurons relax the extensor muscles that would otherwise oppose that movement. The crossed extensor reflex occurs simultaneously because collaterals of the excitatory and inhibitory interneurons cross to the other side of the spinal cord to motor neurons controlling muscles in the uninjured leg. The excitatory interneurons stimulate motor neurons controlling extensor muscles while the inhibitory interneurons relax the flexor muscles. As a result, your opposite leg straightens to support the shifting weight.

3 **Polysynaptic reflexes** are responsible for the automatic actions involved in complex movements such as walking and running. All polysynaptic reflexes share the basic characteristics summarized in this table.

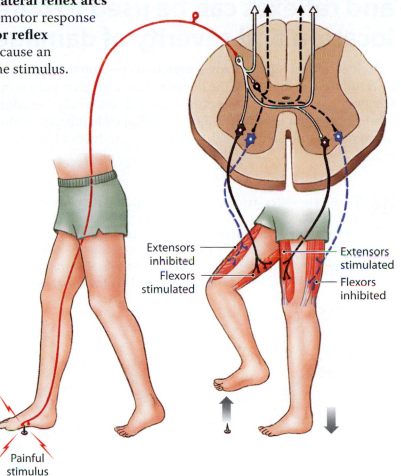

To motor neurons in other segments of the spinal cord

Extensors inhibited
Flexors stimulated

Extensors stimulated
Flexors inhibited

Painful stimulus

Properties of Polysynaptic Reflexes

- **They Involve Pools of Interneurons**. Processing occurs in pools of interneurons before motor neurons are activated. The result may be excitation or inhibition.

- **They Are Intersegmental in Distribution**. The interneuron pools extend across spinal segments and may activate muscle groups in many parts of the body.

- **They Involve Reciprocal Inhibition**. Reciprocal inhibition coordinates muscular contractions and reduces resistance to movement.

- **They Have Reverberating Circuits, Which Prolong the Reflexive Motor Response**. Positive feedback between interneurons that innervate motor neurons and the processing pool maintains the stimulation even after the initial stimulus has faded.

- **Several Reflexes May Cooperate to Produce a Coordinated, Controlled Response**. As a reflex movement gets under way, antagonistic reflexes are inhibited. In complex polysynaptic reflexes, commands may be distributed along the length of the spinal cord, producing a well-coordinated response.

Module 12.11 Review

a. Identify the basic characteristics of polysynaptic reflexes.

b. Describe the flexor reflex.

c. During a withdrawal reflex of the foot, what happens to the limb on the side opposite the stimulus? What is this response called?

The brain can inhibit or facilitate spinal reflexes, and reflexes can be used to determine the location and severity of damage to the CNS

Activities in the brain can have a profound effect on the performance of a reflex by facilitating or inhibiting the motor neurons or interneurons involved. The facilitation of motor neurons involved in reflexes is called **reinforcement**. For example, a voluntary effort to pull apart clasped hands can produce reinforcement of stretch reflexes so that a light patellar tap produces a big kick rather than a twitch.

1 The **biceps reflex**, **triceps reflex**, and **ankle-jerk reflex** are stretch reflexes often tested during a physical exam. Each reflex is controlled by specific segments of the spinal cord. As a result, testing these reflexes provides information about the status of the corresponding spinal segments. The table on the right presents information on several reflexes commonly used in physical exams.

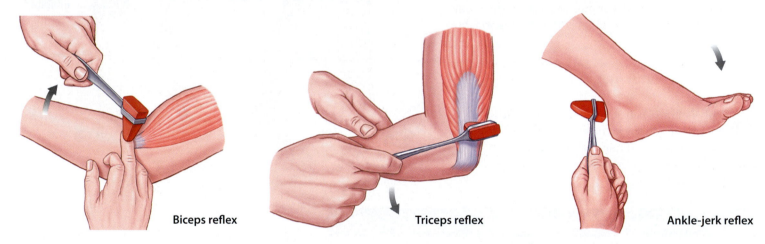

Biceps reflex   Triceps reflex   Ankle-jerk reflex

2 Descending fibers may also have an inhibitory effect on spinal reflexes. Stroking an infant's foot on the lateral side of the sole produces a fanning of the toes known as the **Babinski sign**, or positive Babinski reflex. This response disappears as descending motor pathways develop, because those pathways inhibit this reflex response.

3 In normal adults, stroking the lateral side of the sole produces a curling of the toes, called a **plantar reflex** (or negative Babinski reflex) after about a 1-second delay. If either the higher centers or the descending tracts are damaged, the Babinski sign will reappear in an adult. As a result, this reflex is often tested if CNS injury is suspected.

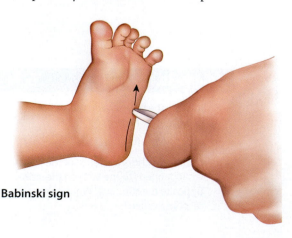

Babinski sign

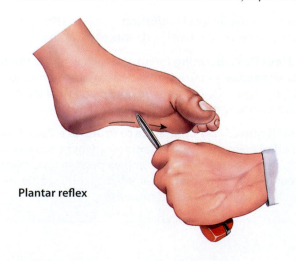

Plantar reflex

Reflexes Used in Diagnostic Testing

Reflex	Stimulus	Afferent Nerve(s)	Spinal Segment	Efferent Nerve(s)	Normal Response
Superficial Reflexes					
Abdominal reflex	Light stroking of skin of abdomen	T_7–T_{12} depending on region stroked	T_7–T_{12} at level of arrival	Same as afferent	Contractions of abdominal muscles that pull navel toward the stimulus
Cremasteric reflex	Stroking of skin of upper thigh	Femoral nerve	L_1	Genitofemoral nerve	Contraction of cremaster, elevation of scrotum
Plantar reflex	Longitudinal stroking of lateral side of sole of foot	Tibial nerve	S_1, S_2	Tibial nerve	Flexion at toe joints
Anal reflex	Stroking of region around the anus	Pudendal nerve	S_4, S_5	Pudendal nerve	Constriction of external anal sphincter
Stretch Reflexes					
Biceps reflex	Tap to tendon of biceps brachii muscle near its insertion	Musculocutaneous nerve	C_5, C_6	Musculocutaneous nerve	Flexion at elbow
Triceps reflex	Tap to tendon of triceps brachii muscle near its insertion	Radial nerve	C_6, C_7	Radial nerve	Extension at elbow
Brachioradialis reflex	Tap to forearm near styloid process of the radius	Radial nerve	C_5, C_6	Radial nerve	Flexion at elbow, supination, and flexion at finger joints
Patellar reflex	Tap to patellar tendon	Femoral nerve	L_2–L_4	Femoral nerve	Extension at knee
Ankle-jerk reflex	Tap to calcaneal tendon	Tibial nerve	S_1, S_2	Tibial nerve	Extension (plantar flexion) at ankle

4 The **abdominal reflex** depends on descending facilitation, rather than descending inhibition. In this reflex, seen in normal adults, a light stroking of the skin produces a reflexive twitch in the abdominal muscles that moves the navel toward the stimulus. The absence of an abdominal reflex may indicate damage to the descending tracts.

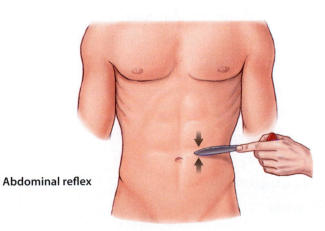

Abdominal reflex

Module 12.12 Review

a. Define reinforcement as it pertains to spinal reflexes.

b. What purpose does reflex testing serve?

c. After injuring her back, 22-year-old Tina exhibits a positive Babinski reflex. What does this imply about her injury?

1. Labeling

Label the neural circuit patterns in the following diagrams.

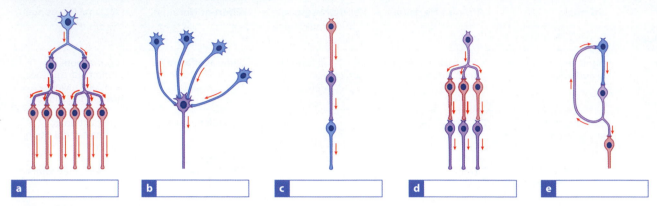

a		b		c		d		e	

2. Labeling

Label the indicated structures in the accompanying diagram of a reflex arc.

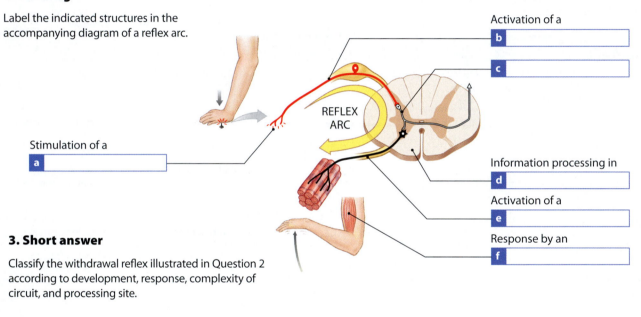

REFLEX ARC

Activation of a
b	

c	

Stimulation of a
a	

Information processing in
d	

Activation of a
e	

Response by an
f	

3. Short answer

Classify the withdrawal reflex illustrated in Question 2 according to development, response, complexity of circuit, and processing site.

_____

_____

4. Vocabulary

Write the boldfaced term introduced in this section that matches each of the following definitions.

a	_____	A reflex in which the sensory stimulus and motor response occur on the same side of the body
b	_____	Reflexes that move affected parts of the body away from a stimulus
c	_____	Controls the sensitivity of a muscle spindle
d	_____	Withdrawal reflex that affects the muscles of a limb
e	_____	Autonomic reflexes that adjust nonskeletal muscles, glands, and adipose tissue
f	_____	Type of reflexes enhanced by repetition
g	_____	A reflex in which the motor response occurs on the side opposite to the stimulus
h	_____	Process involved in preventing opposing muscles from contracting during a reflex
i	_____	The enhancement of reflexes due to facilitation of motor neurons

Visual Outline with Key Terms   Summarize the content of each module using the terms in the order provided.

The Functional Organization of the Spinal Cord

- spinal cord

12.1

The spinal cord contains gray matter and white matter

- cervical enlargement
- lumbar enlargement
- conus medullaris
- cauda equina
- filum terminale
- posterior median sulcus
- dorsal root
- dorsal root ganglion
- spinal nerve
- ventral root
- anterior median fissure
- white matter
- gray matter

12.2

The spinal cord is surrounded by the meninges, which consist of the dura mater, arachnoid mater, and pia mater

- spinal meninges
- pia mater
- arachnoid mater
- subarachnoid space
- dura mater
- cerebrospinal fluid (CSF)
- epidural space
- denticulate ligaments
- lumbar puncture

12.3

Gray matter is the region of integration, and white matter carries information

- horns
- posterior gray horn
- lateral gray horn
- anterior gray horn
- gray commissures
- nuclei
- sensory nuclei
- motor nuclei
- columns
- tracts
- ascending tracts
- descending tracts
- posterior white column
- lateral white column
- anterior white column
- anterior white commissure

12.4

Spinal nerves have a relatively consistent anatomical structure and pattern of distribution

- epineurium
- perineurium
- fascicles
- endoneurium
- rami
- sympathetic division
- dorsal ramus
- ventral ramus
- communicating rami
- dermatome
- shingles

12.5

Each ramus (branch) of a spinal nerve provides motor and sensory innervation to a specific region

- dorsal ramus
- ventral ramus
- white ramus
- gray ramus
- rami communicantes
- sympathetic nerve

12.6

Spinal nerves form nerve plexuses that innervate the skin and skeletal muscles; the cervical plexus is the smallest of these nerve plexuses

- nerve plexus
- cervical plexus
- brachial plexus
- lumbar plexus
- sacral plexus
- phrenic nerve
- ansa cervicalis
- lesser occipital nerve
- transverse cervical nerve
- supraclavicular nerve
- great auricular nerve
- cervical nerves

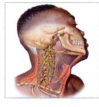

12.7

The brachial plexus innervates the pectoral girdle and upper limbs

- trunks
- cords
- dorsal scapular nerve
- long thoracic nerve
- suprascapular nerve
- pectoral nerves
- subscapular nerves
- thoracodorsal nerve
- axillary nerve
- medial antebrachial cutaneous nerve
- radial nerve
- musculocutaneous nerve
- median nerve
- ulnar nerve

• = *Term boldfaced in this module*

12.8

The lumbar and sacral plexuses innervate the skin and skeletal muscles of the trunk and lower limbs

- lumbar plexus
- sacral plexus
- sciatic nerve
- iliohypogastric nerve
- ilioinguinal nerve
- genitofemoral nerve
- lateral femoral cutaneous nerve
- femoral nerve
- obturator nerve
- saphenous nerve
- gluteal nerves
- posterior femoral cutaneous nerve
- tibial nerve
- fibular nerve
- pudendal nerve
- sural nerve

12.10

The stretch reflex is a monosynaptic reflex involving muscle spindles

- stretch reflex
- muscle spindles
- intrafusal muscle fibers
- gamma motor neurons
- postural reflexes

SECTION 2

An Introduction to Reflexes

- neuronal pools
- neural circuit
- divergence
- parallel processing
- serial processing
- convergence
- reverberation
- reflexes

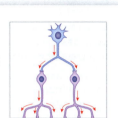

12.11

Withdrawal reflexes and crossed extensor reflexes are polysynaptic reflexes

- withdrawal reflexes
- flexor reflex
- reciprocal inhibition
- ipsilateral reflex arcs
- crossed extensor reflex
- contralateral reflex arc
- polysynaptic reflexes

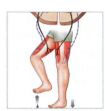

12.9

Reflexes are rapid, automatic responses to stimuli

- reflexes
- reflex arc
- innate reflexes
- acquired reflexes
- somatic reflexes
- visceral reflexes
- polysynaptic reflexes
- monosynaptic reflexes
- spinal reflexes
- intersegmental reflexes
- cranial reflexes

12.12

The brain can inhibit or facilitate spinal reflexes, and reflexes can be used to determine the location and severity of damage to the CNS

- reinforcement
- biceps reflex
- triceps reflex
- ankle-jerk reflex
- Babinski sign
- plantar reflex
- abdominal reflex

• = *Term boldfaced in this module*

Chapter Integration: Applying what you've learned

Karen had just finished her A&P midterm exam and was hurrying out of the classroom to get home and finish packing for her first-ever spring break trip to Cancun, Mexico. She was excited about her upcoming trip, and equally excited about how well she was doing in her first year of nursing school. She thought back to how she had struggled with every high-school science class she'd had, but her experience in her A&P class was completely different. Her professor was wonderful at presenting the information in an understandable manner and in guiding students to numerous ancillary tools to help them comprehend a seemingly overwhelming subject.

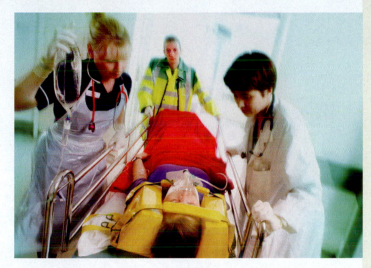

Seeing the crowd of students waiting for the elevator, Karen chose to take the stairs. In her haste to get home, she slipped on the edge of a step and fell down a flight of stairs. During the fall she suffered lumbar and sacral spinal cord damage due to a hyperextension of her back. The injury resulted in edema around the central canal that compressed the anterior horn of the lumbar region.

An additional consequence of her injury was the loss of the ability to control her bowels and urinary bladder, because control of these functions involves spinal reflex arcs located in the sacral region of the spinal cord. In both instances, two sphincter muscles—an inner sphincter of smooth muscle and an outer sphincter of skeletal muscle—control the passage of wastes (whether feces or urine) out of the body.

1. What signs and symptoms would you expect Karen to exhibit during her neurological evaluation?

2. How would a transection of the spinal cord at the L_1 level affect an individual's bowel and urinary bladder control?

3. Compare the effects of a transection of the spinal cord at L_1 in an adult with a newborn's inability to control urination.

Access more review material online in the Study Area at **www.masteringaandp.com.**

There, you'll find:

- **Chapter guides**
- **Chapter quizzes**
- **Practice tests**
- **Labeling activities**
- **MP3 Tutor Sessions**
- **Animations**
- **Flashcards**
- **A glossary with pronunciations**

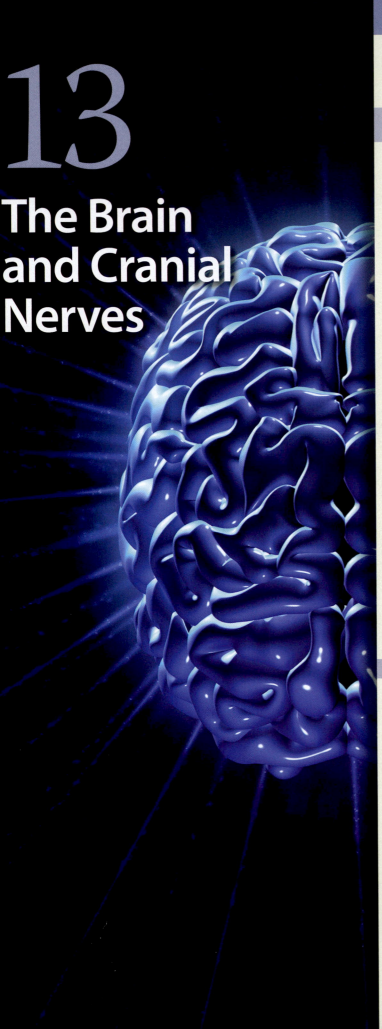

13

The Brain and Cranial Nerves

The Functional Anatomy of the Brain and Cranial Nerves

The first section of this chapter introduces the functional anatomy of the brain and cranial nerves. In adult humans, the brain contains almost 97 percent of the body's neural tissue. A "typical" brain weighs 1.4 kg (3 lb) and has a volume of 1200 mL (71 in.^{3}). Brain size varies considerably among individuals. On average, the brains of males are about 10 percent larger than those of females, owing to differences in average body size. No correlation exists between brain size and intelligence. Individuals with the smallest brains (750 mL) and the largest brains (2100 mL) are functionally normal.

1 This lateral view of the brain of an embryo after 4 weeks of development shows the **neural tube**, the hollow cylinder that is the beginning of the central nervous system. The internal passageway is called the **neurocoel** (NOOR-ō-sēl). In the cephalic portion of the neural tube, three areas enlarge rapidly through expansion of the internal cavity. This enlargement creates three prominent divisions called **primary brain vesicles**. The primary brain vesicles are named for their relative positions.

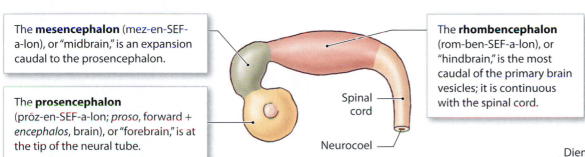

The **mesencephalon** (mez-en-SEF-a-lon), or "midbrain," is an expansion caudal to the prosencephalon.

The **prosencephalon** (prōz-en-SEF-a-lon; *proso*, forward + *encephalos*, brain), or "forebrain," is at the tip of the neural tube.

The **rhombencephalon** (rom-ben-SEF-a-lon), or "hindbrain," is the most caudal of the primary brain vesicles; it is continuous with the spinal cord.

Spinal cord

Neurocoel

2 By week 5 of development, the primary brain vesicles have changed position and the prosencephalon and rhombencephalon have subdivided, forming **secondary brain vesicles.**

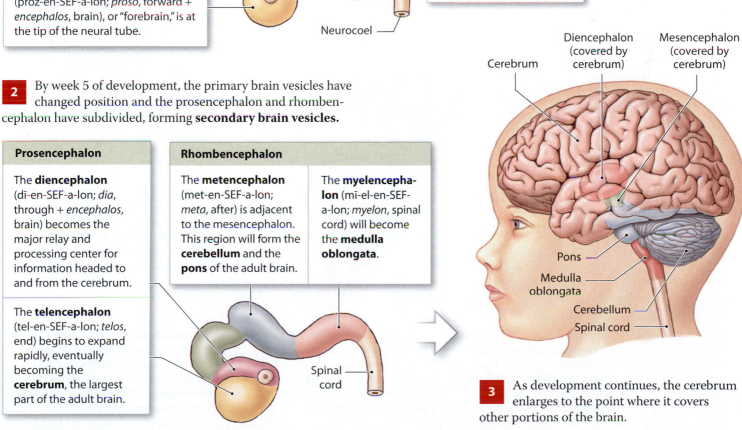

Prosencephalon	Rhombencephalon	
The **diencephalon** (dī-en-SEF-a-lon; *dia*, through + *encephalos*, brain) becomes the major relay and processing center for information headed to and from the cerebrum.	The **metencephalon** (met-en-SEF-a-lon; *meta*, after) is adjacent to the mesencephalon. This region will form the **cerebellum** and the **pons** of the adult brain.	The **myelencephalon** (mī-el-en-SEF-a-lon; *myelon*, spinal cord) will become the **medulla oblongata**.
The **telencephalon** (tel-en-SEF-a-lon; *telos*, end) begins to expand rapidly, eventually becoming the **cerebrum**, the largest part of the adult brain.		

Spinal cord

Cerebrum

Diencephalon (covered by cerebrum)

Mesencephalon (covered by cerebrum)

Pons

Medulla oblongata

Cerebellum

Spinal cord

3 As development continues, the cerebrum enlarges to the point where it covers other portions of the brain.

Each region of the brain has distinct structural and functional characteristics

1 This diagrammatic view of the brain introduces the major regions of the brain and their general functions. The regions are color coded, and these colors will be used as appropriate in illustrations throughout the chapter.

Cerebrum

The **cerebrum** (se-RĒ-brum) of the adult brain is divided into a pair of large **cerebral hemispheres.** The surfaces of the cerebral hemispheres are highly folded and covered by a superficial layer of gray matter called the **cerebral cortex** (*cortex,* rind or bark). Functions include conscious thought, memory storage and processing, sensory processing, and the regulation of skeletal muscle contractions.

Fissures are deep grooves that subdivide the cerebral hemisphere.

Gyri (JĪ-rī; singular, *gyrus*) are folds in the cerebral cortex that increase its surface area.

Sulci (SUL-sī; singular, *sulcus*) are shallow depressions in the cerebral cortex that separate adjacent gyri.

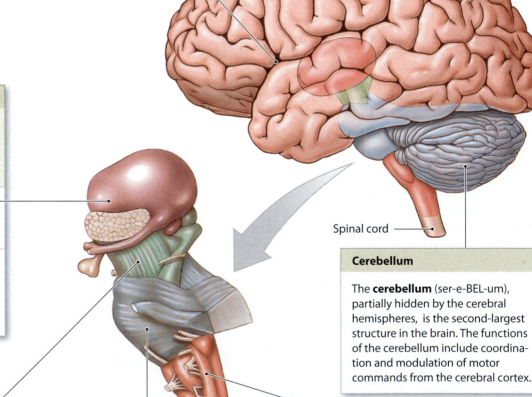

Diencephalon

The **diencephalon** is the structural and functional link between the cerebral hemispheres and the rest of the CNS.

The **thalamus** (THAL-a-mus) contains relay and processing centers for sensory information.

The **hypothalamus** (*hypo-,* below), or floor of the diencephalon, contains centers involved with emotions, autonomic function, and hormone production.

Spinal cord

Cerebellum

The **cerebellum** (ser-e-BEL-um), partially hidden by the cerebral hemispheres, is the second-largest structure in the brain. The functions of the cerebellum include coordination and modulation of motor commands from the cerebral cortex.

Brain stem

The **brain stem** includes the midbrain, pons, and medulla oblongata.

The **midbrain** contains nuclei that process visual and auditory information and control reflexes triggered by these stimuli. It also contains centers that help maintain consciousness.

The **pons** (*pons,* bridge) connects the cerebellum to the brain stem. In addition to tracts and relay centers, the pons also contains nuclei that function in somatic and visceral motor control.

The **medulla oblongata** relays sensory information to other portions of the brain stem, and to the thalamus. The medulla oblongata also contains major centers that regulate autonomic function, such as heart rate and blood pressure.

2 During development, the neurocoel within the cerebral hemispheres, diencephalon, metencephalon, and medulla oblongata expands to form chambers called **ventricles** (VEN-tri-kls). The ventricles are filled with cerebrospinal fluid and lined by ependymal cells.

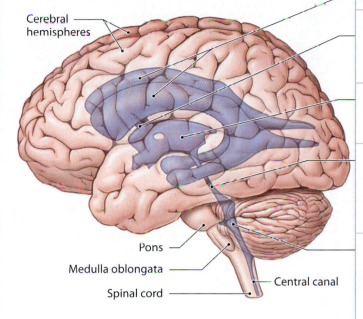

Cerebral hemispheres

Pons

Medulla oblongata

Spinal cord

Central canal

Ventricular system, lateral view

Ventricles of the Brain

Each cerebral hemisphere contains a large **lateral ventricle**.

Each lateral ventricle communicates with the third ventricle through an **interventricular foramen**.

The **third ventricle** is located in the diencephalon.

The **aqueduct of the midbrain** is a slender canal within the midbrain that connects the third ventricle to the fourth ventricle.

The **fourth ventricle** begins in the metencephalon and extends into the superior portion of the medulla oblongata. It then narrows and becomes the central canal of the spinal cord.

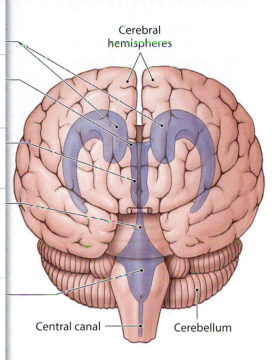

Cerebral hemispheres

Central canal

Cerebellum

Ventricular system, anterior view

3 The interconnections between the ventricles can be seen in this coronal section of the brain.

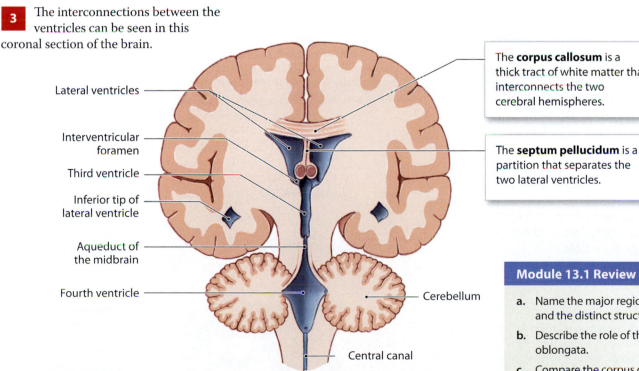

Lateral ventricles

Interventricular foramen

Third ventricle

Inferior tip of lateral ventricle

Aqueduct of the midbrain

Fourth ventricle

Cerebellum

Central canal

The **corpus callosum** is a thick tract of white matter that interconnects the two cerebral hemispheres.

The **septum pellucidum** is a partition that separates the two lateral ventricles.

Module 13.1 Review

a. Name the major regions of the brain and the distinct structures of each.

b. Describe the role of the medulla oblongata.

c. Compare the corpus callosum to the septum pellucidum.

The brain is protected and supported by the cranial meninges and the cerebrospinal fluid

The delicate tissues of the brain are protected from mechanical forces by the bones of the cranium, the tissue structures of the **cranial meninges**, and the cerebrospinal fluid. In addition, the neural tissue of the brain is biochemically isolated from the general circulation by the blood–brain barrier.

1 The three layers that make up the cranial meninges—the cranial **dura mater**, **arachnoid mater**, and **pia mater**—are continuous with those of the spinal meninges.

1 Dura mater

The cranial dura mater consists of outer and inner fibrous layers. The outer layer is fused to the periosteum of the cranial bones. As a result, there is no epidural space. The outer (endosteal) and inner (meningeal) layers of the cranial dura mater are typically separated by a slender gap that contains tissue fluids and blood vessels, including several large dural sinuses, which collect blood from the veins of the brain.

2 Arachnoid mater

The cranial arachnoid mater consists of the arachnoid membrane and the arachnoid trabeculae (which connect to the pia mater). The arachnoid membrane provides a smooth covering that does not follow the brain's underlying folds. The subarachnoid space lies between the arachnoid membrane and the pia mater.

Subdural space

Cranium (skull)

Arachnoid membrane

Subarachnoid space

Arachnoid trabeculae

Dura mater (endosteal layer)

Dural sinus

Dura mater (meningeal layer)

Cerebral cortex

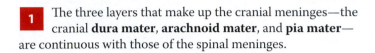

3 Pia mater

Astrocyte processes bind the pia mater to the surface of the brain. The pia mater sticks to the surface of the brain. It extends into every fold and accompanies the branches of cerebral blood vessels as they penetrate the surface of the brain to reach internal structures.

Inferior sagittal sinus

The **superior sagittal sinus** is the largest dural sinus.

2 At several sites, the inner layer of the dura mater extends into the cranial cavity, forming **dural folds**—sheets that dip inward and then return. These provide additional stabilization and support to the brain. **Dural sinuses** are large collecting veins located within the dural folds. There are three large dural folds.

The **tentorium cerebelli** (ten-TŌ-rē-um ser-e-BEL-ī; *tentorium*, a tent) separates the cerebral hemispheres from the cerebellum.

The **falx cerebri** (FALKS SER-e-brī; *falx*, sickle shaped) is a fold of dura mater that projects between the cerebral hemispheres. Its inferior portions attach anteriorly to the crista galli and posteriorly to the internal occipital crest of the occipital bone. The superior and inferior sagittal sinuses lie within this dural fold.

The **falx cerebelli** separates the two cerebellar hemispheres along the midsagittal line inferior to the tentorium cerebelli.

3 **Cerebrospinal fluid (CSF)** completely surrounds and bathes the exposed surfaces of the CNS. Each of the ventricles contains an area of **choroid plexus** (*choroid*, vascular coat; *plexus*, network), which consists of a combination of specialized ependymal cells and capillaries involved in the production and maintenance of cerebrospinal fluid.

The CSF circulates from the choroid plexuses through the ventricles and fills the central canal of the spinal cord. As it circulates, materials diffuse between the CSF and the interstitial fluid of the CNS across the ependymal cells.

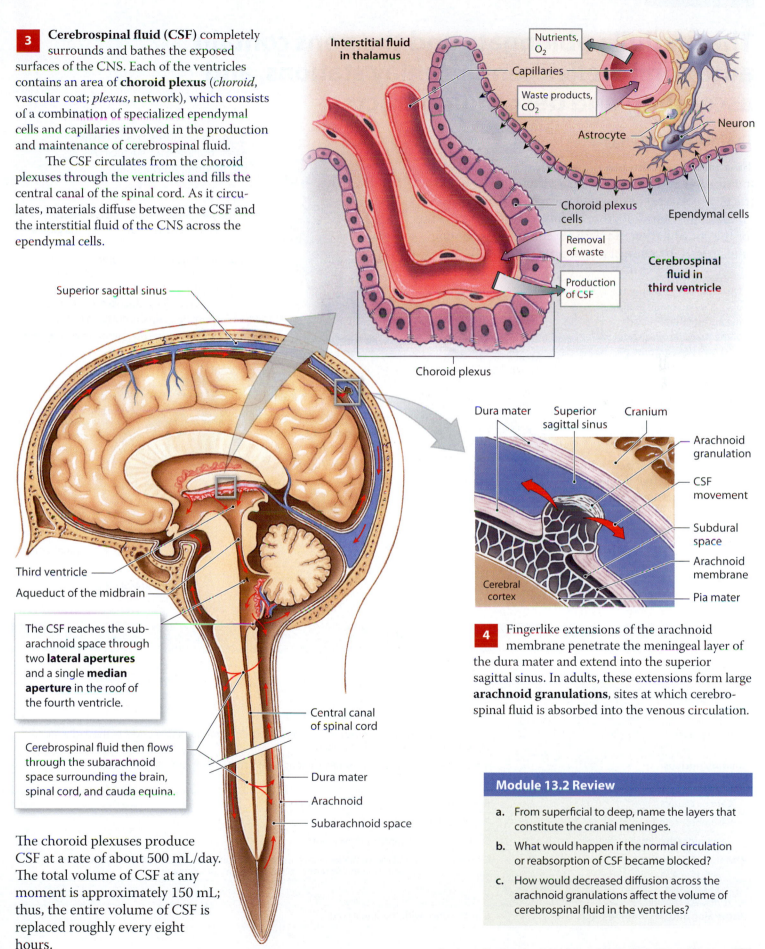

Interstitial fluid in thalamus

Nutrients, O_2

Capillaries

Waste products, CO_2

Neuron

Astrocyte

Choroid plexus cells

Ependymal cells

Removal of waste

Production of CSF

Cerebrospinal fluid in third ventricle

Choroid plexus

Superior sagittal sinus

Third ventricle

Aqueduct of the midbrain

The CSF reaches the subarachnoid space through two **lateral apertures** and a single **median aperture** in the roof of the fourth ventricle.

Cerebrospinal fluid then flows through the subarachnoid space surrounding the brain, spinal cord, and cauda equina.

Central canal of spinal cord

Dura mater

Arachnoid

Subarachnoid space

The choroid plexuses produce CSF at a rate of about 500 mL/day. The total volume of CSF at any moment is approximately 150 mL; thus, the entire volume of CSF is replaced roughly every eight hours.

Dura mater   Superior sagittal sinus   Cranium

Arachnoid granulation

CSF movement

Subdural space

Arachnoid membrane

Cerebral cortex

Pia mater

4 Fingerlike extensions of the arachnoid membrane penetrate the meningeal layer of the dura mater and extend into the superior sagittal sinus. In adults, these extensions form large **arachnoid granulations**, sites at which cerebrospinal fluid is absorbed into the venous circulation.

Module 13.2 Review

a. From superficial to deep, name the layers that constitute the cranial meninges.

b. What would happen if the normal circulation or reabsorption of CSF became blocked?

c. How would decreased diffusion across the arachnoid granulations affect the volume of cerebrospinal fluid in the ventricles?

The medulla oblongata and the pons contain autonomic reflex centers, relay stations, and ascending and descending tracts

The Medulla Oblongata

The **medulla oblongata** is a very busy place—all communication between the brain and spinal cord involves tracts that ascend or descend through the medulla oblongata. In addition, the medulla oblongata is a center for the coordination of relatively complex autonomic reflexes and the control of visceral functions.

The **olive** is a prominent olive-shaped bulge along the anterolateral surface of the medulla oblongata. It follows the contours of the olivary nucleus.

2 Landmarks and structures of the medulla oblongata are shown in these illustrations, and major components and functions are summarized in the table below. The medulla oblongata contains **autonomic centers** controlling vital functions, and **relay stations** along sensory and motor pathways. The medulla oblongata also contains the nuclei associated with five cranial nerves, although those nuclei are not illustrated here.

1 Descending tracts cover the anterior surface of the medulla oblongata.

The **pyramids** contain tracts of motor fibers that originate at the cerebral cortex.

Some of the pyramidal fibers cross over to the opposite side of the medulla as they descend into the spinal cord. That crossing is called a **decussation** (de-kuh-SĀ-shun; *decussation*, crossing over).

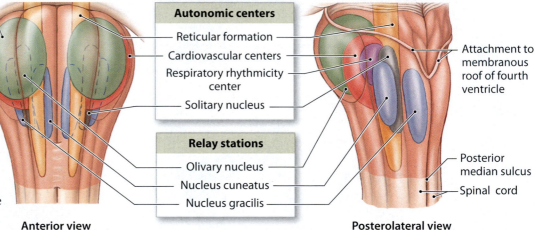

Pons

Autonomic centers
- Reticular formation
- Cardiovascular centers
- Respiratory rhythmicity center
- Solitary nucleus

Relay stations
- Olivary nucleus
- Nucleus cuneatus
- Nucleus gracilis

Attachment to membranous roof of fourth ventricle

Posterior median sulcus

Spinal cord

Anterior view

Posterolateral view

Components and Functions of the Medulla Oblongata	
Components	**Functions**
Gray Matter	
Nucleus gracilis, nucleus cuneatus	Relay somatic sensory information to the thalamus
Olivary nuclei	Located within the olives; relay information from the red nucleus, other nuclei of the midbrain, and the cerebral cortex to the cerebellum
Solitary nucleus	Integrates and relays visceral sensory information to autonomic processing centers
Autonomic reflex centers	
Cardiac centers	Regulate heart rate and force of contraction
Vasomotor centers	Regulate distribution of blood flow
Respiratory rhythmicity centers	Set the pace of respiratory movements
Other nuclei/centers	Contain sensory and motor nuclei of cranial nerves VIII (in part), IX, X, XI (in part), and XII; relay ascending sensory information from the spinal cord to higher centers
White Matter	
Ascending and descending tracts	Link the brain with the spinal cord

The Pons

3 The **pons** links the cerebellum with the midbrain, diencephalon, cerebrum, medulla oblongata, and spinal cord.

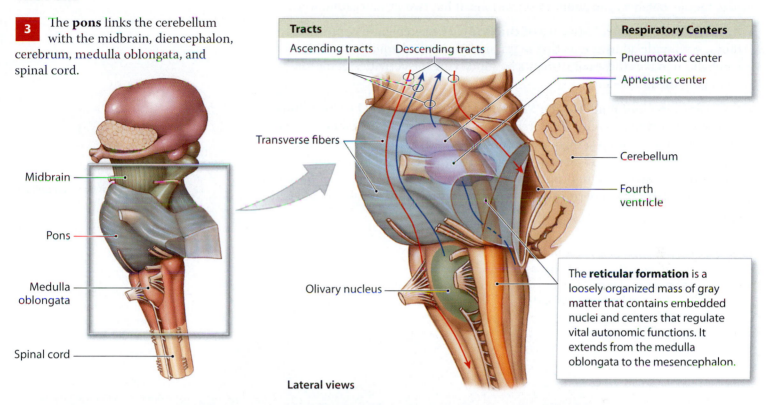

Lateral views

Midbrain

Pons

Medulla oblongata

Spinal cord

Tracts
Ascending tracts   Descending tracts

Transverse fibers

Olivary nucleus

Respiratory Centers
Pneumotaxic center
Apneustic center

Cerebellum

Fourth ventricle

The **reticular formation** is a loosely organized mass of gray matter that contains embedded nuclei and centers that regulate vital autonomic functions. It extends from the medulla oblongata to the mesencephalon.

Components and Functions of the Pons	
Components	**Functions**
Gray Matter	
Nuclei associated with cranial nerves V, VI, VII, and VIII (in part)	Relay sensory information and issue somatic motor commands
Apneustic and pneumotaxic centers	Adjust activities of the respiratory rhythmicity centers in the medulla oblongata
Relay centers	Relay sensory and motor information to the cerebellum
White Matter	
Ascending tracts	Carry sensory information from the nucleus cuneatus and nucleus gracilis to the thalamus
Descending tracts	Carry motor commands from higher centers to motor nuclei of cranial or spinal nerves
Transverse fibers	Interconnect processing centers in the cerebellar hemispheres

4 The table above summarizes the components and functions of the pons.

Module 13.3 Review

a. What is the function of the ascending and descending tracts in the medulla oblongata?

b. Name the medulla oblongata parts that relay somatic sensory information to the thalamus.

c. Describe the pyramids of the medulla oblongata and the result of decussation.

The cerebellum coordinates learned and reflexive patterns of muscular activity at the subconscious level

The **cerebellum** is an automatic processing center that monitors proprioceptive, visual, tactile, balance, and auditory sensations. It has two primary functions:

- **Adjusting the Postural Muscles of the Body.** The cerebellum coordinates rapid, automatic adjustments that maintain balance and equilibrium. The cerebellum makes these adjustments by modifying the activities of motor centers in the brain stem.

- **Programming and Fine-Tuning Movements Controlled at the Conscious and Subconscious Levels.** The cerebellum refines learned movement patterns indirectly by regulating activity along motor pathways at the cerebral cortex, basal nuclei, and motor centers in the brain stem. The cerebellum compares the motor commands with proprioceptive information (perception of muscle and joint position) and performs any adjustments needed to make movements smooth.

Cerebellum

Posterior view

1 As evident in this view of the posterior, superior surface, the cerebellum has large **anterior** and **posterior lobes**. Like the cerebrum, the cerebellum has two hemispheres and the surface is covered with a layer of gray matter—the **cerebellar cortex**.

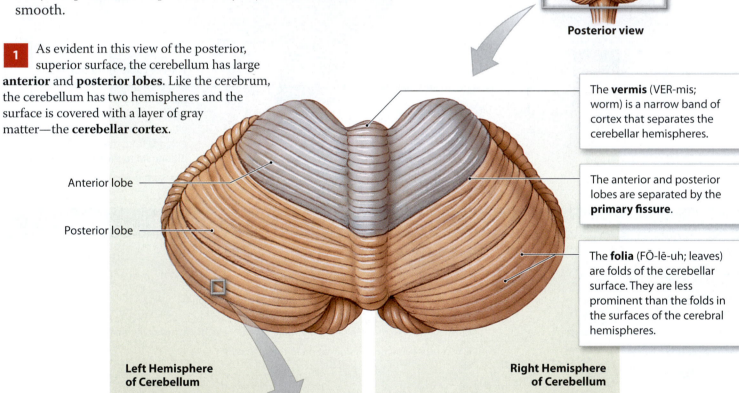

The **vermis** (VER-mis; worm) is a narrow band of cortex that separates the cerebellar hemispheres.

The anterior and posterior lobes are separated by the **primary fissure**.

The **folia** (FŌ-lē-uh; leaves) are folds of the cerebellar surface. They are less prominent than the folds in the surfaces of the cerebral hemispheres.

Anterior lobe

Posterior lobe

Left Hemisphere of Cerebellum

Right Hemisphere of Cerebellum

The extensive dendrites of each Purkinje cell receive input from up to 200,000 synapses.

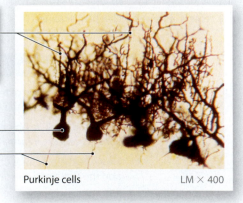

Cell body of Purkinje cell

Purkinje cell axons project into the white matter of the cerebellum.

Purkinje cells   LM × 400

2 Like the cerebrum, the cerebellum has a superficial layer of neural cortex. The cerebellar cortex contains huge, highly branched **Purkinje** (pur-KIN-jē) **cells**. Most axons that carry sensory information do not synapse in the cerebellar nuclei but pass through the deeper layers of the cerebellum on their way to the Purkinje cells. Information about the motor commands issued at the conscious and subconscious levels reaches the Purkinje cells, after being relayed by nuclei in the pons or by the cerebellar nuclei.

3 This sagittal section through the vermis shows the internal organization of the cerebellum. Also shown are the locations of the three cerebellar peduncles.

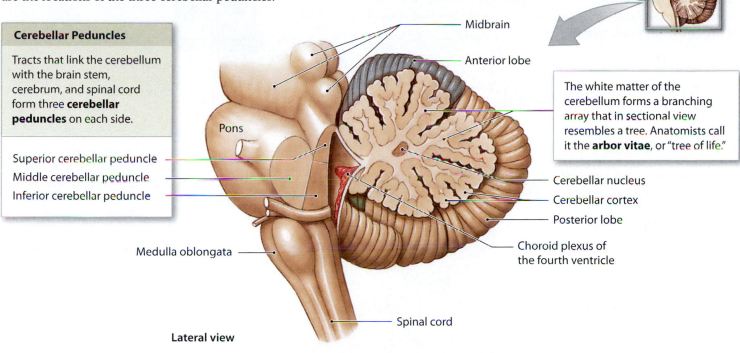

Cerebellar Peduncles

Tracts that link the cerebellum with the brain stem, cerebrum, and spinal cord form three **cerebellar peduncles** on each side.

Superior cerebellar peduncle
Middle cerebellar peduncle
Inferior cerebellar peduncle

Midbrain

Anterior lobe

The white matter of the cerebellum forms a branching array that in sectional view resembles a tree. Anatomists call it the **arbor vitae**, or "tree of life."

Pons

Cerebellar nucleus
Cerebellar cortex
Posterior lobe

Medulla oblongata

Choroid plexus of the fourth ventricle

Spinal cord

Lateral view

Components of the Cerebellum and Their Functions

Region/Nuclei	Function(s)
Gray Matter	
Cerebellar cortex	Involuntary coordination and control of ongoing body movements
Cerebellar nuclei	Involuntary coordination and control of ongoing body movements
White Matter	
Arbor vitae	Connects cerebellar cortex and nuclei with cerebellar peduncles
Cerebellar peduncles	
Superior	Link the cerebellum with midbrain, diencephalon, and cerebrum
Middle	Contain transverse fibers and carry communications between the cerebellum and pons
Inferior	Link the cerebellum with the medulla oblongata and spinal cord
Transverse fibers	Interconnect pontine nuclei with the cerebellar hemisphere on the opposite side

The cerebellum can be permanently damaged by trauma or stroke, or temporarily affected by drugs such as alcohol. The result is **ataxia** (a-TAK-sē-uh; *ataxia*, lack of order), a disturbance in muscular coordination. In severe ataxia, the individual cannot sit or stand without assistance.

Module 13.4 Review

a. Identify the components of the cerebellar gray matter.

b. Describe the arbor vitae, including its makeup, location, and function.

c. Describe ataxia.

The midbrain regulates auditory and visual reflexes and controls alertness

The **midbrain** is the most complex and integrative portion of the brain stem. Working independently or with the cerebellum, the midbrain can direct complex motor patterns at the subconscious level. It also influences the level of activity in the entire nervous system.

1 This posterior view of the midbrain shows the major superficial landmarks as well as underlying nuclei.

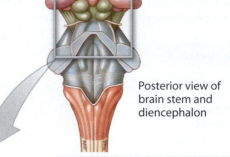

Posterior view of brain stem and diencephalon

Corpora Quadrigemina

The **corpora quadrigemina** (KOR-por-uh qua-dri-JEM-i-nuh) are two pairs of sensory nuclei located in the roof of the midbrain (see **2** below).

Each **superior colliculus** (ko-LIK-ū-lus; *colliculus*, hill) receives visual inputs from the thalamus and controls the reflex movements of the eyes, head, and neck in response to these visual stimuli.

Each **inferior colliculus** receives auditory data from nuclei in the medulla oblongata and pons and controls reflex movements of the head, neck, and trunk in response to these auditory stimuli.

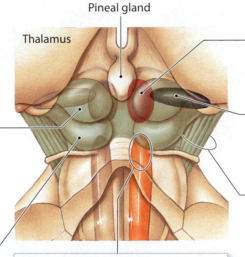

Pineal gland

Thalamus

The **reticular activating system (RAS)** is a specialized part of the reticular formation. Stimulation of the RAS makes you more alert and attentive; damage to the RAS produces unconsciousness.

The **red nucleus** receives information from the cerebrum and cerebellum and issues subconscious motor commands that affect upper limb position and background muscle tone.

The **substantia nigra** (NĪ-gruh; *nigra*, black) contains darkly pigmented cells that adjust activity in the basal nuclei of the cerebrum.

The **cerebral peduncles** (*peduncles*, little feet) are nerve fiber bundles on the ventrolateral surfaces of the midbrain. They contain (1) descending fibers that reach the cerebellum by way of the pons and (2) descending fibers that carry voluntary motor commands issued by the cerebral hemispheres.

2 This superior view of a horizontal section through the midbrain shows the internal subdivisions relative to the aqueduct of the midbrain.

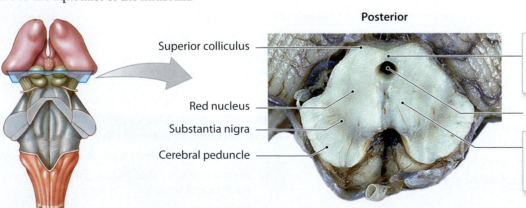

Posterior

Superior colliculus

Red nucleus

Substantia nigra

Cerebral peduncle

The **tectum**, or roof of the midbrain, is the region posterior to the aqueduct of the midbrain.

Aqueduct of the midbrain

The **tegmentum** is the area anterior to the aqueduct of the midbrain.

Anterior

Components and Functions of the Midbrain

Subdivision	Region/Nuclei	Function(s)
Gray Matter		
Tectum (roof)	Superior colliculi	Integrate visual information with other sensory inputs; initiate reflex responses to visual stimuli
	Inferior colliculi	Relay auditory information to medial geniculate nuclei; initiate reflex responses to auditory stimuli
Walls and floor	Red nuclei	Provide subconscious control of upper limb position and background muscle tone
	Substantia nigra	Regulates activity in the basal nuclei
	Reticular formation (headquarters)	Processes incoming sensations and outgoing motor commands automatically; can initiate involuntary motor responses to stimuli; helps maintain consciousness (RAS)
	Other nuclei/centers	Are associated with cranial nerves III and IV
White Matter		
	Cerebral peduncles	Connect primary motor cortex with motor neurons in brain and spinal cord; carry ascending sensory information to thalamus

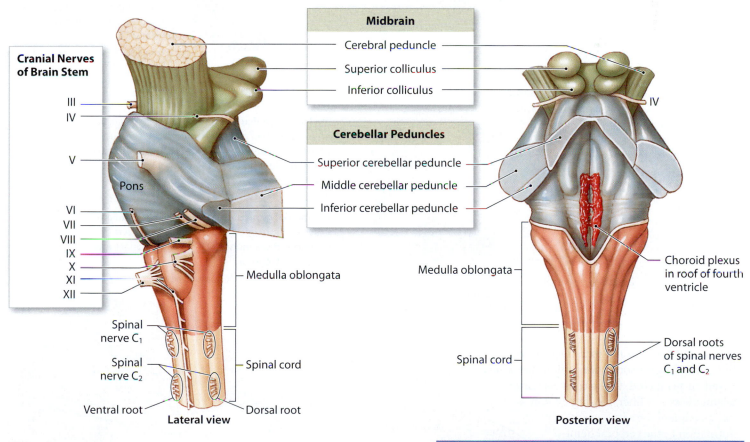

Cranial Nerves of Brain Stem

III
IV
V
Pons
VI
VII
VIII
IX
X
XI
XII

Spinal nerve C₁
Spinal nerve C₂
Ventral root
Lateral view
Dorsal root
Medulla oblongata
Spinal cord

Midbrain
Cerebral peduncle
Superior colliculus
Inferior colliculus

Cerebellar Peduncles
Superior cerebellar peduncle
Middle cerebellar peduncle
Inferior cerebellar peduncle

IV
Medulla oblongata
Choroid plexus in roof of fourth ventricle
Spinal cord
Dorsal roots of spinal nerves C₁ and C₂
Posterior view

3 These views of the brain stem provide reviews of the anatomy of the midbrain in relation to the brain stem as a whole.

Module 13.5 Review

a. Cranial nerves III to XII arise from which structure?

b. Identify the sensory nuclei contained within the corpora quadrigemina.

c. Which area(s) of the midbrain control reflexive movements of the eyes, head, and neck?

The diencephalon consists of the epithalamus, thalamus (left and right), and hypothalamus

Epithalamus

1 As seen in this sagittal section of the brain, the **epithalamus** is the roof of the diencephalon superior to the third ventricle. The anterior portion of the epithalamus contains an extensive area of choroid plexus that extends through the interventricular foramina.

The anterior limit of the diencephalon is marked by the **anterior commissure**, a tract that interconnects the cerebral hemispheres, and the **optic chiasm**, where the optic nerves connect to the brain.

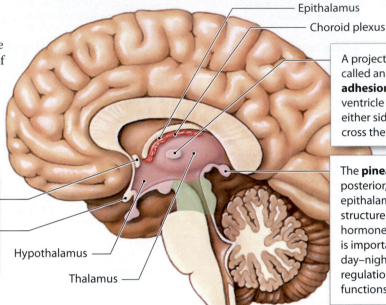

Epithalamus

Choroid plexus

A projection of gray matter called an **interthalamic adhesion** extends into the ventricle from the thalamus on either side, although no fibers cross the midline.

The **pineal gland** lies in the posterior, inferior portion of the epithalamus. It is an endocrine structure that secretes the hormone **melatonin**. Melatonin is important in the regulation of day–night cycles and also in the regulation of reproductive functions.

Hypothalamus

Thalamus

Thalamus

2 On each side of the brain, a **thalamus** (plural, *thalami*) sits superior to the midbrain. The important landmarks in this lateral view can be seen only after the cerebral hemispheres and cerebral peduncles have been removed.

Thalamus

The **lateral geniculate** (je-NIK-ū-lāt; *genicula*, little knee) **nucleus** of each thalamus receives visual information over the **optic tract** and sends signals to both the midbrain and the occipital lobe of the cerebral hemisphere on that side.

The **medial geniculate nucleus** of each thalamus relays auditory information from specialized receptors of the inner ear to the appropriate area of the cerebral cortex.

Optic chiasm

Optic tract

Cerebral peduncle (midbrain)

Lateral view of the left thalamus and midbrain

3 The thalamus is the final relay point for ascending sensory information that will be relayed, or **projected**, to the cerebral cortex. The thalamus acts as a filter, passing on only a small portion of the arriving sensory information to the cerebral hemispheres. Each region of the thalamus contains nuclei or groups of nuclei that connect to specific regions of the cerebral cortex (see table at the upper right).

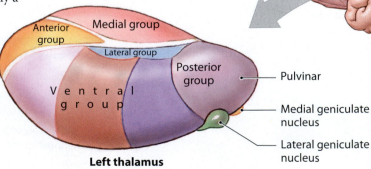

Anterior group

Medial group

Lateral group

Posterior group

Ventral group

Pulvinar

Medial geniculate nucleus

Lateral geniculate nucleus

Left thalamus

Note: colors indicate the associated areas of the cerebral cortex

Components of the Thalamus and Their Functions

Group/Nuclei	Function(s)
Anterior group	Part of the limbic system (see Module 13.7)
Medial group	Integrates sensory information for projection to the frontal lobes of the cerebral hemispheres
Ventral group	Projects sensory information to the primary sensory cortex; relays information from the cerebellum and basal nuclei to the motor area of the cerebral cortex
Posterior group	
Pulvinar	Integrates sensory information for projection to association areas of the cerebral cortex
Lateral geniculate nuclei	Project visual information to the visual cortex
Medial geniculate nuclei	Project auditory information to the auditory cortex
Lateral group	Integrates sensory information and influences emotional states

Hypothalamus

4 The **hypothalamus** contains important control and integrative centers that are shown in this sagittal section. Hypothalamic centers may be stimulated by (1) sensory information from the cerebrum, brain stem, and spinal cord; (2) changes in the composition of the CSF and interstitial fluid; or (3) chemical stimuli in the circulating blood that rapidly enter the hypothalamus because this region lacks a blood–brain barrier.

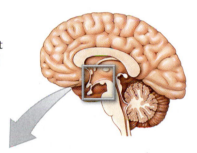

Hypothalamic Nuclei

Autonomic centers control the cardiovascular and vasomotor centers of the medulla oblongata.

The **preoptic area** regulates body temperature by coordinated adjustments in blood flow and sweat gland activity.

The **suprachiasmatic nucleus** coordinates day–night cycles of activity/inactivity.

Hormonal centers secrete chemical messengers that control endocrine cells of the anterior pituitary gland and secrete two hormones at the posterior pituitary gland.

A narrow stalk called the **infundibulum** (in-fun-DIB-ū-lum; *infundibulum*, funnel) extends inferiorly, connecting the floor of the hypothalamus to the pituitary gland.

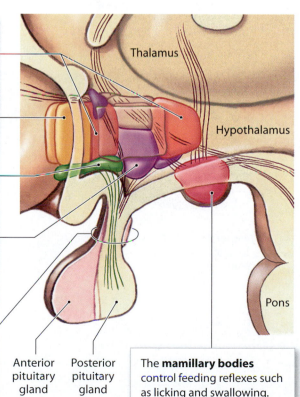

Thalamus

Hypothalamus

Pons

Anterior pituitary gland

Posterior pituitary gland

The **mamillary bodies** control feeding reflexes such as licking and swallowing.

Module 13.6 Review

a. Name the main components of the diencephalon.

b. Damage to the lateral geniculate nuclei of the thalami would interfere with what particular function?

c. Which component of the diencephalon is stimulated by changes in body temperature?

The limbic system is a functional group of tracts and nuclei located in the cerebrum and diencephalon

The **limbic system** (*limbus*, border) includes nuclei and tracts along the border between the cerebrum and diencephalon. This system is a functional grouping rather than an anatomical one. Functions of the limbic system include (1) establishing emotional states; (2) linking the conscious, intellectual functions of the cerebral cortex with the unconscious and autonomic functions of the brain stem; and (3) facilitating memory storage and retrieval.

1 This diagrammatic sagittal section shows the position and orientation of the major components of the limbic system.

Corpus callosum

Fornix

Central sulcus

Pineal gland

Components of the Limbic System in the Cerebrum

The region of the cerebral hemisphere shown in green is known as the **limbic lobe**.

Cingulate gyrus (superior portion of limbic lobe)

Parahippocampal gyrus (inferior portion of limbic lobe)

Hippocampus (see **2**)

Components of the Limbic System in the Diencephalon

Anterior group of thalamic nuclei

Hypothalamus

Mamillary body

Temporal lobe of cerebrum

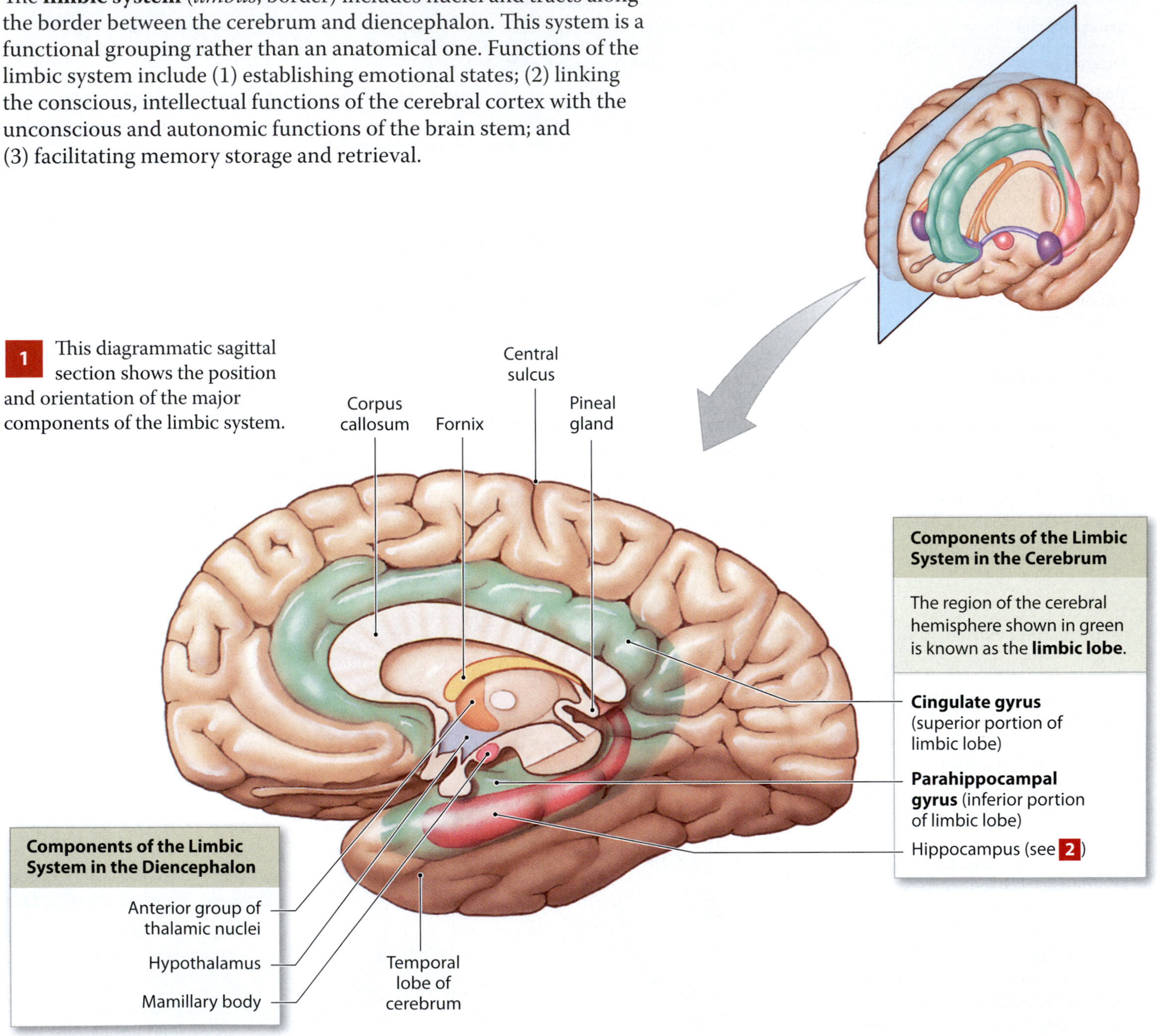

2 The specific functions of important limbic system components and nuclei are indicated in this sectional view.

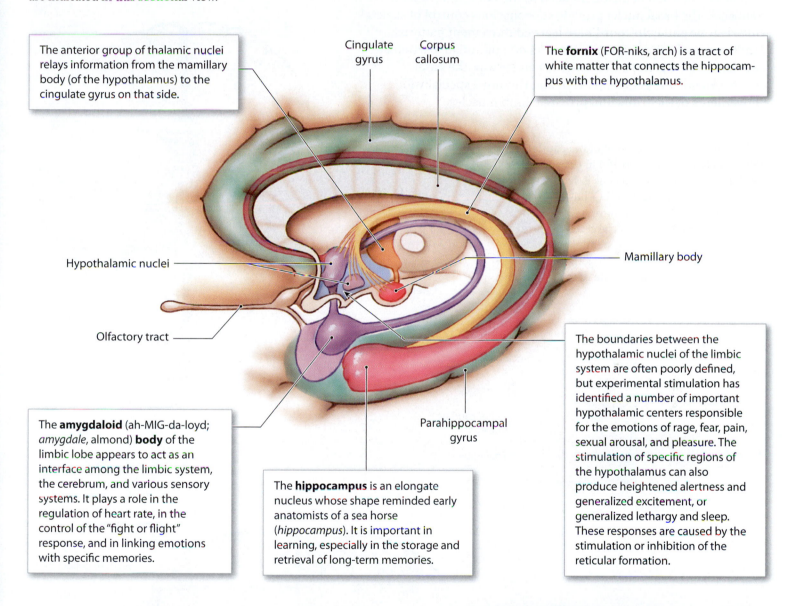

Cingulate gyrus

Corpus callosum

The anterior group of thalamic nuclei relays information from the mamillary body (of the hypothalamus) to the cingulate gyrus on that side.

The **fornix** (FOR-niks, arch) is a tract of white matter that connects the hippocampus with the hypothalamus.

Hypothalamic nuclei

Mamillary body

Olfactory tract

The boundaries between the hypothalamic nuclei of the limbic system are often poorly defined, but experimental stimulation has identified a number of important hypothalamic centers responsible for the emotions of rage, fear, pain, sexual arousal, and pleasure. The stimulation of specific regions of the hypothalamus can also produce heightened alertness and generalized excitement, or generalized lethargy and sleep. These responses are caused by the stimulation or inhibition of the reticular formation.

Parahippocampal gyrus

The **amygdaloid** (ah-MIG-da-loyd; *amygdale*, almond) **body** of the limbic lobe appears to act as an interface among the limbic system, the cerebrum, and various sensory systems. It plays a role in the regulation of heart rate, in the control of the "fight or flight" response, and in linking emotions with specific memories.

The **hippocampus** is an elongate nucleus whose shape reminded early anatomists of a sea horse (*hippocampus*). It is important in learning, especially in the storage and retrieval of long-term memories.

Whereas the sensory cortex, motor cortex, and association areas of the cerebral cortex enable you to perform complex tasks, it is largely the limbic system that makes you want to do them. For this reason, the limbic system is also known as the motivational system.

Module 13.7 Review

a. List the primary functions of the limbic system.

b. Which region of the limbic system is particularly important for the storage and retrieval of long-term memories?

c. Damage to the amygdaloid body would interfere with the regulation of which division of the autonomic nervous system?

The basal nuclei of the cerebrum perform subconscious adjustment and refinement of ongoing voluntary movements

The **basal nuclei**, or basal ganglia, are masses of gray matter that lie within each hemisphere deep to the floor of the lateral ventricle. The basal nuclei provide subconscious control of skeletal muscle tone and help coordinate learned movement patterns. Under normal conditions, these nuclei do not initiate particular movements. But once a movement is under way, the basal nuclei provide the general pattern and rhythm, especially for movements of the trunk and proximal limb muscles.

1 The basal nuclei consist of the **caudate nucleus** and the **lentiform** (lens-shaped) **nucleus**. The lentiform nucleus is subdivided into a medial **globus pallidus** (GLŌ-bus PAL-i-dus; pale globe) and a lateral **putamen** (pū-TĀ-men). The axon bundles that link the cerebral cortex to the diencephalon and brain stem pass between and around the basal nuclei. Together these fibers constitute the **internal capsule** of the cerebrum.

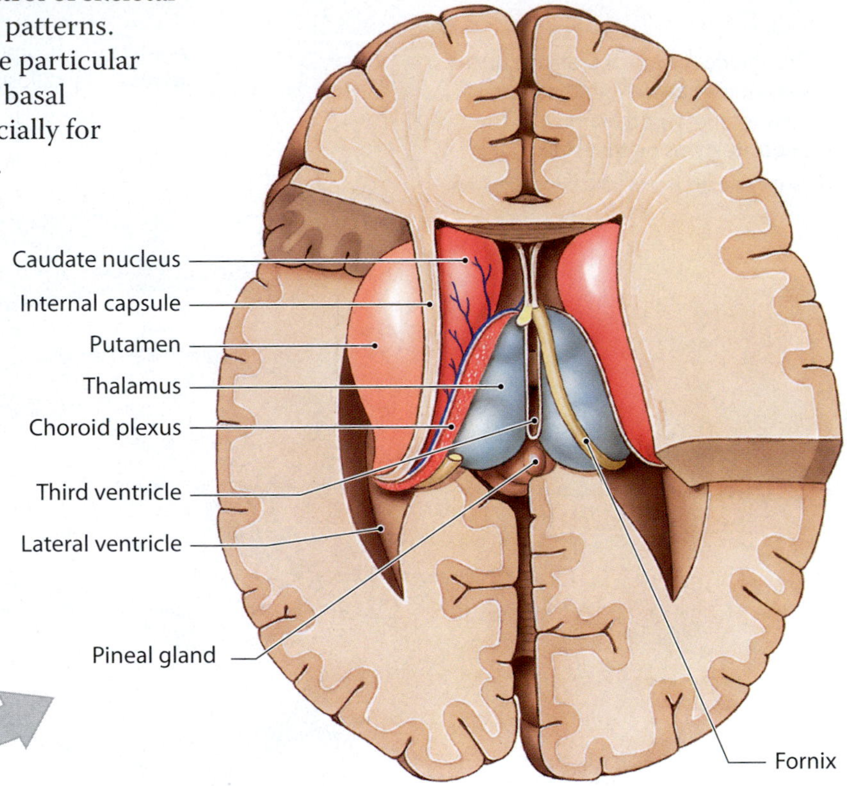

Caudate nucleus
Internal capsule
Putamen
Thalamus
Choroid plexus
Third ventricle
Lateral ventricle
Pineal gland
Fornix

Horizontal section, dissected

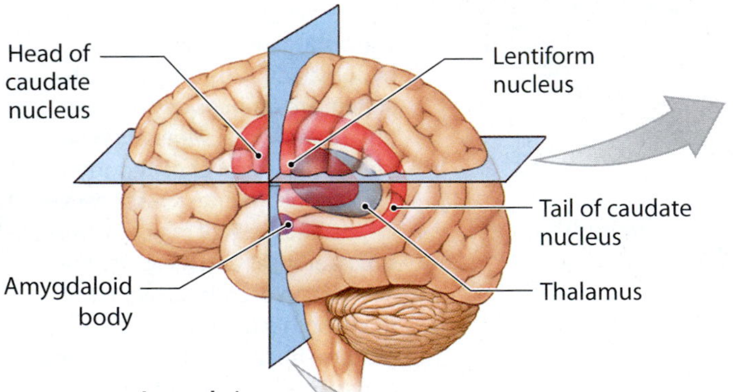

Head of caudate nucleus
Lentiform nucleus
Tail of caudate nucleus
Amygdaloid body
Thalamus

Lateral view

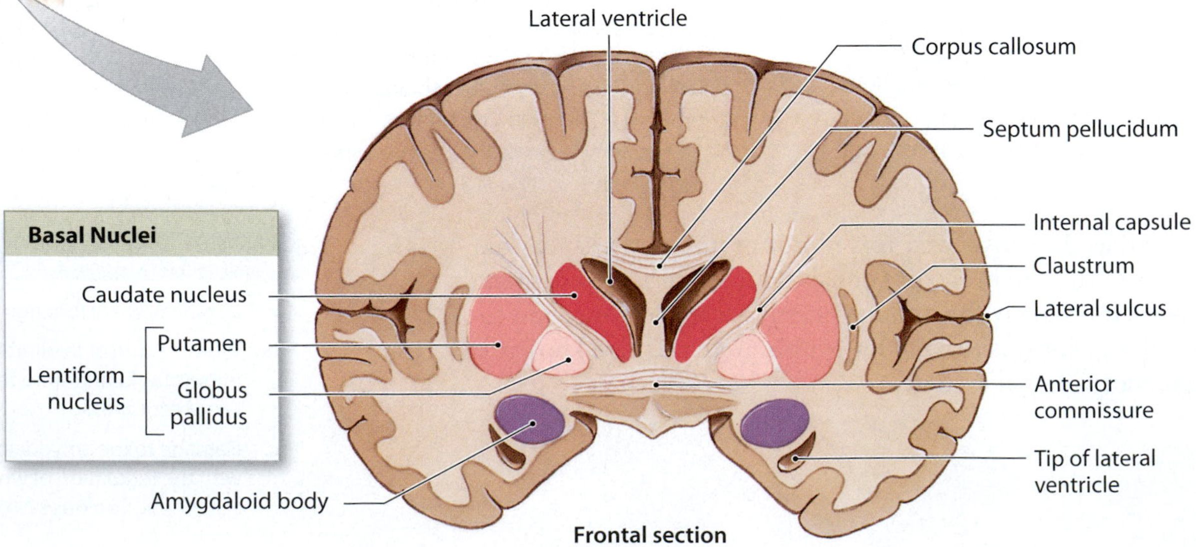

Basal Nuclei	
Caudate nucleus	
Lentiform nucleus	Putamen
	Globus pallidus

Lateral ventricle
Corpus callosum
Septum pellucidum
Internal capsule
Claustrum
Lateral sulcus
Anterior commissure
Tip of lateral ventricle
Amygdaloid body

Frontal section

2 This diagram indicates the roles of the basal nuclei in modifying ongoing movements.

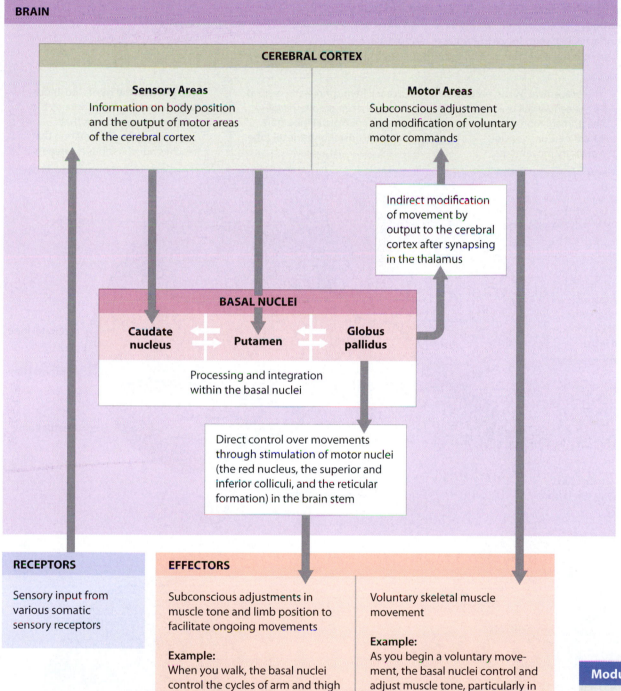

BRAIN

CEREBRAL CORTEX

Sensory Areas
Information on body position and the output of motor areas of the cerebral cortex

Motor Areas
Subconscious adjustment and modification of voluntary motor commands

Indirect modification of movement by output to the cerebral cortex after synapsing in the thalamus

BASAL NUCLEI

Caudate nucleus   Putamen   Globus pallidus

Processing and integration within the basal nuclei

Direct control over movements through stimulation of motor nuclei (the red nucleus, the superior and inferior colliculi, and the reticular formation) in the brain stem

RECEPTORS

Sensory input from various somatic sensory receptors

EFFECTORS

Subconscious adjustments in muscle tone and limb position to facilitate ongoing movements

Example:
When you walk, the basal nuclei control the cycles of arm and thigh movements that occur between the time you decide to "start" walking and the time you give the "stop" order.

Voluntary skeletal muscle movement

Example:
As you begin a voluntary movement, the basal nuclei control and adjust muscle tone, particularly in the appendicular muscles, to set your body position. When you decide to pick up a pencil, you consciously reach and grasp with your forearm, wrist, and hand while the basal nuclei operate at the subconscious level to position your shoulder and stabilize your arm.

Module 13.8 Review

a. Define the basal nuclei.

b. Describe the caudate nucleus.

c. What clinical signs would you expect to observe in an individual who has damage to the basal nuclei?

Superficial landmarks can be used to divide the surface of the cerebral cortex into lobes

1 Each cerebral hemisphere can be divided into regions called **lobes**. Your brain has a unique pattern of sulci and gyri, as individual as a fingerprint, but the boundaries between lobes are reliable landmarks. Lobes on the external surfaces are named after the overlying bones of the skull.

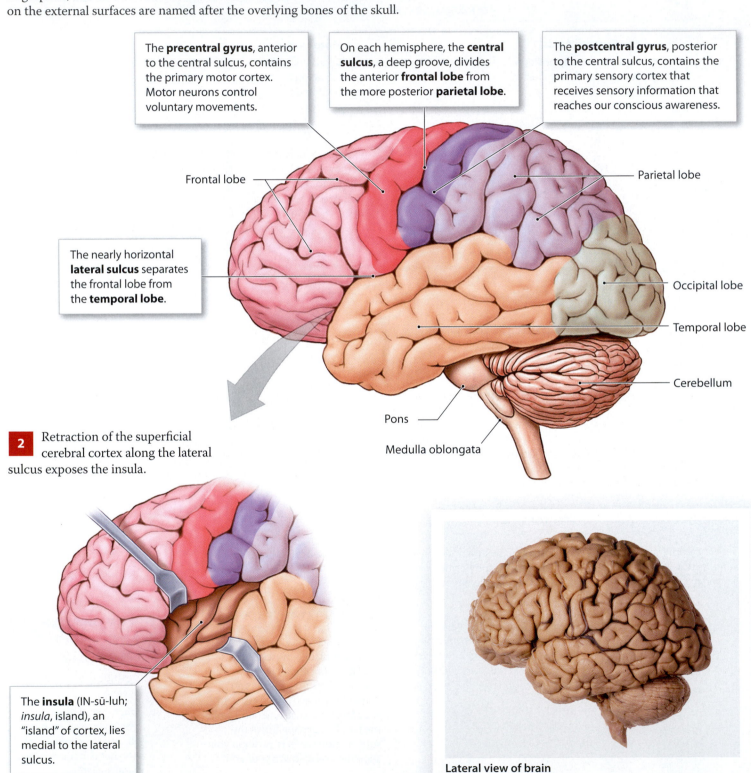

The **precentral gyrus**, anterior to the central sulcus, contains the primary motor cortex. Motor neurons control voluntary movements.

On each hemisphere, the **central sulcus**, a deep groove, divides the anterior **frontal lobe** from the more posterior **parietal lobe**.

The **postcentral gyrus**, posterior to the central sulcus, contains the primary sensory cortex that receives sensory information that reaches our conscious awareness.

Frontal lobe

Parietal lobe

The nearly horizontal **lateral sulcus** separates the frontal lobe from the **temporal lobe**.

Occipital lobe

Temporal lobe

Cerebellum

Pons

Medulla oblongata

2 Retraction of the superficial cerebral cortex along the lateral sulcus exposes the insula.

The **insula** (IN-sū-luh; *insula*, island), an "island" of cortex, lies medial to the lateral sulcus.

Lateral view of brain

3 This midsagittal view indicates the inner boundaries of the lobes and highlights the way the cerebral hemispheres cover the rest of the brain. For clarity, structures and regions outside of the cerebrum are labeled in italics.

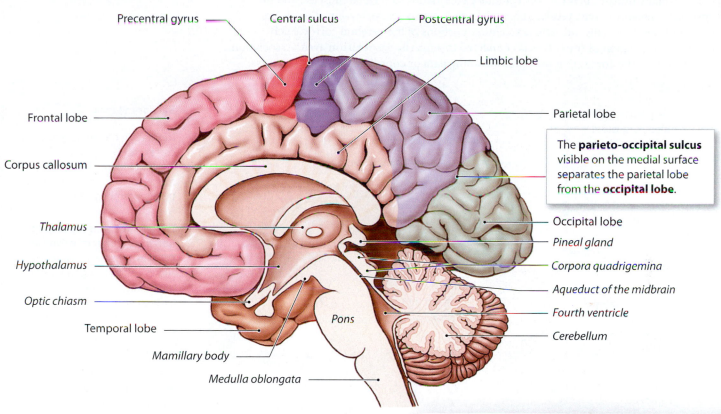

Precentral gyrus — Central sulcus — Postcentral gyrus

Limbic lobe

Frontal lobe

Parietal lobe

Corpus callosum

The **parieto-occipital sulcus** visible on the medial surface separates the parietal lobe from the **occipital lobe**.

Thalamus

Occipital lobe

Pineal gland

Hypothalamus

Corpora quadrigemina

Aqueduct of the midbrain

Optic chiasm

Fourth ventricle

Temporal lobe

Pons

Cerebellum

Mamillary body

Medulla oblongata

As you proceed, you should remember the following general facts about the cerebral hemispheres:

- Each cerebral hemisphere receives sensory information from, and sends motor commands to, the opposite side of the body. Thus the motor areas of the left cerebral hemisphere control muscles on the right side, and motor areas of the right cerebral hemisphere control muscles on the left side. This crossing over, which occurs in the brain stem and spinal cord, has no known functional significance.

- Even though the two hemispheres may look identical and have many similar functions, important differences exist.

- The correspondence between a specific function and a specific region of the cerebral cortex is imprecise. The boundaries are indistinct and have considerable overlap, and some cortical functions, such as consciousness, cannot easily be assigned to any single region. However, we know that normal individuals use all portions of the brain.

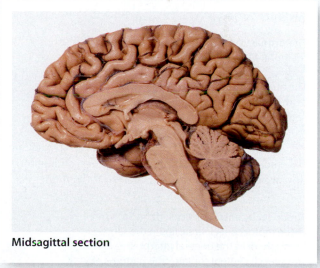

Midsagittal section

Module 13.9 Review

a. Identify the lobes of the cerebrum and indicate the basis for their names.

b. Describe the insula.

c. What effect would damage to the left postcentral gyrus produce?

The lobes of the cerebral cortex contain regions with specific functions

1 The **primary motor cortex** issues voluntary commands to skeletal muscles, and the **primary sensory cortex** receives general somatic sensory information. The special senses of sight, sound, smell, and taste reach other portions of the cerebral cortex. Each sensory and motor region of the cortex is connected to a nearby **association area**. Association areas are regions of the cortex that interpret incoming data or coordinate a motor response.

Motor Cortex

Neurons of the primary motor cortex are called **pyramidal cells**, because their cell bodies resemble little pyramids.

The **somatic motor association area** is responsible for the coordination of learned movements.

Gustatory Cortex

The **gustatory cortex** of the insula receives information from taste receptors.

Olfactory Cortex

The **olfactory cortex** receives sensory information from the olfactory receptors.

Auditory Cortex

The **primary auditory cortex** is responsible for monitoring auditory (sound) information.

The **auditory association area** monitors sensory activity in the auditory cortex and recognizes sounds, such as spoken words.

Sensory Cortex

Neurons in the primary sensory cortex receive somatic sensory information from receptors for touch, pressure, pain, vibration, taste, or temperature.

The **somatic sensory association area** monitors activity in the primary sensory cortex. It allows you to recognize a light touch, such as a mosquito landing on your arm.

Visual Cortex

The **primary visual cortex** receives information from the lateral geniculate nuclei.

The **visual association area** monitors the patterns of activity in the visual cortex and interprets the results. When you see the symbols c, a, and r, your visual association area recognizes that they form the word "car."

Central sulcus

PARIETAL LOBE

FRONTAL LOBE

OCCIPITAL LOBE

Lateral sulcus

TEMPORAL LOBE

2 **Integrative centers** concerned with the performance of complex processes, such as speech, writing, mathematics, and understanding spatial relationships, are restricted to either the left or the right hemisphere.

The **speech center,** also called the Broca area or the motor speech area, lies in the same hemisphere as the general interpretive area. The speech center regulates the patterns of breathing and vocalization needed for normal speech.

The **prefrontal cortex** coordinates information relayed from the association areas of the cortex. In the process, it performs such abstract intellectual functions as predicting the consequences of events or actions.

The **frontal eye field** controls learned eye movements, such as when you scan these lines of text.

The **general interpretive area**, or the Wernicke area, receives information from all the sensory association areas. This analytical center is present in only one hemisphere (typically the left). This region plays an essential role in your personality by integrating sensory information and coordinating access to complex visual and auditory memories.

3 Each of the two cerebral hemispheres is responsible for specific functions that are not ordinarily performed by the opposite hemisphere. This regional specialization is called **hemispheric lateralization**.

Left Cerebral Hemisphere

In most people, the left hemisphere contains the general interpretive and speech centers and is responsible for language-based skills. Reading, writing, and speaking, for example, depend on processing done in the left cerebral hemisphere. In addition, the premotor cortex that functions in the control of hand movements is larger on the left side for right-handed individuals than for left-handed individuals. The left hemisphere is also important in performing analytical tasks, such as mathematics and logic.

Right Cerebral Hemisphere

The right cerebral hemisphere analyzes sensory information and relates the body to the sensory environment. Interpretive centers in this hemisphere enable you to identify familiar objects by touch, smell, sight, taste, or feel. For example, the right hemisphere plays a dominant role in recognizing faces and in understanding three-dimensional relationships. It is also important in analyzing the emotional context of a conversation—for instance, distinguishing between the threat "Get lost!" and the question "Get lost?"

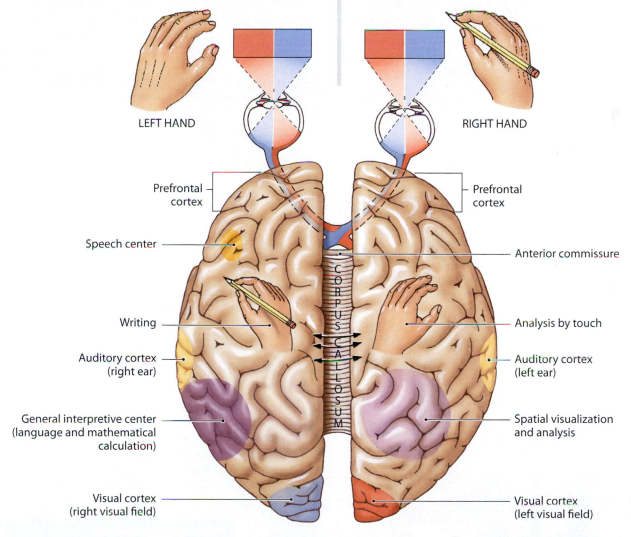

LEFT HAND

RIGHT HAND

Prefrontal cortex

Prefrontal cortex

Speech center

Anterior commissure

CORPUS CALLOSUM

Writing

Analysis by touch

Auditory cortex (right ear)

Auditory cortex (left ear)

General interpretive center (language and mathematical calculation)

Spatial visualization and analysis

Visual cortex (right visual field)

Visual cortex (left visual field)

Left-handed people represent about 9 percent of the human population. Although in most cases the primary motor cortex of the right hemisphere controls motor function for the dominant left hand, the centers involved with speech and analytical function are in the left hemisphere. Interestingly, an unusually high percentage of musicians and artists are left-handed, and the primary motor cortex and association areas on the right cerebral hemisphere are near the association areas involved with spatial visualization and emotions.

Module 13.10 Review

a. Where is the primary motor cortex located?

b. Which senses are affected by damage to the temporal lobes?

c. Which brain part has been affected in a stroke victim who is unable to speak?

White matter interconnects the cerebral hemispheres, the lobes of each hemisphere, and links the cerebrum to the rest of the brain

The interior of the cerebral hemispheres consists primarily of white matter organized into groups that share common functions.

1 **Association fibers** interconnect areas of neural cortex within a single cerebral hemisphere.

The shortest association fibers are called **arcuate** (AR-kū-āt) **fibers**, because they curve in an arc to pass from one gyrus to another.

Longer association fibers are organized into discrete bundles, or fasciculi. The **longitudinal fasciculi** connect the frontal lobe to the other lobes of the same cerebral hemisphere.

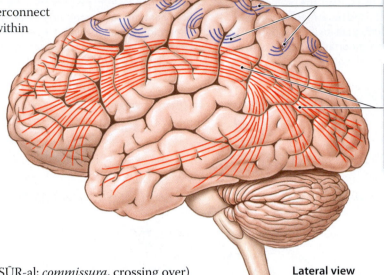

Lateral view

2 **Commissural** (kom-i-SŪR-al; *commissura*, crossing over) **fibers** interconnect the cerebral hemispheres. **Projection fibers** link the cerebral cortex to the diencephalon, brain stem, cerebellum, and spinal cord. All projection fibers must pass through the diencephalon, where axons heading to sensory areas of the cerebral cortex pass among the axons descending from motor areas of the cortex. In gross dissection, the ascending fibers and descending fibers look alike, and the entire mass is known as the **internal capsule**.

Projection fibers of internal capsule

Longitudinal fissure

The **corpus callosum** is the most important band of commissural fibers because it allows communication and coordination between the left and right cerebral hemispheres. It contains more than 200 million axons carrying some 4 billion impulses per second.

The **anterior commissure** is a smaller tract of commissural fibers that provides another route for communication between the cerebral hemispheres. Its importance increases if the corpus callosum is damaged.

Anterior view

Module 13.11 Review

a. What special names are given to axons in the white matter of the cerebral hemispheres?

b. What is the function of the longitudinal fasciculi?

c. What are fibers carrying information between the brain and spinal cord called, and through which brain regions do they pass?

Brain activity can be monitored using external electrodes; the record is called an electroencephalogram, or EEG

1 Neural function depends on electrical impulses, and the brain contains billions of neurons and the axons of the central white matter. The activity under way at any given moment generates an electrical field that can be measured by placing electrodes on the scalp. The electrical activity changes constantly, as nuclei and cortical areas are stimulated or quieted down. A printed report of the electrical activity of the brain is called an **electroencephalogram (EEG)**. The electrical patterns observed are called **brain waves**.

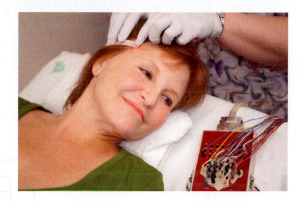

Alpha waves occur in the brains of healthy, awake adults who are resting with their eyes closed. Alpha waves disappear during sleep, but they also vanish when the individual begins to concentrate on some specific task.

Beta waves are higher-frequency waves typical of individuals who are either concentrating on a task, under stress, or in a state of psychological tension.

Theta waves may appear transiently during sleep in normal adults but are most often observed in children and in intensely frustrated adults. The presence of theta waves under other circumstances may indicate the presence of a brain disorder, such as a tumor.

Delta waves are very-large-amplitude, low-frequency waves. They are normally seen during deep sleep in individuals of all ages. Delta waves are also seen in the brains of infants (in whom cortical development is still incomplete) and in awake adults when a tumor, vascular blockage, or inflammation has damaged portions of the brain.

Seconds   1   2   3   4

Electrical activity in the two hemispheres is generally synchronized by a "pacemaker" mechanism that appears to involve the thalamus. Asynchrony between the hemispheres can therefore indicate localized damage or other cerebral abnormalities. A tumor or injury affecting one hemisphere, for example, typically changes the pattern in that hemisphere, and the patterns of the two hemispheres are no longer aligned. A **seizure** is a temporary cerebral disorder accompanied by abnormal movements, unusual sensations, inappropriate behavior, or some combination of these signs and symptoms. Clinical conditions characterized by seizures are known as seizure disorders, or **epilepsies**. Seizures of all kinds are accompanied by a marked change in the pattern of electrical activity recorded in an electroencephalogram. The change begins in one portion of the cerebral cortex but may subsequently spread across the entire cortical surface, like a wave on the surface of a pond.

Module 13.12 Review

a. Define electroencephalogram (EEG).

b. Describe the four wave types associated with an EEG.

c. Differentiate between a seizure and epilepsy.

The twelve pairs of cranial nerves can be classified as sensory, special sensory, motor, or mixed nerves

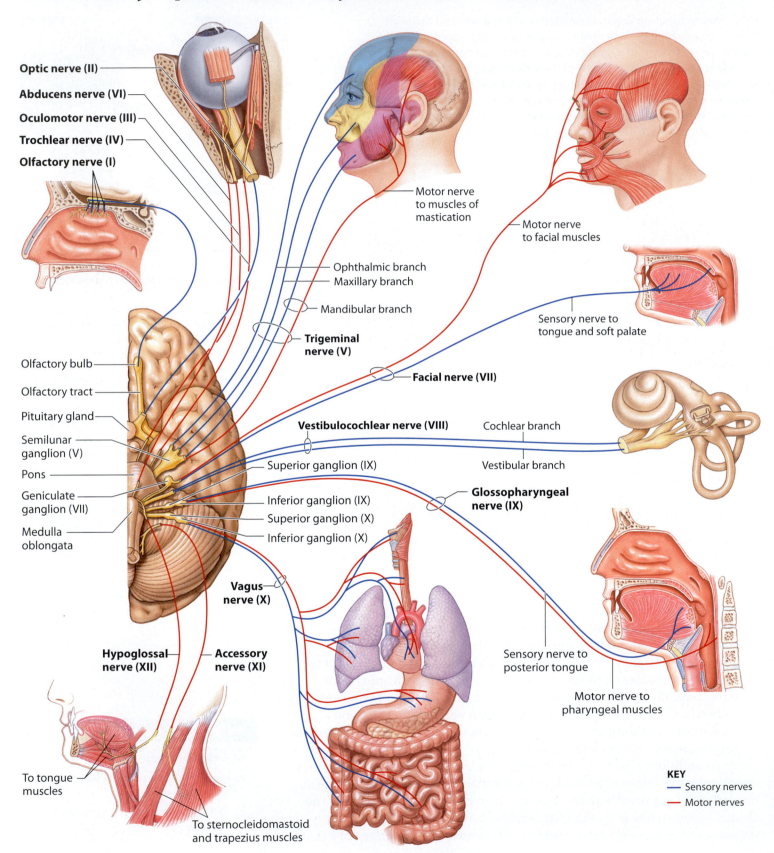

Optic nerve (II)

Abducens nerve (VI)

Oculomotor nerve (III)

Trochlear nerve (IV)

Olfactory nerve (I)

Olfactory bulb

Olfactory tract

Pituitary gland

Semilunar ganglion (V)

Pons

Geniculate ganglion (VII)

Medulla oblongata

Hypoglossal nerve (XII)

Accessory nerve (XI)

To tongue muscles

To sternocleidomastoid and trapezius muscles

Motor nerve to muscles of mastication

Ophthalmic branch

Maxillary branch

Mandibular branch

Trigeminal nerve (V)

Facial nerve (VII)

Vestibulocochlear nerve (VIII)

Superior ganglion (IX)

Inferior ganglion (IX)

Superior ganglion (X)

Inferior ganglion (X)

Vagus nerve (X)

Motor nerve to facial muscles

Sensory nerve to tongue and soft palate

Cochlear branch

Vestibular branch

Glossopharyngeal nerve (IX)

Sensory nerve to posterior tongue

Motor nerve to pharyngeal muscles

KEY
— Sensory nerves
— Motor nerves

The Branches and Functions of Cranial Nerves

Cranial Nerve (Number)	Sensory Ganglion	Branch	Primary Function	Foramen	Innervation
Olfactory (I)			Special sensory	Olfactory foramina of ethmoid	Olfactory epithelium
Optic (II)			Special sensory	Optic canal	Retina of eye
Oculomotor (III)			Motor	Superior orbital fissure	Inferior, medial, superior rectus, inferior oblique, and levator palpebrae superioris muscles; intrinsic eye muscles
Trochlear (IV)			Motor	Superior orbital fissure	Superior oblique muscle
Trigeminal (V)	Semilunar		Mixed	Superior orbital fissure	Areas associated with the jaws
		Ophthalmic	Sensory	Superior orbital fissure	Orbital structures, nasal cavity, skin of forehead, upper eyelid, eyebrows, and part of nose
		Maxillary	Sensory	Foramen rotundum	Lower eyelid; superior lip, gums, and teeth; cheek, part of nose, palate, and part of pharynx
		Mandibular	Mixed	Foramen ovale	*Sensory:* inferior gums, teeth, lips, part of palate, and part of tongue *Motor:* muscles of mastication
Abducens (VI)			Motor	Superior orbital fissure	Lateral rectus muscle
Facial (VII)	Geniculate		Mixed	Internal acoustic meatus to facial canal; exits at stylomastoid foramen	*Sensory:* taste receptors on anterior two-thirds of tongue *Motor:* muscles of facial expression, lacrimal gland, submandibular gland, and sublingual salivary glands
Vestibulocochlear (Acoustic) (VIII)		Cochlear Vestibular	Special sensory	Internal acoustic meatus	Cochlea (receptors for hearing) Vestibule (receptors for motion and balance)
Glossopharyngeal (IX)	Superior and inferior		Mixed	Jugular foramen	*Sensory:* posterior third of tongue; pharynx and part of palate; receptors for blood pressure, pH, oxygen, and carbon dioxide concentrations *Motor:* pharyngeal muscles and parotid salivary gland
Vagus (X)	Superior and inferior		Mixed	Jugular foramen	*Sensory:* pharynx; auricle and external acoustic canal; diaphragm; visceral organs in thoracic and abdominopelvic cavities *Motor:* palatal and pharyngeal muscles and visceral organs in thoracic and abdominopelvic cavities
Accessory (XI)		Internal	Motor	Jugular foramen	Skeletal muscles of palate, pharynx, and larynx (with vagus nerve)
		External	Motor	Jugular foramen	Sternocleidomastoid and trapezius muscles
Hypoglossal (XII)			Motor	Hypoglossal canal	Tongue musculature

Module 13.13 Review

a. Identify the cranial nerves by name and number.

b. Which cranial nerves have motor functions only?

c. Which cranial nerves are mixed nerves?

1. Labeling

Label the structures in the accompanying figure of a lateral view of the human brain.

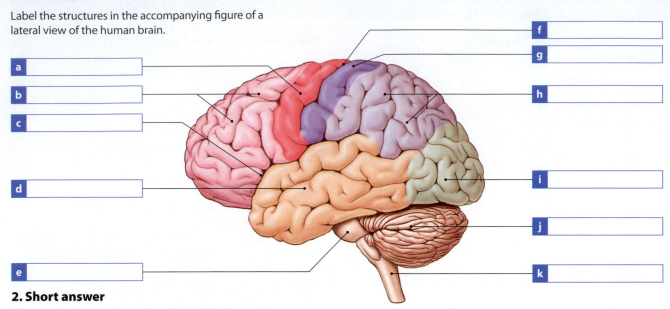

a ___
b ___
c ___
d ___
e ___
f ___
g ___
h ___
i ___
j ___
k ___

2. Short answer

Identify the cranial nerves in the accompanying figure, and indicate the function of each: M = motor, S = sensory, or B = both motor and sensory.

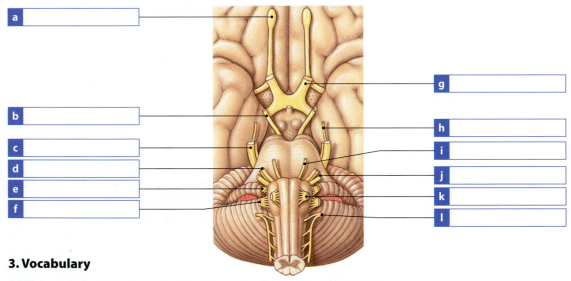

a ___
b ___
c ___
d ___
e ___
f ___
g ___
h ___
i ___
j ___
k ___
l ___

3. Vocabulary

Write the boldfaced term introduced in this section that matches the definition.

a _____   Forms the walls of the diencephalon

b _____   The shortest association fibers in the CNS white matter

c _____   The tract of white matter that connects the hippocampus with the hypothalamus

d _____   The fibers that permit communication between the two cerebral hemispheres

e _____   The nuclei made up of the caudate nucleus and the lentiform nucleus

4. Section integration

Smelling salts may restore consciousness after a person has fainted. The active ingredient of smelling salts is ammonia, and it acts by irritating the lining of the nasal cavity. Propose a mechanism by which smelling salts would raise a person from the unconscious state to the conscious state.

___

___

Sensory and Motor Pathways

Sensory receptors are specialized cells or cell processes that provide your central nervous system with information about conditions inside or outside the body. The term **general senses** is used to describe our sensitivity to temperature, pain, touch, pressure, vibration, and proprioception. General sensory receptors are distributed throughout the body, and they are relatively simple in structure. Sensory pathways begin at peripheral receptors and end within the CNS, often at the diencephalon and/or cerebral hemispheres. Much of the information carried by a sensory pathway never reaches the primary sensory cortex and our awareness. The information carried by a sensory pathway is called a **sensation**, and the conscious awareness of a sensation is called a **perception**.

1 The area monitored by a single receptor cell is its **receptive field**. The larger the receptive field, the poorer your ability to localize a stimulus. A touch receptor on the general body surface, for example, may have a receptive field 7 cm (2.5 in.) in diameter. As a result, you can describe a light touch there as affecting only a general area, not an exact spot. On the tongue or fingertips, where the receptive fields are less than a millimeter in diameter, you can be very precise about the location of a stimulus.

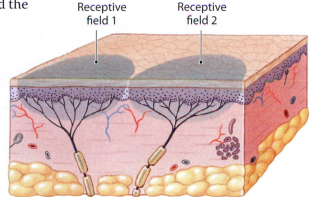

Receptive field 1   Receptive field 2

2 This diagram presents the basic events that occur along sensory and motor pathways.

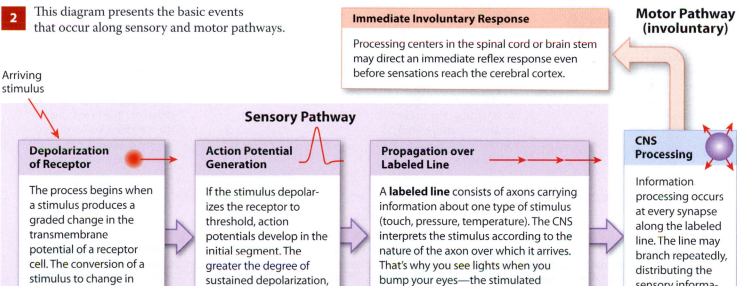

Motor Pathway (involuntary)

Immediate Involuntary Response

Processing centers in the spinal cord or brain stem may direct an immediate reflex response even before sensations reach the cerebral cortex.

Arriving stimulus

Sensory Pathway

Depolarization of Receptor

The process begins when a stimulus produces a graded change in the transmembrane potential of a receptor cell. The conversion of a stimulus to change in membrane potential is called **transduction**.

Action Potential Generation

If the stimulus depolarizes the receptor to threshold, action potentials develop in the initial segment. The greater the degree of sustained depolarization, the higher the frequency of action potentials.

Propagation over Labeled Line

A **labeled line** consists of axons carrying information about one type of stimulus (touch, pressure, temperature). The CNS interprets the stimulus according to the nature of the axon over which it arrives. That's why you see lights when you bump your eyes—the stimulated receptor sends information over axons that normally carry visual information.

CNS Processing

Information processing occurs at every synapse along the labeled line. The line may branch repeatedly, distributing the sensory information to multiple nuclei and centers in the spinal cord and brain.

Voluntary Response

The voluntary response, which isn't immediate, can moderate, enhance, or supplement the relatively simple reflexive response.

Motor Pathway (voluntary)

Perception

Only about 1 percent of arriving sensations are relayed to the primary sensory cortex, and often those are sensations that merit a voluntary response.

453

Receptors for the general senses can be classified by function and by sensitivity

1 The simplest receptors are the dendrites of sensory neurons. As shown at the right, the branching tips of these dendrites, called free nerve endings, are not protected by accessory structures. Free nerve endings extend through a tissue the way grass roots extend into the soil. They can be stimulated by many different stimuli and therefore exhibit little receptor specificity. For example, free nerve endings that respond to tissue damage by providing pain sensations may be stimulated by chemical stimulation, pressure, temperature changes, or trauma. The sensitivity and specificity of a free nerve ending may be altered by its location and the presence of accessory structures.

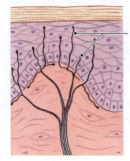

Free nerve endings

2 As indicated in this concept map, general sensory receptors can be classified according to the nature of the primary stimulus.

A Functional Classification of General Sensory Receptors

Nociceptors

Nociceptors are pain receptors. They are free nerve endings with large receptive fields and broad sensitivity. Two types of axons—Type A and Type C fibers—carry pain sensations.

Thermoreceptors

Thermoreceptors, or temperature receptors, are free nerve endings located in the dermis, in skeletal muscles, in the liver, and in the hypothalamus. Cold receptors are three or four times more numerous than warm receptors. No structural differences between warm and cold thermoreceptors have been identified.

Chemoreceptors

Chemoreceptors respond to water-soluble and lipid-soluble substances that are dissolved in body fluids (interstitial fluid, plasma, and CSF).

Mechanoreceptors

Mechanoreceptors are sensitive to stimuli that distort their plasma membranes. These membranes contain mechanically gated ion channels whose gates open or close in response to stretching, compression, twisting, or other distortions of the membrane.

Myelinated **Type A fibers** carry sensations of **fast pain**, or prickling pain. An injection or a deep cut produces this type of pain. These sensations very quickly reach the CNS, where they often trigger somatic reflexes. They are also relayed to the primary sensory cortex and so receive conscious attention. In most cases, the arriving information permits the stimulus to be localized to an area several centimeters in diameter.

Slower, unmyelinated **Type C fibers** carry sensations of **slow pain**, or burning and aching pain. These sensations cause a generalized activation of the reticular formation and thalamus. The individual becomes aware of the pain but has only a general idea of the area affected.

Proprioceptors monitor the positions of joints and muscles. They are the most structurally and functionally complex of the general sensory receptors. One example is the muscle spindle, considered in Module 12.10.

Baroreceptors (bar-ō-rē-SEP-torz; *baro-*, pressure) detect pressure changes in the walls of blood vessels and in portions of the digestive, respiratory, and urinary tracts.

Tactile receptors provide the sensations of touch, pressure, and vibration. Touch sensations provide information about shape or texture, whereas pressure sensations indicate the degree and frequency of mechanical distortion. Extremely sensitive **fine touch and pressure receptors** provide detailed information about a stimulation, whereas **crude touch and pressure receptors** provide poor localization and give little information.

3 Receptors can also be categorized based on the nature of their response to stimulation.

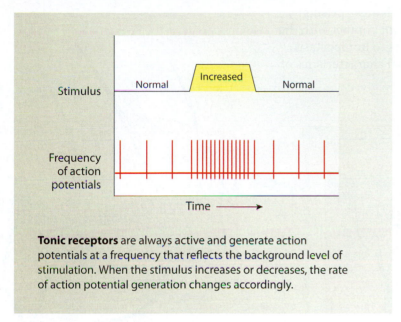

Tonic receptors are always active and generate action potentials at a frequency that reflects the background level of stimulation. When the stimulus increases or decreases, the rate of action potential generation changes accordingly.

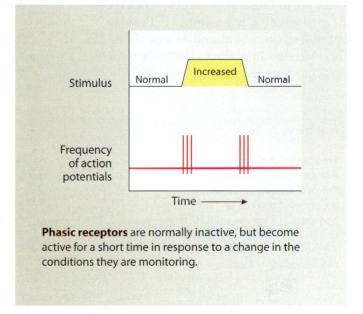

Phasic receptors are normally inactive, but become active for a short time in response to a change in the conditions they are monitoring.

4 A reduction in sensitivity in the presence of a constant stimulus is called **adaptation**. You seldom notice the rumble of the tires when you ride in a car, or the background noise of the air conditioner, because your nervous system quickly adapts to stimuli that are painless and constant. Adaptation may be peripheral or central.

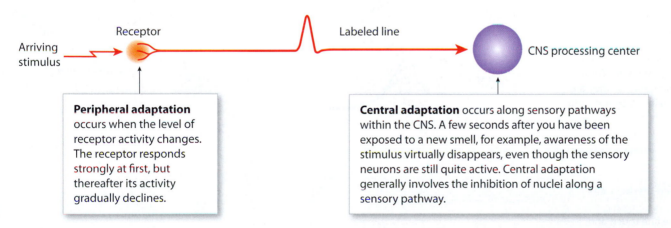

Peripheral adaptation occurs when the level of receptor activity changes. The receptor responds strongly at first, but thereafter its activity gradually declines.

Central adaptation occurs along sensory pathways within the CNS. A few seconds after you have been exposed to a new smell, for example, awareness of the stimulus virtually disappears, even though the sensory neurons are still quite active. Central adaptation generally involves the inhibition of nuclei along a sensory pathway.

Module 13.14 Review

a. List the four types of general sensory receptors based on function, and identify the type of stimulus that excites each type.

b. Describe the three classes of mechanoreceptors.

c. Explain adaptation, and differentiate between peripheral adaptation and central adaptation.

General sensory receptors are relatively simple in structure and widely distributed in the body

There are millions of general sensory receptors in the body. Not surprisingly, the greatest diversity is found in the skin, which is in constant contact with the external environment and its associated hazards and threats to homeostasis. This overview introduces the basic structural and functional characteristics of the important receptor types in the skin.

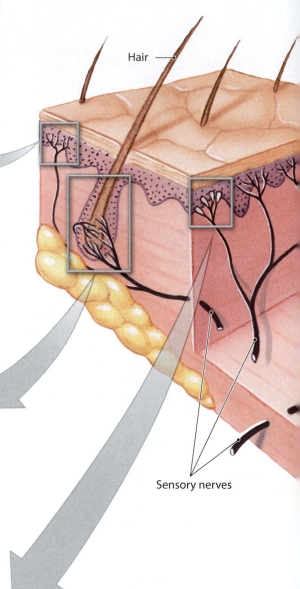

Hair

Sensory nerves

Free Nerve Endings

Free nerve endings are the branching tips of sensory neurons. They are not protected by any accessory structures, and they are nonspecific: They can respond to tactile, pain, and temperature stimuli. Free nerve endings are the most common receptors in the skin.

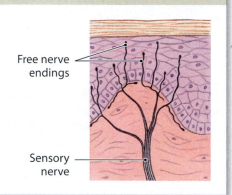

Free nerve endings

Sensory nerve

Root Hair Plexus

Wherever hairs are located, the nerve endings of the **root hair plexus** monitor distortions and movements across the body surface. When a hair is displaced, the movement of the follicle distorts the sensory dendrites and produces action potentials. These receptors adapt rapidly, so they are best at detecting initial contact and subsequent movements.

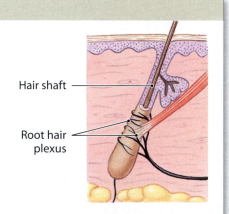

Hair shaft

Root hair plexus

Tactile Discs and Merkel Cells

Tactile discs are fine touch and pressure receptors. They are extremely sensitive tonic receptors with very small receptive fields. The dendritic processes of a single myelinated afferent fiber make close contact with **Merkel cells**, unusually large epithelial cells in the stratum basale of the skin.

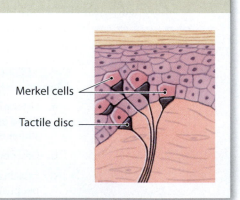

Merkel cells

Tactile disc

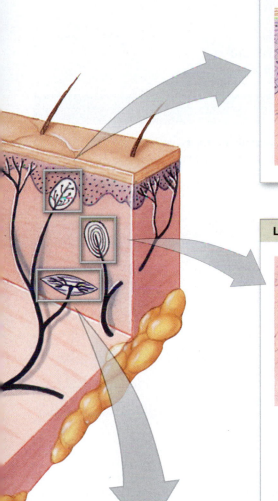

Tactile Corpuscles

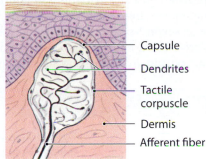

Capsule
Dendrites
Tactile corpuscle
Dermis
Afferent fiber

Tactile corpuscles, or Meissner (MĪS-ner) corpuscles, provide sensations of fine touch and pressure and low-frequency vibration. They adapt to stimulation within a second after contact. Tactile corpuscles are fairly large structures, measuring roughly 100 μm in length and 50 μm in width. These receptors are most abundant in the eyelids, lips, fingertips, nipples, and external genitalia. The dendrites are highly coiled and interwoven, and they are surrounded by modified Schwann cells. A fibrous capsule surrounds the entire complex and anchors it within the dermis.

Lamellated Corpuscles

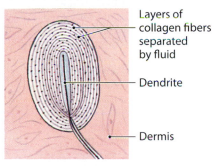

Layers of collagen fibers separated by fluid
Dendrite
Dermis

Lamellated (LAM-e-lāt-ed; *lamella*, thin plate) **corpuscles**, or pacinian (pa-SIN-ē-an) corpuscles, are sensitive to deep pressure. Because they are fast-adapting receptors, they are most sensitive to pulsing or high-frequency vibrating stimuli. A single dendrite lies within a series of concentric layers of collagen fibers and specialized fibroblasts. The entire corpuscle may reach 4 mm in length and 1 mm in diameter. The concentric layers, separated by interstitial fluid, shield the dendrite from virtually every source of stimulation other than direct pressure. Somatic sensory information is provided by lamellated corpuscles located throughout the dermis, notably in the fingers, mammary glands, and external genitalia; in the superficial and deep fasciae; and in joint capsules. Visceral sensory information is provided by lamellated corpuscles in mesenteries, in the pancreas, and in the walls of the urethra and urinary bladder.

Ruffini Corpuscles

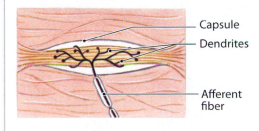

Capsule
Dendrites
Afferent fiber

Ruffini (roo-FĒ-nē) **corpuscles** are sensitive to pressure and distortion of the reticular (deep) dermis. These receptors are tonic and show little if any adaptation. A capsule surrounds a core of collagen fibers that are continuous with those of the surrounding dermis. Within the capsule, a network of dendrites is intertwined with the collagen fibers. Any tension or distortion of the dermis tugs or twists the capsular fibers, stretching or compressing the attached dendrites and altering the activity in the myelinated afferent fiber.

Module 13.15 Review

a. Identify the six types of tactile receptors located in the skin, and describe their sensitivities.

b. Which types of tactile receptors are located only in the dermis?

c. Which is likely to be more sensitive to continuous deep pressure: a lamellated corpuscle or a Ruffini corpuscle?

Three major somatic sensory pathways carry information from the skin and musculature to the CNS

1 Two tracts within the **spinothalamic pathway** provide conscious sensations of poorly localized ("crude") touch, pressure, pain, and temperature. In this pathway, axons of **first-order neurons** enter the spinal cord and synapse on **second-order neurons** within the posterior gray horns. The axons of these interneurons cross to the opposite side of the spinal cord before ascending to the thalamus. **Third-order neurons** synapse in the primary sensory cortex.

Spinothalamic Pathway

The **anterior spinothalamic tracts** of the spinothalamic pathway carry crude touch and pressure sensations.

The **lateral spinothalamic tracts** of the spinothalamic pathway carry pain and temperature sensations.

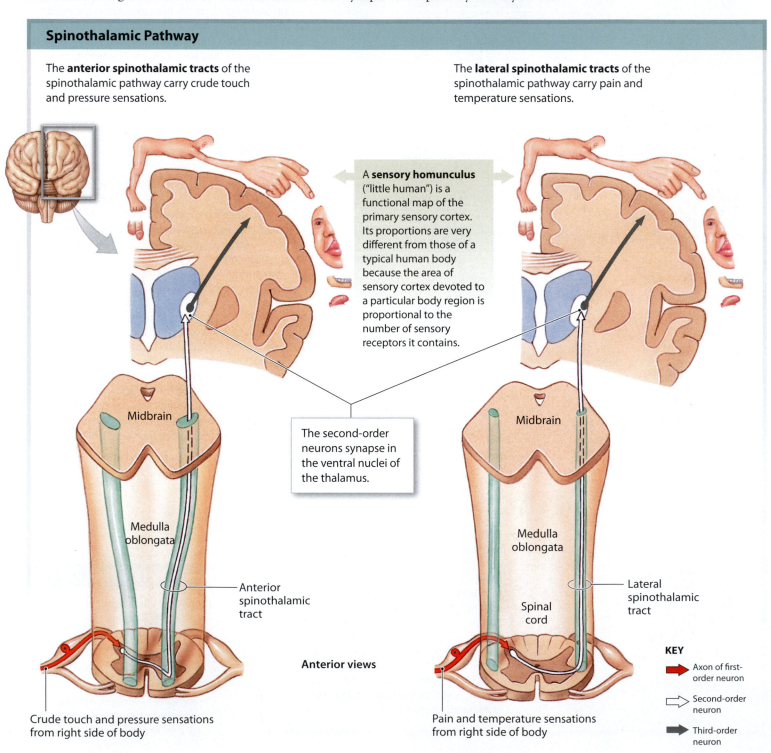

A **sensory homunculus** ("little human") is a functional map of the primary sensory cortex. Its proportions are very different from those of a typical human body because the area of sensory cortex devoted to a particular body region is proportional to the number of sensory receptors it contains.

The second-order neurons synapse in the ventral nuclei of the thalamus.

Midbrain

Midbrain

Medulla oblongata

Medulla oblongata

Anterior spinothalamic tract

Lateral spinothalamic tract

Spinal cord

Anterior views

Crude touch and pressure sensations from right side of body

Pain and temperature sensations from right side of body

KEY

→ Axon of first-order neuron

⇨ Second-order neuron

→ Third-order neuron

2 The **posterior column pathway** carries sensations of highly localized ("fine") touch, pressure, vibration, and proprioception. This pathway begins at a peripheral receptor and ends at the primary sensory cortex of the cerebral hemispheres.

Posterior Column Pathway

Ventral nuclei in thalamus

Midbrain

The **medial lemniscus** is a tract leading from the nucleus gracilis/ nucleus cuneatus to the thalamus.

Nucleus gracilis and nucleus cuneatus

The sensory axons ascend in the fasciculus gracilis and fasciculus cuneatus.

Medulla oblongata

Spinal cord

Dorsal root ganglion

Fine-touch, vibration, pressure, and proprioception sensations from right side of body

3 The cerebellum receives proprioceptive information about the position of skeletal muscles, tendons, and joints along the **spinocerebellar pathway**. The posterior spinocerebellar tracts contain axons that do not cross over to the opposite side of the spinal cord. These axons reach the cerebellar cortex via the inferior cerebellar peduncle of that side. The anterior spinocerebellar tracts are dominated by axons that have crossed over to the opposite side of the spinal cord.

Spinocerebellar Pathway

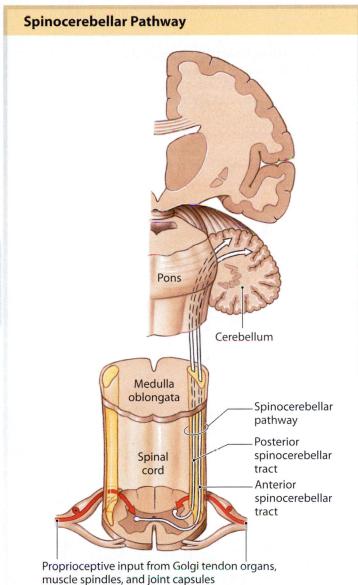

Pons

Cerebellum

Medulla oblongata

Spinal cord

Spinocerebellar pathway

Posterior spinocerebellar tract

Anterior spinocerebellar tract

Proprioceptive input from Golgi tendon organs, muscle spindles, and joint capsules

4 This cross section through the spinal cord shows the locations of the somatic sensory pathways.

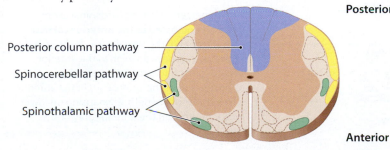

Posterior

Posterior column pathway

Spinocerebellar pathway

Spinothalamic pathway

Anterior

Module 13.16 Review

a. Define sensory homunculus.

b. Which spinal tracts carry action potentials generated by nociceptors?

c. Which cerebral hemisphere receives impulses conducted by the right fasciculus gracilis of the spinal cord?

The somatic nervous system controls skeletal muscles through upper and lower motor neurons

Somatic motor pathways always involve at least two motor neurons: an **upper motor neuron**, whose cell body lies in a CNS processing center, and a **lower motor neuron**, whose cell body lies in a nucleus of the brain stem or spinal cord. The upper motor neuron synapses on the lower motor neuron, which in turn innervates a single motor unit in a skeletal muscle.

1 The **corticospinal pathway** provides voluntary control over skeletal muscles. This pathway is sometimes called the pyramidal system because it begins at the pyramidal cells of the primary motor cortex. The axons of these upper motor neurons descend into the brain stem and spinal cord to synapse on lower motor neurons that control skeletal muscles.

Each region of the primary motor cortex corresponds with a specific region of the body. A functional map of the cortical areas is a **motor homunculus.** The proportions of the motor homunculus differ from those of the body because the motor area devoted to a specific region of the cortex is proportional to the number of motor units innervated and the degree of fine motor control available.

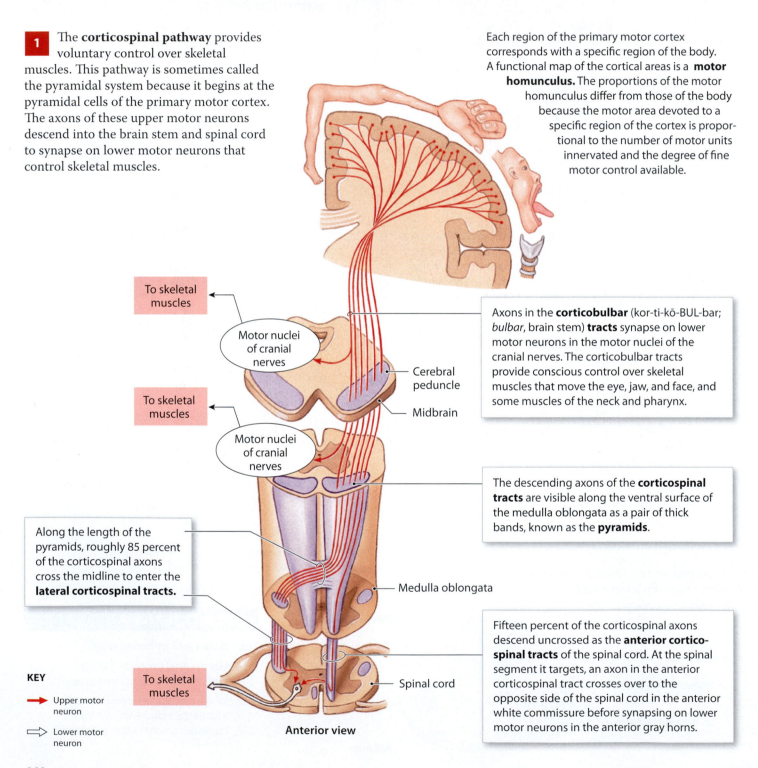

To skeletal muscles

Motor nuclei of cranial nerves

Cerebral peduncle

Midbrain

To skeletal muscles

Motor nuclei of cranial nerves

Axons in the **corticobulbar** (kor-ti-kō-BUL-bar; *bulbar*, brain stem) **tracts** synapse on lower motor neurons in the motor nuclei of the cranial nerves. The corticobulbar tracts provide conscious control over skeletal muscles that move the eye, jaw, and face, and some muscles of the neck and pharynx.

The descending axons of the **corticospinal tracts** are visible along the ventral surface of the medulla oblongata as a pair of thick bands, known as the **pyramids.**

Along the length of the pyramids, roughly 85 percent of the corticospinal axons cross the midline to enter the **lateral corticospinal tracts.**

Medulla oblongata

KEY

➡ Upper motor neuron

⇨ Lower motor neuron

To skeletal muscles

Spinal cord

Anterior view

Fifteen percent of the corticospinal axons descend uncrossed as the **anterior corticospinal tracts** of the spinal cord. At the spinal segment it targets, an axon in the anterior corticospinal tract crosses over to the opposite side of the spinal cord in the anterior white commissure before synapsing on lower motor neurons in the anterior gray horns.

2 Several centers in the cerebrum, diencephalon, and brain stem may issue somatic motor commands as a result of processing performed at a subconscious level. The locations of those centers are indicated in the diagram to the right. The components of the **medial pathway** help control gross movements of the trunk and proximal limb muscles, whereas those of the **lateral pathway** help control the distal limb muscles that perform more precise movements.

Motor cortex

Thalamus

Basal nuclei

The **red nucleus**, the primary nucleus of the lateral pathway, receives information from the cerebrum and cerebellum and adjusts upper limb position and background muscle tone.

Cerebellar nuclei

Nuclei of the Medial Pathway

Superior and inferior colliculi

Reticular formation

Vestibular nucleus

Medulla oblongata

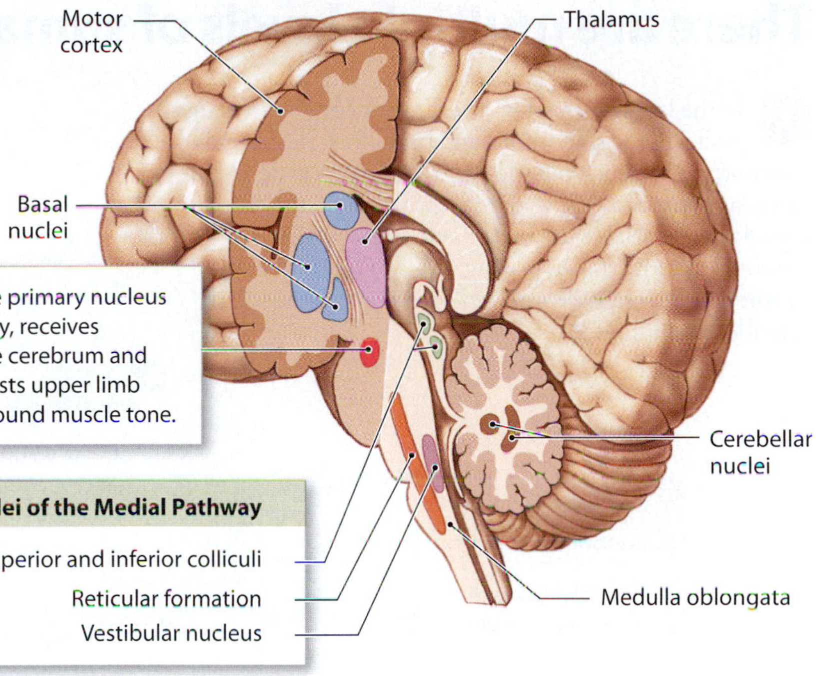

3 The locations of the medial and lateral pathways are indicated in this cross section of the spinal cord.

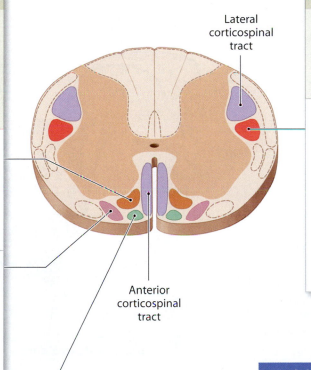

Lateral corticospinal tract

Anterior corticospinal tract

Medial Pathway

The medial pathway is primarily concerned with the control of muscle tone and gross movements of the neck, trunk, and proximal limb muscles. The upper motor neurons of the medial pathway are located in the vestibular nuclei, the superior and inferior colliculi, and the reticular formation.

The **reticulospinal tracts** contain the axons of upper motor neurons in the reticular formation. The reticular formation receives input from almost every ascending and descending pathway. It also has extensive interconnections with the cerebrum, the cerebellum, and brain stem nuclei.

The **vestibulospinal tracts** begin at the vestibular nuclei of N VIII. These nuclei receive sensory information from the inner ear about the position and movement of the head. These nuclei respond to changes in the orientation of the head by issuing motor commands that alter the muscle tone and position of the neck, eyes, head, and limbs.

The **tectospinal tracts** contain the axons of upper motor neurons in the superior and inferior colliculi of the midbrain. Axons in the tectospinal tracts direct reflexive changes in the position of the head, neck, and upper limbs in response to bright lights, sudden movements, or loud noises.

Lateral Pathway

The lateral pathway is primarily concerned with the control of muscle tone and the more precise movements of the distal parts of the limbs.

The upper motor neurons of the lateral pathway lie within the red nuclei of the midbrain. Axons of these motor neurons cross to the opposite side of the brain and descend into the spinal cord in the **rubrospinal tracts** (*ruber*, red). In humans, the rubrospinal tracts are small and extend only to the cervical spinal cord, where they provide motor control over distal muscles of the upper limbs.

There are multiple levels of somatic motor control

1 In the last two chapters you have encountered many nuclei in the spinal cord and brain that play roles in the control of skeletal muscle contractions. In general, the closer a motor center is to the cerebral cortex, the more complex and variable the motor activities will be. The numbers indicate the increasing levels of motor complexity; the cerebellum is involved in coordinating the motor activities at multiple levels.

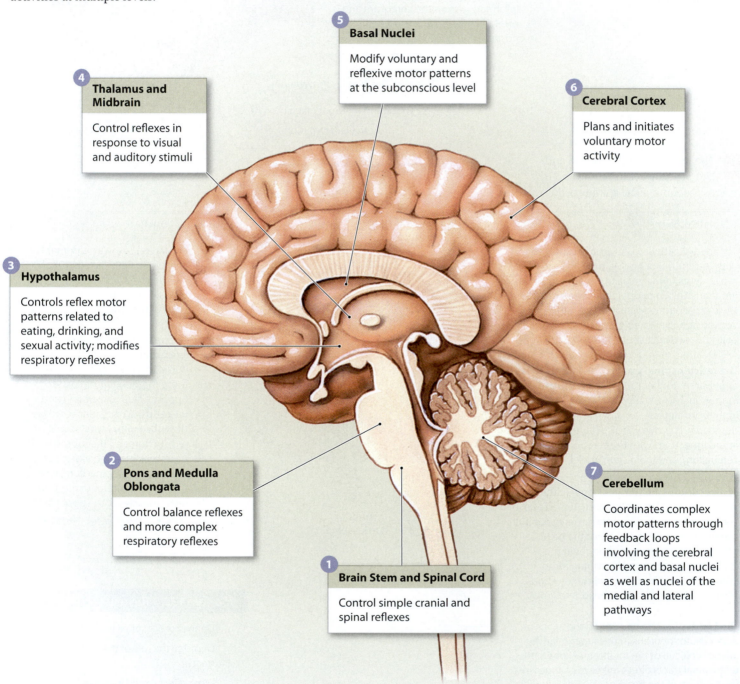

5 Basal Nuclei

Modify voluntary and reflexive motor patterns at the subconscious level

4 Thalamus and Midbrain

Control reflexes in response to visual and auditory stimuli

6 Cerebral Cortex

Plans and initiates voluntary motor activity

3 Hypothalamus

Controls reflex motor patterns related to eating, drinking, and sexual activity; modifies respiratory reflexes

2 Pons and Medulla Oblongata

Control balance reflexes and more complex respiratory reflexes

7 Cerebellum

Coordinates complex motor patterns through feedback loops involving the cerebral cortex and basal nuclei as well as nuclei of the medial and lateral pathways

1 Brain Stem and Spinal Cord

Control simple cranial and spinal reflexes

Preparing for Movement

2 When you make a conscious decision to perform a specific movement, information is relayed from the frontal lobes to motor association areas. These areas in turn relay the information to the cerebellum and basal nuclei.

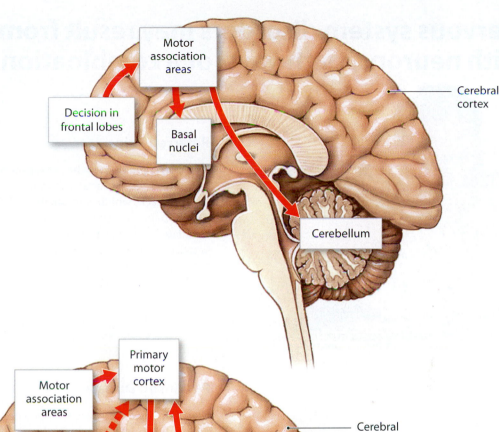

Performing a Movement

3 As the movement begins, the motor association areas send instructions to the primary motor cortex. Feedback from the basal nuclei and cerebellum modifies those commands, and output along the medial and lateral pathways directs involuntary adjustments in position and muscle tone.

The basal nuclei adjust patterns of movement in two ways:
1. They alter the sensitivity of the pyramidal cells to adjust the output along the corticospinal tract.
2. They change the excitatory or inhibitory output of the medial and lateral pathways.

As the movement proceeds, the cerebellum monitors proprioceptive and vestibular information and compares the arriving sensations with those experienced during previous movements. It then adjusts the activities of the upper motor neurons involved.

If the primary motor cortex is damaged, the individual loses the ability to exert fine control over skeletal muscles. However, some voluntary movements can still be controlled by the basal nuclei based on information from the prefrontal cortex concerning planned movements. However, because the corticospinal pathway is inoperative, the cerebellar feedback cannot fine-tune the ongoing movements. An individual in this condition can stand, maintain balance, and even walk, but all movements are hesitant, awkward, and poorly controlled.

Module 13.18 Review

a. The basic motor patterns related to eating and drinking are controlled by what region of the brain?

b. Which brain regions control reflexes in response to visual and auditory stimuli that are experienced while viewing a movie?

c. During a tennis match, you decide how and where to hit the ball. Explain how the motor association areas are involved.

Nervous system disorders may result from problems with neurons, pathways, or a combination of the two

Heart

Referred Pain

1 **Referred pain** is the sensation of pain in a part of the body other than its actual source. A familiar example is the pain of a heart attack, which is frequently felt in the left arm. Strong visceral pain sensations arriving at a segment of the spinal cord can stimulate interneurons that are part of the spinothalamic pathway. Activity in these interneurons leads to the stimulation of the primary sensory cortex, so the individual feels pain in a specific part of the body surface.

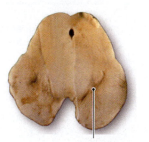

Normal substantia nigra   Diminished substantia nigra in Parkinson patient

Parkinson Disease

2 **Parkinson disease** results when neurons of the substantia nigra are damaged or secrete less dopamine. The basal nuclei become more active, which raises skeletal muscle tone and produces rigidity and stiffness. Individuals with Parkinson disease have difficulty starting voluntary movements, because opposing muscle groups do not relax; they must be overpowered. Once a movement is under way, every aspect must be voluntarily controlled through intense effort and concentration.

Rabies

3 **Rabies** is a dramatic example of a clinical condition directly related to retrograde flow in peripheral axons. A bite from a rabid animal injects the rabies virus into peripheral tissues, where virus particles quickly enter synaptic knobs. Retrograde flow then carries the virus into the CNS, with potentially fatal results. Many toxins (including heavy metals), some pathogenic bacteria, and other viruses also bypass CNS defenses by exploiting axoplasmic transport.

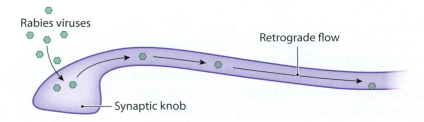

Rabies viruses

Retrograde flow

Synaptic knob

Cerebral Palsy

4 The term **cerebral palsy (CP)** refers to a number of disorders that affect voluntary motor performance; they appear during infancy or childhood and persist throughout the life of affected individuals. The cause may be trauma associated with premature or unusually stressful birth, maternal exposure to drugs (including alcohol), or a genetic defect that causes the improper development of motor pathways.

ALS

5 **Amyotrophic lateral sclerosis (ALS)** is a progressive, degenerative disorder that affects motor neurons in the spinal cord, brain stem, and cerebral hemispheres. The degeneration affects both upper and lower motor neurons. A defect in axonal transport is thought to underlie the disease. Because a motor neuron and its dependent muscle fibers are so intimately related, the destruction of CNS neurons causes atrophy of the associated skeletal muscles. It is commonly known as *Lou Gehrig disease*, named after the famous New York Yankees player who died of the disorder. Noted physicist Stephen Hawking is also afflicted with this condition.

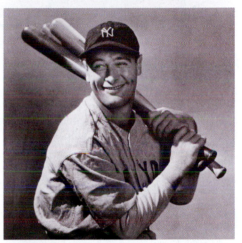

Alzheimer Disease

6 **Alzheimer disease (AD)** is a progressive disorder characterized by the loss of higher-order cerebral functions. It is the most common cause of **senile dementia**, or senility. Symptoms may appear at 50–60 years of age or later, although the disease occasionally affects younger individuals. An estimated 2 million people in the United States—including roughly 15 percent of those over age 65, and nearly half of those over age 85—have some form of the condition, and it causes approximately 100,000 deaths each year. Microscopic examination of the brains of AD patients reveals intracellular and extracellular abnormalities in brain regions such as the hippocampus, specifically associated with memory processing.

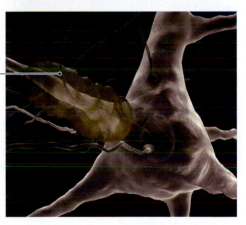

Abnormal dendrites, axons, and extracellular proteins form complexes known as Alzheimer plaques.

Multiple Sclerosis

7 **Multiple sclerosis** (skler-Ō-sis; *sklerosis*; hardness), or MS, is a disease characterized by recurrent incidents of demyelination that affects axons in the optic nerve, brain, and spinal cord. Common signs and symptoms include partial loss of vision and problems with speech, balance, and general motor coordination, including bowel and urinary bladder control. The time between incidents and the degree of recovery vary from case to case. In about one-third of all cases, the disorder is progressive, and functional impairment increases following each new incident. The first attack typically occurs in individuals 30–40 years old; the incidence among women is 1.5 times that among men.

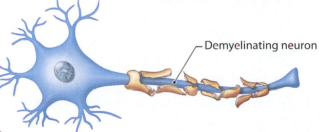
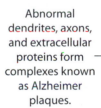

Demyelinating neuron

Module 13.19 Review

a. Define referred pain.

b. Describe how rabies is contracted.

c. Describe amyotrophic lateral sclerosis (ALS).

1. Labeling

Label each type of tactile receptor found in the skin.

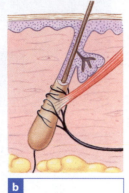
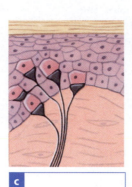
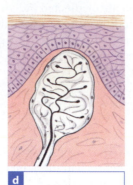
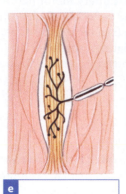
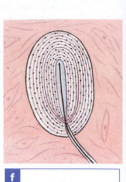

a _____
b _____
c _____
d _____
e _____
f _____

2. Short answer

Identify the descending and ascending tracts and pathways in the accompanying sectional diagram of the spinal cord, and then describe the general functions of the tracts of each pathway.

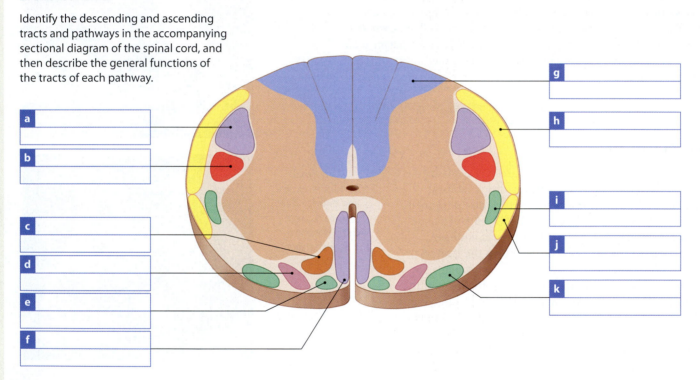

a _____

b _____

c _____

d _____

e _____

f _____

g _____

h _____

i _____

j _____

k _____

3. Short answer

The general organization of the spinal cord is such that motor tracts are a _____ (anterior or posterior), and sensory tracts are b _____ (anterior or posterior).

4. Section integration

An individual whose primary cortex has been injured retains the ability to walk, maintain balance, and perform other voluntary and involuntary movements. Even though the movements lack precision and are awkward and poorly controlled, why is the ability to walk and maintain balance possible?

_____

_____

Visual Outline with Key Terms
Summarize the content of each module using the terms in the order provided.

SECTION 1

The Functional Anatomy of the Brain and Cranial Nerves

- neural tube
- neurocoel
- primary brain vesicles
- mesencephalon
- prosencephalon
- rhombencephalon
- secondary brain vesicles
- diencephalon
- telencephalon
- cerebrum
- metencephalon
- cerebellum
- pons
- myelencephalon
- medulla oblongata

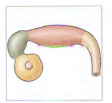

13.1

Each region of the brain has distinct structural and functional characteristics

- cerebrum
- cerebral hemispheres
- cerebral cortex
- fissures
- gyri
- sulci
- diencephalon
- thalamus
- hypothalamus
- brain stem
- midbrain
- pons
- medulla oblongata
- cerebellum
- ventricles
- lateral ventricle
- interventricular foramen
- third ventricle
- aqueduct of the midbrain
- fourth ventricle
- corpus callosum
- septum pellucidum

13.2

The brain is protected and supported by the cranial meninges and the cerebrospinal fluid

- cranial meninges
- dura mater
- arachnoid mater
- pia mater
- dural folds
- dural sinuses
- falx cerebri
- superior sagittal sinus
- tentorium cerebelli
- falx cerebelli
- cerebrospinal fluid (CSF)
- choroid plexus
- lateral apertures
- median aperture
- arachnoid granulations

13.3

The medulla oblongata and the pons contain autonomic reflex centers, relay stations, and ascending and descending tracts

- medulla oblongata
- pyramids
- decussation
- autonomic centers
- relay stations
- olive
- nucleus gracilis
- nucleus cuneatus
- olivary nuclei
- solitary nucleus
- ascending and descending tracts
- pons
- tracts
- respiratory centers
- reticular formation
- nuclei associated with cranial nerves (V, VI, VII, VIII)
- apneustic and pneumotaxic centers
- transverse fibers

13.4

The cerebellum coordinates learned and reflexive patterns of muscular activity at the subconscious level

- cerebellum
- anterior lobe
- posterior lobe
- cerebellar cortex
- vermis
- primary fissure
- folia
- Purkinje cells
- cerebellar peduncles
- arbor vitae
- transverse fibers
- ataxia

13.5

The midbrain regulates auditory and visual reflexes and controls alertness

- midbrain
- corpora quadrigemina
- superior colliculus
- inferior colliculus
- reticular activating system (RAS)
- red nucleus
- substantia nigra
- cerebral peduncles
- tectum
- tegmentum

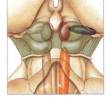

13.6

The diencephalon consists of the epithalamus, thalamus (left and right), and hypothalamus

- epithalamus
- anterior commissure
- optic chiasm
- interthalamic adhesion
- pineal gland
- melatonin
- thalamus
- lateral geniculate nucleus
- optic tract
- medial geniculate nucleus
- projected
- hypothalamus
- preoptic area
- suprachiasmatic nucleus
- infundibulum
- mamillary bodies

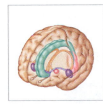

13.7

The limbic system is a functional group of tracts and nuclei located in the cerebrum and diencephalon

- limbic system
- limbic lobe
- cingulate gyrus
- parahippocampal gyrus
- fornix
- amygdaloid body
- hippocampus

● = *Term boldfaced in this module*

13.8

The basal nuclei of the cerebrum perform subconscious adjustment and refinement of ongoing voluntary movements

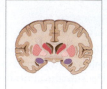

- basal nuclei
- caudate nucleus
- lentiform nucleus
- globus pallidus
- putamen
- internal capsule

13.9

Superficial landmarks can be used to divide the surface of the cerebral cortex into lobes

- lobes
- precentral gyrus
- central sulcus
- frontal lobe
- parietal lobe
- postcentral gyrus
- lateral sulcus
- temporal lobe
- insula
- parieto-occipital sulcus
- occiptal lobe

13.10

The lobes of the cerebral cortex contain regions with specific functions

- primary motor cortex
- primary sensory cortex
- association area
- motor cortex
- pyramidal cells
- somatic motor association area
- gustatory cortex
- olfactory cortex
- auditory cortex
- primary auditory cortex
- auditory association area
- sensory cortex
- somatic sensory association area
- visual cortex
- primary visual cortex
- visual association area
- integrative centers
- speech center
- prefrontal cortex
- frontal eye field
- general interpretive area
- hemispheric lateralization

13.11

White matter interconnects the cerebral hemispheres, the lobes of each hemisphere, and links the cerebrum to the rest of the brain

- association fibers
- arcuate fibers
- longitudinal fasciculi
- commissural fibers
- projection fibers
- internal capsule
- corpus callosum
- anterior commissure

13.12

Brain activity can be monitored using external electrodes; the record is called an electroencephalogram, or EEG

- electroencephalogram (EEG)
- brain waves
- alpha waves
- beta waves
- theta waves
- delta waves
- seizure
- epilepsies

13.13

The twelve pairs of cranial nerves can be classified as sensory, special sensory, motor, or mixed nerves

- olfactory nerves (I)
- optic nerves (II)
- oculomotor nerves (III)
- trochlear nerves (IV)
- trigeminal nerves (V)
- abducens nerves (VI)
- facial nerves (VII)
- vestibulocochlear nerves (VIII)
- glossopharyngeal nerves (IX)
- vagus nerves (X)
- accessory nerves (XI)
- hypoglossal nerves (XII)

SECTION 2

Sensory and Motor Pathways

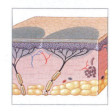

- general senses
- sensation
- perception
- receptive field
- transduction
- labeled line

13.14

Receptors for the general senses can be classified by function and by sensitivity

- nociceptors
- Type A fibers
- fast pain
- Type C fibers
- slow pain
- thermoreceptors
- chemoreceptors
- mechanoreceptors
- proprioceptors
- baroreceptors
- tactile receptors
- fine touch and pressure receptors
- crude touch and pressure receptors
- tonic receptors
- phasic receptors
- adaptation
- peripheral adaptation
- central adaptation

13.15

General sensory receptors are relatively simple in structure and widely distributed in the body

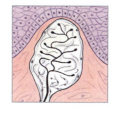

- free nerve endings
- root hair plexus
- tactile discs
- Merkel cells
- tactile corpuscles
- ○ Meissner corpuscles
- lamellated corpuscles
- ○ pacinian corpuscles
- Ruffini corpuscles

13.16

Three major somatic sensory pathways carry information from the skin and musculature to the CNS

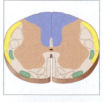

- spinothalamic pathway
- first-order neurons
- second-order neurons
- third-order neurons
- anterior spinothalamic tracts
- lateral spinothalamic tracts
- sensory homunculus
- posterior column pathway
- medial lemniscus
- spinocerebellar pathway

• = *Term boldfaced in this module*

13.17

The somatic nervous system controls skeletal muscles through upper and lower motor neurons

- upper motor neuron
- lower motor neuron
- corticospinal pathway
- motor homunculus
- corticobulbar tracts
- corticospinal tracts
- pyramids
- anterior corticospinal tracts

- lateral corticospinal tracts
- medial pathway
- lateral pathway
- red nucleus
- reticulospinal tracts

- vestibulospinal tracts
- tectospinal tracts
- rubrospinal tracts

13.19

Nervous system disorders may result from problems with neurons, pathways, or a combination of the two

- referred pain
- Parkinson disease
- rabies
- cerebral palsy (CP)
- amyotrophic lateral sclerosis (ALS)

- Alzheimer disease (AD)
- senile dementia
- multiple sclerosis (MS)

13.18

There are multiple levels of somatic motor control

- brain stem and spinal cord
- pons and medulla oblongata
- hypothalamus

- thalamus and midbrain
- basal nuclei
- cerebral cortex
- cerebellum

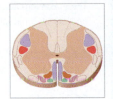

● = *Term boldfaced in this module*

Chapter Integration: Applying what you've learned

Infants have little to no control of the movements of their head, so they are susceptible to shaken-baby syndrome (SBS), caused by vigorous shaking of an infant or young child held by the arms, legs, chest, or shoulders. Dr. John Caffey, a radiologist, and Dr. Norman Guthkelch, a neurosurgeon, first described the syndrome around 1970.

Considered to be a form of child abuse, such forceful shaking can cause brain damage that may lead to mental retardation, speech and learning disabilities, paralysis, seizures, hearing loss, and even death. In many instances, parents or caregivers are angry or frustrated; only in *rare* instances may the injury be caused by tossing a baby in the air or jogging with a baby in a backpack. It is estimated that 1200–1600 children are affected each year in the United States.

1. Damage to which areas of the brain would account for the clinical signs and symptoms observed in shaken-baby syndrome?

2. Describe the physical events that occur when a baby is shaken forcefully.

3. Structures damaged by SBS can include the medulla oblongata. Explain why injury to the medulla oblongata is often fatal.

Access more review material online in the Study Area at **www.masteringaandp.com**.

There, you'll find:
- **Chapter guides**
- **Chapter quizzes**
- **Practice tests**
- **Labeling activities**
- **Flashcards**
- **A glossary with pronunciations**

14 The Autonomic Nervous System

LEARNING OUTCOMES

These Learning Outcomes correspond by number to this chapter's modules and indicate what you should be able to do after completing the chapter.

SECTION 1 • The Functional Anatomy and Organization of the Autonomic Nervous System (ANS)

14.1 List the divisions of the ANS and the general functions of each.

14.2 Describe the structures and functions of the sympathetic and parasympathetic divisions of the autonomic nervous system.

14.3 Describe the innervation patterns of the sympathetic and parasympathetic divisions of the autonomic nervous system.

14.4 Describe the various types of sympathetic and parasympathetic receptors and their associated neurotransmitters.

14.5 Describe the mechanisms of neurotransmitter release in the ANS, and explain the effects of neurotransmitters on target organs and tissues.

SECTION 2 • Autonomic Regulation and Control Mechanisms

14.6 Discuss the relationship between the two divisions of the ANS and the significance of dual innervation.

14.7 Define a visceral reflex, and explain the significance of such reflexes.

14.8 Explain the roles of baroreceptors and chemoreceptors in homeostasis.

14.9 Describe the hierarchy of interacting levels of control in the autonomic nervous system beginning with the hypothalamus.

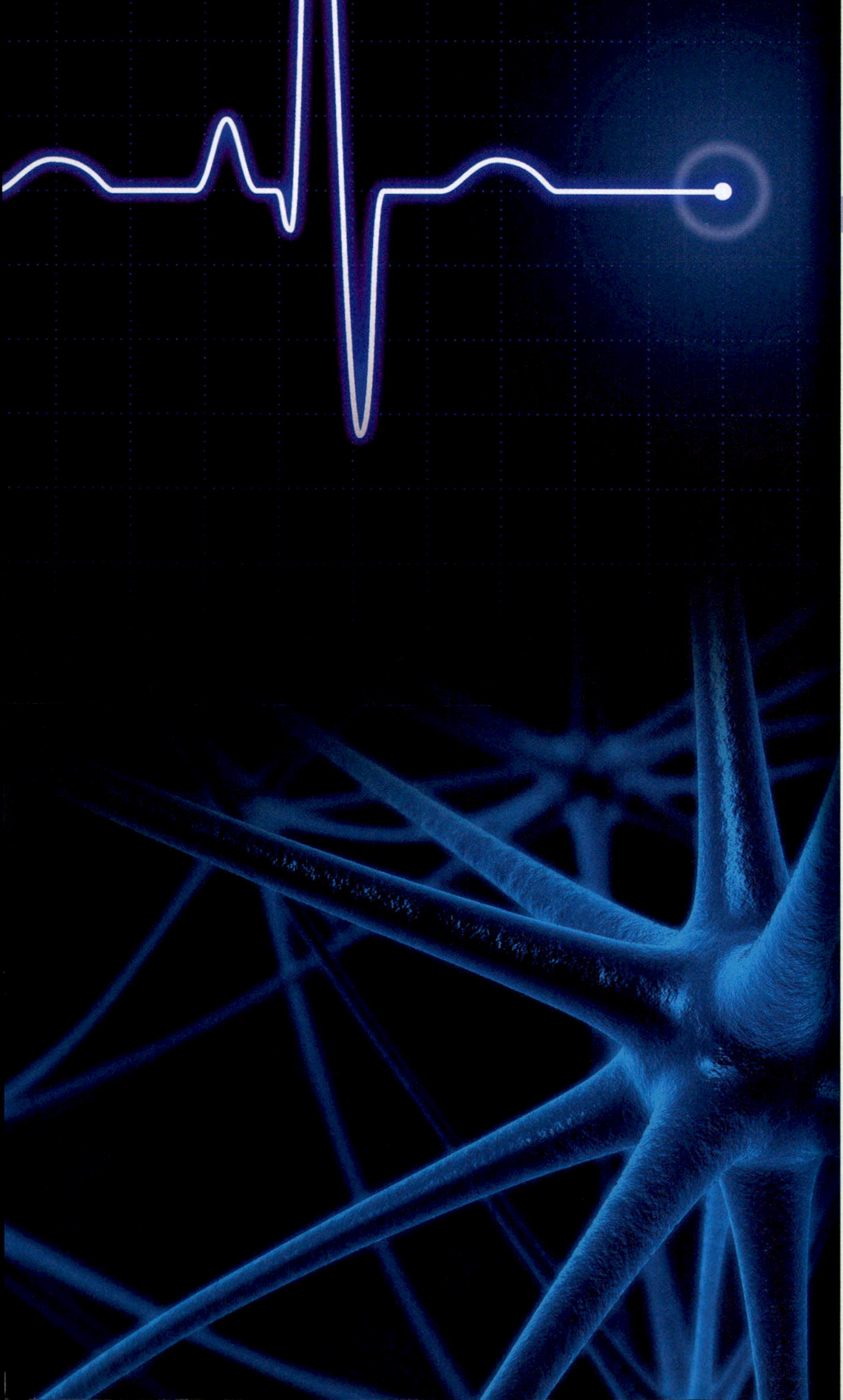

The Functional Anatomy and Organization of the Autonomic Nervous System (ANS)

In Chapters 12 and 13 we considered the organization of the **somatic nervous system**, or **SNS**. The SNS provides conscious and subconscious control over the skeletal muscles of the body. In this section we consider the **autonomic nervous system**, or **ANS**, which controls visceral function largely outside of our awareness.

1 In the SNS, motor neurons of the central nervous system exert direct control over skeletal muscles. The lower motor neurons may be controlled by reflexes based in the spinal cord or brain, or by upper motor neurons whose cell bodies lie within nuclei of the brain or at the primary motor cortex.

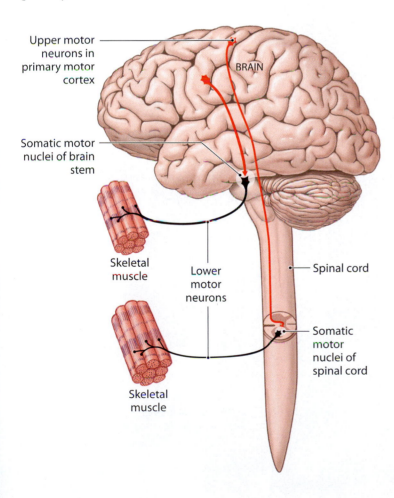

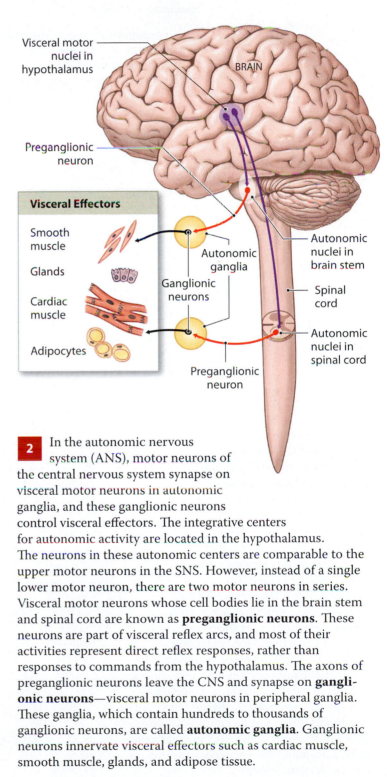

2 In the autonomic nervous system (ANS), motor neurons of the central nervous system synapse on visceral motor neurons in autonomic ganglia, and these ganglionic neurons control visceral effectors. The integrative centers for autonomic activity are located in the hypothalamus. The neurons in these autonomic centers are comparable to the upper motor neurons in the SNS. However, instead of a single lower motor neuron, there are two motor neurons in series. Visceral motor neurons whose cell bodies lie in the brain stem and spinal cord are known as **preganglionic neurons**. These neurons are part of visceral reflex arcs, and most of their activities represent direct reflex responses, rather than responses to commands from the hypothalamus. The axons of preganglionic neurons leave the CNS and synapse on **ganglionic neurons**—visceral motor neurons in peripheral ganglia. These ganglia, which contain hundreds to thousands of ganglionic neurons, are called **autonomic ganglia**. Ganglionic neurons innervate visceral effectors such as cardiac muscle, smooth muscle, glands, and adipose tissue.

The ANS consists of sympathetic, parasympathetic, and enteric divisions

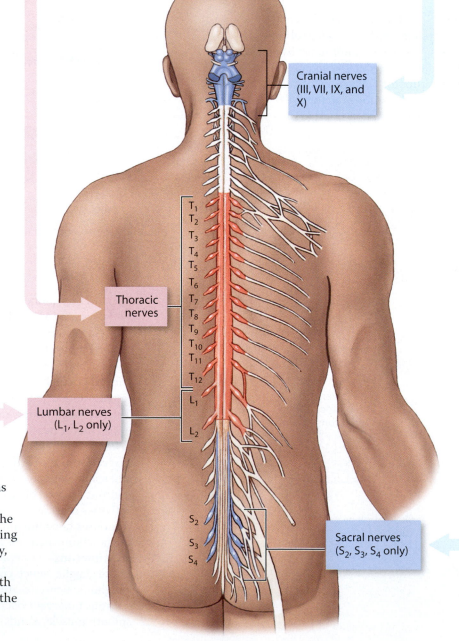

Autonomic Nervous System

Sympathetic Division

In the sympathetic division, or **thoracolumbar** (thor-a-kō-LUM-bar) **division**, axons emerge from the thoracic and superior lumbar segments of the spinal cord and innervate ganglia relatively close to the spinal cord.

Parasympathetic Division

In the parasympathetic division, or **cranio-sacral** (krā-nē-ō-SĀ-krul) **division**, axons emerge from the brain stem and the sacral segments of the spinal cord, and they innervate ganglia very close to (or within) target organs.

Cranial nerves (III, VII, IX, and X)

Thoracic nerves

T$_1$
T$_2$
T$_3$
T$_4$
T$_5$
T$_6$
T$_7$
T$_8$
T$_9$
T$_{10}$
T$_{11}$
T$_{12}$

Lumbar nerves (L$_1$, L$_2$ only)

L$_1$
L$_2$

S$_2$
S$_3$
S$_4$

Sacral nerves (S$_2$, S$_3$, S$_4$ only)

1 The ANS contains two well-known divisions whose names are probably already familiar to you: the **sympathetic division** and the **parasympathetic division**. Most often, these two divisions have opposing effects; if the sympathetic division causes excitation, the parasympathetic causes inhibition. However, this is not always the case. The two divisions may work independently, because some structures are innervated by only one division; or the two divisions may work together, each controlling a stage of a complex process. In general, the sympathetic division "kicks in" only during periods of exertion, stress, or emergency, and the parasympathetic division predominates under resting conditions. Both primary divisions of the ANS influence the third autonomic division, known as the enteric nervous system.

2 The **enteric nervous system (ENS)** is an extensive network of neurons and nerve networks located in the walls of the digestive tract. Although the sympathetic and parasympathetic divisions influence the activities of the enteric nervous system, many complex visceral reflexes are initiated and coordinated locally, without instructions from the CNS. Altogether, the ENS has roughly 100 million neurons—at least as many as the spinal cord—and all of the neurotransmitters found in the brain. The ENS will be discussed in our examination of the digestive system in Chapter 21.

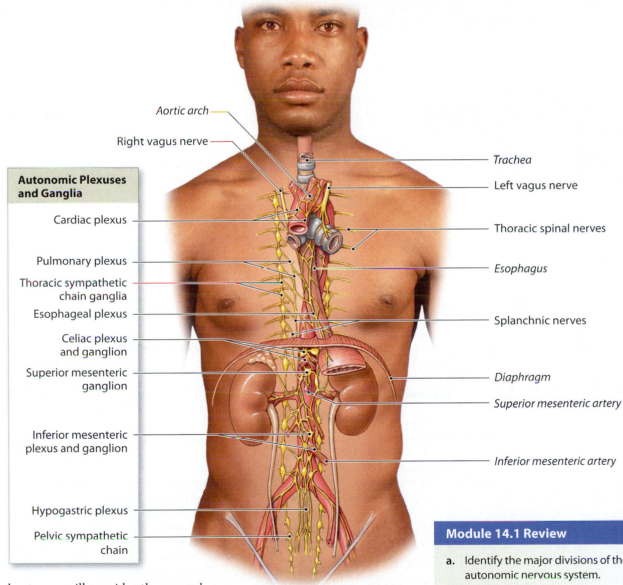

Esophagus

Stomach

Large intestine

Small intestine

Autonomic Plexuses and Ganglia

Cardiac plexus

Pulmonary plexus

Thoracic sympathetic chain ganglia

Esophageal plexus

Celiac plexus and ganglion

Superior mesenteric ganglion

Inferior mesenteric plexus and ganglion

Hypogastric plexus

Pelvic sympathetic chain

Aortic arch

Right vagus nerve

Trachea

Left vagus nerve

Thoracic spinal nerves

Esophagus

Splanchnic nerves

Diaphragm

Superior mesenteric artery

Inferior mesenteric artery

3 In this chapter we will consider the general organization of the sympathetic and parasympathetic divisions. In dissection, the sympathetic and parasympathetic nerves are not clearly segregated. Instead, they form a complex of intertwining nerves and plexuses. To clarify the structure of these divisions, we will rely on diagrammatic views, but you should return to this image from time to time to compare the diagrammatic views with a more anatomically accurate representation.

Module 14.1 Review

a. Identify the major divisions of the autonomic nervous system.

b. What division of the ANS is responsible for the physiological changes you experience when startled by a loud noise?

c. Compare the anatomy of the sympathetic division with the parasympathetic division.

The sympathetic division has chain ganglia, collateral ganglia, and the adrenal medullae …

Organization of the Sympathetic Division

1 In the sympathetic division, preganglionic neurons from the thoracic and superior lumbar segments of the spinal cord synapse on ganglionic neurons that may be located (1) within the sympathetic chain of ganglia near the spinal cord, (2) in collateral ganglia that lie within the thoracic or abdominopelvic cavities, or (3) within modified ganglion cells in the adrenal medullae. Because the ganglionic neurons are relatively close to the vertebral column, the axons of the preganglionic neurons (**preganglionic fibers**) are short, and the axons of the ganglionic neurons (**postganglionic fibers**) are long.

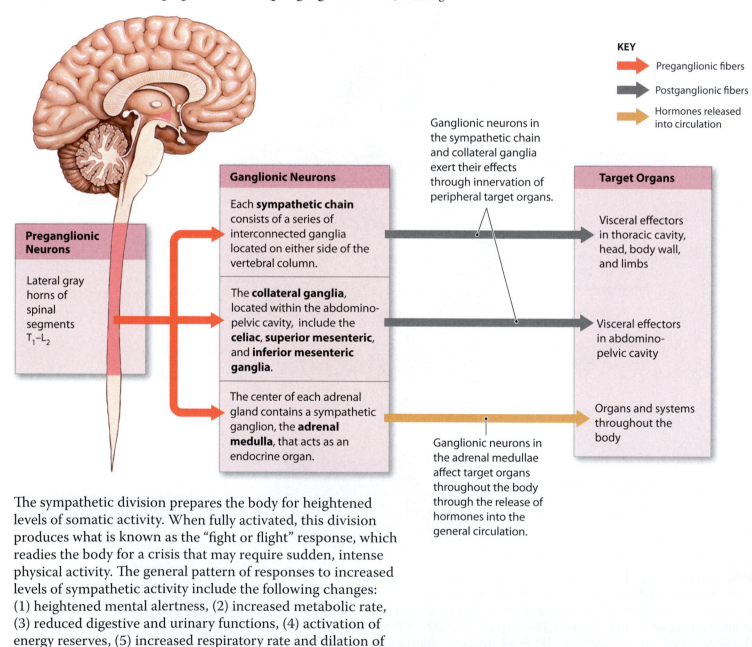

KEY

➤ Preganglionic fibers

➤ Postganglionic fibers

➤ Hormones released into circulation

Preganglionic Neurons

Lateral gray horns of spinal segments T_1–L_2

Ganglionic Neurons

Each **sympathetic chain** consists of a series of interconnected ganglia located on either side of the vertebral column.

The **collateral ganglia**, located within the abdomino-pelvic cavity, include the **celiac**, **superior mesenteric**, and **inferior mesenteric ganglia**.

The center of each adrenal gland contains a sympathetic ganglion, the **adrenal medulla**, that acts as an endocrine organ.

Ganglionic neurons in the sympathetic chain and collateral ganglia exert their effects through innervation of peripheral target organs.

Ganglionic neurons in the adrenal medullae affect target organs throughout the body through the release of hormones into the general circulation.

Target Organs

Visceral effectors in thoracic cavity, head, body wall, and limbs

Visceral effectors in abdomino-pelvic cavity

Organs and systems throughout the body

The sympathetic division prepares the body for heightened levels of somatic activity. When fully activated, this division produces what is known as the "fight or flight" response, which readies the body for a crisis that may require sudden, intense physical activity. The general pattern of responses to increased levels of sympathetic activity include the following changes: (1) heightened mental alertness, (2) increased metabolic rate, (3) reduced digestive and urinary functions, (4) activation of energy reserves, (5) increased respiratory rate and dilation of respiratory passageways, (6) elevated heart rate and blood pressure, and (7) activation of sweat glands.

... whereas the parasympathetic division has terminal or intramural ganglia

Organization of the Parasympathetic Division

2 In the parasympathetic division, a typical preganglionic fiber synapses on six to eight ganglionic neurons. These neurons may be situated in **terminal ganglia**, located near the target organ, or in **intramural ganglia** (*murus*, wall), which are embedded in the tissues of the target organ. Terminal ganglia are usually paired; examples include the parasympathetic ganglia associated with the cranial nerves. Intramural ganglia typically consist of interconnected masses and clusters of ganglion cells.

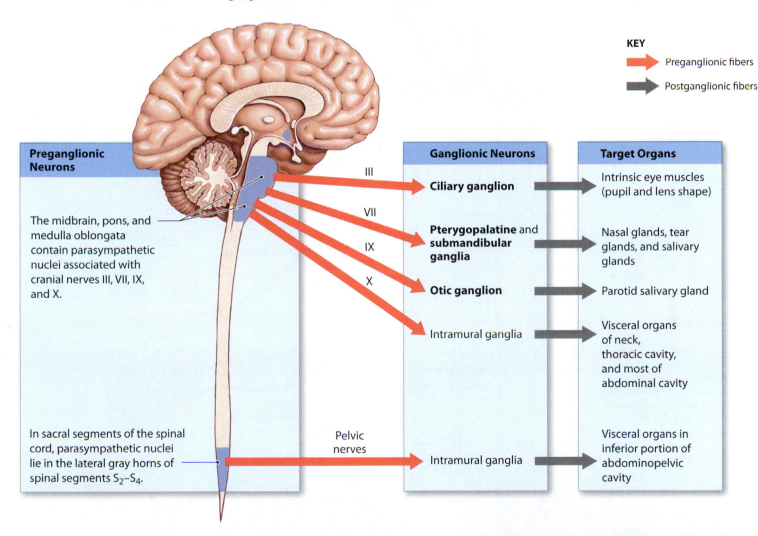

KEY

➡ Preganglionic fibers

➡ Postganglionic fibers

Preganglionic Neurons

The midbrain, pons, and medulla oblongata contain parasympathetic nuclei associated with cranial nerves III, VII, IX, and X.

In sacral segments of the spinal cord, parasympathetic nuclei lie in the lateral gray horns of spinal segments S_2–S_4.

III

VII

IX

X

Pelvic nerves

Ganglionic Neurons

Ciliary ganglion

Pterygopalatine and **submandibular ganglia**

Otic ganglion

Intramural ganglia

Intramural ganglia

Target Organs

Intrinsic eye muscles (pupil and lens shape)

Nasal glands, tear glands, and salivary glands

Parotid salivary gland

Visceral organs of neck, thoracic cavity, and most of abdominal cavity

Visceral organs in inferior portion of abdominopelvic cavity

The parasympathetic division is concerned with the regulation of visceral function and energy conservation. It is known as the "rest and digest" system. The overall pattern of responses to increased levels of parasympathetic activity include the following changes: (1) decreased metabolic rate, (2) decreased heart rate and blood pressure, (3) increased secretion by salivary and digestive glands, (4) increased motility and blood flow in the digestive tract, and (5) stimulation of urination and defecation.

Module 14.2 Review

a. List general responses to increased sympathetic activity and to parasympathetic activity.

b. Describe an intramural ganglion.

c. Starting in the spinal cord, trace the path of a nerve impulse through the sympathetic ANS to its target organ in the abdominopelvic cavity.

The two ANS divisions innervate many of the same structures, but the innervation patterns are different

Innervation in the Sympathetic Division

1 The left side of this image shows the distribution to the skin (and to skeletal muscles and other tissues of the body wall), whereas the right side depicts the innervation of visceral organs.

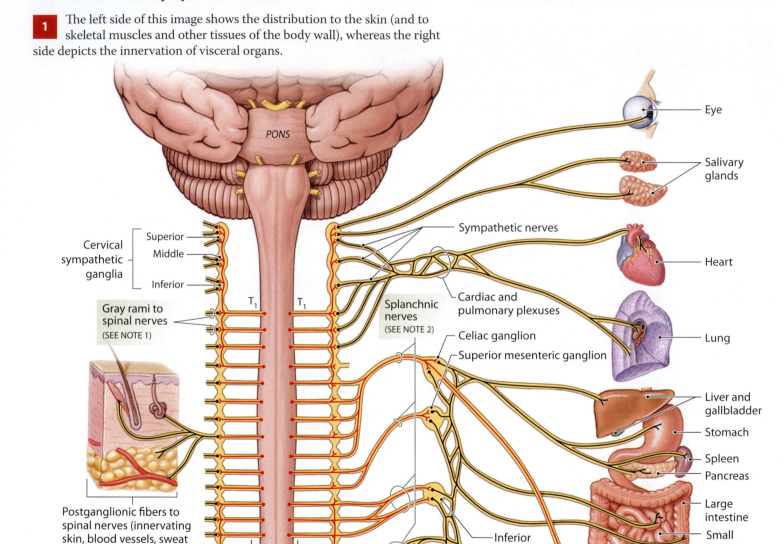

Cervical sympathetic ganglia
- Superior
- Middle
- Inferior

Gray rami to spinal nerves
(SEE NOTE 1)

T_1

Sympathetic nerves

Cardiac and pulmonary plexuses

Splanchnic nerves
(SEE NOTE 2)

Celiac ganglion

Superior mesenteric ganglion

Eye

Salivary glands

Heart

Lung

Liver and gallbladder

Stomach

Spleen

Pancreas

Large intestine

Small intestine

Postganglionic fibers to spinal nerves (innervating skin, blood vessels, sweat glands, arrector pili muscles, adipose tissue)

L_2

Inferior mesenteric ganglion

Adrenal medulla

Kidney

Sympathetic chain ganglia

Spinal cord

Coccygeal ganglia (Co_1) fused together

Ovary

Uterus

Penis

Scrotum

Urinary bladder

NOTE 1

Every spinal nerve has a gray ramus that carries sympathetic postganglionic fibers for distribution in the body wall and limbs. In the head and neck, postganglionic sympathetic fibers leaving the superior cervical sympathetic ganglia supply the regions innervated by cranial nerves III, VII, IX, and X.

NOTE 2

Preganglionic fibers on their way to the collateral ganglia form the **splanchnic** (SPLANK-nik) **nerves.** Postganglionic fibers innervating structures in the thoracic cavity, such as the heart and lungs, form bundles known as **sympathetic nerves.**

KEY
— Preganglionic neurons
— Ganglionic neurons

Innervation in the Parasympathetic Division

2 This image shows the parasympathetic innervation on one side of the body; the innervation on the opposite side (not shown) follows the same pattern.

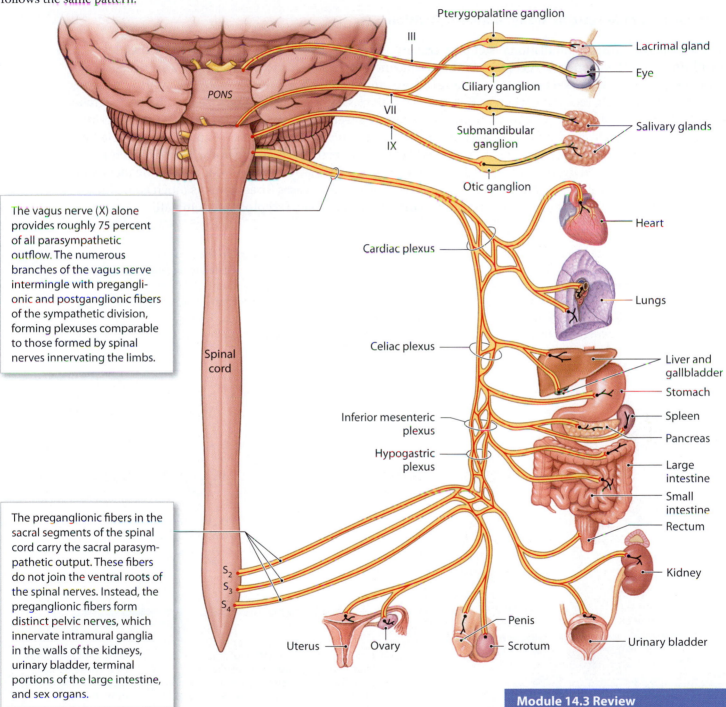

Pterygopalatine ganglion

III

Lacrimal gland

Eye

Ciliary ganglion

PONS

VII

Salivary glands

Submandibular ganglion

IX

Otic ganglion

The vagus nerve (X) alone provides roughly 75 percent of all parasympathetic outflow. The numerous branches of the vagus nerve intermingle with preganglionic and postganglionic fibers of the sympathetic division, forming plexuses comparable to those formed by spinal nerves innervating the limbs.

Cardiac plexus

Heart

Lungs

Spinal cord

Celiac plexus

Liver and gallbladder

Stomach

Spleen

Inferior mesenteric plexus

Pancreas

Hypogastric plexus

Large intestine

Small intestine

Rectum

The preganglionic fibers in the sacral segments of the spinal cord carry the sacral parasympathetic output. These fibers do not join the ventral roots of the spinal nerves. Instead, the preganglionic fibers form distinct pelvic nerves, which innervate intramural ganglia in the walls of the kidneys, urinary bladder, terminal portions of the large intestine, and sex organs.

S_2
S_3
S_4

Kidney

Penis

Uterus

Ovary

Scrotum

Urinary bladder

Module 14.3 Review

a. Define splanchnic nerves.

b. Which nerve carries the majority of the parasympathetic outflow?

c. Describe sympathetic nerves.

The effects of sympathetic and parasympathetic stimulation are mediated by membrane receptors at the target organs

Neurotransmitter Release in the Sympathetic Division

The effects of sympathetic stimulation result primarily from the interactions of norepinephrine (NE) and epinephrine (E) with **adrenergic receptors** in the plasma membrane. There are two classes of sympathetic adrenergic receptors: **alpha receptors** and **beta receptors**. In general, NE stimulates alpha receptors to a greater degree than it does beta receptors, whereas E stimulates both classes of receptors. Localized sympathetic activity involving the release of NE at sympathetic terminals primarily affects alpha receptors located near those terminals. The effects typically persist for a few seconds. By contrast, generalized sympathetic activation and release of E and NE by the adrenal medullae affects alpha and beta receptors throughout the body. Tissue concentrations of E and NE released by the adrenal medullae remain elevated for as long as 30 seconds, and the metabolic effects may persist for several minutes. Because the adrenal medullae releases three times as much epinephrine as it does norepinephrine, during sympathetic activation the effects of beta receptors predominate.

1 The stimulation of alpha (α) receptors activates enzymes on the inside of the plasma membrane. The stimulation of alpha-1 receptors generally has an excitatory effect on the target cell, whereas stimulation of alpha-2 receptors generally has an inhibitory effect.

2 Beta (β) receptors are located on the plasma membranes of cells in many organs, including skeletal muscles, the lungs, the heart, and the liver. The stimulation of beta receptors triggers changes in the metabolic activity of the target cell. There are three major types of beta receptors: beta-1 (β_1), beta-2 (β_2), and beta-3 (β_3).

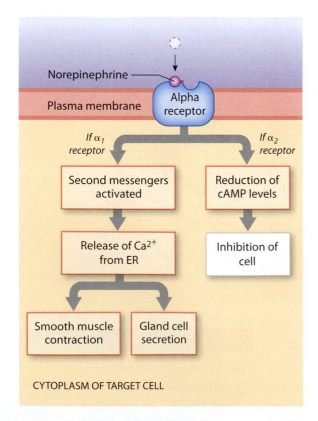

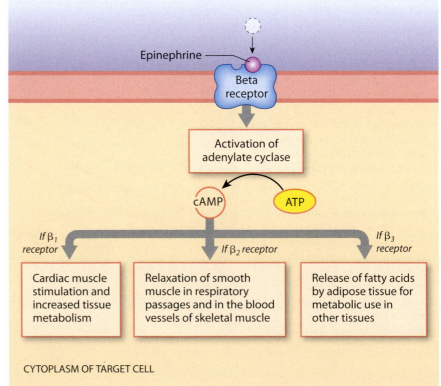

Neurotransmitter Release in the Parasympathetic Division

Although all the synapses and neuromuscular or neuroglandular junctions of the parasympathetic division use the same transmitter—ACh—two types of ACh receptors occur on the postsynaptic membranes: nicotinic receptors and muscarinic receptors.

3 **Nicotinic** (nik-ō-TIN-ik) **receptors** are located on the surfaces of ganglion cells of the parasympathetic division (and also on the surfaces of sympathetic ganglion cells and at neuromuscular junctions of the SNS). Exposure to ACh always causes excitation of the ganglionic neuron or muscle fiber by opening chemically gated channels in the postsynaptic membrane. Nicotinic receptors are also stimulated by nicotine, a powerful toxin that can be obtained from a variety of sources, including tobacco.

4 **Muscarinic** (mus-ka-RIN-ik) **receptors** are located at cholinergic neuromuscular or neuroglandular junctions in the parasympathetic division, as well as at the few cholinergic junctions in the sympathetic division. Muscarinic receptor stimulation produces longer-lasting effects than does the stimulation of nicotinic receptors. The response, which reflects the activation or inactivation of specific enzymes, can be either excitatory or inhibitory. Muscarinic receptors are stimulated by muscarine, a toxin produced by some poisonous mushrooms.

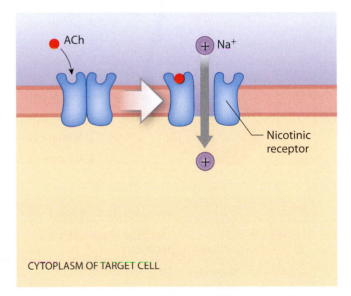

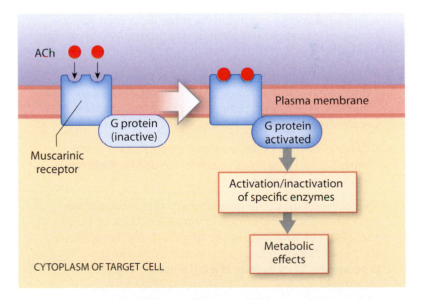

Module 14.4 Review

a. Compare and contrast alpha and beta receptors.

b. Compare nicotinic receptors with muscarinic receptors.

c. An individual with high blood pressure (hypertension) is prescribed a drug that blocks beta receptors. How could this medication alleviate hypertension?

The functional differences between the two ANS divisions reflect their divergent anatomical and physiological characteristics

Functional Characteristics of the Sympathetic Division

1 This table summarizes the effects of sympathetic activation.

2 This illustration and the table below it summarize the anatomical characteristics of the sympathetic division of the ANS.

Effects of Sympathetic Activation

The sympathetic division can change the activities of tissues and organs by releasing NE at peripheral synapses, and by distributing E and NE throughout the body in the bloodstream. The visceral motor fibers that target specific effectors, such as smooth muscle fibers in blood vessels of the skin, can be activated in reflexes that do not involve other visceral effectors. In a crisis, however, the entire division responds. This event, called **sympathetic activation**, is controlled by sympathetic centers in the hypothalamus. The effects are not limited to peripheral tissues; sympathetic activation also alters CNS activity. During sympathetic activation, the following changes occur in an individual:

- Increased alertness via stimulation of the reticular activating system, causing the individual to feel "on edge."

- A feeling of energy and euphoria, often associated with a disregard for danger and a temporary insensitivity to painful stimuli.

- Increased activity in the cardiovascular and respiratory centers of the pons and medulla oblongata, leading to elevations in blood pressure, heart rate, breathing rate, and depth of respiration.

- A general elevation in muscle tone through stimulation of the medial and lateral pathways, so the person looks tense and may begin to shiver.

- The mobilization of energy reserves, through the accelerated breakdown of glycogen in muscle and liver cells and the release of lipids by adipose tissues. These changes, plus the peripheral changes already noted, complete the preparations necessary for the individual to cope with a stressful situation.

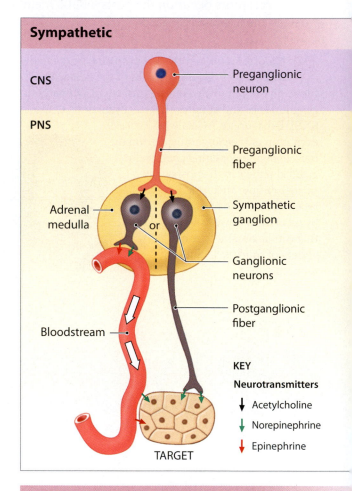

Characteristic	Sympathetic Division
Location of CNS visceral motor neurons	Lateral gray horns of spinal segments T_1–L_2
Location of PNS ganglia	Near vertebral column
Preganglionic fibers Neurotransmitter	Short ACh
Postganglionic fibers Neurotransmitter	Long Normally NE, sometimes NO or ACh
General functions	Stimulates metabolism; increases alertness; prepares for emergency ("fight or flight")

Functional Characteristics of the Parasympathetic Division

3 This illustration and the table below it summarize the anatomical characteristics of the parasympathetic division of the ANS.

4 This table summarizes the effects of **parasympathetic stimulation**.

Parasympathetic

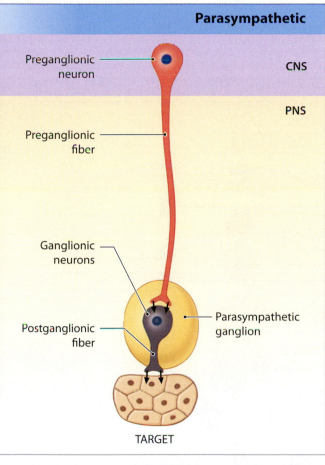

Preganglionic neuron

CNS

PNS

Preganglionic fiber

Ganglionic neurons

Postganglionic fiber

Parasympathetic ganglion

TARGET

Characteristic	Parasympathetic Division
Location of CNS visceral motor neurons	Brain stem and spinal segments S_2–S_4
Location of PNS ganglia	Typically intramural
Preganglionic fibers Neurotransmitter	Relatively long ACh
Postganglionic fibers Neurotransmitter	Relatively short ACh
General functions	Promotes relaxation, nutrient uptake, energy storage ("rest and digest")

Effects of Parasympathetic Activation

Under normal conditions, the entire parasympathetic division—unlike the sympathetic division—is neither controlled nor activated as a whole. Although it is active continuously, the activities are reflex responses to conditions within specific structures or regions. Examples of the major effects produced by the parasympathetic division include the following:

- Constriction of the pupils (to restrict the amount of light that enters the eyes) and focusing of the lenses of the eyes on nearby objects.

- Secretion by digestive glands, including salivary glands, gastric glands, duodenal glands, intestinal glands, the pancreas (exocrine and endocrine), and the liver.

- Secretion of hormones that promote the absorption and utilization of nutrients by peripheral cells.

- Changes in blood flow and glandular activity associated with sexual arousal.

- Increased smooth muscle activity along the digestive tract.

- Stimulation and coordination of defecation.

- Contraction of the urinary bladder during urination.

- Constriction of the respiratory passageways.

- Reduction in heart rate and in the force of contraction.

These functions center on relaxation, food processing, and energy absorption. The parasympathetic division has been called the **anabolic system** (*anabole*, a rising up), because its stimulation leads to a general increase in the nutrient content of the blood. Cells throughout the body respond to this increase by absorbing nutrients and using them to support growth, cell division, and the creation of energy reserves in the form of lipids or glycogen.

Module 14.5 Review

a. What neurotransmitter is released by all parasympathetic neurons?

b. Why is the parasympathetic division called the anabolic system?

c. What physiological changes are typical in tense (anxious) individuals?

1. Concept map

Use each of the following terms once to fill in the blank spaces to correctly complete the autonomic nervous system concept map.

- sympathetic division
- craniosacral division
- cranial nerves III, VII, IX, X
- thoracolumbar division
- parasympathetic division
- sacral nerves
- thoracic nerves
- enteric nervous system
- lumbar nerves

Autonomic Nervous System

contains

a _____ also known as b _____ communicates with | Enteric nervous system | preganglionic fibers within c _____ d _____

e _____ also known as f _____ communicates with i _____ preganglionic fibers within g _____ h _____

2. Labeling

Fill in the missing labels in this diagram of the sympathetic division of the ANS. Also indicate the distribution of the sympathetic innervation using red for the preganglionic fibers and black for the postganglionic fibers.

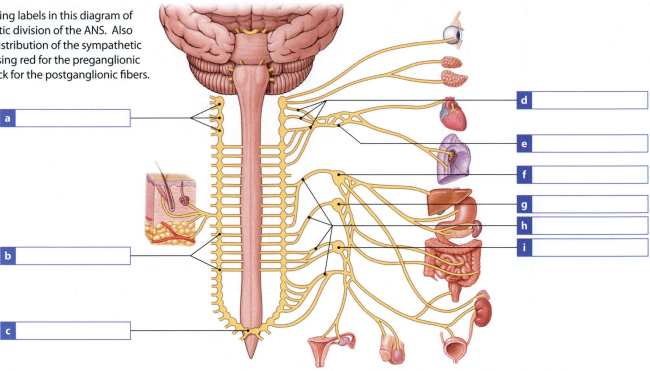

a _____
b _____
c _____

d _____
e _____
f _____
g _____
h _____
i _____

3. Matching

Match the following terms with the most closely related description.

- nicotinic, muscarinic
- secrete norepinephrine
- receptors
- alpha, beta
- cholinergic
- splanchnic nerves
- acetylcholine
- parasympathetic activation

a _____	Collateral ganglia
b _____	Parasympathetic neurotransmitter
c _____	Sexual arousal
d _____	Adrenal medullae
e _____	Cholinergic receptors
f _____	Determines neurotransmitter effects
g _____	All parasympathetic neurons
h _____	Adrenergic receptors

Autonomic Regulation and Control Mechanisms

If all consciousness were eliminated, vital physiological processes would continue virtually unchanged; a night's sleep is not a life-threatening event. Longer, deeper states of unconsciousness are not necessarily more dangerous, as long as nourishment and other basic care is provided. People who have suffered severe brain injuries can survive in a coma for decades. Over that period, the ANS continues to adjust activities of the digestive, cardiovascular, respiratory, and reproductive systems to maintain homeostasis, without instructions or interference from the conscious mind.

1 Here is a general overview of the way information is distributed and motor commands are issued by the nervous system.

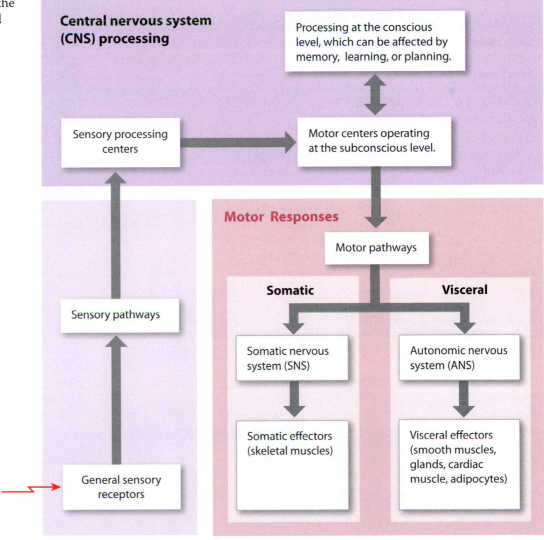

Central nervous system (CNS) processing

Processing at the conscious level, which can be affected by memory, learning, or planning.

Sensory processing centers

Motor centers operating at the subconscious level.

Motor Responses

Motor pathways

Somatic

Visceral

Sensory pathways

Somatic nervous system (SNS)

Autonomic nervous system (ANS)

Somatic effectors (skeletal muscles)

Visceral effectors (smooth muscles, glands, cardiac muscle, adipocytes)

Stimulus ⟶ General sensory receptors

The output of the ANS has an impact on virtually every body system. This section considers those effects and how autonomic output is coordinated and directed.

The ANS provides precise control over visceral functions

1 This table lists the effects of sympathetic and parasympathetic innervation on major organs and systems of the body. Even in the absence of stimuli, autonomic motor neurons show a continuous level of spontaneous activity called **autonomic tone**. Many vital organs receive **dual innervation**, receiving instructions from both the sympathetic and parasympathetic divisions. The effects may be opposing or complementary. In cases where only the sympathetic division provides innervation, the response may still vary widely depending on the type of receptor stimulated.

A Functional Comparison of the Sympathetic and Parasympathetic Divisions of the ANS

Structure	Sympathetic Effects (receptor type)	Parasympathetic Effects (all cholinergic)
Eye		
	Dilation of pupil (α_1); accommodation for distance vision (β_2)	Constriction of pupil; accommodation for close vision
Lacrimal glands	None (not innervated)	Secretion
Skin		
Sweat glands	Increased secretion, palms and soles (α_1); generalized increase in secretion (cholinergic)	None (not innervated)
Arrector pili muscles	Contraction; erection of hairs (α_1)	None (not innervated)
Cardiovascular System		
Blood vessels		
To skin	Dilation (β_2 and cholinergic); constriction (α_1)	None (not innervated)
To skeletal muscles	Dilation (β_2 and cholinergic)	None (not innervated)
To heart	Dilation (β_2); constriction (α_1, α_2)	None (not innervated)
To lungs	Dilation (β_2); constriction (α_2)	None (not innervated)
To digestive viscera	Constriction (α_1); dilation (α_2)	None (not innervated)
To kidneys	Constriction, decreased urine production (α_1, α_2); dilation, increased urine production (β_1, β_2)	None (not innervated)
To brain	Dilation (cholinergic)	None (not innervated)
Veins	Constriction (α_1, β_2)	None (not innervated)
Heart	Increased heart rate, force of contraction, and blood pressure (α_1, β_1)	Decreased heart rate, force of contraction, and blood pressure
Endocrine System		
Adrenal gland	Secretion of epinephrine, norepinephrine by adrenal medulla	None (not innervated)
Neurohypophysis	Secretion of ADH (β_1)	None (not innervated)
Pancreas	Decreased insulin secretion (α_2)	Increased insulin secretion
Pineal gland	Increased melatonin secretion (β)*	Inhibition of melatonin synthesis
Respiratory System		
Airways	Increased airway diameter (β_2)	Decreased airway diameter
Secretory glands	Mucous secretion (α_1)	None (not innervated)
Digestive System		
Salivary glands	Production of viscous secretion (α_1, β_1) containing mucins and enzymes	Production of copious, watery secretion
Sphincters	Constriction (α_1)	Dilation
General level of activity	Decreased (α_2, β_2)	Increased
Secretory glands	Inhibition (α_2)	Stimulation
Liver	Glycogen breakdown, glucose synthesis and release (α_1, β_2)	Glycogen synthesis
Pancreas	Decreased exocrine secretion (α_1)	Increased exocrine secretion

* The type of beta receptor in the pineal gland has not yet been determined.

A Functional Comparison of the Sympathetic and Parasympathetic Divisions of the ANS (Continued)

Structure	Sympathetic Effects (receptor type)	Parasympathetic Effects (all cholinergic)
Muscular System		
	Increased force of contraction, glycogen breakdown (β_2)	None (not innervated)
	Facilitation of ACh release at neuromuscular junction (α_2)	None (not innervated)
Adipose Tissue		
	Lipolysis, fatty acid release ($\alpha_1, \beta_1, \beta_3$)	None (not innervated)
Urinary System		
Kidneys	Secretion of renin (β_1)	Uncertain effects on urine production
Urinary bladder	Constriction of internal sphincter; relaxation of urinary bladder (α_1, β_2)	Tensing of urinary bladder, relaxation of internal sphincter to eliminate urine
Male Reproductive System		
	Increased glandular secretion and ejaculation (α_1)	Erection
Female Reproductive System		
	Increased glandular secretion; contraction of pregnant uterus (α_1)	Variable (depending on hormones present)
	Relaxation of nonpregnant uterus (β_1)	Variable (depending on hormones present)

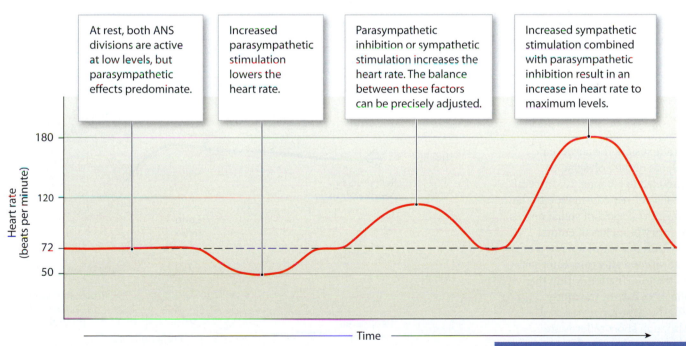

At rest, both ANS divisions are active at low levels, but parasympathetic effects predominate.

Increased parasympathetic stimulation lowers the heart rate.

Parasympathetic inhibition or sympathetic stimulation increases the heart rate. The balance between these factors can be precisely adjusted.

Increased sympathetic stimulation combined with parasympathetic inhibition result in an increase in heart rate to maximum levels.

2 This graph shows the effects of both autonomic divisions on the heart, which receives dual innervation. The heart consists of cardiac muscle tissue, and its contractions are triggered by specialized pacemaker cells. The two autonomic divisions have opposing effects on pacemaker function. ACh released by the parasympathetic division reduces the heart rate, whereas NE released by the sympathetic division accelerates the heart rate. Because autonomic tone exists, small amounts of both of these neurotransmitters are released continuously. However, under resting conditions, the effects of parasympathetic innervation predominate.

Module 14.6 Review

a. Define dual innervation.

b. Discuss autonomic tone and its significance in controlling visceral function.

c. You go outside in winter and blood flow to your skin is reduced, conserving body heat. You become angry, and your face turns red. Explain these changes.

Most visceral functions are controlled by visceral reflexes

Visceral reflexes provide automatic motor responses that can be modified, facilitated, or inhibited by higher centers, especially those of the hypothalamus. All visceral reflexes are polysynaptic. Each visceral reflex arc consists of a receptor, a sensory neuron, a processing center (one or more interneurons), and one or two visceral motor neurons.

1 **Short reflexes** bypass the CNS entirely; they involve sensory neurons and interneurons whose cell bodies are located within autonomic ganglia. The interneurons synapse on ganglionic neurons, and the motor commands are then distributed to effectors by postganglionic fibers. Short reflexes control very simple motor responses with localized effects. In general, short reflexes control patterns of activity in one small part of a target organ, whereas long reflexes coordinate the activities of an entire organ. Short reflexes predominate in the enteric nervous system, which operates largely outside of the awareness and control of the CNS.

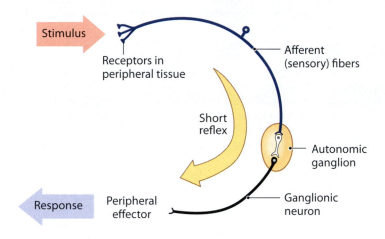

2 **Long reflexes** are the autonomic equivalents of the polysynaptic reflexes introduced in Chapter 12. Visceral sensory neurons deliver information to the CNS along the dorsal roots of spinal nerves, within the sensory branches of cranial nerves, and within the autonomic nerves that innervate visceral effectors. Interneurons process the information within the CNS, and the ANS carries the motor commands to the appropriate visceral effectors. Although short reflexes are present in the sympathetic and parasympathetic divisions, long reflexes predominate, and they are responsible for coordinating responses involving multiple organ systems.

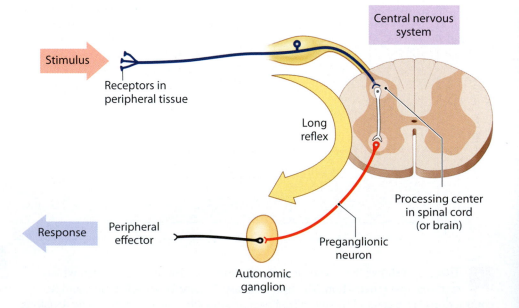

Representative Visceral Reflexes

	Reflex	Stimulus	Response	Comments
SYMPATHETIC	**Cardioacceleratory reflex**	Sudden decline in blood pressure in carotid artery	Increase in heart rate and force of contraction	Coordinated in cardiac center of medulla oblongata
	Vasomotor reflexes	Changes in blood pressure in major arteries	Changes in diameter of peripheral vessels	Coordinated in vasomotor center in medulla oblongata
	Pupillary reflex	Low light level reaching visual receptors	Dilation of pupil	
	Ejaculation (in males)	Erotic stimuli (primarily tactile)	Skeletal muscle contractions ejecting semen	
PARASYMPATHETIC	**Gastric and intestinal reflexes**	Pressure and physical contact	Smooth muscle contractions that propel food materials and mix them with secretions	Via vagus nerve
	Defecation	Distention of rectum	Relaxation of internal anal sphincter	Requires voluntary relaxation of external anal sphincter
	Urination	Distention of urinary bladder	Contraction of walls of urinary bladder; relaxation of internal urethral sphincter	Requires voluntary relaxation of external urethral sphincter
	Direct light and consensual light reflexes	Bright light shining in eye(s)	Constriction of pupils of both eyes	
	Swallowing reflex	Movement of food and liquids into pharynx	Smooth muscle and skeletal muscle contractions	Coordinated by medullary swallowing center
	Coughing reflex	Irritation of respiratory tract	Sudden explosive ejection of air	Coordinated by medullary coughing center
	Baroreceptor reflex	Sudden rise in carotid blood pressure	Reduction in heart rate and force of contraction	Coordinated in cardiac center of medulla oblongata
	Sexual arousal	Erotic stimuli (visual or tactile)	Increased glandular secretions, sensitivity, erection	

Interoceptors provide visceral sensory information.

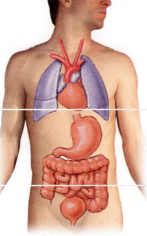

The axons of the visceral sensory neurons usually travel in company with autonomic motor fibers innervating the same visceral structures.

Mouth, palate, pharynx, larynx, trachea, esophagus, and associated vessels and glands → Cranial nerves V, VII, IX, and X

Visceral organs located between the diaphragm and the pelvic cavity → Dorsal roots of spinal nerves T_1–L_2

Organs in the inferior portion of the pelvic cavity → Dorsal roots of spinal nerves S_2–S_4

The visceral sensory information is delivered to the solitary nucleus.

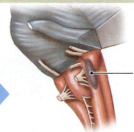

A solitary nucleus is located on each side of the medulla oblongata.

Each solitary nucleus is a major processing and sorting center for visceral sensory information; it has extensive connections with the various cardiovascular and respiratory centers as well as with the reticular formation. The information is seldom relayed to higher centers, so we remain unaware of the sensations or the resulting motor responses.

3 Visceral sensory information is collected by interoceptors monitoring visceral tissues and organs, primarily within the thoracic and abdominopelvic cavities. These interoceptors include nociceptors, thermoreceptors, tactile receptors, baroreceptors, and chemoreceptors, although none of them are as numerous as they are in somatic tissues. Almost all of the processing of this sensory information is done at the subconscious level, in nuclei in the spinal cord and the **solitary nuclei** in the brain stem.

Module 14.7 Review

a. Define visceral reflex.

b. Describe the solitary nucleus.

c. Compare short reflexes with long reflexes.

Baroreceptors and chemoreceptors initiate important autonomic reflexes involving visceral sensory pathways

Baroreceptors and chemoreceptors play key roles in the control of visceral function by the autonomic nervous system.

1 **Baroreceptors** are stretch receptors that monitor changes in pressure. The receptor consists of free nerve endings that branch within the elastic tissues in the walls of hollow organs, blood vessels, and tubes in the respiratory, digestive, or urinary tract. When the pressure changes, the elastic walls of these structures stretch or recoil. These changes in shape distort the receptor's dendritic branches and alter the rate of action-potential generation. Baroreceptors monitor blood pressure in the walls of major vessels, including the carotid artery (at the carotid sinus) and the aorta (at the aortic sinus). The information provided by these baroreceptors plays a major role in regulating cardiac function and adjusting blood flow to vital tissues. Baroreceptors in the lungs monitor the degree of lung expansion. This information is relayed to the respiratory rhythmicity center, which sets the pace of respiration. Baroreceptors in the urinary and digestive tracts trigger a variety of visceral reflexes, such as urination.

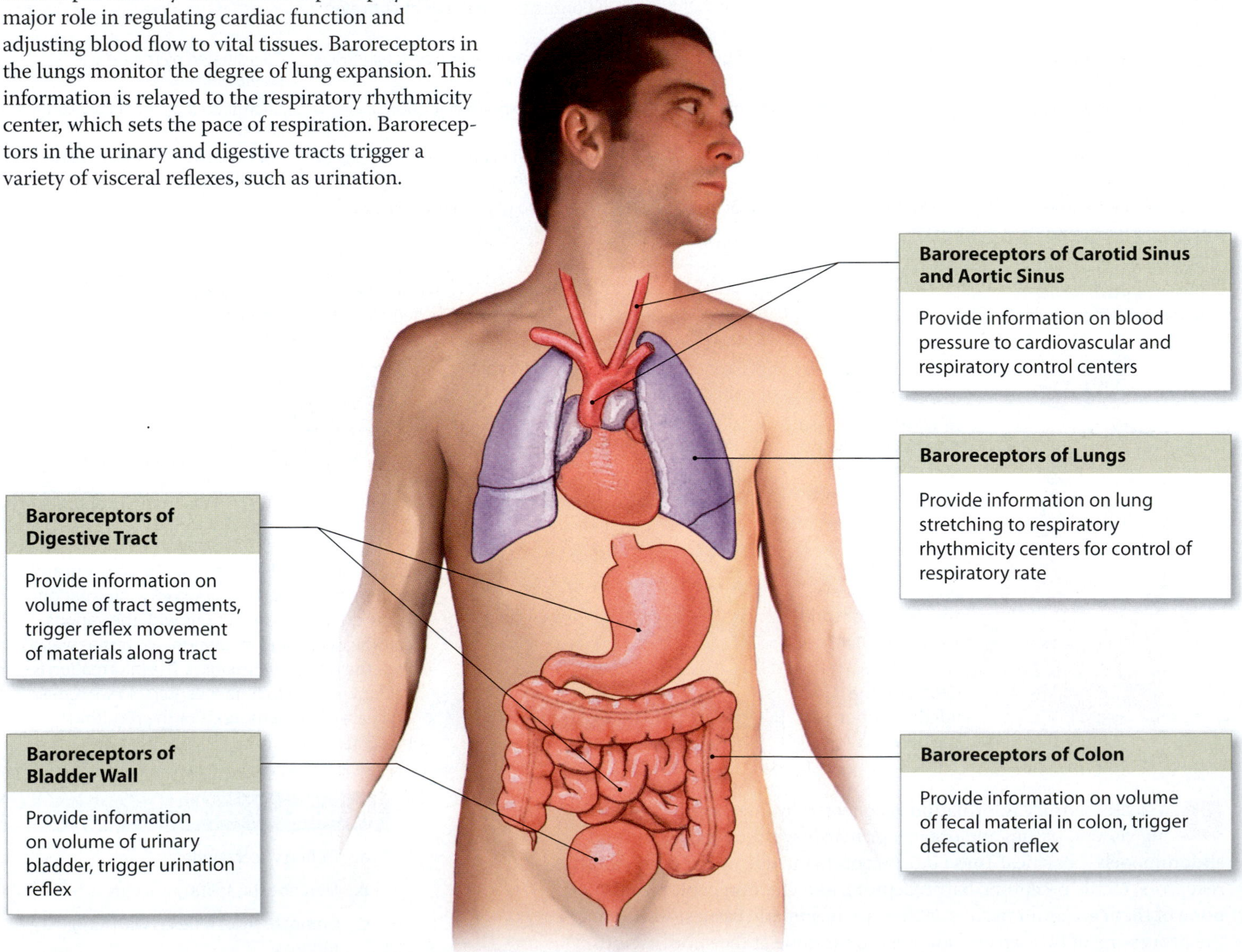

Baroreceptors of Carotid Sinus and Aortic Sinus

Provide information on blood pressure to cardiovascular and respiratory control centers

Baroreceptors of Lungs

Provide information on lung stretching to respiratory rhythmicity centers for control of respiratory rate

Baroreceptors of Digestive Tract

Provide information on volume of tract segments, trigger reflex movement of materials along tract

Baroreceptors of Bladder Wall

Provide information on volume of urinary bladder, trigger urination reflex

Baroreceptors of Colon

Provide information on volume of fecal material in colon, trigger defecation reflex

2   **Chemoreceptors** are specialized neurons that can detect small changes in the concentrations of specific chemicals or compounds. Chemoreceptive neurons are found (1) within the medulla oblongata and elsewhere in the brain, (2) in the **carotid bodies**, near the origin of the internal carotid arteries on each side of the neck, and (3) in the **aortic bodies** between the major branches of the aortic arch. The neurons in the respiratory centers in the medulla oblongata monitor the pH and P_{CO_2} in cerebrospinal fluid (CSF). The carotid and aortic bodies monitor the pH and levels of carbon dioxide (P_{CO_2}) and oxygen (P_{O_2}) in arterial blood. These chemoreceptors play an important role in the reflexive control of respiration and cardiovascular function.

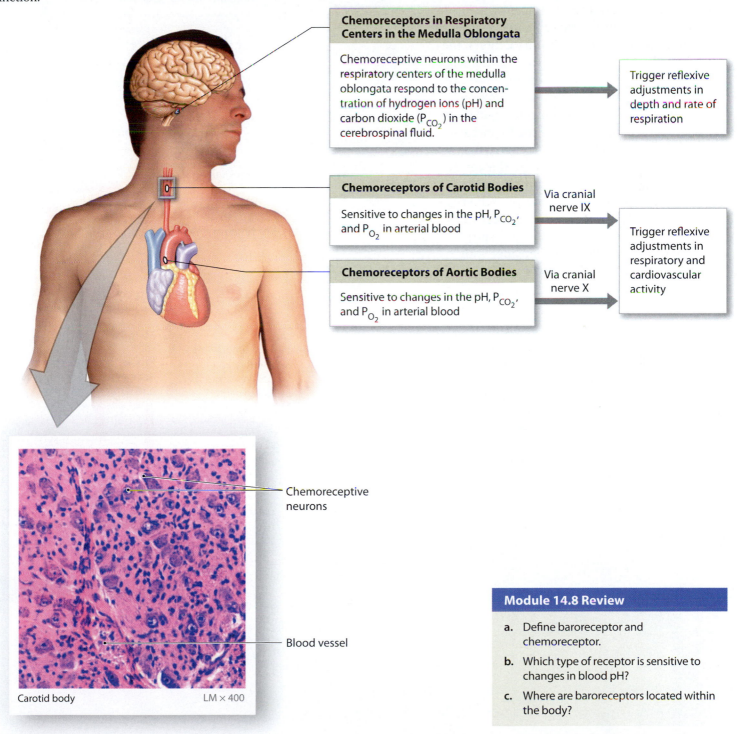

Chemoreceptors in Respiratory Centers in the Medulla Oblongata

Chemoreceptive neurons within the respiratory centers of the medulla oblongata respond to the concentration of hydrogen ions (pH) and carbon dioxide (P_{CO_2}) in the cerebrospinal fluid.

Trigger reflexive adjustments in depth and rate of respiration

Chemoreceptors of Carotid Bodies

Sensitive to changes in the pH, P_{CO_2}, and P_{O_2} in arterial blood

Via cranial nerve IX

Chemoreceptors of Aortic Bodies

Sensitive to changes in the pH, P_{CO_2}, and P_{O_2} in arterial blood

Via cranial nerve X

Trigger reflexive adjustments in respiratory and cardiovascular activity

Chemoreceptive neurons

Blood vessel

Carotid body   LM × 400

Module 14.8 Review

a. Define baroreceptor and chemoreceptor.

b. Which type of receptor is sensitive to changes in blood pH?

c. Where are baroreceptors located within the body?

The autonomic nervous system has multiple levels of motor control

The levels of activity in the sympathetic and parasympathetic divisions of the ANS are controlled by centers in the brain stem that regulate specific visceral functions. As in the SNS, in the ANS simple reflexes based in the autonomic ganglia and the spinal cord provide relatively rapid and automatic responses to stimuli. More complex sympathetic and parasympathetic reflexes are coordinated by processing centers in the medulla oblongata. In addition to the cardiovascular and respiratory centers, the medulla oblongata contains centers and nuclei involved with salivation, swallowing, digestive secretions, peristalsis, and urinary function. These centers are in turn subject to regulation by the hypothalamus. Because the hypothalamus interacts with all other portions of the brain, activity in the limbic system, thalamus, or cerebral cortex can have dramatic effects on autonomic function. For example, when you become angry, your heart rate accelerates, your blood pressure rises, and your respiratory rate increases; when you consider your next meal, your stomach "growls" and your mouth waters.

1 This diagram depicts the pathways and levels of ANS control. The hypothalamus acts as the "headquarters" of both the sympathetic and parasympathetic divisions, and hypothalamic output is indicated by solid red arrows. Nuclei in the spinal cord and brain stem control basic visceral reflex patterns. There is also continual feedback between higher brain centers and the hypothalamus and brain stem; this subconscious communication is indicated by the dashed red arrows.

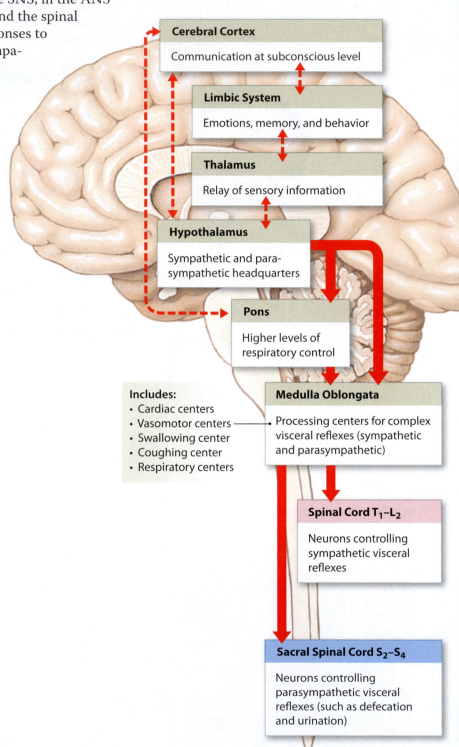

Cerebral Cortex

Communication at subconscious level

Limbic System

Emotions, memory, and behavior

Thalamus

Relay of sensory information

Hypothalamus

Sympathetic and parasympathetic headquarters

Pons

Higher levels of respiratory control

Includes:
- Cardiac centers
- Vasomotor centers
- Swallowing center
- Coughing center
- Respiratory centers

Medulla Oblongata

Processing centers for complex visceral reflexes (sympathetic and parasympathetic)

Spinal Cord T$_1$–L$_2$

Neurons controlling sympathetic visceral reflexes

Sacral Spinal Cord S$_2$–S$_4$

Neurons controlling parasympathetic visceral reflexes (such as defecation and urination)

2 The SNS and ANS are organized in parallel and are integrated at the level of the brain stem. This diagram shows the basic patterns of organization and integration. Blue arrows indicate ascending sensory information; red arrows, descending motor commands; dashed lines indicate pathways of communication and feedback among higher centers. The activities of the somatic nervous system and those of the autonomic nervous system are integrated at many levels. Although we have considered somatic and visceral motor pathways separately, the two have many parallels, in terms of both organization and function. Sensory pathways may carry information distributed to both the SNS and the ANS, triggering integrated and compatible reflexes. Higher levels of integration involve the brain stem, and both systems are influenced, if not controlled, by higher centers.

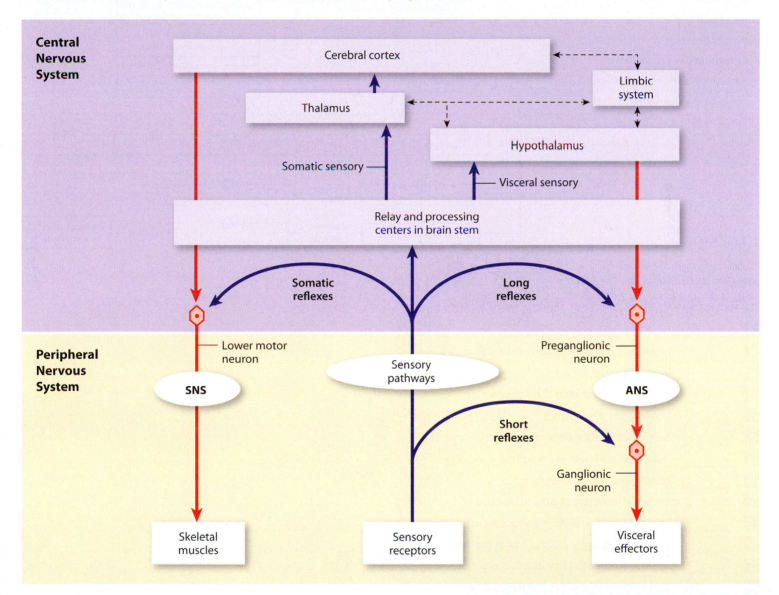

Module 14.9 Review

a. What structure within the brain relays sensory information?

b. Identify the target structures for the SNS and ANS.

c. Harry has a brain tumor that is pressing against his hypothalamus. Would you expect this tumor to interfere with autonomic function? Why or why not?

1. Concept map

Use each of the following terms once to fill in the blank spaces to correctly complete the levels of autonomic control concept map.

- respiratory
- pons
- spinal cord T_1–L_2
- vasomotor
- coughing
- hypothalamus
- sympathetic visceral reflexes
- parasympathetic visceral reflexes
- complex visceral reflexes
- limbic system and thalamus

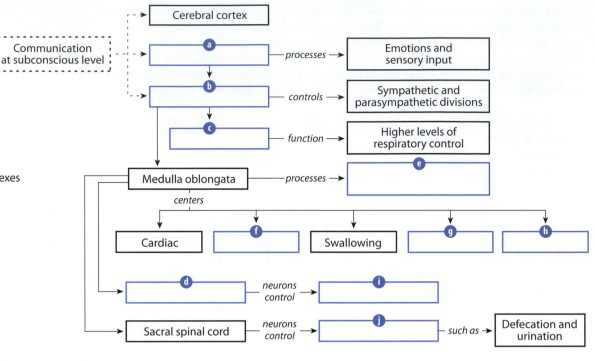

2. Matching

Write S (for sympathetic) or P (for parasympathetic) to indicate the ANS division responsible for each of the following effects.

a _____ decreased metabolic rate

b _____ increased salivary and digestive secretions

c _____ increased metabolic rate

d _____ stimulation of urination and defecation

e _____ activation of sweat glands

f _____ heightened mental alertness

g _____ decreased heart rate and blood pressure

h _____ activation of energy reserves

i _____ increased heart rate and blood pressure

j _____ reduced digestive and urinary functions

k _____ increased motility and blood flow in the digestive tract

l _____ increased respiratory rate and dilation of respiratory passages

m _____ constriction of the pupils and focus of the eyes on nearby objects

3. Section integration

Recent surveys show that about one-third of the American adult population is involved in some type of exercise program. What contributions does sympathetic activation make to help the body adjust to changes that occur during exercise and still maintain homeostasis?

_____

_____

Visual Outline with Key Terms

Summarize the content of each module using the terms in the order provided.

SECTION 1

The Functional Anatomy and Organization of the Autonomic Nervous System (ANS)

- somatic nervous system (SNS)
- autonomic nervous system (ANS)
- preganglionic neurons
- ganglionic neurons
- autonomic ganglia

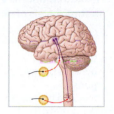

14.1

The ANS consists of sympathetic, parasympathetic, and enteric divisions

- sympathetic division
- parasympathetic division
- thoracolumbar division
- craniosacral division
- enteric nervous system (ENS)

14.2

The sympathetic division has chain ganglia, collateral ganglia, and the adrenal medullae; whereas the parasympathetic division has terminal or intramural ganglia

- preganglionic fibers
- postganglionic fibers
- sympathetic chain
- collateral ganglia
- celiac ganglia
- superior mesenteric ganglia
- inferior mesenteric ganglia
- adrenal medulla
- terminal ganglia
- intramural ganglia
- ciliary ganglion
- pterygopalatine ganglion
- submandibular ganglion
- otic ganglion

14.3

The two ANS divisions innervate many of the same structures, but the innervation patterns are different

- splanchnic nerves
- sympathetic nerves

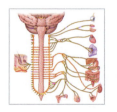

14.4

The effects of sympathetic and parasympathetic stimulation are mediated by membrane receptors at the target organs

- adrenergic receptors
- alpha receptors
- beta receptors
- nicotinic receptors
- muscarinic receptors

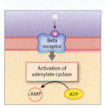

14.5

The functional differences between the two ANS divisions reflect their divergent anatomical and physiological characteristics

- sympathetic activation
- parasympathetic stimulation
- anabolic system

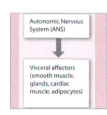

SECTION 2

Autonomic Regulation and Control Mechanisms

- central nervous system (CNS) processing
- sensory processing centers
- motor centers
- motor responses
- somatic effectors
- visceral effectors
- autonomic output

14.6

The ANS provides precise control over visceral functions

- autonomic tone
- dual innervation
- ○ sympathetic effects
- ○ parasympathetic effects

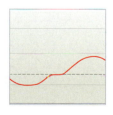

• = *Term boldfaced in this module*

14.7

Most visceral functions are controlled by visceral reflexes

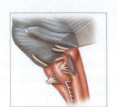

- visceral reflexes
- short reflexes
- long reflexes
- solitary nuclei

14.9

The autonomic nervous system has multiple levels of motor control

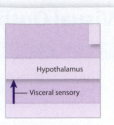

Hypothalamus

Visceral sensory

- ○ cerebral cortex
- ○ limbic system
- ○ thalamus
- ○ hypothalamus
- ○ pons
- ○ medulla oblongata
- ○ spinal cord T_1–L_2
- ○ sacral spinal cord S_2–S_4

14.8

Baroreceptors and chemoreceptors initiate important autonomic reflexes involving visceral sensory pathways

- baroreceptors
- chemoreceptors
- carotid bodies
- aortic bodies

• = *Term boldfaced in this module*

Chapter Integration: Applying what you've learned

John is outside gardening when he is stung on his cheek by a bee. Allergic reactions to flying stinging insects are relatively common, and John is allergic to bee venom. Shortly afterwards, he experiences pain, swelling, redness, and itching in the immediate area of the sting; his throat begins to swell and breathing becomes more difficult.

An ambulance is called, and John is taken to the emergency room. Drugs are administered to prevent anaphylaxis, a severe, life-threatening allergic response. Anaphylaxis is characterized by the contraction of smooth muscles along the respiratory passageways and a peripheral vasodilation that produces a fall in blood pressure, which would affect multiple organ systems.

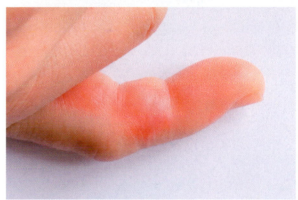

Approximately 40 people in the United States die every year from a venom allergy, most of whom had no known medical history of venom allergy.

1. Would acetylcholine or epinephrine be more helpful in relieving his condition? Why?

2. If the condition progressed to the point where there is a fall in blood pressure and inadequate blood flow to body tissues, describe what ANS responses would attempt to restore normal blood flow and blood pressure.

Mastering A&P™

15
The Special Senses

An Introduction to the Special Senses: Olfaction and Gustation

The special senses provide us with vital information about our environment. Although the sensory information provided is diverse and complex, each special sense originates at receptor cells that may be neurons or specialized receptor cells that communicate with sensory neurons.

1 Olfactory receptors are the dendrites of specialized neurons. When dissolved chemicals contact the dendritic processes, there is a depolarization, called a **generator potential**. This graph shows the action potentials produced by a generator potential.

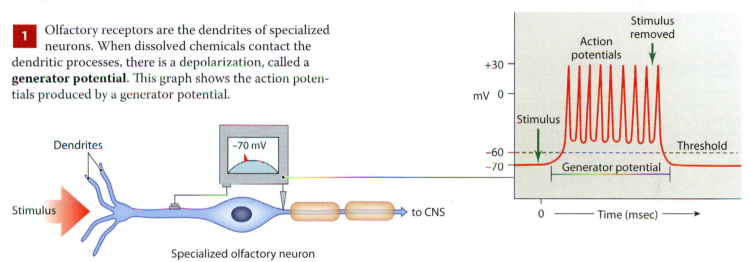

Specialized olfactory neuron

2 The receptors for the senses of taste, vision, equilibrium, and hearing are specialized cells that have inexcitable membranes and form synapses with the processes of sensory neurons. When stimulated, the membrane of the receptor cell undergoes a graded depolarization that triggers the release of chemical transmitters at the synapse. These transmitters then depolarize the sensory neuron, creating a generator potential and action potentials that are propagated to the CNS. Because a synapse is involved, there is a slight synaptic delay. However, this arrangement permits modification of the sensitivity of the receptor cell by presynaptic facilitation or inhibition.

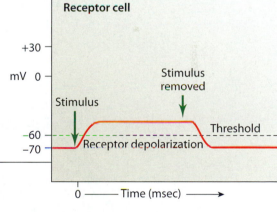

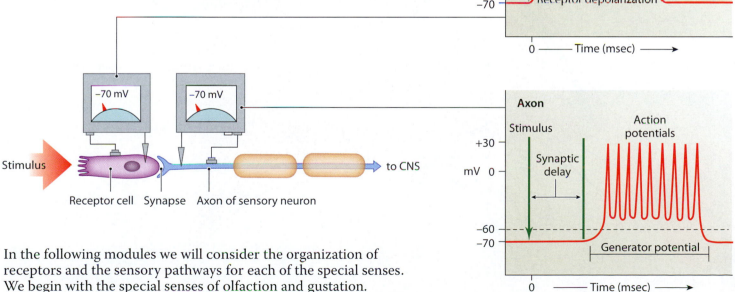

Receptor cell   Synapse   Axon of sensory neuron

In the following modules we will consider the organization of receptors and the sensory pathways for each of the special senses. We begin with the special senses of olfaction and gustation.

Olfaction involves specialized chemoreceptive neurons and delivers sensations directly to the cerebrum

1 The sense of smell, more precisely called olfaction, is provided by paired **olfactory organs**. These organs are located in the nasal cavity on either side of the nasal septum. They cover the inferior surface of the cribriform plate, the superior portion of the perpendicular plate, and the superior nasal conchae of the ethmoid.

Olfactory Pathway to the Cerebrum

The sensory neurons within the olfactory organ are stimulated by chemicals in the air.	Axons leaving the olfactory epithelium collect into 20 or more bundles that penetrate the cribriform plate of the ethmoid.	The first synapse occurs in the **olfactory bulb**, which is located just superior to the cribriform plate.	Axons leaving the olfactory bulb travel along the **olfactory tract** to reach the olfactory cortex, the hypothalamus, and portions of the limbic system.	The distribution of olfactory information to the limbic system and hypothalamus explains the profound emotional and behavioral responses, as well as the memories, that can be triggered by certain smells.

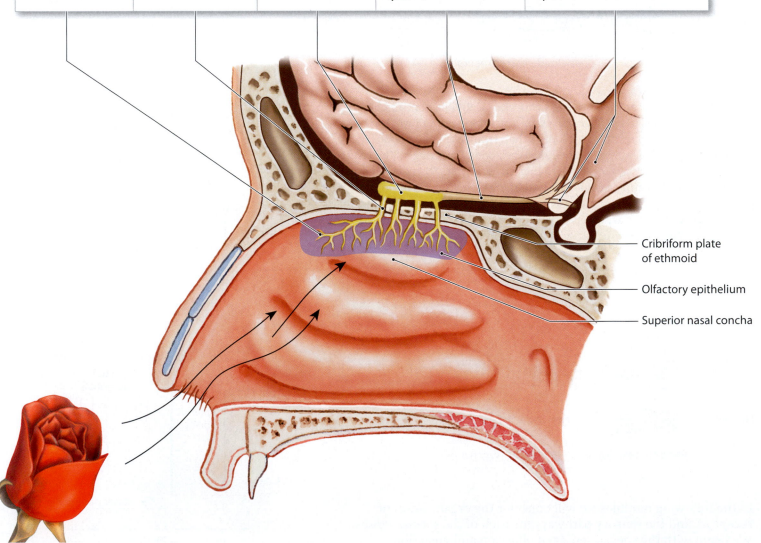

Cribriform plate of ethmoid

Olfactory epithelium

Superior nasal concha

2 The olfactory organs are made up of two layers: the **olfactory epithelium** and the **lamina propria**. The olfactory receptors are part of the epithelium. The exposed tip of each receptor cell forms a knob that projects beyond the epithelial surface. The knob provides a base for up to 20 cilia that extend into the surrounding mucus and lie parallel to the epithelial surface, exposing their considerable surface area to dissolved compounds. Between 10 and 20 million olfactory receptors are packed into an area of roughly 5 cm². If we take into account the exposed ciliary surfaces, the actual sensory area probably approaches that of the entire body surface.

The underlying lamina propria consists of areolar tissue, numerous blood vessels, and nerves. This layer also contains **olfactory glands**, or Bowman glands, whose secretions absorb water and form a thick, pigmented mucus.

The olfactory epithelium contains the **olfactory receptor cells**, **supporting cells**, and regenerative **basal cells** (stem cells). The olfactory receptor population undergoes considerable turnover; new receptor cells are produced by the division and differentiation of basal cells. This turnover is one of the few examples of neuronal replacement in adult humans.

Olfactory (Bowman) gland

To olfactory bulb

Olfactory nerve fibers

Basal cell: divides to replace worn-out olfactory receptor cells

Developing olfactory receptor cell

Olfactory receptor cell

Supporting cell

Mucous layer

Knob

Olfactory cilia: surfaces contain receptor proteins

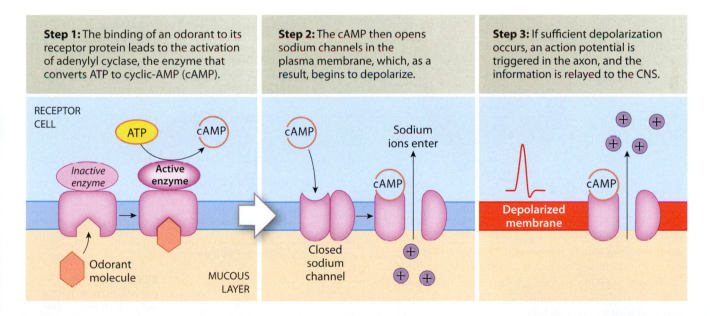

Step 1: The binding of an odorant to its receptor protein leads to the activation of adenylyl cyclase, the enzyme that converts ATP to cyclic-AMP (cAMP).

Step 2: The cAMP then opens sodium channels in the plasma membrane, which, as a result, begins to depolarize.

Step 3: If sufficient depolarization occurs, an action potential is triggered in the axon, and the information is relayed to the CNS.

RECEPTOR CELL

ATP cAMP

Inactive enzyme Active enzyme

Odorant molecule

MUCOUS LAYER

cAMP

Sodium ions enter

cAMP

Closed sodium channel

cAMP

Depolarized membrane

3 Olfactory reception occurs on the surface membranes of the olfactory cilia. **Odorants**—dissolved chemicals that stimulate olfactory receptors—interact with receptors called odorant binding proteins on the membrane surface. In general, odorants are small organic molecules; the strongest smells are associated with molecules of either high water or high lipid solubilities. As few as four odorant molecules can activate an olfactory receptor.

Module 15.1 Review

a. Describe olfaction.

b. Which neurons associated with olfaction are capable of regenerating?

c. Trace the olfactory pathway, beginning at the olfactory epithelium.

Gustation involves epithelial chemoreceptor cells located in taste buds

Gustation, or taste, provides information about the foods and liquids we consume. **Taste receptors**, or gustatory (GUS-ta-tor-ē) receptors, are distributed over the superior surface of the tongue and adjacent portions of the pharynx and larynx. The most important taste receptors are on the tongue; by the time we reach adulthood, the taste receptors on the pharynx, larynx, and epiglottis have decreased in importance and abundance.

1 The superior surface of the tongue bears epithelial projections called **lingual papillae** (pa-PIL-ē; *papilla*, a nipple-shaped mound). The human tongue bears three types of lingual papillae: (1) circumvallate (sir-kum-VAL-āt; *circum-*, around + *vallum*, wall) papillae, (2) fungiform (*fungus*, mushroom) papillae, and (3) filiform (*filum*, thread) papillae. Many of these papillae contain taste receptors and specialized epithelial cells in sensory structures called **taste buds**. An adult has about 5000 taste buds.

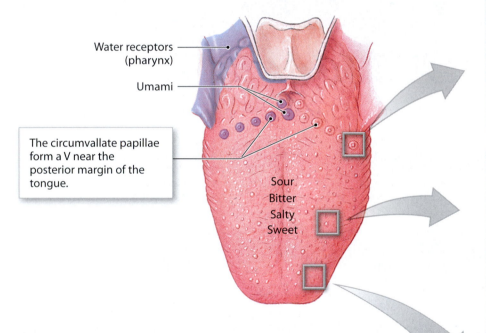

Water receptors (pharynx)

Umami

The circumvallate papillae form a V near the posterior margin of the tongue.

Sour
Bitter
Salty
Sweet

2 There is some evidence that sensitivity to the four primary taste sensations—sweet, salty, sour, and bitter—varies along the long axis of the tongue. However, there are no differences in the structure of the taste buds, and taste buds in all portions of the tongue provide all four primary taste sensations. There are also two other taste sensations. **Umami** (oo-MAH-mē) is a pleasant taste that is characteristic of beef broth, chicken broth, and Parmesan cheese. This taste is detected by receptors sensitive to the presence of amino acids, small peptides, and nucleotides. These receptors are present in taste buds of the circumvallate papillae. **Water receptors** have been demonstrated in humans, and they appear to be especially concentrated in the pharynx. The sensory output of these receptors is processed in the hypothalamus and affects water balance and the regulation of blood volume. For example, minor reductions in antidiuretic hormone (ADH) secretion occur each time you take a long drink.

Circumvallate Papillae

Circumvallate papillae are relatively large, shaped like the tip of a pencil eraser, and surrounded by deep epithelial folds. Each of these papillae contains as many as 100 taste buds.

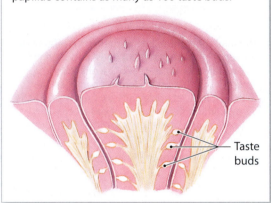

Taste buds

Fungiform Papillae

Fungiform papillae are shaped like small buttons within shallow depressions. Each papilla contains about five taste buds.

Filiform Papillae

Filiform papillae provide friction that helps the tongue move objects around in the mouth but do not contain taste buds. Filiform and fungiform papillae are found on the anterior two-thirds of the superior surface of the tongue.

3 Taste buds are recessed into the surrounding epithelium, isolated from the relatively unprocessed contents of the mouth. Each taste bud contains 40–100 receptor cells and many small stem cells.

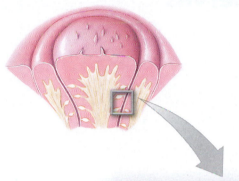

Each **gustatory cell** extends slender microvilli, sometimes called taste hairs, into the surrounding fluids through the **taste pore**, a narrow opening. A typical gustatory cell survives for only about 10 days before it is replaced.

Basal cells are stem cells that divide to produce daughter cells that will mature into gustatory cells.

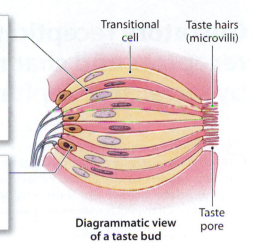

Transitional cell

Taste hairs (microvilli)

Taste pore

Diagrammatic view of a taste bud

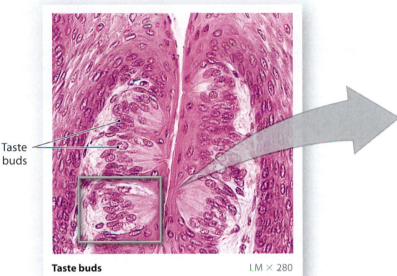

Taste buds

Taste buds   LM × 280

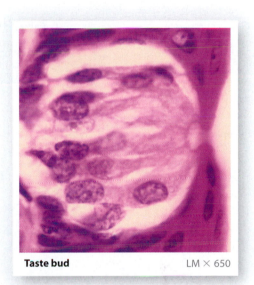

Taste bud   LM × 650

4 The threshold for taste receptor stimulation varies for each of the primary taste sensations, and taste receptors respond more readily to unpleasant than to pleasant stimuli. For example, we are 100,000 times more sensitive to bitter substances and almost 1000 times more sensitive to acids (sour) than to either sweet or salty chemicals. This sensitivity has survival value, because acids can damage the mucous membranes of the mouth and pharynx, and many potent biological toxins have an extremely bitter taste. Our tasting abilities change with age. We begin life with more than 10,000 taste buds, but the number begins declining dramatically by age 50. The sensory loss becomes especially significant because aging individuals also experience a decline in the number of olfactory receptors. As a result, many elderly people find that their food tastes bland and unappetizing, whereas children tend to find the same foods too spicy.

Module 15.2 Review

a. Define gustation.

b. Describe filiform papillae.

c. Relate the adaptive sensitivity of taste receptors for bitter and sour sensations, to sweet and salty sensations.

Gustatory reception relies on membrane receptors and channels, and sensations are carried by facial, glossopharyngeal, and vagus nerves

1 The mechanism involved in gustatory reception resembles that of olfaction. Dissolved chemicals contacting the taste hairs bind to receptor proteins of the gustatory cell. The different tastes involve different receptor mechanisms. Taste receptors adapt slowly, but central adaptation quickly reduces your sensitivity to a new taste.

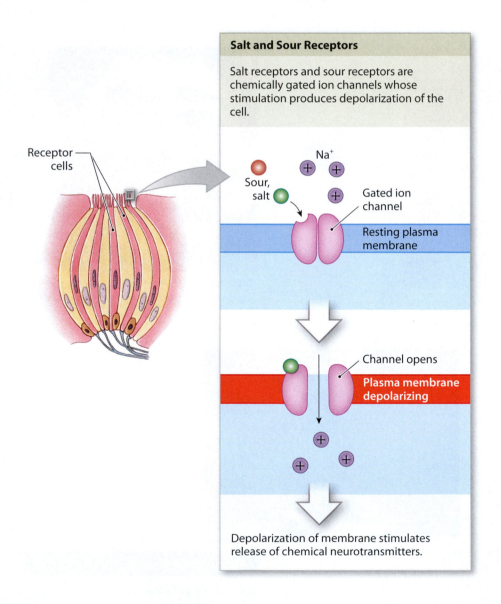

Receptor cells

Salt and Sour Receptors

Salt receptors and sour receptors are chemically gated ion channels whose stimulation produces depolarization of the cell.

Na⁺

Sour, salt

Gated ion channel

Resting plasma membrane

Channel opens

Plasma membrane depolarizing

Depolarization of membrane stimulates release of chemical neurotransmitters.

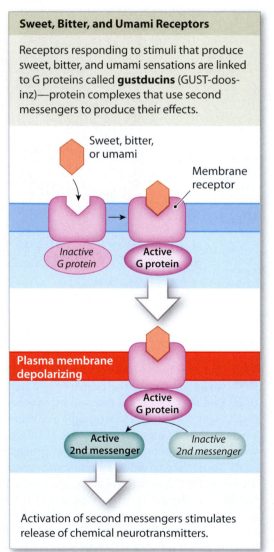

Sweet, Bitter, and Umami Receptors

Receptors responding to stimuli that produce sweet, bitter, and umami sensations are linked to G proteins called **gustducins** (GUST-doos-inz)—protein complexes that use second messengers to produce their effects.

Sweet, bitter, or umami

Membrane receptor

Inactive G protein

Active G protein

Plasma membrane depolarizing

Active G protein

Active 2nd messenger

Inactive 2nd messenger

Activation of second messengers stimulates release of chemical neurotransmitters.

2 This illustration follows the gustatory pathway from the receptors to the cerebral cortex. A conscious perception of taste is produced by processing at the primary sensory cortex, as the information received from the taste buds is correlated with other sensory data. Information about the texture of food, along with taste-related sensations such as "peppery" or "burning hot," is provided by sensory (afferent) fibers in the trigeminal nerve (V). In addition, the level of stimulation from the olfactory receptors plays an overwhelming role in taste perception. Thus, you are several thousand times more sensitive to "tastes" when your olfactory organs are fully functional. By contrast, when you have a cold and your nose is stuffed up, airborne molecules cannot reach your olfactory receptors, so meals taste dull and unappealing. This reduction in taste perception occurs even though the taste buds are responding normally.

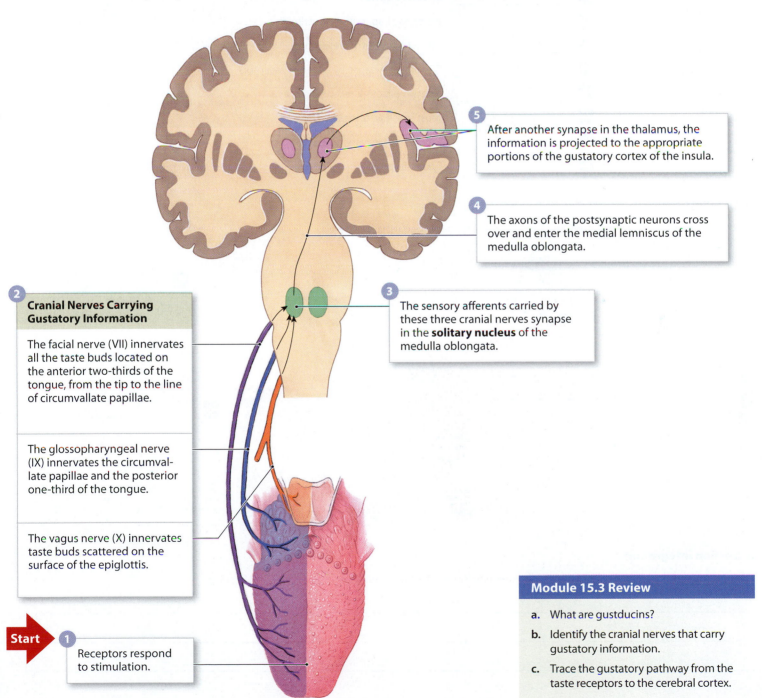

5 After another synapse in the thalamus, the information is projected to the appropriate portions of the gustatory cortex of the insula.

4 The axons of the postsynaptic neurons cross over and enter the medial lemniscus of the medulla oblongata.

2 Cranial Nerves Carrying Gustatory Information

The facial nerve (VII) innervates all the taste buds located on the anterior two-thirds of the tongue, from the tip to the line of circumvallate papillae.

The glossopharyngeal nerve (IX) innervates the circumvallate papillae and the posterior one-third of the tongue.

The vagus nerve (X) innervates taste buds scattered on the surface of the epiglottis.

3 The sensory afferents carried by these three cranial nerves synapse in the **solitary nucleus** of the medulla oblongata.

Start **1** Receptors respond to stimulation.

Module 15.3 Review

a. What are gustducins?

b. Identify the cranial nerves that carry gustatory information.

c. Trace the gustatory pathway from the taste receptors to the cerebral cortex.

1. Matching

Match the following terms with the most closely related description.

- gustation
- depolarization
- Bowman glands
- lingual papillae
- G proteins
- stem cells
- bitter
- olfactory cilia
- olfaction
- taste bud
- cerebral cortex
- olfactory bulb
- odorant

a _____ Sweet, bitter, and umami sensations

b _____ Sense of smell

c _____ Basal cells

d _____ Chemical stimulus

e _____ Sense of taste

f _____ Olfactory glands

g _____ Receives all special senses stimuli

h _____ Cluster of gustatory receptors

i _____ Site of first synapse by olfactory receptors

j _____ Contain olfactory receptor proteins

k _____ Most sensitive taste sensation

l _____ Produces generator potential

m _____ Epithelial projections of tongue

2. Labeling

Label the areas of taste on the tongue (a–e) and the three types of lingual papillae (f–h).

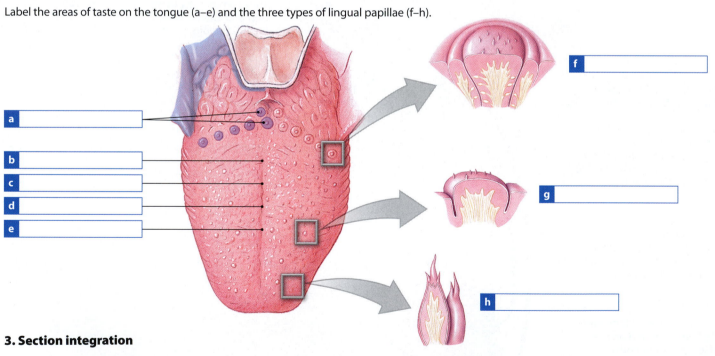

a _____

b _____

c _____

d _____

e _____

f _____

g _____

h _____

3. Section integration

Contrast the sensory receptors for olfaction with the sensory receptors for taste, vision, equilibrium, and hearing.

_____

_____

_____

_____

_____

Equilibrium and Hearing

1 In olfaction, the sensory receptors are modifed neurons; in gustation, the receptors are specialized receptor cells that communicate with sensory neurons. In each case, the sensory receptors are located within epithelia exposed to the external environment, and the information is routed directly to the CNS for processing.

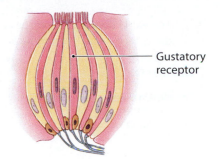

Olfactory receptor

Gustatory receptor

2 In contrast, the receptors responsible for the sensations of equilibrium and hearing are isolated and protected from the external environment. These receptors are located within the **inner ear**, a complex sense organ, and the sensory information may be integrated and organized before it is forwarded to the CNS.

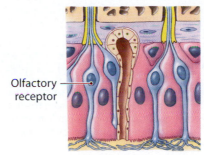

Inner ear

3 The receptors of the inner ear are called **hair cells** because their free surfaces are covered with specialized processes similar to the cilia and microvilli of other cells. They are basically mechanoreceptors sensitive to contact or movement, and they are always surrounded by supporting cells and monitored by the dendrites of sensory neurons. When an external force pushes against the processes of a hair cell, the distortion of the plasma membrane alters the rate at which the hair cell releases chemical transmitters. In this way, hair cells provide information about the direction and strength of mechanical stimuli.

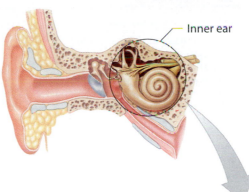

Displacement in this direction stimulates hair cell

Displacement in this direction inhibits hair cell

Each hair cell in the vestibule contains a **kinocilium** (ki-nō-SIL-ē-um), a single large cilium. Hair cells do not actively move their kinocilia (or stereocilia).

The free surface of every hair cell supports 80–100 long **stereocilia**, which resemble very long microvilli.

Hair cell

Dendrite of sensory neuron

Supporting cell

The inner ear has a complex three-dimensional structure that determines what stimuli can reach the hair cells in each region. For example, hair cells in one region can respond only to gravity or acceleration, whereas those in other regions respond only to rotation or only to sound. In this section you will learn how the anatomy of the inner ear enables a single receptor cell type to provide such a diversity of information.

505

The ear is divided into the external ear, the middle ear, and the inner ear

1 The ear is divided into three anatomical regions: the external ear, the middle ear, and the inner ear.

External Ear	Middle Ear	Inner Ear
The **external ear**—the visible portion of the ear—collects and directs sound waves toward the middle ear.	The **middle ear**, or tympanic cavity, is an air-filled chamber separated from the external acoustic meatus by the tympanic membrane. It is connected to the pharynx by the auditory tube.	The **inner ear** contains the sensory organs for hearing and equilibrium. It receives amplified sound waves from the middle ear.

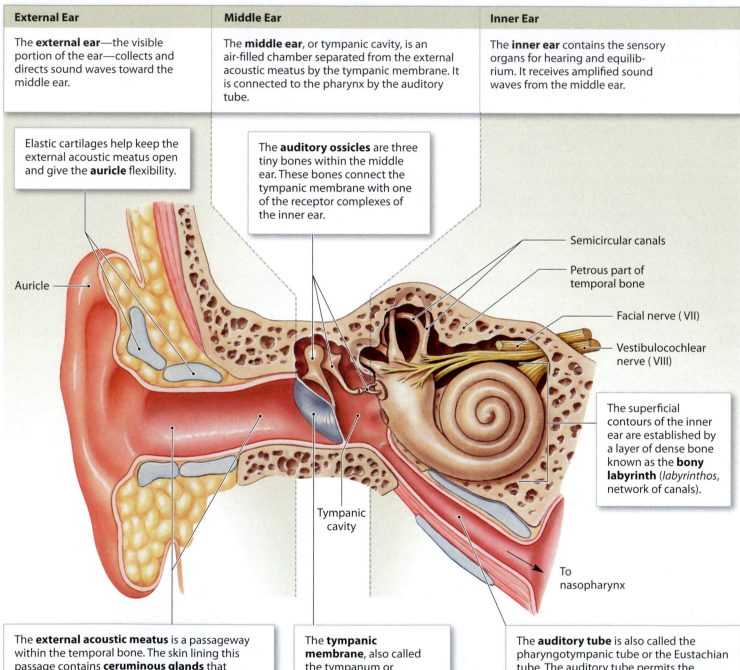

Elastic cartilages help keep the external acoustic meatus open and give the **auricle** flexibility.

The **auditory ossicles** are three tiny bones within the middle ear. These bones connect the tympanic membrane with one of the receptor complexes of the inner ear.

Auricle

Semicircular canals

Petrous part of temporal bone

Facial nerve (VII)

Vestibulocochlear nerve (VIII)

The superficial contours of the inner ear are established by a layer of dense bone known as the **bony labyrinth** (*labyrinthos*, network of canals).

Tympanic cavity

To nasopharynx

The **external acoustic meatus** is a passageway within the temporal bone. The skin lining this passage contains **ceruminous glands** that secrete a waxy material, **cerumen**, and it has many small, outwardly projecting hairs. Together, cerumen and the hairs help keep foreign objects and insects from reaching more delicate internal structures. Cerumen also slows the growth of microorganisms in the external acoustic meatus, reducing the chances for an external ear infection.

The **tympanic membrane**, also called the tympanum or eardrum, lies at the end of the external acoustic meatus. The tympanic membrane is a thin, semitransparent sheet that separates the external ear from the middle ear.

The **auditory tube** is also called the pharyngotympanic tube or the Eustachian tube. The auditory tube permits the equalization of pressures on either side of the tympanic membrane. Unfortunately, the auditory tube can also allow microorganisms to travel from the nasopharynx into the middle ear. Invasion by microorganisms can lead to an unpleasant middle ear infection known as **otitis media**.

2 The middle ear contains the three auditory ossicles and communicates both with the superior portion of the pharynx (the nasopharynx), through the auditory tube, and with the mastoid air cells, through a number of small connections. It contains the three auditory ossicles. The articulations between the auditory ossicles are the smallest synovial joints in the body. Each ossicle has a tiny capsule and supporting extracapsular ligaments.

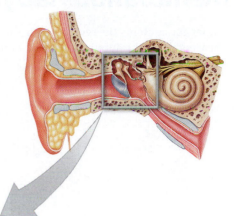

Auditory Ossicles

The **malleus** (*malleus*, hammer) attaches at three points to the interior surface of the tympanic membrane.

The **incus** (*incus*, anvil), the middle ossicle, attaches the malleus to the stapes.

The edges of the base of the **stapes** (*stapes*, stirrup) are bound to the edges of the oval window, an opening in the bone that surrounds the inner ear.

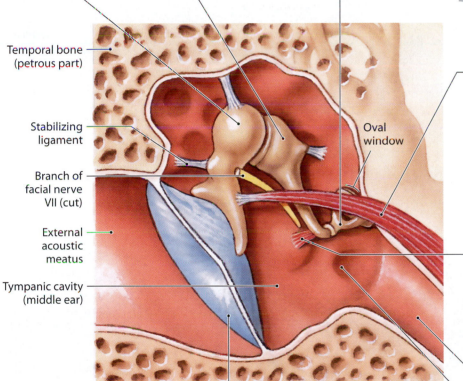

Temporal bone (petrous part)

Stabilizing ligament

Branch of facial nerve VII (cut)

External acoustic meatus

Tympanic cavity (middle ear)

Oval window

Auditory tube

Round window

Muscles of the Middle Ear

The **tensor tympani** (TEN-sor tim-PAN-ē) **muscle** is a short ribbon of muscle whose origin is the petrous portion of the temporal bone and the auditory tube, and whose insertion is on the "handle" of the malleus. When the tensor tympani contracts, the malleus is pulled medially, stiffening the tympanic membrane and reducing the amount it can vibrate in response to a sound. The tensor tympani muscle is innervated by motor fibers of the mandibular branch of the trigeminal nerve (V).

The **stapedius** (sta-PĒ-dē-us) **muscle**, innervated by the facial nerve (VII), originates from the posterior wall of the middle ear and inserts on the stapes. Contraction of the stapedius pulls the stapes, reducing movement of the stapes at the oval window.

Arriving sound waves vibrate the tympanic membrane, thereby converting the waves into mechanical movements. The auditory ossicles conduct those vibrations to the inner ear, because they are connected in such a way that an in–out movement of the tympanic membrane produces a rocking motion of the stapes. The ossicles thus function as a lever system that collects the force applied to the tympanic membrane and focuses it on the oval window. Because the tympanic membrane is 22 times larger and heavier than the oval window, considerable amplification occurs, so we can hear very faint sounds. But that degree of amplification can be a problem when we are exposed to very loud noises. Contractions of the tensor tympani and the stapedius muscles protect the tympanic membrane and ossicles from violent movements under very noisy conditions.

Module 15.4 Review

a. Name the three tiny bones located in the middle ear.

b. What is the function of the auditory tube?

c. Why are external ear infections relatively uncommon?

The bony labyrinth protects the membranous labyrinth

1 The inner ear contains the receptors for the special senses of equilibrium and hearing. It can be subdivided into a bony labyrinth and a membranous labyrinth.

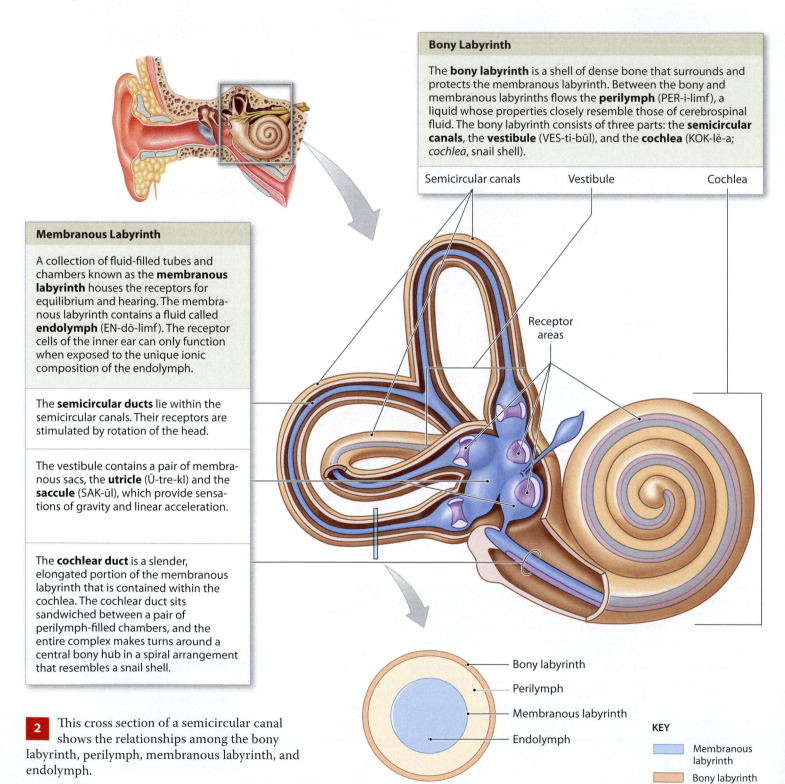

Bony Labyrinth

The **bony labyrinth** is a shell of dense bone that surrounds and protects the membranous labyrinth. Between the bony and membranous labyrinths flows the **perilymph** (PER-i-limf), a liquid whose properties closely resemble those of cerebrospinal fluid. The bony labyrinth consists of three parts: the **semicircular canals**, the **vestibule** (VES-ti-būl), and the **cochlea** (KOK-lē-a; *cochlea*, snail shell).

Semicircular canals   Vestibule   Cochlea

Membranous Labyrinth

A collection of fluid-filled tubes and chambers known as the **membranous labyrinth** houses the receptors for equilibrium and hearing. The membranous labyrinth contains a fluid called **endolymph** (EN-dō-limf). The receptor cells of the inner ear can only function when exposed to the unique ionic composition of the endolymph.

The **semicircular ducts** lie within the semicircular canals. Their receptors are stimulated by rotation of the head.

The vestibule contains a pair of membranous sacs, the **utricle** (Ū-tre-kl) and the **saccule** (SAK-ūl), which provide sensations of gravity and linear acceleration.

The **cochlear duct** is a slender, elongated portion of the membranous labyrinth that is contained within the cochlea. The cochlear duct sits sandwiched between a pair of perilymph-filled chambers, and the entire complex makes turns around a central bony hub in a spiral arrangement that resembles a snail shell.

Receptor areas

Bony labyrinth
Perilymph
Membranous labyrinth
Endolymph

KEY

Membranous labyrinth
Bony labyrinth

2 This cross section of a semicircular canal shows the relationships among the bony labyrinth, perilymph, membranous labyrinth, and endolymph.

3 Here is a simple concept map that summarizes information about the regions and functions of the membranous labyrinth.

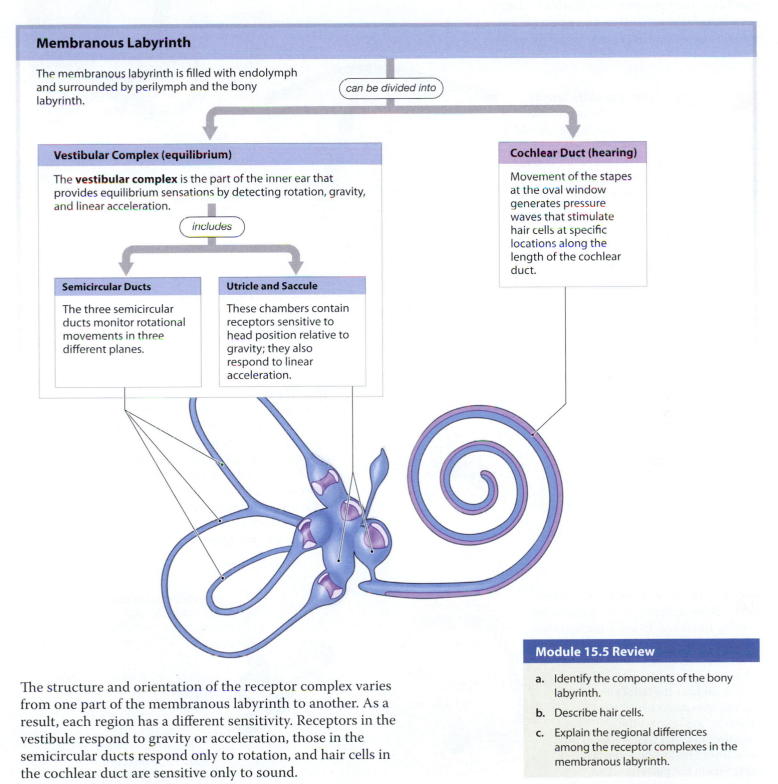

Membranous Labyrinth

The membranous labyrinth is filled with endolymph and surrounded by perilymph and the bony labyrinth.

can be divided into

Vestibular Complex (equilibrium)

The **vestibular complex** is the part of the inner ear that provides equilibrium sensations by detecting rotation, gravity, and linear acceleration.

includes

Semicircular Ducts

The three semicircular ducts monitor rotational movements in three different planes.

Utricle and Saccule

These chambers contain receptors sensitive to head position relative to gravity; they also respond to linear acceleration.

Cochlear Duct (hearing)

Movement of the stapes at the oval window generates pressure waves that stimulate hair cells at specific locations along the length of the cochlear duct.

The structure and orientation of the receptor complex varies from one part of the membranous labyrinth to another. As a result, each region has a different sensitivity. Receptors in the vestibule respond to gravity or acceleration, those in the semicircular ducts respond only to rotation, and hair cells in the cochlear duct are sensitive only to sound.

Module 15.5 Review

a. Identify the components of the bony labyrinth.

b. Describe hair cells.

c. Explain the regional differences among the receptor complexes in the membranous labyrinth.

Hair cells in the semicircular ducts respond to rotation, while those in the utricle and saccule respond to gravity and linear acceleration

The **anterior**, **posterior**, and **lateral semicircular ducts** are continuous with the utricle. Each semicircular duct contains an **ampulla**, an expanded region that contains the receptors.

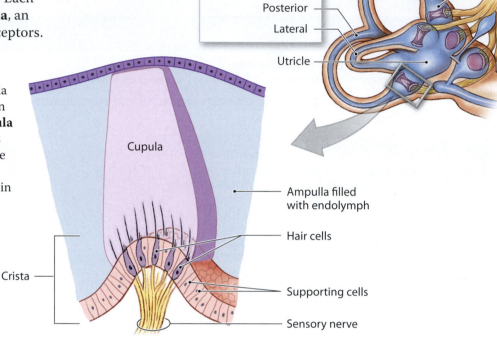

Semicircular Ducts
Anterior
Posterior
Lateral

Ampulla

Utricle

1 The region in the wall of the ampulla that contains the receptors is known as a **crista**. Each crista is bound to a **cupula** (KŪ-pū-luh), a flexible, elastic, gelatinous structure that extends the full width of the ampulla. At a crista, the kinocilia and stereocilia of the hair cells are embedded in the cupula.

Cupula

Ampulla filled with endolymph

Hair cells

Crista

Supporting cells

Sensory nerve

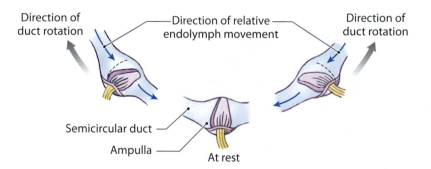

Direction of duct rotation

Direction of relative endolymph movement

Direction of duct rotation

Semicircular duct

Ampulla

At rest

2 The cupula has a density very close to that of the surrounding endolymph, so it essentially floats above the receptor surface. When your head rotates in the plane of a semicircular duct, the movement of endolymph along the length of the duct pushes the cupula to the side, distorting the receptor processes. Movement of fluid in one direction stimulates the hair cells, and movement in the opposite direction inhibits them. When the endolymph stops moving, the elastic cupula rebounds to its normal position.

3 Even the most complex movement can be analyzed in terms of motion in three rotational planes. Each semicircular duct responds to one of these rotational movements. A horizontal rotation, as in shaking your head "no," stimulates the hair cells of the lateral semicircular duct. Nodding "yes" excites the anterior duct, and tilting your head from side to side activates receptors in the posterior duct.

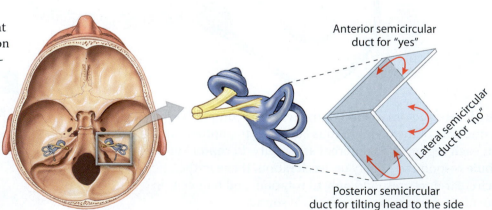

Anterior semicircular duct for "yes"

Lateral semicircular duct for "no"

Posterior semicircular duct for tilting head to the side

4 The utricle and saccule provide equilibrium sensations, whether the body is moving or stationary. These two chambers are connected by a slender passageway that is continuous with the narrow **endolymphatic duct**, which ends in a blind pouch called the **endolymphatic sac**. This sac projects into the subarachnoid space, where it is surrounded by a capillary network. Portions of the cochlear duct secrete endolymph continuously, and endolymph returns to the general circulation at the endolymphatic sac.

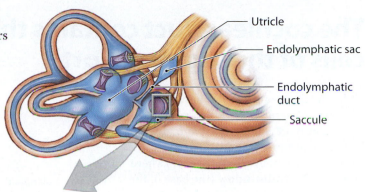

- Utricle
- Endolymphatic sac
- Endolymphatic duct
- Saccule

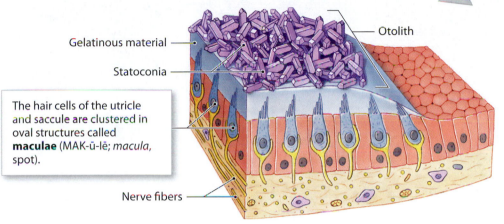

- Gelatinous material
- Statoconia
- Otolith
- Nerve fibers

The hair cells of the utricle and saccule are clustered in oval structures called **maculae** (MAK-ū-lē; *macula*, spot).

5 The hair cell processes are embedded in a gelatinous mass whose surface contains densely packed calcium carbonate crystals known as **statoconia** (*statos*, standing + *conia*, dust). The complex as a whole (gelatinous matrix and statoconia) is called an **otolith** ("ear stone").

6 Changes in the position of the head cause distortion of the hair cell processes in the maculae and send signals to the brain.

When your head is in the normal, upright position, the statoconia sit atop the macula. Their weight presses on the macular surface, pushing the hair cell processes down rather than to one side or another.

When your head is tilted, the pull of gravity on the statoconia shifts them to the side, thereby distorting the hair cell processes and stimulating the macular receptors. A similar mechanism accounts for your perception of linear acceleration, as when your car speeds up suddenly. The statoconia lag behind, and the effect on the hair cells is comparable to tilting your head back.

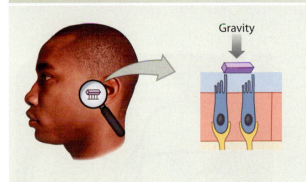

Gravity

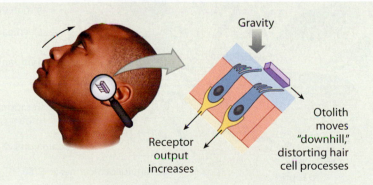

Gravity

Receptor output increases

Otolith moves "downhill," distorting hair cell processes

Module 15.6 Review

a. Define statoconia.

b. Cite the function of receptors in the saccule and utricle.

c. Damage to the cupula of the lateral semicircular duct would interfere with what perception?

The cochlear duct contains the hair cells of the organ of Corti

1 The **cochlear duct** lies between a pair of perilymphatic chambers: the **vestibular duct** and the **tympanic duct**. The outer surfaces of these ducts are encased by the bony labyrinth everywhere except at the **oval window** (the base of the vestibular duct) and the **round window** (the base of the tympanic duct). Because the vestibular and tympanic ducts are interconnected at the tip of the cochlear spiral, they really form one long and continuous perilymphatic chamber. This chamber begins at the oval window, extends through the vestibular duct, and proceeds around the top of the cochlea and along the tympanic duct; it ends at the round window.

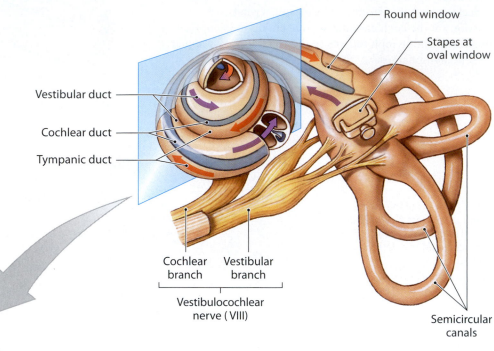

Round window

Stapes at oval window

Vestibular duct

Cochlear duct

Tympanic duct

Cochlear branch

Vestibular branch

Vestibulocochlear nerve (VIII)

Semicircular canals

KEY

→ From oval window to tip of spiral

→ From tip of spiral to round window

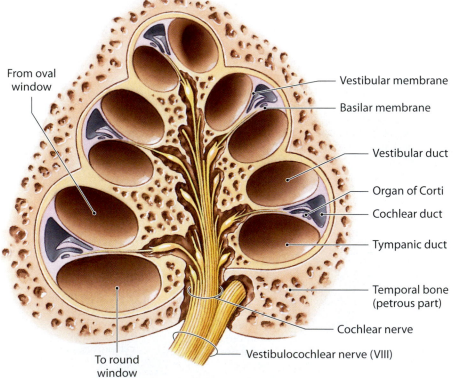

From oval window

Vestibular membrane

Basilar membrane

Vestibular duct

Organ of Corti

Cochlear duct

Tympanic duct

Temporal bone (petrous part)

Cochlear nerve

Vestibulocochlear nerve (VIII)

To round window

2 The cochlear duct is a long, coiled tube suspended between the vestibular duct and the tympanic duct. The **vestibular membrane** separates the cochlear duct from the vestibular duct, and the **basilar membrane** separates the cochlear duct from the tympanic duct. The hair cells of the cochlear duct are located in a structure called the **organ of Corti**, which is located on the basilar membrane.

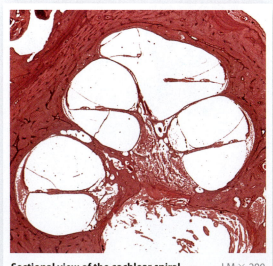

Sectional view of the cochlear spiral   LM × 200

3 This sectional view shows a single turn of the cochlea. The vestibular duct and tympanic duct are filled with perilymph, whereas the cochlear duct, which contains the organ of Corti, is filled with endolymph.

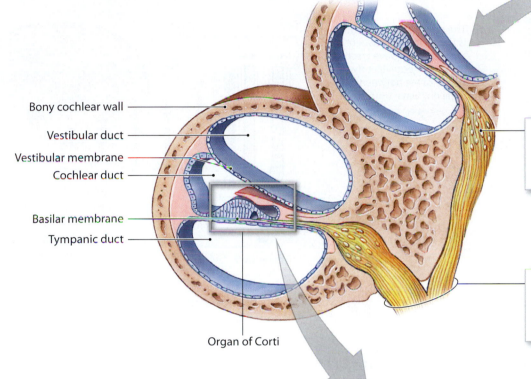

Bony cochlear wall

Vestibular duct

Vestibular membrane

Cochlear duct

Basilar membrane

Tympanic duct

Organ of Corti

The **spiral ganglion** contains the cell bodies of sensory neurons that monitor the adjacent hair cells of the organ of Corti.

The cochlear branch of the vestibulo-cochlear nerve (VIII) contains the axons of the neurons of the spiral ganglion.

4 The hair cells are arranged in a series of longitudinal rows. They lack kinocilia, and their stereocilia are in contact with the overlying **tectorial** (tek-TOR-ē-al; *tectum*, roof) **membrane**. This membrane is firmly attached to the inner wall of the cochlear duct. When a portion of the basilar membrane bounces up and down in response to pressure fluctuations within the perilymph, the stereocilia of the hair cells are pressed against the tectorial membrane and distorted. If the amount of movement increases, more hair cells—and more rows of hair cells—are stimulated. These pressure changes within the perilymph are triggered by sound waves arriving at the tympanic membrane.

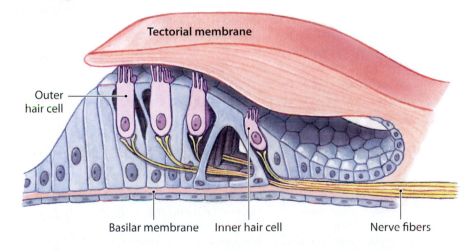

Tectorial membrane

Outer hair cell

Basilar membrane   Inner hair cell   Nerve fibers

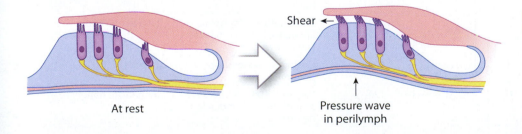

At rest

Shear

Pressure wave in perilymph

Module 15.7 Review

a. Where is the organ of Corti located?

b. Name the fluids found within the vestibular duct, tympanic duct, and cochlear duct.

c. Identify the features visible in the LM sectional view of the cochlear spiral shown in **2**.

The organ of Corti provides sensations of pitch and volume

1 Hearing is the perception of sound, which consists of waves of pressure conducted through a medium such as air or water. In air, each pressure wave consists of a region where the air molecules are crowded together and one where they are farther apart.

The **wavelength** of sound is the distance between two adjacent wave crests (peaks) or, equivalently, the distance between two adjacent wave troughs.

Air molecules

Tympanic membrane

Tuning fork

2 Sound waves can be graphed as S-shaped curves that repeat in a regular pattern. At sea level, sound waves travel through the air at about 1235 km/h (768 mph). The **frequency** is the number of waves that pass a fixed reference point—such as the tympanic membrane—in a given time. Physicists use the term "cycles" rather than waves. Hence, the frequency of a sound is measured in terms of the number of cycles per second (cps), a unit called **hertz (Hz)**. What we perceive as the pitch of a sound is our sensory response to its frequency. A high-frequency sound (high pitch, short wavelength) might have a frequency of 15,000 Hz or more; a very low-frequency sound (low pitch, long wavelength) could have a frequency of 100 Hz or less.

Because all sound waves travel at the same speed, as the frequency increases, the wavelength must become shorter.

The **amplitude** is determined by the amount of energy carried by the wave. The louder the sound, the higher the amplitude.

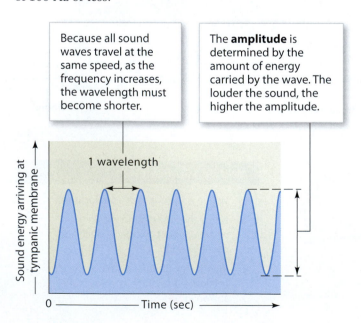

1 wavelength

Sound energy arriving at tympanic membrane

0   Time (sec)

3 This table lists the **intensity**—or energy in sound waves—of familiar sounds. Intensity determines how loud it seems; the greater the energy content, the larger the amplitude, and the louder the sound. Sound energy is reported in **decibels** (DES-i-belz) **(dB)**.

The Intensity of Representative Sounds

Typical Decibel Level	Example	Dangerous Time Exposure
0	Lowest audible sound	
30	Quiet library; soft whisper	
40	Quiet office; living room; bedroom away from traffic	
50	Light traffic at a distance; refrigerator; gentle breeze	
60	Air conditioner from 20 feet; conversation; sewing machine in operation	
70	Busy traffic; noisy restaurant	Some damage if continuous
80	Subway; heavy city traffic; alarm clock at 2 feet; factory noise	More than 8 hours
90	Truck traffic; noisy home appliances; shop tools; gas lawn mower	Less than 8 hours
100	Chain saw; boiler shop; pneumatic drill	2 hours
120	"Heavy metal" rock concert; sandblasting; thunderclap nearby	Immediate danger
140	Gunshot; jet plane	Immediate danger
160	Rocket launching pad	Hearing loss inevitable

4 The energy of sound waves is a physical pressure. When sound waves strike a flexible object, the object responds to that pressure. Given the right combination of frequencies and amplitudes, the object will begin to vibrate at the same frequency as the sound, a phenomenon called **resonance**. The higher the amplitude, the greater the amount of vibration. For you to be able to hear a sound, your tympanic membrane must vibrate in resonance with the sound waves. Pressure waves at the tympanic membrane generate movement of the stapes at the oval window. The flexibility of the basilar membrane varies along its length, so pressure waves of different frequencies affect different parts of the membrane. This diagram depicts the movement produced by sound waves with a frequency of 6000 Hz. When the stapes moves inward, the basilar membrane distorts toward the round window, which bulges into the middle-ear cavity. When the stapes moves outward, the basilar membrane rebounds and distorts toward the oval window. The location of the vibration is interpreted as **pitch**; the number of stimulated hair cells is interpreted as **volume**.

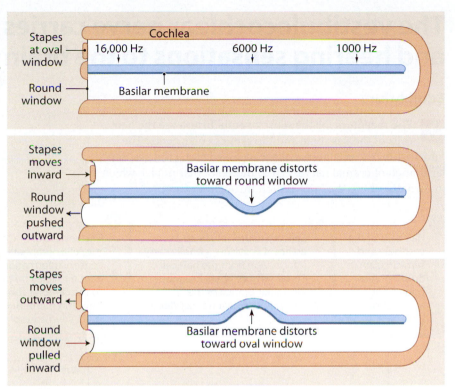

5 This illustration summarizes the events involved in hearing.

Events Involved in Hearing

Sound waves arrive at the tympanic membrane.	Movement of the tympanic membrane causes displacement of the auditory ossicles.	Movement of the stapes at the oval window establishes pressure waves in the perilymph of the vestibular duct.	The pressure waves distort the basilar membrane on their way to the round window of the tympanic duct.	Vibration of the basilar membrane causes vibration of hair cells against the tectorial membrane.	Information about the region and the intensity of stimulation is relayed to the CNS over the cochlear branch of cranial nerve VIII.
①	②	③	④	⑤	⑥

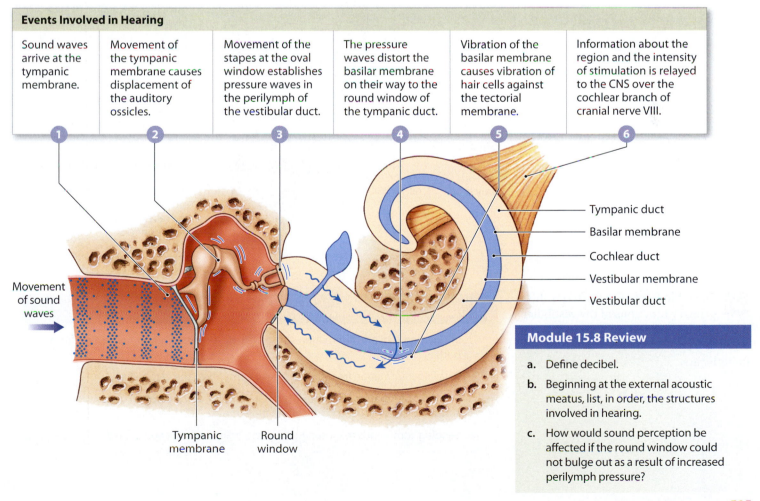

Tympanic duct
Basilar membrane
Cochlear duct
Vestibular membrane
Vestibular duct

Movement of sound waves

Tympanic membrane   Round window

Module 15.8 Review

a. Define decibel.

b. Beginning at the external acoustic meatus, list, in order, the structures involved in hearing.

c. How would sound perception be affected if the round window could not bulge out as a result of increased perilymph pressure?

The vestibulocochlear nerve carries equilibrium and hearing sensations to the brain stem

Equilibrium

1 The receptors for the sense of equilibrium are the hair cells of the vestibule and the semicircular ducts. The information the receptors collect is passed along the vestibular branch of cranial nerve VIII to the vestibular nuclei, which in turn distribute the information throughout the CNS.

1 Hair cells of the vestibule and semicircular ducts monitor body position and motion.

2 Sensory neurons located in adjacent vestibular ganglia carry information from the hair cells. These sensory fibers form the vestibular branch of the vestibulocochlear nerve (VIII).

3 The vestibular nuclei in the medulla oblongata integrate sensory information from both ears, and relay that information to the cerebral cortex, cerebellum, and motor nuclei in the brain stem and spinal cord.

4 The automatic movements of the eyes that occur in response to sensations of motion are directed by the superior colliculi. These movements attempt to keep your gaze focused on a specific point in space, despite changes in body position and orientation.

The reflexive motor commands issued by the vestibular nuclei are distributed to the motor nuclei for the cranial nerves involved with eye, head, and neck movements (N III, N IV, N VI, and N XI).

The vestibular nuclei relay information about position and balance to the cerebellum.

Instructions descending in the vestibulospinal tracts of the spinal cord adjust peripheral muscle tone and complement the reflexive movements of the head or neck.

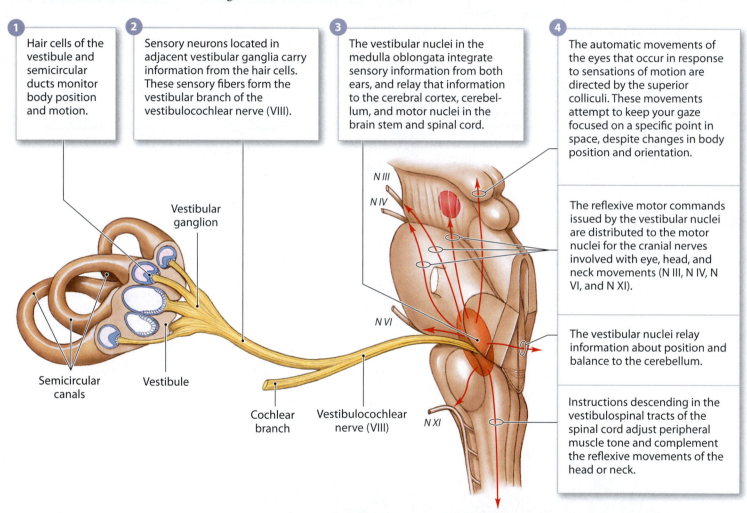

Vestibular ganglion

Semicircular canals

Vestibule

Cochlear branch

Vestibulocochlear nerve (VIII)

N III
N IV
N VI
N XI

2 The table at right summarizes the primary functions of the vestibular nuclei as they process information arriving from the equilibrium receptors in both ears.

Functions of the Vestibular Nuclei

1. Integrating sensory information about balance and equilibrium that arrives from both ears.
2. Relaying information from the vestibular complex to the cerebellum.
3. Relaying information from the vestibular complex to the cerebral cortex, providing a conscious sense of head position and movement.
4. Sending commands to motor nuclei in the brain stem and spinal cord.

Hearing

3 Hair cells along the basilar membrane are the receptors for the sense of hearing. Information they collect is passed along the cochlear branch of the vestibulocochlear nerve (VIII) to the cochlear nuclei, which in turn distribute that information to the inferior colliculus and to the auditory cortex.

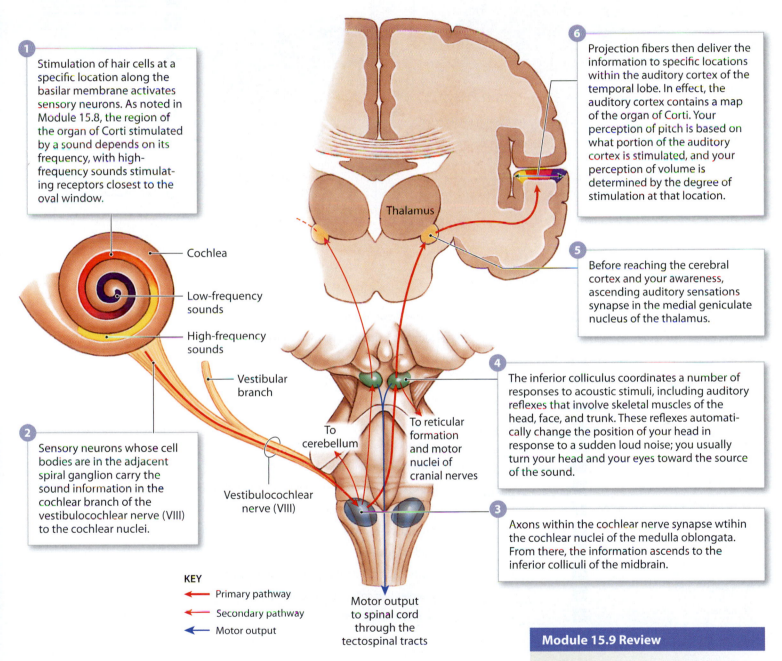

1 Stimulation of hair cells at a specific location along the basilar membrane activates sensory neurons. As noted in Module 15.8, the region of the organ of Corti stimulated by a sound depends on its frequency, with high-frequency sounds stimulating receptors closest to the oval window.

Cochlea

Low-frequency sounds

High-frequency sounds

Vestibular branch

2 Sensory neurons whose cell bodies are in the adjacent spiral ganglion carry the sound information in the cochlear branch of the vestibulocochlear nerve (VIII) to the cochlear nuclei.

Vestibulocochlear nerve (VIII)

Thalamus

To cerebellum

To reticular formation and motor nuclei of cranial nerves

Motor output to spinal cord through the tectospinal tracts

6 Projection fibers then deliver the information to specific locations within the auditory cortex of the temporal lobe. In effect, the auditory cortex contains a map of the organ of Corti. Your perception of pitch is based on what portion of the auditory cortex is stimulated, and your perception of volume is determined by the degree of stimulation at that location.

5 Before reaching the cerebral cortex and your awareness, ascending auditory sensations synapse in the medial geniculate nucleus of the thalamus.

4 The inferior colliculus coordinates a number of responses to acoustic stimuli, including auditory reflexes that involve skeletal muscles of the head, face, and trunk. These reflexes automatically change the position of your head in response to a sudden loud noise; you usually turn your head and your eyes toward the source of the sound.

3 Axons within the cochlear nerve synapse wtihin the cochlear nuclei of the medulla oblongata. From there, the information ascends to the inferior colliculi of the midbrain.

KEY

← Primary pathway
← Secondary pathway
← Motor output

Most of the auditory information from one cochlea is projected to the auditory cortex of the cerebral hemisphere on the opposite side of the brain. However, each auditory cortex also receives information from the cochlea on that side. These interconnections play a role in localizing sounds (left/right); they can also reduce the functional impact of damage to a cochlea or ascending pathway.

Module 15.9 Review

a. Where are the hair cells for equilibrium located?

b. Which cranial nerves are involved with eye, head, and neck movements?

c. What is your reflexive response to hearing a loud noise, such as a firecracker?

1. Labeling

Label the structures in the following diagram of the right ear.

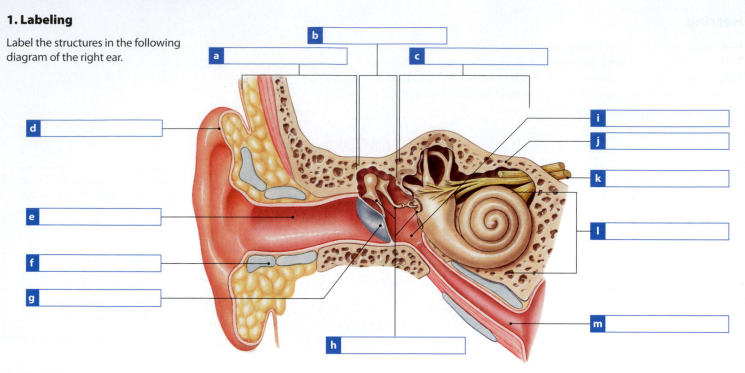

2. Matching

Match the following terms with the most closely related description.

- inner ear
- organ of Corti
- round window
- ampulla
- decibel
- auricle
- temporal lobe
- incus
- cerumen
- tectorial membrane
- tympanic membrane
- endolymph
- stapes
- tympanic cavity
- equilibrium

a _____	Ceruminous glands
b _____	Aids in determining the direction of sound
c _____	Fluid within chambers and canals of inner ear
d _____	Opens into auditory tube
e _____	Membranous labyrinth
f _____	Utricle, saccule
g _____	Lies at end of external acoustic meatus
h _____	Attached to oval window
i _____	Separates perilymph of cochlea from middle ear
j _____	Middle auditory ossicle
k _____	Contains receptors for rotation in semicircular ducts
l _____	Auditory cortex
m _____	Cochlear structure that provides information to the CNS
n _____	Overlies hair cells in the cochlear duct
o _____	Unit of sound intensity

3. Section integration

For a few seconds after you ride the express elevator from the 25th floor to the ground floor, you still feel as if you are descending, even though you have come to a stop. Why?

_____

_____

Vision

This section will examine the anatomical and physiological properties of the adult eyes. The eyes, which provide us with the sense of vision, are our most complex sense organs. A tremendous amount of sensory integration, feedback, and processing occurs within the eye before any sensations are relayed to the CNS. This functional complexity far exceeds that of the receptor complexes responsible for olfaction, gustation, equilibrium, and hearing. That complexity is directly related to the way the eyes form during embryological development. In many ways the eyes resemble detached CNS nuclei that are totally devoted to processing visual stimuli.

Week 4 embryo

1 The first indication of eye development appears as a pair of bulges called **optic vesicles** form in the lateral walls of the prosencephalon. These bulges extend to either side like a pair of dumbbells, each containing a cavity continuous with the neurocoel.

Optic vesicle

2 The lateral bulges become indented, forming a pair of **optic cups** that remain connected to the diencephalon by slender stalks. The epidermis overlying the optic cup responds by forming a pocket that later pinches off and develops into the lens of the eye.

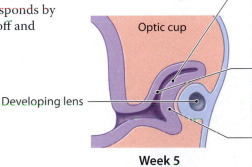

Optic cup

Developing lens

Week 5

Layers of Developing Retina

The ependymal cells on the outer wall of the optic cup develop into the photoreceptors.

The ependymal cells on the inner wall of the optic cup develop into pigment cells that absorb light that has passed through the photoreceptor layer. The separation gradually decreases until the photoreceptors and pigment cell layers are in contact.

The neural tissue of the outer wall of the optic cup forms layers of neurons, ganglion cells, and specialized glial cells that are responsible for preliminary processing and integration of visual information.

3 Mesoderm aggregating around the optic cup forms supporting layers of connective tissue that isolate the neural tissues from the rest of the body. The eye develops interior chambers filled with fluid that is continuously generated and reabsorbed. Modules within this section will examine each of these components in greater detail.

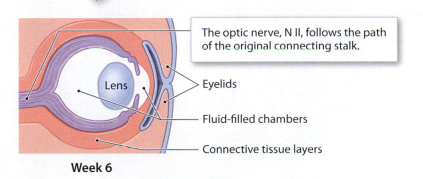

The optic nerve, N II, follows the path of the original connecting stalk.

Lens

Eyelids

Fluid-filled chambers

Connective tissue layers

Week 6

The accessory structures of the eye provide protection while allowing light to reach the interior of the eye

1 The **accessory structures** of the eye include the eyelids, eyelashes, the superficial epithelium of the eye, and the structures associated with the production, secretion, and removal of tears.

The **cornea** is a transparent area on the anterior surface of the eye.

Laterally, the two eyelids are connected at the **lateral canthus**.

Light enters the eye by passing through the cornea and then through the **pupil**, an opening at the center of the colored **iris**.

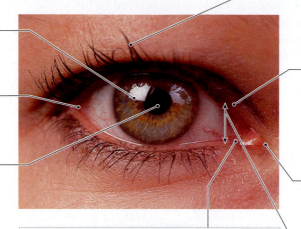

Glands in the **lacrimal caruncle** (KAR-ung-kul) produce the thick secretions that contribute to the gritty deposits that sometimes appear after a good night's sleep.

Eyelids and Eyelashes

The **eyelashes**, along the margins of the eyelids, are very robust hairs that help prevent foreign matter from reaching the surface of the eye.

The eyelid, or **palpebra** (pal-PĒ-bra), is a continuation of the skin. The continual blinking of the palpebrae keeps the surface of the eye lubricated, and removes dust and debris. The eyelids can also close firmly to protect the delicate surface of the eye.

Medially, the two eyelids are connected at the **medial canthus** (KAN-thus).

The **palpebral fissure** is the gap that separates the free margins of the upper and lower eyelids.

2 The epithelium covering the inner surfaces of the eyelids and the outer surface of the eye is called the **conjunctiva** (kon-junk-TĪ-vuh). It is a mucous membrane covered by a specialized stratified squamous epithelium.

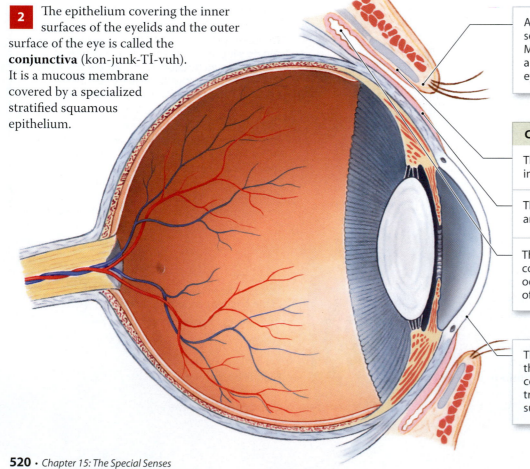

Along the inner margin of the lid, modified sebaceous glands called **tarsal glands**, or Meibomian (mī-BŌ-mē-an) glands, secrete a lipid-rich product that helps keep the eyelids from sticking together.

Conjunctiva

The **palpebral conjunctiva** covers the inner surface of the eyelids.

The **ocular conjunctiva** covers the anterior surface of the eye.

The pocket created where the palpebral conjunctiva becomes continuous with the ocular conjunctiva is known as the **fornix** of the eye.

The ocular conjunctiva is continuous with the very delicate corneal epithelium that covers the surface of the cornea, a transparent portion of the eye's anterior surface.

3 A constant flow of tears keeps conjunctival surfaces moist and clean. Tears reduce friction, remove debris, prevent bacterial infection, and provide nutrients and oxygen to portions of the conjunctival epithelium. The **lacrimal apparatus** produces, distributes, and removes tears. The lacrimal apparatus of each eye consists of (1) a lacrimal gland (or tear gland) with associated ducts, (2) paired lacrimal canaliculi, (3) a lacrimal sac, and (4) a nasolacrimal duct.

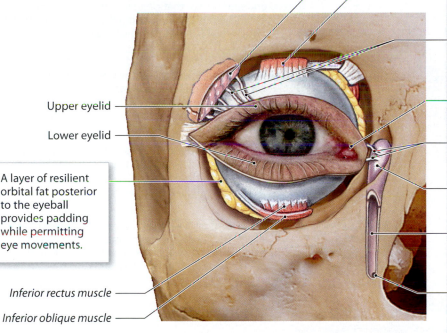

Superior rectus muscle

Upper eyelid

Lower eyelid

A layer of resilient orbital fat posterior to the eyeball provides padding while permitting eye movements.

Inferior rectus muscle

Inferior oblique muscle

Components of the Lacrimal Apparatus

The almond-shaped **lacrimal gland** is about 12–20 mm (0.5–0.75 in.) long. Each day it produces about 1 mL of watery, slightly alkaline tears. The tears provide lubrication and supply the nutrient and oxygen demands of the corneal cells. Lacrimal gland secretions contain the antibacterial enzyme **lysozyme** and antibodies that attack pathogens before they enter the body.

There are 10–12 **tear ducts** that deliver tears from the lacrimal gland to the space behind the upper eyelid.

The **lacrimal puncta** (singular, punctum) are two small pores that drain the lacrimal lake.

The **lacrimal canaliculi** are small canals that connect the lacrimal puncta to the lacrimal sac.

The **lacrimal sac** is a small chamber that nestles within the lacrimal sulcus of the orbit.

The **nasolacrimal duct** originates at the inferior tip of the lacrimal sac. It passes through the nasolacrimal canal to deliver tears to the nasal cavity.

The nasolacrimal duct empties into the inferior meatus, a narrow passageway inferior and lateral to the inferior nasal concha.

4 **Conjunctivitis**, or pinkeye, results from damage to, and irritation of, the conjunctival surface. The most obvious sign, redness, is due to the dilation of blood vessels deep to the conjunctival epithelium. This condition may be caused by pathogenic infection or by physical, allergic, or chemical irritation of the conjunctival surface.

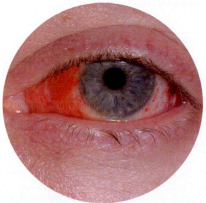

Module 15.10 Review

a. List the accessory structures associated with the eye.

b. Explain conjunctivitis.

c. Which layer of the eye would be the first affected by inadequate tear production?

The eye has a layered wall; it is hollow, with fluid-filled anterior and posterior cavities

1 The wall of the eye has three layers, also known as **tunics**.

Fibrous Tunic

The **fibrous tunic**, the outermost layer of the eye, consists of the cornea and the **sclera** (SKLER-uh). These two components are continuous, and the border between the two is called the **corneal limbus**. The fibrous tunic (1) provides mechanical support and some degree of physical protection, (2) serves as an attachment site for the extrinsic eye muscles, and (3) contains the cornea, which permits the passage of light and whose curvature aids in the focusing process.

Vascular Tunic

The **vascular tunic**, or **uvea** (Ū-vē-uh), contains numerous blood vessels, lymphatic vessels, and the intrinsic (smooth) muscles of the eye. The functions of this middle layer include (1) providing a route for blood vessels and lymphatics that supply tissues of the eye; (2) regulating the amount of light that enters the eye; (3) secreting and reabsorbing the aqueous humor that circulates within the chambers of the eye; and (4) controlling the shape of the lens, an essential part of the focusing process.

The **iris** is visible through the transparent corneal surface. It contains blood vessels, pigment cells, and layers of smooth muscle fibers. When these muscles contract, they change the diameter of the pupil.

The **ciliary body** is a thickened region that bulges into the interior of the eye. Ligaments extend from the ciliary body to the lens, holding it in position posterior to the pupil.

The **choroid** is a vascular layer that is covered by the sclera. The choroid contains an extensive capillary network that delivers oxygen and nutrients to the neural tissue within the neural tunic.

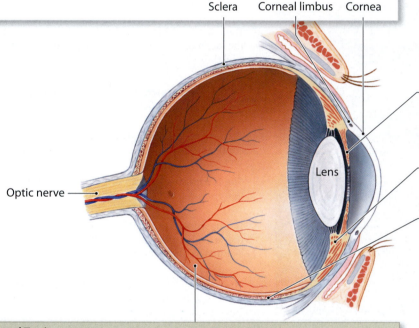

Sclera Corneal limbus Cornea

Lens

Optic nerve

Neural Tunic

The **neural tunic**, or **retina**, is the innermost layer of the eye. The retina consists of a thin outer layer (the pigmented layer) that absorbs light, and a thick inner layer (the neural layer) that contains the **photoreceptors**—the cells that are sensitive to light.

2 This sectional view reveals that the ciliary body and the lens divide the interior of the eye into a small **anterior cavity** and a large **posterior cavity**.

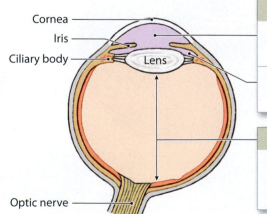

Cornea
Iris
Ciliary body
Lens
Optic nerve

Anterior Cavity

The **anterior chamber** extends from the cornea to the iris.

The **posterior chamber** extends between the iris and the ciliary body and lens.

Posterior Cavity

Most of the posterior cavity's volume is taken up by a gelatinous substance known as the **vitreous body**, or vitreous humor.

3 A view of the anterior cavity at greater magnification reveals additional details about the structure of the iris and ciliary body. Eye color is determined by (1) genes that influence the density and distribution of melanocytes on the anterior surface and interior of the iris, and (2) the density of the pigmented epithelium. When the connective tissue of the iris contains few melanocytes, light passes through it and bounces off the pigmented epithelium; the eye then appears blue. Individuals with green, brown, or black eyes have increasing numbers of melanocytes in the body and surface of the iris. The eyes of human albinos appear a very pale gray or blue-gray.

4 **Aqueous humor** is a fluid that circulates within the anterior cavity, passing from the posterior to the anterior chamber through the pupil. It also freely diffuses through the vitreous body and across the surface of the retina. Circulation of the aqueous humor provides an important route for nutrient and waste transport, in addition to forming a fluid cushion. The pressure exerted by the fluid aqueous humor helps retain the eye's shape; it also stabilizes the position of the retina, pressing the neural part against the pigmented part. A procedure called **tonometry** measures the eye's intraocular pressure—the fluid pressure within the anterior chamber. Normal intraocular pressure ranges from 12 to 21 mm Hg.

The body of the iris consists of a highly vascular, pigmented, loose connective tissue. The anterior surface has no epithelial covering; instead, it has an incomplete layer of fibroblasts and melanocytes. The posterior surface is covered by a pigmented epithelium that is part of the neural tunic.

Aqueous humor forms through active secretion by epithelial cells of the ciliary body's ciliary processes at a rate of 1–2 µL per minute. The epithelial cells regulate its composition, which resembles that of cerebrospinal fluid.

Aqueous humor leaves the eye through the **canal of Schlemm** (or scleral venous sinus), a passageway that extends completely around the eye at the level of the corneal limbus. Collecting channels deliver the aqueous humor from this canal to veins in the sclera. The rate of removal normally keeps pace with the rate of secretion, and aqueous humor is removed and recycled within a few hours of its formation.

The bulk of the ciliary body consists of the **ciliary muscle**, a smooth muscular ring that projects into the interior of the eye.

The **ciliary processes** are folds of epithelium covering the ciliary muscle.

Cornea

Anterior chamber

Lens

Conjunctiva

Posterior cavity (vitreous chamber)

The **ora serrata** (Ō-ra ser-RA-tuh; serrated mouth) is the jagged anterior edge of the thick, inner portion of the neural tunic. The pigmented outer layer of the neural tunic continues anteriorly across the posterior surface of the iris.

The **suspensory ligaments** attach to the tips of the ciliary processes. The connective-tissue fibers of these ligaments hold the lens posterior to the iris and centered on the pupil. As a result, any light passing through the pupil will also pass through the lens.

Module 15.11 Review

a. Name the three tunics of the eye.

b. What give eyes their characteristic color?

c. Where in the eye is aqueous humor located?

The eye is highly organized and has a consistent visual axis that directs light to the fovea of the retina

1 The sectional view below presents key aspects of eye anatomy that are associated with positioning the eye and allowing light to reach the photoreceptors of the retina.

The cornea permits the entry of light into the eye, so its transparency and clarity are vital to eye function. The cornea consists primarily of a dense matrix containing multiple layers of collagen fibers, organized so as not to interfere with the passage of light. The cornea has no blood vessels; the superficial epithelial cells must obtain oxygen and nutrients from the tears that flow across their free surfaces.

The lens lies posterior to the cornea, held in place by the suspensory ligaments that originate on the ciliary body. The lens consists of concentric layers of cells surrounded by a dense fibrous capsule. The cells are slender, elongate, and filled with transparent proteins called **crystallins**, which are responsible for both the clarity and the focusing power of the lens. Around the edges of the lens, its capsular fibers intermingle with those of the suspensory ligaments. The primary function of the lens is to focus the visual image on the photoreceptors, and the lens accomplishes this by changing shape.

Tension in the suspensory ligaments resists the tendency of the lens to assume a spherical shape.

The ciliary body supports the lens and controls its shape.

The retina contains the photoreceptors, pigment cells, supporting cells, and neurons.

The blood vessels of the choroid directly or indirectly provide nutrients to all structures within the eye.

The sclera, or "white of the eye," consists of dense fibrous connective tissue containing both collagen and elastic fibers. This layer is thickest over the posterior surface of the eye, near the exit of the optic nerve, and thinnest over the anterior surface. The sclera stabilizes the shape of the eye during eye movements; the six extrinsic eye muscles insert on the sclera, blending their collagen fibers with those of the sclera.

The **optic nerve** (N II) carries visual information to the brain.

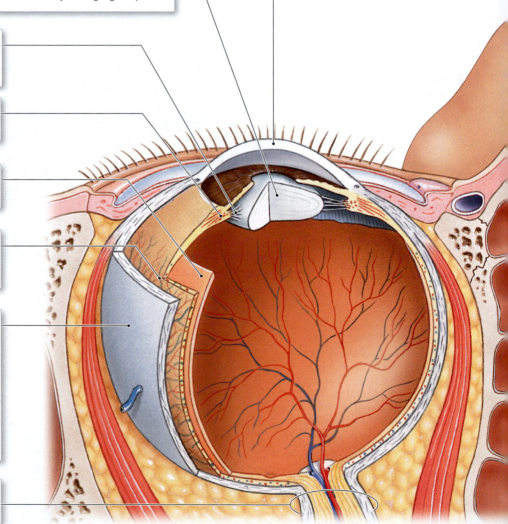

2 The amount of light entering the eye and passing through the lens is controlled by the two layers of the pupillary muscles of the iris. When these smooth muscles contract, they change the diameter of the pupil. Both muscle layers are controlled by the autonomic nervous system. Parasympathetic activation in response to bright light causes the pupils to constrict, whereas sympathetic activation in response to dim light causes the pupils to dilate.

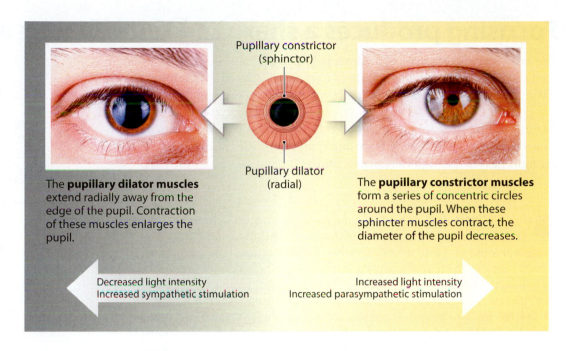

Pupillary constrictor (sphincter)

Pupillary dilator (radial)

The **pupillary dilator muscles** extend radially away from the edge of the pupil. Contraction of these muscles enlarges the pupil.

The **pupillary constrictor muscles** form a series of concentric circles around the pupil. When these sphincter muscles contract, the diameter of the pupil decreases.

Decreased light intensity
Increased sympathetic stimulation

Increased light intensity
Increased parasympathetic stimulation

3 Light passing through the center of the cornea and the center of the lens strikes a specific location that contains the highest density of photoreceptors anywhere in the eye.

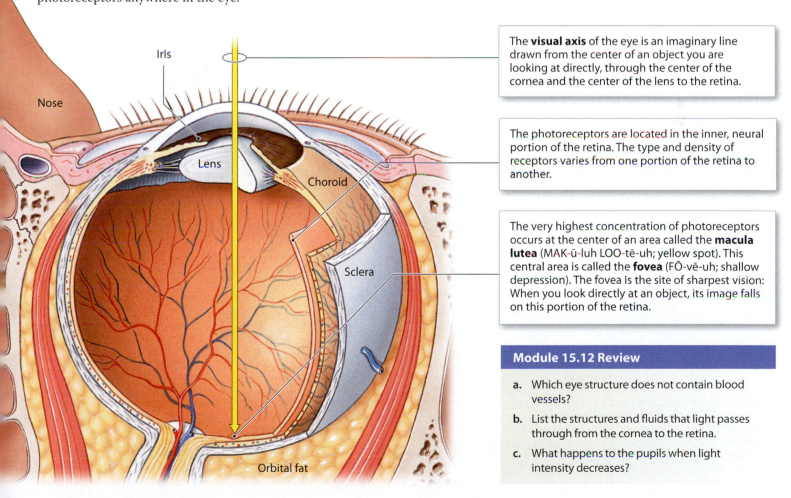

Iris

Nose

Lens

Choroid

Sclera

Orbital fat

The **visual axis** of the eye is an imaginary line drawn from the center of an object you are looking at directly, through the center of the cornea and the center of the lens to the retina.

The photoreceptors are located in the inner, neural portion of the retina. The type and density of receptors varies from one portion of the retina to another.

The very highest concentration of photoreceptors occurs at the center of an area called the **macula lutea** (MAK-ū-luh LOO-tē-uh; yellow spot). This central area is called the **fovea** (FŌ-vē-uh; shallow depression). The fovea is the site of sharpest vision: When you look directly at an object, its image falls on this portion of the retina.

Module 15.12 Review

a. Which eye structure does not contain blood vessels?

b. List the structures and fluids that light passes through from the cornea to the retina.

c. What happens to the pupils when light intensity decreases?

Focusing produces a sharply defined image at the retina

The eye is often compared to a camera. To provide useful information, the lens of the eye, like a camera lens, must focus the arriving image. For an image to be "in focus," the rays of light arriving from an object must strike the sensitive surface of the retina in precise order, forming a miniature image of the object. If the rays are not perfectly focused, the image is blurry. Focusing typically occurs in two steps: as the light passes through the cornea, and as it passes through the lens.

1 Light is **refracted**, or bent, when it passes from one medium to another medium with a different density. In the human eye, the greatest amount of refraction occurs when light passes from the air into the corneal tissues, which have a density close to that of water. You cannot vary the amount of refraction that occurs at the cornea. Additional refraction takes place when the light passes from the aqueous humor into the relatively dense lens. The lens provides the extra refraction needed to focus the light rays from an object toward a **focal point**—a specific point of intersection on the retina. The distance between the center of the lens and its focal point is the **focal distance** of the lens. Whether in the eye or in a camera, the focal distance is determined by the distance from the object to the lens, and the shape of the lens.

2 A camera focuses an image by moving the lens toward or away from the digital image sensor. This method of focusing cannot work in our eyes, because the distance from the lens to the macula lutea cannot change. We focus images on the retina by changing the shape of the lens to keep the focal distance constant, a process called **accommodation**.

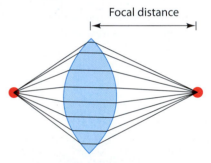

The closer the light source, the longer the focal distance

The rounder the lens, the shorter the focal distance

For Close Vision: Ciliary Muscle Contracted, Lens Rounded

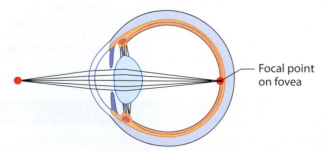

Focal point on fovea

When the ciliary muscle contracts, the ciliary body moves toward the lens, thereby reducing the tension in the suspensory ligaments. The elastic capsule of the lens then pulls it into a more spherical shape that increases the refractive power of the lens, enabling it to bring light from nearby objects into focus on the retina.

For Distant Vision: Ciliary Muscle Relaxed, Lens Flattened

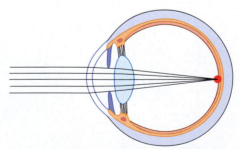

When the ciliary muscle relaxes, the suspensory ligaments pull at the circumference of the lens, making the lens flatter and bringing the image of a distant object into focus on the retina.

3 An object in view isn't a single point; the image consists of a large number of individual points, like the pixels on a computer screen. In the eye, light from each point is focused on the retina, creating a miniature image of the original. However, the image is inverted and reversed. The brain compensates for this image reversal, and we are not aware of any difference between the orientation of the image on the retina and that of the object. The compensation is learned by experience—a person wearing glasses that reverse and invert the visual image can adapt to the change relatively quickly.

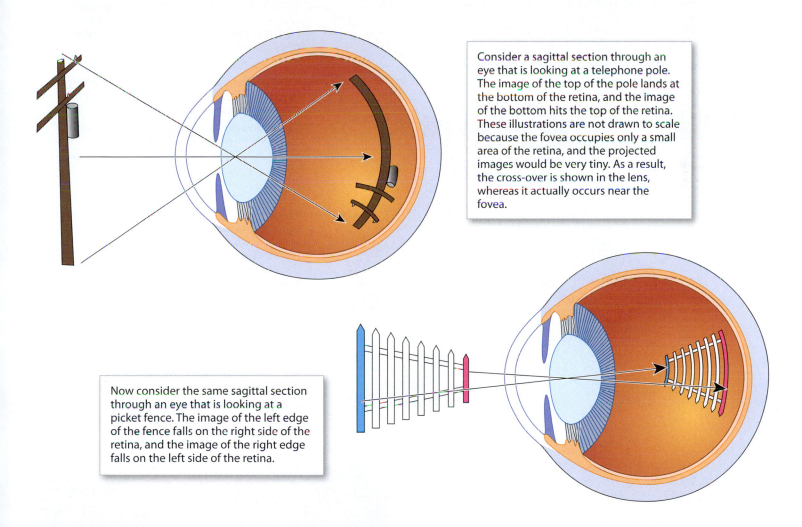

Consider a sagittal section through an eye that is looking at a telephone pole. The image of the top of the pole lands at the bottom of the retina, and the image of the bottom hits the top of the retina. These illustrations are not drawn to scale because the fovea occupies only a small area of the retina, and the projected images would be very tiny. As a result, the cross-over is shown in the lens, whereas it actually occurs near the fovea.

Now consider the same sagittal section through an eye that is looking at a picket fence. The image of the left edge of the fence falls on the right side of the retina, and the image of the right edge falls on the left side of the retina.

The greatest amount of refraction is required to view objects that are very close to the lens. The inner limit of clear vision, known as the **near point of vision**, is determined by the degree of elasticity in the lens. Children can usually focus on something 7–9 cm from the eye, but over time the lens tends to become stiffer and less responsive. A young adult can usually focus on objects 15–20 cm away. As aging proceeds, this distance gradually increases; the near point at age 60 is typically about 83 cm.

Module 15.13 Review

a. Define focal point.

b. When the ciliary muscles are relaxed, are you viewing something close up or something in the distance?

c. Why does the near point of vision typically increase with age?

The neural tissue of the retina contains multiple layers of specialized photoreceptors, neurons, and supporting cells

1 This diagrammatic sectional view through the eye shows the retina near the origin of the optic nerve.

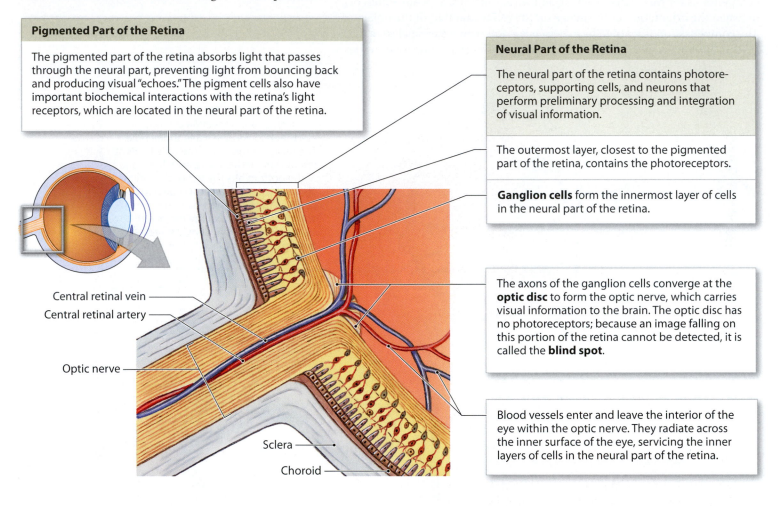

Pigmented Part of the Retina

The pigmented part of the retina absorbs light that passes through the neural part, preventing light from bouncing back and producing visual "echoes." The pigment cells also have important biochemical interactions with the retina's light receptors, which are located in the neural part of the retina.

Neural Part of the Retina

The neural part of the retina contains photoreceptors, supporting cells, and neurons that perform preliminary processing and integration of visual information.

The outermost layer, closest to the pigmented part of the retina, contains the photoreceptors.

Ganglion cells form the innermost layer of cells in the neural part of the retina.

The axons of the ganglion cells converge at the **optic disc** to form the optic nerve, which carries visual information to the brain. The optic disc has no photoreceptors; because an image falling on this portion of the retina cannot be detected, it is called the **blind spot**.

Blood vessels enter and leave the interior of the eye within the optic nerve. They radiate across the inner surface of the eye, servicing the inner layers of cells in the neural part of the retina.

Central retinal vein

Central retinal artery

Optic nerve

Sclera

Choroid

2 This is a photograph of the retinal surface, taken through the cornea, pupil, and lens of the right eye.

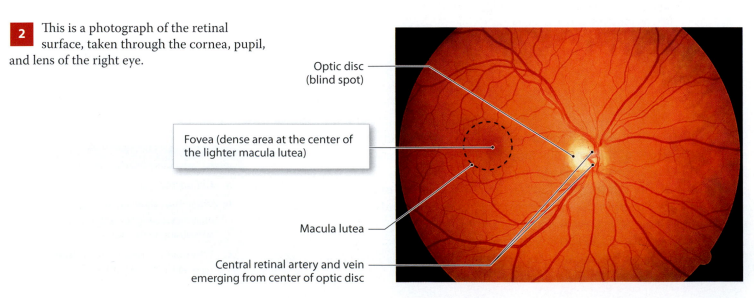

Optic disc (blind spot)

Fovea (dense area at the center of the lighter macula lutea)

Macula lutea

Central retinal artery and vein emerging from center of optic disc

3 This sectional view shows that the retina contains multiple layers of specialized cells, including two types of photoreceptors: rods and cones.

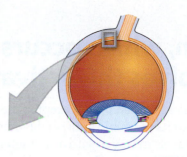

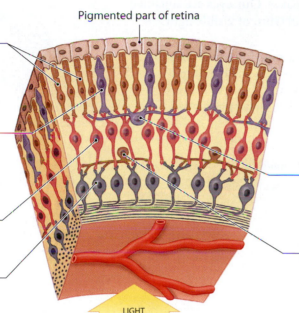

Pigmented part of retina

Photoreceptors of the Retina

Rods do not discriminate among colors of light. Highly sensitive, they enable us to see in dimly lit rooms, at twilight, and in pale moonlight.

Cones provide us with color vision. They give us sharper, clearer images than rods do, but cones require more intense light.

Rods and cones synapse with neurons called **bipolar cells**.

Bipolar cells synapse on ganglion cells.

Horizontal and Amacrine Cells

These cells can facilitate or inhibit communication between photoreceptors and ganglion cells, thereby altering the sensitivity of the retina. The effect is comparable to adjusting the contrast on a television set. These cells play an important role in the eye's adjustment to dim or brightly lit environments.

A network of **horizontal cells** extends across the outer portion of the retina at the level of the synapses between photoreceptors and bipolar cells.

A layer of **amacrine** (AM-a-krin) **cells** occurs where bipolar cells synapse with ganglion cells.

LIGHT

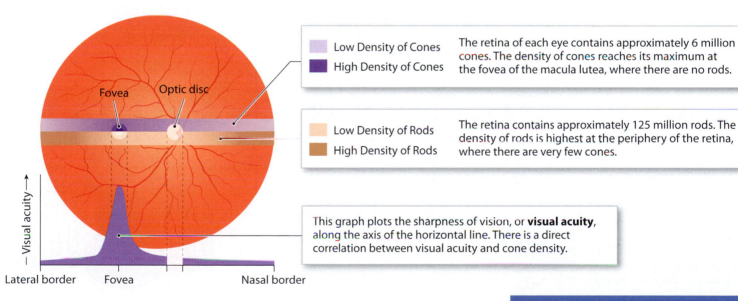

Fovea Optic disc

Low Density of Cones
High Density of Cones

The retina of each eye contains approximately 6 million cones. The density of cones reaches its maximum at the fovea of the macula lutea, where there are no rods.

Low Density of Rods
High Density of Rods

The retina contains approximately 125 million rods. The density of rods is highest at the periphery of the retina, where there are very few cones.

This graph plots the sharpness of vision, or **visual acuity**, along the axis of the horizontal line. There is a direct correlation between visual acuity and cone density.

— Visual acuity →

Lateral border   Fovea   Nasal border

4 The two color bands in this illustration of the retinal surface indicate the relative densities of cones and rods on either side of a horizontal line passing through the fovea and optic disc of the right eye. When you look directly at an object, the image falls on the fovea, the center of color vision and image sharpness. However, in very dim light, cones cannot function. That is why you can't see a dim star if you stare directly at it, but you can see it if you shift your gaze to one side or the other so that the image falls on the more sensitive rods.

Module 15.14 Review

a. Define rods and cones.

b. If you enter a dimly lit room, will you be able to see clearly? Why or why not?

c. If you had been born without cones in your eyes, explain why you would or would not be able to see.

Photoreception, which occurs in the outer segment of rods and cones, involves the activation of visual pigments

The rods and cones of the retina are called photoreceptors because they detect photons, the basic units of visible light. Light energy is a form of radiant energy that travels in waves with a characteristic wavelength (distance between wave peaks). Our eyes are sensitive to wavelengths of 400–700 nm, the spectrum of visible light.

1 This illustration shows the major structural features of rods and cones, and the adjacent pigment epithelium and bipolar cells. The outer segments of both rods and cones have membranous plates, or discs, that contain special organic compounds called **visual pigments**.

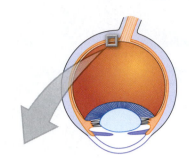

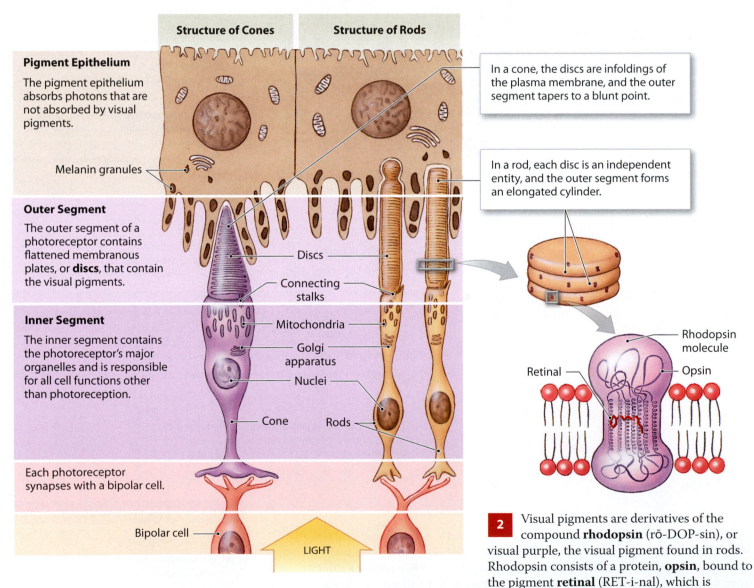

Structure of Cones

Structure of Rods

Pigment Epithelium

The pigment epithelium absorbs photons that are not absorbed by visual pigments.

Melanin granules

In a cone, the discs are infoldings of the plasma membrane, and the outer segment tapers to a blunt point.

In a rod, each disc is an independent entity, and the outer segment forms an elongated cylinder.

Outer Segment

The outer segment of a photoreceptor contains flattened membranous plates, or **discs**, that contain the visual pigments.

Discs

Connecting stalks

Inner Segment

The inner segment contains the photoreceptor's major organelles and is responsible for all cell functions other than photoreception.

Mitochondria

Golgi apparatus

Nuclei

Cone

Rods

Each photoreceptor synapses with a bipolar cell.

Bipolar cell

LIGHT

Rhodopsin molecule

Retinal

Opsin

2 Visual pigments are derivatives of the compound **rhodopsin** (rō-DOP-sin), or visual purple, the visual pigment found in rods. Rhodopsin consists of a protein, **opsin**, bound to the pigment **retinal** (RET-i-nal), which is synthesized from vitamin A. The type of opsin present determines the wavelength of light that can be absorbed by retinal.

3 When light strikes a visual pigment, the retinal molecule changes shape, which changes the permeability of the outer segment. This permeability change is the key to transduction in the eye, as it converts light energy into a nerve impulse.

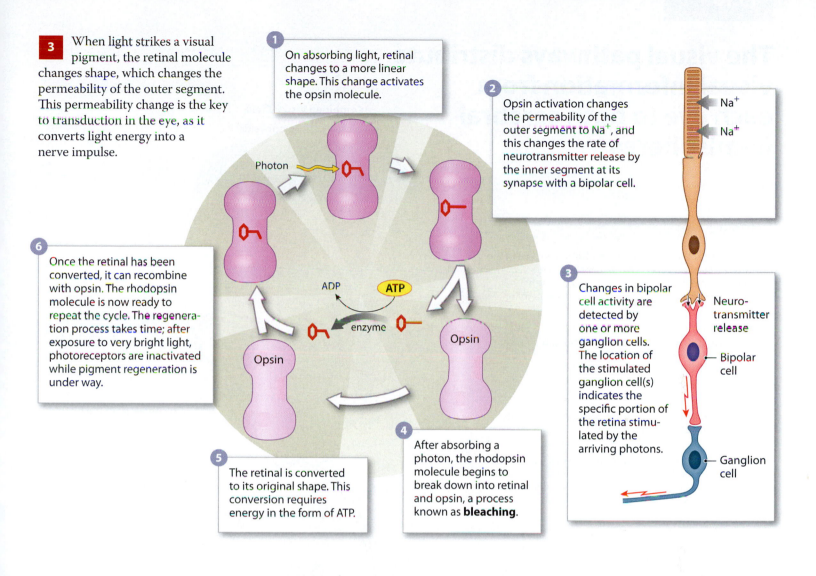

1 On absorbing light, retinal changes to a more linear shape. This change activates the opsin molecule.

2 Opsin activation changes the permeability of the outer segment to Na⁺, and this changes the rate of neurotransmitter release by the inner segment at its synapse with a bipolar cell.

Na⁺
Na⁺

3 Changes in bipolar cell activity are detected by one or more ganglion cells. The location of the stimulated ganglion cell(s) indicates the specific portion of the retina stimulated by the arriving photons.

Neuro-transmitter release

Bipolar cell

Ganglion cell

6 Once the retinal has been converted, it can recombine with opsin. The rhodopsin molecule is now ready to repeat the cycle. The regeneration process takes time; after exposure to very bright light, photoreceptors are inactivated while pigment regeneration is under way.

ADP
ATP
enzyme
Opsin
Opsin

5 The retinal is converted to its original shape. This conversion requires energy in the form of ATP.

4 After absorbing a photon, the rhodopsin molecule begins to break down into retinal and opsin, a process known as **bleaching**.

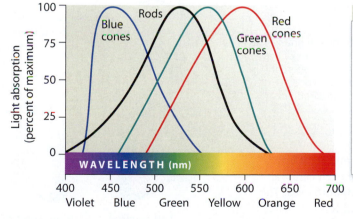

4 There are three types of cones: **blue cones**, **green cones**, and **red cones**. Each type has a different form of opsin that is sensitive to a different range of wavelengths. Their stimulation in various combinations is the basis for color vision. In an individual with normal vision, the cone population consists of 16 percent blue cones, 10 percent green cones, and 74 percent red cones. Although their wavelength sensitivities overlap, each type is most sensitive to a specific portion of the visual spectrum. If all three cone populations are stimulated, we perceive the color as white. We also perceive white if rods (but not cones) are stimulated, which is why everything appears black-and-white when we enter dimly lit surroundings or walk by starlight.

NOTE

If a person lacks one or more cone pigments, **color blindness** results. The common forms are sex-linked and the genetic basis will be considered in a later chapter. Total color blindness (no cone pigments) is very rare, but 2% of males lack either red or green cone pigments and are partially color blind.

Module 15.15 Review

a. Identify the three types of cones.

b. Compare rods with cones.

c. How could a diet deficient in vitamin A affect vision?

The visual pathways distribute visual information from each eye to both cerebral hemispheres

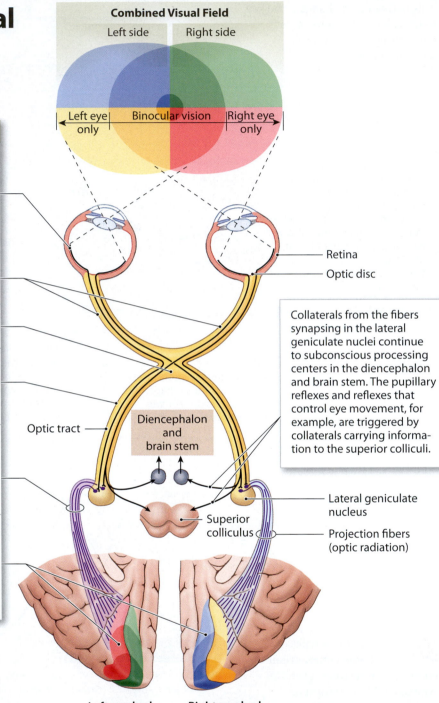

Combined Visual Field

Left side | Right side

Left eye only | Binocular vision | Right eye only

The Visual Pathways

The visual pathways begin at the photoreceptors in the retina. Each photoreceptor monitors a specific receptive field, and when stimulated, passes the information through a bipolar cell and to a ganglion cell.

Axons from the approximately 1 million ganglion cells converge on the optic disc, penetrate the wall of the eye, and proceed toward the diencephalon as the optic nerve (II).

The two optic nerves, one from each eye, reach the diencephalon at the optic chiasm.

From that point, approximately half the fibers proceed toward the lateral geniculate nucleus of the same side of the brain, whereas the other half cross over to reach the lateral geniculate nucleus of the opposite side.

From each lateral geniculate nucleus, visual information travels to the occipital cortex of the cerebral hemisphere on that side. The bundle of projection fibers linking each lateral geniculate nucleus with the visual cortex is known as the **optic radiation**.

The perception of a visual image reflects the integration of information that arrives at the visual cortex of the occipital lobes. Each eye receives a slightly different visual image, because (1) the foveae are 5–7.5 cm (2–3.0 in.) apart, and (2) the nose and eye socket block the view of the opposite side.

Retina

Optic disc

Collaterals from the fibers synapsing in the lateral geniculate nuclei continue to subconscious processing centers in the diencephalon and brain stem. The pupillary reflexes and reflexes that control eye movement, for example, are triggered by collaterals carrying information to the superior colliculi.

Optic tract

Diencephalon and brain stem

Lateral geniculate nucleus

Superior colliculus

Projection fibers (optic radiation)

Left cerebral hemisphere

Right cerebral hemisphere

1 The visual images from the left and right eyes overlap, and the visual cortex of each cerebral hemisphere receives information from both eyes. The information is sorted, however, so that the left visual cortex gets information on the right half of the visual field, and the right visual cortex receives information on the left half of the visual field. **Depth perception**, an interpretation of the three-dimensional relationships among objects in view, is obtained by comparing the relative positions of objects within the images received by the two eyes. The map in the visual cortex is upside down and backward, duplicating the orientation of the visual image at the retina.

Module 15.16 Review

a. Define optic radiation.

b. Where are visual images perceived?

c. Trace the visual pathway, beginning at the photoreceptors in the retina.

Accommodation problems result from abnormalities in the cornea or lens, or in the shape of the eye

Emmetropia

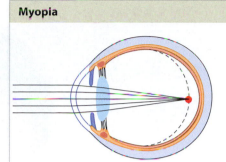

In the normal healthy eye, when the ciliary muscle is relaxed and the lens is flattened, the image of a distant object will be focused on the retina's surface. This condition is called **emmetropia** (*emmetro-*, proper + *opia*, vision), or normal vision.

1 Small variations in the performance of the lens or the structure of the eye can be corrected with external lenses (glasses or contact lenses).

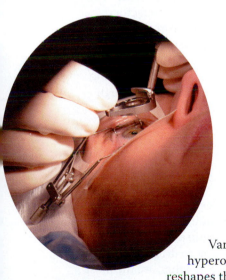

Myopia

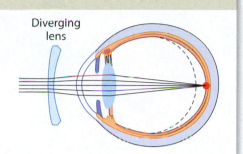

If the eyeball is too deep or the resting curvature of the lens is too great, the image of a distant object is projected in front of the retina. Such individuals are said to be nearsighted because vision at close range is clear but distant objects are blurry and out of focus. Their condition is more formally termed **myopia** (*myein*, to shut + *ops*, eye).

Myopia can be treated by placing a diverging lens in front of the eye. Diverging lenses have at least one concave surface and spread the light rays apart as if the object were closer to the viewer.

Hyperopia

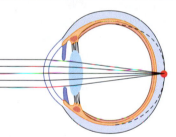 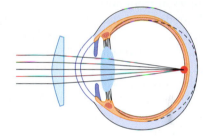

If the eyeball is too shallow or the lens is too flat, **hyperopia** results. The ciliary muscle must contract to focus even a distant object on the retina, and at close range the lens cannot provide enough refraction to focus an image on the retina. Individuals with this problem are said to be farsighted, because they can see distant objects most clearly.

Hyperopia can be corrected by placing a converging lens in front of the eye. Converging lenses have at least one convex surface and provide the additional refraction needed to bring nearby objects into focus.

Variable success at correcting myopia and hyperopia has been achieved by surgery that reshapes the cornea. In **photorefractive keratectomy (PRK)** a computer-guided laser shapes the cornea to exact specifications. Tissue is removed only to a depth of 10–20 μm—no more than about 10 percent of the cornea's thickness. The entire procedure can be done in less than a minute. A variation on PRK is called **laser-assisted in-situ keratomileusis (LASIK)**. In this procedure the interior layers of the cornea are reshaped and covered by a flap of the normal corneal epithelium. Roughly 70 percent of LASIK patients achieve normal vision; it is now the most common form of refractive surgery. Each year, an estimated 100,000 people undergo PRK therapy in the United States. Corneal scarring is rare, and approximately 10 million Americans have had corneal refractive surgery. However, many still need reading glasses, and both immediate and long-term visual problems can occur.

Module 15.17 Review

a. Define emmetropia.

b. Discuss two surgical procedures for correcting myopia and hyperopia.

c. Which type of lens would correct hyperopia?

Aging is associated with many disorders of the special senses; trauma, infection, and abnormal stimuli may cause problems at any age

Olfaction

1 Disorders of the sense of smell may result from a head injury or from normal, age-related changes. If an injury to the head damages the olfactory nerves (N I), then the sense of smell may be impaired. Unlike other populations of neurons, olfactory receptor cells are regularly replaced by the division of stem cells. Despite this process, the total number of receptors declines with age, and the remaining receptors become less sensitive. As a result, the elderly have difficulty detecting odors in low concentrations. This explains why your grandmother may overdo her perfume, and your grandfather's aftershave may seem so strong—they must use more of the odorous solution to be able to smell it themselves.

Gustation

2 Disorders of the sense of taste can be caused by problems with olfactory receptors, damage to taste buds, damage to cranial nerves, and age-related changes. The sense of smell also makes a large contribution to our sense of taste, so conditions that affect the olfactory receptors—such as the common cold—can also dull your sense of taste. A reduced sense of taste may also result from damage to taste buds by inflammation or infections of the mouth. Alternatively, the cranial nerves carrying taste sensations (VII, IX, X) may be damaged through trauma or compression by a tumor.

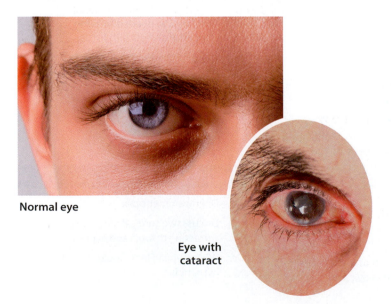

Normal eye

Eye with cataract

Vision

3 Several common eye problems were introduced earlier in the chapter. A condition in which the lens loses its transparency—called a **cataract**—can result from injuries, radiation, or reaction to drugs, but **senile cataracts**, a natural consequence of aging, are the most common form. As the condition advances, the individual needs brighter and brighter light for reading, and visual acuity may eventually decrease to the point of blindness. A damaged or nonfunctional lens can be replaced by an artificial substitute, and vision fine-tuned with glasses or contact lenses.

Equilibrium

4   **Vertigo** is an illusion of movement (within oneself or within one's surroundings). Vertigo is caused by conditions that alter the function of the inner ear receptor complex, the vestibular branch of the vestibulocochlear nerve, or sensory nuclei and pathways in the central nervous system. Any event that sets endolymph into motion can stimulate the equilibrium receptors and produce vertigo; flushing the external auditory canal with cold water may chill the endolymph in the outermost portions of the labyrinth and establish a temperature-related circulation of fluid that produces mild and temporary vertigo. Excessive consumption of alcohol or exposure to certain drugs can also produce vertigo by changing endolymph composition or disturbing hair cells. Perhaps the most common cause of vertigo is **motion sickness**. Its unpleasant signs and symptoms include headache, sweating, flushing of the face, nausea, and vomiting. The drugs commonly administered to prevent motion sickness appear to depress activity in the brain stem.

Hearing

5   An estimated 6 million Americans have at least a partial hearing deficit, or deafness. **Conductive deafness** results from interference with the normal transfer of vibrations from the tympanic membrane to the oval window. Causes include excess wax or trapped water in the external acoustic meatus, scarring or perforation of the tympanic membrane, or immobilization of one or more auditory ossicles by fluid or a tumor. In **nerve deafness**, the problem lies within the cochlea or somewhere along the auditory pathway. Vibrations reach the oval window, but the receptors either cannot respond or their response cannot reach the central nervous system. Very loud noises can cause nerve deafness by damaging the sensory cilia on the receptor cells. Bacterial or viral infections may also kill receptor cells and damage sensory nerves. Young children have the greatest hearing range: They can detect sounds ranging from a 20-Hz buzz to a 20,000-Hz whine. With age, damage due to loud noises or other injuries accumulates: The tympanic membrane gets less flexible, the articulations between the auditory ossicles stiffen, and the round window may begin to ossify. As a result, older individuals typically show some degree of hearing loss.

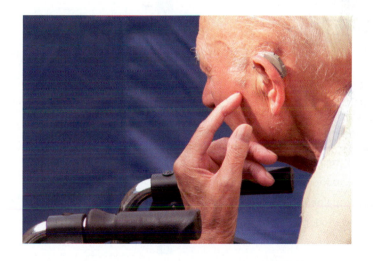

Module 15.18 Review

a. Which cranial nerves provide taste sensations from the tongue?

b. Identify two common classes of hearing-related disorders.

c. What causes vertigo?

1. Labeling

Label the structures in the following
diagram of a sagittal section of the left eye.

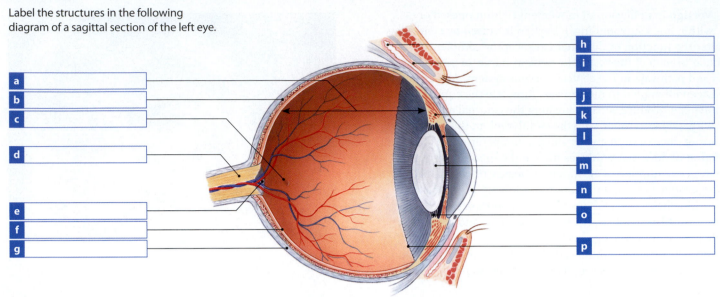

a

b

c

d

e

f

g

h

i

j

k

l

m

n

o

p

2. Matching

Match the following terms with the most closely related description.

- ganglion cells
- cones
- vascular tunic
- rods
- optic disc
- posterior chamber
- crystallins
- sclera
- retina
- occipital lobe
- palpebrae
- rhodopsin
- fovea
- posterior cavity
- pupil

a _____ Visual pigment

b _____ Eyelids

c _____ Transparent proteins in the cells of a lens

d _____ White of the eye

e _____ Opening surrounded by the iris

f _____ Neural tunic

g _____ Site of vitreous body

h _____ Photoreceptors that enable vision in dim light

i _____ Extends between the iris and the ciliary body and lens

j _____ Sharpest vision

k _____ Visual cortex

l _____ Photoreceptors that provide the perception of color

m _____ Their axons form the optic nerves

n _____ Iris, ciliary body, and choroid

o _____ Region of retina called the "blind spot"

3. Section integration

A bright flash of light from nearby exploding fireworks blinds Rachel's eyes. The result is a "ghost" image
that temporarily remains on her retinas. What might account for the images and their subsequent disappearance?

_____

_____

_____

Visual Outline with Key Terms

Summarize the content of each module using the terms in the order provided.

SECTION 1

An Introduction to the Special Senses: Olfaction and Gustation

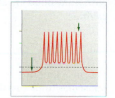

- generator potential
- synaptic delay

SECTION 2

Equilibrium and Hearing

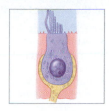

- inner ear
- hair cells
- kinocilium
- stereocilia

15.1

Olfaction involves specialized chemoreceptive neurons and delivers sensations directly to the cerebrum

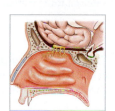

- olfactory organs
- olfactory bulb
- olfactory tract
- olfactory epithelium
- lamina propria
- olfactory glands
- olfactory receptor cells
- supporting cells
- basal cells
- odorants

15.4

The ear is divided into the external ear, the middle ear, and the inner ear

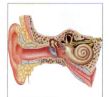

- external ear
- middle ear
- inner ear
- auricle
- external acoustic meatus
- ceruminous glands
- cerumen
- auditory ossicles
- tympanic membrane
- bony labyrinth
- auditory tube
- otitis media
- malleus
- incus
- stapes

- tensor tympani muscle
- stapedius muscle

15.2

Gustation involves epithelial chemoreceptor cells located in taste buds

- gustation
- taste receptors
- lingual papillae
- taste buds
- circumvallate papillae
- fungiform papillae
- filiform papillae
- umami
- water receptors
- gustatory cells
- taste pore
- basal cells

15.5

The bony labyrinth protects the membranous labyrinth

- inner ear
- bony labyrinth
- perilymph
- semicircular canals
- vestibule
- cochlea
- membranous labyrinth
- endolymph
- semicircular ducts
- utricle
- saccule
- cochlear duct
- vestibular complex

15.3

Gustatory reception relies on membrane receptors and channels, and sensations are carried by facial, glossopharyngeal, and vagus nerves

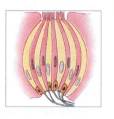

- central adaptation
- gustducins
- solitary nucleus
- thalamus
- primary sensory cortex

15.6

Hair cells in the semicircular ducts respond to rotation, while those in the utricle and saccule respond to gravity and linear acceleration

- anterior semicircular duct
- posterior semicircular duct
- lateral semicircular duct
- ampulla
- crista
- cupula
- endolymphatic duct
- endolymphatic sac
- statoconia
- otolith
- maculae

• = Term boldfaced in this module

15.7

The cochlear duct contains the hair cells of the organ of Corti

- cochlear duct
- vestibular duct
- tympanic duct
- oval window
- round window
- vestibular membrane
- basilar membrane
- organ of Corti
- spiral ganglion
- tectorial membrane

15.8

The organ of Corti provides sensations of pitch and volume

- wavelength
- frequency
- hertz (Hz)
- amplitude
- intensity
- decibels (dB)
- resonance
- pitch
- volume

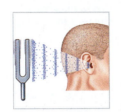

15.9

The vestibulocochlear nerve carries equilibrium and hearing sensations to the brain stem

- vestibular ganglion
- vestibular nuclei
- spiral ganglion
- cochlear nuclei
- inferior colliculus
- thalamus
- auditory cortex

SECTION 3

Vision

- optic vesicles
- optic cups
- ependymal cells

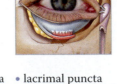

15.10

The accessory structures of the eye provide protection while allowing light to reach the interior of the eye

- accessory structures
- cornea
- lateral canthus
- pupil
- iris
- lacrimal caruncle
- eyelashes
- palpebra
- medial canthus
- palpebral fissure
- conjunctiva
- tarsal glands
- palpebral conjunctiva
- ocular conjunctiva
- fornix
- lacrimal apparatus
- lacrimal gland
- lysozyme
- tear ducts
- lacrimal puncta
- lacrimal canaliculi
- lacrimal sac
- nasolacrimal duct
- conjunctivitis

15.11

The eye has a layered wall; it is hollow, with fluid-filled anterior and posterior cavities

- tunics
- fibrous tunic
- sclera
- corneal limbus
- vascular tunic (uvea)
- iris
- ciliary body
- choroid
- neural tunic (retina)
- photoreceptors
- anterior cavity
- anterior chamber
- posterior chamber
- posterior cavity
- vitreous body
- ciliary muscle
- ciliary processes
- ora serrata
- suspensory ligaments
- aqueous humor
- canal of Schlemm

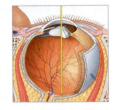

15.12

The eye is highly organized and has a consistent visual axis that directs light to the fovea of the retina

- crystallins
- optic nerve
- pupillary dilator muscles
- papillary constrictor muscles
- visual axis
- macula lutea
- fovea

15.13

Focusing produces a sharply defined image at the retina

- refracted
- focal point
- focal distance
- accommodation
- near point of vision

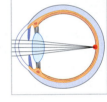

15.14

The neural tissue of the retina contains multiple layers of specialized photoreceptors, neurons, and supporting cells

- ganglion cells
- optic disc
- blind spot
- rods
- cones
- bipolar cells
- horizontal cells
- amacrine cells
- visual acuity

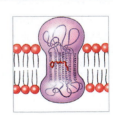

15.15

Photoreception, which occurs in the outer segment of rods and cones, involves the activation of visual pigments

- visual pigments
- pigment epithelium
- photoreceptor
- outer segment
- discs
- inner segment
- rhodopsin
- opsin
- retinal
- bleaching
- blue cones
- green cones
- red cones
- color blindness

• = *Term boldfaced in this module*

15.16

The visual pathways distribute visual information from each eye to both cerebral hemispheres

- depth perception
- optic radiation
- ○ lateral geniculate nucleus
- ○ superior colliculus

15.18

Aging is associated with many disorders of the special senses; trauma, infection, and abnormal stimuli may cause problems at any age

- cataract
- senile cataracts
- vertigo
- motion sickness
- conductive deafness
- nerve deafness

15.17

Accommodation problems result from abnormalities in the cornea or lens, or in the shape of the eye

- emmetropia
- myopia
- hyperopia
- photorefractive keratectomy (PRK)
- laser-assisted in-situ keratomileusis (LASIK)

● = *Term boldfaced in this module*

Chapter Integration: Applying what you've learned

John Drummond, 75 years old, visits his physician for a routine examination. During the history and physical, Mr. Drummond states that he has been experiencing difficulty seeing at a distance while driving and that at times he feels dizzy, especially when he closes his eyes. Additionally, Mr. Drummond tells his physician that he has a slight "cold." The physician asks Mr. Drummond to stand with his feet together and arms extended forward. The doctor discovers that as long as Mr. Drummond keeps his eyes open, he exhibits very little swaying movement. However, when he closes his eyes, his body begins to sway a great deal, and his arms tend to drift together toward the left side of his body. After a complete workup, which includes a neurological evaluation and a basic vision-screening test, the preliminary diagnosis is a slight fever, myopia, and vertigo.

It is estimated that 20–25 percent of all adults in the United States have myopia, while 40 percent of people in the United States experience vertigo at least once during their lifetime. The prevalence for both conditions increases with age.

1. Explain the diagnosis of myopia.

2. What treatment is available for persons with myopia?

3. Based on the information given, provide an explanation for Mr. Drummond's vertigo.

4. Why is Mr. Drummond's vertigo worse when his eyes are closed?

5. Why might Mr. Drummond's arms move toward his left side?

Access more review material online in the Study Area at **www.masteringaandp.com.**

There, you'll find:
- **Chapter guides**
- **Chapter quizzes**
- **Practice tests**
- **Labeling activities**
- **MP3 Tutor Sessions**
- **Tutorials**
- **Animations**
- **Flashcards**
- **A glossary with pronunciations**

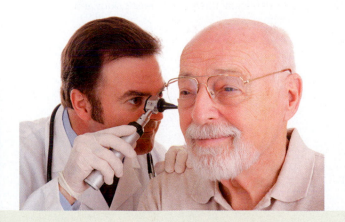

16

The Endocrine System

These Learning Outcomes correspond by number to this chapter's modules and indicate what you should be able to do after completing the chapter.

SECTION 1 • Hormones and Intercellular Communication

16.1 Explain the classification of hormones, and identify key functions of hormones secreted by organs and tissues of the endocrine system.

16.2 Explain the general mechanisms of hormonal action.

16.3 Describe how the hypothalamus controls endocrine organs.

16.4 Describe the location and structure of the pituitary gland, and identify pituitary hormones and their functions.

16.5 Describe the role of negative feedback in the functional relationship between the hypothalamus and pituitary gland.

16.6 Describe the location and structure of the thyroid gland, identify the hormones it produces, and specify the functions of those hormones.

16.7 Describe the location of the parathyroid glands, and identify the functions of the hormone they produce.

16.8 Describe the location, structure, and general functions of the adrenal glands, identify the hormones produced by the adrenal cortex and the adrenal medulla, and specify the functions of each hormone.

16.9 Describe the location and structure of the pancreas, identify the hormones it produces, and specify the functions of those hormones.

16.10 Describe the location of the pineal gland, and identify the functions of the hormone that it produces.

16.11 ✚ **CLINICAL MODULE** Explain diabetes mellitus: its types, clinical manifestations, and treatments.

SECTION 2 • Hormones and System Integration

16.12 Describe the functions of the hormones produced by the kidneys and the heart, and explain the roles of other endocrine organs and hormones in normal growth and development.

16.13 Define the general adaptation syndrome, and compare homeostatic responses with stress responses.

16.14 ✚ **CLINICAL MODULE** Describe key endocrine disorders, citing their characteristic signs and symptoms.

Hormones and Intercellular Communication

To preserve homeostasis, cellular activities must be coordinated throughout the body. The table below introduces the various ways by which cells communicate and coordinate their activities using chemical messengers.

Mechanisms of Intercellular Communication

Mechanism	Transmission	Chemical Mediators	Distribution of Effects
Direct communication 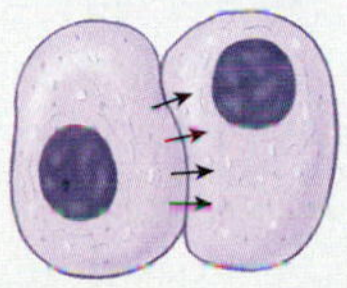	Through gap junctions	Ions, small solutes, lipid-soluble materials	Usually limited to adjacent cells of the same type that are interconnected by connexons
Paracrine communication 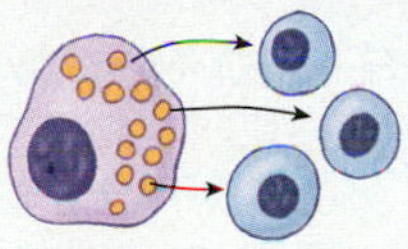	Through extracellular fluid	**Paracrine factors**	Primarily limited to the local area, where paracrine factor concentrations are relatively high; target cells must have appropriate receptors
Endocrine communication 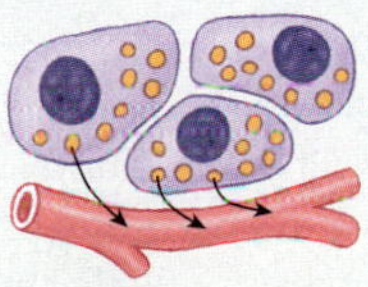	Through the bloodstream	**Hormones**	Target cells are primarily in other tissues and organs and must have appropriate receptors
Synaptic communication 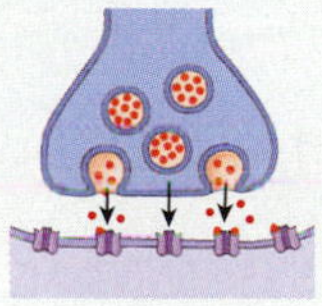	Across synaptic clefts	Neurotransmitters	Limited to very specific area; target cells must have appropriate receptors

Viewed from a general perspective, the differences between the nervous and endocrine systems seem relatively clear. In fact, these broad organizational and functional distinctions are the basis for treating them as two separate systems. Yet when we consider them in detail, we see similarities in the ways they are organized:

- Both systems rely on the release of chemicals that bind to specific receptors on their target cells.
- The two systems share many chemical messengers; for example, norepinephrine and epinephrine are called hormones when released into the bloodstream, but they are called neurotransmitters when released across synapses.
- Both systems are regulated primarily by negative feedback control mechanisms.
- The two systems share a common goal: to preserve homeostasis by coordinating and regulating the activities of other cells, tissues, organs, and systems.

In this section we will consider the structure and functions of the endocrine system and its products, as well as the integration of neural and endocrine activities.

Hormones may be amino acid derivatives, peptides, or lipid derivatives

1 The hormones and paracrine factors of the body can be divided into three groups on the basis of their chemical structure: (1) **amino acid derivatives**, (2) **peptide hormones**, and (3) **lipid derivatives**.

Amino Acid Derivatives

This group of hormones includes (1) thyroid hormones, produced by the thyroid gland; (2) the compounds epinephrine (E), norepinephrine (NE), and dopamine, which are sometimes called **catecholamines** (kat-e-KŌ-la-mēnz); and (3) melatonin (mel-a-TŌ-nin), a derivative of tryptophan that is secreted by the pineal gland.

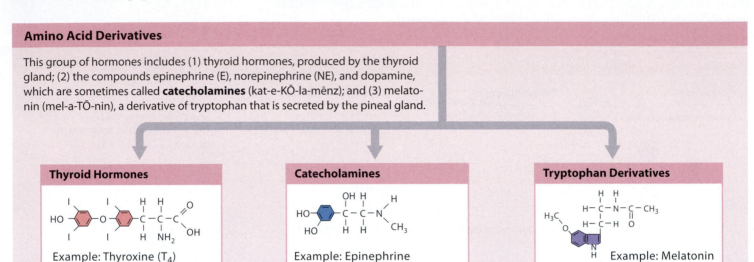

Thyroid Hormones

Example: Thyroxine (T_4)

Catecholamines

Example: Epinephrine

Tryptophan Derivatives

Example: Melatonin

Peptide Hormones

In general, peptide hormones are synthesized as **prohormones**—inactive molecules that are converted to active hormones either before or after they are secreted. Peptide hormones range from short polypeptide chains, such as antidiuretic hormone (ADH) and oxytocin (9 amino acids apiece), to small proteins, such as growth hormone (GH; 191 amino acids) and prolactin (PRL; 198 amino acids). This group includes all the hormones secreted by the hypothalamus, heart, thymus, digestive tract, and pancreas (including insulin), and most of the hormones of the pituitary gland. Glycoproteins— polypeptides that have carbohydrate side chains—may also function as hormones. Examples include thyroid-stimulating hormone (TSH), luteinizing hormone (LH), and follicle-stimulating hormone (FSH) from the pituitary gland, as well as several hormones produced in other organs.

Lipid Derivatives

These hormones consist of carbon rings and side chains built either from fatty acids (eicosanoids) or cholesterol (steroid hormones).

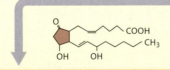

Eicosanoids   Example: Prostaglandin E

Eicosanoids (Ī-kō-sa-noydz) are important paracrine factors that coordinate cellular activities and affect enzymatic processes (such as blood clotting) in extracellular fluids. Some eicosanoids, such as **leukotrienes** (loo-kō-TRĪ-ēns), have secondary roles as hormones. A second group of eicosanoids—**prostaglandins**—is involved primarily in coordinating local cellular activities.

Steroid Hormones   Example: Estrogen

Steroid hormones are released by the reproductive organs (androgens by the testes in males, estrogens and progestins by the ovaries in females), by the cortex of the adrenal glands (corticosteroids), and by the kidneys (calcitriol). Because circulating steroid hormones are bound to specific transport proteins in the plasma, they remain in circulation longer than do secreted peptide hormones.

2 Hormones and paracrine factors are secreted by many different organs, but not all of those organs are part of the endocrine system. The **endocrine system** includes only those organs (indicated in purple in this figure) whose primary function is the production of hormones or paracrine factors. Many other organs contain tissues that secrete hormones, but their endocrine functions are secondary. Examples include the heart, the kidneys, the intestines, the thymus, and the reproductive organs. The endocrine functions of these organs will be considered primarily in later chapters.

Hypothalamus

The **hypothalamus** secretes hormones involved with fluid balance, smooth muscle contraction, and the control of hormone secretion by the anterior pituitary gland.

Pituitary Gland

The **pituitary gland** secretes multiple hormones that regulate the endocrine activities of the adrenal cortex, thyroid gland, and reproductive organs, and a hormone that stimulates melanin production.

Thyroid Gland

The **thyroid gland** secretes hormones that affect metabolic rate and calcium levels in body fluids.

Adrenal Glands

The **adrenal glands** secrete hormones involved with mineral balance, metabolic control, and resistance to stress; the adrenal medullae release E and NE during sympathetic activation.

Pancreas (Pancreatic Islets)

The **pancreatic islets** secrete hormones regulating the rate of glucose uptake and utilization by body tissues.

Pineal Gland

The **pineal gland** secretes melatonin, which affects reproductive function and helps establish circadian (day/night) rhythms.

Parathyroid Glands

The **parathyroid glands** secrete a hormone important to the regulation of calcium ion concentrations in body fluids.

Organs with Secondary Endocrine Functions

Heart: Secretes hormones involved in the regulation of blood volume	See Chapter 18
Thymus: Secretes hormones involved in the stimulation and coordination of the immune response	See Chapter 19
Digestive Tract: Secretes numerous hormones involved in the coordination of system functions, glucose metabolism, and appetite	See Chapter 21
Kidneys: Secrete hormones that regulate blood cell production and the rates of calcium and phosphate absorption by the intestinal tract	See Chapter 23
Gonads: Secrete hormones affecting growth, metabolism, and sexual characteristics, as well as hormones coordinating the activities of organs in the reproductive system	See Chapter 25

Testis

Ovary

Module 16.1 Review

a. Define endocrine system.

b. Name the organs of the endocrine system.

c. Describe the structural classification of hormones.

Hormones affect target cells after binding to receptors in the plasma membrane, cytoplasm, or nucleus

To affect a target cell, a hormone must first interact with an appropriate receptor—a protein molecule to which a particular molecule binds strongly. Each cell has receptors for responding to several different hormones, but cells in different tissues have different combinations of receptors. This arrangement is one reason hormones have differential effects on specific tissues. For every cell, the presence or absence of a specific receptor determines the cell's hormonal sensitivities. If a cell has a receptor that can bind a particular hormone, that cell will respond to the hormone's presence. If a cell lacks the proper receptor for that hormone, the hormone will have no effect on that cell. Hormone receptors are located either in the plasma membrane or inside the cell.

1 A hormone that binds to receptors in the plasma membrane cannot have a direct effect on the activities under way inside the target cell. Such a hormone cannot, for example, begin building a protein or catalyze a specific reaction. Instead, the hormone, or **first messenger**, uses an intracellular intermediary, or **second messenger**, to exert the hormone's effects in the cytoplasm. Thus hormone binding gives rise to a second messenger, which may act as an enzyme activator, inhibitor, or cofactor. Regardless, the net result is a change in the rates of various metabolic reactions. The two most important second messengers are **cyclic-AMP (cAMP)**, a derivative of ATP, and calcium ions.

The link between the first messenger and the second messenger generally involves a **G protein**, an enzyme complex coupled to a membrane receptor.

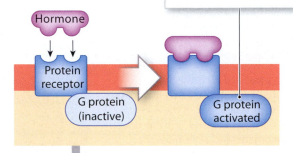

Effects on cAMP Levels

Many G proteins, once activated, exert their effects by changing the concentration of cyclic-AMP, which acts as the second messenger within the cell.

If levels of cAMP increase, enzymes may be activated or ion channels may be opened, accelerating the metabolic activity of the cell.

In some instances, G protein activation results in decreased levels of cAMP in the cytoplasm. This decrease has an inhibitory effect on the cell.

Effects on Ca²⁺ Levels

Some G proteins use Ca²⁺ as a second messenger.

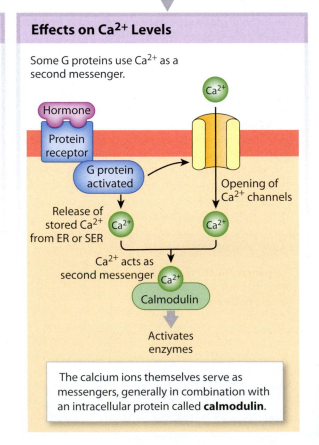

The calcium ions themselves serve as messengers, generally in combination with an intracellular protein called **calmodulin**.

2 Steroid hormones diffuse across the phospholipid bilayer of the plasma membrane and bind to receptors in the cytoplasm or nucleus. The hormone-receptor complexes then alter the activity of specific genes. By this mechanism, steroid hormones can alter the rate of DNA transcription in the nucleus, changing the pattern of protein synthesis. The resulting changes in the synthesis of enzymes or structural proteins directly affect the target cell's metabolic activity and structure. For example, in response to the sex hormone testosterone, skeletal muscle fibers increase production of enzymes and structural proteins, causing increases in muscle size and strength.

3 Thyroid hormones are primarily transported across the plasma membrane by carrier-mediated processes. Once in the cells, these hormones bind to receptors on mitochondria and within the nucleus. Thyroid hormones bound to mitochondria increase the mitochondrial rates of ATP production. Hormone-receptor complexes in the nucleus activate specific genes or change the rate of transcription, which affects the metabolic activities of the cell by increasing or decreasing the concentrations of specific enzymes.

Steroid hormone

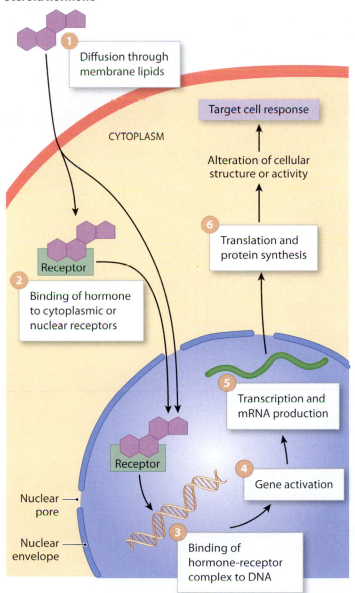

Thyroid hormone

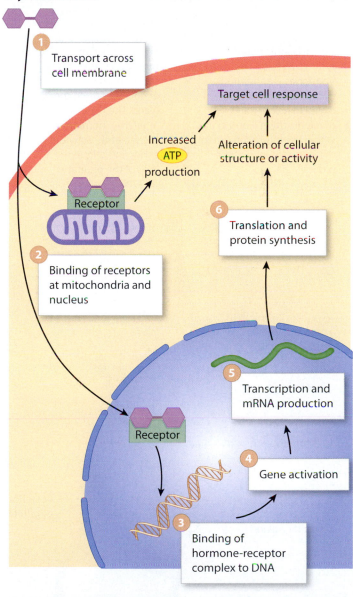

Module 16.2 Review

a. Define hormone receptor.

b. Differentiate between a first messenger and a second messenger.

c. Which type of hormone diffuses across the plasma membrane and binds to receptors in the cytoplasm?

The hypothalamus exerts direct or indirect control over the activities of many different endocrine organs

1 The **hypothalamus** provides the highest level of endocrine control; it integrates the activities of the nervous and endocrine systems. The hypothalamus accomplishes this integration through three mechanisms.

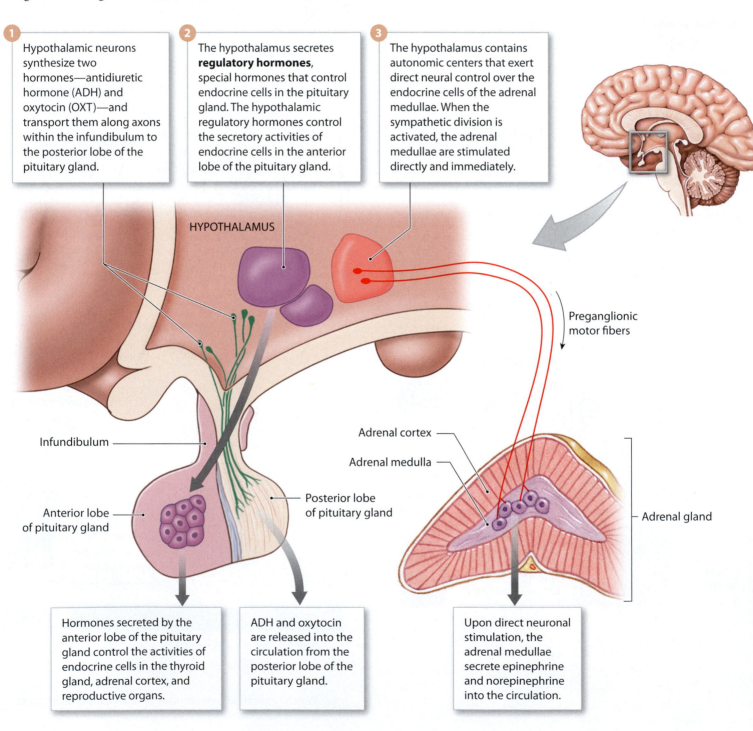

1 Hypothalamic neurons synthesize two hormones—antidiuretic hormone (ADH) and oxytocin (OXT)—and transport them along axons within the infundibulum to the posterior lobe of the pituitary gland.

2 The hypothalamus secretes **regulatory hormones**, special hormones that control endocrine cells in the pituitary gland. The hypothalamic regulatory hormones control the secretory activities of endocrine cells in the anterior lobe of the pituitary gland.

3 The hypothalamus contains autonomic centers that exert direct neural control over the endocrine cells of the adrenal medullae. When the sympathetic division is activated, the adrenal medullae are stimulated directly and immediately.

HYPOTHALAMUS

Preganglionic motor fibers

Infundibulum

Adrenal cortex

Adrenal medulla

Anterior lobe of pituitary gland

Posterior lobe of pituitary gland

Adrenal gland

Hormones secreted by the anterior lobe of the pituitary gland control the activities of endocrine cells in the thyroid gland, adrenal cortex, and reproductive organs.

ADH and oxytocin are released into the circulation from the posterior lobe of the pituitary gland.

Upon direct neuronal stimulation, the adrenal medullae secrete epinephrine and norepinephrine into the circulation.

2 By secreting specific regulatory hormones, the hypothalamus controls the production of hormones in the anterior lobe of the pituitary gland. At the **median eminence**, a swelling near the attachment of the infundibulum, hypothalamic neurons release regulatory factors into the surrounding interstitial fluids. These secretions enter the bloodstream quite easily, because the endothelial cells lining the capillaries in this region are unusually permeable. These **fenestrated** (FEN-es-trā-ted; *fenestra*, window) **capillaries** allow relatively large molecules to enter or leave the bloodstream.

Hypophyseal Portal System

The capillary networks and the interconnecting vessels constitute a portal system. Portal systems are named after their destinations; hence, this particular system is known as the **hypophyseal** (hī-po-FI-sē-al) **portal system** (*hypophysis* is the Latin name for the pituitary gland). This portal system provides an efficient means of chemical communication by ensuring that all the hypothalamic hormones entering the portal vessels will reach their target cells in the anterior lobe before being diluted through mixing with the general circulation. The communication is strictly one way, however, because any chemicals released by the cells "downstream" must do a complete circuit of the cardiovascular system before they reach the capillaries of the portal system.

The capillary networks in the median eminence are supplied by the superior hypophyseal artery. Before leaving the hypothalamus, the capillary networks unite to form a series of larger vessels that spiral around the infundibulum to reach the anterior lobe.

The vessels between the median eminence and the anterior lobe carry blood from one capillary network to another. Blood vessels that link two capillary networks are called **portal vessels**; in this case, they have the histological structure of veins, so they are called portal veins.

Once within the anterior lobe, these vessels form a second capillary network that branches among the endocrine cells.

Neurons of the supraoptic and paraventricular nuclei manufacture **antidiuretic hormone (ADH)** and **oxytocin (OXT)**, respectively. These hormones are released by synaptic terminals at fenestrated capillaries in the posterior lobe of the pituitary gland.

Supraoptic nuclei

Paraventricular nuclei

Neurosecretory neurons

HYPOTHALAMUS

MEDIAN EMINENCE

Superior hypophyseal artery

Infundibulum

Inferior hypophyseal artery

Posterior lobe of pituitary gland

Endocrine cells

Anterior lobe of pituitary gland

Hypophyseal veins

The regulatory hormones secreted at the hypothalamus are transported directly to the anterior lobe by the hypophyseal portal system. Hypothalamic regulatory hormones are of two types: **Releasing hormones (RH)** stimulate the synthesis and secretion of one or more hormones at the anterior lobe; **inhibiting hormones (IH)** prevent the synthesis and secretion of hormones from the anterior lobe. A given endocrine cell may be controlled by releasing hormones, by inhibiting hormones, or by some combination of the two.

Module 16.3 Review

a. Define regulatory hormone.

b. Identify the three mechanisms by which the hypothalamus integrates neural and endocrine function.

c. Name and describe the characteristics and functions of the blood vessels that link the hypothalamus with the anterior lobe of the pituitary gland.

The pituitary gland consists of an anterior lobe and a posterior lobe

1 The pituitary gland, or **hypophysis** (hī-POF-i-sis), is a small, oval gland that lies nestled within the sella turcica, a depression in the sphenoid bone. Nine important peptide hormones are released by the pituitary gland—seven by the anterior lobe and two by the posterior lobe. All nine hormones bind to membrane receptors, and all nine use cAMP as a second messenger. The hormones of the anterior lobe are also called **tropic hormones** (*trope*, a turning), because they "turn on" endocrine glands or support the functions of other organs.

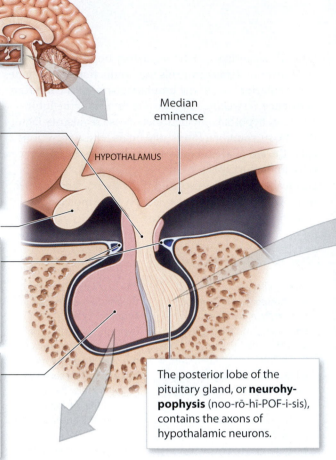

The **infundibulum** (in-fun-DIB-ū-lum; funnel) is a funnel-shaped stalk that connects the pituitary gland to the inferior surface of the hypothalamus.

Median eminence

HYPOTHALAMUS

Optic chiasm

A fold of the dura mater encircles the base of the infundibulum, locking the pituitary gland in position and isolating it from the cranial cavity.

The anterior lobe of the pituitary gland, or **adeno-hypophysis** (ad-e-nō-hī-POF-i-sis), contains a variety of endocrine cells.

The posterior lobe of the pituitary gland, or **neurohy-pophysis** (noo-rō-hī-POF-i-sis), contains the axons of hypothalamic neurons.

Hormones of the Anterior Lobe

TSH

Thyroid-stimulating hormone (TSH) targets the thyroid gland, where it triggers the release of thyroid hormones. TSH is released in response to **thyrotropin-releasing hormone (TRH)** from the hypothalamus. As circulating concentrations of thyroid hormones rise, the rates of TRH and TSH production decline—another example of negative feedback.

Thyroid gland

ACTH

Adrenocorticotropic hormone (ACTH), also known as corticotropin, stimulates the release of steroid hormones by the adrenal cortex, the outer portion of the adrenal gland. ACTH specifically targets cells that produce hormones that affect glucose metabolism. ACTH release occurs under the stimulation of **corticotropin-releasing hormone (CRH)** from the hypothalamus.

Adrenal gland

Gonadotropins (FSH and LH)

The hormones called **gonadotropins** (gō-nad-ō-TRŌ-pinz) regulate the activities of the gonads. (These organs—the testes in males and ovaries in females—produce reproductive cells as well as hormones.) The production of gonadotropins occurs under stimulation by **gonadotropin-releasing hormone (GnRH)** from the hypothalamus.

Follicle-stimulating hormone (FSH) promotes ovarian follicle development in females and, in combination with luteinizing hormone, stimulates the secretion of **estrogens** (ES-trō-jenz) by ovarian cells. In males, FSH promotes the physical maturation of developing sperm. FSH production is inhibited by **inhibin**, a peptide hormone released by cells in the testes and ovaries.

Luteinizing (LOO-tē-in-ī-zing) **hormone (LH)** induces ovulation, the release of reproductive cells in females. It also promotes the secretion, by the ovaries, of estrogens and progestins (such as progesterone), which prepare the body for possible pregnancy. In males, this gonadotropin stimulates the production of sex hormones by the interstitial cells of the testes. These sex hormones are called **androgens** (AN-drō-jenz; *andros*, man), the most important of which is testosterone.

Ovary

Testis

Hormones of the Posterior Lobe

ADH

Antidiuretic hormone (ADH), also known as **arginine vasopressin (AVP)**, is released in response to a variety of stimuli, most notably a rise in the solute concentration in the blood or a fall in blood volume or blood pressure. A rise in the solute concentration stimulates specialized hypothalamic neurons. Because they respond to a change in the osmotic concentration of body fluids, these neurons are called **osmoreceptors.** The osmoreceptors then stimulate the neurosecretory neurons that release ADH. The primary function of ADH is to decrease the amount of water lost at the kidneys. With losses minimized, any water absorbed from the digestive tract will be retained, reducing the concentrations of electrolytes in the extracellular fluid. In high concentrations, ADH also causes vasoconstriction, a constriction of peripheral blood vessels that helps elevate blood pressure. ADH release is inhibited by alcohol, which explains the increased fluid excretion that follows the consumption of alcoholic beverages.

Kidney

OXT

In women, **oxytocin** (*okytokos,* swift birth), or **OXT**, stimulates smooth muscle contraction in the wall of the uterus, promoting labor and delivery. After delivery, oxytocin stimulates the contraction of myoepithelial cells around the secretory alveoli and the ducts of the mammary glands, promoting the ejection of milk. Although the functions of oxytocin in sexual activity remain unclear, it is known that circulating concentrations of oxytocin rise during sexual arousal and peak at orgasm in both sexes. Oxytocin release is triggered by sensory input, so it is an example of a **neuroendocrine reflex**.

Uterus

GH

Growth hormone (GH) stimulates cell growth and reproduction by accelerating the rate of protein synthesis. Skeletal muscle cells and chondrocytes are particularly sensitive to GH. The production of GH is regulated by **growth hormone–releasing hormone (GH–RH)** and **growth hormone–inhibiting hormone (GH–IH)** from the hypothalamus. The stimulation of growth by GH involves two different mechanisms of control:

Musculo-skeletal system

The primary control mechanism is indirect. Liver cells respond to GH by synthesizing and releasing **somatome-dins**, which are peptide hormones that bind to receptor sites on a variety of plasma membranes and increase the rate of uptake of amino acids and their incorporation into new proteins.

The direct actions of GH are more selective:
- In epithelia and connective tissues, GH stimulates stem cell divisions and the differentiation of daughter cells.
- In adipose tissue, GH stimulates the breakdown of stored triglycerides by adipocytes (fat cells), which then release fatty acids into the blood. Many tissues then stop breaking down glucose and use fatty acids instead to generate ATP. This is termed a **glucose-sparing effect**.
- In the liver, GH stimulates the breakdown of glycogen reserves, leading to the release of glucose into the bloodstream.

PRL

Prolactin (*pro-*, before + *lac*, milk) **(PRL)** works with other hormones to stimulate mammary gland development. In pregnancy and during the nursing period that follows delivery, PRL also stimulates milk production by the mammary glands. Prolactin production is inhibited by **prolactin-inhibiting hormone (PIH)** and stimulated by several prolactin-releasing factors.

MSH

The pars intermedia, a narrow portion of the anterior lobe closest to the posterior lobe, may secrete **melanocyte-stimulating hormone (MSH)**. MSH stimulates the melanocytes of the skin to increase their production of melanin. In adults, this portion of the anterior lobe is virtually nonfunctional, and the circulating blood usually does not contain MSH.

Module 16.4 Review

a. Name the two lobes of the pituitary gland.

b. Identify the nine pituitary hormones and their target tissues.

c. In a dehydrated person, how would the amount of ADH released by the posterior pituitary change?

Negative feedback mechanisms control the secretion rates of the hypothalamus and pituitary gland

1 Hormone secretion is typically controlled through negative feedback; the relationships among the the hypothalamus, the pituitary, and the endocrine target organ provide many useful examples. In the common arrangement depicted here, the hypothalamus produces a releasing hormone or factor that triggers the release of a hormone by the anterior lobe of the pituitary gland. The pituitary hormone stimulates release of a second hormone by the target organ. This second hormone suppresses secretion of both the hypothalamic releasing hormone and the pituitary hormone. The table at right lists major hormones whose secretion is controlled in this way.

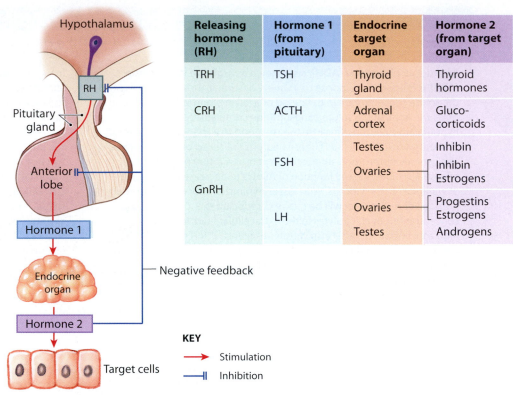

Releasing hormone (RH)	Hormone 1 (from pituitary)	Endocrine target organ	Hormone 2 (from target organ)
TRH	TSH	Thyroid gland	Thyroid hormones
CRH	ACTH	Adrenal cortex	Gluco-corticoids
GnRH	FSH	Testes	Inhibin
		Ovaries	Inhibin / Estrogens
	LH	Ovaries	Progestins / Estrogens
		Testes	Androgens

KEY

→ Stimulation

—I Inhibition

2 The secretion of growth hormone involves both releasing and inhibiting hormones. The hypothalamic neurons releasing these hormones are sensitive to levels of **somatomedins** (compounds that stimulate tissue growth by increasing amino acid uptake), which function like Hormone 2 in the general pattern shown above. However, the somatomedins not only suppress secretion of GH-RH but stimulate secretion of GH-IH. This combination provides more rapid and precise regulation of GH levels.

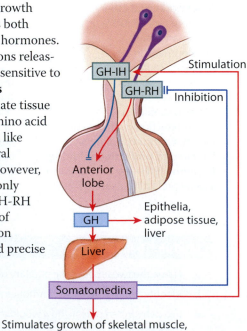

Stimulates growth of skeletal muscle, cartilage, and many other tissues

3 Secretion of prolactin (PRL) is also controlled by a pair of regulatory hormones. Secretion of prolactin-releasing factor (PRF) is inhibited by PRL, whereas secretion of prolactin-inhibiting hormone (PIH) is stimulated by PRL.

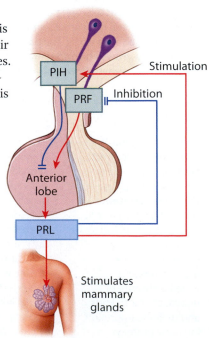

Stimulates mammary glands

4 This figure provides an overview of the hormonal relationships we have considered in Modules 16.3 through 16.5.

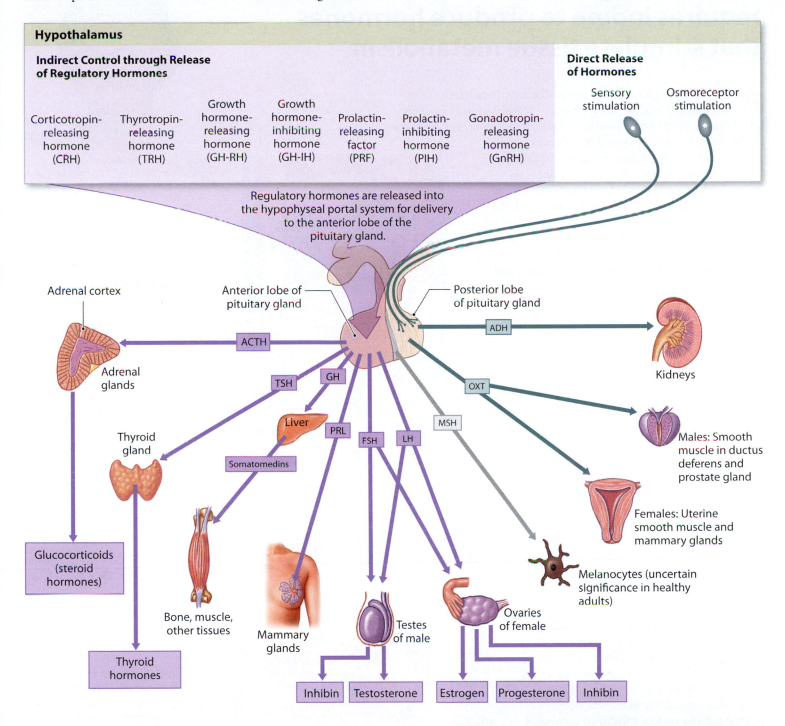

Hypothalamus

Indirect Control through Release of Regulatory Hormones

Corticotropin-releasing hormone (CRH)

Thyrotropin-releasing hormone (TRH)

Growth hormone-releasing hormone (GH-RH)

Growth hormone-inhibiting hormone (GH-IH)

Prolactin-releasing factor (PRF)

Prolactin-inhibiting hormone (PIH)

Gonadotropin-releasing hormone (GnRH)

Direct Release of Hormones

Sensory stimulation

Osmoreceptor stimulation

Regulatory hormones are released into the hypophyseal portal system for delivery to the anterior lobe of the pituitary gland.

Adrenal cortex

Anterior lobe of pituitary gland

Posterior lobe of pituitary gland

ACTH

ADH

Kidneys

Adrenal glands

TSH

GH

OXT

Thyroid gland

Liver

PRL

FSH

LH

MSH

Somatomedins

Males: Smooth muscle in ductus deferens and prostate gland

Females: Uterine smooth muscle and mammary glands

Glucocorticoids (steroid hormones)

Melanocytes (uncertain significance in healthy adults)

Bone, muscle, other tissues

Mammary glands

Testes of male

Ovaries of female

Thyroid hormones

Inhibin

Testosterone

Estrogen

Progesterone

Inhibin

Module 16.5 Review

a. List the hypothalamic releasing hormones.

b. If a blood sample contained elevated levels of somatomedins, which pituitary hormone would you expect to be elevated?

c. What effects would elevated circulating levels of glucocorticoids have on the pituitary secretion of ACTH?

The thyroid gland contains follicles and requires iodine to produce hormones that stimulate tissue metabolism

1 The **thyroid gland** curves across the anterior surface of the trachea just inferior to the thyroid ("shield-shaped") cartilage, which forms most of the anterior surface of the larynx. The two lobes of the thyroid gland are united by a slender connection, the isthmus (IS-mus). You can easily feel the gland with your fingers. When something goes wrong with it, the thyroid gland typically becomes visible as it enlarges and distorts the surface of the neck. The size of the gland is quite variable, depending on heredity and environmental and nutritional factors, but its average weight is about 34 g (1.2 oz). An extensive blood supply gives the thyroid gland a deep red color.

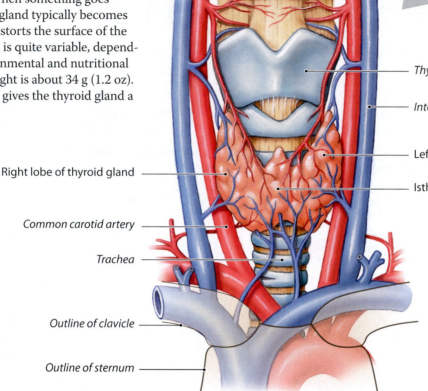

Thyroid cartilage

Internal jugular vein

Left lobe of thyroid gland

Isthmus of thyroid gland

Right lobe of thyroid gland

Common carotid artery

Trachea

Outline of clavicle

Outline of sternum

2 The **thyroid gland** contains large numbers of thyroid follicles, hollow spheres lined by a simple cuboidal epithelium. The follicle cells surround a follicle cavity that holds a viscous colloid, a fluid containing large quantities of dissolved proteins. A network of capillaries surrounds each follicle, delivering nutrients and regulatory hormones to the glandular cells and accepting their secretory products and metabolic wastes. The follicle cells synthesize a globular protein called **thyroglobulin** (thī-rō-GLOB-ū-lin) and secrete it into the colloid of the thyroid follicles. Thyroglobulin molecules contain the amino acid tyrosine, the building block of thyroid hormones.

Simple cuboidal epithelium of follicle

Thyroid follicle

Thyroglobulin in colloid

A second population of endocrine cells lies sandwiched between the basal laminae of the follicle cells. These large, pale cells—called **C (clear) cells**—produce the hormone **calcitonin (CT)**, which aids in the regulation of calcium ion concentrations in body fluids.

Section of thyroid gland   LM × 260

3 This figure depicts the continuous process by which thyroid hormones are produced and then stored within thyroglobulin in thyroid follicles. Thyroid hormones are released into and removed from the circulation as needed.

Follicle cavity

3 The hormone **thyroxine** (thī-ROKS-ēn), or **T₄**, contains four iodide ions. A related molecule called T₃ contains three iodide ions. Eventually, each molecule of thyroglobulin contains four to eight molecules of T₃, T₄, or both.

2 The iodide ions diffuse to the apical surface of each follicle cell, where they are enzymatically converted to an activated form of iodide (I⁺). This reaction also attaches one or two activated iodide ions to the tyrosine portions of a thyroglobulin molecule within the follicle lumen.

FOLLICLE CAVITY

Thyroglobulin
(contains T₃ and T₄)

Thyroglobulin

Iodide
(I⁺)

Other amino acids

Tyrosine

Endocytosis

T₄
T₃

FOLLICLE CELL

Diffusion

TSH-
sensitive
ion pump

Diffusion

Start ▶ **1** Iodide ions are absorbed from the diet and are delivered to the thyroid gland by the bloodstream. Carrier proteins in the basal membrane of the follicle cells actively transport iodide ions (I⁻) into the cytoplasm.

Iodide (I⁻)

CAPILLARY

T₄ & T₃

4 Follicle cells remove thyroglobulin from the follicle cavity by endocytosis.

5 Lysosomal enzymes break the thyroglobulin down, and the released amino acids and thyroid hormones enter the cytoplasm. The amino acids are then recycled and used to synthesize more thyroglobulin.

6 The released molecules of T₃ and T₄ diffuse across the basement membrane and enter the bloodstream. About 90 percent of all thyroid secretions is T₄; T₃ is secreted in comparatively small amounts although its metabolic effects are much stronger than those of T₄.

7 Roughly 75 percent of the T₄ molecules and 70 percent of the T₃ molecules entering the bloodstream become attached to transport proteins called **thyroid-binding globulins (TBGs)**. The transport proteins release thyroid hormones only gradually, and the bound thyroid hormones represent a substantial reserve: The bloodstream normally contains more than a week's supply of thyroid hormones.

Effects of Thyroid Hormones on Peripheral Tissues

- Elevated rates of oxygen consumption and energy consumption; in children, may cause a rise in body temperature
- Increased heart rate and force of contraction; generally results in a rise in blood pressure
- Increased sensitivity to sympathetic stimulation
- Maintenance of normal sensitivity of respiratory centers to changes in oxygen and carbon dioxide concentrations
- Stimulation of red blood cell formation and thus enhanced oxygen delivery
- Stimulation of activity in other endocrine tissues
- Accelerated turnover of minerals in bone

Module 16.6 Review

a. Name the hormones of the thyroid gland.

b. What thyroid hormone aids in the regulation of calcium?

c. Why do signs and symptoms of decreased thyroxine concentrations not appear until about a week after a thyroidectomy (surgical removal of the thyroid gland)?

Parathyroid hormone, produced by the parathyroid glands, is the primary regulator of calcium ion levels in body fluids

1 Two pairs of **parathyroid glands** are embedded in the posterior surfaces of the thyroid gland. Altogether, the four parathyroid glands weigh a mere 1.6 g (0.06 oz).

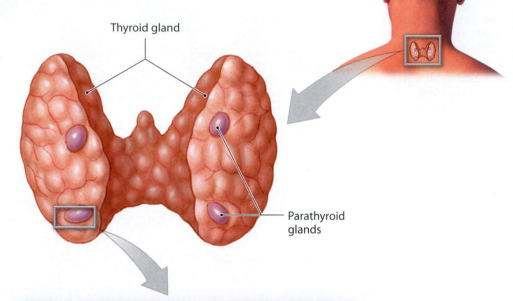

Thyroid gland

Parathyroid glands

2 The parathyroid glands have at least two cell populations: **Parathyroid cells** produce parathyroid hormone. The other cells, called oxyphils, have no known function. Like the C cells of the thyroid gland, the parathyroid cells monitor the circulating concentration of calcium ions. When the Ca^{2+} concentration of the blood falls below normal, the parathyroid cells secrete **parathyroid hormone (PTH)**. The net result of PTH secretion is an increase in Ca^{2+} concentration in body fluids.

Blood vessel

A dense fibrous capsule separates the cells of the parathyroid gland from those of the thyroid gland.

Thyroid follicles

Parathyroid gland — LM × 94

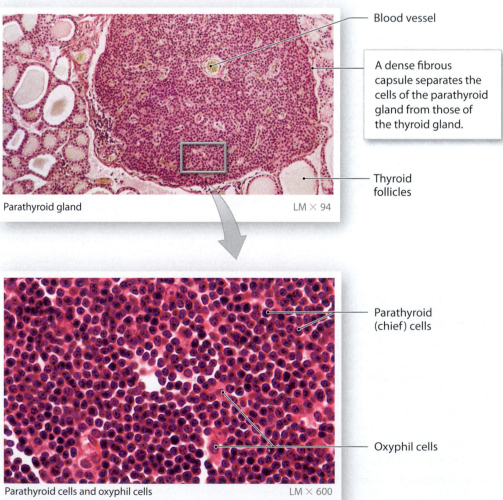

Parathyroid (chief) cells

Oxyphil cells

Parathyroid cells and oxyphil cells — LM × 600

3 Parathyroid hormone and calcitonin (from the thyroid) have opposing effects on levels of calcium ions in body fluids. However, in healthy adults PTH, aided by calcitriol secreted by the kidneys, is the primary regulator of circulating calcium ion concentrations, and removal of the thyroid gland seldom affects calcium ion homeostasis because dietary intake and metabolic demand are so closely balanced that elevated blood calcium levels are very rare. However, calcitonin can be administered clinically to treat several metabolic disorders that cause elevated calcium levels and excessive bone formation.

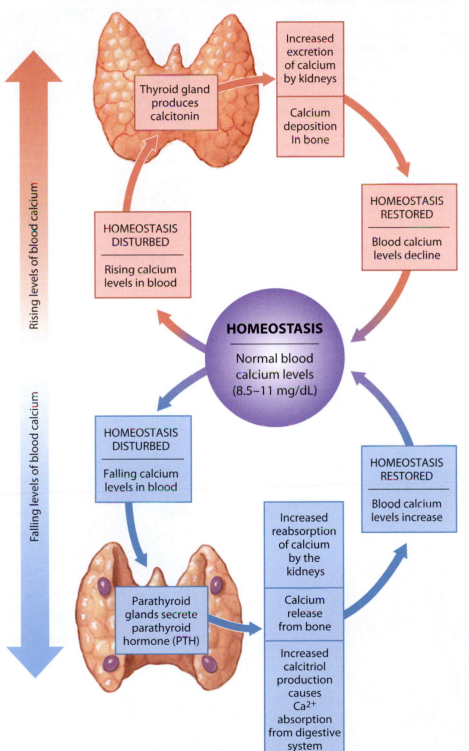

Rising levels of blood calcium

Thyroid gland produces calcitonin

Increased excretion of calcium by kidneys

Calcium deposition in bone

HOMEOSTASIS RESTORED
Blood calcium levels decline

HOMEOSTASIS DISTURBED
Rising calcium levels in blood

HOMEOSTASIS
Normal blood calcium levels (8.5–11 mg/dL)

Falling levels of blood calcium

HOMEOSTASIS DISTURBED
Falling calcium levels in blood

Parathyroid glands secrete parathyroid hormone (PTH)

Increased reabsorption of calcium by the kidneys

Calcium release from bone

Increased calcitriol production causes Ca^{2+} absorption from digestive system

HOMEOSTASIS RESTORED
Blood calcium levels increase

Effects of Parathyroid Hormone on Peripheral Tissues

- PTH mobilizes calcium from bone by affecting osteoblast and osteoclast activity. PTH inhibits osteoblasts, thereby reducing the rate of calcium deposition in bone. Although osteoclasts have no PTH receptors, PTH triggers the release of a growth factor that increases osteoclast numbers. Because osteoclasts are more numerous, osteoclast activity predominates, and as bone matrix erodes, plasma Ca^{2+} levels rise.

- PTH enhances the reabsorption of Ca^{2+} by the kidneys, reducing urinary losses.

- PTH stimulates the formation and secretion of calcitriol at the kidneys. In general, the effects of calcitriol complement or enhance those of PTH, but one major effect of calcitriol is the enhancement of Ca^{2+} and PO$_4^{3-}$ absorption by the digestive tract.

Module 16.7 Review

a. Describe the locations of the parathyroid glands.

b. Explain how parathyroid hormone raises the blood calcium levels.

c. Decreased blood calcium levels would result in increased secretion of which hormone?

The adrenal glands produce hormones involved in metabolic regulation

1 A yellow, pyramid-shaped **adrenal gland,** or suprarenal (soo-pra-RĒ-nal; *supra-*, above + *renes*, kidneys) gland, sits on the superior border of each kidney. The adrenal glands are retroperitoneal, as are the kidneys, and only their anterior surfaces are covered by a layer of parietal peritoneum. Like other endocrine glands, the adrenal glands are richly supplied with blood vessels.

2 The adrenal cortex has a yellow color due to stored lipids, especially cholesterol and various fatty acids. The adrenal cortex produces more than two dozen steroid hormones, collectively called **adrenocortical steroids**, or simply **corticosteroids**. Like other steroid hormones, corticosteroids exert their effects by determining which genes in the nuclei of their target cells are transcribed, and at what rate. The resulting changes in the nature and concentration of enzymes in the cytoplasm affect cellular metabolism. Corticosteroids are vital: If the adrenal glands are destroyed or removed, the individual will die unless corticosteroids are administered.

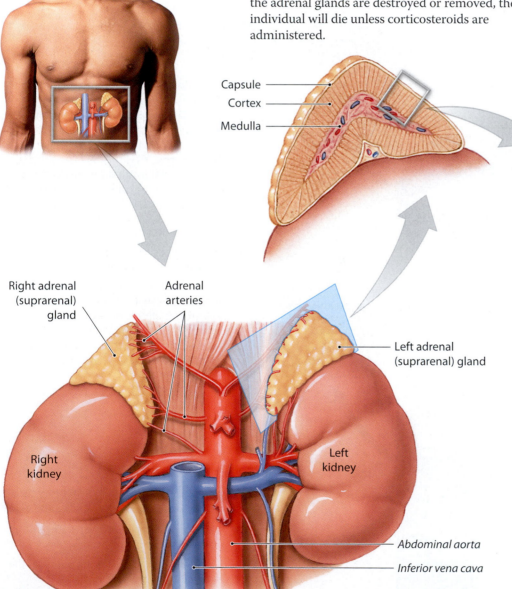

Capsule

Cortex

Medulla

Right adrenal (suprarenal) gland

Adrenal arteries

Left adrenal (suprarenal) gland

Right kidney

Left kidney

Abdominal aorta

Inferior vena cava

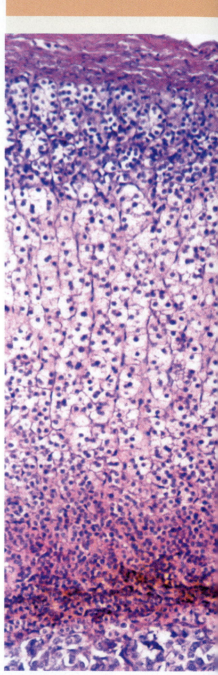

Adrenal gland LM × 250

3 Deep to the adrenal capsule are three distinct regions, or zones, in the adrenal cortex. Each zone synthesizes specific steroid hormones. Deep to the cortex lies the adrenal medulla, which synthesizes epinephrine and norepinephrine.

The Adrenal Hormones

Region/Zone	Hormones	Primary Targets	Hormonal Effects	Regulatory Control
Capsule				
ADRENAL CORTEX				
The **zona glomerulosa** (glō-mer-ū-LŌ-suh) is the outer region of the adrenal cortex.	**Mineralocorticoids (MCs)**, primarily **aldosterone**	Kidneys	Aldosterone increases renal reabsorption of Na$^+$ and water, especially in the presence of ADH. It also accelerates urinary loss of K$^+$.	Mineralocorticoid secretion is stimulated by the activation of the renin-angiotensin system (Module 16.12) and is inhibited by hormones opposing that system.
The **zona fascicu-lata** (fa-sik-ū-LA-tuh; *fasciculus*, little bundle) is the large, central portion of the adrenal cortex.	**Glucocorticoids (GCs)** are steroid hormones that affect glucose metabolism. The primary hormones are **cortisol** (KOR-ti-sol), also called hydrocorti-sone, and smaller amounts of the related steroid **corticosterone** (kor-ti-KOS-te-rōn). The liver converts some of the circulating cortisol to **cortisone**, another metabolically active glucocorticoid.	Most cells	Glucocorticoids increase rates of glucose and glycogen formation by the liver. They also stimulate the release of amino acids from skeletal muscles, and lipids from adipose tissues, and they promote lipid catabolism within peripheral cells. These actions supplement the glucose-sparing effect of growth hormone (noted in Module 16.4). Cortisol also reduces inflammation (an **anti-inflammatory effect**).	Glucocorticoid secretion is stimu-lated by ACTH from the anterior lobe of the pituitary gland.
The **zona reticularis** (re-tik-ū-LAR-is; *reticulum*, network) forms a narrow band bordering each adrenal medulla.	Small quantities of androgens (male sex hormones) that may be converted to estrogens in the bloodstream	Skin, bones, and other tissues, but minimal effects in normal adults	Adrenal androgens stimulate the development of pubic hair in boys and girls before puberty.	Androgen secretion is stimulated by ACTH.
ADRENAL MEDULLA	Epinephrine (E), norepinephrine (NE)	Most cells	Epinephrine and norepinephrine increase cardiac activity, blood pressure, glycogen breakdown, and blood glucose levels.	Epinephrine and norepinephrine secretion is stimulated by sympa-thetic preganglionic fibers during sympathetic activation.

Module 16.8 Review

a. Identify the two regions of an adrenal gland, and cite the hormones secreted by each.

b. List the three zones of the adrenal cortex.

c. What effect would elevated cortisol levels have on blood glucose levels?

The pancreatic islets secrete insulin and glucagon and regulate glucose utilization by most cells

1 The **pancreas** lies within the abdominopelvic cavity in the loop formed between the inferior border of the stomach and the proximal portion of the small intestine. It is a slender, pale organ with a nodular (lumpy) consistency. In adults, the pancreas is 20–25 cm (8–10 in.) long and weighs about 80 g (2.8 oz). The **exocrine pancreas**, roughly 99 percent of the organ's volume, consists of clusters of gland cells and their attached ducts. Together, the gland and duct cells secrete large quantities of an alkaline, enzyme-rich fluid that reaches the lumen of the digestive tract through one or two pancreatic ducts.

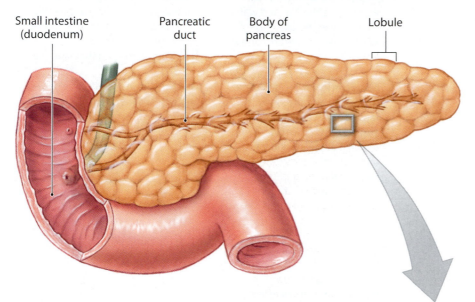

Small intestine (duodenum)   Pancreatic duct   Body of pancreas   Lobule

2 The **endocrine pancreas** consists of small groups of cells scattered among the exocrine cells. The endocrine clusters are known as **pancreatic islets**, or the islets of Langerhans (LAN-ger-hanz). Pancreatic islets account for only about 1 percent of all cells in the pancreas. Nevertheless, a typical pancreas contains roughly 2 million pancreatic islets, and their secretions are vital to our survival.

The exocrine cells form small clusters that secrete into a lumen continuous with a pancreatic duct. The glandular clusters are called **pancreatic acini** (singular, *acinus*).

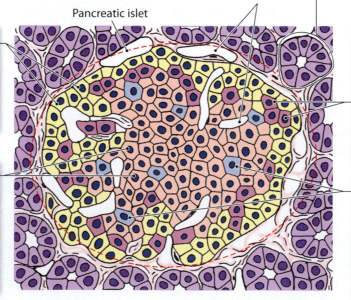

Capillaries

Pancreatic islet

Alpha cells produce the hormone **glucagon** (GLOO-ka-gon). Glucagon raises blood glucose levels by increasing the rates of glycogen breakdown and glucose release by the liver.

Beta cells produce the hormone **insulin** (IN-suh-lin). Insulin lowers blood glucose levels by increasing the rate of glucose uptake and utilization by cells, and by increasing glycogen synthesis in skeletal muscles and the liver.

Delta cells produce a peptide hormone identical to growth hormone–inhibiting hormone (GH–IH). GH–IH suppresses the release of glucagon and insulin by other islet cells and slows the rates of food absorption and enzyme secretion along the digestive tract.

F cells produce the hormone **pancreatic polypeptide (PP)**. PP inhibits gallbladder contractions and regulates the production of some pancreatic enzymes, and it may help control the rate of nutrient absorption by the digestive tract.

3 Insulin and glucagon are the primary hormones responsible for the regulation of blood glucose levels. When blood glucose levels rise, beta cells secrete insulin, which then stimulates the transport of glucose across plasma membranes and into target cells. When blood glucose levels decline, alpha cells secrete glucagon, which stimulates glycogen breakdown and glucose release by the liver.

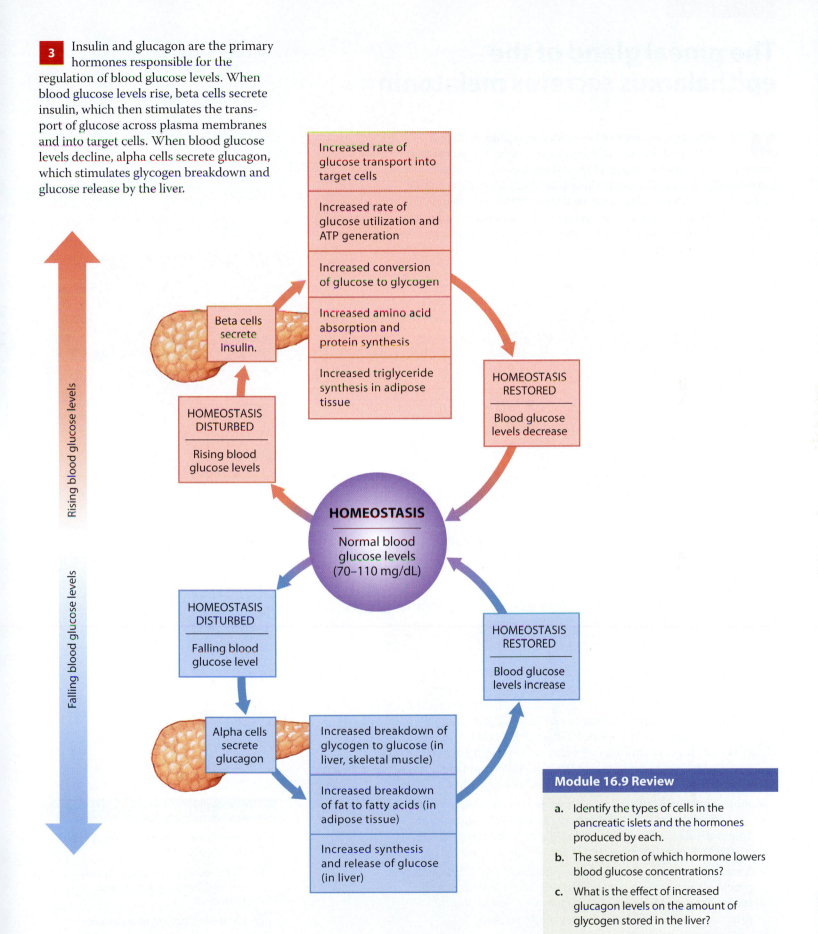

Rising blood glucose levels

Falling blood glucose levels

Increased rate of glucose transport into target cells

Increased rate of glucose utilization and ATP generation

Increased conversion of glucose to glycogen

Increased amino acid absorption and protein synthesis

Increased triglyceride synthesis in adipose tissue

Beta cells secrete Insulin.

HOMEOSTASIS DISTURBED
Rising blood glucose levels

HOMEOSTASIS RESTORED
Blood glucose levels decrease

HOMEOSTASIS
Normal blood glucose levels (70–110 mg/dL)

HOMEOSTASIS DISTURBED
Falling blood glucose level

HOMEOSTASIS RESTORED
Blood glucose levels increase

Alpha cells secrete glucagon

Increased breakdown of glycogen to glucose (in liver, skeletal muscle)

Increased breakdown of fat to fatty acids (in adipose tissue)

Increased synthesis and release of glucose (in liver)

Module 16.9 Review

a. Identify the types of cells in the pancreatic islets and the hormones produced by each.

b. The secretion of which hormone lowers blood glucose concentrations?

c. What is the effect of increased glucagon levels on the amount of glycogen stored in the liver?

The pineal gland of the epithalamus secretes melatonin

1 The **pineal gland**, part of the epithalamus, lies in the posterior portion of the roof of the third ventricle. The pineal gland contains neurons, neuroglia, and special secretory cells called **pinealocytes** (pin-Ē-al-ō-sīts). These cells synthesize the hormone **melatonin** from molecules of the neurotransmitter serotonin. Collaterals from the visual pathways enter the pineal gland and affect the rate of melatonin production, which is lowest during daylight hours and highest at night.

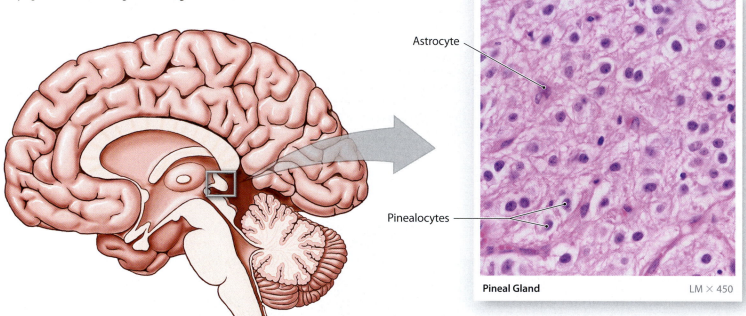

Astrocyte

Pinealocytes

Pineal Gland　　　　　　　LM × 450

Functions of Melatonin in Humans

- *Inhibiting reproductive functions*. In some mammals, melatonin slows the maturation of sperm, oocytes, and reproductive organs by reducing the rate of GnRH secretion. The significance of this effect in humans remains unclear, but circumstantial evidence suggests that melatonin may play a role in the timing of human sexual maturation. Melatonin levels in the blood decline at puberty, and pineal tumors that eliminate melatonin production cause premature puberty in young children.

- *Protecting against tissue damage by free radicals*. Melatonin is a very effective antioxidant that may protect CNS neurons from **free radicals**, such as nitric oxide (NO) or hydrogen peroxide (H_2O_2), that may be generated in active neural tissue.

- *Setting circadian rhythms*. Because pineal activity is cyclical, the pineal gland may also be involved with the maintenance of basic **circadian rhythms**—daily changes in physiological processes that follow a regular day–night pattern.

Module 16.10 Review

a. Identify the hormone-secreting cells of the pineal gland.

b. Increased amounts of light would inhibit the production of which hormone?

c. List three functions of melatonin.

Diabetes mellitus is an endocrine disorder characterized by excessively high blood glucose levels

Diabetes Mellitus

Diabetes mellitus (mel-ĭ-tus; *mellitum*, honey) is characterized by glucose concentrations that are high enough to overwhelm the reabsorption capabilities of the kidneys. (The presence of abnormally high blood glucose levels is called **hyperglycemia** [hĭ-per-glī-SĒ-mē-ah].) In diabetes mellitus, glucose appears in the urine (**glycosuria**; glī-kō-SOO-rē-a), and urine volume generally becomes excessive (**polyuria**). Diabetes mellitus can be caused by genetic abnormalities or mutations that result in inadequate insulin production, the synthesis of abnormal insulin molecules, or the production of defective insulin-receptor proteins.

Type 1 (insulin dependent) Diabetes

Type 1 (insulin dependent) diabetes is characterized by inadequate insulin production by the pancreatic beta cells. Individuals with type 1 diabetes must receive insulin to live—typically multiple injections daily, or continuous infusion through an insulin pump or other device. Type 1 diabetes accounts for only about 5–10 percent of diabetes cases and often develops in childhood.

Type 2 (non-insulin dependent) Diabetes

Type 2 (non-insulin dependent) diabetes is the most common form of diabetes mellitus. Most individuals with this form of diabetes produce normal amounts of insulin, at least initially, but their tissues do not respond properly—a condition known as insulin resistance. Type 2 diabetes is associated with obesity, and weight loss through diet and exercise can be an effective treatment, especially when coupled with drugs that alter rates of glucose synthesis and release by the liver.

1 Untreated diabetes mellitus disrupts metabolic activities throughout the body. Clinical problems arise because the tissues involved are experiencing an energy crisis—in essence, most of the tissues are responding as they would during chronic starvation, breaking down lipids and even proteins because they are unable to absorb glucose from their surroundings. Problems involving abnormal changes in blood vessel structure are particularly dangerous. An estimated 23.6 million people in the United States have some form of diabetes.

Clinical Problems Caused by Diabetes Mellitus

The proliferation of capillaries and hemorrhaging at the retina may cause partial or complete blindness. This condition is called **diabetic retinopathy.**

Degenerative blockages in cardiac circulation can lead to early heart attacks. For a given age group, heart attacks are three to five times more likely in individuals with diabetes than in individuals that do not have the condition.

Degenerative changes in the kidneys, a condition called **diabetic nephropathy**, can lead to kidney failure.

Abnormal blood flow to neural tissues is probably responsible for a variety of problems with peripheral nerves, including abnormal autonomic function. As a group, these disorders are termed **diabetic neuropathy**.

Blood flow to the distal portions of the limbs is reduced, and peripheral tissues may be damaged as a result. A reduction in blood flow to the feet, for example, can lead to tissue death, ulceration, infection, and the loss of toes or a major portion of one or both feet.

Module 16.11 Review

a. Define diabetes mellitus.

b. Identify and describe the two types of diabetes mellitus.

c. Why does a person with diabetes mellitus urinate frequently?

1. Concept map

Place each of the following terms in the appropriate box to correctly complete the hormones concept map.

- steroid hormones
- tryptophan derivatives
- glycoproteins
- short polypeptides
- catecholamines
- peptide hormones
- thyroid hormones
- transport proteins
- lipid derivatives
- small proteins
- eicosanoids

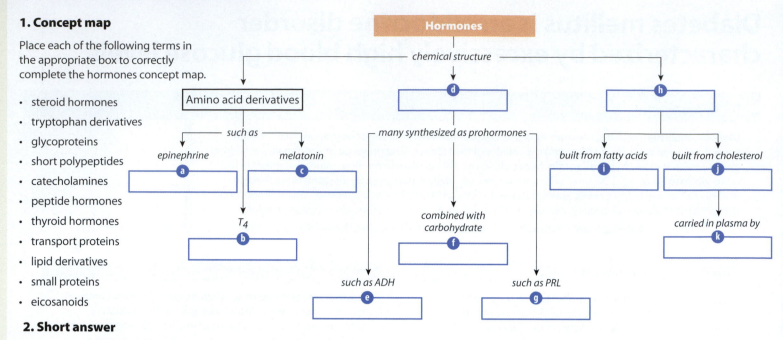

2. Short answer

Identify the endocrine gland—or organ containing endocrine cells—based on the major effects produced by their secreted hormone(s).

a _____ stimulates and coordinates the immune response

b _____ establishes day/night cycles

c _____ secretes insulin and regulates glucose uptake and utilization

d _____ controls hormone secretion of the pituitary gland

e _____ regulates red blood cell production and the absorption of calcium and phosphate by the intestinal tract

f _____ regulates mineral balance, metabolic control, and resistance to stress

g _____ regulates secretions of adrenal cortex, thyroid gland, and reproductive organs

h _____ affects growth, metabolism, and sexual characteristics

i _____ maintains circulation and has a role in regulating blood volume

j _____ coordinates its system functions, glucose metabolism, and appetite

k _____ regulates metabolic rate and calcium levels in body fluids

l _____ glands embedded in posterior surface of thyroid that play an important role in the response to decreasing calcium levels in body fluids

3. Matching

Match the following terms with the most closely related description.

- FSH
- androgens
- F cells
- parathyroid glands
- epinephrine
- direct communication
- tropic hormones
- secretes releasing hormones
- prostaglandins
- cyclic-AMP

a _____ pancreatic polypeptide

b _____ adrenal medulla

c _____ gap junctions

d _____ pituitary gland

e _____ second messenger

f _____ hypothalamus

g _____ zona reticularis

h _____ eicosanoids

i _____ gonadotropin

j _____ chief cells

Hormones and System Integration

Although hormones are usually studied (and memorized) individually, extracellular fluids contain a mixture of hormones whose concentrations change daily or even hourly. As a result, cells never respond to only one hormone; instead, they respond to multiple hormones simultaneously. When a cell receives instructions from two hormones at the same time, there are four possible outcomes.

Antagonistic Effects

The two hormones may have **antagonistic** (opposing) **effects**, as in the case of PTH and calcitonin, or insulin and glucagon. The net result depends on the balance between the two hormones. In general, when antagonistic hormones are present, the observed effects are weaker than those produced by either hormone acting unopposed.

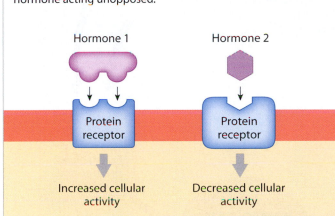

Additive Effects

The two hormones may have **additive effects**, so that the net result is greater than the effect that each would produce acting alone. In some cases, the net result is greater than the sum of the hormones' individual effects. This phenomenon is called a **synergistic effect** (sin-er-JIS-tik; *synairesis*, a drawing together). An example is the enhancement of the glucose-sparing action of GH in the presence of glucocorticoids.

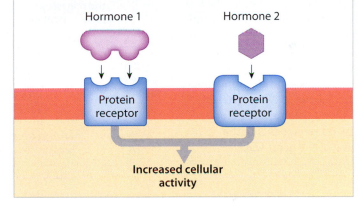

Permissive Effects

One hormone can have a **permissive effect** on another. In such cases, the first hormone is needed for the second to produce its effect. For example, epinephrine does not change the rate of energy consumption in a tissue unless thyroid hormones are also present in normal concentrations.

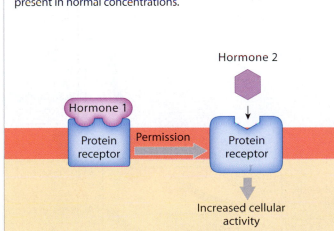

Integrative Effects

Hormones may produce different, but complementary, effects in specific tissues and organs. These **integrative effects** are important in coordinating the activities of diverse physiological systems. One example is the differing effects of calcitriol and parathyroid hormone on tissues involved in calcium metabolism.

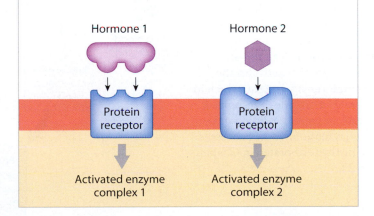

In this section we will consider the ways hormones interact to preserve homeostasis and control short-term or long-term processes.

Long-term regulation of blood pressure, blood volume, and growth involves hormones produced by the endocrine system and by endocrine tissues in other systems

The long-term regulation of blood pressure and blood volume involves not only the pituitary and adrenal glands, but also endocrine cells in the heart and kidneys.

1 The endocrine cells of the heart are located in the heart walls. If blood volume becomes too great, these cells are stretched excessively and begin to secrete **natriuretic peptides** (nā-trē-ū-RET-ik; *natrium*, sodium + *ouresis*, making water). Natriuretic peptides promote the loss of Na⁺ and water at the kidneys, and inhibit renin release and the secretion of ADH and aldosterone. They also suppress thirst and prevent antagonistic hormones from elevating blood pressure. The net result is a reduction in both blood volume and blood pressure, thereby reducing the stretching of the heart walls.

2 If blood volume or blood pressure falls below the normal range, blood flow to the kidneys decreases. Endocrine cells in the kidneys then release a hormone, **erythropoietin** (**EPO**), and an enzyme, **renin**, that activates the **renin-angiotensin system,** which in turn creates a cascade of responses that leads to increased fluid intake and fluid retention. These processes will be further elaborated in Chapters 19 and 23.

Dilation of blood vessels → Reduced blood pressure

Supression of thirst → Reduced fluid intake

Inhibition of antagonistic hormones

Na⁺ and water loss from kidneys → Increased fluid loss

Stretch receptors in the heart cause the release of natriuretic peptides

Rising blood pressure and volume

HOMEOSTASIS DISTURBED — Rising blood pressure and volume

HOMEOSTASIS RESTORED — Falling blood pressure and volume

HOMEOSTASIS — Normal blood pressure and volume

Falling blood pressure and volume

HOMEOSTASIS DISTURBED — Falling blood pressure and volume

HOMEOSTASIS RESTORED — Rising blood pressure and volume

Falling renal blood flow and O₂

Erythropoietin released

Renin released

Activation of renin-angiotensin system

Increased red blood cell production

Increased fluid intake and retention

Aldosterone secreted

ADH secreted

Stimulation of thirst

3 Normal growth requires the cooperation of many endocrine organs. Several hormones—GH, thyroid hormones, insulin, PTH, calcitriol, and reproductive hormones—are especially important, although many others have secondary effects on growth. The circulating concentrations of these hormones are regulated independently. Every time the hormonal mixture changes, metabolic operations are modified to some degree. The modifications vary in duration and intensity, producing unique growth patterns in different individuals.

Insulin

Growing cells need adequate supplies of energy and nutrients. Without insulin, the passage of glucose and amino acids across plasma membranes is drastically reduced or eliminated.

Parathyroid Hormone, Calcitriol, and Calcitonin

Parathyroid hormone (PTH) and calcitriol promote the absorption of calcium salts for subsequent deposition in bone; calcitonin accelerates the rate of deposition. Without adequate levels of both hormones, bones can still enlarge, but they will be poorly mineralized, weak, and flexible.

Thyroid Hormones

Normal growth requires appropriate levels of thyroid hormones. If these hormones are absent during fetal development or the first year of life, the nervous system will not develop normally, and mental retardation will result. If thyroid hormone concentrations decline before puberty, normal skeletal development will not continue.

Reproductive Hormones

The activity of osteoblasts in key locations and the growth of specific cell populations are affected by the presence or absence of reproductive hormones (androgens in males, estrogens in females). These sex hormones stimulate cell growth and differentiation in their target tissues. The targets differ for androgens and estrogens, and the differential growth induced by each accounts for gender-related differences in skeletal proportions and secondary sex characteristics.

5 ft
4 ft
3 ft
2 ft
1 ft

Newborn   Child (6 years)   Adult

Growth Hormone

In Children	In Adults
The effects of GH on protein synthesis and cellular growth are most apparent in children, in whom GH supports muscular and skeletal development.	In adults, growth hormone assists in the maintenance of normal blood glucose concentrations and in the mobilization of lipid reserves stored in adipose tissue. It is not the primary hormone involved, however, and an adult with a GH deficiency but normal levels of thyroid hormones, insulin, and glucocorticoids will have no physiological problems.

Module 16.12 Review

a. Name a peptide secreted by the heart and a hormone released by the kidneys.

b. Identify several hormones necessary for normal growth and development.

c. Explain the action of renin in the bloodstream.

The stress response is a predictable response to any significant threat to homeostasis

Any condition—whether it is physical or emotional—that threatens homeostasis is a form of **stress**. Many stresses are opposed by specific homeostatic adjustments. For example, a decline in body temperature leads to shivering or changes in the pattern of blood flow, which can restore body temperature to normal. In addition, the body has a general response to stress that can occur while other, more specific responses are under way. Exposure to a wide variety of stress-causing factors will produce the same general pattern of hormonal and physiological adjustments. These responses are part of the **stress response**, also known as the **general adaptation syndrome (GAS)**. The stress response is divided into three phases.

1 Alarm Phase ("Fight or Flight")

During the **alarm phase**, an immediate response to the stress occurs. This response is directed by the sympathetic division of the autonomic nervous system. In the alarm phase, (1) energy reserves are mobilized, mainly in the form of glucose, and (2) the body prepares to deal with the stress-causing factor through "fight or flight" responses. Epinephrine is the dominant hormone of the alarm phase, and its secretion accompanies a generalized sympathetic activation.

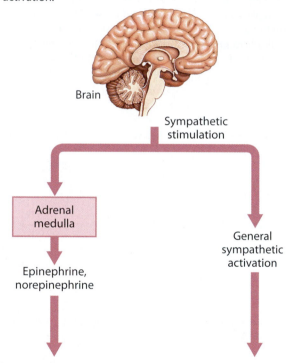

Brain

Sympathetic stimulation

Adrenal medulla

Epinephrine, norepinephrine

General sympathetic activation

Immediate Short-Term Responses to Crises

- Increased mental alertness
- Increased energy use by all cells
- Mobilization of glycogen and lipid reserves
- Changes in circulation
- Reduction in digestive activity and urine production
- Increased sweat gland secretion
- Increased heart rate and respiratory rate

2 Resistance Phase

If a stress lasts longer than a few hours, the individual enters the **resistance phase** of the stress response. Glucocorticoids are the dominant hormones of the resistance phase. Epinephrine, GH, and thyroid hormones are also involved. Energy demands in the resistance phase remain higher than normal, due to the combined effects of these hormones. Neural tissue has a high demand for energy, and neurons must have a reliable supply of glucose. Glycogen reserves are adequate to maintain normal glucose concentrations during the alarm phase but are nearly exhausted after several hours. The hormones of the resistance phase mobilize the body's metabolic reserves while shifting tissue metabolism away from glucose, so that whatever glucose becomes available can be used by neural tissues.

Brain

Sympathetic stimulation

ACTH

Renin-angiotensin system

Adrenal cortex

Pancreas

Mineralocorticoids (with ADH)

Glucocorticoids

Glucagon

Growth hormone

Long-Term Metabolic Adjustments

- Mobilization of remaining energy reserves: Lipids are released by adipose tissue; amino acids are released by skeletal muscle
- Conservation of glucose: Peripheral tissues (except neural) break down lipids to obtain energy
- Elevation of blood glucose concentrations: Liver synthesizes glucose from other carbohydrates, amino acids, and lipids
- Conservation of salts and water, loss of K^+ and H^+

3 Exhaustion Phase

The body's lipid reserves are sufficient to maintain the resistance phase for a period of weeks or even months. But the resistance phase cannot be sustained indefinitely. When the resistance phase ends, homeostatic regulation breaks down and the **exhaustion phase** begins. Unless corrective actions are taken almost immediately, the failure of one or more organ systems will prove fatal. Mineral imbalances contribute to the existing problems with major systems. The production of aldosterone throughout the resistance phase results in a conservation of Na^+ at the expense of K^+. As the body's K^+ content declines, a variety of cells—notably neurons and muscle fibers—begin to malfunction. Although a single cause (such as heart failure) may be listed as the cause of death, the underlying problem is the body's inability to sustain the endocrine and metabolic adjustments of the resistance phase.

Factors That Can Trigger the Exhaustion Phase

- Exhaustion of lipid reserves and the breakdown of structural proteins as the body's primary energy source, damaging vital organs
- Infections that develop due to suppression of inflammation and of the immune response, a secondary effect of the glucocorticoids that are essential to the metabolic activities of the resistance phase
- Cardiovascular damage and complications that are related to the ADH and aldosterone-related elevations in blood pressure and blood volume
- Inability of the adrenal cortex to continue producing glucocorticoids, which results in a failure to maintain acceptable blood glucose concentrations
- Failure to maintain adequate fluid and electrolyte balance

Module 16.13 Review

a. List the three phases of the stress response.

b. Describe the resistance phase.

c. During which phase of the general adaptation syndrome is there a collapse of vital systems?

Endocrine disorders may result from overproduction or underproduction of hormones

Endocrine disorders may develop for a variety of reasons, including abnormalities in the endocrine gland, the endocrine or neural regulatory mechanisms, or the target tissues. For example, a hormone level may rise because its target organs are becoming less responsive, because a tumor has formed among the gland cells, or because something has interfered with the normal feedback control mechanism. When naming endocrine disorders, clinicians use the prefix *hyper-* when referring to excessive hormone production and *hypo-* when referring to inadequate hormone production. Endocrine tumors can result in **hypersecretion**, but its incidence is relatively rare. Most endocrine disorders are the result of problems within the endocrine gland that result in **hyposecretion**, the production of inadequate levels of a particular hormone.

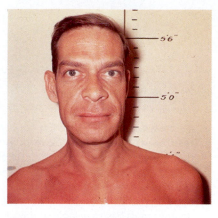

1 **Acromegaly** results from the overproduction of growth hormone after the epiphyseal plates have fused. Bone shapes change and cartilaginous areas of the skeleton enlarge. Note the broad facial features and the enlarged lower jaw.

Common Causes of Hormone Hyposecretion

Metabolic Factors	Physical Damage	Congenital Disorders
Hyposecretion may result from a deficiency in some key substrate needed to synthesize the hormone in question. For example, hypothyroidism can be caused by inadequate levels of iodine in the diet.	Any condition that interrupts the normal circulatory supply to endocrine cells or that physically damages those cells may cause them to become inactive immediately or after an initial surge of hormone release.	An individual may be unable to produce normal amounts of a particular hormone because (1) the gland is too small, (2) the required enzymes are abnormal, (3) the receptors that trigger secretion are relatively insensitive, or (4) the gland cells lack the receptors normally involved in stimulating secretory activity.

2 An enlarged thyroid gland, or **goiter**, is usually associated with thyroid hyposecretion due to nutritional iodine deficiency.

Endocrine abnormalities can also be caused by the presence of abnormal hormonal receptors in target tissues. In such a case, the gland and the regulatory mechanisms involved are normal, but the peripheral cells are unable to respond to the circulating hormone. The best example of this type of abnormality is type 2 diabetes, in which peripheral cells do not respond normally to insulin.

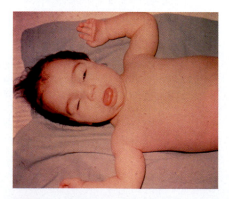

3 **Cretinism** results from thyroid hormone insufficiency in infancy.

An Overview of Endocrine Disorders

Hormone	Results of Under-production or Tissue Insensitivity	Principal Signs and Symptoms	Results of Over-production or Tissue Hypersensitivity	Principal Signs and Symptoms
Growth hormone (GH)	Pituitary growth failure	Retarded growth, abnormal fat distribution, low blood glucose hours after a meal	Gigantism, acromegaly	Excessive growth
Antidiuretic hormone (ADH)	Diabetes insipidus	Polyuria, dehydration, thirst	SIADH (syndrome of inappropriate ADH secretion)	Increased body weight and water content
T$_4$ (thyroxine) and T$_3$	Myxedema, cretinism	Low metabolic rate; low body temperature; impaired physical and mental development	Hyperthyroidism, Grave disease	High metabolic rate and body temperature
Parathyroid hormone (PTH)	Hypoparathyroidism	Muscular weakness, neurological problems, formation of dense bones, tetany due to low blood Ca^{2+} concentrations	Hyperparathyroidism	Neurological, mental, and muscular problems due to high blood Ca^{2+} concentrations; weak and brittle bones
Insulin	Diabetes mellitus (type 1)	High blood glucose, impaired glucose utilization, dependence on lipids for energy; glycosuria	Excess insulin production (also caused by administering too much insulin)	Low blood glucose levels, possibly causing coma
Mineralocorticoids (MCs) Example: aldosterone	Hypoaldosteronism	Polyuria, low blood volume, high blood K$^+$, and low blood Na$^+$ concentrations	Aldosteronism	Increased body weight due to Na$^+$ and water retention; low blood K$^+$ concentration
Glucocorticoids (GCs) Example: cortisol	Addison disease	Inability to tolerate stress, mobilize energy reserves, or maintain normal blood glucose concentrations	Cushing disease	Excessive breakdown of tissue proteins and lipid reserves; impaired glucose metabolism
Epinephrine (E), norepinephrine (NE)	None identified	None identified	Pheochromocytoma	High metabolic rate, body temperature, and heart rate; elevated blood glucose levels
Estrogens (females)	Hypogonadism	Sterility, lack of secondary sex characteristics	Adrenogenital syndrome	Overproduction of androgens by zona reticularis of adrenal cortex; leads to masculinization
			Precocious puberty	Premature sexual maturation and related behavioral changes
Androgens (males)	Hypogonadism	Sterility, lack of secondary sex characteristics	Adrenogenital syndrome (gynecomastia)	Abnormal production of estrogens, sometimes due to adrenal or interstitial cell tumors; leads to breast enlargement
			Precocious puberty	Premature sexual maturation and related behavioral changes

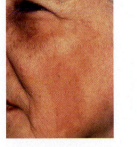

 4 **Addison disease** is caused by hyposecretion of corticosteroids, especially glucocorticoids. Pigment changes result from stimulation of melanocytes by ACTH, which is similar in structure to MSH.

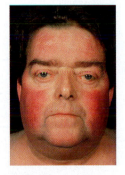

5 **Cushing disease** is caused by hypersecretion of glucocorticoids. Lipid reserves are mobilized, and adipose tissue accumulates in the cheeks and at the base of the neck.

Module 16.14 Review

a. Define the prefixes *hyper-* and *hypo-* in the context of endocrine disorders.

b. Identify three common causes of hormone hyposecretion.

c. What condition is characterized by increased body weight due to Na$^+$ and water retention and a low blood K$^+$ concentration?

1. Matching

Match the following terms with the most closely related description.

- sympathetic activation
- reduce blood pressure and volume
- PTH and calcitonin
- GH and glucocorticoids
- increase blood pressure and volume
- homeostasis threat
- glucocorticoids
- PTH and calcitriol
- gigantism
- protein synthesis

a _____ antagonistic effect

b _____ resistance phase

c _____ alarm phase

d _____ renin and EPO effect

e _____ growth hormone (GH)

f _____ additive effect

g _____ excessive GH in children

h _____ natriuretic peptides effect

i _____ stress

j _____ integrative effect

2. Short answer

Write the phrases at left in the boxes to complete the diagram of the homeostatic regulation of blood pressure and volume.

- increased fluid loss
- erythropoietin released
- reduced blood pressure
- aldosterone secreted
- suppression of thirst
- falling blood pressure and volume
- rising blood pressure and volume
- renin released
- ADH secreted
- release of natriuretic peptides
- Na$^+$ and H$_2$O loss from kidneys
- increased red blood cell production

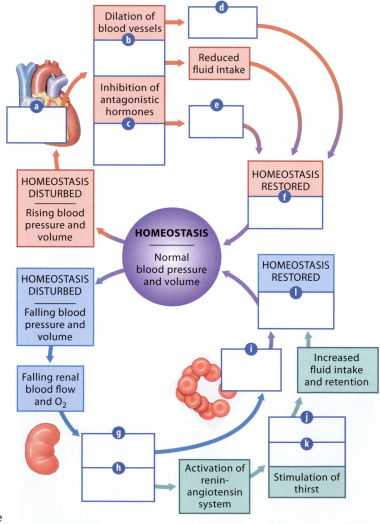

3. Section integration

Describe, and give an example of, the four possible effects that may occur when a cell receives instructions from two different hormones.

Visual Outline with Key Terms

Summarize the content of each module using the terms in the order provided.

SECTION 1

Hormones and Intercellular Communication

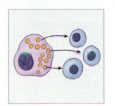

- ○ direct communication
- ○ paracrine communication
- ● paracrine factors
- ○ endocrine communication
- ● hormones
- ○ synaptic communication
- ○ neurotransmitters

16.1

Hormones may be amino acid derivatives, peptides, or lipid derivatives

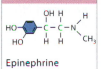

Catecholamines

Epinephrine

- ● amino acid derivatives
- ● peptide hormones
- ● lipid derivatives
- ● catecholamines
- ● prohormones
- ● eicosanoids
- ● leukotrienes
- ● prostaglandins
- ● steroid hormones
- ● endocrine system
- ● hypothalamus
- ● pituitary gland
- ● thyroid gland
- ● adrenal glands
- ● pancreatic islets
- ● pineal gland
- ● parathyroid glands

16.2

Hormones affect target cells after binding to receptors in the plasma membrane, cytoplasm, or nucleus

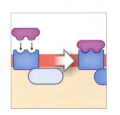

- ● first messenger
- ● second messenger
- ● cyclic-AMP (cAMP)
- ○ calcium ions
- ● G protein
- ● calmodulin
- ○ steroid hormones
- ○ thyroid hormones

16.3

The hypothalamus exerts direct or indirect control over the activities of many different endocrine organs

- ● hypothalamus
- ● regulatory hormones
- ● median eminence
- ● fenestrated capillaries
- ● hypophyseal portal system
- ● portal vessels
- ● antidiuretic hormone (ADH)
- ● oxytocin (OXT)
- ● releasing hormones (RH)
- ● inhibiting hormones (IH)

16.4

The pituitary gland consists of an anterior lobe and a posterior lobe

- ● hypophysis
- ● tropic hormones
- ● infundibulum
- ● adenohypophysis
- ● neurohypophysis
- ● antidiuretic hormone (ADH)
- ● arginine vasopressin (AVP)
- ○ osmoreceptors
- ○ oxytocin (OXT)
- ● neuroendocrine reflex
- ● thyroid-stimulating hormone (TSH)
- ● thyrotropin-releasing hormone (TRH)
- ● adrenocorticotropic hormone (ACTH)
- ● corticotropin-releasing hormone (CRH)
- ● gonadotropins
- ● gonadotropin-releasing hormone (GnRH)
- ● follicle-stimulating hormone (FSH)
- ● estrogens
- ● inhibin
- ● luteinizing hormone (LH)
- ● androgens
- ● growth hormone (GH)
- ● growth hormone-releasing hormone (GH–RH)
- ● growth hormone-inhibiting hormone (GH–IH)
- ● somatomedins
- ● glucose-sparing effect
- ● prolactin (PRL)
- ● prolactin-inhibiting hormone (PIH)
- ● melanocyte-stimulating hormone (MSH)

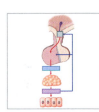

16.5

Negative feedback mechanisms control the secretion rates of the hypothalamus and pituitary gland

- ○ negative feedback
- ● somatomedins

16.6

The thyroid gland contains follicles and requires iodine to produce hormones that stimulate tissue metabolism

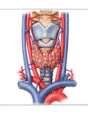

- ● thyroid gland
- ● isthmus
- ● thyroid follicles
- ● thyroglobulin
- ● C (clear) cells
- ● calcitonin (CT)
- ● thyroid hormones (T_4 and T_3)
- ● thyroid-binding globulins (TBGs)

● = *Term boldfaced in this module*

16.7

Parathyroid hormone, produced by the parathyroid glands, is the primary regulator of calcium ion levels in body fluids

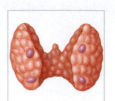

- parathyroid glands
- parathyroid cells
- parathyroid hormone (PTH)
- calcitriol
- kidneys

16.8

The adrenal glands produce hormones involved in metabolic regulation

- adrenal gland
- retroperitoneal
- adrenal cortex
- adrenocortical steroids
- corticosteroids
- zona glomerulosa
- mineralocorticoids (MCs)
- aldosterone
- zona fasciculata
- glucocorticoids (GCs)
- cortisol
- corticosterone
- cortisone
- glucose-sparing effect
- anti-inflammatory effect
- zona reticularis
- androgens
- adrenal medulla
- epinephrine (E)
- norepinephrine (NE)

16.9

The pancreatic islets secrete insulin and glucagon and regulate glucose utilization by most cells

- pancreas
- exocrine pancreas
- endocrine pancreas
- pancreatic islets
- alpha cells
- glucagon
- beta cells
- insulin
- pancreatic acini
- delta cells
- F cells
- GH-IH
- pancreatic polypeptide (PP)

16.10

The pineal gland of the epithalamus secretes melatonin

- pineal gland
- pinealocytes
- melatonin
- free radicals
- circadian rhythms

16.11

Diabetes mellitus is an endocrine disorder characterized by excessively high blood glucose levels

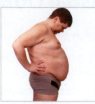

- diabetes mellitus
- hyperglycemia
- glycosuria
- polyuria
- type 1 (insulin dependent) diabetes
- type 2 (non-insulin dependent) diabetes
- diabetic retinopathy
- diabetic nephropathy
- diabetic neuropathy

SECTION 2

Hormones and System Integration

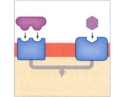

- antagonistic effects
- additive effects
- synergistic effect
- permissive effects
- integrative effects

16.12

Long-term regulation of blood pressure, blood volume, and growth involves hormones produced by the endocrine system and by endocrine tissues in other systems

- natriuretic peptides
- erythropoietin (EPO)
- renin
- renin-angiotensin system
- insulin
- parathyroid hormone
- calcitriol
- thyroid hormones
- reproductive hormones
- growth hormone

16.13

The stress response is a predictable response to any significant threat to homeostasis

- stress
- stress response
- general adaptation syndrome (GAS)
- alarm phase
- resistance phase
- exhaustion phase

16.14

Endocrine disorders may result from overproduction or underproduction of hormones

- hypersecretion
- hyposecretion
- acromegaly
- goiter
- cretinism
- Addison disease
- Cushing disease

• = *Term boldfaced in this module*

Chapter Integration: Applying what you've learned

Sherry, a 43-year-old mother of two, had been experiencing some slightly troublesome physical and mental problems that began about eight months ago. She had been a physically fit 120-lb person with a bright, happy, and cheerful disposition all her adult life. Within the past few months, however, her husband has repeatedly commented about her overreactions to otherwise normal family situations, and that she seemed irritated by the slightest problems. She had also lost a considerable amount of weight, even though her diet and exercise levels had not changed. Because Sherry just wasn't feeling "right," she decided to make an appointment with her family physician.

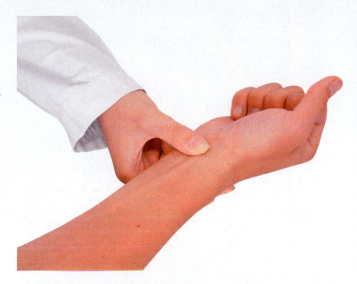

Sherry tells her physician that she has been restless, anxious, and quite irritable lately. She is having a hard time sleeping, complains of diarrhea and weight loss, and has an increased appetite without weight gain. Visual examination of her neck reveals an enlargement of the right anterior neck. During the examination, her physician notices a higher-than-normal heart rate and a fine tremor in her outstretched fingers. Although further clinical tests are necessary, the doctor suspects hyperthyroidism.

1. Why did the physician suspect hyperthyroidism?

2. Explain hyperthyroidism.

3. What tests could the physician perform to make a positive diagnosis of Sherry's condition?

4. Which of Sherry's signs and symptoms indicate nervous system involvement?

Access more review material online in the Study Area at **www.masteringaandp.com.**

There, you'll find:
- **Chapter guides**
- **Chapter quizzes**
- **Practice tests**
- **Labeling activities**
- **MP3 Tutor Sessions**
- **Tutorials**
- **Animations**
- **Flashcards**
- **A glossary with pronunciations**

iP® Use *Interactive Physiology*® (IP) to help you understand difficult physiological concepts in this chapter. Go to **Endocrine System** and find the following topics:
- **Orientation**
- **Endocrine System Review**
- **Biochemistry, Secretion, and Transport of Hormones**
- **The Actions of Hormones on Target Cells**
- **The Hypothalamic-Pituitary Axis**
- **Response to Stress**

17

Blood and Blood Vessels

LEARNING OUTCOMES

These Learning Outcomes correspond by number to this chapter's modules and indicate what you should be able to do after completing the chapter.

SECTION 1 · Blood

17.1 Describe the important components and major properties of blood.

17.2 List the characteristics and functions of red blood cells, and describe the structure and functions of hemoglobin.

17.3 Describe how the components of aged or damaged red blood cells are recycled.

17.4 Explain the importance of blood typing and the basis for ABO and Rh incompatibilities.

17.5 ✚ **CLINICAL MODULE** Describe hemolytic disease of the newborn, explain the clinical significance of the cross-reaction between fetal and maternal blood types, and cite preventive measures.

17.6 Categorize the various types of white blood cells on the basis of their structures and functions.

17.7 Explain the origins and differentiation of the formed elements.

17.8 Discuss the mechanisms that control blood loss after an injury, and describe the reaction sequences responsible for blood clotting.

17.9 ✚ **CLINICAL MODULE** Explain how blood disorders are detected, and describe examples of the various categories of blood disorders.

SECTION 2 · The Functional Anatomy of Blood Vessels

17.10 Distinguish among the types of blood vessels on the basis of their structure and function.

17.11 Describe the structures of capillaries and their functions in the exchange of dissolved materials between blood and interstitial fluid.

17.12 Describe the venous system, and indicate the distribution of blood within the cardiovascular system.

17.13 Identify the major arteries and veins of the pulmonary circuit, and name the areas each serves.

17.14 Identify the major arteries and veins of the systemic circuit, and name the areas each serves.

17.15 Identify the branches of the aortic arch and the tributaries of the superior vena cava, and name the areas each serves.

17.16 Identify the branches of the carotid arteries and the tributaries of the external jugular veins, and name the areas each serves.

17.17 Identify the branches of the internal carotid and vertebral arteries, and the tributaries of the internal jugular veins, and name the areas each serves.

17.18 Identify the branches of the descending aorta and the tributaries of the venae cavae, and name the areas each serves.

17.19 Identify the branches of the visceral arterial vessels and the venous tributaries of the hepatic portal system, and name the areas each serves.

17.20 Identify the branches of the common iliac artery and the tributaries of the common iliac vein, and name the areas each serves.

17.21 ✚ **CLINICAL MODULE** Identify the differences between fetal and adult circulation patterns, and describe the changes in blood flow patterns that occur at birth.

Blood

1 The **cardiovascular system** includes a fluid (blood), a series of conducting tubes (the blood vessels) that distribute the fluid throughout the body, and a pump (the heart) that keeps the fluid in motion.

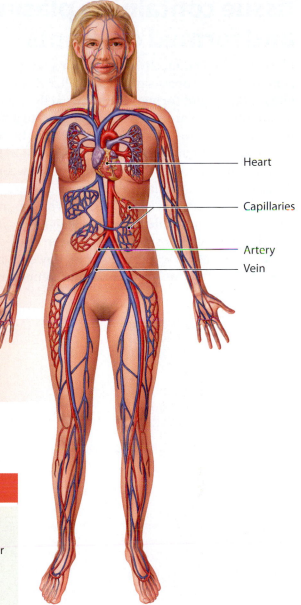

The Components of the Cardiovascular System

THE HEART propels blood and maintains blood pressure

BLOOD VESSELS distribute blood around the body

Capillaries permit diffusion between blood and interstitial fluids

Arteries carry blood away from the heart to the capillaries

Veins return blood from capillaries to the heart

BLOOD distributes oxygen, carbon dioxide, and blood cells; delivers nutrients and hormones; transports waste products; and assists in temperature regulation and defense against disease.

Heart

Capillaries

Artery

Vein

Functions of the Blood

- **Transportation of Dissolved Gases, Nutrients, Hormones, and Metabolic Wastes.** Blood carries oxygen from the lungs to peripheral tissues, and carbon dioxide from those tissues to the lungs. Blood distributes nutrients absorbed at the digestive tract or released from storage in adipose tissue or in the liver. It carries hormones from endocrine glands toward their target cells, and it absorbs and carries the wastes produced by tissue cells to the kidneys for excretion.

- **Regulation of the pH and Ion Composition of Interstitial Fluids.** Diffusion between interstitial fluids and blood eliminates local deficiencies or excesses of ions such as calcium or potassium. Blood also absorbs and neutralizes acids generated by active tissues, such as lactic acid produced by skeletal muscles.

- **Restriction of Fluid Losses at Injury Sites.** Blood contains enzymes and other substances that respond to breaks in vessel walls by initiating the process of clotting. A blood clot acts as a temporary patch that prevents further blood loss.

- **Defense against Toxins and Pathogens.** Blood transports white blood cells, specialized cells that migrate into peripheral tissues to fight infections or remove debris. Blood also transports antibodies, proteins that specifically attack invading organisms or foreign compounds.

- **Stabilization of Body Temperature.** Blood absorbs the heat generated by active skeletal muscles and redistributes it to other tissues. If body temperature is already high, that heat will be lost across the surface of the skin. If body temperature is too low, the warm blood is directed to the brain and to other temperature-sensitive organs.

2 This section examines the structure and function of blood, a fluid with multiple functions (see the table at left) and remarkable properties. In adults, circulating blood provides each of the body's roughly 75 trillion cells a source of nutrients, oxygen, and chemical instructions, and a way of transporting wastes to their sites of removal. The blood also transports specialized cells that defend peripheral tissues from infection and disease. These services are so essential that a body region deprived of circulation dies in a matter of minutes.

Blood is a fluid connective tissue containing plasma and formed elements

Blood is a fluid connective tissue with a unique composition. It consists of **plasma** (PLAZ-muh), a liquid matrix, and **formed elements** (cells and cell fragments). The cardiovascular system of an adult male contains 5–6 liters (5.3–6.4 quarts) of blood; that of an adult female contains 4–5 liters (4.2–5.3 quarts). The difference in blood volume between the sexes primarily reflects differences in average body size. After blood is removed for analysis or storage, the term **whole blood** is used to indicate that the blood composition has not been altered. The components of whole blood can, however, be separated, or fractionated, if only one component is of interest.

1 Plasma forms 46–63 percent of the volume of whole blood. In many respects, the composition of plasma resembles that of interstitial fluid. This similarity exists because water, ions, and small solutes are continuously exchanged between plasma and interstitial fluids across the walls of capillaries. The primary differences between plasma and interstitial fluid involve (1) the levels of respiratory gases (oxygen and carbon dioxide), due to the respiratory activities of tissue cells, and (2) the concentrations and types of dissolved proteins (because plasma proteins cannot cross capillary walls).

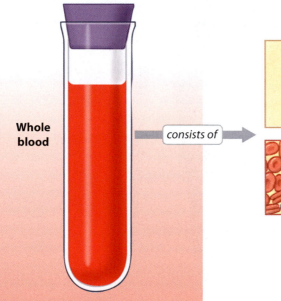

Whole blood — consists of — Plasma (46–63%) + Formed elements (37–54%)

PLASMA COMPOSITION	
Plasma proteins	7%
Other solutes	1%
Water	92%

Transports organic and inorganic molecules, formed elements, and heat

The **hematocrit** (he-MAT-ō-krit) is the percentage of whole blood volume contributed by formed elements, 99.9 percent of which are red blood cells. In adult males, the normal hematocrit, or **packed cell volume (PCV)**, averages 47 (range: 40–54); the average for adult females is 42 (range: 37–47). The difference in hematocrit between the sexes primarily reflects the fact that androgens (male hormones) stimulate red blood cell production, whereas estrogens (female hormones) do not.

FORMED ELEMENTS	
Platelets	< .1%
White blood cells	< .1%
Red blood cells	99.9%

2 Formed elements are blood cells and cell fragments suspended in plasma, and they account for 37–54 percent of the volume of whole blood.

Properties of Whole Blood

- Blood temperature is roughly 38°C (100.4°F), slightly above normal body temperature.
- Blood is five times as viscous as water—that is, five times as resistant to flow. Blood's high viscosity results from interactions among dissolved proteins, formed elements, and water molecules in the plasma.
- Blood is slightly alkaline, with a pH between 7.35 and 7.45 (average: 7.4).

Plasma Proteins

Plasma proteins are in solution rather than forming insoluble fibers like those in other connective tissues, such as loose connective tissue or cartilage. On average, each 100 mL of plasma contains 7.6 g of protein, almost five times the concentration in interstitial fluid. The large size and globular shapes of most blood proteins usually prevent them from leaving the bloodstream. The liver synthesizes and releases more than 90 percent of all plasma proteins.

Albumins (al-BŪ-minz) constitute roughly 60 percent of the plasma proteins. As the most abundant plasma proteins, they are major contributors to the osmotic pressure of plasma.

Globulins (GLOB-ū-linz) account for approximately 35 percent of the proteins in plasma. Important plasma globulins include antibodies and transport globulins. Antibodies, also called **immunoglobulins** (i-mū-nō-GLOB-ū-linz), attack foreign proteins and pathogens. **Transport globulins** bind small ions, hormones, lipids, and other compounds.

Fibrinogen (fi-BRIN-ō-jen) functions in clotting and normally accounts for roughly 4 percent of plasma proteins. Under certain conditions, fibrinogen molecules interact to form large, insoluble strands of **fibrin** (FĪ-brin) that form the basic framework for a blood clot.

The plasma also contains active and inactive enzymes and hormones whose concentrations vary widely.

Other Solutes

Other solutes are generally present in plasma in concentrations similar to those in the interstitial fluids. However, differences in the concentrations of nutrients and waste products can exist between arterial blood and venous blood.

Electrolytes: Normal extracellular ion composition is essential for vital cellular activities. The major plasma electrolytes are Na^+, K^+, Ca^{2+}, Mg^{2+}, Cl^-, HCO_3^-, HPO_4^-, and SO_4^{2-}.

Organic nutrients: Organic nutrients are used for ATP production, growth, and maintenance of cells. This category includes lipids (fatty acids, cholesterol, and glycerides), carbohydrates (primarily glucose), and amino acids.

Organic wastes: Waste products are carried to sites of breakdown or excretion. Examples of organic wastes include urea, uric acid, creatinine, bilirubin, and ammonium ions.

Platelets

Platelets (PLĀT-lets) are small, membrane-bound cell fragments that contain enzymes and other substances important to the process of clotting.

White Blood Cells

White blood cells (**WBCs**), or leukocytes (LOO-kō-sīts; *leukos*, white + *-cyte*, cell), participate in the body's defense mechanisms. There are five classes of leukocytes, each with slightly different functions that will be explored later in the chapter.

Neutrophils

Eosinophils

Basophils

Lymphocytes

Monocytes

Red Blood Cells

Red blood cells (**RBCs**), or erythrocytes (e-RITH-rō-sits; *erythros*, red + *-cyte*, cell), are the most abundant blood cells. These specialized cells are essential for the transport of oxygen in the blood.

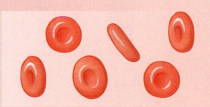

Module 17.1 Review

a. Define hematocrit.

b. Identify the two components constituting whole blood, and list the composition of each.

c. Which specific plasma proteins would you expect to be elevated during an infection?

Red blood cells, the most common formed elements, contain hemoglobin

1 The human body contains an enormous number of red blood cells (RBCs). A standard blood test—the **red blood cell count**—reports the number of RBCs per microliter (µL) of whole blood. In adult males, 1 µL, or 1 cubic millimeter (mm^3), of whole blood contains 4.5–6.3 million RBCs; in adult females, 1 µL contains 4.2–5.5 million. A single drop of whole blood contains approximately 260 million RBCs, and the blood of an average adult has 25 trillion RBCs. RBCs thus account for roughly one-third of all cells in the human body.

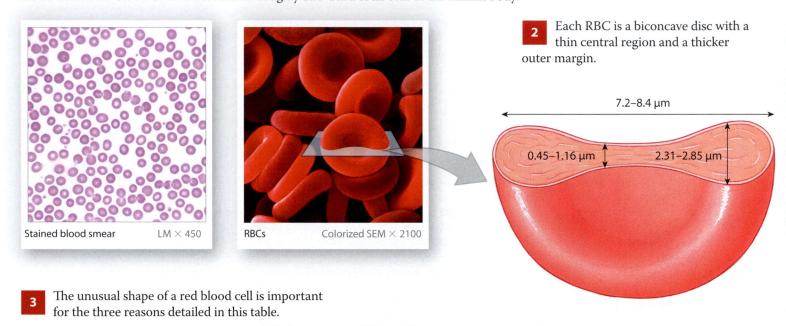

Stained blood smear — LM × 450

RBCs — Colorized SEM × 2100

2 Each RBC is a biconcave disc with a thin central region and a thicker outer margin.

7.2–8.4 µm

0.45–1.16 µm   2.31–2.85 µm

3 The unusual shape of a red blood cell is important for the three reasons detailed in this table.

Functional Aspects of Red Blood Cells

- **Large surface area-to-volume ratio.** Each RBC carries oxygen bound to hemoglobin, an intracellular protein, and that oxygen must be absorbed or released quickly as the RBC passes through the capillaries. The greater the surface area per unit volume, the faster the exchange between the RBC's interior and the surrounding plasma. The total surface area of all the RBCs in the blood of a typical adult is about 3800 square meters, roughly 2000 times the total surface area of the body.

- **RBCs can form stacks.** Like dinner plates, RBCs can form stacks that ease the flow through narrow blood vessels. An entire stack can pass along a blood vessel only slightly larger than the diameter of a single RBC, whereas individual cells would bump the walls, bang together, and form logjams that could restrict or prevent blood flow.

- **Flexibility.** Red blood cells are very flexible and can bend and flex when entering small capillaries and branches. By changing shape, individual RBCs can squeeze through capillaries as narrow as 4 µm.

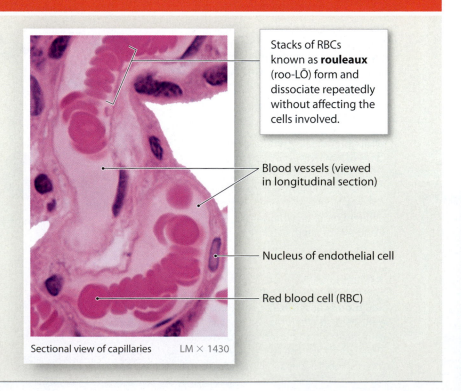

Stacks of RBCs known as **rouleaux** (roo-LŌ) form and dissociate repeatedly without affecting the cells involved.

Blood vessels (viewed in longitudinal section)

Nucleus of endothelial cell

Red blood cell (RBC)

Sectional view of capillaries — LM × 1430

4 A red blood cell is very different from the "typical cell" we discussed in Chapter 3. As our RBCs develop, they lose most of their organelles, including nuclei; they retain only the cytoskeleton. Because mature RBCs lack nuclei and ribosomes, they cannot divide or synthesize structural proteins or enzymes. As a result, RBCs cannot repair themselves, and their life span is normally less than 120 days. In effect, a developing RBC loses any organelle not directly associated with this primary function: the transport of respiratory gases. That function is performed by molecules of **hemoglobin (Hb)**, which account for more than 95 percent of an RBC's intracellular proteins. The hemoglobin content of whole blood is reported in grams of Hb per deciliter (100 mL) of whole blood (g/dL). Normal ranges are 14–18 g/dL in males and 12–16 g/dL in females.

5 Hemoglobin has a complex quaternary structure. Each Hb molecule has two **alpha (α) chains** and two **beta (β) chains** of polypeptides. Each chain is a globular protein subunit that resembles the myoglobin in skeletal and cardiac muscle cells. Like myoglobin, each Hb chain contains a single molecule of **heme**, a non-protein pigment complex.

α chain 1

β chain 1

β chain 2

Heme

α chain 2

6 Each heme unit holds an iron ion in such a way that the iron can interact with an oxygen molecule, forming **oxyhemoglobin, HbO₂**. Blood containing RBCs filled with oxyhemoglobin is bright red. The iron–oxygen interaction is very weak; the two can easily dissociate without damaging the heme unit or the oxygen molecule. The binding of an oxygen molecule to the iron in a heme unit is therefore completely reversible. A hemoglobin molecule whose iron is not bound to oxygen is called **deoxyhemoglobin**. Blood containing RBCs filled with deoxyhemoglobin is dark red—almost burgundy.

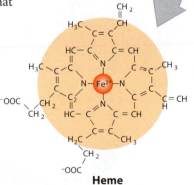

Heme

Each red blood cell contains about 280 million Hb molecules. Because a Hb molecule contains four heme units, each RBC can potentially carry more than a billion molecules of oxygen at a time. Roughly 98.5 percent of the oxygen carried by the blood travels through the bloodstream bound to Hb molecules inside RBCs; the rest is dissolved in the plasma.

Module 17.2 Review

a. Define rouleaux.

b. Describe hemoglobin.

c. Compare oxyhemoglobin with deoxyhemoglobin.

Red blood cells are continuously produced and recycled

About 1 percent of the circulating RBCs are replaced each day, and in the process approximately 3 million new RBCs enter the bloodstream each second! Such a rapid rate of replacement is necessary because a typical RBC has a relatively short life span. After it travels about 700 miles in 120 days, either its plasma membrane ruptures or it is engulfed by macrophages in the liver, spleen, or bone marrow. The continuous elimination of RBCs usually goes unnoticed, as long as new ones enter the bloodstream at a comparable rate.

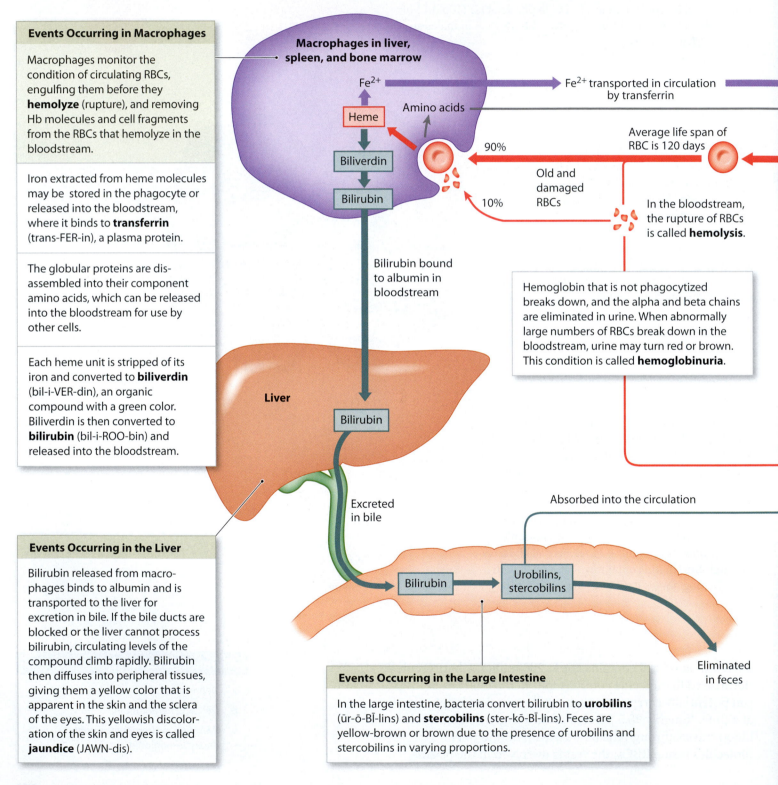

Events Occurring in Macrophages

Macrophages monitor the condition of circulating RBCs, engulfing them before they **hemolyze** (rupture), and removing Hb molecules and cell fragments from the RBCs that hemolyze in the bloodstream.

Iron extracted from heme molecules may be stored in the phagocyte or released into the bloodstream, where it binds to **transferrin** (trans-FER-in), a plasma protein.

The globular proteins are dis-assembled into their component amino acids, which can be released into the bloodstream for use by other cells.

Each heme unit is stripped of its iron and converted to **biliverdin** (bil-i-VER-din), an organic compound with a green color. Biliverdin is then converted to **bilirubin** (bil-i-ROO-bin) and released into the bloodstream.

Macrophages in liver, spleen, and bone marrow

Fe^{2+}

Heme

Amino acids

Biliverdin

Bilirubin

Fe^{2+} transported in circulation by transferrin

90%

Average life span of RBC is 120 days

Old and damaged RBCs

10%

In the bloodstream, the rupture of RBCs is called **hemolysis**.

Hemoglobin that is not phagocytized breaks down, and the alpha and beta chains are eliminated in urine. When abnormally large numbers of RBCs break down in the bloodstream, urine may turn red or brown. This condition is called **hemoglobinuria**.

Bilirubin bound to albumin in bloodstream

Liver

Bilirubin

Excreted in bile

Absorbed into the circulation

Bilirubin

Urobilins, stercobilins

Eliminated in feces

Events Occurring in the Liver

Bilirubin released from macro-phages binds to albumin and is transported to the liver for excretion in bile. If the bile ducts are blocked or the liver cannot process bilirubin, circulating levels of the compound climb rapidly. Bilirubin then diffuses into peripheral tissues, giving them a yellow color that is apparent in the skin and the sclera of the eyes. This yellowish discolor-ation of the skin and eyes is called **jaundice** (JAWN-dis).

Events Occurring in the Large Intestine

In the large intestine, bacteria convert bilirubin to **urobilins** (ūr-ō-BĪ-lins) and **stercobilins** (ster-kō-BĪ-lins). Feces are yellow-brown or brown due to the presence of urobilins and stercobilins in varying proportions.

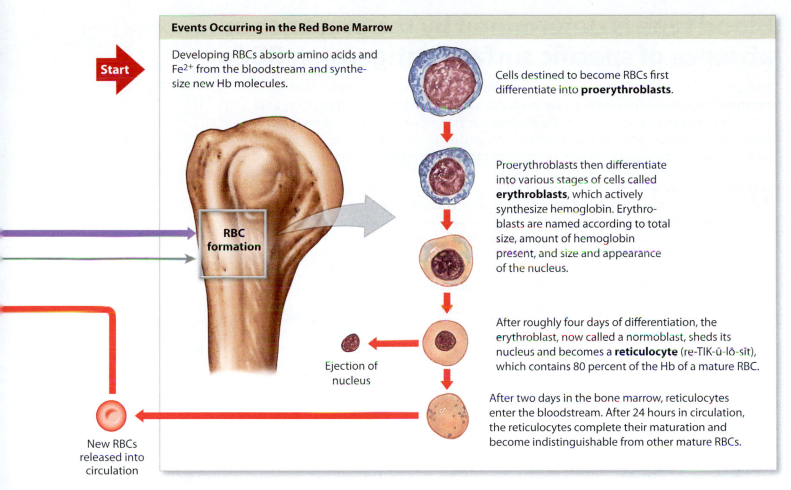

Events Occurring in the Red Bone Marrow

Start

Developing RBCs absorb amino acids and Fe²⁺ from the bloodstream and synthesize new Hb molecules.

RBC formation

Ejection of nucleus

New RBCs released into circulation

Cells destined to become RBCs first differentiate into **proerythroblasts**.

Proerythroblasts then differentiate into various stages of cells called **erythroblasts**, which actively synthesize hemoglobin. Erythroblasts are named according to total size, amount of hemoglobin present, and size and appearance of the nucleus.

After roughly four days of differentiation, the erythroblast, now called a normoblast, sheds its nucleus and becomes a **reticulocyte** (re-TIK-ū-lō-sīt), which contains 80 percent of the Hb of a mature RBC.

After two days in the bone marrow, reticulocytes enter the bloodstream. After 24 hours in circulation, the reticulocytes complete their maturation and become indistinguishable from other mature RBCs.

In adults, blood cell formation, or **erythropoiesis** (e-rith-rō-poy-Ē-sis), occurs only in red bone marrow, or **myeloid** (MĪ-e-loyd) **tissue** (*myelos*, marrow). This tissue is located in portions of the vertebrae, sternum, ribs, skull, scapulae, pelvis, and proximal limb bones. Other marrow areas contain a fatty tissue known as yellow bone marrow. Under extreme stimulation, such as severe and sustained blood loss, areas of yellow marrow can convert to red marrow, increasing the rate of RBC formation.

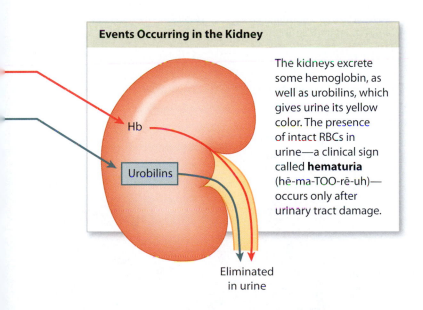

Events Occurring in the Kidney

The kidneys excrete some hemoglobin, as well as urobilins, which gives urine its yellow color. The presence of intact RBCs in urine—a clinical sign called **hematuria** (hē-ma-TOO-rē-uh)—occurs only after urinary tract damage.

Hb

Urobilins

Eliminated in urine

Module 17.3 Review

a. Define hemolysis.

b. Identify the products formed during the breakdown of heme.

c. In what way would a liver disease affect the level of bilirubin in the blood?

Blood type is determined by the presence or absence of specific surface antigens on RBCs

Antigens are substances that can trigger a protective defense mechanism called an **immune response**. Most antigens are proteins, although some other types of organic molecules are antigens as well. The plasma membranes of your cells contain **surface antigens**, substances your immune system recognizes as "normal" or "self." In other words, your immune system ignores these substances rather than attacking them as "foreign."

1 Your **blood type** is a classification determined by specific surface antigens in RBC plasma membranes. These surface antigens are genetically determined membrane glycoproteins or glycolipids. Your immune system ignores the surface antigens on your own RBCs, However, your plasma may contain antibodies that will attack the antigens on "foreign" RBCs. Although red blood cells have at least 50 kinds of surface antigens, three surface antigens are of particular importance: A, B, and Rh (or D). There are four blood types based on the presence or absence of the A and B surface antigens.

Type A	Type B	Type AB	Type O
Type A blood has RBCs with surface antigen A only.	**Type B** blood has RBCs with surface antigen B only.	**Type AB** blood has RBCs with both A and B surface antigens.	**Type O** blood has RBCs lacking both A and B surface antigens.
Surface antigen A	Surface antigen B		
If you have Type A blood, your plasma contains anti-B antibodies, which will attack Type B surface antigens.	If you have Type B blood, your plasma contains anti-A antibodies.	If you have Type AB blood, your plasma has neither anti-A nor anti-B antibodies.	If you have Type O blood, your plasma contains both anti-A and anti-B antibodies.

2 If surface antigens on RBCs of one blood type are exposed to the corresponding antibodies from another blood type, the RBCs will **agglutinate**, or clump together; this process is called **agglutination**. The cells may also undergo hemolysis. Such a **cross-reaction** is very dangerous because clumps and fragments of RBCs can plug small blood vessels in the kidneys, lungs, heart, or brain, damaging or destroying affected tissues. Accidental cross-reactions may occur if a person being treated for severe blood loss is accidentally given a transfusion of the wrong blood type.

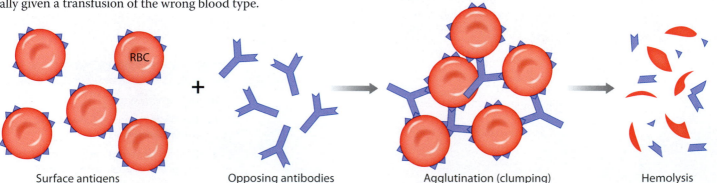

Surface antigens   +   Opposing antibodies   →   Agglutination (clumping)   →   Hemolysis

RBC

3 As this table indicates, the various blood types are not evenly distributed throughout the population. The term **Rh positive** (**Rh⁺**) indicates the presence of the **Rh** surface antigen, also called the **D** antigen. The absence of this antigen is indicated as Rh negative (Rh⁻). When the complete blood type is recorded, "Rh" is usually omitted and blood type is reported as O negative (O^-), A positive (A^+), and so on.

Differences in Blood Type Distribution

Population	Percentage with Each Blood Type				
	O	**A**	**B**	**AB**	**Rh⁺**
U.S. (Average)	46	40	10	4	85
African American	49	27	20	4	95
Caucasian	45	40	11	4	85
Chinese American	42	27	25	6	100
Filipino American	44	22	29	6	100
Hawaiian	46	46	5	3	100
Japanese American	31	39	21	10	100
Korean American	32	28	30	10	100
Native North American	79	16	4	<1	100
Native South American	100	0	0	0	100
Australian Aborigine	44	56	0	0	100

4 Shown here are the results of blood typing tests on blood samples from four individuals. Drops taken from the sample at the left are mixed with solutions containing antibodies to the surface antigens A, B, and D (Rh). Clumping (agglutination) occurs when the sample contains the corresponding surface antigen or antigens. The blood type for each individual is shown at right. Because cross-reactions, or **transfusion reactions**, are so dangerous, care must be taken to ensure that the blood types of a blood donor and the recipient are **compatible**—that is, the donor's blood cells and the recipient's plasma will not cross-react.

Anti-A	Anti-B	Anti-D	Blood type
			A⁺
			B⁺
			AB⁺
			O⁻

The presence of anti-A and/or anti-B antibodies is genetically determined and remains constant throughout life, regardless of whether the individual has ever been exposed to foreign RBCs. In contrast, the plasma of an Rh-negative individual does not necessarily contain anti-Rh antibodies. These antibodies are present only if the individual has been sensitized by previous exposure to Rh-positive RBCs.

Module 17.4 Review

a. What is determined by the surface antigens on RBCs?

b. Which blood type(s) can be safely transfused into a person with Type O blood?

c. Why can't a person with Type A blood safely receive blood from a person with Type B blood?

Hemolytic disease of the newborn is an RBC-related disorder caused by a cross-reaction between fetal and maternal blood types

Genes controlling the presence or absence of any surface antigen in the plasma membrane of a red blood cell are provided by both parents, so a child can have a blood type different from that of either parent. During pregnancy, when fetal and maternal vascular systems are closely intertwined, the mother's antibodies against RBC surface antigens may cross the placenta, attacking and destroying fetal RBCs. The resulting condition, called **hemolytic disease of the newborn (HDN)**, has many forms, some so mild as to remain undetected. Those involving the Rh surface antigen are quite dangerous, because unlike anti-A and anti-B antibodies, anti-Rh antibodies are able to cross the placenta and enter the fetal bloodstream.

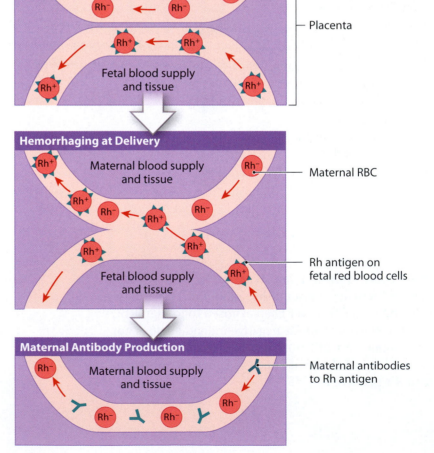

First Pregnancy of an Rh⁻ Mother with an Rh⁺ Infant

The most common form of hemolytic disease of the newborn develops after an Rh⁻ woman has carried an Rh⁺ fetus.

1 Problems seldom develop during a first pregnancy, because very few fetal cells enter the maternal circulation then, and thus the mother's immune system is not stimulated to produce anti-Rh antibodies.

2 Exposure to fetal red blood cell antigens generally occurs during delivery, when bleeding takes place at the placenta and uterus. Such mixing of fetal and maternal blood can stimulate the mother's immune system to produce anti-Rh antibodies, leading to **sensitization**.

3 Roughly 20 percent of Rh⁻ mothers who carried Rh⁺ children become sensitized within 6 months of delivery. Because the anti-Rh antibodies are not produced in significant amounts until after delivery, a woman's first infant is not affected.

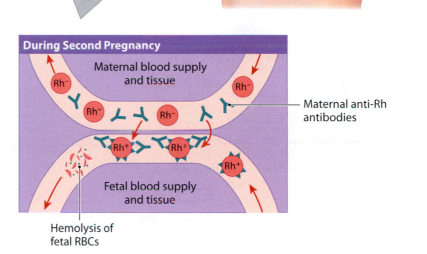

Rh⁻
mother

Rh⁺
fetus

Second Pregnancy of an Rh⁻ Mother with an Rh⁺ Infant

4 If a subsequent pregnancy involves an Rh⁺ fetus, maternal anti-Rh antibodies produced after the first delivery cross the placenta and enter the fetal bloodstream. These antibodies destroy fetal RBCs, producing a dangerous anemia. The fetal demand for blood cells increases, and they begin leaving the bone marrow and entering the bloodstream before completing their development. Because these immature RBCs are erythroblasts, HDN is also known as **erythroblastosis fetalis** (e-rith-rō-blas-TŌ-sis fē-TAL-is). Without treatment, the fetus will probably die before delivery or shortly thereafter. A newborn with severe HDN is anemic, and the high concentration of circulating bilirubin produces jaundice. Because the maternal antibodies remain active in the newborn for one to two months after delivery, the infant's entire blood volume may require replacement to remove the maternal anti-Rh antibodies, as well as the damaged RBCs. Fortunately, the mother's anti-Rh antibody production can be prevented if anti-Rh antibodies (available under the name RhoGAM) are administered to the mother in weeks 26–28 of pregnancy and during and after delivery. These antibodies destroy any fetal RBCs that cross the placenta before they can stimulate a maternal immune response. Because maternal sensitization does not occur, no anti-Rh antibodies are produced. In the United States, this relatively simple procedure has almost entirely eliminated HDN mortality caused by Rh incompatibilities.

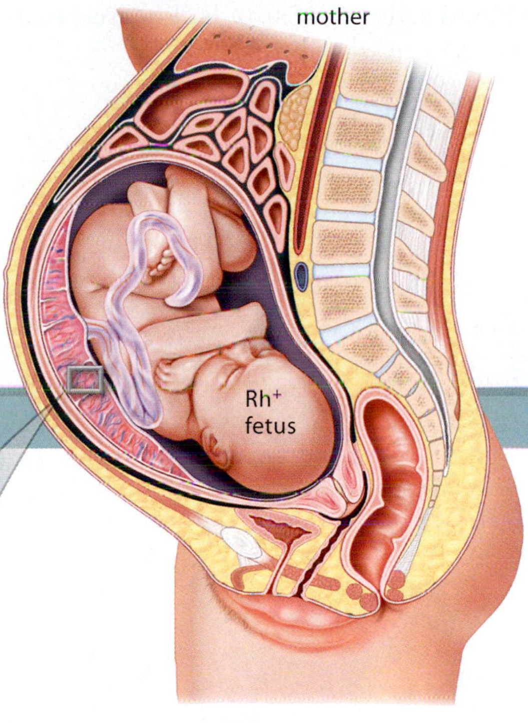

During Second Pregnancy

Maternal blood supply
and tissue

Rh⁻

Rh⁻

Rh⁻

Maternal anti-Rh
antibodies

Rh⁺

Rh⁺

Rh⁺

Fetal blood supply
and tissue

Hemolysis of
fetal RBCs

Module 17.5 Review

a. Define hemolytic disease of the newborn (HDN).

b. Why is RhoGAM administered to Rh⁻ mothers?

c. Does an Rh⁺ mother carrying an Rh⁻ fetus require a RhoGAM injection? Explain your answer.

White blood cells defend the body against pathogens, toxins, cellular debris, and abnormal or damaged cells

White blood cells (WBCs), or **leukocytes,** share the properties described below.

1 **Granular leukocytes** have abundant cytoplasmic granules that absorb histological stains, such as Wright stain or Giemsa stain.

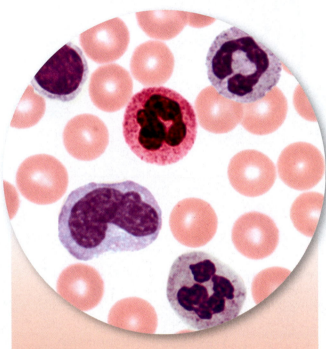

WBCs can be divided into two classes

GRANULAR LEUKOCYTES
Neutrophil
Eosinophil
Basophil

AGRANULAR LEUKOCYTES
Monocyte
Lymphocyte

Shared Properties of WBCs

- WBCs circulate for only a short portion of their life span, using the bloodstream primarily to travel between organs and to rapidly reach areas of infection or injury. White blood cells spend most of their time migrating through loose and dense connective tissues throughout the body.

- All WBCs can migrate out of the bloodstream. When circulating white blood cells in the bloodstream become activated, they contact and adhere to the vessel walls and squeeze between adjacent endothelial cells to enter the surrounding tissue. This process is called **emigration,** or **diapedesis** (*dia*, through; *pedesis*, a leaping).

- All WBCs are attracted to specific chemical stimuli. This characteristic, called **positive chemotaxis** (kē-mō-TAK-sis), guides WBCs to invading pathogens, damaged tissues, and other active WBCs.

- Neutrophils, eosinophils, and monocytes are capable of phagocytosis. These phagocytes can engulf pathogens, cell debris, or other materials. Macrophages are monocytes that have moved out of the bloodstream and have become actively phagocytic.

2 **Agranular leukocytes** have few, if any, cytoplasmic granules that absorb histological stain.

White Blood Cells

Cell	Abundance (Average Number per µL)	Appearance in a Stained Blood Smear	Functions	Remarks
Neutrophils	4150 (range: 1800–7300) Differential count: 50–70%	Round cell; nucleus lobed and may resemble a string of beads; cytoplasm contains large, pale inclusions	Phagocytic: engulf pathogens or debris in injured or infected tissues; release cytotoxic enzymes and chemicals	Move into tissues after several hours; may survive for minutes to days, depending on tissue activity; produced in red bone marrow
Eosinophils	165 (range: 0–700) Differential count: 2–4%	Round cell; nucleus generally has two lobes; cytoplasm contains large granules that generally stain bright red	Phagocytic: engulf antibody-labeled materials; release cytotoxic enzymes; reduce inflammation; increase in abundance in allergies and parasitic infections	Move into tissues after several hours; may survive for minutes to days, depending on activity in tissues; produced in red bone marrow
Basophils	44 (range: 0–150) Differential count: <1%	Round cell; nucleus generally cannot be seen through dense, blue-stained granules in cytoplasm	Enter damaged tissues and release histamine and other chemicals that promote inflammation	Survival time unknown; assist mast cells of tissues in producing inflammation; produced in red bone marrow
Monocytes	456 (range: 200–950) Differential count: 2–8%	Very large cell; nucleus kidney bean–shaped; abundant cytoplasm	Enter tissues and become macrophages; engulf pathogens or debris	Move into tissues after 1–2 days; survive for months or longer; produced primarily in red bone marrow
Lymphocytes	2185 (range: 1500–4000) Differential count: 20–30%	Generally round cell, slightly larger than RBC; round nucleus; very little cytoplasm	Cells of lymphatic system; provide defense against specific pathogens or toxins	Survive for months to decades; circulate from blood to tissues and back; produced in red bone marrow and lymphoid tissues

A variety of conditions, including pathogenic infections, inflammation, and allergic reactions, cause characteristic changes in the populations of circulating WBCs. Examining a stained blood smear can provide a **differential count** of the WBC population. The values reported indicate the number of each type of cell in a sample of 100 WBCs.

Module 17.6 Review

a. Identify the five types of white blood cells.

b. Which type of white blood cell would you find in the greatest numbers in an infected cut?

c. How do basophils respond during inflammation?

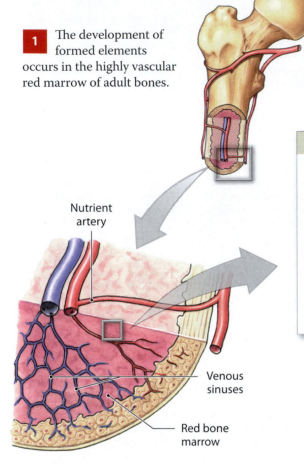

Formed elements are produced by stem cells derived from hemocytoblasts

1 The development of formed elements occurs in the highly vascular red marrow of adult bones.

Nutrient artery

Venous sinuses

Red bone marrow

Hemocytoblasts

Hemocytoblasts (*hemo-*, blood + *cyte*, cell + *blastos*, precursor), or multipotent stem cells, are found in the red bone marrow of adults. Their divisions give rise to two types of stem cells responsible for producing all formed elements.

Lymphoid Stem Cells

Lymphoid stem cells, which are responsible for the production of lymphocytes, originate in the red bone marrow. Some remain there, while others migrate from the bone marrow to other **lymphoid tissues,** including the thymus, spleen, and lymph nodes. As a result, lymphocytes are produced in these organs as well as in the red bone marrow.

Myeloid Stem Cells

Myeloid stem cells are stem cells in red bone marrow that divide to give rise to all types of formed elements other than lymphocytes.

2 The term "formed elements" is appropriate because platelets are cell fragments, rather than specialized cells. This table summarizes important information about platelets.

Structure and Function of Platelets

Appearance in a Stained Blood Smear	Abundance (Average Number per µL)	Function	Remarks
Platelets (PLĀT-lets) are flattened discs that appear round when viewed from above, and spindle-shaped in section or in a blood smear.	350,000 (range: 150,000–500,000)	Platelets clump together and stick to damaged vessel walls, and they release chemicals that stimulate blood clotting.	Platelets are continuously replaced. Each platelet circulates for 9–12 days before being removed by phagocytes, mainly in the spleen.

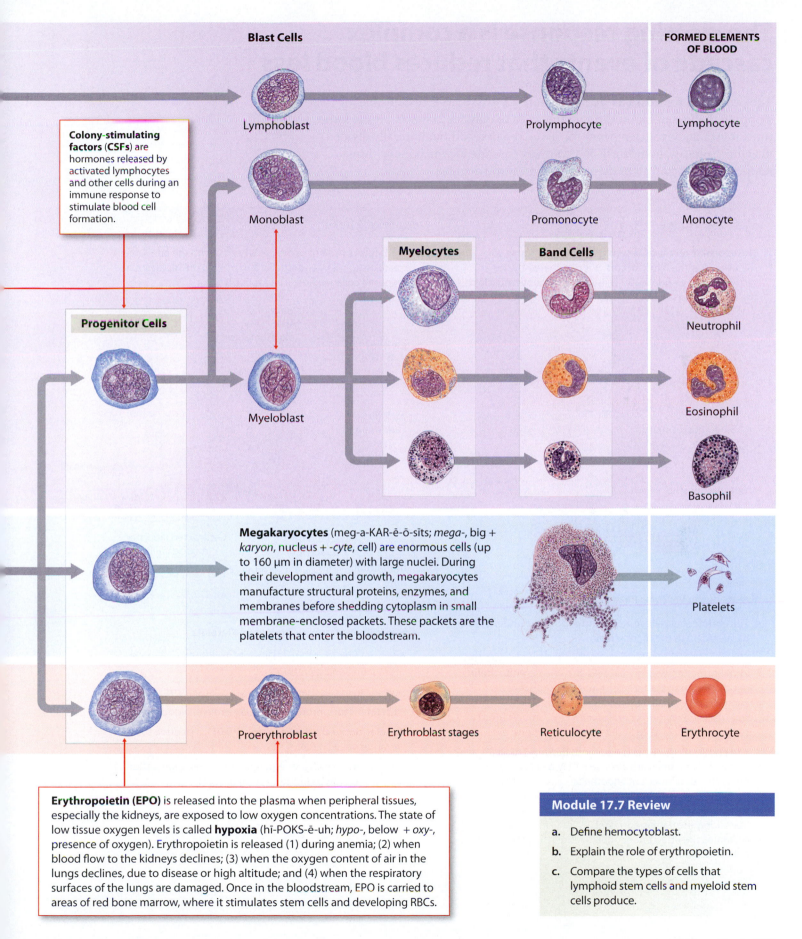

Blast Cells

FORMED ELEMENTS OF BLOOD

Lymphoblast → Prolymphocyte → Lymphocyte

Colony-stimulating factors (CSFs) are hormones released by activated lymphocytes and other cells during an immune response to stimulate blood cell formation.

Monoblast → Promonocyte → Monocyte

Myelocytes

Band Cells

Progenitor Cells

Myeloblast → Neutrophil

→ Eosinophil

→ Basophil

Megakaryocytes (meg-a-KAR-ē-ō-sīts; *mega-*, big + *karyon*, nucleus + -*cyte*, cell) are enormous cells (up to 160 μm in diameter) with large nuclei. During their development and growth, megakaryocytes manufacture structural proteins, enzymes, and membranes before shedding cytoplasm in small membrane-enclosed packets. These packets are the platelets that enter the bloodstream.

Platelets

Proerythroblast → Erythroblast stages → Reticulocyte → Erythrocyte

Erythropoietin (EPO) is released into the plasma when peripheral tissues, especially the kidneys, are exposed to low oxygen concentrations. The state of low tissue oxygen levels is called **hypoxia** (hī-POKS-ē-uh; *hypo-*, below + *oxy-*, presence of oxygen). Erythropoietin is released (1) during anemia; (2) when blood flow to the kidneys declines; (3) when the oxygen content of air in the lungs declines, due to disease or high altitude; and (4) when the respiratory surfaces of the lungs are damaged. Once in the bloodstream, EPO is carried to areas of red bone marrow, where it stimulates stem cells and developing RBCs.

Module 17.7 Review

a. Define hemocytoblast.

b. Explain the role of erythropoietin.

c. Compare the types of cells that lymphoid stem cells and myeloid stem cells produce.

The clotting response is a complex cascade of events that reduces blood loss

The process of **hemostasis** (*haima*, blood + *stasis*, halt) is responsible for stopping the loss of blood through the walls of damaged vessels. At the same time, it establishes a framework for tissue repairs. Although usually divided into three phases, hemostasis is a complex cascade in which many things happen at once, and all of them interact to some degree.

Vascular Phase

The **vascular phase** of hemostasis lasts for roughly 30 minutes after the injury occurs. It is dominated by the response of the endothelial cells and the smooth muscle of the vessel walls.

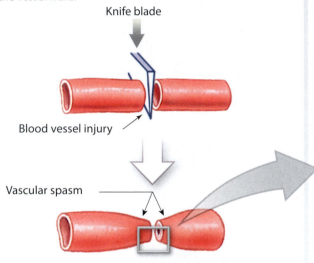

Events of the Vascular Phase

- The endothelial cells contract and expose the underlying basal lamina to the bloodstream.

- The endothelial cells begin releasing chemical factors and local hormones. Endothelial cells also release **endothelins**, peptide hormones that (1) stimulate smooth muscle contraction and promote vascular spasms and (2) stimulate the division of endothelial cells, smooth muscle cells, and fibroblasts to accelerate the repair process.

- The endothelial plasma membranes become "sticky." A tear in the wall of a small artery or vein may be partially sealed off by the attachment of endothelial cells on either side of the break. In small capillaries, endothelial cells on opposite sides of the vessel may stick together and prevent blood flow along the damaged vessel. The stickiness also facilitates the attachment of platelets as the platelet phase gets under way.

Platelet Phase

The **platelet phase** of hemostasis begins with the attachment of platelets to sticky endothelial surfaces, to the basal lamina, to exposed collagen fibers, and to each other.

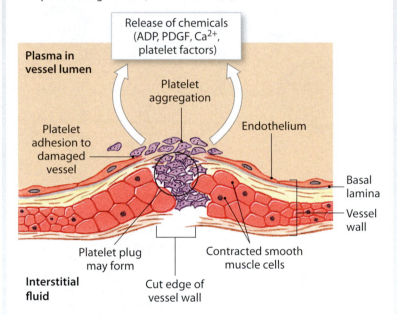

Chemicals Released by Activated Platelets

- Adenosine diphosphate (ADP), which stimulates platelet aggregation and secretion

- Several chemicals that stimulate vascular spasms

- **Platelet factors**, proteins that play a role in blood clotting

- **Platelet-derived growth factor (PDGF)**, a peptide that promotes vessel repair

- Calcium ions, which are required for platelet aggregation and in several steps in the clotting process

Coagulation Phase

The **coagulation** (cō-ag-ū-LĀ-shun) **phase** of hemostasis does not start until 30 seconds or more after the vessel has been damaged. **Coagulation**, or blood clotting, involves a complex sequence of steps leading to the conversion of circulating fibrinogen into the insoluble protein **fibrin**. As the fibrin network grows, blood cells and additional platelets are trapped in the fibrous tangle, forming a blood clot that seals off the damaged portion of the vessel. **Procoagulants** (clotting factors) in the plasma play a key role in this phase. Important clotting factors include Ca^{2+} and 11 different proteins (identified by Roman numerals). Many clotting factors are proenzymes, which, when converted to active enzymes, direct essential reactions in the clotting response. The activation of one proenzyme commonly creates a chain reaction, or **cascade**.

Common Pathway

The **common pathway** begins when enzymes from either the extrinsic or intrinsic pathway activate Factor X, forming the enzyme **prothrombinase**. Prothrombinase converts the proenzyme **prothrombin** into the enzyme **thrombin** (THROM-bin). Thrombin then completes the clotting process by converting fibrinogen, a soluble plasma protein, to insoluble strands of fibrin.

Extrinsic Pathway

The **extrinsic pathway** begins with the release of **tissue factor** (Factor III) by damaged endothelial cells or peripheral tissues. The greater the damage, the more tissue factor is released and the faster clotting occurs. Tissue factor then combines with Ca^{2+} and another clotting factor to form an enzyme complex capable of activating Factor X, the first step in the common pathway.

Intrinsic Pathway

The **intrinsic pathway** begins with the activation of proenzymes exposed to collagen fibers at the injury site. This pathway proceeds with the assistance of **PF-3**, a platelet factor released by aggregating platelets. After a series of linked reactions, activated clotting factors combine to form an enzyme complex capable of activating Factor X.

Factor X — Prothrombinase — Prothrombin → Thrombin — Fibrinogen → Fibrin

Tissue factor complex — Clotting factor (VII) — Ca^{2+} — Tissue factor (Factor III)

Factor X activator complex — Clotting factors (VIII, IX) — Ca^{2+} — Platelet factor (PF-3) — Activated proenzymes (usually Factor XII)

Tissue damage

Contracted smooth muscle cells

Blood clot containing trapped RBCs SEM × 1200

Clot Retraction

Once the fibrin meshwork has formed, platelets and red blood cells stick to the fibrin strands. The platelets then contract, and the entire clot begins to undergo **clot retraction**, a process that continues over a period of 30–60 minutes and pulls the cut edges together.

As repairs proceed, the clot gradually dissolves. This process, called **fibrinolysis** (fī-bri-NOL-i-sis), begins with the activation of the proenzyme **plasminogen** by thrombin, produced by the common pathway, and **tissue plasminogen activator (t-PA)**, released by damaged tissues. The activation of plasminogen produces the enzyme **plasmin** (PLAZ-min), which erodes the foundation of the clot.

Module 17.8 Review

a. Define hemostasis.

b. Briefly describe the vascular, platelet, and coagulation phases of hemostasis.

c. Provide the correct sequence for the following list of events involved in the process of hemostasis: 1. coagulation; 2. fibrinolysis; 3. vascular spasm; 4. retraction; 5. platelet phase.

Blood disorders can be classified by their origins and the changes in blood characteristics

1 The procedure called **venipuncture** (VĒN-i-punk-chur; *vena*, vein + *punctura*, a piercing) is crucial in diagnosing blood disorders. Fresh whole blood is generally collected from a superficial vein, such as the median cubital vein on the anterior surface of the elbow. Venipuncture is commonly used because (1) superficial veins are easy to locate, (2) the walls of veins are thinner than those of comparably sized arteries, and (3) blood pressure in the venous system is relatively low, so the puncture wound seals quickly. The most common clinical procedures examine venous blood.

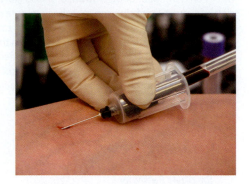

2 Blood disorders can have a variety of causes. The major categories of blood disorders are introduced here.

Nutritional Blood Disorders

In **iron deficiency anemia**, normal hemoglobin synthesis cannot occur because iron reserves or the dietary intake of iron are inadequate. Because developing RBCs cannot synthesize functional hemoglobin, as a result they are unusually small (**microcytic**). Women are especially dependent on a normal dietary supply of iron because their iron reserves are about one-half that of a typical man.

A deficiency in **vitamin B_{12}** prevents normal stem cell divisions in the bone marrow, which can result in **pernicious** (per-NISH-us) **anemia**. Fewer red blood cells are produced, and those that are produced are abnormally large (**macrocytic**) and may develop bizarre shapes.

Calcium ions and **vitamin K** affect almost every aspect of the clotting process. All three pathways (intrinsic, extrinsic, and common) require Ca^{2+}, so any disorder that lowers plasma Ca^{2+} concentrations can impair blood clotting. Additionally, the liver requires adequate amounts of vitamin K for synthesizing four of the clotting factors, including prothrombin.

Congenital Blood Disorders

Sickle cell anemia results from a mutation affecting the amino acid sequence of the beta chains of the Hb molecule. Affected RBCs take on a sickled shape when they release bound oxygen. This makes the RBCs fragile and easily damaged. Moreover, a sickled RBC can become stuck in a narrow capillary. A circulatory blockage results, and nearby tissues become starved for oxygen. To develop sickle cell anemia, an individual must have two copies of the sickling gene— one from each parent. If only one sickling gene is present, the individual has the **sickling trait**. In such cases, most of the hemoglobin is of the normal form, and the RBCs function normally. However, having the sickling trait gives an individual some resistance to malaria, a deadly mosquito-borne parasitic disease. Infection of RBCs by the parasites induces sickling, and sickled cells containing the parasites are engulfed and destroyed by macrophages.

Normal and sickled RBCs

Hemophilia (hē-mō-FĒ-lē-a) is an inherited bleeding disorder. About 1 person in 10,000 is a hemophiliac, and of those, 80–90 percent are males. In most cases, hemophilia is caused by the reduced production of a single clotting factor. The severity of hemophilia varies, depending on how little clotting factor is produced. In severe cases, extensive bleeding occurs with relatively minor contact, and bleeding occurs at joints and around muscles.

The **thalassemias** (thal-ah-SĒ-mē-uhs) are a diverse group of inherited blood disorders caused by an inability to produce adequate amounts of normal protein subunits of hemoglobin. The severity of different types of thalassemias depends on which and how many protein subunits are abnormal.

Infections of the Blood

Blood is normally free of microorganisms, but they can enter the blood through a wound or infection. **Bacteremia** is a condition in which bacteria circulate in blood, but do not multiply there; **viremia** is a similar condition associated with viruses. **Sepsis** (SEP-sis) is a widespread pathogenic infection of body tissues. Sepsis of the blood, or **septicemia** (formerly known as "blood poisoning"), results if pathogens are present and multiplying in the blood, and spreading throughout the body.

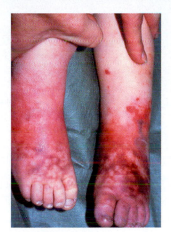

Infant with septicemia from a meningococcal bacteria

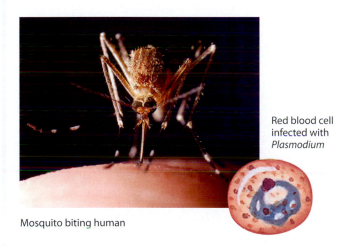

Mosquito biting human

Red blood cell infected with *Plasmodium*

Malaria is a parasitic disease caused by several species of the protozoan *Plasmodium*. It is one of the most severe diseases in tropical countries, killing 1.5–3 million people per year, of which up to half are children under the age of 5. Malaria is transmitted from person to person by a mosquito. The parasite initially infects liver cells, then enlarges and fragments into smaller forms that infect red blood cells. Periodically, at intervals of 2–3 days, all of the infected RBCs rupture simultaneously and release more parasites that infect additional RBCs. The release and reinfection of RBCs corresponds to the cycles of fever and chills that characterize malaria. Dead RBCs can block blood vessels leading to vital organs, such as the kidney and brain, resulting in tissue death.

Cancers of the Blood

The **leukemias** are cancers of blood-forming tissues. The cancerous cells of leukemia do not form a compact tumor, but instead are spread throughout the body from their origin in red bone marrow. Both of the two types of leukemia—myeloid and lymphoid—are characterized by elevated levels of circulating WBCs. **Myeloid leukemia** is characterized by the presence of abnormal granulocytes (neutrophils, eosinophils, and basophils) or other cells of the bone marrow. **Lymphoid leukemia** involves lymphocytes and their stem cells. The first symptoms appear when immature and abnormal white blood cells appear in the circulation. Untreated leukemia is invariably fatal.

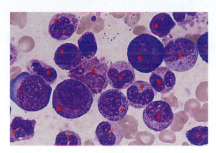

Abnormal WBCs (✱) seen in a blood smear of a patient with myeloid leukemia

Degenerative Blood Disorders

In **disseminated intravascular coagulation (DIC)**, bacterial toxins activate several steps in the coagulation process that converts fibrinogen to fibrin within the circulating blood. Although much of the fibrin is removed by phagocytes or is dissolved by plasmin, small clots may block small vessels and damage nearby tissues. If the liver cannot produce enough circulating fibrinogen to keep pace with the rate at which fibrinogen is being removed, clotting abilities decline and uncontrolled bleeding may occur.

Module 17.9 Review

a. Define venipuncture.

b. Identify the two types of leukemia.

c. Compare pernicious anemia with iron deficiency anemia.

1. Concept map

Use the terms in the following list to complete the whole blood concept map.

- globulins
- electrolytes, glucose, urea
- protein
- albumins
- leukocytes
- fibrinogen
- solutes
- basophils
- erythrocytes
- formed elements
- plasma
- eosinophils
- lymphocytes
- water
- platelets
- monocytes
- neutrophils

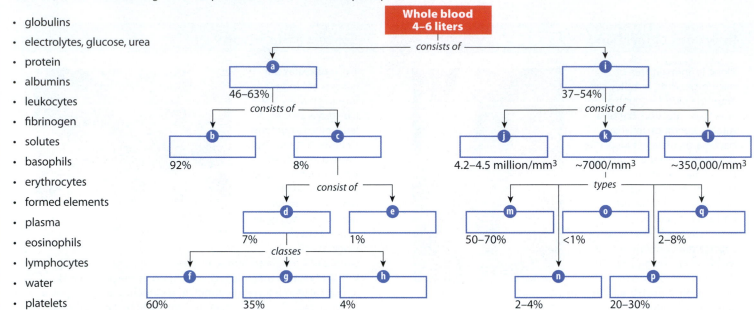

2. Matching

Match the following terms with the most closely related description.

- matrix
- transport protein
- jaundice
- venipuncture
- red bone marrow
- mature RBCs
- pigment complex
- platelets
- erythropoietin
- cross-reaction
- monocytes
- lymphocytes

a	_____	myeloid tissue
b	_____	anucleated
c	_____	plasma
d	_____	macrophages
e	_____	globulin
f	_____	agglutination
g	_____	specific immunity
h	_____	bilirubin
i	_____	median cubital vein
j	_____	heme
k	_____	hormone
l	_____	blood clotting

3. Section integration

Why are mature red blood cells in humans incapable of protein synthesis and mitosis?

_____

_____

The Functional Anatomy of Blood Vessels

Blood flows through a network of blood vessels that extend between the heart and peripheral tissues. Those blood vessels can be organized into a **pulmonary circuit**, which carries blood to and from the gas exchange surfaces of the lungs, and a **systemic circuit**, which transports blood to and from the rest of the body. Each circuit begins and ends at the heart, and blood travels through these circuits in sequence. Thus, blood returning to the heart from the systemic circuit must complete the pulmonary circuit before reentering the systemic circuit. Blood is carried away from the heart by **arteries**, or efferent vessels, and returns to the heart by way of **veins**, or afferent vessels. Microscopic thin-walled vessels called **capillaries** interconnect the smallest arteries and the smallest veins. Capillaries are called exchange vessels, because their thin walls permit the exchange of nutrients, dissolved gases, and waste products between the blood and surrounding tissues.

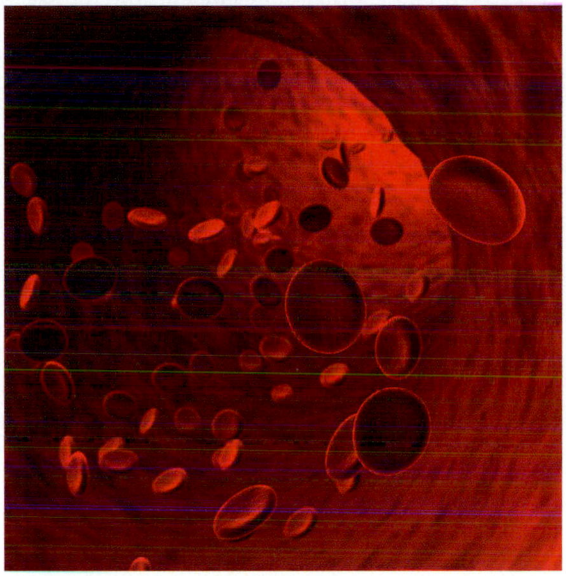

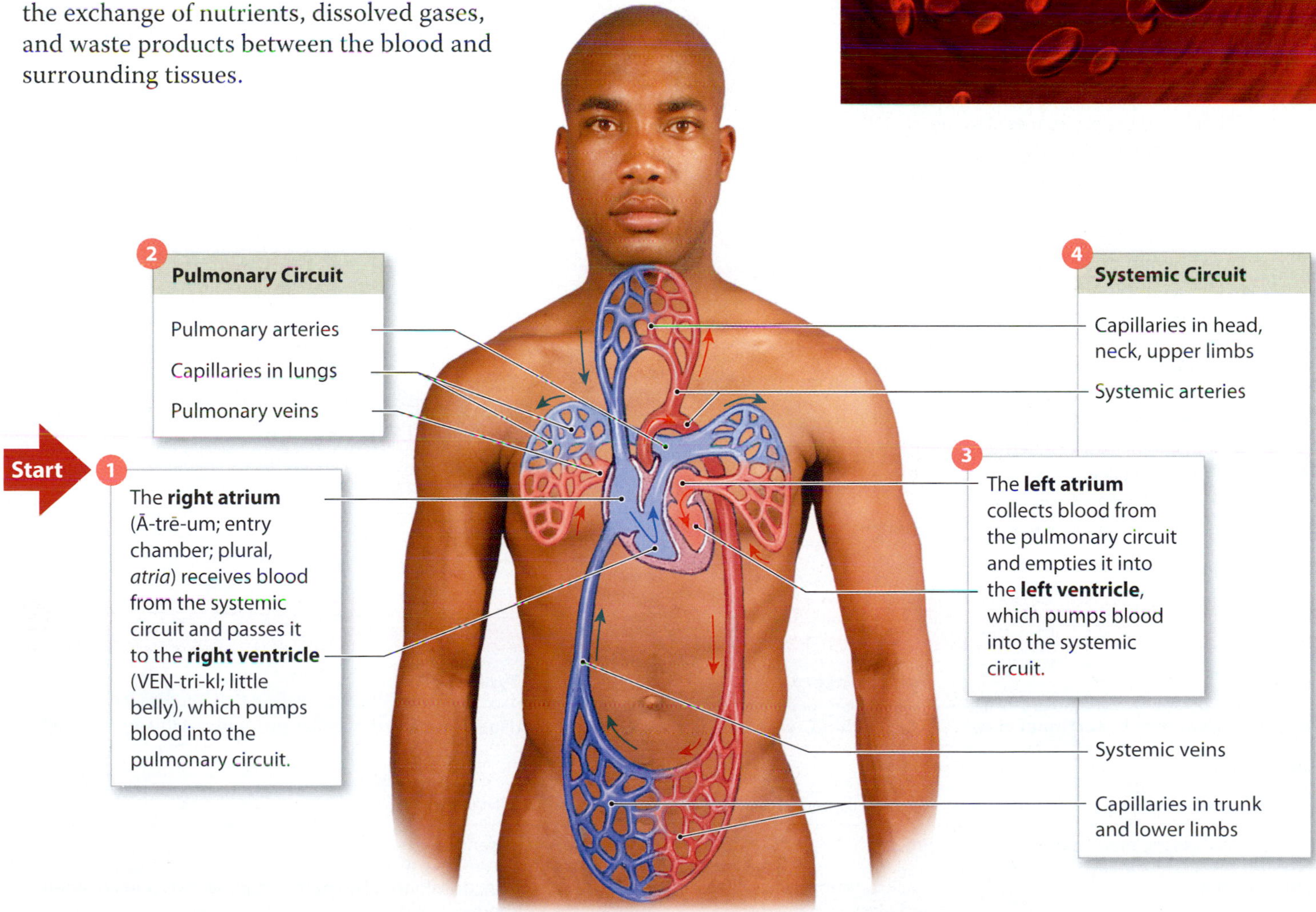

2 Pulmonary Circuit

Pulmonary arteries

Capillaries in lungs

Pulmonary veins

Start

1 The **right atrium** (Ā-trē-um; entry chamber; plural, *atria*) receives blood from the systemic circuit and passes it to the **right ventricle** (VEN-tri-kl; little belly), which pumps blood into the pulmonary circuit.

4 Systemic Circuit

Capillaries in head, neck, upper limbs

Systemic arteries

3 The **left atrium** collects blood from the pulmonary circuit and empties it into the **left ventricle**, which pumps blood into the systemic circuit.

Systemic veins

Capillaries in trunk and lower limbs

In this section we will take a close look at the structure and distribution of the major blood vessels in the body.

17.10

Arteries and veins differ in the structure and thickness of their walls

1 The walls of arteries and veins contain three distinct layers: the tunica intima, tunica media, and tunica externa.

The **tunica intima** (IN-ti-muh), or tunica interna, is the innermost layer of a blood vessel. This layer includes the endothelial lining and an underlying layer of connective tissue containing elastic fibers. In arteries, the outer margin of the tunica intima contains a thick layer of elastic fibers called the **internal elastic membrane**.

The **tunica media**, the middle layer, contains concentric sheets of smooth muscle tissue in a framework of loose connective tissue. When these smooth muscles contract, the vessel decreases in diameter; this is called **vasoconstriction**. When the smooth muscles relax, the diameter increases; this is called **vasodilation**. Collagen fibers bind the tunica media to the tunica intima and tunica externa.

The **tunica externa** (eks-TER-nuh) or tunica adventitia, the outermost layer of a blood vessel, is a connective tissue sheath. In arteries, this layer contains collagen fibers with scattered bands of elastic fibers. In veins, it is generally thicker than the tunica media and contains networks of elastic fibers and bundles of smooth muscle cells. The connective tissue fibers of the tunica externa typically blend into those of adjacent tissues, stabilizing and anchoring the blood vessel.

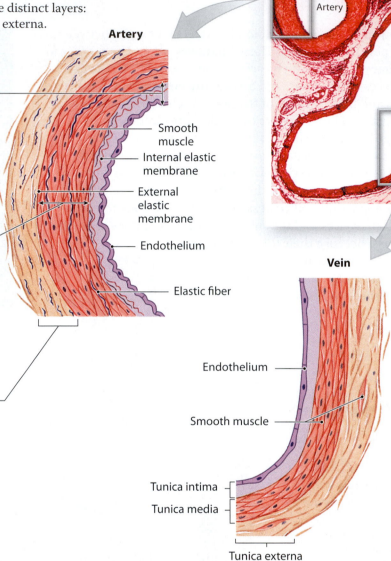

Artery

- Smooth muscle
- Internal elastic membrane
- External elastic membrane
- Endothelium
- Elastic fiber

Artery

Vein

LM × 60

Vein

- Endothelium
- Smooth muscle
- Tunica intima
- Tunica media
- Tunica externa

Features of Typical (Medium-Sized) Arteries and Veins

Feature	Typical Artery	Typical Vein
General appearance in sectional view	Usually round, with relatively thick wall	Usually flattened or collapsed, with relatively thin wall
Tunica Intima		
Endothelium	Usually rippled, due to vessel constriction	Often smooth
Internal elastic membrane	Present	Absent
Tunica Media	Thick, dominated by smooth muscle cells and elastic fibers	Thin, dominated by smooth muscle cells and collagen fibers
External elastic membrane	Present	Absent
Tunica Externa	Collagen and elastic fibers	Collagen, elastic fibers, and smooth muscle cells

Large Vein

Large veins include the superior and inferior venae cavae and their tributaries. All three vessel wall layers are present in all large veins. The slender tunica media is surrounded by a thick tunica externa composed of a mixture of elastic and collagen fibers.

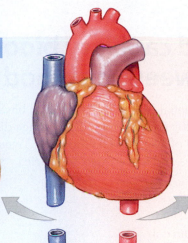

- Tunica externa
- Tunica media
- Tunica intima

Elastic Artery

Elastic arteries are large vessels that transport blood away from the heart. The pulmonary trunk and aorta, as well as their major arterial branches, are elastic arteries. These are resilient, elastic vessels capable of stretching and recoiling as the heart beats and arterial pressures change.

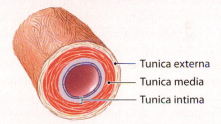

- Internal elastic layer
- Tunica intima
- Tunica media
- Tunica externa

Medium-sized Vein

Medium-sized veins range from 2 to 9 mm in internal diameter. In these veins, the tunica media is thin and contains smooth muscle cells and collagen fibers. The thickest layer is the tunica externa, which contains smooth muscle cells and longitudinal bundles of elastic and collagen fibers.

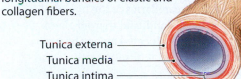

- Tunica externa
- Tunica media
- Tunica intima

Muscular Artery

Muscular arteries, or medium-sized arteries, distribute blood to the body's skeletal muscles and internal organs.

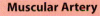

- Tunica externa
- Tunica media
- Tunica intima

Venule

Venules collect blood from capillary beds and are the smallest venous vessels. Venules smaller than 50 µm lack a tunica media and resemble expanded capillaries.

- Tunica externa
- Endothelium

Arteriole

Arterioles have a poorly defined tunica externa, and the tunica media consists of only one or two layers of smooth muscle cells.

- Smooth muscle cells
- Endothelium

Capillaries

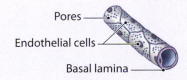

- Pores
- Endothelial cells
- Basal lamina

Capillaries are the only blood vessels whose walls permit exchange between the blood and the surrounding interstitial fluids. Because capillary walls are thin, diffusion distances are short, so exchange can occur quickly. Capillary structure is examined in greater detail in Module 17.11.

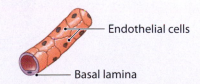

- Endothelial cells
- Basal lamina

---

2 There are five general classes of blood vessels in the cardiovascular system. **Arteries** carry blood away from the heart. As arteries enter peripheral tissues they branch repeatedly, and the branches decrease in diameter. The smallest arterial branches are called arterioles (ar-TĒR-ē-ōls). From the arterioles, blood moves into capillaries, where diffusion occurs between blood and interstitial fluid. From the capillaries, blood enters small venules (VEN-ūls), which unite to form larger **veins** that return blood to the heart.

Module 17.10 Review

a. List the five general classes of blood vessels.

b. Describe a capillary.

c. A cross section of tissue shows several small, thin-walled vessels with very little smooth muscle tissue in the tunica media. Which type of vessels are these?

Capillary structure and capillary blood flow affect the rates of exchange between the blood and interstitial fluid

A typical capillary consists of a tube of endothelial cells within a delicate basal lamina; neither a tunica media nor a tunica externa is present. The average diameter of a capillary is a mere 8 µm, very close to that of a single red blood cell. Two major types of capillaries exist: continuous capillaries and fenestrated capillaries.

Continuous Capillary

1 In a **continuous capillary**, the endothelium is a complete lining. A cross section through a large continuous capillary cuts across several endothelial cells. In a small continuous capillary, a single endothelial cell may completely encircle the lumen. Continuous capillaries are located throughout the body, in all tissues except epithelia and cartilage. Continuous capillaries permit the diffusion of water, small solutes, and lipid-soluble materials into the surrounding interstitial fluid, but prevent the loss of blood cells and plasma proteins. In addition, some exchange may occur between blood and interstitial fluid through selective vesicular transport. In specialized continuous capillaries throughout most of the central nervous system and in the thymus, the endothelial cells are bound together by tight junctions. Permeability in these capillaries is both restricted and precisely regulated.

Fenestrated Capillary

2 **Fenestrated** (FEN-es-trā-ted) **capillaries** (*fenestra*, window) are capillaries that contain "windows," or pores, that penetrate the endothelial lining. The pores permit the rapid exchange of water and solutes as large as small peptides between plasma and interstitial fluid. Examples of fenestrated capillaries include the choroid plexus of the brain and the capillaries of the hypothalamus, pituitary gland, pineal gland, and thyroid gland. Fenestrated capillaries are also located along absorptive areas of the intestinal tract and at filtration sites in the kidneys.

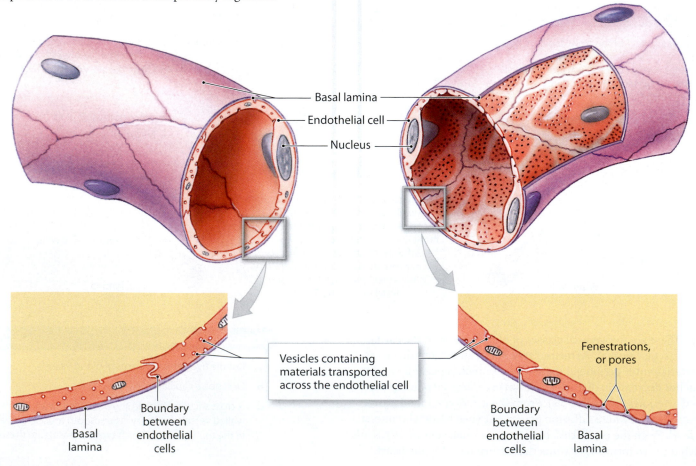

Basal lamina — Endothelial cell — Nucleus

Vesicles containing materials transported across the endothelial cell

Fenestrations, or pores

Basal lamina — Boundary between endothelial cells

Boundary between endothelial cells — Basal lamina

3 **Sinusoids** (SĪ-nuh-soydz) resemble fenestrated capillaries that are flattened and irregularly shaped. In contrast to fenestrated capillaries, sinusoids commonly have gaps between adjacent endothelial cells, and the basal lamina is either thinner or absent. As a result, sinusoids permit the free exchange of water and solutes as large as plasma proteins between the slow-moving blood and interstitial fluid. Sinusoids occur in the liver, bone marrow, spleen, and many endocrine organs, including the pituitary and adrenal glands.

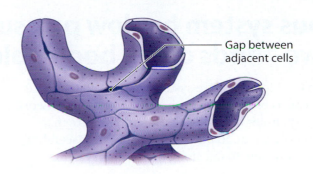

Gap between adjacent cells

A capillary bed may receive blood from more than one artery. These **collateral arteries** enter the region and fuse before giving rise to arterioles. The fusion of two collateral arteries that supply a capillary bed is an example of an **arterial anastomosis**. An arterial anastomosis is in effect an insurance policy: If one artery is compressed or blocked, capillary circulation will continue.

A single arteriole generally gives rise to dozens of capillaries that empty into several venules.

The entrance to each capillary is guarded by a band of smooth muscle called a **precapillary sphincter**. Contraction or relaxation of the smooth muscle cells changes the diameter of the capillary entrance, thereby controlling the flow of blood through the sphincter.

An **arteriovenous** (ar-tēr-ē-ō-VĒ-nus) **anastomosis** is a direct connection between an arteriole and a venule. When this anastomosis is dilated, blood will bypass the capillary bed and flow directly into the venous circulation. The pattern of blood flow through these anastomoses is regulated primarily by sympathetic innervation under the control of the cardiovascular centers of the medulla oblongata.

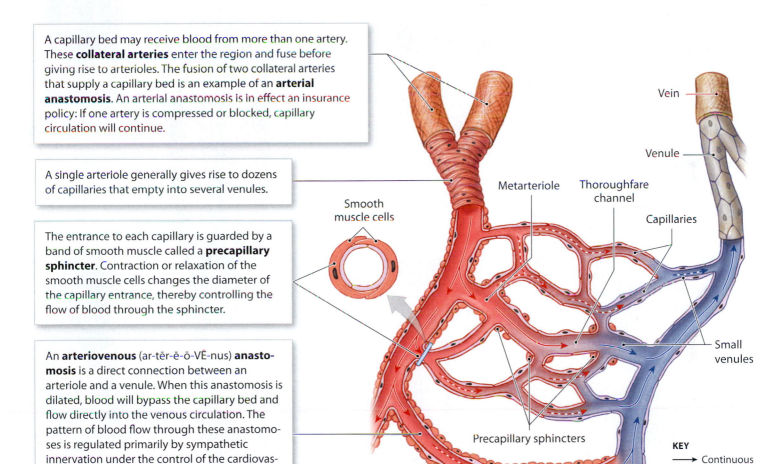

Vein

Venule

Smooth muscle cells

Metarteriole

Thoroughfare channel

Capillaries

Small venules

Precapillary sphincters

KEY

→ Continuous blood flow

---> Variable blood flow

4 Capillaries function as part of an interconnected network called a **capillary bed.** A capillary bed contains several relatively direct connections between arterioles and venules. The wall in the initial part of such a passageway possesses smooth muscle capable of changing its diameter. This segment is called a **metarteriole** (met-ar-TĒR-ē-ōl). The rest of the passageway, which resembles a typical capillary in structure, is called a **thoroughfare channel**. Although blood normally flows at a constant rate from the arteriole to the venule across the capillary bed, the flow within each capillary is quite variable. Bands of smooth muscle at the entrance to each capillary—called precapillary sphincters—alternately contract and relax, perhaps a dozen times per minute. As a result, the blood flow within the associated capillary occurs in pulses rather than as a steady and constant stream. The cycling of contraction and relaxation of smooth muscles that changes blood flow through capillary beds is called **vasomotion**.

Module 17.11 Review

a. Identify the two types of capillaries.

b. At what sites in the body are fenestrated capillaries located?

c. Why do capillaries permit the diffusion of materials, whereas arteries and veins do not?

The venous system has low pressures and contains almost two-thirds of the body's blood volume

The arterial system is a high-pressure system: Almost all the force developed by the heart is required to push blood along the network of arteries and through miles of capillaries. Blood pressure in a peripheral venule is only about 10 percent of that in the ascending aorta, and pressures continue to fall along the venous system.

1 The blood pressure in venules and medium-sized veins is so low that it cannot overcome the force of gravity. In the limbs, veins of this size contain **valves**, folds of the tunica intima that project from the vessel wall and point in the direction of blood flow. Venous valves (like valves in the heart) permit blood flow in one direction only, thereby preventing the backflow of blood toward the capillaries. If the walls of the veins near the valves weaken or become stretched and distorted, the valves may no longer work properly. Blood then pools in the veins, which become grossly distended. The effects range from mild discomfort and a cosmetic problem, as in superficial **varicose veins** in the thighs and legs, to painful distortion of adjacent tissues, as in the **hemorrhoids** that form in venous networks of the anal canal.

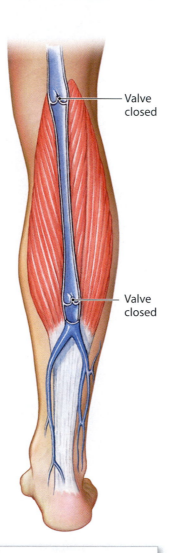

Valve closed

Valve closed

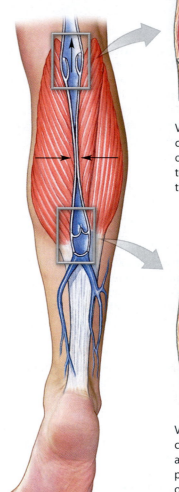

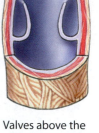

Valves above the contracting muscle open, allowing blood to move toward the heart.

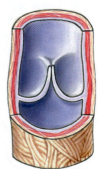

Valves below the contracting muscle are forced closed, preventing backflow of blood to the capillaries.

When you are standing, venous blood from your feet must overcome the pull of gravity to ascend to the heart. Valves compartmentalize the blood within the veins, thereby dividing the weight of the blood between the compartments.

Any contraction of the surrounding skeletal muscles squeezes the blood toward the heart. Although you are probably not aware of it, when you stand, rapid cycles of contraction and relaxation are occurring within your leg muscles, helping to push blood toward the trunk.

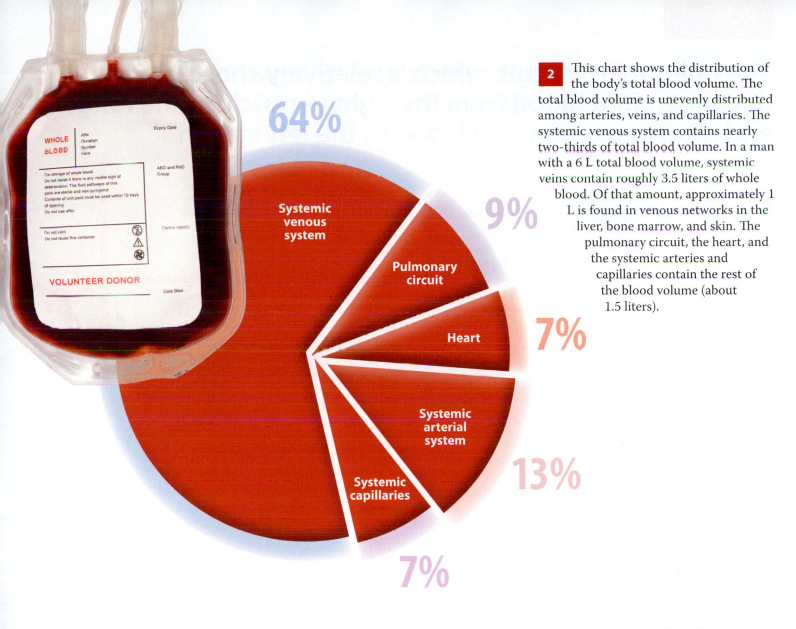

64%

Systemic venous system

9%

Pulmonary circuit

7%

Heart

Systemic arterial system

13%

Systemic capillaries

7%

2 This chart shows the distribution of the body's total blood volume. The total blood volume is unevenly distributed among arteries, veins, and capillaries. The systemic venous system contains nearly two-thirds of total blood volume. In a man with a 6 L total blood volume, systemic veins contain roughly 3.5 liters of whole blood. Of that amount, approximately 1 L is found in venous networks in the liver, bone marrow, and skin. The pulmonary circuit, the heart, and the systemic arteries and capillaries contain the rest of the blood volume (about 1.5 liters).

3 If serious hemorrhaging occurs, the body maintains blood volume within the arterial system at near-normal levels by reducing the volume of blood in the venous system. In the mechanism involved, the vasomotor center in the medulla oblongata stimulates sympathetic nerves innervating smooth muscle cells in the walls of medium-sized veins. The resulting contraction produces **venoconstriction**, reducing the diameter of the veins and the amount of blood contained in the venous system. In addition, blood enters the general circulation from venous networks in the liver, bone marrow, and skin. Reducing the amount of blood in the venous system can maintain the volume within the arterial system at near-normal levels despite a significant blood loss.

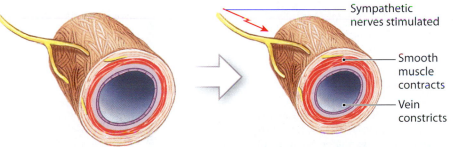

Sympathetic nerves stimulated

Smooth muscle contracts

Vein constricts

Module 17.12 Review

a. Define varicose veins.

b. Why are valves located in veins, but not in arteries?

c. How is blood pressure maintained in veins to counter the force of gravity?

The pulmonary circuit, which is relatively short, carries deoxygenated blood from the right ventricle to the lungs and returns oxygenated blood to the left atrium

1 This flow chart provides an overview of the organization of the cardiovascular system. The pulmonary circuit is composed of arteries and veins that transport blood between the heart and the lungs. This circuit begins at the right ventricle and ends at the left atrium. From the left ventricle, the arteries of the systemic circuit transport oxygenated blood and nutrients to all organs and tissues, ultimately returning deoxygenated blood to the right atrium.

2 The table below summarizes the overall functional patterns of the organization of the blood vessels in both the pulmonary and systemic circuits.

Major Patterns of Blood Vessel Organization

1. The peripheral distributions of arteries and veins on the body's left and right sides are generally identical, except near the heart, where the largest vessels connect to the atria or ventricles.

2. A single vessel may have several names as it crosses specific anatomical boundaries, making accurate anatomical descriptions possible when the vessel extends far into the periphery. For example, the external iliac artery becomes the femoral artery as it leaves the trunk and enters the lower limb.

3. Tissues and organs are usually serviced by several arteries and veins. Often, anastomoses between adjacent arteries or veins reduce the impact of a temporary or even permanent **occlusion** (blockage) of a single blood vessel.

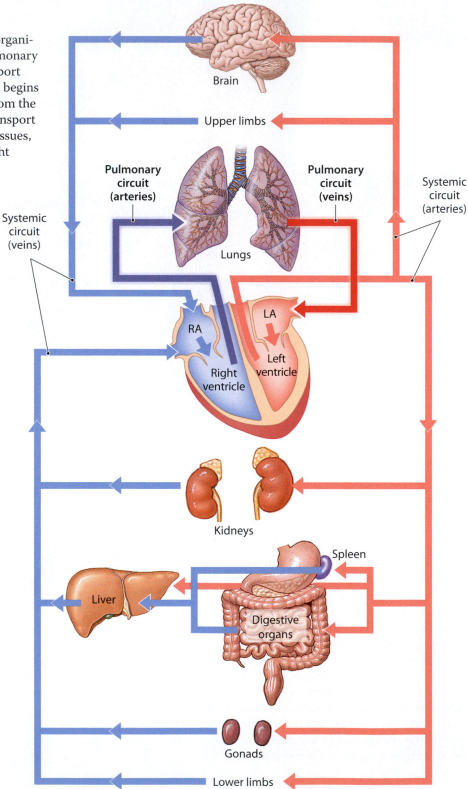

Brain

Upper limbs

Pulmonary circuit (arteries)

Pulmonary circuit (veins)

Systemic circuit (arteries)

Systemic circuit (veins)

Lungs

LA

RA

Left ventricle

Right ventricle

Kidneys

Spleen

Liver

Digestive organs

Gonads

Lower limbs

3 Arteries of the pulmonary circuit differ from those of the systemic circuit in that they carry deoxygenated blood. (This is why most color-coded diagrams show the pulmonary arteries in blue, the same color as systemic veins.) By convention, several large arteries are called **trunks**; the **pulmonary trunk** is one important example. As the pulmonary trunk curves over the superior border of the heart, it gives rise to the left and right **pulmonary arteries**. These large arteries enter the lungs before branching repeatedly, giving rise to smaller and smaller arteries. The smallest branches, the **pulmonary arterioles**, provide blood to **alveolar capillaries** that surround small air pockets called **alveoli** (al-VĒ-ō-lī; singular, *alveolus*). The walls of the alveoli are thin enough for gas to be exchanged between the capillary blood and inspired air; the blood absorbs oxygen and eliminates carbon dioxide. Oxygenated blood leaving the alveolar capillaries enters venules that in turn unite to form larger vessels carrying blood toward the **pulmonary veins**. These four veins, two from each lung, deliver oxygenated blood into the left atrium, completing the pulmonary circuit.

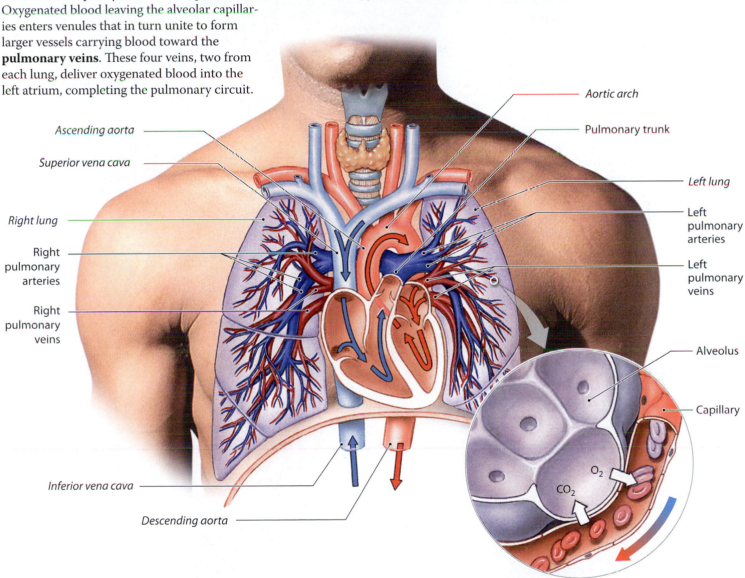

Aortic arch

Ascending aorta

Pulmonary trunk

Superior vena cava

Left lung

Right lung

Left pulmonary arteries

Right pulmonary arteries

Left pulmonary veins

Right pulmonary veins

Alveolus

Capillary

O_2

CO_2

Inferior vena cava

Descending aorta

Module 17.13 Review

a. Identify the two circulatory circuits of the cardiovascular system.

b. Briefly describe the three major patterns of blood vessel organization.

c. Trace a drop of blood through the lungs, beginning at the right ventricle and ending at the left atrium.

The systemic arterial and venous systems operate in parallel, and the major vessels often have similar names

1 This figure provides an overview of the systemic **arterial system**. Note that all the vessels of the systemic arterial system originate from the aorta, the large elastic artery extending from the left ventricle of the heart. In each of the six modules that follow, the left-hand page will illustrate and discuss the systemic arterial vessels and their branches. To reduce clutter, "artery" will not be repeated in every label. Because most of the major arteries are paired, with one artery of each pair on either side of the body, the terms "right" and "left" will appear in figure labels only when the arteries on both sides are labeled.

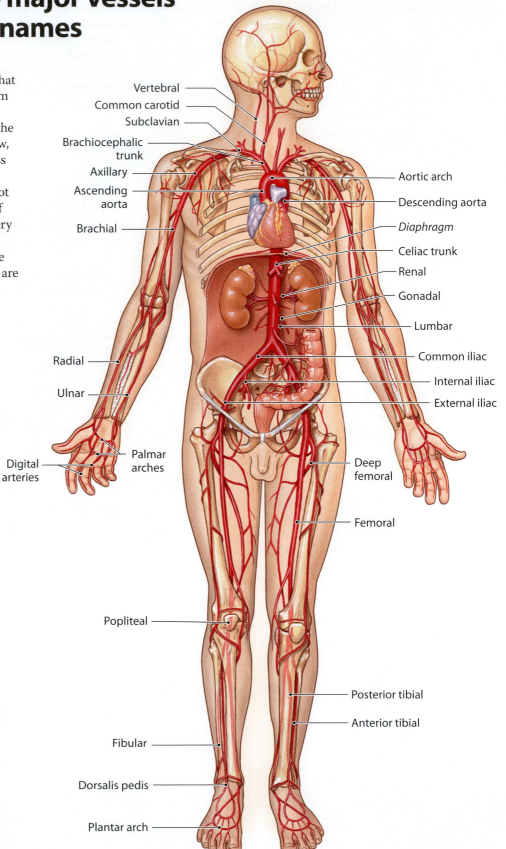

Vertebral

Common carotid

Subclavian

Brachiocephalic trunk

Axillary

Ascending aorta

Brachial

Radial

Ulnar

Digital arteries

Palmar arches

Popliteal

Fibular

Dorsalis pedis

Plantar arch

Aortic arch

Descending aorta

Diaphragm

Celiac trunk

Renal

Gonadal

Lumbar

Common iliac

Internal iliac

External iliac

Deep femoral

Femoral

Posterior tibial

Anterior tibial

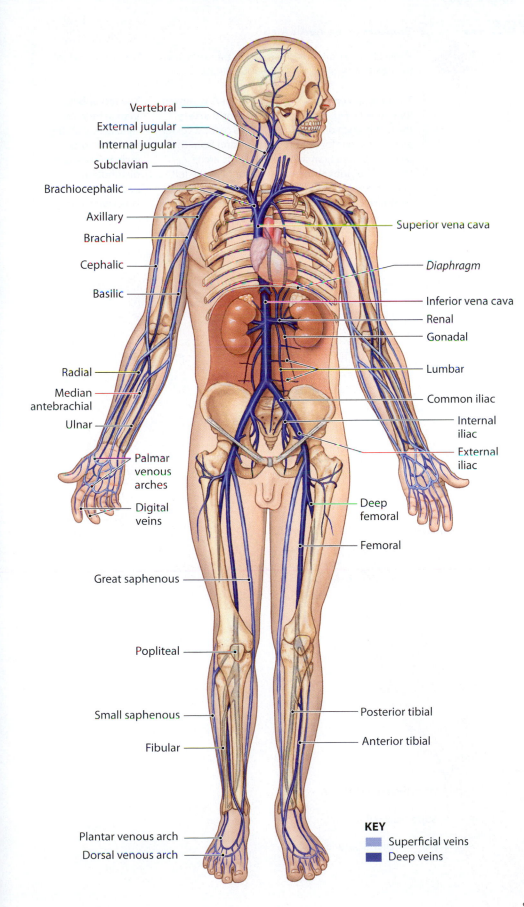

Vertebral
External jugular
Internal jugular
Subclavian
Brachiocephalic
Axillary
Brachial
Cephalic
Basilic
Radial
Median antebrachial
Ulnar
Palmar venous arches
Digital veins
Great saphenous
Popliteal
Small saphenous
Fibular
Plantar venous arch
Dorsal venous arch

Superior vena cava
Diaphragm
Inferior vena cava
Renal
Gonadal
Lumbar
Common iliac
Internal iliac
External iliac
Deep femoral
Femoral
Posterior tibial
Anterior tibial

KEY
Superficial veins
Deep veins

2 This figure provides an overview of the systemic **venous system**. Note that all of the vessels of the systemic venous system merge into two large veins: the **superior vena cava**, which collects systemic blood from the head, chest, and upper limbs, and the **inferior vena cava**, which collects systemic blood from all structures inferior to the diaphragm. In each of the six modules that follow, the right-hand page will illustrate and discuss the systemic venous vessels and their tributaries. To reduce clutter, "vein" will not be repeated in every label; the same name often applies to both the artery and the vein servicing a particular structure or region. Additionally, "right" and "left" will appear in figure labels only when the veins on both sides are shown.

One significant difference between the arterial and venous systems concerns the distribution of major veins in the neck and limbs. Arteries in these areas are located deep beneath the skin, protected by bones and surrounding soft tissues. In contrast, the neck and limbs generally have two sets of peripheral veins, one superficial and the other deep. This dual venous drainage is important for controlling body temperature. In hot weather, venous blood flows through superficial veins, where heat loss can occur; in cold weather, blood is routed to the deep veins to minimize heat loss.

Module 17.14 Review

a. Name the two large veins that collect blood from the systemic circuit.

b. Identify the largest artery in the body.

c. Besides containing valves, cite another major difference between the arterial and venous systems.

The branches of the aortic arch supply structures ...

Start

Branches of the Aortic Arch

Three elastic arteries originate along the aortic arch and deliver blood to the head, neck, shoulders, and upper limbs:

The **brachiocephalic** (brā-kē-ō-se-FAL-ik) **trunk** is the first branch off the aortic arch. It ascends for a short distance before branching to form the **right subclavian artery** and the **right common carotid artery**.

The **left common carotid artery**, the second vessel arising from the aortic arch, supplies the left side of the head and neck.

The **left subclavian artery**, the third artery arising from the aortic arch, has the same distribution pattern as the right subclavian artery.

The Right Subclavian Artery

Two major branches arise before a subclavian artery leaves the thoracic cavity: (1) the **internal thoracic artery**, supplying the pericardium and anterior wall of the chest; and (2) the **vertebral artery**, which provides blood to the brain and spinal cord.

Arteries of the Arm

After leaving the thoracic cavity and passing across the superior border of the first rib, the subclavian is called the **axillary artery**. This artery crosses the axilla to enter the arm, where it becomes the **brachial artery**, which supplies blood to the upper limb. The brachial artery gives rise to the deep brachial artery, which supplies deep structures on the posterior aspect of the arm, and the ulnar collateral arteries, which supply the area around the elbow.

Arteries of the Forearm

As it approaches the coronoid fossa of the humerus, the brachial artery divides into the **radial artery**, which follows the radius, and the **ulnar artery**, which follows the ulna to the wrist. At the wrist, the radial and ulnar arteries fuse to form the superficial and deep **palmar arches**, which supply blood to the hand and to the **digital arteries** of the thumb and fingers.

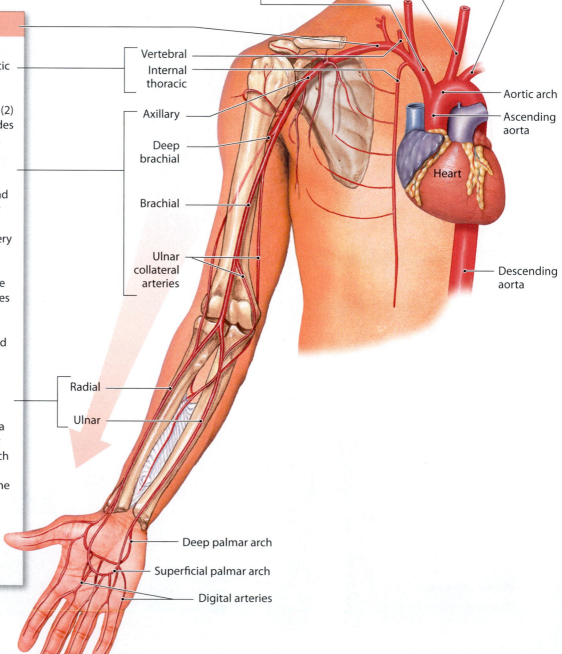

Vertebral

Internal thoracic

Axillary

Deep brachial

Brachial

Ulnar collateral arteries

Radial

Ulnar

Aortic arch

Ascending aorta

Heart

Descending aorta

Deep palmar arch

Superficial palmar arch

Digital arteries

... that are drained by the superior vena cava

Veins of the Neck

The **external jugular vein** drains superficial structures of the head and neck.

The **internal jugular vein** drains deep structures of the head and neck.

The **vertebral vein** drains the cervical spinal cord and the posterior surface of the skull.

The Right Subclavian Vein

The axillary vein is joined by the cephalic vein on the lateral surface of the first rib, forming the **subclavian vein**, which continues into the chest.

Veins of the Arm

As the **brachial vein** merges with the **basilic vein** it becomes the **axillary vein**, which enters the axilla.

The **cephalic vein** extends along the lateral side of the arm.

Veins of the Forearm

The **median cubital vein** interconnects the cephalic and basilic veins. (The median cubital is the vein from which venous blood samples are typically collected.)

The **ulnar vein** and the **radial vein** drain the **deep palmar arch**. After crossing the elbow, these veins fuse to form the brachial vein.

The cephalic vein, the **median antebrachial vein**, and the basilic vein drain the **superficial palmar arch**.

The **digital veins** empty into superficial and deep veins of the hand, which are interconnected to form the palmar venous arches.

Start

The **brachiocephalic vein** forms as the jugular veins empty into the axillary vein. It receives blood from the vertebral vein and the internal thoracic vein draining the anterior chest wall.

The **superior vena cava (SVC)** carries blood from the two brachiocephalic veins to the right atrium of the heart.

The **internal thoracic vein** collects blood from the intercostal veins and delivers it to the brachiocephalic vein.

Brachial
Basilic

KEY

Superficial veins
Deep veins

Cephalic

Radial

Basilic

Ulnar

Deep palmar arch

Superficial palmar arch

Module 17.15 Review

a. Name the two arteries formed by the division of the brachiocephalic trunk.

b. A blockage of which branch from the aortic arch would interfere with blood flow to the left arm?

c. Whenever Thor gets angry, a large vein bulges in the lateral region of his neck. Which vein is this?

The external carotid arteries supply the neck, lower jaw, and face, and the internal carotid and vertebral arteries supply the brain ...

1 The **common carotid arteries** ascend deep in the tissues of the neck and supply blood to the structures of the face, neck, and brain. To locate the carotid artery, gently press a finger along either side of the windpipe (trachea) until you feel a strong pulse. Details of the arterial supply of the brain are covered in Module 17.17.

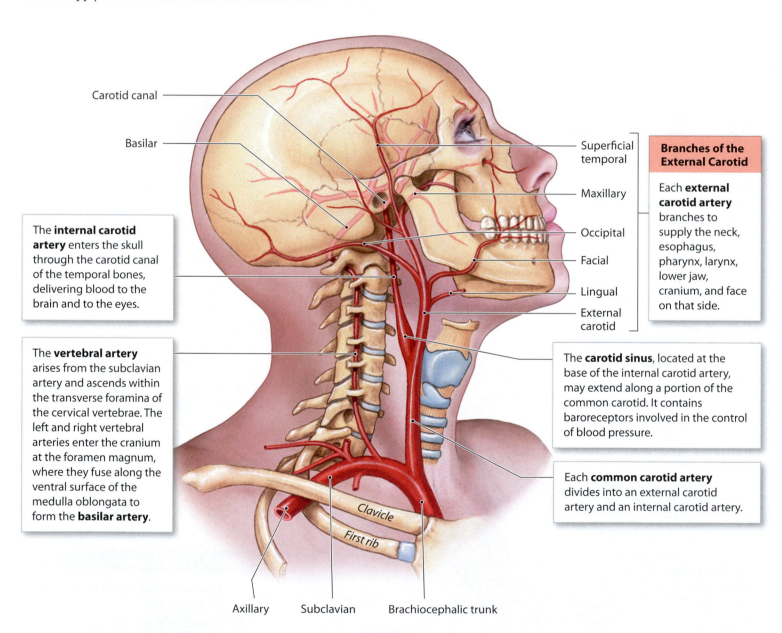

Carotid canal

Basilar

The **internal carotid artery** enters the skull through the carotid canal of the temporal bones, delivering blood to the brain and to the eyes.

The **vertebral artery** arises from the subclavian artery and ascends within the transverse foramina of the cervical vertebrae. The left and right vertebral arteries enter the cranium at the foramen magnum, where they fuse along the ventral surface of the medulla oblongata to form the **basilar artery**.

Superficial temporal

Maxillary

Occipital

Facial

Lingual

External carotid

Branches of the External Carotid

Each **external carotid artery** branches to supply the neck, esophagus, pharynx, larynx, lower jaw, cranium, and face on that side.

The **carotid sinus**, located at the base of the internal carotid artery, may extend along a portion of the common carotid. It contains baroreceptors involved in the control of blood pressure.

Each **common carotid artery** divides into an external carotid artery and an internal carotid artery.

Clavicle

First rib

Axillary   Subclavian   Brachiocephalic trunk

... while the external jugular veins drain the regions supplied by the external carotids, and the internal jugular veins drain the brain

2 The **external jugular veins** are formed by the maxillary and temporal veins, while the **internal jugular veins** drain the blood from the various venous sinuses within the cranium (see Module 17.17 for additional details). The external and internal jugular veins combine with the vertebral and subclavian to form the **brachiocephalic vein**.

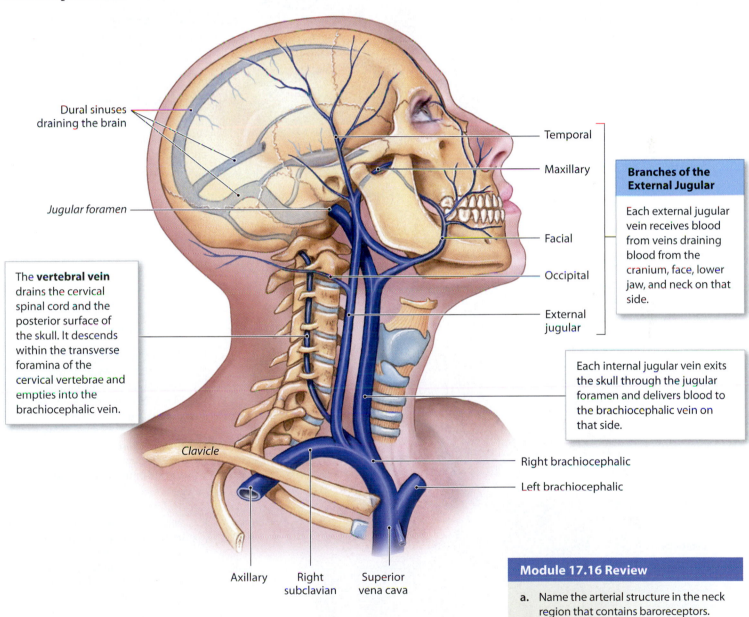

Dural sinuses draining the brain

Jugular foramen

The **vertebral vein** drains the cervical spinal cord and the posterior surface of the skull. It descends within the transverse foramina of the cervical vertebrae and empties into the brachiocephalic vein.

Temporal

Maxillary

Facial

Occipital

External jugular

Branches of the External Jugular

Each external jugular vein receives blood from veins draining blood from the cranium, face, lower jaw, and neck on that side.

Each internal jugular vein exits the skull through the jugular foramen and delivers blood to the brachiocephalic vein on that side.

Clavicle

Right brachiocephalic

Left brachiocephalic

Axillary   Right subclavian   Superior vena cava

Module 17.16 Review

a. Name the arterial structure in the neck region that contains baroreceptors.

b. Identify branches of the external carotid artery.

c. Identify the veins that combine to form the brachiocephalic vein.

The internal carotid arteries and the vertebral arteries supply the brain ...

1 This lateral view shows the major arteries supplying the brain. The internal carotid arteries normally supply the arteries of the anterior half of the cerebrum, and the rest of the brain receives blood from the vertebral and basilar arteries. The internal carotid artery ascends to the level of the optic nerves, where each artery divides into three branches: (1) an **ophthalmic artery**, which supplies the eyes; (2) an **anterior cerebral artery**, which supplies the frontal and parietal lobes of the brain; and (3) a **middle cerebral artery**, which supplies the midbrain and the lateral surfaces of the cerebral hemispheres.

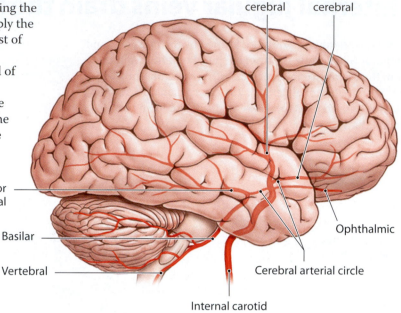

2 The internal carotid arteries and the basilar artery are interconnected in a ring-shaped anastomosis called the **cerebral arterial circle**.

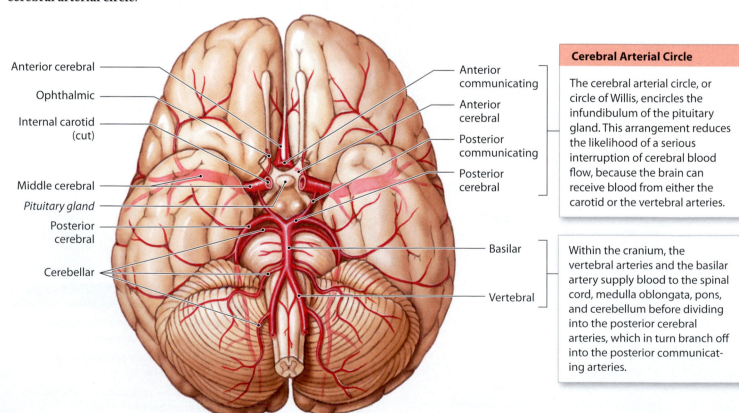

Cerebral Arterial Circle

The cerebral arterial circle, or circle of Willis, encircles the infundibulum of the pituitary gland. This arrangement reduces the likelihood of a serious interruption of cerebral blood flow, because the brain can receive blood from either the carotid or the vertebral arteries.

Within the cranium, the vertebral arteries and the basilar artery supply blood to the spinal cord, medulla oblongata, pons, and cerebellum before dividing into the posterior cerebral arteries, which in turn branch off into the posterior communicating arteries.

... which is drained by the dural sinuses and the internal jugular veins

3 The superficial cerebral veins and small veins of the brain stem empty into a network of dural sinuses. Most of the deep cerebral veins converge within the brain to form the **great cerebral vein**, which delivers blood from the interior of the cerebral hemispheres and the choroid plexus to the **straight sinus**. Numerous small veins from the orbit and other cerebral veins drain into the **cavernous sinus**.

The **superior sagittal sinus**, in the falx cerebri, is the largest dural sinus.

Unpaired Median Sinuses

Superior sagittal sinus
Inferior sagittal sinus
Straight sinus
Cavernous sinus
Occipital sinus

Paired Lateral Sinuses

Right transverse sinus
Right sigmoid sinus
Petrosal sinuses

Great cerebral vein

Internal jugular

The **vertebral vein** on each side receives blood from the transverse sinus and occipital sinus as well as superficial veins of the skull and veins draining the cervical vertebrae.

4 The cavernous sinus empties into the two **petrosal sinuses**, which in turn drain into the **transverse sinuses**. The transverse sinuses, the straight sinus, and the superior sagittal sinus converge to form the **sigmoid sinuses**, which penetrate the jugular foramina and leave the skull as the internal jugular veins.

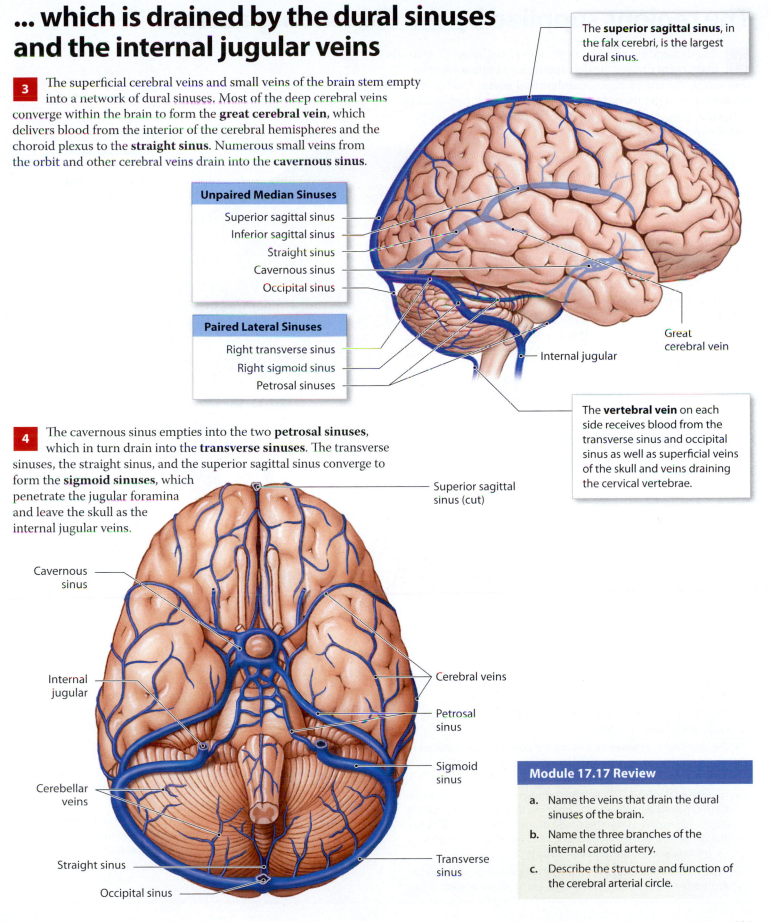

Superior sagittal sinus (cut)

Cavernous sinus

Internal jugular

Cerebral veins

Petrosal sinus

Sigmoid sinus

Cerebellar veins

Straight sinus

Occipital sinus

Transverse sinus

Module 17.17 Review

a. Name the veins that drain the dural sinuses of the brain.

b. Name the three branches of the internal carotid artery.

c. Describe the structure and function of the cerebral arterial circle.

The regions supplied by the descending aorta ...

1 The descending aorta is continuous with the aortic arch. The diaphragm divides the descending aorta into a superior **thoracic aorta** and an inferior **abdominal aorta.** The figure details the branches of the thoracic aorta and introduces the branches of the abdominal aorta that are detailed further in the table below.

Visceral Branches of the Thoracic Aorta

Visceral branches of the thoracic aorta supply the organs of the chest.

Bronchial arteries supply the tissues of the lungs not involved in gas exchange.

Esophageal arteries supply the esophagus.

Mediastinal arteries supply the tissues of the mediastinum.

Pericardial arteries supply the pericardium.

Somatic Branches of the Thoracic Aorta

Intercostal arteries supply the chest muscles and the vertebral column area.

Superior phrenic (FREN-ik) **arteries** deliver blood to the superior surface of the diaphragm.

Aortic arch
Internal thoracic
Thoracic aorta

Diaphragm
Inferior phrenic
Adrenal
Renal
Gonadal
Lumbar
Common iliac

Celiac trunk

Left gastric
Splenic   Branches of
Common   the celiac
hepatic   trunk

Superior mesenteric
Abdomial aorta
Inferior mesenteric

Major Paired Branches of the Abdominal Aorta

- The **inferior phrenic arteries** supply the inferior surface of the diaphragm and the inferior portion of the esophagus.
- The **adrenal arteries** supply the adrenal glands, which cap the superior part of each kidney.
- The short **renal arteries** arise along the posterolateral surface of the abdominal aorta, just inferior to the superior mesenteric artery. We will consider the branches of the renal arteries in Chapter 26.
- The **gonadal** (gō-NAD-al) **arteries** originate between the superior and inferior mesenteric arteries. In males, they are called testicular arteries; in females, they are termed ovarian arteries. The distribution of gonadal vessels (both arteries and veins) differs by gender; we will examine the differences in Chapter 25.
- Small **lumbar arteries** arise on the posterior surface of the aorta and supply the vertebrae, spinal cord, and abdominal wall.

Major Unpaired Branches of the Abdominal Aorta

- The **celiac** (SĒ-lē-ak) **trunk** divides into three branches: (1) the left gastric artery, which supplies the stomach and the inferior portion of the esophagus; (2) the splenic artery, which supplies the spleen and arteries to the stomach; and (3) the common hepatic artery, which supplies arteries to the liver, stomach, gallbladder, and the proximal portion of the small intestine.
- The **superior mesenteric** (mez-en-TER-ik) **artery** arises inferior to the celiac trunk to supply arteries to the pancreas and duodenum, and to most of the large intestine.
- The **inferior mesenteric artery** arises just superior to the bifurcation of the aorta, and it delivers blood to the terminal portions of the colon and the rectum. We will examine the branches of these vessels in Module 17.19.

... are drained by the superior and inferior venae cavae

2 The superior vena cava drains blood from the head, neck, shoulders, chest, and upper limbs. The chief collecting vessels of the thorax are the **azygos** (AZ-i-gos) **vein** and the **hemiazygos vein**. These veins receive blood from (1) **intercostal veins**, which in turn receive blood from the chest muscles; (2) **esophageal veins**, which drain blood from the inferior portion of the esophagus; (3) **bronchial veins** draining the passageways of the lungs; and (4) **mediastinal veins** draining other mediastinal structures. The inferior vena cava collects most of the blood inferior to the diaphragm.

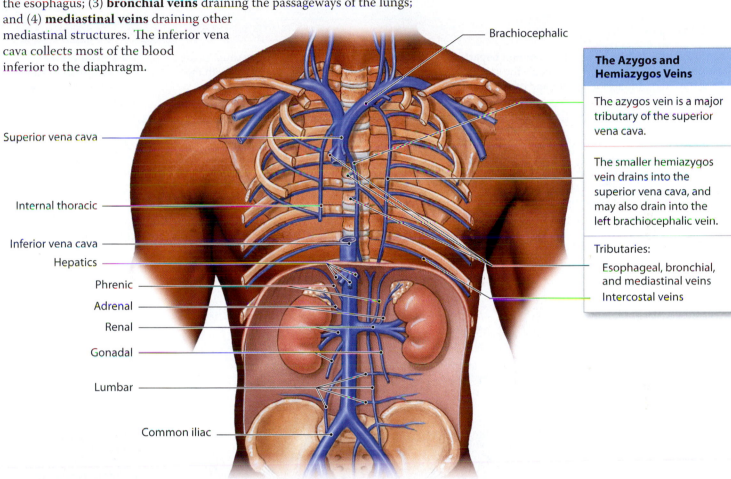

Brachiocephalic

Superior vena cava

Internal thoracic

Inferior vena cava

Hepatics

Phrenic

Adrenal

Renal

Gonadal

Lumbar

Common iliac

The Azygos and Hemiazygos Veins

The azygos vein is a major tributary of the superior vena cava.

The smaller hemiazygos vein drains into the superior vena cava, and may also drain into the left brachiocephalic vein.

Tributaries:
  Esophageal, bronchial, and mediastinal veins
  Intercostal veins

Major Tributaries of the Inferior Vena Cava

- **Lumbar veins** drain the lumbar portion of the abdomen, including the spinal cord and muscles of the body wall.
- **Gonadal** (ovarian or testicular) **veins** drain the ovaries or testes. The right gonadal vein empties into the inferior vena cava; the left gonadal vein generally drains into the left renal vein.
- **Hepatic veins** drain the sinusoids of the liver.
- **Renal veins**, the largest tributaries of the inferior vena cava, collect blood from the kidneys.
- **Adrenal veins** drain the adrenal glands. In most individuals, only the right adrenal vein drains into the inferior vena cava; the left adrenal vein drains into the left renal vein.
- **Phrenic veins** drain the diaphragm. Only the right phrenic vein drains into the inferior vena cava; the left drains into the left renal vein.

Module 17.18 Review

a. Which vessel collects most of the venous blood inferior to the diaphragm?

b. Identify the major tributaries of the inferior vena cava.

c. Grace is in an automobile accident, and her celiac trunk is ruptured. Which organs will be affected most directly by this injury?

The viscera supplied by the celiac trunk and mesenteric arteries ...

1 The abdominal aorta begins immediately inferior to the diaphragm. Three unpaired branches supply the abdominal viscera: the celiac trunk, the superior mesenteric artery, and the inferior mesenteric artery.

The Celiac Trunk

The **celiac trunk** divides into the common hepatic artery, the left gastric artery, and the splenic artery.

The **common hepatic artery** branches to supply the liver, stomach, gallbladder, and the duodenum (the proximal segment of the small intestine).

The **left gastric artery** supplies the stomach. Its anastomosis with the right gastric artery ensures the stomach a continuous blood supply.

The **splenic artery** supplies the spleen and sends branches to the stomach and pancreas.

Branches of the Common Hepatic Artery

Hepatic artery proper (liver)

Cystic (gallbladder)

Gastroduodenal (stomach and duodenum)

Right gastric (stomach)

Right gastroepiploic (stomach and duodenum)

Superior pancreatico-duodenal (duodenum)

Ascending colon

Superior Mesenteric Artery

The **superior mesenteric artery** branches to supply the pancreas and duodenum, small intestine, and most of the large intestine.

Inferior pancreatico-duodenal (pancreas and duodenum)

Right colic (large intestine)

Ileocolic (large intestine)

Middle colic (cut) (large intestine)

Intestinal arteries (small intestine)

Branches of the Splenic Artery

Left gastroepiploic (stomach)

Pancreatic (pancreas)

Spleen

Inferior Mesenteric Artery

The **inferior mesenteric artery** delivers blood to the terminal portions of the colon and the rectum.

Left colic (colon)

Sigmoid (colon)

Rectal (rectum)

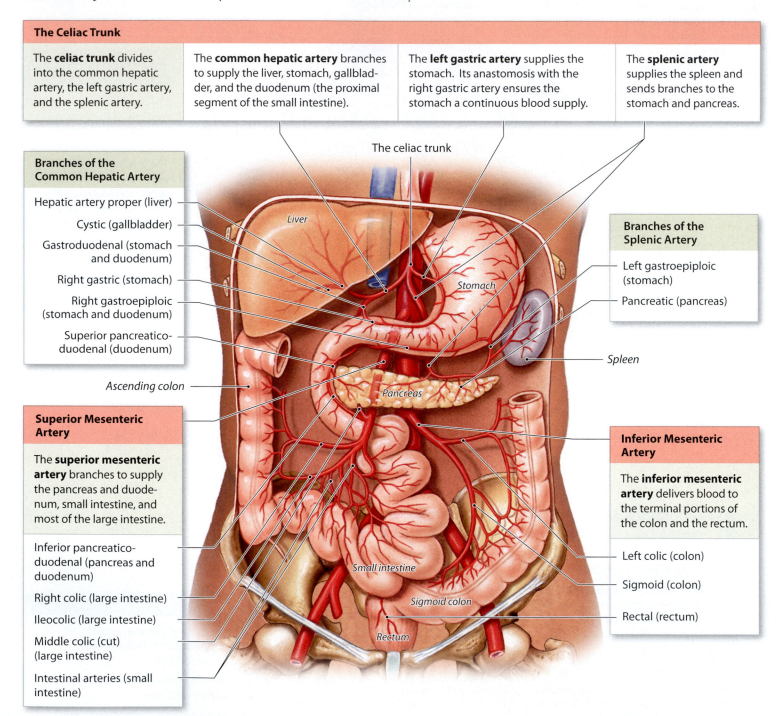

The celiac trunk

Liver

Stomach

Pancreas

Small intestine

Sigmoid colon

Rectum

... are drained by the tributaries of the hepatic portal vein

2 The **hepatic portal vein** forms through the fusion of the **superior mesenteric**, **inferior mesenteric**, and **splenic veins**. The largest volume of blood (and most of the nutrients) flows through the superior mesenteric vein. The hepatic portal vein receives blood from the left and right **gastric veins**, which drain the medial border of the stomach, and from the **cystic vein**, emanating from the gallbladder.

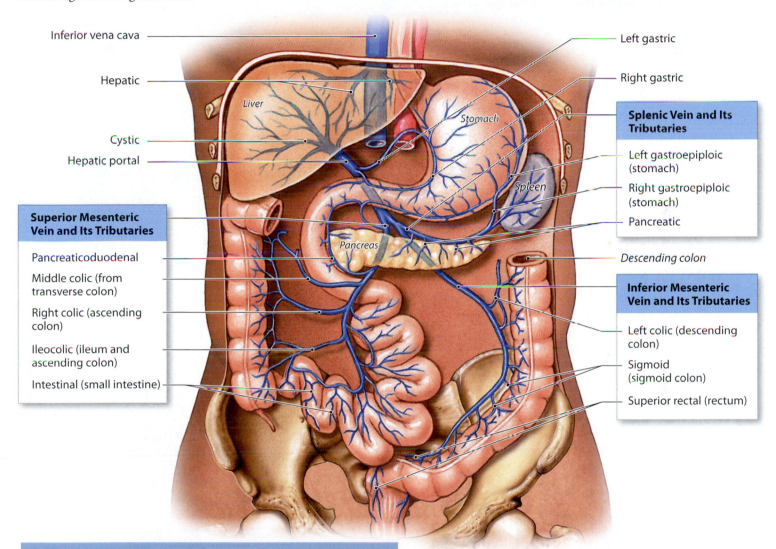

Inferior vena cava

Hepatic

Liver

Cystic

Hepatic portal

Left gastric

Right gastric

Stomach

Splenic Vein and Its Tributaries

Left gastroepiploic (stomach)

Spleen

Right gastroepiploic (stomach)

Pancreatic

Superior Mesenteric Vein and Its Tributaries

Pancreaticoduodenal

Middle colic (from transverse colon)

Right colic (ascending colon)

Ileocolic (ileum and ascending colon)

Intestinal (small intestine)

Pancreas

Descending colon

Inferior Mesenteric Vein and Its Tributaries

Left colic (descending colon)

Sigmoid (sigmoid colon)

Superior rectal (rectum)

Tributaries of the Hepatic Portal Vein

- The inferior mesenteric vein collects blood from capillaries along the inferior portion of the large intestine. It drains the left colic vein and the superior rectal veins, which collect venous blood from the descending colon, sigmoid colon, and rectum.
- The splenic vein is formed by the union of the inferior mesenteric vein and veins from the spleen, the lateral border of the stomach (left gastroepiploic vein), and the pancreas (pancreatic veins).
- The superior mesenteric vein collects blood from veins draining the stomach (right gastroepiploic vein), the small intestine (intestinal and pancreaticoduodenal veins), and two-thirds of the large intestine (ileocolic, right colic, and middle colic veins).

Module 17.19 Review

a. List the unpaired branches of the abdominal aorta that supply blood to the visceral organs.

b. Identify the three veins that merge to form the hepatic portal vein.

c. Identify two veins that carry blood away from the stomach.

The pelvis and lower limbs are supplied by branches of the common iliac arteries ...

1 Near the level of vertebra L_4, the abdominal aorta divides to form a pair of elastic arteries: the **right** and **left common iliac** (IL-ē-ak) **arteries**. At the level of the lumbosacral joint, each common iliac divides to form an **internal iliac artery** and an **external iliac artery**.

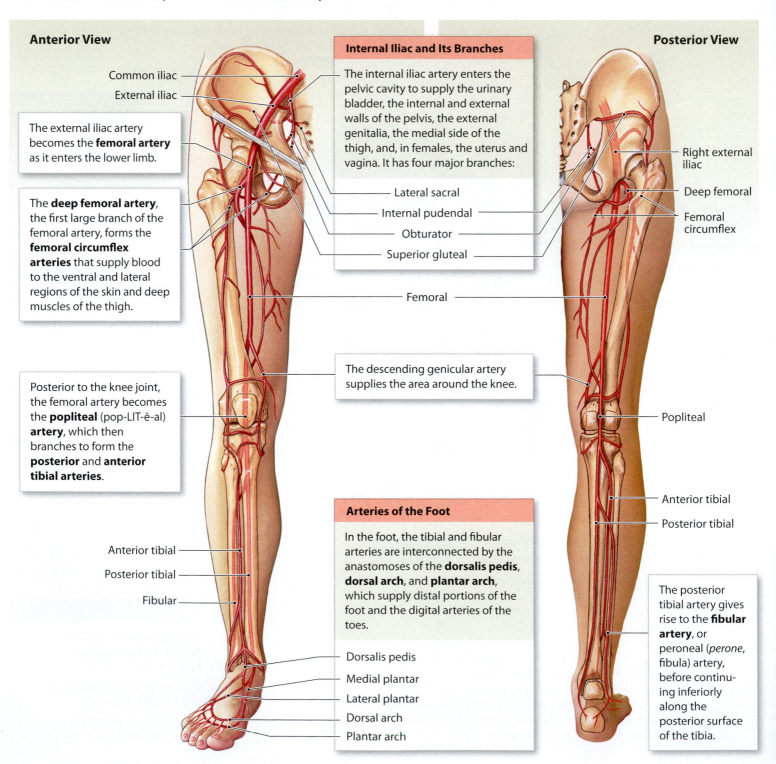

Anterior View

Common iliac

External iliac

Posterior View

The external iliac artery becomes the **femoral artery** as it enters the lower limb.

Internal Iliac and Its Branches

The internal iliac artery enters the pelvic cavity to supply the urinary bladder, the internal and external walls of the pelvis, the external genitalia, the medial side of the thigh, and, in females, the uterus and vagina. It has four major branches:

Right external iliac

The **deep femoral artery**, the first large branch of the femoral artery, forms the **femoral circumflex arteries** that supply blood to the ventral and lateral regions of the skin and deep muscles of the thigh.

Lateral sacral

Internal pudendal

Obturator

Superior gluteal

Deep femoral

Femoral circumflex

Femoral

The descending genicular artery supplies the area around the knee.

Posterior to the knee joint, the femoral artery becomes the **popliteal** (pop-LIT-ē-al) **artery**, which then branches to form the **posterior** and **anterior tibial arteries**.

Popliteal

Arteries of the Foot

In the foot, the tibial and fibular arteries are interconnected by the anastomoses of the **dorsalis pedis**, **dorsal arch**, and **plantar arch**, which supply distal portions of the foot and the digital arteries of the toes.

Anterior tibial

Posterior tibial

Anterior tibial

Posterior tibial

Fibular

The posterior tibial artery gives rise to the **fibular artery**, or peroneal (*perone*, fibula) artery, before continuing inferiorly along the posterior surface of the tibia.

Dorsalis pedis

Medial plantar

Lateral plantar

Dorsal arch

Plantar arch

... and drained by tributaries of the common iliac veins

2 The external iliac veins receive blood from the lower limbs, the pelvis, and the lower abdomen. As the left and right external iliac veins cross the inner surface of the ilium, they are joined by the internal iliac veins, which drain the pelvic organs. The internal iliac veins are formed by the fusion of the gluteal, internal pudendal, obturator, and lateral sacral veins. The union of external and internal iliac veins forms the **common iliac vein**.

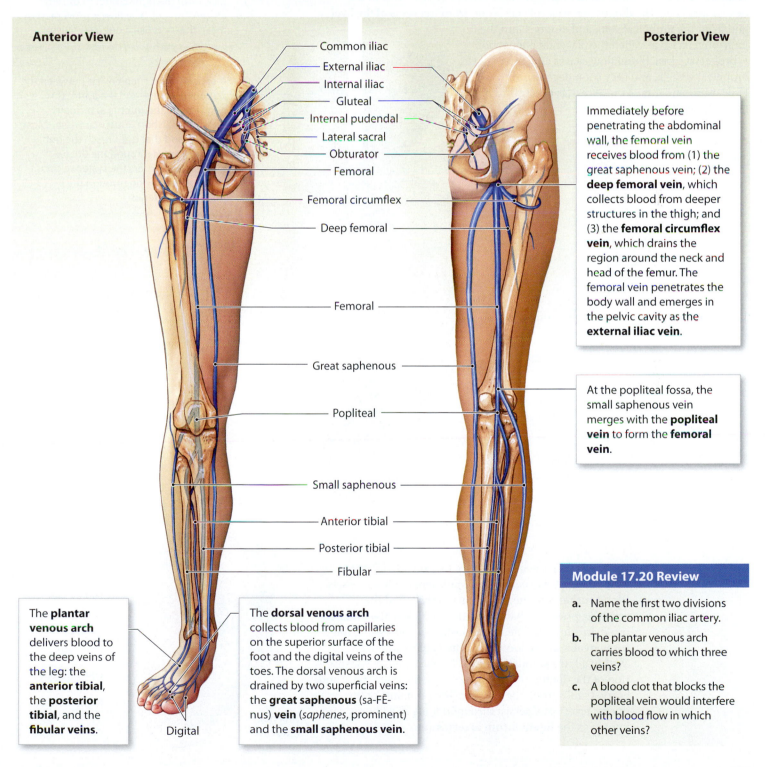

Anterior View

Posterior View

Common iliac
External iliac
Internal iliac
Gluteal
Internal pudendal
Lateral sacral
Obturator
Femoral
Femoral circumflex
Deep femoral
Femoral
Great saphenous
Popliteal
Small saphenous
Anterior tibial
Posterior tibial
Fibular
Digital

Immediately before penetrating the abdominal wall, the femoral vein receives blood from (1) the great saphenous vein; (2) the **deep femoral vein**, which collects blood from deeper structures in the thigh; and (3) the **femoral circumflex vein**, which drains the region around the neck and head of the femur. The femoral vein penetrates the body wall and emerges in the pelvic cavity as the **external iliac vein**.

At the popliteal fossa, the small saphenous vein merges with the **popliteal vein** to form the **femoral vein**.

The **plantar venous arch** delivers blood to the deep veins of the leg: the **anterior tibial**, the **posterior tibial**, and the **fibular veins**.

The **dorsal venous arch** collects blood from capillaries on the superior surface of the foot and the digital veins of the toes. The dorsal venous arch is drained by two superficial veins: the **great saphenous** (sa-FĒ-nus) **vein** (*saphenes*, prominent) and the **small saphenous vein**.

Module 17.20 Review

a. Name the first two divisions of the common iliac artery.

b. The plantar venous arch carries blood to which three veins?

c. A blood clot that blocks the popliteal vein would interfere with blood flow in which other veins?

The pattern of blood flow through the fetal heart and the systemic circuit must change at birth

1 Fetal blood flows to the placenta through a pair of **umbilical arteries**, which arise from the internal iliac arteries and enter the umbilical cord. Blood returns from the placenta in the single **umbilical vein**, bringing oxygen and nutrients to the developing fetus. The umbilical vein drains into the **ductus venosus**, a vascular connection to an intricate network of veins within the developing liver. The ductus venosus collects blood from the veins of the liver and from the umbilical vein, and empties into the inferior vena cava. When the placental connection is broken at birth, blood flow ceases along the umbilical vessels, and they soon degenerate. However, remnants of these vessels persist throughout life as fibrous cords.

Prior to birth, the lungs are collapsed and most blood bypasses the pulmonary circuit completely. The **foramen ovale**, or interatrial opening, allows blood to pass from the right atrium to the left atrium, but any backflow is prevented by a flap that acts like a one-way valve.

A second short circuit exists between the pulmonary trunk and the aorta. This connection, the **ductus arteriosus**, consists of a short, muscular vessel. Most of the blood that does reach the right ventricle flows through the ductus venosus and enters the systemic circuit rather than flowing through the pulmonary arteries.

Aorta

Placenta

Liver

Umbilical vein

Pulmonary trunk

Inferior vena cava

Ductus venosus

Umbilical cord

Umbilical arteries

Full-term fetus (before birth)

After delivery

2 At birth, the infant takes a first breath, inflating the lungs and expanding the pulmonary blood vessels. Blood rushes into the pulmonary vessels, and the resulting pressure changes at the heart close the foramen ovale. In adults, the heart has a shallow depression, the **fossa ovalis**, at the location of the fetal passageway. Within a few seconds, rising O_2 levels stimulate constriction of the ductus arteriosus, isolating the pulmonary and aortic trunks from one another. The remnants of the ductus arteriosus persist throughout life as a fibrous cord known as the **ligamentum arteriosum**.

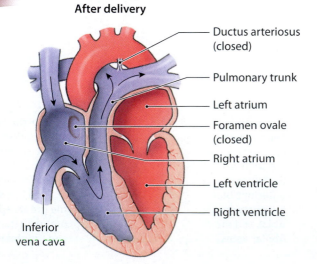

Ductus arteriosus (closed)

Pulmonary trunk

Left atrium

Foramen ovale (closed)

Right atrium

Left ventricle

Right ventricle

Inferior vena cava

Ventricular Septal Defect

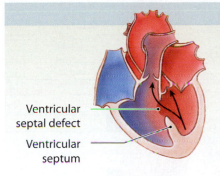

Ventricular septal defect

Ventricular septum

Ventricular septal defects are openings in the interventricular septum that separate the right and left ventricles. These defects are the most common congenital heart problems, affecting 0.12 percent of newborns. The opening between the two ventricles has an effect similar to a connection between the atria: When the more powerful left ventricle beats, it ejects blood into the right ventricle and pulmonary circuit.

3 Although minor individual variations in the vascular network are quite common, congenital cardio-vascular problems serious enough to threaten homeostasis are relatively rare. Most problems reflect abnormal formation of the heart or problems with the interconnections between the heart and the great vessels. If diagnosed early, most can be surgically corrected—sometimes prior to delivery.

Patent Foramen Ovale and Patent Ductus Arteriosus

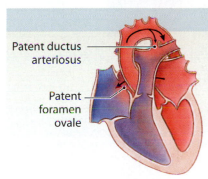

Patent ductus arteriosus

Patent foramen ovale

If the foramen ovale remains open, or **patent**, blood recirculates through the pulmonary circuit instead of entering the left ventricle. The movement, driven by the relatively high systemic pressure, is called a "left-to-right shunt." Arterial oxygen content is normal, but the left ventricle must work much harder than usual to provide adequate blood flow through the systemic circuit. Hence, pressures rise in the pulmonary circuit. If the pulmonary pressures rise enough, they may force blood into the systemic circuit through the ductus arteriosus. This condition—a **patent ductus arteriosus**—creates a "right-to-left shunt." Because the circulating blood is not adequately oxygenated, it develops a deep red color. The skin then develops the blue tones typical of cyanosis and the infant is known as a "blue baby."

Tetralogy of Fallot

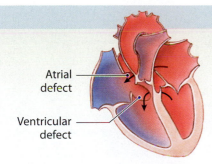

Patent ductus arteriosus

Pulmonary stenosis

Ventricular septal defect

Enlarged right ventricle

The **tetralogy of Fallot** (fa-LŌ) is a complex group of heart and circulatory defects that affect 0.10 percent of newborn infants. In this condition, (1) the pulmonary trunk is abnormally narrow (pulmonary stenosis), (2) the interventricular septum is incomplete, (3) the aorta originates where the interventricular septum normally ends, and (4) the right ventricle is enlarged and both ventricles thicken in response to the increased work-load.

Atrioventricular Septal Defect

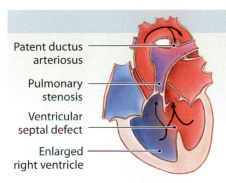

Atrial defect

Ventricular defect

In an **atrioventricular septal defect**, both the atria and ventricles are incompletely separated. The results are quite variable, depending on the extent of the defect and the effects on the atrioventricular valves. This type of defect most commonly affects infants with Down syndrome, a disorder caused by the presence of an extra copy of chromosome 21.

Transposition of the Great Vessels

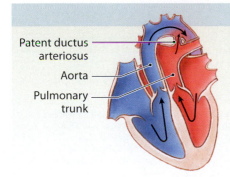

Patent ductus arteriosus

Aorta

Pulmonary trunk

In the **transposition of great vessels**, the aorta is connected to the right ventricle instead of to the left ventricle, and the pulmonary artery is connected to the left ventricle instead of the right ventricle. This malformation affects 0.05 percent of newborn infants.

Module 17.21 Review

a. Describe the pattern of fetal blood flow to and from the placenta.

b. Identify the six structures that are necessary in the fetal circulation but cease to function at birth, and describe what becomes of these structures.

c. Compare a ventricular septal defect with tetralogy of Fallot.

1. Labeling

Label the major arteries in the diagram at right.

a

b

c

d

e

f

g

h

i

j

k

l

m

n

2. Labeling

Label the major veins in the accompanying diagram.

a

b

c

d

e

f

g

h

i

j

k

l

m

n

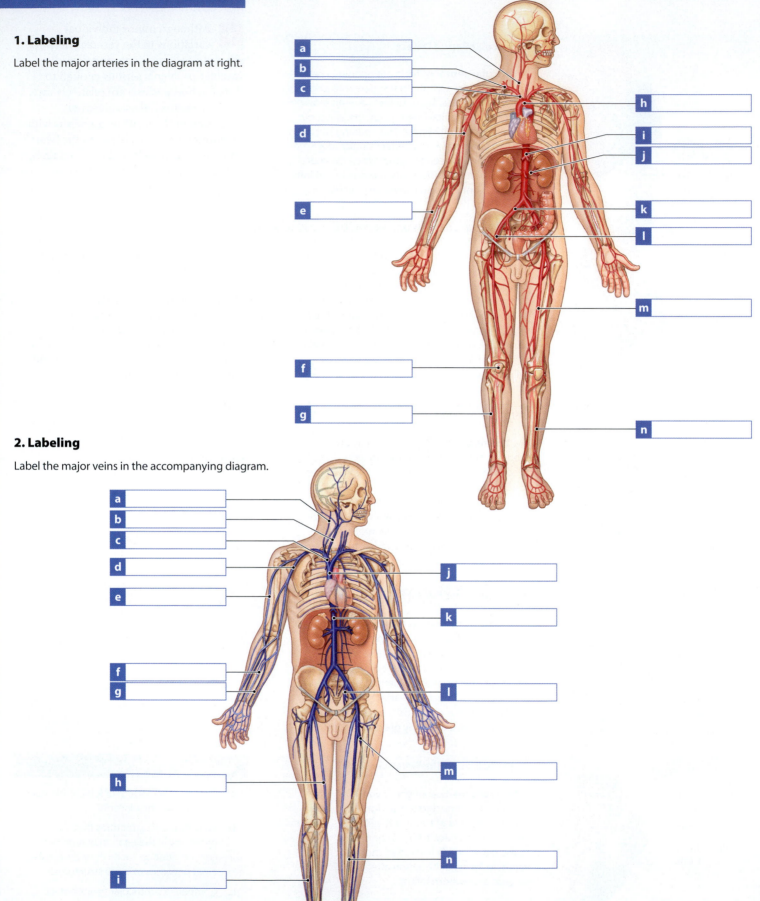

Visual Outline with Key Terms

Summarize the content of each module using the terms in the order provided.

SECTION 1

Blood

- cardiovascular system
  - heart
  - blood vessels
  - blood

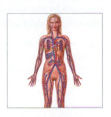

17.1

Blood is a fluid connective tissue containing plasma and formed elements

- plasma
- formed elements
- whole blood
- albumin
- globulins
- immunoglobulins
- transport globulins
- fibrinogen
- fibrin
- electrolytes
- organic nutrients
- organic wastes
- platelets
- white blood cells (WBCs)
- red blood cells (RBCs)
- hematocrit
- packed cell volume (PCV)

17.2

Red blood cells, the most common formed elements, contain hemoglobin

- red blood cell count
- rouleaux
- hemoglobin (Hb)
- alpha (α) chains
- beta (β) chains
- heme
- oxyhemoglobin (HbO$_2$)
- deoxyhemoglobin

17.3

Red blood cells are continuously produced and recycled

- proerythroblasts
- erythroblasts
- reticulocyte
- erythropoiesis
- myeloid tissue
- hemolysis
- hemoglobinuria
- hemolyze
- transferrin
- biliverdin
- bilirubin
- jaundice
- urobilins
- stercobilins
- hematuria

17.4

Blood type is determined by the presence or absence of specific surface antigens on RBCs

- antigens
- immune response
- surface antigens
- blood type
- Type A
- Type B
- Type AB
- Type O
- agglutinate
- agglutination
- cross-reaction
- Rh positive (Rh$^+$)
- Rh negative (Rh$^-$)
- transfusion reactions
- compatible

17.5

Hemolytic disease of the newborn is an RBC-related disorder caused by a cross-reaction between fetal and maternal blood types

- hemolytic disease of the newborn (HDN)
- sensitization
- erythroblastosis fetalis

17.6

White blood cells defend the body against pathogens, toxins, cellular debris, and abnormal or damaged cells

- white blood cells (WBCs)
- leukocytes
- emigration (diapedesis)
- positive chemotaxis
- granular leukocytes
- neutrophils
- eosinophils
- basophils
- agranular leukocytes
- monocytes
- lymphocytes
- differential count

17.7

Formed elements are produced by stem cells derived from hemocytoblasts

- hemocytoblasts
- lymphoid stem cells
- lymphoid tissues
- colony-stimulating factors (CSFs)
- blast cells
- myeloid stem cells
- progenitor cells
- myelocytes
- band cells
- megakaryocytes
- erythropoietin (EPO)
- hypoxia
- platelets

17.8

The clotting response is a complex cascade of events that reduces blood loss

- hemostasis
- vascular phase
- endothelins
- platelet phase
- platelet factors
- platelet-derived growth factor (PDGF)
- coagulation phase
- coagulation
  - fibrinogen
- fibrin
- procoagulants
- cascade
- common pathway
- prothrombinase
- prothrombin
- thrombin
- extrinsic pathway
- tissue factor
- intrinsic pathway
- PF-3
- clot retraction
- fibrinolysis
- plasminogen
- tissue plasminogen activator (t-PA)
- plasmin

17.9

Blood disorders can be classified by their origins and the changes in blood characteristics

- venipuncture
- iron deficiency anemia
- microcytic
- vitamin B$_{12}$
- pernicious anemia
- macrocytic
- vitamin K
- sickle cell anemia
- sickling trait
- hemophilia
- thalassemias
- bacteremia
- viremia
- sepsis
- septicemia
- malaria
- leukemias
- myeloid leukemia
- lymphoid leukemia
- disseminated intravascular coagulation (DIC)

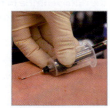

SECTION 2

The Functional Anatomy of Blood Vessels

- pulmonary circuit
- systemic circuit
- arteries
- veins
- capillaries
- right atrium
- right ventricle
- left atrium
- left ventricle

17.10

Arteries and veins differ in the structure and thickness of their walls

- tunica intima
- internal elastic membrane
- tunica media
- vasoconstriction
- vasodilation
- tunica externa
- arteries
- veins
- large veins
- medium-sized veins
- venules
- elastic arteries
- muscular arteries
- arterioles
- capillaries

17.11

Capillary structure and capillary blood flow affect the rates of exchange between the blood and interstitial fluid

- continuous capillary
- fenestrated capillary
- sinusoids
- capillary bed
- metarteriole
- thoroughfare channel
- vasomotion
- collateral arteries
- arterial anastomosis
- precapillary sphincter
- arteriovenous anastomosis

17.12

The venous system has low pressures and contains almost two-thirds of the body's blood volume

- valves
- varicose veins
- hemorrhoids
- venoconstriction

17.13

The pulmonary circuit, which is relatively short, carries deoxygenated blood from the right ventricle to the lungs and returns oxygenated blood to the left atrium

- occlusion
- trunks
- pulmonary trunk
- pulmonary arteries
- pulmonary arterioles
- alveolar capillaries
- alveoli
- pulmonary veins

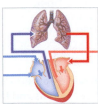

17.14

The systemic arterial and venous systems operate in parallel, and the major vessels often have similar names

- arterial system
- venous system
- superior vena cava
- inferior vena cava
- dual venous drainage

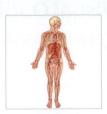

17.15

The branches of the aortic arch supply structures that are drained by the superior vena cava

- brachiocephalic trunk
- right subclavian artery
- right common carotid artery
- left common carotid artery
- left subclavian artery
- internal thoracic artery
- vertebral artery
- axillary artery
- brachial artery
- radial artery
- ulnar artery
- palmar arches
- digital arteries
- digital veins
- superficial palmar arch
- median antebrachial vein
- deep palmar arch
- radial vein
- ulnar vein
- median cubital vein
- cephalic vein
- axillary vein
- basilic vein
- brachial vein
- subclavian vein
- external jugular vein
- internal jugular vein
- vertebral vein
- brachiocephalic vein
- superior vena cava (SVC)
- internal thoracic vein

17.16

The external carotid arteries supply the neck, lower jaw, and face, and the internal carotid and vertebral arteries supply the brain, while the external jugular veins drain the regions supplied by the external carotids, and the internal jugular veins drain the brain

- common carotid arteries
- internal carotid artery
- vertebral artery
- basilar artery
- external carotid artery
- carotid sinus
- common carotid artery
- external jugular veins
- internal jugular vein
- brachiocephalic vein
- vertebral vein

17.17

The internal carotid arteries and the vertebral arteries supply the brain, which is drained by the dural sinuses and the internal jugular veins

- ophthalmic artery
- anterior cerebral artery
- middle cerebral artery
- cerebral arterial circle
- great cerebral vein
- straight sinus
- cavernous sinus
- superior sagittal sinus
- vertebral vein
- petrosal sinuses
- transverse sinuses
- sigmoid sinuses

• = *Term boldfaced in this module*

17.18

The regions supplied by the descending aorta are drained by the superior and inferior venae cavae

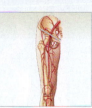

- thoracic aorta
- abdominal aorta
- • intercostal arteries
- superior phrenic arteries
- bronchial arteries
- esophageal arteries
- mediastinal arteries
- pericardial arteries
- inferior phrenic arteries
- adrenal arteries
- renal arteries
- gonadal arteries
- lumbar arteries
- • celiac trunk
- • superior mesenteric artery
- inferior mesenteric artery
- azygos vein
- hemiazygos vein
- intercostal veins
- esophageal veins
- bronchial veins
- mediastinal veins
- lumbar veins
- gonadal veins
- hepatic veins
- renal veins
- adrenal veins
- phrenic veins

17.19

The viscera supplied by the celiac trunk and mesenteric arteries are drained by the tributaries of the hepatic portal vein

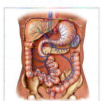

- • celiac trunk
- • common hepatic artery
- • left gastric artery
- • splenic artery
- • superior mesenteric artery
- • inferior mesenteric artery
- • hepatic portal vein
- • superior mesenteric vein
- • inferior mesenteric vein
- • splenic vein
- • gastric veins
- • cystic vein

17.20

The pelvis and lower limbs are supplied by branches of the common iliac arteries and drained by tributaries of the common iliac veins

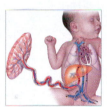

- right common iliac artery
- left common iliac artery
- internal iliac artery
- external iliac artery
- femoral artery
- deep femoral artery
- femoral circumflex arteries
- popliteal artery
- posterior tibial artery
- anterior tibial artery
- dorsalis pedis
- dorsal arch
- plantar arch
- fibular artery
- common iliac vein
- plantar venous arch
- anterior tibial vein
- posterior tibial vein
- fibular vein
- dorsal venous arch
- great saphenous vein
- small saphenous vein
- deep femoral vein
- femoral circumflex vein
- external iliac vein
- popliteal vein
- femoral vein

17.21

The pattern of blood flow through the fetal heart and the systemic circuit must change at birth

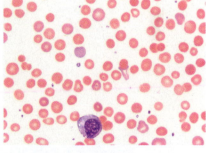
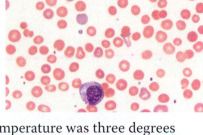

- umbilical arteries
- umbilical vein
- ductus venosus
- foramen ovale
- ductus arteriosus
- fossa ovalis
- ligamentum arteriosum
- ventricular septal defects
- patent
- ○ patent foramen ovale
- patent ductus arteriosus
- tetralogy of Fallot
- atrioventricular septal defect
- transposition of great vessels

• = Term boldfaced in this module

Chapter Integration: Applying what you've learned

Ursula is a 27-year-old, single female with no known medical history. A normally active person, with a well-balanced diet, she rarely was ill. One weekend, Ursula felt as though she was "coming down with something." On Saturday, she was fatigued, but had no trouble eating, but by Sunday, she was extremely tired and simply could not eat. Monday afternoon, her lethargy was so profound, she was unable to get out of bed, so she called a family member to take her to the physician's office.

While her physician was performing a physical exam, her blood pressure plummeted to 60/20, her heart rate increased to 190 beats per minute, and her body temperature was three degrees below normal (95.6 °F). She was immediately rushed to the hospital, where blood was drawn and tested. Obtaining blood using a venipuncture was nearly impossible, because Ursula was in the beginning stages of vascular collapse. Her gums (gingivae) and skin were pale.

When viewed through a microscope, Ursula's red blood cells were smaller than normal and spherically shaped, and thus were characterized as "spherocytes." Antibodies in her plasma were attacking her red blood cells, causing them to hemolyze and transform into "spherocytes." Additionally, her blood's hemoglobin levels were well below normal. Ultimately, Ursula was diagnosed with immune-mediated hemolytic anemia, and crisis treatment—a blood transfusion and the administration of corticosteroids and chemotherapeutic agents—was begun.

1. Describe the shape of normal red blood cells, and define hemolysis.

2. Explain why Ursula's body temperature had dropped significantly.

3. Provide a plausible explanation for Ursula's fatigue.

MasteringA&P™

Access more review material online in the Study Area at **www.masteringaandp.com**.

There, you'll find:
- **Chapter guides**
- **Chapter quizzes**
- **Practice tests**
- **Labeling activities**
- **MP3 Tutor Sessions**
- **Animations**
- **Flashcards**
- **A glossary with pronunciations**

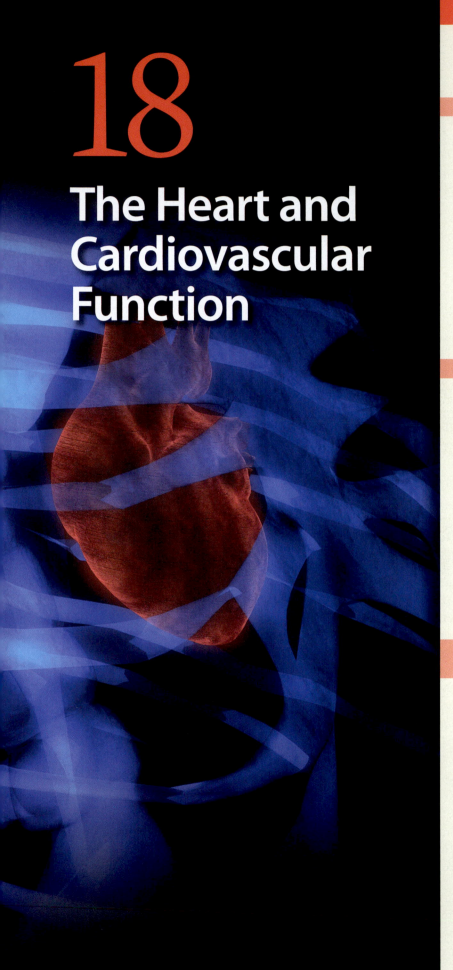

18

The Heart and Cardiovascular Function

The Structure of the Heart

1 The heart is located near the anterior chest wall, directly posterior to the sternum. A midsagittal section through the trunk does not divide the heart into two equal halves, because the center lies slightly to the left of the midline, and the entire heart is rotated to the left, so that the right atrium and right ventricle dominate an anterior view of the heart.

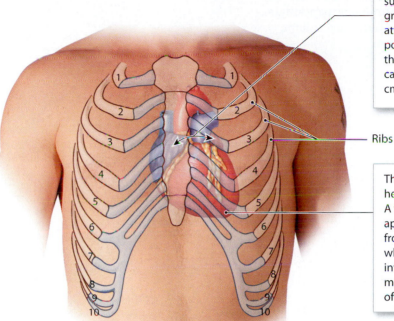

The **base** of the heart is at the superior surface, where the great veins and arteries are attached. The base sits posterior to the sternum at the level of the third costal cartilage, centered about 1.2 cm (0.5 in.) to the left side.

Ribs

The inferior, pointed tip of the heart is the free **apex** (Ā-peks). A typical adult heart measures approximately 12.5 cm (5 in.) from the base to the apex, which reaches the fifth intercostal space approximately 7.5 cm (3 in.) to the left of the midline.

2 This anterior view illustrates the borders of the heart. The base forms the **superior border**. The **right border** of the heart is formed by the right atrium; the **left border** is formed by the left ventricle and a small portion of the left atrium. The left border extends to the apex, where it meets the inferior border. The **inferior border** is formed mainly by the inferior wall of the right ventricle.

This section examines the anatomy of the heart. We will then consider the regulation of cardiac function (Section 2) before considering how cardiac and vasomotor activities are coordinated.

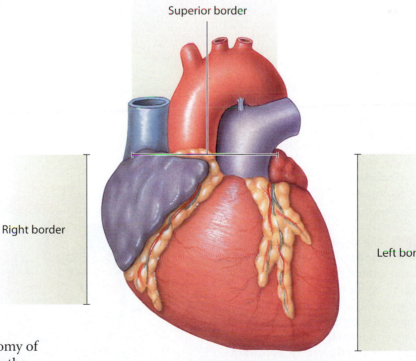

Superior border

Right border

Left border

Inferior border

The wall of the heart contains concentric layers of cardiac muscle tissue

1 This is a view of a section taken from the wall of the heart and the surrounding pericardium. The heart wall contains three layers: epicardium, myocardium, and endocardium.

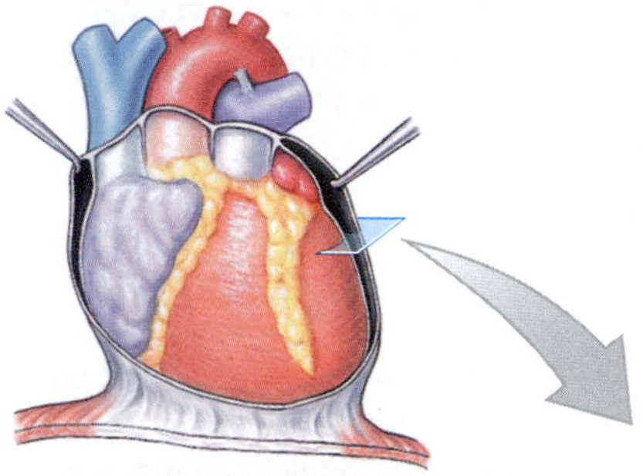

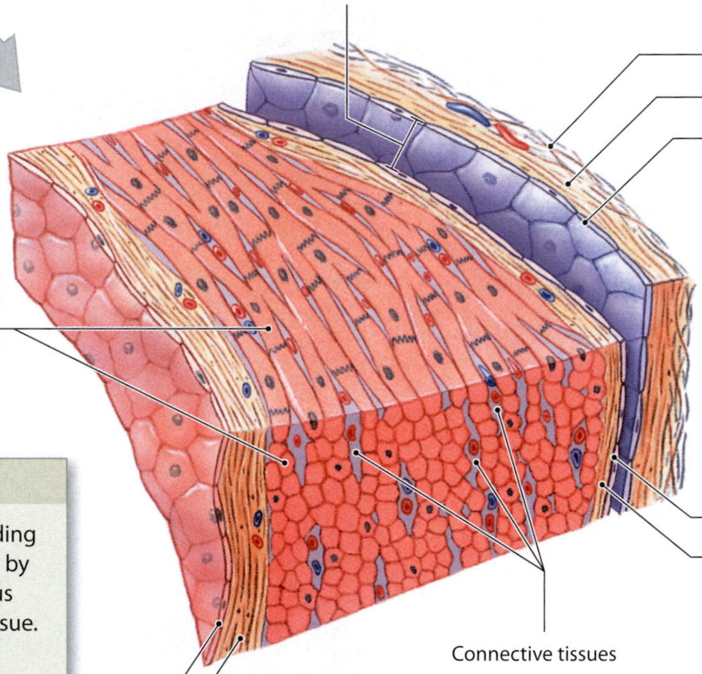

The Pericardium

The relationship between the heart and the pericardium was detailed in Module 1.10, p.28.

Parietal Pericardium

The **parietal pericardium** is the serous membrane that forms the outer wall of the pericardial cavity. The parietal pericardium is reinforced by a dense fibrous layer; together they form the **pericardial sac** that surrounds the heart.

Pericardial cavity (contains serous fluid)

Dense fibrous layer
Areolar tissue
Mesothelium

Epicardium

The **epicardium**, or visceral pericardium, covers the outer surface of the heart. This serous membrane consists of an exposed mesothelium and an underlying layer of areolar tissue that is attached to the myocardium.

Mesothelium
Areolar tissue

Myocardium

The **myocardium**, or muscular wall of the heart, forms both atria and ventricles. This middle layer contains cardiac muscle tissue, blood vessels, and nerves. The myocardium consists of concentric layers of cardiac muscle tissue.

Endocardium

The inner surfaces of the heart, including those of the heart valves, are covered by the **endocardium**, a simple squamous epithelium and underlying areolar tissue. The squamous epithelial lining of the cardiovascular system is called an endothelium; the endothelium of the heart is continuous with the endothelium of the attached great vessels.

Connective tissues

Endothelium
Areolar tissue

2 The atrial myocardium contains muscle bundles that wrap around the atria and form figure-eights that encircle the great vessels. Superficial ventricular muscles wrap around both ventricles; deeper muscle layers spiral around and between the ventricles toward the apex in a figure-eight pattern.

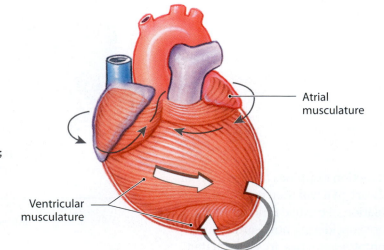

Atrial musculature

Ventricular musculature

3 This light micrograph shows the histological characteristics that distinguish cardiac muscle tissue from skeletal muscle tissue: (1) small cell size, (2) a single, centrally located nucleus, (3) branching interconnections between cells, and (4) specialized intercellular connections.

Each cardiac muscle cell is connected to several others at specialized sites known as **intercalated** (in-TER-ka-lā-ted) **discs**.

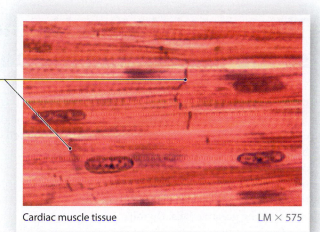

Cardiac muscle tissue   LM × 575

4 Cardiac muscle cells are found only in the heart. Like skeletal muscle fibers, cardiac muscle cells contain organized myofibrils, and the presence of many aligned sarcomeres gives the cells a striated appearance. They are almost totally dependent on aerobic metabolism to obtain the energy they need to continue contracting. The sarcoplasm of a cardiac muscle cell contains large numbers of mitochondria and abundant reserves of myoglobin that store oxygen. Because these cells are metabolically very active and have a high demand for oxygen and nutrients, cardiac tissues are richly supplied with capillaries.

Cardiac muscle cells are relatively small, averaging 10–20 μm in diameter and 50–100 μm in length.

Intercalated disc (sectioned)

Nucleus

Mitochondria

Bundles of myofibrils

Intercalated disc

5 At an intercalated disc, the plasma membranes of two adjacent cardiac muscle cells are extensively intertwined and bound together by gap junctions and desmosomes. These connections help stabilize the relative positions of adjacent cells. The gap junctions allow ions and small molecules to move from one cell to another. This creates a direct electrical connection between the two muscle cells. An action potential can travel across an intercalated disc, moving quickly from one cardiac muscle cell to another. Myofibrils in the two interlocking muscle cells are firmly anchored to the membrane at the intercalated disc and can "pull together" with maximum efficiency. Because the cardiac muscle cells are mechanically, chemically, and electrically connected to one another, the entire tissue resembles a single, enormous muscle cell. For this reason, cardiac muscle has been called a **functional syncytium** (sin-SISH-ē-um; a fused mass of cells).

Gap junction

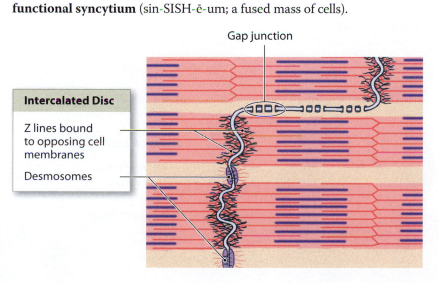

Intercalated Disc

Z lines bound to opposing cell membranes

Desmosomes

Module 18.1 Review

a. From superficial to deep, name the layers of the heart wall.

b. Describe the tissue layers of the epicardium.

c. Why is it important that cardiac tissue be richly supplied with mitochondria and capillaries?

The heart is located in the mediastinum, suspended within the pericardial cavity

1 The position and orientation of the heart relative to the major vessels and the ribs, sternum, and lungs can be seen in this anterior view. The heart, surrounded by the pericardial sac, sits in the anterior portion of the **mediastinum** (mē-dē-AS-ti-num or mē-dē-a-STĪ-num), the region between the two pleural cavities. The mediastinum also contains the great vessels, thymus, esophagus, and trachea.

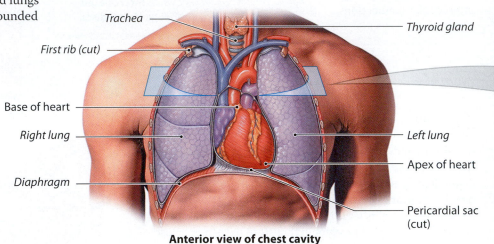

Trachea

Thyroid gland

First rib (cut)

Base of heart

Right lung

Left lung

Apex of heart

Diaphragm

Pericardial sac (cut)

Anterior view of chest cavity

2 To visualize the relationship between the heart and the pericardial cavity, imagine pushing your fist toward the center of a large, partially inflated balloon. The balloon represents the pericardium, and your fist is the heart. Your wrist, where the balloon folds back on itself, corresponds to the base of the heart, to which the **great vessels**, the largest veins and arteries in the body, are attached. The air space inside the balloon corresponds to the pericardial cavity.

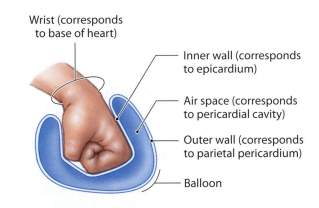

Wrist (corresponds to base of heart)

Inner wall (corresponds to epicardium)

Air space (corresponds to pericardial cavity)

Outer wall (corresponds to parietal pericardium)

Balloon

The pericardial cavity contains 15–50 mL of pericardial fluid, secreted by the pericardial membranes. This fluid acts as a lubricant that reduces friction between the opposing surfaces as the heart beats. Pathogens can infect the pericardium, producing the condition **pericarditis**. The inflamed pericardial surfaces rub against one another, producing a distinctive scratching sound that can be heard through a stethoscope.

Base of heart

Cut edge of parietal pericardium

Fibrous tissue of pericardial sac

Parietal Pericardium

Areolar tissue

Mesothelium

Cut edge of epicardium

Fibrous attachment to diaphragm

Apex of heart

3 This is a diagrammatic superior view of a partial dissection of the thoracic cavity. This three-dimensional image illustrates the position of the pericardial cavity and the physical relationships among the components in the mediastinum.

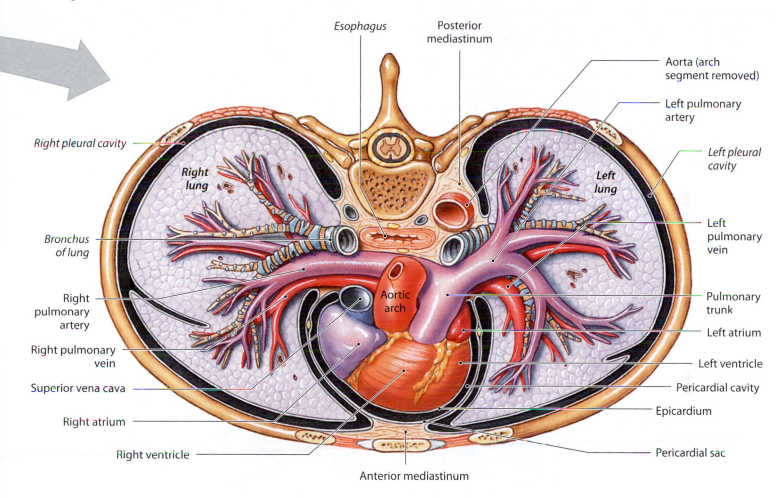

- Esophagus
- Posterior mediastinum
- Aorta (arch segment removed)
- Left pulmonary artery
- Right pleural cavity
- Left pleural cavity
- Right lung
- Left lung
- Bronchus of lung
- Left pulmonary vein
- Right pulmonary artery
- Aortic arch
- Pulmonary trunk
- Left atrium
- Right pulmonary vein
- Left ventricle
- Superior vena cava
- Pericardial cavity
- Epicardium
- Right atrium
- Pericardial sac
- Right ventricle
- Anterior mediastinum

As you see in this section, the heart does not have a lot of empty space around it—the thoracic cavity is very crowded. Traumatic injuries that damage the pericardium or chest wall can result in fluid accumulation within the pericardial cavity, which can restrict the movement of the heart. This condition, called **cardiac tamponade** (tam-po-NĀD; *tampon*, plug), can also be caused by acute pericarditis.

Module 18.2 Review

a. Define mediastinum.

b. Describe the heart's location.

c. Why can cardiac tamponade be a life-threatening condition?

The boundaries between the chambers of the heart can be distinguished on its external surface

1 The four cardiac chambers can easily be identified in a superficial view of the **anterior surface** of the heart. The two atria have relatively thin muscular walls and are highly expandable. When not filled with blood, the outer portion of each atrium deflates and becomes a lumpy, wrinkled flap. Shallow grooves, or **sulci** (singular: *sulcus*), mark the boundaries between the atria and ventricles and between the left and right ventricles. The connective tissue of the epicardium generally contains substantial amounts of fat, especially along the sulci. In fresh or preserved hearts, this fat must be stripped away to expose the underlying grooves. These sulci also contain the arteries and veins that carry blood to and from the cardiac muscle.

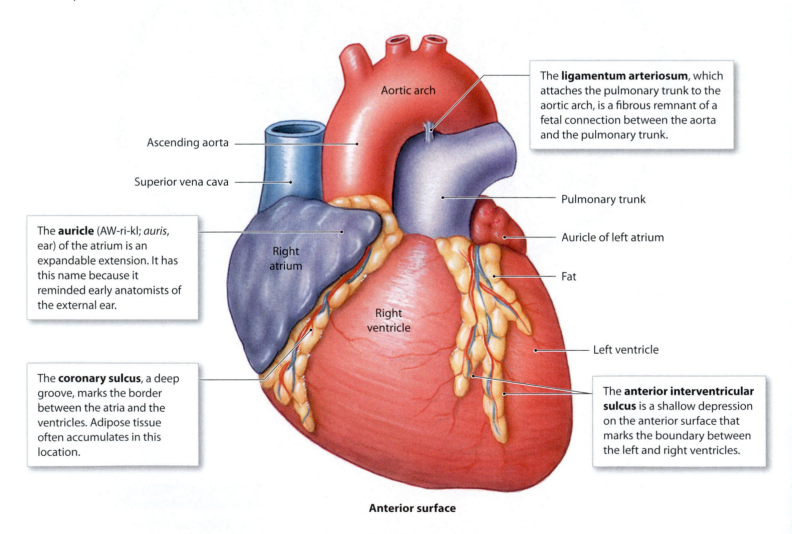

Aortic arch

Ascending aorta

Superior vena cava

The **ligamentum arteriosum**, which attaches the pulmonary trunk to the aortic arch, is a fibrous remnant of a fetal connection between the aorta and the pulmonary trunk.

Pulmonary trunk

The **auricle** (AW-ri-kl; *auris*, ear) of the atrium is an expandable extension. It has this name because it reminded early anatomists of the external ear.

Right atrium

Auricle of left atrium

Fat

Right ventricle

Left ventricle

The **coronary sulcus**, a deep groove, marks the border between the atria and the ventricles. Adipose tissue often accumulates in this location.

The **anterior interventricular sulcus** is a shallow depression on the anterior surface that marks the boundary between the left and right ventricles.

Anterior surface

2 This view of the **posterior surface** of the heart shows the left atrium and its connection to the pulmonary veins, as well as the right atrium and its connection to the coronary veins and the venae cavae.

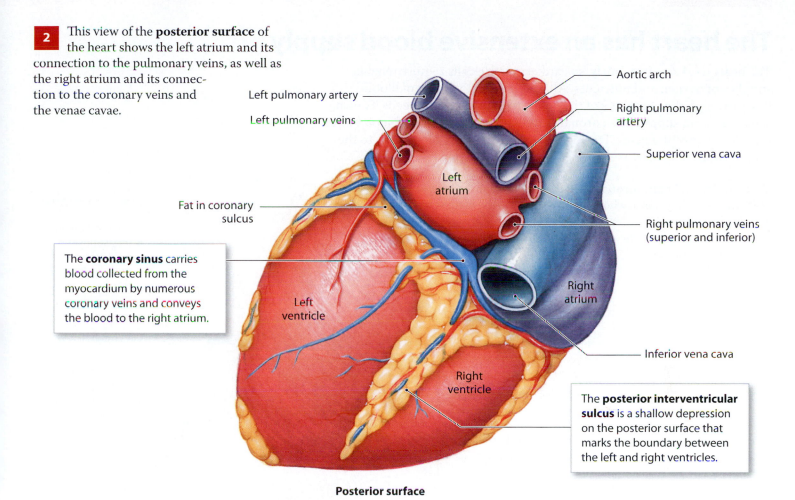

Left pulmonary artery

Left pulmonary veins

Aortic arch

Right pulmonary artery

Superior vena cava

Left atrium

Fat in coronary sulcus

The **coronary sinus** carries blood collected from the myocardium by numerous coronary veins and conveys the blood to the right atrium.

Right pulmonary veins (superior and inferior)

Right atrium

Left ventricle

Inferior vena cava

Right ventricle

The **posterior interventricular sulcus** is a shallow depression on the posterior surface that marks the boundary between the left and right ventricles.

Posterior surface

3 As you see in this anterior view, a dissected heart from a preserved cadaver is not conveniently color-coded.

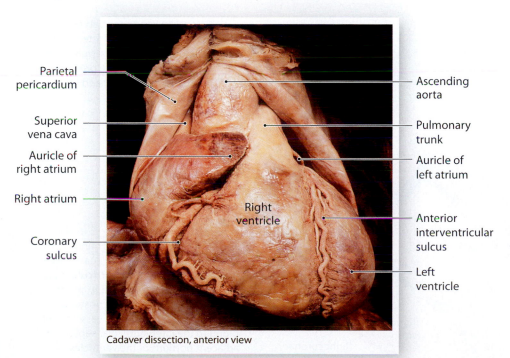

Parietal pericardium

Superior vena cava

Auricle of right atrium

Right atrium

Coronary sulcus

Ascending aorta

Pulmonary trunk

Auricle of left atrium

Right ventricle

Anterior interventricular sulcus

Left ventricle

Cadaver dissection, anterior view

Module 18.3 Review

a. Name the four cardiac chambers.

b. Name and describe the shallow depressions and grooves found on the heart's external surface.

c. Which structures collect blood from the myocardium, and into which heart chamber does this blood flow?

The heart has an extensive blood supply

The heart works continuously, so cardiac muscle cells require reliable supplies of oxygen and nutrients. Although a great volume of blood flows through the chambers of the heart, the myocardium needs its own, separate blood supply. The coronary circulation supplies that blood to the muscle tissue of the heart. During maximum exertion, blood flow to the myocardium may increase to nine times that of resting levels.

1 The left and right **coronary arteries** originate at the base of the ascending aorta, where blood pressure is the highest in the systemic circuit. However, myocardial blood flow is not steady; it peaks while the heart muscle is relaxed, and almost ceases while it contracts.

Left Coronary Artery

The **left coronary artery** supplies blood to the left ventricle, left atrium, and interventricular septum.

Circumflex artery

The large **anterior interventricular artery** runs along the surface within the anterior interventricular sulcus.

Right Coronary Artery

The **right coronary artery**, which follows the coronary sulcus around the heart, supplies blood to the right atrium, portions of both ventricles, and portions of the conducting system of the heart, which contols and coordinates the heartbeat.

Marginal arteries from the right coronary artery supply the surface of the right ventricle.

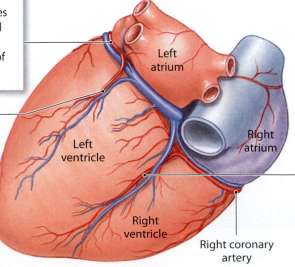

Anterior view

2 Branches of the left and right coronary arteries continue onto the posterior surface of the heart.

The **circumflex artery** is a branch of the left coronary artery that curves to the left around the coronary sulcus, eventually meeting and fusing with small branches of the right coronary artery. A marginal artery branches from the circumflex artery to supply the posterior surface of the left ventricle.

Arterial anastomoses between the anterior and posterior interventricular arteries maintain a relatively continuous blood flow despite pressure fluctuations in the left and right coronary arteries.

The right coronary artery continues across the posterior surface of the heart, supplying the **posterior interventricular artery**, or posterior descending artery, which runs toward the apex within the posterior interventricular sulcus. This vessel supplies blood to the interventricular septum and adjacent portions of the ventricles.

Posterior view

3 This view identifies the major collecting vessels on the anterior surface of the heart.

Aortic arch

Left atrium

Right atrium

Right ventricle

Left ventricle

The **anterior cardiac veins**, which drain the anterior surface of the right ventricle, empty directly into the right atrium.

The **great cardiac vein** begins on the anterior surface of the ventricles, along the interventricular sulcus. This vein drains blood from the region supplied by the anterior interventricular artery. The great cardiac vein reaches the level of the atria and then curves around the left side of the heart within the coronary sulcus to empty into the coronary sinus.

Anterior view

4 This view identifies the major collecting vessels on the posterior surface of the heart.

Great cardiac vein

Left atrium

The **coronary sinus** is an expanded vein that opens into the right atrium near the base of the inferior vena cava.

Left ventricle

Right atrium

The **posterior cardiac vein** drains the area supplied by the circumflex artery.

Right ventricle

The **small cardiac vein** receives blood from the posterior surfaces of the right atrium and ventricle. It empties into the coronary sinus in company with the middle cardiac vein.

The **middle cardiac vein**, draining the area supplied by the posterior interventricular artery, empties into the coronary sinus.

Posterior view

Each time the left ventricle contracts, it forces blood into the aorta. The arrival of additional blood at elevated pressures stretches the elastic walls of the aorta. When the left ventricle relaxes, pressure declines, and the walls of the aorta recoil. This recoil, called **elastic rebound**, pushes blood both forward, into the systemic circuit, and backward, into the coronary arteries. Thus, the combination of blood pressure and elastic rebound ensures a continuous flow of blood to meet the demands of active cardiac muscle tissue.

Module 18.4 Review

a. List the arteries and veins of the heart.

b. Describe what happens to blood flow during elastic rebound.

c. Identify the main vessel that collects venous blood from the myocardium.

Internal valves control the direction of blood flow between the heart chambers

1 In a sectional view, you can see that the right atrium communicates with the right ventricle, and the left atrium with the left ventricle. The atria are separated by the **interatrial septum** (*septum*, wall); the ventricles are separated by the much thicker **interventricular septum**. Each septum is a muscular partition. **Atrioventricular (AV) valves**, folds of fibrous tissue, extend into the openings between the atria and ventricles. These valves permit blood flow in one direction only: from the atria to the ventricles.

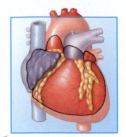

Right Atrium

The right atrium receives blood from the superior and inferior venae cavae. It also receives blood from the cardiac veins through the coronary sinus.

The **fossa ovalis** marks the location of the foramen ovale in the embryo. The foramen ovale closes at birth, and this opening between the atria is permanently sealed off during the next 3 months.

The anterior atrial wall and the inner surface of the auricle contain prominent muscular ridges called the **pectinate muscles** (*pectin*, comb).

The opening of the coronary sinus carries blood from the cardiac veins.

Right Ventricle

Blood travels from the right atrium into the right ventricle through a broad opening bounded by the **right atrioventricular (AV) valve**, also known as the **tricuspid** (trī- KUS-pid; *tri*, three) **valve**.

The free edge of each valve consists of three flaps, or **cusps**, attached to tendinous connective tissue fibers called the **chordae tendineae** (KOR-dē TEN-di-nē-ē; tendinous cords).

The fibers originate at conical muscular projections called the **papillary** (PAP-i-ler-ē) **muscles**.

The superior portion of the right ventricle tapers toward the **pulmonary valve**, or pulmonary semilunar valve. Blood leaving the right ventricle passes through this valve to enter the pulmonary trunk.

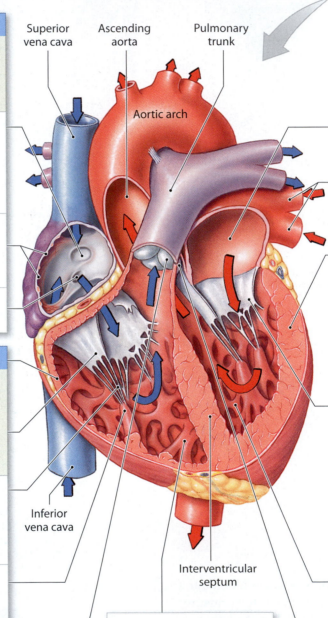

Superior vena cava

Ascending aorta

Pulmonary trunk

Aortic arch

Inferior vena cava

Interventricular septum

Left Atrium

The left atrium receives blood from the pulmonary veins.

Left pulmonary veins

Left Ventricle

The left ventricle is much larger than the right ventricle. Its thick, muscular wall enables the left ventricle to develop pressure sufficient to push blood through the large systemic circuit, whereas the right ventricle needs to pump blood, at lower pressure, through the nearby lungs.

The **left atrioventricular (AV) valve**, or **bicuspid** (bī-KUS-pid) **valve**, permits the flow of blood from the left atrium into the left ventricle but prevents backflow during ventricular contraction. As the name bicuspid implies, the left AV valve contains a pair, not a trio, of cusps. Clinicians often call this valve the **mitral** (MĪ-tral; *mitre*, a bishop's hat) **valve**.

The **trabeculae carneae** (tra-BEK-ū-lē KAR-nē-ē; *carneus*, fleshy) are a series of muscular ridges on the inner surfaces of the right and left ventricles.

Blood leaves the left ventricle by passing through the **aortic valve** (aortic semilunar valve) and into the ascending aorta.

The **moderator band** provides a rapid-conduction path that tenses the papillary muscles before the ventricular myocardium contracts; this prevents "slamming" of the right AV cusps.

2 The function of an atrium is to collect blood that is returning to the heart and convey it to the attached ventricle. The functional demands on the right and left atria are similar, and the two chambers look almost identical. The demands on the right and left ventricles, however, are very different, and the two have significant structural differences. These differences are best seen in sectional view.

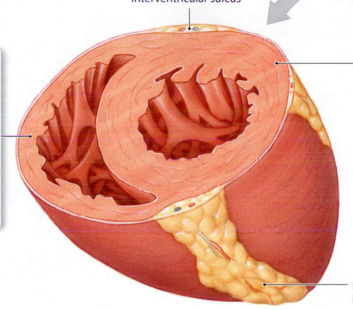

Posterior interventricular sulcus

In sectional view, the relatively thin wall of the right ventricle resembles a pouch attached to the massive wall of the left ventricle. Such a thin wall is adequate because the right ventricle normally does not need to work very hard to push blood through the pulmonary circuit—the lungs are close to the heart, and the pulmonary vessels are relatively short and wide.

The left ventricle has an extremely thick muscular wall and is round in cross section. It must develop four to six times as much pressure to push blood around the systemic circuit as the right ventricle develops to push blood around the pulmonary circuit.

Fat in anterior interventricular sulcus

3 These sections indicate the changes in ventricular shape that occur when the ventricles contract.

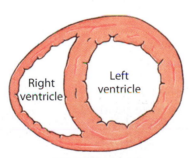

Right ventricle

Left ventricle

Dilated (relaxed)

When the right ventricle contracts, it acts like a bellows, squeezing the blood against the thick wall of the left ventricle. This mechanism moves blood very efficiently with minimal effort, but it develops relatively low pressures.

When the left ventricle contracts, (1) the diameter of the ventricular chamber decreases, and (2) the distance between the base and apex decreases. The effect is similar to simultaneously squeezing and rolling up the end of a toothpaste tube. This motion also reduces the volume of the right ventricle, and individuals whose right ventricular musculature has been severely damaged may still survive, because the contraction of the left ventricle helps push blood into the pulmonary circuit.

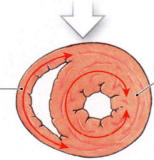

Contracted

Module 18.5 Review

a. Damage to the semilunar valves on the right side of the heart would affect blood flow to which vessel?

b. What prevents the AV valves from swinging into the atria?

c. Why is the left ventricle more muscular than the right ventricle?

When the heart beats, the AV valves close before the semilunar valves open, and the semilunar valves close before the AV valves open

1 When the ventricles are relaxed, the chordae tendineae are loose, and the AV valves offer no resistance to the flow of blood from the atria into the ventricles. Blood pressure in the pulmonary and systemic circuits keeps the aortic and pulmonary valves closed until the ventricles contract.

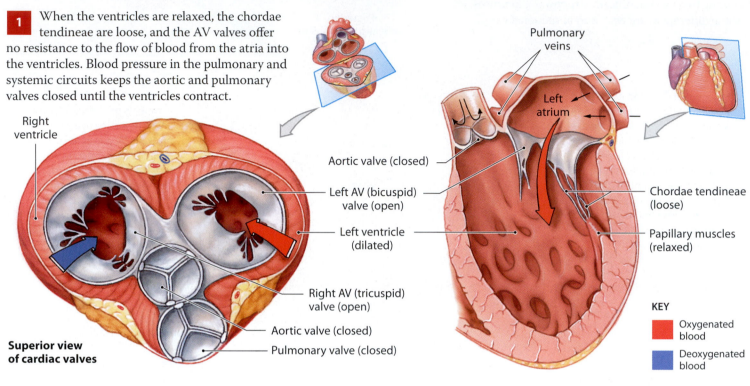

Pulmonary veins

Left atrium

Aortic valve (closed)

Left AV (bicuspid) valve (open)

Left ventricle (dilated)

Chordae tendineae (loose)

Papillary muscles (relaxed)

Right ventricle

Right AV (tricuspid) valve (open)

Aortic valve (closed)

Pulmonary valve (closed)

Superior view of cardiac valves

KEY

- Oxygenated blood
- Deoxygenated blood

2 When the ventricles contract, blood moving back toward the atria pushes the cusps of the AV valves together, closing them and preventing backflow. At the same time, the contraction of the papillary muscles tenses the chordae tendineae, stopping the cusps before they swing into the atria. If the chordae tendineae are cut or the papillary muscles are damaged, backflow (**regurgitation**) of blood into the atria occurs each time the ventricles contract. As the ventricular pressures rise above those in the pulmonary and systemic circuits, the aortic and pulmonary valves open and blood flows out of the ventricles.

The **aortic sinuses** are saclike dilations adjacent to each cusp of the aortic valve. The right and left coronary arteries originate at the aortic sinuses.

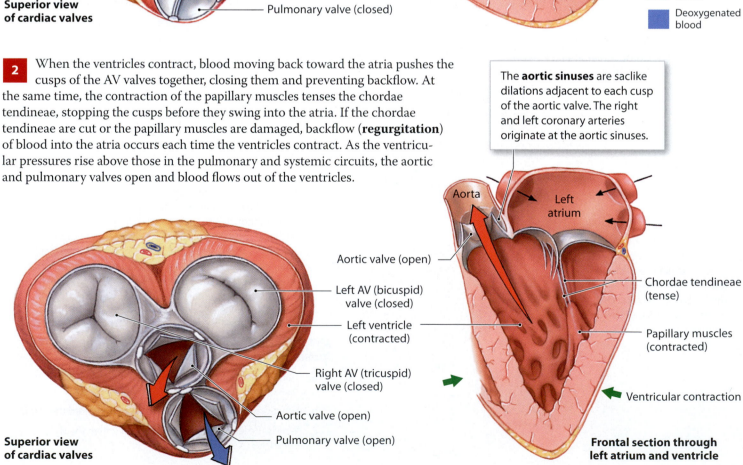

Aorta

Left atrium

Aortic valve (open)

Left AV (bicuspid) valve (closed)

Left ventricle (contracted)

Right AV (tricuspid) valve (closed)

Aortic valve (open)

Pulmonary valve (open)

Superior view of cardiac valves

Chordae tendineae (tense)

Papillary muscles (contracted)

Ventricular contraction

Frontal section through left atrium and ventricle

3 The heart valves are encircled and supported by flexible connective tissues known as the **cardiac skeleton**. The cardiac skeleton consists of dense bands of tough elastic tissue that encircle the heart valves and the bases of the pulmonary trunk and aorta. These bands of connective tissue stabilize the positions of the heart valves and ventricular muscle cells and isolate the ventricular myocardium from the atrial myocardium.

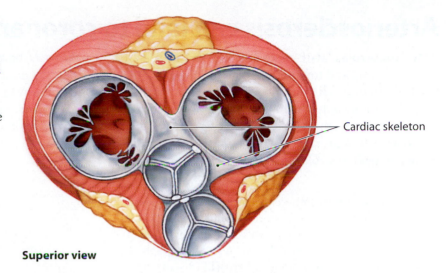

Cardiac skeleton

Superior view

4 The pulmonary and aortic valves each consist of three semilunar (half-moon shaped) cusps of thick connective tissue. Unlike the AV valves, the **semilunar valves** do not require muscular braces, because the cusps are stable. When the semilunar valves close, the three symmetrical cusps support one another like the legs of a tripod.

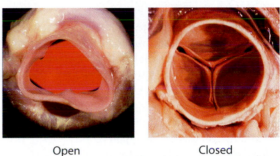

Open   Closed

Superior views

5 Serious valve problems can interfere with cardiac function. If valve function deteriorates to the point at which the heart cannot maintain adequate blood flow, symptoms of **valvular heart disease (VHD)** appear. Congenital malformations may be responsible, but in many cases the condition develops as a consequence of **carditis**, an inflammation of the heart. In severe cases, the only option may be to replace the damaged valve with a prosthetic valve.

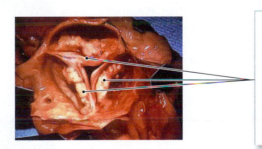

This is a superior view of a damaged aortic valve. The cusps are irregular in shape; they are also stiff and relatively inflexible. Such a valve would not open properly and cannot close completely.

This bioprosthetic valve is an example of an artificial valve that uses the cusps from a pig's heart. Pig or cow valves do not stimulate the clotting system, but they may wear out after roughly 10 years of service.

Module 18.6 Review

a. Define cardiac regurgitation.

b. Compare the structure of the tricuspid valve with that of the pulmonary valve.

c. What do semilunar valves prevent?

Arteriosclerosis can lead to coronary artery disease

Arteriosclerosis (ar-tē-rē-ō-skle-RŌ-sis; *arterio-*, artery + *sklerosis,* hardness) is a thickening and toughening of arterial walls. This condition may not sound life-threatening, but complications related to arteriosclerosis account for roughly half of all deaths in the United States. The effects of arteriosclerosis are varied; for example, arteriosclerosis of coronary vessels is responsible for coronary artery disease (CAD), and arteriosclerosis of arteries supplying the brain can lead to strokes.

1 **Atherosclerosis** (ath-er-ō-skler-Ō-sis; *athero-*, fatty degeneration) is the formation of lipid deposits in the arterial tunica media associated with damage to the endothelial lining. Here is a histological view and a photograph of a dissected atherosclerotic artery. Atherosclerosis, the most common form of arteriosclerosis, tends to develop in people whose blood contains elevated levels of plasma lipids—specifically, cholesterol. The circulating lipids are removed and deposited within arterial walls, which become increasingly abnormal in structure. The result is an atherosclerotic **plaque**, a fatty mass of tissue that projects into the lumen of the vessel and restricts blood flow. Elderly individuals—especially elderly men—are most likely to develop atherosclerotic plaques. In addition to being older and male, other important risk factors for atherosclerosis include high blood cholesterol levels, high blood pressure, and cigarette smoking. Roughly 20 percent of middle-aged men have all three of these risk factors; these individuals are four times as likely to experience a heart attack as other men in their age group. Although fewer women develop atherosclerotic plaques, elderly female smokers with high blood cholesterol and high blood pressure are at much greater risk than other women.

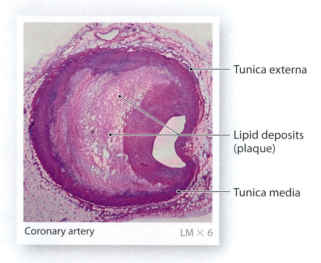

Tunica externa

Lipid deposits (plaque)

Tunica media

Coronary artery   LM × 6

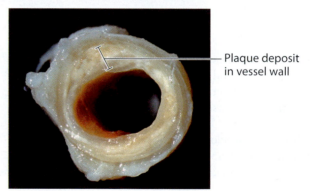

Plaque deposit in vessel wall

2 Plaques can be treated by removing the damaged segment of the vessel and replacing it (often with a superficial vein removed from the leg), but such surgery can be difficult and dangerous. In **balloon angioplasty** (AN-jē-ō-plas-tē; *angeion*, vessel), the tip of a catheter contains an inflatable balloon. Once in position, the balloon is inflated, pressing the plaque against the vessel walls. This photo shows the catheter within an artery of a cadaver; the artery has been opened to show the orientation and relative sizes of the catheter and artery. Balloon angioplasty is most effective in treating small, soft plaques. Several factors make this a highly attractive treatment: (1) the mortality rate during surgery is only about 1 percent; (2) the success rate is over 90 percent; and (3) the procedure can be performed on an outpatient basis.

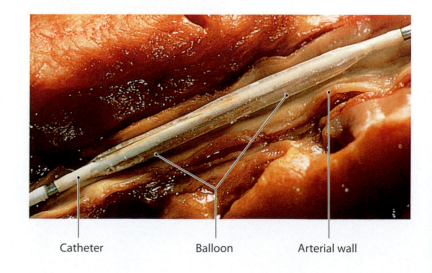

Catheter   Balloon   Arterial wall

3 The term **coronary artery disease (CAD)** refers to areas of partial or complete blockage of coronary circulation. Cardiac muscle cells need a constant supply of oxygen and nutrients, so any reduction in blood flow to the heart muscle produces a corresponding reduction in cardiac performance. Such reduced circulatory supply, known as **coronary ischemia** (is-KĒ-mē-uh), generally results from partial or complete blockage of the coronary arteries. The usual cause is the formation of an atherosclerotic plaque in a coronary artery. The plaque, or an associated **thrombus** (blood clot), then narrows the passageway and reduces blood flow. Spasms in the smooth muscles of the vessel wall can further decrease or even stop blood flow. Plaques may be visible in clinical scans or high-resolution ultrasound images.

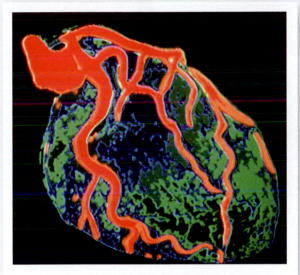

This is a color-enhanced **digital subtraction angiography (DSA)** scan of a normal heart. The major branches of the left and right coronary arteries are clearly visible.

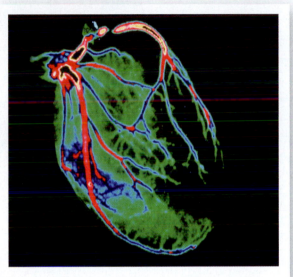

This is a color-enhanced DSA scan of the heart of an individual with advanced CAD. Blood flow to the ventricular myocardium is severely restricted.

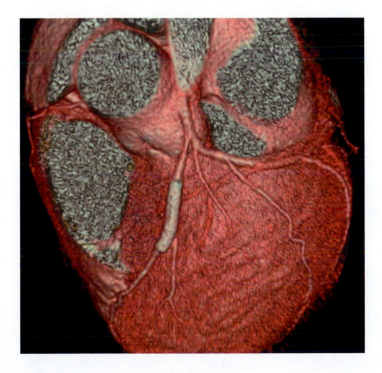

4 Because plaques commonly rebound or redevelop after angioplasty, a fine wire-mesh tube called a **stent** may be inserted into the vessel. The stent pushes against the vessel wall, holding it open. This is a scan of a stent that has been placed in the anterior interventricular artery. (This imaging technique reveals the lumen of the blood vessels, rather than the vessel, so the stent appears to surround the vessel.) Stents are routinely used by many cardiac specialists because their long-term success rate and incidence of complications are significantly lower than those for balloon angioplasty alone. If the circulatory blockage is extensive, mutiple stents can be inserted along the length of the vessel.

Module 18.7 Review

a. Compare arteriosclerosis with atherosclerosis.

b. What is coronary ischemia, and what danger does it pose?

c. Describe the purpose of a stent.

1. Labeling

Label each of the structures in the accompanying figure.

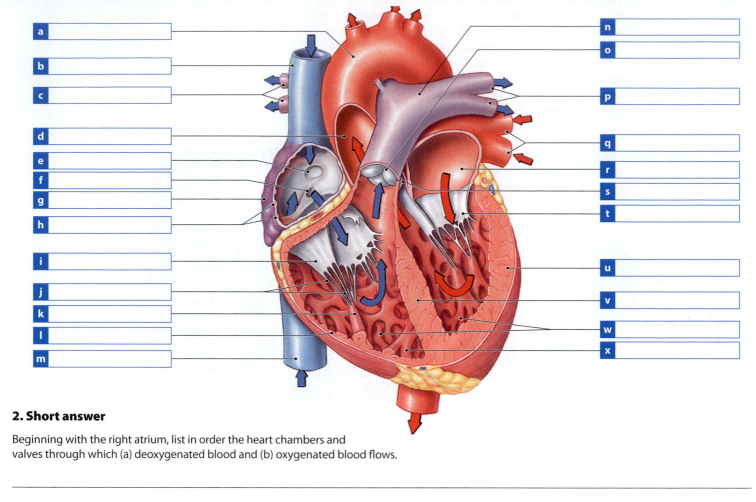

a

b

c

d

e

f

g

h

i

j

k

l

m

n

o

p

q

r

s

t

u

v

w

x

2. Short answer

Beginning with the right atrium, list in order the heart chambers and valves through which (a) deoxygenated blood and (b) oxygenated blood flows.

3. Matching

Match the following terms with the most closely related description.

- cardiac skeleton
- fossa ovalis
- intercalated discs
- serous membrane
- tricuspid valve
- aortic valve
- endocardium
- aorta
- myocardium
- anastomoses
- coronary sinus
- calcium ions

a _____ Blood to systemic arteries

b _____ Activates contraction

c _____ Muscular wall of heart

d _____ Returns blood from heart back to heart

e _____ Depression in interatrial septum

f _____ Right atrioventricular valve

g _____ Interconnections betwen blood vessels

h _____ Cardiac muscle fiber connections

i _____ Pericardium

j _____ Inner surface of heart

k _____ Stabilizing connective tissue

l _____ Semilunar valve

The Cardiac Cycle

1   Each heartbeat is followed by a brief resting phase, which allows time for the chambers to relax and prepare for the next heartbeat. The period between the start of one heartbeat and the beginning of the next is a single **cardiac cycle**. The cardiac cycle, therefore, includes alternating periods of contraction and relaxation.

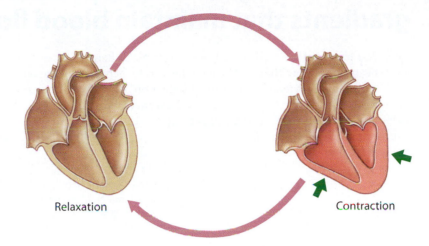

Relaxation   Contraction

2   Although we think of the heart as a pump, it is really four pumps that work in pairs, and thus a heartbeat is a complicated event. If all four chambers contracted at once, normal blood flow couldn't occur. Instead, the two atria contract first, pushing blood into the ventricles, and then the two ventricles contract, pushing blood through the pulmonary and systemic circuits and into the atria. The elaborate pacemaking and conducting systems within the heart normally provide the required spacing between atrial and ventricular contractions.

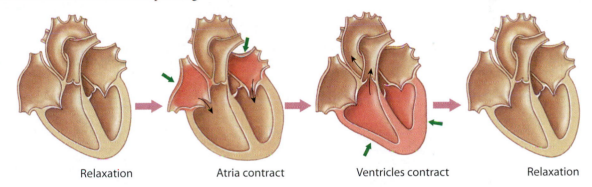

Relaxation   Atria contract   Ventricles contract   Relaxation

3   For any one chamber in the heart, the cardiac cycle can be divided into two phases: (1) systole and (2) diastole. During **systole** (SIS-tō-lē), or contraction, the chamber contracts and pushes blood into an adjacent chamber or into an arterial trunk. Systole is followed by **diastole** (dī-AS-tō-lē), or relaxation. During diastole, the chamber fills with blood and prepares for the next contraction. At a representative heart rate of 75 beats per minute (bpm), a sequence of systole and diastole in either the atria or the ventricles lasts 800 msec. For convenience, we will assume that the cardiac cycle is determined by the atria, and that it includes one cycle of atrial systole and atrial diastole. This section will examine the details of a representative cardiac cycle—how it is initiated and coordinated and how the pressures generated by the contracting chambers result in directional blood flow.

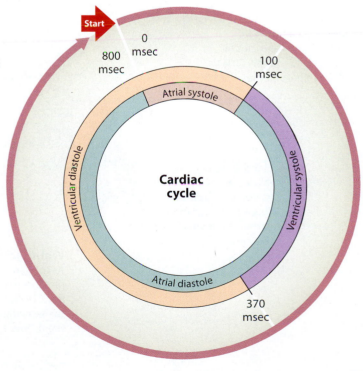

Start

0 msec

800 msec

100 msec

Atrial systole

Ventricular diastole

Ventricular systole

Cardiac cycle

Atrial diastole

370 msec

641

The cardiac cycle creates pressure gradients that maintain blood flow

1 The phases of the cardiac cycle are diagrammed here for a heart rate of 75 beats per minute (bpm). When the heart rate increases, all the phases of the cardiac cycle are shortened. The greatest reduction occurs in the length of time spent in diastole. When the heart rate climbs from 75 bpm to 200 bpm, the time spent in systole drops by less than 40 percent, but the duration of diastole is reduced by almost 75 percent.

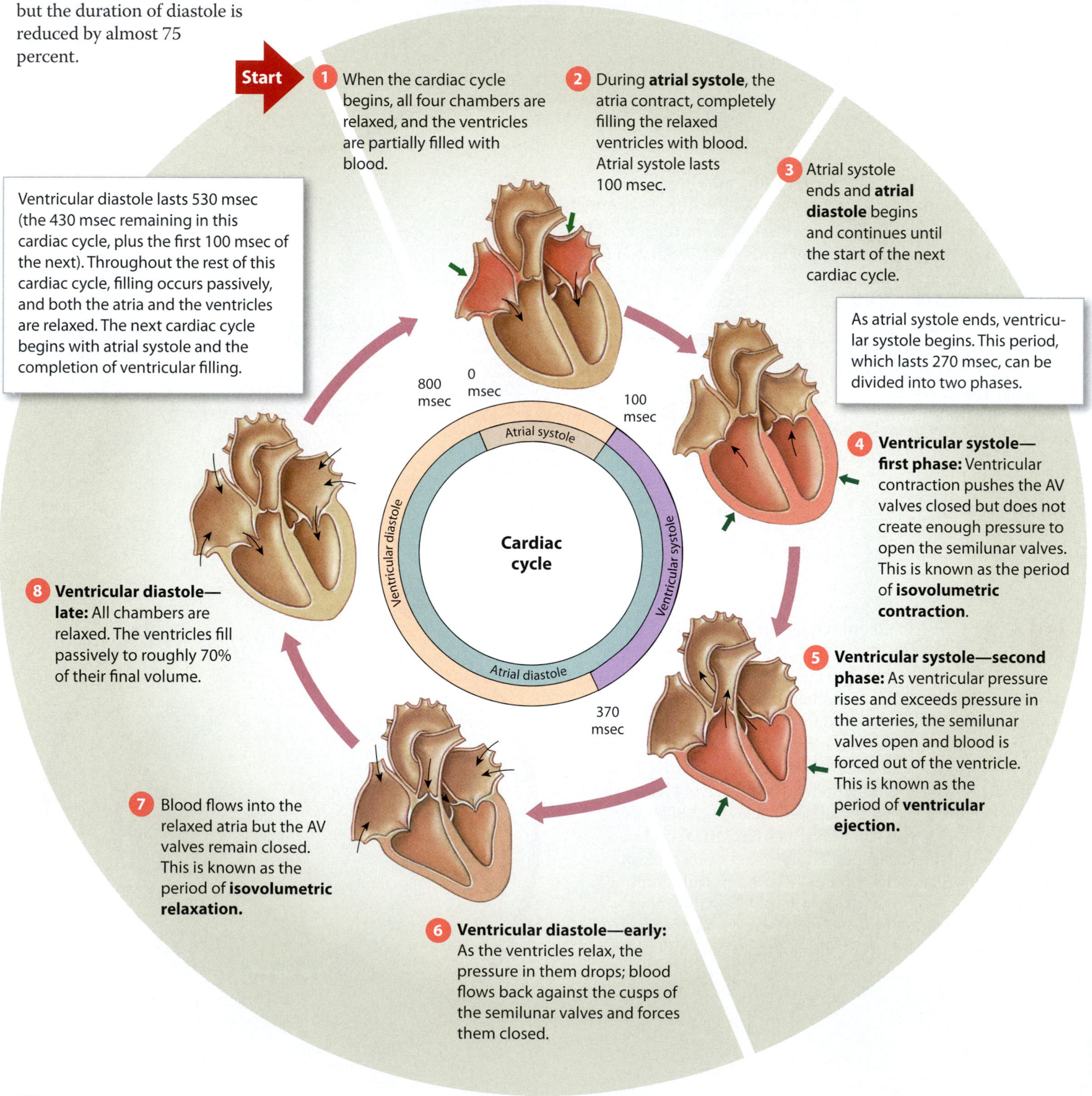

Start

1 When the cardiac cycle begins, all four chambers are relaxed, and the ventricles are partially filled with blood.

2 During **atrial systole**, the atria contract, completely filling the relaxed ventricles with blood. Atrial systole lasts 100 msec.

3 Atrial systole ends and **atrial diastole** begins and continues until the start of the next cardiac cycle.

Ventricular diastole lasts 530 msec (the 430 msec remaining in this cardiac cycle, plus the first 100 msec of the next). Throughout the rest of this cardiac cycle, filling occurs passively, and both the atria and the ventricles are relaxed. The next cardiac cycle begins with atrial systole and the completion of ventricular filling.

As atrial systole ends, ventricular systole begins. This period, which lasts 270 msec, can be divided into two phases.

4 **Ventricular systole— first phase:** Ventricular contraction pushes the AV valves closed but does not create enough pressure to open the semilunar valves. This is known as the period of **isovolumetric contraction**.

5 **Ventricular systole—second phase:** As ventricular pressure rises and exceeds pressure in the arteries, the semilunar valves open and blood is forced out of the ventricle. This is known as the period of **ventricular ejection**.

8 **Ventricular diastole— late:** All chambers are relaxed. The ventricles fill passively to roughly 70% of their final volume.

7 Blood flows into the relaxed atria but the AV valves remain closed. This is known as the period of **isovolumetric relaxation**.

6 **Ventricular diastole—early:** As the ventricles relax, the pressure in them drops; blood flows back against the cusps of the semilunar valves and forces them closed.

800 msec   0 msec   100 msec

370 msec

Atrial systole

Ventricular diastole

Ventricular systole

Cardiac cycle

Atrial diastole

2 This graph plots the pressure changes within the aorta, the left atrium, and the left ventricle during the cardiac cycle. Although pressures are lower in the right atrium and right ventricle, the same principles apply; both sides of the heart contract at the same time, and they eject equal volumes of blood.

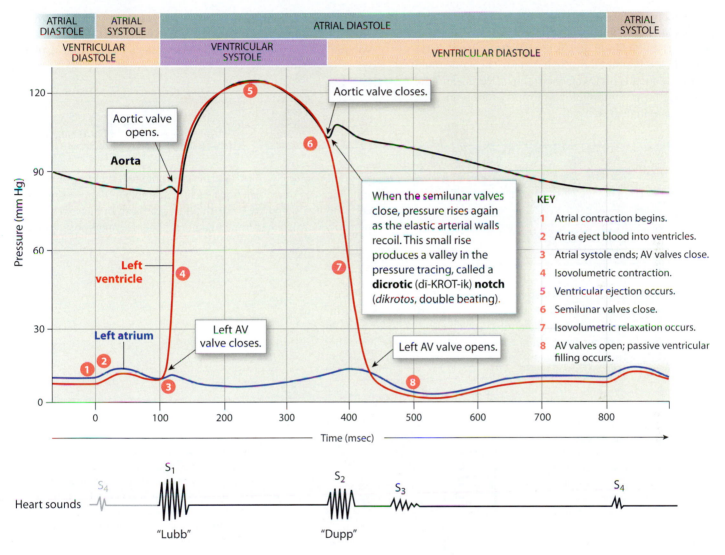

When the semilunar valves close, pressure rises again as the elastic arterial walls recoil. This small rise produces a valley in the pressure tracing, called a **dicrotic** (dī-KROT-ik) **notch** (*dikrotos*, double beating).

KEY

1 Atrial contraction begins.
2 Atria eject blood into ventricles.
3 Atrial systole ends; AV valves close.
4 Isovolumetric contraction.
5 Ventricular ejection occurs.
6 Semilunar valves close.
7 Isovolumetric relaxation occurs.
8 AV valves open; passive ventricular filling occurs.

3 There are four heart sounds, designated as **S₁** through **S₄**. If you listen to your own heart with a stethoscope, you will clearly hear the first and second heart sounds. The first heart sound, known as "lubb" (S_1), lasts a little longer than the second, called "dupp" (S_2). S_1, which marks the start of ventricular contraction, is produced as the AV valves close; S_2 occurs when the semilunar valves close. Third and fourth heart sounds are usually very faint and seldom are audible in healthy adults. These sounds are associated with blood flowing into the ventricles (S_3) and atrial contraction (S_4), rather than with valve action.

Module 18.8 Review

a. Provide the alternate terms for heart contraction and heart relaxation.

b. List the phases of the cardiac cycle.

c. Is the heart always pumping blood when pressure in the left ventricle is rising? Explain.

The heart rate, a key factor in cardiac output, is established by the SA node and distributed by the conducting system

1 The goal of cardiovascular regulation is the maintenance of adequate blood flow to vital tissues. The best overall indicator of peripheral blood flow is the **cardiac output (CO)**, the amount of blood pumped by the left ventricle in one minute. Cardiac output depends on two factors: the heart rate and the **stroke volume**, the amount of blood pumped out of the ventricle during a single heartbeat.

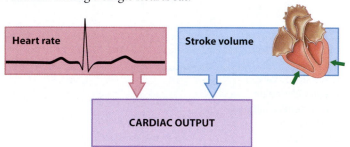

2 The calculation of cardiac output is very straightforward: you multiply the heart rate (HR) by the average stroke volume (SV). For example, if the heart rate is 75 bpm and stroke volume is 80 mL/beat:

The body precisely adjusts cardiac output such that peripheral tissues receive an adequate circulatory supply under a variety of conditions. When necessary, the heart rate can increase by 250 percent, and stroke volume in a normal heart can almost double.

3 Cardiac muscle tissue contracts on its own, in the absence of neural or hormonal stimulation. This property is called **automaticity**. The **conducting system** is a network of specialized cardiac muscle cells responsible for initiating and distributing the stimulus to contract. The following illustration introduces the components of the conducting system and their specific functions.

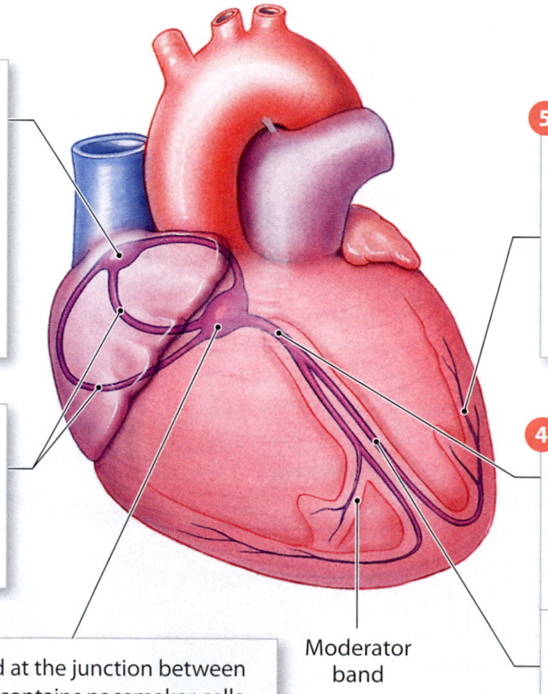

1 Each heartbeat begins with an action potential generated at the **sinoatrial** (sī-nō-Ā-trē-al) **node**, or simply the **SA node**. The SA node is embedded in the posterior wall of the right atrium, near the entrance of the superior vena cava. The electrical impulse generated by this cardiac pacemaker is then distributed by other cells of the conducting system.

2 In the atria, conducting cells are found in **internodal pathways**, which distribute the contractile stimulus to atrial muscle cells as the impulse travels toward the ventricles.

3 The **atrioventricular (AV) node** is located at the junction between the atria and ventricles. The AV node also contains pacemaker cells, but they do not ordinarily affect the heart rate. However, if the SA node or internodal pathways are damaged, the heart will continue to beat because in the absence of commands from the SA node, the AV node will generate impulses at a rate of 40–60 beats per minute.

5 **Purkinje fibers** are large-diameter conducting cells that propagate action potentials very rapidly—as fast as small myelinated axons. Purkinje cells are the final link in the distribution network, and they are responsible for the depolarization of the ventricular myocardial cells that triggers ventricular systole.

4 The AV node delivers the stimulus to the **AV bundle**, located within the interventricular septum. The AV bundle is normally the only electrical connection between the atria and the ventricles.

The AV bundle leads to the right and left **bundle branches**. The left bundle branch, which supplies the massive left ventricle, is much larger than the right bundle branch. Both branches extend toward the apex of the heart, turn, and fan out deep to the endocardial surface.

Moderator band

1 An action potential is generated at the SA node, and atrial activation begins.

SA node

Time = 0

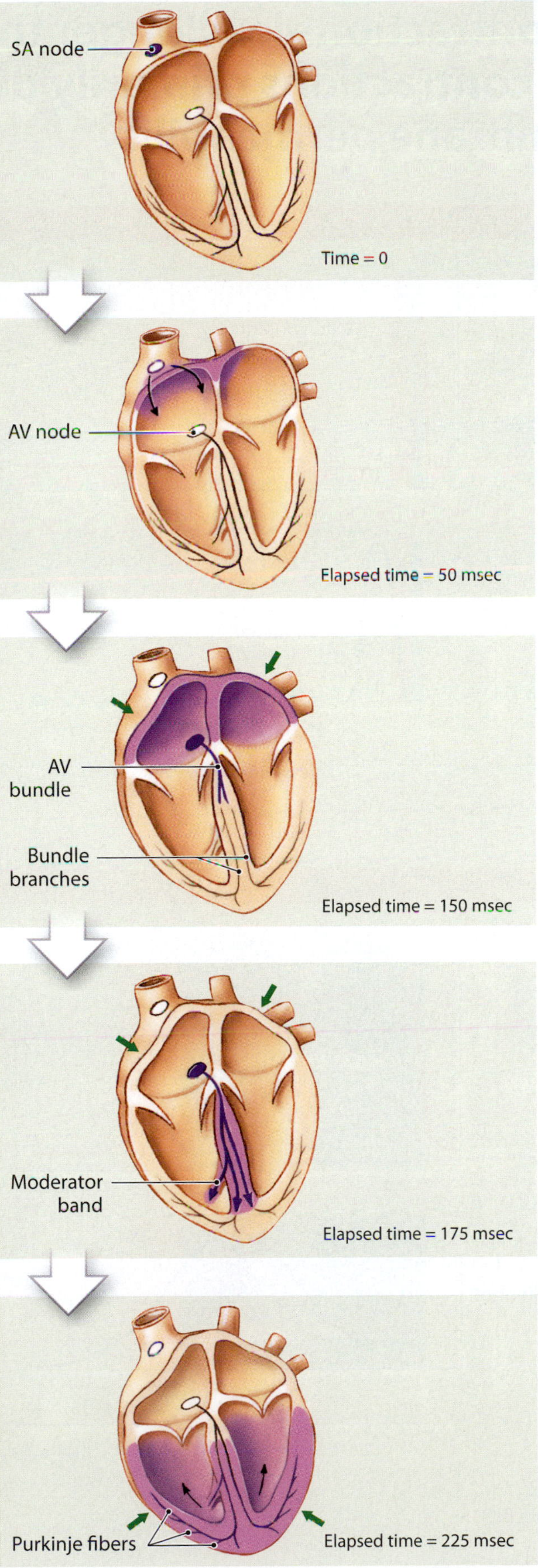

4 This sequence illustrates the distribution of the contractile stimulus and shows how the conducting system coordinates the contractions of the cardiac cycle. The P wave begins at time zero in this figure, and atrial contraction—the start of the cardiac cycle, detailed in Module 18.8—begins 50 msec after an impulse is generated at the SA node.

2 The stimulus spreads across the atrial surfaces by cell-to-cell contact within the internodal pathways and soon reaches the AV node.

AV node

Elapsed time = 50 msec

3 A 100-msec delay occurs at the AV node. During this delay, atrial contraction occurs.

AV bundle

Bundle branches

Elapsed time = 150 msec

The cells of the AV node can conduct impulses at a maximum rate of 230 per minute. Because each impulse results in a ventricular contraction, this value is the maximum normal heart rate. Even if the SA node generates impulses at a faster rate, the ventricles will still contract at 230 bpm. Higher heart rates occur only when the heart or the conducting system has been damaged or stimulated by drugs.

4 As atrial contraction is completed, the impulse travels along the interventricular septum within the AV bundle and the bundle branches to the Purkinje fibers and, via the moderator band, to the papillary muscles of the right ventricle. Ventricular contraction begins.

Moderator band

Elapsed time = 175 msec

5 The impulse is distributed by Purkinje fibers and relayed throughout the ventricular myocardium. Ventricular contraction reaches full force and proceeds to completion.

Purkinje fibers

Elapsed time = 225 msec

Module 18.9 Review

a. Define automaticity.

b. If the cells of the SA node failed to function, how would the heart rate be affected?

c. Why is it important for impulses from the atria to be delayed at the AV node before they pass into the ventricles?

Cardiac muscle cell contractions last longer than skeletal muscle fiber contractions primarily due to differences in membrane permeability

Skeletal Muscle

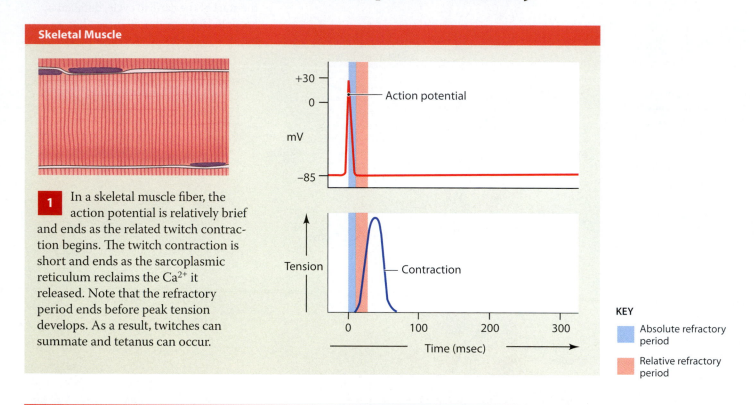

1 In a skeletal muscle fiber, the action potential is relatively brief and ends as the related twitch contraction begins. The twitch contraction is short and ends as the sarcoplasmic reticulum reclaims the Ca^{2+} it released. Note that the refractory period ends before peak tension develops. As a result, twitches can summate and tetanus can occur.

KEY

- Absolute refractory period
- Relative refractory period

Cardiac Muscle

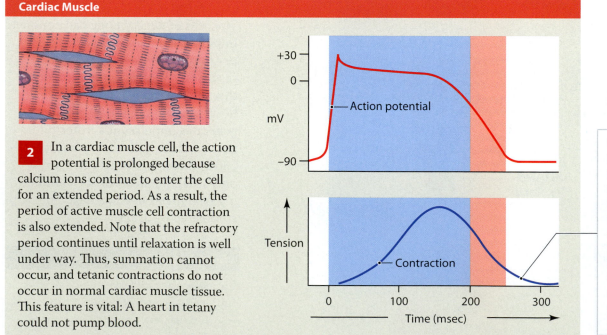

2 In a cardiac muscle cell, the action potential is prolonged because calcium ions continue to enter the cell for an extended period. As a result, the period of active muscle cell contraction is also extended. Note that the refractory period continues until relaxation is well under way. Thus, summation cannot occur, and tetanic contractions do not occur in normal cardiac muscle tissue. This feature is vital: A heart in tetany could not pump blood.

With a single twitch lasting 250 msec or longer, a normal cardiac muscle cell could reach 300–400 contractions per minute under maximum stimulation. The normal heart rate never gets that high because the stimulus for contraction cannot spread that quickly through the heart muscle.

3 The action potential in a cardiac muscle cell can be divided into three stages.

Rapid Depolarization

The stage of rapid depolarization in a cardiac muscle cell resembles that in a skeletal muscle fiber. At threshold, voltage-gated sodium channels open, and the membrane suddenly becomes permeable to Na^+. The result is a massive influx of sodium ions and the rapid depolarization of the sarcolemma. The channels involved are called **fast sodium channels**, because they open quickly and remain open for only a few milliseconds.

Plateau

During the plateau, the transmembrane potential remains near 0 mV. Two opposing factors are involved. As the transmembrane potential approaches +30 mV, the voltage-gated sodium channels close and the cell begins actively pumping Na^+ out of the cell. However, as the sodium channels are closing, voltage-gated calcium channels are opening. These channels are called **slow calcium channels**, because they open slowly and remain open for a relatively long period—roughly 175 msec. While the slow calcium channels are open, the entry of Ca^{2+} roughly balances the loss of Na^+, and the transmembrane potential hovers near 0 mV.

Repolarization

After approximately 175 msec, slow calcium channels begin closing, and slow potassium channels begin opening. As the channels open, potassium ions (K^+) rush out of the cell, and the net result is a period of rapid repolarization that restores the resting potential.

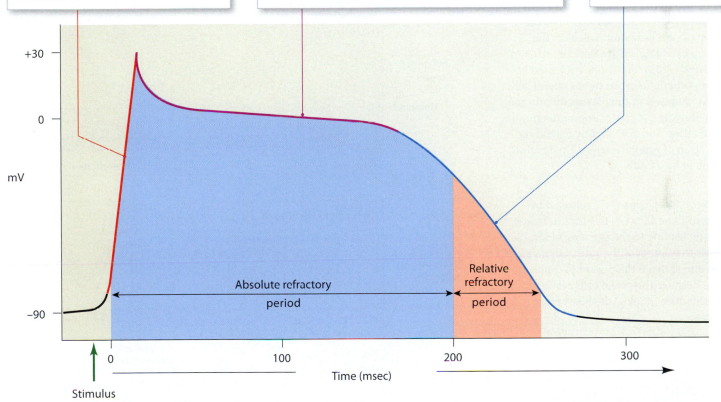

Module 18.10 Review

a. List the three stages of an action potential in a cardiac muscle cell.

b. Describe slow calcium channels and the significance of their activity.

c. Why does tetany not occur in cardiac muscle?

The intrinsic heart rate can be altered by autonomic activity

1 Cells of the SA and AV nodes cannot maintain a stable resting potential. After each repolarization, the membrane gradually drifts toward threshold. This gradual depolarization is called a **prepotential** or **pacemaker potential**. The rate of spontaneous depolarization is fastest at the SA node, which in the absence of neural or hormonal stimulation generates action potentials at a rate of 80–100 per minute. Because the SA node reaches threshold first, it establishes the heart rate—the impulse generated by the SA node brings the AV nodal cells to threshold before the prepotential of the AV nodal cells can do so.

2 Any factor that changes the rate of spontaneous depolarization or the duration of repolarization will alter the heart rate by changing the time required to reach threshold. Acetylcholine released by parasympathetic neurons opens chemically gated K^+ channels in the plasma membrane, thereby slowing the rate of spontaneous depolarization and also slightly extending the duration of repolarization. The result is a decline in heart rate.

3 Norepinephrine (NE) released by sympathetic neurons binds to beta-1 receptors, leading to the opening of ion channels that increase the rate of depolarization and shorten the period of repolarization. Because the nodal cells reach threshold more quickly, the heart rate increases.

Normal (resting)

Membrane potential (mV)

+20
0
−30 Threshold
−60

Prepotential (spontaneous depolarization)

Heart rate: 75 bpm

0.8   1.6   2.4

Parasympathetic stimulation

Membrane potential (mV)

+20
0
−30 Threshold
−60

Hyperpolarization

Slower depolarization

Heart rate: 40 bpm

0.8   1.6   2.4

Sympathetic stimulation

Membrane potential (mV)

+20
0
−30 Threshold
−60

Reduced repolarization

More rapid depolarization

Heart rate: 120 bpm

0.8   1.6   2.4

Time (sec)

4 Each individual has a characteristic resting heart rate that varies with age, general health, and physical conditioning. However, there is a normal range of heart rates; the American Heart Association considers 60–100 bpm to be the normal range of resting heart rates.

Bradycardia (brād-ē-KAR-dē-uh; *bradys*, slow) is a condition in which the heart rate is slower than normal.

← 60 bpm

Normal range of resting heart rates

100 bpm →

Tachycardia (tak-ē-KAR-dē-uh; *tachys*, swift) indicates a faster-than-normal heart rate.

5 The cardiac centers of the medulla oblongata contain the autonomic headquarters for cardiac control. These centers innervate the heart by means of the cardiac plexus.

The **cardioinhibitory center** controls the parasympathetic neurons that slow the heart rate.

The **cardioacceleratory center** controls sympathetic neurons that increase the heart rate.

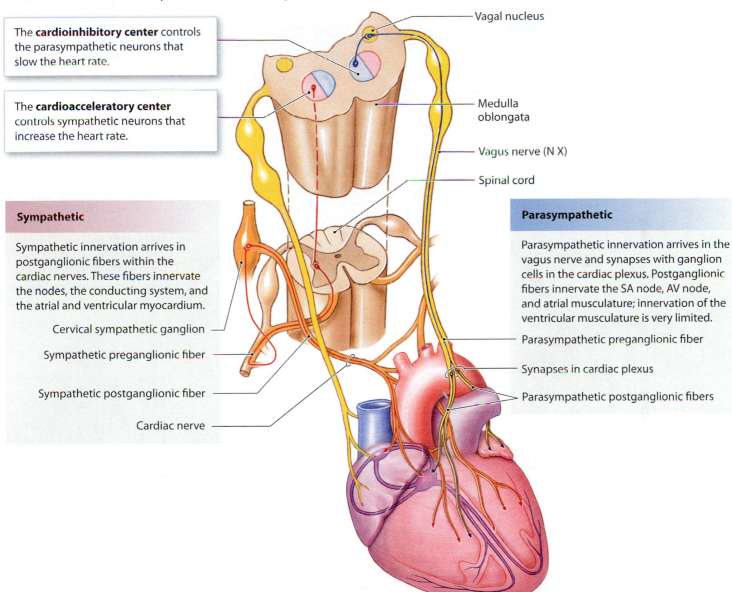

Vagal nucleus

Medulla oblongata

Vagus nerve (N X)

Spinal cord

Sympathetic

Sympathetic innervation arrives in postganglionic fibers within the cardiac nerves. These fibers innervate the nodes, the conducting system, and the atrial and ventricular myocardium.

Cervical sympathetic ganglion

Sympathetic preganglionic fiber

Sympathetic postganglionic fiber

Cardiac nerve

Parasympathetic

Parasympathetic innervation arrives in the vagus nerve and synapses with ganglion cells in the cardiac plexus. Postganglionic fibers innervate the SA node, AV node, and atrial musculature; innervation of the ventricular musculature is very limited.

Parasympathetic preganglionic fiber

Synapses in cardiac plexus

Parasympathetic postganglionic fibers

The activities of the cardiac centers are regulated through reflex pathways and by input from higher centers, especially from the parasympathetic and sympathetic headquarters in the hypothalamus. Autonomic tone exists at the heart, as both divisions are chronically active. The normal resting heart rate is somewhat slower than the intrinsic SA nodal stimulation rate of 80–100 bpm, because at rest the effects of parasympathetic innervation predominate.

Module 18.11 Review

a. Compare bradycardia with tachycardia.

b. Describe the sites and actions of the cardio-inhibitory and cardioacceleratory centers.

c. Caffeine has effects on conducting cells and contractile cells that are similar to those of NE. What effect would drinking large amounts of caffeinated beverages have on the heart rate?

Stroke volume depends on the relationship between EDV and ESV

1 The **stroke volume** of the heart can be compared to pumping water with a manual pump. The amount pumped varies with the amount of movement of the pump handle. Although the heart contains two large pumps—the right and left ventricles—we can use a single pump as a model because the two work in the same way and pump equal amounts of blood.

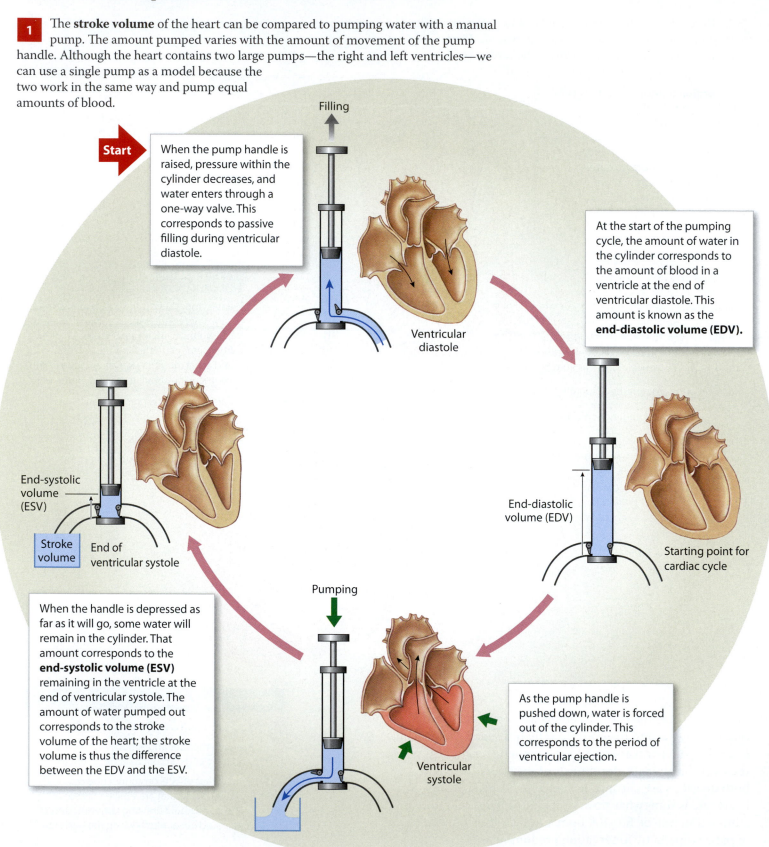

Start

When the pump handle is raised, pressure within the cylinder decreases, and water enters through a one-way valve. This corresponds to passive filling during ventricular diastole.

Filling

At the start of the pumping cycle, the amount of water in the cylinder corresponds to the amount of blood in a ventricle at the end of ventricular diastole. This amount is known as the **end-diastolic volume (EDV).**

Ventricular diastole

End-systolic volume (ESV)

Stroke volume

End of ventricular systole

End-diastolic volume (EDV)

Starting point for cardiac cycle

When the handle is depressed as far as it will go, some water will remain in the cylinder. That amount corresponds to the **end-systolic volume (ESV)** remaining in the ventricle at the end of ventricular systole. The amount of water pumped out corresponds to the stroke volume of the heart; the stroke volume is thus the difference between the EDV and the ESV.

Pumping

Ventricular systole

As the pump handle is pushed down, water is forced out of the cylinder. This corresponds to the period of ventricular ejection.

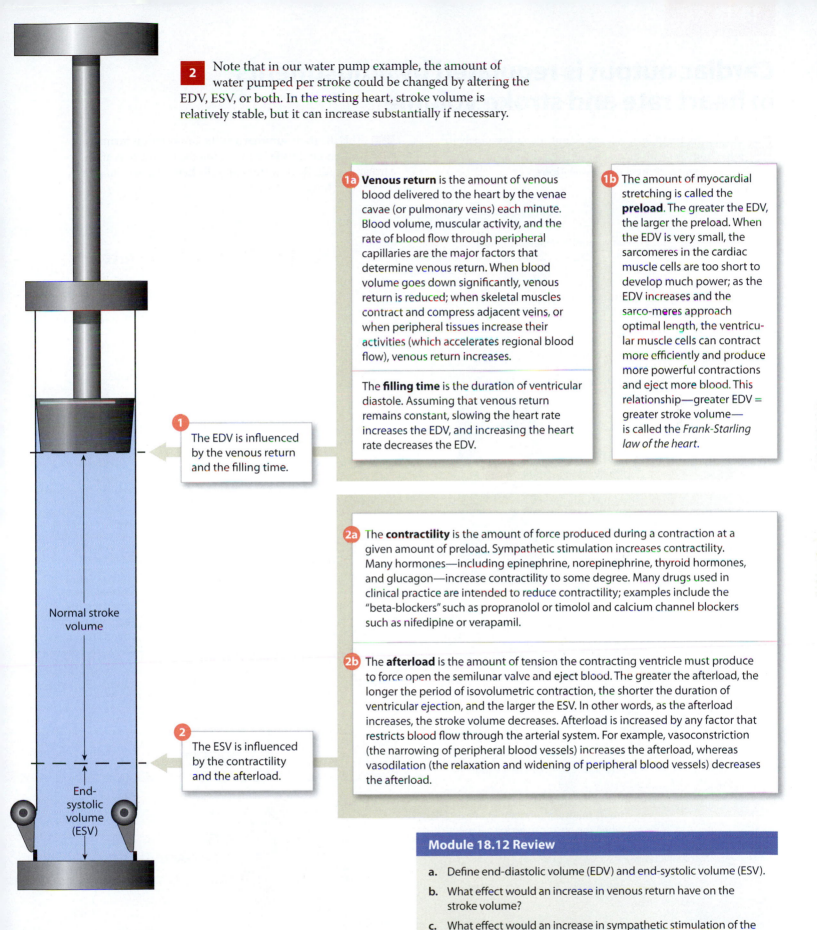

2 Note that in our water pump example, the amount of water pumped per stroke could be changed by altering the EDV, ESV, or both. In the resting heart, stroke volume is relatively stable, but it can increase substantially if necessary.

1a **Venous return** is the amount of venous blood delivered to the heart by the venae cavae (or pulmonary veins) each minute. Blood volume, muscular activity, and the rate of blood flow through peripheral capillaries are the major factors that determine venous return. When blood volume goes down significantly, venous return is reduced; when skeletal muscles contract and compress adjacent veins, or when peripheral tissues increase their activities (which accelerates regional blood flow), venous return increases.

The **filling time** is the duration of ventricular diastole. Assuming that venous return remains constant, slowing the heart rate increases the EDV, and increasing the heart rate decreases the EDV.

1b The amount of myocardial stretching is called the **preload**. The greater the EDV, the larger the preload. When the EDV is very small, the sarcomeres in the cardiac muscle cells are too short to develop much power; as the EDV increases and the sarco-meres approach optimal length, the ventricular muscle cells can contract more efficiently and produce more powerful contractions and eject more blood. This relationship—greater EDV = greater stroke volume— is called the *Frank-Starling law of the heart*.

1 The EDV is influenced by the venous return and the filling time.

2a The **contractility** is the amount of force produced during a contraction at a given amount of preload. Sympathetic stimulation increases contractility. Many hormones—including epinephrine, norepinephrine, thyroid hormones, and glucagon—increase contractility to some degree. Many drugs used in clinical practice are intended to reduce contractility; examples include the "beta-blockers" such as propranolol or timolol and calcium channel blockers such as nifedipine or verapamil.

2b The **afterload** is the amount of tension the contracting ventricle must produce to force open the semilunar valve and eject blood. The greater the afterload, the longer the period of isovolumetric contraction, the shorter the duration of ventricular ejection, and the larger the ESV. In other words, as the afterload increases, the stroke volume decreases. Afterload is increased by any factor that restricts blood flow through the arterial system. For example, vasoconstriction (the narrowing of peripheral blood vessels) increases the afterload, whereas vasodilation (the relaxation and widening of peripheral blood vessels) decreases the afterload.

2 The ESV is influenced by the contractility and the afterload.

Module 18.12 Review

a. Define end-diastolic volume (EDV) and end-systolic volume (ESV).

b. What effect would an increase in venous return have on the stroke volume?

c. What effect would an increase in sympathetic stimulation of the heart have on the end-systolic volume (ESV)?

Cardiac output is regulated by adjustments in heart rate and stroke volume

1 As this figure indicates, cardiac output varies widely to meet metabolic demands, and the body must closely monitor and tightly control those variations.

2 This diagram summarizes the important information presented thus far concerning the factors that affect cardiac output. Review this carefully before proceeding to the next section.

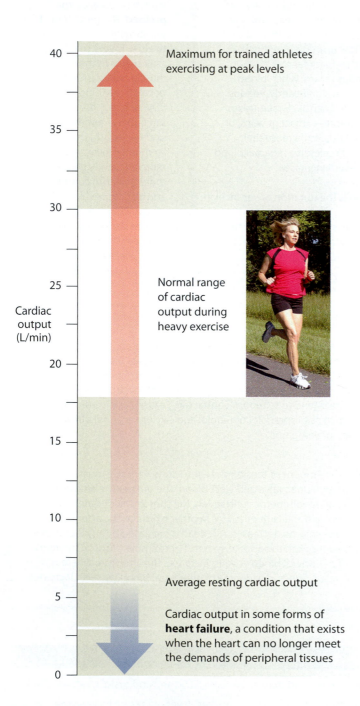

40 — Maximum for trained athletes exercising at peak levels

Cardiac output (L/min)

Normal range of cardiac output during heavy exercise

Average resting cardiac output

Cardiac output in some forms of **heart failure**, a condition that exists when the heart can no longer meet the demands of peripheral tissues

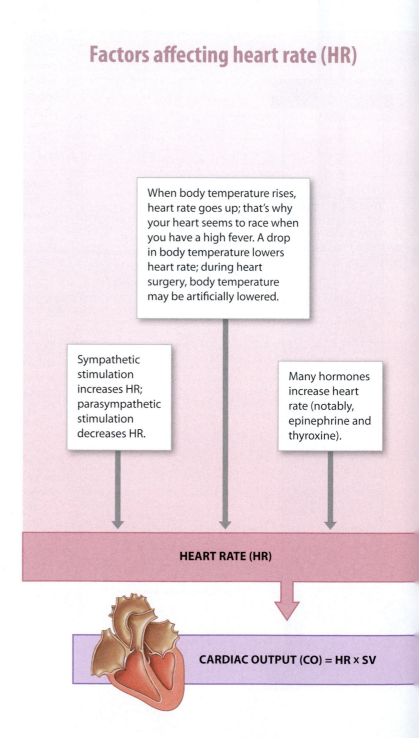

Factors affecting heart rate (HR)

When body temperature rises, heart rate goes up; that's why your heart seems to race when you have a high fever. A drop in body temperature lowers heart rate; during heart surgery, body temperature may be artificially lowered.

Sympathetic stimulation increases HR; parasympathetic stimulation decreases HR.

Many hormones increase heart rate (notably, epinephrine and thyroxine).

HEART RATE (HR)

CARDIAC OUTPUT (CO) = HR × SV

Factors affecting stroke volume (SV)

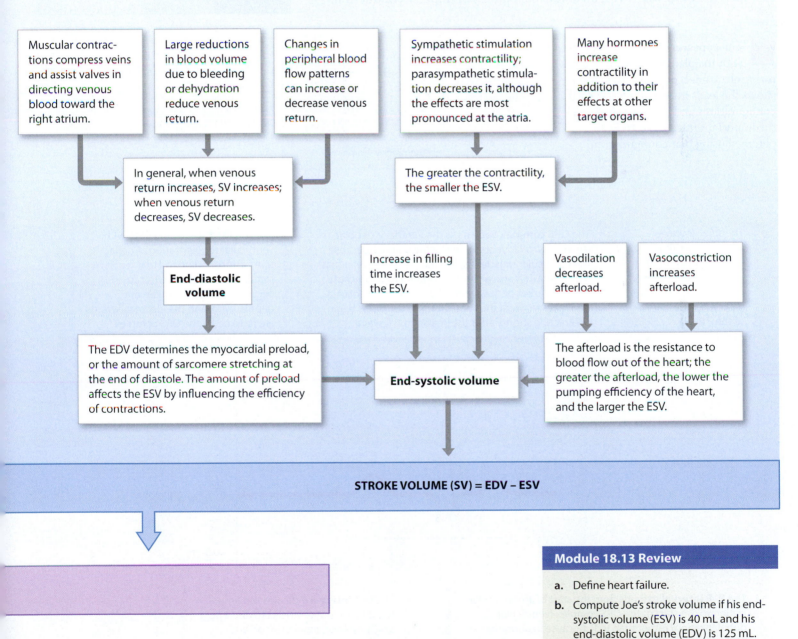

Muscular contractions compress veins and assist valves in directing venous blood toward the right atrium.

Large reductions in blood volume due to bleeding or dehydration reduce venous return.

Changes in peripheral blood flow patterns can increase or decrease venous return.

Sympathetic stimulation increases contractility; parasympathetic stimulation decreases it, although the effects are most pronounced at the atria.

Many hormones increase contractility in addition to their effects at other target organs.

In general, when venous return increases, SV increases; when venous return decreases, SV decreases.

The greater the contractility, the smaller the ESV.

End-diastolic volume

Increase in filling time increases the ESV.

Vasodilation decreases afterload.

Vasoconstriction increases afterload.

The EDV determines the myocardial preload, or the amount of sarcomere stretching at the end of diastole. The amount of preload affects the ESV by influencing the efficiency of contractions.

End-systolic volume

The afterload is the resistance to blood flow out of the heart; the greater the afterload, the lower the pumping efficiency of the heart, and the larger the ESV.

STROKE VOLUME (SV) = EDV – ESV

Normal and abnormal cardiac activity can be detected in an electrocardiogram

The electrical events occurring in the heart are powerful enough to be detected by electrodes on the surface of the body. A recording of these events over time is an **electrocardiogram** (e-lek-trō-KAR-dē-ō-gram), also called an **ECG** or **EKG**. Clinicians can use an ECG to assess the performance of specific nodal, conducting, and contractile components. When a portion of the heart has been damaged by a heart attack, for example, the ECG will reveal an abnormal pattern of impulse conduction.

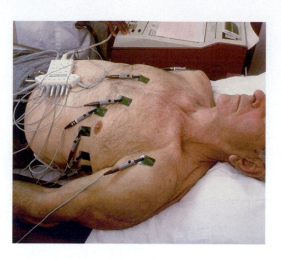

1 The appearance of the ECG varies with the placement of the monitoring electrodes, or leads. The photo shows the leads in one of the standard configurations and the graph indicates the important features of an ECG obtained using that configuration.

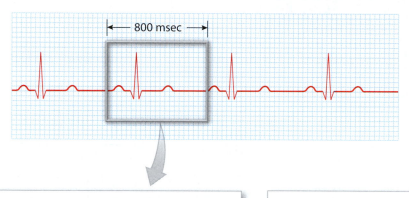

The small **P wave** accompanies the depolarization of the atria. The atria begin contracting about 25 msec after the start of the P wave.

The **QRS complex** appears as the ventricles depolarize. This is a relatively strong electrical signal, because the ventricular muscle is much more massive than that of the atria. It is also a complex signal, largely because of the complex pathway that the spread of depolarization takes through the ventricles. The ventricles begin contracting shortly after the peak of the **R wave**.

The smaller **T wave** coincides with ventricular repolarization. A deflection corresponding to atrial repolarization is not apparent, because it occurs while the ventricles are depolarizing, and the electrical events are masked by the QRS complex.

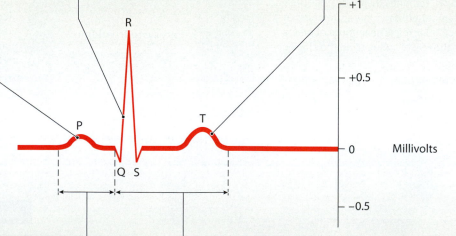

The **P–R interval** extends from the start of atrial depolarization to the start of the QRS complex (ventricular depolarization) rather than to R, because in abnormal ECGs the peak at R can be difficult to determine. Extension of the P–R interval to more than 200 msec can indicate damage to the conducting pathways or AV node.

The **Q–T interval** indicates the time required for the ventricles to undergo a single cycle of depolarization and repolarization. It is usually measured from the end of the P–R interval rather than from the bottom of the Q wave.

2 Despite the variety of sophisticated equipment available to assess or visualize cardiac function, in the majority of cases the ECG provides the most important diagnostic information. ECG analysis is especially useful in detecting and diagnosing **cardiac arrhythmias** (ā-RITH-mē-az)—abnormal patterns of cardiac electrical activity. Momentary arrhythmias are not inherently dangerous; about 5 percent of healthy individuals experience a few abnormal heartbeats each day. Clinical problems appear when arrhythmias reduce the pumping efficiency of the heart. Several important examples of arrhythmias are described here.

Premature Atrial Contractions (PACs)

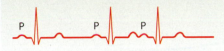

Premature atrial contractions (PACs) often occur in healthy individuals. In a PAC, the normal atrial rhythm is momentarily interrupted by a "surprise" atrial contraction. Stress, caffeine, and various drugs may increase the incidence of PACs, presumably by increasing the permeabilities of the SA pacemakers. The impulse spreads along the conduction pathway, and a normal ventricular contraction follows the atrial beat.

Paroxysmal Atrial Tachycardia (PAT)

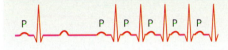

In **paroxysmal** (par-ok-SIZ-mal) **atrial tachycardia**, or **PAT**, a premature atrial contraction triggers a flurry of atrial activity. The ventricles are still able to keep pace, and the heart rate jumps to about 180 beats per minute.

Atrial Fibrillation (AF)

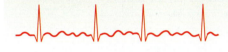

During **atrial fibrillation** (fib-ri-LĀ-shun), the impulses move over the atrial surface at rates of perhaps 500 beats per minute. The atrial wall quivers instead of producing an organized contraction. The ventricular rate cannot follow the atrial rate and may remain within normal limits. Even though the atria are now nonfunctional, their contribution to ventricular end-diastolic volume is so small that the condition may go unnoticed in older individuals.

Premature Ventricular Contractions (PVCs)

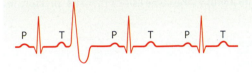

Premature ventricular contractions (PVCs) occur when a Purkinje cell or ventricular myocardial cell depolarizes to threshold and triggers a premature contraction. Single PVCs are common and not dangerous. The cell responsible is called an **ectopic pacemaker**. The frequency of PVCs can be increased by exposure to epinephrine, to other stimulatory drugs, or to ionic changes that depolarize cardiac muscle cell membranes.

Ventricular Tachycardia (VT)

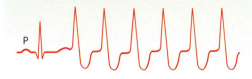

Ventricular tachycardia is defined as four or more PVCs without intervening normal beats. It is also known as **VT** or **V-tach**. Multiple PVCs and VT may indicate that serious cardiac problems exist.

Ventricular Fibrillation (VF)

Ventricular fibrillation (VF) is responsible for the condition known as **cardiac arrest**. VF is rapidly fatal, because the ventricles quiver and stop pumping blood.

Module 18.14 Review

a. Define electrocardiogram.

b. List the important features of the ECG, and indicate what each represents.

c. Why is ventricular fibrillation fatal?

1. Short answer

Refer to the accompanying graph of the cardiac cycle in answering questions a through f below.

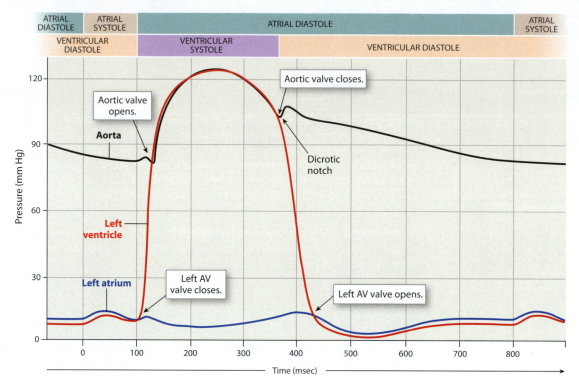

a What event occurs when the pressure in the left ventricle rises above that in the left atrium? _____

b During ventricular systole, the blood volume in the atria is _____ (increasing/decreasing), and the volume in the ventricle is _____ (increasing/decreasing).

c During most of ventricular diastole, the pressure in the left ventricle is _____ (greater than/the same as/less than) the pressure in the left atrium.

d What event occurs when the pressure within the left ventricle becomes greater than the pressure within the aorta? _____

e During isovolumetric contraction, pressure is highest in the _____ .

f During what part of the cardiac cycle is blood pressure highest in the large systemic arteries? _____

2. Matching

Match the following terms with the most closely related description.

- P wave
- cardiac output
- automaticity
- "lubb" sound
- "dubb" sound
- sympathetic neurons
- stroke volume
- tachycardia
- bradycardia
- parasympathetic neurons

a _____ AV valves close

b _____ Self-stimulated cardiac muscle contractions

c _____ Atrial depolarization

d _____ Amount of blood ejected by ventricle during a single beat

e _____ HR x SV

f _____ Decrease the heart rate

g _____ Term for slower-than-normal HR

h _____ Increase the heart rate

i _____ Semilunar valve closes

j _____ Term for faster-than-normal HR

The Coordination of Cardiac Output and Peripheral Blood Flow

You are now familiar with the factors involved in the regulation of cardiac output. Both cardiac output and the distribution of blood within the pulmonary and systemic circuits must constantly be adjusted to meet the demands of active tissues. The sites and mechanisms of cardiovascular regulation are summarized here.

Cardiac Output

To maintain **cardiac output**, the heart must generate enough pressure to force blood through thousands of miles of peripheral capillaries, most scarcely larger in diameter than a single red blood cell.

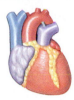

Venous Return

The **venous return** is the amount of blood arriving at the right ventricle each minute; on average, it is equal to the cardiac output.

Arterial Blood Pressure

Blood pressure is the pressure within the cardiovascular system as a whole. **Arterial pressure** is much higher than **venous pressure** because it must push blood through smaller and smaller arteries and then through innumerable capillary networks.

Regulation (Neural and Hormonal)

Neural and hormonal regulation make coordinated adjustments in heart rate, stroke volume, peripheral resistance, and venous pressure so that cardiac output is sufficient to meet the demands of peripheral tissues.

Venous Pressure

Valves and muscular compression of peripheral veins play an important role in maintaining venous pressure and venous blood flow. As blood moves toward the heart, the vessels get larger and larger in diameter, and resistance is continuously decreasing.

Peripheral Resistance

Resistance is a force that opposes movement. The **peripheral resistance** is the resistance of the arterial system as a whole; resistance increases as the arterial branches get smaller and smaller.

Capillary Pressure

Because capillary pressures are very low, blood flows slowly, allowing plenty of time for diffusion between the blood and the surrounding interstitial fluid. This interplay, called **capillary exchange**, is the primary focus of the entire cardiovascular system.

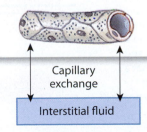

Capillary exchange

Interstitial fluid

In this section we will consider the functioning of the cardiovascular system as a whole, and learn how neural and hormonal mechanisms regulate cardiovascular function.

There are multiple sources of resistance within the cardiovascular system, but vessel diameter is the primary factor under normal conditions

For circulation to occur, the heart must develop sufficient pressure to overcome the **total peripheral resistance**—the resistance of the entire cardiovascular system. The total peripheral resistance of the cardiovascular system depends on three factors: vascular resistance, viscosity, and turbulence.

1 **Vascular resistance**, the resistance of the blood vessels, is the largest component of total peripheral resistance. Vascular resistance primarily results from friction between blood and the vessel walls. The amount of friction depends on two factors: vessel length and vessel diameter.

Friction and Vessel Length

Friction occurs between the moving blood and the walls of the vessel. The longer the vessel, the greater the surface area in contact with the blood, and the greater the resistance. You can easily blow the water out of a snorkel that is 25 cm (10 in.) long, but you cannot blow the water out of a 15 m garden hose of the same diameter, because the total frictional resistance is too great. The most dramatic changes in blood vessel length occur between birth and maturity, as individuals grow to adult size. In adults, vessel length can be considered constant.

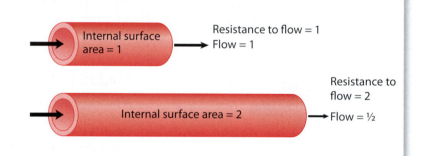

Friction and Vessel Diameter

Friction also occurs between layers of fluid moving at different speeds. The layer of blood closest to the vessel wall is slowed down by friction with the endothelial surface. The adjacent layer of blood is slowed down by friction with the more superficial layer. This effect gradually diminishes as the distance from the wall increases. In a small-diameter vessel, all the blood is slowed to some degree, and resistance is relatively high. In a large-diameter vessel the central region is unaffected by events at the periphery, so the resistance is relatively low.

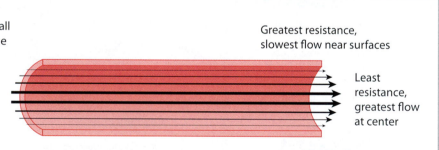

Vessel Length versus Vessel Diameter

Differences in diameter have much more significant effects on resistance than do differences in length. As shown above, if two vessels are equal in diameter but one is twice as long as the other, the longer vessel offers twice as much resistance to blood flow. But for two vessels of equal length, one twice the diameter of the other, the narrower one offers 16 times as much resistance to blood flow. This relationship, expressed in terms of the vessel radius (r) and resistance (R), can be summarized as $R = 1/r^4$. This means that a small change in vessel diameter produces a large change in resistance. The vasomotor centers control peripheral resistance and blood flow primarily by altering the diameters of arterioles.

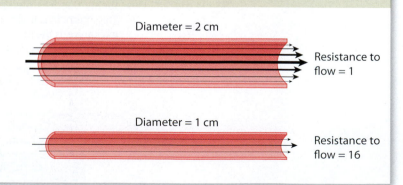

2 **Viscosity** is the resistance to flow caused by interactions among molecules and suspended materials in a liquid. Liquids of low viscosity, such as water (viscosity 1.0), flow at low pressures; thick, syrupy fluids, such as molasses (viscosity 300), flow only under higher pressures. Whole blood has a viscosity about five times that of water, due to the presence of plasma proteins and blood cells. Under normal conditions, the viscosity of blood remains stable, but disorders that affect the hematocrit or plasma composition also change blood viscosity, and thus peripheral resistance.

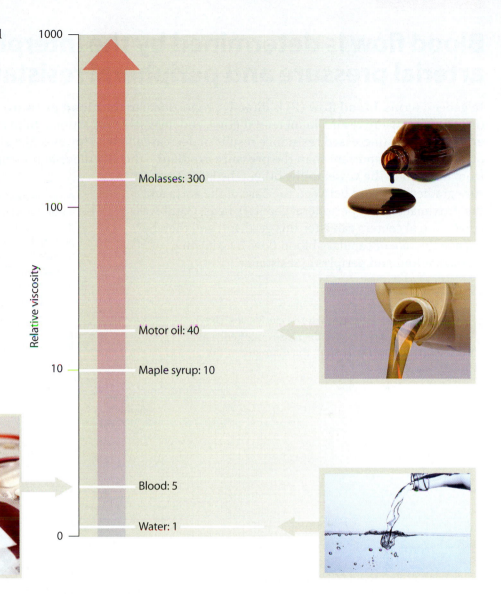

Relative viscosity

1000

100

Molasses: 300

Motor oil: 40

10

Maple syrup: 10

Blood: 5

Water: 1

0

3 High flow rates, irregular surfaces, and sudden changes in vessel diameter upset the smooth flow of blood, creating eddies and swirls. This phenomenon, called **turbulence**, increases resistance and slows blood flow. Here you see the turbulence created by the presence of a plaque within a small artery (Module 18.7). Turbulence also occurs in normal individuals when blood flows between the atria and the ventricles, and between the ventricles and the aortic and pulmonary trunks. Turbulent flow is responsible for the production of the third and fourth heart sounds (Module 18.8).

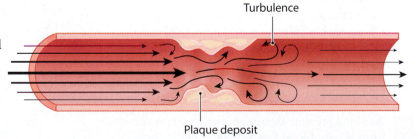

Turbulence

Plaque deposit

Module 18.15 Review

a. List the factors that contribute to total peripheral resistance.

b. Which would reduce peripheral resistance: an increase in vessel length or an increase in vessel diameter?

c. Explain the formula $R = 1/r^4$.

Blood flow is determined by the interplay between arterial pressure and peripheral resistance

In general terms, **blood flow (F)** is directly proportional to the **blood pressure** (increased pressure results in increased flow), and inversely proportional to **peripheral resistance** (increased resistance results in decreased flow). However, the absolute pressure is less important than the **pressure gradient**—the difference in pressure from one end of the vessel to the other. The largest pressure gradient is found between the base of the aorta and the proximal ends of peripheral capillary beds. Cardiovascular control centers can alter this gradient, and thereby change the rate of capillary blood flow, by adjusting cardiac output and peripheral resistance.

1 As blood proceeds toward the capillaries, the diameter of the arteries decreases markedly; as blood returns toward the heart, the diameter of the veins increases.

As blood proceeds from the aorta toward the capillaries, divergence of blood vessels occurs; the arteries branch repeatedly, and each branch is smaller in diameter than the preceding one.

As blood proceeds from the capillaries toward the venae cavae, convergence of blood vessels occurs; vessel diameters increase as venules combine to form small and medium-sized veins.

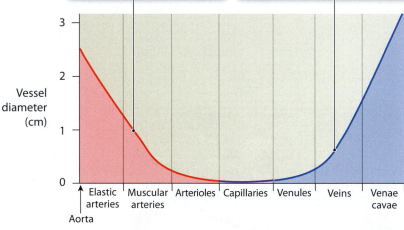

2 The heart generates a pressure of around 100 mm Hg as it pushes blood into the aorta, which has a cross-sectional area of roughly 4.5 cm². At each branching of the arterial system, the arterial pressure drops as blood is pushed into ever-increasing numbers of smaller and smaller branches. At the start of peripheral capillaries, arterial pressure has fallen to approximately 35 mm Hg, and by the time blood reaches the venules, it has fallen to approximately 18 mm Hg.

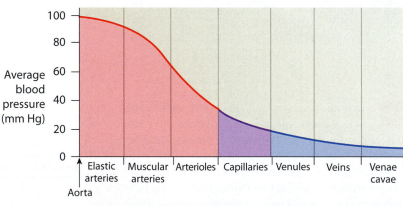

3 As blood pressure decreases, the blood moves more slowly. Blood flow is highest in the aorta, where blood pressure is high and the large diameter means that resistance is low. Blood flow is slowest in the capillaries, whose diameter is roughly the same as that of a single red blood cell. This is important because exchange with interstitial fluid occurs only in the capillaries, and diffusion is a relatively slow process. Blood flow then accelerates in the venous system because although pressures are falling, the vessels are merging into large-diameter passages with very low resistance.

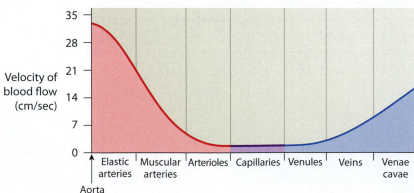

4 Arterial pressure is not constant; it rises during ventricular systole and falls during ventricular diastole as the elastic arterial walls stretch and recoil. The peak blood pressure measured during ventricular systole is called **systolic pressure**, and the minimum blood pressure at the end of ventricular diastole is called **diastolic pressure**. In recording blood pressure, we separate systolic and diastolic pressures with a slash. Thus the blood pressure illustrated here is written as 120/90 (read as "one-twenty over ninety").

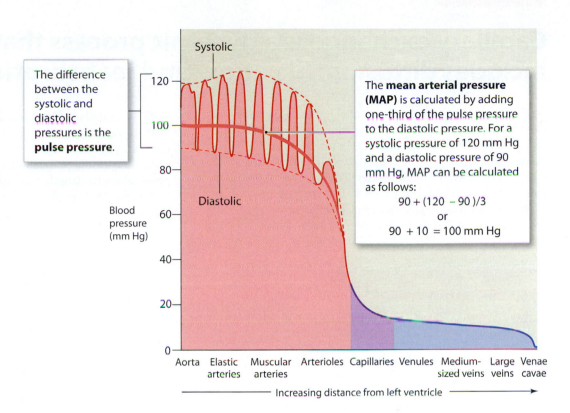

The difference between the systolic and diastolic pressures is the **pulse pressure**.

The **mean arterial pressure (MAP)** is calculated by adding one-third of the pulse pressure to the diastolic pressure. For a systolic pressure of 120 mm Hg and a diastolic pressure of 90 mm Hg, MAP can be calculated as follows:

$$90 + (120 - 90)/3$$
or
$$90 + 10 = 100 \text{ mm Hg}$$

5 **Capillary exchange** involves a combination of filtration, diffusion, and osmosis. **Capillary hydrostatic pressure (CHP)** is the blood pressure within capillary beds, and it provides the driving force for filtration. Along the length of a typical capillary, CHP pushes water and soluble molecules out of the bloodstream and into the interstitial fluid. Only small solutes can cross the endothelium; larger molecules, such as plasma proteins, must stay in the bloodstream. This size-selective process is called **filtration**. We will consider further details about capillary exchange in the next module.

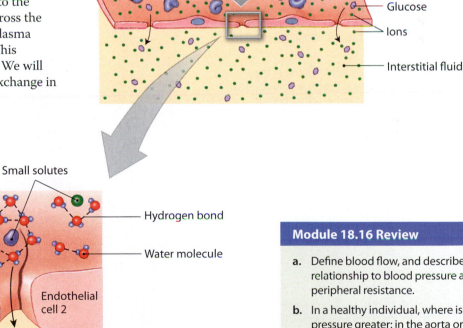

Relatively small solutes are carried along as CHP pushes water into the interstitial spaces.

Module 18.16 Review

a. Define blood flow, and describe its relationship to blood pressure and peripheral resistance.

b. In a healthy individual, where is blood pressure greater: in the aorta or in the inferior vena cava? Explain.

c. For an individual with a blood pressure of 125/70, calculate the mean arterial pressure (MAP).

Capillary exchange is a dynamic process that includes diffusion, filtration, and reabsorption

1 Diffusion is the net movement of ions or molecules from an area where their concentration is higher to an area where their concentration is lower. Diffusion occurs most rapidly when (1) the distances involved are short, (2) the concentration gradient is large, and (3) the ions or molecules involved are small. Diffusion occurs continuously across capillary walls, but different substances use different routes, as noted in the table below.

2 In a capillary, blood pressure declines as blood flows from the arterial end to the venous end. As a result, the rates of filtration and reabsorption gradually change as blood passes along the length of a capillary. The factors involved are diagrammed in the figure below.

Routes of Diffusion across Capillary Walls

- Water, ions, and small organic molecules, such as glucose, amino acids, and urea, can usually enter or leave the bloodstream by diffusion either between adjacent endothelial cells or through the pores of fenestrated capillaries.

- Many ions, including sodium, potassium, calcium, and chloride, can diffuse across endothelial cells by passing through channels in plasma membranes.

- Large water-soluble compounds are unable to enter or leave the bloodstream except at fenestrated capillaries, such as those in the hypothalamus, the kidneys, many endocrine organs, and the intestinal tract.

- Lipids, such as fatty acids and steroids, and lipid-soluble materials, including soluble gases such as oxygen and carbon dioxide, can cross capillary walls by diffusion through the endothelial plasma membranes.

- Plasma proteins are normally unable to cross the endothelial lining anywhere except in sinusoids, such as those of the liver, where plasma proteins enter the bloodstream.

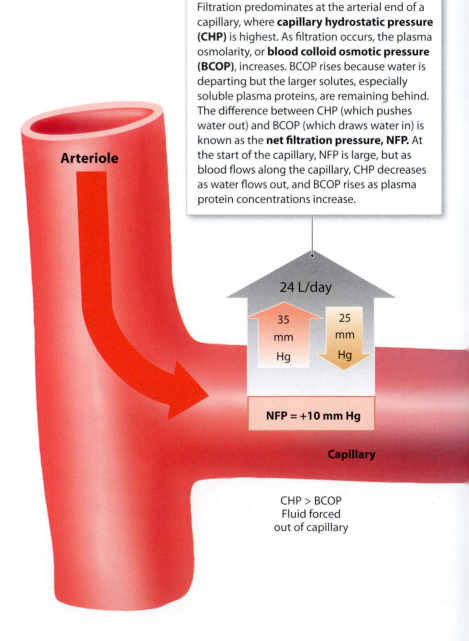

Filtration Predominates

Filtration predominates at the arterial end of a capillary, where **capillary hydrostatic pressure (CHP)** is highest. As filtration occurs, the plasma osmolarity, or **blood colloid osmotic pressure (BCOP)**, increases. BCOP rises because water is departing but the larger solutes, especially soluble plasma proteins, are remaining behind. The difference between CHP (which pushes water out) and BCOP (which draws water in) is known as the **net filtration pressure, NFP.** At the start of the capillary, NFP is large, but as blood flows along the capillary, CHP decreases as water flows out, and BCOP rises as plasma protein concentrations increase.

Arteriole

24 L/day

35 mm Hg

25 mm Hg

NFP = +10 mm Hg

Capillary

CHP > BCOP
Fluid forced
out of capillary

3 Any conditions that affect either blood pressure or the osmotic pressures of the blood or tissues will shift the balance between hydrostatic and osmotic forces. We can then predict the effects on the basis of an understanding of capillary dynamics; three examples are presented in the table to the right.

Representative Variations in Capillary Exchange

- If hemorrhaging occurs, both blood volume and blood pressure decline. This reduction in CHP lowers the NFP and increases the amount of capillary reabsorption. The result is a reduction in the volume of interstitial fluid and an increase in the circulating plasma volume. This process is known as a recall of fluids.

- If dehydration occurs, the plasma volume decreases due to water loss, and the concentration of plasma proteins increases. The increase in BCOP accelerates reabsorption and a recall of fluids that delays the onset and severity of clinical problems caused by low blood volume and blood pressure.

- If the CHP rises or the BCOP declines, fluid moves out of the blood in capillaries and builds up in peripheral tissues, an abnormal condition called edema.

No Net Movement

At a point roughly two-thirds of the way along the capillary, NFP is zero and there is no net movement of fluid into or out of the capillary.

Reabsorption Predominates

In the final segment of the capillary, CHP falls below BCOP, and water flows back into the capillary. Note that more water leaves the bloodstream during filtration than gets retrieved through reabsorption. The difference, roughly 3.6 L/day, flows through peripheral tissues and enters lymphatic vessels that eventually return it to the venous system.

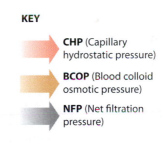

Venule

No net fluid movement

20.4 L/day

25 mm Hg   25 mm Hg

18 mm Hg   25 mm Hg

NFP = 0

NFP = –7 mm Hg

Capillary

CHP = BCOP
No net movement of fluid

BCOP > CHP
Fluid moves into capillary

KEY

CHP (Capillary hydrostatic pressure)

BCOP (Blood colloid osmotic pressure)

NFP (Net filtration pressure)

Module 18.17 Review

a. Define edema.

b. Identify the conditions that would shift the balance between hydrostatic and osmotic forces.

c. Under what general conditions would fluid move into a capillary?

18.18

Cardiovascular regulatory mechanisms respond to changes in blood pressure or blood composition

1 Homeostatic mechanisms regulate cardiovascular activity to ensure that **tissue perfusion**—the blood flow through the tissues—meets the tissue demands for oxygen and nutrients. There are two regulatory pathways: one that involves local autoregulation and, if that is ineffective, a second that involves central regulation by neural and endocrine mechanisms.

Central Regulation

Central regulation involves both neural and endocrine mechanisms. Activation of the cardiovascular centers involves both the **cardioacceleratory centers** (which stimulate the heart) and the vasomotor centers (which control the degree of peripheral vasoconstriction). Neural mechanisms elevate cardiac output and reduce blood flow to nonessential or inactive tissues. The primary **vasoconstrictor** involved in neural regulation is norepinephrine (NE). Endocrine mechanisms involve long-term increases in blood volume and blood pressure.

Autoregulation

Autoregulation involves changes in the pattern of blood flow within capillary beds as precapillary sphincters open and close in response to chemical changes in the interstitial fluid. Factors that promote the dilation of blood vessels are called vasodilators. Local **vasodilators** such as lactate accelerate blood flow through their tissue of origin.

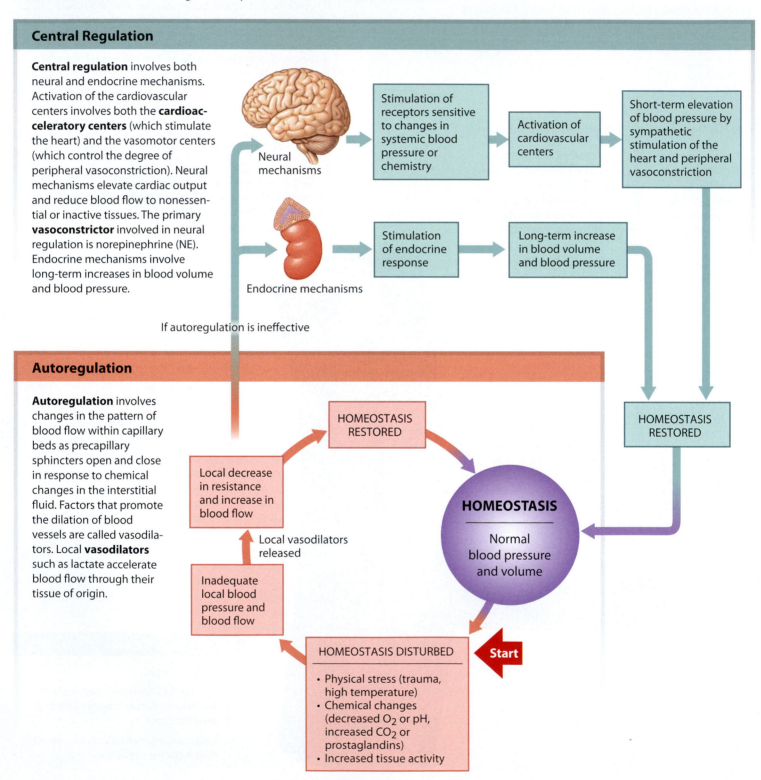

If autoregulation is ineffective

continued

2 The **baroreceptor reflexes** (*baro-*, pressure) respond to changes in blood pressure. As noted in Module 14.8, p.488, the receptors are located in the walls of (1) the carotid sinuses, expanded chambers near the bases of the internal carotid arteries of the neck; (2) the aortic sinuses, pockets in the walls of the ascending aorta adjacent to the heart; and (3) the right atrium.

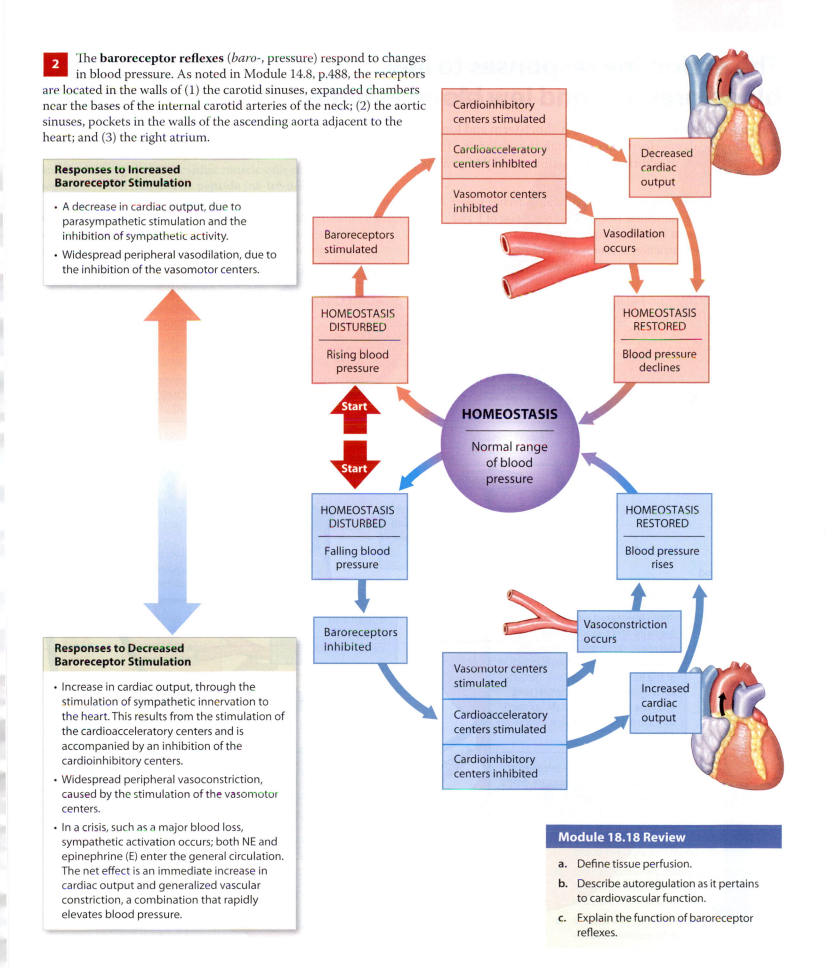

Responses to Increased Baroreceptor Stimulation

- A decrease in cardiac output, due to parasympathetic stimulation and the inhibition of sympathetic activity.
- Widespread peripheral vasodilation, due to the inhibition of the vasomotor centers.

Cardioinhibitory centers stimulated
Cardioaccelleratory centers inhibited
Vasomotor centers inhibited

Decreased cardiac output

Baroreceptors stimulated

Vasodilation occurs

HOMEOSTASIS DISTURBED

Rising blood pressure

HOMEOSTASIS RESTORED

Blood pressure declines

Start

Start

HOMEOSTASIS

Normal range of blood pressure

HOMEOSTASIS DISTURBED

Falling blood pressure

HOMEOSTASIS RESTORED

Blood pressure rises

Baroreceptors inhibited

Vasoconstriction occurs

Vasomotor centers stimulated
Cardioaccelleratory centers stimulated
Cardioinhibitory centers inhibited

Increased cardiac output

Responses to Decreased Baroreceptor Stimulation

- Increase in cardiac output, through the stimulation of sympathetic innervation to the heart. This results from the stimulation of the cardioaccelleratory centers and is accompanied by an inhibition of the cardioinhibitory centers.
- Widespread peripheral vasoconstriction, caused by the stimulation of the vasomotor centers.
- In a crisis, such as a major blood loss, sympathetic activation occurs; both NE and epinephrine (E) enter the general circulation. The net effect is an immediate increase in cardiac output and generalized vascular constriction, a combination that rapidly elevates blood pressure.

Module 18.18 Review

a. Define tissue perfusion.

b. Describe autoregulation as it pertains to cardiovascular function.

c. Explain the function of baroreceptor reflexes.

Short-term and long-term mechanisms compensate for a reduction in blood volume

1 When hemostasis fails to prevent significant blood loss, the entire cardiovascular system makes adjustments to maintain blood pressure and restore blood volume. The immediate problem is the maintenance of adequate blood pressure and peripheral blood flow. The long-term problem is the restoration of normal blood volume. These mechanisms can cope with blood losses equivalent to approximately 30 percent of total blood volume.

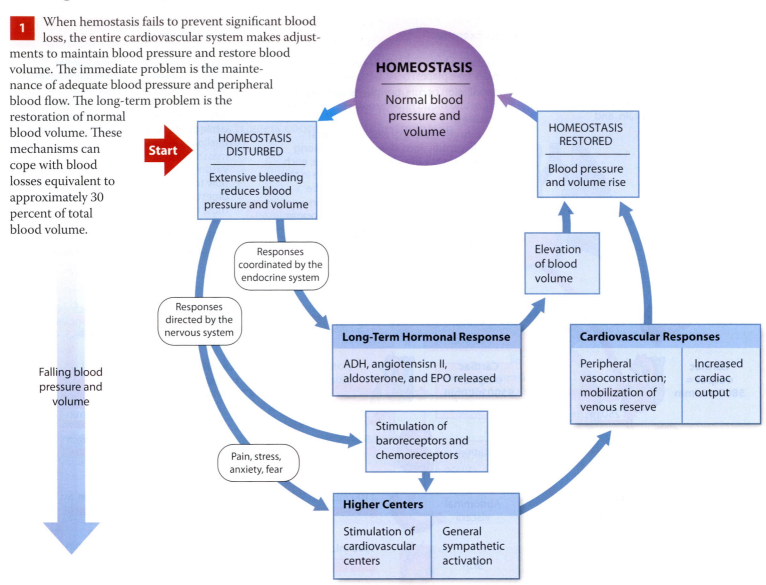

HOMEOSTASIS

Normal blood pressure and volume

Start

HOMEOSTASIS DISTURBED

Extensive bleeding reduces blood pressure and volume

Responses coordinated by the endocrine system

Responses directed by the nervous system

Falling blood pressure and volume

Pain, stress, anxiety, fear

HOMEOSTASIS RESTORED

Blood pressure and volume rise

Elevation of blood volume

Long-Term Hormonal Response

ADH, angiotensisn II, aldosterone, and EPO released

Cardiovascular Responses

| Peripheral vasoconstriction; mobilization of venous reserve | Increased cardiac output |

Stimulation of baroreceptors and chemoreceptors

Higher Centers

| Stimulation of cardiovascular centers | General sympathetic activation |

Short-Term Responses to Blood Loss

- Carotid and aortic reflexes increase cardiac output and cause peripheral vasoconstriction. Cardiac output is maintained by increasing the heart rate, typically to 180–200 bpm.
- The combination of stress and anxiety stimulates the sympathetic nervous system, which causes a further increase in vasomotor tone, constricting the arterioles and elevating blood pressure. At the same time, venoconstriction, the constriction of large and medium-sized veins, improves venous return and shifts blood into the arterial system.
- Sympathetic activation causes the secretion of E and NE by the adrenal medullae, increasing cardiac output and extending peripheral vasoconstriction. The release of ADH by the neurohypophysis and the production of angiotensin II enhance vasoconstriction; ADH also participates in the long-term response.

Long-Term Responses to Blood Loss

- The decline in capillary blood pressure triggers a recall of fluids from the interstitial spaces.
- Aldosterone and ADH promote fluid retention and reabsorption by the kidneys.
- Thirst increases, and additional water is obtained by absorption across the digestive tract.
- Erythropoietin stimulates the maturation of red blood cells in the red bone marrow, resulting in increased blood volume and improved oxygen delivery to peripheral tissues.

2 **Shock** is an acute cardiovascular crisis marked by low blood pressure (**hypotension**) and inadequate peripheral blood flow. Severe and potentially fatal signs and symptoms develop as vital tissues become starved for oxygen and nutrients. The most common cardiovascular causes of shock are a drop in cardiac output after hemorrhaging or damage to the heart, as in a heart attack. Shock develops when normal compensation mechanisms fail to maintain adequate cardiac output and peripheral blood flow. **Circulatory shock** involves a series of interlocking positive feedback loops that begin when blood losses exceed roughly 35 percent of total blood volume. **Progressive shock** is an initial stage characterized by positive feedback loops that accelerate tissue damage. Unless the positive feedback loops are broken, the individual will eventually enter the fatal stage of **irreversible shock**.

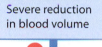

HOMEOSTATIC FAILURE

Severe reduction in blood volume ❶

Damage to heart muscle

Decreased cardiac output

Positive feedback loop ❷

Increased capillary permeability

Decreased venous return

Decreased cardiac blood flow

Positive feedback loop ❹

Clotting within vessels

Decreased arterial pressure

If feedback loops at 2–4 can be broken, the individual may survive. If the mean arterial blood pressure (MAP) falls below 50 mm Hg, damage soon becomes irreversible

Positive feedback loop ❸

Increased PCO₂ and lactate, decreased pH

Decreased peripheral blood flow

Maximum vasoconstriction ❺

Decreased sympathetic activity

Irreversible cardiac muscle damage

Irreversible CNS damage

Positive feedback loop ❻

Falling arterial pressures ❼

Decreased blood flow to CNS

Circulatory collapse

DEATH

Progressive Shock

1 When blood volume declines by more than 35 percent, blood pressure remains abnormally low, venous return is reduced, and cardiac output is inadequate despite sustained vasoconstriction and the mobilization of the venous reserve.

2 The first of several dangerous positive feedback loops begins when low cardiac output reduces blood flow to the heart; this damages the myocardium, which leads to a further reduction in cardiac output.

3 Reduced cardiac output accelerates oxygen starvation in peripheral tissues, and the resulting chemical changes promote intravascular clotting that further restricts peripheral blood flow.

4 Local pH changes increase capillary permeability, and this further reduces blood volume.

Irreversible Shock

5 Carotid sinus baroreceptors trigger a massive activation of the vasomotor centers. To preserve blood flow to the brain at any cost, the sympathetic output causes a sustained and maximal vasoconstriction. This reduces peripheral circulation to an absolute minimum, but it elevates blood pressure to about 70 mm Hg and temporarily improves blood flow to cerebral vessels.

6 Unless prompt treatment is provided, blood pressure will again decline. By now the heart muscle will be seriously damaged as blood flow to the brain decreases.

7 **Circulatory collapse** occurs when arteriolar smooth muscles and precapillary sphincters become unable to contract, despite the commands of the vasomotor centers. The result is widespread peripheral vasodilation, an immediate and fatal decline in blood pressure, and the cessation of blood flow.

Module 18.22 Review

a. Identify the compensatory mechanisms that respond to blood loss.

b. Explain the role of aldosterone and ADH in the long-term restoration of blood volume.

c. Name the immediate and long-term problems related to profuse blood loss.

1. Short answer

What three primary factors influence blood pressure and blood flow? _____

2. Matching

Match the following terms with the most closely related description.

- autoregulation
- venous return
- local vasodilators
- natriuretic peptides
- chemoreceptors
- baroreceptors
- turbulence
- net hydrostatic pressure
- medulla oblongata
- edema
- viscosity
- osmotic pressure

a	_____	Detect changes in pressure
b	_____	Vasomotor center
c	_____	Aided by thoracic pressure changes due to breathing and skeletal muscle activity
d	_____	Causes immediate, local homeostatic responses
e	_____	Decreased tissue O_2 and increased CO_2
f	_____	Carotid bodies
g	_____	Forces water into a capillary
h	_____	Opposite of smooth blood flow
i	_____	Resistance to flow
j	_____	Forces water out of a capillary
k	_____	Peripheral vasodilation
l	_____	Excess interstitial fluid accumulation

3. Multiple choice

Choose the bulleted item that best completes each statement.

a Of the following blood vessels, the greatest drop in blood pressure occurs in the _____ .

- capillaries
- veins
- venules
- arterioles

b The central regulation of cardiac output primarily involves the _____ .

- somatic nervous system
- central nervous system
- autonomic nervous system
- all of these

c Hormonal regulation by ADH, epinephrine, angiotensin II, and norepinephrine results in _____ .

- decreasing peripheral vasoconstriction
- increasing peripheral vasoconstriction
- increasing peripheral vasodilation
- all of these

d The three primary interrelated changes that occur as exercise begins are _____ .

- decreased vasodilation, increased venous return, increased cardiac output
- increased vasodilation, decreased venous return, increased cardiac output
- decreased vasodilation, decreased venous return, decreased cardiac output
- increased vasodilation, increased venous return, increased cardiac output

e The only part of the body where the blood supply is unchanged during exercise at maximal levels is the _____ .

- heart
- brain
- kidney
- skin

f The systems responsible for modifying heart rate and regulating blood pressure are the _____ systems.

- respiratory and muscular
- respiratory and nervous
- muscular and urinary
- nervous and endocrine

Visual Outline with Key Terms

Summarize the content of each module using the terms in the order provided.

SECTION 1

The Structure of the Heart

- base
- apex
- superior border
- right border
- left border
- inferior border

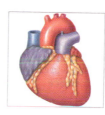

18.1

The wall of the heart contains concentric layers of cardiac muscle tissue

- myocardium
- endocardium
- parietal pericardium
- pericardial sac
- epicardium

- intercalated disc
- functional syncytium

18.2

The heart is located in the mediastinum, suspended within the pericardial cavity

- mediastinum
- great vessels
- pericarditis
- cardiac tamponade

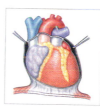

18.3

The boundaries between the chambers of the heart can be distinguished on its external surface

- anterior surface
- sulci
- auricle
- coronary sulcus
- ligamentum arteriosum

- anterior interventricular sulcus
- posterior surface
- coronary sinus

- posterior interventricular sulcus

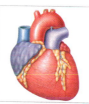

18.4

The heart has an extensive blood supply

- coronary arteries
- right coronary artery
- marginal arteries
- left coronary artery
- anterior interventricular artery
- circumflex artery
- posterior interventricular artery

- anterior cardiac veins
- great cardiac vein
- coronary sinus
- posterior cardiac vein
- small cardiac vein

- middle cardiac vein
- elastic rebound

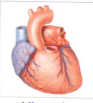

• = *Term boldfaced in this module*

18.5

Internal valves control the direction of blood flow between the heart chambers

- interatrial septum
- interventricular septum
- atrioventricular (AV) valves
- fossa ovalis
- pectinate muscles
- right atrioventricular (AV) valve
- tricuspid valve

- cusps
- chordae tendineae
- papillary muscles
- pulmonary valve
- moderator band
- left atrioventricular (AV) valve

- bicuspid valve
- mitral valve
- trabeculae carneae
- aortic valve

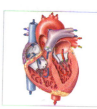

18.6

When the heart beats, the AV valves close before the semilunar valves open, and the semilunar valves close before the AV valves open

- regurgitation
- aortic sinuses
- cardiac skeleton
- semilunar valves

- valvular heart disease (VHD)
- carditis

18.7

Arteriosclerosis can lead to coronary artery disease

- arteriosclerosis
- atherosclerosis
- plaque
- balloon angioplasty
- coronary artery disease (CAD)
- coronary ischemia

- thrombus
- digital subtraction angiography (DSA)
- stent

SECTION 2

The Cardiac Cycle

- cardiac cycle
- systole
- diastole

18.8

The cardiac cycle creates pressure gradients that maintain blood flow

- atrial systole
- atrial diastole
- ventricular systole—first phase
- isovolumetric contraction
- ventricular systole—second phase

- ventricular ejection
- ventricular diastole—early
- isovolumetric relaxation
- ventricular diastole—late

- dicrotic notch
- heart sounds (S_1–S_4)

Physical barriers prevent pathogens and toxins from entering body tissues ...

To cause trouble, an antigenic compound or a pathogen must enter body tissues, which requires crossing an epithelium—either at the skin or across a mucous membrane.

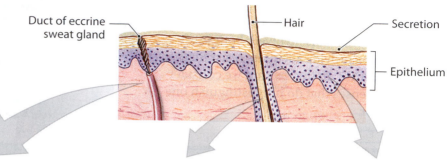

Duct of eccrine sweat gland — Hair — Secretion — Epithelium

1 The integumentary system provides the major physical barrier to the external environment.

Sebaceous gland

Most epithelia are protected by specialized accessory structures and secretions. The epidermal surface also receives the secretions of sebaceous and sweat glands. These secretions, which flush the surface to wash away microorganisms and chemical agents, may also contain bactericidal chemicals, destructive enzymes (lysozymes), and antibodies.

The hairs on most areas of your body's surface provide some protection against mechanical abrasion (especially on the scalp), and they often prevent hazardous materials or insects from contacting your skin.

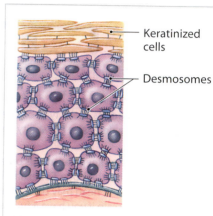

Keratinized cells — Desmosomes

The epithelial covering of the skin has multiple layers, a coating of keratinized cells, and a network of desmosomes that lock adjacent cells together.

2 The epithelia lining the digestive, respiratory, urinary, and reproductive tracts also provide an important barrier that protects against antigenic compounds and pathogens. As noted in Module 19.4, MALT in these locations provides a secondary line of nonspecific defense.

Mucus coating — Secretory cell — Tight junctions — Basal lamina

Mucus bathes most surfaces of your digestive tract, and your stomach contains a powerful acid that can destroy many pathogens. Mucus moves across the lining of the respiratory tract, urine flushes the urinary passageways, and glandular secretions do the same for the reproductive tract. Special enzymes, antibodies, and an acidic pH add to the effectiveness of these secretions.

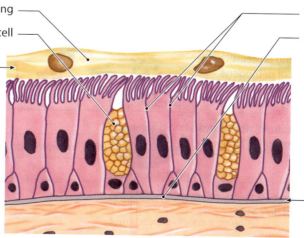

Along the more delicate internal passageways, epithelial cells are usually tied together by tight junctions and supported by a fibrous basal lamina.

... and phagocytes provide the next line of defense

3 Phagocytes perform janitorial and police services in peripheral tissues, removing cellular debris and responding to invasion by foreign compounds or pathogens. Phagocytes represent the "first line of cellular defense" against pathogenic invasion. Many phagocytes attack and remove microorganisms even before lymphocytes detect their presence. All phagocytic cells function in much the same way, although the target of phagocytosis may differ from one type of phagocyte to another.

Types of Phagocytes

12 μm

8–10 μm

Neutrophils are abundant, mobile, and quick to phagocytize cellular debris or invading bacteria. They circulate in the bloodstream and roam through peripheral tissues, especially at sites of injury or infection.

Eosinophils, which are less abundant than neutrophils, phagocytize foreign compounds or pathogens that have been coated with antibodies.

There are two major classes of **macrophages** derived from the monocytes of the circulating blood. This collection of phagocytic cells is called the **monocyte–macrophage system**, or the reticuloendothelial system.

Fixed macrophages are permanent residents of specific tissues and organs and are scattered among connective tissues. They normally do not move within these tissues.

Free macrophages travel throughout the body, arriving at the site of an injury by migrating through adjacent tissues or by recruitment from the circulating blood.

4 All phagocytes share several capabilities, which are summarized in the table below.

Functional Characteristics of Phagocytes

- Phagocytes can leave capillaries by squeezing between adjacent endothelial cells, a process known as **emigration**, or **diapedesis**.

- Phagocytes may be attracted to or repelled by chemicals in the surrounding fluids, a phenomenon called **chemotaxis**. They are particularly sensitive to chemicals released by either body cells or pathogens.

- Phagocytosis always begins with the attachment of the phagocyte to its target. In this process, receptors on the plasma membrane of the phagocyte bind to the surface of the target.

- After attachment, the phagocyte may either destroy the target itself or promote its destruction by activating specific defenses.

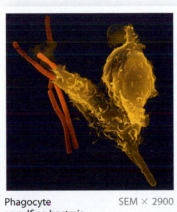

Phagocyte engulfing bacteria   SEM × 2900

Module 19.7 Review

a. Define chemotaxis.

b. How does the integumentary system protect the body?

c. Identify the types of phagocytes in the body, and differentiate between fixed macrophages and free macrophages.

NK cells perform immunological surveillance, detecting and destroying abnormal cells

The immune system generally ignores the body's own cells unless they become abnormal in some way. Natural killer (NK) cells are responsible for recognizing and destroying abnormal cells when they appear in peripheral tissues. The constant monitoring of normal tissues by NK cells is called **immunological surveillance**. The plasma membranes of cancer cells generally contain unusual proteins called **tumor-specific antigens**, which NK cells recognize as abnormal. The affected cells are then destroyed, preserving tissue integrity.

1 NK cells recognize bacteria, foreign cells, cells infected by viruses, and cancer cells. In each case, the steps leading to target cell destruction are similar.

Step 1: If a cell has unusual components in its plasma membrane, an NK cell recognizes that other cell as abnormal. Such recognition activates the NK cell, which then adheres to its target cell.	**Step 2:** The Golgi apparatus moves around the nucleus until the maturing face points directly toward the abnormal cell. A flood of secretory vesicles is then produced at the Golgi apparatus. These vesicles, which contain proteins called **perforins**, travel through the cytoplasm toward the cell surface.	**Step 3:** The perforins are released at the cell surface by exocytosis and diffuse across the narrow gap separating the NK cell from its target.	**Step 4:** As a result of the pores made of perforin molecules, the target cell can no longer maintain its internal environment, and it quickly disintegrates.

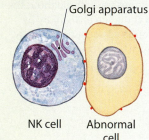

Golgi apparatus

NK cell   Abnormal cell

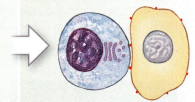

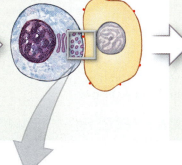

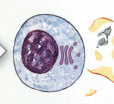

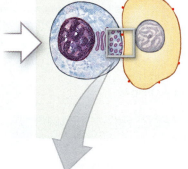

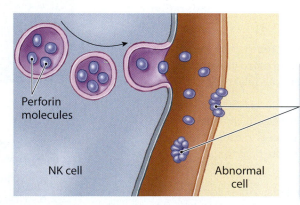

Perforin molecules

NK cell

Abnormal cell

On reaching the opposing plasma membrane, perforin molecules interact to create a network of pores. These pores are large enough to permit the free passage of ions, proteins, and other intracellular materials.

2 Cell division is a complicated process, and mistakes occasionally occur. The abnormal daughter cells usually cannot survive, and NK cells detect and destroy the few that can still function. That detection is important because a small number of those abnormal cells are potentially lethal cancer cells.

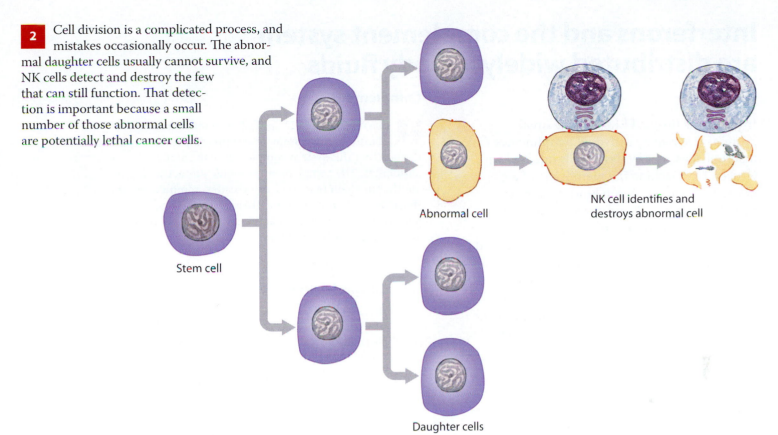

Stem cell

Abnormal cell

NK cell identifies and destroys abnormal cell

Daughter cells

3 Cancer cells often mutate and can sometimes avoid detection by NK cells. This process of avoiding detection or neutralizing body defenses is called **immunological escape**.

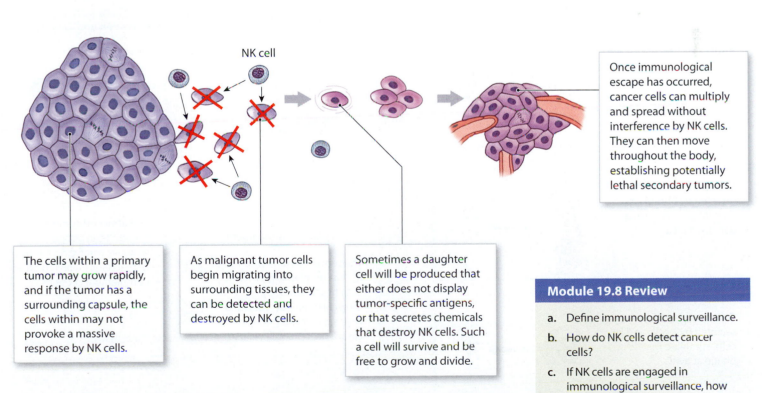

NK cell

Once immunological escape has occurred, cancer cells can multiply and spread without interference by NK cells. They can then move throughout the body, establishing potentially lethal secondary tumors.

The cells within a primary tumor may grow rapidly, and if the tumor has a surrounding capsule, the cells within may not provoke a massive response by NK cells.

As malignant tumor cells begin migrating into surrounding tissues, they can be detected and destroyed by NK cells.

Sometimes a daughter cell will be produced that either does not display tumor-specific antigens, or that secretes chemicals that destroy NK cells. Such a cell will survive and be free to grow and divide.

Module 19.8 Review

a. Define immunological surveillance.

b. How do NK cells detect cancer cells?

c. If NK cells are engaged in immunological surveillance, how do cancer cells spread?

Interferons and the complement system are distributed widely in body fluids

Interferons

Interferons (in-ter-FĒR-onz) are small proteins released by activated lymphocytes and macrophages, and by tissue cells infected with viruses. On reaching the membrane of a normal cell, an interferon binds to surface receptors on the cell and, via second messengers, triggers the production of antiviral proteins in the cytoplasm. Antiviral proteins do not interfere with the entry of viruses, but they do interfere with viral replication inside the cell. In addition to their role in slowing the spread of viral infections, interferons stimulate the activities of macrophages and NK cells.

1 At least three types of interferons exist, each of which has additional specialized functions. Interferons are examples of **cytokines** (SĪ-tō-kīnz)—chemical messengers released by tissue cells to coordinate local activities. Cytokines produced by most cells are used only for cell-to-cell communication within a given tissue. However, cytokines released by phagocytes and lymphocytes can also act as hormones, affecting cells and tissues throughout the body.

1 **Alpha (α)-interferons** are produced by cells infected with viruses. They attract and stimulate NK cells and enhance resistance to viral infection.

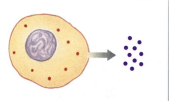

2 **Beta (β)-interferons**, secreted by fibroblasts, slow inflammation in a damaged area.

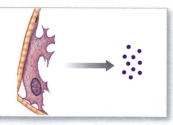

3 **Gamma (γ)-interferons**, secreted by T cells and NK cells, stimulate macrophage activity.

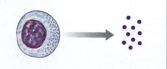

Complement

2 Plasma contains 11 special proteins that form the **complement system**. The term *complement* refers to the fact that this system complements the action of antibodies. The complement proteins interact with one another in chain reactions, or cascades, reminiscent of those of the clotting system. There are two possible pathways for complement action: the **classical pathway** and the **alternative pathway**.

Classical Pathway

The most rapid and effective activation of the complement system occurs through the classical pathway.

Antibody Binding and Complement Attachment

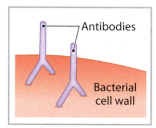

Antibodies

Bacterial cell wall

Antibody binding

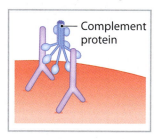

Complement protein

Complement attachment

The process begins when one of the complement proteins attaches to antibody molecules already bound to their specific antigen—in this case, a bacterial cell wall.

Activation and Cascade

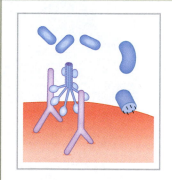

The attached complement protein then acts as an enzyme, catalyzing a series of reactions involving other complement proteins.

Alternative Pathway

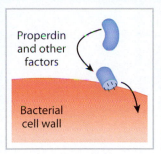

Properdin and other factors

Bacterial cell wall

The alternative pathway begins when several complement proteins, notably **properdin**, interact in the plasma. This interaction can be triggered by exposure to foreign materials, such as the capsule of a bacterium. The end result is the attachment of an activated complement protein to the bacterial cell wall.

The alternative pathway is important in the defense against bacteria, some parasites, and virus-infected cells.

Complement Attachment (alternate pathway)

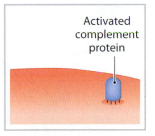

Activated complement protein

Complement Attachment (classical pathway)

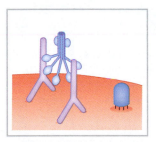

The classical pathway ends with the conversion of an inactive complement protein to an activated form that attaches to the cell wall.

Pore Formation

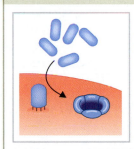

Once an activated complement protein has attached to the cell wall, additional complement proteins form a pore in the membrane that destroys the integrity of the target cell.

Multiple pores in bacterium

Cell lysis

Enhanced Phagocytosis

A coating of complement proteins and antibodies both attracts phagocytes and makes the target cell easier to engulf. This enhancement of phagocytosis, a process called **opsonization**, occurs because macrophage membranes contain receptors that detect and bind to complement proteins and bound antibodies.

Histamine Release

Release of histamine by mast cells and basophils in tissues increases the degree of local inflammation and accelerates blood flow to the region.

Module 19.9 Review

a. Define interferons.

b. Briefly explain the role of complement proteins.

c. What is the effect of histamine released by complement system activation?

Inflammation is a localized tissue response to injury; fever is a generalized response to tissue damage and infection

Inflammation

1 **Inflammation**, or the **inflammatory response**, is a localized tissue response to injury. Inflammation produces local swelling, redness, heat, and pain. Many stimuli, including impact, abrasion, distortion, chemical irritation, infection by pathogens, and extreme temperatures (hot or cold), can produce inflammation. Each of these stimuli kills cells, damages connective tissue fibers, or injures the tissue in some other way. The changes alter the chemical composition of the interstitial fluid. Damaged cells release prostaglandins, proteins, and potassium ions, and the injury itself may have introduced foreign proteins or pathogens. The changes in the interstitial environment trigger the complex process of inflammation.

Fever

2 **Fever** is the maintenance of body temperature greater than 37.2°C (99°F). Circulating proteins called **pyrogens** (PĪ-rō-jenz; *pyro-*, fever or heat + *-gen*, substance) can reset the temperature thermostat in the hypothalamus and raise body temperature. Within limits, a fever can be beneficial. High body temperatures may inhibit some viruses and bacteria, but the most likely beneficial effect is on body metabolism. For each 1°C rise in body temperature, metabolic rate jumps by 10 percent. The net results may be the quicker mobilization of tissue defenses and an accelerated repair process.

Tissue Damage

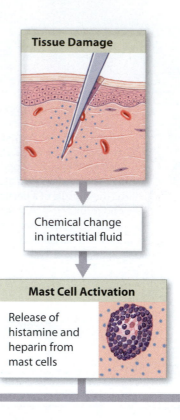

↓

Chemical change in interstitial fluid

↓

Mast Cell Activation

Release of histamine and heparin from mast cells

Redness, Swelling, Warmth, and Pain

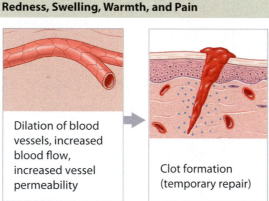 

Dilation of blood vessels, increased blood flow, increased vessel permeability

Clot formation (temporary repair)

Phagocyte Attraction

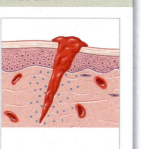

Attraction of phagocytes, especially neutrophils

Release of cytokines

Removal of debris by neutrophils and macrophages; stimulation of fibroblasts

Activation of specific defenses

Tissue Repair

Pathogen removal, clot erosion, scar tissue formation

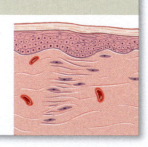

Summary of Nonspecific Defenses

3 This table summarizes the information on nonspecific defenses presented in this section.

Physical Barriers

Prevent approach of and deny access to pathogens

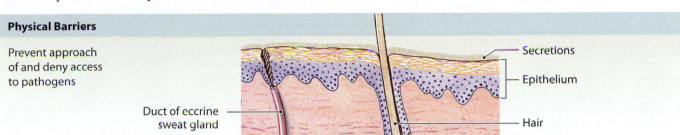

- Secretions
- Epithelium
- Hair
- Duct of eccrine sweat gland

Phagocytes

Remove debris and pathogens

Neutrophil   Eosinophil   Monocyte   Free macrophage   Fixed macrophage

Immunological Surveillance

Destroys abnormal cells

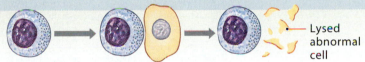

Natural killer cell

Lysed abnormal cell

Interferons

Increase resistance of cells to viral infection; slow the spread of disease

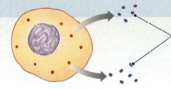

Interferons released by activated lymphocytes, macrophages, or virus-infected cells

Complement System

Attacks and breaks down the surfaces of cells, bacteria, and viruses; attracts phagocytes; stimulates inflammation

Complement

Lysed pathogen

Inflammatory Response

Multiple effects

Mast cell

- Blood flow increased
- Phagocytes activated
- Damaged area isolated by clotting reaction
- Capillary permeability increased
- Complement activated
- Regional temperature increased
- Specific defenses activated

Fever

Mobilizes defenses; accelerates repairs; inhibits pathogens

Body temperature rises above 37.2°C in response to pyrogens

Module 19.10 Review

a. List the body's nonspecific defenses.

b. A rise in the level of interferons in the body suggests what kind of infection?

c. What effects do pyrogens have in the body?

1. Labeling

Fill in the spaces with the name of the nonspecific defense described at right.

a _____ _____

b _____

c _____ _____

d _____

e _____

f _____ _____

g _____

a keep hazardous organisms and materials outside the body. For example, a mosquito that lands on your head may be unable to reach the surface of the scalp if you have a full head of hair.

b are cells that engulf pathogens and cell debris. Examples are the macrophages of peripheral tissues and the eosinophils and neutrophils of blood.

c is the destruction of abnormal cells by NK cells in peripheral tissues.

 Destruction of abnormal cells

d are chemical messengers that coordinate the defenses against viral infections.

e is a system of circulating proteins that assists antibodies in the destruction of pathogens.

f is a localized, tissue-level response that tends to limit the spread of an injury or infection.

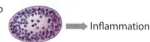 Inflammation

g is an elevation of body temperature that accelerates tissue metabolism and the activity of defenses.

2. Multiple choice

Choose the bulleted item that best completes each statement.

a A physical barrier such as the skin provides a nonspecific body defense due to its makeup, which includes _____.

- multiple layers
- a coating of keratinized cells
- a network of desmosomes locking adjacent cells together
- all of these

b NK (natural killer) cells sensitive to the presence of abnormal plasma membranes are primarily involved in _____.

- defenses against specific threats
- phagocytic activity for defense
- complex, time-consuming defense mechanisms
- immunological surveillance

c The nonspecific defense that breaks down cells, attracts phagocytes, and stimulates inflammation is _____.

- the inflammatory response
- the action of interferons
- the complement system
- immunological surveillance

d The protein(s) that interfere with the replication of viruses is (are) _____.

- complement proteins
- heparin
- pyrogens
- interferons

e Circulating proteins that reset the thermostat in the hypothalamus, causing a rise in body temperature, are called _____

- pyrogens
- interferons
- lysosomes
- complement proteins

f The "first line of cellular defense" against pathogenic invasion is _____.

- phagocytes
- mucus
- hair
- interferon

3. Short answer

We usually associate a fever with illness or disease. How may a fever be beneficial?

_____

Specific Defenses

1 **Specific defenses** are provided by the coordinated activities of T cells and B cells. These activities produce **immunity**, a specific resistance against potentially dangerous antigens. In general, T cells are responsible for cell-mediated immunity, which defends against abnormal cells and pathogens inside cells, and B cells provide antibody-mediated immunity, which defends against antigens and pathogens in body fluids. However, as indicated in this flowchart, there are many different forms of immunity.

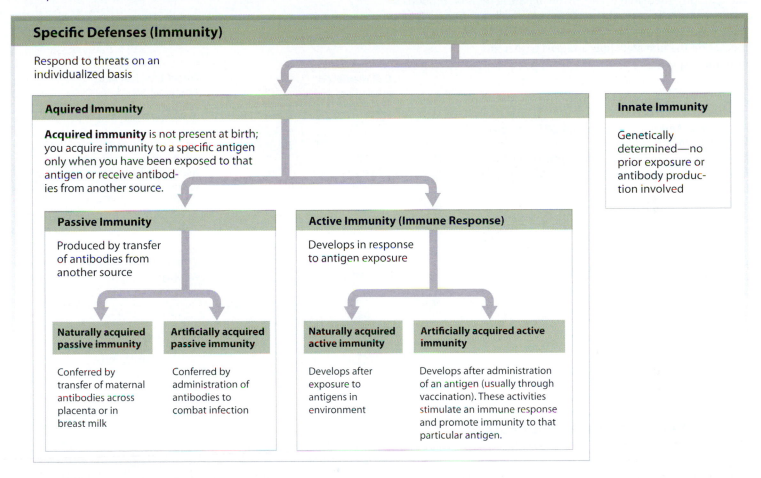

Specific Defenses (Immunity)

Respond to threats on an individualized basis

Aquired Immunity

Acquired immunity is not present at birth; you acquire immunity to a specific antigen only when you have been exposed to that antigen or receive antibodies from another source.

Innate Immunity

Genetically determined—no prior exposure or antibody production involved

Passive Immunity

Produced by transfer of antibodies from another source

Active Immunity (Immune Response)

Develops in response to antigen exposure

Naturally acquired passive immunity

Conferred by transfer of maternal antibodies across placenta or in breast milk

Artificially acquired passive immunity

Conferred by administration of antibodies to combat infection

Naturally acquired active immunity

Develops after exposure to antigens in environment

Artificially acquired active immunity

Develops after administration of an antigen (usually through vaccination). These activities stimulate an immune response and promote immunity to that particular antigen.

2 Regardless of the form, immunity exhibits four general properties, which are summarized in the following table.

Properties of Immunity

- **Specificity** results from the activation of appropriate lymphocytes and the production of antibodies with targeted effects. Each T cell or B cell has receptors that bind to one specific antigen, but ignore all others. The response of an activated T cell or B cell is equally specific, and leaves other antigens unaffected.
- **Versatility** results from the large diversity of lymphocytes present in the body. There are millions of different lymphocyte populations, each sensitive to a different antigen. When activated by an appropriate antigen, a lymphocyte divides, producing more lymphocytes with the same specificity. All the cells produced by the division of an activated lymphocyte constitute a **clone**.

- **Immunologic memory** exists because cell divisions of activated lymphocytes produce two groups of cells: one group that attacks the invader immediately, and another that remains inactive unless it is exposed to the same antigen at a later date. These inactive **memory cells** enable your immune system to "remember" antigens it has previously encountered, and to launch a faster, stronger, and longer-lasting counterattack if such an antigen reappears.
- **Tolerance** exists because the immune response ignores normal ("self") tissues but targets abnormal and foreign cells ("non-self") as well as toxins. Tolerance can also develop over time in response to chronic exposure to an antigen in the environment. Such tolerance generally lasts only as long as the exposure continues.

Specific defenses are triggered by exposure to antigenic fragments

1 This figure provides a general "big picture" overview of the immune response: Phagocytes activated by exposure to an antigen stimulate cell-mediated events involving attacks by T cells, and antibody-mediated events involving the actions of antibodies produced by cells derived from B cells. We will examine each of these processes more closely in later modules; this module focuses on the first step: how antigens trigger an immune response.

Antigens or Antigenic Fragments in Body Fluids

Most antigens must either infect cells or be "processed" by phagocytes before specific defenses are activated. The trigger is the appearance of antigens or antigenic fragments in plasma membranes; this is called **antigen presentation**.

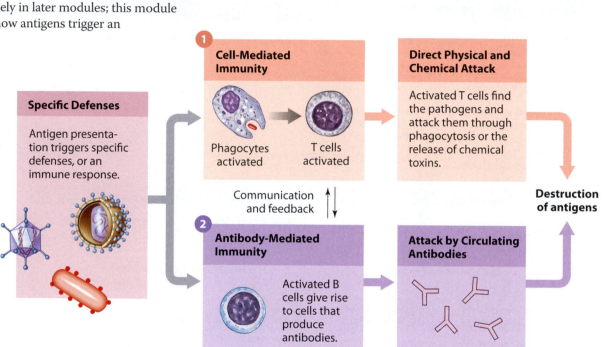

MHC Proteins and Antigen Presentation

2 Antigen presentation occurs when an antigen-glycoprotein combination capable of activating T cells appears in a plasma membrane. The structure of these glycoproteins is genetically determined. The genes controlling their synthesis are located along one portion of chromosome 6, in a region called the **major histocompatibility complex (MHC)**. These membrane glycoproteins are called **MHC proteins**. **Class I MHC proteins** are always present in the membranes of all nucleated cells.

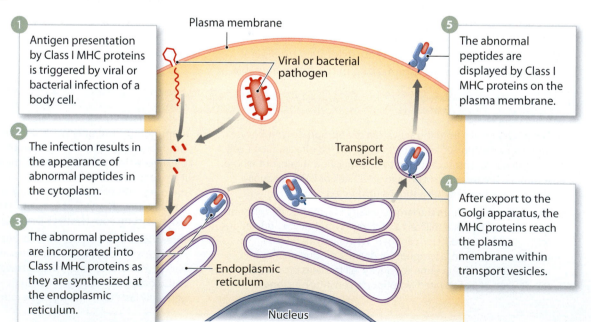

3 **Class II MHC proteins** are present only in the membranes of antigen-presenting cells and lymphocytes. **Antigen-presenting cells (APCs)** are specialized cells that include all the phagocytic cells of the monocyte–macrophage group and the dendritic cells of the skin and lymphoid organs. Class II MHC proteins appear in the plasma membrane only when the cell is processing antigens. The process of antigen presentation is shown here for a phagocytic cell.

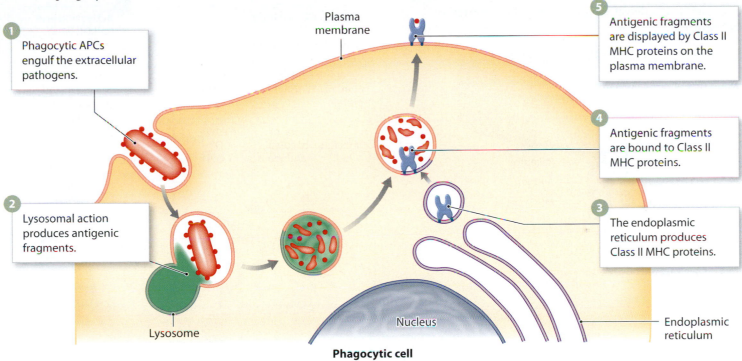

1 Phagocytic APCs engulf the extracellular pathogens.

2 Lysosomal action produces antigenic fragments.

Plasma membrane

5 Antigenic fragments are displayed by Class II MHC proteins on the plasma membrane.

4 Antigenic fragments are bound to Class II MHC proteins.

3 The endoplasmic reticulum produces Class II MHC proteins.

Lysosome

Nucleus

Endoplasmic reticulum

Phagocytic cell

4 In this preparation, cells were exposed to fluorescent-tagged antibodies that bind to specific structures and which glow when the cells are illuminated with fluorescent light. The green cell is an APC and the red cell is a lymphocyte; the bright yellow areas indicate where labeled antigens are being displayed on MHC proteins.

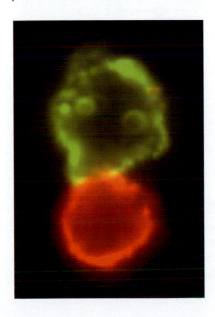

Module 19.11 Review

a. Describe antigen presentation.

b. What is the major histocompatibility complex (MHC)?

c. Where are Class I MHC proteins and Class II MHC proteins found?

Infected cells stimulate the formation and division of cytotoxic T cells, memory T_C cells, and suppressor T cells

1 Inactive T cells have receptors that recognize either Class I or Class II MHC proteins on other cells when those proteins are bound to specific antigens. T cell binding will occur only if the MHC protein contains the antigen that the T cell is programmed to detect. This process is called **antigen recognition**.

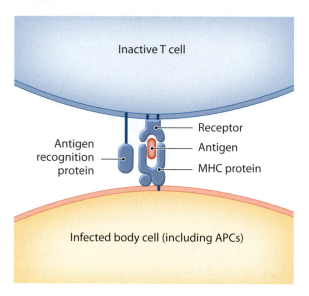

Inactive T cell

Antigen recognition protein — Receptor

Antigen

MHC protein

Infected body cell (including APCs)

3 Antigen recognition prepares a T cell for activation. A CD8 T cell can recognize antigens bound to Class I MHC proteins **1**. Full activation requires exposure of the cell to specific physical or chemical stimuli **2**. This results in activation and cell division **3**.

1
Antigen Recognition

Antigen recognition occurs when a CD8 T cell encounters an appropriate antigen on the surface of another cell, bound to a Class I MHC protein.

Infected cell

Viral or bacterial antigen

Inactive CD8 T cell

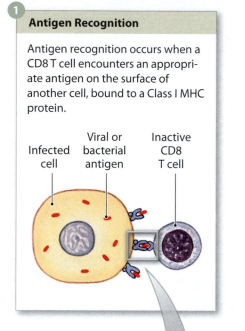

3
Activation and Cell Division

Antigen recognition results in T cell activation and cell division, producing three different types of CD8 T cells.

2 The membrane proteins involved in antigen recognition are members of a class of proteins called **CD** (cluster of differentiation) **markers**.

CD Markers

There are at least 70 different CD markers, but only two associated with T cells are important to our discussion.

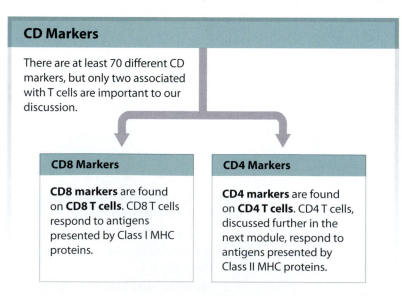

CD8 Markers

CD8 markers are found on **CD8 T cells**. CD8 T cells respond to antigens presented by Class I MHC proteins.

CD4 Markers

CD4 markers are found on **CD4 T cells**. CD4 T cells, discussed further in the next module, respond to antigens presented by Class II MHC proteins.

2
Costimulation

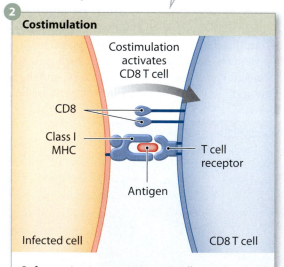

Costimulation activates CD8 T cell

CD8

Class I MHC

T cell receptor

Antigen

Infected cell

CD8 T cell

Before activation can occur, a T cell must be chemically or physically stimulated by the abnormal target cell. This vital secondary binding process, called **costimulation**, confirms the activation signal. Costimulation is like the safety on a gun: It helps prevent T cells from mistakenly attacking normal (self) tissues.

4 Following activation, the cell divides repeatedly, and daughter cells differentiate into three different types of CD8 T cells.

Cytotoxic T Cells Seek Out Antigen-Bearing Cells

Cytotoxic T cells, also called **T$_C$ cells**, seek out and destroy abnormal and infected cells. Cytotoxic T cells are highly mobile cells that roam throughout injured tissues. When a T$_C$ cell encounters its target antigens bound to Class I MHC proteins, it attacks the target cell.

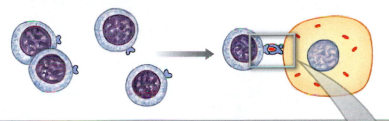

Destruction of Target Cells

The T$_C$ cell destroys the antigen-bearing cell. It may use several different mechanisms to kill the target cell.

Memory T$_C$ Cells Are Produced

Memory T$_C$ cells are produced by the same cell divisions that produce cytotoxic T cells. Thousands of these cells are produced, but they do not differentiate further the first time the antigen triggers an immune response.

Memory T$_C$ cells (inactive)

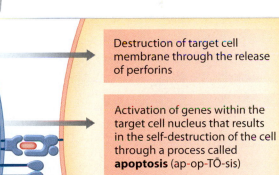

Destruction of target cell membrane through the release of perforins

Activation of genes within the target cell nucleus that results in the self-destruction of the cell through a process called **apoptosis** (ap-op-TŌ-sis)

Disruption of cell metabolism through the release of **lymphotoxin** (lim-fō-TOK-sin)

Suppressor T Cells Provide a Delayed Suppression

Suppressor T cells (T$_S$ cells) suppress the responses of other T cells and of B cells by secreting **suppression factors** that limit the degree of immune system activation. Suppression does not occur immediately, because suppressor T cell activation takes much longer than the activation of other types of T cells, and so suppressor T cells act only after the initial immune response.

Suppressor T cells

The first time CD8 T cells encounter a specific antigen, the entire sequence of events—from the appearance of the antigen in a tissue to widespread cell destruction by cytotoxic T cells—takes two days or more. Over this period, the damage or infection may spread, making it more difficult to control. However, if the same antigen appears a second time, memory T$_C$ cells will immediately differentiate into cytotoxic T cells, producing a prompt, effective cellular response that can overwhelm an invading organism before it becomes well established in the tissues.

Module 19.12 Review

a. Identify the three major types of T cells activated by Class I MHC proteins.

b. Describe CD markers.

c. How can the presence of an abnormal antigen in the cytoplasm of a cell initiate an immune response?

APCs can stimulate the activation of CD4 T cells; this produces helper T cells that promote B cell activation and antibody production

1 Before they can initiate antibody-mediated immunity, inactive CD4 T cells must be exposed to appropriate antigens bound to Class II MHC proteins. Costimulation then completes their activation. Upon activation, CD4 T cells undergo a series of divisions, and daughter cells differentiate into active **helper T cells (T_H cells)** and **memory T_H cells**.

2 When a B cell encounters its specific antigen, it prepares to undergo activation. This preparatory process is called **sensitization**. During sensitization, antigens brought into the cell by endocytosis subsequently appear on the surface of the B cell, bound to Class II MHC proteins.

Antigen Recognition by CD4 T Cell

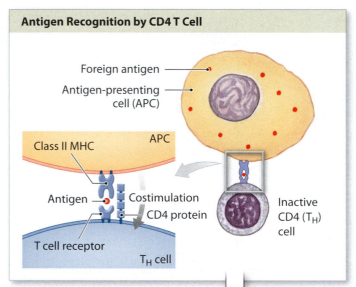

Foreign antigen
Antigen-presenting cell (APC)
APC
Class II MHC
Antigen
Costimulation
CD4 protein
T cell receptor
T_H cell
Inactive CD4 (T_H) cell

Sensitization

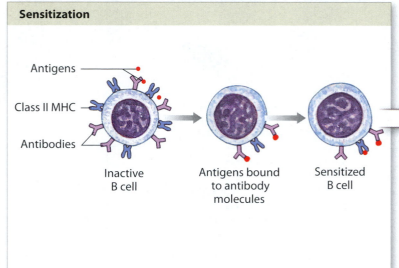

Antigens
Class II MHC
Antibodies
Inactive B cell
Antigens bound to antibody molecules
Sensitized B cell

CD4 T Cell Activation and Cell Division

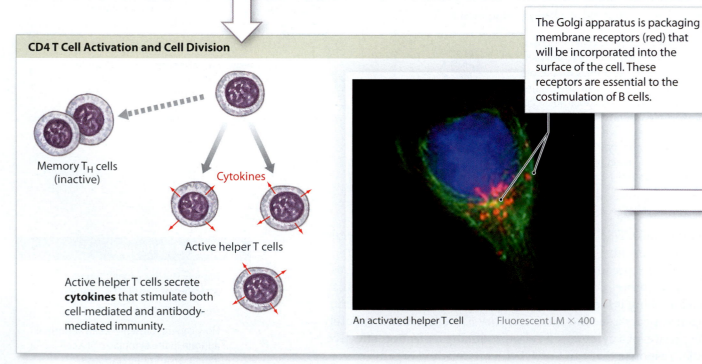

Memory T_H cells (inactive)

Cytokines

Active helper T cells

Active helper T cells secrete **cytokines** that stimulate both cell-mediated and antibody-mediated immunity.

The Golgi apparatus is packaging membrane receptors (red) that will be incorporated into the surface of the cell. These receptors are essential to the costimulation of B cells.

An activated helper T cell   Fluorescent LM × 400

3 To become fully activated, a sensitized B cell must encounter a helper T cell that was activated by exposure to the same antigen. When that happens, the helper T cell binds to the MHC complex of the sensitized B cell, recognizes the presence of the antigen, and begins secreting cytokines that promote B cell activation.

4 Under the continued stimulation of cytokines released by active helper T cells, an activated B cell undergoes a series of cell divisions. These divisions yield daughter cells with two different fates.

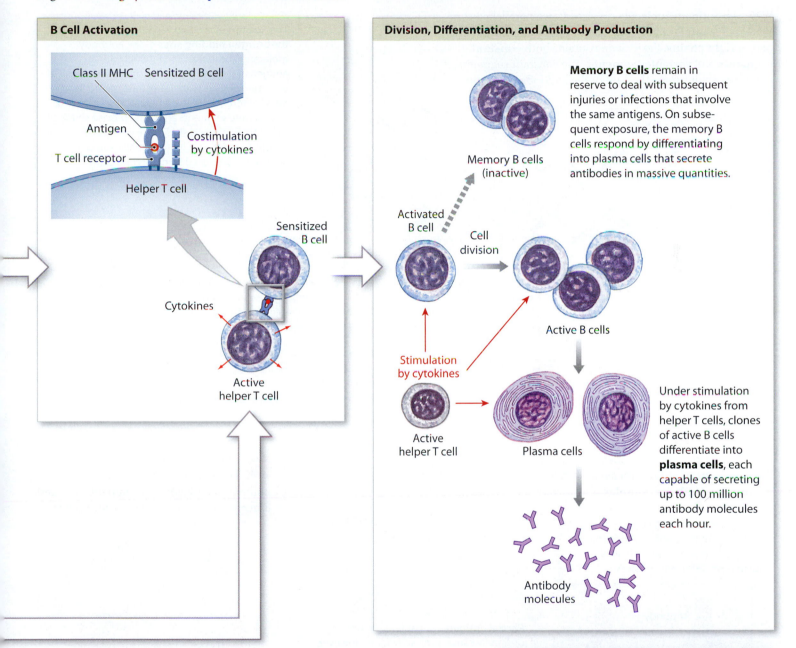

B Cell Activation

Class II MHC   Sensitized B cell

Antigen

Costimulation by cytokines

T cell receptor

Helper T cell

Cytokines

Sensitized B cell

Active helper T cell

Division, Differentiation, and Antibody Production

Memory B cells remain in reserve to deal with subsequent injuries or infections that involve the same antigens. On subsequent exposure, the memory B cells respond by differentiating into plasma cells that secrete antibodies in massive quantities.

Memory B cells (inactive)

Activated B cell

Cell division

Active B cells

Stimulation by cytokines

Active helper T cell

Plasma cells

Under stimulation by cytokines from helper T cells, clones of active B cells differentiate into **plasma cells**, each capable of secreting up to 100 million antibody molecules each hour.

Antibody molecules

Module 19.13 Review

a. Define sensitization.

b. Explain the function of cytokines secreted by helper T cells.

c. If you observed a higher-than-normal number of plasma cells in a sample of lymph, would you expect antibody levels in the blood to be higher or lower than normal?

Antibodies are small soluble proteins that bind to specific antigens; they may inactivate the antigens or trigger another defensive process

1 An antibody molecule consists of two parallel pairs of polypeptide chains: one pair of **heavy chains** and one pair of **light chains**. Each chain contains both **constant segments** and **variable segments**. The constant segments of the heavy chains form the base of the antibody molecule.

The free tips of the two variable segments form the **antigen binding sites** of the antibody molecule. These sites can interact with an antigen in the same way that the active site of an enzyme interacts with a substrate molecule. Small differences in the amino acid sequence of the variable segments affect the precise shape of the antigen binding site. These differences account for differences in specificity among the antibodies produced by different B cells.

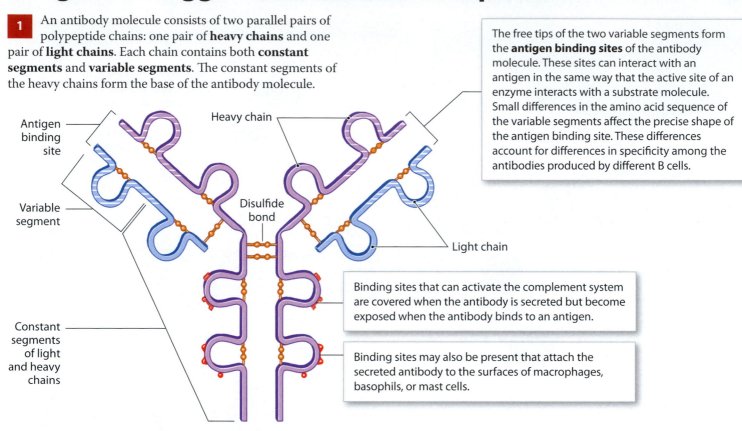

Antigen binding site

Heavy chain

Variable segment

Disulfide bond

Light chain

Constant segments of light and heavy chains

Binding sites that can activate the complement system are covered when the antibody is secreted but become exposed when the antibody binds to an antigen.

Binding sites may also be present that attach the secreted antibody to the surfaces of macrophages, basophils, or mast cells.

2 When an antibody molecule binds to its corresponding antigen molecule, an **antigen-antibody complex** is formed.

1 Antibodies bind not to the entire antigen, but to specific portions of its exposed surface—regions called **antigenic determinant sites**.

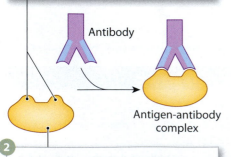

Antibody

Antigen-antibody complex

2 A **complete antigen** is an antigen with at least two antigenic determinant sites, one for each of the antigen binding sites on an antibody molecule.

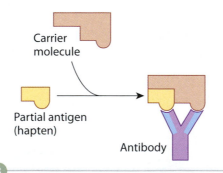

Carrier molecule

Partial antigen (hapten)

Antibody

3 **Partial antigens**, or haptens, do not ordinarily cause B cell activation. However, they may become attached to carrier molecules, forming combinations that can function as complete antigens. The antibodies produced will attack both the hapten and the carrier molecule. If the carrier molecule is normally present in the tissues, the antibodies may begin attacking and destroying normal cells. This is the basis for several drug reactions, including allergies to penicillin.

3 The exposed surface of something as large as a bacterium contains millions of antigenic determinant sites, and it may become carpeted with antibodies.

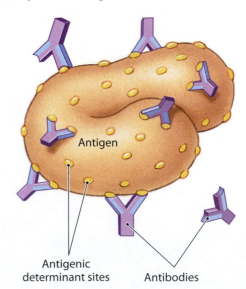

Antigen

Antigenic determinant sites

Antibodies

4 There are five different classes of antibodies, or **immunoglobulins** (**Igs**). The classes are determined by differences in the structure of the heavy-chain constant segments and so have no effect on the antibody's specificity, which is determined by the antigen binding sites.

Classes of Antibodies

   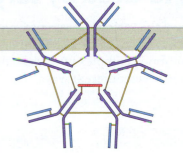 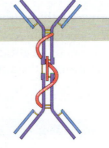

IgG antibodies account for 80 percent of all antibodies. IgG antibodies are responsible for resistance against many viruses, bacteria, and bacterial toxins.

IgE attaches as an individual molecule to the exposed surfaces of basophils and mast cells.

IgD is an individual molecule on the surfaces of B cells, where it can bind antigens in the extracellular fluid. This binding can play a role in the sensitization of the B cell involved.

IgM is the first class of antibody secreted after an antigen is encountered. IgM concentration declines as IgG production accelerates. The anti-A and anti-B antibodies responsible for the agglutination of incompatible blood types are IgM antibodies.

IgA is found primarily in glandular secretions such as mucus, tears, saliva, and semen. These antibodies attack pathogens before they gain access to internal tissues.

5 The initial response to exposure to an antigen is called the **primary response**. Because the antigen must activate the appropriate B cells, which must then differentiate into antibody-secreting plasma cells, the primary response takes time to develop. During the primary response, the **antibody titer**, or level of antibody activity in the plasma, does not peak until one to two weeks after the initial exposure. If the individual is no longer exposed to the antigen, the antibody titer and concentration declines thereafter.

6 When an antigen is encountered a second time, it triggers a more extensive and prolonged **secondary response**. During the secondary response, antibody titers increase more rapidly and reach levels many times higher than they did in the primary response. This reflects the presence of large numbers of memory cells that are already primed for the arrival of the antigen. These memory B cells respond immediately—much faster than the B cells stimulated during the initial exposure. The secondary response appears even if the second exposure occurs years after the first, because memory cells may survive for 20 years or more.

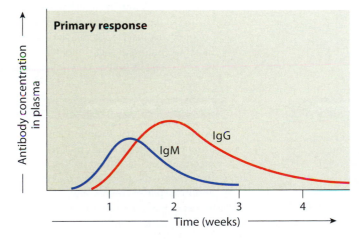

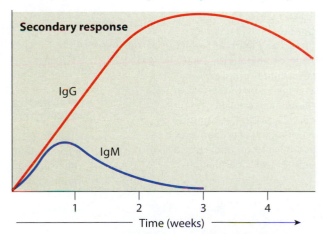

Module 19.14 Review

a. Define antigenic determinant site.

b. Describe the structure of an antibody.

c. Which would be more affected by a lack of memory B cells and memory T cells: the primary response or the secondary response?

Antibodies use many different mechanisms to destroy target antigens

As a group, the antibodies of the various classes provide a versatile and effective defense against a variety of threats. The formation of an antigen-antibody complex may cause the elimination of the antigen in seven different ways.

Neutralization

Both viruses and bacterial toxins must bind to the plasma membranes of body cells before they can enter or injure those cells. Binding occurs at superficial sites on the bacteria or toxins. Antibodies may bind to those sites, making the virus or toxin incapable of attaching itself to a cell. This mechanism is known as **neutralization**.

Prevention of Pathogen Adhesion

Antibodies dissolved in saliva, mucus, tears, and perspiration coat epithelia, providing an additional layer of defense. A covering of antibodies makes it difficult for bacteria or viruses to adhere to and penetrate body surfaces.

Activation of Complement

Upon binding to an antigen, portions of the antibody molecule change shape, exposing areas that bind complement proteins. The bound complement molecules then activate the complement system, which destroys the antigen.

Potential Beneficial Effects of Antigen-Antibody Complex Formation

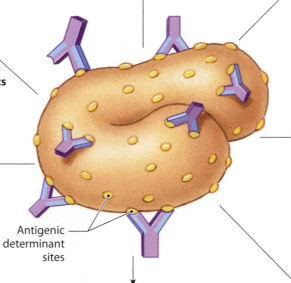

Antigenic determinant sites

Opsonization

Phagocytes can bind more easily to antibodies and complement proteins on the surface of a pathogen than they can to an antibody-free and complement-free surface. As a result, a coating of antibodies and complement proteins increases the effectiveness of phagocytosis. This effect, called **opsonization**, was introduced in Module 19.9. Some bacteria have slick plasma membranes or capsules, and phagocytes must be able to hang onto their prey before they can engulf it.

Precipitation and Agglutination

If antigens are close together, an antibody can bind to antigenic determinant sites on two different antigens. In this way, antibodies can tie large numbers of antigens together, creating an **immune complex**. When the target antigen is on the surface of a cell or a virus, the formation of immune complexes is called **agglutination**. The clumping of erythrocytes that occurs when incompatible blood types are mixed is an agglutination reaction.

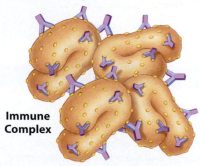

Immune Complex

Stimulation of Inflammation

Antibodies may promote inflammation by stimulating the release of chemicals from basophils and mast cells.

Attraction of Phagocytes

Antigens covered with antibodies attract eosinophils, neutrophils, and macrophages—cells that phagocytize pathogens and destroy foreign or abnormal plasma membranes.

Module 19.15 Review

a. Define opsonization.

b. List the ways that antigen-antibody complexes can destroy target antigens.

c. Which cells are involved in the inflammatory response?

Allergies and anaphylaxis are caused by antibody responses

Allergies are inappropriate or excessive immune responses to antigens. The sudden increase in cellular activity or antibody titers can have several unpleasant side effects. For example, neutrophils or cytotoxic T cells may destroy normal cells while attacking the antigen, or the antigen-antibody complex may trigger a massive inflammatory response. Antigens that trigger allergic reactions are often called **allergens**.

1 Sensitization to an allergen during the initial exposure leads to the production of large quantities of IgE. The tendency to produce IgE antibodies in response to specific allergens may be genetically determined.

First Exposure

Allergen fragment

Allergens

Macrophage

T$_H$ cell activation

B cell sensitization and activation

Plasma cell

IgE antibodies

2 **Immediate hypersensitivity** is a rapid and especially severe response to the presence of an antigen. One form, **allergic rhinitis**, includes hay fever and other environmental allergies and may affect 15 percent of the U.S. population. Allergic rhinitis, characterized by inflammation of the nasal membranes, is one example of a **hypersensitivity reaction** that occurs at and is restricted to the body surface. If the allergen enters the bloodstream, however, the response could be lethal. In **anaphylaxis** (an-a-fi-LAK-sis; *ana-*, again + *phylaxis*, protection), a circulating allergen affects mast cells throughout the body. In severe cases extensive peripheral vasodilation occurs, producing a fall in blood pressure that can lead to a circulatory collapse. This response is called **anaphylactic shock**.

Subsequent Exposure

Allergen

IgE

Granules

Sensitization of mast cells and basophils

Massive stimulation of mast cells and basophils

Release of histamines, leukotrienes, and other chemicals that cause pain and inflammation

Localized Allergic Reactions	Systemic Allergic Reactions
If the allergen is at the body surface: localized inflammation, pain, and itching	If the allergen is in the bloodstream: itching, swelling, and difficulty breathing (due to airway constriction)
Example: allergic rhinitis	Example: anaphylaxis

Module 19.16 Review

a. Define allergy and allergen.

b. What is anaphylaxis?

c. Which chemicals do mast cells and basophils release when stimulated in an allergic reaction?

Specific and nonspecific defenses work together to defeat pathogens

1 This flowchart depicts the relationships among the elements of the nonspecific defenses and those of the specific defenses (the immune response).

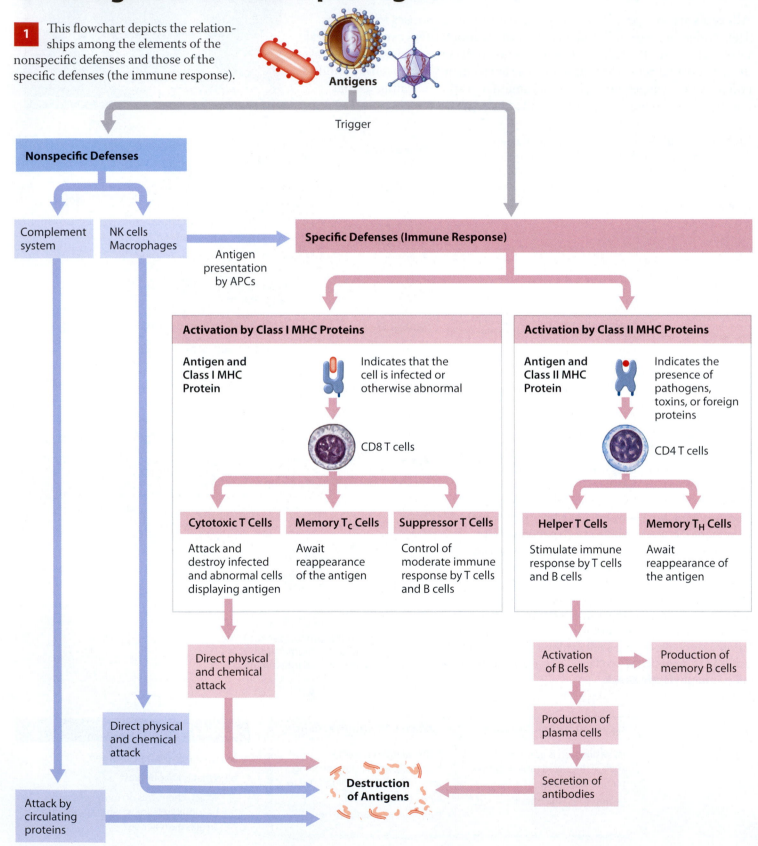

Antigens

Trigger

Nonspecific Defenses

Complement system

NK cells Macrophages

Antigen presentation by APCs

Specific Defenses (Immune Response)

Activation by Class I MHC Proteins

Antigen and Class I MHC Protein

Indicates that the cell is infected or otherwise abnormal

CD8 T cells

Cytotoxic T Cells

Attack and destroy infected and abnormal cells displaying antigen

Memory T$_C$ Cells

Await reappearance of the antigen

Suppressor T Cells

Control of moderate immune response by T cells and B cells

Activation by Class II MHC Proteins

Antigen and Class II MHC Protein

Indicates the presence of pathogens, toxins, or foreign proteins

CD4 T cells

Helper T Cells

Stimulate immune response by T cells and B cells

Memory T$_H$ Cells

Await reappearance of the antigen

Direct physical and chemical attack

Direct physical and chemical attack

Activation of B cells

Production of memory B cells

Production of plasma cells

Destruction of Antigens

Secretion of antibodies

Attack by circulating proteins

2 This illustration provides an overview of the course of events responsible for overcoming a bacterial infection. The most effective defenses against bacteria involve phagocytosis and antigen presentation by APCs.

3 The basic sequence of events following a viral infection differs from those of a bacterial infection because cytotoxic T cells and NK cells can be activated by direct contact with virus-infected cells.

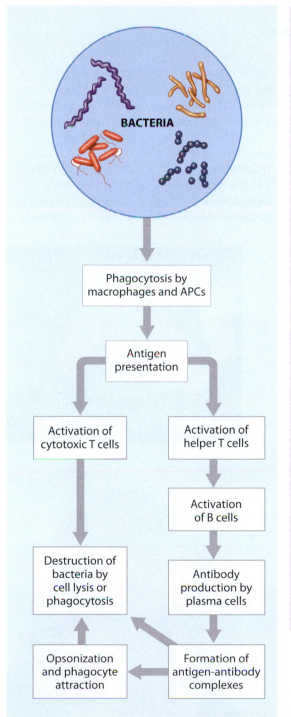

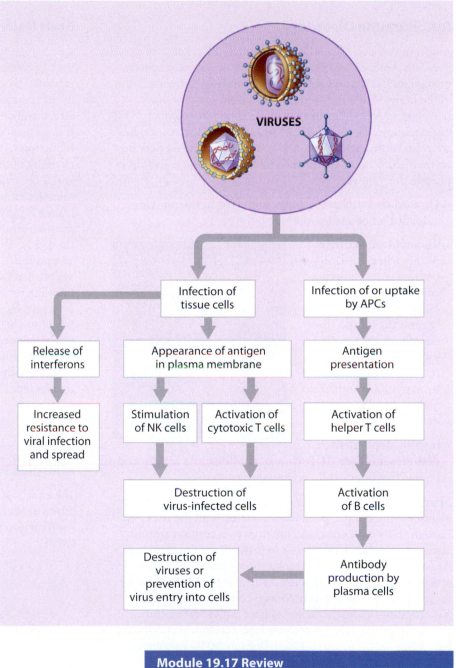

Module 19.17 Review

a. Identify the type of T cell whose plasma membrane contains CD8 markers and the type with CD4 markers.

b. Which cells can be activated by direct contact with virus-infected cells?

c. Which cells produce antibodies?

Immune disorders involving both overactivity and underactivity can be harmful

Excessive or Misdirected Immune Response

Autoimmune Disorders

Autoimmune disorders affect an estimated 5 percent of adults in North America and Europe. The immune system usually recognizes but ignores self-antigens—normal antigens found in the body. When the recognition system malfunctions, however, activated B cells may make antibodies against normal body cells and tissues. These "misguided" antibodies are called **autoantibodies**. The condition produced depends on the specific antigen attacked by autoantibodies. Examples include the following:

- **Thyroiditis** is inflammation resulting from the release of autoantibodies against thyroglobulin.

- **Rheumatoid arthritis** occurs when autoantibodies form immune complexes within connective tissues around the joints. These complexes lead to an excessive immune response characterized by marked inflammation and eventually joint destruction.

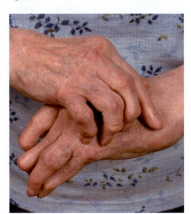

- **Insulin-dependent diabetes mellitus (IDDM)** is generally caused by autoantibodies that attack cells in the pancreatic islets.

Many autoimmune disorders appear to be cases of mistaken identity. For example, proteins associated with the measles, Epstein–Barr, influenza, and other viruses contain amino acid sequences that are similar to those of myelin proteins. As a result, antibodies that target these viruses may also attack myelin sheaths. This mechanism is likely responsible for multiple sclerosis. For unknown reasons, the risk of autoimmune disorders increases if an individual has an unusual type of MHC protein. At least 50 clinical conditions have been linked to specific variations in MHC structure.

Graft Rejection

After organ transplantation surgery has been performed, the major problem is **graft rejection**. In graft rejection, T cells are activated by contact with MHC proteins on plasma membranes in the donated tissues. The cytotoxic T cells that develop then attack and destroy the foreign cells. Significant improvements in transplant success can be made by **immunosuppression**, a reduction in the sensitivity of the immune system. An understanding of the communication among T cells, macrophages, and B cells has led to the development of drugs with more selective effects. **Cyclosporin A (CsA)**, a compound derived from a fungus, was the most important immunosuppressive drug developed in the 1980s. This compound suppresses the immune response primarily by inhibiting helper T cell activity while leaving suppressor T cells relatively unaffected.

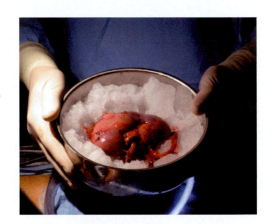

Allergies

The effects of the many forms of allergies range from mild to potentially lethal.

Immunodeficiency Diseases

Immunodeficiency diseases result from (1) problems with the embryological development of lymphoid organs and tissues; (2) an infection with a virus that depresses immune function; or (3) treatment with, or exposure to, immunosuppressive agents, such as radiation or drugs. **Acquired immune deficiency syndrome (AIDS)**, the most common immunodeficiency disease, is caused by the **human immunodeficiency virus (HIV)**. The virus binds to CD4 proteins and infects helper T cells. The infected cells begin synthesizing viral proteins, and these new viruses are then shed from the cell surface. Cells infected with HIV are ultimately destroyed either by the virus or immune defenses. The gradual destruction of helper T cells impairs both cell-mediated and antibody-mediated responses to antigens. Making matters worse, suppressor T cells are relatively unaffected by the virus, and over time the excess of suppressing factors "turns off" the normal immune response. Circulating antibody levels decline, cell-mediated immunity is reduced, and the body is left vulnerable to microbial invaders. With immune function suppressed, ordinarily harmless microorganisms can initiate lethal **opportunistic infections**. Because immune surveillance is also depressed, the risk of cancer increases. Infection with HIV occurs through intimate contact with the body fluids of infected individuals. An estimated 33 million people are infected worldwide; 22 million of them are in sub-Saharan Africa, and 1.2 million in North America. AIDS causes about 17,000 deaths each year in the United States, and 2 million deaths worldwide.

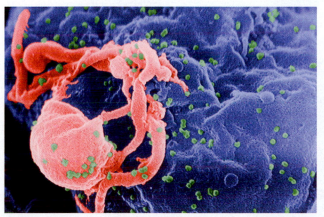

HIV (green) budding from an infected T$_H$ cell   SEM × 40,000

Age-Related Reductions in Immune Activity

With advancing age, the immune system becomes less effective at combating disease. T cells become less responsive to antigens, so fewer cytotoxic T cells respond to an infection. This effect may, at least in part, be associated with the gradual involution of the thymus and a reduction in circulating levels of thymic hormones. Because the number of helper T cells is also reduced, B cells are less responsive, so antibody levels do not rise as quickly after antigen exposure. The net result is an increased susceptibility to viral and bacterial infections. For this reason, vaccinations for acute viral diseases such as the flu (influenza), and for pneumococcal pneumonia, are strongly recommended for elderly individuals. The increased incidence of cancer in the elderly reflects the fact that immune surveillance declines, so tumor cells are not eliminated as effectively.

Module 19.18 Review

a. Define autoimmune disorders.

b. Describe immunosuppression.

c. Provide a plausible explanation for the increased incidence of cancer in the elderly.

1. Matching

Match the following terms with the most closely related description.

- opsonization
- helper T cells
- antibody
- Class II MHC
- costimulation
- IgM
- Class I MHC
- IgG
- passive immunity
- anaphylaxis
- CD4 markers
- acquired immunity
- B lymphocytes

a _____ Two parallel pairs of polypeptide chains

b _____ Found on helper T cells

c _____ Active and passive

d _____ Transfer of antibodies

e _____ Attacked by HIV

f _____ Enhances phagocytosis

g _____ MHC proteins present in the plasma membranes of all nucleated cells

h _____ Differentiate into memory and plasma cells

i _____ MHC proteins present in the plasma membranes of all APCs and lymphocytes

j _____ Antibodies used to determine blood type

k _____ Secondary binding process required for T cell activation

l _____ Accounts for 80 percent of all immunoglobulins

m _____ Circulating allergen stimulates mast cells throughout body

2. Matching

Match the following terms with the most closely related description.

- cytotoxic T cells
- viruses
- B cells
- antibodies
- helper T cells
- macrophages
- natural killer (NK) cells
- suppressor T cells
- memory T cells and B cells

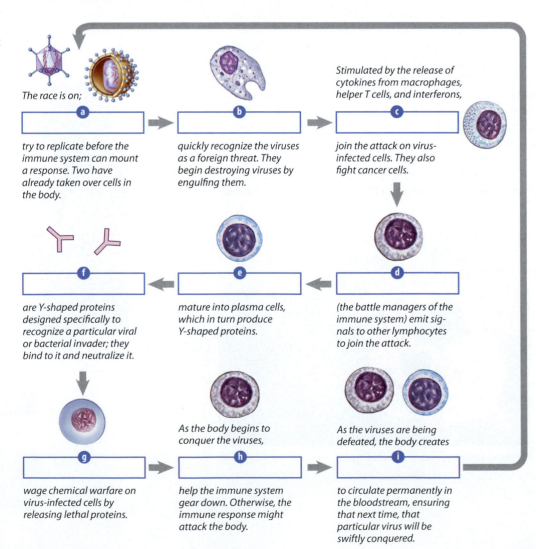

The race is on;

a ☐

try to replicate before the immune system can mount a response. Two have already taken over cells in the body.

b ☐

quickly recognize the viruses as a foreign threat. They begin destroying viruses by engulfing them.

Stimulated by the release of cytokines from macrophages, helper T cells, and interferons,

c ☐

join the attack on virus-infected cells. They also fight cancer cells.

f ☐

are Y-shaped proteins designed specifically to recognize a particular viral or bacterial invader; they bind to it and neutralize it.

e ☐

mature into plasma cells, which in turn produce Y-shaped proteins.

d ☐

(the battle managers of the immune system) emit signals to other lymphocytes to join the attack.

g ☐

wage chemical warfare on virus-infected cells by releasing lethal proteins.

As the body begins to conquer the viruses,

h ☐

help the immune system gear down. Otherwise, the immune response might attack the body.

As the viruses are being defeated, the body creates

i ☐

to circulate permanently in the bloodstream, ensuring that next time, that particular virus will be swiftly conquered.

Visual Outline with Key Terms   Summarize the content of each module using the terms in the order provided.

Anatomy of the Lymphatic System

- ○ lymphatic system
- • lymphocytes
- • lymph
- • lymphatics

19.1

Interstitial fluid continuously flows into lymphatic capillaries, and leaves tissue as lymph within lymphatic vessels

- • lymphatic vessels
- • lymphatic capillaries

19.2

Small lymphatic vessels collect to form lymphatic ducts that empty into the subclavian veins

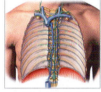

- • superficial lymphatics
- • deep lymphatics
- • lymphatic trunks
- • thoracic duct
- • right lymphatic duct
- • right and left jugular trunks
- • right and left subclavian trunks
- • right and left broncho-mediastinal trunks
- ○ right and left subclavian veins
- • cisterna chyli
- • lumbar trunks
- • intestinal trunk
- • lymphedema

19.3

Lymphocytes are responsible for the immune functions of the lymphatic system

- • antigens
- • T cells
- • cytotoxic T cells
- • cell-mediated immunity
- • helper T cells
- • suppressor T cells
- • B cells
- • plasma cells
- • antibody-mediated immunity
- • NK (natural killer) cells
- • immunological surveillance
- • lymphopoiesis
- ○ red bone marrow
- • lymphoid stem cells
- ○ thymus
- • blood–thymus barrier

19.4

Lymphocytes aggregate within lymphoid tissues and lymphoid organs

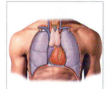

- • lymphoid tissues
- • lymphoid nodule
- • lymphoid organs
- • aggregated lymphoid nodules
- • germinal center
- • mucosa-associated lymphoid tissue (MALT)
- • tonsils
- • palatine tonsils
- • pharyngeal tonsil
- • lingual tonsils
- • tonsillitis
- • lymph nodes
- • lymph glands
- • afferent lymphatics
- • dendritic cells
- • outer cortex
- • deep cortex
- • medullary sinus
- • efferent lymphatics
- • hilum
- • trabeculae
- • appendicitis

19.5

The thymus is a lymphoid organ that produces functional T cells

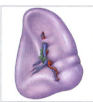

- ○ thymus
- • thymosins
- • involution
- • lobes
- • septa
- • lobules
- • cortex
- • medulla
- • reticular epithelial cells
- • blood–thymus barrier
- • thymic corpuscles

19.6

The spleen, the largest lymphoid organ, responds to antigens in the bloodstream

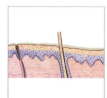

- ○ spleen
- • gastrosplenic ligament
- • diaphragmatic surface
- • splenectomy
- • visceral surface
- • gastric area
- • renal area
- • hilum
- • pulp
- • red pulp
- • white pulp
- • trabeculae
- • trabecular arteries
- • central arteries
- • trabecular veins

Nonspecific Defenses

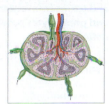

- ○ resistance
- • nonspecific defenses
- • nonspecific resistance
- • physical barriers
- • phagocytes
- • immunological surveillance
- • interferons
- • complement
- • inflammatory response
- • fever
- • specific defenses
- • immunity
- • specific resistance

• = *Term boldfaced in this module*

19.7

Physical barriers prevent pathogens and toxins from entering body tissues and phagocytes provide the next line of defense

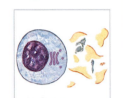

- ○ integumentary system
- ○ epithelia
- ○ phagocytes
- neutrophils
- eosinophils
- macrophages
- monocyte–macrophage system
- fixed macrophages
- free macrophages
- emigration
- diapedesis
- chemotaxis

19.8

NK cells perform immunological surveillance, detecting and destroying abnormal cells

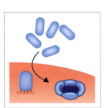

- immunological surveillance
- tumor-specific antigens
- perforins
- immunological escape

19.9

Interferons and the complement system are distributed widely in body fluids

- interferons
- cytokines
- alpha (α)-interferons
- beta (β)-interferons
- gamma (γ)-interferons
- complement system
- classical pathway
- alternative pathway
- properdin
- opsonization

19.10

Inflammation is a localized tissue response to injury; fever is a generalized response to tissue damage and infection

- inflammation
- inflammatory response
- fever
- pyrogens

SECTION 3

Specific Defenses

- specific defenses
- immunity
- acquired immunity
- ○ passive immunity
- ○ naturally acquired passive immunity
- ○ artificially acquired passive immunity
- ○ active immunity (immune response)
- ○ naturally acquired active immunity
- ○ artificially acquired active immunity
- ○ innate immunity
- specificity
- versatility
- clone
- immunologic memory

Specific Defenses (Immunity)

Respond to threats on an individualized basis

- memory cells
- tolerance

19.11

Specific defenses are triggered by exposure to antigenic fragments

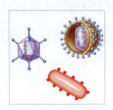

- antigen presentation
- major histocompatibility complex (MHC)
- MHC proteins
- Class I MHC proteins
- Class II MHC proteins
- antigen-presenting cells (APCs)

19.12

Infected cells stimulate the formation and division of cytotoxic T cells, memory T_C cells, and suppressor T cells

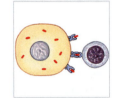

- antigen recognition
- CD markers
- CD8 markers
- CD8 T cells
- CD4 markers
- CD4 T cells
- costimulation
- cytotoxic T cells (T_C cells)
- memory T_C cells
- suppressor T cells (T_S cells)
- suppression factors
- apoptosis
- lymphotoxin

19.13

APCs can stimulate the activation of CD4 T cells; this produces helper T cells that promote B cell activation and antibody production

- helper T cells (T_H cells)
- memory T_H cells
- cytokines
- sensitization
- memory B cells
- plasma cells

19.14

Antibodies are small soluble proteins that bind to specific antigens; they may inactivate the antigens or trigger another defensive process

- ○ antibody molecule
- heavy chains
- light chains
- constant segments
- variable segments
- antigen binding sites
- antigen-antibody complex
- antigenic determinant sites
- complete antigen
- partial antigens
- immunoglobulins (Igs)
- IgG, IgE, IgD, IgM, IgA
- primary response
- antibody titer
- secondary response

19.15

Antibodies use many different mechanisms to destroy target antigens

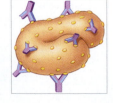

- neutralization
- opsonization
- immune complex
- agglutination

● = *Term boldfaced in this module*

19.16

Allergies and anaphylaxis are caused by antibody responses

- allergies
- allergens
- immediate hypersensitivity
- allergic rhinitis
- hypersensitivity reaction
- anaphylaxis
- anaphylactic shock
- localized allergic reactions

- systemic allergic reactions

19.17

Specific and nonspecific defenses work together to defeat pathogens

- nonspecific defenses
- specific defenses

BACTERIA

19.18

Immune disorders involving both overactivity and underactivity can be harmful

- autoimmune disorders
- autoantibodies
- thyroiditis
- rheumatoid arthritis
- insulin-dependent diabetes mellitus (IDDM)
- graft rejection
- immunosuppression
- cyclosporin A (CsA)
- allergies
- immuno-deficiency diseases
- acquired immune deficiency syndrome (AIDS)

- human immunodeficiency virus (HIV)
- opportunistic infections

● = *Term boldfaced in this module*

Chapter Integration: Applying what you've learned

The varicella-zoster virus (VZV) is responsible for two significant diseases: chickenpox (varicella) and shingles (herpes zoster). Chickenpox is a common, contagious, childhood disease characterized by groups of small, itchy blisters, inflamed skin, and a rash that spreads across the body surface. Signs and symptoms, including a low-grade fever, fatigue, headache, and flu-like symptoms, appear a few days after initial exposure. The blisters then develop over the next four days. Subsequently these blisters dry out, crust over, and form scabs that slough off 9–13 days later. Shingles is caused by a reactivation of the varicella-zoster virus later in the individual's life.

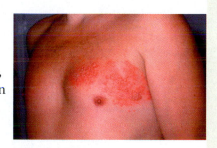

A live chickenpox vaccine was developed in 1994, and the Centers for Disease Control and Prevention (CDC) recommends that people aged 19–49 who had chickenpox earlier in life receive the chickenpox vaccine to prevent shingles; the CDC also recommends the shingles vaccine, Zostavax, for people 60 years old and older to prevent shingles.

An anatomy and physiology student discovers that she has been exposed to chickenpox after spending the weekend at home baby-sitting her twin nieces, who had been coughing and sneezing a considerable amount. Because the virus is spread via droplet transmission, she is concerned that she might have contracted the disease. She goes to see her physician, who takes a blood sample and sends it to a lab for antibody titers. The results reveal an elevated level of IgM antibodies to the varicella (chickenpox) virus but very few IgG antibodies to the virus.

1. Has this student come down with chickenpox?

2. Who is susceptible for contracting chickenpox?

3. What is the causative agent for chickenpox and for shingles?

4. Distinguish between IgM and IgG antibodies.

5. Explain how antigenic specificity of antibodies is possible.

6. Compare the primary and secondary immune responses as they pertain to IgM and IgG.

7. Administration of the chickenpox vaccine confers what type of immunity?

Access more review material online in the Study Area at **www.masteringaandp.com.**

There, you'll find:
- **Chapter guides**
- **Chapter quizzes**
- **Practice tests**
- **Labeling activities**
- **MP3 Tutor Sessions**
- **Tutorials**
- **Animations**
- **Flashcards**
- **A glossary with pronunciations**

iP® Use *Interactive Physiology*® (IP) to help you understand difficult physiological concepts in this chapter. Go to Immune System and find the following topics:

- **Immune System Overview**
- **Anatomy Review**
- **Innate Host Defenses**
- **Common Characteristics of B and T Lymphocytes**
- **Humoral Immunity**
- **Cellular Immunity**

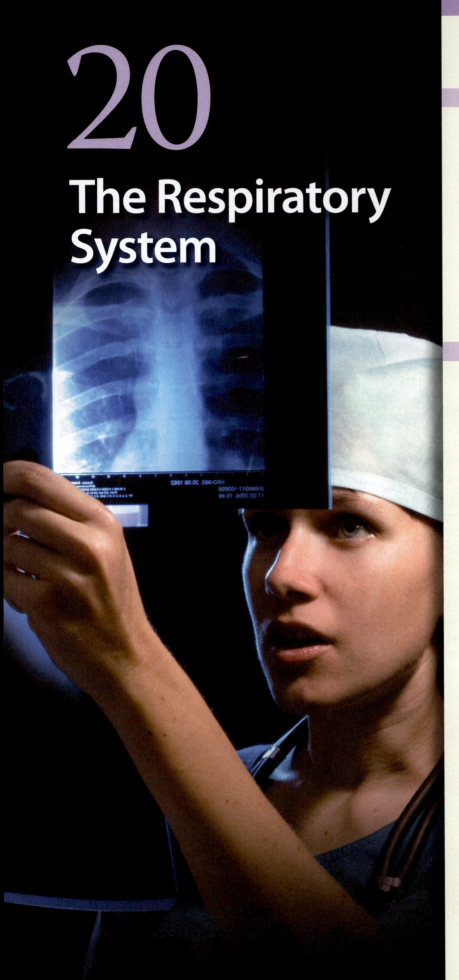

20

The Respiratory System

Functional Anatomy of the Respiratory System

The **respiratory system** is composed of structures involved in ventilation (airflow to and from the lungs) and gas exchange. In this section we consider this body system's functions and structural organization.

1 This illustration introduces the two anatomical divisions of the respiratory system. The passageways of both divisions that carry air to and from the gas exchange surfaces of the lungs make up the **respiratory tract**. The tract is composed of a **conducting portion**—which begins at the entrance to the nasal cavity and extends to fine passageways called **bronchioles**—and a **respiratory portion**, which includes the smallest, most delicate bronchioles and the associated air-filled pockets (**alveoli**) where gas exchange occurs between air and blood.

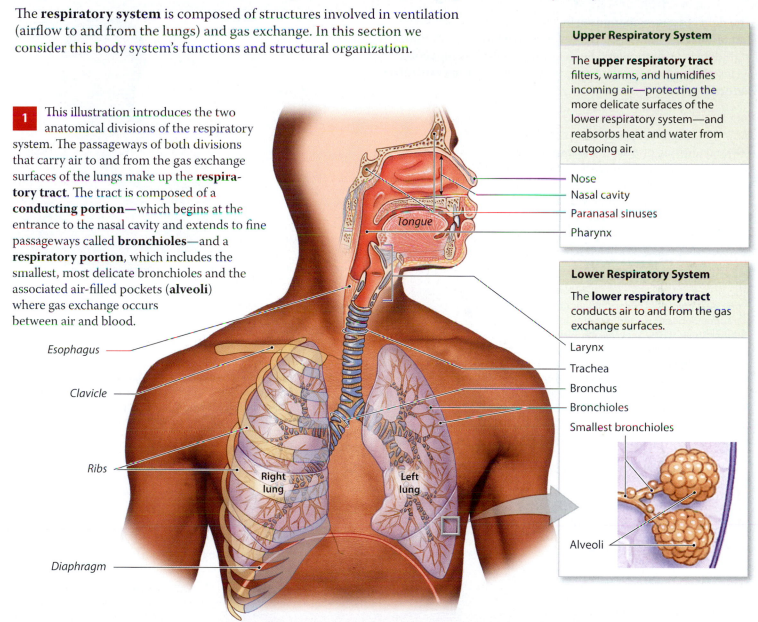

Upper Respiratory System

The **upper respiratory tract** filters, warms, and humidifies incoming air—protecting the more delicate surfaces of the lower respiratory system—and reabsorbs heat and water from outgoing air.

- Nose
- Nasal cavity
- Paranasal sinuses
- Pharynx

Lower Respiratory System

The **lower respiratory tract** conducts air to and from the gas exchange surfaces.

- Larynx
- Trachea
- Bronchus
- Bronchioles
- Smallest bronchioles
- Alveoli

Tongue

Esophagus

Clavicle

Ribs

Right lung

Left lung

Diaphragm

2 This table summarizes the major functions of the respiratory system.

Major Functions of the Respiratory System

- Providing an extensive surface area for gas exchange between air and circulating blood.
- Moving air to and from the exchange surfaces of the lungs along the respiratory passageways.
- Protecting respiratory surfaces from dehydration, temperature changes, or other environmental variations, and defending the respiratory system and other tissues from invasion by pathogens.
- Producing sounds involved in speaking, singing, and other forms of communication.
- Facilitating the detection of olfactory stimuli by olfactory receptors in the superior portions of the nasal cavity.

The respiratory mucosa is protected by the respiratory defense system

The delicate gas exchange surfaces of the respiratory system can be severely damaged if inhaled air becomes contaminated with debris or pathogens. Such contamination is prevented by a series of filtration mechanisms that constitute the **respiratory defense system**.

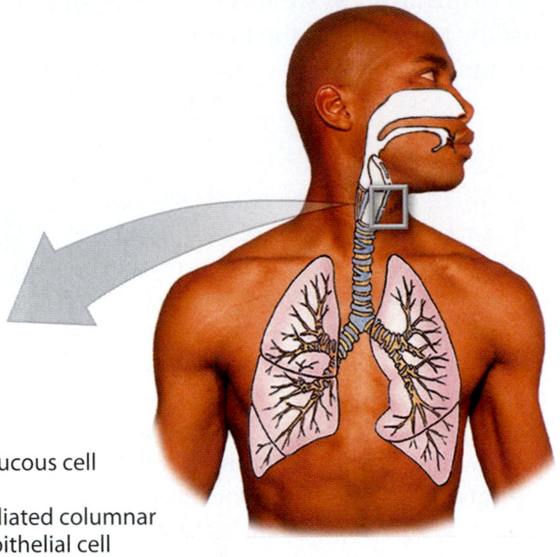

1 The **respiratory mucosa** (mū-KŌ-suh) lines the conducting portion of the respiratory tract. The structure of the respiratory epithelium changes along the respiratory tract. A pseudostratified ciliated columnar epithelium with numerous mucous cells lines the nasal cavity, the superior portion of the pharynx, and the trachea, bronchi, and large bronchioles.

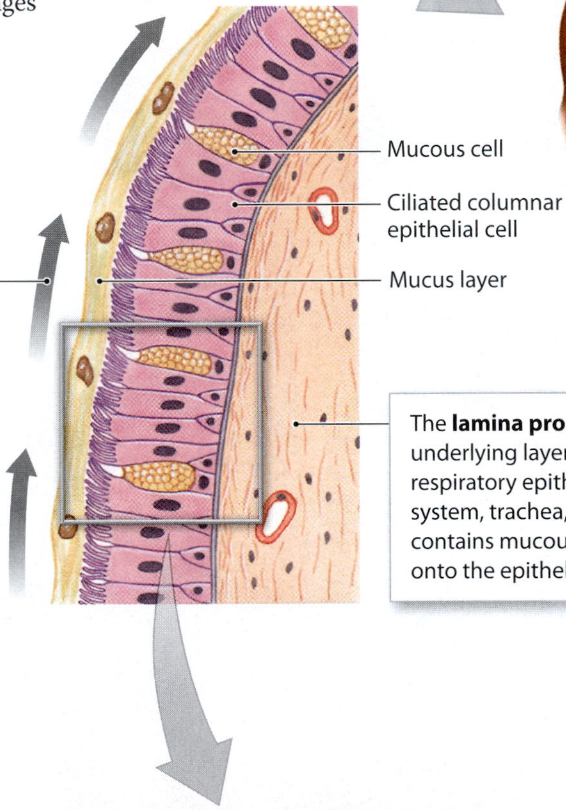

The beating of cilia sweeps mucus and any trapped debris or microorganisms toward the pharynx, where they will be swallowed and exposed to the acids and enzymes of the stomach. This ciliary movement continuously cleans and protects the respiratory surfaces. The flow of mucus is often described as a **mucus escalator**. By the time air reaches the respiratory portion of the tract, particles larger than around 5 μm have been trapped and removed.

Mucous cell

Ciliated columnar epithelial cell

Mucus layer

The **lamina propria** (LAM-in-nuh PRŌ-prē-uh) is the underlying layer of areolar tissue that supports the respiratory epithelium. In the upper respiratory system, trachea, and bronchi, the lamina propria contains mucous glands that discharge secretions onto the epithelial surface.

2 This sectional view of the respiratory mucosa shows the histological appearance of the respiratory epithelium.

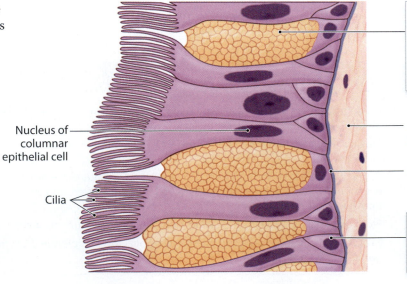

Mucous cells in the epithelium and mucous glands in the lamina propria produce a sticky mucus that bathes exposed surfaces.

Nucleus of columnar epithelial cell

Lamina propria

Basal lamina

Cilia

Stem cells within the epithelium divide to replace damaged or aged cells.

3 The structure of the respiratory epithelium changes along the respiratory tract.

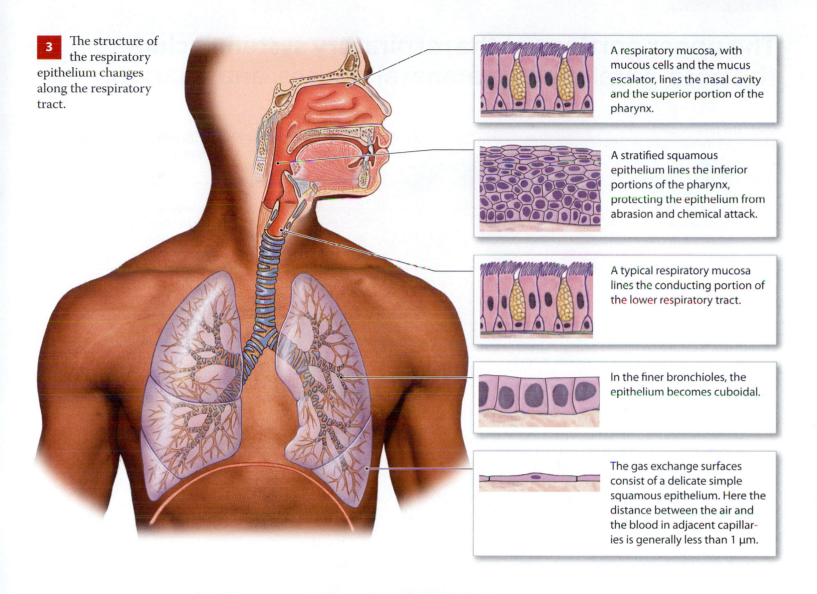

A respiratory mucosa, with mucous cells and the mucus escalator, lines the nasal cavity and the superior portion of the pharynx.

A stratified squamous epithelium lines the inferior portions of the pharynx, protecting the epithelium from abrasion and chemical attack.

A typical respiratory mucosa lines the conducting portion of the lower respiratory tract.

In the finer bronchioles, the epithelium becomes cuboidal.

The gas exchange surfaces consist of a delicate simple squamous epithelium. Here the distance between the air and the blood in adjacent capillaries is generally less than 1 μm.

4 **Cystic fibrosis (CF)** is the most common lethal inherited disease among Caucasians of Northern European descent; it occurs at a frequency of 1 in 2500 births. The most dangerous signs and symptoms result from the production of abnormally thick and sticky mucus in conducting portions of the respiratory tract. The respiratory defense system cannot transport such dense mucus, and it accumulates, restricting airflow and possibly blocking the smaller respiratory passageways. Potentially lethal infections may develop in the respiratory passageways and lungs if bacteria such as *Pseudomonas aeruginosa* colonize the stagnant mucus.

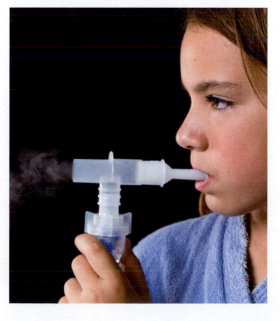

Module 20.1 Review

a. Define respiratory defense system.

b. What membrane lines the conducting portion of the respiratory tract?

c. Why can cystic fibrosis become lethal?

The upper portions of the respiratory system include the nose, nasal cavity, paranasal sinuses, and pharynx

1 The nose is the primary passageway for air entering the respiratory system when you are resting and breathing quietly. This anterior view shows the nasal cartilages and external landmarks.

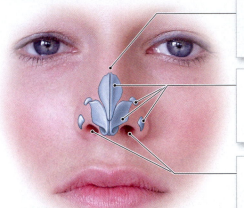

The **bridge of the nose** is supported by the anterior portion of the nasal septum, which is formed of hyaline cartilage.

Small, elastic **nasal cartilages** extend laterally from the bridge of the nose. These cartilages help to keep the external nares open and prevent their collapse during a strong inhalation.

Air normally enters through the paired **external nares** (NĀ-res), or nostrils, which open into the nasal cavity.

2 Several important features of the nasal cavity can be best seen in frontal section. The superior, middle, and inferior nasal conchae (singular: *concha*) project toward the nasal septum from the lateral walls of the nasal cavity. To pass from the external nares to the internal nares, air flows between adjacent conchae, through the **superior**, **middle**, and **inferior meatuses** (mē-Ā-tus-ez; *meatus*, a passage). The incoming air bounces off the conchal surfaces and churns like a stream flowing over rocks. As the air swirls, small airborne particles contact and are trapped in the mucus that coats the lining of the nasal cavity.

Paranasal Sinuses

The maxillary, frontal, ethmoid, and sphenoid bones that form the lateral and superior walls of the nasal cavity contain paranasal sinuses. The mucous secretions produced in these sinuses, aided by tears draining through the nasolacrimal ducts, keep the surfaces of the nasal cavity moist and clean.

Right eye

Superior nasal concha

Superior meatus

Middle nasal concha

Middle meatus

Inferior nasal concha

Inferior meatus

Frontal sinus

Ethmoidal air cell

Maxillary sinus

Hard palate

Tongue

Mandible

Nasal Septum

The **nasal septum**, formed by the fusion of the perpendicular plate of the ethmoid and the vomer, divides the nasal cavity into left and right portions.

Perpendicular plate of ethmoid

Vomer

3 Other features of the upper respiratory system are best seen in sagittal section.

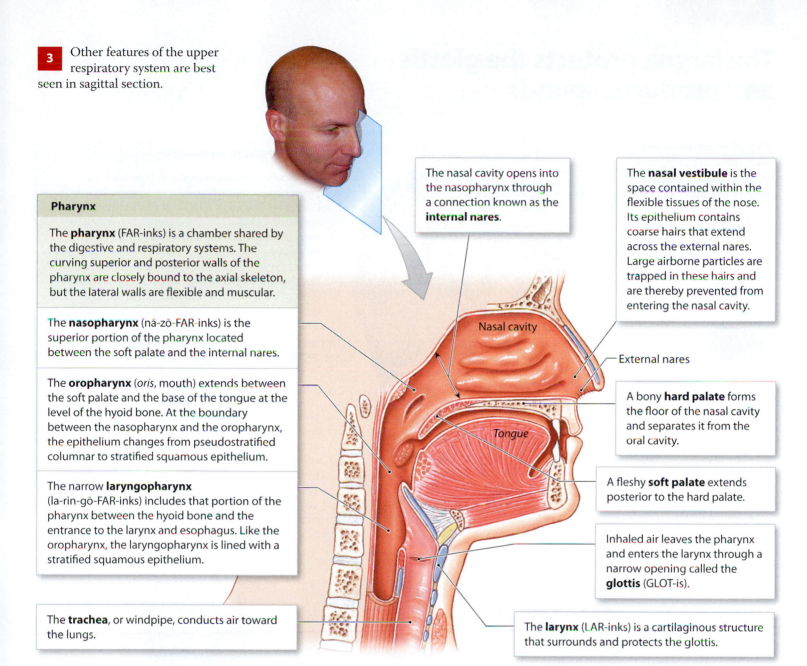

The nasal cavity opens into the nasopharynx through a connection known as the **internal nares**.

The **nasal vestibule** is the space contained within the flexible tissues of the nose. Its epithelium contains coarse hairs that extend across the external nares. Large airborne particles are trapped in these hairs and are thereby prevented from entering the nasal cavity.

Pharynx

The **pharynx** (FAR-inks) is a chamber shared by the digestive and respiratory systems. The curving superior and posterior walls of the pharynx are closely bound to the axial skeleton, but the lateral walls are flexible and muscular.

The **nasopharynx** (nā-zō-FAR-inks) is the superior portion of the pharynx located between the soft palate and the internal nares.

The **oropharynx** (*oris*, mouth) extends between the soft palate and the base of the tongue at the level of the hyoid bone. At the boundary between the nasopharynx and the oropharynx, the epithelium changes from pseudostratified columnar to stratified squamous epithelium.

The narrow **laryngopharynx** (la-rin-gō-FAR-inks) includes that portion of the pharynx between the hyoid bone and the entrance to the larynx and esophagus. Like the oropharynx, the laryngopharynx is lined with a stratified squamous epithelium.

The **trachea**, or windpipe, conducts air toward the lungs.

Nasal cavity

External nares

A bony **hard palate** forms the floor of the nasal cavity and separates it from the oral cavity.

Tongue

A fleshy **soft palate** extends posterior to the hard palate.

Inhaled air leaves the pharynx and enters the larynx through a narrow opening called the **glottis** (GLOT-is).

The **larynx** (LAR-inks) is a cartilaginous structure that surrounds and protects the glottis.

Throughout much of the nasal cavity, the lamina propria contains an extensive network of large and highly expandable veins that can release heat like a radiator. As cool, dry air passes inward over the exposed surfaces of the nasal cavity, the air warms and water in the mucus evaporates. Air moving from your nasal cavity to your lungs is thus heated almost to body temperature, and it is nearly saturated with water vapor. This protects more delicate respiratory surfaces from chilling or drying out. As air moves out of the respiratory tract, it again passes over the epithelium of the nasal cavity. The air now is warmer and more humid than it was when it entered; the air cools as it transfers heat to the nasal mucosa, and moisture condenses on the epithelial surfaces. This reduces both heat loss and water loss to the environment. Breathing through your mouth eliminates much of the conditioning of inhaled air and increases heat and water loss at every exhalation.

Module 20.2 Review

a. List the components of the upper respiratory system.

b. Trace the pathway of air as it enters the upper respiratory system.

c. Why is the vascularization of the nasal cavity important?

The larynx protects the glottis and produces sounds

1 The **larynx** consists of three large unpaired cartilages and three small paired cartilages.

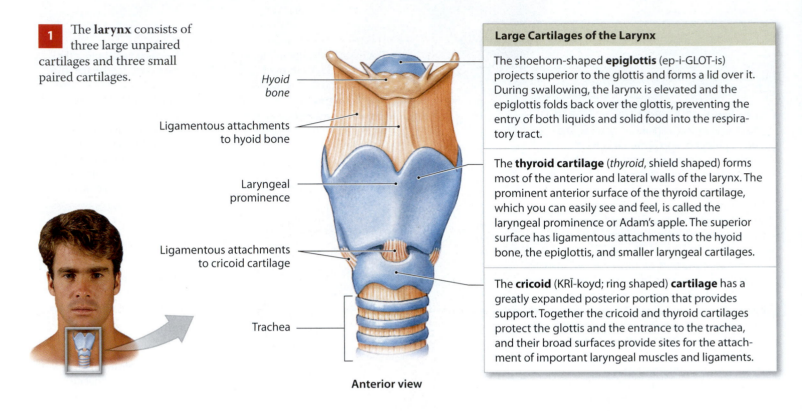

Hyoid bone

Ligamentous attachments to hyoid bone

Laryngeal prominence

Ligamentous attachments to cricoid cartilage

Trachea

Anterior view

Large Cartilages of the Larynx

The shoehorn-shaped **epiglottis** (ep-i-GLOT-is) projects superior to the glottis and forms a lid over it. During swallowing, the larynx is elevated and the epiglottis folds back over the glottis, preventing the entry of both liquids and solid food into the respiratory tract.

The **thyroid cartilage** (*thyroid*, shield shaped) forms most of the anterior and lateral walls of the larynx. The prominent anterior surface of the thyroid cartilage, which you can easily see and feel, is called the laryngeal prominence or Adam's apple. The superior surface has ligamentous attachments to the hyoid bone, the epiglottis, and smaller laryngeal cartilages.

The **cricoid** (KRĪ-koyd; ring shaped) **cartilage** has a greatly expanded posterior portion that provides support. Together the cricoid and thyroid cartilages protect the glottis and the entrance to the trachea, and their broad surfaces provide sites for the attachment of important laryngeal muscles and ligaments.

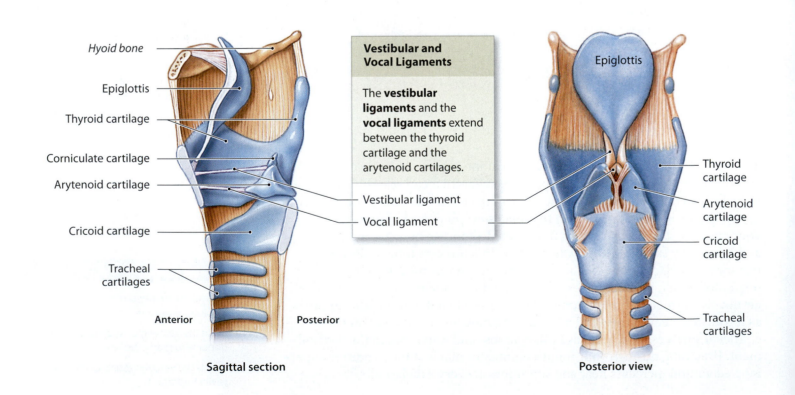

Hyoid bone

Epiglottis

Thyroid cartilage

Corniculate cartilage

Arytenoid cartilage

Cricoid cartilage

Tracheal cartilages

Anterior   Posterior

Sagittal section

Vestibular and Vocal Ligaments

The **vestibular ligaments** and the **vocal ligaments** extend between the thyroid cartilage and the arytenoid cartilages.

Vestibular ligament

Vocal ligament

Epiglottis

Thyroid cartilage

Arytenoid cartilage

Cricoid cartilage

Tracheal cartilages

Posterior view

2 The cartilages of the larynx are isolated in this posterior view. Now the small laryngeal cartilages can be seen in relationship to the larger, supporting cartilages.

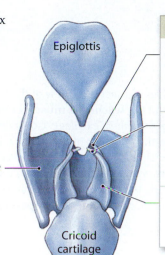

Epiglottis

Thyroid cartilage

Cricoid cartilage

Small Laryneal Cartilages

The **cuneiform** (kū-NĒ-i-form; wedge shaped) **cartilages** are long and curved, and they lie within folds of tissue that extend between the lateral surface of each arytenoid cartilage and the epiglottis.

The **corniculate** (kor-NIK-ū-lat; horn shaped) **cartilages** articulate with the arytenoid cartilages. The corniculate and arytenoid cartilages function in the opening and closing of the glottis and the production of sound.

The small, paired **arytenoid** (ar-i-TĒ-noyd; ladle shaped) **cartilages** articulate with the superior surface of the cricoid cartilage.

3 These diagrammatic superior views show the glottis in the open and closed positions. The opening or closing of the glottis involves rotational movements of the arytenoid cartilages. When the glottis is open, air passing through it vibrates the **vocal folds**, tissue folds that contain the elastic **vocal ligaments**. The vibration of the vocal folds produces sound waves, and the pitch of the sound produced depends on the diameter, length, and tension in the vocal folds. The tension is controlled by the contraction of voluntary muscles that reposition the arytenoid cartilages relative to the thyroid cartilage.

The vocal folds, which contain the vocal ligaments, lie inferior to the vestibular folds. The vocal folds are involved with the production of sound and are also known as the **vocal cords.**

The vestibular ligaments lie within a pair of relatively inelastic **vestibular folds**. These folds help prevent foreign objects from entering the glottis and contacting the more delicate vocal folds.

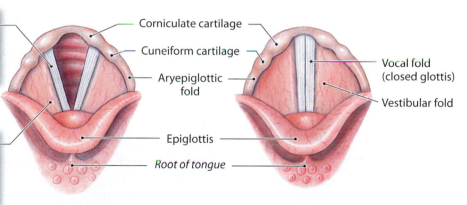

Corniculate cartilage

Cuneiform cartilage

Aryepiglottic fold

Vocal fold (closed glottis)

Vestibular fold

Epiglottis

Root of tongue

Glottis (open)

Glottis (closed)

4 Sound production at the larynx is called **phonation** (fō-NĀ-shun; *phone*, voice). Phonation is one component of speech production. However, clear speech also requires **articulation**, the modification of those sounds by other structures, such as the tongue, teeth, and lips. In a stringed instrument, such as a guitar, the quality of the sound produced does not depend solely on the nature of the vibrating string. Rather, the entire instrument becomes involved as the walls vibrate and the composite sound echoes within the hollow body. Similar amplification and resonance occur within your pharynx, oral cavity, nasal cavity, and paranasal sinuses. The combination determines the particular and distinctive sound of your voice.

Module 20.3 Review

a. Identify the paired and unpaired cartilages that compose the larynx.

b. What are the highly elastic vocal folds of the larynx also called?

c. Distinguish between phonation and articulation.

The trachea and primary bronchi convey air to and from the lungs

1 The **trachea** (TRĀ-kē-uh), or windpipe, is a tough, flexible tube with a diameter of about 2.5 cm (1 in.). Within the mediastinum, the trachea branches to form the right and left **primary bronchi** (BRONG-kī; singular, *bronchus*).

2 This is a sectional view of the trachea. An elastic ligament and the **trachealis muscle** connect the ends of each C-shaped tracheal cartilage. Contraction of the trachealis muscle reduces the diameter of the trachea, increasing the resistance to airflow. The diameter of the trachea changes from moment to moment; sympathetic stimulation relaxes the trachealis muscle, increasing the diameter of the trachea and making it easier to move air along the respiratory passageways.

The trachea contains 15–20 **tracheal cartilages**, which stiffen the tracheal walls and protect the airway. They also prevent its collapse or overexpansion as pressures change within the respiratory system.

The right primary bronchus is larger in diameter than the left, and descends toward the lung at a steeper angle. Thus, most foreign objects that enter the trachea find their way into the right bronchus.

Because the tracheal cartilages are incomplete posteriorly, the posterior tracheal wall can easily distort when large masses of food pass along the esophagus.

The mucosa of the trachea resembles that of the nasal cavity and the nasopharynx.

- Hyoid bone
- Larynx
- Trachea
- Left primary bronchus
- Secondary bronchi
- Esophagus
- Trachealis muscle
- Lumen of trachea
- *Thyroid gland*
- Tracheal cartilage

Right lung

Left lung

Sectional view of trachea   LM × 3

3 This flowchart indicates the general pattern of airway distribution in the conducting portion of the respiratory tract.

Air Conduction Passageways in the Lower Respiratory Tract

The trachea is a single conducting tube, extending from the larynx to the mediastinum.	The trachea branches to form two primary bronchi, one for each lung. Like the trachea, the primary bronchi have cartilaginous C-shaped rings, but the ends of the C overlap.	After entering the lung, each primary bronchus divides to form **secondary bronchi**. The right lung has three and the left lung has two. The secondary bronchi are supported by small cartilage plates rather than rings.	Each secondary bronchus divides to form **tertiary bronchi**. The cartilages in the walls of tertiary bronchi resemble those of secondary bronchi.	Each tertiary bronchus branches several times, giving rise to multiple **bronchioles**.

4 This diagrammatic view highlights important structural features of bronchi and bronchioles, and the general pattern of airway distribution by the respiratory tract. At each new branch, the diameter decreases; the trachea has a diameter of approximately 2.5 cm, whereas the terminal bronchioles have a diameter of 0.3–0.5 mm.

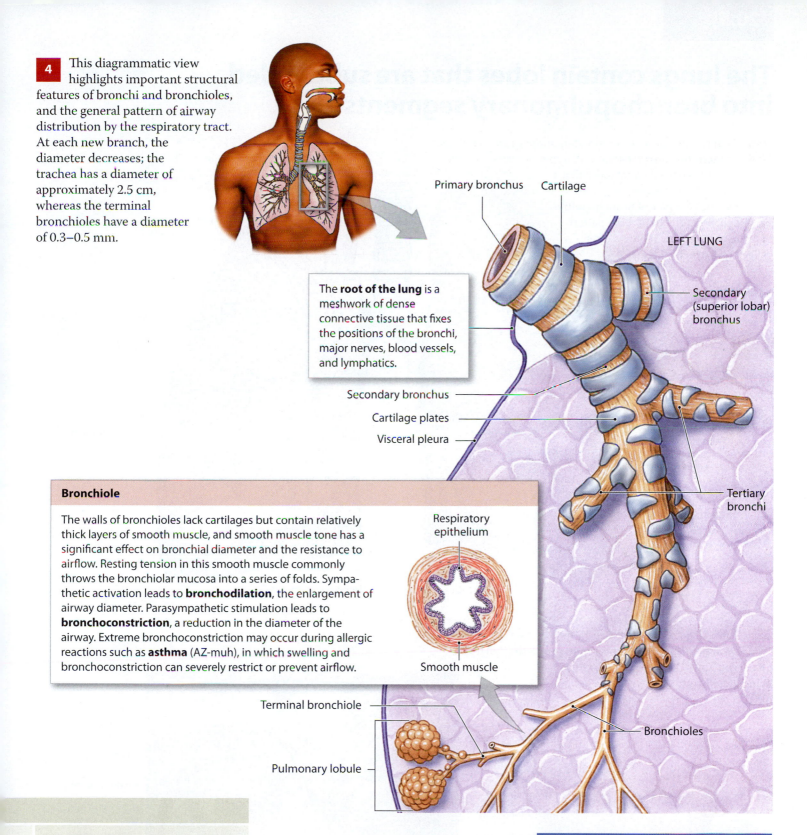

Primary bronchus   Cartilage

LEFT LUNG

The **root of the lung** is a meshwork of dense connective tissue that fixes the positions of the bronchi, major nerves, blood vessels, and lymphatics.

Secondary (superior lobar) bronchus

Secondary bronchus

Cartilage plates

Visceral pleura

Tertiary bronchi

Bronchiole

The walls of bronchioles lack cartilages but contain relatively thick layers of smooth muscle, and smooth muscle tone has a significant effect on bronchial diameter and the resistance to airflow. Resting tension in this smooth muscle commonly throws the bronchiolar mucosa into a series of folds. Sympathetic activation leads to **bronchodilation**, the enlargement of airway diameter. Parasympathetic stimulation leads to **bronchoconstriction**, a reduction in the diameter of the airway. Extreme bronchoconstriction may occur during allergic reactions such as **asthma** (AZ-muh), in which swelling and bronchoconstriction can severely restrict or prevent airflow.

Respiratory epithelium

Smooth muscle

Terminal bronchiole

Bronchioles

Pulmonary lobule

Each bronchiole branches further to form **terminal bronchioles**. Roughly 6500 terminal bronchioles arise from each tertiary bronchus, and each terminal bronchiole supplies a single **pulmonary lobule** where gas exchange occurs.

Module 20.4 Review

a. List the functions of the trachea.

b. What function do the C-shaped tracheal cartilages allow?

c. Trace the pattern of airflow along the passages of the lower respiratory tract.

The lungs contain lobes that are subdivided into bronchopulmonary segments

1 Each tertiary bronchus ultimately supplies air to a single **bronchopulmonary segment**, a specific region of one lung. The lungs have distinct lobes that are separated by deep fissures; each lobe contains at least a pair of bronchopulmonary segments. The branching pattern of bronchi and bronchioles is often called the **bronchial tree**.

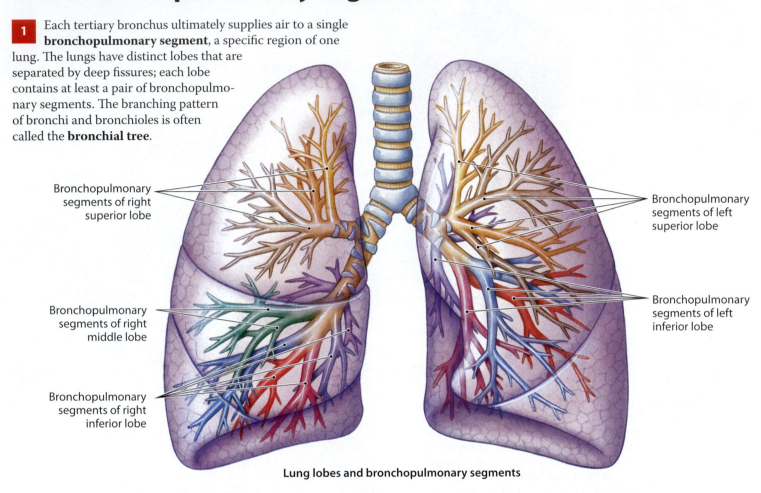

Bronchopulmonary segments of right superior lobe

Bronchopulmonary segments of left superior lobe

Bronchopulmonary segments of right middle lobe

Bronchopulmonary segments of left inferior lobe

Bronchopulmonary segments of right inferior lobe

Lung lobes and bronchopulmonary segments

2 The left and right lungs are in the left and right pleural cavities, respectively. This photo shows the collapsed, preserved lungs in a cadaver; the lungs in a living, healthy nonsmoker have a lighter, pinkish color.

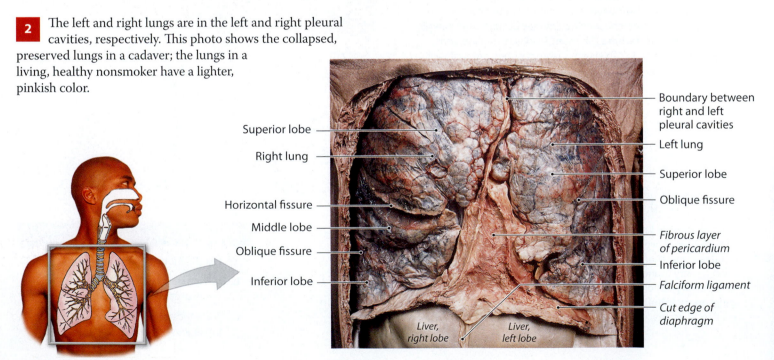

Superior lobe

Right lung

Horizontal fissure

Middle lobe

Oblique fissure

Inferior lobe

Boundary between right and left pleural cavities

Left lung

Superior lobe

Oblique fissure

Fibrous layer of pericardium

Inferior lobe

Falciform ligament

Cut edge of diaphragm

Liver, right lobe

Liver, left lobe

3   Each lung is a blunt cone, the tip of which extends superior to the first rib. The broad concave inferior portion of each lung rests on the superior surface of the diaphragm. Major superficial landmarks on the left and right lungs can be seen in these lateral and medial views.

Lateral Surfaces

The curving anterior and lateral surfaces of each lung follow the inner contours of the rib cage.

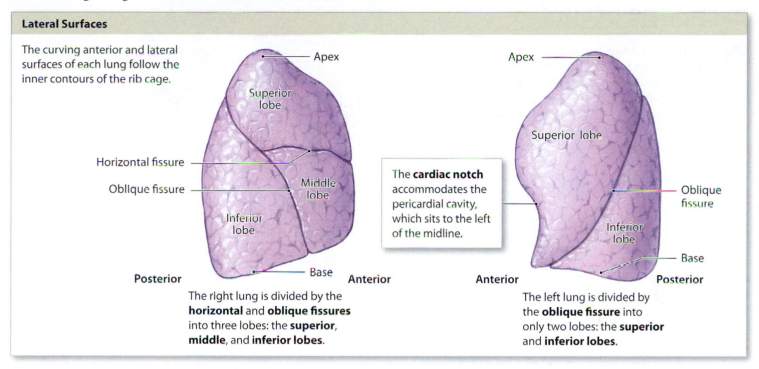

The **cardiac notch** accommodates the pericardial cavity, which sits to the left of the midline.

The right lung is divided by the **horizontal** and **oblique fissures** into three lobes: the **superior**, **middle**, and **inferior lobes**.

The left lung is divided by the **oblique fissure** into only two lobes: the **superior** and **inferior lobes**.

Medial Surfaces

The medial surfaces, which contain the hilum, have more irregular shapes. The medial surfaces of both lungs bear grooves that mark the positions of the great vessels and the heart.

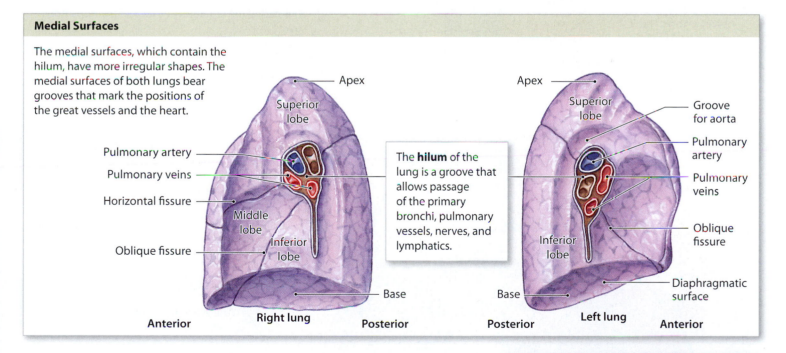

The **hilum** of the lung is a groove that allows passage of the primary bronchi, pulmonary vessels, nerves, and lymphatics.

Module 20.5 Review

a. Define bronchopulmonary segment.

b. Name the lobes and fissures of each lung.

c. Describe the location of the lungs within the thoracic cavity.

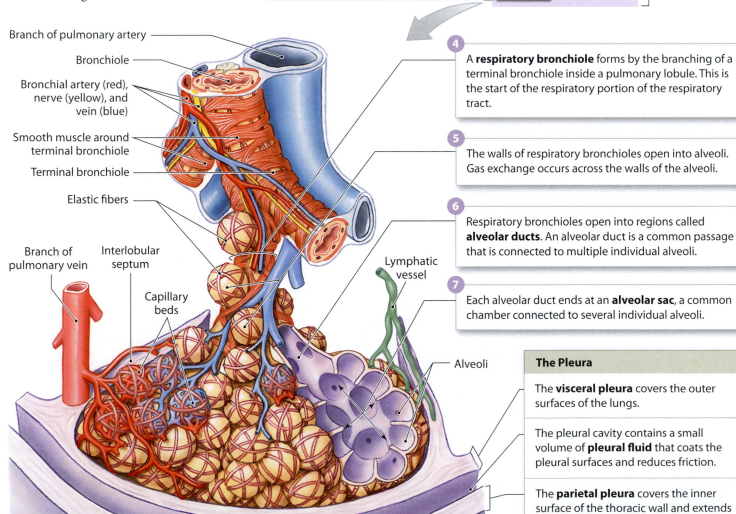

Pulmonary lobules contain alveoli, where gas exchange occurs

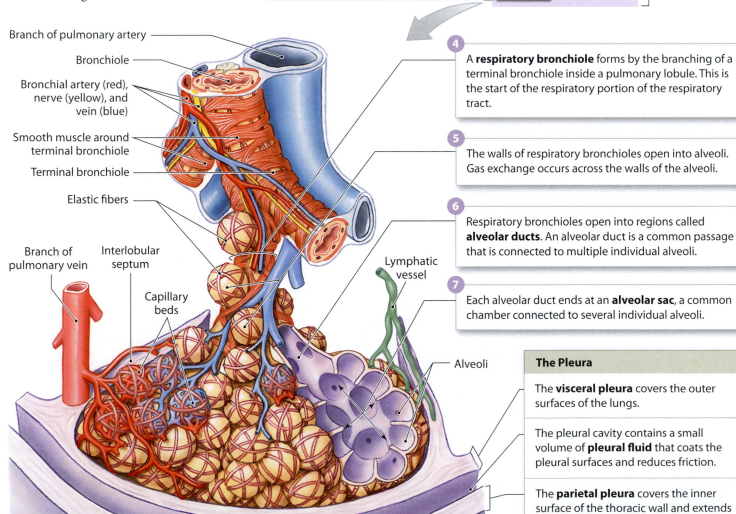

1 The conducting and respiratory portions of the respiratory tract are continuous and end in air sacs called **alveoli** (al-VĒ-ō-lī; singular, alveolus). Each lung contains about 150 million alveoli, and their abundance gives the lung an open, spongy appearance. Each alveolus is surrounded by an extensive capillary network that receives blood from a branch of a pulmonary artery and discharges blood into a tributary of a pulmonary vein. A network of elastic fibers surrounds these capillaries. Recoil of these fibers during exhalation reduces the size of the alveoli and helps push air out of the lungs.

Trachea

Left primary bronchus

Secondary bronchus

1 Each tertiary bronchus delivers air to a single bronchopulmonary segment.

2 Within a bronchopulmonary segment, bronchioles branch repeatedly, forming roughly 6500 terminal bronchioles.

3 Each terminal bronchiole supplies a single **pulmonary lobule**.

Broncho-pulmonary segment

Bronchioles

Respiratory bronchiole

Alveoli

Branch of pulmonary artery

Bronchiole

Bronchial artery (red), nerve (yellow), and vein (blue)

Smooth muscle around terminal bronchiole

Terminal bronchiole

Elastic fibers

Branch of pulmonary vein

Interlobular septum

Capillary beds

Lymphatic vessel

Alveoli

4 A **respiratory bronchiole** forms by the branching of a terminal bronchiole inside a pulmonary lobule. This is the start of the respiratory portion of the respiratory tract.

5 The walls of respiratory bronchioles open into alveoli. Gas exchange occurs across the walls of the alveoli.

6 Respiratory bronchioles open into regions called **alveolar ducts**. An alveolar duct is a common passage that is connected to multiple individual alveoli.

7 Each alveolar duct ends at an **alveolar sac**, a common chamber connected to several individual alveoli.

The Pleura

The **visceral pleura** covers the outer surfaces of the lungs.

The pleural cavity contains a small volume of **pleural fluid** that coats the pleural surfaces and reduces friction.

The **parietal pleura** covers the inner surface of the thoracic wall and extends over the diaphragm and mediastinum.

2 The alveolar epithelium is primarily a simple squamous epithelium. The surrounding capillaries differ functionally from other capillaries; they dilate when alveolar oxygen levels are high, and constrict when alveolar oxygen levels are low. This response directs blood flow to the alveoli containing the most oxygen.

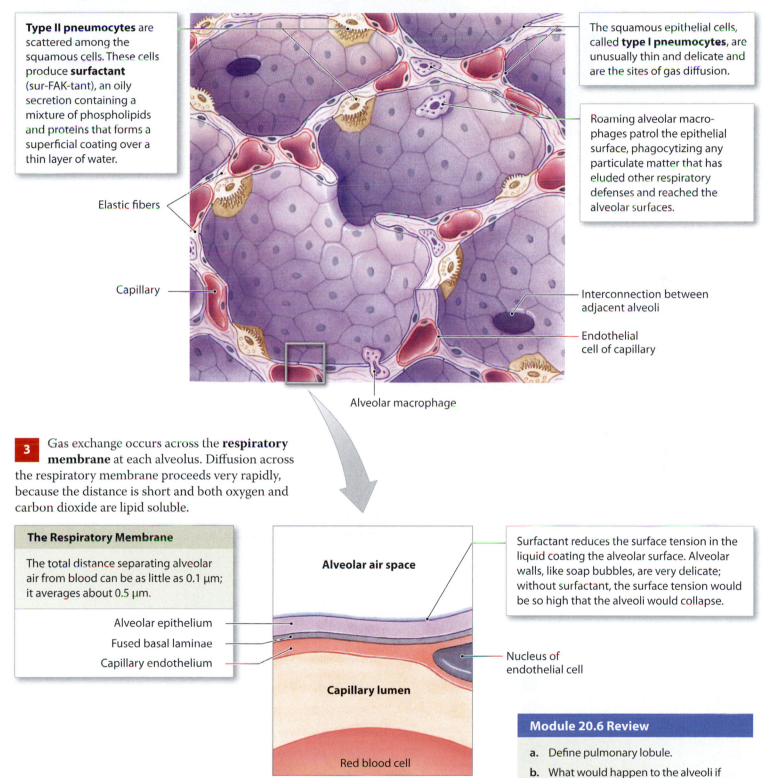

Type II pneumocytes are scattered among the squamous cells. These cells produce **surfactant** (sur-FAK-tant), an oily secretion containing a mixture of phospholipids and proteins that forms a superficial coating over a thin layer of water.

The squamous epithelial cells, called **type I pneumocytes**, are unusually thin and delicate and are the sites of gas diffusion.

Roaming alveolar macrophages patrol the epithelial surface, phagocytizing any particulate matter that has eluded other respiratory defenses and reached the alveolar surfaces.

Elastic fibers

Capillary

Interconnection between adjacent alveoli

Endothelial cell of capillary

Alveolar macrophage

3 Gas exchange occurs across the **respiratory membrane** at each alveolus. Diffusion across the respiratory membrane proceeds very rapidly, because the distance is short and both oxygen and carbon dioxide are lipid soluble.

The Respiratory Membrane

The total distance separating alveolar air from blood can be as little as 0.1 μm; it averages about 0.5 μm.

Alveolar epithelium
Fused basal laminae
Capillary endothelium

Alveolar air space

Surfactant reduces the surface tension in the liquid coating the alveolar surface. Alveolar walls, like soap bubbles, are very delicate; without surfactant, the surface tension would be so high that the alveoli would collapse.

Nucleus of endothelial cell

Capillary lumen

Red blood cell

Module 20.6 Review

a. Define pulmonary lobule.

b. What would happen to the alveoli if surfactant were not produced?

c. Trace the path of airflow from the glottis to the respiratory membrane.

1. Labeling

Label each of the respiratory system structures in the following figure.

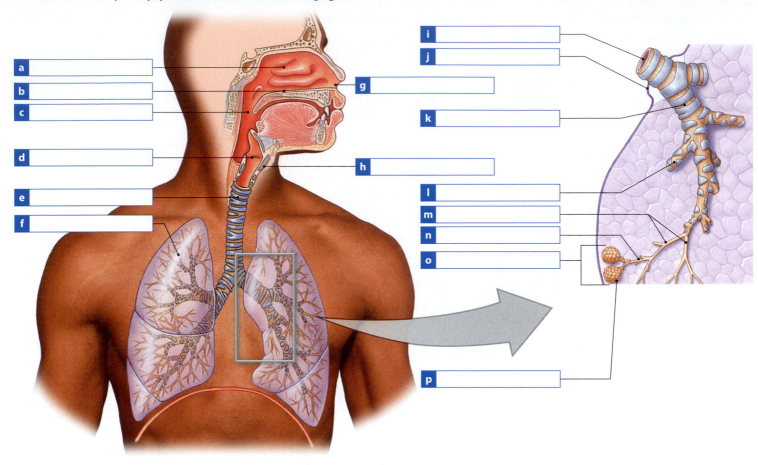

a _____

b _____

c _____

d _____

e _____

f _____

g _____

h _____

i _____

j _____

k _____

l _____

m _____

n _____

o _____

p _____

2. Matching

Match the following terms with the most closely related description.

- respiratory bronchiole
- respiratory mucosa
- phonation
- bronchodilation
- terminal bronchiole
- laryngeal prominence
- type I pneumocytes
- type II pneumocytes
- cystic fibrosis
- trachea
- pharynx
- respiratory membrane
- larynx
- bronchoconstriction

a _____ Produce surfactant

b _____ Windpipe

c _____ Simple squamous epithelial cells

d _____ Sympathetic activation

e _____ Supplies a pulmonary lobule

f _____ Parasympathetic activation

g _____ Start of respiratory portion of respiratory tract

h _____ Gas exchange

i _____ Sound production at the larynx

j _____ Chamber shared by respiratory and digestive systems

k _____ Surrounds and protects the glottis

l _____ Lethal inherited respiratory disease

m _____ Lines conducting portion of respiratory tract

n _____ Anterior surface of thyroid cartilage

Respiratory Physiology

The general term **respiration** refers to two integrated processes: external respiration and internal respiration.

1 This illustration provides an overview of respiration and shows relationships among external respiration, gas diffusion and transport, and internal respiration.

Respiration

External Respiration

External respiration includes all the processes involved in the exchange of oxygen and carbon dioxide between the body's interstitial fluids and the external environment. The purpose of external respiration, and the primary function of the respiratory system, is meeting the respiratory demands of cells.

Gas diffusion occurs across the respiratory membrane between alveoli and capillaries, and across capillary walls between blood and other tissues.

Internal Respiration

Internal respiration is the absorption of O_2 and the release of CO_2 by tissue cells. We will consider the biochemical pathways responsible for O_2 consumption and CO_2 generation in Chapter 22.

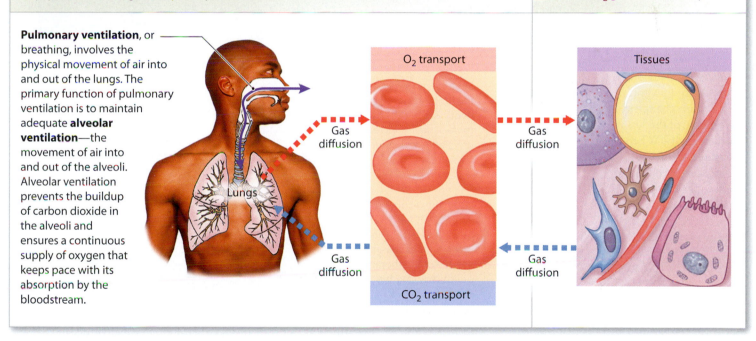

Pulmonary ventilation, or breathing, involves the physical movement of air into and out of the lungs. The primary function of pulmonary ventilation is to maintain adequate **alveolar ventilation**—the movement of air into and out of the alveoli. Alveolar ventilation prevents the buildup of carbon dioxide in the alveoli and ensures a continuous supply of oxygen that keeps pace with its absorption by the bloodstream.

Lungs

O_2 transport

Gas diffusion

Gas diffusion

CO_2 transport

Gas diffusion

Tissues

Gas diffusion

In this section, we will examine the integrated steps involved in external respiration. Abnormalities affecting any of the steps involved in external respiration will ultimately affect the gas concentrations of interstitial fluids, and thus cellular activities as well. If the oxygen content declines, the affected tissues will become starved for oxygen. **Hypoxia**, or low tissue oxygen levels, places severe limits on the metabolic activities of the affected area. If the supply of oxygen is cut off completely, the condition called **anoxia** (an-OK-sē-a; *a-*, without *ox-*, oxygen) results. Much of the damage caused by strokes and heart attacks is the result of localized anoxia.

Pulmonary ventilation is driven by pressure changes within the pleural cavities

In a gas, such as air, the molecules bounce around as independent entities. At normal atmospheric pressures, gas molecules are much farther apart than the molecules in a liquid, so the density of air is relatively low. The pressure exerted by the enclosed gas results from the collision of gas molecules with the walls of the container. The greater the number of collisions, the higher the pressure.

1 These diagrams should help you understand the relationships between pressure and volume in a gas. For a gas in a closed container and at a constant temperature, pressure (P) is inversely proportional to volume (V). That is, if you decrease the volume of a gas, its pressure will rise; if you increase the volume of a gas, its pressure will fall. The relationship between pressure and volume is reciprocal: If you reduce the volume of a flexible container by half, the pressure within it will double; if you double the volume of the container, the pressure inside it will decline by half. This relationship, first recognized by Robert Boyle in the 1600s, is called **Boyle's law**.

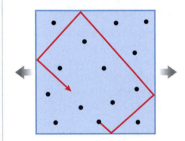

If you decrease the volume of the container, collisions occur more frequently per unit time, elevating the pressure of the gas.

If you increase the volume, fewer collisions occur per unit time, because it takes longer for a gas molecule to travel from one wall to another. As a result, the gas pressure inside the container declines.

2 Movements of the diaphragm and rib cage cause a change in the volume of the thoracic cavity. When the shape of the thoracic cavity changes, it expands or compresses the lungs, changing the air pressure within the respiratory tract.

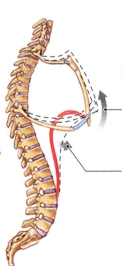

Superior movement of the rib cage increases the depth and width of the thoracic cavity, increasing its volume and reducing pressure within it.

When the diaphragm contracts, it tenses and moves inferiorly. This movement increases the volume of the thoracic cavity, reducing the pressure within it.

3 At the start of a breath, pressures inside and outside the thoracic cavity are identical, and no air moves into or out of the lungs.

Thoracic wall

Parietal pleura

Pleural fluid

Visceral pleura

Lung

Pleural cavity

Diaphragm

$$P_{outside} = P_{inside}$$

Pressure outside and inside are equal, so no air movement occurs

Although separated by the pleural cavity, the layer of pleural fluid makes the lungs stick to the inner walls of the thorax. This kind of fluid bond is responsible for making a coaster stick to the bottom of a wet glass. The elastic tissues of the lungs are always trying to recoil and reduce lung volume to about 5 percent of its normal size. This collapse is prevented by the fluid bond between the parietal and visceral pleura. If an injury allows air into the pleural cavity, this bond is broken and the lung collapses. This condition is called **atelectasis** (a-te-LEK-ta-sis).

4 Air will flow from an area of higher pressure to an area of lower pressure. When the thoracic cavity enlarges during inhalation, pressure falls inside the lungs and air flows in. When the thoracic cavity decreases in volume during exhalation, pressure rises inside the lungs, forcing air out of the respiratory tract.

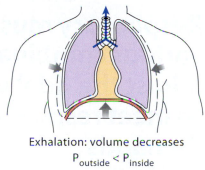

Inhalation: volume increases
$P_{outside} > P_{inside}$
Pressure inside falls, so air flows in

Exhalation: volume decreases
$P_{outside} < P_{inside}$
Pressure inside rises, so air flows out

5 The direction of airflow is determined by the difference between atmospheric pressure and **intrapulmonary** (in-tra-PUL-mo-nār-ē) **pressure**, the pressure inside the respiratory tract (usually measured at the alveoli).

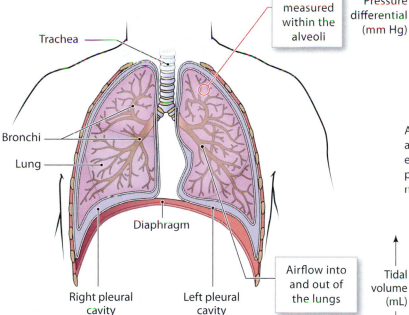

Trachea

Bronchi

Lung

Diaphragm

Right pleural cavity

Left pleural cavity

Pressures measured within the alveoli

Airflow into and out of the lungs

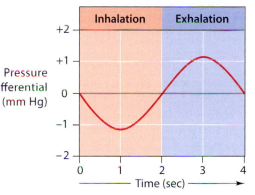

A pressure differential of 0 mm Hg exists when atmospheric and intrapulmonary pressures are equal. Positive intrapulmonary pressures will push air out of the lungs; negative intrapulmonary pressures will pull air into the lungs.

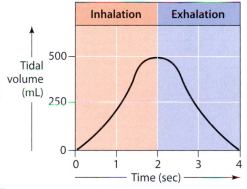

The **tidal volume (V_T)** is the amount of air moved into the lungs during inhalation and out of the lungs during exhalation. At rest, the tidal volume is approximately 500 mL.

6 Although we will use mm Hg to report pressures, other units are also used in clinical practice. The table below compares several units used to report gas pressures.

Four Common Units for Reporting Gas Pressures

- **Millimeters of mercury (mm Hg):** This is the most common unit for reporting blood pressure and gas pressures. Normal atmospheric pressure is approximately 760 mm Hg.

- **Torr:** This unit of measurement is preferred by many respiratory therapists; it is also commonly used in Europe and in some technical journals. One torr is equivalent to 1 mm Hg; in other words, normal atmospheric pressure is equal to 760 torr.

- **Centimeters of water (cm H_2O):** In a hospital setting, anesthetic gas pressures and oxygen pressures are commonly measured in centimeters of water. One cm H_2O is equivalent to 0.735 mm Hg; normal atmospheric pressure is 1033.6 cm H_2O.

- **Pounds per square inch (psi):** Pressures in compressed gas cylinders and other industrial applications are generally reported in psi. Normal atmospheric pressure at sea level is approximately 15 psi.

Module 20.7 Review

a. Define Boyle's law.

b. What physical changes affect the volume of the lungs?

c. What pressures determine the direction of airflow within the respiratory tract?

Respiratory muscles in various combinations adjust the tidal volume to meet respiratory demands

Respiratory muscles may be involved with either inhalation or exhalation. Those involved with inhalation are often called **inspiratory muscles**; those involved in exhalation are often called **expiratory muscles**.

1 This anterior view introduces the **primary** and **accessory respiratory muscles**. The primary muscles, the diaphragm and external intercostal muscles, are both involved in inhalation. When you are breathing quietly, inhalation is active but exhalation is passive—elastic forces and gravity are sufficient to reduce the volume of the lungs.

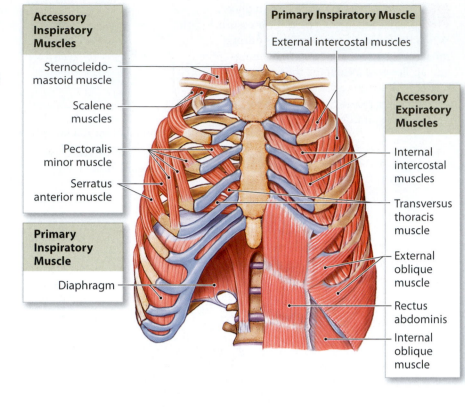

Accessory Inspiratory Muscles

Sternocleido-mastoid muscle

Scalene muscles

Pectoralis minor muscle

Serratus anterior muscle

Primary Inspiratory Muscle

Diaphragm

Primary Inspiratory Muscle

External intercostal muscles

Accessory Expiratory Muscles

Internal intercostal muscles

Transversus thoracis muscle

External oblique muscle

Rectus abdominis

Internal oblique muscle

2 This lateral view during inhalation shows the inspiratory muscles that elevate the ribs and depress the diaphragm to enlarge the thoracic cavity.

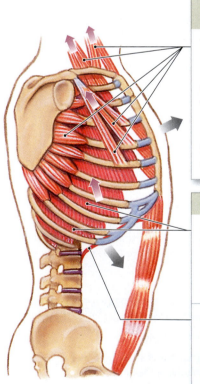

Accessory Inspiratory Muscles (active when needed)

The contraction of accessory muscles assists the external intercostal muscles in elevating the ribs. The muscles increase the speed and amount of rib movement when the primary respiratory muscles are unable to move enough air to meet the oxygen demands of tissues.

Primary Inspiratory Muscles

Contraction of the external intercostal muscles elevates the ribs. This action contributes roughly 25 percent to the volume of air in the lungs at rest.

Contraction of the diaphragm flattens the floor of the thoracic cavity, increasing its volume and drawing air into the lungs. This is responsible for roughly 75 percent of the air movement in normal breathing at rest.

3 This corresponding lateral view during active exhalation shows the accessory expiratory muscles that depress the ribs and push the relaxed diaphragm into the thoracic cavity. The abdominal muscles that assist in exhalation are represented by a single muscle (the rectus abdominis).

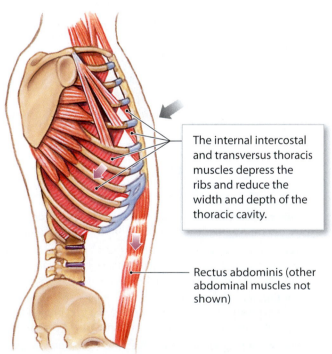

The internal intercostal and transversus thoracis muscles depress the ribs and reduce the width and depth of the thoracic cavity.

Rectus abdominis (other abdominal muscles not shown)

4 Only a small proportion of the air in the lungs is exchanged during a single quiet respiratory cycle (consisting of an inhalation and an exhalation); the tidal volume can be increased by inhaling more vigorously and exhaling more completely. We can divide the total volume of the lungs into a series of **volumes** and **capacities** (each the sum of various volumes), as indicated in this graph. The red line indicates the volume of air within the lungs as respiratory movements are performed.

Pulmonary Volumes and Capacities (adult male)

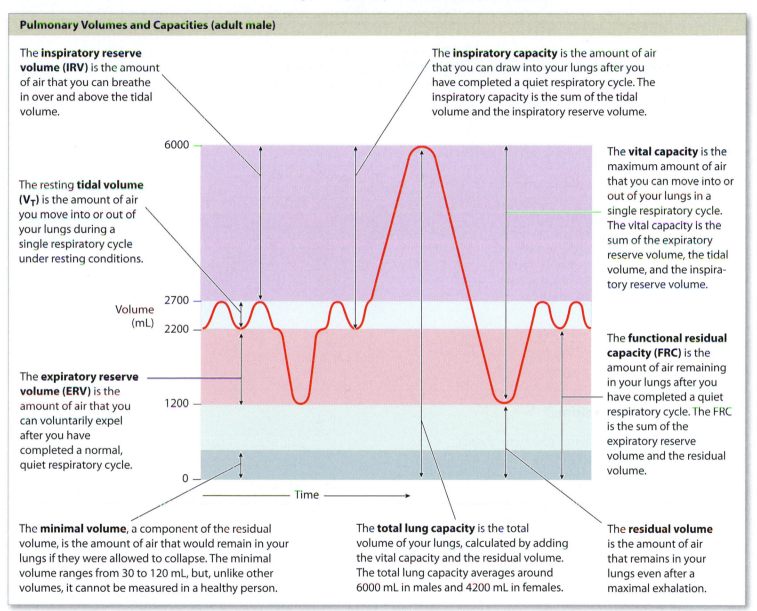

The **inspiratory reserve volume (IRV)** is the amount of air that you can breathe in over and above the tidal volume.

The resting **tidal volume (V$_T$)** is the amount of air you move into or out of your lungs during a single respiratory cycle under resting conditions.

The **expiratory reserve volume (ERV)** is the amount of air that you can voluntarily expel after you have completed a normal, quiet respiratory cycle.

The **inspiratory capacity** is the amount of air that you can draw into your lungs after you have completed a quiet respiratory cycle. The inspiratory capacity is the sum of the tidal volume and the inspiratory reserve volume.

The **vital capacity** is the maximum amount of air that you can move into or out of your lungs in a single respiratory cycle. The vital capacity is the sum of the expiratory reserve volume, the tidal volume, and the inspiratory reserve volume.

The **functional residual capacity (FRC)** is the amount of air remaining in your lungs after you have completed a quiet respiratory cycle. The FRC is the sum of the expiratory reserve volume and the residual volume.

The **minimal volume**, a component of the residual volume, is the amount of air that would remain in your lungs if they were allowed to collapse. The minimal volume ranges from 30 to 120 mL, but, unlike other volumes, it cannot be measured in a healthy person.

The **total lung capacity** is the total volume of your lungs, calculated by adding the vital capacity and the residual volume. The total lung capacity averages around 6000 mL in males and 4200 mL in females.

The **residual volume** is the amount of air that remains in your lungs even after a maximal exhalation.

Pulmonary Volumes

		Males	Females	
Vital capacity	IRV	3300	1900	Inspiratory capacity
	V$_T$	500	500	
	ERV	1000	700	Functional residual capacity
Residual volume		1200	1100	
Total lung capacity		6000 mL	4200 mL	

Module 20.8 Review

a. Name the various measurable pulmonary volumes.

b. Identify the primary inspiratory muscles.

c. When do the accessory respiratory muscles become active?

Pulmonary ventilation must be closely regulated to meet tissue oxygen demands

The respiratory system adjusts pulmonary ventilation over a broad range to meet the oxygen demands of the body. These adjustments involve varying both the number of breaths per minute and the amount of air moved per breath. When you are exercising at peak levels, the amount of air moving into and out of the respiratory tract can be 50 times the amount moved at rest. The factors involved in pulmonary ventilation should remind you of the factors involved in regulating cardiovascular function.

Respiratory Rate

1 Your **respiratory rate** is the number of breaths you take each minute. As you read this, you are probably breathing quietly, with a low respiratory rate. The normal respiratory rate of a resting adult ranges from 12 to 18 breaths each minute, roughly one for every four heartbeats. Children breathe more rapidly, at rates of about 18–20 breaths per minute.

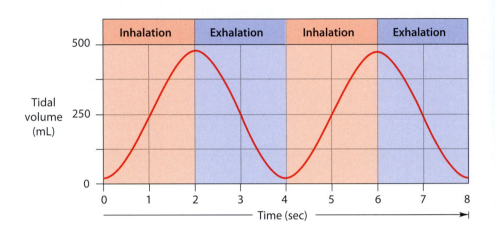

Respiratory Minute Volume

2 We can calculate the volume of air moved each minute, symbolized V_E, by multiplying the respiratory rate, f, by the tidal volume, V_T. This value is called the **respiratory minute volume**. The respiratory rate at rest averages 12 breaths per minute, and the tidal volume at rest averages around 500 mL per breath. On that basis, we can calculate that respiratory minute volume at rest is approximately 6 L/min.

$$V_E \quad = \quad f \quad \times \quad V_T$$

$$\left(\begin{array}{c} \text{Volume of air moved} \\ \text{each minute} \end{array} \right) = \left(\begin{array}{c} \text{Breaths per} \\ \text{minute} \end{array} \right) \times \left(\begin{array}{c} \text{Tidal} \\ \text{volume} \end{array} \right)$$

= 12 × 500 mL per minute

= 6000 mL per minute

= 6.0 liters per minute

In other words, the respiratory minute volume at rest is approximately 6 liters per minute.

3 The factors involved in respiratory minute volume are easily diagrammed, but their functional relationships are complex. Increasing either respiratory rate or tidal volume will increase the respiratory minute volume. But what if the respiratory rate goes up, and the tidal volume goes down? The respiratory minute volume may remain the same, but the effects on respiratory performance are very different; we will now consider the differences and their significance.

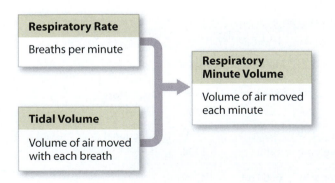

Alveolar Ventilation

4 **Alveolar ventilation,** symbolized V_A, is the amount of air reaching the alveoli each minute. The alveolar ventilation is less than the respiratory minute volume, because some of the air never reaches the alveoli, but remains in the conducting portion of the lungs. This is known as the **anatomic dead space (V_D)**, and at rest it amounts to roughly 150 mL of the 500 mL of tidal air.

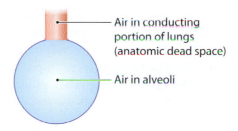

Air in conducting portion of lungs (anatomic dead space)

Air in alveoli

5 We can calculate alveolar ventilation by subtracting the dead space from the tidal volume:

V_A		f		V_T		V_D	
Alveolar ventilation	=	Breaths per minute	× (	Tidal volume	−	Anatomic dead space	)

V_A	=	12	×	(500 mL − 150 mL)
	=	12	×	350 mL
	=	4200 mL		

However, the composition of the gas arriving in the alveoli is significantly different from that of the surrounding atmosphere, because inhaled air always mixes with "used" air in the conducting passageways (the anatomic dead space) on its way to the exchange surfaces. The air in alveoli thus contains less oxygen and more carbon dioxide than does atmospheric air.

6 Let's return to the effects of altering tidal volume and respiratory rate. If the respiratory rate jumps to 20 per minute but the tidal volume drops to 300 mL, the respiratory minute volume will remain unchanged. However, the alveolar ventilation rate drops dramatically, falling from 4.2 L/min to 3 L/min.

$$V_E = f \times V_T$$
$$= 20 \times 300 \text{ mL per minute}$$
$$= 6.0 \text{ liters per minute}$$

$$V_A = f \times (V_T - V_D)$$
$$= 20 \times (300 \text{ mL} - 150 \text{ mL}) \text{ per minute}$$
$$= 3.0 \text{ liters per minute}$$

This alveolar ventilation rate is almost 30 percent below its original value of 4.2 L/min, and if tissue demands are elevated as well, widespread tissue hypoxia could result. Thus, whenever the demand for oxygen increases, both the tidal volume *and* the respiratory rate must be increased.

Module 20.9 Review

a. Define respiratory rate.

b. How does the respiratory minute volume differ from alveolar ventilation?

c. Which ventilates alveoli more effectively: slow, deep breaths or rapid, shallow breaths? Explain why.

Gas diffusion depends on the partial pressures and solubilities of gases

The principles that govern the movement and diffusion of gas molecules are relatively straightforward. These principles, known as **gas laws**, have been understood for roughly 250 years. You have already heard about Boyle's law, which determines the direction of air movement in pulmonary ventilation. Now you will learn about other gas laws and factors that determine the rate of oxygen and carbon dioxide diffusion across the respiratory membrane.

1 The **partial pressure (P)** of a gas is the pressure contributed by a single gas in a mixture of gases. The partial pressures for atmospheric gases are included in this table. All the partial pressures added together equal the total pressure exerted by the gas mixture; this is known as **Dalton's law**. As soon as air enters the respiratory tract, its characteristics begin to change. In passing through the nasal cavity, inhaled air becomes warmer, and the amount of water vapor increases. On reaching the alveoli, the incoming air mixes with air remaining in the alveoli from the previous respiratory cycle. During the subsequent exhalation, the departing alveolar air mixes with air in the anatomic dead space, producing yet another mixture that differs from both atmospheric and alveolar air samples.

Partial Pressures (mm Hg) and Normal Gas Concentrations (%) in Air				
Source of Sample	**Nitrogen (N$_2$)**	**Oxygen (O$_2$)**	**Carbon Dioxide (CO$_2$)**	**Water Vapor (H$_2$O)**
Inhaled air (dry)	597 (78.6%)	159 (20.9%)	0.3 (0.04%)	3.7 (0.5%)
Alveolar air (saturated)	573 (75.4%)	100 (13.2%)	40 (5.2%)	47 (6.2%)
Exhaled air (saturated)	569 (74.8%)	116 (15.3%)	28 (3.7%)	47 (6.2%)

2 At a given temperature, the amount of a particular gas in solution is directly proportional to the partial pressure of that gas. This principle is known as **Henry's law**.

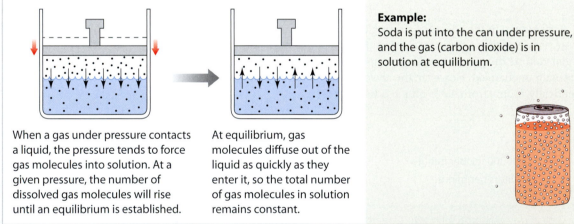

When a gas under pressure contacts a liquid, the pressure tends to force gas molecules into solution. At a given pressure, the number of dissolved gas molecules will rise until an equilibrium is established.

At equilibrium, gas molecules diffuse out of the liquid as quickly as they enter it, so the total number of gas molecules in solution remains constant.

Example:
Soda is put into the can under pressure, and the gas (carbon dioxide) is in solution at equilibrium.

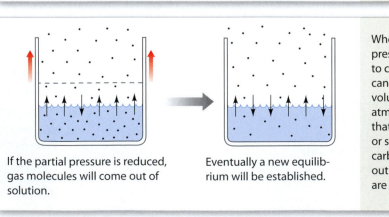

If the partial pressure is reduced, gas molecules will come out of solution.

Eventually a new equilibrium will be established.

When you open a soda can, the internal pressure falls and the gas molecules begin to come out of solution. The volume of the can is so small, and the volume of the atmosphere so great, that within a half hour or so virtually all the carbon dioxide comes out of solution, and you are left with "flat" soda.

3 This illustration depicts the partial pressures of oxygen and carbon dioxide during external respiration in the pulmonary circuit and during internal respiration in the systemic circuit.

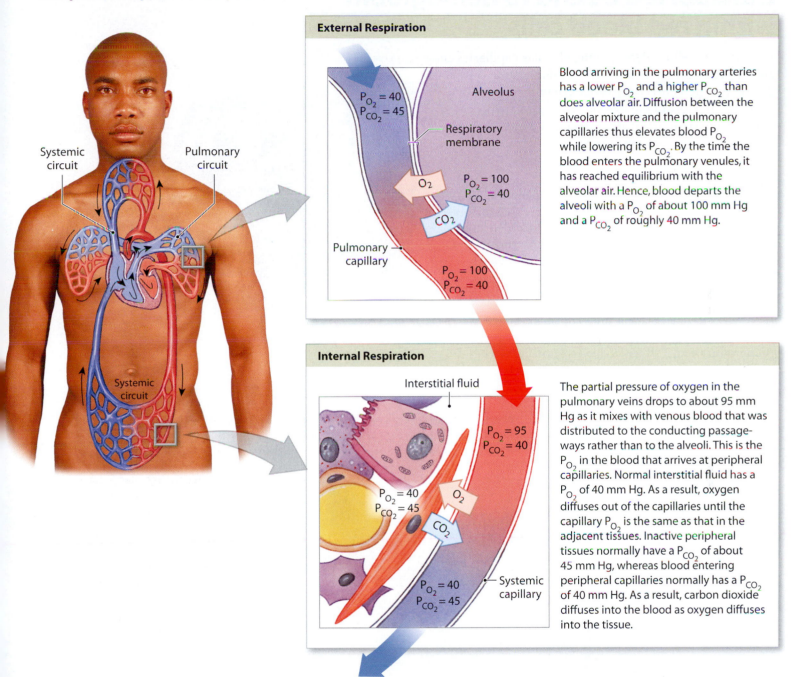

Systemic circuit

Pulmonary circuit

Systemic circuit

External Respiration

$P_{O_2} = 40$
$P_{CO_2} = 45$

Alveolus

Respiratory membrane

O_2

$P_{O_2} = 100$
$P_{CO_2} = 40$

CO_2

Pulmonary capillary

$P_{O_2} = 100$
$P_{CO_2} = 40$

Blood arriving in the pulmonary arteries has a lower P_{O_2} and a higher P_{CO_2} than does alveolar air. Diffusion between the alveolar mixture and the pulmonary capillaries thus elevates blood P_{O_2} while lowering its P_{CO_2}. By the time the blood enters the pulmonary venules, it has reached equilibrium with the alveolar air. Hence, blood departs the alveoli with a P_{O_2} of about 100 mm Hg and a P_{CO_2} of roughly 40 mm Hg.

Internal Respiration

Interstitial fluid

$P_{O_2} = 95$
$P_{CO_2} = 40$

$P_{O_2} = 40$
$P_{CO_2} = 45$

O_2

CO_2

$P_{O_2} = 40$
$P_{CO_2} = 45$

Systemic capillary

The partial pressure of oxygen in the pulmonary veins drops to about 95 mm Hg as it mixes with venous blood that was distributed to the conducting passageways rather than to the alveoli. This is the P_{O_2} in the blood that arrives at peripheral capillaries. Normal interstitial fluid has a P_{O_2} of 40 mm Hg. As a result, oxygen diffuses out of the capillaries until the capillary P_{O_2} is the same as that in the adjacent tissues. Inactive peripheral tissues normally have a P_{CO_2} of about 45 mm Hg, whereas blood entering peripheral capillaries normally has a P_{CO_2} of 40 mm Hg. As a result, carbon dioxide diffuses into the blood as oxygen diffuses into the tissue.

The system just described is at equilibrium, and the P_{O_2} and P_{CO_2} are stable in the alveoli and in the tissues. Every oxygen molecule entering peripheral tissues is balanced by an oxygen molecule absorbed at the alveoli, and the absorbed oxygen molecule will be replaced in the next respiratory cycle. But if tissue oxygen demand accelerates, that equilibrium is disturbed; the respiratory rate and tidal volume must then increase.

Module 20.10 Review

a. Define Dalton's law.

b. Explain the decrease in P_{O_2} from the pulmonary venules to the blood arriving in the peripheral capillaries of the systemic circuit.

c. What is the significance of Henry's law to the process of respiration?

Almost all of the oxygen in the blood is transported bound to hemoglobin within red blood cells

Each 100 mL of blood leaving the alveolar capillaries carries away roughly 20 mL of oxygen. Of this amount, only about 0.3 mL (1.5 percent) consists of oxygen molecules in solution. The rest of the oxygen molecules are bound to hemoglobin (Hb) molecules—specifically, to the iron ions in the center of **heme units**.

1 A hemoglobin molecule consists of four globular protein subunits, each containing a heme unit. Thus, each hemoglobin molecule can reversibly bind up to four molecules of oxygen, forming **oxyhemoglobin (HbO$_2$)**. Carbon monoxide, CO, is a gas released by petroleum-burning engines and heaters. It is dangerous because it will irreversibly bind to heme units, making them unavailable for oxygen transport.

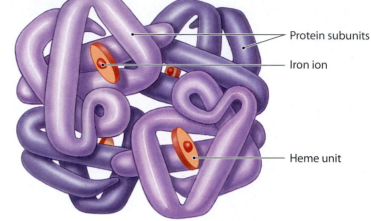

Protein subunits

Iron ion

Heme unit

2 The percentage of heme units containing bound oxygen at any given moment is called the **hemoglobin saturation**. If all the Hb molecules in the blood are fully loaded with oxygen, saturation is 100 percent. If, on average, each Hb molecule carries two O$_2$ molecules, saturation is 50 percent. This graph is an **oxygen-hemoglobin saturation curve**, which shows the saturation of hemoglobin at different partial pressures of oxygen. Notice that hemoglobin will be more than 90 percent saturated if exposed to a P$_{O_2}$ above 60 mm Hg.

Where the slope is steep, a very small change in plasma P$_{O_2}$ will result in a large change in the amount of oxygen bound to Hb or released from HbO$_2$.

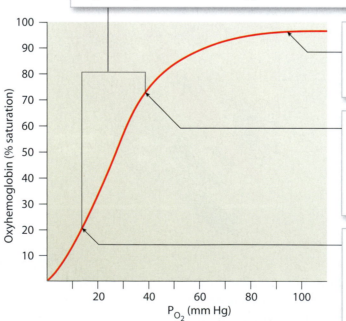

Blood entering the systemic circuit has a P$_{O_2}$ of 95 mm Hg, and the hemoglobin is about 97% saturated with oxygen.

Blood leaving peripheral tissues has an average P$_{O_2}$ of 40 mm Hg, so the hemoglobin drops from 97% to 75% saturation. It releases only 22% of its stored oxygen, so even venous blood contains substantial oxygen reserves.

The P$_{O_2}$ in active muscle tissue may drop to 15–20 mm Hg. Hemoglobin passing through these capillaries will go from 97% saturation to about 20% saturation. This means that an active tissue can receive 3.5 times as much oxygen as an inactive tissue, even if the rate of blood flow remains the same.

Note: The curve has this shape because each arriving oxygen molecule increases the affinity of hemoglobin for the next oxygen molecule. Once the first oxygen molecule binds to the hemoglobin, the slope rises rapidly until reaching a plateau near 100 percent saturation.

3 Blood pH has a direct effect on the oxygen-hemoglobin saturation curve because the shape of hemoglobin molecules changes as the number of bound O_2 molecules increases and these changes affect its affinity for oxygen. This change is called the **Bohr effect**. The curves in this figure show how the slope of the oxygen-hemoglobin saturation curve changes when the pH of the blood shifts away from the normal average value.

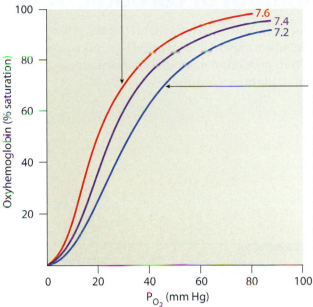

If the pH increases, the saturation curve shifts to the left, and hemoglobin releases less oxygen.

If the pH decreases, the saturation curve shifts to the right, and hemoglobin releases more oxygen. At a pH of 7.4, hemoglobin saturation would be 75% at a P_{O_2} of 40 mm Hg; if the pH shifts to 7.2, at the same P_{O_2} hemoglobin saturation would be 60%. Thus the pH shift triggered the release of an additional 15% of the bound oxygen.

4 Temperature changes also affect the saturation curve. The higher the temperature, the more readily hemoglobin gives up its oxygen reserves.

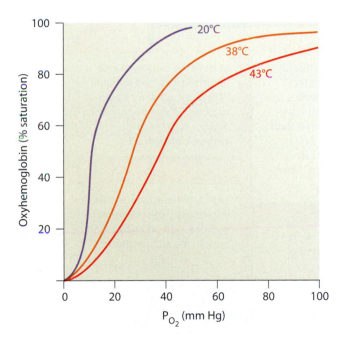

Red blood cells (RBCs) do not contain mitochondria, so they can only generate ATP through glycolysis. The metabolic pathways involved in glycolysis also generate the compound **2,3-bisphosphoglycerate** (biz-fos-fō-GLIS-er-āt), or **BPG**. For any partial pressure of oxygen, the higher the concentration of BPG, the more oxygen will be released by the Hb molecules. BPG production decreases as RBCs age, and levels of BPG can determine how long a blood bank can store fresh whole blood. When BPG levels get too low, hemoglobin becomes firmly bound to the available oxygen. The blood is then useless for transfusions, because the RBCs will no longer release oxygen to peripheral tissues, even at a disastrously low P_{O_2}.

Module 20.11 Review

a. Define oxyhemoglobin.

b. Explain the relationship among BPG, oxygen, and hemoglobin.

c. During exercise, hemoglobin releases more oxygen to active skeletal muscles than it does when those muscles are at rest. Why?

Most carbon dioxide transport occurs through the reversible formation of carbonic acid

Carbon dioxide is generated by aerobic metabolism in peripheral tissues. After entering the bloodstream, a CO_2 molecule is either (1) converted to a molecule of carbonic acid, (2) bound to the protein portion of hemoglobin molecules within red blood cells, or (3) dissolved in plasma. All three reactions are completely reversible.

1 We will consider the events that occur as blood enters peripheral tissues in which the P_{CO_2} is 45 mm Hg.

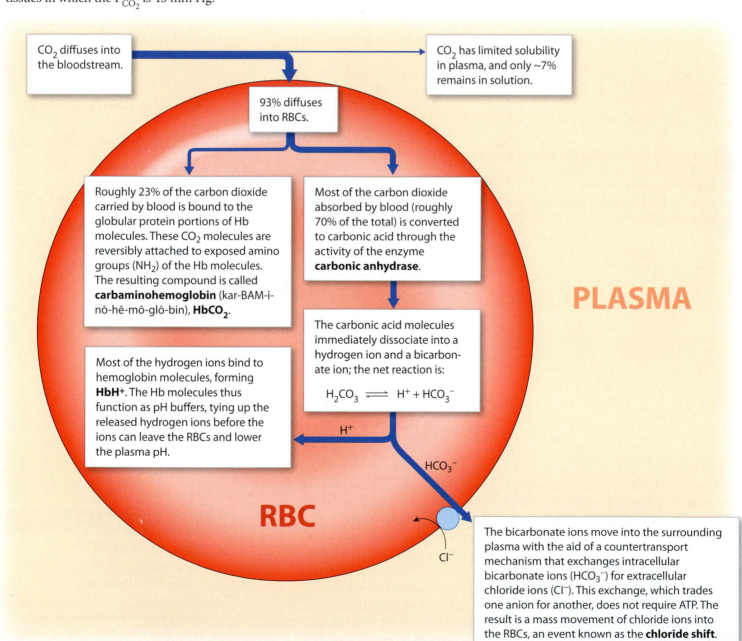

CO_2 diffuses into the bloodstream.

CO_2 has limited solubility in plasma, and only ~7% remains in solution.

93% diffuses into RBCs.

Roughly 23% of the carbon dioxide carried by blood is bound to the globular protein portions of Hb molecules. These CO_2 molecules are reversibly attached to exposed amino groups (NH_2) of the Hb molecules. The resulting compound is called **carbaminohemoglobin** (kar-BAM-i-nō-hē-mō-glō-bin), **HbCO$_2$**.

Most of the carbon dioxide absorbed by blood (roughly 70% of the total) is converted to carbonic acid through the activity of the enzyme **carbonic anhydrase**.

The carbonic acid molecules immediately dissociate into a hydrogen ion and a bicarbonate ion; the net reaction is:

$$H_2CO_3 \rightleftharpoons H^+ + HCO_3^-$$

Most of the hydrogen ions bind to hemoglobin molecules, forming **HbH$^+$**. The Hb molecules thus function as pH buffers, tying up the released hydrogen ions before the ions can leave the RBCs and lower the plasma pH.

H^+

HCO_3^-

PLASMA

RBC

Cl^-

The bicarbonate ions move into the surrounding plasma with the aid of a countertransport mechanism that exchanges intracellular bicarbonate ions (HCO_3^-) for extracellular chloride ions (Cl^-). This exchange, which trades one anion for another, does not require ATP. The result is a mass movement of chloride ions into the RBCs, an event known as the **chloride shift**.

2 This illustration summarizes the transportation of oxygen and carbon dioxide in the lungs (at left) and in peripheral tissues (at right). This system is at equilibrium, and the P_{O_2} and P_{CO_2} are stable in the alveoli and in the tissues. Every oxygen molecule entering peripheral tissues is balanced by an oxygen molecule absorbed at the alveoli, and the absorbed oxygen molecule will be replaced in the next respiratory cycle.

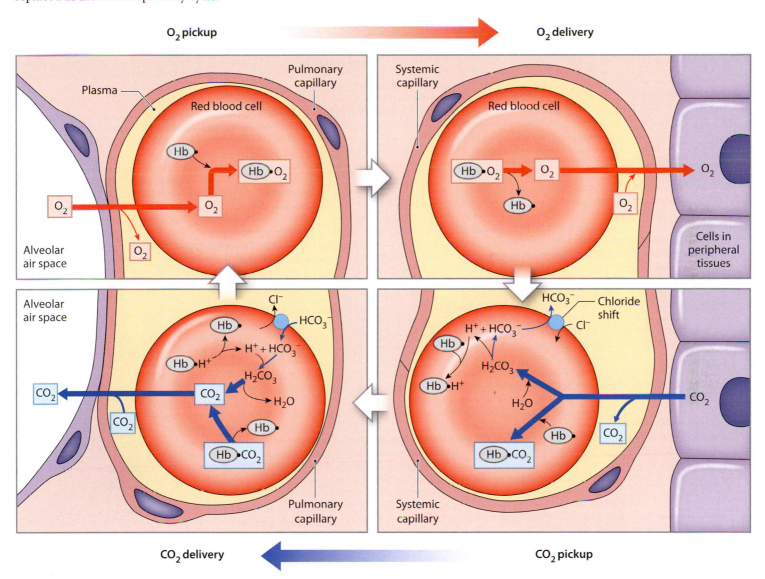

The equilibrium between oxygen absorption and oxygen use is disturbed when tissue oxygen demand increases. If the respiratory rate and tidal volume do not increase, the alveolar P_{O_2} will steadily decline, and the alveolar, blood, and tissue P_{CO_2} will steadily rise. This is a particulary unpleasant combination that can lead to widespread hypoxia and a dangerous decrease in the pH of body fluids. In a clinical setting, pulmonary ventilation is therefore closely regulated by monitoring the P_{O_2}, P_{CO_2}, and pH of body fluids.

Module 20.12 Review

a. Identify three ways that carbon dioxide is transported in the bloodstream.

b. Describe the forces that drive oxygen and carbon dioxide transport between the blood and peripheral tissues.

c. How would blockage of the trachea affect blood pH?

Pulmonary disease can affect both lung elasticity and airflow

1 The **compliance** of the lungs is an indication of their expandability, how easily the lungs expand. It is influenced by the internal structure of the lungs (elasticity and resilience) and the flexibility of the chest wall. Compliance is a static measurement determined by monitoring the intrapulmonary pressure at different lung volumes.

2 The **resistance** of the lungs is an indication of how much force is required to inflate or deflate them. At rest, the muscular activity involved in pulmonary ventilation accounts for 3–5 percent of the resting energy demand. If resistance increases, that figure climbs dramatically, and an individual may become exhausted simply trying to continue breathing.

Compliance

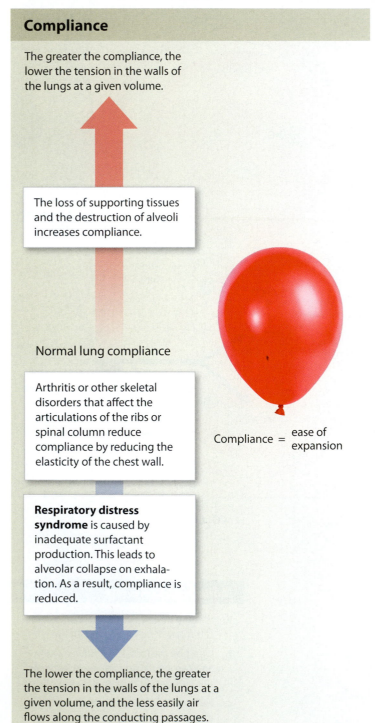

The greater the compliance, the lower the tension in the walls of the lungs at a given volume.

The loss of supporting tissues and the destruction of alveoli increases compliance.

Normal lung compliance

Arthritis or other skeletal disorders that affect the articulations of the ribs or spinal column reduce compliance by reducing the elasticity of the chest wall.

Respiratory distress syndrome is caused by inadequate surfactant production. This leads to alveolar collapse on exhalation. As a result, compliance is reduced.

Compliance = ease of expansion

The lower the compliance, the greater the tension in the walls of the lungs at a given volume, and the less easily air flows along the conducting passages.

Resistance

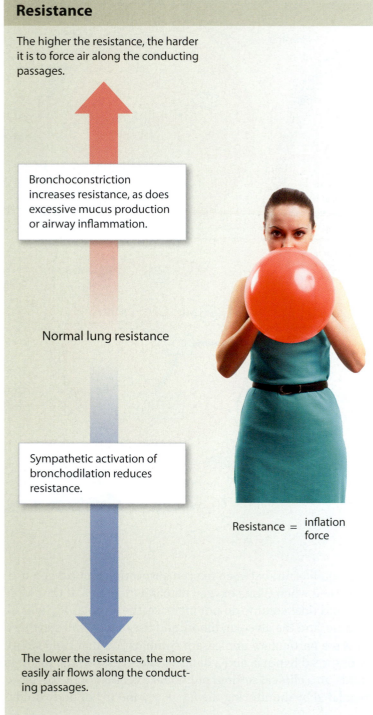

The higher the resistance, the harder it is to force air along the conducting passages.

Bronchoconstriction increases resistance, as does excessive mucus production or airway inflammation.

Normal lung resistance

Sympathetic activation of bronchodilation reduces resistance.

Resistance = inflation force

The lower the resistance, the more easily air flows along the conducting passages.

Chronic Obstructive Pulmonary Disease

3 **Chronic obstructive pulmonary disease (COPD)** is a general term indicating a progressive disorder of the airways that restricts airflow and reduces alveolar ventilation. Three different conditions are included under this heading. Asthma, or asthmatic bronchitis, is the term used when symptoms are acute and intermittent. The terms chronic bronchitis and emphysema are usually applied when signs and symptoms are chronic and progressive with occasional crises due to infections.

Asthma

Asthma (AZ-muh) is a condition characterized by conducting passageways that are extremely sensitive to irritation. The airways respond to irritation by constricting smooth muscles all along the bronchial tree. This is accompanied by edema and swelling of the mucosa of the respiratory passageways, and the accelerated production of mucus. The combination makes breathing very difficult, and resistance is markedly increased. This can be caused by allergies, toxins, or exercise.

Chronic Bronchitis

Chronic bronchitis (brong-KĪ-tis) is a long-term inflammation and swelling of the bronchial lining, leading to overproduction of mucous secretions. The characteristic sign is frequent coughing with copious sputum production. This condition is most commonly related to cigarette smoking but also results from other environmental irritants, such as chemical vapors. Over time, the increased mucus production can block smaller airways, increasing resistance and reducing respiratory efficiency. Chronic bacterial infections leading to more lung damage are common. Individuals with chronic bronchitis may have symptoms of heart failure, including widespread edema. Their blood oxygenation is low, and their skin may have a bluish color. The combination of widespread edema and bluish coloration has led to the descriptive term **blue bloaters** for individuals with this condition.

Emphysema

Emphysema (em-fi-ZĒ-muh) is a chronic, progressive condition characterized by shortness of breath and an inability to tolerate physical exertion. The underlying problem is the destruction of alveolar surfaces and inadequate surface area for oxygen and carbon dioxide exchange. The alveoli gradually expand, and adjacent alveoli merge to form larger air spaces supported by fibrous tissue without alveolar capillary networks. As elastic connective tissues are lost, compliance increases, but the loss of respiratory surface area restricts oxygen absorption, so the individual becomes short of breath. The respiratory muscles work hard, and these individuals, who use a lot of energy just breathing, tend to be thin. On chest x-rays, such as the one at right, the lungs appear overexpanded. Because their exaggerated respiratory movements typically maintain near normal blood oxygenation, the skin of pale-skinned emphysema patients is usually pink. The combination of heavy breathing and pink coloration has led to the descriptive term **pink puffers** for these individuals.

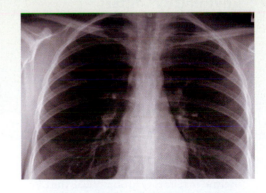

Module 20.13 Review

a. Define compliance and resistance.

b. Identify three chronic obstructive pulmonary diseases (COPDs).

c. Compare chronic bronchitis with emphysema.

Respiratory control mechanisms involve interacting centers in the brain stem

1 Respiratory control involves multiple levels of regulation. Most of the regulatory activities occur outside of our awareness.

Level 3: Higher Centers

Higher centers in the hypothalamus, limbic system, and cerebral cortex can alter the activity of the pneumotaxic centers, but essentially normal respiratory cycles continue even if the brain stem superior to the pons has been severely damaged.

Level 2: Apneustic and Pneumotaxic Centers

The **apneustic** (ap-NOO-stik) **centers** and the **pneumotaxic** (noo-mō-TAKS-ik) **centers** of the pons are paired nuclei that adjust the output of the respiratory rhythmicity centers.

Level 1: Respiratory Rhythmicity Centers

Start

The most basic level of respiratory control involves pacemaker cells in the medulla oblongata. These neurons generate cycles of contraction and relaxation in the diaphragm. The **respiratory rhythmicity centers** set the pace of respiration by adjusting the activities of these pacemakers and coordinating the activities of additional respiratory muscles. Each rhythmicity center can be subdivided into a **dorsal respiratory group** (**DRG**) and a **ventral respiratory group** (**VRG**). The DRG modifies its activities in response to input from chemoreceptors and baroreceptors that monitor O_2, CO_2, and pH in the blood and CSF and from stretch receptors that monitor the degree of stretching in the walls of the lungs.

Higher Centers
- Cerebral cortex
- Limbic system
- Hypothalamus

The pneumotaxic centers inhibit the apneustic centers and thereby promote passive or active exhalation. An increase in pneumotaxic output quickens the pace of respiration by shortening the duration of each inhalation; a decrease in pneumotaxic output slows the respiratory pace but increases the depth of respiration, because the apneustic centers are more active.

The apneustic centers promote inhalation by stimulating the DRG. During forced breathing, the apneustic centers adjust the degree of stimulation in response to sensory information from N X (the vagus nerve) concerning the amount of lung inflation.

Pons

Medulla oblongata

To diaphragm ←

The **inspiratory center** of the DRG contains neurons that control lower motor neurons innervating the external intercostal muscles and the diaphragm. This center functions in every respiratory cycle.

To external intercostal muscles ←

To accessory inspiratory muscles ←

The VRG has inspiratory and expiratory centers that function only when ventilation demands increase and accessory respiratory muscles become involved.

To accessory expiratory muscles ←

2 This flowchart shows the events involved in **quiet breathing** in a resting individual.

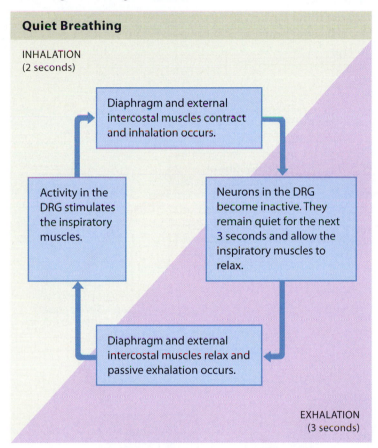

Quiet Breathing

INHALATION
(2 seconds)

Diaphragm and external intercostal muscles contract and inhalation occurs.

Activity in the DRG stimulates the inspiratory muscles.

Neurons in the DRG become inactive. They remain quiet for the next 3 seconds and allow the inspiratory muscles to relax.

Diaphragm and external intercostal muscles relax and passive exhalation occurs.

EXHALATION
(3 seconds)

3 During **forced breathing**, pulmonary ventilation is increased toward maximal levels through the involvement of accessory respiratory muscles.

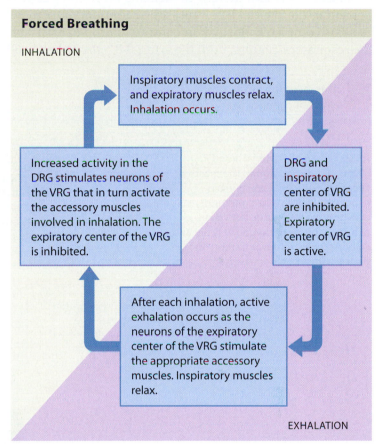

Forced Breathing

INHALATION

Inspiratory muscles contract, and expiratory muscles relax. Inhalation occurs.

Increased activity in the DRG stimulates neurons of the VRG that in turn activate the accessory muscles involved in inhalation. The expiratory center of the VRG is inhibited.

DRG and inspiratory center of VRG are inhibited. Expiratory center of VRG is active.

After each inhalation, active exhalation occurs as the neurons of the expiratory center of the VRG stimulate the appropriate accessory muscles. Inspiratory muscles relax.

EXHALATION

4 This table lists the various sensory stimuli that can modify the activities of the respiratory centers. These stimuli can trigger stereotyped, automatic responses known as **respiratory reflexes**, which are the focus of the next module.

Representative Respiratory Reflexes

- Chemoreceptors sensitive to the pH, P_{O_2}, or P_{CO_2} of the blood or cerebrospinal fluid alter the activities of the respiratory centers.
- Baroreceptors in the aortic or carotid sinuses sensitive to changes in blood pressure alter the activities of the respiratory centers.
- Stretch receptors that respond to changes in the volume of the lungs are responsible for inflation and deflation reflexes.
- Irritating physical or chemical stimuli in the nasal cavity, larynx, or bronchial tree initiate protective reflexes, such as coughing or sneezing.

Module 20.14 Review

a. Name the paired CNS nuclei that adjust the pace of respiration.

b. Which brain stem centers generate the respiratory pace?

c. Which chemical factors in blood or cerebrospinal fluid stimulate the respiratory centers?

Respiratory reflexes provide rapid automatic adustments in pulmonary ventilation

Chemoreceptor Reflexes

Under normal conditions, the P_{CO_2} is the most important factor stimulating chemoreceptors and thereby influencing respiratory activity. A rise of just 10 percent in the arterial P_{CO_2} causes the respiratory rate to double, even if the P_{O_2} remains completely normal. In contrast, a drop in arterial P_{O_2} has little effect on the respiratory centers, until the arterial P_{O_2} drops below 60 mm Hg.

1 This diagram shows the effect of changes in P_{CO_2} on the respiratory rate.

An increase in the arterial blood P_{CO_2} constitutes **hypercapnia**. The most common cause of hypercapnia is hypoventilation—when respiratory activity is insufficient to meet the demands for tissue oxygen delivery and carbon dioxide removal. Carbon dioxide then accumulates in the blood.

Hyperventilation, when the rate and depth of respiration exceed the demands for oxygen delivery and carbon dioxide removal, gradually leads to **hypocapnia**, an abnormally low P_{CO_2}. Snorkelers sometimes hyperventilate to extend their time underwater; this works because it is the P_{CO_2} that stimulates respiratory activity. If the P_{CO_2} is driven down too far, a snorkeler may become unconscious from oxygen starvation in the brain without ever feeling the urge to breathe. This condition is called **shallow water blackout**.

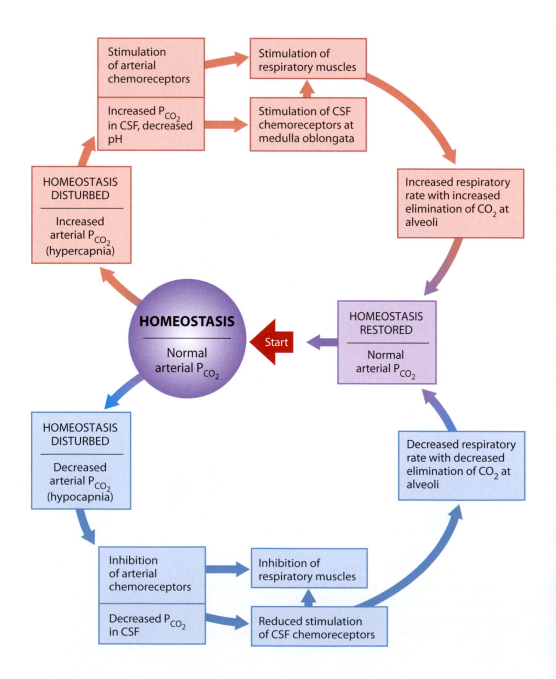

Baroreceptor Reflexes

2 Baroreceptors in the carotid and aortic sinuses are monitored by sensory nerves within the glossopharyngeal and vagus nerves, respectively. The cardiovascular effects of their stimulation were considered in Chapter 18. The sensory information is also distributed to the respiratory centers, triggering adjustments in the respiratory minute volume.

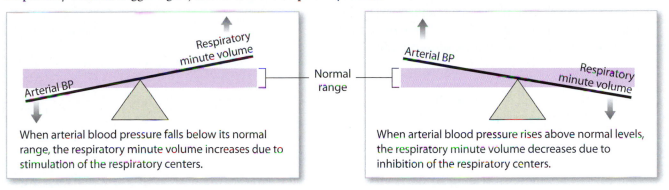

When arterial blood pressure falls below its normal range, the respiratory minute volume increases due to stimulation of the respiratory centers.

When arterial blood pressure rises above normal levels, the respiratory minute volume decreases due to inhibition of the respiratory centers.

Inflation/Deflation Reflexes

3 Inflation and deflation reflexes are activated by stretch receptors in the lungs during forced breathing when the tidal volume is large (1000 mL or more). This sensory information is distributed to the apneustic centers and the VRG.

The **inflation reflex** prevents overexpansion of the lungs during forced breathing.

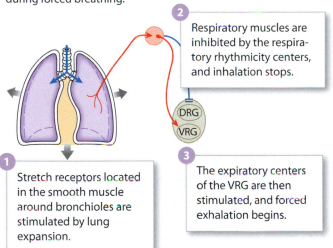

2 Respiratory muscles are inhibited by the respiratory rhythmicity centers, and inhalation stops.

1 Stretch receptors located in the smooth muscle around bronchioles are stimulated by lung expansion.

3 The expiratory centers of the VRG are then stimulated, and forced exhalation begins.

The **deflation reflex** inhibits the expiratory centers and stimulates the inspiratory centers when the lungs are deflating.

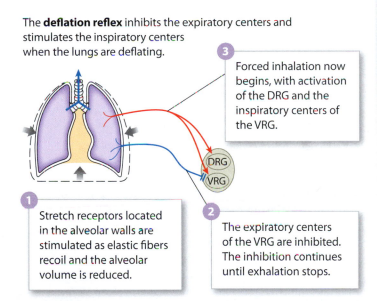

3 Forced inhalation now begins, with activation of the DRG and the inspiratory centers of the VRG.

1 Stretch receptors located in the alveolar walls are stimulated as elastic fibers recoil and the alveolar volume is reduced.

2 The expiratory centers of the VRG are inhibited. The inhibition continues until exhalation stops.

Protective Reflexes

4 **Protective reflexes** include sneezing and coughing. Sneezing is triggered by an irritation of the nasal cavity wall. Coughing is triggered by an irritation of the larynx, trachea, or bronchi. Both reflexes involve **apnea** (AP-nē-uh), a period in which respiration is suspended, usually followed by a forceful expulsion of air to remove the offending stimulus. Air leaving the larynx can travel at 160 kph (99 mph), carrying mucus, foreign particles, and irritating gases out of the respiratory tract via the nose or mouth.

Module 20.15 Review

a. Define hypercapnia and hypocapnia.

b. Are chemoreceptors more sensitive or less sensitive to plasma levels of CO_2 than they are to plasma levels of O_2?

c. Little Johnny is angry with his mother, so he tells her that he will hold his breath until he turns blue and dies. Should Johnny's mother worry that this will happen?

Respiratory function decreases with age; smoking makes matters worse

All aspects of respiratory function are affected by aging. As elastic tissue deteriorates throughout the body, vital capacity decreases; and as arthritic changes stiffen rib articulations, compliance is reduced, as is the maximum respiratory minute volume. These respiratory limitations make strenuous exercise difficult or impossible.

1 In addition to skeletal and connective tissue changes associated with age, some degree of emphysema is normal in individuals over age 50. However, the extent varies widely with the lifetime exposure to cigarette smoke and other respiratory irritants. This graph compares the respiratory performance of individuals who have never smoked with individuals who have smoked for various periods of time. The message is quite clear: Although some decrease in respiratory performance is inevitable, you can prevent serious respiratory deterioration by stopping smoking or never starting.

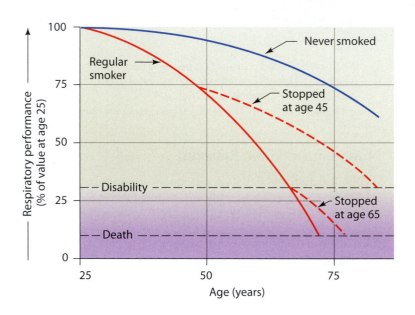

Lung Cancer

2 **Lung cancers** now account for 12.6 percent of new cancer cases in both men and women, and it kills more people each year than colon, breast, and prostate cancer combined. Over 50 percent of lung cancer patients die within a year of diagnosis. Detailed statistical and experimental evidence has shown that 85–90 percent of all lung cancers are the direct result of cigarette smoking. Before about 1970, this disease affected primarily middle-aged men, but as the number of women smokers has increased (a trend that started in the 1940s), so has the number of women who develop lung cancer. This graph shows that while the incidence of lung cancer among men has gradually declined since 1992, the incidence among women has been increasing.

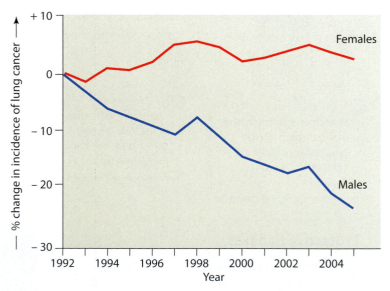

3 Smoke is an irritant, and the chemicals contained in cigarette smoke include several carcinogens (cancer-causing substances). This figure shows the progression of events leading to lung cancer. The fact that cigarette smoking causes cancer is not surprising, given that the smoke contains carcinogens. What is surprising is that more smokers do not develop lung cancer. Evidence suggests that some smokers have a genetic predisposition to developing at least one form of lung cancer.

Normal Respiratory Epithelium

The normal respiratory epithelium consists of ciliated columnar epithelium with an abundance of mucous cells, which produce mucus that helps clean inhaled air. The irritants and inhaled carcinogens contained in cigarette smoke begin a series of changes to this epithelium.

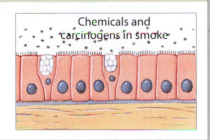

Chemicals and carcinogens in smoke

Reversible

Dysplasia

In **dysplasia**, cells are damaged and the functional characteristics change. The cilia of respiratory epithelial cells are damaged and paralyzed by exposure to cigarette smoke. These changes cause the local buildup of mucus and reduce the effectiveness of the epithelium in protecting deeper, more delicate portions of the respiratory tract.

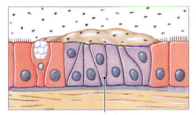

Abnormal cells

Reversible

Metaplasia

In **metaplasia**, a tissue changes its structure, in response to injury or chemical stresses. In this case the stressed respiratory surface converts to a stratified epithelium that protects underlying connective tissue but does nothing for other areas of the respiratory tract. Metaplasia may be reversed if the stressful stimulus is removed before further damage occurs.

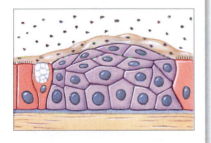

Irreversible

Neoplasia and Anaplasia

In **neoplasia**, the abnormal cells form a cancerous tumor, or **neoplasm**. In **anaplasia**, the most dangerous stage, the cells become malignant and spread, or metastasize, to other parts of the body. Neither neoplasia nor anaplasia are reversible, although the cancers may be treated by chemicals, radiation, and/or surgery.

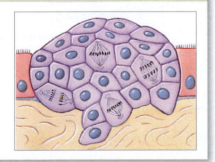

Module 20.16 Review

a. Name several age-related factors that affect the respiratory system.

b. Compare dysplasia, metaplasia, neoplasia, and anaplasia.

c. Is the incidence of lung cancer in males increasing or decreasing?

1. Short answer

Identify and describe the various pulmonary volumes and capacities indicated in the following graph.

a _____

b _____

c _____

d _____

e _____

f _____

g _____

h _____

i _____

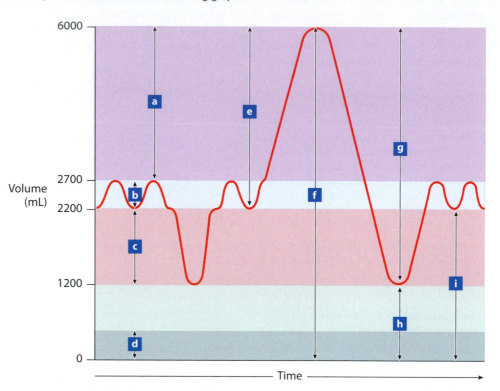

2. Matching

Match the following terms with the most closely related description.

- apnea
- hemoglobin releases more O_2
- lowers vital capacity
- external intercostals
- iron ion
- Boyle's law
- compliance
- bicarbonate ion
- anoxia
- atelectasis
- hypocapnia
- pneumotaxic centers
- apneustic centers
- partial pressure

a _____ Single gas in a mixture

b _____ A cause of tissue death

c _____ Inverse pressure/volume relationship

d _____ Expandability of lungs

e _____ CO_2 transport

f _____ Elastic tissue deterioration

g _____ Act to elevate ribs

h _____ Heme unit

i _____ Blood pH decreases

j _____ Promotes passive or active exhalation

k _____ Stimulates DRG and promotes inhalation

l _____ Hyperventilation

m _____ Collapsed lung

n _____ Period of suspended respiration

3. Section integration

Compare and contrast external respiration, pulmonary ventilation, and internal respiration. _____

_____

_____

Visual Outline with Key Terms

Summarize the content of each module using the terms in the order provided.

SECTION 1

Functional Anatomy of the Respiratory System

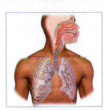

- respiratory system
- respiratory tract
- conducting portion
- bronchioles
- respiratory portion
- alveoli
- upper respiratory system
- lower respiratory system

20.1

The respiratory mucosa is protected by the respiratory defense system

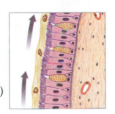

- respiratory defense system
- respiratory mucosa
- mucus escalator
- lamina propria
- cystic fibrosis (CF)

20.2

The upper portions of the respiratory system include the nose, nasal cavity, paranasal sinuses, and pharynx

- bridge of the nose
- nasal cartilages
- external nares
- superior, middle, and inferior meatuses
- nasal septum
- paranasal sinuses
- pharynx
- nasopharynx
- oropharynx
- laryngopharynx
- trachea
- internal nares
- nasal vestibule
- hard palate
- soft palate
- glottis
- larynx

20.3

The larynx protects the glottis and produces sounds

- larynx
- epiglottis
- thyroid cartilage
- cricoid cartilage
- vestibular ligaments
- vocal ligaments
- cuneiform cartilages
- corniculate cartilages
- arytenoid cartilages
- vocal folds
- vocal ligaments
- vocal cords
- vestibular folds
- phonation
- articulation

20.4

The trachea and primary bronchi convey air to and from the lungs

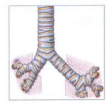

- trachea
- primary bronchi
- tracheal cartilages
- trachealis muscle
- secondary bronchi
- tertiary bronchi
- bronchioles
- terminal bronchioles
- pulmonary lobule
- root of the lung
- bronchodilation
- bronchoconstriction
- asthma

20.5

The lungs contain lobes that are subdivided into bronchopulmonary segments

- bronchopulmonary segment
- bronchial tree
- right lung
- horizontal and oblique fissures
- superior, middle, and inferior lobes
- cardiac notch
- left lung
- oblique fissure
- superior and inferior lobes
- hilum

20.6

Pulmonary lobules contain alveoli, where gas exchange occurs

- alveoli
- pulmonary lobule
- respiratory bronchiole
- alveolar ducts
- alveolar sac
- visceral pleura
- pleural fluid
- parietal pleura
- type II pneumocytes
- surfactant
- type I pneumocytes
- respiratory membrane

SECTION 2

Respiratory Physiology

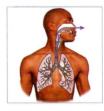

- respiration
- external respiration
- pulmonary ventilation
- alveolar ventilation
- gas diffusion
- internal respiration
- hypoxia
- anoxia

• = *Term boldfaced in this module*

20.7

Pulmonary ventilation is driven by pressure changes within the pleural cavities

- Boyle's law
- atelectasis
- intrapulmonary pressure
- tidal volume (V_T)
- ○ gas pressures
- millimeters of mercury (mm Hg)
- torr
- centimeters of water (cm H_2O)
- pounds per square inch (psi)

20.11

Almost all of the oxygen in the blood is transported bound to hemoglobin within red blood cells

- heme units
- oxyhemoglobin (HbO_2)
- hemoglobin saturation
- oxygen-hemoglobin saturation curve
- Bohr effect
- 2,3-bisphosphoglycerate (BPG)

20.8

Respiratory muscles in various combinations adjust the tidal volume to meet respiratory demands

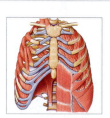

- inspiratory muscles
- expiratory muscles
- primary respiratory muscles
- accessory respiratory muscles
- ○ diaphragm
- ○ external intercostal muscles
- volumes
- capacities
- inspiratory reserve volume (IRV)
- tidal volume (V_T)
- expiratory reserve volume (ERV)
- inspiratory capacity
- vital capacity
- functional residual capacity (FRC)
- minimal volume
- total lung capacity
- residual volume

20.12

Most carbon dioxide transport occurs through the reversible formation of carbonic acid

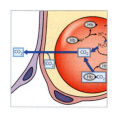

- carbaminohemoglobin ($HbCO_2$)
- carbonic anhydrase
- HbH^+
- chloride shift

20.9

Pulmonary ventilation must be closely regulated to meet tissue oxygen demands

- respiratory rate (f)
- respiratory minute volume (V_E)
- alveolar ventilation (V_A)
- anatomic dead space (V_D)

$$\mathbf{V_E} = f \times \mathbf{V_T}$$

20.13

Pulmonary disease can affect both lung elasticity and airflow

- compliance
- respiratory distress syndrome
- resistance
- chronic obstructive pulmonary disease (COPD)
- asthma
- chronic bronchitis
- blue bloaters
- emphysema
- pink puffers

20.10

Gas diffusion depends on the partial pressures and solubilities of gases

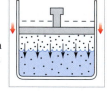

- gas laws
- partial pressure (P)
- Dalton's law
- Henry's law
- ○ external respiration
- ○ internal respiration

20.14

Respiratory control mechanisms involve interacting centers in the brain stem

- respiratory rhythmicity centers
- dorsal respiratory group (DRG)
- ventral respiratory group (VRG)
- inspiratory center
- apneustic centers
- pneumotaxic centers
- higher centers
- quiet breathing
- forced breathing
- respiratory reflexes

• = *Term boldfaced in this module*

20.15

20.15

Respiratory reflexes provide rapid automatic adjustments in pulmonary ventilation

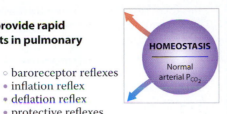

- chemoreceptor reflexes
- hypercapnia
- hyperventilation
- hypocapnia
- shallow water blackout
- baroreceptor reflexes
- inflation reflex
- deflation reflex
- protective reflexes
- apnea

HOMEOSTASIS

Normal arterial P_{CO_2}

20.16

Respiratory function decreases with age; smoking makes matters worse

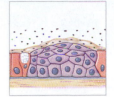

- lung cancers
- dysplasia
- metaplasia
- neoplasia
- neoplasm
- anaplasia

• = *Term boldfaced in this module*

Chapter Integration: Applying what you've learned

Jake is an active 22-year-old college student who enjoys water sports. In fact, he has an athletic scholarship to swim on the university swim team. While on spring break at an ocean-front hotel with both beach and pool access, Jake decides he'll take advantage of his prime location.

While at the beach one day, he goes snorkeling. The sun is shining, the air temperature is 80°F, and the ocean is a balmy 72°F. His normal alveolar ventilation rate (AVR) during mild exercise is 6.0 L/min, and his snorkel has a volume of 50 mL.

After snorkeling in the ocean all afternoon, Jake heads to the pool to meet up with some friends. Before diving in, he hyperventilates for several minutes. After he enters and begins swimming underwater, he blacks out and almost drowns. Fortunately, friends notice that something's wrong and pull him out of the pool to safety. He recovered within 15 seconds and did not require pulmonary resuscitation or medical treatment.

1. Assuming a constant tidal volume of 500 mL and an anatomic dead space of 150 mL, what would Jake's respiratory rate have to be for him to maintain an AVR of 6.0 L/min while snorkeling?

2. What caused Jake to black out and nearly drown?

MasteringA&P™

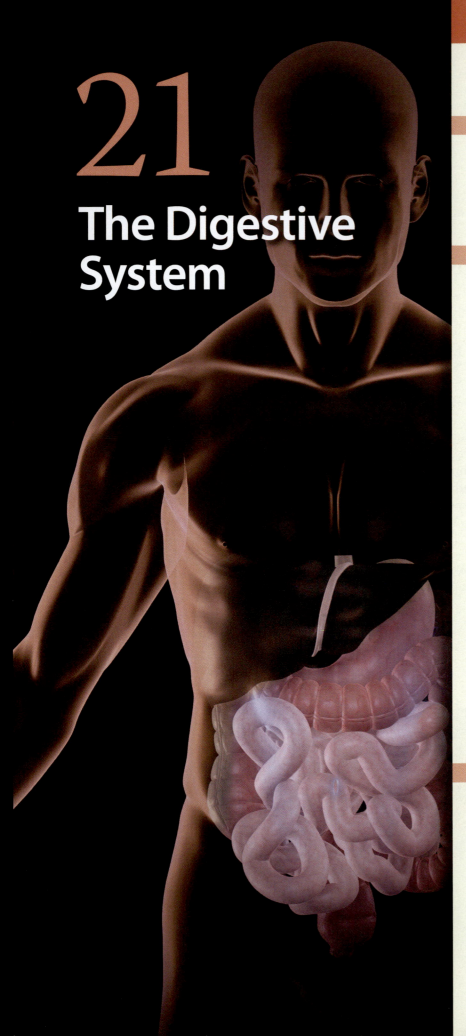

21

The Digestive System

General Organization of the Digestive System

The **digestive system** consists of a muscular tube, the digestive tract—also called the gastrointestinal (GI) tract—plus various accessory organs. Food enters the mouth and passes along the length of the digestive tract. On the way, accessory organs produce secretions containing water, enzymes, buffers and other components that assist in preparing organic and inorganic nutrients for absorption.

1 The digestive system works with other systems to support tissues that have no direct connection with the outside environment and no other means of obtaining nutrients.

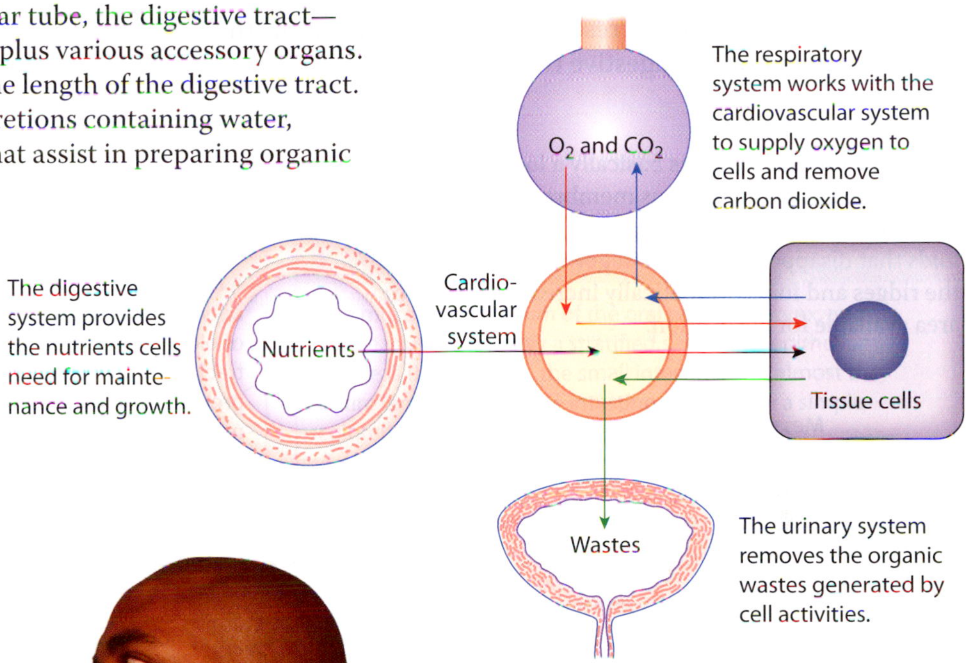

The digestive system provides the nutrients cells need for maintenance and growth.

The respiratory system works with the cardiovascular system to supply oxygen to cells and remove carbon dioxide.

O_2 and CO_2

Nutrients

Cardio-vascular system

Tissue cells

Wastes

The urinary system removes the organic wastes generated by cell activities.

2 The **digestive tract** begins at the mouth and continues through the oral cavity, pharynx, esophagus, stomach, small intestine, and large intestine, which opens to the exterior at the anus. These structures have distinctive structural and functional characteristics, but all share an underlying pattern of histological organization that will be considered in this section.

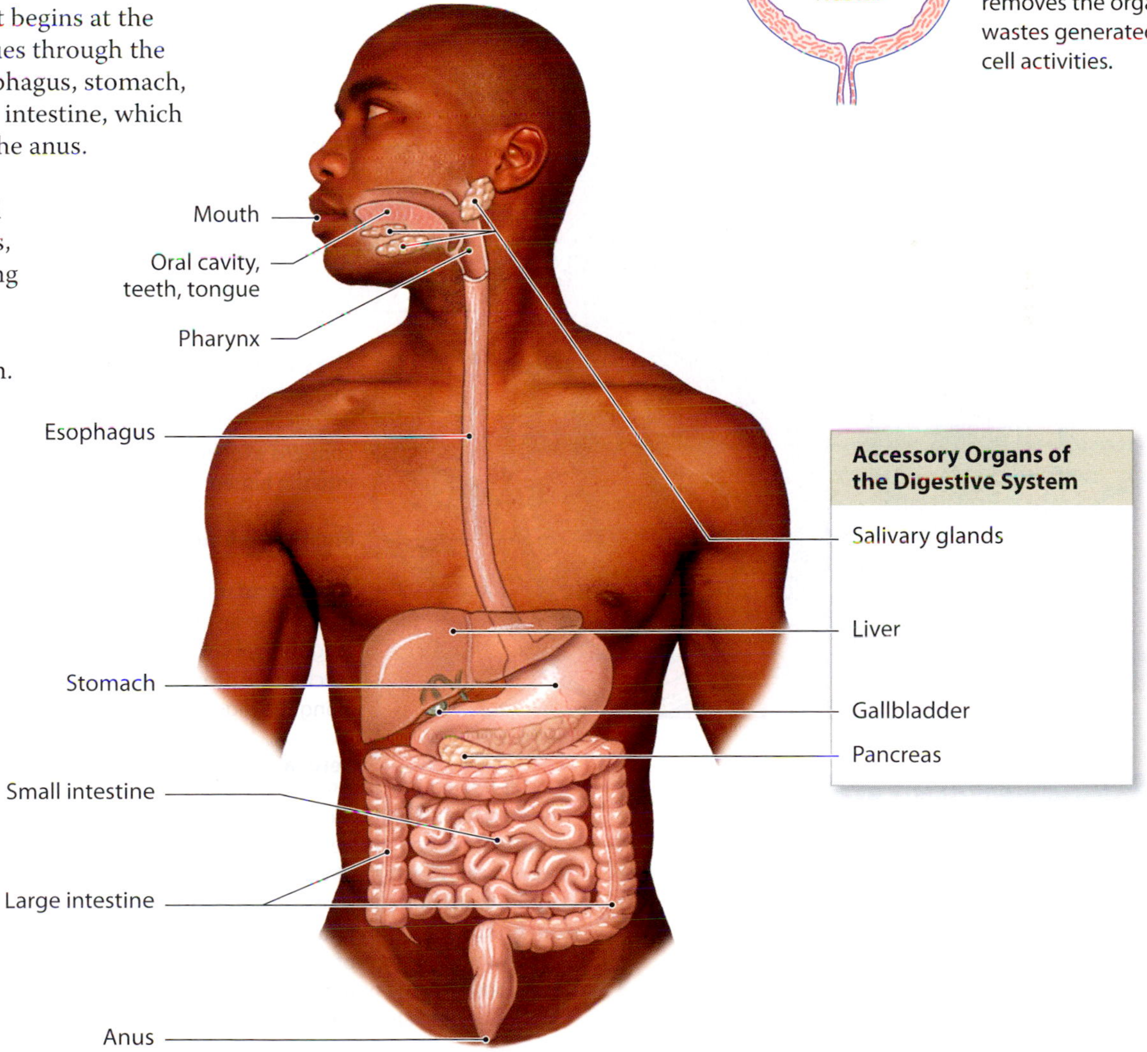

Mouth

Oral cavity, teeth, tongue

Pharynx

Esophagus

Stomach

Small intestine

Large intestine

Anus

Accessory Organs of the Digestive System

Salivary glands

Liver

Gallbladder

Pancreas

Smooth muscle contractions mix the contents of the digestive tract and propel materials along its length

The coordinated contractions of the smooth muscle layers in the muscularis externa play a vital role in both moving material along the digestive tract (through peristalsis) and in mechanical processing (through segmentation).

Peristalsis

1 Food enters the digestive tract as a moist, compact mass known as a **bolus** (BŌ-lus). The muscularis externa propels materials from one portion of the digestive tract to another by contractions known as **peristalsis** (per-i-STAL-sis). During a peristaltic movement, a wave of contraction in the circular muscles forces the bolus forward.

Segmentation

2 Most areas of the small intestine and some portions of the large intestine undergo cycles of contraction that churn and fragment the bolus, mixing the contents with intestinal secretions. This activity is called **segmentation**. These rhythmic cycles of contraction do not follow a set pattern and thus do not push materials along the tract in any one direction.

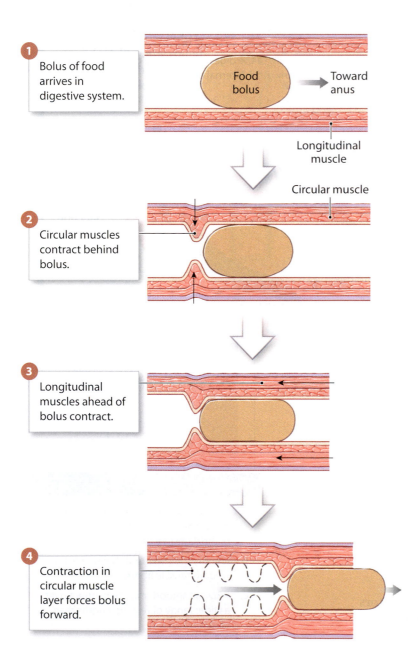

1 Bolus of food arrives in digestive system.

Food bolus

Toward anus

Longitudinal muscle

Circular muscle

2 Circular muscles contract behind bolus.

3 Longitudinal muscles ahead of bolus contract.

4 Contraction in circular muscle layer forces bolus forward.

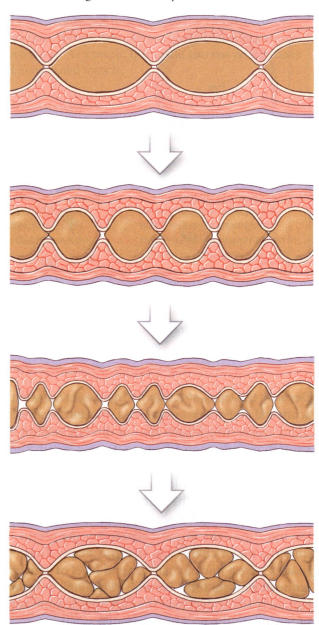

3 Three major mechanisms regulate and control digestive activities. The primary stimuli involved are local factors that may in turn activate neural or hormonal control mechanisms.

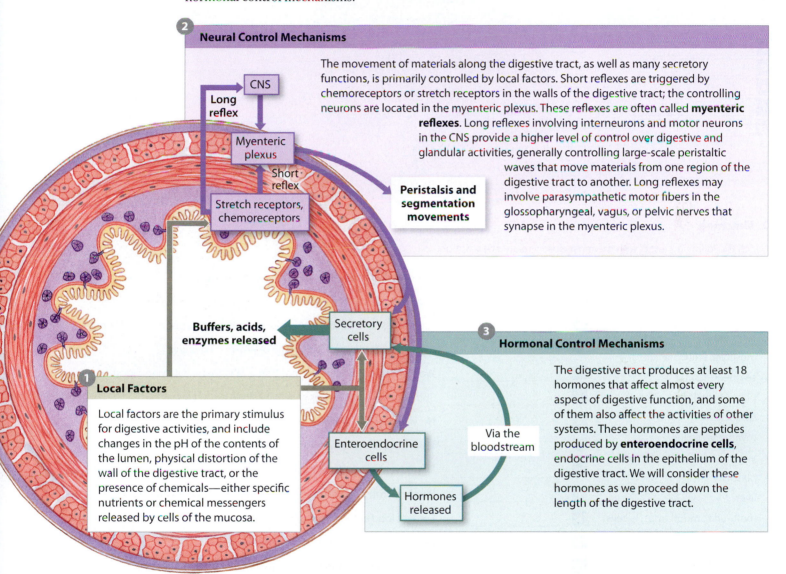

② Neural Control Mechanisms

The movement of materials along the digestive tract, as well as many secretory functions, is primarily controlled by local factors. Short reflexes are triggered by chemoreceptors or stretch receptors in the walls of the digestive tract; the controlling neurons are located in the myenteric plexus. These reflexes are often called **myenteric reflexes**. Long reflexes involving interneurons and motor neurons in the CNS provide a higher level of control over digestive and glandular activities, generally controlling large-scale peristaltic waves that move materials from one region of the digestive tract to another. Long reflexes may involve parasympathetic motor fibers in the glossopharyngeal, vagus, or pelvic nerves that synapse in the myenteric plexus.

CNS

Long reflex

Myenteric plexus

Short reflex

Stretch receptors, chemoreceptors

Peristalsis and segmentation movements

Buffers, acids, enzymes released

Secretory cells

③ Hormonal Control Mechanisms

The digestive tract produces at least 18 hormones that affect almost every aspect of digestive function, and some of them also affect the activities of other systems. These hormones are peptides produced by **enteroendocrine cells**, endocrine cells in the epithelium of the digestive tract. We will consider these hormones as we proceed down the length of the digestive tract.

Via the bloodstream

Enteroendocrine cells

Hormones released

① Local Factors

Local factors are the primary stimulus for digestive activities, and include changes in the pH of the contents of the lumen, physical distortion of the wall of the digestive tract, or the presence of chemicals—either specific nutrients or chemical messengers released by cells of the mucosa.

Module 21.3 Review

a. Describe enteroendocrine cells.

b. Cite the major mechanisms that regulate and control digestive activities.

c. Which is more efficient in propelling intestinal contents along the digestive tract: peristalsis or segmentation? Why?

1. Labeling

Label each of the structures of the digestive tract in the following figure.

a _____

b _____

c _____

d _____

e _____

f _____

g _____

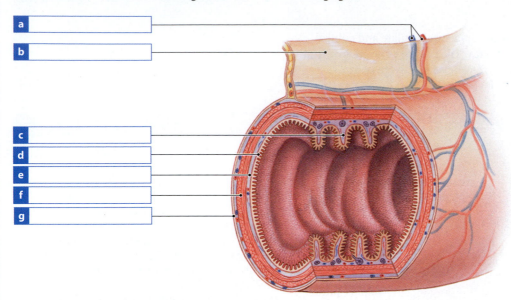

2. Matching

Match the following terms with the most closely related description.

- lamina propria
- peristalsis
- pacesetter cells
- esophagus
- muscularis mucosa
- segmentation
- plicae circulares
- sphincter
- myenteric plexus
- visceral smooth muscle cells
- plasticity
- liver
- multi-unit smooth muscle cells
- bolus

a _____ Digestive tube between the pharynx and stomach

b _____ Moves plicae circulares and villi

c _____ Areolar tissue layer containing blood vessels, nerve endings, and lymphatics

d _____ Permanent transverse folds in the digestive tract lining

e _____ Waves of muscular contractions that propel materials along digestive tract

f _____ Stimulate rhythmic cycles of activity along digestive tract

g _____ Nerve network within the muscularis externa

h _____ Have direct contact with motor neurons

i _____ Digestive system accessory organ

j _____ Rhythmic muscular contractions that mix materials in digestive tract

k _____ Form of food entering the digestive tract

l _____ Lack direct contact with motor neurons

m _____ Ability of smooth muscle cells to function over varied lengths

n _____ Ring of muscle tissue

3. Section integration

How would a decrease in smooth muscle tone affect the digestive processes and possibly promote constipation (infrequent bowel movement)?

_____

_____

_____

The Digestive Tract

The digestive tract is a muscular tube with a length of approximately 10 m (33 ft). It can be divided into regions that differ in histological structure and functional properties. In this section we consider each of the major subdivisions of the digestive tract, which are summarized below.

Mouth

Major Subdivisions of the Digestive Tract

Oral Cavity, Teeth, Tongue

Mechanical processing, moistening, mixing with salivary secretions

Pharynx

Muscular propulsion of materials into the esophagus

Esophagus

Transport of materials to the stomach

Stomach

Chemical breakdown of materials via acid and enzymes; mechanical processing through muscular contractions

Small Intestine

Enzymatic digestion and absorption of water, organic substrates, vitamins, and ions

Large Intestine

Dehydration and compaction of indigestible materials in preparation for elimination

Accessory Organs of the Digestive System

Salivary glands

Liver

Gallbladder

Pancreas

Anus

1 This table summarizes the general functions of the digestive tract. The lining of the digestive tract also protects surrounding tissues against digestive acids and enzymes, abrasion, and pathogens within the digestive tract.

General Functions of the Digestive Tract

- **Ingestion** occurs when foods and liquids enter the digestive tract via the mouth.
- **Mechanical processing** occurs to most ingested solids either before they are swallowed or in the proximal portions of the digestive tract.
- **Digestion** is the chemical and enzymatic breakdown of food into small organic molecules that can be absorbed by the digestive epithelium.

- **Secretion** is performed along most of the digestive tract (however, most of the acids, enzymes, and buffers required for digestion are provided by the accessory digestive organs).
- **Absorption** is the movement of organic molecules, electrolytes, vitamins, and water across the digestive epithelium and into the interstitial fluid of the digestive tract.
- **Compaction** is the progressive dehydration of indigestible materials and organic wastes prior to elimination from the body. The compacted material is called **feces**; the elimination of feces from the body is called **defecation** (def-e-KĀ-shun).

The stomach breaks down the organic nutrients in ingested materials

1 The wall of the stomach is relatively thick and muscular, and its mucosa has deep folds that form gastric glands.

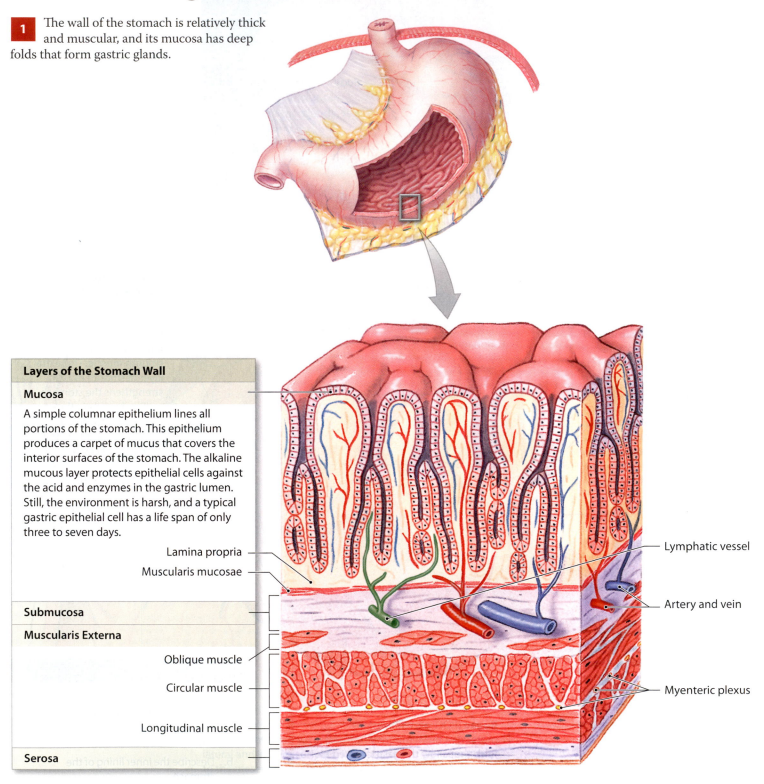

Layers of the Stomach Wall

Mucosa

A simple columnar epithelium lines all portions of the stomach. This epithelium produces a carpet of mucus that covers the interior surfaces of the stomach. The alkaline mucous layer protects epithelial cells against the acid and enzymes in the gastric lumen. Still, the environment is harsh, and a typical gastric epithelial cell has a life span of only three to seven days.

Lamina propria

Muscularis mucosae

Submucosa

Muscularis Externa

Oblique muscle

Circular muscle

Longitudinal muscle

Serosa

Lymphatic vessel

Artery and vein

Myenteric plexus

2 **Gastric glands** in the fundus and body secrete most of the acid and enzymes involved in gastric digestion. The gastric glands in these areas are dominated by **parietal cells** and **chief cells**. Together, they secrete about 1500 mL of gastric juice each day. Gastric glands in the pylorus secrete mucus and hormones involved in the coordination and control of digestive activity.

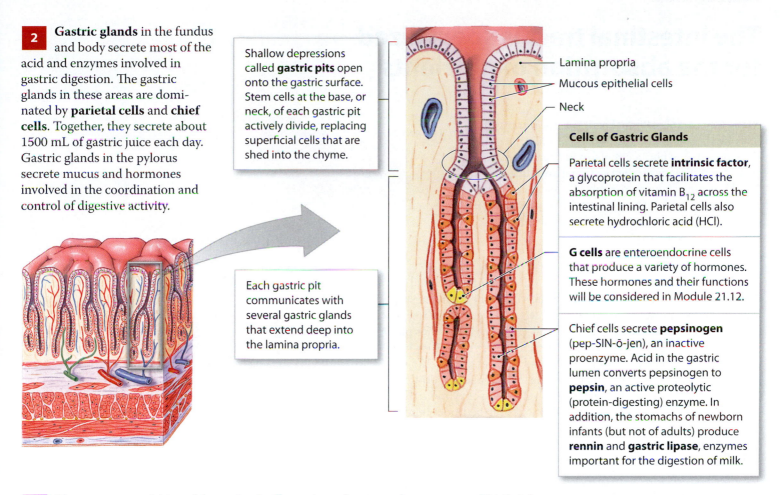

Shallow depressions called **gastric pits** open onto the gastric surface. Stem cells at the base, or neck, of each gastric pit actively divide, replacing superficial cells that are shed into the chyme.

Each gastric pit communicates with several gastric glands that extend deep into the lamina propria.

- Lamina propria
- Mucous epithelial cells
- Neck

Cells of Gastric Glands

Parietal cells secrete **intrinsic factor**, a glycoprotein that facilitates the absorption of vitamin B_{12} across the intestinal lining. Parietal cells also secrete hydrochloric acid (HCl).

G cells are enteroendocrine cells that produce a variety of hormones. These hormones and their functions will be considered in Module 21.12.

Chief cells secrete **pepsinogen** (pep-SIN-ō-jen), an inactive proenzyme. Acid in the gastric lumen converts pepsinogen to **pepsin**, an active proteolytic (protein-digesting) enzyme. In addition, the stomachs of newborn infants (but not of adults) produce **rennin** and **gastric lipase**, enzymes important for the digestion of milk.

3 The secretory activities of the parietal cells can keep the stomach contents at pH 1.5–2.0. However, parietal cells do not produce HCl in the cytoplasm, because it is such a strong acid that it would erode secretory vesicles and destroy the cell. Instead, H^+ and Cl^- are transported and secreted independently, as diagrammed below. When gastric glands are actively secreting, enough bicarbonate ions enter the bloodstream to increase the pH of the blood significantly. This sudden influx of bicarbonate ions has been called the **alkaline tide**.

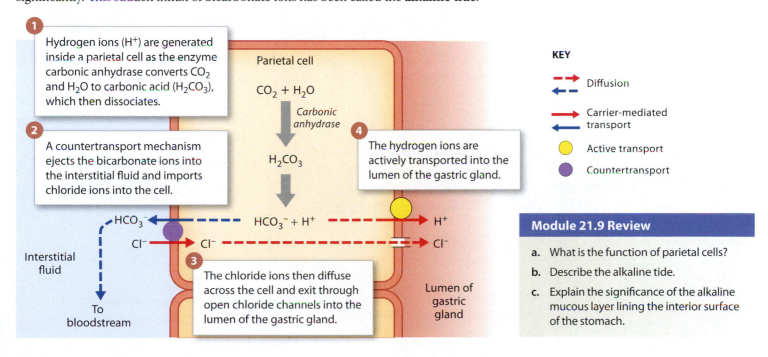

1 Hydrogen ions (H^+) are generated inside a parietal cell as the enzyme carbonic anhydrase converts CO_2 and H_2O to carbonic acid (H_2CO_3), which then dissociates.

2 A countertransport mechanism ejects the bicarbonate ions into the interstitial fluid and imports chloride ions into the cell.

4 The hydrogen ions are actively transported into the lumen of the gastric gland.

3 The chloride ions then diffuse across the cell and exit through open chloride channels into the lumen of the gastric gland.

Parietal cell

$CO_2 + H_2O$

Carbonic anhydrase

H_2CO_3

$HCO_3^- + H^+$

HCO_3^-

Cl^- → Cl^-

Interstitial fluid

To bloodstream

Lumen of gastric gland

KEY

- Diffusion
- Carrier-mediated transport
- Active transport
- Countertransport

Module 21.9 Review

a. What is the function of parietal cells?

b. Describe the alkaline tide.

c. Explain the significance of the alkaline mucous layer lining the interior surface of the stomach.

The intestinal tract is specialized for the absorption of nutrients

1 The intestinal lining bears a series of transverse folds called **plicae circulares**. Unlike the rugae in the stomach, the plicae circulares are permanent features that do not disappear when the small intestine fills. The small intestine contains roughly 800 plicae circulares, most of them within the jejunum. Their presence greatly increases the surface area available for absorption.

Plica circulares

Villi

2 The mucosa of the small intestine is thrown into a series of fingerlike projections, the **intestinal villi**. If the small intestine were a simple tube with smooth walls, it would have a total absorptive area of roughly 3300 cm^2 (3.6 ft^2). Instead, the mucosa contains plicae circulares; each plica circulares supports a forest of villi, and each villus is covered by epithelial cells whose exposed surfaces are covered with microvilli. This arrangement increases the total area for absorption by a factor of more than 600, to approximately 2 million cm^2 (more than 2200 ft^2, roughly the floor space of a spacious four-bedroom home).

3 This diagrammatic sectional view of the intestinal wall shows features common to all segments of the small intestine.

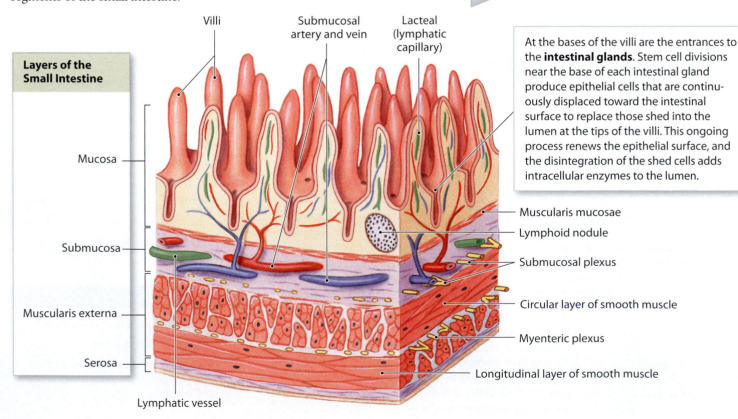

Villi

Submucosal artery and vein

Lacteal (lymphatic capillary)

At the bases of the villi are the entrances to the **intestinal glands**. Stem cell divisions near the base of each intestinal gland produce epithelial cells that are continuously displaced toward the intestinal surface to replace those shed into the lumen at the tips of the villi. This ongoing process renews the epithelial surface, and the disintegration of the shed cells adds intracellular enzymes to the lumen.

Layers of the Small Intestine

Mucosa

Submucosa

Muscularis externa

Serosa

Muscularis mucosae

Lymphoid nodule

Submucosal plexus

Circular layer of smooth muscle

Myenteric plexus

Longitudinal layer of smooth muscle

Lymphatic vessel

4 Each villus has a complex internal structure. The lamina propria of each villus contains an extensive network of capillaries that originate in a vascular network within the submucosa. These capillaries carry absorbed nutrients to the hepatic portal circulation for delivery to the liver, which adjusts the nutrient concentrations in blood before the blood reaches the general systemic circulation.

5 The surface of each villus consists of a simple columnar epithelium that is carpeted with microvilli. Because the microvilli project from the epithelium like the bristles on a brush, these cells are said to have a **brush border**. Brush border enzymes are integral membrane proteins located on the surfaces of intestinal microvilli. These enzymes break down materials that come in contact with the brush border. The epithelial cells then absorb the breakdown products.

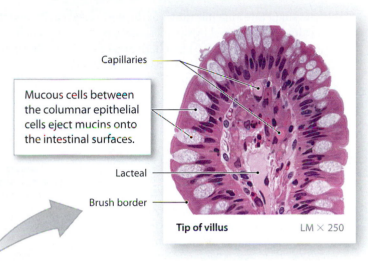

Capillaries

Mucous cells between the columnar epithelial cells eject mucins onto the intestinal surfaces.

Lacteal

Brush border

Tip of villus   LM × 250

Each villus contains a lymphatic capillary called a **lacteal** (LAK-tē-ul; *lacteus*, milky). Lacteals transport materials that cannot enter blood capillaries. For example, absorbed fatty acids are assembled into protein–lipid packages that are too large to diffuse into the bloodstream. These lipid-rich packets reach the venous circulation as the thoracic duct delivers lymph to the left subclavian vein.

Columnar epithelial cell

Mucous cell

Nerve

Capillary network

Lamina propria

Arteriole

Lymphatic vessel

Venule

Contractions of the muscularis mucosae and smooth muscle cells within the intestinal villi move the villi back and forth, exposing the epithelial surfaces to the liquefied intestinal contents. This movement improves the efficiency of absorption by quickly eliminating local differences in nutrient concentration. Movements of the villi also squeeze the lacteals, thereby assisting in the movement of lymph out of the villi.

Muscularis mucosae

Module 21.10 Review

a. Name the layers of the small intestine from superficial to deep.

b. Describe the anatomy of the intestinal mucosa.

c. Explain the function of lacteals.

The small intestine is divided into the duodenum, jejunum, and ileum

The small intestine plays the key role in the digestion and absorption of nutrients. Ninety percent of nutrient absorption occurs in the small intestine; most of the rest occurs in the large intestine.

1 The small intestine fills much of the peritoneal cavity, and its position is stabilized by the mesentery proper. The small intestine averages 6 m (19.7 ft) in length and has a diameter ranging from 4 cm (1.6 in.) at the stomach to about 2.5 cm (1 in.) at the junction with the large intestine.

Regions of the Small Intestine

The **duodenum** (doo-AH-de-num or doo-ō-DĒ-num), 25 cm (10 in.) in length, is the segment closest to the stomach. This portion of the small intestine receives chyme from the stomach and digestive secretions from the pancreas and liver. Except for the proximal 2.5 cm (1 in.), the duodenum is in a retroperitoneal position, firmly attached to the posterior body wall.

A rather abrupt bend marks the boundary between the duodenum and the **jejunum** (je-JOO-num). At this junction, the small intestine reenters the peritoneal cavity, supported by a sheet of mesentery. The jejunum is about 2.5 meters (8.2 ft) long. The bulk of chemical digestion and nutrient absorption occurs in the jejunum.

The **ileum** (IL-ē-um), the final segment of the small intestine, is also the longest, averaging 3.5 meters (11.5 ft) in length. The ileum ends at the **ileocecal** (il-ē-o-SĒ-kal) **valve**, a sphincter that controls the flow of material from the ileum into the cecum of the large intestine.

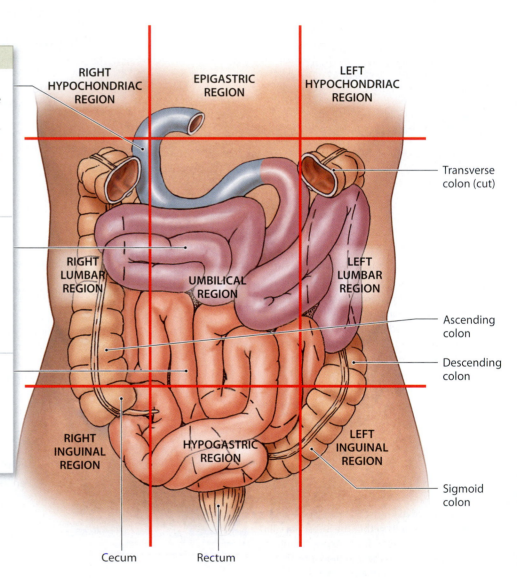

RIGHT HYPOCHONDRIAC REGION

EPIGASTRIC REGION

LEFT HYPOCHONDRIAC REGION

Transverse colon (cut)

RIGHT LUMBAR REGION

UMBILICAL REGION

LEFT LUMBAR REGION

Ascending colon

Descending colon

RIGHT INGUINAL REGION

HYPOGASTRIC REGION

LEFT INGUINAL REGION

Sigmoid colon

Cecum

Rectum

2 Each segment of the small intestine has characteristic features related to its primary functions. The transition from one region to another is gradual, and the boundaries are indistinct.

Duodenum

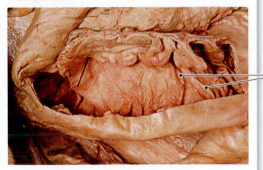

Plicae circulares

The duodenum has few plicae circulares, and their villi are small. The primary function of the duodenum is to receive chyme from the stomach and neutralize its acids before they can damage the absorptive surfaces of the small intestine.

Jejunum

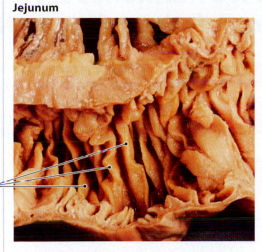

Plicae circulares

The jejunum has numerous plicae circulares and the villi are abundant and very long.

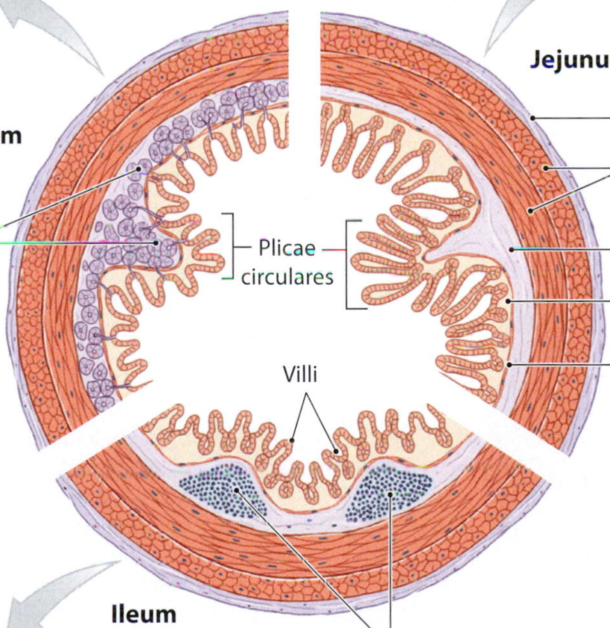

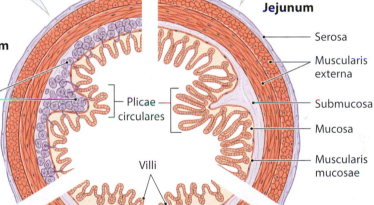

Duodenum

The submucosa of the duodenum is dominated by **duodenal glands** that produce mucous secretions.

Plicae circulares

Villi

Jejunum

Serosa

Muscularis externa

Submucosa

Mucosa

Muscularis mucosae

Ileum

Aggregated lymphoid nodules

Ileum

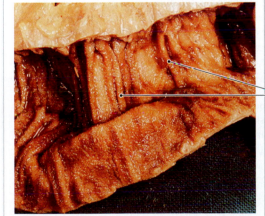

Plicae circulares

The ileum has few if any plicae circulares, and the villi are relatively stumpy. The submucosa contains aggregated lymphoid nodules.

Module 21.11 Review

a. Name the three regions of the small intestine from proximal to distal.

b. Identify the region of the small intestine found within the epigastric region.

c. What is the primary function of the duodenum?

Five hormones are involved in the regulation of digestive activities

1 Four of the five major hormones involved in regulating digestive function are produced by the duodenum, which receives
partially digested materials from the stomach. The duodenum adjusts
gastric activity and coordinates the secretions of accessory digestive
organs according to the characteristics of the arriving chyme.

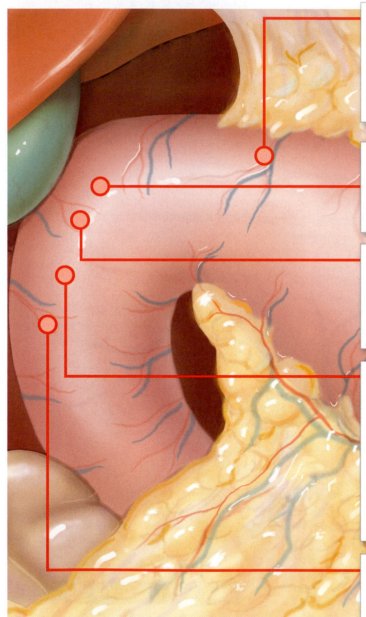

Gastrin is secreted by G cells in the pyloric antrum and enteroendocrine
cells in the duodenum. The pyloric antrum secretes gastrin when
stimulated by the vagus nerves or when food arrives in the stomach.
Duodenal cells release gastrin when they are exposed to large quantities
of incompletely digested proteins. The functions of gastrin include
promoting increased stomach motility and stimulating the production of
gastric acids and enzymes.

Secretin is released when chyme arrives in the duodenum. Secretin's
primary effect is an increase in the secretion of bile (by the liver) and
buffers (by the pancreas) which in turn act to increase the pH of the
chyme. Among its secondary effects, secretin reduces gastric motility
and secretory rates.

Gastric inhibitory peptide (GIP) is secreted when fats and
carbohydrates—especially glucose—enter the small intestine. The
inhibition of gastric activity is accompanied by the stimulation of insulin
release at the pancreatic islets. GIP has several secondary effects,
including stimulating duodenal gland activity, stimulating lipid synthesis
in adipose tissue, and increasing glucose use by skeletal muscles.

Cholecystokinin (CCK) is secreted when chyme arrives in the duodenum, especially when the chyme contains lipids and partially digested
proteins. In the pancreas, CCK accelerates the production and secretion
of all types of digestive enzymes. It also causes a relaxation of the
hepatopancreatic sphincter and contraction of the gallbladder, resulting
in the ejection of bile and pancreatic juice into the duodenum. Thus, the
net effects of CCK are to increase the secretion of pancreatic enzymes
and to push pancreatic secretions and bile into the duodenum. The
presence of CCK in high concentrations has two additional effects: It
inhibits gastric activity, and it appears to have CNS effects that reduce
the sensation of hunger.

Vasoactive intestinal peptide (VIP) stimulates the secretion of
intestinal glands, dilates regional capillaries, and inhibits acid production
in the stomach. By dilating capillaries in active areas of the intestinal tract,
VIP provides an efficient mechanism for removing absorbed nutrients.

2 This flowchart summarizes the pattern of hormone release and the effects of those hormones within the digestive system.

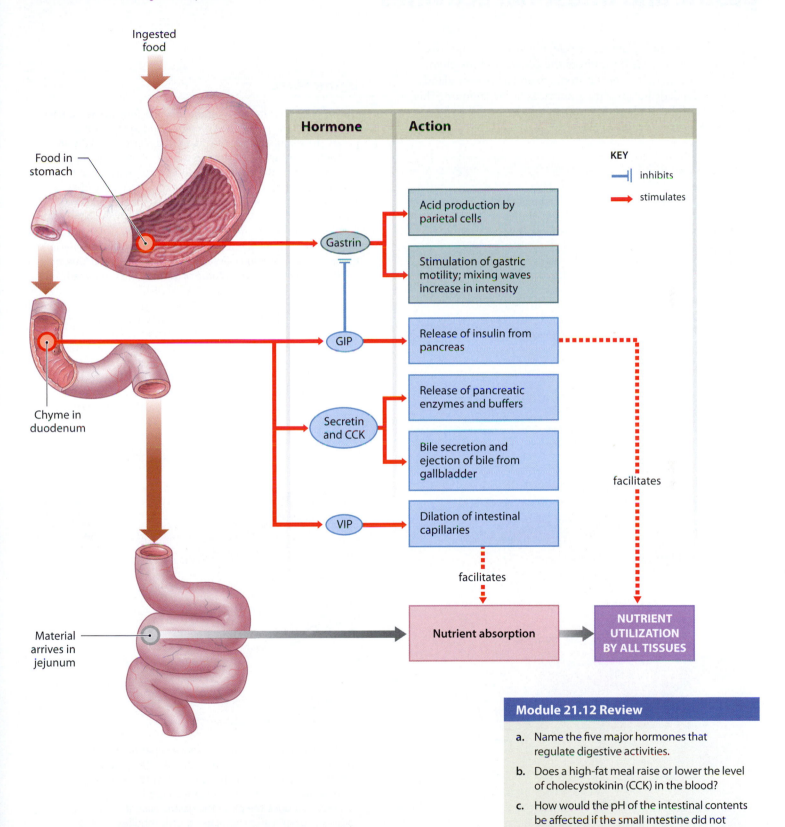

Ingested food

Food in stomach

Chyme in duodenum

Material arrives in jejunum

Hormone	Action

KEY

⊣ inhibits

→ stimulates

Gastrin
- Acid production by parietal cells
- Stimulation of gastric motility; mixing waves increase in intensity

GIP
- Release of insulin from pancreas

Secretin and CCK
- Release of pancreatic enzymes and buffers
- Bile secretion and ejection of bile from gallbladder

VIP
- Dilation of intestinal capillaries

facilitates

facilitates

Nutrient absorption

NUTRIENT UTILIZATION BY ALL TISSUES

Module 21.12 Review

a. Name the five major hormones that regulate digestive activities.

b. Does a high-fat meal raise or lower the level of cholecystokinin (CCK) in the blood?

c. How would the pH of the intestinal contents be affected if the small intestine did not produce secretin?

Central and local mechanisms coordinate gastric and intestinal activities

1 The duodenum plays a key role in controlling digestive function because it monitors the contents of the chyme and adjusts the activities of the stomach and accessory glands to protect the delicate absorptive surfaces of the jejunum. This pivotal role of the duodenum is apparent when you consider the three **phases of gastric secretion**.

Cephalic Phase

The **cephalic phase** of gastric secretion begins when you see, smell, taste, or think of food. This phase, which is directed by the CNS, prepares the stomach to receive food. The neural output proceeds by way of the parasympathetic division of the autonomic nervous system, and the vagus nerves innervate the submucosal plexus of the stomach. Next, postganglionic parasympathetic fibers innervate mucous cells, chief cells, parietal cells, and G cells of the stomach. In response to stimulation, the production of gastric juice accelerates, reaching rates of about 500 mL/h. This phase generally lasts only minutes.

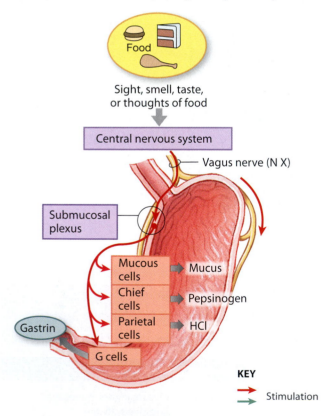

Gastric Phase

The **gastric phase** begins with the arrival of food in the stomach and builds on the stimulation provided during the cephalic phase. The stimuli that initiate the gastric phase are (1) distension of the stomach, (2) an increase in the pH of the gastric contents, and (3) the presence of undigested materials in the stomach, especially proteins and peptides. The gastric phase may continue for three to four hours while the acid and enzymes process the ingested materials. During this period, gastrin stimulates contractions in the muscularis externa of the stomach and intestinal tract. After the first hour, the material in the stomach is churning like clothing in a washing machine. As mixing continues, a large volume of gastric juice is secreted.

The stimulation of stretch receptors and chemoreceptors triggers short reflexes coordinated in the submucosal and myenteric plexuses. This in turn activates the stomach's secretory cells. The stimulation of the myenteric plexus produces powerful contractions called **mixing waves** in the muscularis externa.

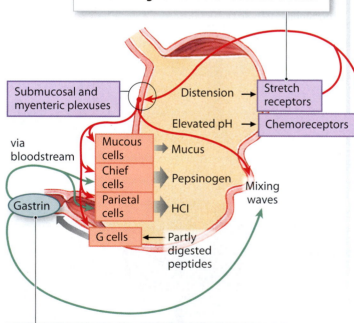

Neural stimulation and the presence of peptides and amino acids in chyme stimulate the secretion of the hormone gastrin. Gastrin travels via the bloodstream to parietal and chief cells, whose increased secretions reduce the pH of the gastric juice. In addition, gastrin also stimulates gastric motility.

Intestinal Phase

The **intestinal phase** of gastric secretion begins when chyme first enters the small intestine, usually after several hours of mixing contractions. The function of the intestinal phase is controlling the rate of gastric emptying to ensure that the secretory, digestive, and absorptive functions of the small intestine can proceed with reasonable efficiency. Although here we consider the intestinal phase as it affects stomach activity, the arrival of chyme in the small intestine also triggers other neural and hormonal events that coordinate the activities of the intestinal tract and the pancreas, liver, and gallbladder.

Chyme leaving the stomach decreases the distension in the stomach, thereby reducing the stimulation of stretch receptors. Distension of the duodenum by chyme stimulates stretch receptors and chemoreceptors that trigger the **enterogastric reflex**. This reflex inhibits both gastrin production and gastric contractions and stimulates the contraction of the pyloric sphincter, which prevents further discharge of chyme. At the same time, local reflexes at the duodenum stimulate mucus production, which helps protect the duodenal lining from the arriving acid and enzymes.

2 Two central reflexes are triggered by the stimulation of stretch receptors in the stomach wall as it fills. These reflexes accelerate movement along the small intestine while the enterogastric reflex controls the rate of chyme entry into the duodenum.

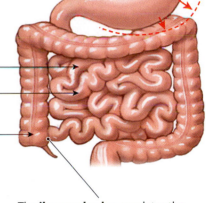

Central Gastric Reflexes

The **gastroenteric reflex** stimulates motility and secretion along the entire small intestine.

The **gastroileal** (gas-trō-IL-ē-al) **reflex** triggers the opening of the ileocecal valve, allowing materials to pass from the small intestine into the large intestine.

The **ileocecal valve** regulates the passage of materials into the large intestine.

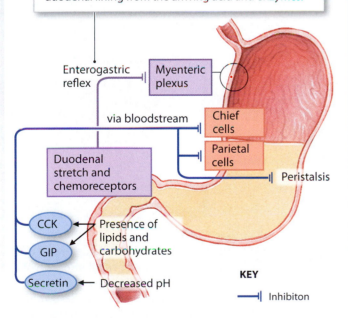

KEY

⊣ Inhibiton

Module 21.13 Review

a. Name and briefly describe an important characteristic of each of the three overlapping phases of gastric secretion.

b. Describe two central reflexes triggered by stimulation of the stretch receptors in the stomach wall.

c. Why might severing the branches of the vagus nerves that supply the stomach provide relief for a person who suffers from chronic gastric ulcers (sores on the stomach lining)?

The large intestine stores and concentrates fecal material

The **large intestine**, also known as the large bowel, has an average length of about 1.5 meters (4.9 ft) and a width of 7.5 cm (3 in.). The major functions of the large intestine include (1) the reabsorption of water and the compaction of the intestinal contents into feces, (2) the absorption of important vitamins liberated by bacterial action, and (3) the storage of fecal material prior to defecation. The large intestine consists of three segments: the cecum, the colon, and the rectum.

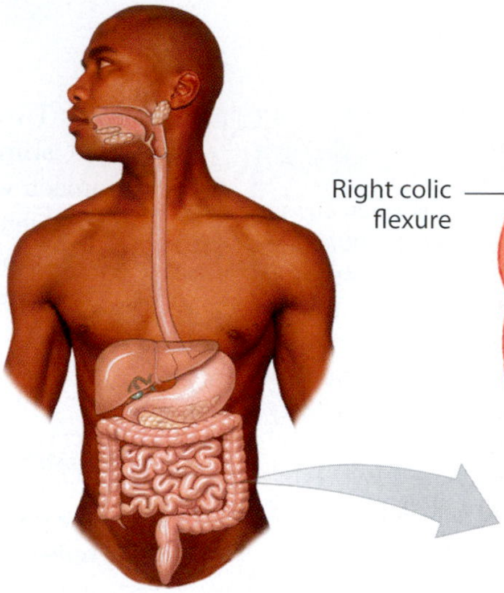

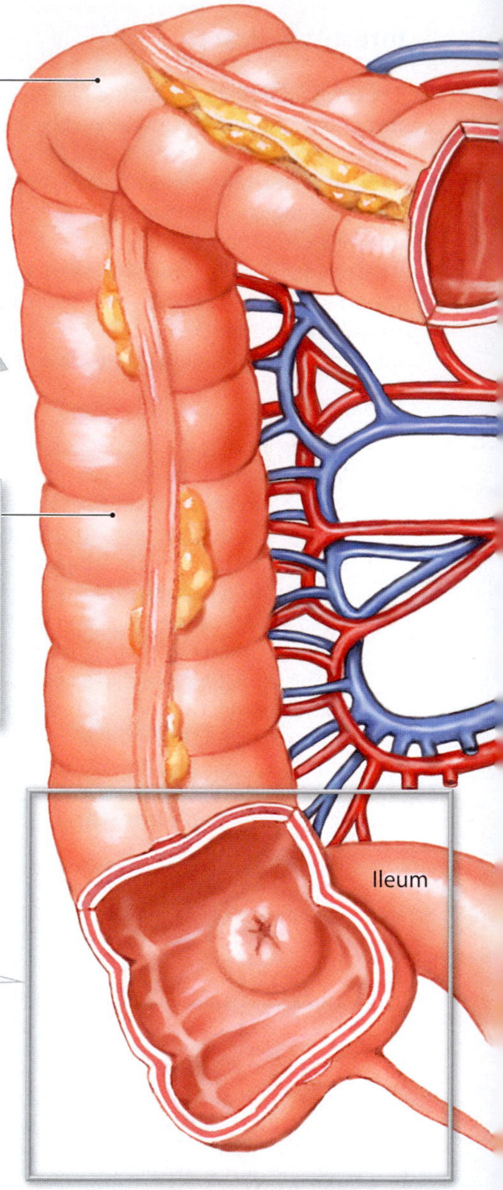

1 Material arriving from the ileum first enters an expanded pouch called the **cecum** (SĒ-kum). The cecum collects and stores materials from the ileum and begins the process of **compaction** (the forming of feces by compression).

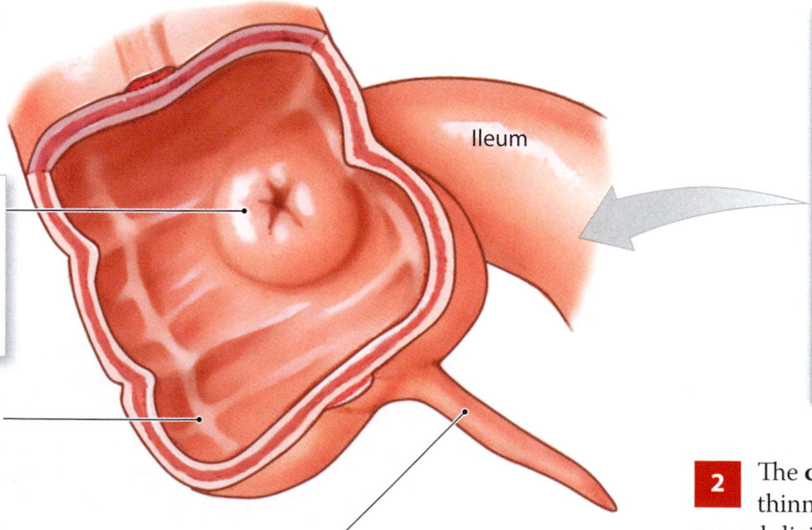

Right colic flexure

The **ascending colon** begins at the cecum and ascends along the right margin of the peritoneal cavity to the inferior surface of the liver. There, the colon bends sharply to the left at the **right colic flexure**, and this marks the end of the ascending colon.

Ileum

The ileum attaches to the medial surface of the cecum and opens into the cecum at the **ileocecal valve**.

Ileum

Cecum

The slender, hollow **appendix** is attached to the cecum. The appendix is generally about 9 cm (3.6 in.) long, but its size and shape are quite variable. The mucosa and submucosa of the appendix are dominated by lymphoid nodules, and the appendix functions primarily as an organ of the lymphatic system. Inflammation of the appendix is known as **appendicitis**.

2 The **colon** has a larger diameter and a thinner wall than the small intestine. We can subdivide the colon into four regions: the ascending colon, transverse colon, descending colon, and sigmoid colon. The ascending and descending colon are retroperitoneal and firmly attached to the abdominal wall. The transverse colon and sigmoid colon are suspended by remnants of the embryonic mesocolon.

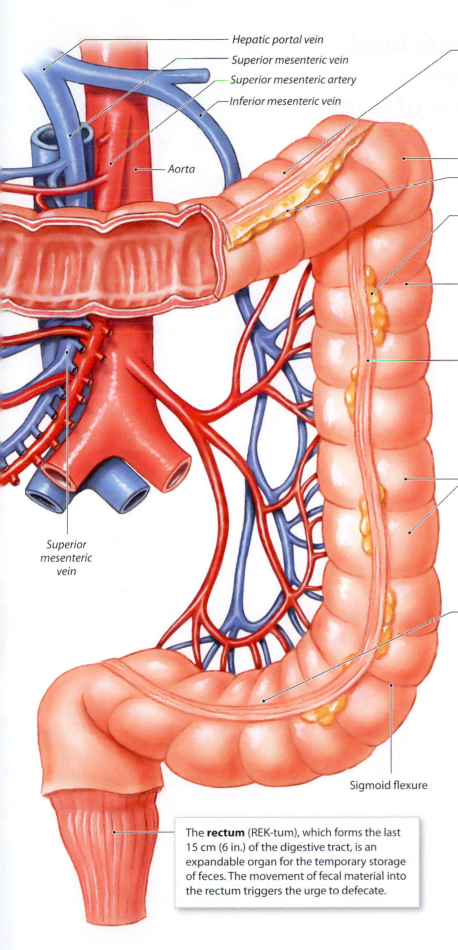

Hepatic portal vein

Superior mesenteric vein

Superior mesenteric artery

Inferior mesenteric vein

Aorta

Superior
mesenteric
vein

The **transverse colon** crosses the abdomen from right to left. It is supported by the transverse mesocolon and is separated from the anterior abdominal wall by the layers of the greater omentum. As the transverse colon reaches the left side of the body, the colon makes a 90° turn at the **left colic flexure**.

Left colic flexure

Greater omentum (cut)

The serosa of the colon contains numerous teardrop-shaped sacs of fat called **fatty appendices**.

The **descending colon** proceeds inferiorly along the body's left side until reaching the iliac fossa. At the iliac fossa, the descending colon ends at the **sigmoid flexure**.

Three separate longitudinal bands of smooth muscle—called the **taeniae coli** (TĒ-nē-ē KŌ-lē; singular, *taenia*)—run along the outer surfaces of the colon just deep to the serosa. These bands correspond to the outer layer of the muscularis externa in other portions of the digestive tract.

Muscle tone within the taeniae coli is what creates **haustra** (HAWS-truh), a series of pouches in the wall of the colon. Cutting into the intestinal lumen reveals that the creases between the haustra affect the mucosal lining as well, producing a series of internal folds. Haustra permit the expansion and elongation of the colon, rather like the bellows that allow an accordion to lengthen.

The sigmoid flexure is the start of the **sigmoid** (SIG-moyd; *sigmeidos*, Greek letter S) **colon**, an S-shaped segment that is about 15 cm (6 in.) long and empties into the rectum.

Powerful peristaltic contractions called **mass movements** occur a few times each day in response to distension of the stomach and duodenum. These contractions begin at the transverse colon and push materials along the distal portion of the large intestine.

Sigmoid flexure

The **rectum** (REK-tum), which forms the last 15 cm (6 in.) of the digestive tract, is an expandable organ for the temporary storage of feces. The movement of fecal material into the rectum triggers the urge to defecate.

Module 21.14 Review

a. Identify the segments of the large intestine and the four regions of the colon.

b. Name the major functions of the large intestine.

c. Describe mass movements.

The large intestine compacts fecal material; the defecation reflex coordinates the elimination of feces

1 Although the diameter of the large intestine is roughly three times that of the small intestine, its wall is much thinner. The major characteristics of the wall of the large intestine are the lack of villi, the abundance of mucous cells, and the presence of distinctive intestinal glands dominated by mucous cells. The mucosa of the large intestine does not produce enzymes; any digestion that occurs results from enzymes secreted into the small intestine or released by bacterial action. The mucus provides lubrication as the fecal material becomes drier and more compact.

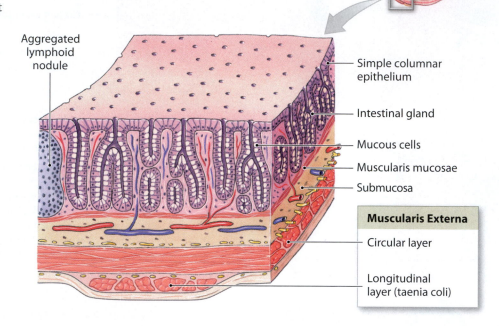

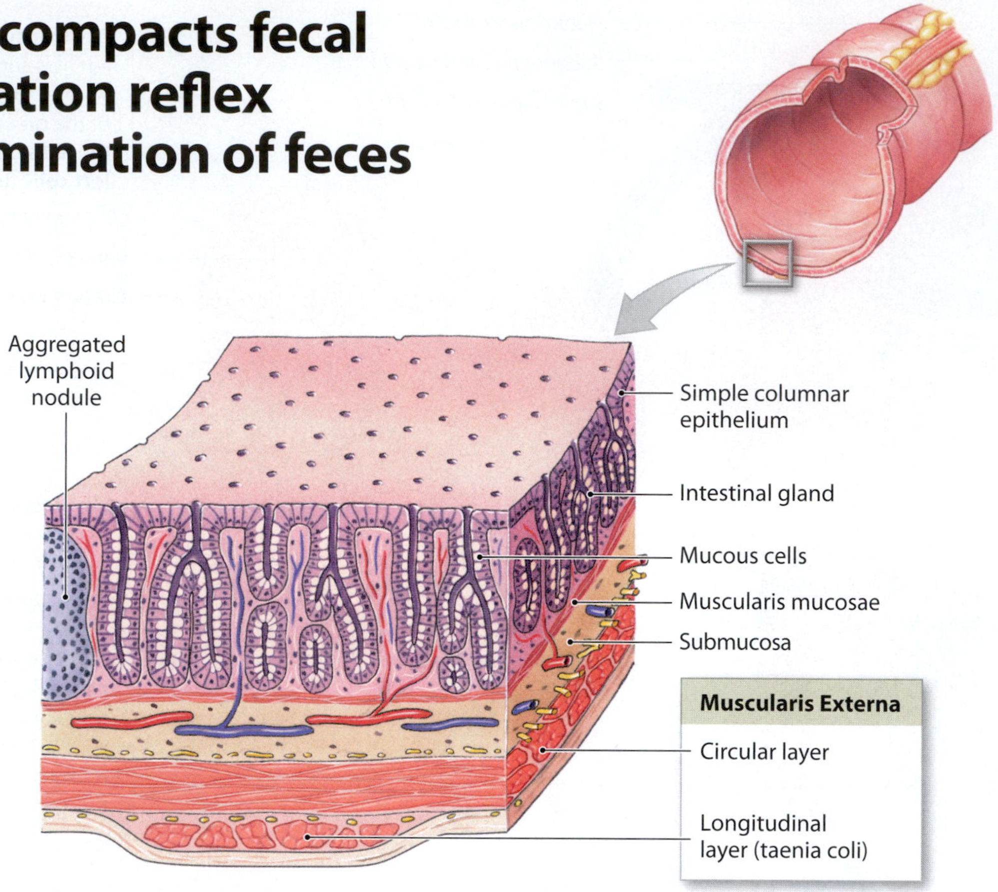

Aggregated lymphoid nodule

Simple columnar epithelium

Intestinal gland

Mucous cells

Muscularis mucosae

Submucosa

Muscularis Externa

Circular layer

Longitudinal layer (taenia coli)

2 This illustration shows the characteristic features of the rectum, the last segment of the digestive tract. The lamina propria and submucosa of the distal portion of the rectum contain a network of veins. If venous pressures there rise too high due to straining during defecation, the veins can become distended, producing **hemorrhoids**.

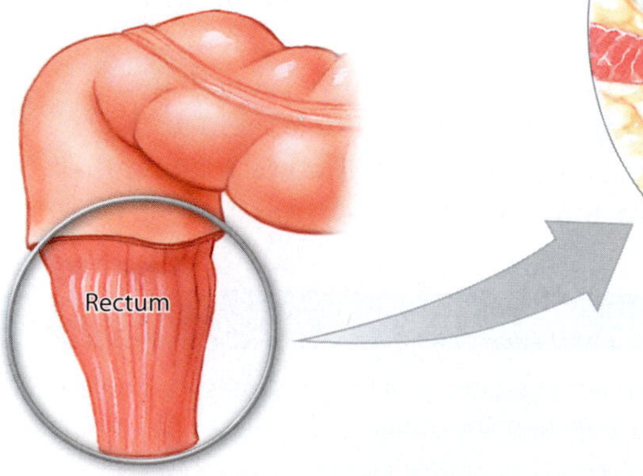

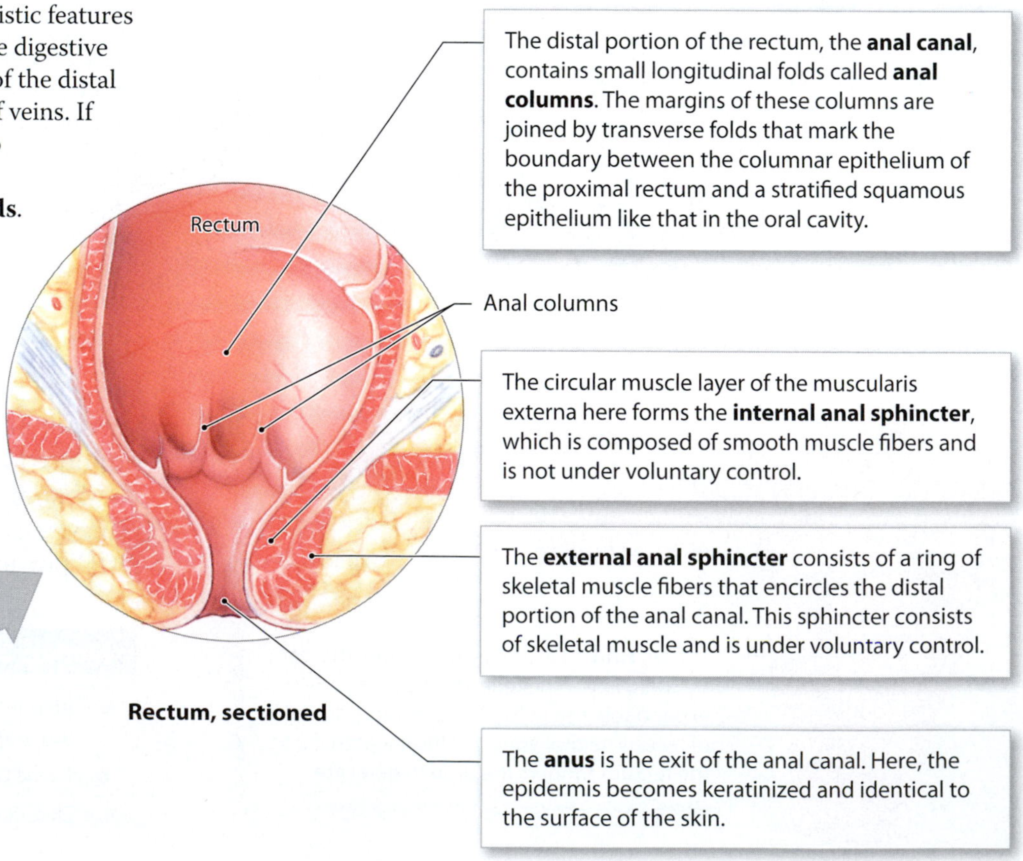

The distal portion of the rectum, the **anal canal**, contains small longitudinal folds called **anal columns**. The margins of these columns are joined by transverse folds that mark the boundary between the columnar epithelium of the proximal rectum and a stratified squamous epithelium like that in the oral cavity.

Rectum

Anal columns

The circular muscle layer of the muscularis externa here forms the **internal anal sphincter**, which is composed of smooth muscle fibers and is not under voluntary control.

The **external anal sphincter** consists of a ring of skeletal muscle fibers that encircles the distal portion of the anal canal. This sphincter consists of skeletal muscle and is under voluntary control.

Rectum

Rectum, sectioned

The **anus** is the exit of the anal canal. Here, the epidermis becomes keratinized and identical to the surface of the skin.

3 Less than 10 percent of the nutrient absorption under way in the digestive tract occurs in the large intestine. Nevertheless, the absorptive operations in this segment of the digestive tract are very important. In addition to preventing dehydration by reabsorbing water, the epithelium absorbs three vitamins produced by the normal bacterial residents of the colon.

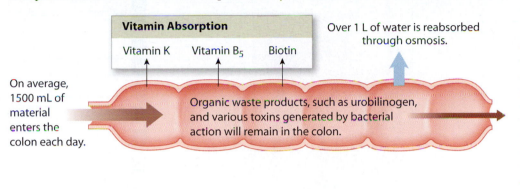

On average, 1500 mL of material enters the colon each day.

Vitamin Absorption

Vitamin K   Vitamin B$_5$ Biotin

Organic waste products, such as urobilinogen, and various toxins generated by bacterial action will remain in the colon.

Over 1 L of water is reabsorbed through osmosis.

Only 200 mL of feces is ejected. Fecal material is 75% water, 5% bacteria, and the rest a mixture of indigestible materials, inorganic matter, and the remains of epithelial cells. Bacterial action produces several compounds that contribute to the odor of feces, including ammonia, **indole** and **skatole** (two nitrogen-containing compounds), and **hydrogen sulfide** (H_2S), a gas that produces a "rotten egg" odor.

4 The rectal chamber is usually empty, except when a powerful peristaltic contraction forces feces out of the sigmoid colon. Distension of the rectal wall then starts the **defecation reflex**, which involves two positive feedback loops, triggered by the stimulation of stretch receptors in the walls of the rectum.

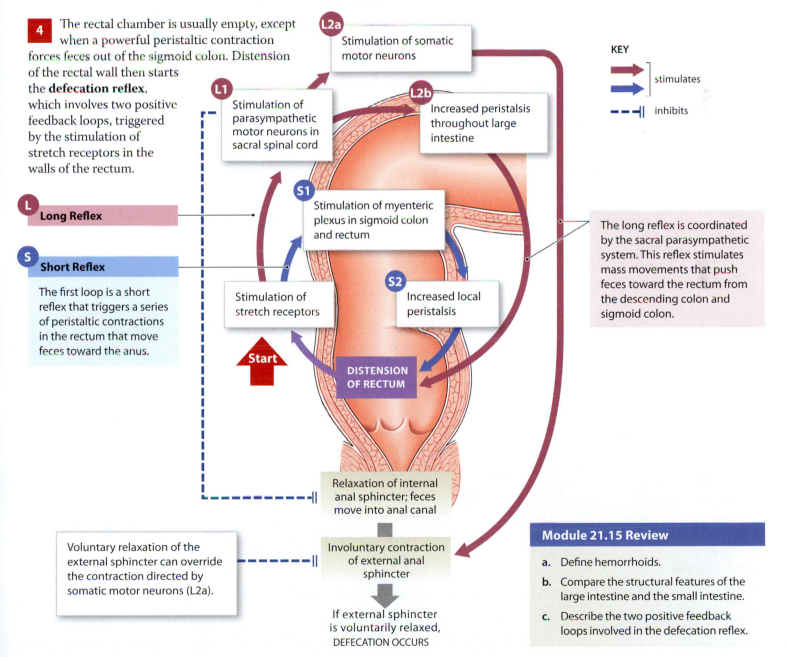

L2a Stimulation of somatic motor neurons

L1 Stimulation of parasympathetic motor neurons in sacral spinal cord

L2b Increased peristalsis throughout large intestine

KEY

stimulates

inhibits

L Long Reflex

S Short Reflex

The first loop is a short reflex that triggers a series of peristaltic contractions in the rectum that move feces toward the anus.

S1 Stimulation of myenteric plexus in sigmoid colon and rectum

S2 Increased local peristalsis

Stimulation of stretch receptors

Start

DISTENSION OF RECTUM

The long reflex is coordinated by the sacral parasympathetic system. This reflex stimulates mass movements that push feces toward the rectum from the descending colon and sigmoid colon.

Relaxation of internal anal sphincter; feces move into anal canal

Voluntary relaxation of the external sphincter can override the contraction directed by somatic motor neurons (L2a).

Involuntary contraction of external anal sphincter

If external sphincter is voluntarily relaxed, DEFECATION OCCURS

Module 21.15 Review

a. Define hemorrhoids.

b. Compare the structural features of the large intestine and the small intestine.

c. Describe the two positive feedback loops involved in the defecation reflex.

1. Labeling

Label the structures of a typical tooth in the following figure.

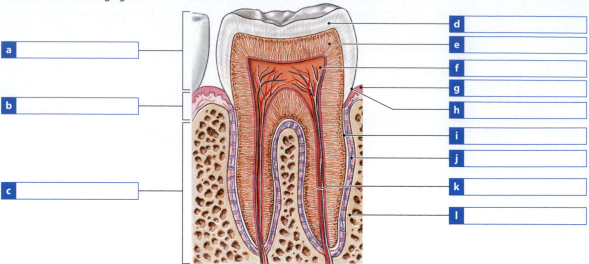

a	
b	
c	

d	
e	
f	
g	
h	
i	
j	
k	
l	

2 . Concept map

Using the following terms, fill in the blank boxes to complete the hormones of digestive activity concept map.

- acid production
- gastrin
- VIP
- insulin
- intestinal capillaries
- GIP

- material in jejunum
- gallbladder
- bile
- inhibits
- secretin and CCK
- nutrient utilization by tissues

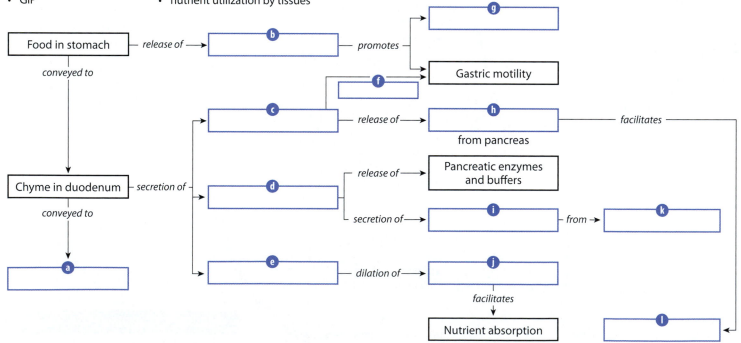

3. Short answer

Briefly describe the similarities and differences between parietal cells and chief cells in the stomach wall. _____

_____

Accessory Digestive Organs

The major accessory digestive organs are the salivary glands, the gallbladder, the pancreas, and the liver. The salivary glands and pancreas produce and store enzymes and buffers that are essential to normal digestive function. In addition to their roles in digestion, the salivary glands, liver, and pancreas have vital metabolic and endocrine functions.

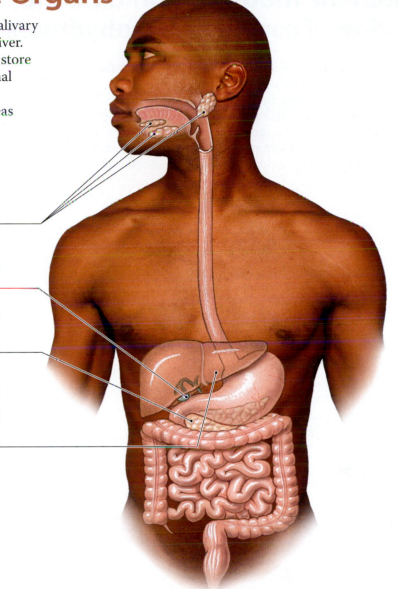

Accessory Digestive Organs

Salivary Glands

Three pairs of salivary glands produce secretions that contain mucins and enzymes.

Gallbladder

The gallbladder stores and concentrates bile secreted by the liver.

Pancreas

Exocrine cells secrete buffers and digestive enzymes; endocrine cells secrete insulin, glucagon, pancreatic polypeptide, and GH–IH, hormones introduced in Module 16.9.

Liver

The liver has almost 200 known functions. The most important are listed in the table below.

Digestive and Metabolic Functions of the Liver

- Synthesis and secretion of bile
- Storage of glycogen and lipid reserves
- Maintenance of normal concentrations of glucose, amino acids, and fatty acids in the bloodstream
- Synthesis and interconversion of nutrient types (such as the conversion of carbohydrates to lipids)
- Synthesis and release of cholesterol bound to transport proteins
- Inactivation of toxins
- Storage of iron reserves
- Storage of fat-soluble vitamins

Other Major Functions

- Synthesis of plasma proteins
- Synthesis of clotting factors
- Phagocytosis of damaged red blood cells (by Kupffer cells)
- Storage of blood
- Absorption and breakdown of circulating hormones and immunoglobulins
- Absorption and inactivation of lipid-soluble drugs

The salivary glands lubricate and moisten the mouth and initiate the digestion of complex carbohydrates

1 Three pairs of salivary glands secrete into the oral cavity. Each pair has a distinctive cellular organization and produces saliva with slightly different properties. Any object in your mouth can trigger a salivary reflex by stimulating receptors monitored by the trigeminal nerve (V) or taste buds innervated by cranial nerves VII, IX, or X. Parasympathetic stimulation accelerates secretion by all the salivary glands, resulting in the production of large amounts of saliva.

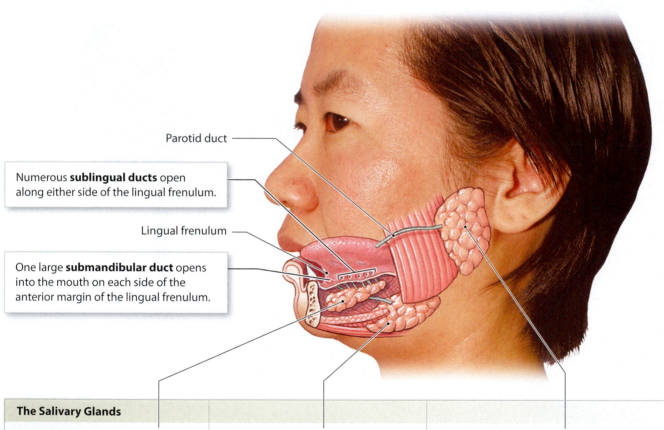

Parotid duct

Numerous **sublingual ducts** open along either side of the lingual frenulum.

Lingual frenulum

One large **submandibular duct** opens into the mouth on each side of the anterior margin of the lingual frenulum.

The Salivary Glands

The **sublingual** (sub-LING-gwal) **salivary glands** are covered by the mucous membrane of the floor of the mouth. These glands produce a mucous secretion that acts as a buffer and lubricant.

The **submandibular salivary glands** are in the floor of the mouth along the inner surfaces of the mandible within the mandibular groove. Cells of the submandibular glands secrete a mixture of buffers, mucins, and **salivary amylase**, an enzyme that breaks down starches (complex carbohydrates). The gland cells also transport antibodies (IgA) into the saliva, to provide additional protection against pathogens in food.

The large **parotid** (pa-ROT-id) **salivary glands** lie inferior to the zygomatic arch deep to the skin covering the lateral and posterior surface of the mandible. Each gland has an irregular shape, extending from the mastoid process of the temporal bone across the outer surface of the masseter muscle. The parotid salivary glands produce a secretion containing large amounts of salivary amylase. The secretions of each parotid gland are drained by a **parotid duct**, which empties into the vestibule at the level of the second upper molar.

2 Each submandibular salivary gland contains a mixture of secretory cells, some specialized for mucous secretion and others specialized for enzyme production. The **saliva** in the mouth is a mixture of glandular secretions; about 70 percent of the saliva originates in the submandibular salivary glands, 25 percent from the parotid salivary glands, and 5 percent from the sublingual salivary glands. Collectively the salivary glands produce 1.0–1.5 L of saliva each day; 99.4 percent of that volume is water.

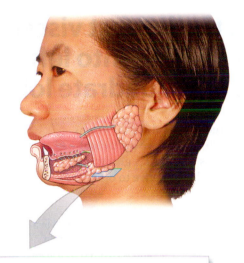

Ducts collect the secretions, and the duct cells assist in the secretion of buffers and antibodies.

Mucous cells secrete mucins, water, and buffers.

Serous cells secrete salivary amylase and **lysozyme**, an antibacterial enzyme. They also transport antibodies from the interstitial fluid into the saliva.

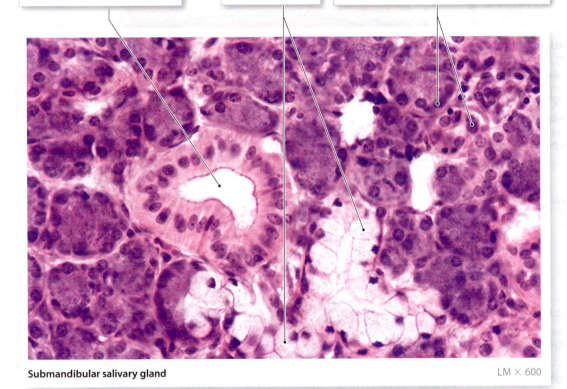

Submandibular salivary gland   LM × 600

A continuous background level of saliva secretion flushes the oral surfaces, helping keep them clean. Buffers in the saliva keep the pH of your mouth near 7.0 and prevent the buildup of acids produced by bacterial action. In addition, saliva contains antibodies (IgA) and lysozyme, which help control populations of oral bacteria. Food usually remains in the mouth long enough for chewing to mix it with saliva and break the combination into a relatively homogeneous, pulpy mass. This is compacted by the tongue to form a bolus that can be easily swallowed.

Module 21.16 Review

a. Name the three pairs of salivary glands.

b. Which glandular secretions contribute least to saliva production?

c. The digestion of which nutrient would be affected by damage to the parotid salivary glands?

1. Labeling

Label each of the structures of a liver lobule in the following diagram.

a _____

b _____

c _____

d _____

e _____

f _____

g _____

h _____

i _____

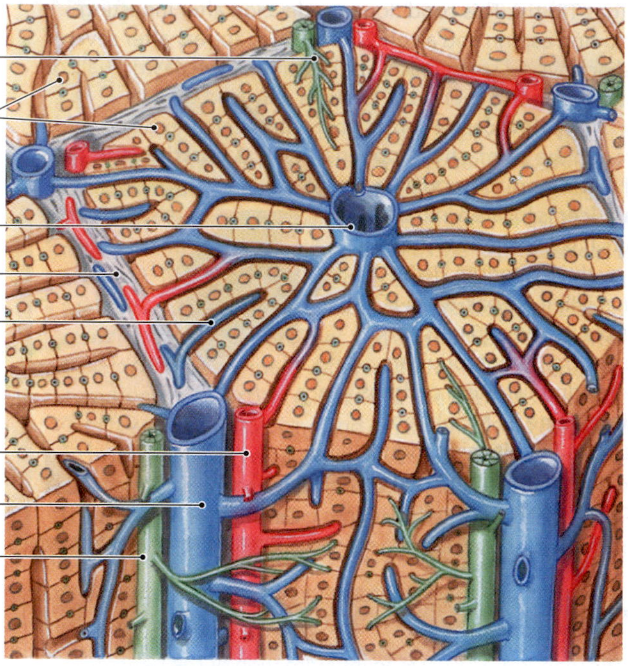

2. Matching

Match the following terms with the most closely related description.

- lysozyme
- emulsification
- gallstones
- Kupffer cells
- pancreatic lipase
- liver
- starch
- pancreas
- submandibular glands
- hepatocytes
- gallbladder
- mumps
- common bile duct
- peptic ulcer

a _____ Pancreatic alpha-amylase

b _____ Retroperitoneal organ

c _____ Drains liver and gallbladder

d _____ Bile-secreting cells

e _____ Viral infection of salivary glands

f _____ Digestive epithelial damage by acids

g _____ Process of breaking lipid droplets apart

h _____ Pancreatic enzyme that breaks down complex lipids

i _____ Organ that secretes bile continuously

j _____ Antibacterial enzyme

k _____ Greatest producer of saliva

l _____ Phagocytize and store iron

m _____ Stores bile

n _____ Cholecystitis

3. Short answer

Describe the beneficial roles of saliva. _____

_____

4. Section integration

Predict the consequences of a blockage of the duodenal ampulla by a tumor. _____

_____

Visual Outline with Key Terms

Summarize the content of each module using the terms in the order provided.

SECTION 1

General Organization of the Digestive System

- digestive system
- digestive tract
- gastrointestinal (GI) tract

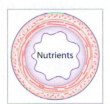

21.1

The digestive tract is a muscular tube lined by a mucous epithelium

- mesentery
- mucosa
- submucosa
- muscularis externa
- serosa
- adventitia
- secretory glands
- plicae circulares
- villi
- muscularis mucosae
- submucosal plexus
- myenteric plexus

21.2

Smooth muscle tissue is found throughout the body, but it plays a particularly prominent role in the digestive tract

- dense bodies
- multi-unit smooth muscle cells
- visceral smooth muscle cells
- pacesetter cells
- plasticity
- smooth muscle tone

21.3

Smooth muscle contractions mix the contents of the digestive tract and propel materials along its length

- bolus
- peristalsis
- segmentation
- myenteric reflexes
- enteroendocrine cells

SECTION 2

The Digestive Tract

- oral cavity, teeth, tongue
- pharynx
- esophagus
- stomach
- small intestine
- large intestine
- ingestion
- mechanical processing
- digestion
- secretion
- absorption
- compaction
- feces
- defecation

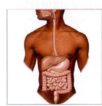

21.4

The oral cavity contains the tongue, salivary glands, and teeth, and receives secretions of the salivary glands

- oral cavity
- oral mucosa
- hard palate
- soft palate
- uvula
- root (of the tongue)
- labia
- cheeks
- body (of the tongue)
- vestibule
- labial frenulum
- lingual frenulum
- pharyngeal arches
- fauces
- tongue
- lingual lipase
- gingivae

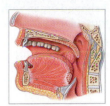

21.5

Teeth in different regions of the jaws vary in size, shape, and function

- dentin
- pulp cavity
- crown
- neck
- root
- alveolus
- occlusal surface
- enamel
- gingival sulcus
- cementum
- periodontal ligament
- gomphosis
- root canal
- apical foramen
- incisors
- cuspids
- bicuspids
- molars
- deciduous teeth
- primary dentition
- secondary dentition
- wisdom teeth
- dental arcades
- gingivitis
- tooth decay
- dental plaque

21.6

The muscular walls of the pharynx and esophagus play a key role in swallowing

- pharynx
- esophagus
- adventitia
- deglutition
- buccal phase
- pharyngeal phase
- esophageal phase
- secondary peristaltic waves
- esophageal hiatus
- upper esophageal sphincter
- lower esophageal sphincter

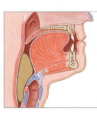

21.7

The stomach and most of the intestinal tract are suspended by mesenteries within the peritoneal cavity

- peritoneal cavity
- visceral peritoneum
- parietal peritoneum
- dorsal and ventral mesenteries
- greater omentum
- lesser omentum
- falciform ligament
- mesentery proper
- mesocolon
- ascites

• = *Term boldfaced in this module*

21.8

The stomach is a muscular, expandable, J-shaped organ with three layers in the muscularis externa

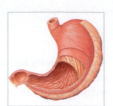

- lesser curvature
- greater curvature
- fundus
- cardia
- body
- pylorus
- pyloric antrum
- pyloric canal
- pyloric sphincter
- rugae

21.9

The stomach breaks down the organic nutrients in ingested materials

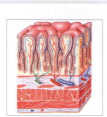

- mucosa
- submucosa
- muscularis externa
- serosa
- gastric glands
- parietal cells
- chief cells
- gastric pits
- intrinsic factor
- G cells
- pepsinogen
- pepsin
- rennin
- gastric lipase
- alkaline tide

21.10

The intestinal tract is specialized for the absorption of nutrients

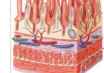

- plicae circulares
- intestinal villi
- intestinal glands
- lacteal
- brush border

21.11

The small intestine is divided into the duodenum, jejunum, and ileum

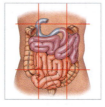

- duodenum
- jejunum
- ileum
- ileocecal valve
- plicae circulares
- duodenal glands

21.12

Five hormones are involved in the regulation of digestive activities

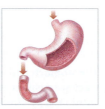

- gastrin
- secretin
- gastric inhibitory peptide (GIP)
- cholecystokinin (CCK)
- vasoactive intestinal peptide (VIP)

21.13

Central and local mechanisms coordinate gastric and intestinal activities

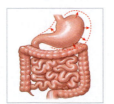

- phases of gastric secretion
- cephalic phase
- gastric phase
- mixing waves
- intestinal phase
- enterogastric reflex
- gastroenteric reflex
- gastroileal reflex
- ileocecal valve

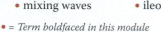
• = Term boldfaced in this module

21.14

The large intestine stores and concentrates fecal material

- large intestine
- cecum
- compaction
- ileocecal valve
- appendix
- appendicitis
- colon
- ascending colon
- right colic flexure
- transverse colon
- left colic flexure
- fatty appendices
- descending colon
- sigmoid flexure
- taeniae coli
- haustra
- sigmoid colon
- rectum
- mass movements

21.15

The large intestine compacts fecal material; the defecation reflex coordinates the elimination of feces

- hemorrhoids
- anal canal
- anal columns
- internal anal sphincter
- external anal sphincter
- anus
- indole
- skatole
- hydrogen sulfide
- defecation reflex
- long reflex
- short reflex

SECTION 3

Accessory Digestive Organs

- salivary glands
- gallbladder
- pancreas
- liver

21.16

The salivary glands lubricate and moisten the mouth and initiate the digestion of complex carbohydrates

- sublingual salivary glands
- submandibular salivary glands
- salivary amylase
- parotid salivary glands
- parotid duct
- sublingual ducts
- submandibular duct
- saliva
- serous cells
- lysozyme
- mucous cells

21.17

The liver, the largest visceral organ, is divided into left, right, caudate, and quadrate lobes

- liver
- falciform ligament
- left lobe
- right lobe
- porta hepatis
- coronary ligament
- bare area
- round ligament
- caudate lobe
- quadrate lobe
- gallbladder
- common bile duct

21.18

The liver tissues have an extensive and complex blood supply

- liver lobules
- hepatocytes
- portal areas
- liver sinusoids
- interlobular septum
- portal triad
- Kupffer cells
- central vein
- bile
- bile canaliculi
- bile ductules
- bile ducts
- viral hepatitis
- portal hypertension

21.19

The gallbladder stores and concentrates bile, and the pancreas has vital endocrine and exocrine functions

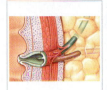

- gallbladder
- fundus, body, and neck of gallbladder
- right and left hepatic ducts
- common hepatic duct
- cystic duct
- common bile duct
- duodenal ampulla
- duodenal papilla
- hepatopancreatic sphincter
- emulsification
- pancreas
- head (of the pancreas)
- pancreatic duct
- pancreatic juice
- pancreatic lobules
- accessory pancreatic duct
- pancreatic acini
- pancreatic acinar cells
- pancreatic alpha-amylase
- pancreatic lipase
- nucleases
- proteolytic enzymes

• = *Term boldfaced in this module*

21.20

Disorders of the digestive system are diverse and relatively common

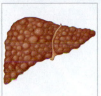

- periodontal disease
- gingivitis
- mumps virus
- mumps
- esophagitis
- gastroesophageal reflux
- hepatitis
- cirrhosis
- hepatitis A, B, and C
- jaundice
- gallstones
- cholecystitis
- gastritis
- peptic ulcer
- gastric ulcer
- duodenal ulcer
- cimetidine
- pancreatitis
- enteritis
- diarrhea
- dysentery
- gastroenteritis
- colitis
- constipation
- colorectal cancer
- polyps

Chapter Integration: Applying what you've learned

Obesity is a medical condition in which excess body fat adversely affects the quality of life, leading to increased health problems and decreased life expectancy. According to 2007 statistics reported by the Centers for Disease Control and Prevention (CDC), 25.6 percent of Americans are obese. Moreover, despite public awareness and health initiatives, not one of the 50 states met the government-sponsored Healthy People 2010 goal of reducing the percentage of the population that is obese to 15 percent.

Because obesity is a leading preventable cause of death, more people have turned to surgery to help them lose weight. One form of weight control surgery involves gastric stapling. In this procedure, a large portion of the gastric lumen is stapled shut, leaving only a small pouch in contact with the esophagus and duodenum.

Gastric bypass is another surgical procedure to induce weight loss. In this surgery, the proximal small intestine (duodenum) is connected to a small pouch formed by a superior portion of the stomach. Although this procedure seems more effective than gastric stapling, it involves more complicated surgery.

A third procedure, adjustable gastric band surgery (also known as lap-band), involves placing an inflatable silicone device around the superior aspect of the stomach via laparoscopic surgery. This procedure is performed on obese patients with a body mass index (BMI) of 35–40 or greater. (The body mass index is a ratio calculated by dividing one's weight in kilograms by the square of one's height in meters; it is used to identify individuals who are overweight or underweight.)

After each type of surgery, the stomach will hold approximately 110–220 grams of food at each meal, compared to 1500 grams for a normal, distended stomach. Although surgical intervention is successful in many cases, the risks for potential complications are also high.

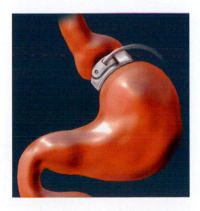

1. How would gastric stapling result in weight loss?

2. Explain the roles of the stretch receptors and the feeling of fullness after the gastric stapling procedure.

3. Using your knowledge of smooth muscle tissue, what will happen to the gastric muscularis externa over time after gastric stapling?

4. How do gastric bypass surgery and lap-band surgery help achieve the goal of weight loss?

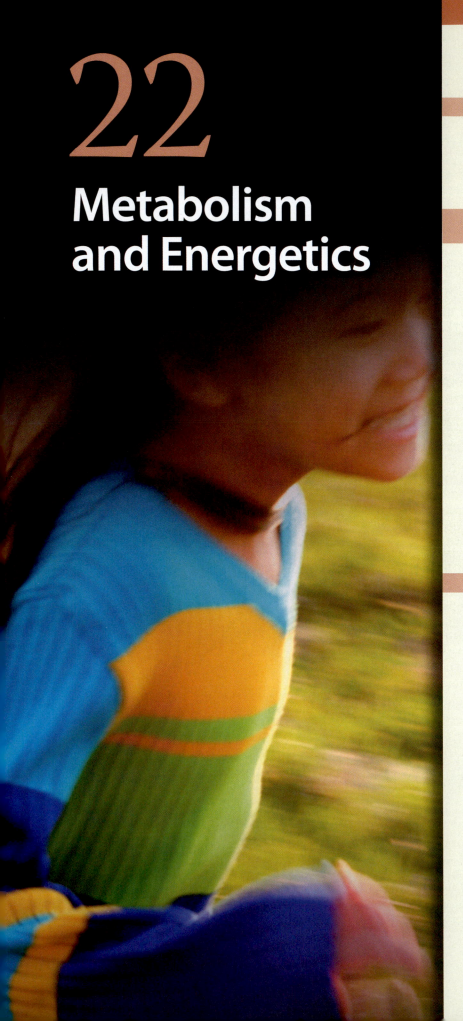

22

Metabolism and Energetics

An Introduction to Cellular Metabolism

The term **metabolism** (me-TAB-ō-lizm) refers to all the chemical reactions that occur in an organism. Chemical reactions within cells, collectively known as **cellular metabolism**, provide the energy needed to maintain homeostasis and to perform essential functions. As noted in Chapter 2, **catabolism** is the breakdown of organic substrates in the body, whereas **anabolism** is the synthesis of new organic molecules.

An Overview of Cellular Metabolism

In the process of **metabolic turnover**, cells continuously break down and replace all their organic components except DNA. The catabolic reactions involved provide negligible amounts of ATP to the cell. All of the cell's organic building blocks form a **nutrient pool**—an accessible reserve of organic substrates that can be used for metabolic turnover or energy production.

Cells continuously absorb organic molecules from the surrounding interstitial fluids, supplementing those released through catabolic reactions in metabolic turnover. The components of the nutrient pool can be either used for anabolism or broken down further for ATP production. This section considers the origins and fate of the nutrient pool, beginning with an overview of the catabolic pathways that provide ATP.

In mitochondria, the catabolic reactions of aerobic metabolism release significant amounts of energy. Roughly 40 percent of the energy released in these catabolic reactions can be captured and used to convert ADP to ATP. The other 60 percent escapes as heat that warms the interior of the cell and the surrounding tissues.

The ATP produced by mitochondria provides energy to support both anabolism and other cell functions.

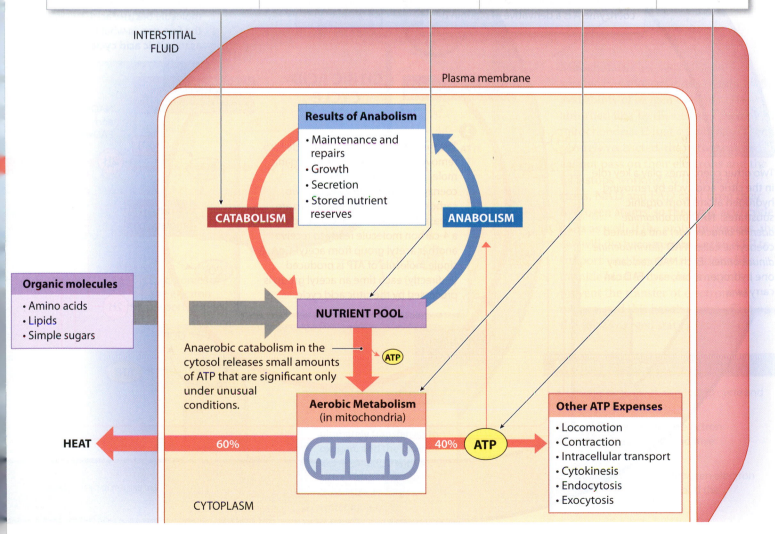

Cells can break down any available substrate from the nutrient pool to obtain the energy they need

The nutrient pool is the source of the substrates for both catabolism and anabolism. Cells tend to conserve the materials needed to build new compounds and break down the excess.

1 The nutrient pool is the key to the cell's survival because it contains the organic materials required for both anabolism and catabolism. Cells are continuously replacing membranes, organelles, enzymes, and structural proteins as well as performing their specialized functions. All of these functions require ATP, so the cell's mitochondria must have a continuous supply of two-carbon substrate molecules. In this diagram, the thickness of the arrow indicates the relative importance of that pathway in an inactive cell.

Organic compounds that can be absorbed by cells are distributed to cells throughout the body by the bloodstream.

Structural, functional, and storage components

Triglycerides | Glycogen | Proteins

Nutrient pool

Fatty acids | Glucose | Amino acids

Three-carbon chains

Two-carbon chains

MITOCHONDRIA

ATP

Citric acid cycle → Coenzymes → Electron transport system

O₂

H₂O

CO₂

KEY

= Catabolic pathway

= Anabolic pathway

2 The nutrient pool of each cell contributes to the metabolic reserves of the body as a whole. When absorption across the digestive tract is insufficient to maintain normal nutrient levels in the blood, those reserves can be mobilized.

Liver cells store triglycerides and glycogen reserves. If absorption by the digestive tract fails to maintain normal nutrient levels, the triglycerides and glycogen are broken down and the fatty acids and glucose are released.

Adipocytes convert excess fatty acids to triglycerides for storage. If absorption by the digestive tract and reserves in the liver fail to maintain normal nutrient levels, the triglycerides are broken down and the fatty acids released.

Nutrients obtained through digestion and absorption → Nutrients distributed in the blood

Neural tissue requires a continuous supply of glucose. During starvation, other tissues shift to fatty acid or amino acid catabolism, conserving glucose for neural tissue.

Cells in most tissues continuously absorb and catabolize glucose.

Skeletal muscles at rest metabolize fatty acids and use glucose to build glycogen reserves. Amino acids are used to increase the number of myofibrils. If the digestive tract, adipocytes, and liver are unable to maintain normal nutrient levels, the contractile proteins can be broken down and amino acids released into the circulation for use by other tissues.

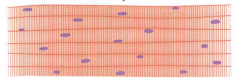

3 Neither the diet nor the nutrient pool provides everything needed to build every protein, carbohydrate, and lipid a cell might require. With few exceptions, this is not a problem because the cell also contains the enzymes necessary to synthesize what it needs from available substrates. This diagram is a general overview of the catabolic and anabolic pathways involved; these pathways will be the focus of the next section. Several modules in earlier chapters provided an overview of some of the key steps and pathways. Those modules are noted here, and you should take the time to go back and review that material before proceeding.

KEY

= Catabolic pathway

= Anabolic pathway

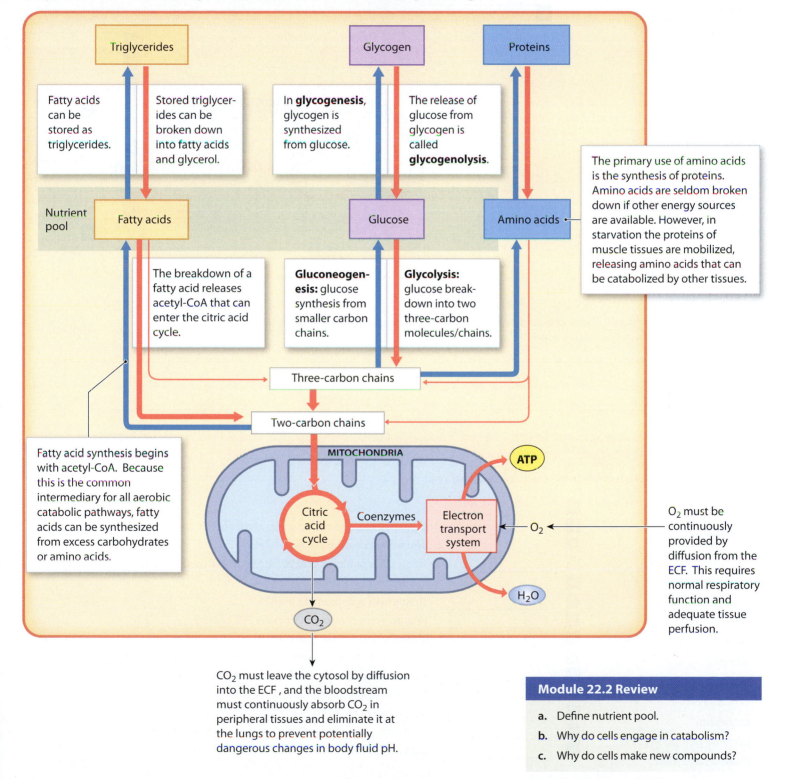

Triglycerides   Glycogen   Proteins

Fatty acids can be stored as triglycerides.

Stored triglycerides can be broken down into fatty acids and glycerol.

In **glycogenesis**, glycogen is synthesized from glucose.

The release of glucose from glycogen is called **glycogenolysis**.

The primary use of amino acids is the synthesis of proteins. Amino acids are seldom broken down if other energy sources are available. However, in starvation the proteins of muscle tissues are mobilized, releasing amino acids that can be catabolized by other tissues.

Nutrient pool   Fatty acids   Glucose   Amino acids

The breakdown of a fatty acid releases acetyl-CoA that can enter the citric acid cycle.

Gluconeogenesis: glucose synthesis from smaller carbon chains.

Glycolysis: glucose breakdown into two three-carbon molecules/chains.

Three-carbon chains

Two-carbon chains

Fatty acid synthesis begins with acetyl-CoA. Because this is the common intermediary for all aerobic catabolic pathways, fatty acids can be synthesized from excess carbohydrates or amino acids.

MITOCHONDRIA

ATP

Citric acid cycle   Coenzymes   Electron transport system

O_2

O_2 must be continuously provided by diffusion from the ECF. This requires normal respiratory function and adequate tissue perfusion.

H_2O

CO_2

CO_2 must leave the cytosol by diffusion into the ECF , and the bloodstream must continuously absorb CO_2 in peripheral tissues and eliminate it at the lungs to prevent potentially dangerous changes in body fluid pH.

Module 22.2 Review

a. Define nutrient pool.

b. Why do cells engage in catabolism?

c. Why do cells make new compounds?

Section 1: An Introduction to Cellular Metabolism · **815**

1. Matching

Use the following terms to fill in the blanks in the cellular metabolism figure to the right.

- glucose
- electron transport system
- O_2
- fatty acids
- proteins
- citric acid cycle
- ATP
- CO_2
- H_2O
- coenzymes
- two-carbon chains

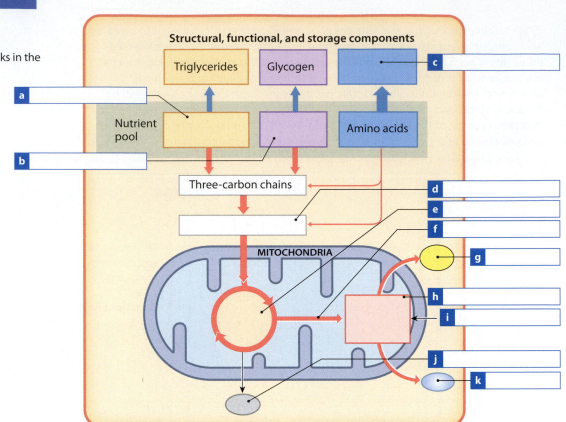

2. Short answer

Neural tissue requires a constant supply of glucose. What general shifts in cellular metabolism occur during fasting or starvation to meet that requirement?

_____

_____

_____

3. Matching

Match each of the following terms with the most closely related item.

- coenzymes
- cytochromes
- citric acid
- nutrients scarce
- nutrient pool
- anabolism
- ATP
- water
- acetate
- oxygen
- citric acid cycle
- catabolism
- oxidative phosphorylation
- nutrients abundant

a	_____	6-carbon molecule
b	_____	Collection of all the cell's organic substances
c	_____	Synthesis of new organic molecules
d	_____	Process that produces over 90 percent of ATP used by body cells
e	_____	ETS proteins
f	_____	Shuttle hydrogen atoms to the ETS
g	_____	Final acceptor of electrons from the ETS
h	_____	Breakdown of organic molecules
i	_____	Condition when cells preferentially break down carbohydrates
j	_____	Product of hydrogen ion diffusion within mitochondria
k	_____	Source of mitochondrial CO_2 production
l	_____	By-product of the ETS
m	_____	Condition when cells preferentially break down lipids
n	_____	Common substrate for mitochondrial ATP production

The Digestion and Metabolism of Organic Nutrients

The food we eat has an organized physical structure, and the organic compounds it contains are large and often insoluble. During digestion the physical structure is broken down, and a combination of chemical and enzymatic attack breaks the complex organic compounds into simpler components that can be absorbed by the digestive tract, and subsequently distributed by the bloodstream. Cells throughout the body rely on the availability of these organic molecules for energy production and to replenish the intracellular nutrient pool.

This section will provide an overview of the digestion, absorption, and fates of carbohydrates, lipids, and proteins. The modules will emphasize general patterns that will make it easier for you to understand the specifics presented in biochemistry or nutrition courses. We will focus attention on how nutrients are absorbed, stored or interconverted, or catabolized to yield the energy needed to support vital activities. We will not consider the major anabolic pathways here because Chapter 2 discussed the synthesis of carbohydrates, lipids, and proteins in sufficient detail.

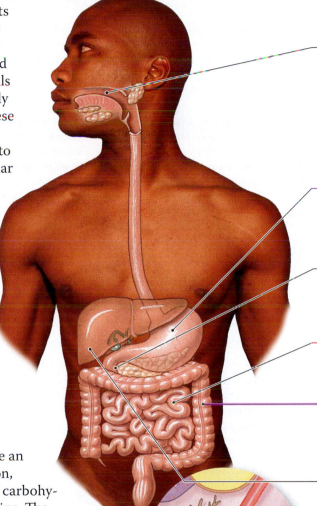

Steps in the Process of Digestion

In the oral cavity, saliva dissolves some organic nutrients, and mechanical processing with the teeth and tongue disrupts the physical structure of the material and provides access for digestive enzymes. Those enzymes begin the digestion of complex carbohydrates (polysaccharides) and lipids.

In the stomach, the material is further broken down physically and chemically by stomach acid and by enzymes that can operate at an extremely low pH.

In the duodenum, buffers from the pancreas and liver moderate the pH of the arriving chyme, and various digestive enzymes are secreted by the pancreas that catalyze the catabolism of carbohydrates, lipids, proteins, and nucleic acids.

Nutrient absorption then occurs in the small intestine, primarily in the jejunum, and the nutrients enter the bloodstream.

Indigestible materials and wastes enter the large intestine, where water is reabsorbed and bacterial action generates both organic nutrients and vitamins. These organic products are absorbed before the residue is ejected at the anus.

Most of the nutrients absorbed by the digestive tract end up in a tributary of the hepatic portal vein that ends at the liver. The liver absorbs nutrients as needed to maintain normal levels in the systemic circuit.

Within peripheral tissues, cells absorb the nutrients needed to maintain their nutrient pool and ongoing operations.

Carbohydrates are usually the preferred substrates for catabolism and ATP production under resting conditions

Carbohydrates ingested

1 In the mouth, chewing saturates the bolus with the secretions of the salivary glands. **Salivary amylase** is a salivary enzyme that breaks down complex carbohydrates into a mixture of disaccharides and trisaccharides.

2 Salivary amylase continues to digest carbohydrates from the meal until the pH throughout the contents of the stomach falls below 4.5. This enzyme generally remains active for one to two hours after a meal.

3 When chyme arrives in the duodenum, secretin stimulates the release of buffers that shift the duodenal pH from acidic to alkaline (all intestinal enzymes require an alkaline pH). At the same time, cholecystokinin (CCK) release triggers the secretion of pancreatic buffers and enzymes, including **pancreatic alpha-amylase**. This enzyme has the same functions as salivary amylase, which was deactivated by denaturation in the stomach.

Hepatic portal vein

4 The arrival of chyme containing large amounts of carbohydrates triggers the release of gastric inhibitory peptide (GIP), which stimulates insulin release by the pancreas (Module 21.12).

5 The epithelial cells lining the jejunum finish the digestion of carbohydrates. The plasma membrane at the brush border contains the enzymes **maltase** (glucose + glucose), **sucrase** (glucose + fructose), and **lactase** (glucose + galactose), which break the disaccharides maltose, sucrose, and lactose into simple sugars that are then absorbed. As these enzymes function, they transport the monosaccharides across the plasma membrane and release them into the cytosol.

The simple sugars that are transported into the cell at its apical surface diffuse through the cytoplasm and reach the interstitial fluid by facilitated diffusion across the basolateral surfaces. These monosaccharides then diffuse into the capillaries of the intestinal villi for eventual transport to the liver in the hepatic portal vein.

6 Intestinal enzymes do not alter indigestible carbohydrates such as cellulose, so they arrive in the colon virtually intact. This provides a reliable nutrient source for colonic bacteria, whose metabolic activities are responsible for generating small quantities of **flatus**, or intestinal gas. Foods containing large amounts of indigestible carbohydrates (such as beans) stimulate bacterial gas production, leading to distension of the colon, cramps, and the frequent discharge of intestinal gases.

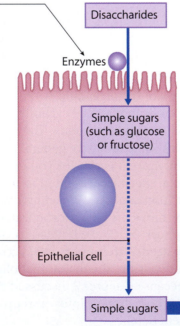

Disaccharides

Enzymes

Simple sugars (such as glucose or fructose)

Epithelial cell

Simple sugars

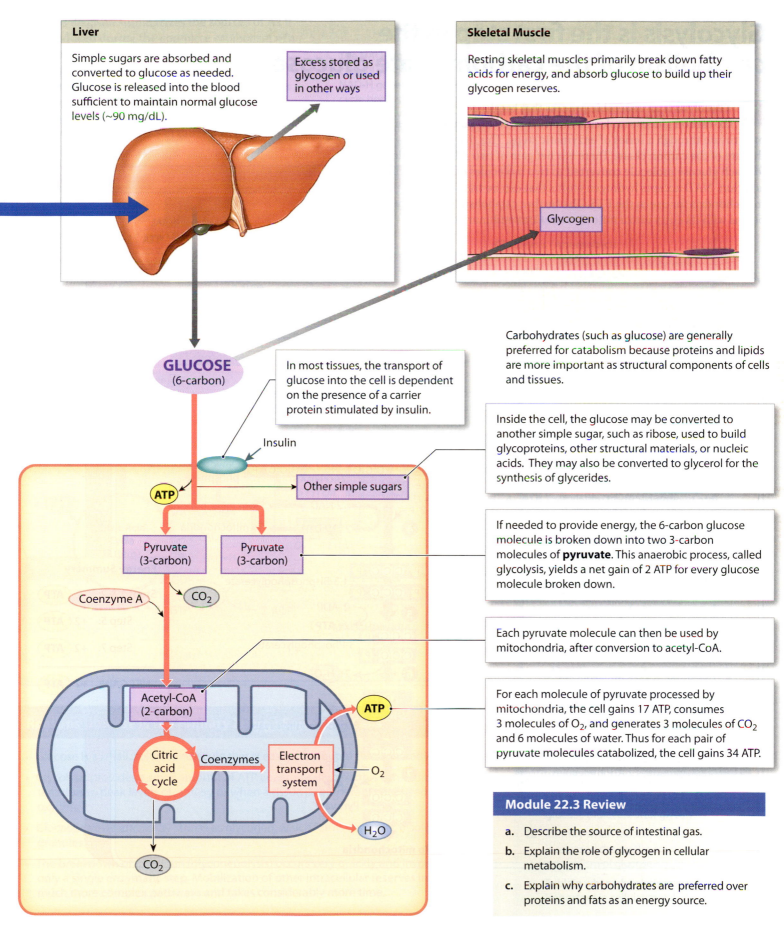

Liver

Simple sugars are absorbed and converted to glucose as needed. Glucose is released into the blood sufficient to maintain normal glucose levels (~90 mg/dL).

Excess stored as glycogen or used in other ways

Skeletal Muscle

Resting skeletal muscles primarily break down fatty acids for energy, and absorb glucose to build up their glycogen reserves.

Glycogen

GLUCOSE
(6-carbon)

In most tissues, the transport of glucose into the cell is dependent on the presence of a carrier protein stimulated by insulin.

Insulin

ATP

Other simple sugars

Carbohydrates (such as glucose) are generally preferred for catabolism because proteins and lipids are more important as structural components of cells and tissues.

Inside the cell, the glucose may be converted to another simple sugar, such as ribose, used to build glycoproteins, other structural materials, or nucleic acids. They may also be converted to glycerol for the synthesis of glycerides.

Pyruvate
(3-carbon)

Pyruvate
(3-carbon)

Coenzyme A

CO_2

If needed to provide energy, the 6-carbon glucose molecule is broken down into two 3-carbon molecules of **pyruvate**. This anaerobic process, called glycolysis, yields a net gain of 2 ATP for every glucose molecule broken down.

Acetyl-CoA
(2-carbon)

ATP

Each pyruvate molecule can then be used by mitochondria, after conversion to acetyl-CoA.

Citric acid cycle

Coenzymes

Electron transport system

O_2

H_2O

CO_2

For each molecule of pyruvate processed by mitochondria, the cell gains 17 ATP, consumes 3 molecules of O_2, and generates 3 molecules of CO_2 and 6 molecules of water. Thus for each pair of pyruvate molecules catabolized, the cell gains 34 ATP.

Module 22.3 Review

a. Describe the source of intestinal gas.

b. Explain the role of glycogen in cellular metabolism.

c. Explain why carbohydrates are preferred over proteins and fats as an energy source.

An amino acid not needed for protein synthesis may be broken down or converted to a different amino acid

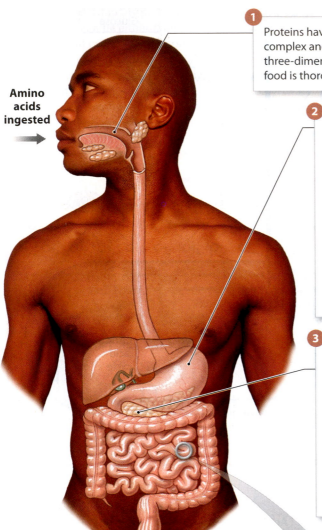

Amino acids ingested

1 Proteins have very complex structures, so protein digestion is both complex and time-consuming. The first step, disrupting the tough three-dimensional organization of the food, begins in the mouth as the food is thoroughly chewed and mixed with saliva.

2 Additional mechanical processing occurs in the stomach through churning and mixing. Exposure of the bolus to a strongly acidic environment kills pathogens and breaks down connective tissues and plant cell walls. Stomach acids also denature most proteins and disrupt tertiary and secondary protein structure, exposing peptide bonds to enzymatic attack by the proteolytic enzyme **pepsin** secreted by the parietal cells as the inactive proenzyme pepsinogen (Module 21.9). Protein digestion is not completed in the stomach, because time is limited and pepsin attacks only specific types of peptide bonds, not all of them. However, pepsin generally has enough time to break down complex proteins into smaller peptide and polypeptide chains before the chyme enters the duodenum.

Hepatic portal vein

3 When acid chyme arrives in the duodenum, CCK stimulates production and release of pancreatic enzymes. These enzymes are secreted as inactive proenzymes that are then activated within the duodenum. **Enteropeptidase**, an enzyme released by the duodenal epithelium, begins the process, converting the proenzyme trypsinogen to the proteolytic enzyme **trypsin**. Trypsin then converts the other proenzymes to yield **chymotrypsin, carboxypeptidase**, and **elastase**. Each enzyme attacks peptide bonds linking specific amino acids and ignores others. Together, they break down proteins into a mixture of dipeptides, tripeptides, and amino acids.

4 The epithelial surfaces of the small intestine contain several **peptidases**—enzymes that break peptide bonds—notably **dipeptidases** that break dipeptides apart and release individual amino acids. These amino acids, as well as those released by the action of pancreatic enzymes, are absorbed through both facilitated diffusion and cotransport mechanisms.

After diffusing to the basal surface of the cell, the amino acids are released into interstitial fluid by facilitated diffusion and cotransport. Once in the interstitial fluid, the amino acids enter intestinal capillaries for transport to the liver in the hepatic portal vein.

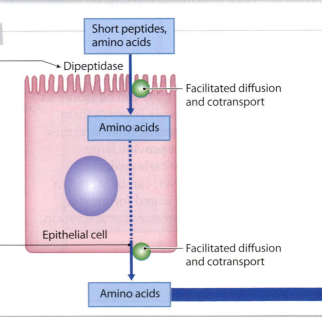

Short peptides, amino acids

Dipeptidase

Facilitated diffusion and cotransport

Amino acids

Epithelial cell

Facilitated diffusion and cotransport

Amino acids

⑤

The liver does not control circulating levels of amino acids as precisely as it does glucose concentrations. Plasma amino acid levels normally range between 35 and 65 mg/dL, but they may become elevated after a protein-rich meal. The liver itself uses many amino acids for synthesizing plasma proteins, and it has all of the enzymes needed to synthesize, convert, or catabolize amino acids. In addition, amino acids that can be broken down to 3-carbon molecules can be used for gluconeogenesis when other sources of glucose are unavailable.

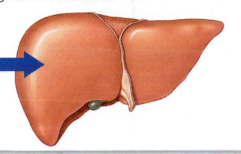

Note: Alanine aminotransferase (ALT) and aspartate aminotransferase (AST) are transaminases that leak from injured hepatocytes into the circulation. High levels in a blood test indicate liver disease or inflammation.

Amino Acid Synthesis

Liver cells and other body cells can readily synthesize the carbon frameworks of roughly half of the amino acids needed to synthesize proteins. There are 10 **essential amino acids** that the body either cannot synthesize or that cannot be produced in amounts sufficient for growing children.

In an **amination** reaction, an ammonium ion (NH_4^+) is used to form an amino group that is attached to a molecule, yielding an amino acid.

α–Ketoglutarate → Glutamic acid

In a **transamination**, the amino group of one amino acid gets transferred to another molecule, yielding a different amino acid. The remaining carbon chain can then be broken down or used in other ways.

Glutamic acid + Organic acid 1 ⇌ (Transaminase) Organic acid 2 + Tyrosine

Amino Acid Catabolism

The first step in amino acid catabolism is the removal of the amino group, leaving a carbon chain that can usually be converted to pyruvate, acetyl-CoA, or an intermediary product in the citric acid cycle.

In **deamination**, the amino group is removed and an ammonium ion is released.

Glutamic acid → (Deaminase, H_2O, H^+) Organic acid + NH_4^+ Ammonium ion

Ammonium ions are highly toxic, even in low concentrations. Liver cells, the primary sites of deamination, have enzymes that use ammonium ions to synthesize **urea**, a relatively harmless water-soluble compound that is excreted in urine. The **urea cycle** is the reaction sequence responsible for the production of urea.

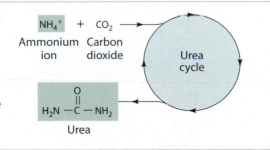

NH_4^+ + CO_2
Ammonium ion Carbon dioxide
→ Urea cycle →
$H_2N-\overset{\overset{\displaystyle O}{\|}}{C}-NH_2$
Urea

When broken down in the mitochondria, the energy yield of amino acids is comparable to that of carbohydrates. However, a cell surviving by amino acid catabolism is like someone in a winter cabin burning the walls to stay warm. It can help temporarily, but it isn't a permanent solution, and in the long run it only makes matters worse.

Module 22.7 Review

a. Name the enzyme secreted by parietal cells that is necessary for protein digestion.

b. Identify the processes by which the amino group is removed.

c. What happens to the ammonium ions that are removed from amino acids during deamination?

There are two general patterns of metabolic activity: the absorptive and postabsorptive states

Metabolic activity changes from moment to moment, but there are broad patterns that follow a daily cycle. These cycles are maintained by many different hormones. The degree of response differs from tissue to tissue, but we will consider a representative cell that shares characteristics with all body cells.

1 The **absorptive state** is the period following a meal, when nutrient absorption is under way. After a typical meal, the absorptive state continues for about 4 hours. If you are fortunate enough to eat three meals a day, you spend 12 out of every 24 hours in the absorptive state. Insulin is the primary hormone of the absorptive state, although various other hormones stimulate amino acid uptake (growth hormone, or GH) and protein synthesis (GH, androgens, and estrogens).

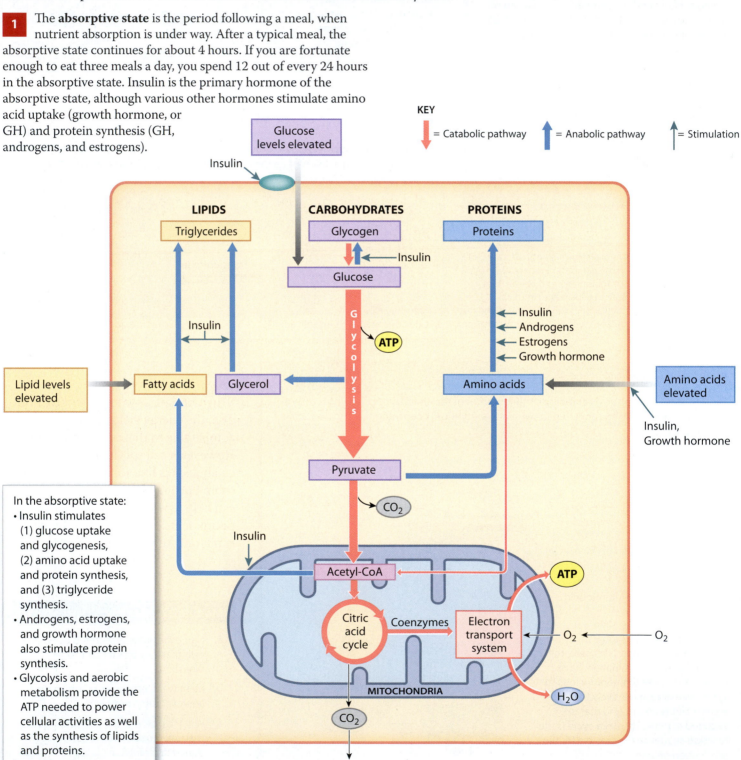

KEY

↓ = Catabolic pathway   ↑ = Anabolic pathway   ↑ = Stimulation

In the absorptive state:
- Insulin stimulates
  (1) glucose uptake and glycogenesis,
  (2) amino acid uptake and protein synthesis,
  and (3) triglyceride synthesis.
- Androgens, estrogens, and growth hormone also stimulate protein synthesis.
- Glycolysis and aerobic metabolism provide the ATP needed to power cellular activities as well as the synthesis of lipids and proteins.

2 The **postabsorptive state** is the period when nutrient absorption is not under way and your body must rely on internal energy reserves to continue meeting its energy demands. You spend roughly 12 hours each day in the postabsorptive state, although a person who is skipping meals can extend that time considerably. Metabolic activity in the postabsorptive state is focused on the mobilization of energy reserves and the maintenance of normal blood glucose levels. These activities are coordinated by several hormones, including glucagon, epinephrine, glucocorticoids, and growth hormone.

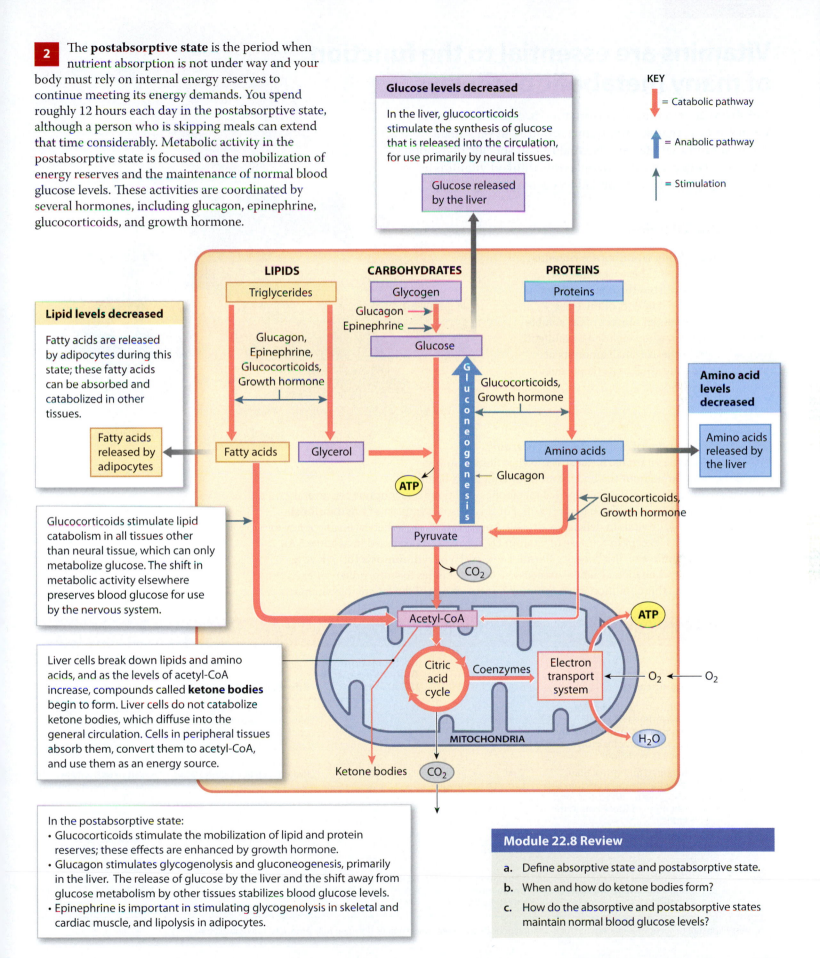

KEY

= Catabolic pathway

= Anabolic pathway

= Stimulation

Glucose levels decreased

In the liver, glucocorticoids stimulate the synthesis of glucose that is released into the circulation, for use primarily by neural tissues.

Glucose released by the liver

LIPIDS

Triglycerides

CARBOHYDRATES

Glycogen

PROTEINS

Proteins

Glucagon →
Epinephrine →

Glucose

Lipid levels decreased

Fatty acids are released by adipocytes during this state; these fatty acids can be absorbed and catabolized in other tissues.

Glucagon, Epinephrine, Glucocorticoids, Growth hormone

Glucocorticoids, Growth hormone

Amino acid levels decreased

Fatty acids released by adipocytes

Fatty acids

Glycerol

Gluconeogenesis

Amino acids

Amino acids released by the liver

ATP

← Glucagon

Glucocorticoids stimulate lipid catabolism in all tissues other than neural tissue, which can only metabolize glucose. The shift in metabolic activity elsewhere preserves blood glucose for use by the nervous system.

Pyruvate

Glucocorticoids, Growth hormone

CO$_2$

Acetyl-CoA

ATP

Liver cells break down lipids and amino acids, and as the levels of acetyl-CoA increase, compounds called **ketone bodies** begin to form. Liver cells do not catabolize ketone bodies, which diffuse into the general circulation. Cells in peripheral tissues absorb them, convert them to acetyl-CoA, and use them as an energy source.

Citric acid cycle

Coenzymes

Electron transport system

O$_2$ ← O$_2$

H$_2$O

MITOCHONDRIA

Ketone bodies

CO$_2$

In the postabsorptive state:
- Glucocorticoids stimulate the mobilization of lipid and protein reserves; these effects are enhanced by growth hormone.
- Glucagon stimulates glycogenolysis and gluconeogenesis, primarily in the liver. The release of glucose by the liver and the shift away from glucose metabolism by other tissues stabilizes blood glucose levels.
- Epinephrine is important in stimulating glycogenolysis in skeletal and cardiac muscle, and lipolysis in adipocytes.

Module 22.8 Review

a. Define absorptive state and postabsorptive state.

b. When and how do ketone bodies form?

c. How do the absorptive and postabsorptive states maintain normal blood glucose levels?

Vitamins are essential to the function of many metabolic pathways

The absorption of nutrients from food is called **nutrition**. **Vitamins** are organic compounds required in very small quantities but that play an essential role in specific metabolic pathways. However, many people only think of vitamins as something they take with breakfast each morning.

Fat-Soluble Vitamins

1 The **fat-soluble vitamins** are vitamins A, D₃, E, and K. These vitamins are absorbed primarily from the digestive tract along with the lipid contents of micelles. Vegetables are potential sources of fat-soluble vitamins. However, when exposed to sunlight, your skin can synthesize small amounts of vitamin D₃, and intestinal bacteria produce some vitamin K.

2 This table summarizes information concerning the fat-soluble vitamins. Because these vitamins are stored in lipid deposits throughout the body, your body contains a significant reserve of these vitamins, and normal metabolic operations can continue for several months after dietary sources have been cut off. For this reason, symptoms of **avitaminosis** (ā-vī-ta-min-Ō-sis), or vitamin deficiency disease, rarely result from a dietary insufficiency of fat-soluble vitamins (except in the case of vitamin D₃, for reasons discussed in Module 5.8). Too much of a vitamin can also produce harmful effects. **Hypervitaminosis** (hī-per-vī-ta-min-Ō-sis) occurs when dietary intake exceeds the body's abilities to store, utilize, or excrete a particular vitamin. This condition most commonly involves one of the fat-soluble vitamins.

The Fat-Soluble Vitamins

Vitamin	Significance	Sources	Recommended Daily Allowance (RDA) in mg	Effects of Deficiency	Effects of Excess
A	Maintains epithelia; required for synthesis of visual pigments; supports immune system; promotes growth and bone remodeling	Leafy green and yellow vegetables	0.7–0.9	Retarded growth, night blindness, deterioration of epithelial membranes	Liver damage, skin paling, CNS effects (nausea, anorexia)
D₃	Required for normal bone growth, intestinal calcium and phosphorus absorption, and retention of these ions at the kidneys	Synthesized in skin exposed to sunlight	0.005–0.015*	Rickets, skeletal deterioration	Calcium deposits in many tissues, disrupting functions
E	Prevents breakdown of vitamin A and fatty acids	Meat, milk, vegetables	15	Anemia, other problems suspected	Nausea, stomach cramps, blurred vision, fatigue
K	Essential for liver synthesis of prothrombin and other clotting factors	Vegetables; production by intestinal bacteria	0.09–0.12	Bleeding disorders	Liver dysfunction, jaundice

*Unless exposure to sunlight is inadequate for extended periods and alternative sources (fortified milk products) are unavailable.

Water-Soluble Vitamins

3 The **water-soluble vitamins** are the B vitamins and vitamin C. Most of them are components of coenzymes. The B vitamins are found in meat, eggs, and dairy products, while vitamin C is found in citrus fruits.

4 This table summarizes information about water-soluble vitamins. These water-soluble vitamins are rapidly exchanged between the fluid compartments of the digestive tract and the circulating blood, and excessive amounts are readily excreted in urine. For this reason, hypervitaminosis involving water-soluble vitamins is unlikely unless you are taking megadoses of vitamin supplements.

The Water-Soluble Vitamins

Vitamin	Component or Precursor of	Sources	Recommended Daily Allowance (RDA) in mg	Effects of Deficiency	Effects of Excess
B_1 (thiamine)	Coenzyme in many pathways	Milk, meat, bread	1.1–1.2	Muscle weakness, CNS and cardiovascular problems, including heart disease; called *beriberi*	Hypotension
B_2 (riboflavin)	Part of FAD, involved in multiple pathways, including glycolysis and citric acid cycle	Milk, meat, eggs and cheese	1.1–1.3	Epithelial and mucosal deterioration	Itching, tingling
B_3 (niacin)	Part of NAD, involved in multiple pathways	Meat, bread, potatoes	14–16	CNS, GI, epithelial, and mucosal deterioration; called *pellagra*	Itching, burning; vasodilation; death after large dose
B_5 (pantothenic acid)	Coenzyme A, in multiple pathways	Milk, meat	10	Retarded growth, CNS disturbances	None reported
B_6 (pyridoxine)	Coenzyme in amino acid and lipid metabolism	Meat, whole grains, vegetables, orange juice, cheese and milk	1.3–1.7	Retarded growth, anemia, convulsions, epithelial changes	CNS alterations, perhaps fatal
B_9 (folic acid)	Coenzyme in amino acid and nucleic acid metabolism	Leafy vegetables, some fruits, liver, cereal and bread	0.2–0.4	Retarded growth, anemia, gastrointestinal disorders, developmental abnormalities	Few noted, except at massive doses
B_{12} (cobalamin)	Coenzyme in nucleic acid metabolism	Milk, meat	0.0024	Impaired RBC production, causing pernicious anemia	Polycythemia
B_7 (biotin)	Coenzyme in many pathways	Eggs, meat, vegetables	0.03	Fatigue, muscular pain, nausea, dermatitis	None reported
C (ascorbic acid)	Coenzyme in many pathways	Citrus fruits	75–90; Smokers add 35 mg	Epithelial and mucosal deterioration; called *scurvy*	Kidney stones

Only vitamins B_{12} and C are stored in significant quantities, so insufficient intake of other water-soluble vitamins can lead to initial signs and symptoms of vitamin deficiency within a period of days to weeks. The bacterial inhabitants of the intestines help prevent deficiency diseases by producing small amounts of four of the nine water-soluble vitamins (B_5, B_7, B_9, and B_{12}). The intestinal epithelium can easily absorb all the water-soluble vitamins except B_{12}. The B_{12} molecule is large, and it must be bound to **intrinsic factor** synthesized by the gastric mucosa before absorption can occur.

Module 22.9 Review

a. Define nutrition.

b. Identify the two classes of vitamins.

c. If vitamins do not provide a source of energy, what is their role in nutrition?

Proper nutrition depends on eating a balanced diet

A **balanced diet** contains all the ingredients needed to maintain homeostasis, including adequate substrates for ATP production, essential amino acids and fatty acids, and vitamins. In addition, the diet must include electrolytes and enough water to replace losses in urine, feces, and evaporation. A balanced diet prevents **malnutrition**, an unhealthy state resulting from inadequate or excessive absorption of one or more nutrients.

1 One method of avoiding malnutrition is to consume a diet based on the updated food pyramid, called the **MyPyramid.gov Steps to a Healthier You**. The United States Department of Agriculture created the food pyramid to offer personalized eating plans and food assessments based on the current Dietary Guidelines for Americans. The color-coded vertical food group bands running from the apex to the base indicate the proportions of food we should consume from each of the five basic food groups: grains (orange), vegetables (green), fruits (red), milk products (blue), and meat and beans (purple). Oils (yellow) are to be used sparingly in addition to the five basic food groups. Twelve different pyramids have been developed based on level of physical activity and general health; all aim to increase healthful eating habits.

Know the Limits on Fats, Sugars, and Salt (Sodium)

- Make most of your fat sources from fish, nuts, and vegetable oils, potential sources of essential fatty acids.

- Limit solid fats like butter, shortening, and lard that are high in saturated fats and cholesterol.

- Check the Nutrition Facts label to keep saturated fats, trans fats, and sodium low.

- Choose food and beverages low in added sugars. Added sugars contribute to excessive caloric intake.

Find Your Balance between Food and Physical Activity

- Be sure to stay within your daily calorie needs.

- Be physically active for at least 30 minutes most days of the week.

- About 60 minutes a day of physical activity may be needed to prevent weight gain.

- For sustaining weight loss, at least 60 to 90 minutes a day of physical activity may be required.

- Children and teenagers should be physically active for 60 minutes most days.

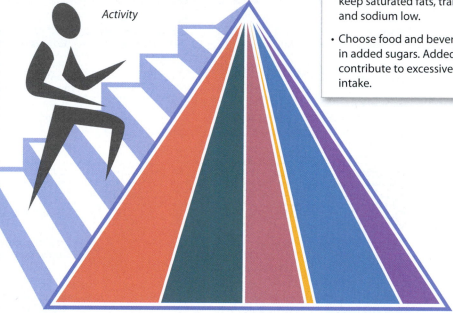

Activity

GRAINS	VEGETABLES	FRUITS	O I L S	MILK	MEAT & BEANS
Make half your grains whole	Vary your veggies	Focus on fruits		Get your calcium-rich foods	Go lean with proteins

2 You must be concerned with not only what types of food, but how much energy that food contains relative to how much you use on a daily basis. The energy content of food may be expressed in **calories** or **joules** (one joule equals 0.239 calories). One calorie is the energy needed to raise the temperature of 1 g of water by 1°C. The **kilocalorie** (KIL-ō-kal-o-rē) (kcal or Calorie) or **kilojoule** (**kJ**), or **Calorie** is used when talking about the metabolism of the entire body. One Calorie is the amount of energy needed to raise the temperature of 1 kilogram of water 1°C. The energy yield of the components of your diet differ; the values are 4.18 Cal/g for carbohydrates, 4.32 Cal/g for proteins, and 9.46 Cal/g for lipids. Depending on the level of activity, the average adult needs between 2000 and 3000 Cal each day to maintain stable weight. That's a lot of mitochondrial activity—1 Cal is roughly equivalent to 83 million trillion ATP.

Food name	Serving	Cal	kJ
Nutrigrain bar	1 bar	368	1540
Long-grain rice	1.5 cup	308	1289
Bread, whole wheat	4 slices	277	1159
Butter	0.5 tbsp	51	213
Beer, regular	12 fl oz	160	669
Cola, regular	12 fl oz	140	586

3 Here is additional information about the various food groups. The food groups in themselves are less important than making intelligent choices about what (and how much) you eat. Poor choices can lead to malnutrition even if all five groups are represented.

Basic Food Groups of the 2005 Dietary Guidelines and Their General Effects on Health

Nutrient Group	Provides	Health Effects
Grains (recommended: at least half of the total eaten as whole grains)	Carbohydrates; vitamins E, thiamine, niacin, folate; calcium; phosphorus; iron; sodium; dietary fiber	Whole grains prevent rapid rise in blood glucose levels, and consequent rapid rise in insulin levels
Vegetables (recommended: especially dark-green and orange vegetables)	Carbohydrates; vitamins A, C, E, folate; dietary fiber; potassium	Reduce risk of cardiovascular disease; protect against colon cancer (folate) and prostate cancer (lycopene in tomatoes)
Fruits (recommended: a variety of fruit each day)	Carbohydrates; vitamins A, C, E, folate; dietary fiber; potassium	Reduce risk of cardiovascular disease; protect against colon cancer (folate)
Milk (recommended: low-fat or fat-free milk, yogurt, and cheese)	Complete proteins; fats; carbohydrates; calcium; potassium; magnesium; sodium; phosphorus; vitamins A, B_{12}, pantothenic acid, thiamine, riboflavin	Whole milk: High in calories, may cause weight gain; saturated fats correlated with heart disease
Meat and Beans (recommended: lean meats, fish, poultry, eggs, dry beans, nuts, legumes)	Complete proteins; fats; calcium; potassium; phosphorus; iron; zinc; vitamins E, thiamine, B_6	Fish and poultry lower risk of heart disease and colon cancer (compared to red meat). Consumption of up to one egg per day does not appear to increase incidence of heart disease; nuts and legumes improve blood cholesterol ratios, lower risk of heart disease and diabetes

4 Some members of the milk products and meat and beans groups—specifically, beef, fish, poultry, eggs, and milk—provide all the essential amino acids in sufficient quantities. They are said to contain **complete proteins**.

5 Many plants supply adequate *amounts* of protein but they contain **incomplete proteins**, which are deficient in one or more of the essential amino acids. Vegetarians, who largely restrict themselves to the grains, vegetables, and fruits groups (with or without the milk products group), must become adept at varying their food choices to include combinations of ingredients that meet all their amino acid requirements. Even with a proper balance of amino acids, vegans, who avoid all animal products, face a significant problem, because vitamin B_{12} is obtained only from animal products or from fortified cereals or tofu.

Module 22.10 Review

a. Define balanced diet.

b. Distinguish between a complete protein and an incomplete protein.

c. Of these three —carbohydrates, lipids, or proteins—which one releases the greatest number of Calories per gram during catabolism?

Metabolic disorders may result from nutritional or biochemical problems

Disorders Related to Diet and Digestion

Eating Disorders

Eating disorders are psychological problems that result in either inadequate or excessive food consumption. There are many different forms; a common thread in these conditions is an obsessive concern about food and body weight.

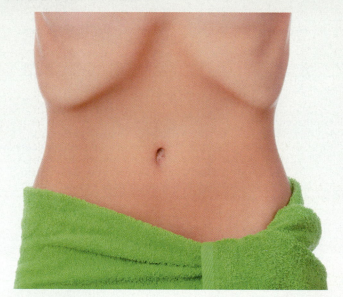

Anorexia Nervosa
Anorexia (an-o-REK-sē-ah) is the lack or loss of appetite. It may also accompany disorders that involve other systems. **Anorexia nervosa** is a form of self-induced starvation that appears to be the result of severe psychological problems. It is most common in adolescent Caucasian females whose weight is roughly 30 percent below normal levels. Although very obviously underweight, patients are convinced that they are too fat and refuse to eat normal amounts of food. Death rates from severe cases range from 10 to 15 percent.

Bulimia
In **bulimia** (bu-LĒM-ē-ah) an individual goes on an "eating binge" that may involve a meal that lasts 1–2 hours and may include 20,000 or more calories. The meal is followed by induced vomiting, usually along with the use of laxatives (to promote the movement of material through the digestive tract), and diuretics (drugs that promote fluid loss in the urine). Bulimia is more common than anorexia nervosa, and generally affects adolescent females.

Obesity

Obesity is defined as a condition of being 20 percent over ideal weight, because it is at this point that serious health risks appear. On that basis, the U.S. Centers for Disease Control estimate that 32 percent of men and 35 percent of women in the United States can be considered obese. Basically, obese individuals take in more food energy than they are using. There are two major categories of obesity: regulatory obesity and metabolic obesity. **Regulatory obesity**, the most common form, results from a failure to regulate food intake so that appetite, diet, and activity are in balance. In **metabolic obesity**, the condition is secondary to some underlying bodily malfunction that affects cell and tissue metabolism. Cases of metabolic obesity are relatively rare.

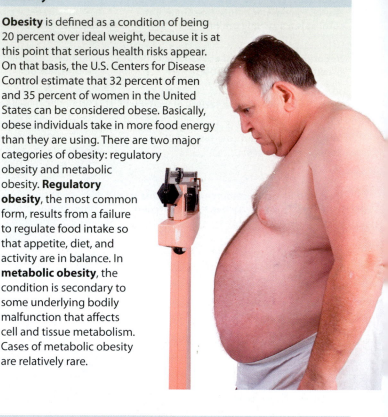

Elevated Cholesterol Levels

Earlier chapters have noted that elevated cholesterol levels are associated with the development of atherosclerosis and coronary artery disease. Nutritionists currently recommend that you reduce cholesterol intake to under 300 mg per day. This amount represents a 40 percent reduction for the average American adult. In fasting individuals, triglycerides are usually present at levels of 40–150 mg/dL. A high total cholesterol value linked to a high LDL level spells trouble. In effect, an excessive amount of cholesterol is being exported to peripheral tissues. Problems can also exist in individuals with low HDL levels (below 35 mg/dL). In such cases, excess cholesterol delivered to the tissues cannot easily be returned to the liver for excretion. In either event, the amount of cholesterol in peripheral tissues—and especially in arterial walls—is likely to increase.

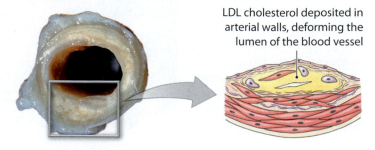

LDL cholesterol deposited in arterial walls, deforming the lumen of the blood vessel

Nutritional/Metabolic Disorders

Phenylketonuria

Several inherited metabolic disorders result from an inability to produce specific enzymes involved in amino acid metabolism. Individuals with **phenylketonuria** (fen-il-kē-tō-NOO-rē-uh), or **PKU**, for example, cannot convert phenylalanine to tyrosine. This reaction is an essential step in the synthesis of norepinephrine, epinephrine, dopamine, and melanin. If PKU is not detected in infancy, central nervous system development is inhibited, and severe brain damage results. The condition is common enough that a warning is printed on the packaging of products, such as diet drinks, that contain phenylalanine.

Protein Deficiency Diseases

Regardless of the energy content of the diet, if it is deficient in essential amino acids, the individual will be malnourished to some degree. In a **protein deficiency disease**, protein synthesis decreases throughout the body. As protein synthesis in the liver fails to keep pace with the breakdown of plasma proteins, plasma osmolarity falls. This reduced osmolarity results in a fluid shift as more water moves out of the capillaries and into interstitial spaces, the peritoneal cavity, or both. The longer the individual remains in this state, the more severe the ascites and edema that result. It is estimated that more than 100 million children worldwide suffer from protein deficiency diseases.

Kwashiorkor (kwash-ē-OR-kor) occurs in children whose protein intake is inadequate, even if the caloric intake is acceptable. In each case, additional complications include damage to the developing brain. The term is from the Ghana language and its literal translation means "first-second", describing the development of the disease in an older child who had been weaned from his mother's breast when the younger sibling was born.

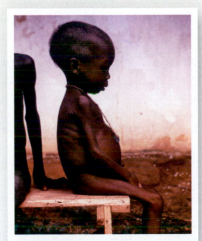

Kwashiorkor

Ketoacidosis

When glucose supplies are limited, the breakdown of fatty acids and some amino acids in liver cells elevates acetyl-CoA levels and results in the production of small organic acids called **ketone bodies**. Most of these compounds diffuse out of the liver and accumulate in the bloodstream. One of the ketone bodies, acetone, can often be smelled in the breath of someone who has skipped one or more meals. This condition is called **ketosis** (ke-TŌ-sis), and over time ketone bodies can lower the pH of blood. This acidification of the blood is called **ketoacidosis** (kē-tō-as-i-DŌ-sis). In severe cases, the blood pH may drop below 7.05, and this may cause coma, cardiac arrhythmias, and death. An individual with poorly controlled or undiagnosed diabetes mellitus is at serious risk of developing ketoacidosis because the liver responds as if the person is starving, catabolizing proteins and lipids and dumping ketone bodies into the circulation.

Gout

When RNA is recycled as part of metabolic turnover, the purines (adenine and guanine) cannot be catabolized. Instead, they are deaminated and excreted as **uric acid**. Like urea, uric acid is a relatively nontoxic waste product, but it is far less soluble than urea. Urea and uric acid are called **nitrogenous wastes**, because they are waste products that contain nitrogen atoms. Normal uric acid concentrations in plasma average 2.7–7.4 mg/dL, depending on gender and age. At concentrations over 7.4 mg/dL, body fluids become saturated with uric acid and insoluble uric acid crystals may begin to form. The condition that then develops is called **gout**. Initially, the joints of the limbs, especially the metatarsal–phalangeal joint of the great toe, are likely to be affected. This intensely painful condition, called **gouty arthritis**, may persist for several days and then disappear for a period of days to years.

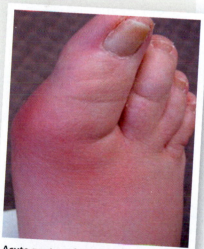

Acute gouty arthritis of the great toe

Module 22.11 Review

a. Identify and briefly define two eating disorders.

b. Define protein deficiency disease and cite an example.

c. Briefly describe phenylketonuria (PKU).

1. Matching

Match each of the following terms with the most closely related item.

- lipogenesis
- anorexia
- lipolysis
- A, D, E, K
- absorptive state
- deamination
- ketone bodies
- calorie
- uric acid
- B complex and C
- urea formation
- lipoproteins
- insulin
- skeleton muscle

a	_____	Absorptive state hormone
b	_____	Glycogen reserves
c	_____	Water-soluble vitamins
d	_____	Fat catabolism
e	_____	Lipid synthesis
f	_____	Amino acid catabolism
g	_____	Fat-soluble vitamins
h	_____	Lipid transport
i	_____	Removal of amino group
j	_____	Lipid breakdown
k	_____	Gout
l	_____	Unit of energy
m	_____	Period following a meal
n	_____	Lack or loss of appetite

2. Multiple choice

Choose the bulleted item that best completes each statement.

a Intestinal absorption of nutrients occurs in the _____ .

- duodenum
- ileocecum
- ileum
- jejunum

b When blood glucose concentrations are elevated, the glucose molecules are _____ .

- catabolized for energy
- used to build proteins
- used for tissue repair
- all of these

c Most of the lipids absorbed by the digestive tract are immediately transferred to the _____ .

- liver
- red blood cells
- hepatocytes for storage
- venous circulation by the thoracic duct

d Hypervitaminosis involving water-soluble vitamins is relatively uncommon because _____ .

- the excess amount is stored in adipose tissue
- the excess amount is readily excreted in the urine
- the excess amount is stored in the bones
- excess amounts are readily absorbed by skeletal muscle tissue

3. Short answer

a What is the difference between an essential amino acid and a non-essential amino acid? _____
_____
_____
_____

b Describe four reasons why protein catabolism is an impractical source of quick energy. _____
_____
_____
_____

c What is the primary difference between the absorptive and postabsorptive states? _____
_____
_____
_____

d Why is the liver the focal point for metabolic regulation and control? _____
_____
_____
_____

4. Section integration

Darla suffers from anorexia nervosa. One afternoon she is rushed to the emergency room because of cardiac arrhythmias. Her breath has the smell of an aromatic hydrocarbon, and blood and urine samples contain high levels of ketone bodies. Why do you think she is having the arrhythmias?

_____
_____
_____

Energetics and Thermoregulation

The amount of energy needed to support ongoing activities varies from moment to moment. The study of the flow of energy and its change(s) from one form to another is called **energetics**. A common benchmark in energetics studies is the **basal metabolic rate (BMR)**, the minimum resting energy expenditure of an awake, alert person.

1 A direct method of determining the BMR involves monitoring respiratory activity. If we assume that average amounts of carbohydrates, lipids, and proteins are being catabolized, 4.825 Calories are expended per liter of oxygen consumed.

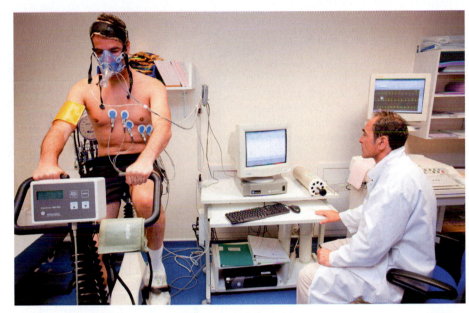

2 An average individual has a BMR of 70 Cal per hour, or about 1680 Cal per day; the actual energy consumption per day may be several times that amount, depending on a person's size, weight, and level of physical activity. This graph shows the approximate number of Calories expended per hour at various levels of physical exertion.

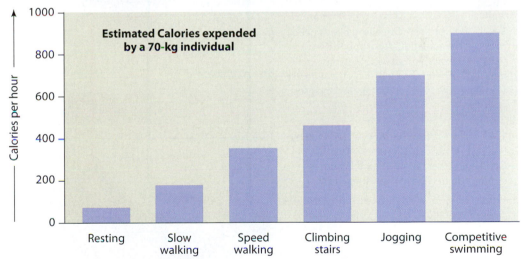

Estimated Calories expended by a 70-kg individual

To maintain energy balance, food intake must be adequate to support the activities under way. This section thus begins with a look at the control of appetite. All of the reactions that generate ATP also generate heat; only about 40 percent of the energy released through catabolism can be used to form ATP, and the rest warms the surrounding cytoplasm. When activity levels increase, ATP production accelerates, and more heat is generated. Enzymes will only function normally over a relatively narrow range of temperatures, and metabolic pathways are at risk unless heat is lost as quickly as it is produced. So this section ends with the topic of **thermoregulation**—the homeostatic control of body temperature.

The control of appetite is complex and involves both short-term and long-term mechanisms

The **feeding center** and the **satiety center** are hypothalamic nuclei involved with the control of appetite. Multiple factors influence these centers, and the most important satiety factors are diagrammed here. In addition, social factors, psychological pressures, and dietary habits can play a role, although the mechanisms and pathways involved remain to be determined.

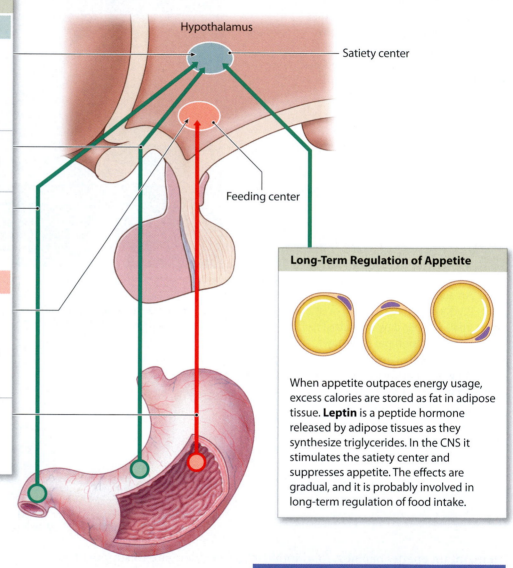

Short-Term Regulation of Appetite

Stimulation of Satiety Center ⟶

Elevated blood glucose levels depress appetite, and low blood glucose stimulates appetite. The likely mechanism is glucose entry stimulating the neurons of the satiety center.

Several hormones of the digestive tract, including CCK , suppress appetite during the absorptive state.

Stimulation of stretch receptors along the digestive tract, especially in the stomach, causes a sense of satiation and suppresses appetite.

Stimulation of Feeding Center ⟶

Several neurotransmitters have been linked to appetite regulation. **Neuropeptide Y (NPY)**, for example, is a hypothalamic neurotransmitter that (among other effects) stimulates the feeding center and increases appetite.

The hormone **ghrelin** (GREL-in), secreted by the gastric mucosa, stimulates appetite. Ghrelin levels are high when the stomach is empty, and decline as the stomach fills.

Hypothalamus

Satiety center

Feeding center

Long-Term Regulation of Appetite

When appetite outpaces energy usage, excess calories are stored as fat in adipose tissue. **Leptin** is a peptide hormone released by adipose tissues as they synthesize triglycerides. In the CNS it stimulates the satiety center and suppresses appetite. The effects are gradual, and it is probably involved in long-term regulation of food intake.

In general, activation of the satiety center causes inhibition of the feeding center, and vice versa. In addition, factors that primarily stimulate one center often have a secondary, inhibitory effect on the other.

Module 22.12 Review

a. What hormone inhibits the satiety center and stimulates appetite in the short-term?

b. Describe leptin and its effect on appetite.

c. How might a lack of Neuropeptide Y in the hypothalamus affect the control of appetite?

To maintain a constant body temperature, heat gain and heat loss must be in balance

Only about 40 percent of energy released by catabolism can be captured as ATP, and the rest is lost as heat that warms surrounding tissues. If body temperature is to remain constant, heat production and heat loss must be kept in balance despite wide variations in activity levels and environmental conditions.

1 This illustration introduces the primary mechanisms of heat transfer between the body and the surrounding environment.

2 The effects of a failure to control body temperature are indicated here.

Primary Mechanisms of Heat Transfer

Radiation: Warm objects lose heat energy as infrared radiation. When you feel the heat from the sun, you are detecting that radiation. Your body loses heat the same way, but in proportionately smaller amounts. More than 50 percent of the heat you lose indoors is attributable to radiation; the exact amount varies with both body temperature and skin temperature.

Convection: Convection is the result of conductive heat loss to the air that overlies the surface of the body. As your body conducts heat to the air next to your skin, that air warms and rises, moving away from the surface of the skin. Cooler air replaces it, and as this air in turn becomes warmed, the pattern repeats. Convection accounts for roughly 15 percent of the body's heat loss indoors.

Evaporation: When water evaporates, it changes from a liquid to a vapor. Evaporation absorbs energy—roughly 0.58 Cal per gram of water evaporated—and cools the surface where evaporation occurs. Each hour, 20–25 mL of water crosses epithelia and evaporates from the alveolar surfaces and the surface of the skin. This **insensible perspiration** remains relatively constant; at rest, it accounts for roughly 20 percent of your body's average indoor heat loss. The sweat glands responsible for **sensible perspiration** have a tremendous scope of activity, ranging from virtual inactivity to secretory rates of 2–4 liters per hour.

Conduction: Conduction is the direct transfer of energy through physical contact. Conduction is generally not an effective mechanism for gaining or losing heat, and its impact depends on the temperature of the object and the amount of skin area it contacts. When you are standing, conductive losses are negligible.

Underlying physical or environmental condition	°F	°C	Thermoregulatory capabilities	Major physiological effects
	114	44	Severely impaired	Death
CNS damage				Proteins denature
	110			Convulsions
Heat stroke		42	Impaired	Cell damage
	106	40		
Disease-related fevers	102			Disorientation
Severe exercise		38	Effective	
Active children				
Normal range (oral)	98	36		Systems normal
Early mornings in cold weather	94	34	Impaired	Disorientation
	90	32		Loss of muscle control
Severe exposure	86	30	Severely impaired	Loss of consciousness
Hypothermia for open heart surgery	82	28		Cardiac arrest
	78	26	Lost	Death
	74	24		

Module 22.13 Review

a. Define insensible perspiration.

b. What heat transfer process accounts for about one-half of an individual's heat loss when indoors?

c. How is heat loss different between conduction and convection?

Hypothalamic thermoregulatory centers adjust rates of heat gain and heat loss

Heat loss and heat gain involve the activities of many systems. Those activities are coordinated by the **heat-loss center** and **heat-gain center**, respectively, in the preoptic area of the hypothalamus.

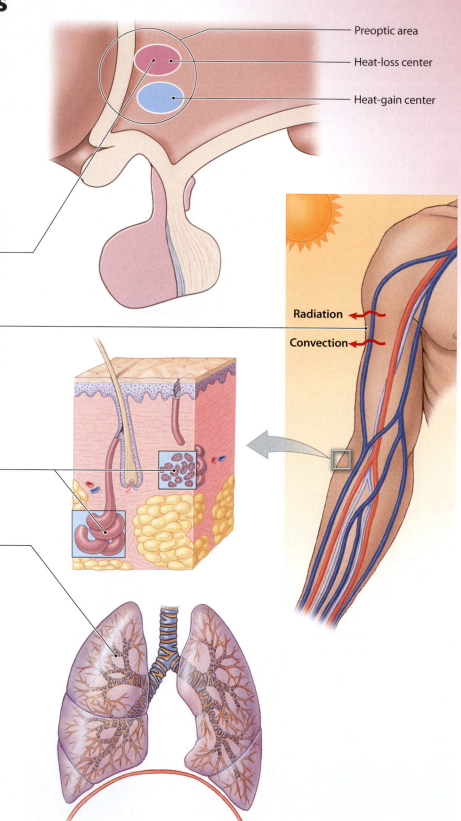

Preoptic area

Heat-loss center

Heat-gain center

Radiation

Convection

Responses to High Body Temperature Coordinated by the Heat-Loss Center

Behavioral Changes: A sense of discomfort leads to behavioral responses—getting into the shade, going into the water, or taking other steps that reduce body temperature.

Vasodilation and Shunting of Blood to Skin Surface: The inhibition of the vasomotor center causes peripheral vasodilation, and warm blood flows to the surface of the body. The skin takes on a reddish color, skin temperatures rise, and radiational and convective losses increase.

Sweat Production: As blood flow to the skin increases, sweat glands are stimulated to increase their secretory output. The perspiration flows across the body surface, and evaporative heat losses accelerate. Maximal secretion, if completely evaporated, would remove 2320 Cal per hour.

Respiratory Heat Loss: The respiratory centers are stimulated, and the depth of respiration increases. Often, the individual begins respiring through an open mouth rather than through the nasal passageways, increasing evaporative heat losses through the lungs.

When Body Temperature Falls

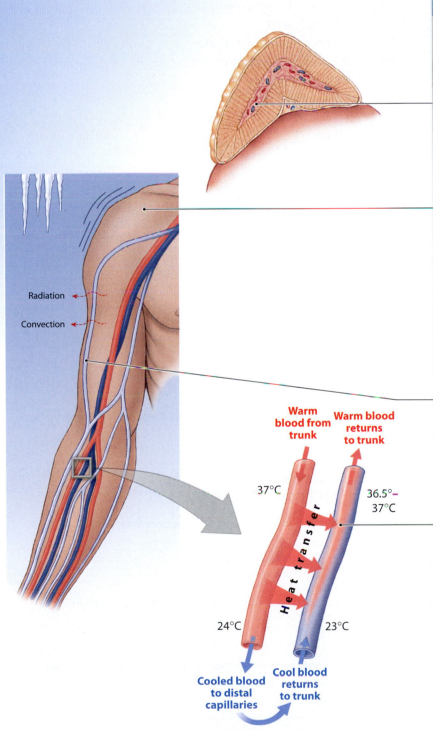

Radiation

Convection

Warm blood from trunk **Warm blood returns to trunk**

37°C   36.5°– 37°C

Heat transfer

24°C   23°C

Cooled blood to distal capillaries **Cool blood returns to trunk**

Responses Coordinated by Heat-Gain Center

The heat-gain center responds to low body temperature in two ways:

Increased Generation of Body Heat

Nonshivering thermogenesis (ther-mō-JEN-e-sis) involves the release of hormones that increase the metabolic activity of all tissues. Sympathetic stimulation of the adrenal medullae releases epinephrine, which quickly increases the rates of glycogenolysis in liver and skeletal muscle and the metabolic rate of most tissues.*

In **shivering thermogenesis**, a gradual increase in muscle tone increases the energy consumption of skeletal muscle tissue throughout your body. Both agonists and antagonists are involved, and muscle tone gradually increases to the point at which stretch receptor stimulation will produce brief, oscillatory contractions of antagonistic muscles. In other words, you begin to shiver. Shivering can elevate body temperature quite effectively, increasing the rate of heat generation by as much as 400 percent.

Conservation of Body Heat

The vasomotor center decreases blood flow to the dermis, thereby reducing losses by radiation and convection. The skin cools, and with blood flow restricted, it may take on a bluish or pale color. The epithelial cells are not damaged, because they can tolerate extended periods at temperatures as low as 25°C (77°F) or as high as 49°C (120°F).

The deep veins lie alongside the deep arteries, and heat is conducted from the warm blood flowing outward to the limbs to the cooler blood returning from the periphery. This arrangement traps the heat close to the body core and dramatically reduces heat loss. The transfer of heat, water, or solutes between fluids moving in opposite directions is called **countercurrent exchange**.

Module 22.14 Review

a. Name the heat conservation mechanism that results in the conduction of heat from deep arteries to adjacent deep veins in the limbs.

b. Describe the role of nonshivering thermogenesis in regulating body temperature.

c. Predict the effect of peripheral vasodilation on an individual's body temperature.

* Note: In children, but not usually in adults, the heat-gain center can increase the rate of TRH release by the hypothalamus when body temperature is below normal. This stimulates the release of thyroid-stimulating hormone (TSH) by the anterior lobe of the pituitary gland, and as the rate of thyroid hormone release increases, so do the rates of catabolism throughout the body.

1. Matching

Match each of the following terms with the most closely related item.

- ghrelin
- basal metabolic rate
- 40 percent
- leptin
- insensible perspiration
- inhibits feeding center
- shivering thermogenesis
- peripheral vasoconstriction
- thermoregulation
- sensible perspiration
- neuropeptide Y
- 60 percent
- nonshivering thermogenesis
- peripheral vasodilation

a _____ Sweat gland activity; heat loss

b _____ Adipose tissue hormone

c _____ Homeostatic control of body temperature

d _____ General role of satiety center

e _____ Release of hormones; increased metabolism

f _____ Percent of catabolic energy released as heat

g _____ Stimulation of vasomotor center

h _____ Appetite-regulating neurotransmitter

i _____ Resting energy expenditure

j _____ Stomach hormone

k _____ Percent of catabolic energy captured as ATP

l _____ Inhibition of vasomotor center

m _____ Epithelial water loss

n _____ Result of increased skeletal muscle tone

2. Multiple choice

Choose the bulleted item that best completes each statement.

a An individual's BMR is influenced by their _____ .

- gender
- body weight
- age
- all of these

b The four processes involved in heat exchange with the environment are _____ .

- sensible, insensible, heat loss, and heat gain
- radiation, conduction, convection, and evaporation
- physiological responses and behavioral modifications
- sensible, insensible, hormones, and heat conservation

c The primary mechanisms for increasing heat loss from the body include _____ .

- vasomotor and respiratory
- sensible and insensible
- physiological responses and behavioral modifications
- acclimatization and vasomotor

d All of the following are responses to an increase in body temperature, except _____ .

- stimulation of the respiratory centers
- stimulation of sweat glands
- peripheral vasoconstriction
- peripheral vasodilation

e If daily intake exceeds total energy demands, the excess energy is stored primarily as _____ .

- triglycerides in adipose tissue
- lipoproteins in the liver
- glycogen in the liver
- glucose in the bloodstream

f All of the following factors suppress appetite, except _____ .

- low blood glucose levels
- high blood glucose levels
- leptin
- stimulation of stretch receptors along the digestive tract

3. Short answer

a Why can energy consumption at rest be estimated by monitoring oxygen utilization? _____

b Describe the responses generated by the heat-gain center. _____

c Describe the heat-gain mechanisms involved in nonshivering thermogenesis. _____

Visual Outline with Key Terms

Summarize the content of each module using the terms in the order provided.

SECTION 1

An Introduction to Cellular Metabolism

- metabolism
- cellular metabolism
- catabolism
- anabolism
- metabolic turnover
- nutrient pool

NUTRIENT POOL

ATP

Aerobic Metabolism (in mitochondria)

22.1

Cells obtain most of their ATP from the electron transport system, which is linked to the citric acid cycle

- acetate
- acetyl-CoA
- acetyl group
- citric acid
- citric acid cycle
- NAD
- FAD
- oxidative phosphorylation
- electron transport system (ETS)
- cytochromes
- ATP synthase

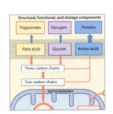

22.2

Cells can break down any available substrate from the nutrient pool to obtain the energy they need

- nutrient pool
- catabolic pathway
- anabolic pathway
- glycogenesis
- glycogenolysis
- gluconeogenesis
- glycolysis

SECTION 2

The Digestion and Metabolism of Organic Nutrients

- oral cavity
- stomach
- duodenum
- jejunum
- hepatic portal vein
- liver

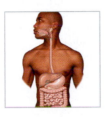

22.3

Carbohydrates are usually the preferred substrates for catabolism and ATP production under resting conditions

- salivary amylase
- pancreatic alpha-amylase
- maltase
- sucrase
- lactase
- flatus
- liver
- skeletal muscle
- glucose
- glycogen
- insulin
- pyruvate

22.4

Glycolysis is the first step in the catabolism of the carbohydrate glucose

- anaerobic
- cytosol
- pyruvate
- glycogenolysis

22.5

Lipids reach the bloodstream in chylomicrons; the cholesterol is then extracted and released as lipoproteins

- lingual lipase
- pancreatic lipase
- bile salts
- emulsification
- micelles
- chylomicrons
- lipoproteins
- lacteals
- thoracic duct
- lipoprotein lipase
- cholesterol
- low-density lipoproteins (LDLs)
- high-density lipoproteins (HDLs)

22.6

Fatty acids can be broken down to provide energy or converted to other lipids

- lipolysis
- beta-oxidation
- lipogenesis
- linolenic acid
- omega-3
- linoleic acid
- omega-6
- essential fatty acids

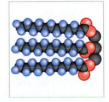

• = *Term boldfaced in this module*

22.7

An amino acid not needed for protein synthesis may be broken down or converted to a different amino acid

- pepsin
- enteropeptidase
- trypsin
- chymotrypsin
- carboxypeptidase
- elastase
- peptidases
- dipeptidases
- ○ hepatic portal vein
- ○ liver
- essential amino acids
- amination
- ○ ammonium ion
- transamination
- deamination
- urea
- urea cycle

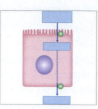

SECTION 3

Energetics and Thermoregulation

- energetics
- basal metabolic rate (BMR)
- thermoregulation

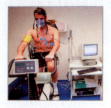

22.8

There are two general patterns of metabolic activity: the absorptive and postabsorptive states

- absorptive state
- ○ insulin
- ○ growth hormone
- ○ androgens
- ○ estrogens
- postabsorptive state
- ketone bodies
- ○ glucocorticoids
- ○ growth hormone
- ○ glucagon
- ○ epinephrine

Glucose levels elevated

Lipid levels elevated

Amino acids elevated

22.12

The control of appetite is complex and involves both short-term and long-term mechanisms

- feeding center
- satiety center
- neuropeptide Y (NPY)
- ghrelin
- leptin

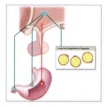

22.9

Vitamins are essential to the function of many metabolic pathways

- nutrition
- vitamins
- fat-soluble vitamins
- avitaminosis
- hypervitaminosis
- water-soluble vitamins
- intrinsic factor

22.13

To maintain a constant body temperature, heat gain and heat loss must be in balance

- ○ heat transfer
- radiation
- convection
- evaporation
- insensible perspiration
- sensible perspiration
- conduction

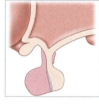

22.10

Proper nutrition depends on eating a balanced diet

- balanced diet
- malnutrition
- ○ food pyramid
- MyPyramid.gov Steps to a Healthier You
- calories
- joules
- kilocalorie
- kilojoule (kJ)
- Calorie
- complete proteins
- incomplete proteins

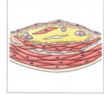

22.14

Hypothalamic thermoregulatory centers adjust rates of heat gain and heat loss

- heat-loss center
- heat-gain center
- ○ preoptic area
- behavioral changes
- vasodilation and shunting of blood to skin surface
- ○ vasomotor center
- ○ radiation
- ○ convection
- sweat production
- respiratory heat loss
- nonshivering thermogenesis
- shivering thermogenesis
- countercurrent exchange

22.11

Metabolic disorders may result from nutritional or biochemical problems

- eating disorders
- anorexia
- anorexia nervosa
- bulimia
- obesity
- regulatory obesity
- metabolic obesity
- ○ cholesterol
- ○ atherosclerosis
- ○ coronary artery disease
- phenylketonuria (PKU)
- protein deficiency disease
- kwashiorkor
- ketone bodies
- ketosis
- ketoacidosis
- uric acid
- nitrogenous wastes
- gout
- gouty arthritis

● = *Term boldfaced in this module*

Chapter Integration: Applying what you've learned

Diet has a profound influence on a person's general health. Too many or too few nutrients, hypervitaminosis (too many vitamins) or avitaminosis (too few vitamins), and above-normal or below-normal concentrations of minerals can adversely affect health. Subtle, long-term problems can occur when the diet includes the wrong proportions or combinations of nutrients. The average diet in the United States contains too much sodium and too many calories in general, and lipids (particularly saturated fats) provide too great a proportion of those calories. Poor diets increase the incidence of obesity, heart disease, atherosclerosis, hypertension, and diabetes in the U.S. population.

Patient X, who has a family history of cardiovascular disease, is concerned about his health. A comprehensive blood test reveals elevated levels of LDL (so-called "bad cholesterol") and decreased levels of HDL (so-called "good cholesterol"), normal glucose, and adequate vitamin levels except vitamin B_6. The physician orders a consultation with the registered dietician, who discovers that Patient X's semi-vegetarian diet is high in carbohydrates and fats.

1. Why are vitamins and minerals essential components of the diet?

2. Explain the labeling of HDL as "good cholesterol" and LDL as "bad cholesterol."

3. Why are high-density lipoproteins (HDLs) considered beneficial?

4. What process in the liver increases after a high-carbohydrate meal?

5. What is the significance of a pyridoxine (vitamin B_6) deficiency?

Access more review material online in the Study Area at www.masteringaandp.com.

There, you'll find:
- Chapter guides
- Chapter quizzes
- Practice tests
- Labeling activities
- Animations
- Tutorials
- MP3 Tutor Sessions
- Flashcards
- A glossary with pronunciations

The kidneys are paired retroperitoneal organs

In this module you will learn the location of the kidneys and their placement in relation to the axial skeleton and the organs of the abdominopelvic cavity. You will also see how the kidneys are supported, protected, and stabilized in position.

1 The kidneys, ureters, urinary bladder, and the associated blood vessels are shown in this anterior view. Because the kidneys are in a retroperitoneal position, they are clearly visible in an anterior view only after other abdominal organs have been removed.

A typical adult **kidney** is reddish-brown and about 10 cm (4 in.) long, 5.5 cm (2.2 in.) wide, and 3 cm (1.2 in.) thick. Each kidney weighs about 150 g (5.25 oz).

The **hilum**, a prominent medial indentation, is the point of entry for the renal artery and renal nerves, and the point of exit for the renal vein and the ureter.

The **ureters** pass inferiorly and cross the anterior surfaces of the external iliac artery and vein before emptying into the posterior, inferior surface of the urinary bladder.

This is the cut edge of the posterior peritoneum, which has been removed. The kidneys, adrenal glands, and ureters lie between the muscles of the posterior body wall and the parietal peritoneum, in a retroperitoneal position.

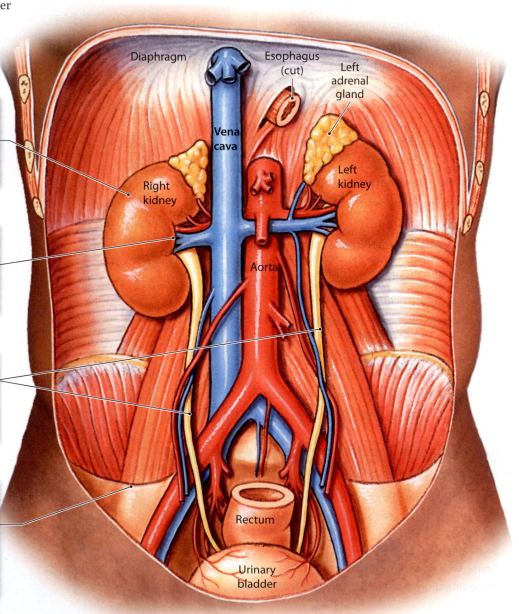

Diaphragm · Esophagus (cut) · Left adrenal gland · Vena cava · Right kidney · Left kidney · Aorta · Rectum · Urinary bladder

2 In this posterior view, you can see that the kidneys are located on either side of the vertebral column, between vertebrae T$_{12}$ and L$_3$. In this position the kidneys are protected by the visceral organs (anteriorly), the musculature of the body wall, and the 11th and 12th ribs (posteriorly and laterally). The left kidney lies slightly superior to the right kidney.

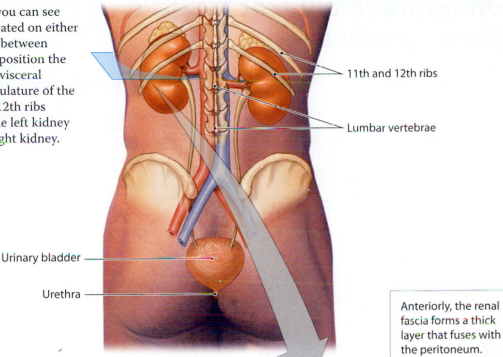

11th and 12th ribs

Lumbar vertebrae

Urinary bladder

Urethra

Anteriorly, the renal fascia forms a thick layer that fuses with the peritoneum.

3 The position of the kidneys in the abdominal cavity is maintained by (1) the overlying peritoneum, (2) contact with adjacent visceral organs, and (3) supporting connective tissues. As seen in this sectional view, each kidney is protected and stabilized by three concentric layers of connective tissue.

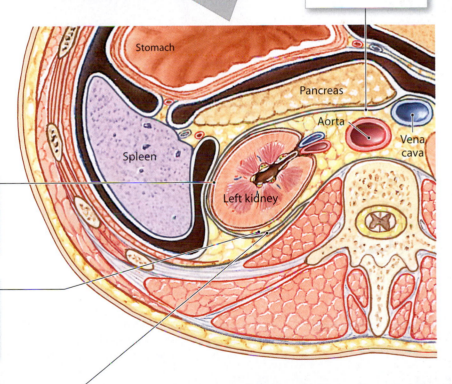

Stomach

Pancreas

Aorta

Vena cava

Spleen

Left kidney

Fibrous Capsule

The **fibrous capsule** is a layer of collagen fibers that covers the outer surface of the entire organ.

Perinephric Fat Capsule

The **perinephric fat capsule** is a thick layer of adipose tissue that surrounds the fibrous capsule.

Renal Fascia

The **renal fascia** is a dense, fibrous outer layer that anchors the kidney to surrounding structures. Collagen fibers extend outward from the fibrous capsule through the perinephric fat to this layer. Posteriorly, the renal fascia fuses with the deep fascia surrounding the muscles of the body wall.

Module 23.1 Review

a. List the main structures composing the urinary system.

b. Describe the concentric layers of connective tissue that protect and anchor the kidney.

c. What would happen to a kidney's position if the perinephric fat layer were depleted and the collagen fibers of the fibrous capsule were to become detached?

The kidneys are complex at the gross and microscopic levels

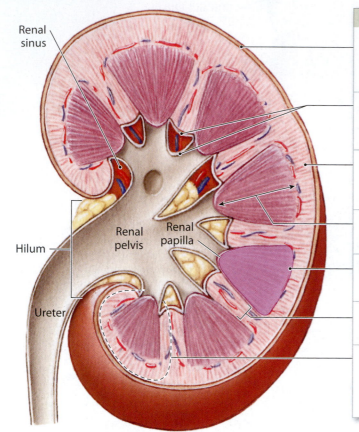

Major Structural Landmarks of the Kidney
The **fibrous capsule** covering the outer surface of the kidney also lines the **renal sinus**, an internal cavity within the kidney. The outer and inner linings are continuous at the hilum.
Within the renal sinus, the fibrous capsule stabilizes the positions of the ureter, the renal blood vessels, and renal nerves.
The **renal cortex** is the superficial portion of the kidney, in contact with the fibrous capsule. The cortex is reddish brown and granular.
The **renal medulla** extends from the renal cortex to the renal sinus.
A **renal pyramid** is a conical structure extending from the cortex to a tip called the **renal papilla**.
A **renal column** is a band of granular tissue that separates adjacent pyramids.
A **kidney lobe** consists of a renal pyramid, the overlying area of renal cortex, and adjacent tissues of the renal columns. Each kidney contains 6–18 kidney lobes. Urine production occurs in the kidney lobes.

Labels on diagram: Renal sinus, Hilum, Ureter, Renal pelvis, Renal papilla

1 Here is a diagrammatic view of a sectioned kidney, showing the major landmarks and features. The blood vessels servicing the kidney enter through the **hilum.**

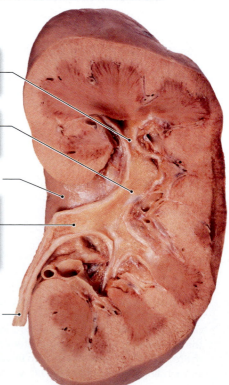

A **minor calyx** collects the urine produced by a single kidney lobe.

A **major calyx** forms through the fusion of 4–5 minor calyces [KĀ-li-sēz].

Hilum

The **renal pelvis** is a large, funnel-shaped structure that collects urine from the major calyces. It is continuous with the ureter.

Ureter

2 Here is a frontal section of a human kidney. The labels follow the path of urine within the collecting system of the renal sinus, which communicates with the ureter at the hilum.

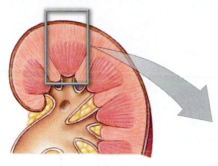

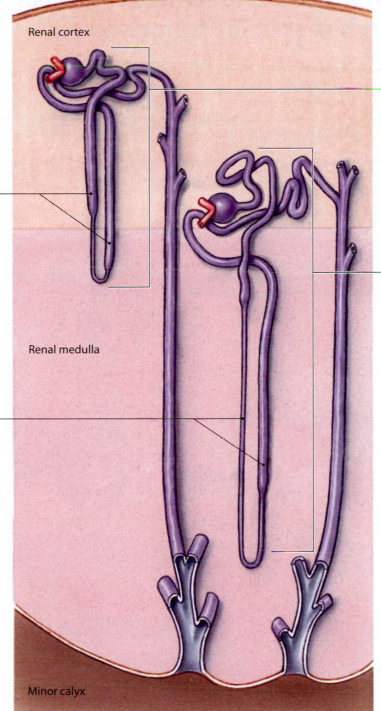

Renal cortex

Cortical nephrons are the most abundant, and they are responsible for most of the regulatory functions of the kidneys.

Nephron loop of cortical nephron

Juxtamedullary nephrons form a small percentage of the total number of nephrons but play a crucial role in establishing conditions in the renal medulla that are essential to water conservation and the production of concentrated urine.

Renal medulla

Nephron loop of juxtamedullary nephron

3   A **nephron** is a microscopic structure that performs the essential functions of the kidney. Nephrons from different locations differ slightly in structure. Roughly 85 percent of all nephrons are **cortical nephrons**, located almost entirely within the superficial cortex of the kidney. The remaining 15 percent of nephrons, termed **juxta-medullary** (juks-tuh-MED-u-lar-ē; *juxta*, near) **nephrons**, have long **nephron loops** that extend deep into the renal medulla.

Minor calyx

Module 23.2 Review

a. Which structure is a conical mass within the renal medulla that ends at the papilla?

b. Describe the renal papilla.

c. Which type of nephron is essential for the conservation of water and the production of concentrated urine?

A nephron can be divided into regions; each region has specific functions

1   The nephron consists of a **renal corpuscle** and a **renal tubule**. At the renal corpuscle, blood pressure forces water and dissolved solutes out of the glomerular capillaries and into a chamber—the **capsular space**—that is continuous with the lumen of the renal tubule. Filtration produces an essentially protein-free solution, known as a **filtrate**, that is otherwise similar to blood plasma. After modification by the renal tubule and collecting system, the filtrate leaves the kidneys as urine.

Nephron

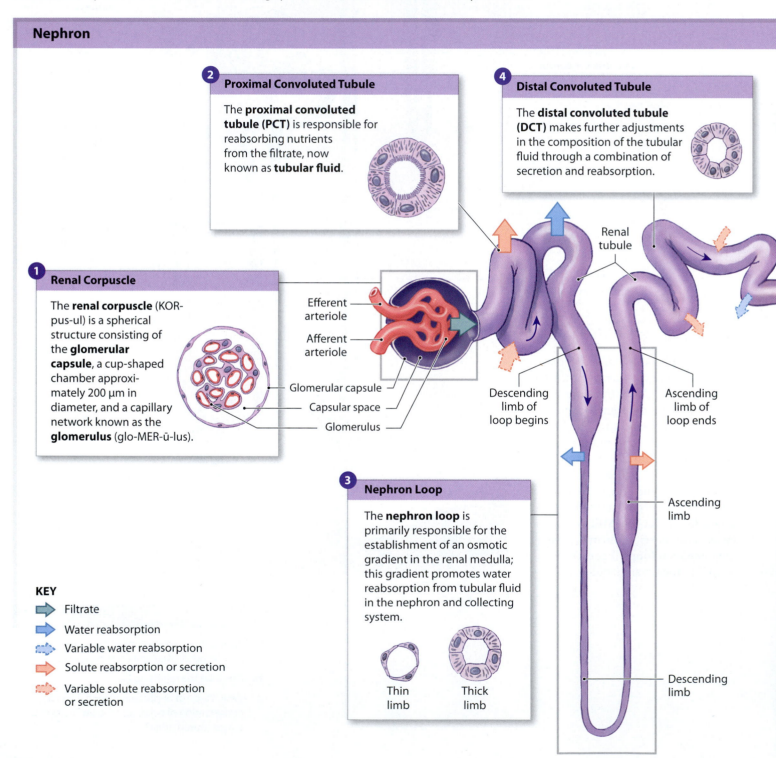

2 **Proximal Convoluted Tubule**

The **proximal convoluted tubule (PCT)** is responsible for reabsorbing nutrients from the filtrate, now known as **tubular fluid**.

4 **Distal Convoluted Tubule**

The **distal convoluted tubule (DCT)** makes further adjustments in the composition of the tubular fluid through a combination of secretion and reabsorption.

1 **Renal Corpuscle**

The **renal corpuscle** (KOR-pus-ul) is a spherical structure consisting of the **glomerular capsule**, a cup-shaped chamber approximately 200 μm in diameter, and a capillary network known as the **glomerulus** (glo-MER-ū-lus).

Efferent arteriole

Afferent arteriole

Glomerular capsule

Capsular space

Glomerulus

Renal tubule

Descending limb of loop begins

Ascending limb of loop ends

Ascending limb

3 **Nephron Loop**

The **nephron loop** is primarily responsible for the establishment of an osmotic gradient in the renal medulla; this gradient promotes water reabsorption from tubular fluid in the nephron and collecting system.

Thin limb

Thick limb

Descending limb

KEY

Filtrate

Water reabsorption

Variable water reabsorption

Solute reabsorption or secretion

Variable solute reabsorption or secretion

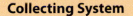

2 Each nephron empties into the **collecting system**, a series of tubes that carry tubular fluid away from the nephron. Collecting ducts receive this fluid from many nephrons. Each collecting duct begins in the renal cortex and descends into the renal medulla, carrying fluid to a papillary duct that drains into a minor calyx.

Collecting System

The collecting system receives the urine from individual nephrons and performs final adjustments in urine volume and composition before delivering it to a minor calyx.

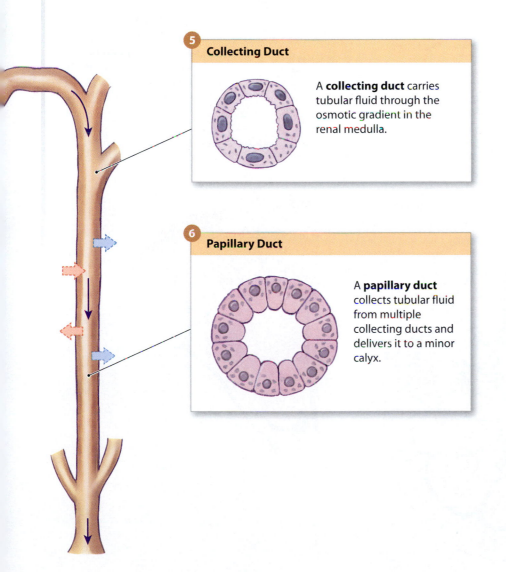

5
Collecting Duct

A **collecting duct** carries tubular fluid through the osmotic gradient in the renal medulla.

6
Papillary Duct

A **papillary duct** collects tubular fluid from multiple collecting ducts and delivers it to a minor calyx.

As it travels along the renal tubule, tubular fluid gradually changes in composition. The characteristics of the urine that enters the minor calyx vary from moment to moment depending on the activities under way in each segment of the nephron and collecting system.

Module 23.3 Review

a. List the primary structures of the nephron and collecting system.

b. Identify the components of the renal corpuscle.

c. What is the difference between fluid formed in the glomerulus and blood plasma?

The kidneys are highly vascular, and the circulation patterns are complex

1 Here are diagrammatic views of the **arterial** system supplying the kidney (top) and the **venous** system draining the kidney (bottom).

Interlobar arteries branch from the segmental arteries and radiate outward within the renal columns.

Segmental arteries form through the branching of the renal artery inside the renal sinus.

Each kidney receives blood through a **renal artery**, which originates at the aorta near the origin of the superior mesenteric artery.

Arcuate arteries originate at interlobar arteries and arch along the boundary between the renal cortex and renal medulla.

Cortical radiate arteries supply the cortical portions of adjacent kidney lobes.

Afferent arterioles that branch off the cortical radiate arteries supply blood to individual nephrons.

Each afferent arteriole delivers blood to a capillary knot called a **glomerulus**. Blood is then distributed to the capillaries of the nephron as detailed in **2**.

Cortical radiate veins collect blood from the capillaries of the nephrons.

Arcuate veins collect blood from associated cortical radiate veins.

Interlobar veins collect blood from arcuate veins. They drain directly into the renal vein because there are no segmental veins.

The **renal vein** returns the blood to the inferior vena cava.

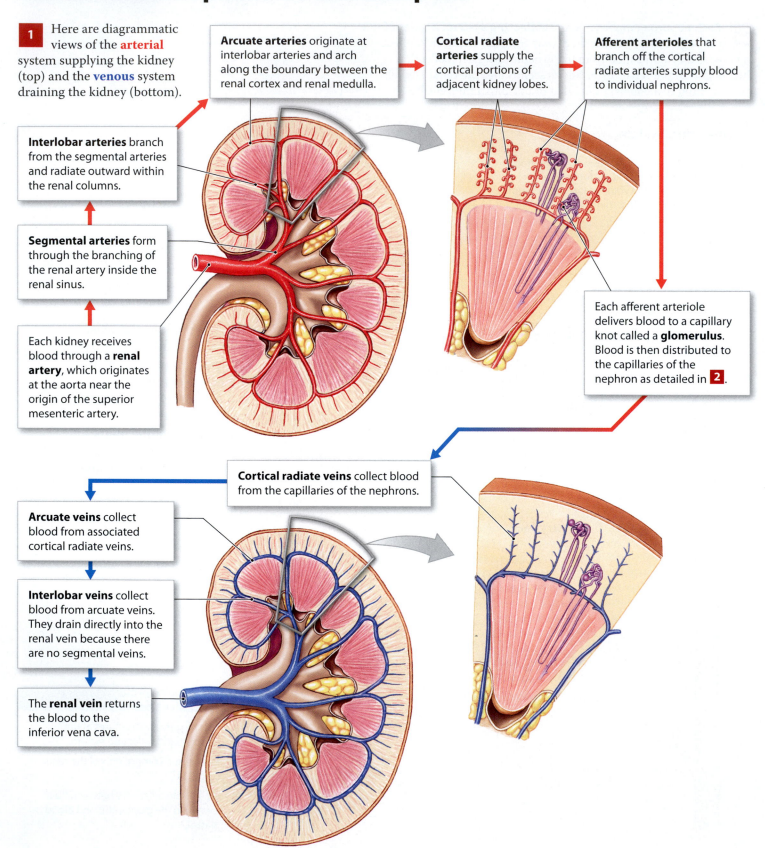

2 In a cortical nephron, the nephron loop is relatively short, and the **efferent arteriole** delivers blood to a network of **peritubular capillaries**, which surround the entire renal tubule. Both the nephrons and the peritubular capillaries are surrounded by interstitial fluid called peritubular fluid. These capillaries drain into small venules that carry blood to the cortical radiate veins.

3 In a juxtamedullary nephron, the peritubular capillaries are connected to the **vasa recta** (*vasa*, vessel + *recta*, straight)—long, straight capillaries that parallel the nephron loop.

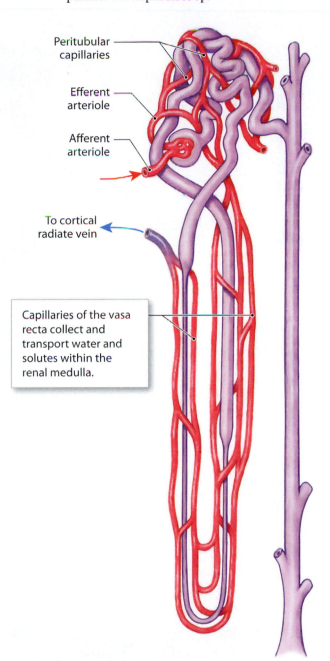

3 The peritubular capillaries collect water and solutes reabsorbed by the nephron, and deliver other solutes to the nephron for secretion.

Glomerulus in renal corpuscle

2 The efferent arteriole carries blood from the glomerulus to the peritubular capillaries.

Start **1** The afferent arteriole delivers blood to the glomerulus, where filtration occurs.

To the cortical radiate vein

Peritubular capillaries

Efferent arteriole

Afferent arteriole

To cortical radiate vein

Capillaries of the vasa recta collect and transport water and solutes within the renal medulla.

Each kidney has approximately 1.25 million nephrons, with a combined length of about 145 km (85 miles). Both the cortical and the juxtamedullary nephrons are innervated by renal nerves that enter at the hilum and follow the branches of the renal arteries. Most of the nerve fibers involved are sympathetic postganglionic fibers from the celiac plexus and the inferior splanchnic nerves. Sympathetic innervation adjusts blood flow and blood pressure at the glomeruli, and stimulates the release of renin.

Module 23.4 Review

a. Trace the pathway of blood from the renal artery to the renal vein.

b. Describe how blood enters and leaves the glomerulus.

c. What are the vasa recta?

1. Short answer

Label the kidney structures in the following diagram, and then provide a brief functional/anatomical description of each.

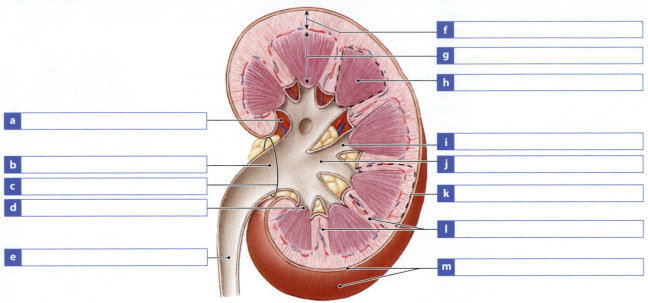

a _____
b _____
c _____
d _____
e _____
f _____
g _____
h _____
i _____
j _____
k _____
l _____
m _____

2. Concept map

Use each of the following terms once to fill in the blank boxes to correctly complete the urinary system concept map.

- ureter
- proximal convoluted tubule
- glomerulus
- urinary bladder
- renal tubules
- papillary duct
- major calyces
- renal medulla
- renal sinus
- nephrons

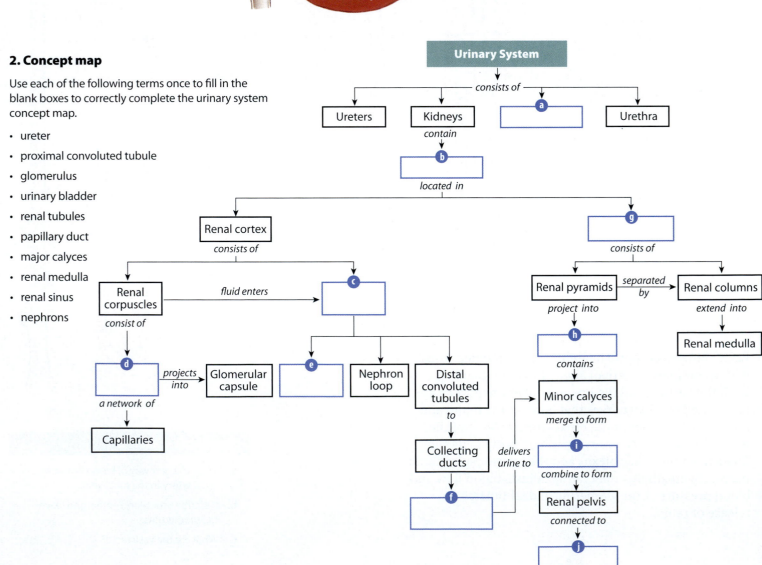

Overview of Renal Physiology

The goal of urine production is to maintain homeostasis by regulating the volume and composition of blood. This process involves the excretion of solutes—specifically, metabolic waste products. Three organic waste products are noteworthy:

- **Urea** is the most abundant organic waste. Urea is a by-product of the breakdown of amino acids in the liver.
- **Creatinine** is generated in skeletal muscle tissue through the breakdown of creatine phosphate, a high-energy compound that plays an important role in muscle contraction.
- **Uric acid** is a waste product formed during the recycling of the nitrogenous bases of RNA molecules.

The kidneys are usually capable of producing concentrated urine with an osmotic concentration of 1200–1400 mOsm/L, more than four times that of plasma.

1 The kidneys produce a fluid that is very different from other body fluids. This table demonstrates renal efficiency by comparing the concentrations of representative substances in urine and plasma.

Normal Laboratory Values for Solutes in Plasma and Urine		
Solute	**Plasma**	**Urine**
Ions (mEq/L)		
Sodium (Na$^+$)	135–145	40–220
Potassium (K$^+$)	3.5–5.0	25–100
Chloride (Cl$^-$)	100–108	110–250
Bicarbonate (HCO$_3^-$)	20–28	1.9
Metabolites and Nutrients (mg/dl)		
Glucose	70–110	0.009
Lipids	450–1000	0.002
Amino acids	40	0.188
Proteins	6.0–8.0 g/dL	0.000
Nitrogenous Wastes (mg/dL)		
Urea	8–25	1800
Creatinine	0.6–1.5	150
Ammonia	<0.1	60
Uric acid	2–6	40

Note: The values indicated are typical ranges; specific numbers vary depending on the laboratory and methods used.

2 To perform their functions, the kidneys rely on three distinct physiological processes.

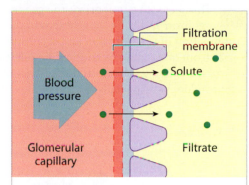

In **filtration**, blood pressure forces water and solutes across the membranes of the glomerular capillaries and into the capsular space. Solute molecules small enough to pass through the filtration membrane are carried by the surrounding water molecules.

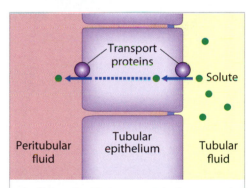

Reabsorption is the removal of water and solutes from the tubular fluid and their movement across the tubular epithelium and into the peritubular fluid.

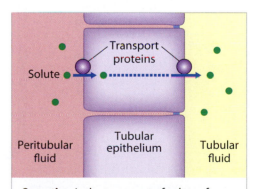

Secretion is the transport of solutes from the peritubular fluid, across the tubular epithelium, and into the tubular fluid.

Filtration, reabsorption, and secretion occur in specific regions of the nephron and collecting system

1 This diagram summarizes the general functions of the various segments of the nephron and collecting system in the formation of urine. Most regions perform a combination of reabsorption and secretion, but the balance between the two processes varies from one region to another. Regulation of the final volume and solute concentration of the urine results from the interaction between the collecting system and the nephron loops—especially the long loops of the juxtamedullary nephrons.

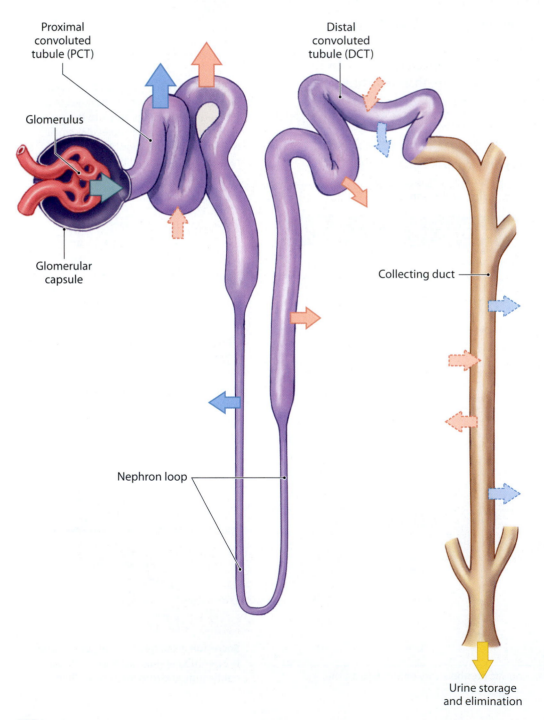

Proximal convoluted tubule (PCT)

Distal convoluted tubule (DCT)

Glomerulus

Glomerular capsule

Collecting duct

Nephron loop

Urine storage and elimination

KEY

Filtration occurs exclusively in the renal corpuscle, across the filtration membrane.

Water reabsorption occurs primarily along the PCT and the descending limb of the nephron loop, but also to a variable degree in the DCT and collecting system.

Variable water reabsorption occurs in the DCT and collecting system.

Solute reabsorption occurs along the PCT, the ascending limb of the nephron loop, the DCT, and the collecting system.

Variable solute reabsorption or secretion occurs at the PCT, the DCT, and the collecting system.

Renal Structures and Their Functions

Segment	General Functions	Specific Functions
Renal corpuscle	*Filtration* of plasma; generates approximately 180 L/day of filtrate similar in composition to blood plasma but without plasma proteins	*Filtration:* water and inorganic and organic solutes from plasma *Retention:* plasma proteins and blood cells
Proximal convoluted tubule (PCT)	*Reabsorption* of 60–70% of the water (108–116 L/day), 99–100% of the organic substrates, and 60–70% of the sodium and chloride ions in the original filtrate	*Active reabsorption:* glucose, other simple sugars, amino acids, vitamins, ions (including sodium, potassium, calcium, magnesium, phosphate, and bicarbonate) *Passive reabsorption:* urea, chloride ions, lipid-soluble materials, water *Secretion:* hydrogen ions, ammonium ions, creatinine, drugs, and toxins
Nephron loop	*Reabsorption* of 25% of the water (45 L/day) and 20–25% of the sodium and chloride ions present in the original filtrate; creation of the concentration gradient in the renal medulla	*Reabsorption:* sodium and chloride ions, water
Distal convoluted tubule (DCT)	*Reabsorption* of a variable amount of water (usually 5%, or 9 L/day) under ADH stimulation, and a variable amount of sodium ions under aldosterone stimulation	*Reabsorption:* sodium and chloride ions, sodium ions (variable), calcium ions (variable), water (variable) *Secretion:* hydrogen ions, ammonium ions, creatinine, drugs, and toxins
Collecting system	*Reabsorption* of a variable amount of water (usually 9.3%, or 16.8 L/day) under ADH stimulation, and a variable amount of sodium ions under aldosterone stimulation	*Reabsorption:* sodium ions (variable), bicarbonate ions (variable), water (variable) *Secretion:* potassium and hydrogen ions (variable)
Peritubular capillaries	*Redistribution* of water and solutes reabsorbed in the renal cortex	Return of water and solutes from the peritubular fluid to the general circulation
Vasa recta	*Redistribution* of water and solutes reabsorbed in the renal medulla, and stabilization of the concentration gradient of the renal medulla	Return of water and solutes from the peritubular fluid to the general circulation

2 The table above provides a brief overview of the functions of the various parts of the nephron and collecting system. We will consider these segments and their roles in greater detail in subsequent modules.

Module 23.5 Review

a. Identify the three distinct processes of urine formation in the kidney.

b. Where does filtration exclusively occur in the kidney?

c. What occurs when the plasma concentration of a substance exceeds the kidney's capacity for transporting that substance?

Filtration occurs at the renal corpuscle

The renal corpuscle, the start of the nephron, is responsible for the filtration of blood. This is the vital first step in the formation of urine.

1 At the renal corpuscle, the capillary knot of the glomerulus projects into the capsular space like the heart projects into the pericardial cavity. Like the pericardium, the glomerular capsule has an outer parietal layer and an inner visceral layer.

The glomerular capsule forms the outer wall of the renal corpuscle and covers the glomerular capillaries.

The **capsular space** separates the parietal and visceral layers of the glomerular capsule.

Initial segment of renal tubule

The **efferent arteriole** delivers blood to peritubular capillaries. It has a smaller diameter than the afferent arteriole; this elevates the blood pressure within the glomerulus.

DCT

The **juxtaglomerular complex** consists of specialized cells that secrete renin when glomerular blood pressure falls.

Parietal layer

Visceral layer

The **afferent arteriole** delivers blood from a cortical radiate artery.

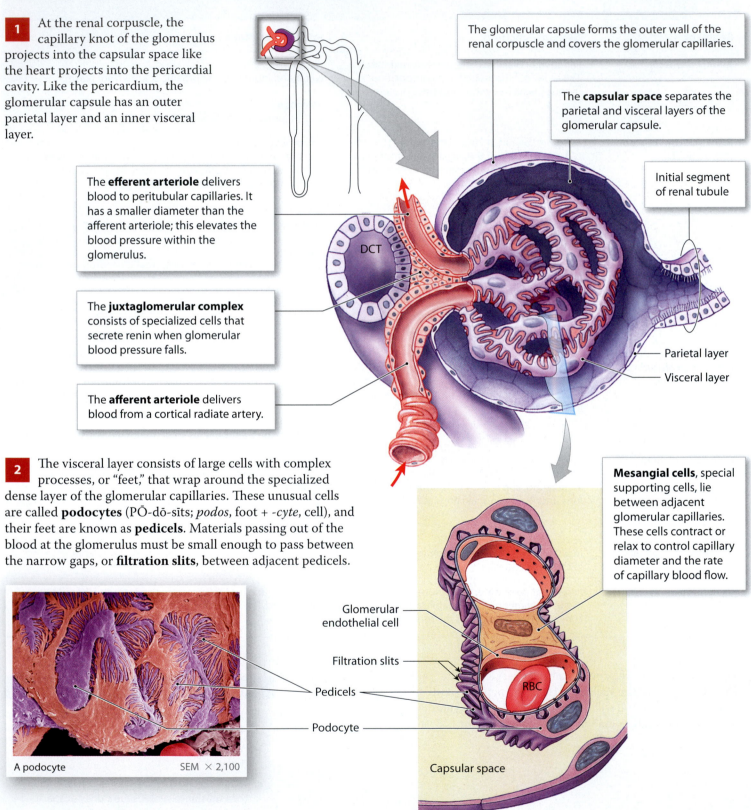

2 The visceral layer consists of large cells with complex processes, or "feet," that wrap around the specialized dense layer of the glomerular capillaries. These unusual cells are called **podocytes** (PŌ-dō-sīts; *podos*, foot + *-cyte*, cell), and their feet are known as **pedicels**. Materials passing out of the blood at the glomerulus must be small enough to pass between the narrow gaps, or **filtration slits**, between adjacent pedicels.

Mesangial cells, special supporting cells, lie between adjacent glomerular capillaries. These cells contract or relax to control capillary diameter and the rate of capillary blood flow.

Glomerular endothelial cell

Filtration slits

Pedicels

Podocyte

RBC

A podocyte   SEM × 2,100

Capsular space

3 The glomerular capillaries are fenestrated capillaries containing large-diameter pores. The endothelium covers a **dense layer**, a specialized type of basal lamina. Together, the fenestrated endothelium, the dense layer, and the filtration slits form the **filtration membrane**. Under normal circumstances only a few plasma proteins—such as albumin molecules, with an average diameter of 7 nm—can cross the filtration membrane and enter the capsular space.

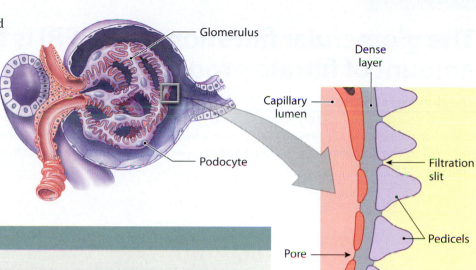

Glomerulus

Podocyte

Dense layer

Capillary lumen

Filtration slit

Pedicels

Pore

Capsular space

Filtration membrane

Factors Controlling Glomerular Filtration

The **glomerular hydrostatic pressure (GHP)** is the blood pressure in the glomerular capillaries. This pressure tends to push water and solute molecules out of the plasma and into the filtrate. The GHP, which averages 50 mm Hg, is significantly higher than capillary pressures elsewhere in the systemic circuit, due to the different diameters of the afferent and efferent capillaries.

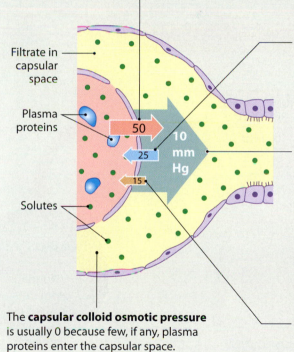

Filtrate in capsular space

Plasma proteins

Solutes

50

25

15

10 mm Hg

The **blood colloid osmotic pressure (BCOP)** tends to draw water out of the filtrate and into the plasma; it thus opposes filtration. Over the entire length of the glomerular capillary bed, the BCOP averages about 25 mm Hg.

The **net filtration pressure (NFP)** is the pressure acting across the glomerular capillaries. It represents the sum of the hydrostatic pressures and the colloid osmotic pressures. Under normal circumstances, the net filtration pressure is approximately 10 mm Hg. This is the average pressure forcing water and dissolved materials out of the glomerular capillaries and into the capsular space.

The **capsular colloid osmotic pressure** is usually 0 because few, if any, plasma proteins enter the capsular space.

Capsular hydrostatic pressure (CsHP) opposes GHP. CsHP, which tends to push water and solutes out of the filtrate and into the plasma, results from the resistance of filtrate already present in the nephron that must be pushed toward the renal pelvis.

4 The primary factor involved in glomerular filtration is basically the same as that governing fluid and solute movement across capillaries throughout the body: the balance between **hydrostatic pressure** (fluid pressure) and **colloid osmotic pressure** (pressure due to materials in solution) on either side of the capillary membrane.

Module 23.6 Review

a. The capsular space separates which layers of the glomerular capsule?

b. Explain why blood pressure is higher in glomerular capillaries than in other systemic capillaries.

c. Blood colloidal osmotic pressure tends to draw water out of the filtrate and into the plasma. Why does this occur?

The glomerular filtration rate (GFR) is the amount of filtrate produced each minute

The **glomerular filtration rate (GFR)** is the amount of filtrate the kidneys produce each minute. Two interacting levels of control stabilize GFR: (1) autoregulation at the local level, and (2) central regulation, which has an endocrine component initiated by the kidneys and an autonomic component involving the sympathetic division of the ANS.

1 Through autoregulation the kidneys adjust GFR in response to changes in the local environment at and around the nephrons.

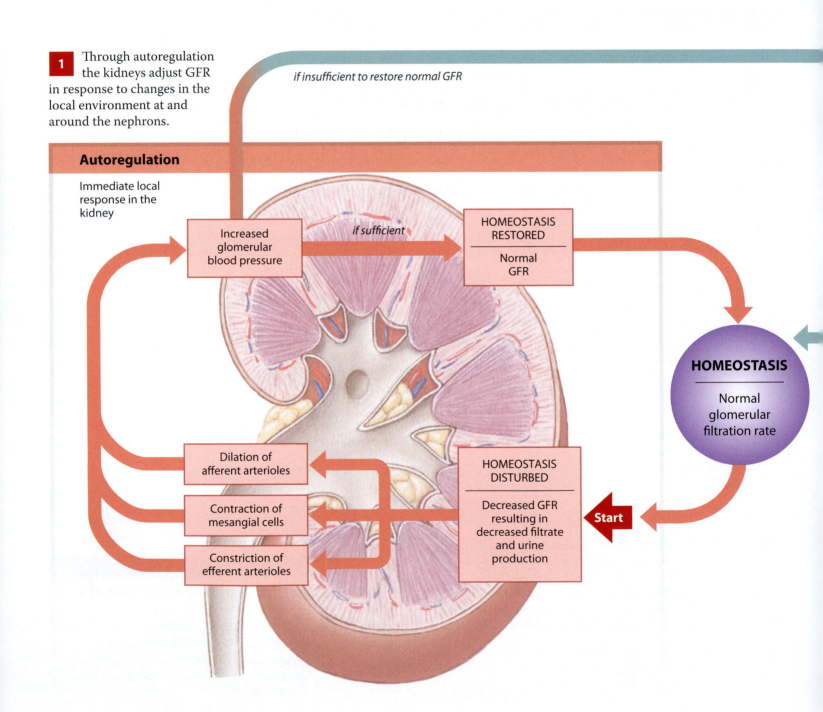

if insufficient to restore normal GFR

Autoregulation

Immediate local response in the kidney

Increased glomerular blood pressure

if sufficient

HOMEOSTASIS RESTORED

Normal GFR

Dilation of afferent arterioles

Contraction of mesangial cells

Constriction of efferent arterioles

HOMEOSTASIS DISTURBED

Decreased GFR resulting in decreased filtrate and urine production

Start

HOMEOSTASIS

Normal glomerular filtration rate

2 In essence, the kidneys call for a central response that elevates GFR if autoregulation is ineffective. This response involves multiple systems and mechanisms.

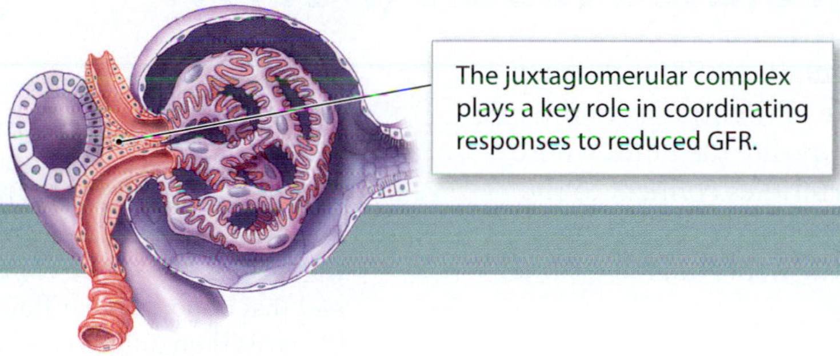

The juxtaglomerular complex plays a key role in coordinating responses to reduced GFR.

Central Regulation

Integrated endocrine and neural mechanisms activated

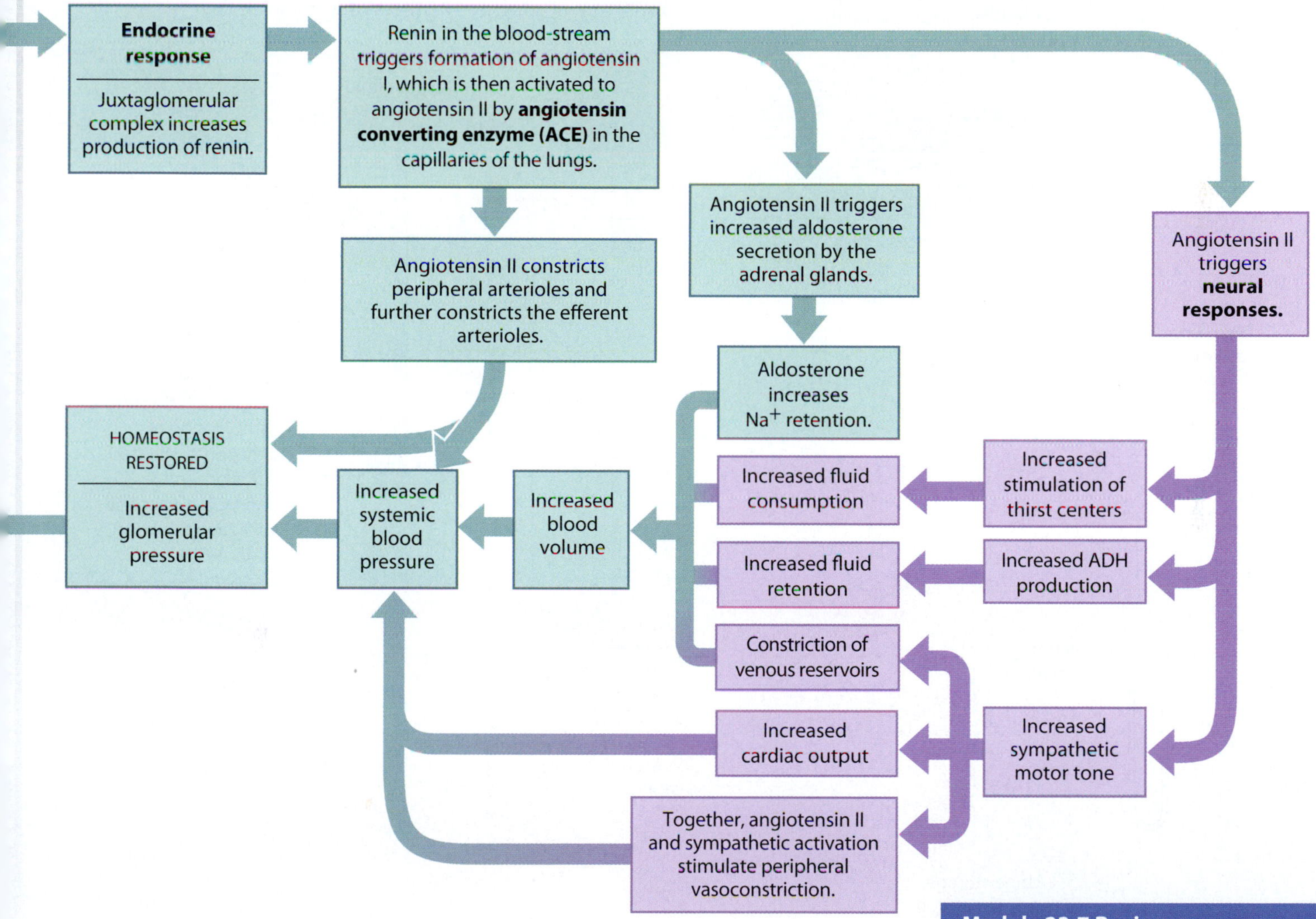

Endocrine response

Juxtaglomerular complex increases production of renin.

Renin in the blood-stream triggers formation of angiotensin I, which is then activated to angiotensin II by **angiotensin converting enzyme (ACE)** in the capillaries of the lungs.

Angiotensin II constricts peripheral arterioles and further constricts the efferent arterioles.

Angiotensin II triggers increased aldosterone secretion by the adrenal glands.

Angiotensin II triggers **neural responses.**

Aldosterone increases Na^+ retention.

HOMEOSTASIS RESTORED
_____
Increased glomerular pressure

Increased systemic blood pressure

Increased blood volume

Increased fluid consumption

Increased stimulation of thirst centers

Increased fluid retention

Increased ADH production

Constriction of venous reservoirs

Increased cardiac output

Increased sympathetic motor tone

Together, angiotensin II and sympathetic activation stimulate peripheral vasoconstriction.

Each kidney contains about 6 m² — some 64 square feet — of filtration surface, and the GFR averages an astounding 125 mL per minute. This means that roughly 10 percent of the fluid delivered to the kidneys by the renal arteries leaves the bloodstream and enters the capsular spaces. In the course of a single day, the glomeruli generate about 180 liters (48 gal) of filtrate, about 70 times the total plasma volume. But as filtrate passes through the renal tubules, about 99 percent of it is reabsorbed.

Module 23.7 Review

a. In response to decreased filtration pressure, the juxtaglomerular complex does what?

b. Angiotensin II has what effect on nephrons?

c. Angiotensin II has what effect on the CNS?

Reabsorption predominates along the proximal convoluted tubule (PCT)...

The proximal convoluted tubule (PCT) is the first segment of the renal tubule. The entrance to the PCT lies almost directly opposite the point where the afferent and efferent arterioles connect to the glomerulus. The distal convoluted tubule, which forms the last segment of the nephron, makes final adjustments in the solute composition of the tubular fluid.

1 Under normal circumstances, before the tubular fluid enters the nephron loop, the PCT reabsorbs more than 99 percent of the glucose, amino acids, and other organic nutrients in the fluid. The PCT also reabsorbs sodium, potassium, bicarbonate, magnesium, phosphate, and sulfate ions. As reabsorption occurs, the solute concentration of tubular fluid decreases, and that of peritubular fluid and adjacent capillaries increases. Osmosis then pulls water out of the tubular fluid and into the peritubular fluid. Along the PCT, this mechanism results in the reabsorption of roughly 108 liters of water each day.

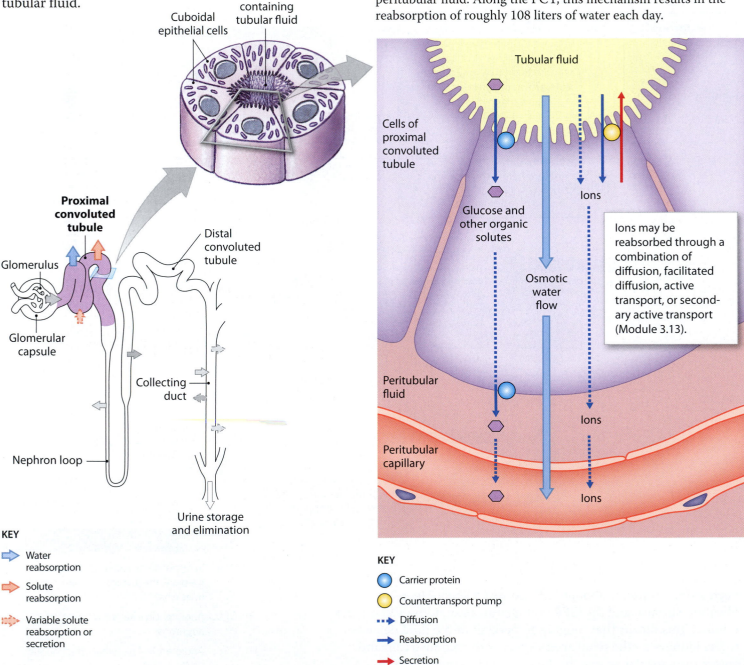

Lumen containing tubular fluid

Cuboidal epithelial cells

Proximal convoluted tubule

Glomerulus

Glomerular capsule

Distal convoluted tubule

Nephron loop

Collecting duct

Urine storage and elimination

Tubular fluid

Cells of proximal convoluted tubule

Glucose and other organic solutes

Ions

Ions may be reabsorbed through a combination of diffusion, facilitated diffusion, active transport, or secondary active transport (Module 3.13).

Osmotic water flow

Peritubular fluid

Peritubular capillary

Ions

Ions

KEY

⇨ Water reabsorption

⇨ Solute reabsorption

⇢⇢ Variable solute reabsorption or secretion

KEY

🔵 Carrier protein

🟡 Countertransport pump

┅▶ Diffusion

──▶ Reabsorption

──▶ Secretion

...whereas reabsorption and secretion are often linked along the distal convoluted tubule (DCT)

2 Only 15–20 percent of the initial filtrate volume reaches the DCT, and the concentrations of electrolytes and organic wastes in the arriving tubular fluid no longer resemble the concentrations in blood plasma. In the DCT, a combination of secretion and reabsorption further alters the solute composition of the tubular fluid.

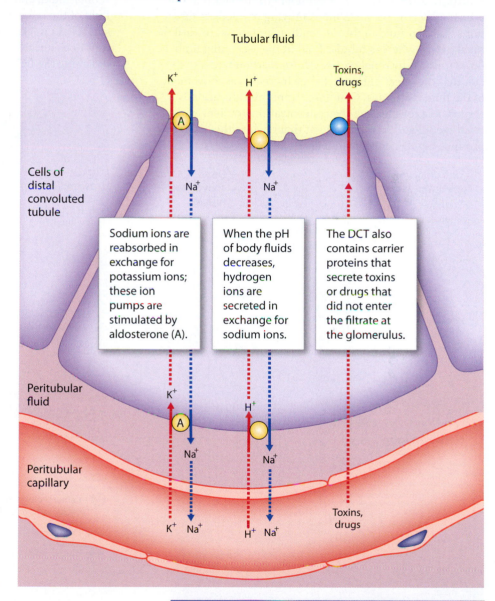

Tubular fluid

Cells of distal convoluted tubule

Sodium ions are reabsorbed in exchange for potassium ions; these ion pumps are stimulated by aldosterone (A).

When the pH of body fluids decreases, hydrogen ions are secreted in exchange for sodium ions.

The DCT also contains carrier proteins that secrete toxins or drugs that did not enter the filtrate at the glomerulus.

Peritubular fluid

Peritubular capillary

Toxins, drugs

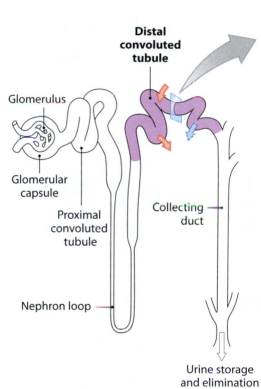

Distal convoluted tubule

Glomerulus

Glomerular capsule

Proximal convoluted tubule

Collecting duct

Nephron loop

Urine storage and elimination

Module 23.8 Review

a. Identify the segment of the nephron that makes final adjustments to the composition of tubular fluid.

b. What effect would increased amounts of aldosterone have on the K^+ concentration in urine?

c. What effect would a decrease in the Na^+ concentration of filtrate have on the pH of tubular fluid?

Feedback between the nephron loop and the collecting duct creates the osmotic gradient in the renal medulla

1 The thin descending limb and the thick ascending limb of the nephron loop are very close together, separated only by peritubular fluid. The exchange that occurs between these segments is called **countercurrent multiplication**. *Countercurrent* refers to the fact that the exchange occurs between fluids moving in opposite directions: Tubular fluid in the descending limb flows toward the renal pelvis, whereas tubular fluid in the ascending limb flows toward the renal cortex. *Multiplication* refers to the fact that the effect of the exchange increases as movement of the fluid continues. Countercurrent multiplication is responsible for creating the concentration gradient in the renal medulla. It is this gradient that enables the kidney to produce highly concentrated urine.

2 The key to this process is the transport activity performed by the cells of the thick ascending limb. Active transport at the apical surface moves sodium and chloride ions out of the tubular fluid, and they then enter the peritubular fluid of the renal medulla. Because the apical surface is impermeable to water, the transport of ions does not result in an osmotic water flow as it did along the PCT.

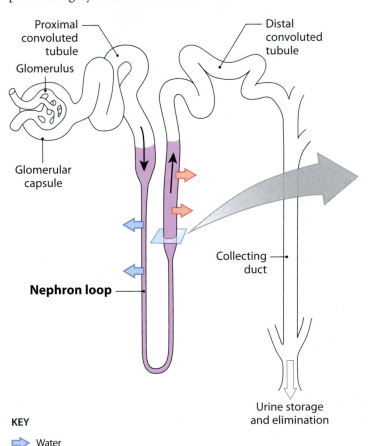

KEY

Water reabsorption

Solute reabsorption

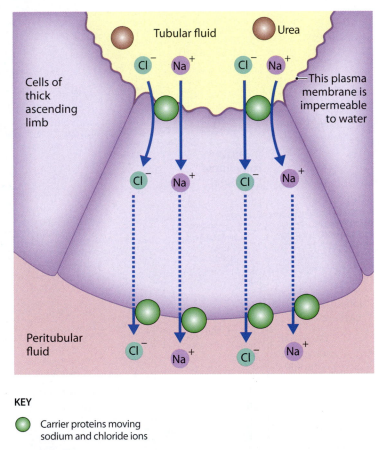

KEY

Carrier proteins moving sodium and chloride ions

Diffusion

Reabsorption

3 The removal of sodium and chloride ions from the tubular fluid in the ascending limb elevates the osmotic concentration of the peritubular fluid around the thin descending limb. Because the thin descending limb is permeable to water but impermeable to solutes, as tubular fluid travels deeper into the renal medulla along the thin descending limb, osmosis moves water into the peritubular fluid.

4 Urea is an important solute that is eliminated in the urine. As water is reabsorbed along the DCT and collecting duct, the concentration of urea gradually rises in the tubular fluid. The tubular fluid reaching the papillary duct typically contains urea at a concentration of about 450 mOsm/L.

The thin descending limb is permeable to water, but impermeable to solutes. Solutes remain behind as water departs, so the tubular fluid reaching the turn of the nephron loop has a higher osmotic concentration than it did at the start.

1 The thin descending limb is permeable to water, but impermeable to solutes, including urea.

2 The thick ascending limb of the nephron loop is impermeable to water and solutes.

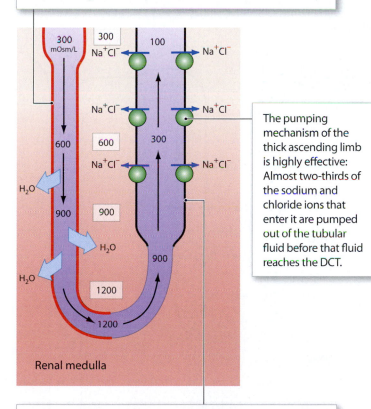

The pumping mechanism of the thick ascending limb is highly effective: Almost two-thirds of the sodium and chloride ions that enter it are pumped out of the tubular fluid before that fluid reaches the DCT.

The thick ascending limb is impermeable to water and other solutes. In other tissues, differences in solute concentration are quickly resolved by osmosis. But since osmosis cannot occur across the impermeable wall of the thick ascending limb, the solute concentration in the tubular fluid declines as Na^+ and Cl^- are removed.

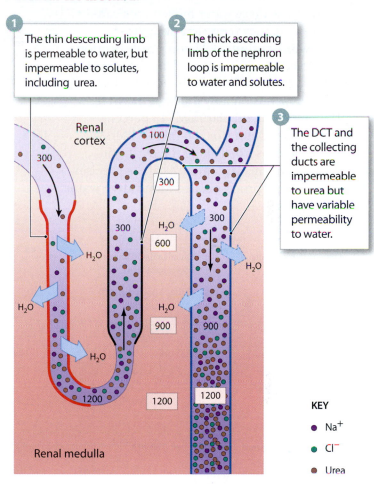

3 The DCT and the collecting ducts are impermeable to urea but have variable permeability to water.

KEY

- Na^+
- Cl^-
- Urea

Module 23.9 Review

a. Define countercurrent multiplication as it occurs in the kidneys.

b. The thick ascending limb of the nephron loop actively pumps what substances into the peritubular fluid?

c. An increase in sodium and chloride ions in the peritubular fluid affects the descending limb in what way?

Urine volume and concentration are hormonally regulated

Urine volume and osmotic concentration are regulated through the control of water reabsorption. The water permeabilities of the PCT and descending limb of the nephron loop cannot be adjusted, and water reabsorption occurs whenever the osmotic concentration of the peritubular fluid exceeds that of the tubular fluid. Because these water movements cannot be prevented, they represent **obligatory water reabsorption**. Obligatory reabsorption usually recovers 85 percent of the volume of filtrate produced. The volume of water lost in urine depends on how much of the water in the remaining tubular fluid (15 percent of the filtrate volume, or roughly 27 liters per day) is reabsorbed along the DCT and collecting system. The amount can be precisely controlled by a process called **facultative water reabsorption**.

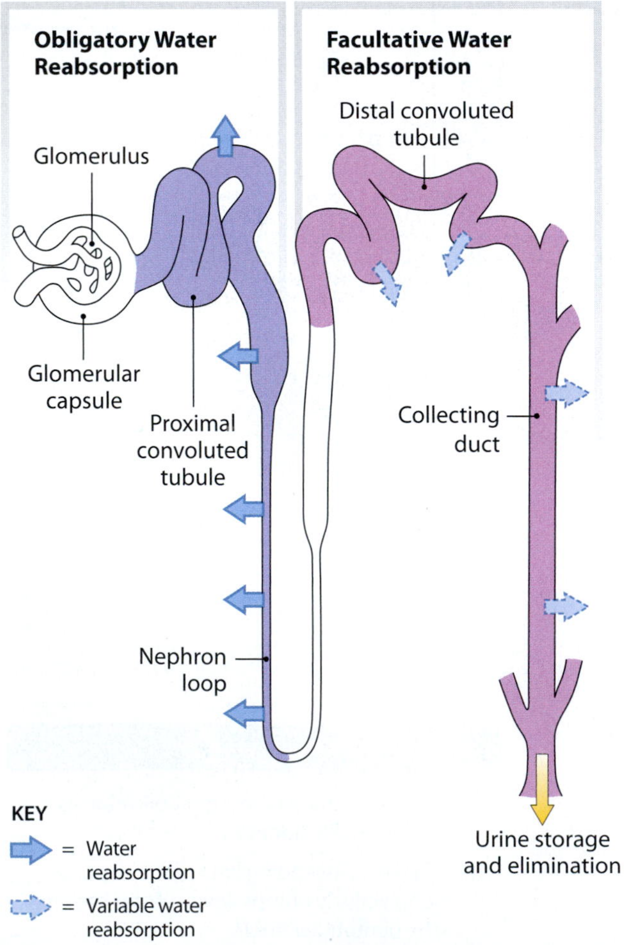

1 In the absence of ADH , water is not reabsorbed in the distal collecting tubule or the collecting duct, so all the fluid reaching the DCT is lost in the urine. No facultative water reabsorption occurs, and the individual then produces large amounts of very dilute urine.

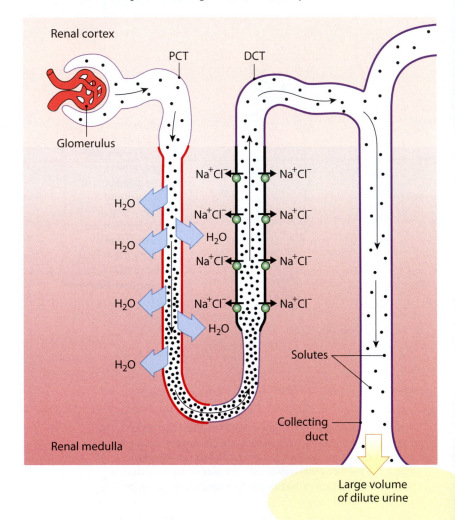

Obligatory Water Reabsorption

Glomerulus

Glomerular capsule

Proximal convoluted tubule

Nephron loop

Facultative Water Reabsorption

Distal convoluted tubule

Collecting duct

Urine storage and elimination

KEY

➡ = Water reabsorption

⇢ = Variable water reabsorption

Renal cortex

PCT   DCT

Glomerulus

Na^+Cl^- Na^+Cl^-

H_2O Na^+Cl^- Na^+Cl^-

H_2O H_2O

Na^+Cl^- Na^+Cl^-

H_2O Na^+Cl^- Na^+Cl^-

H_2O

H_2O

Renal medulla

Solutes

Collecting duct

Large volume of dilute urine

2 ADH causes the appearance of special water channels, called **aquaporins**, in the apical plasma membranes lining the DCT and collecting duct. This dramatically enhances the rate of osmotic water movement. As ADH levels rise, the DCT and collecting system thus become more permeable to water, the amount of water reabsorbed increases, and the urine osmotic concentration climbs. Under maximum ADH stimulation, the DCT and collecting system become so permeable to water that the osmotic concentration of the urine is equal to that of the deepest portion of the renal medulla.

3 A healthy adult typically produces 1200 mL of urine per day (about 0.6 percent of the filtrate volume), with an osmotic concentration of roughly 1000 mOsm/L. However, normal values differ from individual to individual and from day to day, as the kidneys alter their function to maintain homeostatic conditions within body fluids.

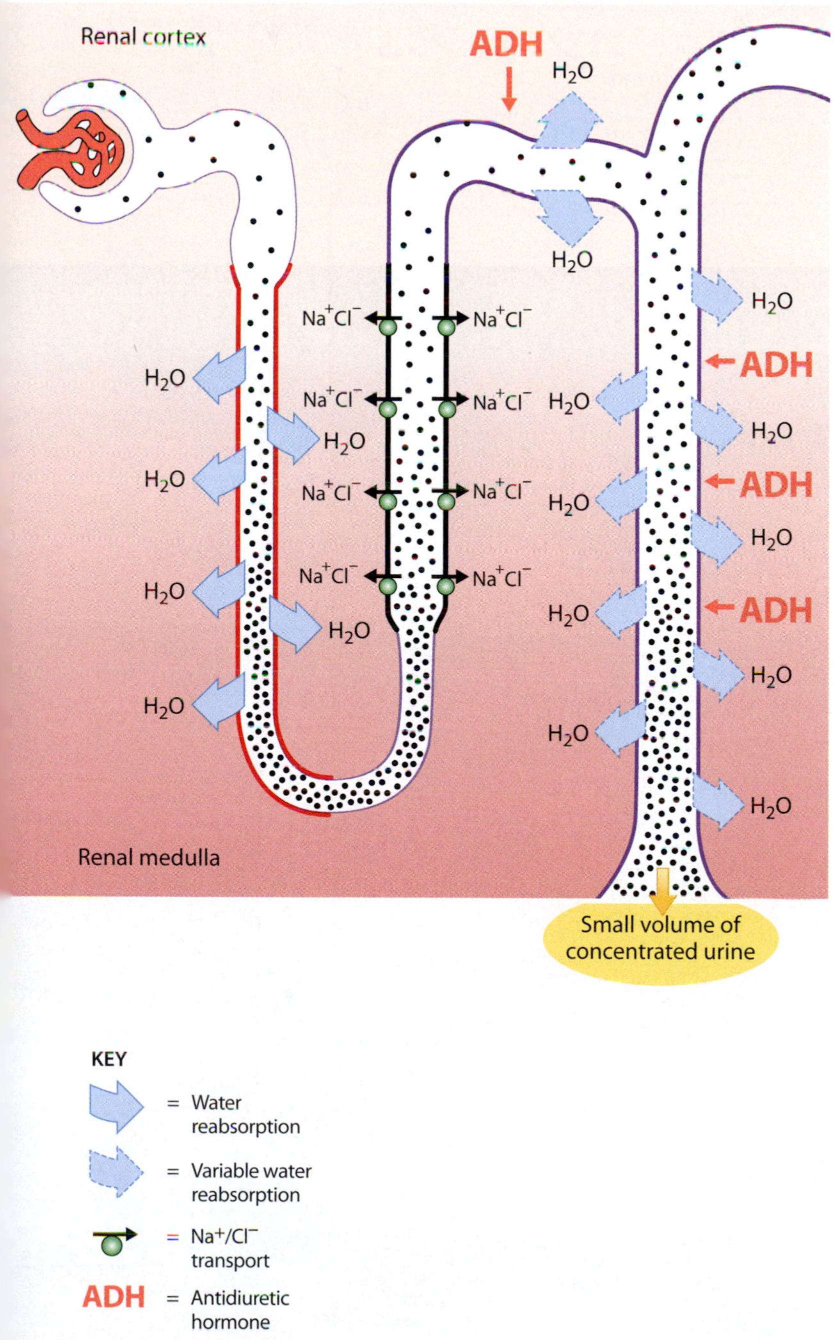

Renal cortex

ADH
H_2O

H_2O

H_2O

Na^+Cl^- ⟷ Na^+Cl^-

H_2O ⟵ ADH

H_2O Na^+Cl^- ⟷ Na^+Cl^- H_2O H_2O

H_2O

H_2O Na^+Cl^- ⟷ Na^+Cl^- H_2O ⟵ ADH

H_2O

H_2O Na^+Cl^- ⟷ Na^+Cl^-

H_2O ⟵ ADH

H_2O

H_2O

H_2O

Renal medulla

H_2O

Small volume of concentrated urine

KEY

⬇ = Water reabsorption

⬇ = Variable water reabsorption

● = Na^+/Cl^- transport

ADH = Antidiuretic hormone

General Characteristics of Normal Urine	
Characteristic	**Normal Range**
pH	4.5–8 (average: 6.0)
Specific gravity	1.003–1.030
Osmotic concentration (osmolarity)	855–1335 mOsm/L
Water content	93–97%
Volume	700–2000 mL/day
Color	Clear yellow
Odor	Varies with composition
Bacterial content	None (sterile)

Module 23.10 Review

a. Can the permeability of the PCT or DCT ever change? Why or why not?

b. What effect does an increase in ADH levels have on the DCT?

c. When ADH levels in the DCT decrease, what happens to the urine osmotic concentration?

Renal function is an integrative process

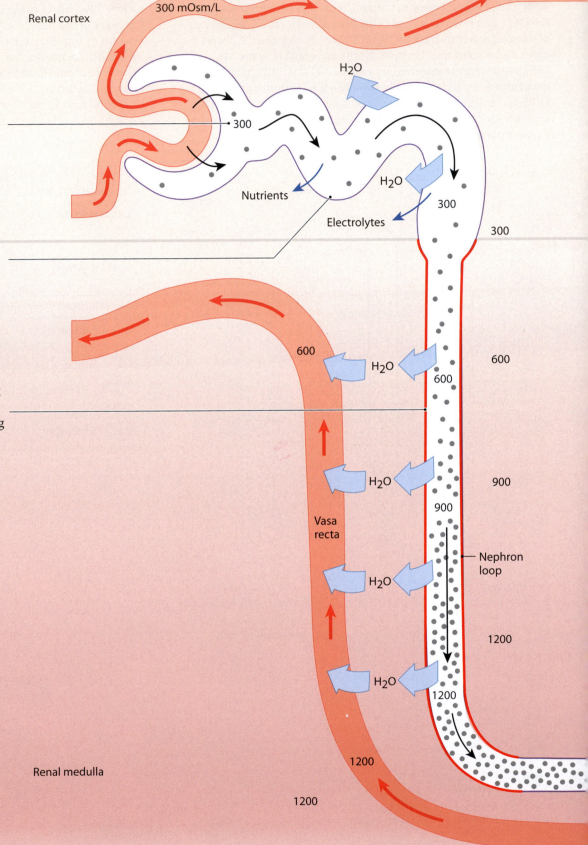

1 The filtrate produced at the renal corpuscle has the same osmotic concentration as plasma—about 300 mOsm/L. It has the same composition as plasma but does not contain plasma proteins.

2 In the proximal convoluted tubule (PCT), the active removal of ions and organic nutrients produces a continuous osmotic flow of water out of the tubular fluid. This reduces the volume of filtrate but keeps the solutions inside and outside the tubule isotonic.

3 In the PCT and descending limb of the nephron loop, water moves into the surrounding peritubular fluids, leaving a small volume of highly concentrated tubular fluid. This reduction occurs by obligatory water reabsorption.

Renal cortex

300 mOsm/L

H_2O

300

Nutrients

H_2O

300

Electrolytes

300

600

H_2O

600

600

H_2O

900

900

Vasa recta

H_2O

Nephron loop

1200

H_2O

1200

1200

Renal medulla

1200

KEY

= Water reabsorption

= Variable water reabsorption

= Na^+/Cl^- transport

Ⓐ = Aldosterone-regulated pump

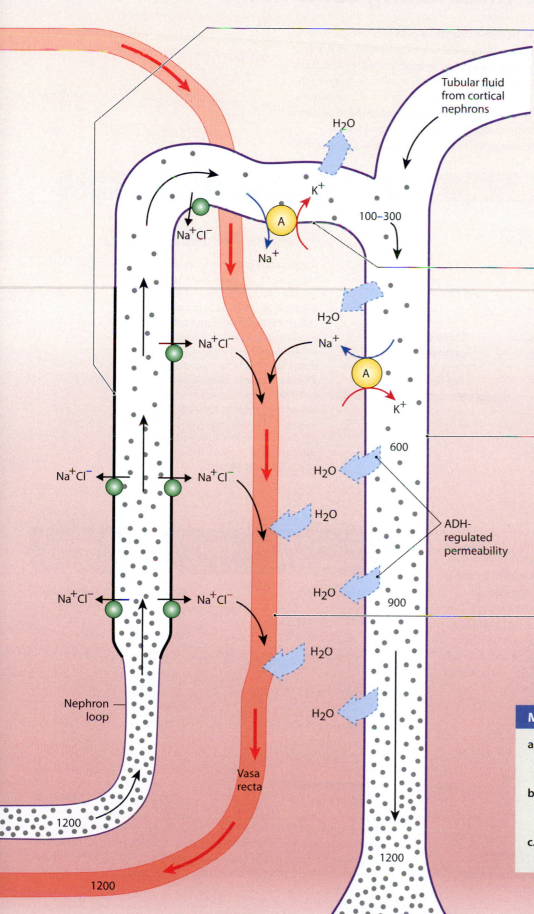

4 The thick ascending limb is impermeable to water and solutes. The tubule cells actively transport Na⁺ and Cl⁻ out of the tubule, thereby lowering the osmotic concentration of the tubular fluid. Because just Na⁺ and Cl⁻ are removed, urea accounts for a higher proportion of the total osmotic concentration at the end of the nephron loop.

5 The final adjustments in the composition of the tubular fluid occur in the DCT and the collecting system. The osmotic concentration of the tubular fluid can be adjusted through active transport (reabsorption or secretion).

6 The final adjustments in the volume and osmotic concentration of the tubular fluid are made by controlling the water permeabilities of the distal portions of the DCT and the collecting system. The level of exposure to ADH determines the final urine concentration.

7 The vasa recta absorbs the solutes and water reabsorbed by the nephron loop and the collecting ducts. By transporting these solutes and water into the main circulatory system, the vasa recta maintains the concentration gradient of the renal medulla.

Module 23.11 Review

a. The filtrate produced at the renal corpuscle has the same osmotic pressure as _____ ?

b. In the PCT, ions and organic substrates are actively removed, thus causing what to occur?

c. How is the concentration gradient of the renal medulla maintained?

Section 2: Overview of Renal Physiology · **871**

Renal failure is a life-threatening condition

1 **Renal failure** occurs when the kidneys become unable to perform the excretory functions needed to maintain homeostasis. When kidney filtration slows for any reason, urine production declines. As the decline continues, symptoms of renal failure appear because water, ions, and metabolic wastes are retained. Virtually all systems in the body are affected. For example, fluid balance, pH, muscular contraction, metabolism, and digestive function are disturbed. The individual generally becomes hypertensive, anemia develops due to a decline in erythropoietin production, and central nervous system problems can lead to sleeplessness, seizures, delirium, and even coma.

Renal Failure

Chronic Renal Failure

In **chronic renal failure**, kidney function deteriorates gradually, and the associated problems accumulate over time. The management of chronic renal failure typically involves restricting water and salt intake and minimizing protein intake. This combination reduces strain on the urinary system by (1) minimizing the volume of urine produced and (2) preventing the generation of large quantities of nitrogenous wastes. Acidosis, a common problem in persons with renal failure, can be countered by the ingestion of bicarbonate ions.

Chronic renal failure generally cannot be reversed; its progression can only be slowed. ⟶

Acute Renal Failure

Acute renal failure occurs when exposure to toxic drugs, renal ischemia, urinary obstruction, or trauma causes filtration to slow suddenly or stop. The reduction in kidney function occurs over a period of a few days and persists for weeks. Sensitized individuals can also develop acute renal failure after an allergic response to antibiotics or anesthetics. Individuals in acute renal failure may recover if they survive the incident. The kidneys may then regain partial or complete function. (With supportive treatment, the survival rate is approximately 50 percent.)

2 In **hemodialysis** (hē-mō-dī-AL-i-sis), an artificial membrane is used to regulate the composition of blood by means of a dialysis machine. The basic principle involved in this process, called **dialysis**, is passive diffusion across a selectively permeable membrane. The patient's blood flows past an artificial dialysis membrane, which contains pores large enough to permit the diffusion of ions, nutrients, and organic wastes, but small enough to prevent the loss of plasma proteins. A special dialysis fluid flows on the other side of the membrane.

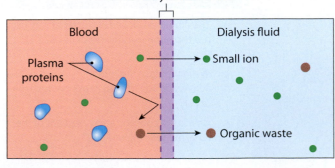

Artificial dialysis membrane

Blood — Dialysis fluid

Plasma proteins

Small ion

Organic waste

Comparative Composition of Plasma and Dialysis Fluid

Component	Plasma	Dialysis Fluid
Electrolytes (mEq/L)		
Sodium (Na^+)	135–145	136–140
Potassium (K^+)	3.5–5.0	0–3.0
Calcium (Ca^{2+})	4.3–5.3	1.5
Magnesium (Mg^{2+})	1.4–2.0	0.5–1.0
Chloride (Cl^-)	100–108	99–110
Bicarbonate (HCO_3^-)	21–28	27–39
Phosphate (PO_4^{2-})	3	0
Sulfate (SO_4^{2-})	1	0
Nutrients (mg/dL)		
Glucose	70–110	100

Note: Although these values are representative, the precise dialysis fluid composition can be tailored to meet specific clinical needs. For example, if plasma potassium levels are too low, the dialysis fluid potassium ion concentration can be elevated to remedy the situation.

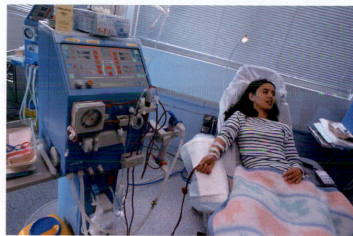

As diffusion takes place across the dialysis membrane, the composition of the blood changes. Potassium ions, phosphate ions, sulfate ions, urea, creatinine, and uric acid diffuse across the membrane into the dialysis fluid. Bicarbonate ions and glucose diffuse into the bloodstream. In effect, diffusion across the dialysis membrane takes the place of normal glomerular filtration, and the characteristics of the dialysis fluid ensure that important metabolites remain in the bloodstream rather than diffusing across the membrane.

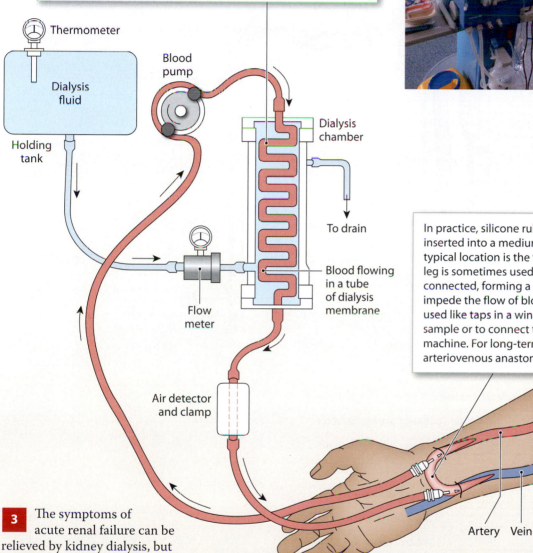

Thermometer

Blood pump

Dialysis fluid

Dialysis chamber

Holding tank

To drain

Blood flowing in a tube of dialysis membrane

Flow meter

Air detector and clamp

In practice, silicone rubber tubes called **shunts** are inserted into a medium-sized artery and vein. (The typical location is the forearm, although the lower leg is sometimes used.) The two shunts are then connected, forming a short circuit that does not impede the flow of blood. The shunts can then be used like taps in a wine barrel, to draw a blood sample or to connect the individual to a dialysis machine. For long-term dialysis, a surgically created arteriovenous anastomosis provides access.

Artery   Vein

3 The symptoms of acute renal failure can be relieved by kidney dialysis, but this treatment is not a cure.

The only real cure for severe renal failure is kidney transplantation. This procedure involves the implantation of a new kidney obtained from a living donor or from a cadaver. In most cases, the damaged kidney is removed and its blood supply is connected to the transplant. The one-year success rate for transplantation is now 85–95 percent. The use of kidneys taken from close relatives significantly improves the chances that the transplant will succeed. Immunosuppressive drugs are administered to reduce tissue rejection, but unfortunately this treatment also lowers the individual's resistance to infection.

Module 23.12 Review

a. Define dialysis.

b. Briefly explain the difference between chronic renal failure and acute renal failure.

c. Explain why patients on dialysis often receive Epogen or Procrit, a synthetic form of erythropoietin.

1. Short answer

Identify the structures of the representative nephron in the following diagram, and describe the functions of each.

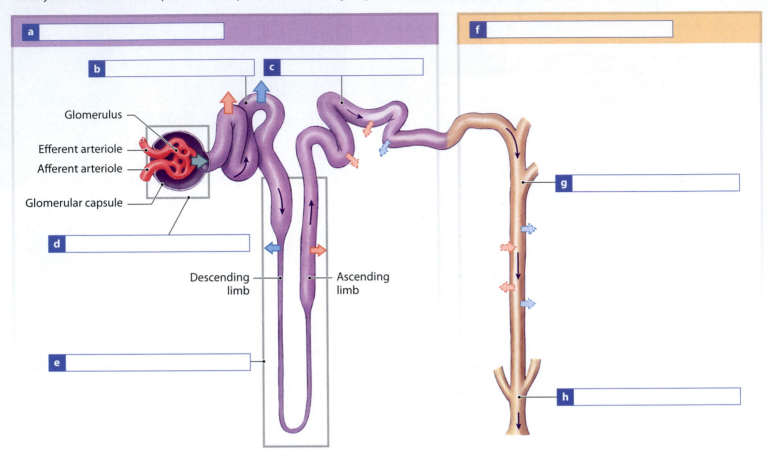

a _____

b _____   **c** _____   **f** _____

Glomerulus

Efferent arteriole

Afferent arteriole

Glomerular capsule

d _____

g _____

Descending limb — Ascending limb

e _____

h _____

2. Matching

Match the following terms with their most closely related item.

- aquaporins
- ADH
- aldosterone
- PCT
- secretion
- renal corpuscle
- nephron loop
- filtrate
- BCOP
- podocytes

a _____ Site of plasma filtration

b _____ Glomerular epithelium

c _____ Protein-free solution

d _____ Opposes filtration

e _____ Countercurrent multiplication

f _____ Water channels

g _____ Primary method for eliminating drugs or toxins

h _____ Ion pump—Na^+ reabsorbed

i _____ Primary site of nutrient reabsorption in the nephron

j _____ Regulates passive reabsorption of water from urine in the collecting system

3. Section integration

Marissa has had a urinalysis that detected large amounts of plasma proteins and white blood cells in her urine. What condition might be responsible, and what effects would it have on her urine output?

_____

_____

Urine Storage and Elimination

Filtrate modification and urine production end when the fluid enters the renal pelvis. The urinary tract (the ureters, urinary bladder, and urethra) is responsible for the transport, storage, and elimination of urine.

1 A **pyelogram** (PĪ-el-ō-gram) is an image of the urinary system, obtained by taking an x-ray of the kidneys after a radiopaque compound has been administered intravenously. Such an image provides an orientation to the relative sizes and positions of the system's main structures.

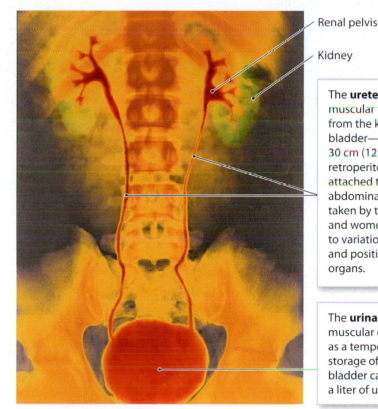

Renal pelvis

Kidney

The **ureters** are a pair of muscular tubes that extend from the kidneys to the urinary bladder—a distance of about 30 cm (12 in.). The ureters are retroperitoneal and are firmly attached to the posterior abdominal wall. The paths taken by the ureters in men and women are different, due to variations in the nature, size, and position of the reproductive organs.

The **urinary bladder** is a hollow, muscular organ that functions as a temporary reservoir for the storage of urine. A full urinary bladder can contain as much as a liter of urine.

2 This is a sectional view of a male pelvis showing the locations of the lower components of the urinary tract.

3 This is a sectional view of a female pelvis showing the locations of the lower components of the urinary tract.

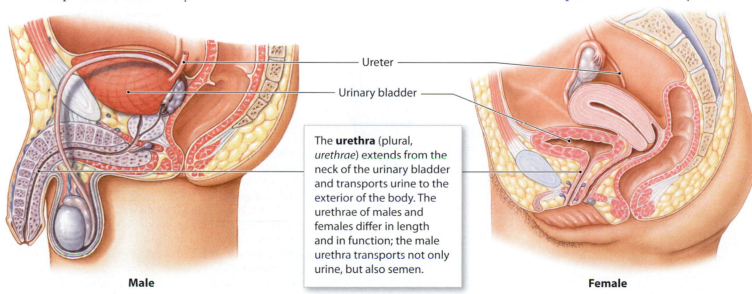

Ureter

Urinary bladder

The **urethra** (plural, *urethrae*) extends from the neck of the urinary bladder and transports urine to the exterior of the body. The urethrae of males and females differ in length and in function; the male urethra transports not only urine, but also semen.

Male

Female

875

The ureters, urinary bladder, and urethra are specialized for the conduction of urine

1 The urinary bladder is a hollow, muscular organ that functions as a temporary reservoir for the storage of urine.

Supporting Ligaments

The **lateral umbilical ligaments** pass along the sides of the bladder to the umbilicus (navel). These fibrous cords are the vestiges of the two umbilical arteries, which supplied blood to the placenta during embryonic and fetal development.

The **middle umbilical ligament** extends from the anterior, superior border toward the umbilicus.

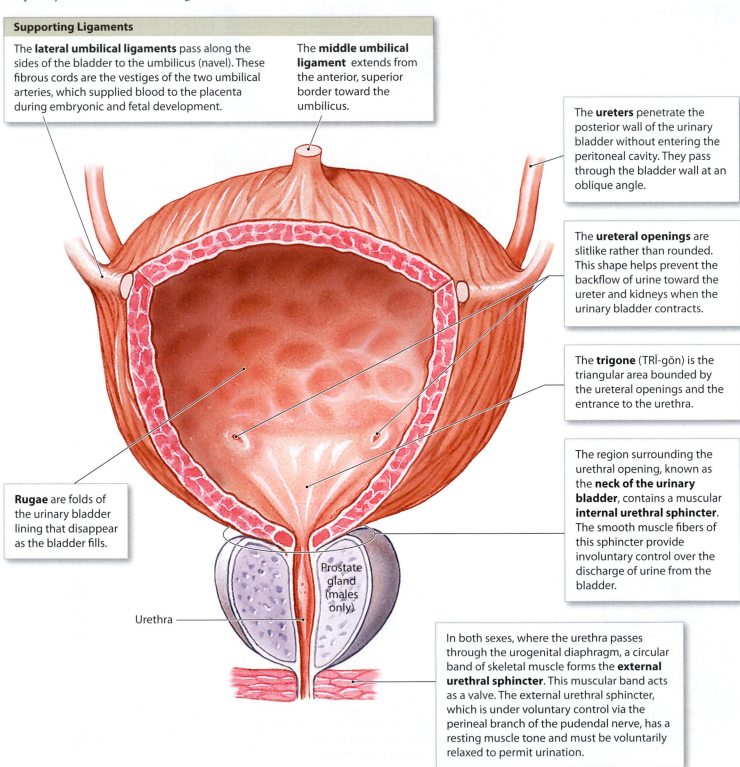

The **ureters** penetrate the posterior wall of the urinary bladder without entering the peritoneal cavity. They pass through the bladder wall at an oblique angle.

The **ureteral openings** are slitlike rather than rounded. This shape helps prevent the backflow of urine toward the ureter and kidneys when the urinary bladder contracts.

The **trigone** (TRĪ-gōn) is the triangular area bounded by the ureteral openings and the entrance to the urethra.

The region surrounding the urethral opening, known as the **neck of the urinary bladder**, contains a muscular **internal urethral sphincter**. The smooth muscle fibers of this sphincter provide involuntary control over the discharge of urine from the bladder.

Rugae are folds of the urinary bladder lining that disappear as the bladder fills.

Prostate gland (males only)

Urethra

In both sexes, where the urethra passes through the urogenital diaphragm, a circular band of skeletal muscle forms the **external urethral sphincter**. This muscular band acts as a valve. The external urethral sphincter, which is under voluntary control via the perineal branch of the pudendal nerve, has a resting muscle tone and must be voluntarily relaxed to permit urination.

2 The wall of each ureter consists of three layers: (1) an inner mucosa, comprising a transitional epithelium and the surrounding lamina propria; (2) a middle muscular layer made up of longitudinal and circular bands of smooth muscle; and (3) an outer connective tissue layer that is continuous with the fibrous capsule and peritoneum. About every 30 seconds, a peristaltic contraction begins at the renal pelvis and sweeps along the ureter, forcing urine toward the urinary bladder.

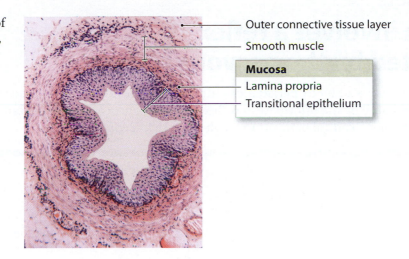

Outer connective tissue layer

Smooth muscle

Mucosa

Lamina propria

Transitional epithelium

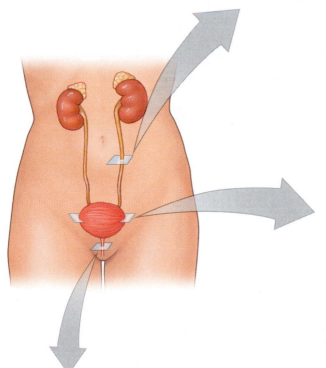

3 The wall of the urinary bladder contains mucosa, submucosa, and muscularis layers. The muscularis layer consists of inner and outer layers of longitudinal smooth muscle, with a circular layer between the two. Collectively, these layers form the powerful **detrusor** (dē-TROO-sor) **muscle** of the urinary bladder. Contraction of this muscle compresses the urinary bladder and expels its contents into the urethra.

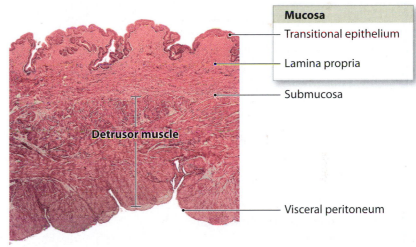

Mucosa

Transitional epithelium

Lamina propria

Submucosa

Detrusor muscle

Visceral peritoneum

4 The urethral lining consists of a stratified epithelium that varies from transitional at the neck of the urinary bladder, to stratified columnar at the midpoint, to stratified squamous near the external urethral orifice. The lamina propria is thick and elastic, and the mucous membrane is thrown into longitudinal folds. Mucin-secreting cells are located in the epithelial pockets. Connective tissues of the lamina propria anchor the urethra to surrounding structures.

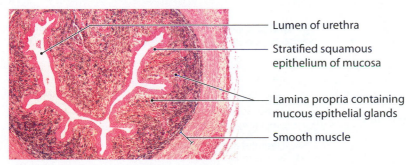

Lumen of urethra

Stratified squamous epithelium of mucosa

Lamina propria containing mucous epithelial glands

Smooth muscle

Module 23.13 Review

a. Urine is transported by the _____ , stored within the _____ , and eliminated through the _____ .

b. The wall of the urinary bladder is composed of a specialized smooth muscle. Name it and describe its physiological role.

c. What has to happen to the external urethral sphincter to allow urination?

23.14

Urination involves a reflex coordinated by the nervous system

Urine reaches the urinary bladder by peristaltic contractions of the ureters. The process of urination is coordinated by the **micturition reflex**, a complex process involving both a local reflex pathway and a central pathway through the cerebral cortex.

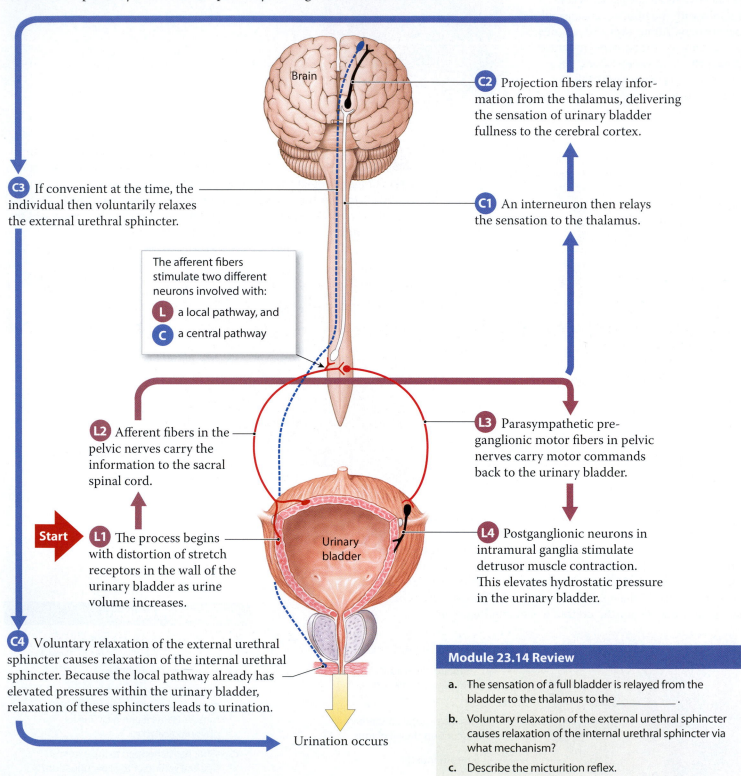

Brain

C2 Projection fibers relay information from the thalamus, delivering the sensation of urinary bladder fullness to the cerebral cortex.

C3 If convenient at the time, the individual then voluntarily relaxes the external urethral sphincter.

C1 An interneuron then relays the sensation to the thalamus.

The afferent fibers stimulate two different neurons involved with:
- **L** a local pathway, and
- **C** a central pathway

L2 Afferent fibers in the pelvic nerves carry the information to the sacral spinal cord.

L3 Parasympathetic preganglionic motor fibers in pelvic nerves carry motor commands back to the urinary bladder.

Start

L1 The process begins with distortion of stretch receptors in the wall of the urinary bladder as urine volume increases.

Urinary bladder

L4 Postganglionic neurons in intramural ganglia stimulate detrusor muscle contraction. This elevates hydrostatic pressure in the urinary bladder.

C4 Voluntary relaxation of the external urethral sphincter causes relaxation of the internal urethral sphincter. Because the local pathway already has elevated pressures within the urinary bladder, relaxation of these sphincters leads to urination.

Urination occurs

Module 23.14 Review

a. The sensation of a full bladder is relayed from the bladder to the thalamus to the _____ .

b. Voluntary relaxation of the external urethral sphincter causes relaxation of the internal urethral sphincter via what mechanism?

c. Describe the micturition reflex.

Urinary disorders can often be detected by physical exams and laboratory tests

1 The primary signs and symptoms of urinary system disorders include changes in volume and appearance of urine, frequency of urination, and pain. The nature and location of the pain can provide clues to the source.

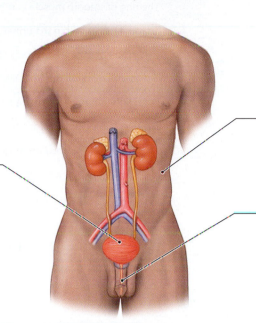

Pain in the superior pubic region may be associated with urinary **bladder** disorders.

Pain in the superior lumbar region or in the flank that radiates to the right upper quadrant or left upper quadrant can be caused by kidney infections such as **pyelonephritis**, or by kidney stones (**renal calculi**).

Dysuria (painful or difficult urination) can occur with cystitis or urethritis, or with urinary obstructions. In males, enlargement of the prostate gland can compress the urethra and lead to dysuria.

2 Characteristic changes in urinary output or frequency give clues to underlying problems with this system or other systems.

Abnormal Urine Output and Frequency

Increased Urgency or Increased Frequency	Changes in Urinary Output	Incontinence	Urinary Retention
An irritation of the lining of the ureters or urinary bladder can lead to the desire to urinate with increased frequency, although the total amount of urine produced each day remains normal.	Changes in the volume of urine produced by a person with average fluid intake indicate problems either at the kidneys or with the control of renal function. **Polyuria**, the production of excessive amounts of urine, results from hormonal or metabolic problems, such as those associated with diabetes or glomerulonephritis. **Oliguria** (a urine volume of 50–500 mL/day) and **anuria** (0–50 mL/day) are conditions that indicate serious kidney problems and potential renal failure.	**Incontinence**, an inability to control urination voluntarily, may involve periodic involuntary leakage (stress incontinence), or inability to delay urination (urge incontinence)—or a continual, slow trickle of urine from a bladder that is always full (overflow incontinence).	In **urinary retention**, renal function is normal, at least initially, but urination does not occur. Urinary retention in males commonly results from enlargement of the prostate and compression of the prostatic urethra.

Important clinical signs of urinary system disorders include the following:

- **Edema**. Renal disorders often lead to protein loss in the urine (**proteinuria**) and, if severe, result in generalized edema in peripheral tissues. Facial swelling, especially around the eyes, is common.

- **Fever**. A fever commonly develops when the urinary system is infected by pathogens. Urinary bladder infections (**cystitis**) may result in a low-grade fever; kidney infections, such as pyelonephritis, can produce very high fevers.

Module 23.15 Review

a. What is the term for painful or difficult urination?

b. Obstruction of a ureter by a kidney stone would interfere with the flow of urine between which two points?

c. Why is urinary obstruction at the urethra more dangerous than at the ureter?

1. Matching

Match the following terms with their descriptions.

- urethra
- external urethral sphincter
- detrusor
- rugae
- internal urethral sphincter
- trigone
- transitional epithelium
- micturition
- stratified squamous epithelium
- external urethral orifice
- middle umbilical ligament

a _____ The ring of smooth muscle in the neck of the urinary bladder

b _____ Triangular area within the urinary bladder

c _____ Relaxation of this muscle leads to urination

d _____ The external opening of the urethra

e _____ Folds lining the surface of the empty urinary bladder

f _____ Superior, supporting fibrous cord of the urinary bladder

g _____ Contraction of this smooth muscle compresses the urinary bladder

h _____ The type of epithelium that lines the ureters

i _____ Tube that transports urine to the exterior

j _____ Term for urination

k _____ Epithelium that lines the urethra

2. Labeling

Use the following descriptions to fill in the boxes in the micturition reflex diagram below.

- sensation relayed to thalamus
- individual relaxes external urethral sphincter
- afferent fibers carry information to sacral spinal cord
- sensation of bladder fullness delivered to cerebral cortex
- stretch receptors stimulated
- detrusor muscle contraction stimulated
- parasympathetic preganglionic fibers carry motor commands
- internal urethral sphincter relaxes

3. Short answer

List four primary signs and symptoms of urinary disorders.

4. Short answer

Briefly describe the similarities and differences in the following pairs of terms.

- cystitis/pyelonephritis
- stress incontinence/ overflow incontinence
- polyuria/proteinuria

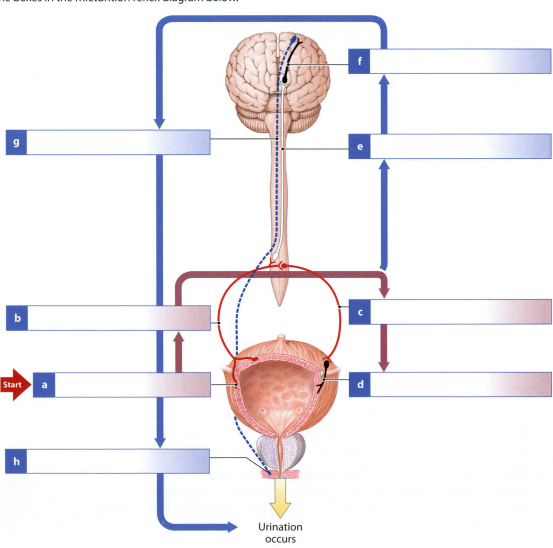

Start

Urination occurs

Visual Outline with Key Terms

Summarize the content of each module using the terms in the order provided.

SECTION 1

Anatomy of the Urinary System

- urinary system
- urinary tract
- kidneys
- urine
- ureters
- urinary bladder
- urethra
- urination

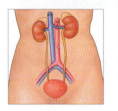

23.1

The kidneys are paired retroperitoneal organs

- kidney
- hilum
- ureters
- ○ retroperitoneal
- fibrous capsule
- perinephric fat capsule
- renal fascia

23.2

The kidneys are complex at the gross and microscopic levels

- hilum
- fibrous capsule
- renal sinus
- renal cortex
- renal medulla
- renal pyramid
- renal papilla
- renal column
- kidney lobe
- minor and major calyces
- renal pelvis
- nephron
- cortical nephron
- juxtamedullary nephron
- nephron loop

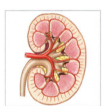

23.3

A nephron can be divided into regions; each region has specific functions

- ○ nephron
- renal corpuscle
- renal tubule
- capsular space
- filtrate
- glomerular capsule
- glomerulus
- proximal convoluted tubule (PCT)
- nephron loop
- distal convoluted tubule (DCT)
- collecting system
- collecting duct
- papillary duct

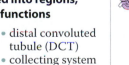

23.4

The kidneys are highly vascular, and the circulation patterns are complex

- renal artery and renal vein
- segmental arteries
- interlobar arteries and veins
- arcuate arteries and veins
- cortical radiate arteries and veins
- afferent and efferent arterioles
- glomerulus
- peritubular capillaries
- vasa recta

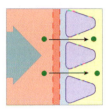

SECTION 2

Overview of Renal Physiology

- urea
- creatinine
- uric acid
- filtration
- reabsorption
- secretion

23.5

Filtration, reabsorption, and secretion occur in specific regions of the nephron and collecting system

- ○ filtration
- ○ water reabsorption
- ○ solute reabsorption
- ○ solute secretion

23.6

Filtration occurs at the renal corpuscle

- efferent arteriole
- juxtaglomerular complex
- afferent arteriole
- capsular space
- podocytes
- pedicels
- filtration slits
- mesangial cells
- dense layer
- filtration membrane
- hydrostatic pressure
- colloid osmotic pressure
- glomerular hydrostatic pressure (GHP)
- blood colloid osmotic pressure (BCOP)
- net filtration pressure (NFP)
- capsular hydrostatic pressure (CsHP)
- capsular colloid osmotic pressure

● = Term boldfaced in this module

23.7

The glomerular filtration rate (GFR) is the amount of filtrate produced each minute

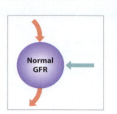

- glomerular filtration rate (GFR)
- ○ filtrate
- autoregulation
- central regulation
- endocrine response
- neural responses
- ○ renin
- ○ angiotensin
- angiotensin converting enzyme (ACE)
- ○ aldosterone
- ○ ADH

23.8

Reabsorption predominates along the PCT, whereas reabsorption and secretion are often linked along the DCT

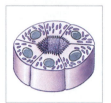

- proximal convoluted tubule (PCT)
- distal convoluted tubule (DCT)
- ○ reabsorption
- ○ secretion
- ○ diffusion
- ○ facilitated diffusion
- ○ active transport
- ○ secondary active transport
- ○ peritubular capillary
- ○ peritubular fluid
- ○ tubular fluid

23.9

Feedback between the nephron loop and the collecting duct creates the osmotic gradient in the renal medulla

- nephron loop
- countercurrent multiplication
- ○ active transport
- ○ impermeable
- ○ osmosis
- ○ thick ascending loop
- ○ thin descending loop
- ○ renal medulla

23.10

Urine volume and concentration are hormonally regulated

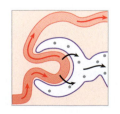

- obligatory water reabsorption
- facultative water reabsorption
- ○ distal convoluted tubule
- ○ collecting duct
- ○ antidiuretic hormone (ADH)
- aquaporins
- ○ urine

23.11

Renal function is an integrative process

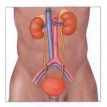

- ○ osmotic concentration
- ○ obligatory water reabsorption
- ○ urea
- ○ aldosterone
- ○ ADH
- ○ vasa recta

• = *Term boldfaced in this module*

23.12

Renal failure is a life-threatening condition

- renal failure
- chronic renal failure
- acute renal failure
- hemodialysis
- dialysis
- ○ dialysis fluid
- shunts

SECTION 3

Urine Storage and Elimination

- pyelogram
- ureters
- urinary bladder
- urethra

23.13

The ureters, urinary bladder, and urethra are specialized for the conduction of urine

- lateral umbilical ligaments
- middle umbilical ligament
- ureters
- ureteral openings
- trigone
- rugae
- neck of the urinary bladder
- internal urethral sphincter
- external urethral sphincter
- detrusor muscle

23.14

Urination involves a reflex coordinated by the nervous system

- micturition reflex
- ○ local pathway
- ○ central pathway
- ○ internal urethral sphincter
- ○ external urethral sphincter

23.15

Urinary disorders can often be detected by physical exams and laboratory tests

- pyelonephritis
- renal calculi
- dysuria
- polyuria
- oliguria
- anuria
- incontinence
- urinary retention
- edema
- proteinuria
- fever
- cystitis

Chapter Integration: Applying what you've learned

For consumers, some recent developments have turned the purity of food and food products, for humans and animals alike, from a matter of trust into a matter of concern.

In 2007, the unintentional adulteration of pet food made in China with the nitrogen-containing industrial chemical melamine caused pet illnesses and deaths in the United States and resulted in massive recalls of contaminated products. In 2008, milk intentionally contaminated with melamine—put there to skew laboratory tests that measure nitrogen as an index of the protein content in a food or drink powder—sickened at least 64,000 children in China, several of whom died. Clearly, vigilance concerning our food supply is crucial, even while standards and regulations are in place.

In the body, melamine contamination has two major effects: (1) the formation of crystalline masses in the filtrate and/or urine, and (2) acidification of the tubular fluid. Resulting clinical problems range from blood in the urine, to acid–base and electrolyte disorders, to urinary obstruction and (in severe cases) kidney failure.

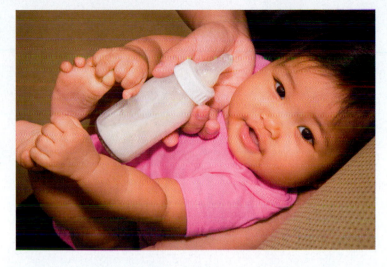

1. Propose a linkage between the stated effects of melamine poisoning and the clinical problems observed.

2. Primary therapeutic options for melamine poisoning include infusion of fluids, dialysis, and medication. Explain how these options address specific problems caused by melamine poisoning and suggest possible follow-up tests.

24

Fluid, Electrolyte, and Acid-Base Balance

Fluid and Electrolyte Balance

Chapter 22 considered the metabolism of the organic components of the body. This chapter takes a broader look at the composition of the body as a whole. We will focus on the inorganic components: water and minerals. **Minerals** are the inorganic substances that dissociate in body fluids to form ions called electrolytes.

1 These pie charts compare the total body composition of adult males and females. The greatest variation is in the intracellular fluid (ICF), or cytosol, as a result of differences in the intracellular water content of fat versus muscle. Less striking differences occur in the extracellular fluid (ECF) values, due to variations in the interstitial fluid volume of various tissues and the larger blood volume in males versus females. The ECF and ICF are called **fluid compartments**, because they commonly behave as distinct entities. Because cells have a plasma membrane and active transport occurs at the membrane surface, cells are able to maintain internal environments quite distinct from that of the ECF.

2 Solid components account for only 40–50 percent of the mass of the body as a whole. This bar graph presents an overview of the solid components of a 70-kg (154-pound) individual with a minimum of body fat. The distribution was obtained by averaging values for males and females ages 18–40 years.

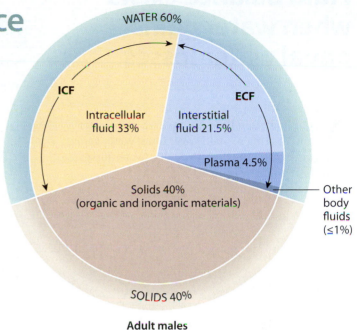

Adult males

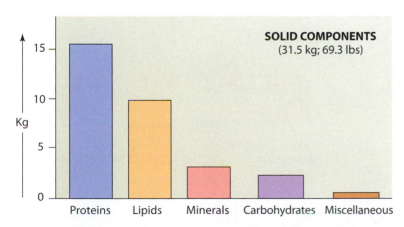

SOLID COMPONENTS
(31.5 kg; 69.3 lbs)

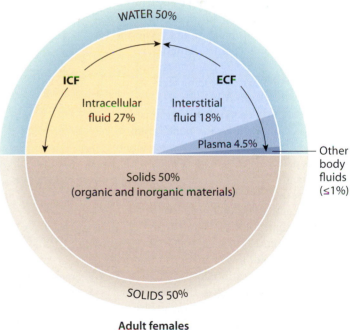

Adult females

In this section we will consider the exchange of water and electrolytes between the ECF and ICF, and between the body and the external environment.

Fluid balance exists when water gains equal water losses

1 Your body is in **fluid balance** when its water content remains stable over time. Water gains occur primarily in the digestive tract. Water losses occur through many routes, but almost half of daily water loss occurs through urination.

Fluid Balance

Source	Daily Input (mL)
Water content of food	1000
Water consumed as liquid	1200
Metabolic water produced during catabolism	300
Total	**2500**

Method of Elimination	Daily Output (mL)
Urination	1200
Evaporation at skin	750
Evaporation at lungs	400
Loss in feces	150
Total	**2500**

2 This diagram indicates where water enters the digestive tract through ingestion or secretion, and where it is reabsorbed; only a small amount leaves the digestive tract in feces. The situation is complicated by the fact that the accessory digestive glands are producing watery secretions that are mixed with arriving food. Most of that secreted water must be recovered along with water gained from food and drink. All of this water movement involves passive water flow down osmotic gradients. Intestinal epithelial cells continuously absorb nutrients and ions, and these activities gradually lower the solute concentration in the lumen and elevate the solute concentration in the interstitial fluid of the lamina propria. As the solute concentration drops in the lumen, water moves across the epithelium and into the interstitial fluid, maintaining osmotic equilibrium. Once within the interstitial fluid, the absorbed water is rapidly distributed throughout the ECF.

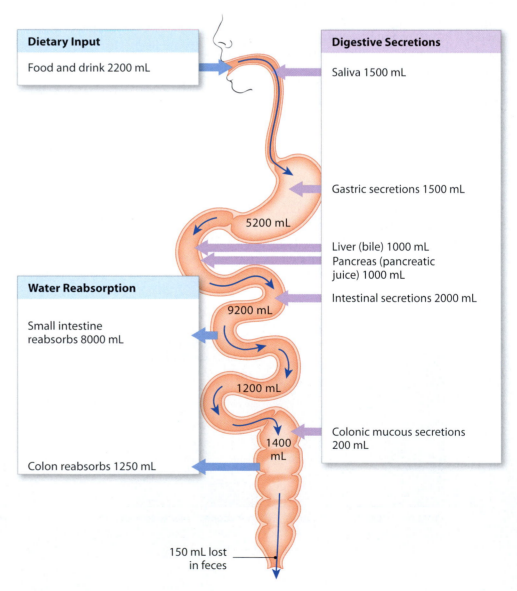

Dietary Input

Food and drink 2200 mL

Digestive Secretions

Saliva 1500 mL

Gastric secretions 1500 mL

5200 mL

Liver (bile) 1000 mL
Pancreas (pancreatic juice) 1000 mL

Intestinal secretions 2000 mL

Water Reabsorption

Small intestine reabsorbs 8000 mL

9200 mL

1200 mL

Colonic mucous secretions 200 mL

1400 mL

Colon reabsorbs 1250 mL

150 mL lost in feces

3 This diagram illustrates the major factors that affect ECF volume. Although the composition of the ECF and ICF are very different, the two are at osmotic equilibrium. Note that the volume of the ICF is larger than that of the ECF. The volume of water held within cells represents a significant reserve that can prevent sudden changes in the solute and water concentrations in the ECF. A rapid water movement between the ECF and the ICF in response to an osmotic gradient is called a **fluid shift**. Fluid shifts occur rapidly in response to changes in the osmotic concentration of the ECF and reach equilibrium within minutes to hours. This can be an important factor when water intake is restricted but water losses are severe.

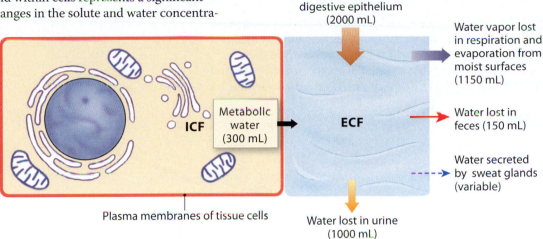

Water absorbed across digestive epithelium (2000 mL)

Water vapor lost in respiration and evaporation from moist surfaces (1150 mL)

Metabolic water (300 mL)

ICF

ECF

Water lost in feces (150 mL)

Water secreted by sweat glands (variable)

Plasma membranes of tissue cells

Water lost in urine (1000 mL)

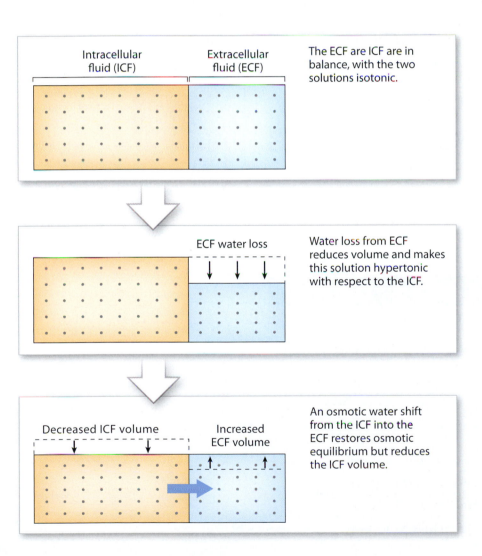

Intracellular fluid (ICF)

Extracellular fluid (ECF)

The ECF are ICF are in balance, with the two solutions isotonic.

ECF water loss

Water loss from ECF reduces volume and makes this solution hypertonic with respect to the ICF.

Decreased ICF volume

Increased ECF volume

An osmotic water shift from the ICF into the ECF restores osmotic equilibrium but reduces the ICF volume.

4 **Dehydration** develops when water losses outpace water gains. When you lose water but retain electrolytes, the osmotic concentration of the ECF rises. Osmosis then moves water out of the ICF and into the ECF until the two solutions are again isotonic. At that point, both the ECF and ICF are somewhat more concentrated than normal, and both volumes are lower than they were before the fluid loss. Because the ICF is considerably larger than the ECF, the ICF is an effective water reserve. However, if the fluid imbalance continues unchecked, the loss of water from the ICF will produce severe thirst, dryness, and wrinkling of the skin. Eventually plasma volume and blood pressure may drop to the point at which circulatory shock develops.

Module 24.1 Review

a. Identify routes of fluid loss from the body.

b. Describe a fluid shift.

c. Explain dehydration and its effect on the osmotic concentration of plasma.

Mineral balance involves balancing electrolyte gains and losses

1 **Mineral balance** is the balance between ion absorption, which occurs across the lining of the small intestine and colon, and ion excretion, which occurs primarily at the kidneys. Sweat glands are a potential source of both water and mineral loss, but the rate of secretion is extremely variable.

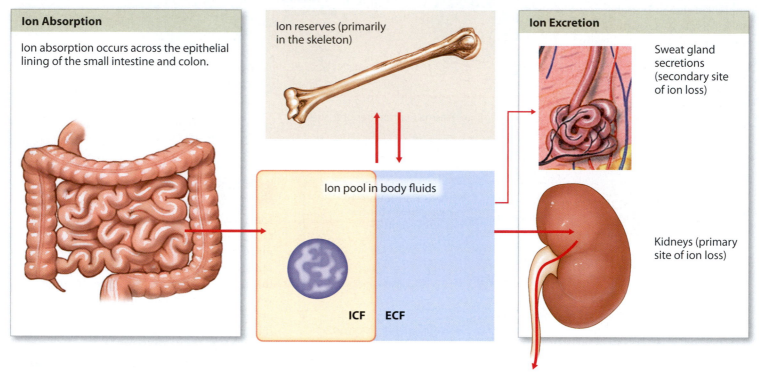

Ion Absorption

Ion absorption occurs across the epithelial lining of the small intestine and colon.

Ion reserves (primarily in the skeleton)

Ion pool in body fluids

ICF   ECF

Ion Excretion

Sweat gland secretions (secondary site of ion loss)

Kidneys (primary site of ion loss)

2 This table lists the mechanisms involved in the absorption of major electrolytes along the digestive tract.

Mineral Absorption

Electrolyte	Mechanism(s)
Na^+	Channel-mediated diffusion, cotransport, or active transport
Ca^{2+}	Active transport
K^+	Channel-mediated diffusion
Mg^{2+}	Active transport
Fe^{2+}	Active transport
Cl^-	Channel-mediated diffusion or carrier-mediated transport
I^-	Channel-mediated diffusion or carrier-mediated transport
HCO_3^-	Channel-mediated diffusion or carrier-mediated transport
NO_3^-	Channel-mediated diffusion or carrier-mediated transport
PO_4^{3-}	Active transport
SO_4^{2-}	Active transport

3   The body contains substantial reserves of key minerals. This table summarizes the functions of the various minerals and indicates the primary routes of ion excretion. The amount lost each day must on average equal the daily intake if the individual is to stay in mineral balance.

Minerals and Mineral Reserves

Mineral	Functions	Total Body Content	Primary Route of Excretion	Recommended Daily Allowance (RDA) (in mg)
Bulk Minerals				
Sodium	Major cation in body fluids; essential for normal membrane function	110 g, primarily in body fluids	Urine, sweat, feces	1500
Potassium	Major cation in cytoplasm; essential for normal membrane function	140 g, primarily in cytoplasm	Urine	4700
Chloride	Major anion in body fluids; functions in forming HCl	89 g, primarily in body fluids	Urine, sweat	2300
Calcium	Essential for normal muscle and neuron function and normal bone structure	1.36 kg, primarily in skeleton	Urine, feces	1000–1200
Phosphorus	In high-energy compounds, nucleic acids, and bone matrix (as phosphate)	744 g, primarily in skeleton	Urine, feces	700
Magnesium	Cofactor of enzymes, required for normal membrane functions	29 g (skeleton, 17 g; cytoplasm and body fluids, 12 g)	Urine	310–400
Trace Minerals				
Iron	Component of hemoglobin, myoglobin, and cytochromes	3.9 g (1.6 g stored as ferritin or hemosiderin)	Urine (traces)	8–18
Zinc	Cofactor of enzyme systems, notably carbonic anhydrase	2 g	Urine, hair (traces)	8–11
Copper	Required as cofactor for hemoglobin synthesis	127 mg	Urine, feces (traces)	0.9
Manganese	Cofactor for some enzymes	11 mg	Feces, urine (traces)	1.8–2.3
Cobalt	Cofactor for transaminations; mineral in vitamin B_{12} (cobalamin)	1.1 g	Feces, urine	0.0001

Module 24.2 Review

a. Define mineral balance.

b. Identify the significance of two important body minerals: sodium and calcium.

c. Identify the ions absorbed by active transport.

Water balance depends on sodium balance, and the two are regulated simultaneously

1 **Sodium balance** exists when sodium gains equal sodium losses. The regulatory mechanism involved changes the ECF volume but keeps the Na^+ concentration relatively stable. When sodium gains exceed losses, the ECF volume increases; when sodium losses exceed gains, the volume of the ECF decreases. The changes in ECF volume occur without a significant change in the osmotic concentration of the ECF. These adjustments cause minor changes in ECF volume that do not cause adverse physiological effects.

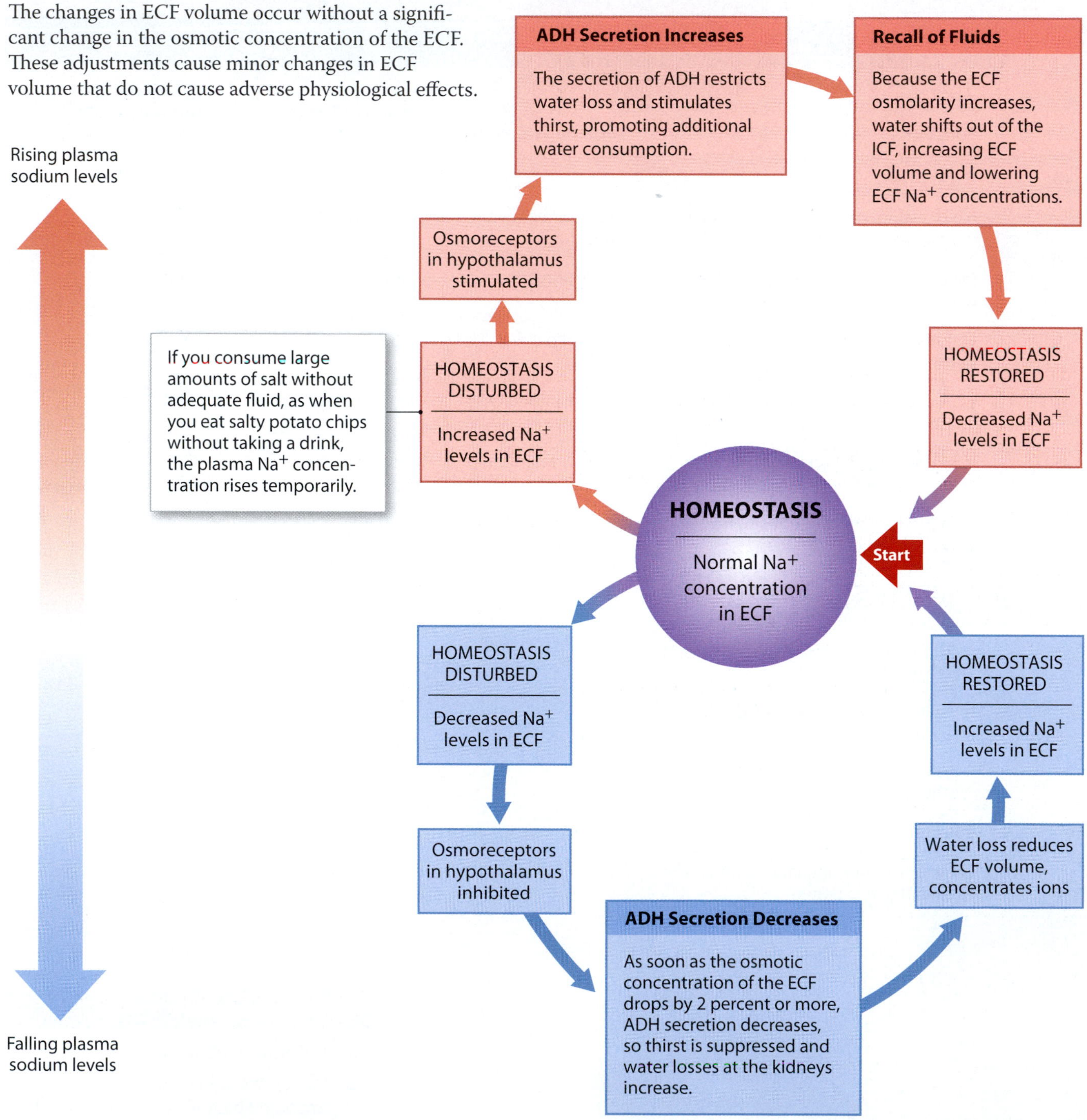

Rising plasma sodium levels

Falling plasma sodium levels

ADH Secretion Increases

The secretion of ADH restricts water loss and stimulates thirst, promoting additional water consumption.

Recall of Fluids

Because the ECF osmolarity increases, water shifts out of the ICF, increasing ECF volume and lowering ECF Na^+ concentrations.

Osmoreceptors in hypothalamus stimulated

If you consume large amounts of salt without adequate fluid, as when you eat salty potato chips without taking a drink, the plasma Na^+ concentration rises temporarily.

HOMEOSTASIS DISTURBED

Increased Na^+ levels in ECF

HOMEOSTASIS RESTORED

Decreased Na^+ levels in ECF

HOMEOSTASIS

Normal Na^+ concentration in ECF

Start

HOMEOSTASIS DISTURBED

Decreased Na^+ levels in ECF

HOMEOSTASIS RESTORED

Increased Na^+ levels in ECF

Osmoreceptors in hypothalamus inhibited

Water loss reduces ECF volume, concentrates ions

ADH Secretion Decreases

As soon as the osmotic concentration of the ECF drops by 2 percent or more, ADH secretion decreases, so thirst is suppressed and water losses at the kidneys increase.

2 If the changes in ECF volume caused by disturbances in sodium balance are extreme, the homeostatic mechanisms responsible for regulating blood volume and blood pressure will be activated. This is the case because when ECF volume changes, so does plasma volume and, in turn, blood volume. If ECF volume rises, blood volume goes up; if ECF volume drops, blood volume goes down.

Rising blood pressure and volume

Falling blood pressure and volume

Natriuretic peptides released by cardiac muscle cells

Increased blood volume and atrial distension

Responses to Natriuretic Peptides

Increased Na$^+$ loss in urine

Increased water loss in urine

Reduced thirst

Inhibition of ADH, aldosterone, epinephrine, and norepinephrine release

Combined Effects

Reduced blood volume

Reduced blood pressure

HOMEOSTASIS DISTURBED

Rising ECF volume by fluid gain or fluid and Na$^+$ gain

HOMEOSTASIS

Normal ECF volume

Start

HOMEOSTASIS RESTORED

Falling ECF volume

HOMEOSTASIS DISTURBED

Falling ECF volume by fluid loss or fluid and Na$^+$ loss

HOMEOSTASIS RESTORED

Rising ECF volume

Decreased blood volume and blood pressure

Endocrine Responses

Increased renin secretion and angiotensin II activation

Increased aldosterone release

Increased ADH release

Combined Effects

Increased urinary Na$^+$ retention

Decreased urinary water loss

Increased thirst

Increased water intake

Sustained abnormalities in the Na$^+$ concentration in the ECF occur only when there are severe problems with fluid balance. When the Na$^+$ concentration of the ECF falls below 136 mEq/L, a state of **hyponatremia** (*natrium*, sodium) exists; this can be caused by excessive water intake (overhydration) or inadequate salt intake. When body water content declines, the Na$^+$ concentration rises; when that concentration exceeds 145 mEq/L, **hypernatremia** exists; dehydration is the most common cause. Both conditions are serious and potentially life-threatening.

Module 24.3 Review

a. What effect does inhibition of osmoreceptors have on ADH secretion and thirst?

b. What effect does aldosterone have on sodium ion concentration in the ECF?

c. Briefly summarize the relationship between sodium ion concentration and the ECF.

Disturbances of potassium balance are uncommon but extremely dangerous

1 This diagram indicates the major factors involved in **potassium balance**. The key factors in the maintenance of potassium balance are (1) the rate of K⁺ entry across the digestive epithelium and (2) the rate of K⁺ loss into urine.

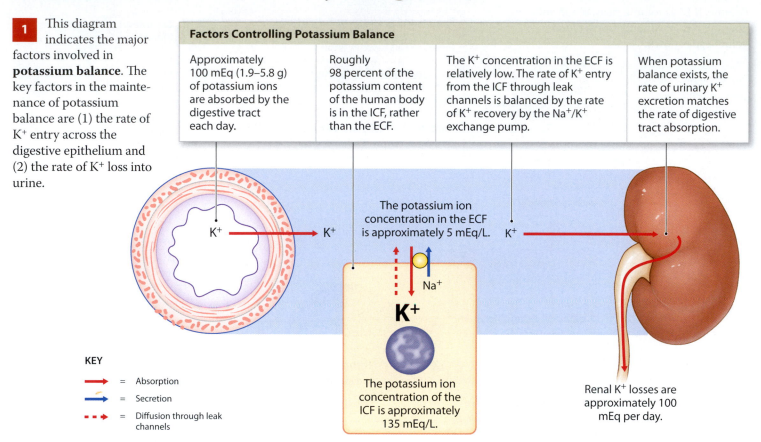

Factors Controlling Potassium Balance

Approximately 100 mEq (1.9–5.8 g) of potassium ions are absorbed by the digestive tract each day.

Roughly 98 percent of the potassium content of the human body is in the ICF, rather than the ECF.

The K⁺ concentration in the ECF is relatively low. The rate of K⁺ entry from the ICF through leak channels is balanced by the rate of K⁺ recovery by the Na⁺/K⁺ exchange pump.

When potassium balance exists, the rate of urinary K⁺ excretion matches the rate of digestive tract absorption.

The potassium ion concentration in the ECF is approximately 5 mEq/L.

The potassium ion concentration of the ICF is approximately 135 mEq/L.

Renal K⁺ losses are approximately 100 mEq per day.

KEY
→ = Absorption
→ = Secretion
▪ ▪ ▸ = Diffusion through leak channels

2 Because dietary intake is relatively constant, the kidneys are the main factor determining the K⁺ concentration in the ECF. Potassium loss at the kidneys is regulated by controlling the activities of ion pumps along the distal portions of the nephron and collecting system. The activity of these pumps is regulated by circulating levels of aldosterone. Aldosterone stimulates Na⁺ reabsorption while simultaneously accelerating K⁺ excretion.

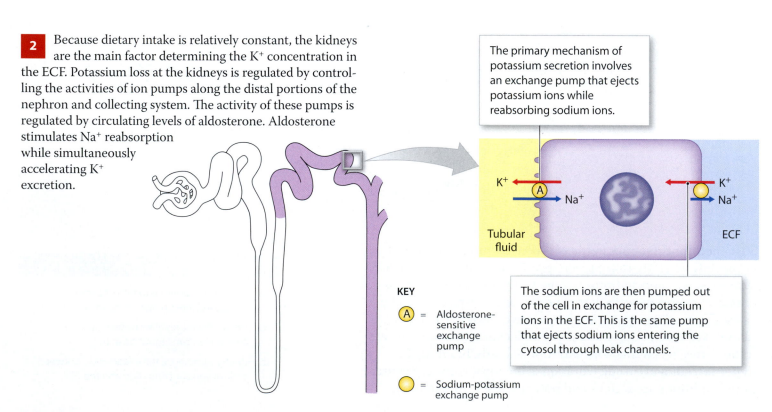

The primary mechanism of potassium secretion involves an exchange pump that ejects potassium ions while reabsorbing sodium ions.

The sodium ions are then pumped out of the cell in exchange for potassium ions in the ECF. This is the same pump that ejects sodium ions entering the cytosol through leak channels.

Tubular fluid

ECF

KEY
Ⓐ = Aldosterone-sensitive exchange pump

○ = Sodium-potassium exchange pump

3 Potassium balance does not occur in isolation. For example, aldosterone secretion increases in response to either high plasma K⁺ concentrations or a decline in ECF sodium levels. However, disturbances of acid-base balance can disrupt potassium balance by affecting the aldosterone-sensitive pumps.

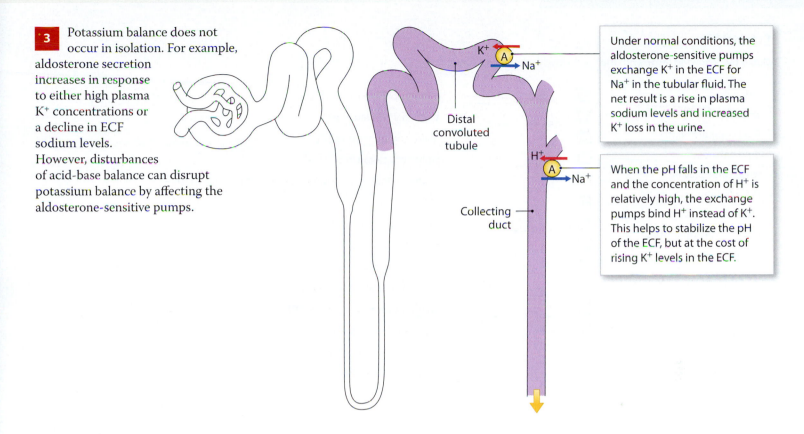

Distal convoluted tubule

Collecting duct

Under normal conditions, the aldosterone-sensitive pumps exchange K⁺ in the ECF for Na⁺ in the tubular fluid. The net result is a rise in plasma sodium levels and increased K⁺ loss in the urine.

When the pH falls in the ECF and the concentration of H⁺ is relatively high, the exchange pumps bind H⁺ instead of K⁺. This helps to stabilize the pH of the ECF, but at the cost of rising K⁺ levels in the ECF.

4 The diagram below indicates major factors involved in disturbances of potassium balance. The normal range of K⁺ concentration in the ECF is 3.5–5.5 mEq/L. Variations outside of that range are unusual and potentially dangerous.

When the plasma concentration of potassium falls below 2 mEq/L, extensive muscular weakness develops, followed by eventual paralysis. This condition, called **hypokalemia** (*kalium*, potassium), is potentially lethal due to its effects on the heart.

Normal potassium levels in serum (3.5–5.5 mEq/L)

High K⁺ concentrations in the ECF produce an equally dangerous condition known as **hyperkalemia**. Severe cardiac arrhythmias appear when the K⁺ concentration exceeds 8 mEq/L.

Factors Promoting Hypokalemia	
Several diuretics, including Lasix, can produce hypokalemia by increasing the volume of urine produced.	The endocrine disorder called **aldosteronism**, characterized by excessive aldosterone secretion, results in hypokalemia by overstimulating sodium retention and potassium loss.

Factors Promoting Hyperkalemia		
Chronically low body fluid pH promotes hyperkalemia by interfering with K⁺ excretion at the kidneys.	Kidney failure due to damage or disease will prevent normal K⁺ secretion and thereby produce hyperkalemia.	Several drugs promote diuresis by blocking Na⁺ reabsorption at the kidneys. When sodium reabsorption slows down, so does potassium secretion, and hyperkalemia can result.

Treatment for hypokalemia generally includes increasing dietary intake of potassium by salting food with potassium salts (KCl) or by taking potassium tablets, such as Slow-K. Treatment for hyperkalemia typically includes diluting the ECF with a solution low in K⁺, stimulating K⁺ loss in urine by using diuretics such as Lasix, adjusting the pH of the ECF, and restricting dietary K⁺ intake. In cases resulting from renal failure, kidney dialysis may also be required.

Module 24.4 Review

a. Define hypokalemia and hyperkalemia.

b. What organs are primarily responsible for regulating the potassium ion concentration of the ECF?

c. Identify factors that cause potassium excretion.

1 . Matching

Match the following terms with the most closely related description.

- kidneys
- potassium
- fluid compartments
- fluid balance
- hypertonic plasma
- dehydration
- aldosterone
- plasma, interstitial fluid
- osmoreceptors
- fluid shift
- hypokalemia
- ADH
- hyponatremia
- sodium

a	_____	Monitor plasma osmotic concentration
b	_____	Water gain = water loss
c	_____	Major components of ECF
d	_____	Dominant cation in ECF
e	_____	Hormone that restricts water loss and stimulates thirst
f	_____	Overhydration
g	_____	Dominant cation in ICF
h	_____	ICF and ECF
i	_____	Most important sites of sodium ion regulation
j	_____	Water movement between ECF and ICF
k	_____	Water moves from cells into ECF
l	_____	Result of aldosteronism
m	_____	Water losses greater than water gains
n	_____	Regulates sodium ion absorption along DCT and collecting system

2. Multiple choice

Choose the bulleted item that best completes each statement.

a Nearly two-thirds of the total body water content is

_____ .

- extracellular fluid (ECF)
- intracellular fluid (ICF)
- tissue fluid
- interstitial fluid

b Electrolyte balance involves balancing the rates of absorption across

the digestive tract with rates of loss at the _____ .

- heart and lungs
- stomach and liver
- kidneys and sweat glands
- pancreas and gallbladder

c If the ECF is hypertonic with respect to the ICF, water will move

_____ .

- from the ECF into the cells until osmotic equilibrium is restored
- from the cells into the ECF until osmotic equilibrium is restored
- in both directions until osmotic equilibrium is restored
- in response to the sodium–potassium exchange pump

d When pure water is consumed, the ECF

_____ .

- becomes hypotonic with respect to the ICF
- becomes hypertonic with respect to the ICF
- becomes isotonic with respect to the ICF
- electrolytes become more concentrated

e Physiological adjustments affecting fluid and electrolyte balance are

mediated primarily by _____ .

- antidiuretic hormone
- aldosterone
- natriuretic peptides
- all of these

f When water is lost but electrolytes are retained, the osmolarity of the

ECF rises, and osmosis then moves water _____ .

- out of the ECF and into the ICF
- back and forth between the ICF and ECF
- out of the ICF and into the ECF
- none of the above

3. Section integration

Malia, a nursing student, has been caring for burn patients. She notices that they consistently show elevated levels
of potassium in their urine and wonders why. What would you tell her? _____

_____

_____

Acid-Base Balance

Your body is in **acid-base balance** when the production of hydrogen ions is precisely offset by their loss, and when the pH of body fluids remains within normal limits.

1 This diagram shows the major factors involved in the maintenance of acid-base balance. The primary challenge to homeostasis is that your body generates a variety of acids during normal metabolic operations, and a significant decline in body fluid pH must be prevented.

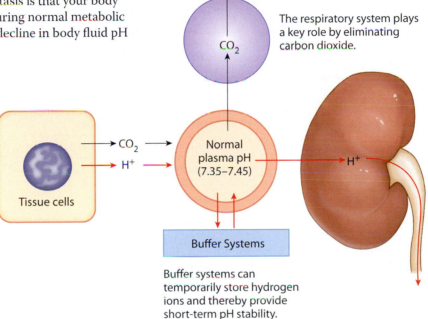

Active tissues continuously generate carbon dioxide, which in solution forms carbonic acid. Additional acids, such as lactic acid, are produced in the course of normal metabolic operations.

The respiratory system plays a key role by eliminating carbon dioxide.

The kidneys play a major role by secreting hydrogen ions into the urine and generating buffers that enter the bloodstream. The rate of excretion rises and falls as needed to maintain normal plasma pH. As a result, the normal pH of urine varies widely but averages 6.0—slightly acidic.

Buffer systems can temporarily store hydrogen ions and thereby provide short-term pH stability.

2 There are three classes of acids that can threaten pH balance.

Classes of Acids

Fixed Acids

Fixed acids are acids that do not leave solution; once produced, they remain in body fluids until they are eliminated at the kidneys. Sulfuric acid and phosphoric acid are the most important fixed acids in the body. They are generated in small amounts during the catabolism of amino acids and compounds that contain phosphate groups, including phospholipids and nucleic acids.

Organic Acids

Organic acids are acid participants in, or by-products of, cellular metabolism. Important organic acids include lactic acid (produced by the anaerobic metabolism of pyruvate) and ketone bodies (synthesized from acetyl-CoA). Under normal conditions, most organic acids are metabolized rapidly, so significant accumulations do not occur.

Volatile Acids

Volatile acids can leave the body by entering the atmosphere at the lungs. Carbonic acid (H_2CO_3) is a volatile acid that forms through the interaction of water and carbon dioxide.

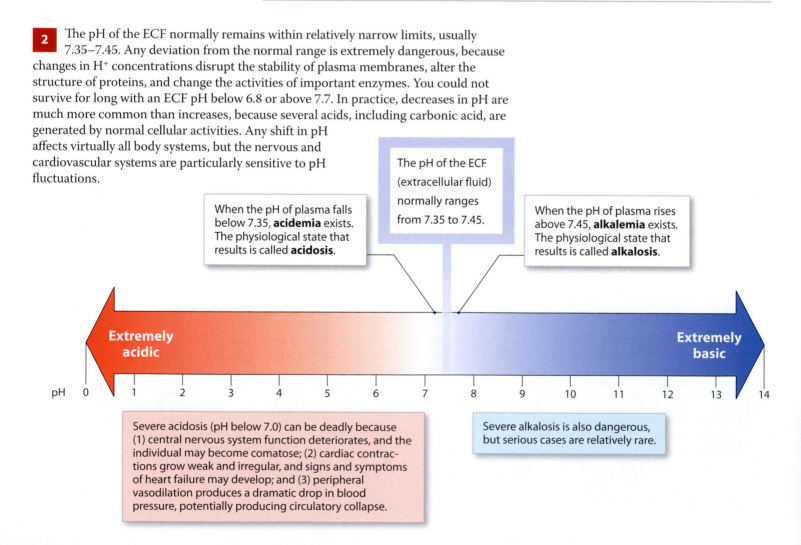

Potentially dangerous disturbances in acid-base balance are opposed by buffer systems

1 The topic of pH and the chemical nature of acids, bases, and buffers was introduced in Module 2.9. This table reviews key terms important to the discussion that follows.

A Review of Important Terms Relating to Acid-Base Balance	
pH	The negative exponent (negative logarithm) of the hydrogen ion concentration [H^+] in a solution
Neutral	A solution with a pH of 7; the solution contains equal numbers of hydrogen ions and hydroxide ions
Acidic	A solution with a pH below 7; in this solution, hydrogen ions (H^+) predominate
Basic, or alkaline	A solution with a pH above 7; in this solution, hydroxide ions (OH^-) predominate
Acid	A substance that dissociates to release hydrogen ions, decreasing pH
Base	A substance that dissociates to release hydroxide ions or to remove hydrogen ions, increasing pH
Salt	An ionic compound consisting of a cation other than a hydrogen ion and an anion other than a hydroxide ion
Buffer	A substance that tends to oppose changes in the pH of a solution by removing or replacing hydrogen ions; in body fluids, buffers maintain blood pH within normal limits (7.35–7.45)

2 The pH of the ECF normally remains within relatively narrow limits, usually 7.35–7.45. Any deviation from the normal range is extremely dangerous, because changes in H^+ concentrations disrupt the stability of plasma membranes, alter the structure of proteins, and change the activities of important enzymes. You could not survive for long with an ECF pH below 6.8 or above 7.7. In practice, decreases in pH are much more common than increases, because several acids, including carbonic acid, are generated by normal cellular activities. Any shift in pH affects virtually all body systems, but the nervous and cardiovascular systems are particularly sensitive to pH fluctuations.

The pH of the ECF (extracellular fluid) normally ranges from 7.35 to 7.45.

When the pH of plasma falls below 7.35, **acidemia** exists. The physiological state that results is called **acidosis**.

When the pH of plasma rises above 7.45, **alkalemia** exists. The physiological state that results is called **alkalosis**.

Extremely acidic

Extremely basic

pH   0   1   2   3   4   5   6   7   8   9   10   11   12   13   14

Severe acidosis (pH below 7.0) can be deadly because (1) central nervous system function deteriorates, and the individual may become comatose; (2) cardiac contractions grow weak and irregular, and signs and symptoms of heart failure may develop; and (3) peripheral vasodilation produces a dramatic drop in blood pressure, potentially producing circulatory collapse.

Severe alkalosis is also dangerous, but serious cases are relatively rare.

3 The partial pressure of carbon dioxide (P_{CO_2}) is the most important factor affecting the pH of body tissues because carbon dioxide combines with water to form **carbonic acid** (H_2CO_3). Because most of the carbon dioxide in solution is converted to carbonic acid by carbonic anhydrase, and most of the carbonic acid dissociates, there is an inverse relationship between the P_{CO_2} and pH.

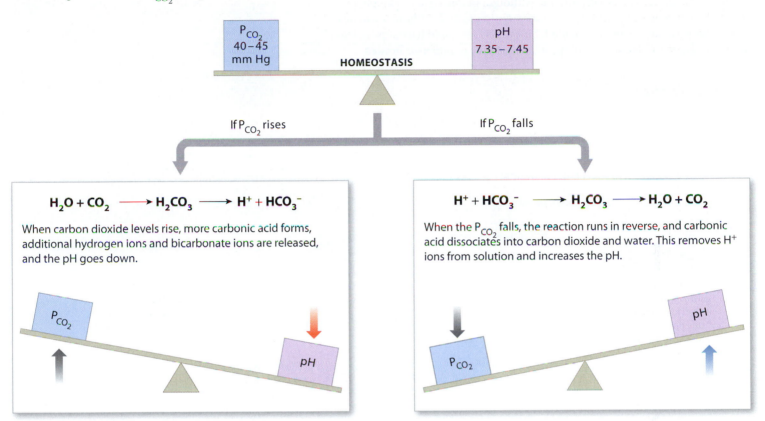

$$H_2O + CO_2 \longrightarrow H_2CO_3 \longrightarrow H^+ + HCO_3^-$$

When carbon dioxide levels rise, more carbonic acid forms, additional hydrogen ions and bicarbonate ions are released, and the pH goes down.

$$H^+ + HCO_3^- \longrightarrow H_2CO_3 \longrightarrow H_2O + CO_2$$

When the P_{CO_2} falls, the reaction runs in reverse, and carbonic acid dissociates into carbon dioxide and water. This removes H^+ ions from solution and increases the pH.

4 A **buffer system** in body fluids generally consists of a combination of a weak acid (HY) and the anion (Y^-) released by its dissociation. The anion functions as a weak base. In solution, molecules of the weak acid exist in equilibrium with its dissociation products. In chemical notation, this relationship is represented as:

$$HY \rightleftharpoons H^+ + Y^-$$

Adding H^+ to the solution upsets the equilibrium and results in the formation of additional molecules of the weak acid.

$$H^+ + Y^- \xrightarrow{H^+} H^+ + HY$$

Removing H^+ from the solution also upsets the equilibrium and results in the dissociation of additional molecules of HY. This releases H^+.

$$H^+ + HY \xrightarrow[H^+]{} H^+ + Y^-$$

Module 24.5 Review

a. Define acidemia and alkalemia.

b. What is the most important factor affecting the pH of the ECF?

c. Summarize the relationship between CO_2 levels and pH.

Buffer systems can delay but not prevent pH shifts in the ICF and ECF

1 The body has three major buffer systems, each with slightly different characteristics and distributions. Although buffer systems can tie up excess H^+, they provide only a temporary solution to an acid-base imbalance. The hydrogen ions are not eliminated, but merely rendered harmless. In the process a buffer molecule is tied up, and the supply of buffers is limited.

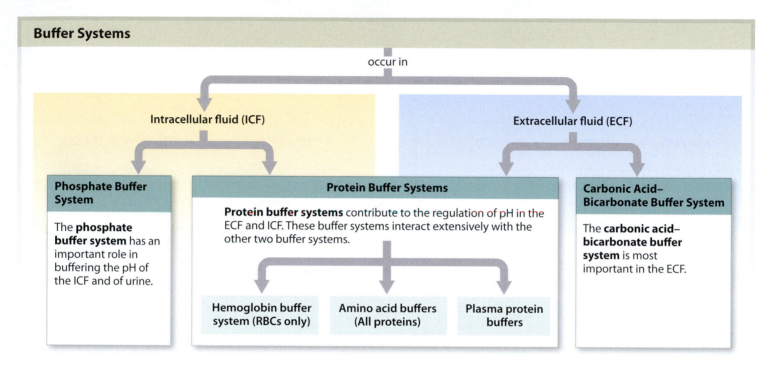

Buffer Systems

occur in

Intracellular fluid (ICF)

Extracellular fluid (ECF)

Phosphate Buffer System

The **phosphate buffer system** has an important role in buffering the pH of the ICF and of urine.

Protein Buffer Systems

Protein buffer systems contribute to the regulation of pH in the ECF and ICF. These buffer systems interact extensively with the other two buffer systems.

Hemoglobin buffer system (RBCs only)

Amino acid buffers (All proteins)

Plasma protein buffers

Carbonic Acid–Bicarbonate Buffer System

The **carbonic acid–bicarbonate buffer system** is most important in the ECF.

2 The **hemoglobin buffer system** is the only intracellular buffer system that can have an immediate effect on the pH of body fluids. In the tissues, red blood cells absorb carbon dioxide from the plasma and convert it to carbonic acid. As the carbonic acid dissociates, the hydrogen ions are buffered by hemoglobin proteins. At the lungs, the entire reaction sequence proceeds in reverse and the CO_2 diffuses into the alveoli for exhalation.

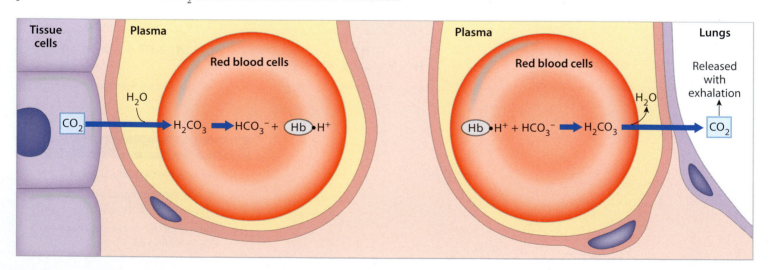

Tissue cells   Plasma   Red blood cells

CO_2 H_2O H_2CO_3 → $HCO_3^- +$ (Hb) • H^+

Plasma   Red blood cells   Lungs

Released with exhalation

(Hb) • $H^+ + HCO_3^-$ → H_2CO_3 H_2O CO_2

3 Protein buffer systems reduce the rate of pH change, usually by binding excess hydrogen ions. These buffer systems depend on the ability of amino acids to respond to pH changes by accepting or releasing H^+. The underlying mechanism is shown here. In a protein, most of the carboxyl and amino groups in the main chain are tied up in peptide bonds, leaving only the $—COO^-$ of the first amino acid and the $—NH_2$ of the last amino acid available as buffers. So most of the buffering capacity of proteins is provided by the R-groups of the component amino acids.

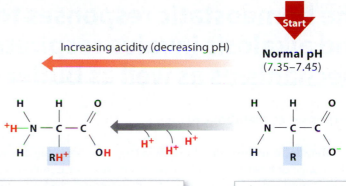

Increasing acidity (decreasing pH)

Start

Normal pH
(7.35–7.45)

If pH drops, the carboxylate ion (COO^-) and the amino group ($—NH_2$) of a free amino acid can act as weak bases and accept additional hydrogen ions, forming a carboxyl group ($—COOH$) and an amino ion ($—NH_3^+$), respectively. Many of the R-groups can also accept hydrogen ions, forming RH^+.

At the normal pH of body fluids (7.35–7.45), the carboxyl groups of most amino acids have released their hydrogen ions.

4 The carbonic acid–bicarbonate buffer system involves freely reversible reactions; a change in the concentration of any participant affects the concentrations of all other participants and shifts the direction of the reactions under way.

BICARBONATE RESERVE

Body fluids contain a large reserve of HCO_3^-, primarily in the form of dissolved molecules of the weak base sodium bicarbonate ($NaHCO_3$). This readily available supply of HCO_3^- is known as the **bicarbonate reserve**.

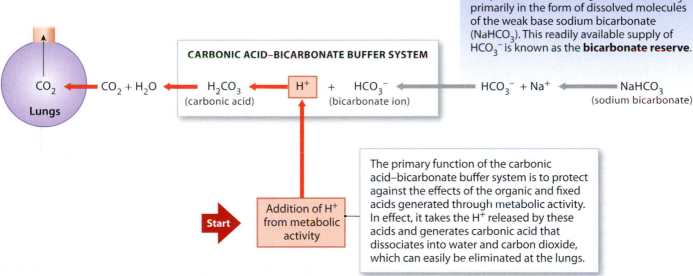

CARBONIC ACID–BICARBONATE BUFFER SYSTEM

Lungs

CO_2 ← $CO_2 + H_2O$ ← H_2CO_3 (carbonic acid) ← H^+ + HCO_3^- (bicarbonate ion) ← $HCO_3^- + Na^+$ ← $NaHCO_3$ (sodium bicarbonate)

Start → Addition of H^+ from metabolic activity

The primary function of the carbonic acid–bicarbonate buffer system is to protect against the effects of the organic and fixed acids generated through metabolic activity. In effect, it takes the H^+ released by these acids and generates carbonic acid that dissociates into water and carbon dioxide, which can easily be eliminated at the lungs.

Metabolic acid-base disorders result from the production or loss of excessive amounts of fixed or organic acids. The primary role of the carbonic acid–bicarbonate buffer system is to protect against such disorders. **Respiratory acid-base disorders** result from an imbalance between the rate of CO_2 generation and the rate of CO_2 elimination at the lungs. The carbonic acid–bicarbonate buffer system cannot provide protection against respiratory disorders; the imbalances must be corrected by reflexive changes in the depth and rate of respiration.

Module 24.6 Review

a. Identify the body's three major buffer systems.

b. Describe the carbonic acid–bicarbonate buffer system.

c. Describe the roles of the phosphate buffer system.

The homeostatic responses to metabolic acidosis and alkalosis involve respiratory and renal mechanisms as well as buffer systems

Responses to Metabolic Acidosis

1 **Metabolic acidosis** develops when large numbers of hydrogen ions are released by organic or fixed acids, and the pH decreases. For homeostasis to be preserved, the excess H^+ must either be permanently tied up through the formation of water (linked to the formation of CO_2 that can be eliminated at the lungs), or removed from body fluids through secretion at the kidneys.

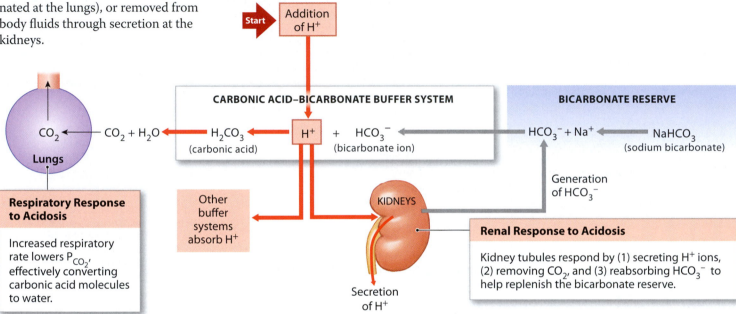

Start → Addition of H^+

CARBONIC ACID–BICARBONATE BUFFER SYSTEM

$$CO_2 \leftarrow CO_2 + H_2O \leftarrow H_2CO_3 \leftarrow H^+ + HCO_3^-$$
$$\text{(carbonic acid)} \quad \text{(bicarbonate ion)}$$

Lungs

BICARBONATE RESERVE

$$HCO_3^- + Na^+ \leftarrow NaHCO_3$$
$$\text{(sodium bicarbonate)}$$

Generation of HCO_3^-

Respiratory Response to Acidosis

Increased respiratory rate lowers P_{CO_2}, effectively converting carbonic acid molecules to water.

Other buffer systems absorb H^+

KIDNEYS

Secretion of H^+

Renal Response to Acidosis

Kidney tubules respond by (1) secreting H^+ ions, (2) removing CO_2, and (3) reabsorbing HCO_3^- to help replenish the bicarbonate reserve.

2 Renal tubule cells secrete H^+ into the tubular fluid along the proximal convoluted tubule (PCT), the distal convoluted tubule (DCT), and the collecting system. They also bolster the capabilities of the carbonic acid–bicarbonate buffer system by increasing the concentration of bicarbonate ions in the ECF, replacing those pulled from the bicarbonate reserve.

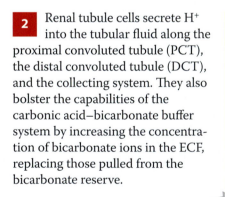
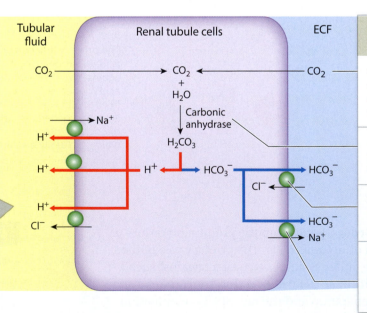

Tubular fluid

Renal tubule cells

ECF

$CO_2 \rightarrow CO_2 + H_2O$

Carbonic anhydrase

H_2CO_3

$CO_2 \leftarrow CO_2$

Na^+

H^+

H^+ ... H^+ → $HCO_3^- \rightarrow HCO_3^-$

Cl^-

H^+

Cl^-

HCO_3^-

Na^+

Steps in CO_2 removal and HCO_3^- production

CO_2 generated by the tubule cell is added to the CO_2 diffusing into the cell from the urine and from the ECF.

Carbonic anhydrase converts CO_2 and water to carbonic acid, which then dissociates.

The chloride ions exchanged for bicarbonate ions are excreted in the tubular fluid.

Bicarbonate ions and sodium ions are transported into the ECF, adding to the bicarbonate reserve.

Responses to Metabolic Alkalosis

3   **Metabolic alkalosis** develops when large numbers of hydrogen ions are removed from body fluids, resulting in a rise in pH. When this occurs, (1) the rate of H^+ secretion at the kidneys declines, (2) tubule cells do not reclaim the bicarbonates in tubular fluid, and (3) the collecting system transports HCO_3^- into tubular fluid while releasing a strong acid (HCl) into the ECF.

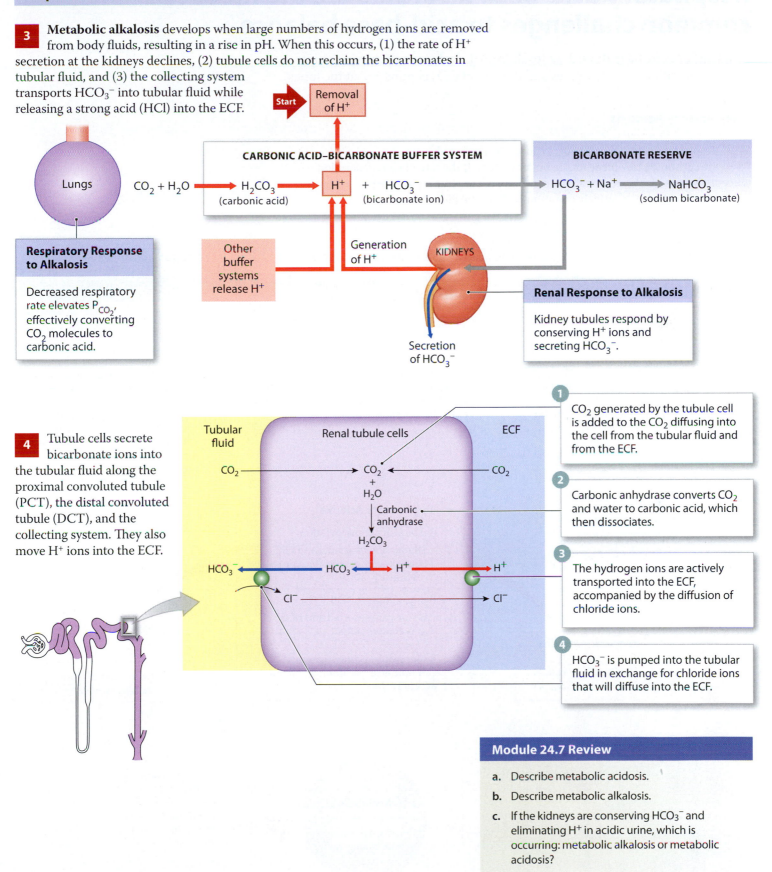

Start   Removal of H^+

CARBONIC ACID–BICARBONATE BUFFER SYSTEM

Lungs

$CO_2 + H_2O \longrightarrow H_2CO_3$ (carbonic acid) $\longrightarrow H^+ \ + \ HCO_3^-$ (bicarbonate ion)

BICARBONATE RESERVE

$HCO_3^- + Na^+ \longrightarrow NaHCO_3$ (sodium bicarbonate)

Respiratory Response to Alkalosis

Decreased respiratory rate elevates P_{CO_2}, effectively converting CO_2 molecules to carbonic acid.

Other buffer systems release H^+

Generation of H^+

KIDNEYS

Secretion of HCO_3^-

Renal Response to Alkalosis

Kidney tubules respond by conserving H^+ ions and secreting HCO_3^-.

4   Tubule cells secrete bicarbonate ions into the tubular fluid along the proximal convoluted tubule (PCT), the distal convoluted tubule (DCT), and the collecting system. They also move H^+ ions into the ECF.

Tubular fluid

Renal tubule cells

ECF

$CO_2 \longrightarrow CO_2 \longleftarrow CO_2$
$+$
H_2O
Carbonic anhydrase
H_2CO_3
$HCO_3^- \longleftarrow HCO_3^- \longrightarrow H^+ \longrightarrow H^+$
$Cl^- \longrightarrow Cl^-$

1 CO_2 generated by the tubule cell is added to the CO_2 diffusing into the cell from the tubular fluid and from the ECF.

2   Carbonic anhydrase converts CO_2 and water to carbonic acid, which then dissociates.

3   The hydrogen ions are actively transported into the ECF, accompanied by the diffusion of chloride ions.

4 HCO_3^- is pumped into the tubular fluid in exchange for chloride ions that will diffuse into the ECF.

Module 24.7 Review

a. Describe metabolic acidosis.

b. Describe metabolic alkalosis.

c. If the kidneys are conserving HCO_3^- and eliminating H^+ in acidic urine, which is occurring: metabolic alkalosis or metabolic acidosis?

Respiratory acid-base disorders are the most common challenges to acid-base balance

Respiratory acid-base disorders result from an imbalance between the rate of CO_2 generation in body tissues and the rate of CO_2 elimination at the lungs.

Respiratory Acidosis

1 Respiratory acid-base disorders cannot be corrected, even temporarily, by the carbonic acid–bicarbonate buffer system. If the rate of CO_2 generation exceeds the rate of CO_2 removal, the condition of **respiratory acidosis** develops. Respiratory acidosis is relatively common and may be life-threatening.

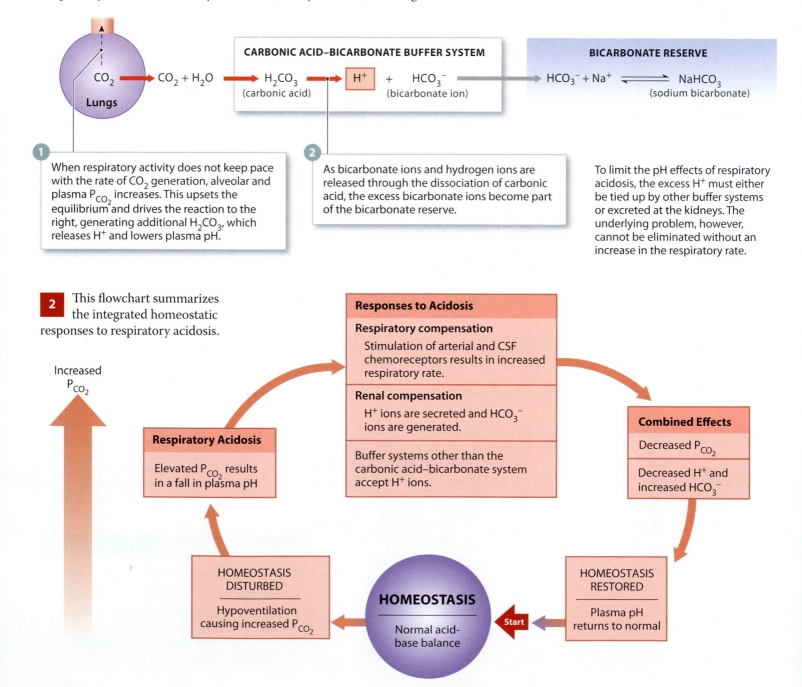

CARBONIC ACID–BICARBONATE BUFFER SYSTEM

CO_2 → $CO_2 + H_2O$ → H_2CO_3 (carbonic acid) → H^+ + HCO_3^- (bicarbonate ion)

Lungs

BICARBONATE RESERVE

$HCO_3^- + Na^+$ ⇌ $NaHCO_3$ (sodium bicarbonate)

1 When respiratory activity does not keep pace with the rate of CO_2 generation, alveolar and plasma P_{CO_2} increases. This upsets the equilibrium and drives the reaction to the right, generating additional H_2CO_3, which releases H^+ and lowers plasma pH.

2 As bicarbonate ions and hydrogen ions are released through the dissociation of carbonic acid, the excess bicarbonate ions become part of the bicarbonate reserve.

To limit the pH effects of respiratory acidosis, the excess H^+ must either be tied up by other buffer systems or excreted at the kidneys. The underlying problem, however, cannot be eliminated without an increase in the respiratory rate.

2 This flowchart summarizes the integrated homeostatic responses to respiratory acidosis.

Increased P_{CO_2}

Responses to Acidosis

Respiratory compensation
Stimulation of arterial and CSF chemoreceptors results in increased respiratory rate.

Renal compensation
H^+ ions are secreted and HCO_3^- ions are generated.

Buffer systems other than the carbonic acid–bicarbonate system accept H^+ ions.

Respiratory Acidosis
Elevated P_{CO_2} results in a fall in plasma pH

Combined Effects
Decreased P_{CO_2}

Decreased H^+ and increased HCO_3^-

HOMEOSTASIS DISTURBED
Hypoventilation causing increased P_{CO_2}

HOMEOSTASIS
Normal acid-base balance

Start

HOMEOSTASIS RESTORED
Plasma pH returns to normal

Respiratory Alkalosis

3 If the rate of CO_2 elimination exceeds the rate of CO_2 generation, the condition of **respiratory alkalosis** develops. Metabolic alkalosis is relatively uncommon and rarely severe.

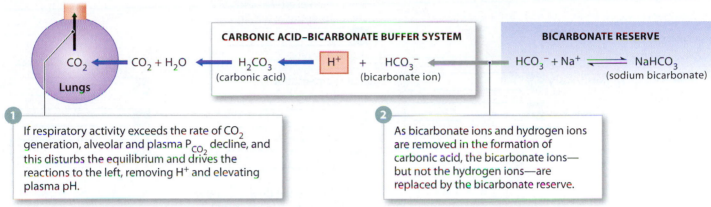

CARBONIC ACID–BICARBONATE BUFFER SYSTEM

$$CO_2 \longleftarrow CO_2 + H_2O \longleftarrow \underset{\text{(carbonic acid)}}{H_2CO_3} \longleftarrow \boxed{H^+} + \underset{\text{(bicarbonate ion)}}{HCO_3^-}$$

Lungs

BICARBONATE RESERVE

$$HCO_3^- + Na^+ \rightleftharpoons \underset{\text{(sodium bicarbonate)}}{NaHCO_3}$$

1 If respiratory activity exceeds the rate of CO_2 generation, alveolar and plasma P_{CO_2} decline, and this disturbs the equilibrium and drives the reactions to the left, removing H^+ and elevating plasma pH.

2 As bicarbonate ions and hydrogen ions are removed in the formation of carbonic acid, the bicarbonate ions—but not the hydrogen ions—are replaced by the bicarbonate reserve.

4 The homeostatic responses to respiratory alkalosis are summarized in this flowchart. Most cases are related to anxiety, and the resulting hyperventilation is self-limiting—the individual often faints and the respiratory rate then drops to normal levels. One common treatment is to have the individual breathe in and out of a paper bag; the rising CO_2 level in the recycled air elevates plasma P_{CO_2} and eliminates the alkalosis.

HOMEOSTASIS
Normal acid-base balance

Start

HOMEOSTASIS DISTURBED

Hyperventilation causing decreased P_{CO_2}

HOMEOSTASIS RESTORED

Plasma pH returns to normal

Decreased P_{CO_2}

Respiratory Alkalosis

Lower P_{CO_2} results in a rise in plasma pH

Responses to Alkalosis

Respiratory compensation

Inhibition of arterial and CSF chemoreceptors results in a decreased respiratory rate.

Renal compensation

H^+ ions are generated and HCO_3^- ions are secreted.

Buffer systems other than the carbonic acid–bicarbonate system release H^+ ions.

Combined Effects

Increased P_{CO_2}

Increased H^+ and decreased HCO_3^-

Module 24.8 Review

a. Define respiratory acidosis and respiratory alkalosis.

b. What would happen to the plasma P_{CO_2} of a patient who has an airway obstruction?

c. How would a decrease in the pH of body fluids affect the respiratory rate?

1. Labeling

Use the following terms to label the boxes in the two flowcharts. Terms may be used more than once.

- plasma pH decrease
- plasma pH increase
- increased P_{CO_2}
- decreased P_{CO_2}
- increased
- decreased
- alkalosis
- acidosis
- generated
- secreted

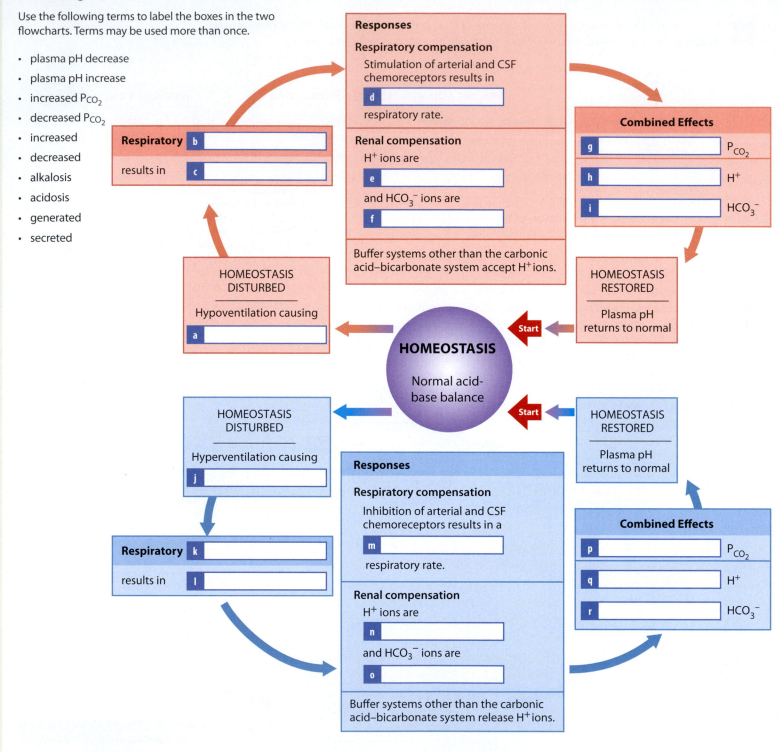

Responses

Respiratory compensation
Stimulation of arterial and CSF chemoreceptors results in
d _____
respiratory rate.

Renal compensation
H^+ ions are
e _____
and HCO_3^- ions are
f _____

Buffer systems other than the carbonic acid–bicarbonate system accept H^+ ions.

Respiratory **b** _____
results in **c** _____

HOMEOSTASIS DISTURBED
_____
Hypoventilation causing
a _____

Combined Effects
g _____ P_{CO_2}
h _____ H^+
i _____ HCO_3^-

HOMEOSTASIS RESTORED
_____
Plasma pH returns to normal

Start

HOMEOSTASIS
Normal acid-base balance

Start

HOMEOSTASIS DISTURBED
_____
Hyperventilation causing
j _____

HOMEOSTASIS RESTORED
_____
Plasma pH returns to normal

Responses

Respiratory compensation
Inhibition of arterial and CSF chemoreceptors results in a
m _____
respiratory rate.

Renal compensation
H^+ ions are
n _____
and HCO_3^- ions are
o _____

Buffer systems other than the carbonic acid–bicarbonate system release H^+ ions.

Respiratory **k** _____
results in **l** _____

Combined Effects
p _____ P_{CO_2}
q _____ H^+
r _____ HCO_3^-

2. Section integration

After falling into an abandoned stone quarry filled with water and nearly drowning, a young boy is rescued. His rescuers assess his condition and find that his body fluids have high P_{CO_2} and lactate levels, and low P_{O_2} levels. Identify the underlying problem and recommend the necessary treatment to restore homeostatic conditions. _____

_____

_____

Visual Outline with Key Terms

Summarize the content of each module using the terms in the order provided.

SECTION 1

Fluid and Electrolyte Balance

- minerals
  - electrolytes
  - intracellular fluid (ICF)
  - extracellular fluid (ECF)
- fluid compartments

24.4

Disturbances of potassium balance are uncommon but extremely dangerous

- potassium balance
- hypokalemia
- hyperkalemia
- aldosteronism

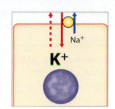

24.1

Fluid balance exists when water gains equal water losses

- fluid balance
- fluid shift
- dehydration

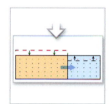

SECTION 2

Acid-Base Balance

- acid-base balance
- fixed acids
- organic acids
- volatile acids

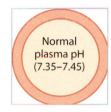

24.2

Mineral balance involves balancing electrolyte gains and losses

- mineral balance
  - ion absorption
  - ion excretion

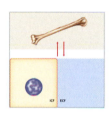

24.5

Potentially dangerous disturbances in acid-base balance are opposed by buffer systems

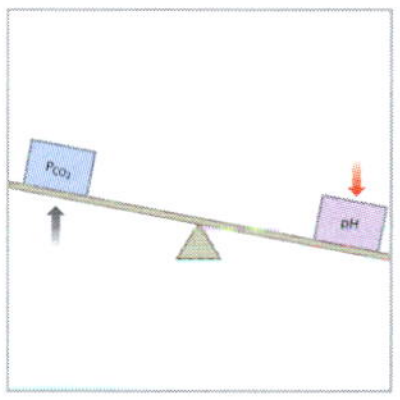

- acidemia
- acidosis
- alkalemia
- alkalosis
- carbonic acid
- buffer system

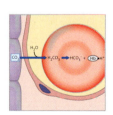

24.3

Water balance depends on sodium balance, and the two are regulated simultaneously

- sodium balance
  - osmoreceptors
- hyponatremia
- hypernatremia

24.6

Buffer systems can delay but not prevent pH shifts in the ICF and ECF

- phosphate buffer system
- protein buffer systems
- carbonic acid–bicarbonate buffer system
- hemoglobin buffer system
- bicarbonate reserve
- metabolic acid-base disorders
- respiratory acid-base disorders

● = *Term boldfaced in this module*

24.7

The homeostatic responses to acidosis and alkalosis involve respiratory and renal mechanisms as well as buffer systems

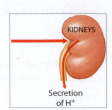

KIDNEYS
Secretion of H^+

- metabolic acidosis
  - carbonic acid–bicarbonate buffer system
  - respiratory response to acidosis
  - renal response to acidosis
- metabolic alkalosis
  - respiratory response to alkalosis
  - renal response to alkalosis

24.8

Respiratory acid-base disorders are the most common challenges to acid-base balance

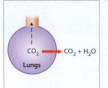

CO_2
$CO_2 + H_2O$
Lungs

- respiratory acid-base disorders
- respiratory acidosis
- respiratory compensation
- renal compensation
- respiratory alkalosis

• = *Term boldfaced in this module*

Chapter Integration: Applying what you've learned

While visiting a foreign country with poor-quality potable water, Beth and Tom each had a drink with ice during dinner. Many travelers avoid drinking the water in foreign countries but forget that ice is often made with local tap water! As a result, both Beth and Tom contracted an intestinal disease that causes severe diarrhea. (Diarrhea is characterized by loose watery bowel movements that, if left untreated, can cause serious health complications.)

It is estimated that each year such "traveler's diarrhea" affects as many as 10 million Americans. High-risk destinations include the lesser-developed countries of Latin America, Africa, the Middle East, and regions of Asia. The Centers for Disease Control and Prevention (CDC) maintains a website link to advise travelers.

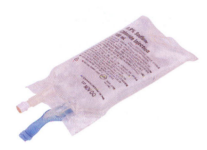

After two days of experiencing uncontrolled bouts of diarrhea, the couple sought medical attention at a nearby medical clinic. Various tests revealed that the culprit was the bacterium *Escherichia coli*. Further tests revealed that both Beth and Tom were dehydrated, so intravenous fluids were prescribed. Unfortunately, the attending nurse became distracted and erroneously gave Tom a hypertonic glucose solution instead of the normal saline that Beth received.

1. How would you expect Beth and Tom's dehydrated states to affect their blood pH, urine pH, and pattern of pulmonary ventilation?

2. What effect will the intravenous hypertonic glucose solution have on Tom's plasma ADH levels and urine volume?

Mastering A&P™

Access more review material online in the Study Area at **www.masteringaandp.com**.

There, you'll find:
- **Chapter guides**
- **Chapter quizzes**
- **Practice tests**
- **Labeling activities**
- **MP3 Tutor Sessions**
- **Flashcards**
- **A glossary with pronunciations**

iP® Animated tutorials in *Interactive Physiology*® (IP) help you understand difficult physiological concepts in this chapter. Go to Fluids and Electrolytes and find the following topics:

- **Introduction to Body Fluids**
- **Water Homeostasis**
- **Electrolyte Homeostasis**
- **Acid/Base Homeostasis**

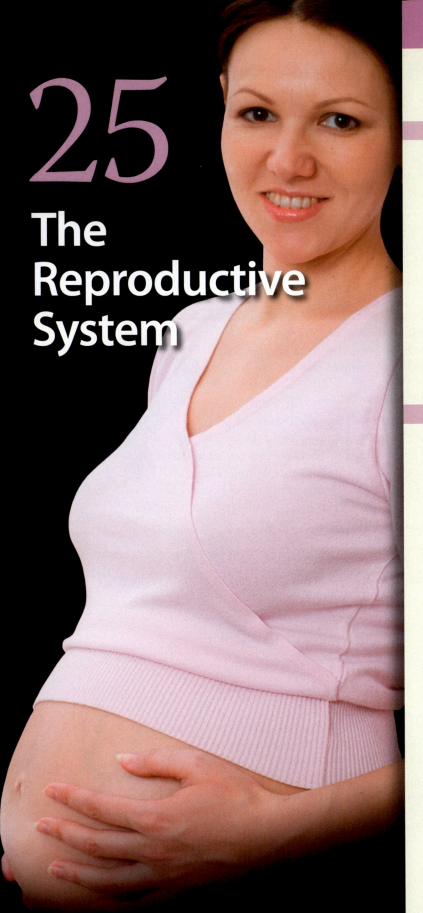

25

The Reproductive System

The Male Reproductive System

The reproductive system of both sexes includes the following basic components:

- **Gonads** (GŌ-nadz; *gone*, seed, generation), or reproductive organs, which produce gametes and hormones.

- Accessory glands and organs that secrete fluids into the ducts of the reproductive system or into other excretory ducts.

- Perineal structures collectively known as the **external genitalia** (jen-i-TĀ-lē-uh).

1 This figure provides an overview of the components of the male reproductive system. The gonad is called a **testis** (plural, *testes*), and it produces male gametes called **spermatozoa** (sper-ma-tō-ZŌ-uh; singular, *spermatozoon*), or sperm. Mature spermatozoa travel along the male duct system, or **male reproductive tract**. As they proceed, they are mixed with the secretions of accessory glands to form a fluid known as **semen** (SĒ-men).

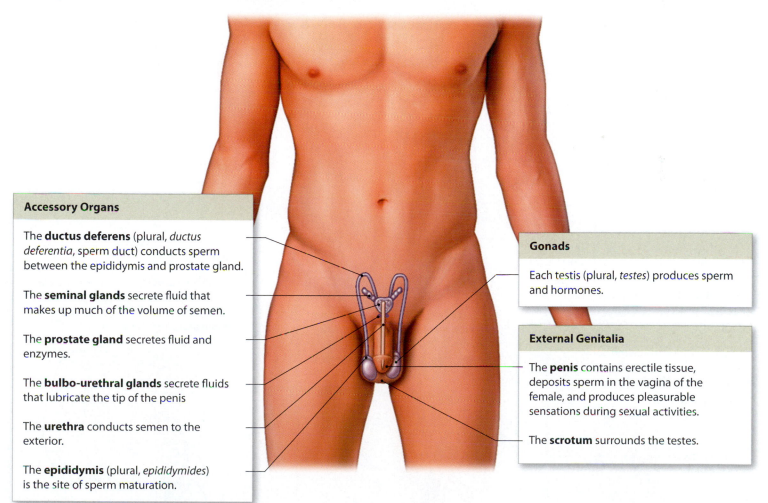

Accessory Organs

The **ductus deferens** (plural, *ductus deferentia*, sperm duct) conducts sperm between the epididymis and prostate gland.

The **seminal glands** secrete fluid that makes up much of the volume of semen.

The **prostate gland** secretes fluid and enzymes.

The **bulbo-urethral glands** secrete fluids that lubricate the tip of the penis

The **urethra** conducts semen to the exterior.

The **epididymis** (plural, *epididymides*) is the site of sperm maturation.

Gonads

Each testis (plural, *testes*) produces sperm and hormones.

External Genitalia

The **penis** contains erectile tissue, deposits sperm in the vagina of the female, and produces pleasurable sensations during sexual activities.

The **scrotum** surrounds the testes.

The coiled seminiferous tubules of the testes are connected to the male reproductive tract

1 The principal structures of the male reproductive system are shown in this sagittal section. Proceeding from a testis, the spermatozoa travel within the **epididymis** (ep-i-DID-i-mus), along the **ductus deferens** (DUK-tus DEF-e-renz), and then along the **ejaculatory duct** and the urethra before leaving the body. Accessory organs—the **seminal** (SEM-i-nal) **glands**, the **prostate** (PROS-tāt) **gland**, and the **bulbo-urethral** (bul-bō-ū-RĒ-thral) **glands**—secrete various fluids into the ejaculatory ducts and urethra. The external genitalia consist of the **scrotum** (SKRŌ-tum), which encloses the testes, and the **penis** (PĒ-nis), an erectile organ through which the distal portion of the urethra passes.

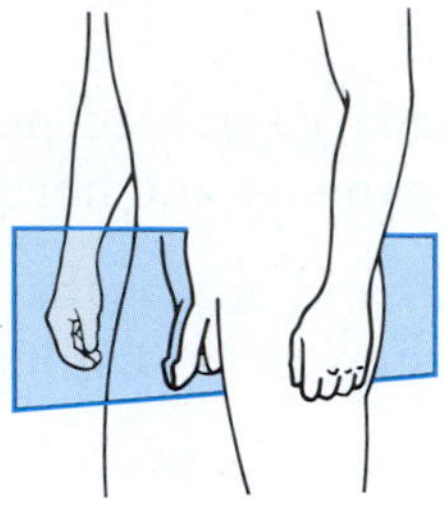

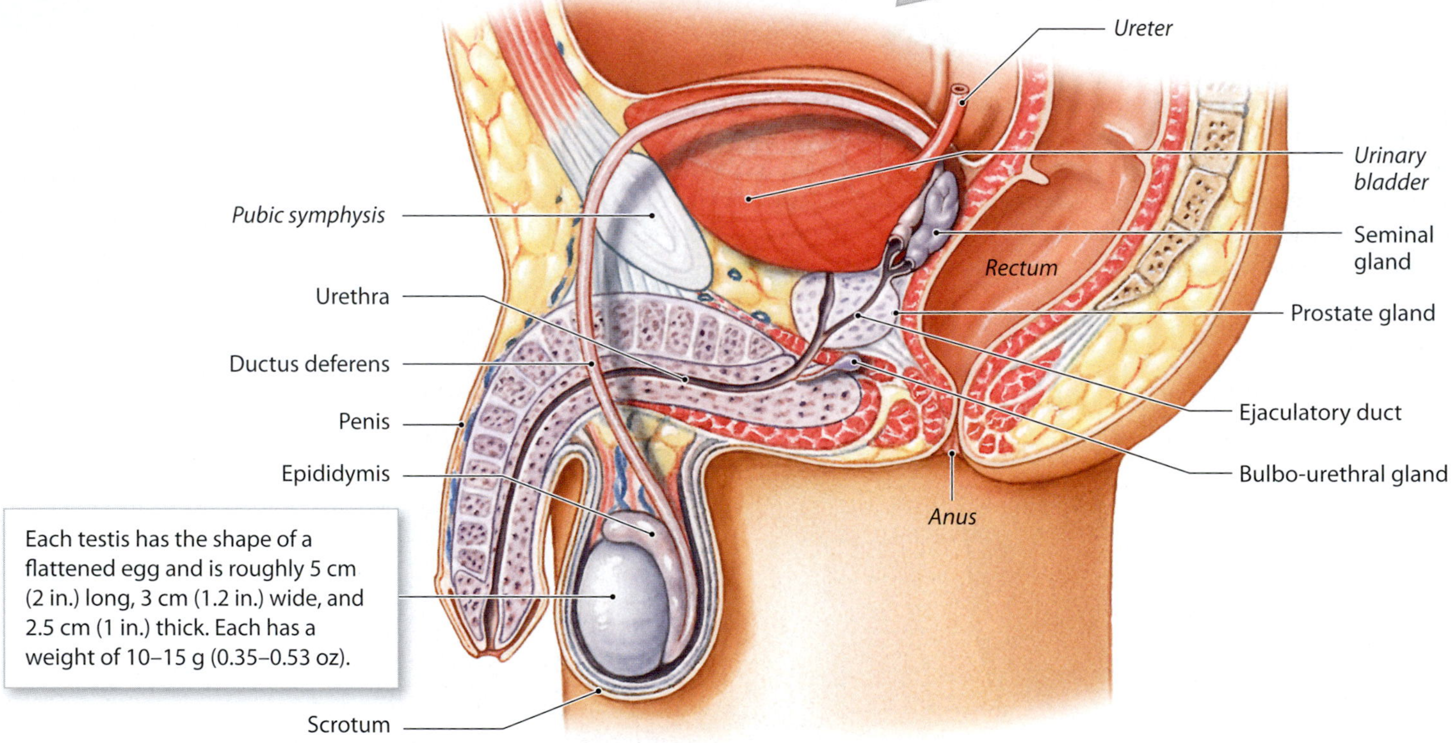

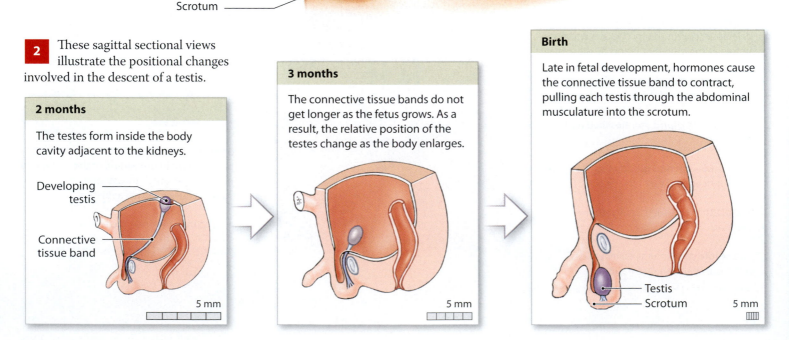

Ureter

Urinary bladder

Seminal gland

Pubic symphysis

Rectum

Urethra

Prostate gland

Ductus deferens

Penis

Ejaculatory duct

Epididymis

Bulbo-urethral gland

Anus

Each testis has the shape of a flattened egg and is roughly 5 cm (2 in.) long, 3 cm (1.2 in.) wide, and 2.5 cm (1 in.) thick. Each has a weight of 10–15 g (0.35–0.53 oz).

Scrotum

2 These sagittal sectional views illustrate the positional changes involved in the descent of a testis.

2 months

The testes form inside the body cavity adjacent to the kidneys.

Developing testis

Connective tissue band

5 mm

3 months

The connective tissue bands do not get longer as the fetus grows. As a result, the relative position of the testes change as the body enlarges.

5 mm

Birth

Late in fetal development, hormones cause the connective tissue band to contract, pulling each testis through the abdominal musculature into the scrotum.

Testis

Scrotum

5 mm

3 This dissection view shows superficial and deeper features of the scrotum, testes, and related structures.

Inguinal ligament

The **inguinal canal** extends from the inguinal ring to the scrotal cavity. In normal adult males, the inguinal canals are closed, but the presence of the spermatic cords creates weak points in the abdominal wall that remain throughout life. As a result, **inguinal hernias**—protrusions of visceral tissues or organs into the inguinal canal—are relatively common in males.

The **superficial inguinal ring** penetrates the layers of abdominal muscles and forms the entrance to the inguinal canal.

The **cremaster** (krē-MAS-ter) **muscle** lies deep to the dermis. Contraction of this muscle during sexual arousal or on exposure to cool temperatures pulls the testes closer to the body.

The scrotum consists of a thin layer of skin and the underlying superficial fascia. The dermis contains a layer of smooth muscle, the **dartos** (DAR-tōs) **muscle**. Resting muscle tone in the dartos muscle elevates the testes and causes the characteristic wrinkling of the scrotal surface.

Nerve
Artery
Venous plexus
Ductus deferens

The **spermatic cords** extend between the abdominopelvic cavity and the testes. Each spermatic cord consists of layers of fascia and muscle enclosing the ductus deferens and the blood vessels, nerves, and lymphatic vessels that supply the testes.

Scrotal septum
Scrotal cavity
Raphe

The left and right **scrotal cavities** are separated by the **scrotal septum**. The partition between the two is marked by a raised thickening in the scrotal surface known as the **raphe** (RĀ-fē).

4 This horizontal section through the scrotum shows the internal organization of the testes.

Epididymis   Efferent ductule   Ductus deferens

A mesothelium lines the scrotal cavity and reduces friction between opposing surfaces.

A tough fibrous capsule, the **tunica albuginea**, (al-bū-JIN-ē-uh) covers the testis. This capsule is continuous with septa that subdivide the interior of the testis into separate lobules.

Scrotum
Skin
Dartos muscle
Superficial fascia
Cremaster muscle
Scrotal cavity

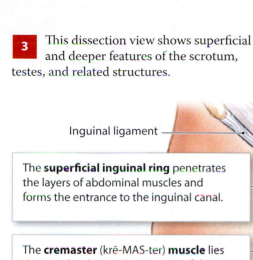

Septa

Each lobule contains a coiled **seminiferous tubule**, which averages about 80 cm (32 in.) in length, and a typical testis contains nearly one-half mile of seminiferous tubules. Sperm production occurs within these tubules.

Each seminiferous tubule is connected to a maze of passageways known as the **rete** (RĒ-tē; *rete*, a net) **testis**. Fifteen to 20 large **efferent ductules** connect the rete testis to the epididymis.

Module 25.1 Review

a. Name the male reproductive structures.

b. Identify the complex network of channels that is connected to the seminiferous tubules.

c. On a warm day, would the cremaster muscle be contracted or relaxed? Why?

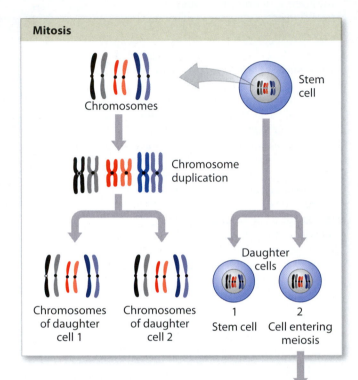

Meiosis in the testes produces haploid spermatids that mature into spermatozoa

Spermatogenesis, or sperm production, involves (1) **mitosis** and cell division, (2) **meiosis** (mī-Ō-sis), a special form of cell division involved in gamete production, and (3) **spermiogenesis**, the differentiation of immature male gametes into physically mature spermatozoa.

1 As you may recall from Chapter 3, somatic cells contain 23 pairs of chromosomes. Mitosis is part of the process of somatic cell division, producing two daughter cells each containing identical pairs of chromosomes. This diagram follows the fates of three representative chromosomes during mitosis and cell division. Because the daughter cells each contain 23 pairs of chromosomes, they are called **diploid** (DIP-loyd; *diplo*, double) cells. Each mitotic division of a stem cell within the seminiferous tubules produces one daughter cell that remains a stem cell, and one daughter cell that enters meiosis.

2 Meiosis involves two cycles of cell division (**meiosis I** and **meiosis II**) and produces four cells, each of which contains 23 individual chromosomes. Because these cells contain only one member of each pair of chromosomes, they are called **haploid** (HAP-loyd; *haplo*, single) cells. The events in the nucleus are the same for the formation of spermatozoa or ova. In **synapsis**, corresponding maternal and paternal chromosomes associate to form 23 chromosome pairs. Each member of a pair consists of two duplicate chromatids, and the set of four chromatids is called a **tetrad** (TET-rad; *tetras*, four).

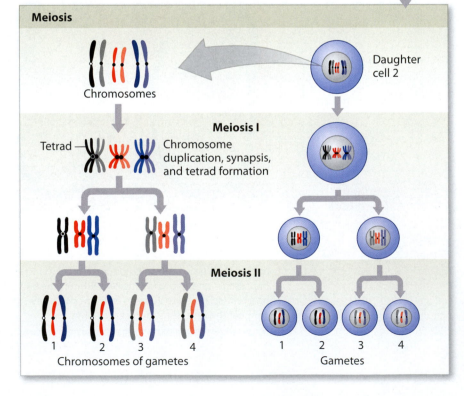

3 **Spermatogonia** (singular, *spermatogonium*) are the stem cells in the seminiferous tubules that divide to produce spermatozoa in a process called spermatogenesis.

Spermatogenesis

Mitosis of spermatogonium

Each division of a diploid spermatogonium produces two daughter cells. One is a spermatogonium that remains in contact with the basal lamina of the tubule, and the other is a **primary spermatocyte** that is displaced toward the lumen.

Primary spermatocyte (diploid)

DNA replication

Primary spermatocyte

Meiosis I

As meiosis I begins, each primary spermatocyte contains 46 individual chromosomes. At the end of meiosis I, the daughter cells are called **secondary spermatocytes**. Every secondary spermatocyte contains 23 chromosomes, each of which consists of a pair of duplicate chromatids.

Synapsis and tetrad formation

Secondary spermatocytes

Meiosis II

The secondary spermatocytes soon enter meiosis II, which yields four haploid **spermatids**, each containing 23 chromosomes. For each primary spermatocyte that enters meiosis, four spermatids are produced.

Spermatids (haploid)

Spermiogenesis (physical maturation)

In spermiogenesis, the last step of spermatogenesis, each spermatid matures into a single spermatozoon, or sperm. The process of spermiogenesis takes roughly 5 weeks to complete.

Spermatozoa (haploid)

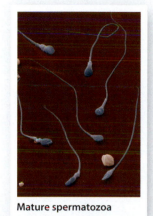

Mature spermatozoa

4 Here you can see the distinctive, specialized features of a **spermatozoon**. Unlike other, less specialized cells, a mature spermatozoon lacks an endoplasmic reticulum, a Golgi apparatus, lysosomes, peroxisomes, inclusions, and many other intracellular structures. The loss of these organelles reduces the cell's size and mass; it is essentially a mobile carrier for the enclosed chromosomes, and extra weight would slow it down.

Structure of a Spermatozoon

The **acrosomal** (ak-rō-SŌ-mal) **cap** is a membranous compartment containing enzymes essential to fertilization.

The **head** is a flattened ellipse containing a nucleus with densely packed chromosomes.

The **neck** contains both centrioles of the original spermatid. The microtubules of the distal centriole are continuous with those of the middle piece and tail.

The **middle piece** contains mitochondria arranged in a spiral around the microtubules. Mitochondrial activity provides the ATP required to move the tail.

The **tail** is a **flagellum**, a whiplike organelle that moves the sperm.

The tail of a spermatozoan is the only flagellum in the human body. Whereas cilia beat in a predictable, wavelike fashion, the flagellum of a spermatozoon has a complex, corkscrew motion.

Module 25.2 Review

a. Define spermatogenesis.

b. How many spermatozoa will eventually be produced from each primary spermatocyte?

c. Describe the functional anatomy of a typical spermatozoon.

Meiosis and early spermiogenesis occur within the seminiferous tubules

1 This is a light micrograph of a horizontal section through a testis. Because the seminiferous tubules are tightly coiled within the lobules, they are often seen in cross section.

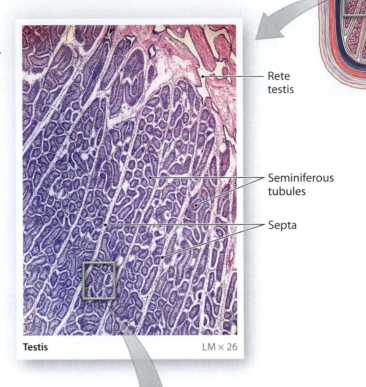

Rete testis

Seminiferous tubules

Septa

Testis LM × 26

2 Here is a higher-magnification view of several sections through a single seminiferous tubule. Spermatogenesis and spermiogenesis together take approximately nine weeks. If spermatogenesis were synchronized along the entire length of a tubule, spermatozoa would be released once every nine weeks. Instead, each segment of a seminiferous tubule is at a different stage of the process of spermatogenesis. As a result, every seminiferous tubule is continuously producing spermatozoa.

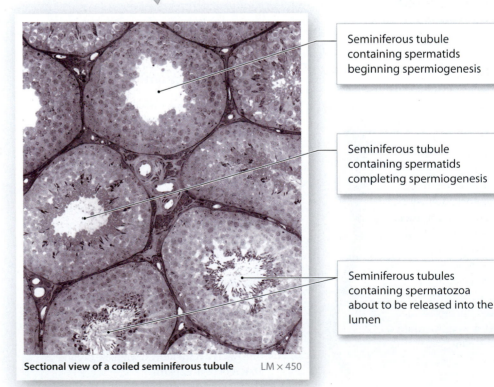

Seminiferous tubule containing spermatids beginning spermiogenesis

Seminiferous tubule containing spermatids completing spermiogenesis

Seminiferous tubules containing spermatozoa about to be released into the lumen

Sectional view of a coiled seminiferous tubule LM × 450

3 Each tubule is surrounded by a delicate connective tissue capsule, and areolar tissue fills the spaces between the tubules. Within those spaces are numerous blood vessels and large **interstitial cells**. Interstitial cells are responsible for the production of androgens, the dominant sex hormones in males. Testosterone is the most important androgen.

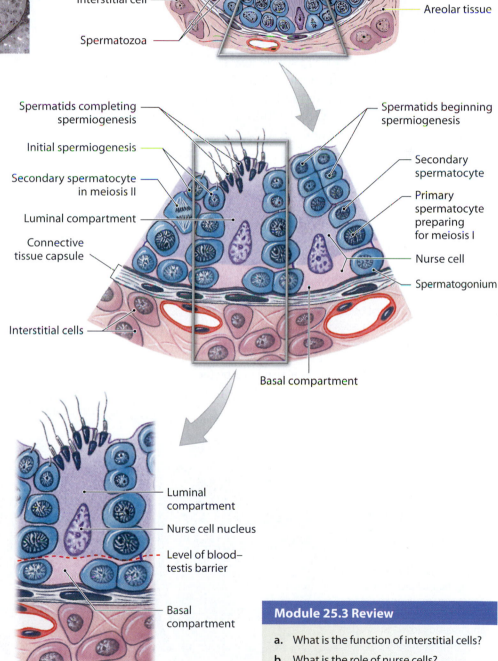

Nurse cell nucleus

Spermatid

Lumen

Capillary

Dividing spermatocytes

Fibroblast

Spermatogonium

Connective tissue capsule

Interstitial cell

Areolar tissue

Spermatozoa

4 Each seminiferous tubule contains spermatogonia, spermatocytes at various stages of meiosis, spermatids, spermatozoa, and large **nurse cells**. Nurse cells are attached to the tubular capsule and extend to the lumen between the other types of cells. Developing spermatocytes undergoing meiosis, and spermatids undergoing spermiogenesis, are not free in the seminiferous tubules. Instead, they are surrounded by the cytoplasm of the nurse cells. As spermiogenesis proceeds, the spermatids gradually develop the appearance of mature spermatozoa. At **spermiation**, a spermatozoon loses its attachment to the nurse cell and enters the lumen of the seminiferous tubule.

Spermatids completing spermiogenesis

Spermatids beginning spermiogenesis

Initial spermiogenesis

Secondary spermatocyte in meiosis II

Secondary spermatocyte

Luminal compartment

Primary spermatocyte preparing for meiosis I

Connective tissue capsule

Nurse cell

Spermatogonium

Interstitial cells

Basal compartment

5 Nurse cells are not only essential to the process of spermatogenesis, but as you will see in Module 25.6, they are involved in its hormonal regulation. One of their many important functions is to isolate the seminiferous tubules from the general circulation by creating a **blood–testis barrier** comparable in function to the blood–brain barrier. Nurse cells are joined by tight junctions, forming a layer that divides the seminiferous tubule into an outer **basal compartment**, which contains the spermatogonia, and an inner **luminal compartment**, where meiosis and spermiogenesis occur.

Luminal compartment

Nurse cell nucleus

Level of blood–testis barrier

Basal compartment

Module 25.3 Review

a. What is the function of interstitial cells?

b. What is the role of nurse cells?

c. List the order of development for a spermatid.

The male reproductive tract receives secretions from the seminal, prostate, and bulbo-urethral glands

The testes produce physically mature spermatozoa; the other portions of the male reproductive system are responsible for the functional maturation, nourishment, storage, and transport of spermatozoa. The spermatozoa leaving the testes are physically mature, but immobile and incapable of fertilizing an oocyte. To become motile (actively swimming) and fully functional, spermatozoa must undergo a process called **capacitation**. Capacitation normally occurs in two steps: (1) Spermatozoa become motile when they are mixed with secretions of the seminal glands, and (2) they become capable of successful fertilization when exposed to conditions in the female reproductive tract.

1 This diagrammatic posterior view shows the urinary bladder, prostate gland, and other structures of the male reproductive system.

The **ampulla** (am-PUL-uh) is the expanded distal portion of the ductus deferens.

The **ejaculatory duct** carries fluid from the seminal gland and ampulla to the urethra.

2 The **epididymis** (ep-i-DID-i-mis; *epi*, on + *didymos*, twin; plural *epididymides*), the start of the male reproductive tract, is a coiled tube bound to the posterior border of each testis. A tubule almost 7 m (23 ft) long, the epididymis is coiled and twisted so as to take up very little space. A spermatozoon passes through the epididymis in about two weeks and completes its functional maturation at that time. Over this period, spermatozoa exist in a sheltered environment that is precisely regulated by the surrounding epithelial cells.

Regions of the Epididymis

The **head** of the epididymis receives spermatozoa from the efferent ductules.

The **body** of the epididymis extends inferiorly along the posterior margin of the testis.

Near the inferior border of the testis, the number of coils decreases, marking the start of the **tail**. The tail recurves and ascends to its connection with the ductus deferens.

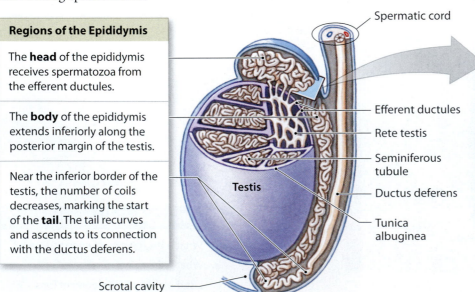

3 The pseudostratified columnar epithelial lining of the epididymis bears extremely long **stereocilia**. These processes increase the surface area available for absorption from, and secretion into, the fluid in the lumen.

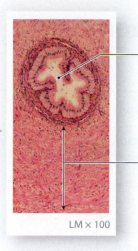

The lumen of the ductus deferens is lined by a pseudostratified ciliated columnar epithelium.

The wall of the ductus deferens contains a thick layer of smooth muscle. Peristaltic contractions in this layer propel spermatozoa and fluid along the duct.

LM × 100

4 Each **ductus deferens**, or *vas deferens*, is 40–45 cm (16–18 in.) long. As part of the spermatic cord, it ascends through the inguinal canal, enters the abdominal cavity, and passes posteriorly to reach the prostate gland. In addition to transporting spermatozoa, the ductus deferens can store spermatozoa for several months. During this time, the spermatozoa remain in a state of suspended animation and have low metabolic rates.

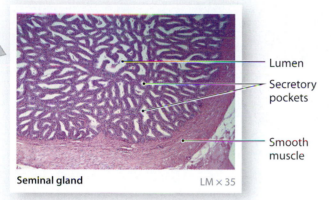

Lumen

Secretory pockets

Smooth muscle

Seminal gland   LM × 35

5 The **seminal glands**, also called the *seminal vesicles*, are embedded in connective tissue on either side of the midline, sandwiched between the posterior wall of the urinary bladder and the rectum. The seminal glands contribute about 60 percent of the volume of **semen**. When mixed with the secretions of the seminal glands, previously inactive but functional spermatozoa undergo the first step in capacitation and begin beating their flagella, becoming highly motile. The secretions are ejected by smooth muscle contractions controlled by the sympathetic division of the ANS.

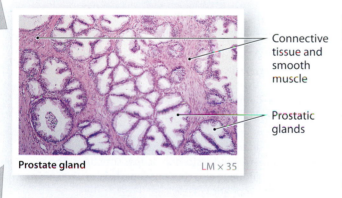

Connective tissue and smooth muscle

Prostatic glands

Prostate gland   LM × 35

6 The **prostate gland** is a small, muscular, rounded organ about 4 cm (1.6 in.) in diameter. The prostate gland encircles the proximal portion of the urethra as it leaves the urinary bladder. The glandular tissue of the prostate is surrounded by and wrapped in a thick blanket of smooth muscle fibers. The prostate gland produces 20–30 percent of the volume of semen. In addition to several other compounds of uncertain significance, prostatic secretions contain **seminalplasmin** (sem-i-nal-PLAZ-min), an antibiotic protein that may help prevent urinary tract infections in males. These secretions are ejected into the prostatic urethra by peristaltic contractions of the muscular prostate wall.

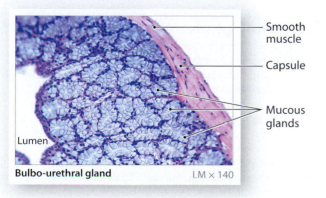

Smooth muscle

Capsule

Lumen

Mucous glands

Bulbo-urethral gland   LM × 140

7 The paired **bulbo-urethral glands** are situated at the base of the penis, covered by the fascia of the urogenital diaphragm. The duct of each gland empties into the urethra. These glands secrete a thick, alkaline mucus that helps neutralize any urinary acids that may remain in the urethra, and it also lubricates the tip of the penis.

Module 25.4 Review

a. Define semen.

b. What are the functions of the secretion of the bulbo-urethral glands?

c. Trace the ductal pathway from the epididymis to the urethra.

The penis conducts semen and urine to the exterior

1 The **penis** is a tubular organ through which the distal portion of the urethra passes. It conducts urine to the exterior and introduces semen into the female's vagina during sexual intercourse.

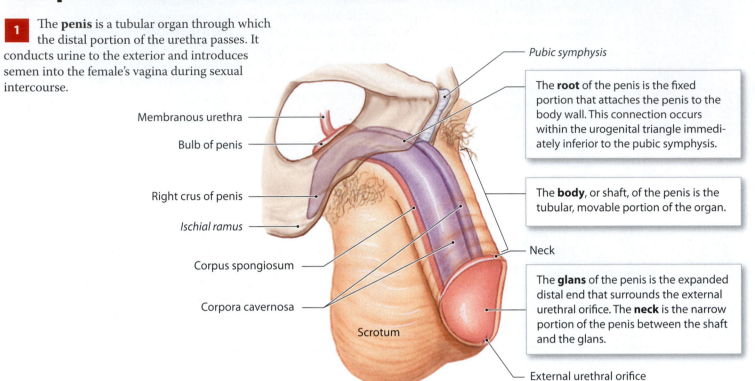

Pubic symphysis

Membranous urethra

Bulb of penis

Right crus of penis

Ischial ramus

Corpus spongiosum

Corpora cavernosa

Scrotum

Neck

External urethral orifice

The **root** of the penis is the fixed portion that attaches the penis to the body wall. This connection occurs within the urogenital triangle immediately inferior to the pubic symphysis.

The **body**, or shaft, of the penis is the tubular, movable portion of the organ.

The **glans** of the penis is the expanded distal end that surrounds the external urethral orifice. The **neck** is the narrow portion of the penis between the shaft and the glans.

2 This cross section shows that most of the body of the penis consists of three cylindrical columns of vascularized **erectile tissue.** Erectile tissue consists of a three-dimensional network with vascular spaces incompletely separated by partitions of elastic connective tissue and smooth muscle fibers. In the resting state, the arterial branches are constricted, and the muscular partitions are tense. This combination restricts blood flow into the erectile tissue.

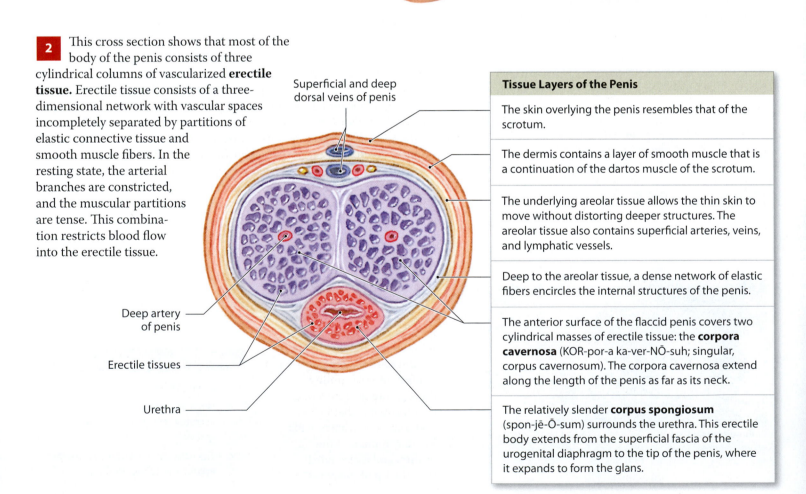

Superficial and deep dorsal veins of penis

Deep artery of penis

Erectile tissues

Urethra

Tissue Layers of the Penis

The skin overlying the penis resembles that of the scrotum.

The dermis contains a layer of smooth muscle that is a continuation of the dartos muscle of the scrotum.

The underlying areolar tissue allows the thin skin to move without distorting deeper structures. The areolar tissue also contains superficial arteries, veins, and lymphatic vessels.

Deep to the areolar tissue, a dense network of elastic fibers encircles the internal structures of the penis.

The anterior surface of the flaccid penis covers two cylindrical masses of erectile tissue: the **corpora cavernosa** (KOR-por-a ka-ver-NŌ-suh; singular, corpus cavernosum). The corpora cavernosa extend along the length of the penis as far as its neck.

The relatively slender **corpus spongiosum** (spon-jē-Ō-sum) surrounds the urethra. This erectile body extends from the superficial fascia of the urogenital diaphragm to the tip of the penis, where it expands to form the glans.

3 This diagrammatic view of the male reproductive tract provides a three-dimensional perspective on the relationships among the glands and passageways. The numbers will be useful as you follow the phases of sexual activity in **4** .

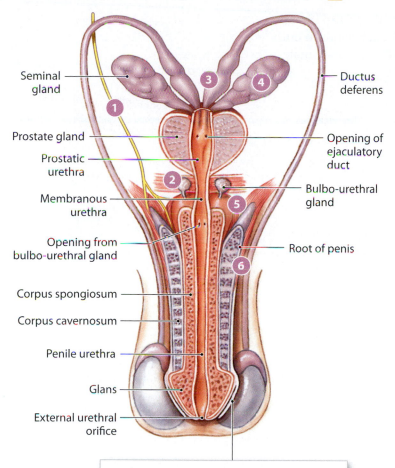

Seminal gland

Ductus deferens

Prostate gland

Opening of ejaculatory duct

Prostatic urethra

Bulbo-urethral gland

Membranous urethra

Opening from bulbo-urethral gland

Root of penis

Corpus spongiosum

Corpus cavernosum

Penile urethra

Glans

External urethral orifice

A fold of skin called the **prepuce** (PRĒ-poos), or foreskin, surrounds the tip of the penis. The prepuce attaches to the relatively narrow neck of the penis and continues over the glans. Glands in the skin of the neck and the inner surface of the prepuce secrete a waxy material known as **smegma** (SMEG-ma).

Impotence is an inability to achieve or maintain an erection. There may be a variety of physical causes, for erection involves vascular changes as well as neural commands. For example, low blood pressure in the arteries of the penis, due to a cardiovascular blockage such as a plaque, can impair the ability to achieve an erection. Psychological factors such as depression or anxiety can also result in impotence. The erection pathway activated by nitric oxide (NO) is opposed by enzymes that cause constriction of the arteries supplying the erectile tissues. The drugs Viagra and Cialis temporarily inactivate these enzymes, so that even small levels of NO are sufficient to produce an erection.

4 The three major phases in male sexual activity are summarized here.

Arousal

During **arousal**, erotic thoughts or stimulation of sensory nerves in the genital region leads to an increase in the parasympathetic outflow over the pelvic nerves.

1 The parasympathetic innervation of the penile arteries involves neurons that release **nitric oxide** (NO) at their synaptic terminals. The smooth muscles in the arterial walls relax when NO is released, at which time the vessels dilate, blood flow increases, the vascular channels become engorged with blood, and **erection** of the penis occurs.

2 Arousal also stimulates the bulbo-urethral glands, whose secretions lubricate the urethra and the tip of the glans.

Emission

Further stimulation leads to sympathetic activation that causes **emission**.

3 Emission begins with peristaltic contractions in the ampullae of the ductus deferens. This pushes spermatozoa into the urethra within the prostate gland.

4 Contractions then begin in the walls of the seminal glands and the prostate gland, and their secretions now enter the urethra and mix with the spermatozoa introduced by ampullary contractions to form **semen**.

Ejaculation

Ejaculation occurs as powerful, rhythmic contractions take place in the ischiocavernosus and bulbocavernosus muscles. These contractions are controlled by somatic motor neurons in the lower lumbar and upper sacral segments of the spinal cord. These contractions are associated with the pleasurable sensations known as **male orgasm**.

5 The bulbocavernosus muscles wrap around the base of the penis, and their contractions push semen toward the external urethral orifice.

6 The ischiocavernosus muscles insert along the sides of the penis, and their contractions serve primarily to stiffen the erect penis.

Module 25.5 Review

a. Name the three columns of erectile tissue in the penis.

b. List the three physiological phases of male sexual activity.

c. An inability to contract the ischiocavernosus and bulbospongiosus muscles would interfere with which phase of male sexual activity?

Testosterone plays a key role in the establishment and maintenance of male sexual function

Testosterone is produced primarily by the interstitial cells of the testes. (Small amounts are also produced by the zona reticularis of the adrenal glands in both sexes.) This flowchart diagrams the hormonal interactions that regulate male reproductive function.

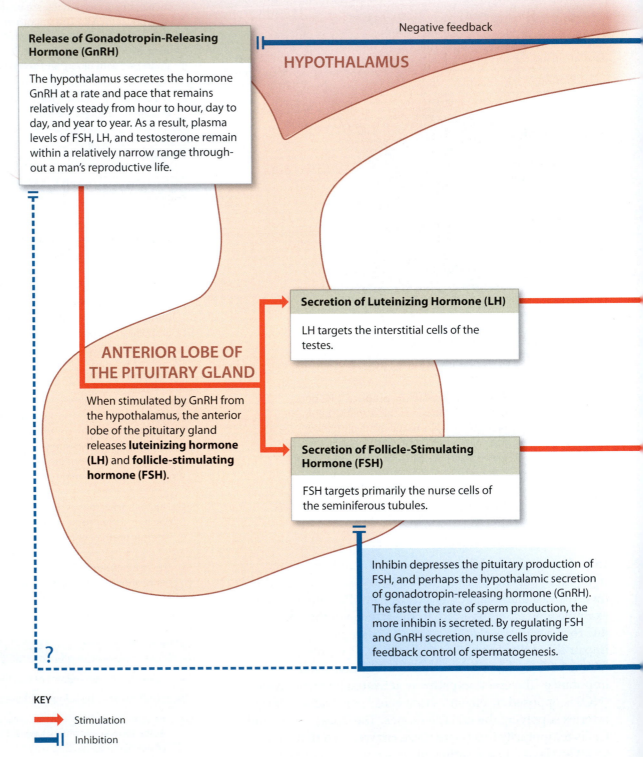

Release of Gonadotropin-Releasing Hormone (GnRH)

The hypothalamus secretes the hormone GnRH at a rate and pace that remains relatively steady from hour to hour, day to day, and year to year. As a result, plasma levels of FSH, LH, and testosterone remain within a relatively narrow range throughout a man's reproductive life.

Negative feedback

HYPOTHALAMUS

Secretion of Luteinizing Hormone (LH)

LH targets the interstitial cells of the testes.

ANTERIOR LOBE OF THE PITUITARY GLAND

When stimulated by GnRH from the hypothalamus, the anterior lobe of the pituitary gland releases **luteinizing hormone (LH)** and **follicle-stimulating hormone (FSH)**.

Secretion of Follicle-Stimulating Hormone (FSH)

FSH targets primarily the nurse cells of the seminiferous tubules.

Inhibin depresses the pituitary production of FSH, and perhaps the hypothalamic secretion of gonadotropin-releasing hormone (GnRH). The faster the rate of sperm production, the more inhibin is secreted. By regulating FSH and GnRH secretion, nurse cells provide feedback control of spermatogenesis.

?

KEY

→ Stimulation

⊣ Inhibition

In many target tissues, some of the arriving testosterone is converted to **dihydrotestosterone (DHT)**. A small amount of DHT diffuses back out of the cell and into the bloodstream, and DHT levels are usually about 10 percent of circulating testosterone levels. Dihydrotestosterone can also enter peripheral cells and bind to the same hormone receptors targeted by testosterone. In addition, some tissues (notably those of the external genitalia) respond to DHT rather than to testosterone, and other tissues (including the prostate gland) are more sensitive to DHT than to testosterone.

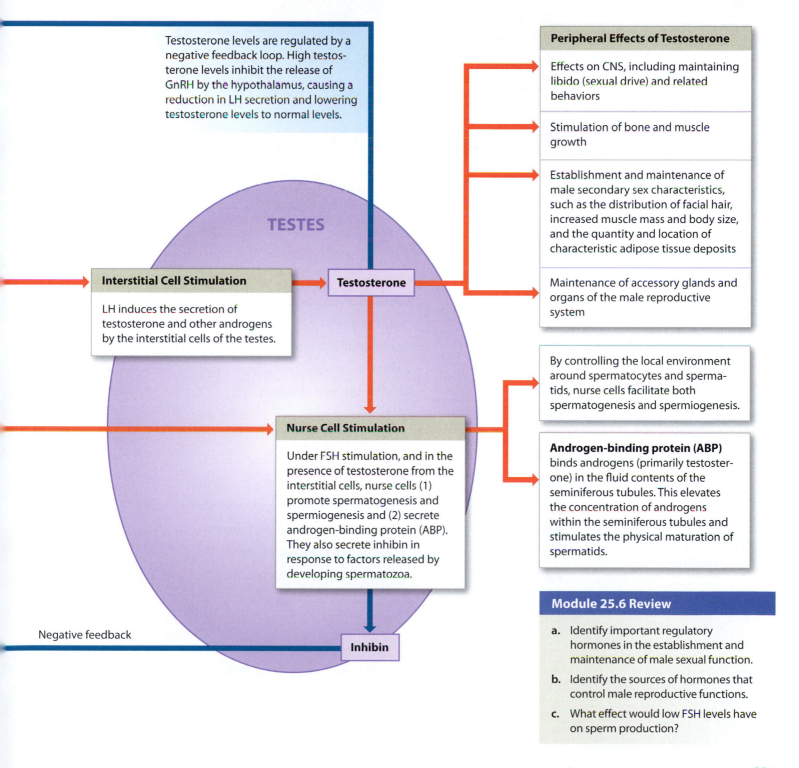

Testosterone levels are regulated by a negative feedback loop. High testosterone levels inhibit the release of GnRH by the hypothalamus, causing a reduction in LH secretion and lowering testosterone levels to normal levels.

TESTES

Interstitial Cell Stimulation

LH induces the secretion of testosterone and other androgens by the interstitial cells of the testes.

Testosterone

Nurse Cell Stimulation

Under FSH stimulation, and in the presence of testosterone from the interstitial cells, nurse cells (1) promote spermatogenesis and spermiogenesis and (2) secrete androgen-binding protein (ABP). They also secrete inhibin in response to factors released by developing spermatozoa.

Negative feedback

Inhibin

Peripheral Effects of Testosterone

Effects on CNS, including maintaining libido (sexual drive) and related behaviors

Stimulation of bone and muscle growth

Establishment and maintenance of male secondary sex characteristics, such as the distribution of facial hair, increased muscle mass and body size, and the quantity and location of characteristic adipose tissue deposits

Maintenance of accessory glands and organs of the male reproductive system

By controlling the local environment around spermatocytes and spermatids, nurse cells facilitate both spermatogenesis and spermiogenesis.

Androgen-binding protein (ABP) binds androgens (primarily testosterone) in the fluid contents of the seminiferous tubules. This elevates the concentration of androgens within the seminiferous tubules and stimulates the physical maturation of spermatids.

Module 25.6 Review

a. Identify important regulatory hormones in the establishment and maintenance of male sexual function.

b. Identify the sources of hormones that control male reproductive functions.

c. What effect would low FSH levels have on sperm production?

1. Labeling

Label the structures of the male reproductive system in the accompanying diagram.

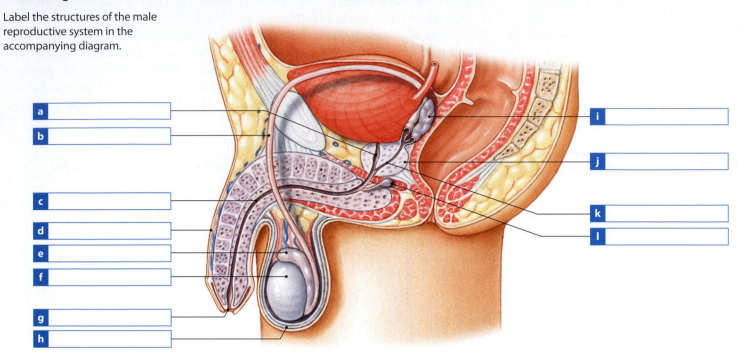

a _____

b _____

c _____

d _____

e _____

f _____

g _____

h _____

i _____

j _____

k _____

l _____

2. Matching

Match the following terms with the most closely related description.

- semen
- epididymis
- nurse cells
- corpus spongiosum
- luteinizing hormone (LH)
- impotence
- follicle-stimulating hormone (FSH)
- spermatogonia
- seminiferous tubules
- dartos muscle
- spermatogenesis
- interstitial cells
- spermiogenesis
- penis and scrotum

a _____ Scrotal smooth muscle

b _____ Sperm stem cells

c _____ Sites of sperm production

d _____ Produce testosterone

e _____ Physical maturation of spermatids

f _____ Sperm production

g _____ External genitalia

h _____ Start of male reproductive tract

i _____ Maintain blood–testis barrier

j _____ Spermatozoa, seminal gland, and other gland secretions

k _____ Inability to achieve or maintain an erection

l _____ Erectile tissue surrounding the urethra

m _____ Induces secretion of androgens

n _____ Hormone that targets nurse cells

3. Section integration

In males, the endocrine disorder hypogonadism is primarily due to the underproduction of testosterone or the lack of tissue sensitivity to testosterone, and results in sterility. What are five primary functions of testosterone in males? _____

_____

_____

The Female Reproductive System

A woman's reproductive system produces sex hormones and functional gametes, and it must also be able to protect and support a developing embryo, maintain a growing fetus, and nourish a newborn infant. The principal organs of the female reproductive system are the ovaries, the uterine tubes, the uterus, the vagina, and the components of the external genitalia. The female reproductive system also includes accessory organs— the mammary glands—and a variety of smaller accessory glands that secrete their products into the female reproductive tract.

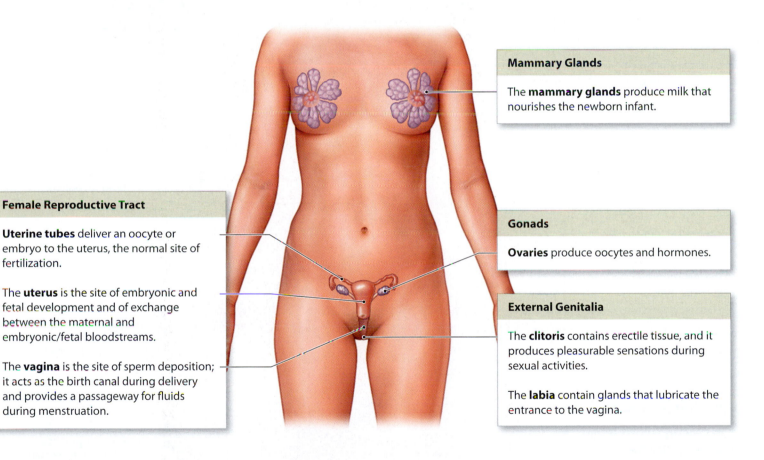

Mammary Glands

The **mammary glands** produce milk that nourishes the newborn infant.

Female Reproductive Tract

Uterine tubes deliver an oocyte or embryo to the uterus, the normal site of fertilization.

The **uterus** is the site of embryonic and fetal development and of exchange between the maternal and embryonic/fetal bloodstreams.

The **vagina** is the site of sperm deposition; it acts as the birth canal during delivery and provides a passageway for fluids during menstruation.

Gonads

Ovaries produce oocytes and hormones.

External Genitalia

The **clitoris** contains erectile tissue, and it produces pleasurable sensations during sexual activities.

The **labia** contain glands that lubricate the entrance to the vagina.

This figure provides an overview of the components of the reproductive system in the female. The gonads in females, called ovaries (singular, *ovary*), produce immature female gametes called **oocytes** (Ŏ-ō-sīts), which later mature into **ova** (singular, *ovum*). Oocytes leave the ovary and then travel along the female duct system, or **female reproductive tract**. If fertilization occurs, it will occur in the uterine tubes and further embryonic development will occur within the uterus.

The ovaries and the female reproductive tract are in close proximity but are not directly connected

1 This sagittal section through the pelvic cavity shows the location and orientation of the female reproductive organs.

The paired **ovaries** are small, lumpy, almond-shaped organs near the lateral walls of the pelvic cavity. The ovaries perform three main functions: (1) production of immature female gametes, or **oocytes**; (2) secretion of female sex hormones, including **estrogens** and **progestins**; and (3) secretion of **inhibin**, involved in the feedback control of pituitary FSH production.

Each **uterine tube** begins with an expanded funnel, called an **infundibulum**, that is open into the pelvic cavity along the medial surface of the ovary. The other end of the uterine tube opens into the uterine cavity.

The **uterus** sits inferior to the ovary, usually angled forward above the urinary bladder.

The pocket formed between the uterus and the posterior wall of the urinary bladder is the **vesicouterine** (ves-i-kō-Ū-ter-in) **pouch**.

External Genitalia

The **clitoris** contains erectile tissue; its stimulation produces pleasurable sensations during sexual activities.

The **labia** contain glands that lubricate the entrance to the vagina.

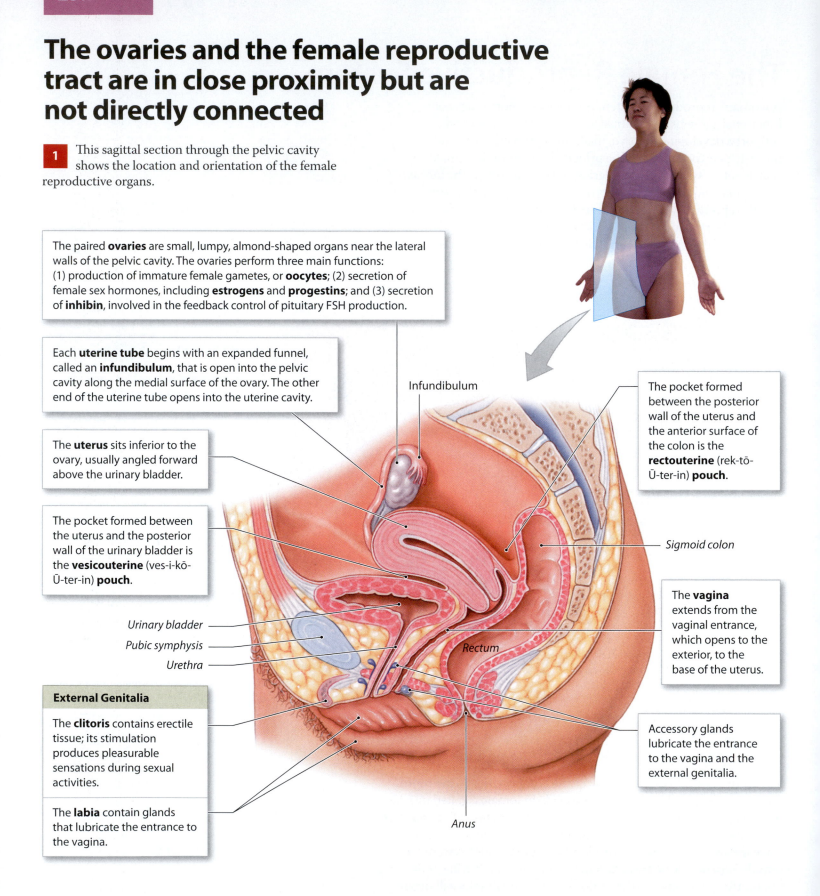

Infundibulum

The pocket formed between the posterior wall of the uterus and the anterior surface of the colon is the **rectouterine** (rek-tō-Ū-ter-in) **pouch**.

Sigmoid colon

The **vagina** extends from the vaginal entrance, which opens to the exterior, to the base of the uterus.

Accessory glands lubricate the entrance to the vagina and the external genitalia.

Urinary bladder

Pubic symphysis

Urethra

Rectum

Anus

2 The position of each ovary is stabilized by several thickened peritoneal folds that are called *ligaments*. This is a view from above and behind, with the left uterine tube pulled away from the ovary to show the ligaments more clearly.

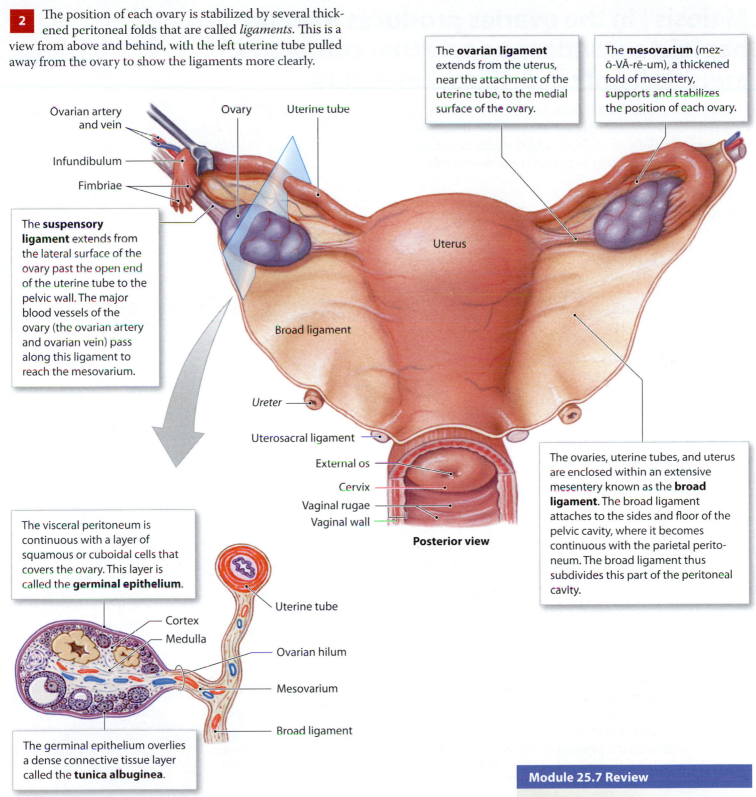

The **ovarian ligament** extends from the uterus, near the attachment of the uterine tube, to the medial surface of the ovary.

The **mesovarium** (mez-ō-VĀ-rē-um), a thickened fold of mesentery, supports and stabilizes the position of each ovary.

The **suspensory ligament** extends from the lateral surface of the ovary past the open end of the uterine tube to the pelvic wall. The major blood vessels of the ovary (the ovarian artery and ovarian vein) pass along this ligament to reach the mesovarium.

Ovarian artery and vein
Infundibulum
Fimbriae
Ovary
Uterine tube
Uterus
Broad ligament
Ureter
Uterosacral ligament
External os
Cervix
Vaginal rugae
Vaginal wall

Posterior view

The ovaries, uterine tubes, and uterus are enclosed within an extensive mesentery known as the **broad ligament**. The broad ligament attaches to the sides and floor of the pelvic cavity, where it becomes continuous with the parietal peritoneum. The broad ligament thus subdivides this part of the peritoneal cavity.

The visceral peritoneum is continuous with a layer of squamous or cuboidal cells that covers the ovary. This layer is called the **germinal epithelium**.

Uterine tube
Cortex
Medulla
Ovarian hilum
Mesovarium
Broad ligament

The germinal epithelium overlies a dense connective tissue layer called the **tunica albuginea**.

3 This is a cross section taken through the mesovarium and the broad ligament. A typical ovary is about 5 cm long, 2.5 cm wide, and 8 mm thick (2 in. by 1 in. by 0.33 in.) and weighs 6–8 g (roughly 0.25 oz). Blood vessels enter and leave the ovary at the ovarian hilum, where the ovary attaches to the mesovarium. The interior of the ovary can be divided into a superficial **cortex** and a deeper **medulla**. Gametes are produced in the cortex.

Module 25.7 Review

a. List the major organs of the female reproductive system.

b. Name the structures enclosed by the broad ligament, and cite the function of the mesovarium.

c. What roles do the ovaries perform?

Meiosis I in the ovaries produces a single haploid secondary oocyte that completes meiosis II only if fertilization occurs

Ovum production, or **oogenesis** (ō-ō-JEN-e-sis; oon, egg), begins before a woman's birth, accelerates at puberty, and ends at menopause. Between puberty and menopause, oogenesis occurs on a monthly basis as part of the ovarian cycle.

1 Although the nuclear events in the ovaries during meiosis are the same as those in the testes, the cytoplasm of the **primary oocyte** is unevenly distributed during the two meiotic divisions. Oogenesis produces one functional **ovum**, which contains most of the original cytoplasm, and two or three **polar bodies**, nonfunctional cells that later disintegrate. Another difference is that the ovary releases a **secondary oocyte** rather than a mature ovum. The secondary oocyte is suspended in metaphase of meiosis II; meiosis will not be completed unless and until fertilization occurs.

Not all primary oocytes produced during development survive until puberty. The ovaries have roughly 2 million primordial follicles at birth, each containing a primary oocyte. Primordial follicles exist at the earliest stages of development. By the time of puberty, the number has dropped to about 400,000. The rest of the primordial follicles degenerate in a process called **atresia** (a-TRĒ-zē-uh).

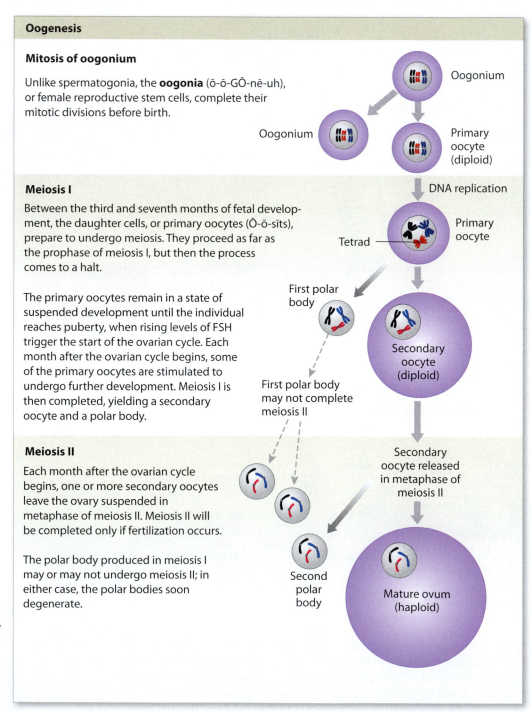

Oogenesis

Mitosis of oogonium

Unlike spermatogonia, the **oogonia** (ō-ō-GŌ-nē-uh), or female reproductive stem cells, complete their mitotic divisions before birth.

Oogonium

Oogonium

Primary oocyte (diploid)

DNA replication

Meiosis I

Between the third and seventh months of fetal development, the daughter cells, or primary oocytes (Ō-ō-sīts), prepare to undergo meiosis. They proceed as far as the prophase of meiosis I, but then the process comes to a halt.

The primary oocytes remain in a state of suspended development until the individual reaches puberty, when rising levels of FSH trigger the start of the ovarian cycle. Each month after the ovarian cycle begins, some of the primary oocytes are stimulated to undergo further development. Meiosis I is then completed, yielding a secondary oocyte and a polar body.

Tetrad

Primary oocyte

First polar body

First polar body may not complete meiosis II

Secondary oocyte (diploid)

Meiosis II

Each month after the ovarian cycle begins, one or more secondary oocytes leave the ovary suspended in metaphase of meiosis II. Meiosis II will be completed only if fertilization occurs.

The polar body produced in meiosis I may or may not undergo meiosis II; in either case, the polar bodies soon degenerate.

Secondary oocyte released in metaphase of meiosis II

Second polar body

Mature ovum (haploid)

2 **Ovarian follicles** are specialized structures in the cortex of the ovaries where both oocyte growth and meiosis I occur. As the ovarian cycle proceeds, the structure of the follicle gradually changes. Important events in the ovarian cycle are summarized here.

Primordial Follicles in Egg Nest

Primary oocyte

Follicle cells

Primary oocytes are located in the outer portion of the ovarian cortex, near the tunica albuginea, in clusters called **egg nests**. An inactive primary oocyte is surrounded by a single squamous layer of follicle cells, forming a **primordial follicle**.

Formation of Primary Follicles

Granulosa cells

Primary oocyte

Thecal cells

Follicular cells enlarge, divide, and form several layers of cells around an activated primary oocyte. These follicle cells are now called **granulosa cells**. A region around the oocyte develops, called the **zona pellucida** (ZŌ-na pe-LOO-si-duh; *pellucidus*, translucent). As the granulosa cells enlarge and multiply, a layer of **thecal cells** (*theca*, a box) forms around the follicle. Thecal cells and granulosa cells work together to produce estrogens.

Formation of Secondary Follicles

Thecal cells

Zona pellucida

Nucleus of primary oocyte

Granulosa cells

Secondary follicles develop as the wall of the follicle thickens and the deeper follicular cells begin secreting fluid that accumulates in small pockets. These pockets gradually expand and separate the inner and outer layers of the follicle.

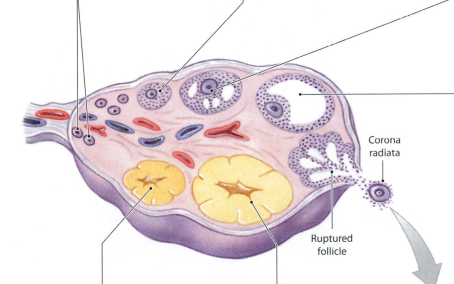

Corona radiata

Ruptured follicle

Formation of a Tertiary Follicle

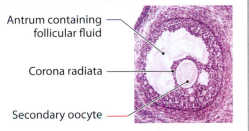

Antrum containing follicular fluid

Corona radiata

Secondary oocyte

By days 10–14 of the cycle, usually only one secondary follicle has become a **tertiary follicle**, or mature graafian (GRAF-ē-an) follicle, roughly 15 mm in diameter. The oocyte projects into the **antrum** (AN-trum), or expanded central chamber of the follicle. The granulosa cells associated with the secondary oocyte form a protective layer known as the **corona radiata** (kō-RŌ-nuh rā-dē-AH-tuh).

Formation of Corpus Albicans

If fertilization does not occur, after 12 days progesterone and estrogen levels fall markedly. Fibroblasts invade the nonfunctional corpus luteum, producing a knot of pale scar tissue called a **corpus albicans** (AL-bi-kanz). The disintegration, or involution, of the corpus luteum marks the end of the ovarian cycle. A new ovarian cycle then begins with the activation of another group of primordial follicles.

Formation of Corpus Luteum

The empty tertiary follicle initially collapses, and under LH stimulation the remaining granulosa cells proliferate to create the **corpus luteum** (LOO-tē-um; *lutea*, yellow), which secretes progesterone (prō-JES-ter-ōn) and estrogens. Progesterone prepares the uterus for pregnancy by stimulating the maturation of the uterine lining and the secretions of uterine glands.

Ovulation

Secondary oocyte

At **ovulation**, the tertiary follicle releases the secondary oocyte and corona radiata into the pelvic cavity. Ovulation marks the end of the **follicular phase** of the ovarian cycle and the start of the **luteal phase**.

Module 25.8 Review

a. Define oocyte.

b. What are the main differences in gamete production between males and females?

c. List the important events in the ovarian cycle.

The uterine tubes are connected to the uterus, a hollow organ with thick muscular walls

1 Each **uterine tube** is a hollow, muscular structure measuring roughly 13 cm (5.2 in.) in length. The distal portion of each uterine tube connects to the uterus. The **uterus** is a hollow muscular organ that is about 7.5 cm (3 in.) long with a maximum diameter of 5 cm (2 in.). It weighs 30–40 g (1–1.4 oz). The sectional illustration below shows the internal structure of the uterine tube and the connection between the lumen of the uterine tube and the large uterine cavity within the uterus.

The thickness of the smooth muscle layers in the wall of the **ampulla**, the middle segment of the uterine tube, gradually increases as the tube approaches the uterus.

The ampulla leads to the **isthmus** (IS-mus) of the uterine tube, a short segment connected to the uterine wall.

The **infundibulum** has numerous fingerlike projections that extend into the pelvic cavity. The projections are called **fimbriae** (FIM-brē-ē). Fimbriae drape over the surface of the ovary, but there is no physical connection between the two structures. The inner surfaces of the infundibulum are lined with cilia that beat toward the lumen of the uterine tube.

Uterine cavity

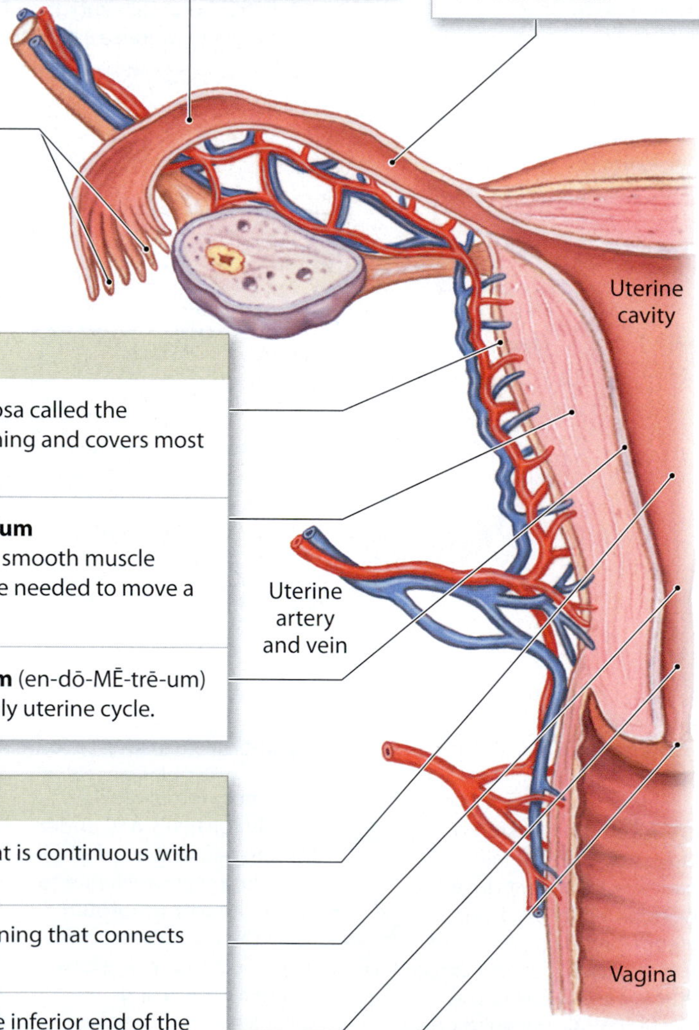

Layers of the Uterine Wall

The outer surface of the uterus is an incomplete serosa called the **perimetrium**. It is continuous with the peritoneal lining and covers most of the uterine surface.

The perimetrium covers a thick, muscular **myometrium** (mī-ō-MĒ-trē-um; *myo-*, muscle + *metra*, uterus). The smooth muscle tissue of the myometrium provides much of the force needed to move a fetus out of the uterus and into the vagina.

The inner lining consists of a glandular **endometrium** (en-dō-MĒ-trē-um) whose character changes in the course of the monthly uterine cycle.

Uterine artery and vein

The Uterine Lumen

The **uterine cavity** is the large, superior chamber that is continuous with the isthmus of the uterine tube on either side.

The **internal os** (*os*, an opening or mouth) is the opening that connects the uterine cavity to the cervical canal.

The **cervical canal** is a constricted passageway at the inferior end of the uterine cavity; it begins at the internal os and ends at the external os.

Within the vagina, the inferior tip of the uterus forms a curving surface that surrounds the **external os**.

Vagina

2 Concentric layers of smooth muscle surround the mucosa of the uterine tube. Oocyte transport along the tube involves a combination of ciliary movement and peristaltic contraction stimulated by autonomic nerves.

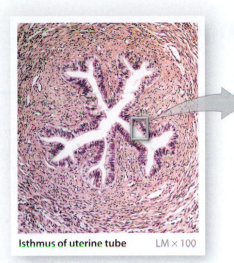

Isthmus of uterine tube   LM × 100

Epithelial surface   SEM × 4000

3 This colorized SEM shows the ciliated epithelium of the uterine tube. The cilia (yellow-green) beat toward the uterine cavity, establishing fluid currents that help collect and transport the secondary oocyte after ovulation.

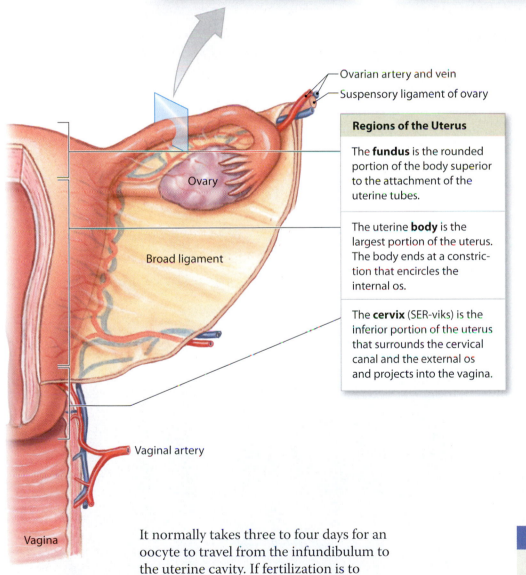

Ovarian artery and vein

Suspensory ligament of ovary

Ovary

Broad ligament

Vaginal artery

Vagina

Regions of the Uterus

The **fundus** is the rounded portion of the body superior to the attachment of the uterine tubes.

The uterine **body** is the largest portion of the uterus. The body ends at a constriction that encircles the internal os.

The **cervix** (SER-viks) is the inferior portion of the uterus that surrounds the cervical canal and the external os and projects into the vagina.

4 The uterus can be divided into three anatomical regions, as indicated here. The uterus, which is capable of great changes in size and shape, provides mechanical protection, nutritional support, and waste removal for the developing **embryo** (weeks 1–8) and **fetus** (week 9 through delivery). In addition, contractions in the muscular wall of the uterus are important in delivering the fetus at birth.

It normally takes three to four days for an oocyte to travel from the infundibulum to the uterine cavity. If fertilization is to occur, the secondary oocyte must encounter spermatozoa during the first 12–24 hours of its passage along the uterine tube.

Module 25.9 Review

a. Name the regions of the uterus.

b. Describe the three layers of the uterine wall.

c. How do recently released secondary oocytes reach the uterine tube?

The uterine cycle involves changes in the functional zone of the endometrium

1 Within the myometrium, branches of the uterine arteries form **arcuate arteries**, which encircle the endometrium. From the arcuate arteries, **radial arteries** supply the endometrium.

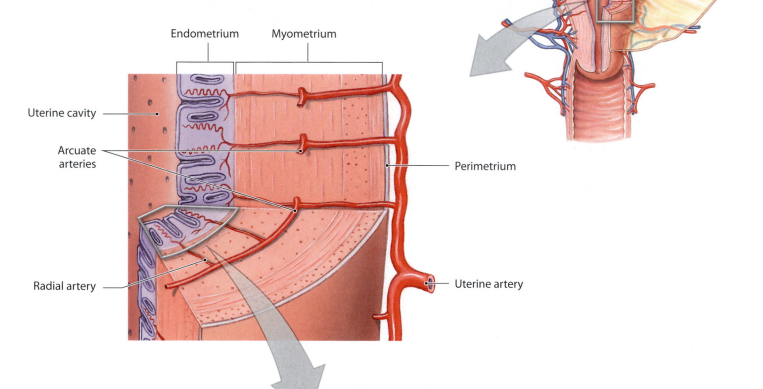

Endometrium   Myometrium

Uterine cavity

Arcuate arteries

Perimetrium

Radial artery

Uterine artery

2 The endometrium contains a **basilar zone** adjacent to the myometrium, and a **functional zone**, the region closest to the uterine cavity. **Straight arteries** deliver blood to the basilar zone, and **spiral arteries** supply the functional zone. The functional zone of the endometrium contains large tubular uterine glands.

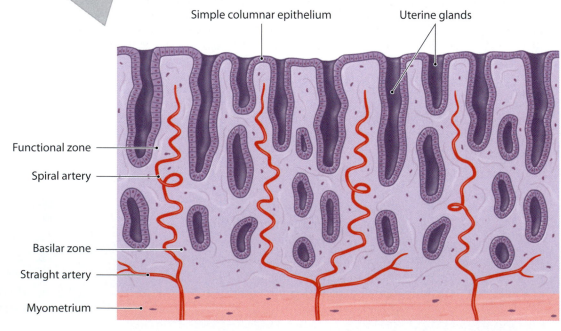

Simple columnar epithelium   Uterine glands

Functional zone

Spiral artery

Basilar zone

Straight artery

Myometrium

3 The structure of the basilar zone remains relatively constant over time, but that of the functional zone undergoes cyclical changes in response to sex hormone levels. These cyclical changes produce the characteristic histological features of the **uterine cycle**. The uterine cycle averages 28 days in length, but it can range from 21 to 35 days in healthy women of reproductive age.

Menses	Proliferative Phase	Secretory Phase

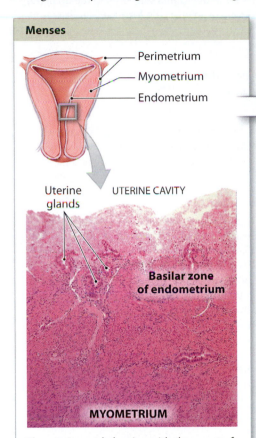

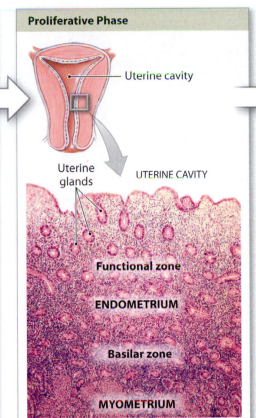

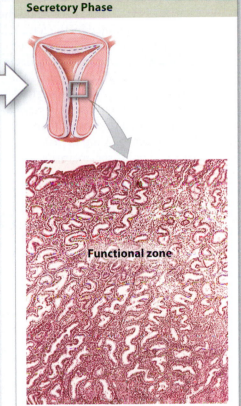

The uterine cycle begins with the onset of **menses** (MEN-sēz), an interval marked by the degeneration of the functional zone of the endometrium. This degeneration is caused by constriction of the spiral arteries, which reduces endometrial blood flow. Eventually, the weakened arterial walls rupture, and blood pours into the connective tissues of the functional zone. Blood cells and degenerating tissues then break away and enter the uterine lumen, to be lost by passage through the external os and into the vagina. The process of endometrial sloughing, called **menstruation** (men-stroo-Ā-shun), generally lasts from one to seven days. Over this period roughly 35 to 50 mL of blood are lost.

The basilar zone and the deepest uterine glands survive menses intact. The epithelial cells of the uterine glands then multiply and spread across the endometrial surface, restoring the integrity of the uterine epithelium. As this reorganization proceeds, the endometrium is in the **proliferative phase**. Restoration is stimulated and sustained by estrogens secreted by the developing ovarian follicles. By the time ovulation occurs, the functional zone is several millimeters thick, and prominent mucous glands extend to the border with the basilar zone. At this time, the uterine glands are manufacturing a glycogen-rich mucus that can be metabolized by an early embryo.

During the **secretory phase**, the uterine glands enlarge, accelerating their rates of secretion, and the arteries that supply the uterine wall elongate and spiral through the tissues of the functional zone. This activity occurs under the combined stimulatory effects of progestins and estrogens from the corpus luteum. The secretory phase begins at the time of ovulation and persists as long as the corpus luteum remains intact. When the corpus luteum stops producing stimulatory hormones, a new uterine cycle begins with the onset of menses and the disintegration of the functional zone.

The uterine cycle, or **menstrual cycle**, begins at puberty. The first cycle, known as **menarche** (me-NAR-kē; *men*, month + *arche*, beginning), typically occurs at age 11–12. The cycles continue until **menopause** (MEN-ō-pawz), the termination of the uterine cycle, at age 45–55. Over the interim, the regular appearance of uterine cycles is interrupted only by circumstances such as illness, stress, starvation, or pregnancy.

Module 25.10 Review

a. Name the zones of the endometrium.

b. Differentiate between menses and menstruation.

c. Describe the phases of the uterine cycle.

The entrance to the vagina is enclosed by external genitalia

1 The **vagina** is an elastic, muscular tube extending between the cervix and the **vestibule**, a space bounded by the female external genitalia. The vagina is typically 7.5–9 cm (3–3.6 in.) long, but its diameter varies because it is highly distensible. The internal passageway is called the **vaginal canal**. The vagina (1) serves as a passageway for the elimination of menstrual fluids; (2) receives the penis during sexual intercourse, and holds spermatozoa prior to their passage into the uterus; and (3) forms the inferior portion of the birth canal, through which the fetus passes during delivery.

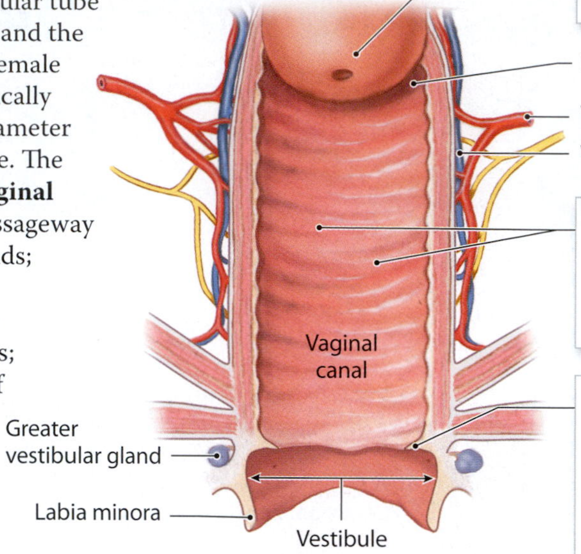

At the proximal end of the vagina, the cervix projects into the vaginal canal. The shallow recess in the vagina surrounding the tip of the cervix is known as the **fornix** (FOR-niks).

Fornix

Vaginal artery

Vaginal vein

In the relaxed state, the vaginal lining forms folds called **rugae**. The vaginal canal is lined by a nonkeratinized stratified squamous epithelium.

Vaginal canal

Greater vestibular gland

Labia minora

Vestibule

Throughout childhood the vagina and vestibule are usually separated by the **hymen** (HĪ-men), an elastic epithelial fold of variable size that partially blocks the entrance to the vagina. An intact hymen is typically stretched or torn during sexual intercourse or tampon use.

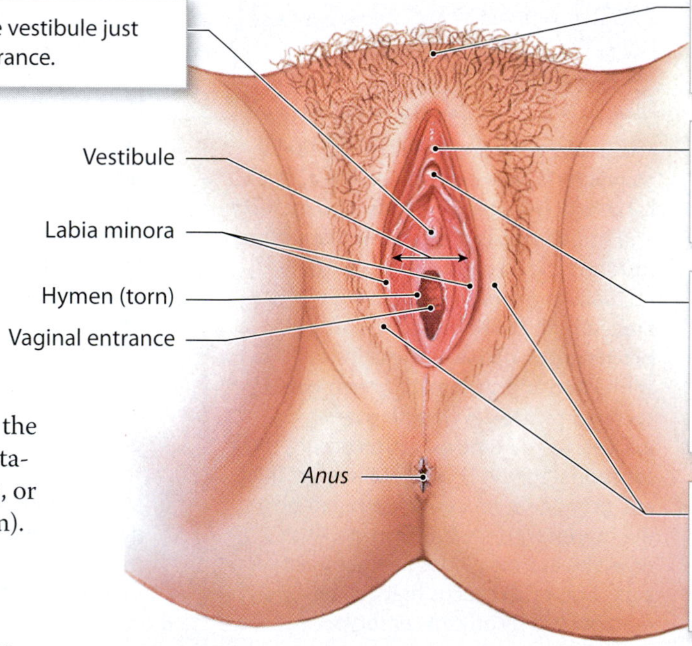

The urethra opens into the vestibule just anterior to the vaginal entrance.

The bulge of the **mons pubis** is created by adipose tissue deep to the skin and superficial to the pubic symphysis.

Vestibule

Labia minora

Hymen (torn)

Vaginal entrance

Extensions of the labia minora encircle the body of the clitoris, forming its **prepuce**, or hood.

The **clitoris** (KLIT-ō-ris or kli-TŌR-is) projects into the vestibule. This small, rounded tissue projection contains erectile tissue comparable to the corpora cavernosa and corpus spongiosum of the penis.

2 The area containing the female external genitalia is the **vulva** (VUL-vuh), or **pudendum** (pū-DEN-dum). The vagina opens into the vestibule, a central space bounded by small folds known as the **labia minora** (LĀ-be-uh mi-NOR-uh; singular, *labium minus*). A variable number of small **lesser vestibular glands** discharge their secretions onto the exposed surface of the vestibule, keeping it moist. During sexual arousal, a pair of ducts discharges the secretions of the **greater vestibular glands** into the vestibule. These mucous glands have the same embryological origins as the bulbo-urethral glands of males.

Anus

The **labia majora** (singular, *labium majus*) are prominent folds of skin that encircle and partially conceal the labia minora and adjacent structures.

Module 25.11 Review

a. List the functions of the vagina.

b. Describe the anatomy of the vagina.

c. Cite the similarities that exist between certain structures in the reproductive systems of females and males.

The mammary glands nourish the infant after delivery

A newborn infant cannot fend for itself, and several of its key systems have yet to complete development. Over the initial period of adjustment to an independent existence, the infant can gain nourishment from the milk secreted by the maternal **mammary glands**. These organs are controlled mainly by hormones released by the reproductive system and the placenta, a temporary structure that provides the embryo and fetus with nutrients. The interaction of these hormones results in milk production, or **lactation** (lak-TĀ-shun).

1 The mammary gland lies directly over the pectoralis major muscle. This dissection shows the internal organization of the mammary tissue and its supporting structures.

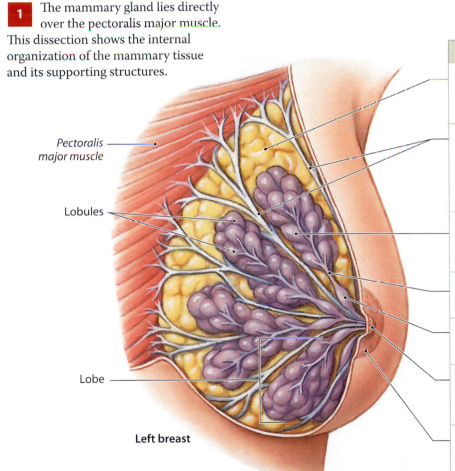

Pectoralis major muscle

Lobules

Lobe

Left breast

The Structure of a Mammary Gland

On each side, a mammary gland lies in the subcutaneous tissue of the **pectoral fat pad** deep to the skin of the chest.

Dense connective tissue surrounds the duct system and forms partitions that extend between the lobes and the lobules. These bands of connective tissue, the **suspensory ligaments of the breast**, originate in the dermis of the overlying skin.

The glandular tissue of the breast consists of separate **lobes**, each containing several secretory **lobules**. Each lobule is composed of many **secretory alveoli**.

Ducts leaving the lobules converge, giving rise to a single **lactiferous** (lak-TIF-er-us) **duct** in each lobe.

Near the nipple, each lactiferous duct enlarges, forming an expanded chamber called a **lactiferous sinus**.

Each breast bears a **nipple**, a small conical projection where 15–20 lactiferous sinuses open onto the body surface.

The reddish-brown skin around each nipple is the **areola** (a-RĒ-ō-luh). Large sebaceous glands deep to the areolar surface give it a grainy texture.

Module 25.12 Review

a. Define lactation.

b. Explain whether the blockage of a single lactiferous sinus would or would not interfere with the delivery of milk to the nipple.

c. Trace the route of milk from its site of production to outside the female.

The ovarian and uterine cycles are regulated by hormones of the hypothalamus, pituitary gland, and ovaries

The ovarian and uterine cycles must operate in synchrony to ensure proper reproductive function. If the two cycles are not properly coordinated, infertility results. A female who doesn't ovulate cannot conceive, even if her uterus is perfectly normal. A female who ovulates normally, but whose uterus is not ready to support an embryo, will also be infertile.

1 As in males, GnRH from the hypothalamus regulates reproductive function in females. However, in females, GnRH levels change throughout the course of the ovarian cycle.

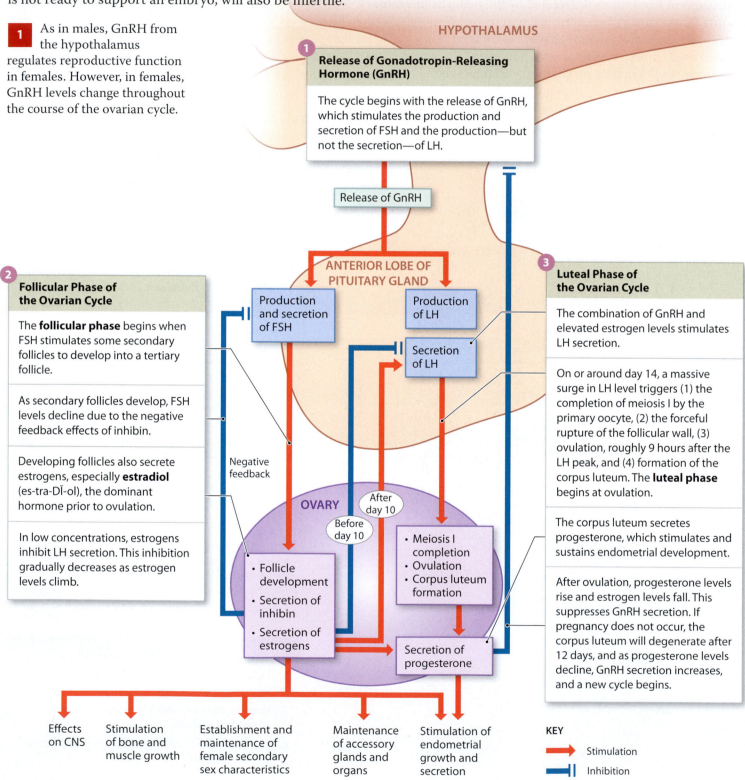

HYPOTHALAMUS

1 **Release of Gonadotropin-Releasing Hormone (GnRH)**

The cycle begins with the release of GnRH, which stimulates the production and secretion of FSH and the production—but not the secretion—of LH.

Release of GnRH

ANTERIOR LOBE OF PITUITARY GLAND

Production and secretion of FSH

Production of LH

Secretion of LH

Negative feedback

OVARY

After day 10

Before day 10

• Follicle development
• Secretion of inhibin
• Secretion of estrogens

• Meiosis I completion
• Ovulation
• Corpus luteum formation

Secretion of progesterone

2 **Follicular Phase of the Ovarian Cycle**

The **follicular phase** begins when FSH stimulates some secondary follicles to develop into a tertiary follicle.

As secondary follicles develop, FSH levels decline due to the negative feedback effects of inhibin.

Developing follicles also secrete estrogens, especially **estradiol** (es-tra-DĪ-ol), the dominant hormone prior to ovulation.

In low concentrations, estrogens inhibit LH secretion. This inhibition gradually decreases as estrogen levels climb.

3 **Luteal Phase of the Ovarian Cycle**

The combination of GnRH and elevated estrogen levels stimulates LH secretion.

On or around day 14, a massive surge in LH level triggers (1) the completion of meiosis I by the primary oocyte, (2) the forceful rupture of the follicular wall, (3) ovulation, roughly 9 hours after the LH peak, and (4) formation of the corpus luteum. The **luteal phase** begins at ovulation.

The corpus luteum secretes progesterone, which stimulates and sustains endometrial development.

After ovulation, progesterone levels rise and estrogen levels fall. This suppresses GnRH secretion. If pregnancy does not occur, the corpus luteum will degenerate after 12 days, and as progesterone levels decline, GnRH secretion increases, and a new cycle begins.

Effects on CNS

Stimulation of bone and muscle growth

Establishment and maintenance of female secondary sex characteristics

Maintenance of accessory glands and organs

Stimulation of endometrial growth and secretion

KEY

→ Stimulation

⊣ Inhibition

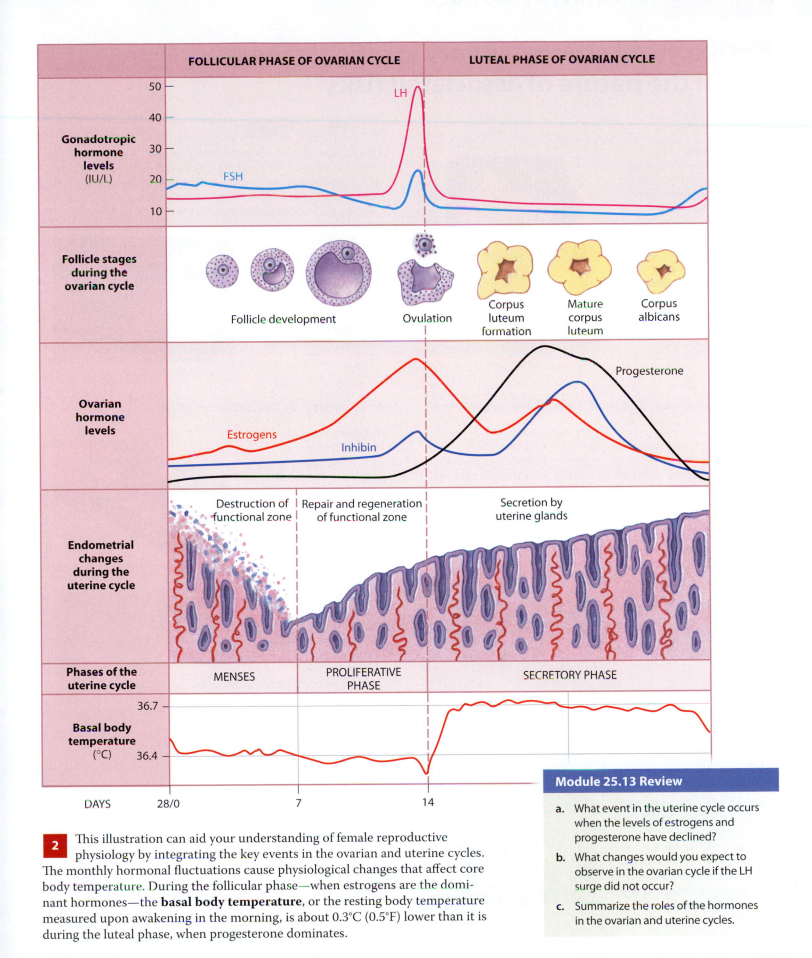

FOLLICULAR PHASE OF OVARIAN CYCLE | **LUTEAL PHASE OF OVARIAN CYCLE**

Gonadotropic hormone levels (IU/L)

LH

FSH

Follicle stages during the ovarian cycle

Follicle development | Ovulation | Corpus luteum formation | Mature corpus luteum | Corpus albicans

Ovarian hormone levels

Progesterone

Estrogens

Inhibin

Endometrial changes during the uterine cycle

Destruction of functional zone | Repair and regeneration of functional zone | Secretion by uterine glands

Phases of the uterine cycle

MENSES | PROLIFERATIVE PHASE | SECRETORY PHASE

Basal body temperature (°C)

DAYS   28/0   7   14

2 This illustration can aid your understanding of female reproductive physiology by integrating the key events in the ovarian and uterine cycles. The monthly hormonal fluctuations cause physiological changes that affect core body temperature. During the follicular phase—when estrogens are the dominant hormones—the **basal body temperature**, or the resting body temperature measured upon awakening in the morning, is about 0.3°C (0.5°F) lower than it is during the luteal phase, when progesterone dominates.

Module 25.13 Review

a. What event in the uterine cycle occurs when the levels of estrogens and progesterone have declined?

b. What changes would you expect to observe in the ovarian cycle if the LH surge did not occur?

c. Summarize the roles of the hormones in the ovarian and uterine cycles.

Birth control strategies vary in effectiveness and in the nature of associated risks

Male Condom

Male condoms (prophylactics or "rubbers") cover the glans and shaft of the penis during intercourse and keep spermatozoa from reaching the female reproductive tract. Of all the strategies described in this module, only latex condoms protect against **sexually transmitted diseases (STDs)**, such as syphilis, gonorrhea, human papilloma virus (HPV), and AIDS.

Diaphragm with Spermicide

A **diaphragm**, the most popular form of vaginal barrier in use today, consists of a dome of latex rubber with a small metal hoop supporting the rim. Because vaginas vary in size, women choosing this method must be individually fitted. Before intercourse, the diaphragm is inserted so that it covers the external os. The diaphragm must be coated with a small amount of spermicidal (sperm-killing) jelly or cream to be an effective contraceptive. The failure rate of a properly fitted and used diaphragm is estimated at 5–6 percent.

Oral Contraceptives—Combined (Estrogen and Progesterone)

At least 20 brands of combination **oral contraceptives** are now available, and more than 200 million women are using them worldwide. In the United States, 33 percent of women under age 45 use a combination pill to prevent conception. The failure rate for combination oral contraceptives, when used as prescribed, is 0.24 percent over a two-year period. (Failure for a birth control method is defined as a pregnancy.) Birth control pills are not risk free: Combination pills can worsen problems associated with severe hypertension, diabetes mellitus, epilepsy, gallbladder disease, heart trouble, and acne. Women taking oral contraceptives are also at increased risk of venous thrombosis, strokes, pulmonary embolism, and (for women over 35) heart disease. However, pregnancy itself has similar or higher risks.

Progesterone-Only Forms of Birth Control

Progesterone-only forms of birth control are now available: The progesterone-only pill and Depo-Provera are examples. The **progesterone-only pill** must be taken daily, and skipping even one pill may result in pregnancy. **Depo-Provera** is injected every 3 months. Uterine cycles are initially irregular and eventually cease in roughly 50 percent of women using this product. The most common problems with this contraceptive method are (1) a tendency to gain weight and (2) a slow return to fertility (up to 18 months) after injections are discontinued.

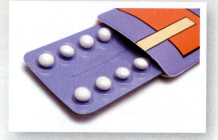

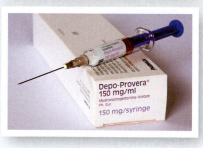

IUD (Intrauterine Device)

An **intrauterine device (IUD)** consists of a small plastic loop or a T that is inserted into the uterine cavity. The mechanism of action remains unclear, but IUDs are known to stimulate prostaglandin production in the uterus, and they are effective for years after insertion.

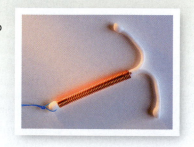

The Rhythm Method

"Natural Family Planning," also called the **rhythm method**, involves abstaining from sexual activity on the days ovulation might be occurring. The timing is estimated on the basis of previous patterns of menstruation; monitoring changes in indications of ovulation, including basal body temperature and cervical mucus texture; and, for some, urine tests for LH. Because of the irregularity of many women's uterine cycles, this method of contraception has a failure rate estimated to be 13–20 percent.

Post-Coital Contraceptives (Preven and Plan B)

Hormonal post-coital contraception, or the emergency "morning after" pill, involves taking either combination estrogen/progesterone birth control pills or progesterone-only pills in two large doses 12 hours apart within 72 hours of unprotected sexual intercourse. Particularly useful when barrier methods malfunction or coerced intercourse occurs, it reduces expected pregnancy rates by up to 89 percent. The progesterone-only version is considered safe for nonprescription use and is available for purchase over the counter for women over 18 years of age. (Purchase by women under age 18 is by prescription only.)

Surgical Sterilization—Male

In a **vasectomy** (vaz-EK-tō-mē), each ductus deferens is cut and either a segment is removed (and the ends tied or cauterized) or silicone plugs are inserted (which makes it relatively easy to reverse the procedure). After a vasectomy, spermatozoa cannot pass from the epididymides to the distal portions of the reproductive tract. The surgery can be performed in a physician's office in a matter of minutes; the failure rate (due to incomplete closure/blockage of either ductus deferens) is 0.3 percent. After vasectomy, men experience normal sexual function, because the secretions of the epididymides and testes normally account for only about 5 percent of the volume of semen. Spermatozoa continue to develop, but they remain inactive and eventually degenerate.

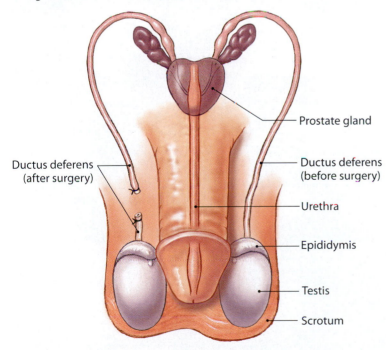

Prostate gland

Ductus deferens (after surgery)

Ductus deferens (before surgery)

Urethra

Epididymis

Testis

Scrotum

Surgical Sterilization—Female

The uterine tubes can be blocked by a surgical procedure known as a **tubal ligation**. The failure rate for this procedure is estimated at 0.45 percent. Because the surgery requires that the abdomino-pelvic cavity be entered, most commonly by laparoscopy, general anesthetic is required and complications are more likely than with a vasectomy.

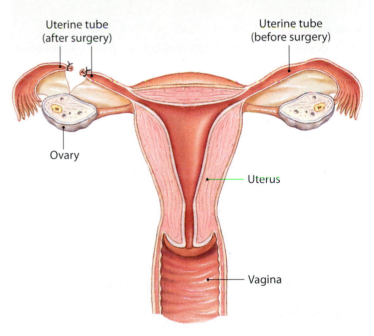

Uterine tube (after surgery)

Uterine tube (before surgery)

Ovary

Uterus

Vagina

Module 25.14 Review

a. Define vasectomy.

b. Which birth control method(s) provide some protection against sexually transmitted diseases?

c. The use of which birth control method often results in the cessation of the uterine cycle?

Reproductive system disorders are relatively common and often deadly

1 Enlargement of the prostate gland, or **benign prostatic hypertrophy (BPH)**, typically occurs spontaneously in men over age 50. The increase in size occurs as testosterone production by the interstitial cells decreases. At the same time, the interstitial cells begin releasing small quantities of estrogens into the bloodstream. The combination of lower testosterone levels and the presence of estrogens probably stimulates prostatic growth. In severe cases, prostatic swelling constricts and blocks the urethra, producing urinary obstruction. **Prostate cancer**, a malignancy of the prostate gland, is the second most common cancer and the second most common cause of cancer deaths in males. The American Cancer Society estimates that approximately 192,000 new prostate cancer cases in 2009 will result in about 27,000 deaths. Blood tests are often used for screening purposes. The most sensitive is a blood test for **prostate-specific antigen (PSA)**. Elevated levels of this antigen, normally present in low concentrations, may indicate the presence of prostate cancer. Screening with periodic PSA tests is now being recommended for men over age 50. Treatment of localized prostate cancer often involves radiation therapy or surgical removal of the prostate gland—a **prostatectomy** (pros-ta-TEK-tō-mē).

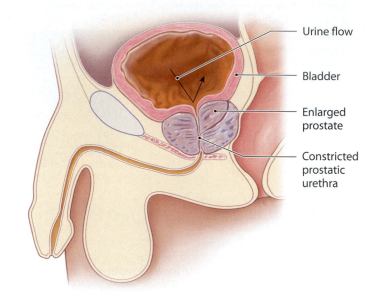

Urine flow

Bladder

Enlarged prostate

Constricted prostatic urethra

2 **Testicular cancer** occurs at a relatively low rate: about 3 cases per 100,000 males per year. Although only about 7900 new cases are reported each year in the United States, with less than 400 deaths, testicular cancer is the most common cancer among males aged 15–35. More than 95 percent of testicular cancers result from abnormal spermatogonia or spermatocytes, rather than abnormal nurse cells, interstitial cells, or other testicular cells. Treatment generally consists of a combination of orchiectomy (removal of the testes) and chemotherapy. The survival rate is now near 95 percent, primarily as a result of earlier diagnosis and improved treatment protocols. Cyclist Lance Armstrong won the grueling Tour de France six consecutive times after successful treatment for advanced testicular cancer.

3 The mammary glands are stimulated by the changing levels of circulating reproductive hormones that accompany the uterine cycle, and, late in the ovarian cycle, occasional discomfort or even inflammation of mammary gland tissues can occur. If inflamed lobules become walled off by scar tissue, **cysts** are created. Clusters of cysts can be felt in the breast as discrete masses, a condition known as **fibrocystic disease**. Biopsies may be needed to distinguish between this benign condition and **breast cancer.** Breast cancer, a malignant, metastasizing tumor of the mammary gland, is the leading cause of death in women between ages 35 and 45, but it is most common in women over age 50. An estimated 12 percent of U.S. women will develop breast cancer at some point in their lifetime. Notable risk factors include (1) a family history of breast cancer, (2) a first pregnancy after age 30, and (3) early menarche or late menopause. Treatment of breast cancer begins with the removal of the tumor. Because in many cases cancer cells begin to spread before the condition is diagnosed, part or all of the affected mammary gland is surgically removed and usually the axillary lymph nodes on that side are biopsied to detect signs of metastasis. A combination of chemotherapy, radiation treatments, and hormone treatments may be used to supplement the surgical procedures.

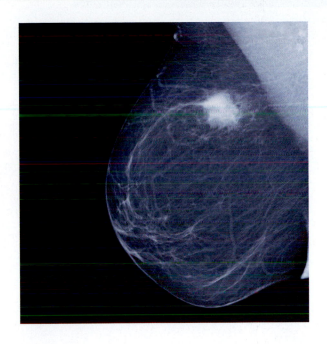

4 A woman in the United States has a 1-in-70 chance of developing **ovarian cancer** in her lifetime. In 2009, there were an estimated 21,500 new cases, and an estimated 14,600 deaths. Although ovarian cancer is the third most common reproductive cancer among women, it is the most dangerous because it is seldom diagnosed in its early stages. The prognosis is relatively good for cancers that originate in the general ovarian tissues or from abnormal oocytes. These cancers respond well to some combination of chemotherapy, radiation, and surgery. However, 85 percent of ovarian cancers are **carcinomas** (cancers derived from epithelial cells), and sustained remission can be obtained in only about one-third of the cases of this type.

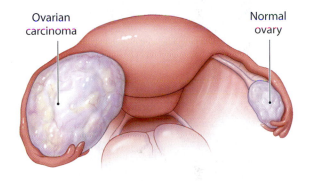

Ovarian carcinoma

Normal ovary

5 **Cervical cancer** is the most common cancer of the reproductive system in women ages 15–34. Each year roughly 13,000 U.S. women are diagnosed with invasive cervical cancer, and approximately one-third of them eventually die from the condition. Another 35,000 women are diagnosed with a less aggressive form of cervical cancer. Gardasil is a new vaccine against the **human papillomaviruses (HPV)** responsible for 70 percent of cervical cancers.

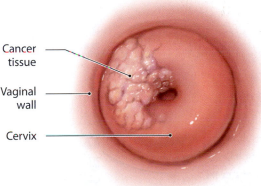

Cancer tissue

Vaginal wall

Cervix

Sexual activity carries with it the risk of infection with a variety of microorganisms. The consequences of such an infection may range from merely inconvenient to potentially lethal. Sexually transmitted diseases (STDs) are transferred from individual to individual, primarily or exclusively by sexual intercourse. At least two dozen bacterial, viral, and fungal infections are currently recognized as STDs. The bacterium *Chlamydia* can cause pelvic inflammatory disease (PID) and infertility; AIDS, caused by HIV, is a deadly viral disease. The incidence of STDs has been increasing in the United States since 1984; an estimated 19 million new cases occur each year; almost 50 percent in persons aged 19–24. Poverty, intravenous drug use, prostitution, and the appearance of drug-resistant pathogens all contribute to the problem.

Module 25.15 Review

a. Define sexually transmitted disease.

b. From which cell type does ovarian cancer usually arise?

c. Which pathogen is associated with most cases of cervical cancer?

1. Labeling

Label the structures of the female reproductive system in the accompanying diagram.

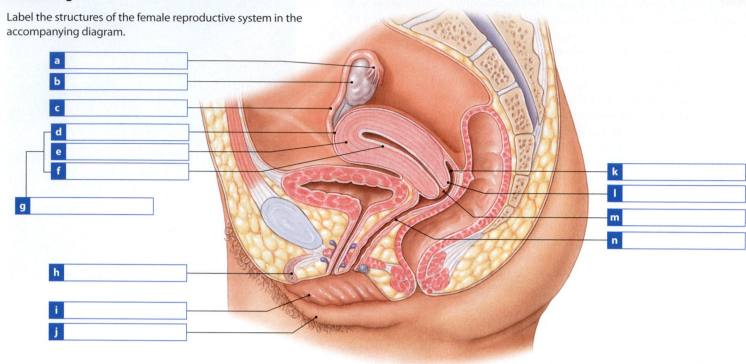

a	
b	
c	
d	
e	
f	
g	
h	
i	
j	
k	
l	
m	
n	

2. Matching

Match the following terms with the most closely related description.

Terms		Descriptions
• LH surge	a _____	Immature female gametes
• rectouterine pouch	b _____	Puberty in female
• tubal ligation	c _____	Pocket anterior to the uterus
• menarche	d _____	Encloses the ovaries, uterine tubes, and uterus
• ovaries	e _____	Pocket posterior to the uterus
• corpus luteum	f _____	Endocrine structure
• vulva	g _____	Averages 28 days
• cervix	h _____	Milk production
• broad ligament	i _____	Oocyte and hormone production
• oocytes	j _____	Triggers ovulation
• GnRH	k _____	Inferior portion of the uterus
• vesicouterine pouch	l _____	Female surgical sterilization
• lactation	m _____	Stimulates FSH production and secretion
• uterine cycle	n _____	Contains female external genitalia

3. Section integration

In a condition known as endometriosis, endometrial cells are believed to migrate from the body of the uterus either into the uterine tubes or through the uterine tubes and into the peritoneal cavity, where they become established. Explain why periodic pain is a major symptom of endometriosis. _____

_____

_____

Visual Outline with Key Terms

Summarize the content of each module using the terms in the order provided.

SECTION 1

The Male Reproductive System

- gonads
- external genitalia
- testis
- spermatozoa
- male reproductive tract
- semen
- ○ accessory organs
- ductus deferens
- seminal glands
- prostate gland
- urethra
- bulbo-urethral glands
- epididymis
- ○ external genitalia
- penis
- scrotum

25.1

The coiled seminiferous tubules of the testes are connected to the male reproductive tract

- epididymis
- ductus deferens
- ejaculatory duct
- seminal glands
- prostate gland
- bulbo-urethral glands
- scrotum
- penis
- superficial inguinal ring
- cremaster muscle
- dartos muscle
- inguinal canal
- inguinal hernias
- spermatic cords
- scrotal cavities
- scrotal septum
- raphe
- tunica albuginea
- seminiferous tubule
- rete testis
- efferent ductules

25.2

Meiosis in the testes produces haploid spermatids that mature into spermatozoa

- spermatogenesis
- mitosis
- meiosis
- spermiogenesis
- diploid
- meiosis I
- meiosis II
- haploid
- synapsis
- tetrad
- spermatogonia
- primary spermatocyte
- secondary spermatocytes
- spermatids
- spermatozoon
- acrosomal cap
- head
- neck
- middle piece
- tail
- flagellum

25.3

Meiosis and early spermiogenesis occur within the seminiferous tubules

- ○ seminiferous tubules
- interstitial cells
- nurse cells
- spermiation
- blood–testis barrier
- basal compartment
- luminal compartment

25.4

The male reproductive tract receives secretions from the seminal, prostate, and bulbo-urethral glands

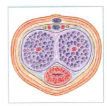

- capacitation
- ampulla
- ejaculatory duct
- epididymis
- head (of epididymis)
- body (of epididymis)
- tail (of epididymis)
- stereocilia
- ductus deferens
- seminal glands
- semen
- prostate gland
- seminalplasmin
- bulbo-urethral glands

25.5

The penis conducts semen and urine to the exterior

- penis
- root
- body
- glans
- neck
- erectile tissue
- corpora cavernosa
- corpus spongiosum
- prepuce
- smegma
- arousal
- nitric oxide
- erection
- emission
- semen
- ejaculation
- male orgasm
- impotence

25.6

Testosterone plays a key role in the establishment and maintenance of male sexual function

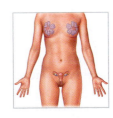

- luteinizing hormone (LH)
- follicle-stimulating hormone (FSH)
- ○ testosterone
- androgen-binding protein (ABP)
- ○ inhibin
- dihydrotestosterone (DHT)

SECTION 2

The Female Reproductive System

- oocytes
- ova
- female reproductive tract
- uterine tubes
- uterus
- vagina
- mammary glands
- ovaries
- ○ external genitalia
- clitoris
- labia

• = *Term boldfaced in this module*

25.7

The ovaries and the female reproductive tract are in close proximity but are not directly connected

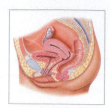

- ovaries
- oocytes
- estrogens
- progestins
- inhibin
- uterine tube
- infundibulum
- uterus
- vesicouterine pouch
- clitoris
- labia
- rectouterine pouch
- vagina
- ovarian ligament
- mesovarium
- broad ligament
- suspensory ligament
- cortex (of ovary)
- medulla (of ovary)
- germinal epithelium
- tunica albuginea

25.8

Meiosis I in the ovaries produces a single haploid secondary oocyte that completes meiosis II only if fertilization occurs

- oogenesis
- primary oocyte
- ovum
- polar bodies
- secondary oocytes
- oogonia
- ovarian follicles
- egg nests
- primordial follicle
- granulosa cells
- zona pellucida
- thecal cells
- secondary follicles
- tertiary follicle
- antrum
- corona radiata
- ovulation
- follicular phase
- luteal phase
- corpus luteum
- corpus albicans
- atresia

25.9

The uterine tubes are connected to the uterus, a hollow organ with thick muscular walls

- uterine tube
- uterus
- infundibulum
- fimbriae
- ampulla (of the uterine tube)
- isthmus (of the uterine tube)
- perimetrium
- myometrium
- endometrium
- uterine cavity
- internal os
- cervical canal
- external os
- embryo
- fetus
- fundus (of the uterus)
- body (of the uterus)
- cervix (of the uterus)

25.10

The uterine cycle involves changes in the functional zone of the endometrium

- arcuate arteries
- radial arteries
- basilar zone
- functional zone
- straight arteries
- spiral arteries
- uterine cycle
- menses
- menstruation
- proliferative phase
- secretory phase
- menstrual cycle
- menarche
- menopause

25.11

The entrance to the vagina is enclosed by external genitalia

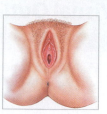

- vagina
- vestibule
- vaginal canal
- fornix
- rugae
- hymen
- vulva
- pudendum
- labia minora
- lesser vestibular glands
- greater vestibular glands
- mons pubis
- prepuce
- clitoris
- labia majora

25.12

The mammary glands nourish the infant after delivery

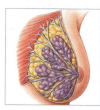

- mammary glands
- lactation
- pectoral fat pad
- suspensory ligaments of the breast
- lobes
- lobules
- secretory alveoli
- lactiferous duct
- lactiferous sinus
- nipple
- areola

25.13

The ovarian and uterine cycles are regulated by hormones of the hypothalamus, pituitary gland, and ovaries

- ovarian cycle
- uterine cycle
- follicular phase
- estrogens
- estradiol
- luteal phase
- progesterone
- basal body temperature

25.14

Birth control strategies vary in effectiveness and in the nature of associated risks

- male condoms
- sexually transmitted diseases (STDs)
- diaphragm
- oral contraceptives
- progesterone-only pill
- Depo-Provera
- intrauterine device (IUD)
- rhythm method
- hormonal post-coital contraception
- vasectomy
- tubal ligation

• = *Term boldfaced in this module*

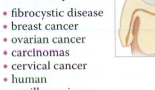

25.15

Reproductive system disorders are relatively common and often deadly

- benign prostatic hypertrophy (BPH)
- prostate cancer
- prostate-specific antigen (PSA)
- prostatectomy
- testicular cancer
- cysts
- fibrocystic disease
- breast cancer
- ovarian cancer
- carcinomas
- cervical cancer
- human papillomaviruses (HPV)

• = *Term boldfaced in this module*

Chapter Integration: Applying what you've learned

Exercise-induced amenorrhea, the abnormal absence of menstruation, occurs in 5–25 percent of women athletes. The variability depends on the level of competition and type of sport. For example, female bodybuilders and ballet dancers, both of whom are likely to have low body fat, commonly experience amenorrhea. Among well-nourished female athletes, hard training and exercise may cause the release of stress hormones, which then interfere with the pituitary gland's production of the hormones necessary for maintaining the menstrual cycle.

Although temporary cessation of menstrual cycles itself is not dangerous, there are long-term health consequences for prolonged exercise-induced amenorrhea in premenopausal women. As a result of long-term exercise-induced amenorrhea, women become estrogen deficient, which can lead to other health-related consequences. Once the diagnosis is confirmed, treatment involves increasing caloric intake and restoring estrogen levels to the normal range. In most cases, exercise-induced amenorrhea is reversible with treatment.

1. What does amenorrhea in female athletes suggest about the relationship between body fat and menstruation?

2. How might exercise-induced amenorrhea be advantageous for survival?

3. What are some health-related consequences of estrogen deficiency?

Access more review material online in the Study Area at **www.masteringaandp.com**.

There, you'll find:
- **Chapter guides**
- **Chapter quizzes**
- **Practice tests**
- **Labeling activities**
- **Animations**
- **MP3 Tutor Sessions**
- **Flashcards**
- **A glossary with pronunciations**

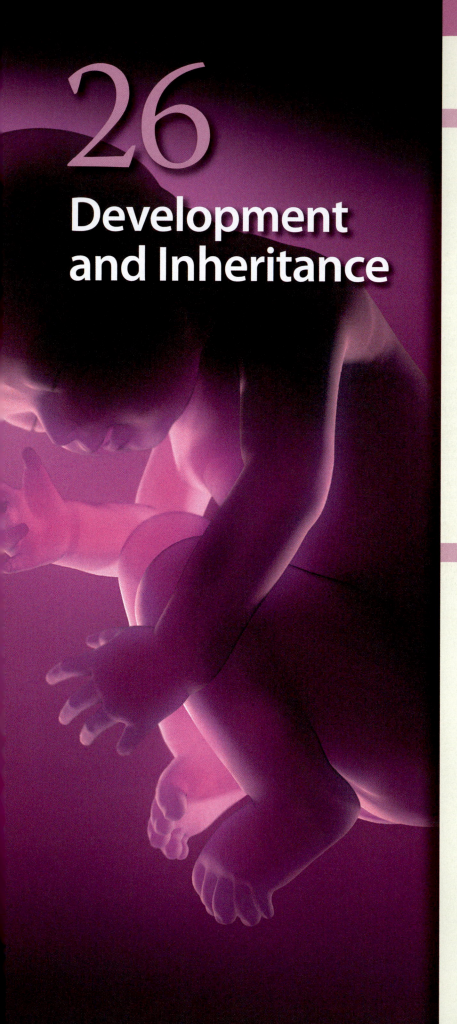

26

Development and Inheritance

An Overview of Development

1   **Development** is the gradual modification of anatomical structures and physiological characteristics during the period from fertilization to maturity. The changes that occur during development are truly remarkable. In a mere 9 months, all the tissues, organs, and organ systems we have studied thus far take shape and begin to function. What begins as a single cell slightly larger than the period at the end of this sentence becomes an individual whose body contains trillions of cells organized into a complex array of highly specialized structures.

Prenatal Development

Embryological development comprises the events that occur during the first two months after fertilization. The study of these events is called **embryology** (em-brē-OL-ō-jē).

Fetal development begins at the start of the ninth week and continues until birth. Embryological and fetal development are sometimes referred to collectively as **prenatal** (*natus*, birth) **development**, the primary focus of this chapter.

4 weeks

8 weeks

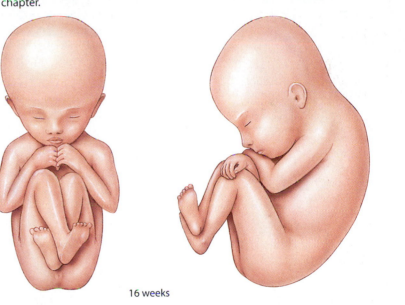

16 weeks

2   The time spent in prenatal development is known as **gestation** (jes-TĀ-shun). For convenience, we usually think of the gestation period as consisting of three integrated trimesters, each three months in duration.

Gestation

First Trimester	Second Trimester	Third Trimester
The **first trimester** is the period of embryological and early fetal development. During this period, the rudiments of all the major organ systems appear.	The **second trimester** is dominated by the development of organs and organ systems, a process that nears completion by the end of the sixth month. During this period, body shape and proportions change; by the end of this trimester, the fetus looks distinctively human.	The **third trimester** is characterized by the largest gain in fetal weight. Early in the third trimester, most of the fetus's major organ systems become fully functional. An infant born one month or even two months prematurely has a reasonable chance of survival if appropriate medical care is available.

Postnatal development begins at birth and continues to **maturity,** the state of full development or completed growth. A basic understanding of prenatal and postnatal development provides important insights into anatomical structures. In addition, many of the mechanisms of development and growth are similar to those responsible for the repair of injuries.

At fertilization, a secondary oocyte and a spermatozoon form a zygote that prepares for cell division

1 **Fertilization** involves the fusion of two haploid gametes, each containing 23 chromosomes, producing a **zygote** (ZĪ-gōt) that contains 46 chromosomes, the normal complement in a somatic cell. Fertilization typically occurs near the junction between the ampulla and isthmus of the uterine tube, generally within a day after ovulation.

Fertilization

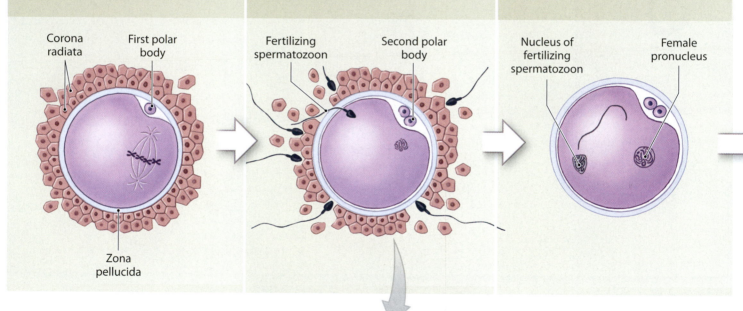

Step 1:
Oocyte at Ovulation
Ovulation releases a secondary oocyte and the first polar body; both are surrounded by the corona radiata. The oocyte is suspended in metaphase of meiosis II.

Corona radiata

First polar body

Zona pellucida

Step 2:
Fertilization and Oocyte Activation
Acrosomal enzymes from multiple sperm create gaps between the cells of the corona radiata. A single sperm then makes contact with the oocyte membrane, and membrane fusion occurs, triggering **oocyte activation** and the completion of meiosis.

Fertilizing spermatozoon

Second polar body

Step 3:
Pronucleus Formation Begins
The sperm is absorbed into the cytoplasm, and the female nuclear material within the ovum reorganizes as the **female pronucleus**.

Nucleus of fertilizing spermatozoon

Female pronucleus

2 This is a photograph of a secondary oocyte surrounded by spermatozoa. The function of a spermatozoon is to deliver the paternal chromosomes to the secondary oocyte and then facilitate the process of fertilization. In contrast, the secondary oocyte provides all the cellular organelles and inclusions, nourishment, and genetic programming necessary to support development of the embryo for nearly a week after conception. The volume of the secondary oocyte is therefore 2000 times greater than that of a spermatozoon.

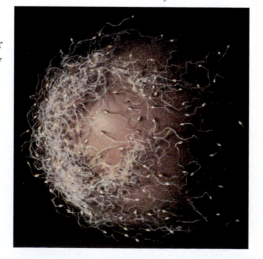

Step 4:
Spindle Formation and Cleavage Preparation
The **male pronucleus** develops, and spindle fibers appear in preparation for cell division. This is the start of the process of **cleavage**, a series of cell divisions that produces an ever-increasing number of smaller and smaller daughter cells.

Step 5:
Amphimixis Occurs and Cleavage Begins
The male pronucleus migrates toward the center of the cell, where the spindle fibers are forming. The two pronuclei then fuse in a process called **amphimixis** (am-fi-MIK-sis). The cell is now a zygote that contains the normal complement of 46 chromosomes, and fertilization is complete.

Step 6:
Cytokinesis Begins
The first cleavage division nears completion roughly 30 hours after fertilization. This division produces two daughter cells, each one-half the size of the original zygote. These cells are called **blastomeres** (BLAS-tō-mērz).

Male pronucleus   Female pronucleus

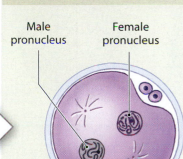

Metaphase of first cleavage division

This is the "moment of conception." Almost immediately the chromosomes line up along a metaphase plate, and the cell prepares to divide.

Blastomeres

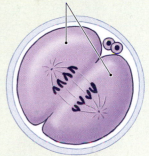

Of the roughly 200 million spermatozoa introduced into the vagina in a typical ejaculation, only about 10,000 enter a uterine tube, and fewer than 100 reach the isthmus. In general, a male with a sperm count below 20 million per milliliter is functionally sterile because too few spermatozoa survive to reach and fertilize an oocyte. While it is true that only one spermatozoon fertilizes an oocyte, dozens of spermatozoa are required for successful fertilization. The additional sperm are essential because one sperm does not contain enough acrosomal enzymes to disrupt the corona radiata that surrounds the secondary oocyte.

Module 26.1 Review

a. Define fertilization.

b. How many chromosomes are contained within a human zygote?

c. Why are numerous spermatozoa required to fertilize a secondary oocyte?

Cleavage continues until the blastocyst implants in the uterine wall

During cleavage, the cytoplasm of the zygote becomes subdivided among an ever-increasing number of progressively smaller blastomeres. A group of blastomeres created by cleavage divisions is called a **pre-embryo**.

1 Cleavage lasts roughly 7 days. Over that period the pre-embryo travels the length of the uterine tube. After three days, the pre-embryo is a solid ball of cells known as a **morula** (MOR-ū-la, blackberry). The morula typically reaches the uterus on day 4. Over the next two days, the blastomeres form a **blastocyst** (blas-tō-sist), a hollow ball with an inner cavity known as the **blastocoele** (BLAS-tō-sēl). At this stage the blastomeres are no longer identical in size and shape.

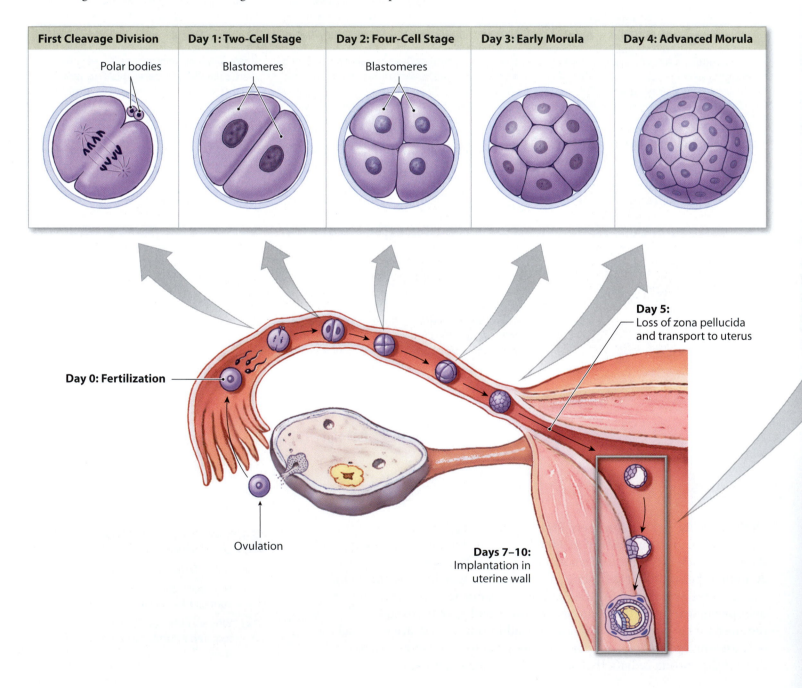

| First Cleavage Division | Day 1: Two-Cell Stage | Day 2: Four-Cell Stage | Day 3: Early Morula | Day 4: Advanced Morula |

Polar bodies

Blastomeres

Blastomeres

Day 0: Fertilization

Ovulation

Day 5:
Loss of zona pellucida and transport to uterus

Days 7–10:
Implantation in uterine wall

Day 6: Blastocyst

The blastocyst is freely exposed to the fluid contents of the uterine cavity, which contains the glycogen-rich secretions of the uterine glands. The rate of growth and cell division now accelerates, and the blastocyst enlarges rapidly.

FUNCTIONAL ZONE OF ENDOMETRIUM

UTERINE CAVITY

Uterine glands

Blastocyst

The outer layer of cells, which separates the outside world from the blastocoele, is called the **trophoblast** (TRŌ-fō-blast, *trophos*, food + *blast*, precursor). The cells in this layer are responsible for providing nutrients to the developing embryo.

Day 7: Implantation

When fully formed, the blastocyst contacts the endometrium. **Implantation** begins with the attachment of the blastocyst to the endometrium of the uterus. Implantation proceeds as the blastocyst erodes the endometrial lining and becomes enclosed within the endometrium by day 10.

Blastocoele

The **inner cell mass** lies clustered at one end of the blastocyst. These cells are exposed to the blastocoele but are insulated from contact with the outside environment by the trophoblast. In time, the inner cell mass will form the embryo.

Day 8: Trophoblast Development

At the point of contact, the trophoblast cells divide rapidly, making the trophoblast several layers thick. The cells closest to the blastocoele remain intact, forming a layer of **cellular trophoblast**. Near the endometrial wall, the plasma membranes separating the trophoblast cells disappear, creating a layer of cytoplasm containing multiple nuclei. This layer is called the **syncytial** (sin-SISH-ul) **trophoblast**.

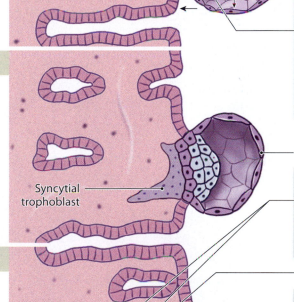

Syncytial trophoblast

Cellular trophoblast

Fingerlike **villi** extend away from the trophoblast into the surrounding endometrium, gradually increasing in size and complexity.

Day 9: Formation of Amniotic Cavity

As implantation proceeds, the syncytial trophoblast continues to enlarge and spread into the surrounding endometrium. The erosion of uterine glands releases nutrients that are absorbed by the syncytial trophoblast and distributed by diffusion through the underlying cellular trophoblast to the inner cell mass. These nutrients provide the energy needed to support the early stages of embryo formation. Trophoblastic extensions (villi) grow around endometrial capillaries. As the capillary walls are destroyed, maternal blood begins to percolate through trophoblastic channels known as **lacunae**.

Lacuna

At the time of implantation, the inner cell mass has separated from the trophoblast. The separation gradually increases, creating a fluid-filled chamber called the **amniotic** (am-nē-OT-ik) **cavity.**

Module 26.2 Review

a. Identify the stage of development that results from cleavage.

b. What developmental stage begins once the zygote arrives in the uterine cavity?

c. Describe the blastocyst and its role in implantation.

Gastrulation produces three germ layers: ectoderm, endoderm, and mesoderm

Day 9: Formation of Amniotic Cavity (continued)

When the amniotic cavity first appears, the cells of the inner cell mass are organized into an oval sheet known as the **blastodisc**. The early blastodisc is two layers thick: a superficial layer that faces the amniotic cavity, and a deeper layer that is exposed to the fluid contents of the blastocoele. Cells of the superficial layer migrate along the walls of the amniotic cavity and separate the amniotic cavity from the tropho-blast. This is the first step in the formation of the **amnion**, one of four **extra-embryonic membranes** we will consider further in Module 26.4. At this stage nutrients released into the amnion and blastocoele by the advancing trophoblast are absorbed directly by the cells of the blastodisc.

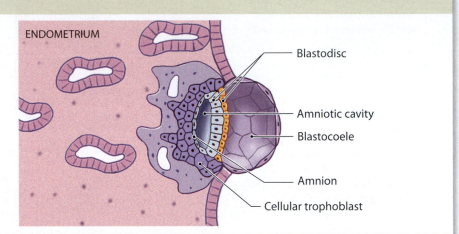

ENDOMETRIUM

- Blastodisc
- Amniotic cavity
- Blastocoele
- Amnion
- Cellular trophoblast

Day 10: Yolk Sac Formation

While cells from the superficial layer of the inner cell mass migrate around the amniotic cavity, forming the amnion, cells from the deeper layer migrate around the outer edges of the blastocoele. This is the first step in the formation of the **yolk sac**, a second extra-embryonic membrane. For roughly the next two weeks, the yolk sac is the primary nutrient source for the inner cell mass; it absorbs and distributes nutrients released into the blastocoele by the trophoblast.

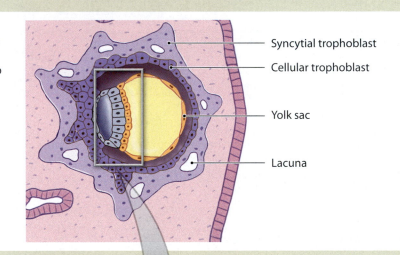

- Syncytial trophoblast
- Cellular trophoblast
- Yolk sac
- Lacuna

Day 12: Gastrulation

By day 12, superficial cells of the blastodisc have begun to migrate toward a central line known as the **primitive streak**. At the primitive streak, the migrat-ing cells leave the surface and move between the two existing layers. This movement creates three distinct embryonic layers: (1) the **ectoderm**, consisting of superficial cells that did not migrate into the interior of the blastodisc; (2) the **endoderm**, consisting of the cells that face the yolk sac; and (3) the **mesoderm**, consisting of the poorly organized layer of migrating cells between the ectoderm and the endoderm. Collectively, these three embryonic layers are called **germ layers**, and the migration process is called **gastrulation** (gas-troo-LĀ-shun). Gastrulation produces an oval, three-layered sheet known as the **embryonic disc**. This disc will form the body of the embryo, whereas all other cells of the blastocyst will be part of the extra-embryonic membranes.

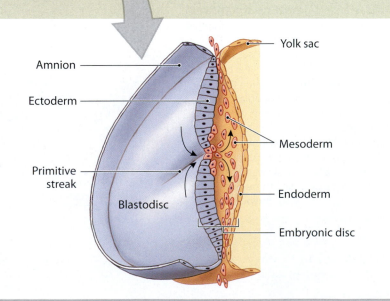

- Yolk sac
- Amnion
- Ectoderm
- Mesoderm
- Primitive streak
- Endoderm
- Blastodisc
- Embryonic disc

The panels to the left illustrate the formation of the three germ layers, and the table below summarizes their importance to later development.

The Fates of the Germ Layers

Body System	Ectodermal Contributions
Integumentary system	Epidermis, hair follicles and hairs, nails, and glands communicating with the skin (sweat glands, mammary glands, and sebaceous glands)
Skeletal system	Pharyngeal cartilages and their derivatives in adults (portion of sphenoid, the auditory ossicles, the styloid processes of the temporal bones, the cornu and superior rim of the hyoid bone)
Nervous system	All neural tissue, including brain and spinal cord
Endocrine system	Pituitary gland and adrenal medullae
Respiratory system	Mucous epithelium of nasal passageways
Digestive system	Mucous epithelium of mouth and anus, salivary glands
Mesodermal Contributions	
Integumentary system	Dermis and hypodermis
Skeletal system	All components except some pharyngeal derivatives
Muscular system	All components
Endocrine system	Adrenal cortex, endocrine tissues of heart, kidneys, and gonads
Cardiovascular system	All components
Lymphatic system	All components
Urinary system	The kidneys, including the nephrons and the initial portions of the collecting system
Reproductive system	The gonads and the adjacent portions of the duct systems
Miscellaneous	The lining of the subdivisions of the ventral body cavity (pleural, pericardial, and peritoneal cavities) and the connective tissues that support all organ systems
Endodermal Contributions	
Endocrine system	Thymus, thyroid gland, and pancreas
Respiratory system	Respiratory epithelium (except nasal passageways) and associated mucous glands
Digestive system	Mucous epithelium (except mouth and anus), exocrine glands (except salivary glands), liver, and pancreas
Urinary system	Urinary bladder and distal portions of the duct system
Reproductive system	Distal portions of the duct system, stem cells that produce gametes

The trophoblast undergoes repeated nuclear divisions, shows extensive and rapid growth, has a very high demand for energy, invades and spreads through adjacent tissues, yet fails to activate the maternal immune system—in short, the trophoblast has many of the characteristics of cancer cells. In about 0.1 percent of pregnancies, something goes wrong with the regulatory mechanisms, and instead of developing normally, the syncytial trophoblast behaves like a tumor. This condition is called **gestational trophoblastic neoplasia**. Approximately 20 percent of gestational trophoblastic neoplasias metastasize to other tissues, with potentially fatal results. Consequently, prompt surgical removal of the mass is essential, and the surgery is sometimes followed by chemotherapy.

Module 26.3 Review

a. Define gestational trophoblastic neoplasia.

b. Describe gastrulation and the formation of the germ layers.

c. What germ layer gives rise to nearly all body systems except the nervous and respiratory systems?

The extra-embryonic membranes form the placenta that supports fetal growth and development

Germ layers participate in the formation of four extra-embryonic membranes: the yolk sac (endoderm + mesoderm), the amnion (ectoderm + mesoderm), the allantois (endoderm + mesoderm), and the chorion (mesoderm + trophoblast). Although these membranes support embryological and fetal development, few traces of their existence remain in adult systems.

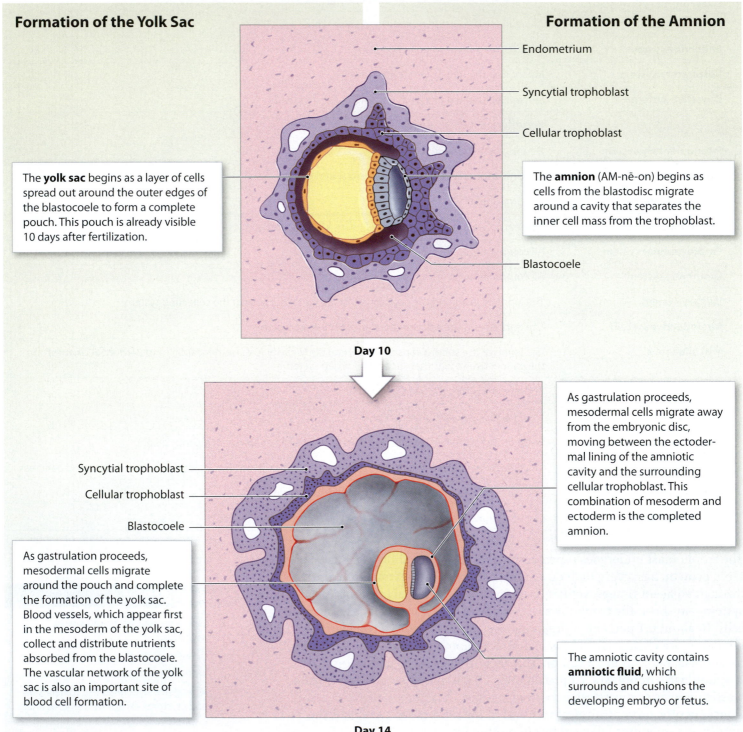

Formation of the Yolk Sac

Formation of the Amnion

Endometrium

Syncytial trophoblast

Cellular trophoblast

The **yolk sac** begins as a layer of cells spread out around the outer edges of the blastocoele to form a complete pouch. This pouch is already visible 10 days after fertilization.

The **amnion** (AM-nē-on) begins as cells from the blastodisc migrate around a cavity that separates the inner cell mass from the trophoblast.

Blastocoele

Day 10

Syncytial trophoblast

Cellular trophoblast

Blastocoele

As gastrulation proceeds, mesodermal cells migrate around the pouch and complete the formation of the yolk sac. Blood vessels, which appear first in the mesoderm of the yolk sac, collect and distribute nutrients absorbed from the blastocoele. The vascular network of the yolk sac is also an important site of blood cell formation.

As gastrulation proceeds, mesodermal cells migrate away from the embryonic disc, moving between the ectodermal lining of the amniotic cavity and the surrounding cellular trophoblast. This combination of mesoderm and ectoderm is the completed amnion.

The amniotic cavity contains **amniotic fluid**, which surrounds and cushions the developing embryo or fetus.

Day 14

Formation of the Allantois

Formation of the Chorion

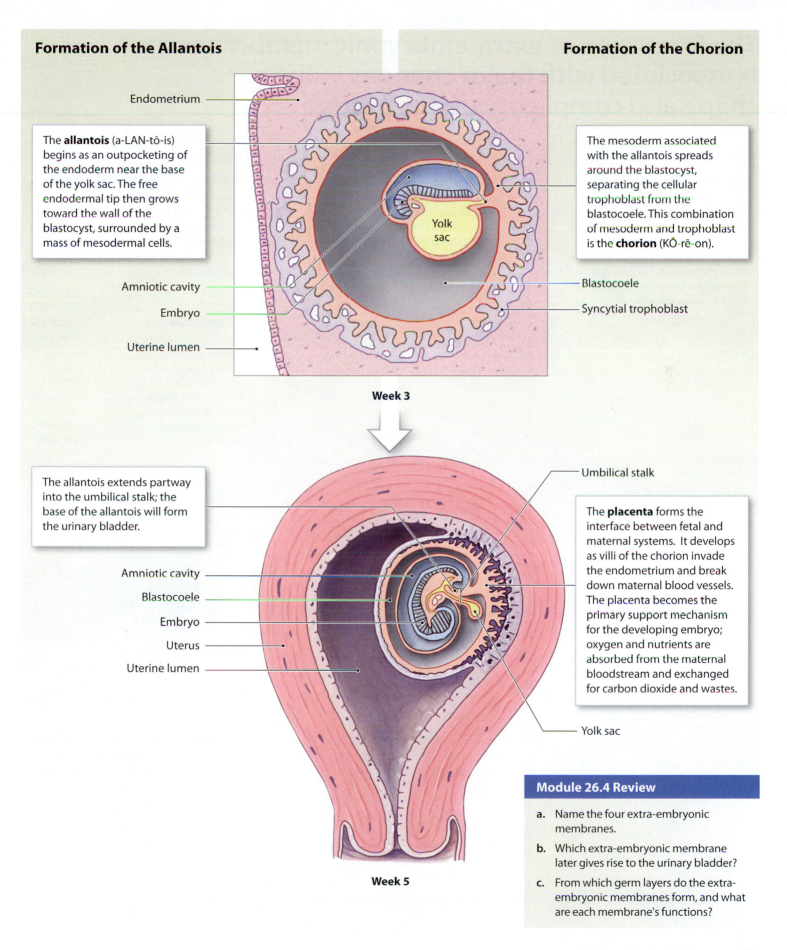

Endometrium

The **allantois** (a-LAN-tō-is) begins as an outpocketing of the endoderm near the base of the yolk sac. The free endodermal tip then grows toward the wall of the blastocyst, surrounded by a mass of mesodermal cells.

The mesoderm associated with the allantois spreads around the blastocyst, separating the cellular trophoblast from the blastocoele. This combination of mesoderm and trophoblast is the **chorion** (KŌ-rē-on).

Yolk sac

Amniotic cavity

Embryo

Uterine lumen

Blastocoele

Syncytial trophoblast

Week 3

The allantois extends partway into the umbilical stalk; the base of the allantois will form the urinary bladder.

Umbilical stalk

The **placenta** forms the interface between fetal and maternal systems. It develops as villi of the chorion invade the endometrium and break down maternal blood vessels. The placenta becomes the primary support mechanism for the developing embryo; oxygen and nutrients are absorbed from the maternal bloodstream and exchanged for carbon dioxide and wastes.

Amniotic cavity

Blastocoele

Embryo

Uterus

Uterine lumen

Yolk sac

Week 5

Module 26.4 Review

a. Name the four extra-embryonic membranes.

b. Which extra-embryonic membrane later gives rise to the urinary bladder?

c. From which germ layers do the extra-embryonic membranes form, and what are each membrane's functions?

26.5

The formation of extra-embryonic membranes is associated with major changes in the shape and complexity of the embryo

Week 2

Migration of mesoderm around the inner surface of the trophoblast creates the chorion. Mesodermal migration around the outside of the amniotic cavity, between the ectodermal cells and the trophoblast, forms the amnion. Mesodermal migration around the endodermal pouch creates the yolk sac.

Syncytial trophoblast   Amnion   Yolk sac

Blastocoele   Cellular trophoblast   Mesoderm

Chorion

Week 3

The embryonic disc bulges into the amniotic cavity at the **head fold**. The allantois, an endodermal extension surrounded by mesoderm, extends toward the trophoblast.

Head fold of embryo   Amniotic cavity (containing amniotic fluid)   Allantois

Yolk sac

Chorion   Syncytial trophoblast

Mesoderm extends along the core of each trophoblastic villus, forming **chorionic villi** in contact with maternal tissues. Embryonic blood vessels develop within each villus. Blood flow through those chorionic vessels begins early in the third week of development, when the embryonic heart starts beating.

Week 4

The embryo now has a **tail fold** as well as a head fold. The anterior/posterior, left/right, and superior/inferior axes of the developing embryo are now clearly established. The connections between the embryo and the surrounding trophoblast begin to constrict.

The **body stalk**, the connection between embryo and chorion, contains the distal portions of the allantois and blood vessels that carry blood to and from the placenta.

The narrow connection between the endoderm of the embryo and the yolk sac is called the **yolk stalk**.

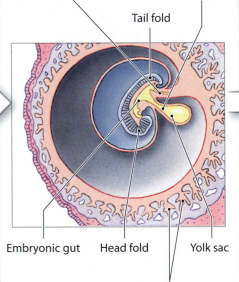

Tail fold

Embryonic gut   Head fold   Yolk sac

As the chorionic villi enlarge, more maternal blood vessels are eroded. Maternal blood now moves slowly through complex lacunae lined by the syncytial trophoblast. Chorionic blood vessels pass close by, and gases and nutrients diffuse between the embryonic and maternal circulations across the layers of the trophoblast.

Week 5

The developing embryo and extra-embryonic membranes bulge into the uterine cavity. The trophoblast pushing out into the uterine lumen remains covered by endometrium but no longer participates in nutrient absorption and embryo support. The embryo moves away from the placenta, and the body stalk and yolk stalk fuse to form an **umbilical stalk**.

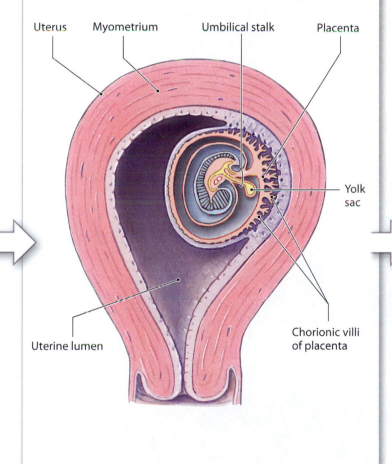

Uterus   Myometrium   Umbilical stalk   Placenta

Yolk sac

Uterine lumen

Chorionic villi of placenta

Week 10

The amnion has expanded greatly, filling the uterine cavity. The fetus is connected to the placenta by an elongated **umbilical cord** that contains a portion of the allantois, blood vessels, and the remnants of the yolk stalk.

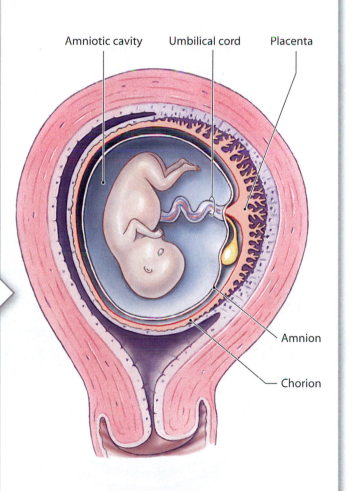

Amniotic cavity   Umbilical cord   Placenta

Amnion

Chorion

Module 26.5 Review

a. Describe the chorionic villi.

b. Compare the body stalk with the yolk stalk.

c. Identify the structure connecting the fetus to the placenta, and name the extra-embryonic membrane from which it is derived.

The placenta performs many vital functions for the duration of prenatal development

1 This illustration provides a closer look at the structure of the placenta. The chorionic villi provide the surface area for active and passive exchanges of gases, nutrients, and waste products between the fetal and maternal bloodstreams. Blood flowing to the placenta through the paired **umbilical arteries** is deoxygenated and contains waste products generated by fetal tissues. At the placenta, oxygen supplies are replenished, organic nutrients are added, and carbon dioxide and other organic waste products are removed. Blood then returns to the fetus within a single **umbilical vein**.

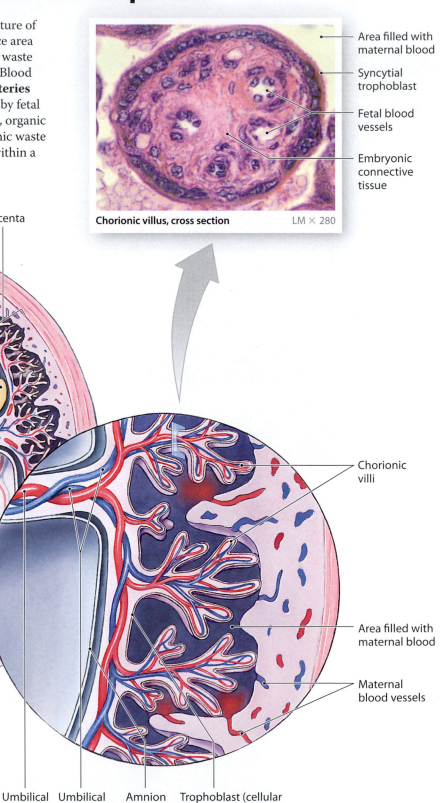

Chorionic villus, cross section   LM × 280

Area filled with maternal blood

Syncytial trophoblast

Fetal blood vessels

Embryonic connective tissue

Umbilical cord (cut)

Yolk sac

Placenta

Amnion

Chorion

Chorionic villi

Area filled with maternal blood

Maternal blood vessels

Myometrium

Uterine cavity

Cervical (mucus) plug in cervical canal

External os

Cervix

Vagina

Umbilical vein

Umbilical arteries

Amnion

Trophoblast (cellular and syncytial layers)

2 In addition to its role in the nutrition of the fetus, the placenta acts as an endocrine organ. Several hormones are synthesized by the syncytial trophoblast and released into the maternal bloodstream.

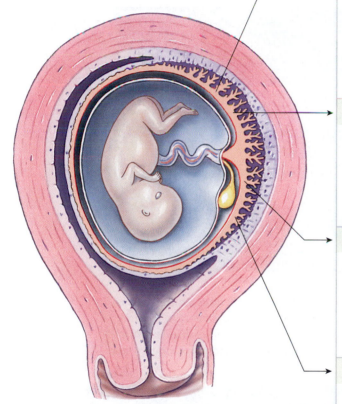

Placental Hormones

Human Chorionic Gonadotropin (hCG)

Human chorionic gonadotropin (hCG) appears in the maternal bloodstream soon after implantation has occurred. The presence of hCG in blood or urine samples provides a reliable indication of pregnancy. Kits sold for the early detection of pregnancy are sensitive to the presence of this hormone. In function, hCG resembles LH, because it maintains the integrity of the corpus luteum and promotes the continued secretion of progesterone. As a result, in pregnancy the endometrial lining remains perfectly functional, and menses does not occur. In the presence of hCG, the corpus luteum persists for three to four months before gradually decreasing in size and secretory function. The decline in luteal function does not trigger the return of uterine cycles, because by the end of the first trimester, the placenta is secreting both estrogens and progesterone.

Human Placental Lactogen (hPL)

Human placental lactogen (hPL) helps prepare the mammary glands for milk production. At the mammary glands, the conversion from inactive to active status requires the presence of placental hormones (hPL, estrogens, and progesterone) as well as several maternal hormones (GH, prolactin, and thyroid hormones).

Relaxin

Relaxin is a peptide hormone that is secreted by the placenta and the corpus luteum during pregnancy. Relaxin (1) increases the flexibility of the pubic symphysis, permitting the pelvis to expand during delivery; (2) causes dilation of the cervix, making it easier for the fetus to enter the vaginal canal; and (3) delays the onset of labor contractions until late in the pregnancy.

Progesterone and Estrogens

After the first trimester, the placenta produces sufficient amounts of progesterone to maintain the endometrial lining and continue the pregnancy. As the end of the third trimester approaches, estrogen production by the placenta accelerates. As we will see in a later module, the rising estrogen levels play a role in stimulating labor and delivery.

Module 26.6 Review

a. Name the hormones synthesized by the syncytial trophoblast.

b. The presence of which hormone in the urine provides a reliable indicator of pregnancy in home pregnancy tests?

c. When does the placenta become sufficiently functional to continue the pregnancy?

Organ systems are established in the first trimester and become functional in the second and third trimesters

The first trimester is a critical period for development, because events in the first 12 weeks establish the basis for **organogenesis**, the process of organ formation. Over the next two trimesters the fetus grows larger and the organ systems increase in complexity to the stage at which they are capable of normal function.

1 This is a scanning electron micrograph of an embryo in the second week of development. The CNS is forming as a deep groove develops in a thick ectodermal band that lies along the posterior midline of the embryo.

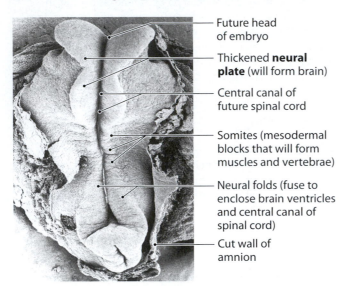

- Future head of embryo
- Thickened **neural plate** (will form brain)
- Central canal of future spinal cord
- Somites (mesodermal blocks that will form muscles and vertebrae)
- Neural folds (fuse to enclose brain ventricles and central canal of spinal cord)
- Cut wall of amnion

2 This photograph of a 4-week embryo shows many features you can probably recognize easily. The heart is beating, pushing blood to and from the placenta, providing the nutrients needed to promote additional growth and development.

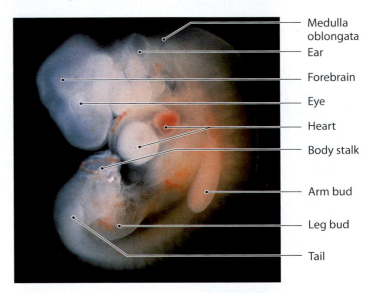

- Medulla oblongata
- Ear
- Forebrain
- Eye
- Heart
- Body stalk
- Arm bud
- Leg bud
- Tail

3 By week 6, the placenta has formed and the embryo floats within the confines of the amniotic cavity. Body proportions are changing, the limbs are growing longer, and the skull bones are beginning to organize around the already-formed brain and eyes.

4 At the end of the first trimester, the fetus is considerably larger and its human features better defined. The axial and appendicular muscles are forming, and fetal movements will soon begin.

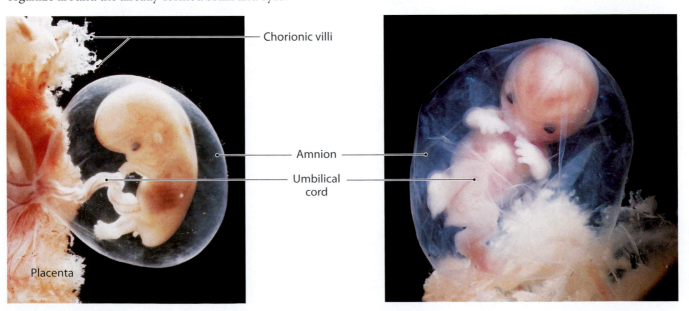

- Chorionic villi
- Amnion
- Umbilical cord
- Placenta

5 This is a fetus after 4 months of gestation. The face and palate have their proper form; the cerebral hemispheres are rapidly enlarging. Hair follicles are present, and hair growth begins. Peripheral nerves have formed, sensory receptors are developing, and the fetus moves frequently. The first 8 weeks of fetal growth is the most rapid, with the fetus increasing in weight some twenty-five-fold. By the end of the second trimester, the fetus will have grown to a weight of about 0.64 kg (1.4 lb).

6 This is an ultrasound of a fetus after 6 months of gestation. During the third trimester, most of the organ systems become ready to perform their normal functions without maternal assistance. The rate of growth starts to slow, but in absolute terms the largest weight gain occurs in this trimester. In the final three months of gestation, the fetus gains about 2.6 kg (5.7 lb), reaching a full-term weight of approximately 3.2 kg (7 lb).

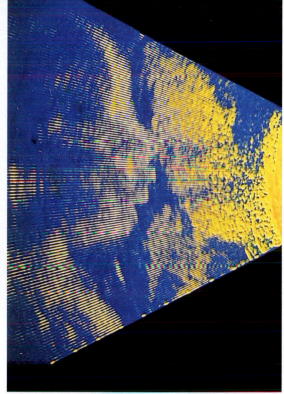

Module 26.7 Review

a. Define organogenesis.

b. Identify the main event in fetal development during the second trimester and third trimester.

c. During which trimester does the fetus undergo its largest absolute weight gain?

Pregnancy places anatomical and physiological stresses on maternal systems

Pregnancy places tremendous strains on the mother. The developing fetus is totally dependent on maternal organ systems for nourishment, respiration, and waste removal. Maternal systems perform these functions in addition to their normal operations. For example, the mother must absorb enough oxygen, nutrients, and vitamins for herself and for her fetus, and she must eliminate all the wastes that are generated. Although this is not a burden over the initial weeks of gestation, the demands placed on the mother become significant as the fetus grows.

1 The physical strains of pregnancy are considerable. It is not unusual for a woman to gain 6–7 kg (13–15 lb) in weight during pregnancy, and the weight is not aligned with the body axis. This means that moving and maintaining balance use additional energy. The sectional views below compare the organ positions in nonpregnant and pregnant women. When the pregnancy is at full term, the uterus and fetus are so large that they push many of the maternal abdominal organs out of their normal positions.

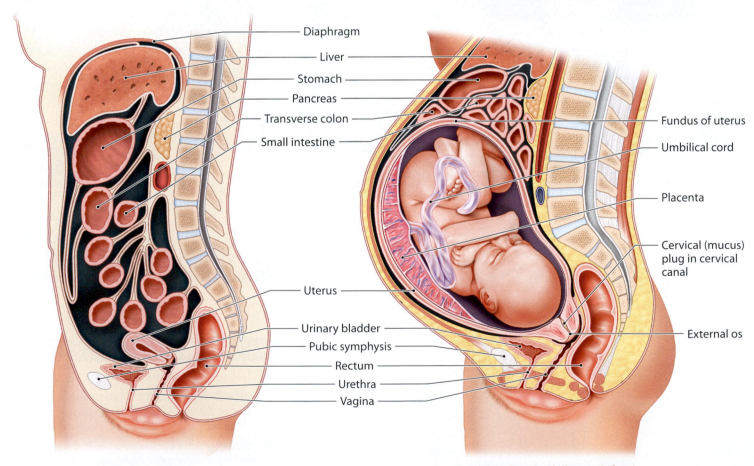

Nonpregnant female

Pregnant female (full-term infant)

Diaphragm
Liver
Stomach
Pancreas
Transverse colon
Small intestine
Fundus of uterus
Umbilical cord
Placenta
Cervical (mucus) plug in cervical canal
Uterus
External os
Urinary bladder
Pubic symphysis
Rectum
Urethra
Vagina

2 The physiological stresses are even more extreme than the physical ones. This is a summary of the physiological changes that have occurred in maternal systems by the end of the third trimester.

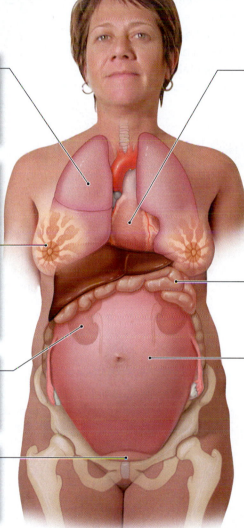

The mother's lungs must deliver the extra oxygen required, and remove the excess carbon dioxide generated, by the fetus. As a result, the maternal respiratory rate goes up and tidal volume increases.

Mammary gland development requires a combination of hormones, including human placental lactogen and placental prolactin (PRL), estrogens, progesterone, GH, and thyroxine from the maternal endocrine organs. By the end of the sixth month of pregnancy, the mammary glands are fully developed and begin to produce clear secretions that are stored in the duct system of those glands and may be expressed from the nipple.

The kidneys must eliminate the wastes produced by maternal and fetal systems. As a result, the maternal glomerular filtration rate (GFR) increases by roughly 50 percent.

Because the volume of urine produced increases and the weight of the uterus presses down on the urinary bladder, pregnant women need to urinate frequently.

The volume of blood flowing into the placenta reduces the volume in the systemic circuit of the mother. At the same time, fetal metabolic activity "steals" maternal oxygen and elevates maternal CO_2 levels. This combination stimulates the production of renin and erythropoietin, leading to an increase in maternal blood volume. By the end of gestation, maternal blood volume has increased by almost 50 percent.

Pregnant women must nourish both themselves and their fetus and so tend to have increased hunger sensations. Maternal requirements for nutrients can climb by up to 30 percent above normal.

At the end of gestation, a typical uterus has grown from 7.5 cm (3 in.) in length and 30–40 g (1–1.4 oz) in weight to 30 cm (12 in.) in length and 1100 g (2.4 lb) in weight. The uterus may then contain 2 liters of fluid, plus fetus and placenta, for a total weight of roughly 6–7 kg (13–15 lb). This remarkable expansion occurs through the enlargement (hypertrophy) of existing cells, especially smooth muscle fibers, rather than by an increase in the total number of cells.

Although pregnancy is a natural phenomenon, the physical and physiological demands on maternal systems make it potentially dangerous. At any age, the risks associated with pregnancy are significantly greater than those associated with the use of oral contraceptives. (The notable exception involves women who both take the pill and smoke.) For pregnant women over age 35 the chances of dying from pregnancy-related complications are almost twice as great as the chances of being killed in an automobile accident.

Module 26.8 Review

a. List the major changes that occur in maternal systems during pregnancy.

b. Why does a mother's blood volume increase during pregnancy?

c. Based on the illustrations showing the locations of the internal organs in nonpregnant and pregnant women, explain why some women experience difficulty breathing while pregnant.

Multiple factors initiate and accelerate the process of labor

The tremendous stretching of the uterus is associated with a gradual increase in the rate of spontaneous smooth muscle contractions in the myometrium. In the early stages of pregnancy, progesterone released by the placenta inhibits the uterine smooth muscle, preventing powerful contractions. Late in pregnancy, some women experience occasional spasms in the uterine musculature, but these contractions are neither regular nor persistent. Such contractions are called **false labor**. **True labor** begins when biochemical and mechanical factors reach a point of no return.

1 This sagittal section shows the position of the fetus at the onset of true labor.

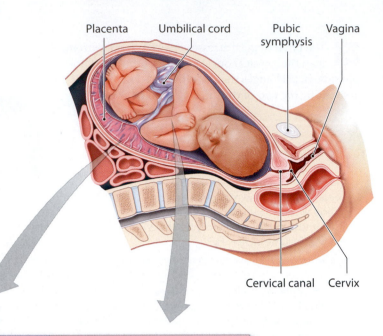

Placenta   Umbilical cord   Pubic symphysis   Vagina

Cervical canal   Cervix

2 After nine months of gestation, multiple factors interact to initiate true labor. Once labor contractions have begun in the myometrium, positive feedback ensures that they will continue until delivery has been completed.

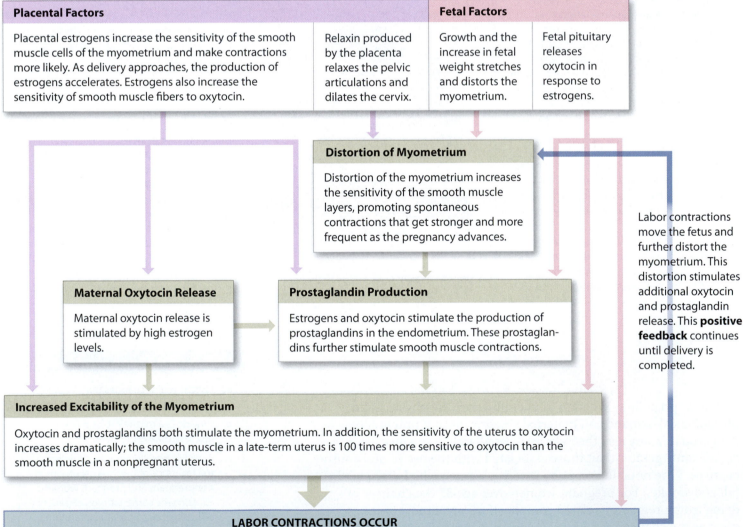

Placental Factors

Placental estrogens increase the sensitivity of the smooth muscle cells of the myometrium and make contractions more likely. As delivery approaches, the production of estrogens accelerates. Estrogens also increase the sensitivity of smooth muscle fibers to oxytocin.

Relaxin produced by the placenta relaxes the pelvic articulations and dilates the cervix.

Fetal Factors

Growth and the increase in fetal weight stretches and distorts the myometrium.

Fetal pituitary releases oxytocin in response to estrogens.

Distortion of Myometrium

Distortion of the myometrium increases the sensitivity of the smooth muscle layers, promoting spontaneous contractions that get stronger and more frequent as the pregnancy advances.

Labor contractions move the fetus and further distort the myometrium. This distortion stimulates additional oxytocin and prostaglandin release. This **positive feedback** continues until delivery is completed.

Maternal Oxytocin Release

Maternal oxytocin release is stimulated by high estrogen levels.

Prostaglandin Production

Estrogens and oxytocin stimulate the production of prostaglandins in the endometrium. These prostaglandins further stimulate smooth muscle contractions.

Increased Excitability of the Myometrium

Oxytocin and prostaglandins both stimulate the myometrium. In addition, the sensitivity of the uterus to oxytocin increases dramatically; the smooth muscle in a late-term uterus is 100 times more sensitive to oxytocin than the smooth muscle in a nonpregnant uterus.

LABOR CONTRACTIONS OCCUR

3 The goal of labor is **parturition** (par-toor-ISH-un), the forcible expulsion of the fetus and placenta. Labor has traditionally been divided into three stages: the dilation stage, the expulsion stage, and the placental stage.

Dilation Stage

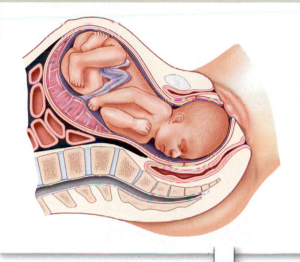

The **dilation stage** begins with the onset of true labor, as the cervix dilates and the fetus begins to shift toward the cervical canal, moved by gravity and uterine contractions. This stage typically lasts eight or more hours. At the start of the dilation stage, labor contractions last up to half a minute and occur once every 10–30 minutes; their frequency increases steadily. Late in this stage, the amnion ruptures, an event sometimes referred to as "having one's water break."

Expulsion Stage

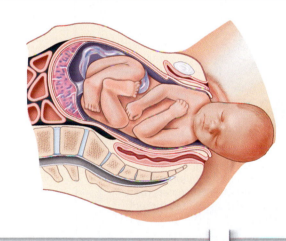

The **expulsion stage** begins as the cervix, pushed open by the approaching fetus, completes its dilation. In this stage, contractions reach maximum intensity, occurring at perhaps two- or three-minute intervals and lasting a full minute. Expulsion continues until the fetus has emerged from the vagina; in most cases, the expulsion stage lasts less than two hours. The arrival of the newborn infant into the outside world is **delivery**, or birth.

Placental Stage

Uterus   Ejection of the placenta

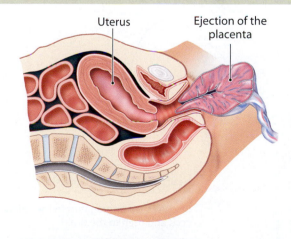

During the **placental stage** of labor, muscle tension builds in the walls of the partially empty uterus, which gradually decreases in size. This uterine contraction tears the connections between the endometrium and the placenta, and the placenta, or **afterbirth**, is ejected. The disruption of the placenta is accompanied by a loss of blood, but associated uterine contractions compress the uterine vessels and usually restrict this flow.

Premature labor occurs when true labor begins before the fetus has completed normal development. The newborn's chances of surviving are directly related to its body weight at delivery. Even with massive supportive efforts, newborns weighing less than 400 g (14 oz) at birth will not survive, primarily because their respiratory, cardiovascular, and urinary systems are unable to support life without aid from maternal systems. Most fetuses born at 25–27 weeks of gestation (a birth weight under 600 g or 21.1 oz) die despite intensive neonatal care, and survivors have a high risk of developmental abnormalities. **Premature delivery** usually refers to birth at 28–36 weeks (a birth weight over 1 kg or 2.2 lb). With care, these newborns have a good chance of surviving and developing normally.

Module 26.9 Review

a. List and describe the factors involved in initiating labor contractions.

b. What chemicals are primarily responsible for initiating contractions of true labor?

c. Name the three stages of labor, and describe the events that characterize each stage.

After delivery, development initially requires nourishment by maternal systems

Developmental processes do not cease at delivery, because newborns have few of the anatomical, functional, or physiological characteristics of mature adults. During an initial **neonatal period**, the newborn is dependent on the mother for nourishment, as well as for transportation and protection from environmental hazards such as extreme changes in temperature.

1 Milk is provided to infants through the **milk let-down reflex**, which is diagrammed here. By the end of the sixth month of pregnancy, the mammary glands are fully developed, and the gland cells begin to produce a secretion known as **colostrum** (kō-LOS-trum). Colostrum contains antibodies that may help the infant ward off infections until the infant's own immune system becomes fully functional. After the first few days of nursing, the mammary gland produces breast milk, which has a much higher fat content but still contains antibodies and also contains **lysozyme**, an enzyme with antibiotic properties.

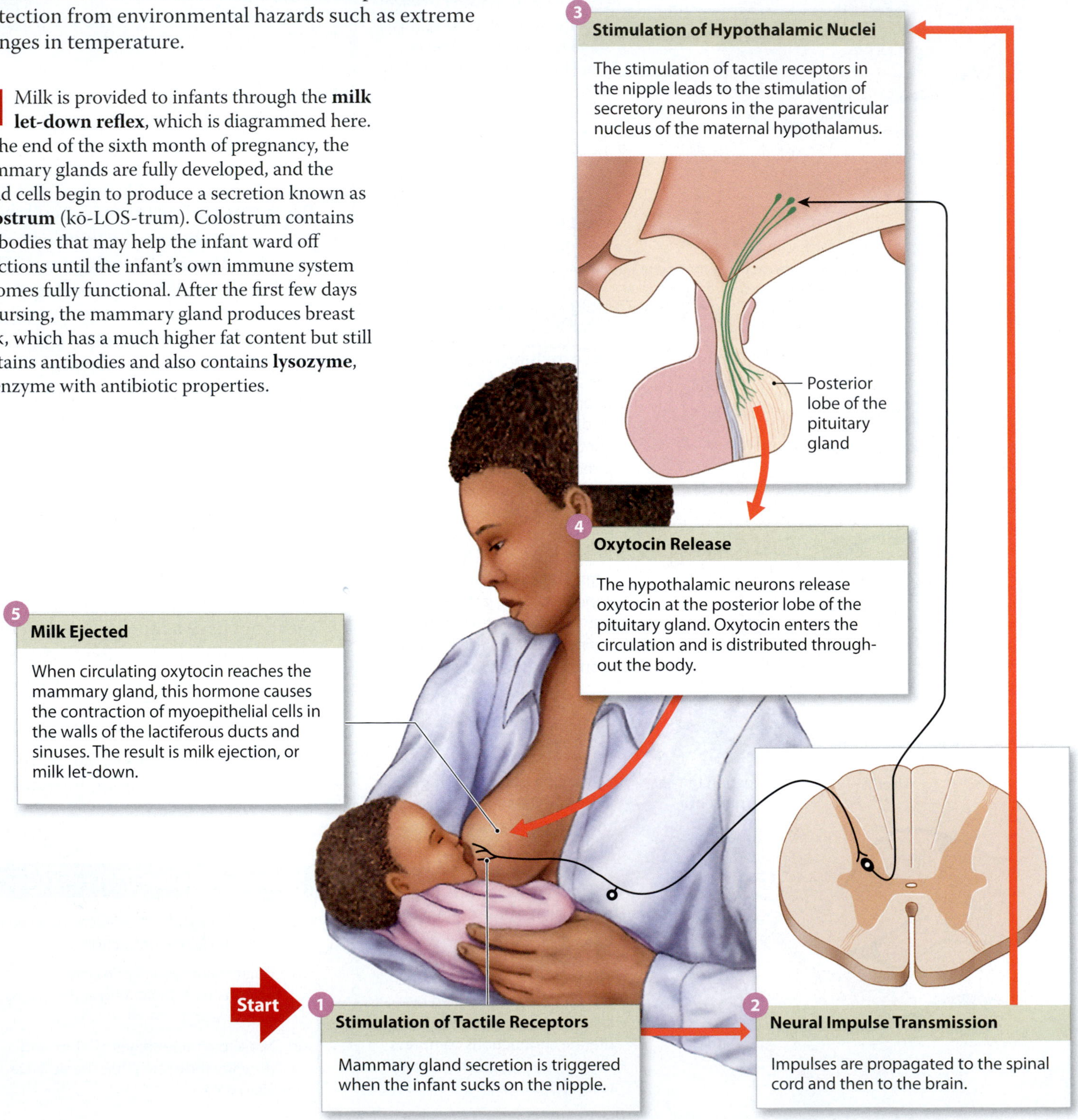

3 Stimulation of Hypothalamic Nuclei

The stimulation of tactile receptors in the nipple leads to the stimulation of secretory neurons in the paraventricular nucleus of the maternal hypothalamus.

Posterior lobe of the pituitary gland

4 Oxytocin Release

The hypothalamic neurons release oxytocin at the posterior lobe of the pituitary gland. Oxytocin enters the circulation and is distributed throughout the body.

5 Milk Ejected

When circulating oxytocin reaches the mammary gland, this hormone causes the contraction of myoepithelial cells in the walls of the lactiferous ducts and sinuses. The result is milk ejection, or milk let-down.

Start

1 Stimulation of Tactile Receptors

Mammary gland secretion is triggered when the infant sucks on the nipple.

2 Neural Impulse Transmission

Impulses are propagated to the spinal cord and then to the brain.

2 Postnatal development includes five **life stages**: (1) the neonatal period, (2) infancy, (3) childhood, (4) adolescence, and (5) maturity. Once maturity has been reached, the individual is subject to the gradual changes that accompany senescence, or aging. Growth during infancy and childhood occurs under the direction of circulating hormones, notably growth hormone, adrenal steroids, and thyroid hormones. These hormones affect each tissue and organ in specific ways, depending on the sensitivities of the individual cells.

Postnatal Development

| Neonatal | Infancy | Childhood | Adolescence | Maturity |

Through the neonatal period and infancy, the newborn is dependent on nutrients contained in milk, typically breast milk secreted by the maternal mammary glands.

In early childhood, the child is weaned from breast milk. Because growth does not occur uniformly, body proportions gradually change. The head, for example, is relatively large at birth but decreases in proportion with the rest of the body as the child grows to adulthood.

Adolescence begins at **puberty**, the period of sexual maturation, and ends when growth is completed.

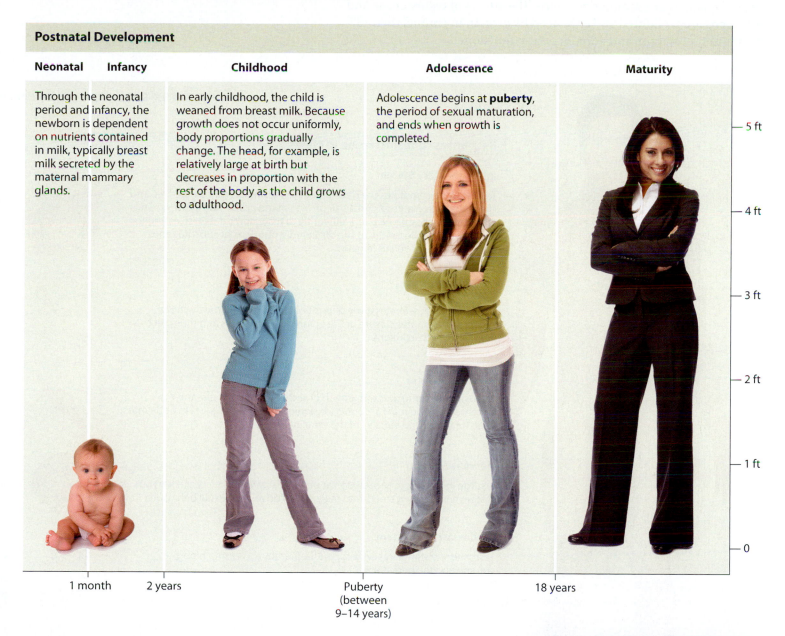

1 month   2 years   Puberty
                             (between
                             9–14 years)   18 years

— 5 ft
— 4 ft
— 3 ft
— 2 ft
— 1 ft
— 0

Module 26.10 Review

a. What hormone causes the milk let-down reflex?

b. Explain the difference between colostrum and breast milk.

c. Name the stages of postnatal development, and describe the time frame involved for each of the stages.

At puberty, male and female sex hormones have differential effects on most body systems

Many body systems alter their activities in response to changes in circulating levels of sex hormones at puberty. The most important sex-related changes are summarized here. The effects of testosterone and estrogens are facilitated and enhanced by growth hormone, thyroid hormones, prolactin, and adrenocortical hormones.

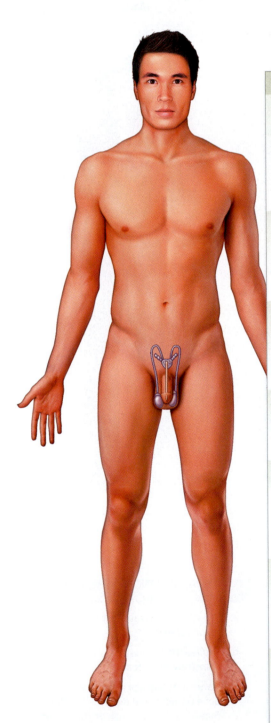

Responses to Testosterone in Males

Integumentary System

Testosterone stimulates the development of terminal hairs on the face and chest, and stimulates terminal hair growth in the axillae and in the genital area. Adipose tissues respond differently to testosterone than to estrogens, and this difference produces the distinct distributions of subcutaneous body fat in males versus females.

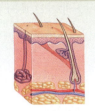

Skeletal System

Testosterone accelerates bone deposition and skeletal growth. In the process, it promotes closure of the epiphyseal cartilages and thus places a limit on growth in height.

Muscular System

Testosterone stimulates the growth of skeletal muscle fibers, and the increased muscle mass accounts for significant sex differences in body mass, even for males and females of the same height.

Nervous System

A surge in testosterone secretion at puberty activates the central nervous system centers concerned with male sexual drive and sexual behaviors.

Cardiovascular System

Testosterone stimulates erythropoiesis, thereby increasing blood volume and the hematocrit.

Respiratory System

Testosterone stimulates disproportionate growth of the larynx and a thickening and lengthening of the vocal cords. These changes cause a gradual deepening of the voice in males.

Reproductive System

Testosterone stimulates the functional development of the accessory reproductive glands, such as the prostate gland and seminal glands, and helps promote spermatogenesis.

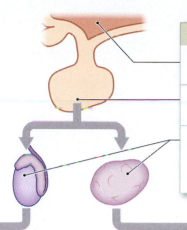

At Puberty

The hypothalamus increases its production of gonadotropin-releasing hormone (GnRH).

Endocrine cells in the anterior lobe of the pituitary gland become more sensitive to the presence of GnRH, and circulating levels of FSH and LH rise rapidly.

Testicular or ovarian cells become more sensitive to FSH and LH, initiating (1) gamete production; (2) the secretion of sex hormones, which stimulate the appearance of secondary sex characteristics and behaviors; and (3) a sudden acceleration in the growth rate, culminating in closure of the epiphyseal cartilages.

Responses to Estrogens in Females

Integumentary System

Estrogens stimulate the hair follicles to continue to produce fine vellus hairs and stimulate terminal hair growth in the axillae and in the genital area. The combination of estrogens, prolactin, growth hormone, and thyroid hormones promotes the initial development of the mammary glands.

Skeletal System

Estrogens cause more rapid epiphyseal closure than does testosterone. In addition, the period of skeletal growth is briefer in females than in males, and so females generally do not grow as tall as males.

Muscular System

Estrogens stimulate the growth of skeletal muscle fibers, increasing strength and endurance, but not to the extent that testosterone does in males.

Nervous System

A surge in estrogen secretion at puberty activates central nervous system centers involved in female sexual drive and sexual behaviors.

Cardiovascular System

The iron loss associated with menses increases the risk of developing iron-deficiency anemia. Estrogens decrease plasma cholesterol levels and slow the formation of plaque within arteries. As a result, premenopausal women have a lower risk of atherosclerosis than do adult men.

Respiratory System

Estrogens do not cause excessive growth of the larynx and vocal cords, so females typically have higher-pitched voices than males.

Reproductive System

Estrogens target the uterus, promoting a thickening of the myometrium and increasing blood flow to the endometrium. Estrogens also promote the functional development of accessory reproductive structures in females.

Module 26.11 Review

a. Name the three major interacting hormonal events associated with the onset of puberty.

b. Why do males have deeper voices and larger larynxes than females?

c. Why are premenopausal women at lesser risk of atherosclerosis than men?

1. Matching

Match the following terms with the most closely related description.

- hCG
- conception
- chorion
- syncytial trophoblast
- colostrum
- amnion
- amphimixis
- morula
- embryonic disc
- neonate
- gestation
- inner cell mass
- relaxin
- blastocyst

a _____ Fertilization

b _____ Newborn infant

c _____ Pronuclei fuse

d _____ Period of prenatal development

e _____ Pregnancy test

f _____ Softens pubic symphysis

g _____ Mammary gland secretion

h _____ Mesoderm and ectoderm

i _____ Hollow ball of cells

j _____ Mesoderm and trophoblast

k _____ Forms the embryo

l _____ Cytoplasm with many nuclei

m _____ Solid ball of cells

n _____ Gastrulation product

2. Multiple choice

Choose the bulleted item that best completes each statement.

a Fertilization typically occurs in the _____ .
- lower part of the uterine tube
- upper part of the uterus
- junction between the ampulla and isthmus of the uterine tube
- cervix

b Fetal development begins at the start of the _____ .
- implantation process
- second month after fertilization
- ninth week after fertilization
- sixth month after fertilization

c Organs and organ systems complete most of their development by the end of the _____ .
- first trimester
- second trimester
- third trimester
- expulsion stage

d The four general processes that occur during the first trimester include _____ .
- blastomere, blastocyst, morula, and trophoblast
- cleavage, implantation, placentation, and embryogenesis
- placentation, dilation, expulsion, and organogenesis
- yolk sac, amnion, allantois, and chorion

e The most dangerous period in prenatal or neonatal life is the _____ .
- first trimester
- second trimester
- third trimester
- expulsion stage

f The systems that were relatively nonfunctional during the fetal period that must become functional at birth are the _____ systems.
- cardiovascular, muscular, and skeletal
- integumentary, reproductive, and nervous
- respiratory, digestive, and urinary
- endocrine, nervous, and digestive

3. Section integration

Tina gives birth to a baby with a congenital deformity of the stomach. Tina believes that her baby's affliction is the result of a viral infection that she suffered during her third trimester. Is this a possibility? Explain. _____

_____

_____

Genetics and Inheritance

Although everyone goes through the same developmental stages, differences in both genetic structure and local environments produce distinctive individual characteristics. The term **inheritance** refers to the transfer of genetically determined characteristics from generation to generation. The study of the mechanisms responsible for inheritance is called **genetics**. We begin our study of genetics by examining two important concepts: *genotype* and *phenotype*.

1 One way to understand genotype and phenotype is to compare them to the architecture of a house: to the house's plans, and to how the finished house looks.

Every nucleated somatic cell in your body carries copies of the original 46 chromosomes present when you were a zygote. Those chromosomes and their component genes constitute your **genotype** (JĒN-ō-tīp; *geno*, gene). In architectural terms, the genotype is a set of plans, like the blueprints for a house.

Through development and differentiation, the instructions contained in the genotype are expressed in many ways; specific genes may be activated or inactivated by interactions with the local environment. Collectively, the pattern of genetic expression within your genotype determines the anatomical and physiological characteristics that make you a unique individual. Those anatomical and physiological characteristics constitute your **phenotype** (FĒ-nō-tīp; *phaino*, to display). In architectual terms, the phenotype is the detailed structure of the completed building.

2 This photograph shows the **karyotype** (*karyon*, nucleus + *typos*, mark), or entire set of chromosomes, of a normal male. At amphimixis, one member of each of the 23 chromosome pairs was contributed by the spermatozoon, and the other by the ovum. The two members of each pair are known as **homologous** (huh-MOL-ō-gus) **chromosomes**. Twenty-two of those pairs are called **autosomal** (aw-tō-SŌ-mul) **chromosomes**. Most of the genes of the autosomal chromosomes affect somatic characteristics, such as hair color and skin pigmentation. The chromosomes of the 23rd pair are called the **sex chromosomes**; one of their functions is to determine whether the individual is genetically male or female. The sex chromosomes of a male consist of an **X chromosome** and a shorter **Y chromosome**, whereas females have two X chromosomes.

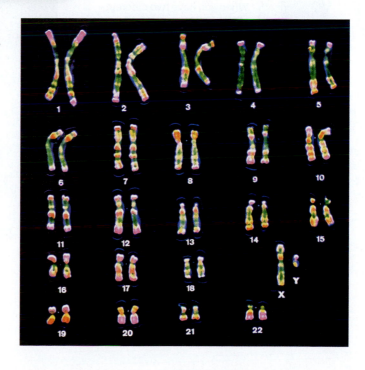

In this section we consider basic genetics as it applies to inherited characteristics, such as sex, hair color, and various clinical conditions.

Genes and chromosomes determine patterns of inheritance

1 The two chromosomes in a homologous autosomal pair have the same structure and carry genes that affect the same traits. The genes are also located at equivalent positions on their respective chromosomes. A gene's position on a chromosome is called a **locus** (LŌ-kus; plural, *loci*). The two chromosomes in a pair may carry the same form or different forms of each gene. The various forms of a given gene are called **alleles** (uh-LĒLZ).

2 The phenotype that results from a heterozygous genotype depends on the nature of the interaction between the corresponding alleles. The most common form of interaction among autosomal genes is called **simple inheritance**. In one kind of simple inheritance—**strict dominance**—any dominant allele will be expressed in the phenotype, regardless of any conflicting instructions carried by the other allele. For instance, a person with only one allele for freckles will have freckles, because that allele is dominant over the "nonfreckle" allele. An allele that is **recessive** will be expressed in the phenotype only if that same allele is present on both chromosomes of a homologous pair.

If the two chromosomes of a homologous pair carry the same allele of a particular gene, you are **homozygous** (hō-mō-ZĪ-gus; *homos*, the same) for the trait affected by that gene. That allele will then be expressed in your phenotype. For example, if you receive a gene for curly hair from your father and a gene for curly hair from your mother, you will be homozygous for curly hair—and you will have curly hair.

When you have two different alleles for the same gene, you are **heterozygous** (het-er-ō-ZĪ-gus; *heteros*, other) for the trait determined by that gene.

The human genome contains roughly 23,000 genes. Thus an "average" autosomal pair contains about 1000 pairs of alleles.

The phenotype "red hair" is determined by a single pair of recessive alleles.

The phenotype "freckles" is an example of strict dominance and requires only one allele for the trait to be expressed.

3 By convention, dominant alleles are indicated by capital letters, and recessive alleles by lowercase letters. Thus for a given trait *T*, the possible genotypes are *TT* (homozygous dominant), *Tt* (heterozygous), and *tt* (homozygous recessive). The **Punnett squares** shown in this figure indicate the possible offspring of an albino mother and a father with normal skin pigmentation. Because albinism is a recessive trait, the maternal alleles are designated *aa*. The father has normal pigmentation, a dominant trait, so his genotype could be either *AA* or *Aa*.

4 Many phenotypic characteristics are determined by interactions among several genes. Such interactions constitute **polygenic inheritance**. Because the resulting phenotype depends not only on the nature of the alleles but how those alleles interact, you cannot predict the presence or absence of phenotypic characters using a simple Punnett square. An example of polygenic inheritance is brown or black hair color.

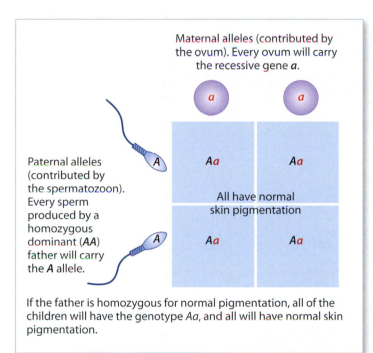

Maternal alleles (contributed by the ovum). Every ovum will carry the recessive gene *a*.

Paternal alleles (contributed by the spermatozoon). Every sperm produced by a homozygous dominant (*AA*) father will carry the *A* allele.

	a	*a*
A	*Aa*	*Aa*
A	*Aa*	*Aa*

All have normal skin pigmentation

If the father is homozygous for normal pigmentation, all of the children will have the genotype *Aa*, and all will have normal skin pigmentation.

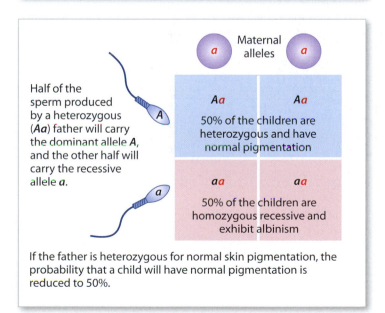

Maternal alleles

Half of the sperm produced by a heterozygous (*Aa*) father will carry the dominant allele *A*, and the other half will carry the recessive allele *a*.

	a	*a*
A	*Aa*	*Aa*
a	*aa*	*aa*

50% of the children are heterozygous and have normal pigmentation

50% of the children are homozygous recessive and exhibit albinism

If the father is heterozygous for normal skin pigmentation, the probability that a child will have normal pigmentation is reduced to 50%.

Module 26.12 Review

a. Describe homozygous and heterozygous.

b. Differentiate between simple inheritance and polygenic inheritance.

c. The trait "curly hair" operates through strict dominance. What would be the phenotype of a person who is heterozygous for this trait?

There are several different patterns of inheritance

1 Inheritance is of two main types: that involving autosomal chromosomes, and that involving the sex chromosomes. This diagram summarizes the major patterns of inheritance.

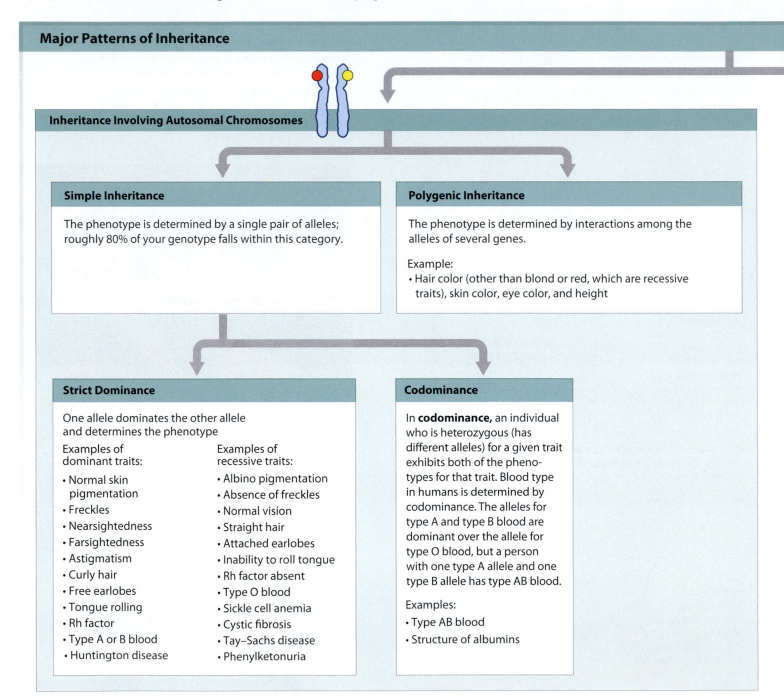

Major Patterns of Inheritance

Inheritance Involving Autosomal Chromosomes

Simple Inheritance

The phenotype is determined by a single pair of alleles; roughly 80% of your genotype falls within this category.

Polygenic Inheritance

The phenotype is determined by interactions among the alleles of several genes.

Example:
• Hair color (other than blond or red, which are recessive traits), skin color, eye color, and height

Strict Dominance

One allele dominates the other allele and determines the phenotype

Examples of dominant traits:

• Normal skin pigmentation
• Freckles
• Nearsightedness
• Farsightedness
• Astigmatism
• Curly hair
• Free earlobes
• Tongue rolling
• Rh factor
• Type A or B blood
• Huntington disease

Examples of recessive traits:

• Albino pigmentation
• Absence of freckles
• Normal vision
• Straight hair
• Attached earlobes
• Inability to roll tongue
• Rh factor absent
• Type O blood
• Sickle cell anemia
• Cystic fibrosis
• Tay–Sachs disease
• Phenylketonuria

Codominance

In **codominance,** an individual who is heterozygous (has different alleles) for a given trait exhibits both of the phenotypes for that trait. Blood type in humans is determined by codominance. The alleles for type A and type B blood are dominant over the allele for type O blood, but a person with one type A allele and one type B allele has type AB blood.

Examples:
• Type AB blood
• Structure of albumins

Inheritance Involving Sex Chromosomes

X-linked inheritance:
The allele on the X chromosome determines the phenotype in the absence of a corresponding allele on the Y chromosome. Most known cases involve alleles that are recessive in females.

Examples:
• Red–green color blindness
• Hemophilia (some forms)
• Duchenne muscular dystrophy

2 The X chromosome is much larger than the Y chromosome, and it carries genes that affect a variety of somatic structures. These characteristics are called **X-linked** (or sex linked), because in most cases there are no corresponding alleles on the Y chromosome. The best known single-allele characteristics are associated with identifiable diseases or functional deficits.

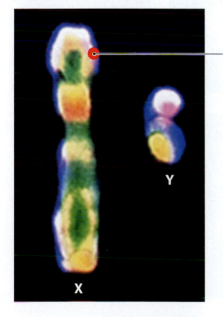

X-linked allele (allele not present on Y chromosome)

3 The inheritance of color blindness exemplifies the differences between sex-linked inheritance and autosomal inheritance. In males, the presence of a dominant allele, *C*, on the X chromosome (X^C) results in normal color vision; a recessive allele, *c*, on the X chromosome (X^c) results in red–green color blindness. This Punnett square reveals that each son of a father with normal vision and a heterozygous (carrier) mother has a 50 percent chance of being red–green color blind, whereas any daughters will have normal color vision.

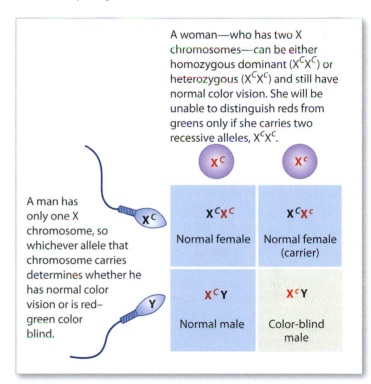

A woman—who has two X chromosomes—can be either homozygous dominant ($X^C X^C$) or heterozygous ($X^C X^c$) and still have normal color vision. She will be unable to distinguish reds from greens only if she carries two recessive alleles, $X^c X^c$.

A man has only one X chromosome, so whichever allele that chromosome carries determines whether he has normal color vision or is red–green color blind.

	X^C	X^c
X^C	$X^C X^C$ Normal female	$X^C X^c$ Normal female (carrier)
Y	$X^C Y$ Normal male	$X^c Y$ Color-blind male

Module 26.13 Review

a. Compare strict dominance with codominance.

b. Why are X-linked traits expressed more frequently in males than females?

c. Indicate the type of inheritance involved in each of the following situations: (1) children who exhibit the trait have at least one parent who also exhibits it; (2) children exhibit the trait even though neither parent exhibits it; and (3) the trait is expressed equally in daughters and sons.

Thousands of clinical disorders have been linked to abnormal chromosomes and/or genes

Chromosomal abnormalities can involve thousands of genes, and as a result they are usually lethal. Most of the time the embryo or fetus dies before delivery. However, there are a few autosomal chromosome abnormalities that do not invariably result in prenatal mortality. In contrast, variations in the structure of individual genes are relatively common. Although more than 99 percent of human nucleotide bases are the same in all people, there are about 1.4 million single-base differences, or **single nucleotide polymorphisms (SNPs)**. Some of these SNPs are inconsequential, but others are associated with specific diseases.

1 **Trisomy 21**, or Down syndrome, is the most common viable chromosomal abnormality. Affected individuals exhibit mental retardation and physical malformations, including a characteristic facial appearance. The degree of mental retardation ranges from moderate to severe, and anatomical problems affecting the cardiovascular system often prove fatal during childhood or early adulthood. Although some individuals survive to moderate old age, many develop Alzheimer disease while still relatively young (before age 40). For unknown reasons, there is a direct correlation between maternal age and the risk of having a child with trisomy 21. For a maternal age below 25, the incidence of Down syndrome approaches 1 in 2000 births, or 0.05 percent. For maternal ages 30–34, the odds increase to 1 in 900, and during ages 35–44 they they go from 1 in 290 to 1 in 46, or more than 2 percent.

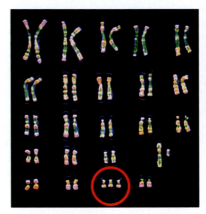

2 In **Klinefelter syndrome**, the individual carries the sex chromosome pattern XXY. The phenotype is male, but the extra X chromosome causes reduced androgen production. As a result, the testes fail to mature so the individuals are sterile, and the breasts are slightly enlarged. The incidence of this condition among newborn males averages 1 in 750 births.

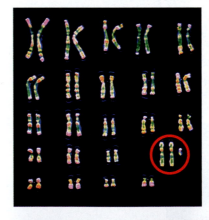

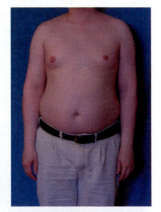

3 Individuals with **Turner syndrome** have only a single, female sex chromosome; their sex chromosome complement is designated XO. This kind of chromosomal deletion is known as **monosomy**. The incidence of this condition at delivery has been estimated as 1 in 10,000 live births. The condition may not be recognized at birth, because the phenotype is normal female. But maturational changes do not appear at puberty. The ovaries are nonfunctional, and estrogen production occurs at negligible levels.

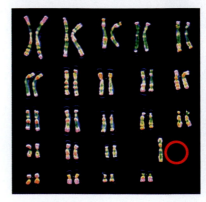

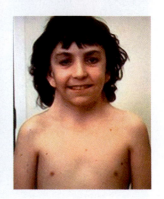

4 The **human genome**—the full set of genetic material (DNA) in our chromosomes, listed nucleotide by nucleotide—contains an estimated 20,000–25,000 genes. The sequence of nucleotides along all of the chromosomes has been determined. Roughly 10,000 different single gene disorders have been described. Most are very rare, but collectively they may affect 1 in every 200 births. Over 900 of these disorders have been mapped on the genome, and several examples are included in this chromosome map. Many of these conditions were noted in earlier chapters.

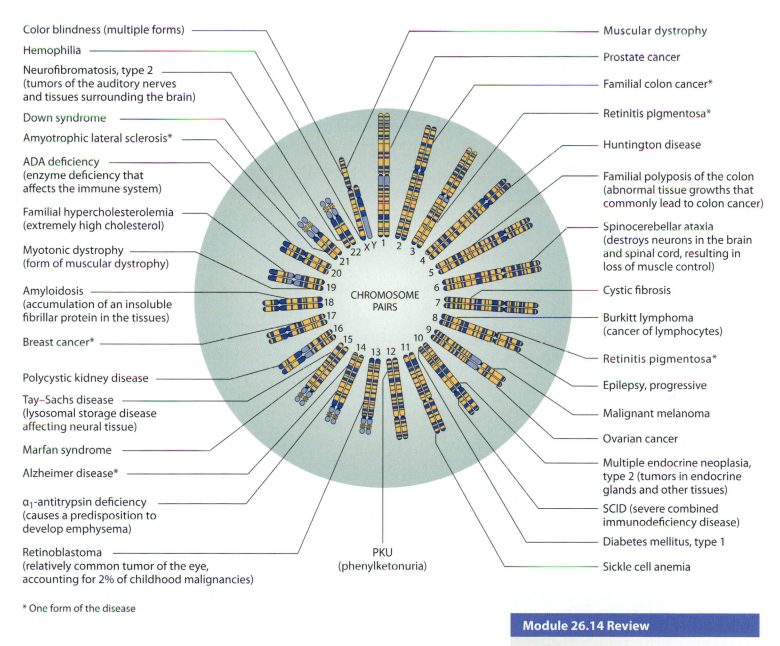

Color blindness (multiple forms)

Hemophilia

Neurofibromatosis, type 2 (tumors of the auditory nerves and tissues surrounding the brain)

Down syndrome

Amyotrophic lateral sclerosis*

ADA deficiency (enzyme deficiency that affects the immune system)

Familial hypercholesterolemia (extremely high cholesterol)

Myotonic dystrophy (form of muscular dystrophy)

Amyloidosis (accumulation of an insoluble fibrillar protein in the tissues)

Breast cancer*

Polycystic kidney disease

Tay–Sachs disease (lysosomal storage disease affecting neural tissue)

Marfan syndrome

Alzheimer disease*

α₁-antitrypsin deficiency (causes a predisposition to develop emphysema)

Retinoblastoma (relatively common tumor of the eye, accounting for 2% of childhood malignancies)

CHROMOSOME PAIRS

PKU (phenylketonuria)

Muscular dystrophy

Prostate cancer

Familial colon cancer*

Retinitis pigmentosa*

Huntington disease

Familial polyposis of the colon (abnormal tissue growths that commonly lead to colon cancer)

Spinocerebellar ataxia (destroys neurons in the brain and spinal cord, resulting in loss of muscle control)

Cystic fibrosis

Burkitt lymphoma (cancer of lymphocytes)

Retinitis pigmentosa*

Epilepsy, progressive

Malignant melanoma

Ovarian cancer

Multiple endocrine neoplasia, type 2 (tumors in endocrine glands and other tissues)

SCID (severe combined immunodeficiency disease)

Diabetes mellitus, type 1

Sickle cell anemia

* One form of the disease

Module 26.14 Review

a. Define single nucleotide polymorphism.

b. Name the disorder characterized by each of the following chromosome patterns: (1) XO and (2) XXY.

c. Identify the chromosome involved in each of the following disorders: (1) ovarian cancer, (2) Tay–Sachs disease, and (3) spinocerebellar ataxia.

1. Matching

Match the following terms with the most closely related description.

- genotype
- heterozygous
- locus
- autosomes
- simple inheritance
- homozygous
- alleles
- polygenic inheritance
- homologous
- genetics
- karyotype
- phenotype

a _____ Visible characteristics

b _____ Alternate forms of a gene

c _____ Refers to the two members of a pair of chromosomes

d _____ Array of the entire set of chromosomes in a cell

e _____ An individual's chromosomes and genes

f _____ Gene's position on a chromosome

g _____ Two different alleles for the same gene

h _____ Study of the mechanisms of inheritance

i _____ Two identical alleles for the same gene

j _____ Interactions between alleles on several genes

k _____ Phenotype determined by a single pair of alleles

l _____ Chromosomes affecting somatic characteristics

2. Section integration

Use the Punnett squares below to answer the questions concerning the following genetic conditions.

a Tongue rolling is inherited as a dominant trait (T). Explain how it is possible for two parents who are tongue rollers to have children who do not have the ability to roll the tongue.

b Achondroplasia dwarfism is an autosomal genetic disorder that results from problems with the replacement of cartilage by bone in the arms and legs. Assume that two dwarf parents have a normal-sized child. Determine the genotype of the parents and predict the probability that a second child would also be normal sized.

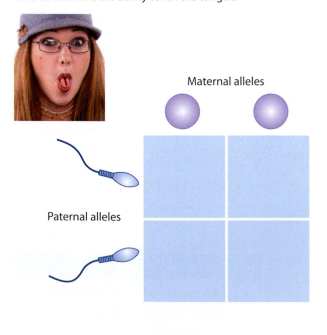

Maternal alleles

Paternal alleles

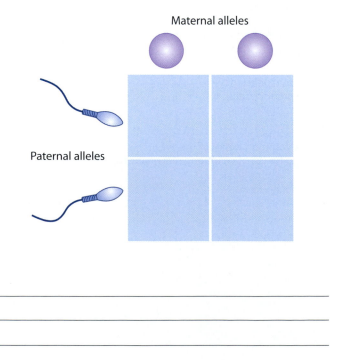

Maternal alleles

Paternal alleles

_____

_____

_____

_____

_____

_____

Visual Outline with Key Terms

Summarize the content of each module using the terms in the order provided.

SECTION 1

An Overview of Development

- development
- embryological development
- embryology
- fetal development
- prenatal development
- gestation
- first trimester
- second trimester
- third trimester
- postnatal development
- maturity

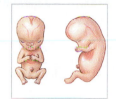

26.1

At fertilization, a secondary oocyte and a spermatozoon form a zygote that prepares for cell division

- fertilization
- zygote
- oocyte activation
- female pronucleus
- male pronucleus
- cleavage
- amphimixis
- blastomeres

26.2

Cleavage continues until the blastocyst implants in the uterine wall

- pre-embryo
- morula
- blastocyst
- blastocoele
- implantation
- trophoblast
- inner cell mass
- cellular trophoblast
- syncytial trophoblast
- lacunae
- villi
- amniotic cavity

26.3

Gastrulation produces three germ layers: ectoderm, endoderm, and mesoderm

- blastodisc
- amnion
- extra-embryonic membranes
- yolk sac
- primitive streak
- ectoderm
- endoderm
- mesoderm
- germ layers
- gastrulation
- embryonic disc
- gestational trophoblastic neoplasia

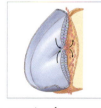

26.4

The extra-embryonic membranes form the placenta that supports fetal growth and development

- yolk sac
- amnion
- amniotic fluid
- allantois
- chorion
- placenta

26.5

The formation of extra-embryonic membranes is associated with major changes in the shape and complexity of the embryo

- head fold
- chorionic villi
- tail fold
- body stalk
- yolk stalk
- umbilical stalk
- umbilical cord

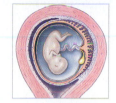

26.6

The placenta performs many vital functions for the duration of prenatal development

- umbilical arteries
- umbilical vein
- human chorionic gonadotropin (hCG)
- human placental lactogen (hPL)
- relaxin
- progesterone
- estrogens

26.7

Organ systems are established in the first trimester and become functional in the second and third trimesters

- organogenesis
- neural plate

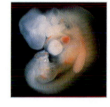

26.8

Pregnancy places anatomical and physiological stresses on maternal systems

- pregnancy
- anatomical changes
- physiological stresses

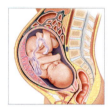

26.9

Multiple factors initiate and accelerate the process of labor

- false labor
- true labor
- positive feedback
- parturition
- dilation stage
- expulsion stage
- delivery
- placental stage
- afterbirth
- premature labor
- premature delivery

• = *Term boldfaced in this module*

26.10

After delivery, development initially requires nourishment by maternal systems

- neonatal period
- milk let-down reflex
- colostrum
- lysozyme
- ○ oxytocin
- life stages
- puberty

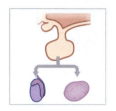

26.11

At puberty, male and female sex hormones have differential effects on most body systems

- ○ responses to testosterone in males
- ○ responses to estrogens in females

SECTION 2

Genetics and Inheritance

- inheritance
- genetics
- genotype
- phenotype
- karyotype
- homologous chromosomes
- autosomal chromosomes
- sex chromosomes
- X chromosome
- Y chromosome

26.12

Genes and chromosomes determine patterns of inheritance

- locus
- alleles
- homozygous
- heterozygous
- simple inheritance
- strict dominance
- recessive
- Punnett squares
- polygenic inheritance

26.13

There are several different patterns of inheritance

- codominance
- X-linked
- ○ carrier

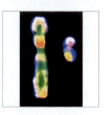

26.14

Thousands of clinical disorders have been linked to abnormal chromosomes and/or genes

- single nucleotide polymorphisms (SNPs)
- trisomy 21
- Klinefelter syndrome
- Turner syndrome
- monosomy
- human genome

• = *Term boldfaced in this module*

Chapter Integration: Applying what you've learned

Joe and Jane desperately want to have children, and although they have tried for two years, they have not been successful. The couple finally consults with a physician and undergoes a battery of tests, including complete physical exams and typical genetics testing. They discover that Joe's genotype is a normal XY and that he suffers from oligospermia (a low sperm count), and that Jane is heterozygous (XX^h) for hemophilia A, a recessive trait in which the blood does not clot properly. The doctor informs Joe and Jane that Joe's low sperm count is interfering with their ability to have children, and that there is a risk that any child who is born will have hemophilia A.

Oligospermia is one cause of male infertility. For most couples, the only sign the male has a low sperm count is the inability to conceive a child after one year of regular intercourse. In most cases, the cause is unknown, but other possible reasons include damaged sperm ducts, infections, anti-sperm antibodies, or an inherited disorder such as Klinefelter syndrome, which results in a complete lack of sperm in the semen (azoospermia).

Hemophilia is a lifelong disease, but many individuals with hemophilia maintain active, productive lives. Signs and symptoms vary depending on the severity of the condition and include unexplained bruising, nosebleeds with unknown cause, and painful headaches. Hemophilia A is the most common type of hemophilia and is caused by insufficient amounts of clotting factor VIII.

1. If it only takes one sperm to fertilize a secondary oocyte, why is Joe's oligospermia problematic for achieving pregnancy?

2. Regarding hemophilia A, what is the probability that the couple will have daughters with hemophilia? What is the probability that the couple will have sons with hemophilia?

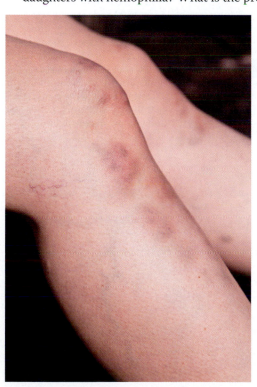

Access more review material online in the Study Area at **www.masteringaandp.com.**

There, you'll find:
- **Chapter guides**
- **Chapter quizzes**
- **Practice tests**
- **Labeling activities**

- **Animations**
- **MP3 Tutor Sessions**
- **Flashcards**
- **A glossary with pronunciations**

Module 1.1 Review

a. Biology is the study of life.

b. The basic functions shared by all living things are responsiveness, adaptability, growth, reproduction, and movement.

c. Most animals have an internal circulation system to transport the inputs and/or products of digestion, respiration, and excretion throughout the body or to excretory sites.

Module 1.2 Review

a. Anatomy is the study of internal and external body structures; physiology is the study of how living organisms perform their vital functions.

b. Gross anatomy (also called macroscopic anatomy) is the study of body structures that can be seen with the unaided eye; microscopic anatomy is the study of body structures that cannot be seen without magnification.

c. The structures of body parts are closely related to their functions—that is, function follows form.

Module 1.3 Review

a. All specific functions are performed by specific structures, and the link between the two is always present, but not always understood.

b. The elbow joint moves in a single plane, like the opening and closing of a door on a hinge.

c. If a structure's anatomy were altered, the structure's function would likely be impaired or perhaps eliminated.

Section 1 Review

1. a. respiration; b. growth and reproduction; c. adaptability; d. circulation; e. excretion; f. digestion; g. movement; h. responsiveness

2. Anatomy: right atrium, myocardium, left ventricle, endocardium, superior vena cava; Physiology: valve to aorta opens, valve between left atrium and left ventricle closes, pressure in left atrium, electrocardiogram

3. Both keys and messenger molecules have specific three-dimensional shapes, so both can "fit" into and function with their complementary structures (a lock or a receptor protein) only if the shapes match closely enough. Because messenger molecules need not be flat in one dimension (as keys tend to be), they can be very complex in shape, and thus may be able to bind with a variety of complementary receptor proteins. Moreover, though a lock has moving parts, it does not actually change shape, whereas a receptor protein bound by a messenger can change shape, which in turn affects its function.

4. Larger organisms simply cannot absorb the amount of needed materials or excrete wastes rapidly enough across their body surfaces as very small organisms can. Their food must be processed, or digested, before being absorbed, and because the processes of absorption, respiration, and excretion occur in different parts of the organism, an efficient, internal circulation network for their materials is necessary for survival.

Module 1.4 Review

a. The cell theory holds that (1) cells are the structural building blocks of all plants and animals, (2) new cells are produced through the division of pre-existing cells, and (3) cells are the smallest structural units that perform all the vital functions of life.

b. The unit used to measure cell size is the micrometer, which is equal to one-millionth of a meter.

c. A fat cell's large volume relative to its surface area makes it ideal for storing fat, whereas the extensive branching of a neuron permits communication with a large number of other cells.

Module 1.5 Review

a. Histology is the study of tissues.

b. The body's four primary tissue types that form all body structures are epithelial, connective, muscle, and neural tissue.

c. Epithelial tissue covers external and internal surfaces and produces secretions; connective tissue fills internal spaces, provides support, and stores energy; muscle tissue is specialized to contract and produce movement; and neural tissue transmits information.

Module 1.6 Review

a. The body's 11 organ systems are the integumentary, skeletal, muscular, nervous, endocrine, cardiovascular, lymphatic, respiratory, digestive, urinary, and reproductive systems.

b. The digestive system provides nutrients and minerals for the skeletal system, which in turn protects soft tissues and organs of the digestive system.

c. Falling down a flight of stairs could cause a compound fracture (a broken bone that protrudes through the skin), which could affect (1) the skeletal system (a broken bone), (2) the integumentary system (disruption of skin integrity), (3) the muscular system (broken bone tearing through a muscle), (4) the cardiovascular system (blood loss at the site of injury), (5) the lymphatic system (mobilization of specialized cells to defend against infection), and (6) the nervous system (pain and nerve injury as a result of the trauma).

Section 2 Review

1. cardiovascular system: blood cells; digestive system: smooth muscle cell; reproductive system: sperm cell or oocyte; skeletal system: bone cell (osteocyte); nervous system: nerve cell (neuron). (Other cells could be listed for other systems.)

2. a. cells; b. organs; c. organ systems; d. epithelial tissue; e. external and internal surfaces; f. glandular secretions; g. connective tissue; h. matrix; i. protein fibers; j. ground substance; k. muscle tissue; l. movement; m. bones of the skeleton; n. blood; o. materials within digestive tract; p. nervous tissue; q. neuroglia

3. tissue: c; cell: b; organ: d; molecule: a; organism: f; organ system: e
4. **a.** skeletal system: support, protection of soft tissues, mineral storage, blood formation; **b.** digestive system: processing of food and absorption of nutrients, minerals, vitamins, and water; **c.** integumentary system: protection from environmental hazards, temperature control; **d.** urinary system: elimination of excess water, salts, and waste products, and control of pH; **e.** nervous system: directing immediate responses to stimuli, usually by coordinating the activities of other organ systems

Module 1.7 Review

a. The components of homeostatic regulation include receptors, a control center, and effectors.
b. Negative feedback systems maintain homeostasis (and provide long-term control over the body's internal conditions and systems) by counteracting any stimulus that moves conditions outside their normal range.
c. Positive feedback is useful in processes such as blood clotting, which, once begun, must move quickly to completion. It is harmful in situations where stable conditions must be maintained, because it tends to exaggerate any departure from the desired condition. Thus positive feedback in the regula-

tion of body temperature would cause a slight fever to spiral out of control, with fatal results.

Section 3 Review

1. **a.** positive feedback; **b.** homeostatic regulation; **c.** homeostasis; **d.** negative feedback; **e.** negative feedback
2. **a.** negative feedback; **b.** positive feedback; **c.** negative feedback; **d.** positive feedback
3. **a.** blood flow to skin increases, sweating increases, body surface cools, temperature declines; **b.** blood flow to skin decreases, shivering occurs, body heat is conserved, temperature rises
4. One reason your body temperature may have dropped is that your body may be losing heat faster than it is being produced. (This is more likely to occur on a cool day.) Perhaps hormones have caused a decrease in your metabolic rate, so your body is not producing as much heat as it normally would. Or you may have an infection that has temporarily reset the set point of the body's "thermostat" to a value higher than normal. The last possibility is the most likely explanation given the circumstances.

Module 1.8 Review

a. A person in the anatomical position is standing erect, facing the observer, arms at the

sides with the palms facing forward, and the feet together.
b. Clinicians base their descriptions on four abdominopelvic quadrants (determined by the intersection of two imaginary perpendicular lines that cross at the navel, or umbilicus), whereas anatomists recognize nine abdominopelvic regions.
c. A person lying face down in the anatomical position is prone.

Module 1.9 Review

a. The purpose of directional and sectional terms is to provide a standardized frame of reference for describing the human body.
b. In the anatomical position, an anterior view shows the subject's face, whereas a posterior view shows the subject's back.
c. A midsagittal section would separate the two eyes.

Module 1.10 Review

a. Body cavities (1) protect internal organs and cushion them from shocks that occur during activity and (2) allow organs within them to change size and shape without disrupting the activities of nearby organs.
b. The ventral body cavity includes the thoracic cavity (which contains the pleural and pericardial cavities) and the abdominopelvic cavity (consisting of the peritoneal, abdominal, and pelvic cavities).

c. The body cavity inferior to the diaphragm is the abdominopelvic or peritoneal cavity.

Section 4 Review

1. **a.** superior; **b.** inferior; **c.** cranial; **d.** caudal; **e.** posterior or dorsal; **f.** anterior or ventral; **g.** lateral; **h.** medial; **i.** proximal; **j.** distal; **k.** proximal; **l.** distal
2. **a.** thoracic cavity; **b.** mediastinum; **c.** left lung; **d.** trachea, esophagus; **e.** heart; **f.** diaphragm; **g.** abdominopelvic cavity; **h.** peritoneal cavity; **i.** digestive glands and organs; **j.** pelvic cavity; **k.** reproductive organs

Chapter Integration

1. The bullet entered the anterior body approximately 2 centimeters inferior to the umbilicus, within the hypogastric (pelvic) abdominopelvic region.
2. Organs found within the abdominal cavity include the liver, stomach, spleen, small intestine, and most of the large intestine. The abdominal aorta and inferior vena cava are two large blood vessels situated posteriorly. If the bullet entered at this location, these structures may have been affected.
3. It is likely that a massive internal hemorrhage led to this man's death.

ANSWERS

Module 2.1 Review

a. An atom is the smallest stable unit of matter.

b. Trace elements, found in very small amounts in the body, are chemical elements required for normal growth and maintenance. Some function as cofactors, but the functions of many have yet to be fully understood.

c. Hydrogen has three isotopes: hydrogen-1, with a mass of 1; deuterium, with a mass of 2; and tritium, with a mass of 3. The heavier sample must contain a higher proportion of one or both of the heavier isotopes.

Module 2.2 Review

a. The maximum number of electrons that can occupy an atom's first three electron shells is 2, 8, and 8, respectively.

b. Atoms of inert elements are unreactive because their outermost electron shell contains the maximum number of electrons possible.

c. A cation is formed when an atom loses one or more electrons from its outermost electron shell; it has an overall positive charge because it contains more protons than electrons. An anion is formed when an atom gains one or more electrons in its outermost electron shell; it has an overall negative charge because it contains more electrons than protons.

Module 2.3 Review

a. The two most common types of chemical bonds are ionic bonds, which result from the attraction of oppositely charged atoms (ions), and covalent bonds, which result from the sharing of electrons.

b. The atoms in a water molecule are held together by polar covalent bonds, in which electrons are shared unequally.

c. The term *molecule* refers only to chemical structures held together by covalent bonds. Table salt is an ionic compound whose components—sodium ions and chloride ions—are held together by ionic bonds.

Module 2.4 Review

a. Solids have a fixed volume and shape, liquids have a constant volume but no fixed shape, and gases have neither a constant volume nor a fixed shape.

b. Water molecules are attracted to each another by hydrogen bonds.

c. The attraction between water molecules at the water's surface creates surface tension, which prevents small objects from penetrating into the water.

Section 1 Review

1. a. 2 b. 4 c. 1 d. 0 e. 6 f. 12
   g. 7 h. 7 i. 20 j. 20

2. a. element; b. compound;
   c. element; d. compound

3. a. ions; b. neutrons; c. compound; d. atomic number;
   e. hydrogen bond; f. covalent bond; g. mass number; h. element; i. protons; j. isotopes;
   k. ionic bond; l. electrons

4. a. Each element consists of atoms containing a characteristic number of protons (atomic number). However, the outer energy level of inert elements is filled with electrons, whereas the outer energy level of reactive elements is not filled with electrons.

   b. Both polar and nonpolar molecules are held together by covalent bonds. However, in a polar molecule the electrons are not shared equally, so it carries small positive and negative charges on its surface; in a nonpolar molecule the electrons are shared equally, so it is electrically neutral.

   c. Both covalent bonds and ionic bonds bind atoms together, but covalent bonds involve the sharing of electrons between atoms, whereas ionic bonds involve the electrical attraction of oppositely charged atoms (ions).

Module 2.5 Review

a. The chemical shorthand used to describe chemical compounds and reactions effectively is known as chemical notation.

b. The molecular formula for glucose, a compound composed of 6 carbon atoms, 12 hydrogen atoms, and 6 oxygen atoms, is $C_6H_{12}O_6$.

c. The weight of one mole of glucose is 180 grams. (Add the atomic weights of 6 C = 72 g, 12 H = 12 g, and 6 O = 96 g.)

Module 2.6 Review

a. Three types of chemical reactions important in human physiology are (1) decomposition reactions, in which a molecule is broken down into smaller fragments; (2) synthesis reactions, in which small molecules are assembled into larger ones; and (3) exchange reactions, in which parts of the reacting molecules are shuffled around to produce new products.

b. In hydrolysis reactions, water is a reactant, whereas in dehydration synthesis reactions water is a product.

c. The source of the released energy was the potential energy stored in the covalent bonds of the glucose molecule. Energy was released when some of the covalent bonds were broken during this catabolic reaction.

Module 2.7 Review

a. An enzyme is a protein that lowers the activation energy, which is the amount of energy required to start a chemical reaction.

b. Metabolites are molecules that can be synthesized or broken down by chemical reactions inside our bodies. Nutrients are essential metabolites normally obtained from the diet.

c. Enzymes promote chemical reactions by lowering the activation energy requirements and making it possible for chemical reactions to proceed under conditions compatible with life.

Section 2 Review

1. a. H_2 b. 2 H c. 6 H_2O
   d. $C_{12}H_{22}O_{11}$

2. $C_6H_{12}O_6 + 6 O_2 \rightarrow 6 CO_2 + 6 H_2O$

3. a. hydrolysis reaction
   b. dehydration synthesis

4. a. enzyme; b. reactants;
   c. hydrolysis; d. endergonic;
   e. exchange reaction; f. organic compounds; g. exergonic;
   h. activation energy

5. A decreased amount of enzyme at the second step would limit the amount of the intermediate products in the next two steps. This would cause a decrease in the amount of the final product.

Module 2.8 Review

a. Electrolytes are soluble inorganic molecules whose ions will conduct an electric current in solution.

b. Hydrophilic molecules are attracted to water molecules, whereas hydrophobic molecules do not interact with water molecules.

c. Sodium chloride dissociates in water as the positive poles of water molecules are attracted to the negatively charged chloride ions, and the negative poles of water molecules are attracted to the positively charged sodium ions. The ions stay dissolved in solution because a layer of surrounding water molecules, or hydration sphere, separates them from each other.

Module 2.9 Review

a. pH is a measure of the hydrogen ion concentration in a solution, defined as the negative logarithm of the hydrogen ion concentration, expressed in moles per liter. On the pH scale, 7 represents neutrality, values below 7 indicate acidic solutions, and values above 7 indicate alkaline (basic) solutions.

b. An acid is a compound whose dissociation in solution releases a hydrogen ion and an anion; a base is a compound whose dissociation releases a hydroxide ion (OH^-) into the solution or removes a hydrogen ion (H^+) from the solution; and a salt is an inorganic compound consisting of a cation other than H^+ and an anion other than OH^-.

c. The pH of various body fluids must remain relatively constant if the body is to maintain homeostasis and remain healthy.

Section 3 Review

1. Four important properties of water in the human body are effective lubrication (as between bony surfaces in a joint), reactivity (participates in

chemical reactions), high heat capacity (readily absorbs and retains heat), and solubility (is a solvent for many substances).

2. **a.** inorganic compounds; **b.** solute; **c.** alkaline; **d.** hydrophilic; **e.** solvent; **f.** salt; **g.** buffers; **h.** acid; **i.** hydrophobic; **j.** water

3. **a.** acidic; **b.** neutral; **c.** alkaline; **d.** The pH 3 solution is 1000 times more acidic than the pH 6 solution—it contains a thousand-fold (10^3) increase in the concentration of hydrogen ions (H^+). **e.** Three negative effects of abnormal fluctuations in pH are cell and tissue damage (due to broken bonds), changes in the shapes of proteins, and altered cellular functions.

4. Table salt dissociates or dissolves in pure water but since it does not release either hydrogen (H^+) ions or (OH^-) ions, no change in pH occurs.

Module 2.10 Review

a. The three structural classes of carbohydrates are monosaccharides (glucose), disaccharides (sucrose), and polysaccharides (starch).

b. Compounds with a C:H:O ratio of 1:2:1 are carbohydrates, used in the body chiefly as energy sources.

c. Muscle cells make (synthesize) glycogen by linking numerous glucose molecules via a series of dehydration synthesis reactions.

Module 2.11 Review

a. Lipids are a diverse group of water-insoluble organic compounds that contain carbon, hydrogen, and oxygen in a ratio that does not approximate 1:2:1 (they contain less oxygen than do carbohydrates). Examples are fatty acids, eicosanoids, glycerides, steroids, phospholipids, and glycolipids.

b. All fatty acids consist of a hydrocarbon chain and a carboxylic acid group. In saturated fatty acids, each carbon atom in the hydrocarbon chain has four single covalent bonds that bind the maximum number of hydrogen atoms possible. In unsaturated fatty acids, one or more of the carbon atoms in the hydrocarbon chain has double covalent bonds, so fewer hydrogen atoms are bonded.

c. In the hydrolysis of a triglyceride, the reactants are a triglyceride and three water molecules; and the products are a glycerol molecule and three fatty acids.

Module 2.12 Review

a. Cholesterol is a component of plasma membranes and is important for cell growth and division.

b. Eicosanoids function primarily as chemical messengers; steroids function as both components of cellular membranes (cholesterol) and sex hormones (androgens and estrogens); and phospholipids and glycolipids are important components of cellular membranes.

c. When phospholipids and glycolipids form a micelle, the hydrophobic tails of both molecules are inside the micelle, and their hydrophilic heads are outside the micelle.

Module 2.13 Review

a. Proteins are organic compounds formed from amino acids that contain a carbon atom, a hydrogen atom, an amino group ($-NH_2$), a carboxylic acid group ($-COOH$), and a variable group, known as an R group or side chain.

b. During the dehydration synthesis of two amino acids, a peptide bond links the amino group of one amino acid with the carboxylic acid group of the other amino acid.

c. The heat of boiling breaks bonds that maintain the protein's tertiary structure, quaternary structure, or both. The resulting change in shape affects the ability of the protein molecule to perform its normal biological functions. These changes are known as denaturation.

Module 2.14 Review

a. The active site is the location on an enzyme where substrate binding occurs.

b. The reactants in an enzymatic reaction are called substrates.

c. An enzyme's specificity results from the unique shape of its active site, which permits only a substrate with a complementary shape to bind.

Module 2.15 Review

a. Cells obtain energy in the high-energy bonds of compounds such as ATP.

b. ATP (adenosine triphosphate) is a compound consisting of adenosine to which three phosphate groups are attached; high-energy bonds attach the second and third phosphates.

c. Adenosine monophosphate (AMP) is a nucleotide consisting of adenosine plus a phosphate group (PO_4^{3-}), and adenosine diphosphate (ADP) is a compound consisting of adenosine with two phosphate groups attached. Adding a phosphate group to AMP creates ADP. Breaking one phosphate linkage of ATP, to form ADP, provides the primary source of energy for physiological processes.

Module 2.16 Review

a. Nucleic acids are large organic molecules, composed of carbon, hydrogen, oxygen, nitrogen, and phosphorus, that regulate the synthesis of proteins and make up the genetic material in cells.

b. The complementary strands of DNA are held together through complementary base pairing between adenine and thymine, and between guanine and cytosine.

c. The nucleic acid RNA (ribonucleic acid) contains the sugar, ribose. The nucleic acid DNA (deoxyribonucleic acid) contains the sugar, deoxyribose; both contain nitrogenous bases and phosphate groups.

Section 4 Review

1. **a.** glycogen; **b.** isomers; **c.** polyunsaturated; **d.** active site; **e.** glycerol; **f.** cholesterol; **g.** ATP; **h.** RNA; **i.** peptide bond; **j.** nucleotide; **k.** monosaccharide

2. **a.** polysaccharide, polyunsaturated, polypeptide; **b.** triglyceride; **c.** disaccharide, diglyceride, dipeptide; **d.** glycogen, glycolipids

3. **a.** carbohydrates; **b.** polysaccharides; **c.** disaccharides; **d.** monosaccharides; **e.** lipids; **f.** fatty acids; **g.** glycerol; **h.** proteins; **i.** amino acids; **j.** nucleic acids; **k.** RNA; **l.** DNA; **m.** nucleotides; **n.** ATP; **o.** phosphate groups

Chapter Integration

1. When in solution, baking soda (or sodium bicarbonate, $NaHCO_3$) reversibly dissociates into a sodium ion (Na^+) and a bicarbonate ion (HCO_3^-), as follows: $NaHCO_3 \leftrightarrow Na^+ + HCO_3^-$. The bicarbonate ions will remove any excess hydrogen ions (H^+) from body fluids and form a weak acid, carbonic acid, as follows: $H^+ + HCO_3^- \leftrightarrow H_2CO_3^-$.

2. Carbohydrates, such as glucose, sucrose, and fructose, are an immediate source of energy for the body, and thus can be metabolized relatively quickly. Carbohydrates can be stored as glycogen, but the muscles and liver have a limited glycogen storage capacity, and carbohydrates in total account for only about 1.5 percent of total body weight. Excess carbohydrates are converted to lipids and stored as fats.

3. It would be slightly alkaline due to the neutralization of hydrochloric acid by the sodium bicarbonate.

4. It is not a good idea to eliminate all fats from the diet, because some important fatty acids cannot be synthesized by the body. Fats are required for normal biological processes and serve as energy reserves, provide insulation, cushion body organs, and are a crucial component of plasma membranes. For fats, the dietary reference intake (DRI)—the established dietary reference value for nutrient intake for Americans and Canadians—is 30 percent of total caloric intake.

Module 3.1 Review

a. Cytoplasm is the material between the plasma membrane and the nuclear membrane; cytosol is the fluid portion of the cytoplasm.

b. The cytoskeleton provides strength and support and enables movement of cellular structures and materials.

c. Membranous organelles (and their functions) include endoplasmic reticulum (synthesis of secretory products, and intracellular storage and transport); nucleus (control of metabolism, storage and processing of genetic information, and control of protein synthesis); rough ER (modification and packaging of newly synthesized proteins); smooth ER (lipid and carbohydrate synthesis); Golgi apparatus (storage, alteration, and packaging of secretory products and lysosomal enzymes); lysosomes (removal of damaged organelles or of intracellular pathogens); mitochondria (production of 95 percent of the ATP required by the cell); and peroxisomes (neutralization of toxic compounds)

Module 3.2 Review

a. The general functions of the plasma membrane include physical isolation of the cell from its environment, regulation of exchange with the environment, sensitivity to the environment, and structural support.

b. The phospholipid bilayer of the plasma membrane is mostly responsible for isolating a cell from its external environment.

c. Channel proteins are integral proteins that allow water and small ions to pass through the plasma membrane.

Module 3.3 Review

a. The three basic components of the cytoskeleton are microfilaments, intermediate filaments, and microtubules.

b. Microtubules are common to both centrioles and cilia.

c. Cilia propel materials across cell surfaces.

Module 3.4 Review

a. Newly synthesized proteins from free ribosomes enter the cytosol; those from fixed ribosomes enter the ER.

b. Smooth endoplasmic reticulum (SER) lacks ribosomes, and its cisternae are tubular.

c. The SER functions in the synthesis of lipids such as steroids. Ovaries and testes produce large amounts of steroid hormones, which are lipids, and thus contain large amounts of SER.

Module 3.5 Review

a. The Golgi apparatus (1) modifies and packages cellular secretions, such as hormones or enzymes; (2) renews or modifies the plasma membrane; and (3) packages special enzymes within vesicles for use within the cell.

b. Lysosomes contain digestive enzymes.

c. Lysosomes may (1) break down other intracellular organelles to enhance recycling and membrane flow; (2) fuse with vesicles containing fluids or solids from the external environment to obtain nutrients or destroy pathogens; and (3) break down and release their digestive enzymes, resulting in the destruction of the cell (autolysis).

Module 3.6 Review

a. A double membrane encloses a mitochondrion; the outer membrane surrounds the organelle, whereas the inner membrane contains folds called cristae and encloses a fluid, enzyme-filled matrix.

b. Mitochondria require oxygen to produce ATP.

c. Mitochondria produce energy, in the form of ATP molecules, for the cell. A large number of mitochondria in a cell indicates a high demand for energy.

Section 1 Review

1. a. microvilli: increased surface area to facilitate absorption of extracellular materials; b. Golgi apparatus: storage, alteration, and packaging of newly synthesized proteins;

c. lysosome: intracellular removal of damaged organelles or of pathogens; d. mitochondrion: production of 95 percent of the ATP required by the cell; e. peroxisome: neutralization of toxic compounds; f. nucleus: control of metabolism, storage and processing of genetic information, and control of protein synthesis; g. endoplasmic reticulum: synthesis of secretory products, and intracellular storage and transport; h. ribosomes: protein synthesis; i. cytoskeleton: provides strength and support, enables movement of cellular structures and materials

2. a. glycolysis; b. aerobic metabolism; c. microvilli; d. lysosome

3. Similar to the role of a plasma membrane around a cell, an organelle membrane physically isolates the organelle's contents from the cytosol, regulates exchange with the cytosol, and provides structural support.

Module 3.7 Review

a. The DNA in the nucleus stores the cell's instructions for synthesizing proteins.

b. The nucleus is a cellular organelle containing DNA, RNA, and proteins. Surrounding the nucleus is the double-membraned nuclear envelope; the gap within this double membrane is the perinuclear space. Nuclear pores allow for chemical communication between the nucleus and the cytosol.

c. The nuclei of human somatic cells contain 23 pairs of chromosomes.

Module 3.8 Review

a. The three types of RNA involved in protein synthesis are ribosomal RNA (rRNA), messenger RNA (mRNA), and transfer RNA (tRNA).

b. A gene is a portion of a DNA strand that functions as a hereditary unit and codes for a specific protein.

c. The genetic code is described as a triplet code because a sequence of three nitrogenous bases—a *triplet*—specifies the identity of a single amino acid.

Module 3.9 Review

a. The DNA template strand is the strand of DNA that will be used to synthesize RNA.

b. Transcription is the encoding of the genetic instructions in a segment of DNA onto a strand of RNA (typically mRNA).

c. A cell that lacked the enzyme RNA polymerase would not be able to transcribe RNA from DNA.

Module 3.10 Review

a. Translation is the synthesis of a protein using the information provided by the sequence of codons along an mRNA strand.

b. The complementary anticodon sequence for GCA is CGU.

c. Deletion of a base during transcription would change the makeup of all subsequent codons in the sequence of mRNA bases; the altered mRNA sequence would then result in the incorporation of a different series of amino acids into the protein during translation. Almost certainly the protein so produced would not be functional.

Section 2 Review

1. a. chromosomes; b. thymine; c. nuclear envelope; d. introns; e. uracil; f. nucleoli; g. nuclear pore; h. transcription; i. gene; j. exons; k. mRNA; l. tRNA; m. nucleus; n. genetic information

2. a. AUG/UUU/UGU/GCC/GCC/UUA b. UAC/AAA/ACA/CGG/CGG/AAU c. Methionine-Phenylalanine-Cysteine-Alanine-Alanine-Leucine

3. The nucleus contains the information for synthesizing proteins in the nucleotide sequence of its DNA. Changes in the extracellular fluid can affect cells through the binding of molecules to plasma membrane receptors or by the diffusion of molecules through membrane channels. Such stimuli may result in alterations of genetic activity in the nucleus. These alterations may change biochemical processes and metabolic pathways through the

synthesis of additional, fewer, or different enzymes. Altered genetic activity may also change the physical structure of the cell by synthesizing additional, fewer, or different structural proteins.

Module 3.11 Review

a. Diffusion is the passive movement of molecules from an area of higher concentration to an area of lower concentration until equilibrium is reached.

b. Factors that influence diffusion rates include distance, molecule size, temperature, concentration gradient size, and electrical forces.

c. Diffusion is driven by a concentration gradient. The larger the concentration gradient, the faster is the rate of diffusion; the smaller the concentration gradient, the slower the rate of diffusion. If the concentration of oxygen in the lungs were to decrease, the concentration gradient between oxygen in the lungs and oxygen in the blood would decrease (as long as the oxygen level of the blood remained constant). Thus, oxygen would diffuse more slowly into the blood.

Module 3.12 Review

a. Osmosis is the passive movement of water across a selectively permeable membrane from a solution with a lower solute concentration to a solution with a higher solute concentration.

b. A hypotonic solution would cause an osmotic flow into the cell, leading to swelling and eventual bursting of the red blood cell (hemolysis). A hypertonic solution would cause an osmotic flow out of the cell, resulting in a shriveled or collapsed red blood cell (crenation).

c. The 10 percent salt solution is hypertonic with respect to

the cells lining the nasal cavity (the solution contains a higher salt concentration than do the cells). The hypertonic solution would draw water out of the cells (which would shrink), adding water to (diluting) the mucus. Thinner mucus would help relieve the congestion.

Module 3.13 Review

a. In carrier-mediated transport, integral proteins bind specific ions or molecules and transport them across the plasma membrane.

b. In both facilitated diffusion and active transport, a carrier molecule transports materials across the plasma membrane.

c. Transporting hydrogen ions against their concentration gradient—that is, from a site where they are less concentrated (the cells lining the stomach, which produce the hydrogen ions) to a region where they are more concentrated (the digesting stomach contents)—requires energy. Thus an active transport process must be involved.

Module 3.14 Review

a. Endocytosis is the process in which a membranous vesicle forms at the cell surface, encloses a relatively large volume of extracellular material, and then moves into the cytoplasm. Types of endocytosis are pinocytosis, the vesicle-mediated movement into the cytoplasm of extracellular fluid and its contents, and phagocytosis, the vesicle-mediated movement into the cytoplasm of extracellular solids, especially bacteria and debris.

b. Exocytosis is the ejection of cytoplasmic materials following the fusion of a membranous vesicle with the plasma membrane.

c. The engulfment of bacteria and their movement into the cell is called phagocytosis.

Section 3 Review

1. a. diffusion; b. facilitated diffusion; c. molecular size; d. net diffusion of water; e. active transport; f. specificity; g. vesicular transport; h. exocytosis; i. pinocytosis; j. "cell eating"

2. a. diffusion; b. diffusion; c. neither; d. diffusion; e. osmosis; f. osmosis

Module 3.15 Review

a. Interphase is the portion of a cell's life cycle during which the chromosomes are uncoiled and all normal cellular functions except mitosis are under way. Its stages are G_1, S, G_2, and G_0.

b. Important enzymes for DNA replication are DNA polymerase and ligases.

c. This cell is likely in the G_1 phase of interphase.

Module 3.16 Review

a. Mitosis is the division of a cell nucleus into two identical daughter cell nuclei, an essential step in cell division. Its four stages are prophase, metaphase, anaphase, and telophase.

b. A chromatid is a copy of a duplicated chromosome. Human cells normally contain 23 pairs of chromosomes, so during mitosis a human cell would contain 92 chromatids.

c. If spindle fibers failed to form during mitosis, the chromosome would not be able to separate into two sets. If cytokinesis occurred, the result would be one cell with two sets of chromosomes, and one cell with none.

Module 3.17 Review

a. Metastasis is the spread of cancer cells from one site to another, leading to the establishment of secondary tumors.

b. A tumor (or neoplasm) is a mass or swelling produced by the abnormal growth and

division of cells. The cells of a benign tumor typically remain within their tissue of origin.

c. Cancer is an illness characterized by mutations that disrupt normal cell control mechanisms and produce malignant cells.

Section 4 Review

1. a. somatic cells; b. G_1 phase; c. G_2 phase; d. DNA replication; e. mitosis; f. metaphase; g. telophase; h. cytokinesis

2. a. telophase; b. prophase; c. centromere

3. A cell that undergoes repeated rounds of the cell cycle without cytokinesis could result in a large, multinucleated cell.

Chapter Integration

1. Mitochondria contain RNA, DNA, and the necessary enzymes to synthesize proteins. Mitochondrial DNA is different from nuclear DNA and allows each mitochrondrion to control its own maintenance, growth, and reproduction. Serious exercisers have high rates of energy consumption and over time their mitochondria respond to this high demand for energy by dividing and increasing in number.

2. Sweating during exercise results in losses of water from the extracellular fluid. As the extracellular fluid osmolarity increases, osmosis moves water out of the cells, which begin to dehydrate. Energy drinks contain a solution of water, electrolytes (especially sodium chloride, which is lost in sweat), and glucose. As the sodium ions and glucose are absorbed by the digestive tract, water follows by osmosis. As the water influx restores normal ECF volume and osmolarity, water moves back into the cells, rehydrating them.

Module 4.1 Review

a. Epithelial tissue provides physical protection, controls permeability, provides sensation, and produces specialized secretions.

b. The three characteristic shapes of epithelial cells are squamous (flat), cuboidal (cube-shaped), and columnar (appear tall and rectangular). A single layer of epithelial cells is a simple epithelium, whereas multiple layers of epithelial cells constitute a stratified epithelium.

c. The presence of many cilia on the free surface of epithelial cells aids the movement of substances over the epithelial surface.

Module 4.2 Review

a. Epithelial intercellular connections include occluding junctions, adhesion belts, gap junctions, desmosomes, and hemidesmosomes.

b. Epithelia rely on blood vessels in underlying tissues to supply needed nutrients.

c. Gap junctions help coordinate the functioning of adjacent cells by allowing small molecules and ions to pass from cell to cell. In epithelial cells, gap junctions help coordinate functions such as the beating of cilia. In cardiac and smooth muscle tissues, they are essential in achieving coordinated muscle cell contractions.

Module 4.3 Review

a. Keratinized epithelia are both tough (have mechanical strength) and water resistant.

b. All these sites are subject to mechanical stresses and abrasion—by food (pharynx and esophagus), by feces (anus), or during intercourse or childbirth (vagina).

c. No. A simple squamous epithelium provides too little protection against infection, abrasion,

or dehydration to function effectively on the skin surface.

Module 4.4 Review

a. The epithelium that lines the urinary bladder and changes in appearance during stretching is a transitional epithelium.

b. In a sectional view, simple cuboidal epithelial cells are square and have central nuclei, and the distance between adjacent nuclei is roughly equal to the height of the epithelium.

c. Stratified cuboidal epithelia are associated with the ducts of sweat glands, mammary glands, and other exocrine glands.

Module 4.5 Review

a. In a sectional view, simple columnar epithelial cells appear as tall slender rectangles containing elongated nuclei close to the basal lamina. Cell height is several times the distance between adjacent nuclei.

b. A pseudostratified columnar epithelium is not truly stratified—even though its nuclei make it appear so—because all of its epithelial cells contact the basal lamina.

c. The columnar epithelium lining the intestine typically has microvilli on its apical surface.

Module 4.6 Review

a. The two primary types of glands are endocrine glands and exocrine glands.

b. The mode of secretion in which secretory cells fill with secretions and then rupture is holocrine secretion.

c. A gland that lacks ducts and releases its secretions directly into the interstitial fluid is an endocrine gland.

Section 1 Review

1. a. simple squamous epithelium;
b. simple cuboidal epithelium;

c. simple columnar epithelium;
d. stratified squamous epithelium; **e.** stratified cuboidal epithelium; **f.** stratified columnar epithelium

2. a. pseudostratified columnar epithelium; **b.** urinary bladder, ureters, urine-collecting chambers in the kidney; **c.** stratified squamous epithelium; **d.** simple columnar epithelium; **e.** lining of the peritoneum and pericardium, exchange surfaces (alveoli) within the lungs; **f.** lining of exocrine glands and ducts, kidney tubules; **g.** stratified cuboidal epithelium

3. a. mucous cells; **b.** mucin; **c.** mucus; **d.** ducts; **e.** epithelial surfaces; **f.** exocrine glands; **g.** merocrine secretion; **h.** apocrine secretion; **i.** holocrine secretion; **j.** interstitial fluid; **k.** endocrine glands

4. a. avascular; **b.** alveolar (acinar) gland; **c.** transitional epithelium; **d.** occluding junction; **e.** basal lamina; **f.** simple gland; **g.** mesothelium (simple squamous epithelium)

Module 4.7 Review

a. Cells found in connective tissue proper are melanocytes, fixed macrophages, mast cells, fibroblasts, adipocytes (fat cells), plasma cells, free macrophages, mesenchymal cells, neutrophils, eosinophils, and lymphocytes.

b. The connective tissue that contains primarily lipids is adipose (fat) tissue.

c. Fibroblasts secrete the proteins that form collagen, elastic fibers, and reticular fibers in the matrix, and fibroblasts also maintain these connective tissue fibers.

Module 4.8 Review

a. The two types of connective tissue that contain a fluid matrix (that is, are fluid connective tissues) are blood and lymph.

b. A vitamin C deficiency would impair the production of collagen fibers, which add strength to connective tissue. The deficiency would result in tissue that is weak and prone to damage.

c. The continuous recirculation of extracellular fluid (plasma, interstitial fluid, and lymph) helps eliminate local differences in the levels of nutrients, wastes, or toxins; maintains blood volume; and alerts the immune system to infectious agents or infections.

Module 4.9 Review

a. Mature cartilage cells are called chondrocytes.

b. The fibers characteristic of cartilage supporting the ear are elastic fibers.

c. The type of cartilage in intervertebral discs is fibrous cartilage.

Module 4.10 Review

a. Mature bone cells in lacunae are called osteocytes.

b. Whereas bone contains only collagen fibers, cartilage may contain collagen, elastic, and reticular fibers.

c. Bone cannot undergo interstitial growth because the matrix of bone is solid and calcified.

Module 4.11 Review

a. The four types of membranes found in the body are mucous membranes, serous membranes, the cutaneous membrane, and synovial membranes.

b. The body cavities lined by serous membranes are the pleural, peritoneal, and pericardial cavities.

c. This tissue is probably fascia, a connective tissue layer and wrapping that supports and surrounds organs.

Section 2 Review

1. **a.** loose connective tissue; **b.** adipose; **c.** regular; **d.** tendons; **e.** ligaments; **f.** fluid connective tissue; **g.** blood; **h.** hyaline; **i.** chondrocytes in lacunae; **j.** bone

2. **a.** perichondrium, periosteum, peritoneum, pericardium; **b.** osteocyte, periosteum; **c.** chondrocyte, perichondrium; **d.** interstitial growth; **e.** lacunae; **f.** chondrocyte; **g.** osseous tissue; **h.** fibrous cartilage; **i.** adipocytes; **j.** synovial membrane; **k.** cutaneous membrane; **l.** perichondrium

Module 4.12 Review

a. The three types of muscle tissue in the body are skeletal muscle tissue, cardiac muscle tissue, and smooth muscle tissue.

b. Muscle tissue containing small, tapering cells with single nuclei and no obvious striations is smooth muscle tissue.

c. These cells are most likely neurons.

Clinical Module 4.13 Review

a. The two processes of the response to tissue injury are inflammation and regeneration.

b. Inflammation produces swelling, redness, warmth, and pain.

c. Inflammation can occur in any organ in the body because all organs have connective tissues.

Section 3 Review

1. **a.** cardiac; **b.** smooth; **c.** non-striated; **d.** multinucleate

2. **a.** cell body; **b.** dendrites; **c.** neuroglia; **d.** maintain physical structure of neural tissue; **e.** repair neural tissue framework after injury; **f.** perform phagocytosis; **g.** provide nutrients to neurons; **h.** regulate the composition of interstitial fluid surrounding neurons

3. **a.** axon; **b.** intercalated disc; **c.** neuroglia; **d.** skeletal muscle tissue; **e.** smooth muscle

tissue; **f.** regeneration; **g.** inflammation

4. Increased blood flow and blood vessel permeability enhance the delivery of oxygen and nutrients and the migration of additional phagocytes into the area, and the removal of toxins and waste products from the area.

Chapter Integration

1. The expected generalized response to tissue injury is inflammation, which is characterized by redness, pain, heat, and swelling. Redness results as blood (fluid connective tissue) flow to the site of injury increases. Pain occurs as sensory nerve endings in nervous tissue respond to chemicals released from damaged tissues and from mast cells. Heat at the injury site results from increased blood flow into the damaged area, and swelling results from the increased permeability of capillary membranes in the affected area. Additionally, a variety of white blood cells, especially phagocytes, aggregate at the injury site to clean up the damage and defend against infection. With time, fibroblasts lay down a network of collagen fibers that aid in the repair and regeneration of the injured tissues.

2. Muscle tissue has an extensive blood supply, so the materials needed to repair damaged muscle tissue arrive relatively quickly and in plentiful supply. In contrast, cartilage lacks a direct blood supply, so the repair of damaged cartilage must rely on the relatively slow process of diffusion.

3. Proteins contain amino acids needed for the synthesis of proteins in tissues undergoing repairs. Vitamin C is required by fibroblasts for the production of collagen, a crucial component of cartilage and other types of connective tissue.

Module 5.1 Review

a. The layers of the epidermis are the stratum basale, stratum spinosum, stratum granulosum, stratum lucidum, and stratum corneum.

b. Dandruff consists of cells from the stratum corneum.

c. A splinter that penetrates to the third epidermal layer of the palm is lodged in the stratum granulosum.

Module 5.2 Review

a. The two pigments in the epidermis are carotene (an orange-yellow pigment) and melanin (a brown, yellow-brown, or black pigment).

b. Exposure to sunlight or sunlamps darkens skin because the ultraviolet radiation they emit stimulates melanocytes in the epidermis and dermis to synthesize the pigment melanin.

c. When skin gets warm, arriving oxygenated blood is diverted to the superficial dermis for the purpose of eliminating heat. The oxygenated blood imparts a reddish coloration to the skin.

Module 5.3 Review

a. The dermis (a connective tissue layer) lies between the epidermis and the hypodermis.

b. The capillaries and sensory neurons that supply the epidermis are located in the papillary layer of the dermis.

c. The presence of elastic fibers allows the dermis to undergo repeated cycles of stretching and recoil (returning to its original shape).

Section 1 Review

1. a. dermis: the connective tissue layer beneath the epidermis; **b.** epidermis: the protective epithelium covering the surface of the skin; **c.** papillary layer: vascularized areolar tissue containing capillaries, lymphatic vessels, and sensory neurons that supply the skin surface; **d.** reticular layer: interwoven meshwork of dense irregular connective tissue containing collagen fibers and elastic fibers; **e.** hypodermis (subcutaneous layer or superficial fascia): layer of loose connective tissue below the dermis

2. a. accessory structures; **b.** epidermis; **c.** granulosum; **d.** papillary layer; **e.** nerves; **f.** reticular layer; **g.** collagen; **h.** hypodermis; **i.** connective; **j.** fat

3. a. Malignant melanoma is often fatal because melanocytes are located close to the dermal layer, so if they become malignant, they can easily metastasize through the blood vessels and lymphatic vessels in nearby connective tissues.

b. Fingers (and toes) swell up because of the hypotonic osmotic flow of water into the dead, keratinized cells of the outer layer of the epidermis, the stratum corneum. Because the underlying strata and dermis do not expand, the larger surface area of the swollen stratum corneum must go somewhere and it forms folds and creases, or wrinkles. Other areas of the body lack a thick stratum corneum, so little swelling and wrinkling result.

Module 5.4 Review

a. A typical hair is a keratinous strand produced by basal (germinative) cells within a hair follicle.

b. When an arrector pili muscle contracts, it pulls the hair follicle erect, depressing the area at the base of the hair and making the surrounding skin appear higher. The overall effect is known as "goose bumps" or "chicken skin."

c. Pulling a hair is painful because its root is attached deep within the hair follicle, the base of which is surrounded by a root hair plexus consisting of sensory nerves. Cutting a hair is painless because a hair shaft contains no sensory nerves.

Module 5.5 Review

a. Two types of exocrine glands found in the skin are sebaceous (oil) glands and sweat glands.

b. Sebaceous secretions (sebum) lubricate and protect the keratin of the hair shaft, lubricate and condition the surrounding skin, and inhibit the growth of bacteria.

c. Deodorants are used to mask the odor of apocrine sweat gland secretions, which contain several kinds of organic compounds. Some of these compounds have an odor, and others produce an odor when metabolized by skin bacteria. Apocrine sweat glands enlarge and increase secretory activity in response to the increase in sex hormones that occurs at puberty.

Module 5.6 Review

a. The hyponychium is the thickened stratum corneum underlying the free edge of a nail.

b. A fingernail is a keratinous structure that is produced by epithelial cells of the nail root and protects the underlying fingertip. Structures of the nail include a distal free edge, lateral nail fold, lunula, proximal nail fold, eponychium, nail root, nail body, and hyponychium.

c. Nail production occurs at the nail root, an epidermal fold that is not visible from the surface.

Module 5.7 Review

a. Common effects of the aging process on the skin include epidermal thinning due to a decline in germinative cell activity; a decrease in melanocyte abundance; a reduction in sebaceous gland secretion; a decline in dendritic cell numbers; a reduction in vitamin D_3 production; a decline in glandular activity; reduced blood flow to the dermis; the stoppage of hair follicle functioning; and a slowing of the pace of skin repair.

b. With advancing age, melanocyte activity decreases, leading to gray or white hair.

c. As a person ages, the blood supply to the dermis decreases, and merocrine sweat glands become less active. Both changes make it more difficult for the elderly to cool themselves in hot weather.

Module 5.8 Review

a. Some hormones that are necessary for maintaining healthy skin are growth hormone, sex hormones, growth factors (including epidermal growth factor), steroid hormones, and thyroid hormones.

b. In the presence of ultraviolet radiation in sunlight, epidermal cells in the stratum spinosum and stratum basale convert a cholesterol-related steroid into cholecalciferol, also known as vitamin D_3.

c. The hormone cholecalciferol (vitamin D_3) is needed to form strong bones and teeth. When the body surface is covered, ultraviolet radiation cannot reach the skin to stimulate cholecalciferol (vitamin D_3) production, so fragile bones can develop.

Module 5.9 Review

a. The first step in the repair of tissue is inflammation.

b. Granulation tissue is the combination of fibrin clots, fibroblasts, and the extensive network of capillaries in healing tissue.

c. Skin can regenerate effectively even after undergoing considerable damage because stem cells persist in both the epithelial and connective tissue components of skin. In response to injury, cells of the stratum basale (germinativum) replace epithelial cells while mesenchymal cells replace lost dermal cells.

Section 2 Review

1. **a.** vitamin D_3; **b.** nail root **c.** apocrine sweat glands **d.** EGF **e.** reticular layer of dermis **f.** wrinkled skin **g.** sebum **h.** malignant melanoma **i.** merocrine sweat glands **j.** eponychium
2. **a.** free edge; **b.** lateral nail fold; **c.** nail body; **d.** lunula; **e.** proximal nail fold; **f.** eponychium; **g.** eponychium; **h.** proximal nail fold; **i.** nail root; **j.** lunula; **k.** nail body; **l.** hyponychium; **m.** phalanx; **n.** dermis; **o.** epidermis
3. **a.** hair shaft; **b.** sebaceous gland; **c.** arrector pili muscle; **d.** connective tissue sheath of hair bulb; **e.** root hair plexus
4. The chemicals in hair dyes break the protective covering of the cortex allowing the dyes to stain the medulla of the shaft. This is not permanent because the cortex remains damaged, allowing shampoo and UV rays from the sun to enter the medulla and affect the color. Also, the viable portion of the hair remains unaffected, so that when the shaft is replaced the color will be lost.

Chapter Integration

1. Year-round sun protection is important because the sun's harmful rays exist during any season, not just summer and it is these rays that cause tissue damage, skin wrinkles, and skin cancer. However, a total lack of exposure to sunlight would reduce the epidermal production of cholecalciferol (vitamin D_3), an essential precursor to calcitriol production in the kidney. Low levels of calcitriol reduce calcium and phosphorus absorption, resulting in fragile bones.

2. Several features of the skin prevent bacteria from reaching deeper tissues. Sebum, the oily secretion of the sebaceous gland, contains a bactericidal agent that kills invading bacteria. Sweat from sweat glands flushes the skin surface to remove bacteria that could establish an infection. The tough stratum corneum and the tightly packed cells of deeper epidermal layers make it difficult for bacteria to penetrate the epidermis. Dendritic cells in the stratum spinosum fight invading bacteria. Finally, the presence of the basal lamina inhibits bacteria from entering the dermis.

3. The hypodermic needle would penetrate (from superficial to deep) the stratum corneum, stratum granulosum, stratum spinosum, stratum basale, papillary layer, and reticular layer.

4. Bedsores in immobile patients are caused by such factors as unrelieved pressure and friction, which irritate the skin and cause blisters, leading to the development of open sores. Pressure sores can extend deep into the body, affecting muscles, tendons, and bones.

Module 6.1 Review

a. A surface feature is a characteristic of a bone's surface that has a certain function, such as forming a joint, serving as a site of muscle attachment, or allowing the passage of nerves and blood vessels.

b. The six broad categories for classifying bones according to shape are flat bones, irregular bones, long bones, sesamoid bones, short bones, and sutural (Wormian) bones.

c. A tubercle is a small, rounded projection on a bone, whereas a tuberosity is a small, rough projection that may occupy a broad area on the bone's surface.

Module 6.2 Review

a. The major parts of a long bone are the epiphysis, diaphysis, metaphysis, and medullary cavity.

b. The medullary cavity—the space within a bone—contains the red bone marrow, the site of blood cell production, and the yellow bone marrow, adipose tissue that is an important site for energy reserves.

c. Articular cartilage is nourished by diffusion from the synovial fluid within the joint.

Module 6.3 Review

a. Osteocytes are cells responsible for the maintenance and turnover of the mineral content of bone; osteoblasts are cells that produce the fibers and matrix of bone; osteoprogenitor cells are stem cells that differentiate into osteoblasts; and osteoclasts are cells that dissolve the fibers and matrix of bone.

b. If the ratio of collagen to hydroxyapatite in a bone increased, the bone's compressive strength would decrease, and it would also become more flexible.

c. If the activity of osteoclasts (which demineralize bone) exceeded osteoblast activity (production of new bone), then the bone's mineral content (and thus its mass) would decline, making it weaker.

Module 6.4 Review

a. An osteon is the basic functional unit of mature compact bone; it consists of osteocytes organized around a central canal and separated by concentric lamellae.

b. Compact bone, which lies over spongy bone and makes up most of a bone's diaphysis, consists of compactly arranged osteons (Haversian systems); it protects, supports, and resists stress. Spongy bone makes up most of the mass of short, flat, and irregular bones and is also found at the epiphyses of long bones; it stores marrow and provides some support.

c. The sample is likely from an epiphysis. The presence of lamellae that are not arranged in osteons is indicative of spongy bone, which occurs in epiphyses.

Module 6.5 Review

a. Appositional growth is enlargement of a bone by the addition of bone matrix at its surface.

b. The periosteum is the layer that surrounds a bone; it consists of an outer fibrous region and an inner cellular region. The endosteum is an incomplete cellular lining on the inner (medullary) surfaces of bones.

c. As a bone increases in diameter, the medullary cavity also increases in diameter.

Module 6.6 Review

a. Endochondral ossification is the replacement of a cartilaginous model to bone.

b. In endochondral ossification, the source of osteoblasts is the differentiation of cells in the inner layer of the perichondrium.

c. X-rays of long bones, such as the femur, can reveal the presence or absence of the epiphyseal cartilage, which separates the epiphysis from the diaphysis so long as the bone is still lengthening. If the epiphyseal cartilage is still present, growth is still occurring; if it is not, the bone has reached its full length.

Module 6.7 Review

a. Intramembranous ossification is the formation of bone within connective tissue without the prior development of a cartilaginous model.

b. During intramembranous ossification, mesenchyme or fibrous connective tissue is replaced by bone.

c. The primary difference between the two types of ossification is that in intramembranous ossification, bone develops from mesenchyme or fibrous connective tissue, whereas in endochondral ossification, bone develops from a cartilage model.

Module 6.8 Review

a. Marfan syndrome is a hereditary disorder of connective tissue, resulting in abnormally long and thin limbs and digits. The condition usually causes life-threatening cardiovascular problems.

b. Gigantism results from overproduction of growth hormone *before* puberty, causing extreme height, whereas acromegaly results from overproduction of growth hormone *after* puberty, causing abnormally thick bones.

c. Pituitary dwarfism is less common today in the United States, because children can be treated with synthetic growth hormone, which will provide adequate or near adequate amounts for normal growth and development.

Section 1 Review

1. a. irregular bones; **b.** epiphyses; **c.** fossa; **d.** medullary cavity; **e.** trabeculae; **f.** osteoclasts; **g.** sesamoid bones; **h.** osteogenesis; **i.** osteon; **j.** appositional growth; **k.** endochondral ossification

2. a. intramembranous ossification; **b.** collagen; **c.** osteocytes; **d.** lacunae; **e.** hyaline cartilage; **f.** periosteum; **g.** compact bone

3. The fracture might have damaged the epiphyseal cartilage in Rebecca's right leg. Even though the bone healed properly, the damaged leg did not produce as much cartilage as did the undamaged leg. The result would be a shorter bone on the side of the injury.

Module 6.9 Review

a. The hormones involved in stimulating and inhibiting the release of calcium ions from bone matrix are parathyroid hormone (PTH), which increases blood calcium levels by causing osteoclasts to release stored calcium from bone; calcitonin, which decreases blood calcium levels by inhibiting osteoclasts and causing osteoblasts to continue depositing calcium in bone; and calcitriol, which causes an increase in intestinal calcium absorption, stimulates calcium resorption from bone, aids the effect of PTH on bone resorption, and increases kidney reabsorption of calcium so fewer calcium ions are lost in urine.

b. Increased PTH secretion would increase blood calcium levels by stimulating osteoclasts to release calcium ions from bone.

c. Calcitonin lowers blood calcium levels by inhibiting osteoclast activity and increasing the rate of calcium excretion at the kidneys.

Clinical Module 6.10 Review

a. An open fracture (also called a compound fracture) is a break in the bone in which bone pierces the skin; a closed fracture (also called a simple fracture) is a break in the bone in which no bone breaks the skin.

b. Immediately following a fracture, extensive bleeding occurs at the injury site. After several hours, a large blood clot called a fracture hematoma develops. Next, an internal callus forms as a network of spongy bone unites the inner edges, and an external callus of cartilage and bone stabilizes the outer edges. The cartilaginous external callus is eventually replaced by bone, and the struts of spongy bone then unite the broken ends. With time, the swelling that initially marked the location of the fracture subsides and the fracture site is remodeled, leaving little evidence that a break occurred.

c. An external callus forms early in the healing process, when cells from the endosteum and periosteum migrate to the area of the fracture. These cells form an enlarged collar (external callus) that encircles the bone in the area of the fracture.

Section 2 Review

1. **a.** calcitonin; **b.** $\downarrow Ca^{2+}$ concentration in body fluids; **c.** $\downarrow$ calcium level; **d.** parathyroid glands; **e.** $\uparrow Ca^{2+}$ concentration in body fluids; **f.** releases stored Ca^{2+} from bone; **g.** homeostasis

2. **a.** spiral fracture; **b.** transverse fracture; **c.** greenstick fracture; **d.** comminuted fracture; **e.** compression fracture; **f.** Colles fracture (typically results from cushioning a fall)

Chapter Integration

1. Astronauts on long space flights should take calcium and vitamin D_3 supplements and should perform any exercise that exerts pressure on the bones; examples include moving on a treadmill and using devices that resist various types of motion.

2. It might be necessary to supplement the child's diet with vitamin D_3, which is required for calcium absorption. Even though the child's diet might contain foods containing adequate amounts of calcium and vitamin D_3, the absence of exposure to sunlight makes vitamin D_3 supplementation a good idea.

3. Bone matrix absorbs traces of various minerals consumed in the diet, and these minerals can be identified hundreds of years later. A diet rich in calcium and vitamin D_3 produces denser bones than will a diet lacking these nutrients. Evidence of cultural practices, such as the binding of appendages or the wrapping of infants' heads, often becomes clear when deformed skeletal remains are discovered. Heavy muscular activity results in large surface features, indicating an active and perhaps athletic lifestyle.

Module 7.1 Review

a. A suture is a fibrous connective tissue joint between the flat bones of the skull.

b. The bones of the cranium are the occipital bone, frontal bone, sphenoid, ethmoid, and the paired parietal and temporal bones.

c. The facial bones protect and support the entrances to the digestive and respiratory tracts. These bones also provide attachment points for the muscles of facial expression.

Module 7.2 Review

a. The facial bones are the paired maxillae, palatine, nasal, inferior nasal conchae, zygomatic, and lacrimal and the unpaired vomer and mandible.

b. The bone that is fractured is the right parietal bone.

c. vomer: facial; ethmoid: cranial; sphenoid: cranial; temporal: cranial; inferior nasal conchae: facial

Module 7.3 Review

a. Both the external acoustic meatus and the internal acoustic meatus are found in the temporal bone.

b. The internal acoustic meatus carries blood vessels and nerves to the inner ear; it also conveys the facial nerve to the stylomastoid foramen.

c. The alveolar processes support the upper teeth in the maxillae, and the lower teeth in the mandible.

Module 7.4 Review

a. The temporal bone contains the carotid canal, through which passes the internal carotid artery supplying the brain.

b. The foramen ovale provides a passageway for nerves innervating the jaw.

c. The foramen magnum is located in the occipital bone. This opening surrounds the connection between the brain and spinal cord.

Module 7.5 Review

a. The optic canal is found in the sphenoid. The optic nerve and ophthalmic artery pass through this opening.

b. The sphenoid bone contains the sella turcica; the fossa within the sella turcica encloses the pituitary gland.

c. The ethmoid contains the cribriform plate. In addition to forming the floor of the cranium and the roof of the nasal cavity, olfactory foramina in this structure serve as passageways for olfactory nerves.

Module 7.6 Review

a. The bones of the orbital complex are the frontal, sphenoid, zygomatic, palatine, maxilla, lacrimal, and ethmoid.

b. The frontal sinuses are cavities within the frontal bone that usually appear after age 6.

c. frontal: both; maxilla: both; palatine: both; nasal: nasal

Module 7.7 Review

a. The foramina of the mandible are the two mental foramina and the two mandibular foramina.

b. Three auditory ossicles are located in each middle ear cavity, found within the petrous portion of the temporal bone. The ossicles play a key role in hearing by conducting vibrations produced by sound waves arriving at the tympanic membrane to the inner ear.

c. Your lab partner is correct: The hyoid bone does not directly attach to any other bone; instead, it supports the larynx and is the attachment site for muscles of the larynx, pharynx, and tongue.

Module 7.8 Review

a. A fontanelle is a relatively soft, flexible, fibrous connective tissue region between two flat bones in the developing skull.

b. The major fontanelles are the anterior fontanelle, occipital fontanelle, sphenoidal fontanelle, and mastoid fontanelle.

c. Because fontanelles are not ossified at birth, they permit flexibility of the skull during childbirth, and they allow for growth of the brain during infancy and early childhood.

Module 7.9 Review

a. The major components of a typical vertebra are the vertebral body, articular processes, and the vertebral arch, the last of which is composed of a spinous process, laminae, transverse processes, and pedicles.

b. The secondary curves of the spine allow us to balance our body weight to permit an upright posture with minimal muscular effort. Without the secondary curves, we would not be able to stand upright for extended periods.

c. The intervertebral discs attach to the body of the vertebra.

Module 7.10 Review

a. The dens is part of the axis, or second cervical vertebra, which is located in the cervical (neck) region of the vertebral column.

b. The presence of transverse foramina indicates that this vertebra is a cervical vertebra.

c. When you run your finger down a person's spine, you can feel the spinous processes of the vertebrae.

Module 7.11 Review

a. There are five vertebrae in the lumbar region, and five fused vertebrae in the sacrum.

b. The sacrum forms the posterior wall of the pelvic girdle.

c. The lumbar vertebrae must support a great deal more weight than do vertebrae that are more superior in the spinal column. The large vertebral bodies allow the weight to be distributed over a larger area.

Module 7.12 Review

a. Vertebrosternal ribs are attached directly to the sternum by their own costal cartilage. Vertebrochondral ribs either do not attach to the sternum (as in the floating ribs) or attach by means of a common costal cartilage (as in the vertebrochondral ribs).

b. Improper chest compressions during CPR can—and commonly do—result in fractures of the sternum or ribs.

c. In addition to the ribs and sternum, the thoracic vertebrae make up the thoracic cage.

Section 1 Review

1. **a.** axial; **b.** longitudinal; **c.** skull; **d.** mandible; **e.** lacrimal; **f.** occipital; **g.** temporal; **h.** hyoid; **i.** vertebral column; **j.** thoracic; **k.** lumbar; **l.** sacral; **m.** floating; **n.** sternum; **o.** xiphoid process

2. **a.** cervical; **b.** thoracic; **c.** lumbar

Module 7.13 Review

a. The bones of the pectoral girdle are two clavicles (collarbones) and two scapulae (shoulder blades).

b. In attaching the scapula to the sternum, the clavicle restricts the scapula's range of movement. A broken clavicle thus gives the scapula a greater range of movement but makes it less stable.

c. The humerus articulates with the scapula at the glenoid cavity.

Module 7.14 Review

a. The bone of the arm is the humerus, and the bones of the forearm are the radius and ulna.

b. The two rounded projections on either side of the elbow are the lateral and medial epicondyles of the humerus.

c. The radius is positioned laterally when the forearm is in the anatomical position.

Module 7.15 Review

a. Phalanges are bones of the fingers (or toes).

b. The carpal bones are the scaphoid, lunate, pisiform, triquetrum, trapezium, trapezoid, capitate, and hamate.

c. Bill has broken the tip of his thumb, also known as the pollex.

Module 7.16 Review

a. The acetabulum is the socket (fossa) on the lateral aspect of the pelvis that articulates with the head of the femur.

b. The three bones that fuse to make a hip bone (coxal bone) are the ilium, ischium, and pubis.

c. When you are seated, your body weight is borne by the ischial tuberosities.

Module 7.17 Review

a. The bones of the pelvis are the two hip (coxal) bones, the sacrum, and the coccyx.

b. The two pubic bones are joined anteriorly by the pubic symphysis.

c. The pelvis of females is adapted for supporting the weight of the developing fetus and enabling the newborn to pass through the pelvic outlet during delivery. Compared to males, the pelvis of females is smoother and lighter; has less-prominent markings; has an enlarged pelvic outlet; has a sacrum and coccyx with less curvature; has a pelvic inlet that is wider and more circular; is relatively broad and low; has ilia that project farther laterally; and has an inferior angle between the pubic bones that is greater than 100° (as opposed to 90° or less for males).

Module 7.18 Review

a. The bones of the lower limb are the femur (thigh), patella (knee-cap), tibia and fibula (leg), tarsal bones, metatarsal bones, and phalanges.

b. The head of the femur articulates with the acetabulum.

c. The fibula both stabilizes the ankle joint and is an important point of attachment for muscles that move the foot and toes. When the fibula is fractured, those muscles cannot function properly, so walking becomes difficult—and painful.

Module 7.19 Review

a. The tarsal bones are the talus, calcaneus, cuboid, navicular, medial cuneiform, intermediate cuneiform, and lateral cuneiform.

b. The talus transmits the weight of the body from the tibia toward the toes.

c. Joey most likely fractured the calcaneus (heel bone).

Section 2 Review

1. a. clavicle; b. scapula; c. humerus; d. radius; e. ulna; f. carpal bones; g. metacarpal bones; h. phalanges; i. hip bone (coxal bone); j. femur; k. patella; l. tibia; m. fibula; n. tarsal bones; o. metatarsal bones; p. phalanges

2. a. anterior view; b. lateral view; c. posterior view; d. acromion; e. coracoid process; f. scapular spine; g. glenoid cavity; h. subscapular fossa; i. supraspinous fossa; j. infraspinous fossa

3. a. sacrum; b. coccyx; c. ilium; d. pubis; e. ischium; f. hip (coxal) bone; g. iliac fossa; h. acetabulum; i. obturator foramen

Chapter Integration

1. Wayne is likely experiencing difficulty breathing because he has two broken ribs. Breathing entails rib movements that change the volume of the thoracic cavity; broken ribs cause difficulty breathing because the pain hampers full expansion of the rib cage.

2. Wayne probably dislocated his shoulder, a common injury due to the weak nature of the shoulder joint (that is, the articulation between the head of the humerus and the glenoid cavity of the scapula).

3. Both the appendicular and axial skeleton have been affected in Wayne's injuries: The ribs are part of the axial skeleton, and the shoulder joint is part of the appendicular skeleton.

Module 8.1 Review

a. In a joint dislocation (luxation), the articulating surfaces of a joint are forced out of position.

b. Components of a synovial joint are a fibrous articular capsule, which surrounds the joint; articular cartilages, which resemble hyaline cartilages and cover the articulating bone surfaces; and synovial fluid, which is located within the joint cavity and provides lubrication, distributes nutrients, and absorbs shocks. Accessory structures include bursae, which are pockets filled with synovial fluid, that reduce friction and absorb shocks; fat pads, which protect the articular cartilages; menisci, which are fibrous cartilage articular discs that allow for variation in the shapes of the articulating surfaces; ligaments, which are cords of fibrous tissue that support, strengthen, and reinforce the joint; and tendons, which pass across or around a joint, limit the range of motion, and provide mechanical support.

c. Articular cartilages lack a blood supply and thus rely on synovial fluid to supply nutrients and remove wastes. If the circulation of synovial fluid were impaired, the cartilages would no longer receive nutrients, and wastes would accumulate. The resulting conditions would cause the cartilages to degenerate, and cells in the tissue may possibly die.

Module 8.2 Review

a. Based on the shapes of the articulating surfaces, synovial joints are classified as gliding, hinge, pivot, ellipsoid, saddle, and ball-and-socket joints.

b. A ball-and-socket joint permits the widest range of motion.

c. shoulder: ball-and socket; elbow: hinge; ankle: gliding; thumb: saddle

Module 8.3 Review

a. When doing jumping jacks, the lower limbs must perform abduction (when the limbs are spread apart) and adduction (when they are brought back together again).

b. Flexion and extension are the movements associated with hinge joints.

c. Dorsiflexion is upward movement of the foot through flexion at the ankle, whereas plantar flexion is ankle extension, as when pointing the toes.

Module 8.4 Review

a. Snapping your fingers involves opposition of the thumb and flexion at the third metacarpophalangeal joint.

b. Pronation and supination of the hand are made possible by the rotation of the radius head.

c. Protraction, supination, and pronation occur while wriggling into tight-fitting gloves.

Section 1 Review

1. a. diarthrosis; **b.** pronation-supination; **c.** shoulder; **d.** articular discs; **e.** synarthrosis; **f.** dislocation; **g.** fluid-filled pouch; **h.** amphiarthrosis

2. a. medullary cavity; **b.** spongy bone; **c.** periosteum; **d.** synovial membrane; **e.** articular cartilage; **f.** joint cavity (containing synovial fluid); **g.** joint capsule; **h.** compact bone

3. a. flexion; **b.** extension; **c.** hyperextension; **d.** flexion; **e.** hyperextension; **f.** abduction; **g.** adduction; **h.** head rotation; **i.** pronation; **j.** abduction; **k.** adduction; **l.** opposition

Module 8.5 Review

a. The primary vertebral ligaments are the ligamentum flavum, posterior longitudinal ligament, interspinous ligament, supraspinous ligament, and anterior longitudinal ligament.

b. The nucleus pulposus is the gelatinous central region of an intervertebral disc. The anulus fibrosus is the tough layer of fibrous cartilage encircling the nucleus pulposus.

c. A slipped disc is a vertebral disc that is displaced or partly protruding as a result of a compressed nucleus pulposus distorting the anulus fibrosus. In a herniated disc, the nucleus pulposus breaks through the anulus fibrosus, causing it to protrude into the vertebral canal.

Module 8.6 Review

a. Ligaments and muscles provide most of the stability for the shoulder joint.

b. The iliofemoral, pubofemoral, and ischiofemoral ligaments are at the hip joint.

c. A shoulder separation is an injury involving partial or complete dislocation of the acromioclavicular joint, thus the bones involved include the clavicle, scapula, and humerus; and the stabilizing ligaments involved are the coracoclavicular, acromioclavicular, coraco-acromial, coracohumeral, and glenohumeral.

Module 8.7 Review

a. Menisci are found in the knee joint.

b. Damage to the menisci of the knee joint decreases the joint's

lateral stability, so the individual would have a difficult time locking the knee in place while standing and would have to use muscle contractions to stabilize the joint. If the person had to stand for a long time, the muscles would fatigue and the knee would "give out." It is also likely that the individual would feel pain.

Module 8.8 Review

a. Rheumatism is a general term describing any painful condition of joints, muscles, or both that is not caused by infection or injury. Osteoarthritis is a form of rheumatism characterized by degeneration of the joint cartilage and the underlying bone. Osteoarthritis results from cumulative wear and tear or genetic factors affecting collagen formation.

b. An arthroscope is an instrument that uses thin, flexible optical fibers to view the interior structures of a joint. This instrument can also be modified to perform surgical procedures without the trauma of major surgery.

c. A person can slow the progression of arthritis by engaging in regular exercise, doing physical therapy, and taking anti-inflammatory drugs.

Section 2 Review

1. **a.** popliteal ligament; **b.** arthritis; **c.** osteoporosis; **d.** reinforce knee joint; **e.** disc outer layer; **f.** acetabulum; **g.** disc inner layer; **h.** dislocation

2. **a.** coracoclavicular ligaments; **b.** acromioclavicular ligament; **c.** tendon of supraspinatus muscle; **d.** acromion; **e.** articular capsule; **f.** subdeltoid bursa; **g.** synovial membrane; **h.** humerus; **i.** clavicle; **j.** coraco-acromial ligament; **k.** coracoid process; **l.** scapula; **m.** articular cartilages; **n.** joint cavity; **o.** glenoid labrum

3. **a.** patellar surface of femur; **b.** fibular collateral ligament; **c.** lateral condyle; **d.** lateral meniscus; **e.** tibia; **f.** fibula; **g.** posterior cruciate ligament (PCL); **h.** medial condyle; **i.** tibial collateral ligament; **j.** medial meniscus; **k.** anterior cruciate ligament (ACL)

4. The sternoclavicular joints are the only articulations between the pectoral girdles and the axial skeleton. The sacro-iliac joints are the articulations between the pelvic girdles and the axial skeleton.

Chapter Integration

1. The bones comprising the knee joint are the femur, patella, and tibia; the fibula does not participate in the articulation.

2. Physical therapy may improve the range of motion at the joint and increase blood flow to the tissue.

3. The knee joint is a diarthrotic synovial joint that functions as a hinge; as such, it allows for flexion and extension at the knee.

4. Individuals who have undergone a total knee arthroplasty may be restricted from running/jogging, high-impact aerobics, and jumping—that is, any activity that places sudden, extreme loads on the artificial joint.

Module 9.1 Review

a. A tendon is collagenous bundle that connects a skeletal muscle to a bone, whereas an aponeurosis is a broad collagenous sheet that takes the place of tendons and connects a skeletal muscle to a wider area of bone or more than one bone.

b. The epimysium is a dense layer of collagen fibers that surrounds the entire muscle; the perimysium divides the skeletal muscle into a series of compartments, each containing a bundle of muscle fibers called a fascicle; and the endomysium surrounds individual skeletal muscle cells (fibers). The collagen fibers of the epimysium, perimysium, and endomysium come together to form either bundles known as tendons, or broad sheets called aponeuroses. Tendons and aponeuroses generally attach skeletal muscles to bones.

c. Because tendons attach muscles to bones, severing the tendon would disconnect the muscle from the bone, and so the muscle could not move a body part.

Module 9.2 Review

a. Transverse tubules are tubular extensions of the sarcolemma that extend deep into the sarcoplasm, contacting cisternae of the sarcoplasmic reticulum.

b. Sarcomeres, the smallest contractile units of a striated muscle cell, are segments of myofibrils. Each sarcomere has dark A bands and light I bands. The A band contains the M line, the H band, and the zone of overlap. Each I band contains thin filaments, but not thick filaments. Z lines mark the boundaries between adjacent sarcomeres.

c. You would expect the greatest concentration of calcium ions in a resting skeletal muscle to be in the cisternae of the sarcoplasmic reticulum.

Module 9.3 Review

a. Thin filaments consist of actin, troponin, and tropomyosin; thick filaments are composed of myosin surrounding a core of titin.

b. The alternating arrangement of A bands (myosin filaments) and I bands (actin filaments) gives skeletal muscle its characteristic striated appearance.

c. The sliding filament theory describes the process of sarcomere shortening caused by the sliding of thin and thick filaments past one another.

Module 9.4 Review

a. The neuromuscular junction is a specialized intercellular connection that enables a motor neuron to communicate with a skeletal muscle fiber.

b. Acetylcholine release is necessary for skeletal muscle contraction, because it serves as the first step in the process. A drug that blocks acetylcholine release would prevent ACh from binding with receptors on the motor end plate, so sodium ions would not rush into the muscle fiber's sarcoplasm, and no action potential would be generated in the sarcolemma. As a result, muscle contraction could not occur.

c. Without AChE (acetylcholinesterase), the motor end plate would be continuously stimulated by acetylcholine, locking the muscle in a state of contraction.

Module 9.5 Review

a. ATP is the molecule that supplies the energy for a muscle contraction.

b. Once the contraction process has begun, the steps that occur are (1) active-site exposure, (2) cross-bridge formation, (3) myosin head pivoting (power stroke), (4) cross-bridge detachment, and (5) myosin reactivation "recocking."

c. The breakdown of ATP into ADP + P enables myosin reactivation, because the energy released during this process is used to "recock" the myosin heads.

Section 1 Review

1. a. mitochondrion; **b.** sarcolemma; **c.** myofibril; **d.** thin filament; **e.** thick filament; **f.** triad; **g.** sarcoplasmic reticulum; **h.** T tubules; **i.** terminal cisterna; **j.** sarcoplasm; **k.** myofibril

2. a. I band; **b.** A band; **c.** H band; **d.** Z line; **e.** titin; **f.** zone of overlap; **g.** M line; **h.** thin filament; **i.** thick filament; **j.** sarcomere

3. a. myoblast, myofibril, myofilament, myosatellite cell; **b.** sarcolemma, sarcoplasm, sarcomere, sarcoplasmic reticulum

Module 9.6 Review

a. The length of individual sarcomeres is a factor that affects the amount of tension produced when a skeletal muscle contracts.

b. According to the length–tension relationship, (1) the greater the zone of overlap in the sarcomere, the greater the tension the muscle can develop; and (2) there is an optimum range of actin and myosin overlap that will produce the greatest amount of tension.

c. In an initial latent period (after the stimulus arrives and before tension begins to increase), an action potential generated in the muscle triggers the release of calcium ions from the SR. In the contraction phase, calcium binds to troponin (cross-bridges form) and tension begins to increase. In the relaxation phase, tension drops because cross-bridges have detached and because calcium levels have fallen; the active sites are once again covered by the troponin–tropomyosin complex.

Module 9.7 Review

a. A motor unit is all of the muscle cells controlled by a single motor neuron.

b. The finer and more precise the movement produced by a particular muscle, the fewer the number of muscle fibers in the motor unit.

c. Incomplete tetanus refers to a muscle producing near-peak tension during rapid cycles of contraction and relaxation, whereas wave summation refers to the addition of one twitch to another.

Module 9.8 Review

a. In an isotonic contraction, tension rises and the skeletal muscle's length changes; in an isometric contraction, tension rises but the muscle's length does not change, and the load does not move.

b. Yes, a skeletal muscle can contract without shortening, as occurs during an isometric contraction. Whether a contracting muscle shortens (a concentric isotonic contraction), elongates (an eccentric isotonic contraction), or remains the same length (an isometric contraction) depends on the relationship between the resistance and the tension produced by actin–myosin interactions.

c. The heavier the load on a muscle, the longer it will take for the muscle to begin to shorten and the less the muscle will shorten.

Module 9.9 Review

a. Three sources of energy utilized by muscle fibers are ATP, CP, and glycogen.

b. Muscle cells continuously synthesize ATP by utilizing creatine phosphate (CP) and by metabolizing glycogen and fatty acids.

c. Muscle fibers produce lactate under conditions of anaerobic metabolism (when there is a lack of oxygen). These conditions occur at peak levels of muscle activity.

Module 9.10 Review

a. Oxygen debt (or excess postexercise oxygen consumption) is the amount of oxygen intake required after strenuous activity to produce the ATP needed to restore normal, pre-exertion conditions in the body.

b. Two processes that are crucial in repaying a muscle's oxygen debt are (1) aerobic respiration by liver cells in making ATP for converting lactate to glucose, and (2) aerobic respiration by muscle cells in restoring ATP, creatine phosphate, and glycogen concentrations to their former levels.

c. The burning sensation in skeletal muscle following strenuous exercise results from the accumulation of lactate.

a. The three types of skeletal muscle fibers are (1) fast fibers—also called white muscle fibers, fast-twitch glycolytic fibers, Type II-B fibers, and fast fatigue (FF) fibers; (2) slow fibers—also called red muscle fibers, slow-twitch oxidative fibers, Type I fibers, and slow oxidative (SO) fibers; and (3) intermediate fibers—also called fast-twitch oxidative fibers, Type II-A fibers, and fast resistant (FR) fibers.

b. A sprinter requires large amounts of energy for a short burst of activity. To supply this energy, the sprinter's muscles switch to anaerobic metabolism. Anaerobic metabolism is less efficient in producing energy than aerobic metabolism, and the process also produces acidic waste products; this combination contributes to muscle fatigue. Conversely, marathon runners derive most of their energy from aerobic metabolism, which is more efficient and produces fewer waste products than anaerobic metabolism does.

c. Individuals who excel at endurance activities have a higher-than-normal percentage of slow fibers. Slow fibers are physiologically better adapted to this type of activity than are fast fibers, which are less vascular and fatigue faster.

a. Muscle hypertrophy is enlargement of fiber size without cell division (and thus of the entire muscle) stemming from increases in myofibrils, mitochondria, glycolytic enzymes, and glycogen reserves, often in response to activities such as bodybuilding. Muscle atrophy is the wasting away of tissues from a lack of use, ischemia, or nutritional abnormalities.

b. While Fred's leg was immobilized, its muscles did not receive sufficient neural stimulation to maintain normal mass, tone, and strength—that is, his leg muscles atrophied—and the muscles were thus unable to support his weight.

c. A murder victim's time of death can be estimated according to the body's flexibility or rigidity because rigor mortis typically begins a few hours after death, reaches maximum rigidity some 2–7 hours after death, and subsides about 1–6 days later or when decomposition begins. Thus, for example, a victim whose body lacks any signs of rigor mortis likely died within the past few hours. At the molecular level, the membranes of the dead cells are no longer selectively permeable and the SR is no longer able to retain calcium ions. As calcium ions enter the sarcoplasm, a sustained contraction develops, making the body extremely stiff. Contraction persists because the dead muscle cells can no longer make the ATP required for cross-bridge detachment from the active sites. Rigor mortis lasts until the lysosomal enzymes released by autolysis break down the myofilaments.

1. **a.** isometric contraction;
   b. isotonic contraction;
   c. eccentric contraction;
   d. concentric contraction

2. **a.** fatty acids; **b.** O_2; **c.** glucose;
   d. glycogen; **e.** CP; **f.** creatine

3. **a.** Small; **b.** Intermediate;
   c. Large; **d.** Red; **e.** White;
   f. Low; **g.** Low; **h.** Dense;
   i. Intermediate; **j.** Many;
   k. Intermediate; **l.** Few;
   m. Prolonged; **n.** Rapid;
   o. Slow; **p.** Fast; **q.** Fast;
   r. Intermediate; **s.** Low;
   t. Low; **u.** High

1. Because organophosphates block the action of acetylcholinesterase, ACh released into the synaptic cleft at neuromuscular junctions would not be removed. The ACh would continue to stimulate the motor end plate and produce action potentials, resulting in a state of extended, uncoordinated contractions. At toxic levels associated with organophosphate poisoning, sustained stimulation leads to muscle fatigue and inexcitability. Twitching, weakness, or paralysis of the diaphragm muscle (the main respiratory muscle) could cause death by suffocation.

2. The chemical could be absorbed through the skin, leading to rashes and localized irritation.

Module 10.1 Review

a. A lever is a rigid structure (such as a bone) that pivots around a fixed point called the fulcrum. In a first-class lever, the fulcrum lies between the applied force and the load; in a second-class lever, the load lies between the fulcrum and the applied force; and in a third-class lever, the applied force is between the fulcrum and the load.

b. The joint between the occipital bone and the first cervical vertebra (atlas) is the fulcrum of a first-class lever; the joint (fulcrum) lies between the skull (the load) and the neck muscles (which supply the applied force).

c. Contraction of a pennate muscle generates more tension than would contraction of a parallel muscle of the same size because pennate muscles contain more muscle fibers—and thus more myofibrils and sarcomeres—than do parallel muscles of the same size.

Module 10.2 Review

a. A synergist is a muscle that helps a larger prime mover (or agonist—a muscle that is responsible for a specific movement) perform its actions more efficiently.

b. Muscles A and B are antagonists, because they perform opposite actions.

c. The name *flexor carpi radialis longus* tells you that this muscle is a long muscle that lies next to the radius and flexes the wrist.

Module 10.3 Review

a. The axial muscles, which arise on the axial skeleton, position the head and spinal column and move the rib cage to make breathing possible.

b. biceps brachii: appendicular; external oblique: axial; temporalis: axial; vastus medialis: appendicular.

c. The following structures labeled in the figures in this module are not muscles: linea alba, flexor retinaculum, iliotibial tract, patella, tibia, clavicle, sternum, superior extensor retinaculum, inferior extensor retinaculum, lateral malleolus

of fibula, medial malleolus of tibia, calcaneal tendon, and calcaneus.

Section 1 Review

1. **a.** deltoid: multipennate; **b.** extensor digitorum: unipennate; **c.** rectus femoris: bipennate

2. **a.** sternocleidomastoid; **b.** deltoid; **c.** biceps brachii; **d.** external oblique; **e.** pronator teres; **f.** brachioradialis; **g.** flexor carpi radialis; **h.** rectus femoris; **i.** vastus lateralis; **j.** vastus medialis; **k.** gastrocnemius; **l.** soleus; **m.** pectoralis major; **n.** rectus abdominis; **o.** iliopsoas; **p.** tensor fasciae latae; **q.** gracilis; **r.** sartorius; **s.** tibialis anterior; **t.** extensor digitorum longus

Module 10.4 Review

a. The muscles associated with the mouth are the buccinator, depressor labii inferioris, levator labii superioris, levator anguli oris, mentalis, orbicularis oris, risorius, depressor anguli oris, zygomaticus major, and zygomaticus minor muscles.

b. buccinator: mouth; corrugator supercilii: eye; mentalis: mouth; nasalis: nose; platysma: neck; procerus: nose; risorius: mouth

c. An individual is able to consciously move the skin on the scalp because the fibers of the epimysium are woven into the fibers of the superficial fascia and the dermis; in the thigh, however, muscle fibers and fibers of the epimysium insert into bone or tendons, not into the skin.

Module 10.5 Review

a. The extrinsic eye muscles are the inferior rectus, medial rectus, superior rectus, lateral rectus, inferior oblique, and superior oblique muscles.

b. The muscles with these origins and insertions are the medial and lateral pterygoid muscles.

c. Contraction of the masseter muscle elevates the mandible, and relaxation of this muscle depresses the mandible, so you would probably be eating or chewing something.

Module 10.6 Review

a. The muscles of the tongue are the genioglossus, hyoglossus, palatoglossus, and styloglossus muscles.

b. The muscles associated with the hyoid that form the floor of the mouth are the mylohyoid, geniohyoid, and digastric muscles.

c. The muscles that elevate the soft palate are the palatal muscles.

Module 10.7 Review

a. The spinal flexor muscles are the longus capitis, longus colli, and quadratus lumborum muscles.

b. The spinalis cervicis muscles enable you to extend the neck.

Module 10.8 Review

a. The rectus abdominis connects the ribs and sternum to the pubic bones.

b. The transversus abdominis forms the deepest layer of the abdominal wall muscles.

c. The external oblique muscle compresses the contents of the abdominal cavity, depresses the ribs, and flexes or bends the spine.

Module 10.9 Review

a. The external urethral sphincter and deep transverse perineal muscles make up the urogenital diaphragm.

b. In females, the bulbospongiosus muscle compresses and stiffens the clitoris and narrows the vaginal opening.

c. The coccygeus muscle extends from the sacrum and coccyx to the ischial spine.

Section 2 Review

1. **a.** occipitofrontalis (frontal belly); **b.** temporalis; **c.** orbicularis oculi; **d.** levator labii superioris; **e.** zygomaticus minor; **f.** zygomaticus major; **g.** buccinator; **h.** orbicularis oris; **i.** risorius; **j.** depressor labii inferioris; **k.** depressor anguli oris; **l.** masseter

2. **a.** mylohyoid; **b.** digastric; **c.** geniohyoid; **d.** omohyoid; **e.** stylohyoid; **f.** thyrohyoid;

g. sternothyroid; **h.** sternohyoid; **i.** sternocleidomastoid

Module 10.10 Review

a. The axial muscle known as the "six-pack" in fit individuals is the rectus abdominis.

b. As you move proximally to distally, appendicular muscles become smaller and more numerous, enabling more precise movements.

c. deltoid: appendicular; external oblique: axial; gluteus maximus: appendicular; pectoralis major: appendicular; platysma: axial; rectus femoris: appendicular

Module 10.11 Review

a. The trapezius is the largest of the superficial muscles that position the pectoral girdle.

b. The levator scapulae muscles enable you to shrug your shoulders.

c. The subclavius originates on the first rib and inserts on the inferior border of the clavicle.

Module 10.12 Review

a. The action line is the line of force produced when a muscle contracts.

b. The deltoid muscle abducts the upper arm.

c. The subscapularis muscle originates on the anterior surface of the scapula and inserts on the lesser tubercle of the humerus.

Module 10.13 Review

a. A retinaculum is a wide band of connective tissue that stabilizes tendons.

b. The wrist extensors are located on the posterior surface.

c. The supinator, pronator quadratus, and pronator teres muscles are involved in turning a doorknob.

Module 10.14 Review

a. The muscles that extend the fingers are the extensor digitorum and extensor digiti minimi muscles.

b. The abductor pollicis longus, extensor pollicis brevis, and extensor pollicis longus abduct the wrist.

c. The names of muscles associated with the thumb, or pollex, frequently include the term pollicis.

Module 10.15 Review

a. The intrinsic muscles of the thumb are the adductor pollicis, flexor pollicis brevis, opponens pollicis, and abductor pollicis brevis muscles.

b. There are no muscles that originate on the phalanges.

c. We are able to move the fingers, even though there are no muscles there, because tendons from forearm muscles extend across the distal finger joints. Contractions of those muscles move the fingers.

Module 10.16 Review

a. The muscles that compose the gluteal group are the gluteus maximus, gluteus medius, gluteus minimus, and tensor fasciae latae muscles.

b. The muscle whose origin is the lateral border of the ischial tuberosity and whose insertion is the intertrochanteric crest of the femur is the quadratus femoris.

c. Injury to the obturator muscles would impair the ability to perform lateral rotation at the hip.

Module 10.17 Review

a. The quadriceps muscles are the rectus femoris, vastus intermedius, vastus lateralis, and vastus medialis muscles.

b. The muscles that flex the knee are the biceps femoris, semimembranosus, semitendinosus, sartorius, and popliteus muscles.

c. The muscle whose origin is on the lateral condyle of the femur is the popliteus.

Module 10.18 Review

a. The muscles involved in flexing the toes are the flexor digitorum longus and flexor hallucis longus muscles.

b. The muscles involved in extending the ankle are the plantaris, gastrocnemius, soleus, tibialis posterior, fibularis longus, and fibularis brevis muscles.

c. A torn calcaneal tendon would make plantar flexion difficult, because this tendon attaches the soleus and gastrocnemius muscles to the calcaneus (heel bone).

Module 10.19 Review

a. The flexor hallucis brevis muscle flexes the great toe.

b. The retinacula stabilize the positions of the tendons descending from the leg.

c. The lumbrical muscles originate on the tendons of the flexor digitorum, and insert on the tendons of the extensor digitorum longus of toes two to five. Their contraction results in flexion at the proximal metatarsophalangeal joints and extension at the distal interphalangeal joints.

Module 10.20 Review

a. The six possible compartments of the muscles of the limbs are the lateral compartment, medial compartment, anterior superficial compartment, anterior deep compartment, posterior superficial compartment, and posterior deep compartment.

b. Compartment syndrome is a condition in which increased pressure within a compartment, a confined anatomical space, adversely affects circulation.

c. Compartment syndrome can be life threatening because the resultant ischemia leads to tissue death.

Section 3 Review

1. a. triceps brachii; b. anconeus; c. extensor carpi ulnaris; d. extensor carpi radialis brevis; e. extensor digitorum; f. flexor carpi ulnaris

2. a. gluteus medius; b. tensor fasciae latae; c. gluteus maximus; d. adductor magnus; e. gracilis; f. biceps femoris; g. semitendinosus; h. semimembranosus; i. sartorius; j. popliteus

3. a. gastrocnemius; b. tibialis anterior; c. fibularis longus; d. soleus; e. extensor digitorum longus; f. fibularis brevis; g. superior extensor retinaculum; h. calcaneal tendon; i. inferior extensor retinaculum

Chapter Integration

1. The triceps brachii and biceps brachii muscles are antagonists; each muscle opposes the movement of the other. To achieve muscle hypertrophy (increased muscle size) and increased definition of both muscles, both exercises need to be done.

2. Sit-ups and trunk-twisting movements will stimulate the abdominal muscles of interest: the rectus abdominis, external and internal obliques, and latissimus dorsi muscles. Placing some weight on the chest while doing sit-ups will produce faster results because the rectus abdominis muscle must then work against a greater load.

Module 11.1 Review

a. Structural components of a typical neuron include a cell body (including the nucleus and perikaryon, plus neurofilaments and neurotubules), an axon (including the axon hillock, axoplasm, axolemma, telodendria, collateral branches, and synaptic terminals), and dendrites (including dendritic spines).

b. A synapse is the site of communication between a neuron and some other cell; if the other cell is not a neuron, the term neuromuscular junction or neuroglandular synapse is often used.

c. Most CNS neurons lack centrioles, which organize the microtubules of the spindle apparatus during mitosis, so these cells cannot divide and replace themselves.

Module 11.2 Review

a. According to structure, neurons are classified as anaxonic, bipolar, unipolar, or multipolar.

b. According to function, neurons are classified as sensory neurons, motor neurons, or interneurons.

c. Most sensory neurons of the PNS are unipolar, so these neurons most likely function as sensory neurons.

Module 11.3 Review

a. Central nervous system neuroglia are ependymal cells, astrocytes, oligodendrocytes, and microglia.

b. Astrocytes protect the CNS from circulating chemicals and hormones by maintaining the blood–brain barrier.

c. Microglia, small phagocytic cells, occur in increased numbers in infected (and damaged) areas of the CNS.

Module 11.4 Review

a. Neuroglia of the peripheral nervous system are satellite cells and Schwann cells.

b. The neurilemma is the outer surface of a Schwann cell encircling an axon.

c. Wallerian degeneration occurs in the PNS, where Schwann cells participate in the repair of damaged nerves.

Section 1 Review

1. a. neuron, neuroglia, neurofilaments, neurofibrils, neurotubules, neurilemma; b. dendrite, dendritic spines, telodendria, oligodendrocytes; c. efferent division, efferent fibers; d. afferent division, afferent fibers
2. a. dendrite; b. Nissl bodies; c. mitochondrion; d. nucleus; e. nucleolus; f. cell body; g. axon hillock; h. axolemma; i. axon; j. telodendrion; k. synaptic terminal
3. a. multipolar; b. unipolar; c. anaxonic; d. bipolar

Module 11.5 Review

a. The resting potential is the transmembrane potential of an undisturbed (nonstimulated) cell.

b. Decreasing the concentration of extracellular potassium ions would cause more potassium to leave the cell, which would make the transmembrane potential of the neuron more negative.

c. The sodium–potassium exchange pump maintains the cell's resting potential by ejecting three sodium ions from the cell for every two potassium ions it recovers from the extracellular fluid.

Module 11.6 Review

a. Gated channels are active channels in the plasma membrane that typically open in response to specific stimuli.

b. Chemically gated channels operate when they bind specific chemicals (such as ACh); voltage-gated channels operate in response to changes in the transmembrane potential; and mechanically gated channels operate in response to physical distortion of the membrane surface.

c. If the voltage-gated sodium channels in a neuron's cell membrane could not open, sodium ions could not rush into the cell, and its transmembrane potential would not change.

Module 11.7 Review

a. A graded potential (also called a local potential) is a change in the transmembrane potential that cannot spread far from the site of stimulation.

b. Depolarization is a shift from the resting potential in which the transmembrane potential becomes less negative. Repolarization is the return of the transmembrane potential to the resting potential after the membrane has been depolarized. Hyperpolarization is a shift from the resting potential in which the transmembrane potential becomes more negative.

c. Movement of sodium ions parallel to the inner and outer surfaces of the plasma membrane—after passing through open chemically gated sodium channels—accounts for the local currents associated with graded potentials.

Module 11.8 Review

a. An action potential is a propagated change in the transmembrane potential of excitable cells, initiated by a change in the plasma membrane's permeability to sodium ions.

b. The events involved in the generation of action potentials are (1) depolarization to threshold, (2) activation of sodium channels and rapid depolarization, (3) inactivation of sodium channels and activation of potassium channels, and (4) closing of potassium channels.

c. The refractory period is the time between the initiation of an action potential and the restoration of the normal resting potential. The absolute refractory period is the portion of the refractory period during which the membrane cannot respond to further stimulation, no matter its magnitude. The relative refractory period is the time during which the membrane can respond only to a larger-than-normal stimulus.

Module 11.9 Review

a. Continuous propagation is the propagation of an action potential along an unmyelinated axon, wherein the action potential affects every portion of the membrane surface. Saltatory propagation is the relatively rapid propagation of an action potential between successive nodes of a myelinated axon.

b. The presence of myelin greatly increases the propagation speed of action potentials.

Module 11.10 Review

a. The parts of a chemical synapse—the site where a neuron communicates with another neuron or with a cell of a different type—are a presynaptic cell and a postsynaptic cell, whose plasma membranes are separated by a narrow gap called the synaptic cleft.

b. In chemical synapses, a neurotransmitter crosses a narrow synaptic cleft, whereas in electrical synapses the membranes of the presynaptic and postsynaptic cells are joined together by gap junctions.

c. Synaptic fatigue occurs in a synaptic knob when intensive stimulation exceeds the knob's ability to keep pace with the demand for neurotransmitter. It is reversed and eliminated by the resynthesis of the neurotransmitter.

a. An excitatory postsynaptic potential (EPSP) is a graded depolarization of a postsynaptic membrane by a chemical neurotransmitter released by a presynaptic cell. An inhibitory postsynaptic potential (IPSP) is a graded hyperpolarization of a postsynaptic membrane after the arrival of a neurotransmitter.

b. Temporal summation is the addition of a rapid succession of stimuli occurring at a single synapse. Spatial summation involves the addition of simultaneous stimuli applied at different locations; that is, it involves multiple synapses that are active simultaneously.

c. No action potential will be generated, because the depolarization did not reach threshold.

a. Regulatory neurons facilitate or inhibit the activities of presynaptic neurons by affecting the plasma membrane of the cell body, or by altering the sensitivity of synaptic knobs.

b. The degree of sustained depolarization at the axon hillock determines the frequency of action potential generation.

c. The greater the degree of sustained depolarization at the axon hillock, the *higher* the frequency of generation of action potentials.

1. a. action potential; b. electrical synapse; c. resting potential; d. gated channels; e. cholinergic synapses; f. hyperpolarization; g. local current; h. depolarization

2. a. acetylcholine (ACh); b. calcium ions (Ca^{2+}); c. synaptic vesicle; d. acetylcholinesterase (AChE); e. ACh receptor; f. sodium ions (Na^+); g. an action potential depolarizes the synaptic knob; h. calcium ions enter the cytoplasm of the synaptic knob; i. ACh is released through exocytosis; j. ACh binds to sodium channel receptors on the postsynaptic membrane, producing a graded depolarization; k. the depolarization ends as ACh is broken down into acetate and choline by AChE; l. the synaptic knob reabsorbs choline from the synaptic cleft and uses it to synthesize new molecules of ACh

3. In myelinated fibers, saltatory propagation transmits nerve impulses to the neuromuscular junctions rapidly enough to initiate muscle contractions and promote normal movements.

In axons that have become demyelinated, nerve impulses cannot be propagated, and so the muscles are not stimulated to contract. Eventually, the muscles atrophy because of the lack of stimulation (a condition termed disuse atrophy).

Chapter Integration

1. Demyelination is the loss or destruction of the myelin sheaths that insulate nerve fibers.

2. The loss of myelin slows nerve impulse propagation, so within the CNS information about limb movement and body position moves slowly; and motor commands move slowly and erratically.

3. The glial cells affected by MS are oligodendrocytes, which form the myelin sheath around axons in the CNS.

Module 12.1 Review

a. A typical spinal cord has 31 pairs of nerves, and the spinal cord ends at the level of lumbar vertebra 1 or 2 (L_1 or L_2).

b. The gray matter in the spinal cord is composed of the cell bodies of neurons, neuroglia, and unmyelinated axons.

c. Gross anatomical features of the cross-sectioned spinal cord include the anterior median fissure (a deep groove along the anterior or ventral surface); the posterior median sulcus (a shallow longitudinal groove); white matter (composed of myelinated and unmyelinated axons); gray matter (composed of cell bodies of neurons, neuroglia, and unmyelinated axons); the central canal (a passageway containing cerebrospinal fluid); a dorsal root of each spinal nerve (axons of neurons whose cell bodies are in the dorsal root ganglion); a ventral root of each spinal nerve (the axons of motor neurons that extend into the periphery to control somatic and visceral effectors); dorsal root ganglia (contain cell bodies of sensory neurons); and spinal nerves (contain the axons of sensory and motor neurons).

Module 12.2 Review

a. The three spinal meninges are the dura mater (the outermost component of the cranial and spinal meninges), arachnoid mater (the middle meninx that encloses cerebrospinal fluid), and pia mater (the innermost layer of the meninges bound to the underlying neural tissue).

b. Cerebrospinal fluid is found in the subarachnoid space, which lies beneath the epithelium of the arachnoid mater and superficial to the pia mater.

c. The lumbar puncture needle would penetrate the epidermis, dermis, subcutaneous layer (hypodermis), and then skeletal muscle before reaching the protective spinal coverings: the dura mater, then the arachnoid matter, and finally the subarachnoid space, which contains cerebrospinal fluid.

Module 12.3 Review

a. Sensory nuclei receive and relay sensory information from peripheral receptors; motor nuclei issue motor commands to peripheral effectors.

b. The poliovirus-infected neurons would be in the anterior gray horns of the spinal cord, where the cell bodies of somatic motor neurons are located.

c. A disease that damages myelin sheaths would affect the white matter columns of the spinal cord, which are composed of bundles of myelinated axons.

Module 12.4 Review

a. The three layers of connective tissue of a spinal nerve are the outer epineurium, middle perineurium, and inner endoneurium; and the major peripheral branches of a spinal nerve are the dorsal root and ventral root.

b. A dermatome is a specific bilateral sensory region monitored by a single pair of spinal nerves.

c. Shingles is caused by a reactivation of the varicella-zoster virus (VZV), the same herpes virus that causes chickenpox. Once reactivated, the virus stimulates painful inflammation of the nerve ganglia and causes skin eruptions in a pattern that reflects the affected dermatome.

Module 12.5 Review

a. The gray ramus is a bundle of postganglionic sympathetic nerve fibers that are distributed to effectors in the body wall, skin, and limbs by way of a spinal nerve; the white ramus is a nerve bundle containing the myelinated preganglionic axons of sympathetic motor neurons en route to sympathetic ganglia.

b. 1) = white rami
2) = gray rami

c. The dorsal ramus of each thoracic or superior lumbar spinal nerve innervates the skin and muscles of the back.

Module 12.6 Review

a. A nerve plexus is a complex, interwoven network of nerves. The major plexuses are the cervical, brachial, lumbar, and sacral plexuses.

b. An anesthetic that blocks the function of the dorsal rami of the cervical spinal nerves would affect the skin and muscles of the back of the neck and of the shoulders.

c. Damage to the cervical plexus—more specifically, to the phrenic nerves, which originate in this plexus and innervate the diaphragm—would interfere with the ability to breathe.

Module 12.7 Review

a. A trunk is a large bundle of axons contributed by several spinal nerves; a cord is a smaller branch of nerves that originates at a trunk.

b. The brachial plexus is a network of nerves formed by branches of spinal nerve segments C_4–T_1, en route to innervating the upper limb.

c. The major nerves associated with the brachial plexus are the dorsal scapular, long thoracic, suprascapular, medial and lateral pectoral, subscapular, thoracodorsal, axillary, medial antebrachial cutaneous, radial, musculocutaneous, median, and ulnar nerves.

Module 12.8 Review

a. The lumbar plexus is a nerve network formed by axons from the ventral rami of spinal nerve segments T_{12}–L_4; the sacral plexus is a nerve network formed by the ventral rami of spinal nerve segments L_4–S_4.

b. The major nerves of the sacral plexus are the superior and inferior gluteal nerves; the posterior femoral cutaneous nerve; the sciatic nerve, which branches into the tibial nerve and the fibular nerve; and the pudendal nerve.

c. Compression of the sciatic nerve produces the sensation that your lower limb has "fallen asleep."

Section 1 Review

1. a. white matter; **b.** dorsal root ganglion; **c.** lateral white column; **d.** posterior gray horn; **e.** lateral gray horn; **f.** anterior gray horn; **g.** posterior median sulcus; **h.** central canal; **i.** sensory nuclei; **j.** motor nuclei; **k.** anterior gray commissure; **l.** ventral root; **m.** anterior white commissure; **n.** anterior median fissure

2. a. anterior view; **b.** radial nerve; **c.** ulnar nerve; **d.** median nerve; **e.** posterior view

3. a. columns; **b.** conus medullaris; **c.** nerves; **d.** meninges; **e.** cauda equina; **f.** brachial plexus; **g.** dura mater; **h.** perineurium; **i.** gray ramus

Module 12.9 Review

a. A reflex is a rapid, automatic response to a specific stimulus. All reflex arcs include a receptor, a sensory neuron, a motor neuron, and a peripheral effector; interneurons may or may not be present as well.

b. All reflexes are rapid, unconscious patterned responses to a physical stimulus, which restore or maintain homeostasis.

c. Reflexes are classified according to their development (innate reflexes vs. acquired reflexes), the nature of the resulting motor response (somatic reflexes vs. visceral reflexes), the complexity of the neural circuit involved (polysynaptic reflexes vs. monosynaptic reflexes), and the site of information processing (spinal reflexes vs. cranial reflexes).

Module 12.10 Review

a. A stretch reflex is a monosynaptic reflex that provides automatic regulation of skeletal muscle length.

b. In the patellar reflex, the response observed is leg extension, and the effectors involved are the quadriceps femoris muscles.

c. When stretch receptors are stimulated by gamma motor neurons, the muscle spindles become more sensitive. As a result, little (if any) stretching stimulus is needed to stimulate the contraction of the quadriceps muscles. Thus, the reflex

response would occur more quickly.

Module 12.11 Review

a. All polysynaptic reflexes involve pools of interneurons, are intersegmental in distribution, involve reciprocal inhibition, and have reverberating circuits.

b. The flexor reflex is an example of a withdrawal reflex that contracts the flexor muscles of a limb in response to a painful stimulus; hence, it has a protective function.

c. During a withdrawal reflex, the limb on the opposite side is extended. This response is called a crossed extensor reflex.

Module 12.12 Review

a. Reinforcement is an enhancement of a spinal reflex through the facilitation of motor neurons involved in reflexes.

b. Reflex testing provides information about the nervous system's functional status.

c. A positive Babinski reflex is abnormal in adults; it indicates possible damage of descending tracts in the spinal cord.

Section 2 Review

1. a. divergence; b. convergence; c. serial processing; d. parallel processing; e. reverberation

2. a. receptor; b. sensory neuron; c. interneuron; d. spinal cord (CNS); e. motor neuron; f. effector

3. The withdrawal reflex illustrated in Question 2 is an innate, somatic, polysynaptic, spinal reflex.

4. a. ipsilateral reflex; b. withdrawal reflexes; c. gamma motor neuron; d. flexor reflex; e. visceral reflexes; f. acquired reflexes; g. contralateral reflex; h. reciprocal inhibition; i. reinforcement

Chapter Integration

1. The anterior horn in the lumbar region of the spinal cord contains somatic motor neurons that direct the activity of skeletal muscles of the hip, lower limb, and foot. As a result of the injury, Karen would be expected to have poor control of most muscles of the lower limbs, causing difficulty walking (if she could walk at all) and

problems maintaining balance (if she could stand).

2. The individual would still exhibit a defecation (bowel) and urination (urinary bladder) reflex because these spinal reflexes are processed at the level of the spinal cord. Afferent impulses from the organs would stimulate specific interneurons in the sacral region that synapse with the motor neurons controlling the sphincters, thus bringing about emptying when the organs began to fill. However, an individual with the spinal cord transection at L_1 would lose voluntary control of the bowel and urinary bladder, because these functions rely on impulses carried by motor neurons in the brain that must travel down the spinal cord and synapse with the interneurons and motor neurons involved in the reflex.

3. The effects of a transection of the spinal cord at L_1 in an adult are the same as the situation in newborns, in whom the descending tracts required for conscious control of urination have not yet fully developed.

Module 13.1 Review

a. The major regions of the brain are the cerebrum (composed of fissures, gyri, and sulci), the diencephalon (thalamus and hypothalamus), the cerebellum, and the brain stem (midbrain, pons, and medulla oblongata). Additionally, the brain contains four ventricles and some connecting passageways (the interventricular foramen and the aqueduct of the midbrain), plus the corpus callosum and the septum pellucidum.

b. The medulla oblongata (the most caudal of the brain regions) relays sensory information to other parts of the brain stem and to the thalamus. It also contains centers that regulate autonomic function, such as heart rate and blood pressure.

c. The corpus callosum is a bundle of axons that links the left and right cerebral hemispheres, whereas the septum pellucidum is a partition that separates the two lateral ventricles.

Module 13.2 Review

a. The layers of the cranial meninges are the outer dura mater, the middle arachnoid mater, and the inner pia mater.

b. If the normal movement of CSF were blocked, CSF would continue to be produced at the choroid plexuses in each ventricle, but the fluid would remain there, causing the ventricles to swell—a condition known as hydrocephalus.

c. If diffusion across the arachnoid granulations decreased, the volume of CSF in the ventricles would increase, because less CSF would reenter the bloodstream. The increased pressure within the brain due to accumulated CSF could damage the brain.

Module 13.3 Review

a. The ascending and descending tracts of the white matter in the medulla oblongata link the brain with the spinal cord.

b. The nucleus gracilis and nucleus cuneatus are parts of the medulla oblongata that relay somatic sensory information to the thalamus.

c. The pyramids contain tracts of motor fibers that originate at the cerebral cortex; the result of decussation (crossing over) is that each side of the motor cortex controls the opposite side of the body.

Module 13.4 Review

a. The components of the cerebellar gray matter are the cerebellar cortex and the cerebellar nuclei.

b. The arbor vitae, which is the white matter of the cerebellum, connects the cerebellar cortex and nuclei with the cerebellar peduncles.

c. Ataxia is the failure of muscular coordination that can result from trauma, stroke, or certain drugs, including alcohol.

Module 13.5 Review

a. Cranial nerves III to XII arise from the brain stem.

b. The two pairs of sensory nuclei contained within the corpora quadrigemina are the superior colliculi and inferior colliculi.

c. The superior colliculi of the midbrain control reflexive movements of the eyes, head, and neck.

Module 13.6 Review

a. The main components of the diencephalon are the epithalamus, thalamus, and hypothalamus.

b. Damage to the lateral geniculate nuclei would interfere with the flow of visual information and thus affect the sense of sight.

c. The preoptic area of the hypothalamus, a component of the diencephalon, is stimulated by changes in body temperature.

Module 13.7 Review

a. The limbic system establishes emotional states; links the conscious, intellectual functions of the cerebral cortex with the unconscious and autonomic functions of the brain stem; and facilitates memory storage and retrieval.

b. The hippocampus is important in the storage and retrieval of long-term memories.

c. Damage to the amygdaloid body would interfere with the sympathetic ("fight or flight") division of the autonomic nervous system.

Module 13.8 Review

a. The basal nuclei (also known as the basal ganglia) are masses of cerebral gray matter that function in the subconscious control of skeletal muscle activity.

b. The caudate nucleus is one of the basal nuclei involved with the subconscious control of skeletal muscular activity.

c. Damage to the basal nuclei would result in decreased muscle tone and the loss of coordination of learned movement patterns.

Module 13.9 Review

a. The lobes of the cerebrum— the frontal lobe, parietal lobe, occipital lobe, and temporal lobe—are named for the overlying bones of the skull.

b. The insula is an island of cortex located medial to the lateral sulcus.

c. Damage to the left postcentral gyrus would interfere with the awareness of sensory information from the right side of the body.

Module 13.10 Review

a. The primary motor cortex is located in the precentral gyrus of the frontal lobe of the cerebrum.

b. Damage to the temporal lobes of the cerebrum would interfere with the processing of olfactory (smell) and auditory (sound) sensations.

c. The stroke has damaged the speech center, located in the frontal lobe.

Module 13.11 Review

a. The axons in the cerebral white matter are called association fibers, commissural fibers, and projection fibers.

b. The longitudinal fasciculi connect the frontal lobe to the other lobes of the same hemisphere.

c. Projection fibers carry information between the cerebral cortex and the spinal cord, in the process passing through the diencephalon, brain stem, and cerebellum.

Module 13.12 Review

a. An electroencephalogram (EEG) is a graph of the electrical activity of the brain.

b. The four wave types associated with an EEG are alpha waves (characteristic of normal resting adults), beta waves (characteristic of a person who is concentrating), theta waves (observed in children and frustrated adults), and delta waves (found in a person who is sleeping deeply or in individuals with certain pathological states).

c. A seizure is a temporary cerebral disorder accompanied by abnormal movements, unusual sensations, inappropriate behavior, or some combination of these signs and symptoms. Epilepsy is a clinical condition characterized by seizures.

Module 13.13 Review

a. The cranial nerves are the olfactory (I), optic (II), oculomotor (III), trochlear (IV), trigeminal (V), abducens (VI), facial (VII), vestibulocochlear (VIII), glossopharyngeal (IX), vagus (X), accessory (XI), and hypoglossal (XII) nerves.

b. The cranial nerves with motor functions only are the oculomotor (III), trochlear (IV), abducens (VI), accessory (XI), and hypoglossal (XII).

c. The cranial nerves with mixed functions are the trigeminal (V), facial (VII), glossopharyngeal (IX), and vagus (X) nerves.

Section 1 Review

1. a. precentral gyrus; **b.** frontal lobe; **c.** lateral sulcus; **d.** temporal lobe; **e.** pons; **f.** central sulcus; **g.** postcentral gyrus; **h.** parietal lobe; **i.** occipital lobe; **j.** cerebellum; **k.** medulla oblongata

2. **a.** olfactory bulb (associated with cranial nerve I, olfactory), S; **b.** oculomotor (III), M; **c.** trigeminal (V), B; **d.** facial (VII), B; **e.** glossopharyngeal (IX), B; **f.** vagus (X), B; **g.** optic (II), S; **h.** trochlear (IV), M; **i.** abducens (VI), M; **j.** vestibulocochlear (VIII), S; **k.** hypoglossal (XII), M; **l.** accessory (XI), M

3. **a.** thalamus; **b.** arcuate fibers; **c.** fornix; **d.** commissural fibers; **e.** basal nuclei

4. The sensory innervation of the nasal lining, or nasal mucosa, is by way of the maxillary branch of the trigeminal nerve (V). Irritation of the nasal lining increases the frequency of action potentials along the maxillary branch of the trigeminal nerve through the semilunar ganglion to reach centers in the midbrain, which in turn excite the neurons of the reticular activating system (RAS). Increased activity by the RAS can raise the cerebrum back to consciousness.

Module 13.14 Review

a. The four types of general sensory receptors (and the stimuli that excite them) are nociceptors (pain), thermoreceptors (temperature), mechanoreceptors (physical distortion), and chemoreceptors (chemicals dissolved in body fluids).

b. The three classes of mechanoreceptors are tactile receptors, which respond to the sense of touch; baroreceptors, which detect changes in pressure; and proprioceptors, which monitor the positions of bones, joints, and muscles.

c. Adaptation is a decrease in receptor sensitivity after chronic stimulation. Peripheral adaptation reduces the amount of information from receptors that reaches the central nervous system. In central adaptation, awareness of a stimulus virtually disappears, even though sensory neurons in the CNS remain active.

Module 13.15 Review

a. The six types of tactile receptors are free nerve endings (are sensitive to touch and pressure), the root hair plexus (monitors distortions of and movements across the body surface), tactile discs and Merkel cells (detect fine touch and pressure), tactile corpuscles (detect fine touch and pressure), lamellated corpuscles (are sensitive to pulsing or vibrating stimuli, such as deep pressure), and Ruffini corpuscles (are sensitive to pressure and distortion of the skin).

b. Tactile receptors found only in the dermis are tactile corpuscles, lamellated corpuscles, and Ruffini corpuscles.

c. A Ruffini corpuscle is more sensitive to continuous deep pressure because, unlike a lamellated corpuscle, it undergoes little adaptation.

Module 13.16 Review

a. A sensory homunculus is a functional map of the primary sensory cortex.

b. The lateral spinothalamic tracts carry action potentials generated by nociceptors.

c. The left cerebral hemisphere (specifically, the primary sensory cortex in that hemisphere) receives impulses conducted by the right fasciculus gracilis.

Module 13.17 Review

a. Corticospinal tracts are descending tracts that carry motor commands from the cerebral cortex to the anterior gray horns of the spinal cord.

b. The corticobulbar tracts (which are descending tracts) carry information or commands from the cerebral cortex to nuclei and centers in the brain stem.

c. Increased stimulation of the motor neurons of the red nucleus would increase stimulation of the skeletal muscles in the upper limbs, thereby increasing their muscle tone.

Module 13.18 Review

a. The basic motor patterns related to eating and drinking are controlled by the hypothalamus.

b. The thalamus and midbrain control reflexes in response to visual and auditory stimuli experienced while viewing a movie.

c. As you decide to hit the ball, the motor association areas receive information from the frontal lobes and then relay that information to the basal nuclei and cerebellum. As the hitting movement begins, the motor association areas send additional information to the primary motor cortex.

Module 13.19 Review

a. Referred pain is a sensation felt in a part of the body other than its actual source.

b. After being bitten by a rabid animal, the rabies virus infects peripheral nerves. Retrograde flow then carries the viral particles into the CNS.

c. Amyotrophic lateral sclerosis (ALS), commonly called Lou Gehrig disease, is a progressive degeneration of the motor neurons of the CNS, leading to muscle atrophy and eventual paralysis.

Section 2 Review

1. **a.** free nerve ending; **b.** root hair plexus; **c.** Merkel cells and tactile discs; **d.** tactile corpuscle; **e.** Ruffini corpuscle; **f.** lamellated corpuscle

2. **a.** lateral corticospinal tract of corticospinal pathway (conscious control of skeletal muscles throughout the body); **b.** rubrospinal tract of lateral pathway (subconscious regulation of muscle tone and movement of distal limb muscles); **c.** reticulospinal tract of medial pathway (subconscious regulation of muscle tone, and movements of the neck, trunk, and proximal limb muscles); **d.** vestibulospinal tract of medial pathway (subconscious regulation of muscle tone, and movements of the neck, trunk, and proximal limb muscles); **e.** tectospinal tract of medial pathway (subconscious regulation of muscle tone, and movements of the neck, trunk, and proximal limb muscles in response to bright lights, sudden movements, and loud noises); **f.** anterior corticospinal tract of corticospinal pathway (conscious control of skeletal muscles throughout the body); **g.** posterior column pathway (carries sensations of "fine" touch, pressure, vibration, and proprioception); **h.** posterior spinocerebellar tract of spinocerebellar pathway (carries proprioceptive information about the position of skeletal muscles, tendons, and joints); **i.** lateral spinothalamic tract of spinothalamic pathway (carries pain and temperature sensations); **j.** anterior spinocerebellar tract of spinocerebellar pathway (carries proprioceptive information about the position of skeletal muscles, tendons, and joints); **k.** anterior spinothalamic tract of spinothalamic pathway (carries "crude" touch and pressure sensations)

3. **a.** anterior; **b.** posterior

4. Injuries to the motor cortex eliminate the ability to exert fine control over motor units, but gross movements may still be produced by cerebral nuclei using the reticulospinal or rubrospinal tracts.

Chapter Integration

1. Most of the functional problems observed in shaken-baby syndrome result from trauma to the cerebral hemispheres due to contact between the brain and the inside of the skull.

2. When a baby is forcefully shaken, the neck muscles are not yet well enough developed to act as shock absorbers and prevent whiplash movements of the head. Such violent movement pitches the infant's brain back and forth within the skull, so that it strikes the inside of the skull. The results can be bruising to and bleeding within the cranial meninges, the brain, and/or the tearing of brain tissue and the nerves within it.

3. Damage to the medulla oblongata can be fatal because these regions contain centers for regulating the cardiovascular and respiratory systems. (Most sensory and motor tracts also pass through the medulla oblongata in communicating between higher and lower centers of the nervous system.)

Module 14.1 Review

a. The major divisions of the autonomic nervous system are the sympathetic division, the parasympathetic division, and the enteric division.

b. The sympathetic division of the ANS is responsible for the physiological changes that occur when you are startled by a loud noise.

c. In the sympathetic division, axons emerge from lumbar and thoracic segments of the spinal cord and innervate ganglia relatively close to the spinal cord, whereas in the parasympathetic division, axons emerge from the brain stem and sacral segments of the spinal cord and innervate ganglia very close to (or within) target organs.

Module 14.2 Review

a. General responses to increased sympathetic activity include heightened mental alertness, increased metabolic rate, reduced digestive and urinary functions, activation of energy reserves, increased respiratory rate and dilation of respiratory passageways, elevated heart rate and blood pressure, and activation of sweat glands. General responses to increased parasympathetic activity include decreased metabolic rate, decreased heart rate and blood pressure, increased secretion by salivary and digestive glands, increased motility and blood flow in the digestive tract, and stimulation of urination and defecation.

b. An intramural ganglion is a group of neurons embedded in the tissue of a target organ.

c. preganglionic neurons (T_1–L_5) → collateral ganglia → ganglionic neurons (postganglionic fibers) → visceral effectors in abdominopelvic cavity

Module 14.3 Review

a. Splanchnic nerves are preganglionic sympathetic fibers on their way to collateral ganglia in the viscera of internal organs, especially those of the abdomen.

b. The vagus nerve (X) carries most of the outflow of the parasympathetic nervous system.

c. Sympathetic nerves are bundles of postganglionic fibers that arise from the cervical sympathetic ganglia and innervate structures in the thoracic cavity.

Module 14.4 Review

a. Both alpha receptors and beta receptors are adrenergic receptors on the membranes of target cells. Alpha receptors are sensitive to norepinephrine (NE) and epinephrine (E), and stimulation typically results in excitation of the target cell; beta receptors are sensitive to epinephrine, and stimulation may result in the excitation or inhibition of the target cell.

b. Nicotinic receptors are acetylcholine (ACh) receptors on the surfaces of sympathetic and parasympathetic ganglionic cells; muscarinic receptors are ACh membrane receptors that are located at all parasympathetic neuromuscular and neuroglandular junctions, and at a few sympathetic neuromuscular and neuroglandular junctions.

c. Blocking the beta-receptors on cells would decrease (or prevent) sympathetic stimulation of tissues containing those cells. As a result, heart rate, force of cardiac muscle contraction, and contraction of the smooth muscle in blood vessel walls would decrease, lowering blood pressure.

Module 14.5 Review

a. Acetylcholine (ACh) is released by all parasympathetic neurons.

b. The parasympathetic division is sometimes referred to as the anabolic system because parasympathetic stimulation leads to a general increase in the nutrient content of the blood. Cells throughout the body respond to the increase by absorbing the nutrients and using them to support growth and other anabolic activities.

c. In tense (anxious) individuals, increased sympathetic stimulation typically causes some or all of the following changes: dry mouth; increased heart rate, blood pressure, and respiration rate; cold sweats; an urge to urinate or defecate; changes in digestive tract motility (that is, "butterflies in the stomach"); and dilated pupils.

Section 1 Review

1. a. sympathetic division; b. thoracolumbar division; c. lumbar nerves; d. thoracic nerves; e. parasympathetic division; f. craniosacral division; g. cranial nerves III, VII, IX, X; h. sacral nerves; i. enteric nervous system

2. a. cervical sympathetic ganglia; b. sympathetic chain ganglia; c. coccygeal ganglia; d. sympathetic nerves; e. cardiac and pulmonary plexuses; f. celiac ganglion; g. superior mesenteric ganglion; h. splanchnic nerves; i. inferior mesenteric ganglion

3. a. splanchnic nerves; b. acetylcholine; c. parasympathetic activation; d. secrete norepi-

nephrine; **e.** nicotinic, muscarinic; **f.** receptors; **g.** cholinergic; **h.** alpha, beta

a. Dual innervation is the situation in which a given body
structure receives instructions
from both the sympathetic and
parasympathetic divisions of
the ANS.

b. Autonomic tone is the background level of activity in sympathetic or parasympathetic
motor neurons under resting
conditions. It provides a
mechanism for fine control of
visceral function because a resting neuron may be *less active* or
more active rather than simply
switching from OFF to ON.
This is particularly important
when only one division innervates a visceral organ, or when
an organ must be precisely
controlled over a broad range of
activity levels (i.e., the heart).

c. The blood vessels of the skin
receive only sympathetic innervation. When you go outside
into the cold, sympathetic
neurons release NE and cause
vasoconstriction of superficial
blood vessels through stimulation of alpha-receptors. When
you get angry, sympathetic
activation occurs and large
amounts of epinephrine enter
the circulation. This stimulates
beta-receptors in the superficial blood vessels, dilating
those vessels, and stimulates
the heart, increasing blood
pressure and blood flow. As
a result, your skin—and most
obviously your face—turns red.

a. A visceral reflex is an autonomic motor response that can be
modified, facilitated, or inhibited by higher centers, especially
those of the hypothalamus.

b. The solitary nucleus is a large
mass of gray matter in the
medulla oblongata that serves
as a processing center and sorting center for visceral sensory
information.

c. Short reflexes are autonomic
responses that bypass the CNS,
whereas long reflexes involve
interneurons within the CNS
and autonomic delivery of motor
commands to the effectors.

a. Baroreceptors are receptors
that detect changes in pressure;
chemoreceptors are receptors
that detect changes in the concentrations of specific chemicals or compounds.

b. Chemoreceptors are sensitive
to changes in blood pH.

c. Baroreceptors are located along
the digestive tract (stomach,
intestines, and colon), within
the walls of the urinary bladder, in the carotid and aortic
sinuses, and in the lungs.

a. The thalamus relays sensory
information.

b. The target structures for the
SNS are skeletal muscles,
whereas those for the ANS are
visceral effectors.

c. A brain tumor pressing on the
hypothalamus could inter-
fere with autonomic function
because the hypothalamus
receives visceral sensory information and controls both sympathetic and parasympathetic
functions.

1. **a.** limbic system and thalamus;
   b. hypothalamus; **c.** pons;
   d. spinal cord T_1–L_2; **e.** complex
   visceral reflexes; **f.** vasomotor;
   g. coughing; **h.** respiratory;
   i. sympathetic visceral reflexes;
   j. parasympathetic visceral
   reflexes
2. **a.** P; **b.** P; **c.** S; **d.** P; **e.** S; **f.** S;
   g. P; **h.** S; **i.** S; **j.** S; **k.** P; **l.** S; **m.** P
3. Even though most sympathetic
   postganglionic fibers are adrenergic, releasing norepinephrine,
   a few are cholinergic, releasing
   acetylcholine. This distribution
   of the cholinergic fibers via the
   sympathetic division provides
   a method of regulating sweat
   gland secretion and selectively
   controlling blood flow to skeletal muscles while reducing the
   flow to other tissues in a body
   wall to maintain homeostasis.

Chapter Integration

1. Epinephrine would be more
   effective because it binds to
   the beta-2 receptors of the
   smooth muscles surrounding
   the airways, resulting in their
   relaxation and airway dilation,
   thus making it easier for John to
   breathe.
2. A fall in blood pressure would
   stimulate baroreceptors in the
   carotid sinus and aortic sinus
   to relay information to the cardiac and vasomotor centers in
   the medulla oblongata. These
   centers respond by increasing sympathetic impulses that
   initiate two complementary
   sympathetic visceral reflexes.
   The cardiac center stimulates
   the cardioacceleratory reflex,
   which increases the heart rate
   and force of contraction, and
   the vasomotor center stimulates vasomotor reflexes, which
   change the diameter of peripheral blood vessels to increase
   blood pressure.

Module 15.1 Review

a. Olfaction—the sense of smell—involves olfactory receptors in paired olfactory organs responding to chemical stimuli.

b. The neurons associated with olfaction that are capable of regenerating are basal cells (stem cells).

c. Axons from the olfactory epithelium collect into bundles that reach the olfactory bulb. Axons leaving the olfactory bulb then travel along the olfactory tract to the olfactory cortex, hypothalamus, and portions of the limbic system.

Module 15.2 Review

a. Gustation is the sense of taste, provided by taste receptors responding to chemical stimuli.

b. Filiform papillae are slender conical projections on the superior surface of the tongue. They help to provide friction for the tongue to move objects in the mouth, but they do not contain taste buds.

c. Taste receptors for bitter and sour sensations are respectively about 100,000 and 1000 times more sensitive than sensations for sweet and salty. Such sensitivity has survival value by helping us avoid substances, such as acids that may harm mucous membranes and bitter-tasting biological toxins.

Module 15.3 Review

a. Gustducins are G proteins—protein complexes that use second messengers to produce effects—that are associated with sweet, bitter, and umami sensations.

b. The cranial nerves that carry gustatory information are the facial nerve (VII), glossopharyngeal nerve (IX), and vagus nerve (X).

c. The pathway for gustation: taste receptors → facial nerve (VII), glossopharyngeal nerve (IX), and vagus nerve (X) → synapse in solitary nucleus of medulla oblongata → medial lemniscus → synapse in thalamus → primary sensory cortex

Section 1 Review

1. a. G proteins b. olfaction c. stem cells d. odorant

e. gustation f. Bowman glands g. cerebral cortex h. taste bud i. olfactory bulb j. olfactory cilia k. bitter l. depolarization m. lingual papillae

2. a. umami b. sour c. bitter d. salty e. sweet f. circumvallate papilla g. fungiform papilla h. filiform papilla

3. The olfactory sensory receptor cells are specialized neurons whose dendrites (also known as olfactory receptor cilia) contain receptor proteins. The binding of odorant molecules to the receptor proteins results in a depolarization of the receptor cell and the production of action potentials. In contrast, the membranes of the sensory receptor cells for taste, vision, equilibrium, and hearing are inexcitable and do not generate action potentials. These cells all form synapses with the processes of sensory neurons, which depolarize and produce action potentials when stimulated by chemical transmitters (neurotransmitters).

Module 15.4 Review

a. The bones in the middle ear are the malleus, incus, and stapes.

b. The auditory tube (also called the Eustachian tube or pharyngotympanic tube) connects the nasopharynx with the middle ear and permits equalization of the pressure on either side of the tympanic membrane.

c. External ear infections are relatively uncommon because the skin in the external acoustic meatus contains glands that secrete cerumen, which inhibits microbial growth.

Module 15.5 Review

a. The bony labyrinth is composed of semicircular canals enclosing the semicircular ducts, the vestibule, and the cochlea.

b. Hair cells are sensory receptors found in the inner ear. As mechanoreceptors, they respond to contact or movement.

c. Within the membranous labyrinth, receptors in the vestibule respond to gravity or linear acceleration, receptors in the semicircular ducts respond only to rotation, and receptors in the cochlear duct respond only to sound.

Module 15.6 Review

a. Statoconia are densely packed calcium carbonate crystals that, together with the gelatinous mass on which the statoconia sit, constitute an otolith.

b. Receptors in the saccule and utricle provide sensations of gravity and linear acceleration.

c. Damage to the cupula of the lateral semicircular duct would interfere with the perception of horizontal rotation of the head.

Module 15.7 Review

a. The organ of Corti is located in the cochlea of the inner ear.

b. Perilymph fills the vestibular duct and the tympanic duct, and endolymph fills the cochlear duct.

c. The features visible in the LM sectional view of the cochlear spiral include the larger tympanic and vestibular ducts, the cochlear duct, and the organ of Corti.

Module 15.8 Review

a. A decibel is the unit of measurement for the intensity of sound.

b. Sound waves enter the external acoustic meatus → tympanic membrane → auditory ossicles → oval window → basilar membrane to tympanic duct → hair cells vibrate against tectorial membrane → information relayed to CNS by cochlear branch of cranial nerve VIII.

c. If the round window could not move, the perilymph would not be moved by the vibration of the stapes at the oval window, reducing or eliminating the perception of sound.

Module 15.9 Review

a. The hair cells for equilibrium are located in the vestibule and the semicircular ducts.

b. Cranial nerves III, IV, VI, and XI are involved with eye, head, and neck movements.

c. The reflexive response to a loud noise is to turn the head and eyes toward the source of the noise.

Section 2 Review

1. a. external ear b. middle ear c. inner ear d. auricle e. external acoustic meatus f. cartilage g. tympanic membrane h. auditory ossicles i. tympanic cavity j. petrous portion of temporal bone k. vestibulocochlear nerve (VIII) l. cochlea m. auditory tube

2. a. cerumen b. auricle c. endolymph d. tympanic cavity e. inner ear f. equilibrium g. tympanic membrane h. stapes i. round window j. incus k. ampulla l. temporal lobe m. organ of Corti n. tectorial membrane o. decibel

3. The rapid descent in the elevator causes the otolith in the macula of the saccule of each vestibule to slide upward, producing the sensation of downward vertical motion. When the elevator abruptly stops, the otoliths do not. It takes a few seconds for them to come to rest in the normal position. As long as the otoliths are displaced, you will perceive movement.

Module 15.10 Review

a. Accessory structures associated with the eye include the eyelids (palpebrae), eyelashes, medial canthus, cornea, lateral canthus, lacrimal caruncle, conjunctiva, tarsal glands, and lacrimal apparatus.

b. Conjunctivitis, or pinkeye, is an irritation (inflammation) of the conjunctiva, generally caused by a pathogenic infection or by physical, allergic, or chemical irritation of the conjunctival surface. The most obvious sign, redness, results from the dilation of blood vessels deep to the conjunctival epithelium.

c. The conjunctiva is the first layer of the eye affected by inadequate tear production.

Module 15.11 Review

a. The tunics of the eye are the fibrous tunic, the vascular tunic (uvea), and the neural tunic (retina).

b. The color of eyes is largely determined by the density and distribution of melanocytes in the iris.

c. Aqueous humor is found in the anterior cavity, between the cornea and the lens.

Module 15.12 Review

a. The cornea does not contain blood vessels.

b. Light passes through the cornea, aqueous humor, pupil of the iris, lens, and vitreous humor before arriving at the retina.

c. When light intensity decreases, sympathetic stimulation causes the pupillary dilator muscles to contract, resulting in dilated, or enlarged, pupils.

Module 15.13 Review

a. The focal point is the point at which the light rays from an object intersect on the retina.

b. When the ciliary muscles are relaxed, you are viewing something in the distance.

c. The near point of vision typically increases with age because the elasticity of the lens tends to decline with age.

Module 15.14 Review

a. Rods are photoreceptors responsible for vision in dim lighting, and cones are photoreceptors requiring more light than rods that provide us with color vision.

b. When you enter a dimly lit room, you are unlikely to be able to see at all. The low-intensity light in the room would be focused on the fovea, which contains only cones, which cannot be stimulated by low-intensity light.

c. If you had been born without cones, you would still be able to see—so long as you had functioning rods—but you would see in black and white only, and with reduced visual acuity.

Module 15.15 Review

a. The three types of cones are blue cones, green cones, and red cones.

b. Rods are more numerous than cones, are responsible for vision in dim light, and provide poorer visual acuity than cones. Several rods are associated with each nerve fiber, whereas the ratio of cones to nerve fibers is 1:1.

c. A dietary vitamin A deficiency would reduce the quantity of retinal the body could produce, thereby interfering with night vision.

Module 15.16 Review

a. The optic radiations are the bundles of projection fibers linking the lateral geniculate nuclei of the thalamus with the visual cortex in each cerebral hemisphere.

b. Visual images are perceived in the visual cortex of the occipital lobes of the cerebrum.

c. The visual pathway is as follows: photoreceptors in the retina → bipolar cell → ganglion cell → axons from the population of ganglion cells converge on optic disc → optic nerve (III) → optic chiasm → optic tract → lateral geniculate nucleus → collateral fibers to diencephalon and brain stem, other collateral fibers to superior colliculus, other collateral fibers to occipital cortex of cerebral hemisphere.

Module 15.17 Review

a. Emmetropia is the term for normal vision.

b. Two surgical procedures for correcting myopia and hyperopia are photorefractive keratectomy (PRK) and laser-assisted in-situ keratomileusis (LASIK). Both procedures—collectively called refractive surgery—use lasers to slice the corneal epithelium, thereby permanently reshaping the cornea.

c. A converging lens (one with at least one convex surface) is used to correct hyperopia.

Module 15.18 Review

a. Cranial nerves VII (facial), IX (glossopharyngeal), and X (vagus) provide taste sensations from the tongue.

b. Two common classes of hearing-related disorders are conductive deafness and nerve deafness.

c. Vertigo is caused by any condition that alters the function of the inner ear receptor complex, the vestibular branch of the vestibulocochlear nerve, or sensory nuclei and pathways in the central nervous system. Common causes include motion sickness, excessive alcohol consumption, and exposure to certain drugs.

Section 3 Review

1. **a.** posterior cavity **b.** choroid **c.** fovea **d.** optic nerve **e.** optic disc **f.** retina **g.** sclera **h.** fornix **i.** palpebral conjunctiva **j.** ocular conjunctiva **k.** ciliary body **l.** iris **m.** lens **n.** cornea **o.** suspensory ligaments **p.** ora serrata

2. **a.** rhodopsin **b.** palpebrae **c.** crystallins **d.** sclera **e.** pupil **f.** retina **g.** posterior cavity **h.** rods **i.** posterior chamber **j.** fovea **k.** occipital lobe **l.** cones **m.** ganglion cells **n.** vascular tunic **o.** optic disc

3. When light falls on the eye, it passes through the cornea and strikes the photoreceptors of the retina, bleaching (breaking down) many molecules of the pigment rhodopsin into retinal and opsin. After an intense exposure to light, a photoreceptor cannot respond to further stimulation until its rhodopsin molecules have been regenerated by the conversion of retinal molecules to their original shape and recombination with opsin molecules. The "ghost" image remains until the rhodopsin molecules are regenerated.

Chapter Integration

1. Myopia is the medical term for nearsightedness; individuals with myopia are able to see nearby objects clearly, but distant objects appear blurry or fuzzy. Hence, Mr. Drummond would have difficulty seeing while driving.

2. Mr. Drummond will likely have a follow-up visit with an optometrist or ophthalmologist, who will prescribe spectacle lenses (glasses) or contanct lenses to alleviate the myopia and restore his distance sight to normal, or near-normal, vision.

3. Vertigo is another term for dizziness and is generally associated with a problem in the inner ear. The movement of Mr. Drummond's arms while standing still provided evidence that there may be a problem with his equilibrium centers (saccule and utricle). Often, inflammation due to infection or a cold affects the inner ear (or N VIII) and causes vertigo.

4. When Mr. Drummond closes his eyes, visual cues are gone, and his brain must rely solely on proprioception (the sense of relative position of body parts) and information from the equilibrium centers of the inner ear to maintain normal posture. Because either the inner ear receptors or the sensory nerves aren't functioning normally, he is unstable.

5. The most likely reason for his drift to the left is that he is getting inappropriate sensations from equilibrium receptors, either at the maculae (affecting his ability to determine which way is "down") or one of the horizontal semicircular ducts (making him attempt to compensate for a perceived roll to the right).

Module 16.1 Review

a. The endocrine system includes organs whose primary function is the production of hormones or paracrine factors, which are chemical secretions that are transported via the extracellular fluid or bloodstream to target cells in other sites within the body.

b. Organs of the endocrine system are the hypothalamus, pituitary gland, thyroid gland, adrenal glands, pancreas (pancreatic islets), pineal gland, and parathyroid glands. Organs of other systems that have endocrine functions are the heart, thymus, digestive tract, kidneys, and gonads.

c. The structural classes of hormones are (1) amino acid derivatives (thyroid hormones, catecholamines, and tryptophan derivatives); (2) peptide hormones (glycoproteins or short polypeptide chains), which are chains of amino acids that are synthesized as prohormones; and (3) lipid derivatives (eicosanoids and steroid hormones), which contain carbon rings and side chains that are built from fatty acids or cholesterol.

Module 16.2 Review

a. A hormone receptor is a protein molecule, located either on the plasma membrane or inside the cell, that binds a specific hormone.

b. A first messenger is a hormone whose binding to a protein receptor in the plasma membrane gives rise to a second messenger in the cytoplasm. The second messenger changes the rate of various metabolic reactions by acting as an enzyme activator, an enzyme inhibitor, or a cofactor.

c. Hormones that diffuse across the plasma membrane and bind to receptors in the cytoplasm are steroid hormones.

Module 16.3 Review

a. A regulatory hormone is a special hormone, secreted by the hypothalamus, that controls endocrine cells in the anterior lobe of the pituitary gland.

b. The three mechanisms of hypothalamic integration of neural and endocrine function are (1) secretion of antidiuretic hormone and oxytocin, (2) secretion of regulatory hormones that control activity of the anterior lobe of the pituitary gland, and (3) neural (sympathetic) control over the endocrine cells of the adrenal medullae.

c. The blood vessels of the hypophyseal portal system link the hypothalamus and anterior lobe. Fenestrated capillary beds in each structure are connected by portal vessels. This arrangement ensures that hypothalamic regulatory hormones reach the "downstream" endocrine cells of the anterior lobe directly, before mixing with, and being diluted by, the general circulation.

Module 16.4 Review

a. The two lobes of the pituitary gland are the anterior lobe and the posterior lobe.

b. The hormones produced and released by the anterior lobe of the pituitary gland are (1) thyroid-stimulating hormone (TSH), which targets the thyroid gland; (2) adrenocorticotropic hormone (ACTH), which targets the adrenal cortex; (3) follicle-stimulating hormone (FSH) and (4) luteinizing hormone (LH), which target the testes in males and the ovaries in females; (5) growth hormone (GH), which targets liver cells (which respond by synthesizing somatomedins); (6) prolactin (PRL), which targets mammary glands in females; and (7) melanocyte-stimulating hormone (MSH), which

targets melanocytes in the skin. Hormones released by the posterior lobe of the pituitary gland are (1) oxytocin (OXT), which targets the mammary glands, and (2) antidiuretic hormone (ADH), which targets the kidneys.

c. In a dehydrated individual, the amount of ADH released by the posterior pituitary increases in response to increased blood osmotic pressure resulting from a rise in solute concentration.

Module 16.5 Review

a. The hypothalamic releasing hormones are corticotropin-releasing hormone (CRH), thyrotropin-releasing hormone (TRH), growth hormone–releasing hormone (GH–RH), prolactin-releasing factor (PRF), and gonadotropin-releasing hormone (GnRH).

b. Somatomedins mediate the action of growth hormone. Elevated levels of growth hormone typically accompany elevated levels of somatomedins.

c. Elevated circulating levels of glucocorticoids inhibit the release of CRH by the hypothalamus. The lack of CRH reduces the secretion of ACTH from the pituitary gland, so ACTH levels would decrease. This is an example of negative feedback.

Module 16.6 Review

a. The hormones of the thyroid gland are thyroxine (T_4), T_3, and calcitonin.

b. Calcitonin aids in calcium regulation.

c. Most of the body's reserves of thyroid hormone, thyroxine, are bound to transport proteins in the bloodstream called thyroid-binding globulins. Because these compounds represent such a large reservoir of thyroxine, it takes several days after removal of the thyroid gland for blood levels of thyroxine to decline.

Module 16.7 Review

a. The parathyroid glands are embedded in the posterior surfaces of the lateral lobes of the thyroid gland.

b. Parathyroid hormone (PTH) raises blood calcium levels by reducing calcium deposition in bones, by increasing reabsorption of calcium from the blood by the kidneys, and by increasing the production of calcitriol by the kidneys.

c. Decreased blood calcium levels result in increased secretion of parathyroid hormone (PTH).

Module 16.8 Review

a. The two regions of an adrenal gland are the cortex and medulla. The cortex secretes mineralocorticoids, primarily aldosterone; glucocorticoids, mainly cortisol (or hydrocortisone) and corticosterone; and androgens. The medulla secretes epinephrine and norepinephrine.

b. The three zones of the adrenal cortex are the zona glomerulosa, zona fasciculata, and zona reticularis.

c. Elevated cortisol levels would result in elevated blood glucose levels, because cortisol reduces the use of glucose by cells while increasing both the available glucose (by promoting the breakdown of glycogen) and the conversion of amino acids to carbohydrates.

Module 16.9 Review

a. The types of cells in the pancreatic islets (and their hormones) are alpha cells (glucagon), beta cells (insulin), delta cells (GH–IH), and F cells (pancreatic polypeptide, or PP).

b. The secretion of insulin lowers blood glucose concentrations.

c. Increased levels of glucagon stimulate the conversion of glycogen to glucose in the liver, which would in turn reduce the amount of glycogen stored in the liver.

Module 16.10 Review

a. The hormone-secreting cells of the pineal gland are pinealocytes.

b. Melatonin secretion is influenced by light–dark cycles. Increased amounts of light would inhibit the production (and release) of melatonin from the pineal gland, which receives neural input concerning the presence of light or darkness from visual pathway collaterals.

c. In humans, melatonin may affect the timing of sexual maturation, protect against damage by free radicals, and set circadian rhythms.

Module 16.11 Review

a. Diabetes mellitus is a condition characterized by elevated blood glucose levels resulting from inadequate insulin production or diminished sensitivity to insulin by cells.

b. The two types of diabetes mellitus are type 1, characterized by inadequate insulin production by the pancreatic beta cells, and type 2, characterized by insulin resistance (failure of the body to use insulin properly).

c. An individual with type 1 or type 2 diabetes has such high blood glucose levels that the kidneys cannot reabsorb all the glucose; some glucose is lost in the urine. Because the urine contains high concentrations of glucose, less water can be reclaimed by osmosis, so the volume of urine production increases, and the individual needs to urinate more often.

Section 1 Review

1. a. catecholamines; **b.** thyroid hormones; **c.** tryptophan derivatives; **d.** peptide hormones; **e.** short polypeptides; **f.** glycoproteins; **g.** small proteins; **h.** lipid derivatives; **i.** eicosanoids; **j.** steroid hormones; **k.** transport proteins

2. a. thymus; **b.** pineal gland; **c.** pancreatic islets; **d.** hypothalamus; **e.** kidneys; **f.** adrenal glands; **g.** pituitary gland; **h.** gonads; **i.** heart; **j.** digestive tract; **k.** thyroid gland; **l.** parathyroid glands

3. a. F cells; **b.** epinephrine; **c.** direct communication; **d.** tropic hormones; **e.** cyclic-AMP; **f.** secretes releasing hormones; **g.** androgens; **h.** prostaglandins; **i.** FSH; **j.** parathyroid glands

Module 16.12 Review

a. The heart secretes natriuretic peptide, and the kidneys release erythropoietin.

b. Hormones necessary for normal growth and development include GH, thyroid hormones, insulin, PTH, calcitriol, and reproductive hormones.

c. Upon its release into the bloodstream, renin functions as an enzyme that activates the renin-angiotensin system, which ultimately causes blood pressure to rise.

Module 16.13 Review

a. The three phases of the stress response are the alarm phase, the resistance phase, and the exhaustion phase.

b. The resistance phase is characterized by long-term metabolic adjustments, including mobilization of remaining energy reserves, conservation of glucose, elevation of blood glucose concentrations, and conservation of salts and water coupled with the loss of K^+ and H^+.

c. The collapse of vital systems occurs during the exhaustion phase of the GAS.

Module 16.14 Review

a. The prefix *hyper-* refers to excessive hormone production, whereas *hypo-* refers to inadequate hormone production.

b. Three common causes of hormone hyposecretion are metabolic factors, physical damage, and congenital disorders.

c. Aldosteronism is characterized by increased body weight due to Na^+ and water retention and a low blood K^+ concentration.

Section 2 Review

1. a. PTH and calcitonin; **b.** glucocorticoids; **c.** sympathetic activation; **d.** increase blood pressure and volume; **e.** protein synthesis; **f.** GH and glucocorticoids; **g.** gigantism; **h.** reduce blood pressure and volume; **i.** homeostasis threat; **j.** PTH and calcitriol

2. a. release of natriuretic peptides; **b.** suppression of thirst; **c.** Na^+ and H_2O loss from kidneys; **d.** reduced blood pressure; **e.** increased fluid loss; **f.** falling blood pressure and volume; **g.** erythropoietin released; **h.** renin released; **i.** increased red blood cell production; **j.** aldosterone secreted; **k.** ADH secreted; **l.** rising blood pressure and volume

3. (1) The two hormones may have opposing or antagonistic effects, such as occurs between insulin (decreases blood glucose levels) and glucagon (increases blood glucose levels). (2) The two hormones may have an additive or synergistic effect, in which the net result is greater than the sum of each acting alone. An example is the enhanced glucose-sparing action of GH in the presence of glucocorticoids. (3) One hormone may have a permissive effect on another, in which the first hormone is needed for the second hormone to produce its effect. For example, epinephrine cannot alter the rate of tissue energy consumption without the presence of thyroid hormones. (4) The hormones may have integrative effects, in which the hormones may produce different but complementary results in specific tissues and organs. An example is the differing effects of calcitriol and parathyroid hormone (PTH) on tissues involved in calcium metabolism; calcitriol increases calcium ion absorption from digestive system, and PTH inhibits osteoblast activity and enhances calcium ion reabsorption by the kidneys.

Chapter Integration

1. Sherry's physician suspected hyperthyroidism because she exhibited the classic signs and symptoms of the condition: restlessness, anxiety, irritability, difficulty sleeping, diarrhea, weight loss, rapid heart rate, and tremors.

2. Hyperthyroidism is the production of excess T_3, T_4, or both, by the thyroid gland. Excess thyroid hormone results in a rapid heartbeat and an increased metabolic rate, among other clinical signs and symptoms.

3. Sherry's physician could order blood tests to assay the levels of TSH, T_3, and T_4. Results confirming hyperthyroidism are lower-than-normal TSH levels and elevated T_3 and T_4 levels. Her physician could also determine whether her condition is primary hyperthyroidism (a problem with the thyroid gland) or secondary hyperthyroidism (a problem with hypothalamo-pituitary control of the thyroid gland).

4. Sherry's signs and symptoms indicating nervous system involvement include tremors and anxiety.

Module 17.1 Review

a. The hematocrit, also called the packed cell volume (PCV), is the percentage of whole blood volume contributed by formed elements.

b. Whole blood is composed of plasma (which contains albumins, globulins, fibrinogen, electrolytes, organic nutrients, and organic wastes) and formed elements (platelets, white blood cells, and red blood cells). White blood cells include neutrophils, eosinophils, basophils, lymphocytes, and monocytes.

c. During an infection, you would expect the level of immunoglobulins (antibodies) in the blood to be elevated.

Module 17.2 Review

a. Rouleaux are stacks of red blood cells.

b. Hemoglobin is a protein—composed of four globular subunits, each bound to a heme molecule—that gives RBCs the ability to transport oxygen in the blood.

c. Oxyhemoglobin is hemoglobin whose iron has bound oxygen; it is bright red. Deoxyhemoglobin is hemoglobin whose iron has not bound oxygen; it is dark red.

Module 17.3 Review

a. Hemolysis is the rupture of red blood cells; it results in the release of hemoglobin.

b. After the removal of iron within macrophages, heme is converted into biliverdin, which is then converted to bilirubin. In the large intestine, bilirubin is converted to either stercobilins, which are eliminated in the feces, or urobilins, which are eliminated in the feces or in urine.

c. Bilirubin would accumulate in the blood, producing jaundice, because diseases that damage the liver impair the liver's ability to excrete bilirubin in the bile.

Module 17.4 Review

a. Surface antigens on RBCs are glycolipids in the plasma membrane; they determine blood type.

b. Only Type O blood can be safely transfused into a person whose blood type is O.

c. A person with type A blood also has anti-B antibodies, so if they received a transfusion of Type B blood, the transfused red blood cells would clump, or agglutinate, potentially blocking blood flow to various organs and tissues.

Module 17.5 Review

a. Hemolytic disease of the newborn (HDN) is a condition in which maternal antibodies attack and destroy fetal red blood cells, resulting in fetal anemia; it occurs in a sensitized Rh− mother who is carrying an Rh+ fetus.

b. When RhoGAM (which contains anti-Rh antibodies) is injected into a pregnant Rh− woman, the anti-Rh antibodies circulate in the mother's bloodstream, where they destroy any fetal RBCs there. This prevents the mother's immune system from making antibodies against the developing fetus's red blood cells.

c. An Rh+ mother carrying an Rh− fetus does not require a RhoGAM injection because the fetus is not at risk. The fetus is not at risk because its RBCs lack Rh surface antigens, and the mother's plasma lacks anti-Rh antibodies.

Module 17.6 Review

a. The five types of white blood cells are neutrophils, eosinophils, basophils, monocytes, and lymphocytes.

b. An infected cut would contain a large number of neutrophils, phagocytic white blood cells that are generally the first to arrive at the site of an injury.

c. Basophils enter damaged tissues and release a variety of chemicals, including histamine, that promote inflammation.

Module 17.7 Review

a. A hemocytoblast is a multipotent stem cell whose divisions produce lymphoid and myeloid stem cells, which divide to form each of the various populations of blood cells.

b. Erythropoietin is a hormone released by tissues (especially the kidneys) exposed to low oxygen concentrations; it stimulates erythropoiesis (red blood cell formation) in red bone marrow.

c. Lymphoid stem cells originate in red bone marrow and give rise to lymphocytes; these stem cells also produce lymphocytes in the thymus, spleen, and lymph nodes. Myeloid stem cells are cells in red bone marrow that give rise to all the formed elements except lymphocytes.

Module 17.8 Review

a. Hemostasis is the stoppage of blood flow involving three phases: the vascular phase, the platelet phase, and the coagulation phase.

b. During the vascular phase, local blood vessel constriction (vascular spasm) occurs at the injury site. In the platelet phase, platelets are activated, aggregate at the site, and adhere to damaged blood vessel surfaces. In the coagulation phase, factors released by platelets and endothelial cells interact with clotting factors (through either the extrinsic pathway, the intrinsic pathway, or the common pathway) to form a blood clot, a process involving the conversion of soluble fibrinogen to insoluble fibers of fibrin.

c. The correct sequence for the events in hemostasis is vascular spasm (3), platelet phase (5), coagulation (1), retraction (4), and fibrinolysis (2).

Module 17.9 Review

a. Venipuncture is the piercing of a vein to obtain a blood sample.

b. The two types of leukemia are myeloid leukemia and lymphoid leukemia.

c. Pernicious anemia is insufficient red blood cell production that results due to a lack of vitamin B_{12}; the blood cells that do develop tend to be macrocytic (abnormally large) and abnormally shaped. Iron deficiency anemia results when the dietary intake or absorption of iron is insufficient, impairing normal hemoglobin synthesis; these blood cells are microcytic (abnormally small).

Section 1 Review

1. a. plasma; b. water; c. solutes; d. proteins; e. electrolytes, glucose, urea; f. albumins; g. globulins; h. fibrinogen; i. formed elements; j. erythrocytes; k. leukocytes; l. platelets; m. neutrophils; n. eosinophils; o. basophils; p. lymphocytes; q. monocytes

2. a. red bone marrow; b. mature RBCs; c. matrix; d. monocytes; e. transport protein; f. cross-reaction; g. lymphocytes; h. jaundice; i. venipuncture; j. pigment complex; k. erythropoietin; l. platelets

3. During differentiation, the red blood cells of humans (and other mammals) lose most of their organelles, including nuclei and ribosomes. As a result, mature circulating RBCs can neither divide nor synthesize the structural proteins and enzymes required for cellular repairs.

Module 17.10 Review

a. The five general classes of blood vessels are arteries, arterioles, capillaries, venules, and veins.

b. A capillary is a small blood vessel, located between an arteriole and a venule, whose thin wall permits the diffusion of gases, nutrients, and wastes between plasma and interstitial fluids.

c. These blood vessels are veins. Arteries and arterioles have a large amount of smooth muscle tissue in a thick, well-developed tunica media.

Module 17.11 Review

a. The two types of capillaries are continuous capillaries and fenestrated capillaries.

b. Fenestrated capillaries are located where solutes as large as small peptides move freely into and out of the blood, including endocrine glands, the choroid plexus of the brain, absorptive areas of the intestine, and filtration areas of the kidneys.

c. Capillary walls are thin, so distances for diffusion are short. Continuous capillaries have small gaps between adjacent endothelial cells that permit the diffusion of water and small solutes into the surrounding interstitial fluid but prevent the loss of blood cells and plasma proteins. Fenestrated capillaries contain pores that permit very rapid exchange of fluids and solutes between interstitial fluid and plasma. The walls of arteries and veins are several cell layers thick and are not specialized for diffusion.

Module 17.12 Review

a. Varicose veins are sagging, swollen veins distorted by the pooling of blood resulting from gravity and the failure of venous valves.

b. In the arterial system, pressures are high enough to keep the blood moving forward. In the venous system, blood pressure is too low to keep the blood moving on toward the heart. Valves in veins prevent blood from flowing backward whenever the venous pressure drops.

c. Assisted by the presence of valves in the veins, which prevent backflow of the blood, the contraction of the surrounding skeletal muscles squeezes venous blood toward the heart.

Module 17.13 Review

a. The two circuits of the cardiovascular system are the pulmonary circuit and the systemic circuit.

b. The three major patterns of blood vessel organization are the following: (1) The peripheral distributions of arteries and veins on the body's left and right sides are generally identical, except near the heart, where the largest vessels connect to the atria or ventricles; (2) a single vessel may have several names as it crosses specific anatomical boundaries, making accurate anatomical descriptions possible; and (3) tissues and organs are usually serviced by several arteries and veins.

c. right ventricle → right and left pulmonary arteries → pulmonary arterioles → alveoli → pulmonary venules → pulmonary veins → left atrium

Module 17.14 Review

a. The two large veins that collect blood from the systemic circuit are the superior vena cava and the inferior vena cava.

b. The largest artery in the body is the aorta.

c. A major anatomical difference between the arterial and venous systems is the existence of dual venous drainage in the neck and limbs.

Module 17.15 Review

a. The two arteries formed by the division of the brachiocephalic trunk are the right common carotid and the right subclavian.

b. A blockage of the left subclavian artery would interfere with blood flow to the left arm.

c. The vein that is bulging is the external jugular vein.

Module 17.16 Review

a. The arterial structure in the neck region that contains baroreceptors is the carotid sinus.

b. Branches of the external carotid artery are the superficial temporal, maxillary, occipital, facial, lingual, and external carotid arteries.

c. The veins that combine to form the brachiocephalic vein are the external jugular, internal jugular, vertebral, and subclavian veins.

Module 17.17 Review

a. The internal jugular veins drain the dural sinuses of the brain.

b. The three branches of the internal carotid artery are the ophthalmic, anterior cerebral, and middle cerebral arteries.

c. The cerebral arterial circle (also known as the circle of Willis) is a ring-shaped anastomosis that encircles the infundibulum of the pituitary gland. Its anatomical arrangement creates alternate pathways in the cerebral circulation, so that if blood flow is interrupted in one area, other blood vessels can continue to perfuse the entire brain with blood.

Module 17.18 Review

a. The inferior vena cava collects most of the venous blood inferior to the diaphragm.

b. The major tributaries of the inferior vena cava are the lumbar, gonadal, hepatic, renal, adrenal, and phrenic veins.

c. Rupture of the celiac trunk would most directly affect the stomach, inferior portion of the esophagus, spleen, liver, gallbladder, and proximal portion of the small intestine.

Module 17.19 Review

a. The unpaired branches of the abdominal aorta that supply blood to the visceral organs are the celiac trunk, superior mesenteric artery, and inferior mesenteric artery.

b. The three veins that merge to form the hepatic portal vein are the superior mesenteric, inferior mesenteric, and splenic veins.

c. The left and right gastroepiploic veins carry blood away from the stomach.

Module 17.20 Review

a. The first two divisions of the common iliac artery are the internal iliac artery and the external iliac artery.

b. The plantar venous arch delivers blood to the anterior tibial, posterior tibial, and fibular veins.

c. A blockage of the popliteal vein would interfere with blood flow in the tibial and fibular veins (which form the popliteal vein) and the small saphenous vein (which joins the popliteal vein).

Module 17.21 Review

a. Deoxygenated blood flows to the placenta through a pair of umbilical arteries, and oxygenated blood returns from the placenta in a single umbilical vein. The umbilical vein then drains into the ductus venosus within the fetal liver.

b. The six necessary structures in the fetal circulation are two umbilical arteries, one umbilical vein, the ductus venosus, the foramen ovale, and the ductus arteriosus. After birth, the foramen ovale closes and persists as the fossa ovalis, a shallow depression; the ductus arteriosus persists as the ligamentum arteriosum, a fibrous cord; and the umbilical vessels and ductus venosus persist throughout life as fibrous cords.

c. Ventricular septal defects are abnormal openings between the left and right ventricles. Tetralogy of Fallot includes a ventricular septum defect plus three other heart defects: a narrowing of the pulmonary trunk, a displaced aorta, and an enlarged right ventricle with corresponding thickened right and left ventricles.

Section 2 Review

1. a. common carotid; **b.** subclavian; **c.** brachiocephalic trunk; **d.** brachial; **e.** radial; **f.** popliteal; **g.** fibular; **h.** aortic arch; **i.** celiac trunk; **j.** renal; **k.** common iliac; **l.** external iliac; **m.** femoral; **n.** anterior tibial

2. a. vertebral; **b.** internal jugular; **c.** brachiocephalic; **d.** axillary; **e.** cephalic; **f.** median antebrachial; **g.** ulnar; **h.** great saphenous; **i.** fibular; **j.** superior vena cava; **k.** inferior vena cava; **l.** internal iliac; **m.** deep femoral; **n.** posterior tibial

Chapter Integration

1. Normal red blood cells are shaped like biconcave discs. Hemolysis is the rupture or destruction of red blood cells. In most cases, hemolysis occurs outside the bloodstream in the spleen, liver, and bone marrow.

2. Some functions of blood include transporting heat throughout the body and oxygen to cells. Because Ursula's oxygen-carrying red blood cells were being actively destroyed, her heart had to pump faster to supply tissues with the oxygen they need. The resulting increase in circulation brought more warm blood to her body surface, where greater-than-normal heat loss lowered her body temperature.

3. Anemia and the accompanying low hemoglobin levels reduce the blood's ability to carry oxygen and nutrients to the body's vital organs and tissues. As a result, Ursula experienced fatigue and weakness (and superficial tissues such as the skin and gingivae became pale).

Module 18.1 Review

a. From superficial to deep, the layers of the heart wall are the epicardium, myocardium, and endocardium.

b. The epicardium consists of an outer mesothelium and an underlying layer of areolar tissue that attaches directly to the myocardium.

c. Cardiac tissue is metabolically active and dependent on mitochondrial activity for ATP, obtaining oxygen and nutrients from local capillaries.

Module 18.2 Review

a. The mediastinum is the region between the two pleural cavities that contains the heart along with the great vessels (large arteries and veins attached to the heart), thymus, esophagus, and trachea.

b. The heart is located within the pericardial sac in the anterior mediastinum, deep to the sternum and superior to the diaphragm.

c. Cardiac tamponade can be a life-threatening condition because the accumulating fluid within the pericardial cavity restricts heart movement.

Module 18.3 Review

a. The four cardiac chambers are the left atrium, right atrium, left ventricle, and right ventricle.

b. The anterior interventricular sulcus marks the boundary between the left and right ventricles on the heart's anterior surface; the shallower posterior interventricular sulcus marks the boundary between the left and right ventricles on the posterior surface; and the coronary sulcus is a deep groove that marks the border between the atria and the ventricles.

c. Coronary veins collect blood from the myocardium and carry it to the right atrium.

Module 18.4 Review

a. Arteries: left coronary artery, anterior interventricular artery, right coronary artery, marginal arteries, circumflex artery, and posterior interventricular artery. Veins: great cardiac vein, anterior cardiac veins, posterior cardiac vein, middle cardiac vein, and small cardiac vein.

b. During elastic rebound, some blood in the aorta is driven forward into the systemic circuit,

and some is forced back toward the left ventricle and into the coronary arteries.

c. The coronary sinus.

Module 18.5 Review

a. Damage to the semilunar valve on the right side of the heart would affect blood flow to the pulmonary trunk.

b. Contraction of the papillary muscles pulls on the chordae tendineae, which prevent the AV valves from swinging into the atria.

c. The more muscular left ventricle must generate enough force to propel blood throughout the body (except the lungs), whereas the right ventricle must generate only enough force to propel blood the short distance to the lungs.

Module 18.6 Review

a. Cardiac regurgitation is the backflow of blood into the atria when the ventricles contract.

b. The tricuspid valve is composed of three relatively large flaps (cusps); the pulmonary valve is made up of three smaller half-moon-shaped cusps.

c. Semilunar valves prevent the backflow of blood into the ventricles.

Module 18.7 Review

a. Arteriosclerosis is any thickening and toughening of arterial walls; atherosclerosis is a type of arteriosclerosis characterized by changes in the endothelial lining and the formation of fatty deposits (plaque) in the tunica media.

b. Coronary ischemia is a condition in which the blood supply of the coronary arteries has been reduced.

c. Stents are artificial mesh "tubes" that prop open the natural blood vessel, creating a conduit to restore blood flow. Without adequate blood flow to the cardiac muscle, the tissue would die.

Section 1 Review

1. a. aortic arch; b. superior vena cava; c. right pulmonary arteries; d. ascending aorta; e. fossa ovalis; f. opening of coronary sinus; g. right atrium; h. pectinate muscles; i. tricuspid valve cusp; j. chordae tendineae; k. papillary muscle; l. right ventricle; m. inferior vena cava; n. pulmonary trunk; o. pulmonary valve; p. left

pulmonary arteries; q. left pulmonary veins; r. left atrium; s. aortic valve; t. bicuspid valve cusp; u. left ventricle; v. interventricular septum; w. trabeculae carneae; x. moderator band

2. a. Deoxygenated blood flow: right atrium → right atrioventricular valve (tricuspid valve) → right ventricle → pulmonary semilunar valve. b. Oxygenated blood flow: left atrium → left atrioventricular valve (bicuspid valve) → left ventricle → aortic semilunar valve.

3. a. aorta; b. calcium ions; c. myocardium; d. coronary sinus; e. fossa ovalis; f. tricuspid valve; g. anastomoses; h. intercalated discs; i. serous membrane; j. endocardium; k. cardiac skeleton; l. aortic valve

Module 18.8 Review

a. The alternate term for heart contraction is systole, and the term for heart relaxation is diastole.

b. Atrial systole, atrial diastole, ventricular systole, and ventricular diastole.

c. No. When pressure in the left ventricle first rises, the heart is contracting but blood is not leaving the heart. During this initial phase of contraction, called the period of isovolumetric contraction, both the AV valves and the semilunar valves are closed. The increase in pressure is the result of the cardiac muscle contracting. When the pressure in the ventricle exceeds that in the aorta, the aortic semilunar valves are forced open, and blood is rapidly ejected from the ventricle.

Module 18.9 Review

a. Automaticity is the ability of cardiac muscle tissue to contract without neural or hormonal stimulation.

b. If the cells of the SA node failed to function, the heart would continue to beat, but at a slower rate; the AV node would act as the pacemaker.

c. If the impulses from the atria were not delayed at the AV node, they would be conducted through the ventricles so quickly by the bundle branches and Purkinje cells that the ventricles would begin contracting before the atria had finished contracting. As a result, the ventricles would not be as full of blood as they could be, and the pumping action of the heart would be less efficient.

Module 18.10 Review

a. Rapid depolarization, plateau, and repolarization.

b. Slow calcium channels are voltage-gated calcium channels that open slowly and remain open for a relatively long period—about 175 msec. When they are open, the entry of calcium ions into the sarcoplasm roughly balances the loss of positive ions through the active transport of sodium ions. As a result, the transmembrane potential remains near 0 mV for an extended period.

c. Cardiac muscle has a long refractory period that continues until relaxation is well under way. As a result, another action potential cannot arrive quickly enough for summation to occur, and thus tetany cannot occur.

Module 18.11 Review

a. Bradycardia is a heart rate below 60 beats per minute; tachycardia is a heart rate above 100 beats per minute.

b. The cardioacceleratory center in the medulla oblongata activates sympathetic neurons to increase heart rate; the cardioinhibitory center (also in the medulla oblongata) controls the parasympathetic neurons that slow heart rate.

c. Like NE, caffeine acts directly on the conducting system and contractile cells of the heart, increasing the rate at which they depolarize. Thus drinking large amounts of caffeinated beverages would increase the heart rate.

Module 18.12 Review

a. The end-diastolic volume (EDV) is the amount of blood a ventricle contains at the end of diastole, just before a contraction begins; the end-systolic volume (ESV) is the amount of blood that remains in the ventricle at the end of ventricular systole.

b. An increase in venous return would stretch the heart muscle. The more the heart muscle is stretched, the more forcefully it will contract (to a point). The more forceful the contraction, the more blood the heart will eject with each beat (stroke volume). Therefore, increased venous return would increase the stroke volume (if all other factors are constant).

c. An increase in sympathetic stimulation of the heart would increase heart rate and force of contraction. The end-systolic volume (ESV) is the amount of blood that remains in a ventricle after a contraction (systole). The more forcefully the heart contracts, the more blood it ejects. Therefore, increased sympathetic stimulation should result in a lower ESV.

Module 18.13 Review

a. Heart failure is a condition in which the heart can no longer meet the oxygen and nutrient demands of peripheral tissues.

b. SV = EDV − ESV, so
SV = 125 mL − 40 mL = 85 mL.

c. The amount of blood the heart pumps is proportional to the amount of blood that enters it. A heart that is beating too rapidly does not have adequate filling time, and it pumps less blood; peripheral tissues can be damaged by inadequate blood flow.

Module 18.14 Review

a. An electrocardiogram (ECG or EKG) is a recording of the electrical activities of the heart over time.

b. The important features of an ECG are the P wave (atrial depolarization), the QRS complex (ventricular depolarization), and the T wave (ventricular repolarization).

c. Ventricular fibrillation, which causes the condition known as cardiac arrest, is fatal because the ventricles merely quiver and do not pump blood into the systemic circulation.

Section 2 Review

1. a. left AV valve closes;
b. increasing, decreasing;
c. less than; **d.** aortic valve is forced open; **e.** aorta;
f. ventricular systole
2. a. "lubb" sound; **b.** automaticity;
c. P wave; **d.** stroke volume; **e.** cardiac output;
f. parasympathetic neurons;
g. bradycardia; **h.** sympathetic neurons; **i.** "dubb" sound;
j. tachycardia

Module 18.15 Review

a. Total peripheral resistance reflects a combination of vascular resistance, vessel length, vessel diameter, blood viscosity, and turbulence.

b. An increase in vessel diameter would reduce peripheral resistance. (An increase in vessel

length increases peripheral resistance.)

c. The formula $R = 1/r^4$ states that resistance (R) is inversely proportional to the fourth power of the vessel radius (r). This means that a small change in vessel diameter results in a large change in resistance.

Module 18.16 Review

a. Blood flow is the volume of blood flowing per unit of time through a vessel or group of vessels; it is directly proportional to blood pressure and inversely proportional to peripheral resistance.

b. In a healthy individual, blood pressure is greater at the aorta than at the inferior vena cava. If the pressure were higher in the inferior vena cava than in the aorta, blood would flow in the reverse direction.

c. Using the formula MAP = diastolic pressure + (pulse pressure)/3, MAP equals 70 + (125 − 70)/3, which equals 70 + 18.3, or 88.3 mmHg.

Module 18.17 Review

a. Edema is an abnormal accumulation of interstitial fluid in peripheral tissues.

b. Any condition that affects either blood pressure or osmotic pressures in the blood or tissues will shift the balance between hydrostatic and osmotic forces.

c. Fluid moves into a capillary whenever blood colloid osmotic pressure (BCOP) is greater than capillary hydrostatic pressure (CHP).

Module 18.18 Review

a. Tissue perfusion is blood flow to tissues that is sufficient to deliver adequate oxygen and nutrients.

b. Cardiovascular autoregulation involves local factors changing the pattern of blood flow within capillary beds in response to chemical changes in interstitial fluids.

c. Baroreceptor reflexes respond to changes in blood pressure. The baroreceptors—located in the walls of the carotid sinuses, aortic sinuses, and right atrium—monitor the degree of stretch at those sites.

Module 18.19 Review

a. Epinephrine and norepinephrine from the adrenal medullae provide short-term regulation of declining blood pressure and blood volume.

b. Antidiuretic hormone (ADH), angiotensin II, erythropoietin (EPO), and aldosterone are hormones involved in the long-term response to declining blood pressure and blood volume.

c. Vasoconstriction of the renal artery would decrease both blood flow and blood pressure at the kidney. In response, the kidney would increase the amount of renin it releases, which in turn would increase the level of angiotensin II. The angiotensin II would bring about increased blood pressure and increased blood volume.

Module 18.20 Review

a. Chemoreceptor reflexes respond to changes in the pH, oxygen, or CO_2 levels in the blood and cerebrospinal fluid (CSF).

b. Chemoreceptors are located in the carotid bodies, in the aortic bodies, and on the ventrolateral surfaces of the medulla oblongata.

c. An increase in the respiratory rate reduces CO_2 levels.

Module 18.21 Review

a. The respiratory pump is a mechanism by which a reduction of pressure in the thoracic cavity during inhalation assists venous return to the heart.

b. During exercise, cardiac output increases, and blood flow to skeletal muscles increases at the expense of blood flow to less essential organs.

c. Unless compensatory vasoconstriction occurs in nonessential organs, vasodilation in skeletal muscles would cause a potentially dangerous fall in blood pressure and blood flow throughout the body.

Module 18.22 Review

a. Compensatory mechanisms that respond to blood loss include an increase in cardiac output, a mobilization of venous reserves, peripheral vasoconstriction, and the release of hormones that promote the retention of fluids and the maturation of erythrocytes.

b. Both aldosterone and ADH promote fluid retention and reabsorption by the kidneys, preventing further reductions in blood volume.

c. The immediate problem during hemorrhaging is the maintenance of adequate blood pressure and

peripheral blood flow; the long-term problem is the restoration of normal blood volume.

Section 3 Review

1. The three primary factors influencing blood pressure and blood flow are cardiac output, blood volume, and peripheral resistance.

2. a. baroreceptors; **b.** medulla oblongata; **c.** venous return; **d.** autoregulation; **e.** local vasodilators; **f.** chemoreceptors; **g.** osmotic pressure; **h.** turbulence; **i.** viscosity; **j.** net hydrostatic pressure; **k.** natriuretic peptides; **l.** edema

3. a. arterioles; **b.** autonomic nervous system; **c.** increasing peripheral vasoconstriction; **d.** increased vasodilation, increased venous return, increased cardiac output; **e.** brain; **f.** nervous and endocrine

Chapter Integration

1. The likely cause of fainting is a reduction in blood flow to the brain due to a sudden drop in blood pressure when moving from a sitting position to a standing position. This condition is called *orthostatic hypotension*. The reduction in cerebral blood flow can cause dizziness, disorientation, and, as in this instance, a temporary loss of consciousness.

2. Normally, when a person moves to a standing position, blood pressure is stabilized by reflexes that elevate the heart rate and cause peripheral vasoconstriction. When he sits up, his heart rate rises but his blood pressure falls. This suggests that the reflexes are functioning normally but his blood volume may be abnormally low.

3. Both the Hct and Hb are lower than normal, which could indicate blood loss from internal bleeding or problems with blood cell production.

4. Mr. Thaddeus was given IV fluids to increase his blood volume and his blood pressure. He was given oxygen to improve oxygen delivery to the tissues and relieve his shortness of breath.

5. The hospital staff performed cross-match testing in case he needed a transfusion of whole blood or packed red blood cells.

Module 19.1 Review

a. Lymphatic vessels transport lymph from peripheral tissues to the venous system.

b. Valves prevent the backflow of lymph in some lymphatic vessels.

c. Overlapping endothelial cells in lymphatic capillaries act as one-way valves that permit the entry of fluids and solutes but prevent their return to the intercellular spaces.

Module 19.2 Review

a. The lymphatic trunks empty into the thoracic duct and the right lymphatic duct.

b. The right lymphatic duct collects lymph from the right side of the body superior to the diaphragm; the thoracic duct collects lymph from the body inferior to the diaphragm and from the left side of the body superior to the diaphragm.

c. Lymphedema, an accumulation of interstitial fluids, results from blocked lymphatic drainage. If the condition does not resolve, connective tissues lose their elasticity, and the swelling becomes permanent.

Module 19.3 Review

a. The three main classes of lymphocytes are T cells, B cells, and natural killer (NK) cells.

b. B cells are responsible for antibody-mediated immunity.

c. The red bone marrow, thymus, and peripheral lymphoid tissues are involved in lymphopoiesis.

Module 19.4 Review

a. Tonsils are large lymphoid nodules in the walls of the pharynx. The five tonsils are the left and right palatine tonsils, a single pharyngeal tonsil, and a pair of lingual tonsils.

b. Mucosa-associated lymphoid tissue (MALT) is the collection of lymphoid tissue that protects epithelia lining the digestive, respiratory, urinary, and reproductive tracts.

c. Lymph flow through a lymph node: afferent lymphatics → subcapsular space → outer cortex → deep cortex → medullary sinus → efferent lymphatics

Module 19.5 Review

a. The thymus is located in the anterior mediastinum (posterior to the sternum).

b. Reticular epithelial cells in the cortex maintain the blood–thymus barrier.

c. The thymus is a pink, grainy organ ranging in weight from 40 g at puberty to less than 12 g at age 50. A capsule covers the thymus and divides it into two lobes, and fibrous partitions called septa divide the lobes into lobules.

Module 19.6 Review

a. The spleen filters the blood; the phagocytes it contains identify and engulf pathogens or infected cells circulating in the blood.

b. Red pulp contains large numbers of red blood cells; white pulp resembles lymphoid nodules.

c. Beginning with the trabecular arteries, blood then flows to the central arteries → capillaries → reticular tissue of red pulp → sinusoids → trabecular veins

Section 1 Review

1. **a.** tonsil; **b.** cervical lymph nodes; **c.** right lymphatic duct; **d.** thymus; **e.** cisterna chyli; **f.** lumbar lymph nodes; **g.** appendix; **h.** lymphatics of lower limb; **i.** lymphatics of upper limb; **j.** axillary lymph nodes; **k.** thoracic duct; **l.** lymphatics of mammary gland; **m.** spleen; **n.** mucosa-associated lymphoid tissue (MALT); **o.** pelvic lymph nodes; **p.** inguinal lymph nodes

2. **a.** lymphatic capillaries; **b.** thymic corpuscles; **c.** right subclavian vein; **d.** thoracic duct; **e.** lymphoid organs; **f.** lymph nodes; **g.** helper T cells and suppressor T cells; **h.** spleen; **i.** reticular epithelial cells; **j.** lymphopoiesis; **k.** cytotoxic T cells; **l.** tonsils; **m.** B cells; **n.** afferent lymphatics

Module 19.7 Review

a. Chemotaxis is movement in response to attraction to or repulsion from chemical stimuli.

b. The integumentary system provides a physical barrier that is the first line of defense in preventing pathogens and toxins from entering body tissues. Skin secretions flush the surface, hair protects against mechanical abrasion, and the multiple layers of the skin's epithelium create an interlocking barrier.

c. The body's phagocytes are neutrophils, eosinophils, and macrophages. Fixed macrophages are scattered among connective tissues and do not move; free macrophages are mobile and reach injury sites by migrating through adjacent tissues or traveling in the bloodstream.

Module 19.8 Review

a. Immunological surveillance is the constant monitoring of normal tissues by NK cells sensitive to abnormal antigens on the surfaces of cells.

b. NK cells recognize unusual proteins, called tumor-specific antigens, on the plasma membranes of cancer cells. When these antigens are detected, the NK cells then destroy the abnormal cells.

c. Cancer cells can mutate such that either they do not display tumor-specific antigens, or they secrete chemicals that can destroy NK cells. This ability to escape detection is referred to as immunological escape.

Module 19.9 Review

a. Interferons are small proteins that are released by activated lymphocytes, macrophages, and cells infected with viruses; they trigger the production of antiviral proteins that interfere with viral replication within tissue cells.

b. The 11 complement proteins of the complement system interact with each other in chain reactions that ultimately produce activated forms that target bacterial cell walls and plasma membranes, stimulate inflammation, attract phagocytes, or enhance phagocytosis.

c. Histamine release by mast cells and basophils in tissues increases local inflammation, thereby accelerating blood flow to the region.

Module 19.10 Review

a. The body's nonspecific defenses include physical barriers, phagocytes, immunological surveillance, interferons, the complement system, the inflammatory response, and fever.

b. A rise in the level of interferons suggests a viral infection.

c. Pyrogens increase body temperature (produce a fever), which can mobilize defenses, accelerate repairs, and inhibit pathogens.

Section 2 Review

1. **a.** physical barriers; **b.** phagocytes; **c.** immunological surveillance; **d.** interferons; **e.** complement; **f.** inflammatory response; **g.** fever

2. **a.** all of these; **b.** immunological surveillance; **c.** the complement system; **d.** interferons; **e.** pyrogens; **f.** phagocytes

3. The high body temperatures of a fever may inhibit some viruses and bacteria or speed their reproductive rates so that the disease runs its course more quickly. High body temperatures also accelerate the body's metabolic processes, which may help to mobilize tissue defenses and speed the repair process.

Module 19.11 Review

a. Antigen presentation occurs when an antigen-glycoprotein combination capable of activating T cells appears in a plasma membrane (typically that of a macrophage). T cells sensitive to this antigen are activated if they contact the antigen on the plasma membrane of the antigen-presenting cell.

b. The major histocompatibility complex (MHC) is a portion of chromosome 6 containing genes that control the synthesis of membrane glycoproteins.

c. Class I MHC proteins are in the plasma membranes of all nucleated body cells. Class II MHC proteins are in the plasma membranes of antigen-presenting cells (APCs) and lymphocytes only.

Module 19.12 Review

a. The three major types of T cells activated by Class I MHC proteins are cytotoxic T cells, memory T_C cells, and suppressor T cells.

b. T cell plasma membranes contain proteins called CD (cluster of differentiation) markers. Cells with CD8 markers respond to antigens presented by Class I MHC proteins and are on cytotoxic T cells, memory T_C cells, and suppressor T cells. Cells with CD4 markers respond to antigens presented by Class II MHC proteins.

c. Abnormal antigens in the cytoplasm of a cell can become attached to MHC proteins and then displayed on the surface of the cell's plasma membrane. The recognition of such antigens by T cells initiates an immune response.

Module 19.13 Review

a. Sensitization is the process by which a B cell prepares to undergo activation, so that it can subsequently react with a specific antigen.

b. Cytokines secreted by activated T cells aid in coordinating specific and nonspecific defenses and regulate cell-mediated and antibody-mediated immunity.

c. Plasma cells produce and secrete antibodies, so observing an elevated number of plasma cells in the lymph would lead us to expect higher than normal antibody levels in the blood.

Module 19.14 Review

a. An antigenic determinant site is the part of an antigen molecule to which an antibody molecule binds.

b. An antibody molecule consists of two parallel pairs of polypeptide chains: a pair of long, heavy chains and a pair of short, light chains. Each chain contains both constant segments and variable segments. The constant segments of the heavy chains form the base of the antibody molecule; the free tips of the two variable segments form the antigen binding sites.

c. The secondary response would be more affected by a lack of memory cells, which are produced in response to an initial exposure to an antigen during the primary response.

Module 19.15 Review

a. Opsonization is the process by which the coating of pathogens with antibodies and complement proteins makes the pathogens more susceptible to phagocytosis.

b. Antigen–antibody complexes help destroy antigens through seven processes: neutralization, prevention of bacterial and viral adhesion, activation of complement, opsonization, attraction of phagocytes, stimulation of inflammation, and precipitation and agglutination.

c. Basophils and mast cells are involved in the inflammatory response.

Module 19.16 Review

a. An allergy is an inappropriate or excessive immune response to an allergen, which is an antigen that triggers an allergic reaction.

b. In anaphylaxis, an immune response to a circulating antigen stimulates mast cells throughout the body to release chemicals that prompt the inflammatory response.

c. Histamines, leukotrienes, and other chemicals that cause pain and inflammation are released when mast cells and basophils are stimulated in an allergic reaction.

Module 19.17 Review

a. CD8 markers are found on cytotoxic T cells, memory T_C cells, and suppressor T cells; CD4 markers are on all helper T cells.

b. Cytotoxic T cells and NK cells can be activated by direct contact with virus-infected cells.

c. Plasma cells produce antibodies.

Module 19.18 Review

a. Autoimmune disorders are diseases that result from the production of antibodies (called autoantibodies) directed against normal substances in the body (self-antigens).

b. Immunosuppression is the partial or complete reduction of the immune response in an individual. It is also induced to enhance the survival of organ transplant recipients.

c. The increased incidence of cancer in the elderly may result from a decline in immunological surveillance, which results in reduced elimination of tumor cells as they arise.

Section 3 Review

1. a. antibody; **b.** CD4 markers; **c.** acquired immunity; **d.** passive immunity; **e.** helper T cells; **f.** opsonization; **g.** Class I MHC; **h.** B lymphocytes; **i.** Class II MHC; **j.** IgM; **k.** costimulation; **l.** IgG; **m.** anaphylaxis

2. a. viruses; **b.** macrophages; **c.** natural killer (NK) cells; **d.** helper T cells; **e.** B cells; **f.** antibodies; **g.** cytotoxic T cells; **h.** suppressor T cells; **i.** memory T cells and B cells

Chapter Integration

1. The student cannot yet know whether she will come down with the chickenpox. Her elevated blood IgM levels indicate that she is in the early stages of a primary response to the chickenpox virus. If her immune response proves unable to control and then eliminate the virus, she will develop chickenpox.

2. Since chickenpox is highly contagious, individuals who have not had chickenpox previously and have not been vaccinated are susceptible to this highly contagious disease.

3. The causative agent for chickenpox and shingles is the varicella-zoster virus (VZV).

4. IgM antibodies are the first class of immunoglobulins secreted after an antigen is encountered. IgG antibodies are the most common class of immunoglobulins.

5. Antigenic specificity results from the arrangement of genes controlling the amino acid sequences of the variable segments of antibodies.

6. Both responses involve the production of antibodies as stimulated by antigens. The primary response develops relatively slowly and levels of IgG rise more slowly than do levels of IgM. The secondary response develops much more rapidly, is more prolonged, and produces much more IgG.

7. Any vaccine, including that for chickenpox, stimulates artificially acquired active immunity.

Module 20.1 Review

a. The respiratory defense system is a series of filtration mechanisms that prevent airway contamination by debris and pathogens.

b. The respiratory mucosa lines the conducting portion of the respiratory tract.

c. Cystic fibrosis is a lethal, inherited disease that results from the production of dense mucus that restricts respiratory passages and accumulates in the lungs. Massive chronic bacterial infection of the lungs inhibits breathing, leading to death.

Module 20.2 Review

a. The components of the upper respiratory system are the nose, nasal cavity, paranasal sinuses, and pharynx.

b. Pathway of entering air: external nares → nasal cavity → nasal vestibule (guarded by hairs that screen out large particles) → superior, middle, and inferior meatuses (air bounces off the conchal surfaces) → internal nares (the connections between the nasal cavity and nasopharynx) → nasopharynx → oropharynx → laryngopharynx → larynx

c. The rich vascularization of the nasal cavity delivers body heat to the nasal cavity, so inhaled air is warmed before it leaves the nasal cavity. The heat also evaporates moisture from the epithelium to humidify the incoming air.

Module 20.3 Review

a. The unpaired laryngeal cartilages are the thyroid cartilage, cricoid cartilage, and epiglottis. The paired cartilages are the arytenoid cartilages, corniculate cartilages, and cuneiform cartilages.

b. The highly elastic vocal folds of the larynx are also called the vocal cords.

c. Phonation is the production of sound; articulation is the clarity of the sound and is provided by the involvement of the tongue, teeth, and lips.

Module 20.4 Review

a. The trachea transports air between the larynx and primary bronchi. The cilia and mucous secretions of tracheal epithelial cells also protect the respiratory tree by trapping inhaled debris and sweeping it toward the pharynx.

b. The C-shaped tracheal cartilages allow room for the esophagus to expand when food or liquids are swallowed.

c. Pathway of airflow along the lower respiratory tract: trachea → primary bronchi → secondary bronchi → tertiary bronchi → terminal bronchioles → pulmonary lobule

Module 20.5 Review

a. A bronchopulmonary segment is a specific region of a lung supplied by a tertiary bronchus.

b. The left lung is divided into a superior lobe and an inferior lobe by the oblique fissure; in the right lung, the horizontal fissure separates the superior lobe from the middle lobe, while the oblique fissure separates the superior and middle lobes from the inferior lobe.

c. Within the thoracic cavity, the left and right lungs are located within the left and right pleural cavities, respectively. The apex of each lung extends superiorly to the first rib, and the base of each lung rests on the superior surface of the diaphragm.

Module 20.6 Review

a. Pulmonary lobules are the smallest subdivisions of the lungs; each is supplied by branches of the pulmonary arteries, pulmonary veins, and tertiary bronchi.

b. Without surfactant, the alveoli would collapse due to the high surface tension in the thin layer of water that moistens the alveolar surfaces.

c. Path of airflow from glottis to the respiratory membrane: glottis → larynx → trachea → primary bronchus → bronchi → bronchioles → terminal bronchiole → respiratory bronchiole → alveolar duct → alveolar sac → alveolus → respiratory membrane

Section 1 Review

1. **a.** nasal cavity; **b.** hard palate; **c.** pharynx; **d.** glottis; **e.** trachea; **f.** right lung; **g.** external nares; **h.** larynx; **i.** primary bronchus; **j.** root of the lung; **k.** secondary bronchus; **l.** tertiary bronchus; **m.** bronchioles; **n.** terminal bronchiole; **o.** pulmonary lobule; **p.** alveolus

2. **a.** type II pneumocytes; **b.** trachea; **c.** type I pneumocytes; **d.** bronchodilation; **e.** terminal bronchiole; **f.** bronchoconstriction; **g.** respiratory bronchiole; **h.** respiratory membrane; **i.** phonation; **j.** pharynx; **k.** larynx; **l.** cystic fibrosis; **m.** respiratory mucosa; **n.** laryngeal prominence

Module 20.7 Review

a. Boyle's law states that at a constant temperature, the pressure of a gas is inversely proportional to its volume.

b. The movements of the diaphragm and ribs affect the volume of the lungs.

c. The intrapulmonary pressure (the pressure inside the respiratory tract) and the atmospheric pressure (the pressure outside the respiratory tract) determine the direction of airflow. Air moves from the area with the higher pressure to the area with the lower pressure.

Module 20.8 Review

a. The measurable pulmonary volumes are the resting tidal volume, expiratory reserve volume (ERV), residual volume, and inspiratory reserve volume (IRV).

b. The primary inspiratory muscles are the diaphragm and the external intercostal muscles.

c. The accessory respiratory muscles become active whenever the primary respiratory muscles are unable to move enough air to meet the oxygen demands of tissues.

Module 20.9 Review

a. The respiratory rate is the number of breaths taken each minute.

b. Respiratory minute volume is the amount of air moved into and out of the respiratory tract each minute, whereas alveolar ventilation is the amount of air reaching the alveoli each minute. Because some of the air never reaches the alveoli but instead remains in the anatomic dead space, alveolar ventilation is smaller than respiratory minute volume.

c. Slow deep breaths ventilate alveoli more effectively, because a smaller amount of the tidal volume of each breath is spent moving air into and out of the anatomic dead space of the lungs.

Module 20.10 Review

a. Dalton's law states that in a mixture of gases, the individual gases exert a pressure proportional to their abundance in the mixture.

b. The P_{O_2} decreases from about 100 mm Hg to 95 mm Hg in the pulmonary veins due to mixing with venous blood from the conducting passageways. The blood arriving at the peripheral capillaries has a P_{O_2} of 95 mm Hg.

c. Henry's law states that, at a given temperature, the amount of a particular gas that dissolves in a liquid is directly proportional to the partial pressure of that gas. Henry's law underlies the diffusion of gases between capillaries and alveoli, and between capillaries and interstitial fluid.

Module 20.11 Review

a. Oxyhemoglobin is hemoglobin to which oxygen molecules have bound.

b. BPG (2,3-bisphosphoglycerate) is a compound that reduces hemoglobin's affinity for oxygen. For any partial pressure of oxygen, if the concentration of BPG increases, the amount of oxygen released by hemoglobin will increase.

c. The combination of increased temperature and lower pH (from heat and acidic waste products generated by active skeletal muscles) causes hemoglobin to release more oxygen than when the muscles are at rest.

Module 20.12 Review

a. Carbon dioxide is transported in the bloodstream as bicarbonate ions, bound to hemoglobin, or dissolved in the plasma.

b. Driven by differences in partial pressure, oxygen enters the blood at the lungs and leaves it in peripheral tissues; similar forces drive carbon dioxide into the blood at the tissues and into the alveoli at the lungs.

c. Blockage of the trachea would interfere with the body's ability to take in oxygen and eliminate carbon dioxide. Because most carbon dioxide is transported in blood as bicarbonate ions formed from the dissociation of carbonic acid, an inability to eliminate carbon dioxide would result in a buildup of excess hydrogen ions, which would lower blood pH.

Module 20.13 Review

a. Compliance is the ease with which the lungs expand and recoil; resistance is an indication of how much force is required to inflate or deflate the lungs.

b. Three COPDs are asthma, chronic bronchitis, and emphysema.

c. Chronic bronchitis is long-standing inflammation of the mucous membranes in the bronchial tubes; emphysema is a condition in which the alveolar surfaces of the lungs are destroyed and alveoli merge, which reduces respiratory surface area and oxygen absorption, causing breathlessness. Individuals with chronic bronchitis are described as blue bloaters, whereas those with emphysema are described as pink puffers.

Module 20.14 Review

a. The pairs of CNS nuclei that adjust the pace of respiration are the apneustic centers and pneumotaxic centers in the pons.

b. The respiratory rhythmicity centers in the medulla oblongata generate the respiratory pace.

c. The pH, P_{O_2}, and P_{CO_2} in blood and cerebrospinal fluid stimulate the respiratory centers.

Module 20.15 Review

a. Hypercapnia is an increase in the P_{CO_2} of arterial blood above the normal range; hypocapnia is an abnormally low arterial P_{CO_2}.

b. Chemoreceptors are more sensitive to carbon dioxide levels than they are to oxygen levels.

c. Johnny's mother should not worry. When Johnny holds his breath, carbon dioxide levels in his blood increase, causing increased stimulation of the inspiratory centers and forcing him to breathe again.

Module 20.16 Review

a. Aging results in deterioration of elastic tissue, arthritic changes that stiffen rib articulations, decreased flexibility at costal cartilages, decreased vital capacity, and some degree of emphysema.

b. Dysplasia is the development of abnormal cells; metaplasia is the development of abnormal changes in tissue structure; neoplasia is the conversion of normal cells to tumor (cancerous) cells, and, in anaplasia, the malignant cells spread (metastasize) throughout the body.

c. The incidence of lung cancer in males is decreasing.

Section 2 Review

1. a. inspiratory reserve volume (IRV): the amount of air that can be taken in above the resting tidal volume; **b.** resting tidal volume (V_T): the amount of air inhaled and exhaled during a single respiratory cycle while resting; **c.** expiratory reserve volume (ERV): the amount of air that can be expelled after a completely normal, quiet respiratory cycle; **d.** minimal volume: the amount of air remaining in the lungs if they were to collapse; **e.** inspiratory capacity: the amount of air that can be drawn into the lungs after completing a quiet respiratory cycle; **f.** total lung capacity: the total volume of the lungs; **g.** vital capacity: the maximum amount of air that can be moved into or out of the lungs in a single respiratory cycle; **h.** residual volume: the amount of air remaining in the lungs after a maximal exhalation; **i.** functional residual capacity (FRC): the amount of air that remains in the lungs after completing a quiet respiratory cycle

2. a. partial pressure; **b.** anoxia; **c.** Boyle's law; **d.** compliance; **e.** bicarbonate ion; **f.** lowers vital capacity; **g.** external intercostals; **h.** iron ion; **i.** hemoglobin releases more O_2; **j.** pneumotaxic centers; **k.** apneustic centers; **l.** hypocapnia; **m.** atelectasis; **n.** apnea

3. External respiration includes all the processes involved in the exchange of oxygen and carbon dioxide between the interstitial fluids and the external environment. Pulmonary ventilation, or breathing, is a process of external respiration that involves the physical movement of air into and out of the lungs. Internal respiration is the absorption of oxygen and the release of carbon dioxide by tissue cells.

Chapter Integration

1. Alveolar ventilation rate (AVR) = respiratory rate × (tidal volume − anatomic dead space). In this case, the dead space is 200 mL (the anatomic dead space plus the volume of the snorkel); therefore, AVR = respiratory rate × (500 − 200). To maintain an AVR of 6.0 L/min, or 6000 mL/minute, the respiratory rate must be 6000/(500 − 200), or 20 breaths per minute.

2. Jake's hyperventilation resulted in abnormally low P_{CO_2}. This reduced his urge to breathe, so he stayed underwater longer, unaware that his P_{O_2} was dropping to the point of loss of consciousness.

Module 21.1 Review

a. The four layers of the digestive tract from superficial to deep are the mucosa (adjacent to the lumen), submucosa, muscularis externa, and serosa.

b. The mesenteries—sheets consisting of two layers of serous membrane separated by loose connective tissue—support and stabilize the organs in the abdominal cavity and provide a route for the passage of associated blood vessels, nerves, and lymphatic vessels.

c. The submucosal plexus is a nerve network that contains sensory neurons, parasympathetic ganglionic neurons, and sympathetic postganglionic fibers that innervate the mucosa and submucosa. The deeper myenteric plexus is a network of parasympathetic neurons, interneurons, and sympathetic postganglionic fibers that lies between the circular and longitudinal muscle layers.

Module 21.2 Review

a. Smooth muscle fibers in either the circular or longitudinal layers of the muscularis externa lie parallel to each other. In a longitudinal section of the digestive tract, the fibers of the superficial circular layer appear as little round balls, whereas the fibers of the deeper longitudinal layer are spindle-shaped.

b. Smooth muscle fibers are spindle shaped; lack T tubules, myofibrils, and sarcomeres; and the sarcoplasmic reticulum forms a loose network throughout the sarcoplasm. Because the tissue lacks sarcomeres, it is nonstriated. Additionally, the thin filaments are anchored to dense bodies.

c. Smooth muscle can contract over a wider range of resting lengths compared to skeletal muscle because of the looser organization of actin and myosin filaments in smooth muscle.

Module 21.3 Review

a. Enteroendocrine cells are endocrine cells that are scattered among the epithelial cells lining the digestive tract and that secrete hormones important to digestion.

b. The major mechanisms that regulate and control digestive activities are local factors (stimuli for digestive activities, such as changes in pH or distortion of the intestinal lumen), neural mechanisms (myenteric reflexes), and hormonal mechanisms (involving neuroendocrine cells).

c. The waves of contractions that constitute peristalsis are more efficient in propelling intestinal contents along the digestive tract than segmentation, which is basically a churning action that mixes intestinal contents with digestive fluids.

Section 1 Review

1. a. mesenteric artery and vein; **b.** mesentery; **c.** plica circulares; **d.** mucosa; **e.** submucosa; **f.** muscularis externa; **g.** serosa

2. a. esophagus; **b.** muscularis mucosae; **c.** lamina propria; **d.** plicae circulares; **e.** peristalsis; **f.** pacesetter cells; **g.** myenteric plexus; **h.** multi-unit smooth muscle cells; **i.** liver; **j.** segmentation; **k.** bolus; **l.** visceral smooth muscle cells; **m.** plasticity; **n.** sphincter

3. With a decrease in smooth muscle tone, general motility along the digestive tract decreases, and peristaltic contractions are weaker.

Module 21.4 Review

a. The hard palate forms the roof of the mouth; the hard palate is supported by bone and the soft palate is composed of muscle with no bony support.

b. The fauces is the arched opening between the oral cavity and the oropharynx.

c. The oral cavity is lined by stratified squamous epithelium, which provides protection against friction and abrasion.

Module 21.5 Review

a. The four types of teeth are incisors, cuspids (canines), bicuspids (premolars), and molars; a typical tooth has a crown, neck, and root.

b. The third set of molars is sometimes called the wisdom teeth.

c. The primary dentition is typically composed of 20 deciduous teeth (primary teeth, milk teeth, or baby teeth), which are temporary and the first teeth to appear in children; the secondary dentition is typically composed of 32 permanent teeth that appear subsequent to the primary dentition.

Module 21.6 Review

a. The pharynx is an anatomical space that serves as a common passageway in the digestive and respiratory tracts; in its digestive function, it receives a food bolus or liquids and passes them to the esophagus as part of the swallowing process.

b. The esophagus is the structure connecting the pharynx to the stomach.

c. During the buccal phase, food is formed into a bolus; during the pharyngeal phase, the bolus contacts the palatal arches and moves into the esophagus; during the esophageal phase, swallowing begins as pharyngeal muscles contract and the bolus is moved toward the stomach via peristaltic waves.

Module 21.7 Review

a. The falciform ligament is a sheet of mesentery that is a remnant of the ventral mesentery between the liver and the anterior wall of the peritoneal cavity.

b. The lesser omentum stabilizes the position of the stomach and provides an access route for blood vessels (and other structures) entering or leaving the liver.

c. Peritoneal fluid separates the parietal and visceral surfaces of the peritoneal cavity and prevents friction and subsequent irritation during sliding movements of organs within the abdominal cavity.

Module 21.8 Review

a. The four regions of the stomach are the cardia, fundus, body, and pylorus.

b. The inner lining of the stomach contains rugae, mucosal folds in the lining of the empty stomach that disappear as gastric distension occurs.

c. The longitudinal, circular, and oblique orientations of the muscle fibers in the three layers of the muscularis externa allow for the mixing and churning actions necessary for chyme formation.

Module 21.9 Review

a. Parietal cells secrete intrinsic factor and hydrochloric acid (HCl).

b. The alkaline tide is a sudden influx of bicarbonate ions into the bloodstream from active parietal cells; it causes a temporary increase in blood pH.

c. The alkaline mucous layer protects epithelial cells against the acid and enzymes in the gastric lumen.

Module 21.10 Review

a. The layers of the small intestine from superficial to deep are the mucosa, submucosa, muscularis externa, and serosa.

b. The intestinal mucosa bears transverse folds called plicae circulares bearing small projections called intestinal villi. These folds and projections increase the surface area available for absorption. Each villus contains a terminal lymphatic capillary called a lacteal. Between the bases of the villi are intestinal glands lined by enteroendocrine, mucous, and stem cells.

c. Lacteals are lymphatic capillaries in the intestinal villi that transport absorbed fatty acids that cannot enter blood capillaries.

Module 21.11 Review

a. The three regions of the small intestine are the duodenum, jejunum, and ileum.

b. The proximal portion of the duodenum is found within the epigastric region.

c. The primary function of the duodenum is to receive chyme from the stomach and neutralize its acids to avoid damaging the absorptive surfaces of the remaining regions of the small intestine.

Module 21.12 Review

a. The five major hormones that regulate digestive activity are gastrin, secretin, gastric inhibitory peptide (GIP), cholecystokinin (CCK), and vasoactive intestinal peptide (VIP).

b. A high-fat meal raises the cholecystokinin (CCK) level in the blood.

c. If the small intestine did not secrete secretin, the pH of the intestinal contents would be lower (more acidic) than normal, because secretin stimulates the pancreas to release a fluid high in buffers that neutralizes the acidic chyme entering the duodenum from the stomach.

Module 21.13 Review

a. The three overlapping phases of gastric secretion (which are named according to the location of the control center involved) are the cephalic phase, which prepares the stomach to receive ingested materials; the gastric phase, which begins with the arrival of food in the stomach; and the intestinal phase, which controls the rate of gastric emptying.

b. The two central reflexes triggered by stimulation of stretch receptors in the stomach wall are the gastroenteric reflex, which stimulates motility and secretion along the entire small intestine, and the gastroileal reflex, which triggers the opening of the ileocecal valve to allow passage of materials from the small intestine into the large intestine.

c. Severing the branches of the vagus nerves that supply the stomach would interrupt parasympathetic stimulation of gastric secretions, and the consequent reduction in acid secretions would provide some relief from gastric ulcers.

Module 21.14 Review

a. The segments of the large intestine are the cecum, colon, and rectum; the four regions of the colon are the ascending colon, transverse colon, descending colon, and sigmoid colon.

b. The major functions of the large intestine are (1) the reabsorption of water and the compaction of material into feces, (2) the absorption of vitamins produced by bacteria, and (3) the storage of fecal material prior to defecation.

c. Mass movements are powerful peristaltic contractions that occur a few times daily in response to distension of the stomach and duodenum.

Module 21.15 Review

a. Hemorrhoids are distended (swollen) veins in the distal portion of the rectum that may result from straining during defecation.

b. The large intestine is larger in diameter and shorter in length than the small intestine, but its relatively thin wall lacks villi and has an abundance of mucous cells

and distinctive intestinal glands (dominated by mucous cells).

c. The two positive feedback loops in the defecation reflex are (1) the short reflex, whereby stretch receptors in the rectal walls promote a series of peristaltic contractions in the colon and rectum, moving feces into the anal canal; and (2) the long reflex, whereby parasympathetic motor neurons in the sacral spinal cord, also activated by stretch receptors, stimulate peristalsis via motor commands distributed by somatic motor neurons.

Section 2 Review

1. a. crown; **b.** neck; **c.** root; **d.** enamel; **e.** dentin; **f.** pulp cavity; **g.** gingiva; **h.** gingival sulcus; **i.** cementum; **j.** periodontal ligament; **k.** root canal; **l.** bone of alveolus

2. a. material in jejunum; **b.** gastrin; **c.** GIP; **d.** secretin and CCK; **e.** VIP; **f.** inhibits; **g.** acid production; **h.** insulin; **i.** bile; **j.** intestinal capillaries; **k.** gallbladder; **l.** nutrient utilization by tissues

3. Both parietal cells and chief cells are secretory cells found in the gastric glands of the wall of the stomach. However, parietal cells secrete intrinsic factor and hydrochloric acid, whereas chief cells secrete pepsinogen, an inactive proenzyme.

Module 21.16 Review

a. The three pairs of salivary glands are the parotid, sublingual, and submandibular salivary glands.

b. The sublingual salivary glands contribute least (about 5 percent) to the secretions that make up saliva.

c. Damage to the parotid salivary glands, which secrete the enzyme salivary amylase, would interfere with the digestion of starches (complex carbohydrates).

Module 21.17 Review

a. The liver is divided into left, right, caudate, and quadrate lobes.

b. The gallbladder temporarily stores bile produced by the liver.

c. The falciform ligament marks the division between the left lobe and the right lobe of the liver.

Module 21.18 Review

a. A hepatocyte is a liver cell.

b. A portal area is located at each of the six corners of a liver lobule; each portal area contains (1) a branch of the hepatic portal vein, (2) a branch of the hepatic artery, and (3) a small branch of the bile duct.

c. Kupffer cells are liver macrophages that engulf pathogens, cell debris, and damaged blood cells.

Module 21.19 Review

a. Emulsification is the breakdown of lipid droplets by bile salts.

b. The gallbladder stores and releases bile, which contains additional buffers and bile salts that facilitate the digestion and absorption of lipids. The pancreas provides several digestive enzymes necessary for the breakdown of starches, lipids, nucleic acids, and proteins.

c. The pathway of a drop of bile: hepatic ducts → common hepatic duct → common bile duct → duodenal ampulla and papilla → duodenal lumen

Module 21.20 Review

a. Cholecystitis is inflammation of the gallbladder, usually resulting from a blockage of the cystic duct or the common bile duct by gallstones.

b. The bacterium responsible for most peptic ulcers is *Helicobacter pylori*.

c. Periodontal disease is characterized by a loosening of the teeth within the alveolar sockets due to erosion of the periodontal ligaments by acids produced by bacterial action.

Section 3 Review

1. a. bile ductule; **b.** hepatocytes; **c.** central vein; **d.** interlobular septum; **e.** sinusoid; **f.** branch of hepatic artery; **g.** branch of hepatic portal vein; **h.** bile duct; **i.** portal area (portal triad)

2. a. starch; **b.** pancreas; **c.** common bile duct; **d.** hepatocytes; **e.** mumps; **f.** peptic ulcer; **g.** emulsification; **h.** pancreatic lipase; **i.** liver; **j.** lysozyme; **k.** submandibular glands;

l. Kupffer cells; **m.** gallbladder; **n.** gallstones

3. Saliva (1) continuously flushes and cleans oral surfaces, (2) contains buffers that prevent the buildup of acids produced by bacterial action, and (3) contains antibodies (IgA) and lysozyme, which help control the growth of oral bacterial populations.

4. Such a blockage would interfere with the release of secretions into the duodenum by the pancreas, gallbladder, and liver. The pancreas normally secretes about 1 liter of pancreatic juice, a mixture of a variety of digestive enzymes and buffer solution. The blockage of pancreatic juice would lead to pancreatitis, an inflammation of the pancreas. Extensive damage to exocrine cells by the blocked digestive enzymes would lead to autolysis that could destroy the pancreas and result in the individual's death. Blockage of bile secretion from the common bile duct could lead to damage of the wall of the gallbladder by the formation of gallstones and to jaundice because bilirubin from the liver would not be excreted in the bile and, instead, would accumulate in body fluids.

Chapter Integration

1. Gastric stapling would result in weight loss because the volume of food (and thus the amount of calories) consumed is reduced because the person feels full after eating only a small amount.

2. After gastric stapling surgery, the individual can eat only a small amount of food before the stretch receptors in the gastric wall become stimulated and a feeling of fullness results.

3. The smooth muscle in the gastric muscularis externa of the functional portion of the stomach gradually becomes increasingly tolerant of distention, and the operation may have to be repeated to achieve the same results.

4. Gastric bypass and lap-band surgeries, like gastric stapling, reduce gastric capacity, thereby decreasing the amount of food a person can ingest before feeling full.

Module 22.1 Review

a. The citric acid cycle is the reaction sequence that occurs in the matrix of mitochondria. In the process, organic molecules are broken down, carbon dioxide molecules are released, and hydrogen atoms are transferred to coenzymes that deliver them to the electron transport system.

b. The electron transport system consists of a series of cytochromes, proteins that transfer electrons along the inner mitochondrial membrane. In the process, hydrogen ions are pumped into the intermembrane space; as those ions diffuse back into the mitochondrial matrix, ATP is generated. The electrons moving along the ETS are ultimately transferred to oxygen atoms, so this is an aerobic process.

c. The primary role of the citric acid cycle in ATP production is to break down organic nutrients and provide hydrogen atoms to the ETS by way of the coenzymes NAD and FAD.

Module 22.2 Review

a. The nutrient pool within a cell includes all of the substrates in the cytoplasm that are available for anabolism or catabolism.

b. Cells carry out catabolism to release energy for use in cell growth, cell division, and tissue-specific activities.

c. Cells make new compounds to maintain and repair structures, to support growth, and to build up nutrient reserves.

Section 1 Review

1. a. fatty acids; b. glucose; c. proteins; d. two-carbon chains; e. citric acid cycle; f. coenzymes; g. ATP; h. electron transport system; i. O_2; j. CO_2; k. H_2O

2. During fasting or starvation, other tissues shift to fatty acid or amino acid catabolism, conserving glucose for neural tissue.

3. a. citric acid; b. nutrient pool; c. anabolism; d. oxidative phosphorylation; e. cytochromes; f. coenzymes; g. oxygen; h. catabolism; i. nutrients abundant; j. ATP; k. citric acid cycle; l. water; m. nutrients scarce; n. acetate

Module 22.3 Review

a. Intestinal gas, or flatus, is a by-product of the breakdown of indigestible carbohydrates by bacteria in the large intestine.

b. Glycogen is synthesized from excess glucose molecules by liver and muscle cells, and serves as an intracellular glucose reserve.

c. Proteins and fats are necessary for building the components of cells and tissues.

Module 22.4 Review

a. Two molecules each of pyruvate, ATP, and NAD•H.

b. Most (two-thirds) of the CO_2 released in the complete catabolism of glucose occurs during the citric acid cycle.

c. ATP produced anaerobically through glycolysis can be important during peak levels of activity or when a tissue is temporarily deprived of oxygen.

Module 22.5 Review

a. Micelles are lipid–bile salt complexes (containing fatty acids, glycerol, and monoglycerides) formed in the intestinal lumen. Chylomicrons are lipoproteins formed in intestinal epithelial cells and contain newly synthesized triglycerides, cholesterol, and other lipids surrounded by phospholipids and proteins.

b. The liver absorbs chylomicrons, removes the triglycerides, combines the cholesterol from the chylomicron with recycled cholesterol, and alters the surface proteins. The new complex is released into the bloodstream as a low-density lipoprotein (LDL).

c. Low-density lipoproteins (LDLs) deliver cholesterol to body tissues, and high-density lipoproteins (HDLs) absorb unused cholesterol from body tissues, returning it to the liver where it may be packaged into new LDLs or excreted with bile salts in bile.

Module 22.6 Review

a. Beta-oxidation is fatty acid catabolism that produces molecules of acetyl-CoA.

b. Acetyl-CoA is a reactant molecule in ATP production and in the synthesis of most types of lipids.

c. Fatty acids may become a source of energy or a component of triglycerides, glycolipids, phospholipids, prostaglandins, cholesterol, and steroids.

Module 22.7 Review

a. Parietal cells secrete the proenzyme pepsinogen, which is then activated to pepsin, the enzyme necessary for protein digestion.

b. The amino group is removed by deamination or transamination.

c. The ammonium ions combine with carbon dioxide to form urea (in the urea cycle), which is ultimately excreted in the urine.

Module 22.8 Review

a. The absorptive state, lasting about 4 hours, is the period following a meal, when nutrient absorption is under way; the postabsorptive state, lasting about 12 hours, is the period when nutrient absorption is not under way and the body relies on internal energy reserves to meet demands.

b. Ketone bodies form during the postabsorptive state when lipids and amino acids are broken down in the liver and the concentration of acetyl-CoA rises.

c. During the absorptive state, insulin prevents a large surge in blood glucose after a meal by stimulating the liver to remove glucose from the circulation. During the postabsorptive state, blood glucose begins to decline, triggering the release of glucagon, which stimulates the liver to release glucose into the circulation.

Module 22.9 Review

a. Nutrition is the absorption of nutrients from food.

b. The two classes of vitamins are fat-soluble vitamins and water-soluble vitamins.

c. Vitamins play an important role in metabolic pathways by serving as coenzymes.

Module 22.10 Review

a. A balanced diet contains all the ingredients needed to maintain homeostasis and prevent malnutrition.

b. A complete protein meets the body's amino acid requirements; an incomplete protein is deficient in one or more amino acids.

c. The catabolism of lipids releases the greatest number of Calories per gram.

Module 22.11 Review

a. Eating disorders are psychological problems that result in inadequate food consumption (anorexia nervosa) or excessive food consumption followed by purging (bulimia).

b. Protein deficiency diseases are nutritional disorders resulting from a lack of one or more essential amino acids; kwashiorkor is an example of a protein deficiency disease.

c. Phenylketonuria (PKU) is an inherited metabolic disorder resulting from an inability to convert phenylalanine to tyrosine.

Section 2 Review

1. a. insulin; b. skeletal muscle; c. B complex and C; d. ketone bodies; e. lipogenesis; f. urea formation; g. A, D, E, K; h. lipoproteins; i. deamination; j. lipolysis; k. uric acid; l. calorie; m. absorptive state; n. anorexia

2. a. jejunum; b. catabolized for energy; c. venous circulation via the thoracic duct; d. the excess amount is readily excreted in the urine

3. a. Essential amino acids are necessary in the diet because they cannot be synthesized by the body. The body can synthesize nonessential amino acids on demand.

b. (1) Proteins are difficult to break apart because of their complex three-dimensional structure. (2) The energy yield of proteins (4.32 Cal/g) is less than that of lipids (9.46 Cal/g). (3) The by-products of protein or amino acid catabolism are ammonium ions, a toxin that can damage cells. (4) Proteins form the most important structural and functional components of cells. Excessive protein catabolism would threaten homeostasis at the cellular to system levels of organization.

c. During the absorptive state, the intestinal mucosa is absorbing nutrients from the digested food. The focus of the postabsorptive state is the mobilization of energy reserves and the maintenance of normal blood glucose levels.

d. Liver cells can break down or synthesize most carbohydrates, lipids, and amino acids. The liver has an extensive blood supply and thus can easily monitor blood composition of these nutrients and regulate accordingly. The liver also stores energy in the form of glycogen.

4. It appears that Darla is suffering from ketoacidosis as a consequence of her anorexia. Because she is literally starving herself, her body is metabolizing large amounts of fatty acids and amino acids to provide energy and in the process is producing large quantities of ketone bodies (normal metabolites from these catabolic processes). One of the ketones that is formed is acetone, which can be eliminated through the lungs. This accounts for the smell of aromatic hydrocarbons on Darla's breath. The ketones are also converted into keto acids. In large amounts this lowers the body's pH and begins to exhaust the alkaline reserves of the buffer system. This is probably the cause of her arrhythmias.

Module 22.12 Review

a. Ghrelin, a hormone secreted by the gastric mucosa when the stomach is not full, inhibits the satiety center and stimulates appetite.

b. Leptin is a peptide hormone produced by adipose tissue during the synthesis of triglycerides. It stimulates the satiety center and suppresses appetite.

c. A lack of this hypothalamic neurotransmitter would probably decrease appetite since it normally stimulates the feeding center.

Module 22.13 Review

a. Insensible perspiration refers to the evaporation of water from the skin and alveolar surfaces of the lungs.

b. Radiation

c. Conduction is the direct transfer of heat through physical contact. Convection is the result of the conduction of heat to the air in contact with the skin. The air warmed by the skin rises and it is repeatedly replaced by cooler air until there is no difference in temperature.

Module 22.14 Review

a. Countercurrent exchange.

b. Nonshivering thermogenesis involves the release of hormones that increase the metabolic activity of all tissues, resulting in an increase in body temperature.

c. The vasodilation of peripheral vessels would increase blood flow to the skin and thus the amount of heat the body can lose. As a result, body temperature would decrease.

Section 3 Review

1. **a.** sensible perspiration; **b.** leptin; **c.** thermoregulation; **d.** inhibits feeding center; **e.** nonshivering thermogenesis; **f.** 60 percent; **g.** peripheral vasoconstriction; **h.** neuropeptide Y; **i.** basal metabolic rate; **j.** ghrelin; **k.** 40 per-

cent; **l.** peripheral vasodilation; **m.** insensible perspiration; **n.** shivering thermogenesis

2. **a.** all of these; **b.** radiation, conduction, convection, and evaporation; **c.** physiological responses and behavioral modifications; **d.** peripheral vasoconstriction; **e.** triglycerides in adipose tissue; **f.** low blood glucose levels

3. **a.** Because energy use at rest is powered by mitochondrial energy production, and mitochondrial energy production is proportional to oxygen consumption.

b. The heat-gain center functions in preventing hypothermia, or below-normal body temperature, by conserving body heat and increasing the rate of heat production by the body.

c. Nonshivering thermogenesis increases the metabolic rate of most tissues through the actions of two hormones, epinephrine and thyroid-stimulating hormone. In the short term, the heat-gain center stimulates the adrenal medullae to release epinephrine via the sympathetic division of the ANS. Epinephrine quickly increases the breakdown of glycogen (glycogenolysis) in liver and skeletal muscle, and the metabolic rate of most tissues. The long-term increase in metabolism occurs primarily in children as the heat-gain center adjusts the rate of thyrotropin-releasing hormone (TRH) release by the hypothalamus. When body temperature is low, additional TRH is released, which stimulates the release of thyroid-stimulating hormone (TSH) by the anterior

lobe of the pituitary gland. The thyroid gland then increases its rate of thyroid hormone release, and these hormones increase rates of catabolism throughout the body.

Chapter Integration

1. Vitamins and minerals are essential components of the diet because the body cannot synthesize most of the vitamins and minerals it requires.

2. These terms refer to the lipoproteins in the blood that transport cholesterol. So-called "good cholesterol" (high-density lipoproteins, or HDLs) transports excess cholesterol to the liver for storage and breakdown, whereas "bad cholesterol" (low-density lipoproteins, or LDLs) transports cholesterol to peripheral tissues, which includes the arteries. The buildup of cholesterol in arterial walls is linked to cardiovascular disease.

3. High-density lipoproteins (HDLs) are considered beneficial because they reduce the amount of cholesterol in the bloodstream by transporting it to the liver for storage or excretion in the bile.

4. Glycogenesis (the formation of glycogen) in the liver increases after a high-carbohydrate meal.

5. A diet deficient in pyridoxine (vitamin B_6) would interfere with the body's ability to metabolize proteins. Subsequent effects of the deficiency include retarded growth, anemia, convulsions, and epithelial changes.

Module 23.1 Review

a. The main structures of the urinary system are the kidneys, ureters, urinary bladder, and urethra.

b. The layers of connective tissue that protect and anchor the kidney are the fibrous capsule, the perinephric fat capsule, and the renal fascia. The fibrous capsule covers the surface of the kidney, the perinephric fat capsule is a thick layer of adipose tissue surrounding the fibrous capsule, and the renal fascia is a dense, fibrous outer layer that anchors the kidney to surrounding structures.

c. If the perinephric fat, which helps hold the kidneys in position against the posterior body wall, were depleted, the kidneys could drop (a condition called renal ptosis) and would no longer be held securely; disruption of the collagen fibers could create a condition called floating kidney, which may cause pain or distortion of other structures.

Module 23.2 Review

a. The renal pyramid is a conical mass within the renal medulla that ends at the papilla.

b. The renal papilla is the tip of the conically shaped renal pyramid; it empties formed urine into the renal pelvis.

c. Juxtamedullary nephrons are essential for the conservation of water and the production of concentrated urine.

Module 23.3 Review

a. The primary structures of the nephron are the renal corpuscle, proximal convoluted tubule, nephron loop, and distal convoluted tubule. The main structures of the collecting system are the collecting duct and papillary duct.

b. The renal corpuscle consists of the glomerular capsule and the glomerulus.

c. Glomerular fluid (also called filtrate) essentially lacks proteins, whereas blood plasma contains a variety of larger protein molecules.

Module 23.4 Review

a. Pathway of renal blood flow: renal artery → segmental arter-ies → interlobar arteries → arcuate arteries → cortical radiate arteries → afferent arterioles → glomerulus → efferent arterioles → peritubular capillaries → cortical radiate veins → arcuate veins → interlobar veins → renal vein

b. Blood enters the glomerulus via the afferent arteriole and leaves via the efferent arteriole.

c. The vasa recta are long, straight peritubular capillaries that parallel the nephron loop of a juxtamedullary nephron.

Section 1 Review

1. **a.** renal sinus: cavity within kidney that contains calyces, pelvis of the ureter, and segmental vessels; **b.** renal pelvis: funnel-shaped expansion of the superior portion of ureter; **c.** hilum: depression on the medial border of kidney, and site of the apex of the renal pelvis and the passage of segmental renal vessels and renal nerves; **d.** renal papilla: tip of the renal pyramid that projects into a minor calyx; **e.** ureter: tube that conducts urine from the renal pelvis to the urinary bladder; **f.** renal cortex: the outer portion of the kidney containing renal lobules, renal columns (extensions between the pyramids), renal corpuscles, and the proximal and distal convoluted tubules; **g.** renal medulla: the inner, darker portion of the kidney that contains the renal pyramids; **h.** renal pyramid: conical mass of the kidney projecting into the medullary region containing part of the secreting tubules and collecting tubules; **i.** minor calyx: subdivision of major calices into which urine enters from the renal papillae; **j.** major calyx: primary subdivision of renal pelvis formed from the merging of four or five minor calyces; **k.** renal lobe: portion of kidney consisting of a renal pyramid and its associated cortical tissue; **l.** renal columns: cortical tissue separating renal pyramids; **m.** fibrous capsule (outer layer): covering of the kidney's outer surface and lining of the renal sinus

2. **a.** urinary bladder; **b.** nephrons; **c.** renal tubules; **d.** glomerulus; **e.** proximal convoluted tubule; **f.** papillary ducts;

g. renal medulla; **h.** renal sinus; **i.** major calyces; **j.** ureter

Module 23.5 Review

a. The three distinct processes of urine formation in the kidney are filtration, reabsorption, and secretion.

b. Filtration exclusively occurs across the filtration membrane in the renal corpuscle.

c. When the plasma concentration of a substance exceeds the kidney's capacity for transporting that substance, the excess is not reabsorbed, so it is excreted in the urine.

Module 23.6 Review

a. The capsular space separates the parietal and visceral layers of the glomerular capsule.

b. The primary factor is the balance between hydrostatic pressure (fluid pressure) and colloid osmotic pressure (pressure due to materials in solution) on either side of the capillary walls.

c. Blood colloidal pressure tends to draw water out of the filtrate and into the plasma because the solute concentration within the blood exceeds that within the filtrate. The "advantage" that results when blood colloidal pressure draws water out of the filtrate and into the plasma is that this action helps conserve body water.

Module 23.7 Review

a. In response to decreased filtration pressure, the juxtaglomerular complex increases renin production and release into the bloodstream.

b. Angiotensin II causes efferent arteriole constriction, which increases glomerular blood pressure.

c. Angiotensin II triggers CNS neural responses, including enhancing the sensation of thirst and increasing ADH production.

Module 23.8 Review

a. The distal convoluted tubule (DCT) makes final adjustments to the composition of tubular fluid.

b. Increased amounts of aldosterone, which promotes Na$^+$

retention and K$^+$ secretion at the kidneys, would elevate the K$^+$ concentration of urine.

c. If the concentration of Na$^+$ in the filtrate decreased, fewer hydrogen ions could be secreted via the countertransport mechanism involving these two ions. As a result, the pH of the tubular fluid would increase.

Module 23.9 Review

a. Countercurrent multiplication in the kidneys is the exchange of substances between two adjacent nephron loop limbs containing fluid moving in opposite directions; the process is responsible for the kidney tubules' ability to concentrate urine.

b. The thick ascending limb actively pumps sodium and chloride into the peritubular fluid; it is impermeable to water.

c. An increase of sodium and chloride ions in the peritubular fluid elevates the osmotic concentrations around the descending limb, resulting in the osmotic flow of water out of it.

Module 23.10 Review

a. No, the permeability of the PCT cannot change, and water reabsorption occurs whenever the osmotic concentration of the peritubular fluid exceeds that of the tubular fluid.

b. Increased ADH levels cause the appearance of more water channels, or aquaporins, in the DCT; as a result, more water is reabsorbed into the peritubular fluid, which reduces the volume of water in the urine.

c. Decreased ADH levels in the DCT reduce the urine osmotic concentration due to the presence of more water in the urine; the result is a larger volume of more dilute urine.

Module 23.11 Review

a. The filtrate produced at the renal corpuscle has the same osmotic pressure as plasma.

b. The active removal of ions and organic substrates from the PCT causes a continuous osmotic flow of water out of the tubular fluid.

c. The concentration gradient of the renal medulla is maintained by the removal of solutes and water from the area by the vasa recta, which transport them to the circulatory system.

Module 23.12 Review

a. Dialysis is the process of using an artificial semipermeable membrane to remove waste products from the blood of a person whose kidneys are not functioning properly.

b. Chronic renal failure involves a gradual loss of renal function, whereas acute renal failure involves a sudden loss of renal function.

c. Patients on dialysis are often given Epogen or Procrit (a synthetic form of erythropoietin) to treat anemia, which occurs because their malfunctioning kidneys produce too little erythropoietin, the hormone that stimulates the development of red blood cells in the bone marrow.

Section 2 Review

1. a. nephron: functional unit of the kidney that filters and excretes waste materials from the blood and forms urine; b. proximal convoluted tubule: reabsorbs water, ions, and all organic nutrients; c. distal convoluted tubule: important site of active secretion; d. renal corpuscle: expanded chamber that encloses the glomerulus; e. nephron loop: portion of the nephron that produces the concentration gradient in the renal medulla; f. collecting system: series of tubes that carry tubular fluid away from the nephron; g. collecting duct: portion of collecting system that receives fluid from many nephrons and performs variable reabsorption of water and reabsorption or secretion of sodium, potassium, hydrogen, and bicarbonate ions;

h. papillary duct: delivers urine to the minor calyx
2. a. renal corpuscle b. podocytes c. filtrate d. BCOP e. nephron loop f. aquaporins g. secretion h. aldosterone i. PCT j. ADH
3. The presence of plasma proteins and numerous WBCs in the urine indicates an increased permeability of the filtration membrane. This condition usually results from inflammation of the filtration membrane within the renal corpuscle. If the condition is temporary, it is probably an acute glomerular nephritis usually associated with a bacterial infection (such as streptococcal sore throat). If the condition is long term, resulting in a nonfunctional kidney, it is referred to as chronic glomerular nephritis. The urine volume would be greater than normal because the plasma proteins increase the osmolarity of the filtrate.

Module 23.13 Review

a. Urine is transported by the ureters, stored within the urinary bladder, and eliminated through the urethra.

b. The specialized smooth muscle in the wall of the urinary bladder is the detrusor muscle; its contraction compresses (squeezes) the urinary bladder and expels urine into the urethra.

c. The external urethral sphincter must be consciously relaxed to allow urination.

Module 23.14 Review

a. The sensation of a full bladder is relayed from the bladder to the thalamus to the cerebral cortex.

b. Voluntary relaxation of the external urethral sphincter causes relaxation of the internal urethral sphincter via the micturition reflex.

c. The urge to urinate usually appears when the urinary blad-

der contains about 200 mL of urine. The micturition reflex begins to function when the stretch receptors have provided adequate stimulation to the parasympathetic motor neurons. The activity in the motor neurons generates action potentials that reach the smooth muscle in the wall of the urinary bladder. These efferent impulses travel over the pelvic nerves, producing a sustained contraction of the urinary bladder.

Module 23.15 Review

a. Dysuria is the term for painful or difficult urination.

b. Obstruction of a ureter by a kidney stone would interfere with the flow of urine between the kidney and the bladder.

c. Urinary obstruction at the urethra is more dangerous than urinary obstruction in the ureter because urine would be prevented from exiting the body and would build up in the urinary bladder, leading to rupture. Obstruction of only one of the two ureters is less hazardous than obstruction of the lone urethra because in the former case, the other kidney could still excrete wastes that could be voided.

Section 3 Review

1. a. internal urethral sphincter; b. trigone; c. external urethral sphincter; d. external urethral orifice; e. rugae; f. middle umbilical ligament; g. detrusor; h. transitional epithelium; i. urethra; j. micturition; k. stratified squamous epithelium
2. a. stretch receptors stimulated; b. afferent fibers carry information to sacral spinal cord; c. parasympathetic preganglionic fibers carry motor commands; d. detrusor muscle contraction stimulated; e. sensation relayed to thala-

mus; f. sensation of bladder fullness delivered to cerebral cortex; g. individual relaxes external urethral sphincter; h. internal urethral sphincter relaxes

3. Four primary signs and symptoms of urinary disorders are (1) changes in the volume of urine, (2) changes in the appearance of urine, (3) changes in the frequency of urination, and (4) pain.

4. cystitis/pyelonephritis: both conditions involve inflammation and infections of the urinary system, but cystitis refers to the urinary bladder, whereas pyelonephritis refers to the kidney; stress incontinence/ overflow incontinence: both conditions involve an inability to control urination, but stress incontinence involves periodic involuntary leakage, whereas overflow incontinence involves a continual, slow trickle of urine; polyuria/proteinuria: both are abnormal urine conditions, but polyuria is the production of excessive amounts of urine, whereas proteinuria refers to the presence of protein in the urine

Chapter Integration

1. The formation of crystalline masses can lead to kidney stones, which can cause urinary obstruction. The kidney stones can cause tiny ruptures along the urinary tract, leading to blood in the urine. The metabolism of nitrogen-containing melamine releases ammonia, which may cause electrolyte imbalance and renal failure.

2. Administration of intravenous fluids and dialysis would flush the body of the harmful chemical. Medication to control blood pressure, nausea, and anemia might also be necessary. Frequent blood tests and urinalysis are also performed to monitor kidney function.

Module 24.1 Review

a. The major routes of fluid loss are urination, evaporation at the skin, evaporation at the lungs, and water loss in feces.

b. A fluid shift is a rapid movement of water between the ECF and ICF in response to an osmotic gradient.

c. Dehydration is a reduction in the water content of the body that develops when water losses outpace water gains. In dehydration, the osmotic concentration of plasma increases.

Module 24.2 Review

a. Mineral balance is the state of the body in which ion gains and losses are equal.

b. Sodium is a major cation that is essential for normal membrane function; calcium is a cation that is essential for normal muscle and neuron function and for normal bone structure.

c. The ions absorbed by active transport are sodium, calcium, magnesium, iron, phosphate, and sulfate.

Module 24.3 Review

a. When osmoreceptors are inhibited, ADH release is decreased, and thirst is suppressed.

b. Aldosterone causes increased urinary sodium retention and thus increases the sodium ion concentration in the ECF.

c. Shifts in sodium balance result in expansion or contraction of the ECF. Large variations in ECF volume are corrected by homeostatic mechanisms triggered by changes in blood volume. If the blood volume becomes too low, ADH and aldosterone are secreted, increasing the sodium ion concentration in the ECF; if the volume becomes too high, natriuretic peptides are secreted.

Module 24.4 Review

a. Hypokalemia is a condition characterized by plasma K^+ levels below 3.5 mEq/L; hyperkalemia is a condition characterized by plasma K^+ levels above 5.5 mEq/L.

b. The kidneys are responsible for regulating the potassium ion concentration of the ECF.

c. Potassium excretion increases as potassium concentrations rise in the ECF, under aldosterone stimulation, and when the ECF pH rises.

Section 1 Review

1. **a.** osmoreceptors; **b.** fluid balance; **c.** plasma, interstitial fluid; **d.** sodium; **e.** ADH; **f.** hyponatremia; **g.** potassium; **h.** fluid compartments; **i.** kidneys; **j.** fluid shift; **k.** hypertonic plasma; **l.** hypokalemia; **m.** dehydration; **n.** aldosterone

2. **a.** intracellular fluid (ICF); **b.** kidneys and sweat glands; **c.** from the cells into the ECF until osmotic equilibrium is restored; **d.** becomes hypotonic with respect to the ICF; **e.** all of these; **f.** out of the ICF and into the ECF

3. When tissues are burned, cells are destroyed and the contents of their cytoplasm leak into the interstitial fluid and then move into the plasma. Since potassium ions are normally found within cells, damage to a large number of cells releases relatively large amounts of potassium ions into the blood. The elevated potassium level would stimulate cells of the adrenal cortex to produce aldosterone, and the cells of the juxtaglomerular complex to produce renin. The renin would activate the angiotensin mechanism. Ultimately, angiotensin II would stimulate still more aldosterone secretion. The elevated levels of aldosterone would promote sodium retention and potassium secretion by the kidneys, thereby accounting for the elevated potassium levels in the patient's urine.

Module 24.5 Review

a. Acidemia is the condition in which plasma pH falls below 7.35; alkalemia exists when the plasma pH is above 7.45.

b. The dissociation of carbonic acid is the most important factor affecting the pH of the ECF.

c. An inverse relationship exists between pH and CO_2 levels.

Module 24.6 Review

a. The body's three major buffer systems are the protein buffer system, the carbonic acid–bicarbonate buffer system, and the phosphate buffer system.

b. The carbonic acid–bicarbonate buffer system prevents pH changes caused by organic acids and fixed acids generated by metabolic activity. It uses the H^+ released by these acids to generate carbonic acid, which dissociates into H_2O and CO_2, the latter of which is exhaled from the lungs.

c. The phosphate buffer system plays an important role in buffering the pH of the ICF and the urine.

Module 24.7 Review

a. Metabolic acidosis results from the depletion of the bicarbonate reserve, caused by an inability to excrete hydrogen ions at the kidneys, the production of large numbers of fixed and organic acids, or bicarbonate loss.

b. Metabolic alkalosis results when bicarbonate ion concentrations become elevated.

c. The kidneys conserve HCO_3^- and eliminate H^+ in the urine during metabolic acidosis.

Module 24.8 Review

a. Respiratory acidosis is lowered blood pH resulting from inadequate respiratory activity and is characterized by elevated levels of carbon dioxide; respiratory alkalosis is elevated blood pH due to excessive respiratory activity, which depresses carbon dioxide levels and elevates the pH of body fluids.

b. The plasma P_{CO_2} of a patient with an airway obstruction would increase, resulting in respiratory acidosis.

c. A decrease in the pH of body fluids would cause an increase in the respiratory rate.

Section 2 Review

1. **a.** increased P_{CO_2}; **b.** acidosis;
   c. plasma pH decrease;
   d. increased; **e.** secreted;
   f. generated; **g.** decreased;
   h. decreased; **i.** increased;
   j. decreased P_{CO_2}; **k.** alkalosis;
   l. plasma pH increase;
   m. decreased; **n.** generated;
   o. secreted; **p.** increased;
   q. increased; **r.** decreased

2. The young boy has metabolic
   and respiratory acidosis. The
   metabolic acidosis resulted primarily from the large amounts of
   lactic acid generated by the boy's
   muscles as he struggled in the
   water. (The dissociation of lactic
   acid releases hydrogen ions and
   lactate ions.) Sustained hypoventilation during drowning contributed to both tissue hypoxia and
   respiratory acidosis. Respiratory
   acidosis developed as the P_{CO_2}
   increased in the ECF, increasing the production of carbonic
   acid and its dissociation into H^+
   and HCO_3^-. Prompt emergency
   treatment is essential; the usual
   procedure involves some form of
   artificial or mechanical respiratory assistance (to increase the
   respiratory rate and decrease
   P_{CO_2} in the ECF) coupled with
   the intravenous infusion of a buffered isotonic solution that would
   absorb the hydrogen ions in the
   ECF and increase body fluid pH.

Chapter Integration

1. Digestive secretions contain high
   levels of bicarbonate, so individuals with diarrhea can lose significant amounts of this important
   ion, leading to acidosis. We
   would expect the blood pH to be
   lower than 7.35, and that of the
   urine to be low (due to increased
   renal excretion of hydrogen ions).
   We would also expect an increase
   in the rate and depth of breathing
   as the respiratory system tries to
   compensate by eliminating carbon dioxide.

2. The hypertonic solution will cause
   fluid to move from the ICF to the
   ECF, further aggravating Tom's
   dehydration. The slight increase
   in pressure and osmolarity of the
   blood should lead to an increase
   in ADH, even though ADH levels
   are probably quite high already.
   Despite the high ADH levels, urine
   volume would probably increase,
   because the kidneys could not
   reabsorb much of the glucose. The
   remaining glucose would increase
   the osmolarity of the tubular fluid,
   decreasing water reabsorption and
   increasing urine volume.

Module 25.1 Review

a. Male reproductive structures are the scrotum, two testes, two epididymides, a pair of ductus deferens, two ejaculatory ducts, a urethra, two seminal glands, a prostate gland, two bulbo-urethral glands, and a penis.

b. The complex network of channels that is connected to the seminiferous tubules is the rete testis.

c. On a warm day, the cremaster muscle (as well as the dartos muscle) would be relaxed so that the scrotal sac could descend away from the warmth of the body, thereby cooling the testes.

Module 25.2 Review

a. Spermatogenesis is the production of spermatozoa and involves mitosis, meiosis, and spermiogenesis.

b. Four haploid spermatozoa will be produced from each diploid primary spermatocyte.

c. A typical spermatozoon has an acrosomal cap that contains enzymes essential to fertilization; a head that is packed with chromosomes; a neck that contains centrioles; a middle piece containing mitochondria to provide ATP for propulsion; and a flagellum, the whiplike structure that moves the cell.

Module 25.3 Review

a. Interstitial cells produce male sex hormones, or androgens, the most important of which is testosterone.

b. Nurse cells provide nutrients to the developing sperm and form the blood–testis barrier that isolates sperm from the blood.

c. Spermatogonia (stem cells) divide by mitosis to produce primary spermatocytes; through meiosis, primary spermatocytes give rise to secondary spermatocytes, which divide and differentiate into spermatids. Each spermatid matures into a spermatozoon.

Module 25.4 Review

a. Semen is sperm plus the secretions of the seminal, prostate, and bulbo-urethral glands.

b. The secretion of the bulbo-urethral glands lubricates the penis tip and neutralizes any urinary acids that may remain in the urethra.

c. The tail of each epididymis connects with the ductus deferens, which passes through the inguinal canal as part of the spermatic cord. Near the prostate gland, each ductus deferens enlarges to form an ampulla. The junction of the base of the seminal gland and the ampulla creates the ejaculatory duct, which empties into the urethra.

Module 25.5 Review

a. The three columns of erectile tissue are the corpus spongiosum and the paired corpora cavernosa.

b. The physiological phases of male sexual activity are arousal, emission, and ejaculation.

c. An inability of a male to contract the ischiocavernosus and bulbospongiosus muscles would interfere with the ejaculation phase, including the contractions that are part of the male orgasm.

Module 25.6 Review

a. Important regulatory hormones in the establishment and maintenance of male sexual function are FSH (follicle-stimulating hormone), LH (luteinizing hormone), and GnRH (gonadotropin-releasing hormone). Testosterone is the most important androgen.

b. The testes, hypothalamus, and the anterior lobe of the pituitary gland secrete the hormones that control male reproductive functions.

c. Low FSH levels would lead to low levels of testosterone in the seminiferous tubules, reducing both the sperm production rate and sperm count.

Section 1 Review

1. a. prostatic urethra; b. ductus deferens; c. penile urethra; d. penis; e. epididymis; f. testis; g. external urethral orifice; h. scrotum; i. seminal gland; j. prostate gland; k. ejaculatory duct; l. bulbo-urethral gland

2. a. dartos muscle; b. spermatogonia; c. seminiferous tubules; d. interstitial cells; e. spermiogenesis; f. spermatogenesis; g. penis and scrotum; h. epididymis; i. nurse cells; j. semen; k. impotence; l. corpus spongiosum; m. luteinizing hormone (LH); n. follicle-stimulating hormone (FSH)

3. Normal levels of testosterone (1) promote the functional maturation of spermatozoa, (2) maintain the accessory organs of the male reproductive tract, (3) are responsible for the establishment and maintenance of male secondary sex characteristics, (4) stimulate bone and muscle growth, and (5) stimulate sexual behaviors and sexual drive (libido).

Module 25.7 Review

a. The major organs of the female reproductive system are the ovaries, uterine tubes, uterus, vagina, and external genitalia.

b. The ovaries, uterine tubes, and uterus are enclosed within the broad ligament. The mesovarium supports and stabilizes each ovary.

c. The ovaries produce immature female gametes called oocytes; secrete female sex hormones, including estrogens and progestins; and secrete inhibin, which is involved in the feedback control of FSH production.

Module 25.8 Review

a. An oocyte is an immature female gamete whose meiotic divisions will produce a single ovum and three polar bodies.

b. Males produce gametes from puberty until death; females produce gametes only from menarche to menopause. Males produce many gametes at a time; females typically produce one or two per 28-day cycle. Males release mature gametes that have completed meiosis; females release secondary oocytes suspended in metaphase of meiosis II.

c. Important events in the ovarian cycle are (1) the formation of primary follicles from primordial follicles in an egg nest, (2) the formation of secondary follicles, (3) the formation of a tertiary follicle, (4) ovulation, and (5) the formation and degeneration of the corpus luteum. A corpus albicans forms if fertilization does not occur.

Module 25.9 Review

a. The regions of the uterus are the fundus, body (ending at the isthmus), and cervix.

b. The perimetrium is the incomplete serosal layer; the myometrium is the outer, muscular layer; and the endometrium is the inner, glandular layer.

c. Recently released secondary oocytes reach the uterine tube with the aid of the beating action of cilia on the inner surfaces of the fimbriae of the infundibulum.

Module 25.10 Review

a. The endometrium contains the deeper basilar zone and the more superficial functional zone.

b. Menses is the first phase of the uterine cycle and is marked by degeneration of the functional zone of the uterus; menstruation is the process of endometrial sloughing that occurs during menses.

c. The uterine cycle begins with menses, the destruction of the functional zone. After menses, the proliferative phase begins, during which the functional zone undergoes repair and it thickens. Following the proliferative phase is the secretory phase, during which uterine (endometrial) glands enlarge. The uterine cycle first begins at menarche and continues until menopause.

Module 25.11 Review

a. The vagina (1) serves as a passageway for the elimination of menstrual fluids; (2) receives the penis during sexual intercourse, and holds spermatozoa prior to their passage into the uterus; and (3) forms the inferior portion of the birth canal, through which the fetus passes during delivery.

b. The vagina is a muscular tube extending between the uterus and external genitalia; its lining forms folds called rugae. The proximal portion of the vagina is marked by the cervix, which dips into the vaginal canal, and the shallow recess known as the fornix. The hymen, a thin epithelial fold, partially blocks the entrance to the vagina until physical distortion ruptures it.

c. The greater vestibular glands in females are similar to the bulbo-urethral glands in males, and both the penis and the clitoris have characteristic erectile tissue (and a glans as well).

Module 25.12 Review

a. Lactation is the secretion of milk by the mammary glands.

b. Blockage of a single lactiferous sinus would not interfere with the delivery of milk to the nipple, because each breast generally has 15–20 lactiferous sinuses.

c. Route of milk flow: secretory alveoli of the secretory lobules → ducts within a lobe → lactiferous duct of the lobe → lactiferous sinus → surface of the nipple

Module 25.13 Review

a. Completion of the decline in the levels of estrogens and progesterone signals the beginning of menses and the start of a new uterine cycle.

b. If the LH surge did not occur during an ovarian cycle, ovulation and corpus luteum formation could not occur.

c. The hypothalamic secretion of GnRH triggers the pituitary secretion of FSH and LH. FSH initiates follicular development, and activated follicles and ovarian interstitial cells produce estrogens. High estrogen levels stimulate LH secretion and increase anterior pituitary gland sensitivity to GnRH, causing the release of LH. Progesterone is the principal hormone of the luteal phase. Changes in estrogen and progesterone levels are responsible for the maintenance of the uterine cycle.

Module 25.14 Review

a. A vasectomy is the surgical removal of a segment of each ductus deferens and the tying or cauterizing of the cut ends, preventing spermatozoa from reaching the distal portions of the male reproductive tract.

b. Condoms provide some protection against sexually transmitted diseases.

c. Depo-Provera injections (a progesterone-only form of birth control) result in the cessation of the uterine cycle in 50 percent of women using this product. (Uterine cycles eventually resume after use of the product is discontinued.)

Module 25.15 Review

a. A sexually transmitted disease is a disease that is transferred from one individual to another primarily or exclusively through sexual contact.

b. Ovarian cancer usually arises from epithelial cells.

c. The human papillomaviruses (HPV) cause most cases of cervical cancer.

Section 2 Review

1. a. infundibulum; b. ovary; c. uterine tube; d. perimetrium; e. myometrium; f. endometrium; g. uterus; h. clitoris; i. labium minus; j. labium majus; k. fornix; l. cervix; m. external os; n. vagina

2. a. oocytes; b. menarche; c. vesicouterine pouch; d. broad ligament; e. rectouterine pouch; f. corpus luteum; g. uterine cycle; h. lactation; i. ovaries; j. LH surge; k. cervix; l. tubal ligation; m. GnRH; n. vulva

3. The endometrial cells have receptors for estrogens and progesterone and respond to these hormones as if the cells were still in the body of the uterus. Under the influence of estrogens, the endometrial cells proliferate at the beginning of the uterine (menstrual) cycle and begin to develop glands and blood vessels, which then further develop under the control of progesterone. This dramatic increase in tissue size exerts pressure on neighboring tissues or in some other way interferes with their function. It is the recurring expansion of tissue in an abnormal location that causes periodic pain.

Chapter Integration

1. The presence of exercise-induced amenorrhea suggests that a certain amount of body fat is necessary for menstrual cycles to occur. If body fat levels fall below some set point, menstruation ceases.

2. Because a woman lacking adequate body fat might not have the energy reserves needed to have a successful pregnancy, compensatory mechanisms in her body prevent pregnancy by shutting down the ovarian cycle, and thus the menstrual cycle. When her body subsequently accumulates sufficient energy reserves in body fat, the cycles begin again.

3. Estrogen deficiency can lead to infertility, vaginal and breast atrophy, and osteoporosis. It may also increase the risk of heart attacks later in life.

Module 26.1 Review

a. Fertilization is the fusion of a secondary oocyte and a spermatozoon to form a zygote.

b. A normal human zygote contains 46 chromosomes.

c. Many spermatozoa are needed to achieve fertilization because one sperm does not contain enough acrosomal enzymes to erode the corona radiata surrounding the secondary oocyte.

Module 26.2 Review

a. The morula is the stage of development that results from cleavage.

b. The blastocyst stage begins once the zygote arrives in the uterine cavity.

c. The blastocyst consists of an outer trophoblast and an inner cell mass. Implantation occurs when the blastocyst adheres to and then becomes enclosed within the uterine lining about 7 days after fertilization.

Module 26.3 Review

a. Gestational trophoblastic neoplasia is a tumor formed by undifferentiated, rapid growth of the syncytial trophoblast; if untreated, the neoplasm may become malignant.

b. Gastrulation is the formation of the primary germ layers—the endoderm, ectoderm, and mesoderm—from the embryonic disc. It is from these germ layers that the body systems differentiate.

c. The mesoderm gives rise to nearly all body systems except the nervous and respiratory systems.

Module 26.4 Review

a. The four extra-embryonic membranes are the yolk sac, amnion, allantois, and chorion.

b. The base of the allantois gives rise to the urinary bladder.

c. The yolk sac forms from endoderm and mesoderm; it is an important site of blood cell formation. The amnion forms from the ectoderm and mesoderm; it encloses the fluid that surrounds and cushions the developing embryo and fetus. The allantois forms from endoderm and mesoderm; its base gives rise to the urinary bladder. The chorion forms from mesoderm and trophoblast; it surrounds the blastocoele.

Module 26.5 Review

a. The chorionic villi are structures that extend outward into the maternal tissues, forming an intricate, branching network through which maternal blood flows. Embryonic blood vessels extend into each chorionic villus.

b. The body stalk is the connection between the embryo and the chorion; it contains portions of the allantois and blood vessels that carry blood to and from the placenta. The yolk stalk is the connection between the endoderm of the embryo and the yolk sac.

c. The umbilical cord connects the fetus to the placenta, and it is derived from the allantois.

Module 26.6 Review

a. The hormones synthesized by the syncytial trophoblast are human chorionic gonadotropin (hCG), human placental lactogen (hPL), relaxin, progesterone, and estrogens.

b. The presence of human chorionic gonadotropin (hCG) in the urine provides a reliable indicator of pregnancy.

c. After the first trimester, the placenta is sufficiently functional to maintain the pregnancy.

Module 26.7 Review

a. Organogenesis is the process of organ formation.

b. In the second trimester, the organ systems increase in complexity. During the third trimester, many of the organ systems become fully functional.

c. During the third trimester, the fetus undergoes its largest weight gain.

Module 26.8 Review

a. The major changes that occur in maternal systems during pregnancy are increases in respiratory rate and tidal volume, blood volume, nutrient requirements, glomerular filtration rate (GFR), and the size of the uterus and mammary glands.

b. A mother's blood volume increases during pregnancy to compensate for the reduction in maternal blood volume resulting from blood flow through the placenta.

c. Difficulty breathing in pregnant women results from the enlarged uterus pressing against the diaphragm and crowding the lungs.

Module 26.9 Review

a. Relaxin, produced by the placenta, softens the pubic symphysis and dilates the cervix, and the weight of the fetus deforms the external os of the uterus. Deformation of the cervix and rising estrogen levels promote the release of oxytocin, and the already stretched smooth muscles of the myometrium become even more excitable.

b. Estrogens and oxytocin stimulate the production of prostaglandins, which are then primarily responsible for the initiation of true labor.

c. The dilation stage begins with the onset of true labor, as the cervix dilates and the fetus begins to move toward the cervical canal; late in this stage, the amnion ruptures. The expulsion stage begins as the cervix dilates completely and continues until the fetus has completely emerged from the vagina (delivery). In the placental stage, the uterus gradually contracts, tearing the connections between the endometrium and the placenta and ejecting the placenta.

Module 26.10 Review

a. Oxytocin causes the milk let-down reflex.

b. Colostrum is produced by the mammary glands from the end of the sixth month of pregnancy until a few days after birth and contains antibodies. After that, the glands begin producing breast milk, which contains antibodies and lysozyme but has a higher fat content than colostrum.

c. The postnatal stages of development are the neonatal period, from birth to one month; infancy, from one month to age 2; childhood, from two until sexual maturation begins; adolescence, which begins with the onset of sexual maturation (puberty) between ages 9–14 and ends when growth in body size ends (around 18 years); and maturity, which includes the rest of the individual's life. A final stage called senescence, or aging, overlaps with maturity.

Module 26.11 Review

a. The three interacting hormonal events associated with the onset of puberty are (1) increased GnRH production by the hypothalamus; (2) increased sensitivity to GnRH by the anterior lobe of the pituitary gland, and a rapid increase in circulating levels of FSH and LH; and (3) increased sensitivity to FSH and LH by ovarian and testicular cells.

b. Males have deeper voices and larger larynxes than do females because testosterone stimulates laryngeal development in males to a greater extent than estrogen stimulates laryngeal development in females.

c. The higher estrogen levels in pre-menopausal women (compared to adult men) cause reductions in plasma cholesterol levels, thereby slowing plaque formation and lowering the risk of atherosclerosis.

Section 1 Review

1. **a.** conception; **b.** neonate; **c.** amphimixis; **d.** gestation; **e.** hCG; **f.** relaxin; **g.** colostrum; **h.** amnion; **i.** blastocyst; **j.** chorion; **k.** inner cell mass; **l.** syncytial trophoblast; **m.** morula; **n.** embryonic disc

2. **a.** junction between the ampulla and isthmus of the uterine tube; **b.** ninth week after fertilization; **c.** second trimester; **d.** cleavage, implantation, placentation, embryogenesis; **e.** first trimester; **f.** respiratory, digestive, and urinary

3. It is very unlikely that the baby's condition is the result of a viral infection contracted during the third trimester. The development of organ systems occurs during the first trimester, and by the end of the second trimester, most organ systems are fully formed. During the third trimester, the fetus undergoes tremendous growth, but very little new organ formation occurs.

Module 26.12 Review

a. Homozygous means that homologous chromosomes carry the same allele of a given gene; heterozygous means that homologous chromosomes carry different alleles of a given gene.

b. In simple inheritance, phenotypic characteristics are determined by interactions between a single pair of alleles; polygenic inheritance involves interactions among alleles of several genes.

c. The phenotype of a person who is heterozygous for curly hair—that is, a person with one dominant allele and one recessive allele for that trait—would be "curly hair."

Module 26.13 Review

a. In strict dominance, one allele dominates the other, so an individual who is heterozygous for a given trait exhibits the dominant phenotype; in codominance, an individual who is heterozygous for a given trait exhibits both of the phenotypes for that trait.

b. X-linked traits are expressed more frequently in males because males have a Y chromosome that has no corresponding allele.

c. (1) simple inheritance (dominant); (2) simple inheritance (recessive); (3) autosomal inheritance

Module 26.14 Review

a. A single nucleotide polymorphism is a variation in a single base pair in a DNA sequence.

b. (1) XO: Turner syndrome; (2) XXY: Klinefelter syndrome

c. (1) ovarian cancer: chromosome 9; (2) Tay–Sachs disease: chromosome 15; (3) spinocerebellar ataxia: chromosome 6

Section 2 Review

1. **a.** phenotype; **b.** alleles; **c.** homologous; **d.** karyotype; **e.** genotype; **f.** locus; **g.** heterozygous; **h.** genetics; **i.** homozygous; **j.** polygenic inheritance; **k.** simple inheritance; **l.** autosomes

2. **a.** Children who cannot roll the tongue must be homozygous recessive for the condition, and the only way a child can receive two recessive alleles from tongue-rolling parents is to have parents who are both heterozygous.

b. The allele for achondroplasia dwarfism must be dominant. In order for two dwarf parents to produce a normal child, the child must be homozygous recessive, and thus both parents must be heterozygous. The probability that their next child (and all subsequent children) will be normal sized is 1 in 4, or 25 percent.

Chapter Integration

1. Although technically it takes only one sperm to fertilize a secondary oocyte, Joe's oligospermia means that because most sperm entering the female reproductive tract are killed or disabled before they reach the uterus, too few sperm reach the secondary oocyte in a uterine tube to produce sufficient acrosomal enzymes to enable one sperm to penetrate the corona radiata.

2. The probability that this couple's daughters will have hemophilia is zero, because each daughter will receive a dominant normal allele from her father. There is a 50 percent chance that a son will have hemophilia, because each son has a 50 percent chance of receiving the mother's normal allele, and a 50 percent chance of receiving the mother's recessive allele.

A

abdomen: The region of the trunk between the inferior margin of the rib cage and the superior margin of the pelvis.

abdominopelvic cavity: The term used to refer to the general region bounded by the abdominal wall and the pelvis; it contains the peritoneal cavity and visceral organs.

abducens: Cranial nerve VI, which innervates the lateral rectus muscle of the eye.

abduction: Movement away from the midline of the body, as viewed in the anatomical position.

abscess: A localized collection of pus within a damaged tissue.

absorption: The active or passive uptake of gases, fluids, or solutes.

accommodation: An alteration in the curvature of the lens of the eye to focus an image on the retina.

acetabulum: The fossa on the lateral aspect of the pelvis that accommodates the head of the femur.

acetylcholine (ACh): A chemical neurotransmitter in the brain and peripheral nervous system; the dominant neurotransmitter in the peripheral nervous system, released at neuromuscular junctions and synapses of the parasympathetic division.

acetylcholinesterase (AChE): An enzyme found in the synaptic cleft, bound to the postsynaptic membrane, and in tissue fluids; breaks down and inactivates acetylcholine molecules.

acetyl-CoA: An acetyl group bound to coenzyme A, a participant in the anabolic and catabolic pathways for carbohydrates, lipids, and many amino acids.

acetyl group: $—CH_3CO$.

acid: A compound whose dissociation in solution releases a hydrogen ion and an anion; an acidic solution has a pH below 7.0 and contains an excess of hydrogen ions.

acidosis: An abnormal physiological state characterized by a plasma pH below 7.35.

acinus/acini: A histological term referring to a blind pocket, pouch, or sac.

acoustic: Pertaining to sound or the sense of hearing.

acromion: A continuation of the scapular spine that projects superior to the capsule of the shoulder joint.

acrosomal cap: A membranous sac at the tip of a spermatozoon that contains hyaluronidase.

actin: The protein component of microfilaments that forms thin filaments in skeletal muscles and produces contractions of all muscles through interaction with thick (myosin) filaments.

action potential: A propagated change in the transmembrane potential of excitable cells, initiated by a change in the membrane permeability to sodium ions.

active transport: The ATP-dependent absorption or secretion of solutes across a plasma membrane.

acute: Sudden in onset, severe in intensity, and brief in duration.

adaptation: A change in pupillary size in response to changes in light intensity; a decrease in receptor sensitivity or perception after chronic stimulation.

Addison disease: A condition resulting from the hyposecretion of glucocorticoids; characterized by lethargy, weakness, hypotension, and increased skin pigmentation.

adduction: Movement toward the axis or midline of the body, as viewed in the anatomical position.

adenine: A purine; one of the nitrogenous bases in the nucleic acids RNA and DNA.

adenohypophysis: The anterior lobe of the pituitary gland.

adenosine: A compound consisting of adenine and ribose.

adenosine diphosphate (ADP): A compound consisting of adenosine with two phosphate groups attached.

adenosine monophosphate (AMP): A nucleotide consisting of adenosine plus a phosphate group ($PO_4{}^{3-}$); also called *adenosine phosphate*.

adenosine triphosphate (ATP): A high-energy compound consisting of adenosine with three phosphate groups attached; the third is attached by a high-energy bond.

adenylate cyclase: An enzyme bound to the inner surfaces of plasma membranes that can convert ATP to cyclic-AMP.

adipocyte: A fat cell.

adipose tissue: Loose connective tissue dominated by adipocytes.

adrenal cortex: The superficial portion of the adrenal gland that produces steroid hormones; also called *suprarenal cortex*.

adrenal gland: A small endocrine gland that secretes steroids and catecholamines and is located superior to each kidney; also called *suprarenal gland*.

adrenal medulla: The core of the adrenal gland; a modified sympathetic ganglion that secretes catecholamines into the blood during sympathetic activation; also called *suprarenal medulla*.

adrenergic: A synaptic terminal that, when stimulated, releases norepinephrine.

adrenocortical hormone: Any steroid produced by the adrenal cortex.

adrenocorticotropic hormone (ACTH): The hormone that stimulates the production and secretion of glucocorticoids by the zona fasciculata of the adrenal cortex; released by the anterior lobe of the pituitary gland in response to corticotropin-releasing hormone.

adventitia: The superficial layer of connective tissue surrounding an internal organ; fibers are continuous with those of surrounding tissues, providing support and stabilization.

aerobic: Requiring the presence of oxygen.

aerobic metabolism: The complete breakdown of organic substrates into carbon dioxide and water, via pyruvate; a process that yields large amounts of ATP but requires mitochondria and oxygen.

afferent: Toward a center.

afferent arteriole: An arteriole that carries blood to a glomerulus of the kidney.

afferent fiber: An axon that carries sensory information to the central nervous system.

agglutination: The aggregation of red blood cells due to interactions between surface antigens and plasma antibodies.

aggregated lymphoid nodules: Lymphoid nodules beneath the epithelium of the small intestine; also called *Peyer patches*.

agonist: A muscle responsible for a specific movement; also called a *prime mover*.

agranular: Without granules; *agranular leukocytes* are monocytes and lymphocytes.

alba: White.

albicans: White.

albuginea: White.

aldosterone: A mineralocorticoid produced by the zona glomerulosa of the adrenal cortex; stimulates sodium and water conservation at the kidneys; secreted in response to the presence of angiotensin II.

alkalosis: The condition characterized by a plasma pH greater than 7.45; associated with a relative deficiency of hydrogen ions or an excess of bicarbonate ions.

alpha receptors: Membrane receptors sensitive to norepinephrine or epinephrine; stimulation normally results in the excitation of the target cell.

alveolar sac: An air-filled chamber that supplies air to several alveoli.

alveolus/alveoli: Blind pockets at the end of the respiratory tree, lined by a simple squamous epithelium and surrounded by a capillary network; sites of gas exchange with the blood; a bony socket that holds the root of a tooth.

Alzheimer disease: A disorder resulting from degenerative changes in populations of neurons in the cerebrum, causing dementia characterized by problems with attention, short-term memory, and emotions.

amination: The attachment of an amino group to a carbon chain; performed by a variety of cells and important in the synthesis of amino acids.

amino acids: Organic compounds whose chemical structure can be summarized as $R—CHNH_2—COOH$.

amino group: $—NH_2$.

amnion: One of the four extraembryonic membranes; surrounds the developing embryo or fetus.

amniotic fluid: Fluid that fills the amniotic cavity; cushions and supports the embryo or fetus.

amphiarthrosis: An articulation that permits a small degree of independent movement; *see* **interosseous membrane** and **pubic symphysis.**

ampulla/ampullae: A localized dilation in the lumen of a canal or passageway.

amygdaloid body: A basal nucleus that is a component of the limbic system and acts as an interface between that system, the cerebrum, and sensory systems.

amylase: An enzyme that breaks down polysaccharides; produced by the salivary glands and pancreas.

anabolism: The synthesis of complex organic compounds from simpler precursors.

anaerobic: Without oxygen.

anal triangle: The posterior subdivision of the perineum.

anaphase: The mitotic stage in which the paired chromatids separate and move toward opposite ends of the spindle apparatus.

anaphylaxis: A hypersensitivity reaction due to the binding of antigens to immunoglobulins (IgE) on the surfaces of mast cells; the release of histamine, serotonin, and prostaglandins by mast cells then causes widespread inflammation; a sudden decline in blood pressure may occur, producing anaphylactic shock.

anastomosis: The joining of two tubes, usually referring to a connection between two peripheral vessels without an intervening capillary bed.

anatomical position: An anatomical reference position; the body viewed from the anterior surface with the palms facing forward.

anatomy: The study of the structure of the body.

androgen: A steroid sex hormone primarily produced by the interstitial cells of the testis and manufactured in small quantities by the adrenal cortex in both sexes.

anemia: The condition marked by a reduction in the hematocrit, the hemoglobin content of the blood, or both.

angiotensin I: The hormone produced by the activation of angiotensinogen by renin; angiotensin-converting enzyme converts angiotensin I into angiotensin II in lung capillaries.

angiotensin II: A hormone that causes an elevation in systemic blood pressure, stimulates the secretion of aldosterone, promotes thirst, and causes the release of antidiuretic hormone; angiotensin-converting enzyme in lung capillaries converts angiotensin I into angiotensin II.

angiotensinogen: The blood protein produced by the liver that is converted to angiotensin I by the enzyme renin.

anion: An ion bearing a negative charge.

anoxia: Tissue oxygen deprivation.

antagonist: A muscle that opposes the movement of an agonist.

antebrachium: The forearm.

anterior: On or near the front, or ventral surface, of the body.

antibiotic: A chemical agent that selectively kills pathogens, primarily bacteria.

antibody: A globular protein produced by plasma cells that will bind to specific antigens and promote their destruction or removal from the body.

antibody-mediated immunity: The form of immunity resulting from the presence of circulating antibodies produced by plasma cells; also called *humoral immunity.*

anticholinesterase: A chemical compound that blocks the action of acetylcholine and causes prolonged and intensive stimulation of postsynaptic membranes.

anticodon: Three nitrogenous bases on a tRNA molecule that interact with a complementary codon on a strand of mRNA.

antidiuretic hormone (ADH): A hormone synthesized in the hypothalamus and secreted at the neurohypophysis (posterior lobe of the pituitary gland); causes water retention by the kidneys and an elevation of blood pressure.

antigen: A substance capable of inducing the production of antibodies.

antigen-antibody complex: The combination of an antigen and a specific antibody.

antigenic determinant site: A portion of an antigen that can interact with an antibody molecule.

antigen-presenting cell (APC): A cell that processes antigens and displays them, bound to MHC proteins; essential to the initiation of a normal immune response.

antihistamines: A chemical agent that blocks the action of histamine on peripheral tissues.

antrum: A chamber or pocket.

anulus: A cartilage or bone shaped like a ring; also spelled *annulus.*

anus: The external opening of the anal canal.

aorta: The large, elastic artery that carries blood away from the left ventricle and into the systemic circuit.

apocrine secretion: A mode of secretion in which the glandular cell sheds portions of its cytoplasm.

aponeurosis/aponeuroses: A broad tendinous sheet that may serve as the origin or insertion of a skeletal muscle.

appendicular: Pertaining to the upper or lower limbs.

appendix: A blind tube connected to the cecum of the large intestine.

appositional growth: The enlargement of a cartilage or bone by the addition of cartilage or bony matrix at its surface.

aqueous humor: A fluid similar to perilymph or cerebrospinal fluid that fills the anterior chamber of the eye.

arachidonic acid: One of the essential fatty acids.

arachnoid granulations: Processes of the arachnoid mater that project into the superior sagittal sinus; sites where cerebrospinal fluid enters the venous circulation.

arachnoid mater: The middle meninx that encloses cerebrospinal fluid and protects the central nervous system.

arbor vitae: The central, branching mass of white matter inside the cerebellum.

arcuate: Curving.

areolar: Containing minute spaces, as in areolar tissue.

areolar tissue: Loose connective tissue with an open framework.

arrector pili: Smooth muscles whose contractions force hairs to stand erect.

arrhythmias: Abnormal patterns of cardiac contractions.

arteriole: A small arterial branch that delivers blood to a capillary network.

artery: A blood vessel that carries blood away from the heart and toward a peripheral capillary.

articular: Pertaining to a joint.

articular capsule: The dense collagen fiber sleeve that surrounds a joint and provides protection and stabilization.

articular cartilage: The cartilage pad that covers the surface of a bone inside a joint cavity.

articulation: A joint; the formation of words.

ascending tract: A tract carrying information from the spinal cord to the brain.

association areas: Cortical areas of the cerebrum that are responsible for the integration of sensory inputs and/or motor commands.

astrocyte: One of the four types of neuroglia in the central nervous system; responsible for maintaining the blood–brain barrier by the stimulation of endothelial cells.

atherosclerosis: The formation of fatty plaques in the walls of arteries, restricting blood flow to deep tissues.

atom: The smallest stable unit of matter.

atomic number: The number of protons in the nucleus of an atom.

atomic weight: Roughly, the average total number of protons and neutrons in the atoms of a particular element.

atria: Thin-walled chambers of the heart that receive venous blood from the pulmonary or systemic circuit.

atrial natriuretic peptide (ANP): *See* **natriuretic peptides.**

atrioventricular (AV) node: Specialized cardiocytes that relay the contractile stimulus to the bundle of His, the bundle branches, the Purkinje fibers, and the ventricular myocardium; located at the boundary between the atria and ventricles.

atrioventricular (AV) valve: One of the valves that prevents backflow into the atria during ventricular systole.

atrophy: The wasting away of tissues from a lack of use, ischemia, or nutritional abnormalities.

auditory: Pertaining to the sense of hearing.

auditory ossicles: The bones of the middle ear: malleus, incus, and stapes.

auditory tube: A passageway that connects the nasopharynx with the middle ear cavity.

auricle: A broad, flattened process that resembles the external ear; in the ear, the expanded, projecting portion that surrounds the external auditory meatus, also called *pinna;* in the heart, the externally visible flap formed by the collapse of the outer wall of a relaxed atrium.

autoantibodies: Antibodies that react with antigens on the surfaces of a person's own cells and tissues.

autoimmunity: The immune system's sensitivity to normal cells and tissues, resulting in the production of autoantibodies.

autolysis: The destruction of a cell due to the rupture of lysosomal membranes in its cytoplasm.

automaticity: The spontaneous depolarization to threshold, characteristic of cardiac pacemaker cells.

autonomic ganglion: A collection of visceral motor neurons outside the central nervous system.

autonomic nerve: A peripheral nerve consisting of preganglionic or postganglionic autonomic fibers.

autonomic nervous system (ANS): Centers, nuclei, tracts, ganglia, and nerves involved in the unconscious regulation of visceral functions; includes components of the central nervous system and the peripheral nervous system.

autopsy: The detailed examination of a body after death.

autoregulation: Changes in activity that maintain homeostasis in direct response to changes in the local environment; does not require neural or endocrine control.

autosomal: Chromosomes other than the X or Y sex chromosome.

avascular: Without blood vessels.

axilla: The armpit.

axolemma: The plasma membrane of an axon, continuous with the plasma membrane of the cell body and dendrites and distinct from any neuroglial coverings.

axon: The elongate extension of a neuron that conducts an action potential.

axon hillock: In a multipolar neuron, the portion of the cell body adjacent to the initial segment.

axoplasm: The cytoplasm within an axon.

B

bacteria: Single-celled microorganisms, some pathogenic, that are common in the environment and in and on the body.

baroreception: The ability to detect changes in pressure.

baroreceptor reflex: A reflexive change in cardiac activity in response to changes in blood pressure.

baroreceptors: The receptors responsible for baroreception.

basal lamina: A layer of filaments and fibers that attach an epithelium to the underlying connective tissue.

basal nuclei: Nuclei of the cerebrum that are important in the subconscious control of skeletal muscle activity.

base: A compound whose dissociation releases a hydroxide ion (OH^-) or removes a hydrogen ion (H^+) from the solution.

basophils: Circulating granulocytes (white blood cells) similar in size and function to tissue mast cells.

B cells: Lymphocytes capable of differentiating into plasmocytes (plasma cells), which produce antibodies.

benign: Not malignant.

beta cells: Cells of the pancreatic islets that secrete insulin in response to elevated blood sugar concentrations.

beta oxidation: Fatty acid catabolism that produces molecules of acetyl-CoA.

beta receptors: Membrane receptors sensitive to epinephrine; stimulation may result in the excitation or inhibition of the target cell.

bicarbonate ions: HCO_3^-; anion components of the carbonic acid–bicarbonate buffer system.

bicuspid: Having two cusps or points; refers to a premolar tooth, which has two roots, or to the left AV valve, which has two cusps.

bicuspid valve: The left atrioventricular (AV) valve, also called *mitral valve.*

bifurcate: To branch into two parts.

bile: The exocrine secretion of the liver; stored in the gallbladder and ejected into the duodenum.

bile salts: Steroid derivatives in bile; responsible for the emulsification of ingested lipids.

bilirubin: A pigment that is the by-product of hemoglobin catabolism.

biopsy: The removal of a small sample of tissue for pathological analysis.

bladder: A muscular sac that distends as fluid is stored and whose contraction ejects the fluid at an appropriate time; used alone, the term usually refers to the urinary bladder.

blastocyst: An early stage in the developing embryo, consisting of an outer trophoblast and an inner cell mass.

blood–brain barrier: The isolation of the central nervous system from the general circulation; primarily the result of astrocyte regulation of capillary permeabilities.

blood–CSF barrier: The isolation of the cerebrospinal fluid from the capillaries of the choroid plexus; primarily the result of specialized ependymal cells.

blood pressure: A force exerted against vessel walls by the blood in the vessels, due to the push exerted by cardiac contraction and the elasticity of the vessel walls; usually measured along one of the muscular arteries, with systolic pressure measured during ventricular systole and diastolic pressure during ventricular diastole.

blood–testis barrier: The isolation of the interior of the seminiferous tubules from the general circulation, due to the activities of the nurse (sustentacular) cells.

Bohr effect: The increased oxygen release by hemoglobin in the presence of elevated carbon dioxide levels.

bolus: A compact mass; usually refers to compacted ingested material on its way to the stomach.

bone: *See* **osseous tissue**.

bowel: The intestinal tract.

brachial: Pertaining to the arm.

brachial plexus: A network formed by branches of spinal nerves C_5–T_1 en route to innervating the upper limb.

brachium: The arm.

bradycardia: An abnormally slow heart rate, usually below 50 bpm.

brain natriuretic peptide (BNP): *See* **natriuretic peptides**.

brain stem: The brain minus the cerebrum, diencephalon, and cerebellum.

brevis: Short.

bronchial tree: The trachea, bronchi, and bronchioles.

bronchodilation: The dilation of the bronchial passages; can be caused by sympathetic stimulation.

bronchus/bronchi: A branch of the bronchial tree between the trachea and bronchioles.

buccal: Pertaining to the cheeks.

buffer: A compound that stabilizes the pH of a solution by removing or releasing hydrogen ions.

buffer system: Interacting compounds that prevent increases or decreases in the pH of body fluids; includes the carbonic acid–bicarbonate buffer system, the phosphate buffer system, and the protein buffer system.

bulbar: Pertaining to the brain stem.

bulbo-urethral glands: Mucous glands at the base of the penis that secrete into the penile urethra; the equivalent of the greater vestibular glands of females; also called *Cowper glands*.

bundle branches: Specialized conducting cells in the ventricles that carry the contractile stimulus from the bundle of His to the Purkinje fibers.

bundle of His: Specialized conducting cells in the interventricular septum that carry the contracting stimulus from the AV node to bundle branches and then to Purkinje fibers.

bursa: A small sac filled with synovial fluid that cushions adjacent structures and reduces friction.

C

calcaneal tendon: The large tendon that inserts on the calcaneus; tension on this tendon produces extension (plantar flexion) of the foot; also called *Achilles tendon*.

calcaneus: The heel bone, the largest of the tarsal bones.

calcification: The deposition of calcium salts within a tissue.

calcitonin: The hormone secreted by C cells of the thyroid when calcium ion concentrations are abnormally high; restores homeostasis by increasing the rate of bone deposition and the rate of calcium loss by the kidneys.

calculus/calculi: A solid mass of insoluble materials that form within body fluids, especially the gallbladder, kidneys, or urinary bladder.

callus: A localized thickening of the epidermis due to chronic mechanical stresses; a thickened area that forms at the site of a bone break as part of the repair process.

canaliculi: Microscopic passageways between cells; bile canaliculi carry bile to bile ducts in the liver; in bone, canaliculi permit the diffusion of nutrients and wastes to and from osteocytes.

cancellous bone: Spongy bone, composed of a network of bony struts.

cancer: An illness caused by mutations leading to the uncontrolled growth and replication of the affected cells.

cannula: A tube that can be inserted into the body; commonly placed in blood vessels prior to transfusion or dialysis.

capacitation: The activation process that must occur before a spermatozoon can successfully fertilize an oocyte; occurs in the vagina after ejaculation.

capillary: A small blood vessel, located between an arteriole and a venule, whose thin wall permits the diffusion of gases, nutrients, and wastes between plasma and interstitial fluids.

capitulum: A general term for a small, elevated articular process; refers to the rounded distal surface of the humerus that articulates with the head of the radius.

caput: The head.

carbaminohemoglobin: Hemoglobin bound to carbon dioxide molecules.

carbohydrase: An enzyme that breaks down carbohydrate molecules.

carbohydrate: An organic compound containing carbon, hydrogen, and oxygen in a ratio that approximates 1:2:1.

carbon dioxide: CO_2; a compound produced by the citric acid cycle reactions of aerobic metabolism.

carbonic anhydrase: An enzyme that catalyzes the reaction $H_2O + CO_2 \rightarrow H_2CO_3$; important in carbon dioxide transport, gastric acid secretion, and renal pH regulation.

carcinogenic: Stimulating cancer formation in affected tissues.

cardia: The area of the stomach surrounding its connection with the esophagus.

cardiac: Pertaining to the heart.

cardiac cycle: One complete heartbeat, including atrial and ventricular systole and diastole.

cardiac output: The amount of blood ejected by the left ventricle each minute.

cardiac reserve: The potential percentage increase in cardiac output above resting levels.

cardiac tamponade: A compression of the heart due to fluid accumulation in the pericardial cavity.

cardiocyte: A cardiac muscle cell.

cardiovascular: Pertaining to the heart, blood, and blood vessels.

cardiovascular centers: Poorly localized centers in the reticular formation of the medulla oblongata of the brain; includes cardioacceleratory, cardioinhibitory, and vasomotor centers.

cardium: The heart.

carotene: A yellow-orange pigment, found in carrots and in green and orange leafy vegetables, that the body can convert to vitamin A.

carotid artery: The principal artery of the neck, servicing cervical and cranial structures; one branch, the internal carotid, provides a major blood supply to the brain.

carotid body: A group of receptors, adjacent to the carotid sinus, that are sensitive to changes in the carbon dioxide levels, pH, and oxygen concentrations of arterial blood.

carotid sinus: A dilated segment at the base of the internal carotid artery whose walls contain baroreceptors sensitive to changes in blood pressure.

carotid sinus reflex: Reflexive changes in blood pressure that maintain homeostatic pressures at the carotid sinus, stabilizing blood flow to the brain.

carpus/carpal: The wrist.

cartilage: A connective tissue with a gelatinous matrix that contains an abundance of fibers.

catabolism: The breakdown of complex organic molecules into simpler components, accompanied by the release of energy.

catalyst: A substance that accelerates a specific chemical reaction but that is not altered by the reaction.

catecholamine: Epinephrine, norepinephrine, dopamine, and related compounds.

catheter: A tube surgically inserted into a body cavity or along a blood vessel or excretory passageway for the collection of body fluids, monitoring of blood pressure, or introduction of medications or radiographic dyes.

cation: An ion that bears a positive charge.

cauda equina: Spinal nerve roots distal to the tip of the adult spinal cord; they extend caudally inside the vertebral canal en route to lumbar and sacral segments.

caudal/caudally: Closest to or toward the tail (coccyx).

caudate nucleus: One of the basal nuclei involved with the subconscious control of skeletal muscular activity.

cavernous tissue: Erectile tissue that can be engorged with blood; located in the penis (males) and clitoris (females).

cell: The smallest living unit in the human body.

cell-mediated immunity: Resistance to disease through the activities of sensitized T cells that destroy antigen-bearing cells by direct contact or through the release of lymphotoxins; also called *cellular immunity*.

center of ossification: The site in a connective tissue where bone formation begins.

central canal: Longitudinal canal in the center of an osteon that contains blood vessels and nerves, a passageway along the longitudinal axis of the spinal cord that contains cerebrospinal fluid.

central nervous system (CNS): The brain and spinal cord.

centriole: A cylindrical intracellular organelle composed of nine groups of microtubules, three in each group; functions in mitosis or meiosis by organizing the microtubules of the spindle apparatus.

centromere: The localized region where two chromatids remain connected after the chromosomes have replicated; site of spindle fiber attachment.

centrosome: A region of cytoplasm that contains a pair of centrioles oriented at right angles to one another.

cephalic: Pertaining to the head.

cerebellum: The posterior portion of the metencephalon, containing the cerebellar hemispheres;

includes the arbor vitae, cerebellar nuclei, and cerebellar cortex.

cerebral cortex: An extensive area of neural cortex covering the surfaces of the cerebral hemispheres.

cerebral hemispheres: A pair of expanded portions of the cerebrum covered in neural cortex.

cerebrospinal fluid (CSF): Fluid bathing the internal and external surfaces of the central nervous system; secreted by the choroid plexus.

cerebrovascular accident (CVA): The occlusion of a blood vessel that supplies a portion of the brain, resulting in damage to the dependent neurons; also called *stroke*.

cerebrum: The largest portion of the brain, composed of the cerebral hemispheres; includes the cerebral cortex, the basal nuclei, and the internal capsule.

cerumen: The waxy secretion of the ceruminous glands along the external acoustic meatus.

ceruminous glands: Integumentary glands that secrete cerumen.

cervix: The inferior portion of the uterus.

chemoreception: The detection of changes in the concentrations of dissolved compounds or gases.

chemotaxis: The attraction of phagocytic cells to the source of abnormal chemicals in tissue fluids.

chloride shift: The movement of plasma chloride ions into red blood cells in exchange for bicarbonate ions generated by the intracellular dissociation of carbonic acid.

cholecystokinin (CCK): A duodenal hormone that stimulates the contraction of the gallbladder and the secretion of enzymes by the exocrine pancreas.

cholesterol: A steroid component of plasma membranes and a substrate for the synthesis of steroid hormones and bile salts.

choline: A breakdown product or precursor of acetylcholine.

cholinergic synapse: A synapse where the presynaptic membrane releases acetylcholine on stimulation.

cholinesterase: The enzyme that breaks down and inactivates acetylcholine.

chondrocyte: A cartilage cell.

chondroitin sulfate: The predominant proteoglycan in cartilage, responsible for the gelatinous consistency of the matrix.

chordae tendineae: Fibrous cords that stabilize the position of the AV valves in the heart, preventing backflow during ventricular systole.

chorion/chorionic: An extraembryonic membrane, consisting of the trophoblast and underlying mesoderm, that forms the placenta.

choroid: The middle, vascular layer in the wall of the eye.

choroid plexus: The vascular complex in the roof of the third and fourth ventricles of the brain, responsible for the production of cerebrospinal fluid.

chromatid: One complete copy of a DNA strand and its associated nucleoproteins.

chromatin: A histological term referring to the grainy material visible in cell nuclei during interphase; the appearance of the DNA content of the nucleus when the chromosomes are uncoiled.

chromosomes: Dense structures, composed of tightly coiled DNA strands and associated histones, that become visible in the nucleus when a cell prepares to undergo mitosis or meiosis; normal human somatic cells each contain 46 chromosomes.

chronic: Habitual or long term.

chylomicrons: Relatively large droplets that may contain triglycerides, phospholipids, and cholesterol in association with proteins; synthesized and released by intestinal cells and transported to the venous blood by the lymphatic system.

ciliary body: A thickened region of the choroid that encircles the lens of the eye; includes the ciliary muscle and the ciliary processes that support the suspensory ligaments of the lens.

cilium/cilia: A slender organelle that extends above the free surface of an epithelial cell and generally undergoes cycles of movement; composed of a basal body and microtubules in a 9 + 2 array.

circulatory system: The network of blood vessels and lymphatic vessels that facilitate the distribution and circulation of extracellular fluid.

circumduction: A movement at a synovial joint in which the distal end of the bone moves in a circular direction, but the shaft does not rotate.

circumvallate papilla: One of the large, dome-shaped papillae on the superior surface of the tongue that forms a V, separating the body of the tongue from the root.

cisterna: An expanded or flattened chamber derived from and associated with the endoplasmc reticulum.

citric acid cycle: The reaction sequence that occurs in the matrix of mitochondria; in the process, organic molecules are broken down, carbon dioxide molecules are released, and hydrogen atoms are transferred to coenzymes that deliver them to the electron transport system.

clot: A network of fibrin fibers and trapped blood cells; also called a *thrombus* if it occurs within the cardiovascular system.

clotting factors: Plasma proteins, synthesized by the liver, that are essential to the clotting response.

clotting response: The series of events that results in the formation of a clot.

coccygeal ligament: The fibrous extension of the dura mater and filum terminale; provides longitudinal stabilization to the spinal cord.

coccyx: The terminal portion of the spinal column, consisting of relatively tiny, fused vertebrae.

cochlea: The spiral portion of the bony labyrinth of the inner ear that surrounds the organ of hearing.

cochlear duct: The central membranous tube within the cochlea that is filled with endolymph and contains the organ of Corti; also called *scala media*.

codon: A sequence of three nitrogenous bases along an mRNA strand that will specify the location of a single amino acid in a peptide chain.

coelom: The ventral body cavity, lined by a serous membrane and subdivided during fetal development into the pleural, pericardial, and abdomino-pelvic (peritoneal) cavities.

coenzymes: Complex organic cofactors; most are structurally related to vitamins.

cofactor: Ions or molecules that must be attached to the active site before an enzyme can function; examples include mineral ions and several vitamins.

collagen: A strong, insoluble protein fiber common in connective tissues.

collateral ganglion: A sympathetic ganglion situated anterior to the spinal column and separate from the sympathetic chain.

colliculus/colliculi: A little mound; in the brain, refers to one of the thickenings in the roof of the mesencephalon; the superior colliculi are associated with the visual system, and the inferior colliculi with the auditory system.

colloid/colloidal suspension: A solution containing large organic molecules in suspension.

colon: The large intestine.

comminuted: Broken or crushed into small pieces.

commissure: A crossing over from one side to another.

common bile duct: The duct formed by the union of the cystic duct from the gallbladder and the bile ducts from the liver; terminates at the duodenal ampulla, where it meets the pancreatic duct.

compact bone: Dense bone that contains parallel osteons.

complement: A system of 11 plasma proteins that interact in a chain reaction after exposure to activated antibodies or the surfaces of certain pathogens; complement proteins promote cell lysis, phagocytosis, and other defense mechanisms.

compliance: Distensibility; the ability of certain organs to tolerate changes in volume; indicates the presence of elastic fibers and smooth muscles.

compound: A molecule containing two or more elements in combination.

concentration: The amount (in grams) or number of atoms, ions, or molecules (in moles) per unit volume.

concentration gradient: Regional differences in the concentration of a particular substance.

conception: Fertilization.

concha/conchae: Three pairs of thin, scroll-like bones that project into the nasal cavities; the superior and middle conchae are part of the ethmoid, and the inferior conchae articulate with the ethmoid, lacrimal, maxilla, and palatine bones.

condyle: A rounded articular projection on the surface of a bone.

congenital: Present at birth.

congestive heart failure (CHF): The failure to maintain adequate cardiac output due to cardiovascular problems or myocardial damage.

conjunctiva: A layer of stratified squamous epithelium that covers the inner surfaces of the eyelids and the anterior surface of the eye to the edges of the cornea.

connective tissue: One of the four primary tissue types; provides a structural framework that stabilizes the relative positions of the other tissue types; includes connective tissue proper, cartilage, bone, and blood; contains cell products, cells, and ground substance.

continuous propagation: The propagation of an action potential along an unmyelinated axon or a muscle plasma membrane, wherein the action potential affects every portion of the membrane surface.

contractility: The ability to contract; possessed by skeletal, smooth, and cardiac muscle cells.

contralateral reflex: A reflex that affects the opposite side of the body from the stimulus.

conus medullaris: The conical tip of the spinal cord that gives rise to the filum terminale.

convergence: In the nervous system, the innervation of a single neuron by axons from several neurons; most common along motor pathways.

coracoid process: A hook-shaped process of the scapula that projects above the anterior surface of the capsule of the shoulder joint.

Cori cycle: The metabolic exchange of lactate from skeletal muscle for glucose from the liver; performed during the recovery period after muscular exertion.

cornea: The transparent portion of the fibrous tunic of the anterior surface of the eye.

corniculate cartilages: A pair of small laryngeal cartilages.

cornu: Horn-shaped.

coronoid: Hooked or curved.

corpora quadrigemina: The superior and inferior colliculi of the mesencephalic tectum (roof) in the brain.

corpus/corpora: Body.

corpus callosum: A large bundle of axons that links centers in the left and right cerebral hemispheres.

corpus luteum: The progestin-secreting mass of follicle cells that develops in the ovary after ovulation.

cortex: The outer layer or portion of an organ.

corticobulbar tracts: Descending tracts that carry information or commands from the cerebral cortex to nuclei and centers in the brain stem.

corticospinal tracts: Descending tracts that carry motor commands from the cerebral cortex to the anterior gray horns of the spinal cord.

corticosteroid: A steroid hormone produced by the adrenal cortex.

corticosterone: A corticosteroid secreted by the zona fasciculata of the adrenal cortex; a glucocorticoid.

corticotropin-releasing hormone (CRH): The releasing hormone, secreted by the hypothalamus, that stimulates secretion of adrenocorticotropic hormone by the anterior lobe of the pituitary gland.

cortisol: A corticosteroid secreted by the zona fasciculata of the adrenal cortex; a glucocorticoid.

costa/costae: A rib/ribs.

cotransport: The membrane transport of a nutrient, such as glucose, in company with the movement of an ion, normally sodium; transport requires a carrier protein but does not involve direct ATP expenditure and can occur regardless of the concentration gradient for the nutrient.

countercurrent exchange: The transfer of heat, water, or solutes between two fluids that travel in opposite directions.

countercurrent multiplication: Active transport between two limbs of a loop that contains a fluid moving in one direction; responsible for the concentration of urine in the kidney tubules.

covalent bond: A chemical bond between atoms that involves the sharing of electrons.

coxal bone: Hip.

cranial: Pertaining to the head.

cranial nerves: Peripheral nerves originating at the brain.

craniosacral division: *See* **parasympathetic division**.

cranium: The braincase; the skull bones that surround and protect the brain.

creatine: A nitrogenous compound, synthesized in the body, that can form a high-energy bond by connecting to a phosphate group and that serves as an energy reserve.

creatine phosphate: A high-energy compound in muscle cells; during muscle activity, the phosphate group is donated to ADP, regenerating ATP; also called *phosphorylcreatine*.

creatinine: A breakdown product of creatine metabolism.

crenation: Cellular shrinkage due to an osmotic movement of water out of the cytoplasm.

cribriform plate: A portion of the ethmoid bone that contains the foramina used by the axons of olfactory receptors en route to the olfactory bulbs of the cerebrum.

cricoid cartilage: A ring-shaped cartilage that forms the inferior margin of the larynx.

crista/cristae: A ridge-shaped collection of hair cells in the ampulla of a semicircular duct; the crista and cupula form a receptor complex sensitive to movement along the plane of the semicircular canal.

cross-bridge: A myosin head that projects from the surface of a thick filament and that can bind to an active site of a thin filament in the presence of calcium ions.

cuneiform cartilages: A pair of small cartilages in the larynx.

cupula: A gelatinous mass that is located in the ampulla of a semicircular duct in the inner ear and whose movement stimulates the hair cells of the crista.

Cushing disease: A condition caused by the oversecretion of adrenal steroids.

cutaneous membrane: The epidermis and papillary layer of the dermis.

cuticle: The layer of dead, keratinized cells that surrounds the shaft of a hair; for nails, *see* **eponychium**.

cyanosis: A bluish coloration of the skin due to the presence of deoxygenated blood in vessels near the body surface.

cystic duct: A duct that carries bile between the gallbladder and the common bile duct.

cytochrome: A pigment component of the electron transport system; a structural relative of heme.

cytokinesis: The cytoplasmic movement that separates two daughter cells at the completion of mitosis.

cytology: The study of cells.

cytoplasm: The material between the plasma membrane and the nuclear membrane; cell contents.

cytosine: A pyrimidine; one of the nitrogenous bases in the nucleic acids RNA and DNA.

cytoskeleton: A network of microtubules and microfilaments in the cytoplasm.

cytosol: The fluid portion of the cytoplasm.

cytotoxic: Poisonous to cells.

cytotoxic T cells: Lymphocytes involved in cell-mediated immunity that kill target cells by direct contact or by the secretion of lymphotoxins; also called *killer T cells* and T_C *cells*.

D

daughter cells: Genetically identical cells produced by somatic cell division.

deamination: The removal of an amino group from an amino acid.

decomposition reaction: A chemical reaction that breaks a molecule into smaller fragments.

decussate: To cross over to the opposite side, usually referring to the crossover of the descending tracts of the corticospinal pathway on the ventral surface of the medulla oblongata.

defecation: The elimination of fecal wastes.

degradation: Breakdown, catabolism.

dehydration: A reduction in the water content of the body that threatens homeostasis.

dehydration synthesis: The joining of two molecules associated with the removal of a water molecule.

demyelination: The loss of the myelin sheath of an axon, normally due to chemical or physical damage to Schwann cells or oligodendrocytes.

denaturation: A temporary or permanent change in the three-dimensional structure of a protein.

dendrite: A sensory process of a neuron.

deoxyribonucleic acid (DNA): A nucleic acid consisting of a double chain of nucleotides that contains the sugar deoxyribose and the nitrogenous bases adenine, guanine, cytosine, and thymine.

deoxyribose: A five-carbon sugar resembling ribose but lacking an oxygen atom.

depolarization: A change in the transmembrane potential from a negative value toward 0 mV.

depression: Inferior (downward) movement of a body part.

dermatome: A sensory region monitored by the dorsal rami of a single spinal segment.

dermis: The connective tissue layer beneath the epidermis of the skin.

detrusor muscle: Collectively, the three layers of smooth muscle in the wall of the urinary bladder.

development: Growth and the acquisition of increasing structural and functional complexity; includes the period from conception to maturity.

diabetes mellitus: Polyuria and glycosuria, most commonly due to the inadequate production or diminished sensitivity to insulin with a resulting elevation of blood glucose levels.

diapedesis: The movement of white blood cells through the walls of blood vessels by migration between adjacent endothelial cells.

diaphragm: Any muscular partition; the respiratory muscle that separates the thoracic cavity from the abdominopelvic cavity.

diaphysis: The shaft of a long bone.

diarthrosis: A synovial joint.

diastolic pressure: Pressure measured in the walls of a muscular artery when the left ventricle is in diastole (relaxation).

diencephalon: A division of the brain that includes the epithalamus, thalamus, and hypothalamus.

differential count: The determination of the relative abundance of each type of white blood cell on the basis of a random sampling of 100 white blood cells.

differentiation: The gradual appearance of characteristic cellular specializations during development as the result of gene activation or repression.

diffusion: Passive molecular movement from an area of higher concentration to an area of lower concentration.

digestion: The chemical breakdown of ingested materials into simple molecules that can be absorbed by the cells of the digestive tract.

digestive system: The digestive tract and associated glands.

digestive tract: An internal passageway that begins at the mouth, ends at the anus, and is lined by a mucous membrane; also called *gastrointestinal tract*.

dilate: To increase in diameter; to enlarge or expand.

disaccharide: A compound formed by the joining of two simple sugars by dehydration synthesis.

dissociation: *See* **ionization**.

distal: A direction away from the point of attachment or origin; for a limb, away from its attachment to the trunk.

distal convoluted tubule (DCT): The portion of the nephron closest to the connecting tubules and collecting duct; an important site of active secretion.

diuresis: Fluid loss at the kidneys; the production of unusually large volumes of urine.

divergence: In neural tissue, the spread of information from one neuron to many neurons; an organizational pattern common along sensory pathways of the central nervous system.

diverticulum: A sac or pouch in the wall of the colon or other organ.

DNA molecule: Two DNA strands wound in a double helix and held together by hydrogen bonds between complementary nitrogenous base pairs.

dopamine: An important neurotransmitter in the central nervous system.

dorsal: Toward the back, posterior.

dorsal root ganglion: A peripheral nervous system ganglion containing the cell bodies of sensory neurons.

dorsiflexion: Upward movement of the foot through flexion at the ankle.

Down syndrome: A genetic abnormality resulting from the presence of three copies of chromosome 21; individuals with this condition have characteristic physical and intellectual deficits.

duct: A passageway that delivers exocrine secretions to an epithelial surface.

ductus arteriosus: A vascular connection between the pulmonary trunk and the aorta that functions throughout fetal life; normally closes at birth or shortly thereafter and persists as the ligamentum arteriosum.

ductus deferens: A passageway that carries spermatozoa from the epididymis to the ejaculatory duct; also called the vas deferens.

duodenal ampulla: A chamber that receives bile from the common bile duct and pancreatic secretions from the pancreatic duct.

duodenal papilla: A conical projection from the inner surface of the duodenum that contains the opening of the duodenal ampulla.

duodenum: The proximal 25 cm (9.8 in.) of the small intestine that contains short villi and submucosal glands.

dura mater: The outermost component of the cranial and spinal meninges.

E

eccrine glands: Sweat glands of the skin that produce a watery secretion.

ectoderm: One of the three primary germ layers; covers the surface of the embryo and gives rise to the nervous system, the epidermis and associated glands, and a variety of other structures.

ectopic: Outside the normal location.

effector: A peripheral gland or muscle cell innervated by a motor neuron.

efferent: Away from an organ or structure.

efferent arteriole: An arteriole carrying blood away from a glomerulus of the kidney.

efferent fiber: An axon that carries impulses away from the central nervous system.

ejaculation: The ejection of semen from the penis as the result of muscular contractions of the bulbospongiosus and ischiocavernosus muscles.

ejaculatory ducts: Short ducts that pass within the walls of the prostate gland and connect the ductus deferens with the prostatic urethra.

elastase: A pancreatic enzyme that breaks down elastin fibers.

elastin: Connective tissue fibers that stretch and recoil, providing elasticity to connective tissues.

electrical coupling: A connection between adjacent cells that permits the movement of ions and the transfer of graded or conducted changes in the transmembrane potential from cell to cell.

electrocardiogram (ECG, EKG): A graphic record of the electrical activities of the heart, as monitored at specific locations on the body surface.

electroencephalogram (EEG): A graphic record of the electrical activities of the brain.

electrolytes: Soluble inorganic compounds whose ions will conduct an electrical current in solution.

electron: One of the three fundamental subatomic particles; bears a negative charge and normally orbits the protons of the nucleus.

electron transport system (ETS): The cytochrome system responsible for aerobic energy production in cells; a complex bound to the inner mitochondrial membrane.

element: All the atoms with the same atomic number.

elevation: Movement in a superior, or upward, direction.

embryo: The developmental stage beginning at fertilization and ending at the start of the third developmental month.

embryology: The study of embryonic development, focusing on the first two months after fertilization.

endocardium: The simple squamous epithelium that lines the heart and is continuous with the endothelium of the great vessels.

endochondral ossification: The replacement of a cartilaginous model by bone; the characteristic mode of formation for skeletal elements other

than the bones of the cranium, the clavicles, and sesamoid bones.

endocrine gland: A gland that secretes hormones into the blood.

endocrine system: The endocrine (ductless) glands/organs of the body.

endocytosis: The movement of relatively large volumes of extracellular material into the cytoplasm via the formation of a membranous vesicle at the cell surface; includes pinocytosis and phagocytosis.

endoderm: One of the three primary germ layers; the layer on the undersurface of the embryonic disc; gives rise to the epithelia and glands of the digestive system, the respiratory system, and portions of the urinary system.

endogenous: Produced within the body.

endolymph: The fluid contents of the membranous labyrinth (the saccule, utricle, semicircular ducts, and cochlear duct) of the inner ear.

endometrium: The mucous membrane lining the uterus.

endomysium: A delicate network of connective tissue fibers that surrounds individual muscle cells.

endoneurium: A delicate network of connective tissue fibers that surrounds individual nerve fibers.

endoplasmic reticulum: A network of membranous channels in the cytoplasm of a cell that function in intracellular transport, synthesis, storage, packaging, and secretion.

endosteum: An incomplete cellular lining on the inner (medullary) surfaces of bones.

endothelium: The simple squamous epithelial cells that line blood and lymphatic vessels.

enteroendocrine cells: Endocrine cells scattered among the epithelial cells that line the digestive tract.

enterogastric reflex: The reflexive inhibition of gastric secretion; initiated by the arrival of chyme in the small intestine.

enterokinase: An enzyme in the lumen of the small intestine that activates the proenzymes secreted by the pancreas.

enzyme: A protein that catalyzes a specific biochemical reaction.

eosinophil: A microphage (white blood cell) with a lobed nucleus and red-staining granules; participates in the immune response and is especially important during allergic reactions.

ependyma: The layer of cells lining the ventricles and central canal of the central nervous system.

epicardium: A serous membrane covering the outer surface of the heart; also called *visceral pericardium*.

epidermis: The epithelium covering the surface of the skin.

epididymis: A coiled duct that connects the rete testis to the ductus deferens; site of functional maturation of spermatozoa.

epidural space: The space between the spinal dura mater and the walls of the vertebral foramen; contains blood vessels and adipose tissue; a common site of injection for regional anesthesia.

epiglottis: A blade-shaped flap of tissue, reinforced by cartilage, that is attached to the dorsal and superior surface of the thyroid cartilage; folds over the entrance to the larynx during swallowing.

epimysium: A dense layer of collagen fibers that surrounds a skeletal muscle and is continuous with the tendons/aponeuroses of the muscle and with the perimysium.

epineurium: A dense layer of collagen fibers that surrounds a peripheral nerve.

epiphyseal cartilage: The cartilaginous region between the epiphysis and diaphysis of a growing bone.

epiphysis: The head of a long bone.

epithelium: One of the four primary tissue types; a layer of cells that forms a superficial covering or an internal lining of a body cavity or vessel.

equilibrium: A dynamic state in which two opposing forces or processes are in balance.

erection: The stiffening of the penis due to the engorgement of the erectile tissues of the corpora cavernosa and corpus spongiosum.

erythema: Redness and inflammation at the surface of the skin.

erythrocyte: A red blood cell; has no nucleus and contains large quantities of hemoglobin.

erythropoietin: A hormone released by most tissues, and especially by the kidneys, when exposed to low oxygen concentrations; stimulates erythropoiesis (red blood cell formation) in bone marrow.

Escherichia coli: A normal bacterial resident of the large intestine.

esophagus: A muscular tube that connects the pharynx to the stomach.

essential amino acids: Amino acids that cannot be synthesized in the body in adequate amounts and must be obtained from the diet.

essential fatty acids: Fatty acids that cannot be synthesized in the body and must be obtained from the diet.

estrogens: A class of steroid sex hormones that includes estradiol.

evaporation: A movement of molecules from the liquid state to the gaseous state.

eversion: A turning outward.

excitable membranes: Membranes that propagate action potentials, a characteristic of muscle cells and nerve cells.

excitatory postsynaptic potential (EPSP): The depolarization of a postsynaptic membrane by a chemical neurotransmitter released by the presynaptic cell.

excretion: The removal of waste products from the blood, tissues, or organs.

exocrine gland: A gland that secretes onto the body surface or into a passageway connected to the exterior.

exocytosis: The ejection of cytoplasmic materials by the fusion of a membranous vesicle with the plasma membrane.

expiration: Exhalation; breathing out.

extension: An increase in the angle between two articulating bones; the opposite of flexion.

external acoustic meatus: A passageway in the temporal bone that leads to the tympanic membrane of the inner ear.

external ear: The auricle, external acoustic meatus, and tympanic membrane.

external nares: The nostrils; the external openings into the nasal cavity.

external respiration: The diffusion of gases between the alveolar air and the alveolar capillaries and between the systemic capillaries and peripheral tissues.

exteroceptors: General sensory receptors in the skin, mucous membranes, and special sense organs that provide information about the external environment and about our position within it.

extracellular fluid: All body fluids other than that contained within cells; includes plasma and interstitial fluid.

extraembryonic membranes: The yolk sac, amnion, chorion, and allantois.

extrinsic pathway: A clotting pathway that begins with damage to blood vessels or surrounding tissues and ends with the formation of tissue thromboplastin.

facilitated: Brought closer to threshold, as in the depolarization of a nerve plasma membrane toward threshold; making the cell more sensitive to depolarizing stimuli.

facilitated diffusion: The passive movement of a substance across a plasma membrane by means of a protein carrier.

falciform ligament: A sheet of mesentery that contains the ligamentum teres, the fibrous remains of the umbilical vein of the fetus.

falx: Sickle-shaped.

falx cerebri: The curving sheet of dura mater that extends between the two cerebral hemispheres; encloses the superior sagittal sinus.

fasciae: Connective tissue fibers, primarily collagenous, that form sheets or bands beneath the skin to attach, stabilize, enclose, and separate muscles and other internal organs.

fasciculus: A small bundle; usually refers to a collection of nerve axons or muscle fibers.

fatty acids: Hydrocarbon chains that end in a carboxylic acid group.

fauces: The passage from the mouth to the pharynx, bounded by the palatal arches, the soft palate, and the uvula.

feces: Waste products eliminated by the digestive tract at the anus; contains indigestible residue, bacteria, mucus, and epithelial cells.

fenestra: An opening.

fertilization: The fusion of a secondary oocyte and a spermatozoon to form a zygote.

fetus: The developmental stage lasting from the start of the third developmental month to delivery.

fibrin: Insoluble protein fibers that form the basic framework of a blood clot.

fibrinogen: A plasma protein that is the soluble precursor of the insoluble protein fibrin.

fibroblasts: Cells of connective tissue proper that are responsible for the production of extracellular fibers and the secretion of the organic compounds of the extracellular matrix.

fibrous cartilage: Cartilage containing an abundance of collagen fibers; located around the edges of joints, in the intervertebral discs, the menisci of the knee, and so on.

fibrous tunic: The outermost layer of the eye, composed of the sclera and cornea.

fibula: The lateral, slender bone of the leg.

filiform papillae: Slender conical projections from the dorsal surface of the anterior two-thirds of the tongue.

filtrate: The fluid produced by filtration at a glomerulus in the kidney.

filtration: The movement of a fluid across a membrane whose pores restrict the passage of solutes on the basis of size.

filum terminale: A fibrous extension of the spinal cord, from the conus medullaris to the coccygeal ligament.

fimbriae: Fringes; the fingerlike processes that surround the entrance to the uterine tube.

fissure: An elongate groove or opening.

flagellum/flagella: An organelle that is structurally similar to a cilium but is used to propel a cell through a fluid; found on spermatozoa.

flatus: Intestinal gas.

flexion: A movement that reduces the angle between two articulating bones; the opposite of extension.

flexor: A muscle that produces flexion.

flexor reflex: A reflex contraction of the flexor muscles of a limb in response to a painful stimulus.

flexure: A bending.

folia: Leaflike folds; the slender folds in the surface of the cerebellar cortex.

follicle: A small secretory sac or gland.

follicle-stimulating hormone (FSH): A hormone secreted by the adenohypophysis (anterior pituitary gland); stimulates oogenesis (female) and spermatogenesis (male).

fontanelle: A relatively soft, flexible, fibrous region between two flat bones in the developing skull; also spelled *fontanel*.

foramen/foramina: An opening or passage through a bone.

forearm: The distal portion of the upper limb between the elbow and wrist.

forebrain: The cerebrum.

fornix: An arch or the space bounded by an arch; in the brain, an arching tract that connects the hippocampus with the mamillary bodies; in the eye, a slender pocket situated where the epithelium of the ocular conjunctiva folds back on itself as the palpebral conjunctiva; in the vagina, the shallow recess surrounding the protrusion of the cervix.

fossa: A shallow depression or furrow in the surface of a bone.

fourth ventricle: An elongate ventricle of the metencephalon (pons and cerebellum) and the myelencephalon (medulla oblongata) of the brain; the roof contains a region of choroid plexus.

fovea: The portion of the retina that provides the sharpest vision because it has the highest concentration of cones; also called *macula lutea*.

fracture: A break or crack in a bone.

frenulum: A bridle; usually referring to a band of tissue that restricts movement (e.g., *lingual frenulum*).

frontal plane: A sectional plane that divides the body into an anterior portion and a posterior portion; also called *coronal plane*.

fructose: A hexose (six-carbon simple sugar) in foods and in semen.

fundus: The base of an organ such as the stomach, uterus, or gallbladder.

gallbladder: The pear-shaped reservoir for bile after it is secreted by the liver.

gametes: Reproductive cells (spermatozoa or oocytes) that contain half the normal chromosome complement.

gametogenesis: The formation of gametes.

gamma aminobutyric acid (GABA): A neurotransmitter of the central nervous system whose effects are generally inhibitory.

gamma motor neurons: Motor neurons that adjust the sensitivities of muscle spindles (intrafusal fibers).

ganglion/ganglia: A collection of neuron cell bodies outside the central nervous system.

gap junctions: Connections between cells that permit electrical coupling.

gaster: The stomach; the body, or belly, of a skeletal muscle.

gastric: Pertaining to the stomach.

gastric glands: The tubular glands of the stomach whose cells produce acid, enzymes, intrinsic factor, and hormones.

gastrointestinal (GI) tract: *See* **digestive tract**.

gene: A portion of a DNA strand that functions as a hereditary unit, is located at a particular site on a specific chromosome, and codes for a specific protein or polypeptide.

genetics: The study of mechanisms of heredity.

geniculate: Like a little knee; the medial geniculates and the lateral geniculates are nuclei in the thalamus of the brain.

genitalia: The reproductive organs.

germinal centers: Pale regions in the interior of lymphoid tissues or lymphoid nodules, where cell divisions occur that produce additional lymphocytes.

gestation: The period of intrauterine development; pregnancy.

gland: Cells that produce exocrine or endocrine secretions.

glenoid cavity: A rounded depression that forms the articular surface of the scapula at the shoulder joint.

glial cells: *See* **neuroglia**.

globular proteins: Proteins whose tertiary structure makes them rounded and compact.

glomerular capsule: The expanded initial portion of the nephron that surrounds the glomerulus.

glomerular filtration rate: The rate of filtrate formation at the glomerulus.

glomerulus: A ball or knot; in the kidneys, a knot of capillaries that projects into the enlarged, proximal end of a nephron; the site of filtration, the first step in the production of urine.

glossopharyngeal nerve: Cranial nerve IX.

glucagon: A hormone secreted by the alpha cells of the pancreatic islets; elevates blood glucose concentrations.

glucocorticoids: Hormones secreted by the zona fasciculata of the adrenal cortex to modify glucose metabolism; cortisol and corticosterone are important examples.

gluconeogenesis: The synthesis of glucose from protein or lipid precursors.

glucose: A six-carbon sugar, $C_6H_{12}O_6$; the preferred energy source for most cells and normally the only energy source for neurons.

glycerides: Lipids composed of glycerol bound to fatty acids.

glycogen: A polysaccharide that is an important energy reserve; a polymer consisting of a long chain of glucose molecules.

glycogenesis: The synthesis of glycogen from glucose molecules.

glycogenolysis: Glycogen breakdown and the liberation of glucose molecules.

glycolipids: Compounds created by the combination of carbohydrate and lipid components.

glycolysis: The anaerobic cytoplasmic breakdown of glucose into two 3-carbon molecules of pyruvate, with a net gain of two ATP molecules.

glycoprotein: A compound containing a relatively small carbohydrate group attached to a large protein.

glycosuria: The presence of glucose in urine.

Golgi apparatus: A cellular organelle consisting of a series of membranous plates that give rise to lysosomes and secretory vesicles.

gomphosis: A fibrous synarthrosis that binds a tooth to the bone of the jaw.

gonadotropin-releasing hormone (GnRH): A hypothalamic releasing hormone that causes the secretion of both follicle-stimulating hormone and luteinizing hormone by the adenohypophysis (anterior pituitary gland).

gonadotropins: Follicle-stimulating hormone and luteinizing hormone, hormones that stimulate gamete development and sex hormone secretion.

gonads: Reproductive organs that produce gametes and hormones.

granulocytes: White blood cells containing granules that are visible with the light microscope; includes eosinophils, basophils, and neutrophils; also called *granular leukocytes*.

gray matter: Areas in the central nervous system that are dominated by neuron cell bodies, neuroglia, and unmyelinated axons.

gray ramus: A bundle of postganglionic sympathetic nerve fibers that are distributed to effectors in the body wall, skin, and limbs by way of a spinal nerve.

greater omentum: A large fold of the dorsal mesentery of the stomach; hangs anterior to the intestines.

groin: The inguinal region.

gross anatomy: The study of the structural features of the body without the aid of a microscope.

growth hormone (GH): An adenohypophysis (anterior pituitary gland) hormone that stimulates tissue growth and anabolism when nutrients are abundant and restricts tissue glucose dependence when nutrients are in short supply.

growth hormone–inhibiting hormone (GH–IH): A hypothalamic regulatory hormone that inhibits growth hormone secretion by the anterior lobe of the pituitary gland.

guanine: A purine; one of the nitrogenous bases in the nucleic acids RNA and DNA.

gustation: Taste.

gyrus: A prominent fold or ridge of neural cortex on the surfaces of the cerebral hemispheres.

H

hair: A keratinous strand produced by epithelial cells of the hair follicle.

hair cells: Sensory cells of the inner ear.

hair follicle: An accessory structure of the integument; a tube lined by a stratified squamous epithelium that begins at the surface of the skin and ends at the hair papilla.

hallux: The big toe.

haploid: Possessing half the normal number of chromosomes; a characteristic of gametes.

hard palate: The bony roof of the oral cavity, formed by the maxillae and palatine bones.

helper T cells: Lymphocytes whose secretions and other activities coordinate cell-mediated and antibody-mediated immunities; also called T_H cells.

hematocrit: The percentage of the volume of whole blood contributed by cells; also called *volume of packed red cells (VPRC)* or *packed cell volume (PCV)*.

hematoma: An abnormal collection of clotted or partially clotted blood outside a blood vessel.

hematuria: The abnormal presence of red blood cells in urine.

heme: A porphyrin ring containing a central iron atom that can reversibly bind oxygen molecules; a component of the hemoglobin molecule.

hemocytoblasts: Stem cells whose divisions produce each of the various populations of blood cells.

hemoglobin: A protein composed of four globular subunits, each bound to a heme molecule; gives red blood cells the ability to transport oxygen in the blood.

hemolysis: The breakdown of red blood cells.

hemopoiesis: Blood cell formation and differentiation.

hemorrhage: Blood loss; to bleed.

hemostasis: The cessation of bleeding.

heparin: An anticoagulant released by activated basophils and mast cells.

hepatic duct: The duct that carries bile away from the liver lobes and toward the union with the cystic duct.

hepatic portal vein: The vessel that carries blood between the intestinal capillaries and the sinusoids of the liver.

hepatocyte: A liver cell.

heterozygous: Possessing two different alleles at corresponding sites on a chromosome pair; the individual's phenotype is determined by one or both of the alleles.

hiatus: A gap, cleft, or opening.

high-density lipoprotein (HDL): A lipoprotein with a relatively small lipid content; responsible for the movement of cholesterol from peripheral tissues to the liver.

hilum: A localized region where blood vessels, lymphatic vessels, nerves, and/or other anatomical structures are attached to an organ.

hippocampus: A region, beneath the floor of a lateral ventricle, involved with emotional states and the conversion of short-term to long-term memories.

histamine: The chemical released by stimulated mast cells or basophils to initiate or enhance an inflammatory response.

histology: The study of tissues.

histones: Proteins associated with the DNA of the nucleus; the DNA strands are wound around them.

holocrine: A form of exocrine secretion in which the secretory cell becomes swollen with vesicles and then ruptures.

homeostasis: The maintenance of a relatively constant internal environment.

hormone: A compound that is secreted by one cell and travels through the bloodstream to affect the activities of cells in another portion of the body.

human chorionic gonadotropin (hCG): The placental hormone that maintains the corpus luteum for the first three months of pregnancy.

human immunodeficiency virus (HIV): The infectious agent that causes acquired immune deficiency syndrome (AIDS).

human leukocyte antigen (HLA): See **MHC protein**.

human placental lactogen (hPL): The placental hormone that stimulates the functional development of the mammary glands.

humoral immunity: See **antibody-mediated immunity**.

hyaluronan: A carbohydrate component of proteoglycans in the matrix of many connective tissues.

hyaluronidase: An enzyme that breaks down the bonds between adjacent follicle cells; produced by some bacteria and found in the acrosomal cap of a spermatozoon.

hydrogen bond: A weak interaction between the hydrogen atom on one molecule and a negatively charged portion of another molecule.

hydrolysis: The breakage of a chemical bond through the addition of a water molecule; the reverse of dehydration synthesis.

hydrophilic: Freely associating with water; readily entering into solution; water-loving.

hydrophobic: Incapable of freely associating with water molecules; insoluble; water-fearing.

hydrostatic pressure: Fluid pressure.

hydroxide ion: OH^-.

hypercapnia: High plasma carbon dioxide concentrations, commonly as a result of hypoventilation or inadequate tissue perfusion.

hyperpolarization: The movement of the transmembrane potential away from the normal resting potential and farther from 0 mV.

hypersecretion: The overactivity of glands that produce exocrine or endocrine secretions.

hypertension: Abnormally high blood pressure.

hypertonic: In comparing two solutions, the solution with the higher osmolarity.

hypertrophy: An increase in tissue size without cell division.

hyperventilation: A rate of respiration sufficient to reduce plasma P_{CO_2} concentrations to levels below normal.

hypocapnia: An abnormally low plasma P_{CO_2} concentration commonly as a result of hyperventilation.

hypodermic needle: A needle inserted through the skin to introduce drugs into the subcutaneous layer.

hypodermis: The layer of loose connective tissue below the dermis; also called *subcutaneous layer* or *superficial fascia*.

hypophyseal portal system: The network of vessels that carries blood from capillaries in the hypothalamus to capillaries in the anterior lobe of the pituitary gland.

hypophysis: The pituitary gland.

hyposecretion: Abnormally low rates of exocrine or endocrine secretion.

hypothalamus: The floor of the diencephalon; the region of the brain containing centers involved with the subconscious regulation of visceral functions, emotions, drives, and the coordination of neural and endocrine functions.

hypotonic: In comparing two solutions, the solution with the lower osmolarity.

hypoventilation: A respiratory rate that is insufficient to keep plasma P_{CO_2} concentrations within normal levels.

hypoxia: A low tissue oxygen concentration.

I

ileum: The distal 2.5 m of the small intestine.

ilium: The largest of the three bones whose fusion creates a coxal bone.

immunity: Resistance to injuries and diseases caused by foreign compounds, toxins, or pathogens.

immunization: The production of immunity by the deliberate exposure to antigens under conditions that prevent the development of illness but stimulate the production of memory B cells.

immunoglobulin: A circulating antibody.

implantation: The attachment of a blastocyst into the endometrium of the uterine wall.

inclusions: Aggregations of insoluble pigments, nutrients, or other materials in cytoplasm.

incus: The central auditory ossicle, situated between the malleus and the stapes in the middle ear cavity.

inexcitable: Incapable of propagating an action potential.

infarct: An area of dead cells that results from an interruption of blood flow.

infection: The invasion and colonization of body tissues by pathogens.

inferior: Below, in reference to a particular structure, with the body in the anatomical position.

inferior vena cava: The vein that carries blood from the parts of the body inferior to the heart to the right atrium.

infertility: The inability to conceive; also called *sterility*.

inflammation: A nonspecific defense mechanism that operates at the tissue level; characterized by swelling, redness, warmth, pain, and some loss of function.

infundibulum: A tapering, funnel-shaped structure; in the brain, the connection between the pituitary gland and the hypothalamus; in the uterine tube, the entrance bounded by fimbriae that receives the oocytes at ovulation.

ingestion: The introduction of materials into the digestive tract by way of the mouth; eating.

inguinal canal: A passage through the abdominal wall that marks the path of testicular descent and that contains the testicular arteries, veins, and ductus deferens.

inguinal region: The area of the abdominal wall near the junction of the trunk and the thighs that contains the external genitalia; the groin.

inhibin: A hormone, produced by nurse (sustentacular) cells of the testes and follicular cells of the ovaries, that inhibits the secretion of follicle-stimulating hormone by the adenohypophysis (anterior lobe of the pituitary gland).

inhibitory postsynaptic potential (IPSP): A hyperpolarization of the postsynaptic membrane after the arrival of a neurotransmitter.

initial segment: The proximal portion of the axon where an action potential first appears.

injection: The forcing of fluid into a body part or organ.

inner cell mass: Cells of the blastocyst that will form the body of the embryo.

inner ear: The membranous labyrinth that contains the organs of hearing and equilibrium.

innervation: The distribution of sensory and motor nerves to a specific region or organ.

insensible perspiration: Evaporative water loss by diffusion across the epithelium of the skin or evaporation across the alveolar surfaces of the lungs.

insertion: A point of attachment of a muscle; the end that is easily movable.

insoluble: Incapable of dissolving in solution.

inspiration: Inhalation; the movement of air into the respiratory system.

insulin: A hormone secreted by beta cells of the pancreatic islets; causes a reduction in plasma glucose concentrations.

integument: The skin.

intercalated discs: Regions where adjacent cardiocytes interlock and where gap junctions permit electrical coupling between the cells.

intercellular cement: Proteoglycans situated between adjacent epithelial cells.

intercellular fluid: *See* **interstitial fluid**.

interferons: Peptides released by virus-infected cells, especially lymphocytes, that slow viral replication and make other cells more resistant to viral infection.

interleukins: Peptides, released by activated monocytes and lymphocytes, that assist in the coordination of cell-mediated and antibody-mediated immunities.

internal capsule: The collection of afferent and efferent fibers of the white matter of the cerebral hemispheres, visible on gross dissection of the brain.

internal nares: The entrance to the nasopharynx from the nasal cavity.

internal respiration: The diffusion of gases between interstitial fluid and cytoplasm.

interneuron: An association neuron; central nervous system neurons that are between sensory and motor neurons.

interoceptors: Sensory receptors monitoring the functions and status of internal organs and systems.

interosseous membrane: The fibrous connective tissue membrane between the shafts of the tibia and fibula and between the radius and ulna; an example of a fibrous amphiarthrosis.

interphase: The stage in the life cycle of a cell during which the chromosomes are uncoiled and all normal cellular functions except mitosis are under way.

intersegmental reflex: A reflex that involves several segments of the spinal cord.

interstitial fluid: The fluid in the tissues that fills the spaces between cells.

interstitial growth: A form of cartilage growth through the growth, mitosis, and secretion of chondrocytes in the matrix.

interventricular foramen: The opening that permits fluid movement between the lateral and third ventricles of the brain.

intervertebral disc: A fibrous cartilage pad between the bodies of successive vertebrae that absorbs shocks.

intestinal crypt: A tubular epithelial pocket that is lined by secretory cells and opens into the lumen of the digestive tract; also called *intestinal gland*.

intestine: The tubular organ of the digestive tract.

intracellular fluid: The cytosol.

intramembranous ossification: The formation of bone within a connective tissue without the prior development of a cartilaginous model.

intrinsic factor: A glycoprotein, secreted by the parietal cells of the stomach, that facilitates the intestinal absorption of vitamin B_{12}.

intrinsic pathway: A pathway of the clotting system that begins with the activation of platelets and ends with the formation of platelet thromboplastin.

inversion: A turning inward.

in vitro: Outside the body, in an artificial environment.

in vivo: In the living body.

involuntary: Not under conscious control.

ion: An atom or molecule bearing a positive or negative charge due to the donation or acceptance, respectively, of one or more electrons.

ionic bond: A molecular bond created by the attraction between ions with opposite charges.

ionization: Dissociation; the breakdown of a molecule in solution to form ions.

ipsilateral: A reflex response that affects the same side as the stimulus.

iris: A contractile structure, made up of smooth muscle, that forms the colored portion of the eye.

ischemia: An inadequate blood supply to a region of the body.

ischium: One of the three bones whose fusion creates a coxal bone.

islets of Langerhans: *See* **pancreatic islets**.

isotonic: A solution with an osmolarity that does not result in water movement across plasma membranes.

isotopes: Forms of an element whose atoms contain the same number of protons but different numbers of neutrons (and thus differ in atomic weight).

isthmus: A narrow band of tissue connecting two larger masses.

J

jejunum: The middle part of the small intestine.

joint: An area where adjacent bones interact; also called *articulation*.

juxtaglomerular complex: Specialized cells in between the walls of the DCT and afferent and efferent arterioles adjacent to the glomerulus; a complex responsible for the release of renin and erythropoietin.

K

keratin: The tough, fibrous protein component of nails, hair, calluses, and the general integumentary surface.

keto acid: A molecule that ends in —COCOOH; the carbon chain that remains after the deamination or transamination of an amino acid.

ketoacidosis: A reduction in the pH of body fluids due to the presence of large numbers of ketone bodies.

ketone bodies: Keto acids produced during the catabolism of lipids and some amino acids; specifically, acetone, acetoacetate, and beta-hydroxybutyrate.

kidney: A component of the urinary system; an organ functioning in the regulation of plasma composition, including the excretion of wastes and the maintenance of normal fluid and electrolyte balances.

killer T cells: *See* **cytotoxic T cells**.

Krebs cycle: *See* **citric acid cycle**.

Kupffer cells: Phagocytic cells of the liver sinusoids.

L

labium/labia: Lip; the labia majora and labia minora are components of the female external genitalia.

labrum: A lip or rim.

labyrinth: A maze of passageways; the structures of the internal ear.

lacrimal gland: A tear gland on the dorsolateral surface of the eye.

lactase: An enzyme that breaks down the milk sugar lactose.

lactation: The production of milk by the mammary glands.

lacteal: A terminal lymphatic within an intestinal villus.

lactate: An anion released by the dissociation of lactic acid, produced from pyruvate under anaerobic conditions.

lacuna: A small pit or cavity.

lambdoid suture: The synarthrosis between the parietal and occipital bones of the cranium.

lamellae: Concentric layers; the concentric layers of bone within an osteon.

lamellated corpuscle: A receptor sensitive to vibration.

lamina: A thin sheet or layer.

lamina propria: The reticular tissue that underlies a mucous epithelium and forms part of a mucous membrane.

Langerhans cells: Cells in the epithelium of the skin and digestive tract that participate in the immune response by presenting antigens to T cells; also called dendritic cells.

large intestine: The terminal portions of the intestinal tract, consisting of the colon, the rectum, and the anal canal.

laryngopharynx: The division of the pharynx that is inferior to the epiglottis and superior to the esophagus.

larynx: A complex cartilaginous structure that surrounds and protects the glottis and vocal cords; the superior margin is bound to the hyoid bone, and the inferior margin is bound to the trachea.

latent period: The time between the stimulation of a muscle and the start of the contraction phase.

lateral: Pertaining to the side.

lateral apertures: Openings in the roof of the fourth ventricle that permit the circulation of cerebrospinal fluid into the subarachnoid space.

lateral ventricle: A fluid-filled chamber within a cerebral hemisphere.

lens: The transparent refractive structure that is between the iris and the vitreous humor.

lesser omentum: A small pocket in the mesentery that connects the lesser curvature of the stomach to the liver.

leukocyte: A white blood cell.

ligament: A dense band of connective tissue fibers that attaches one bone to another.

ligamentum arteriosum: The fibrous strand in adults that is the remnant of the ductus arteriosus of the fetal stage.

ligamentum nuchae: An elastic ligament between the vertebra prominens and the occipital bone.

ligamentum teres: The fibrous strand in the falciform ligament of adults that is the remnant of the umbilical vein of the fetal stage.

ligate: To tie off.

limbic system: The group of nuclei and centers in the cerebrum and diencephalon that are involved with emotional states, memories, and behavioral drives.

lingual: Pertaining to the tongue.

lipid: An organic compound containing carbons, hydrogens, and oxygens in a ratio that does not approximate 1:2:1; includes fats, oils, and waxes.

lipogenesis: The synthesis of lipids from nonlipid precursors.

lipolysis: The catabolism of lipids as a source of energy.

lipoprotein: A compound containing a relatively small lipid bound to a protein.

liver: An organ of the digestive system that has varied and vital functions, including the production of plasma proteins, the excretion of bile, the storage of energy reserves, the detoxification of poisons, and the interconversion of nutrients.

lobule: Histologically, the basic organizational unit of the liver.

local hormone: *See* **prostaglandin**.

loop of Henle: *See* **nephron loop**.

loose connective tissue: A loosely organized, easily distorted connective tissue that contains several fiber types, a varied population of cells, and a viscous ground substance.

lumbar: Pertaining to the lower back.

lumen: The central space within a duct or other internal passageway.

lungs: The paired organs of respiration, situated in the pleural cavities.

luteinizing hormone (LH): A hormone produced by the adenohypophysis (anterior lobe of the pituitary gland). In females, it assists FSH in follicle stimulation, triggers ovulation, and promotes the maintenance and secretion of endometrial glands. In males, it stimulates testosterone secretion by the interstitial cells of the testes.

lymph: The fluid contents of lymphatic vessels, similar in composition to interstitial fluid.

lymphatic vessels: The vessels of the lymphoid system; also called *lymphatics*.

lymph nodes: Lymphoid organs that monitor the composition of lymph.

lymphocyte: A cell of the lymphoid system that participates in the immune response.

lymphokines: Chemicals secreted by activated lymphocytes.

lymphopoiesis: The production of lymphocytes from lymphoid stem cells.

lymphotoxin: A secretion of lymphocytes that kills the target cells.

lysis: The destruction of a cell through the rupture of its plasma membrane.

lysosome: An intracellular vesicle containing digestive enzymes.

lysozyme: An enzyme, present in some exocrine secretions, that has antibiotic properties.

M

macrophage: A phagocytic cell of the monocyte–macrophage system.

macula: A receptor complex, located in the saccule or utricle of the inner ear, that responds to linear acceleration or gravity.

macula lutea: *See* **fovea**.

major histocompatibility complex: *See* **MHC protein**.

malignant tumor: A form of cancer characterized by rapid cell growth and the spread of cancer cells throughout the body.

malleus: The first auditory ossicle, bound to the tympanic membrane and the incus.

malnutrition: An unhealthy state produced by inadequate dietary intake or absorption of nutrients, calories, and/or vitamins.

mamillary bodies: Nuclei in the hypothalamus that affect eating reflexes and behaviors; a component of the limbic system.

mammary glands: Milk-producing glands of the female breast.

manus: The hand.

marrow: A tissue that fills the internal cavities in bone; dominated by hemopoietic cells (red bone marrow) or by adipose tissue (yellow bone marrow).

mast cell: A connective tissue cell that, when stimulated, releases histamine, serotonin, and heparin, initiating the inflammatory response.

mastication: Chewing.

mastoid sinus: Air-filled spaces in the mastoid process of the temporal bone.

matrix: The extracellular fibers and ground substance of a connective tissue.

maxillary sinus: One of the paranasal sinuses; an air-filled chamber lined by a respiratory epithelium that is located in a maxilla and opens into the nasal cavity.

meatus: An opening or entrance into a passageway.

mechanoreception: The detection of mechanical stimuli, such as touch, pressure, or vibration.

medial: Toward the midline of the body.

mediastinum: The central tissue mass that divides the thoracic cavity into two pleural cavities.

medulla: The inner layer or core of an organ.

medulla oblongata: The most caudal of the brain regions, also called the *myelencephalon*.

medullary cavity: The space within a bone that contains the marrow.

medullary rhythmicity center: The center in the medulla oblongata that sets the background pace of respiration; includes inspiratory and expiratory centers.

megakaryocytes: Bone marrow cells responsible for the formation of platelets.

meiosis: Cell division that produces gametes with half the normal somatic chromosome complement.

melanin: The yellow-brown pigment produced by the melanocytes of the skin.

melanocyte: A specialized cell in the deeper layers of the stratified squamous epithelium of the skin; responsible for the production of melanin.

melanocyte-stimulating hormone (MSH): A hormone, produced by the pars intermedia of the adenohypophysis (anterior lobe of the pituitary gland), that stimulates melanin production.

melatonin: A hormone secreted by the pineal gland.

membrane: Any sheet or partition; a layer consisting of an epithelium and the underlying connective tissue.

membrane flow: The movement of sections of membrane surface to and from the cell surface and components of the endoplasmic reticulum, the Golgi apparatus, and vesicles.

membrane potential: *See* **transmembrane potential**.

membranous labyrinth: Endolymph-filled tubes that enclose the receptors of the inner ear.

meninges: Three membranes that surround the surfaces of the central nervous system; the dura mater, the pia mater, and the arachnoid.

meniscus: A fibrous cartilage pad between opposing surfaces in a joint.

menses: The first portion of the uterine cycle in which the endometrial lining sloughs away; menstrual period.

merocrine: A method of secretion in which the cell ejects materials from secretory vesicles through exocytosis.

mesencephalon: The midbrain; the region between the diencephalon and pons.

mesenchyme: Embryonic or fetal connective tissue.

mesentery: A double layer of serous membrane that supports and stabilizes the position of an organ in the abdominopelvic cavity and provides a route for the associated blood vessels, nerves, and lymphatic vessels.

mesoderm: The middle germ layer, between the ectoderm and endoderm of the embryo.

mesothelium: A simple squamous epithelium that lines one of the divisions of the ventral body cavity.

messenger RNA (mRNA): RNA formed at transcription to direct protein synthesis in the cytoplasm.

metabolic turnover: The continuous breakdown and replacement of organic materials within cells.

metabolism: The sum of all biochemical processes under way within the human body at any moment; includes anabolism and catabolism.

metabolites: Compounds produced in the body as a result of metabolic reactions.

metacarpal bones: The five bones of the palm of the hand.

metaphase: The stage of mitosis in which the chromosomes line up along the equatorial plane of the cell.

metaphysis: The region of a long bone between the epiphysis and diaphysis, corresponding to the location of the epiphyseal cartilage of the developing bone.

metarteriole: A vessel that connects an arteriole to a venule and that provides blood to a capillary plexus.

metastasis: The spread of cancer cells from one organ to another, leading to the establishment of secondary tumors.

metatarsal bone: One of the five bones of the foot that articulate with the tarsal bones (proximally) and the phalanges (distally).

metencephalon: The pons and cerebellum of the brain.

MHC protein: A surface antigen that is important to the recognition of foreign antigens and that plays a role in the coordination and activation of the immune response; also called *human leukocyte antigen (HLA)*.

micelle: A droplet with hydrophilic portions on the outside; a spherical aggregation of bile salts, monoglycerides, and fatty acids in the lumen of the intestinal tract.

microfilaments: Fine protein filaments visible with the electron microscope; components of the cytoskeleton.

microglia: Phagocytic neuroglia in the central nervous system.

microtubules: Microscopic tubules that are part of the cytoskeleton and are a component in cilia, flagella, the centrioles, and spindle fibers.

microvilli: Small, fingerlike extensions of the exposed plasma membrane of an epithelial cell.

micturition: Urination.

midbrain: The mesencephalon.

middle ear: The space between the external and internal ears that contains auditory ossicles.

midsagittal plane: A plane passing through the midline of the body that divides it into left and right halves.

mineralocorticoid: Corticosteroids produced by the zona glomerulosa of the adrenal cortex; steroids such as aldosterone that affect mineral metabolism.

mitochondrion: An intracellular organelle responsible for generating most of the ATP required for cellular operations.

mitosis: The division of a single cell nucleus that produces two identical daughter cell nuclei; an essential step in cell division.

mitral valve: *See* **bicuspid valve**.

mixed gland: A gland that contains exocrine and endocrine cells, or an exocrine gland that produces serous and mucous secretions.

mixed nerve: A peripheral nerve that contains sensory and motor fibers.

mole: A quantity of an element or compound having a mass in grams equal to the element's atomic weight or to the compound's molecular weight.

molecular weight: The sum of the atomic weights of all the atoms in a molecule.

molecule: A chemical structure containing two or more atoms that are held together by covalent chemical bonds.

monocytes: Phagocytic agranulocytes (white blood cells) in the circulating blood.

monoglyceride: A lipid consisting of a single fatty acid bound to a molecule of glycerol.

monosaccharide: A simple sugar, such as glucose or ribose.

monosynaptic reflex: A reflex in which the sensory afferent neuron synapses directly on the motor efferent neuron.

motor unit: All of the muscle cells controlled by a single motor neuron.

mucins: Proteoglycans responsible for the lubricating properties of mucus.

mucosa: A mucous membrane; the epithelium plus the lamina propria.

mucosa-associated lymphoid tissue (MALT): The extensive collection of lymphoid tissues linked with the epithelia of the digestive, respiratory, urinary, and reproductive tracts.

mucous (adjective): Indicating the presence or production of mucus.

mucous cell: A goblet-shaped, mucus-producing, unicellular gland in certain epithelia of the digestive and respiratory tracts; also called goblet cells.

mucus (noun): A lubricating fluid that is composed of water and mucins and is produced by unicellular and multicellular glands along the digestive, respiratory, urinary, and reproductive tracts.

multipolar neuron: A neuron with many dendrites and a single axon; the typical form of a motor neuron.

multi-unit smooth muscle: A smooth muscle tissue whose muscle cells are innervated in motor units.

muscarinic receptors: Membrane receptors sensitive to acetylcholine and to muscarine, a toxin produced by certain mushrooms; located at all parasympathetic neuromuscular and neuroglandular junctions and at a few sympathetic neuromuscular and neuroglandular junctions.

muscle: A contractile organ composed of muscle tissue, blood vessels, nerves, connective tissues, and lymphatic vessels.

muscle tissue: A tissue characterized by the presence of cells capable of contraction; includes skeletal, cardiac, and smooth muscle tissues.

muscularis externa: Concentric layers of smooth muscle responsible for peristalsis.

muscularis mucosae: The layer of smooth muscle beneath the lamina propria; responsible for moving the mucosal surface.

mutagens: Chemical agents that induce mutations and may be carcinogenic.

mutation: A change in the nucleotide sequence of the DNA in a cell.

myelencephalon: *See* medulla oblongata.

myelin: An insulating sheath around an axon; consists of multiple layers of neuroglial membrane; significantly increases the impulse propagation rate along the axon.

myelination: The formation of myelin.

myenteric plexus: Parasympathetic motor neurons and sympathetic postganglionic fibers located between the circular and longitudinal layers of the muscularis externa.

myocardial infarction: A heart attack; damage to the heart muscle due to an interruption of regional coronary circulation.

myocardium: The cardiac muscle tissue of the heart.

myofibril: Organized collections of myofilaments in skeletal and cardiac muscle cells.

myofilaments: Fine protein filaments composed primarily of the proteins actin (thin filaments) and myosin (thick filaments).

myoglobin: An oxygen-binding pigment that is especially common in slow skeletal muscle fibers and cardiac muscle tissue.

myogram: A recording of the tension produced by muscle fibers on stimulation.

myometrium: The thick layer of smooth muscle in the wall of the uterus.

myosepta: Connective tissue partitions that separate adjacent skeletal muscles.

myosin: The protein component of thick filaments.

N

nail: A keratinous structure produced by epithelial cells of the nail root.

nares, external: The entrance from the exterior to the nasal cavity.

nares, internal: The entrance from the nasal cavity to the nasopharynx.

nasal cavity: A chamber in the skull that is bounded by the internal and external nares.

nasolacrimal duct: The passageway that transports tears from the nasolacrimal sac to the nasal cavity.

nasolacrimal sac: A chamber that receives tears from the lacrimal ducts.

nasopharynx: A region that is posterior to the internal nares and superior to the soft palate and ends at the oropharynx.

natriuretic peptides (NP): Hormones released by specialized cardiocytes when they are stretched by an abnormally large venous return; promote fluid loss and reductions in blood pressure and in venous return. Include atrial natriuretic peptide (ANP) and brain natriuretic peptide (BNP).

necrosis: The death of cells or tissues from disease or injury.

negative feedback: A corrective mechanism that opposes or negates a variation from normal limits.

neonate: A newborn infant, or baby.

neoplasm: A tumor, or mass of abnormal tissue.

nephron: The basic functional unit of the kidney.

nephron loop: The portion of the nephron that creates the concentration gradient in the renal medulla; also called *loop of Henle.*

nerve impulse: An action potential in a neuron plasma membrane.

net filtration pressure: The hydrostatic pressure responsible for filtration.

neural cortex: An area of gray matter at the surface of the central nervous system.

neurilemma: The outer surface of a neuroglia that encircles an axon.

neurofibrils: Microfibrils in the cytoplasm of a neuron.

neurofilaments: Microfilaments in the cytoplasm of a neuron.

neuroglandular junction: A cell junction at which a neuron controls or regulates the activity of a secretory (gland) cell.

neuroglia: Cells of the central nervous system and peripheral nervous system that support and protect neurons; also called *glial cells.*

neurohypophysis: The posterior lobe of the pituitary gland; contains the axons of hypothalamic neurons, and these axons release OXT and ADH.

neuromuscular junction: A synapse between a neuron and a muscle cell.

neuron: A cell in neural tissue that is specialized for intercellular communication through (1) changes in membrane potential and (2) synaptic connections.

neurotransmitter: A chemical compound released by one neuron to affect the transmembrane potential of another.

neurotubules: Microtubules in the cytoplasm of a neuron.

neurulation: The embryological process responsible for the formation of the central nervous system.

neutron: A fundamental particle that does not carry a positive or a negative charge.

neutrophil: A microphage that is very numerous and normally the first of the mobile phagocytic cells to arrive at an area of injury or infection.

nicotinic receptors: Acetylcholine receptors on the surfaces of sympathetic and parasympathetic ganglion cells; respond to the compound nicotine.

nipple: An elevated epithelial projection on the surface of the breast; contains the openings of the lactiferous sinuses.

Nissl bodies: The ribosomes, Golgi apparatus, rough endoplasmic reticulum, and mitochondria of the perikaryon of a typical neuron.

nitrogenous wastes: Organic waste products of metabolism that contain nitrogen, such as urea, uric acid, and creatinine.

nociception: Pain perception.

node of Ranvier: The area between adjacent neuroglia where the myelin covering of an axon is incomplete.

nodose ganglion: A sensory ganglion of cranial nerve X; also called inferior ganglion.

noradrenaline: *See* norepinephrine.

norepinephrine (NE): A catecholamine neurotransmitter in the peripheral nervous system and central nervous system, released at most sympathetic neuromuscular and neuroglandular junctions, and a hormone secreted by the adrenal medulla; also called *noradrenaline.*

nucleic acid: A polymer of nucleotides that contains a pentose sugar, a phosphate group, and one of four nitrogenous bases that regulate the synthesis of proteins and make up the genetic material in cells.

nucleolus: The dense region in the nucleus that is the site of ribosomal RNA synthesis.

nucleoplasm: The fluid content of the nucleus.

nucleoproteins: Proteins of the nucleus that are generally associated with DNA.

nucleotide: A compound consisting of a nitrogenous base, a simple sugar, and a phosphate group.

nucleus: A cellular organelle that contains DNA, RNA, and proteins; in the central nervous system, a mass of gray matter.

nucleus pulposus: The gelatinous central region of an intervertebral disc.

nurse cells: Supporting cells of the seminiferous tubules of the testis; responsible for the differentiation of spermatids, the maintenance of the blood–testis barrier, and the secretion of inhibin, and androgen-binding protein; also called *sustentacular cells.*

nutrient: An inorganic or organic compound that can be broken down in the body to produce energy.

O

occlusal surface: The opposing surfaces of the teeth that come into contact when chewing food.

ocular: Pertaining to the eye.

oculomotor nerve: Cranial nerve III, which controls the extra-ocular muscles other than the superior oblique and the lateral rectus muscles.

olecranon: The proximal end of the ulna that forms the prominent point of the elbow.

olfaction: The sense of smell.

olfactory bulb: The expanded ends of the olfactory tracts; the sites where the axons of the first cranial nerves (I) synapse on central nervous system interneurons that lie inferior to the frontal lobes of the cerebrum.

oligodendrocytes: Central nervous system neuroglia that maintain cellular organization within gray matter and provide a myelin sheath in areas of white matter.

oocyte: A cell whose meiotic divisions will produce a single ovum and three polar bodies.

oogenesis: Ovum production.

opsonization: An effect of coating an object with antibodies; the attraction and enhancement of phagocytosis.

optic chiasm: The crossing point of the optic nerves.

optic nerve: The second cranial nerve (II), which carries signals from the retina of the eye to the optic chiasm.

optic tract: The tract over which nerve impulses from the retina are transmitted between the optic chiasm and the thalamus.

orbit: The bony recess of the skull that contains the eyeball.

organelle: An intracellular structure that performs a specific function or group of functions.

organic compound: A compound containing carbon, hydrogen, and in most cases oxygen.

organogenesis: The formation of organs during embryological and fetal development.

organs: Combinations of tissues that perform complex functions.

origin: In a skeletal muscle, the point of attachment that does not change position when the muscle contracts; usually defined in terms of movements from the anatomical position.

oropharynx: The middle portion of the pharynx, bounded superiorly by the nasopharynx, anteriorly by the oral cavity, and inferiorly by the laryngopharynx.

osmolarity: The total concentration of dissolved materials in a solution, regardless of their specific identities, expressed in moles; also called *osmotic concentration.*

osmoreceptor: A receptor sensitive to changes in the osmolarity of plasma.

osmosis: The movement of water across a selectively permeable membrane from one solution to another solution that contains a higher solute concentration.

osmotic pressure: The force of osmotic water movement; the pressure that must be applied to prevent osmosis across a membrane.

osseous tissue: A strong connective tissue containing specialized cells and a mineralized matrix of crystalline calcium phosphate and calcium carbonate; also called bone.

ossicles: Small bones.

ossification: The formation of bone; osteogenesis.

osteoblast: A cell that produces the fibers and matrix of bone.

osteoclast: A cell that dissolves the fibers and matrix of bone.

osteocyte: A bone cell responsible for the maintenance and turnover of the mineral content of the surrounding bone.

osteogenic layer: The inner, cellular layer of the periosteum that participates in bone growth and repair.

osteolysis: The breakdown of the mineral matrix of bone.

osteon: The basic histological unit of compact bone, consisting of osteocytes organized around a central canal and separated by concentric lamellae.

otic: Pertaining to the ear.

otolith: A complex formed by the combination of a gelatinous matrix and statoconia, aggregations of calcium carbonate crystals; located above one of the maculae of the vestibule.

oval window: An opening in the bony labyrinth where the stapes attaches to the membranous wall of the vestibular duct.

ovarian cycle: The monthly chain of events that leads to ovulation.

ovary: The female reproductive organ that produces gametes.

ovulation: The release of a secondary oocyte, surrounded by cells of the corona radiata, after the rupture of the wall of a tertiary follicle; in females, the periodic release of an oocyte from an ovary.

ovum/ova: The functional product of meiosis II, produced after the fertilization of a secondary oocyte.

oxytocin (OXT): A hormone produced by hypothalamic cells and secreted into capillaries at the neurohypophysis (posterior lobe of the pituitary gland); stimulates smooth muscle contractions of the uterus or mammary glands in females and the prostate gland in males.

P

pacemaker cells: Cells of the sinoatrial node that set the pace of cardiac contraction.

palate: The horizontal partition separating the oral cavity from the nasal cavity and nasopharynx; divided into an anterior bony (hard) palate and a posterior fleshy (soft) palate.

palatine: Pertaining to the palate.

palpate: To examine by touch.

palpebrae: Eyelids.

pancreas: A digestive organ containing exocrine and endocrine tissues; the exocrine portion secretes pancreatic juice, and the endocrine portion secretes hormones, including insulin and glucagon.

pancreatic duct: A tubular duct that carries pancreatic juice from the pancreas to the duodenum.

pancreatic islets: Aggregations of endocrine cells in the pancreas; also called *islets of Langerhans.*

pancreatic juice: A mixture of buffers and digestive enzymes that is discharged into the duodenum under the stimulation of the enzymes secretin and cholecystokinin.

papilla: A small, conical projection.

paralysis: The loss of voluntary motor control over a portion of the body.

paranasal sinuses: Bony chambers, lined by respiratory epithelium, that open into the nasal cavity; the frontal, ethmoidal, sphenoidal, and maxillary sinuses.

parasagittal: A section or plane that parallels the midsagittal plane but that does not pass along the midline.

parasympathetic division: One of the two divisions of the autonomic nervous system; also called *craniosacral division*; generally responsible for activities that conserve energy and lower the metabolic rate.

parathyroid glands: Four small glands embedded in the posterior surface of the thyroid gland; secrete parathyroid hormone.

parathyroid hormone (PTH): A hormone secreted by the parathyroid glands when plasma calcium levels fall below the normal range; causes increased osteoclast activity, increased intestinal calcium uptake, and decreased calcium ion loss at the kidneys.

parenchyma: The cells of a tissue or organ that are responsible for fulfilling its functional role; distinguished from the stroma of that tissue or organ.

parietal: Relating to the parietal bone; referring to the wall of a cavity.

parietal cells: Cells of the gastric glands that secrete hydrochloric acid and intrinsic factor.

parotid salivary glands: Large salivary glands that secrete a saliva containing high concentrations of salivary (alpha) amylase.

patella: The sesamoid bone of the kneecap.

pathogen: A disease-causing organism.

pathogenic: Disease-causing.

pathologist: A physician specializing in the identification of diseases on the basis of characteristic structural and functional changes in tissues and organs.

pelvic cavity: The inferior subdivision of the abdominopelvic cavity; encloses the urinary bladder, the sigmoid colon and rectum, and male or female reproductive organs.

pelvis: A bony complex created by the articulations among the coxal bones, the sacrum, and the coccyx.

penis: A component of the male external genitalia; a copulatory organ that surrounds the urethra and serves to introduce semen into the female vagina; the developmental equivalent of the female clitoris.

peptide: A chain of amino acids linked by peptide bonds.

peptide bond: A covalent bond between the amino group of one amino acid and the carboxyl group of another.

pericardial cavity: The space between the parietal pericardium and the epicardium (visceral pericardium) that covers the outer surface of the heart.

pericardium: The fibrous sac that surrounds the heart; its inner, serous lining is continuous with the epicardium.

perichondrium: The layer that surrounds a cartilage, consisting of an outer fibrous region and an inner cellular region.

perikaryon: The cytoplasm that surrounds the nucleus in the cell body of a neuron.

perilymph: A fluid similar in composition to cerebrospinal fluid; located in the spaces between the bony labyrinth and the membranous labyrinth of the inner ear.

perimysium: A connective tissue partition that separates adjacent fasciculi in a skeletal muscle.

perineum: The pelvic floor and its associated structures.

perineurium: A connective tissue partition that separates adjacent bundles of nerve fibers in a peripheral nerve.

periodontal ligament: Collagen fibers that bind the cementum of a tooth to the periosteum of the surrounding alveolus.

periosteum: The layer that surrounds a bone, consisting of an outer fibrous region and inner cellular region.

peripheral nervous system (PNS): All neural tissue outside the central nervous system.

peripheral resistance: The resistance to blood flow; primarily caused by friction with the vascular walls.

peristalsis: A wave of smooth muscle contractions that propels materials along the axis of a tube such as the digestive tract, the ureters, or the ductus deferens.

peritoneum: The serous membrane that lines the peritoneal cavity.

peritubular capillaries: A network of capillaries that surrounds the proximal and distal convoluted tubules of the kidneys.

permeability: The ease with which dissolved materials can cross a membrane; if the membrane is freely permeable, any molecule can cross it; if impermeable, nothing can cross; most biological membranes are selectively permeable.

peroxisome: A membranous vesicle containing enzymes that break down hydrogen peroxide (H_2O_2).

pes: The foot.

petrosal ganglion: A sensory ganglion of the glossopharyngeal nerve (N IX).

petrous: Stony; usually refers to the thickened portion of the temporal bone that encloses the internal ear.

pH: The negative exponent of the hydrogen ion concentration, expressed in moles per liter.

phagocyte: A cell that performs phagocytosis.

phagocytosis: The engulfing of extracellular materials or pathogens; the movement of extracellular materials into the cytoplasm by enclosure in a membranous vesicle.

phalanx/phalanges: Bone(s) of the finger(s) or toe(s).

pharmacology: The study of drugs, their physiological effects, and their clinical uses.

pharynx: The throat; a muscular passageway shared by the digestive and respiratory tracts.

phenotype: Physical characteristics that are genetically determined.

phosphate group: PO_4^{3-}; a functional group that can be attached to an organic molecule; required for the formation of high-energy bonds.

phospholipid: An important membrane lipid whose structure includes both hydrophilic and hydrophobic regions.

phosphorylation: The addition of a high-energy phosphate group to a molecule.

photoreception: Sensitivity to light.

physiology: The study of function; deals with the ways organisms perform vital activities.

pia mater: The innermost layer of the meninges bound to the underlying neural tissue.

pineal gland: Neural tissue in the posterior portion of the roof of the diencephalon; secretes melatonin.

pinna: *See* **auricle.**

pinocytosis: The introduction of fluids into the cytoplasm by enclosing them in membranous vesicles at the cell surface.

pituitary gland: An endocrine organ that is situated in the sella turcica of the sphenoid and is connected to the hypothalamus by the infundibulum; includes the posterior lobe (neurohypophysis) and the anterior lobe (adenohypophysis); also called the *hypophysis.*

placenta: A temporary structure in the uterine wall that permits diffusion between the fetal and maternal circulatory systems.

plantar: Referring to the sole of the foot.

plantar flexion: Ankle extension; toe pointing.

plasma: The fluid ground substance of whole blood; what remains after the cells have been removed from a sample of whole blood.

plasma cell: An activated B cell that secretes antibodies; plasmocyte.

plasma membrane: A cell membrane; plasmalemma.

platelets: Small packets of cytoplasm that contain enzymes important in the clotting response; manufactured in bone marrow by megakaryocytes.

pleura: The serous membrane that lines the pleural cavities.

pleural cavities: Subdivisions of the ventral body cavity that contain the lungs.

plexus: A network or braid.

polar body: A nonfunctional packet of cytoplasm that contains chromosomes eliminated from an oocyte during meiosis.

polar bond: A covalent bond in which electrons are shared unequally.

polarized: Referring to cells that have regional differences in organelle distribution or cytoplasmic composition along a specific axis, such as between the basement membrane and free surface of an epithelial cell.

pollex: The thumb.

polypeptide: A chain of amino acids strung together by peptide bonds; those containing more than 100 peptides are called *proteins.*

polyribosome: Several ribosomes linked by their translation of a single mRNA strand.

polysaccharide: A complex sugar, such as glycogen or a starch.

polysynaptic reflex: A reflex in which interneurons are interposed between the sensory fiber and the motor neuron(s).

polyunsaturated fats: Fatty acids containing carbon atoms that are linked by double bonds.

pons: The portion of the metencephalon that is anterior to the cerebellum.

popliteal: Pertaining to the back of the knee.

positive feedback: A mechanism that increases a deviation from normal limits after an initial stimulus.

postcentral gyrus: The primary sensory cortex, where touch, vibration, pain, temperature, and taste sensations arrive and are consciously perceived.

posterior: Toward the back; dorsal.

postganglionic neuron: An autonomic neuron in a peripheral ganglion, whose activities control peripheral effectors.

postsynaptic membrane: The portion of the plasma membrane of a postsynaptic cell that is part of a synapse.

potential difference: The separation of opposite charges; requires a barrier that prevents ion migration.

precentral gyrus: The primary motor cortex of a cerebral hemisphere, located anterior to the central sulcus.

prefrontal cortex: The anterior portion of each cerebral hemisphere; thought to be involved with higher intellectual functions, predictions, calculations, and so forth.

preganglionic neuron: A visceral motor neuron in the central nervous system whose output controls one or more ganglionic motor neurons in the peripheral nervous system.

premotor cortex: The motor association area between the precentral gyrus and the prefrontal area.

preoptic nucleus: The hypothalamic nucleus that coordinates thermoregulatory activities.

presynaptic membrane: The synaptic surface where neurotransmitter release occurs.

prevertebral ganglion: *See* **collateral ganglion.**

prime mover: A muscle that performs a specific action.

proenzyme: An inactive enzyme secreted by an epithelial cell.

progesterone: The most important progestin secreted by the corpus luteum after ovulation.

progestins: Steroid hormones structurally related to cholesterol; progesterone is an example.

prognosis: A prediction about the possible course or outcome from a specific disease.

projection fibers: Axons carrying information from the thalamus to the cerebral cortex.

prolactin: The hormone that stimulates functional development of the mammary glands in females; a secretion of the adenohypophysis (anterior lobe of the pituitary gland).

pronation: The rotation of the forearm that makes the palm face posteriorly.

prone: Lying face down with the palms facing the floor.

pronucleus: An enlarged ovum or spermatozoon nucleus that forms after fertilization but before amphimixis.

prophase: The initial phase of mitosis; characterized by the appearance of chromosomes, the breakdown of the nuclear membrane, and the formation of the spindle apparatus.

proprioception: The awareness of the positions of bones, joints, and muscles.

prostaglandin: A fatty acid secreted by one cell that alters the metabolic activities or sensitivities of adjacent cells; also called *local hormone.*

prostate gland: An accessory gland of the male reproductive tract, contributing roughly one-third of the volume of semen.

prosthesis: An artificial substitute for a body part.

protease: *See* **proteinase.**

protein: A large polypeptide with a complex structure.

proteinase: An enzyme that breaks down proteins into peptides and amino acids.

proteoglycan: A compound containing a large polysaccharide complex attached to a relatively small protein; examples include hyaluronan and chondroitin sulfate.

proton: A fundamental particle bearing a positive charge.

protraction: Movement anteriorly in the horizontal plane.

proximal: Toward the attached base of an organ or structure.

proximal convoluted tubule (PCT): The portion of the nephron that is situated between the glomerular capsule (Bowman capsule) and the nephron loop; the major site of active reabsorption from filtrate.

pseudopodia: Temporary cytoplasmic extensions typical of mobile or phagocytic cells.

pseudostratified epithelium: An epithelium that contains several layers of nuclei but whose cells are all in contact with the underlying basement membrane.

puberty: A period of rapid growth, sexual maturation, and the appearance of secondary sexual characteristics; normally occurs at ages 10–15 years.

pubic symphysis: The fibrocartilaginous amphiarthrosis between the pubic bones of the coxal bones.

pubis: The anterior, inferior component of the hip bone.

pudendum: The external genitalia.

pulmonary circuit: Blood vessels between the pulmonary semilunar valve of the right ventricle and the entrance to the left atrium; the blood flow through the lungs.

pulmonary ventilation: The movement of air into and out of the lungs.

pulvinar: The thalamic nucleus involved in the integration of sensory information prior to projection to the cerebral hemispheres.

pupil: The opening in the center of the iris through which light enters the eye.

purine: A nitrogen compound with a double ring-shaped structure; examples include adenine and guanine, two nitrogenous bases that are common in nucleic acids.

Purkinje cell: A large, branching neuron of the cerebellar cortex.

Purkinje fibers: Specialized conducting cardiocytes in the ventricles of the heart.

pyloric sphincter: A sphincter of smooth muscle that regulates the passage of chyme from the stomach to the duodenum.

pylorus: The gastric region between the body of the stomach and the duodenum; includes the pyloric sphincter.

pyrimidine: A nitrogen compound with a single ring-shaped structure; examples include cytosine, thymine, and uracil, nitrogenous bases that are common in nucleic acids.

pyruvate: The ion formed by the dissociation of pyruvic acid, a three-carbon compound produced by glycolysis.

Q

quaternary structure: The three-dimensional protein structure produced by interactions between protein subunits.

R

rami communicantes: Axon bundles that link the spinal nerves with the ganglia of the sympathetic chain.

ramus/rami: A branch/branches.

raphe: A seam.

receptive field: The area monitored by a single sensory receptor.

rectum: The inferior 15 cm (6 in.) of the digestive tract.

rectus: Straight.

red blood cell (RBC): *See* **erythrocyte.**

reductional division: The first meiotic division, which reduces the chromosome number from 46 to 23.

reflex: A rapid, automatic response to a stimulus.

reflex arc: The receptor, sensory neuron, motor neuron, and effector involved in a particular reflex; interneurons may be present, depending on the reflex considered.

refractory period: The period between the initiation of an action potential and the restoration of the normal resting potential; during this period, the membrane will not respond normally to stimulation.

relaxation phase: The period after a contraction when the tension in the muscle fiber returns to resting levels.

relaxin: A hormone that loosens the pubic symphysis; secreted by the placenta.

renal: Pertaining to the kidneys.

renal corpuscle: The initial portion of the nephron, consisting of an expanded chamber that encloses the glomerulus.

renin: The enzyme released by cells of the juxtaglomerular complex when renal blood flow declines; converts angiotensinogen to angiotensin I.

repolarization: The movement of the transmembrane potential away from a positive value and toward the resting potential.

respiration: The exchange of gases between cells and the environment; includes pulmonary ventilation, external respiration, internal respiration, and cellular respiration.

respiratory minute volume (V_E): The amount of air moved into and out of the respiratory system each minute.

respiratory pump: A mechanism by which changes in the intrapleural pressures during the respiratory cycle assist the venous return to the heart.

resting potential: The transmembrane potential of a normal cell under homeostatic conditions.

rete: An interwoven network of blood vessels or passageways.

reticular activating system (RAS): The mesencephalic portion of the reticular formation; responsible for arousal and the maintenance of consciousness.

reticular formation: A diffuse network of gray matter that extends the entire length of the brain stem.

reticulospinal tracts: Descending tracts of the medial pathway that carry involuntary motor commands issued by neurons of the reticular formation.

retina: The innermost layer of the eye, lining the vitreous chamber; also called *neural tunic.*

retinal: A visual pigment derived from vitamin A.

retraction: Movement posteriorly in the horizontal plane.

retroperitoneal: Behind or outside the peritoneal cavity.

reverberation: A positive feedback along a chain of neurons such that they remain active once stimulated.

rheumatism: A condition characterized by pain in muscles, tendons, bones, or joints.

Rh factor: A surface antigen that may be present (Rh-positive) or absent (Rh-negative) from the surfaces of red blood cells.

rhodopsin: The visual pigment in the membrane disks of the distal segments of rods.

rhythmicity center: A medullary center responsible for the pace of respiration; includes inspiratory and expiratory centers.

ribonucleic acid: A nucleic acid consisting of a chain of nucleotides that contain the sugar ribose and the nitrogenous bases adenine, guanine, cytosine, and uracil.

ribose: A five-carbon sugar that is a structural component of RNA.

ribosome: An organelle that contains rRNA and proteins and is essential to mRNA translation and protein synthesis.

rod: A photoreceptor responsible for vision in dim lighting.

rough endoplasmic reticulum (RER): A membranous organelle that is a site of protein synthesis and storage.

round window: An opening in the bony labyrinth of the inner ear that exposes the membranous wall of the tympanic duct to the air of the middle ear cavity.

rubrospinal tracts: Descending tracts of the lateral pathway that carry involuntary motor commands issued by the red nucleus of the mesencephalon.

rugae: Mucosal folds in the lining of the empty stomach that disappear as gastric distension occurs; folds in the urinary bladder.

S

saccule: A portion of the vestibular apparatus of the internal ear; contains a macula important for static equilibrium.

sagittal plane: A sectional plane that divides the body into left and right portions.

salt: An inorganic compound consisting of a cation other than H^+ and an anion other than OH^-.

saltatory propagation: The relatively rapid propagation of an action potential between successive nodes of a myelinated axon.

sarcomere: The smallest contractile unit of a striated muscle cell.

sarcoplasm: The cytoplasm of a muscle cell.

scala media: *See* **cochlear duct.**

scar tissue: The thick, collagenous tissue that forms at an injury site.

Schwann cells: Neuroglia responsible for the neurilemma that surrounds axons in the peripheral nervous system.

sciatic nerve: A nerve innervating the posteromedial portions of the thigh and leg.

sclera: The fibrous, outer layer of the eye that forms the white area of the anterior surface; a portion of the fibrous tunic of the eye.

sclerosis: A hardening and thickening that commonly occurs secondary to tissue inflammation.

scrotum: The loose-fitting, fleshy pouch that encloses the testes of the male.

sebaceous glands: Glands that secrete sebum; normally associated with hair follicles.

sebum: A waxy secretion that coats the surfaces of hairs.

secondary sex characteristics: Physical characteristics that appear at puberty in response to sex hormones but are not involved in the production of gametes.

secretin: A hormone, secreted by the duodenum, that stimulates the production of buffers by the pancreas and inhibits gastric activity.

semen: The fluid ejaculate that contains spermatozoa and the secretions of accessory glands of the male reproductive tract.

semicircular ducts: The tubular components of the membranous labyrinth of the inner ear; responsible for dynamic equilibrium.

semilunar valve: A three-cusped valve guarding the exit from one of the cardiac ventricles; the pulmonary and aortic valves.

seminal glands: Glands of the male reproductive tract that produce roughly 60 percent of the volume of semen.

seminiferous tubules: Coiled tubules where spermatozoon production occurs in the testes.

senescence: Aging.

sensible perspiration: Water loss due to secretion by sweat glands.

septa: Partitions that subdivide an organ.

serous cell: A cell that produces a serous secretion.

serous membrane: A squamous epithelium and the underlying loose connective tissue; the lining of the pericardial, pleural, and peritoneal cavities.

serous secretion: A watery secretion that contains high concentrations of enzymes.

serum: The ground substance of blood plasma from which clotting agents have been removed.

sesamoid bone: A bone that forms within a tendon.

sigmoid colon: The S-shaped 18-cm (7.1 in.)-long portion of the colon between the descending colon and the rectum.

sign: The visible, objective evidence of the presence of a disease.

simple epithelium: An epithelium containing a single layer of cells above the basal lamina.

sinoatrial (SA) node: The natural pacemaker of the heart; situated in the wall of the right atrium.

sinus: A chamber or hollow in a tissue; a large, dilated vein.

sinusoid: An exchange vessel that is similar in general structure to a fenestrated capillary. The two differ in size (sinusoids are larger and more irregular in cross section), continuity (sinusoids have gaps between endothelial cells), and support (sinusoids have thin basal laminae, if present at all).

skeletal muscle: A contractile organ of the muscular system.

skeletal muscle tissue: A contractile tissue dominated by skeletal muscle fibers; characterized as striated, voluntary muscle.

sliding filament theory: The concept that a sarcomere shortens as the thick and thin filaments slide past one another.

small intestine: The duodenum, jejunum, and ileum; the digestive tract between the stomach and the large intestine.

smooth endoplasmic reticulum (SER): A membranous organelle in which lipid and carbohydrate synthesis and storage occur.

smooth muscle tissue: Muscle tissue in the walls of many visceral organs; characterized as nonstriated, involuntary muscle.

soft palate: The fleshy posterior extension of the hard palate, separating the nasopharynx from the oral cavity.

solute: Any materials dissolved in a solution.

solution: A fluid containing dissolved materials.

somatic: Pertaining to the body.

somatic nervous system (SNS): The efferent division of the nervous system that innervates skeletal muscles.

somatomedins: Compounds stimulating tissue growth; released by the liver after the secretion of growth hormone; also called *insulin-like growth factors*.

sperm: See **spermatozoon**.

spermatic cord: Collectively, the spermatic vessels, nerves, lymphatic vessels, and the ductus deferens, extending between the testes and the proximal end of the inguinal canal.

spermatocyte: A cell of the seminiferous tubules that is engaged in meiosis.

spermatogenesis: Spermatozoon production.

spermatozoon/spermatozoa: A male gamete; also called *sperm*.

sphincter: A muscular ring that contracts to close the entrance or exit of an internal passageway.

spinal nerve: One of 31 pairs of nerves that originate on the spinal cord from anterior and posterior roots.

spindle apparatus: Microtubule-based structure that distributes duplicated chromosomes to opposite ends of a dividing cell during mitosis.

spinocerebellar tracts: Ascending tracts that carry sensory information to the cerebellum.

spinothalamic tracts: Ascending tracts that carry poorly localized touch, pressure, pain, vibration, and temperature sensations to the thalamus.

spinous process: The prominent posterior projection of a vertebra; formed by the fusion of two laminae.

spleen: A lymphoid organ important for the phagocytosis of red blood cells, the immune response, and lymphocyte production.

squama: A broad, flat surface.

squamous: Flattened.

squamous epithelium: An epithelium whose superficial cells are flattened and platelike.

stapes: The auditory ossicle attached to the tympanic membrane.

stenosis: A constriction or narrowing of a passageway.

stereocilia: Elongate microvilli characteristic of the epithelium of the epididymis, portions of the ductus deferens, and the internal ear.

steroid: A ring-shaped lipid structurally related to cholesterol.

stimulus: An environmental change that produces a change in cellular activities; often used to refer to events that alter the transmembrane potentials of excitable cells.

stratified: Containing several layers.

stratum: A layer.

stretch receptors: Sensory receptors that respond to stretching of the surrounding tissues.

stroma: The connective tissue framework of an organ; distinguished from the functional cells (parenchyma) of that organ.

subarachnoid space: A meningeal space containing cerebrospinal fluid; the area between the arachnoid membrane and the pia mater.

subclavian: Pertaining to the region immediately posterior and inferior to the clavicle.

subcutaneous layer: *See* hypodermis.

submucosa: The region between the muscularis mucosae and the muscularis externa.

subserous fascia: The loose connective tissue layer beneath the serous membrane that lines the ventral body cavity.

substrate: A participant (product or reactant) in an enzyme-catalyzed reaction.

sulcus: A groove or furrow.

summation: The temporal or spatial addition of contractile force or neural stimuli.

superior: Above, in reference to a portion of the body in the anatomical position.

superior vena cava (SVC): The vein that carries blood to the right atrium from parts of the body that are superior to the heart.

supination: The rotation of the forearm such that the palm faces anteriorly.

supine: Lying face up, with palms facing anteriorly.

suppressor T cells: Lymphocytes that inhibit B cell activation and the secretion of antibodies by plasma cells.

surfactant: A lipid secretion that coats the alveolar surfaces of the lungs and prevents their collapse.

sustentacular cells: See nurse cells.

sutural bones: Irregular bones that form in fibrous tissue between the flat bones of the developing cranium; also called *Wormian bones*.

suture: A fibrous joint between flat bones of the skull.

sympathetic division: The division of the autonomic nervous system that is responsible for "fight or flight" reactions; primarily concerned with the elevation of metabolic rate and increased alertness.

symphysis: A fibrous amphiarthrosis, such as that between adjacent vertebrae or between the pubic bones of the coxal bones.

symptom: An abnormality of function as a result of disease; subjective experience of patient.

synapse: The site of communication between a nerve cell and some other cell; if the other cell is not a neuron, the term *neuromuscular* or *neuroglandular junction* is often used.

synaptic delay: The period between the arrival of an impulse at the presynaptic membrane and the initiation of an action potential in the postsynaptic membrane.

syncytium: A multinucleate mass of cytoplasm, produced by the fusion of cells or repeated mitoses without cytokinesis.

syndrome: A discrete set of signs and symptoms that occur together.

synergist: A muscle that assists a prime mover in performing its primary action.

synovial cavity: A fluid-filled chamber in a synovial joint.

synovial fluid: The substance secreted by synovial membranes that lubricates joints.

synovial joint: A freely movable joint where the opposing bone surfaces are separated by synovial fluid; a diarthrosis.

synovial membrane: An incomplete layer of fibroblasts facing the synovial cavity, plus the underlying loose connective tissue.

synthesis: Manufacture; anabolism.

system: An interacting group of organs that performs one or more specific functions.

systemic circuit: The vessels between the aortic valve and the entrance to the right atrium; the system other than the vessels of the pulmonary circuit.

systole: A period of contraction in a chamber of the heart, as part of the cardiac cycle.

systolic pressure: The peak arterial pressure measured during ventricular systole.

T

tactile: Pertaining to the sense of touch.

tarsal bones: The bones of the ankle (the talus, calcaneus, navicular, and cuneiform bones).

tarsus: The ankle.

T cells: Lymphocytes responsible for cell-mediated immunity and for the coordination and regulation of the immune response; includes regulatory T cells (helpers and suppressors) and cytotoxic (killer) T cells.

tectospinal tracts: Descending tracts of the medial pathway that carry involuntary motor commands issued by the colliculi.

telodendria: Terminal axonal branches that end in synaptic terminals.

telophase: The final stage of mitosis, characterized by the disappearance of the spindle apparatus, the reappearance of the nuclear membrane, the disappearance of the chromosomes, and the completion of cytokinesis.

temporal: Pertaining to time (temporal summation) or to the temples (temporal bone).

tendon: A collagenous band that connects a skeletal muscle to an element of the skeleton.

teres: Long and round.

terminal: Toward the end.

tertiary structure: The protein structure that results from interactions among distant portions of the same molecule; complex coiling and folding.

testes: The male gonads, sites of gamete production and hormone secretion.

testosterone: The principal androgen produced by the interstitial cells of the testes.

tetraiodothyronine: T_4, or thyroxine, a thyroid hormone.

thalamus: The walls of the diencephalon.

therapy: The treatment of disease.

thermoreception: Sensitivity to temperature changes.

thermoregulation: Homeostatic maintenance of body temperature.

thick filament: A cytoskeletal filament in a skeletal or cardiac muscle cell; composed of myosin, with a core of titin.

thin filament: A cytoskeletal filament in a skeletal or cardiac muscle cell; consists of actin, troponin, and tropomyosin.

thoracolumbar division: The sympathetic division of the autonomic nervous system.

thorax: The chest.

threshold: The transmembrane potential at which an action potential begins.

thrombin: The enzyme that converts fibrinogen to fibrin.

thymine: A pyrimidine; one of the nitrogenous bases in the nucleic acid DNA.

thymosins: Thymic hormones essential to the development and differentiation of T cells.

thymus: A lymphoid organ, the site of T cell formation.

thyroglobulin: A circulating transport globulin that binds thyroid hormones.

thyroid gland: An endocrine gland whose lobes are lateral to the thyroid cartilage of the larynx.

thyroid hormones: Thyroxine (T_4) and T_3, hormones of the thyroid gland; stimulate tissue metabolism, energy utilization, and growth.

thyroid-stimulating hormone (TSH): The hormone, produced by the anterior lobe of the pituitary gland, that triggers the secretion of thyroid hormones by the thyroid gland.

thyroxine: A thyroid hormone; also called T_4 or *tetraiodothyronine*.

tidal volume: The volume of air moved into and out of the lungs during a normal quiet respiratory cycle.

tissue: A collection of specialized cells and cell products that performs a specific function.

tonsil: A lymphoid nodule in the wall of the pharynx; the palatine, pharyngeal, and lingual tonsils.

topical: Applied to the body surface.

toxic: Poisonous.

trabecula: A connective tissue partition that subdivides an organ.

trachea: The windpipe; an airway extending from the larynx to the primary bronchi.

tract: A bundle of axons in the central nervous system.

transcription: The encoding of genetic instructions on a strand of mRNA.

transection: The severing or cutting of an object in the transverse plane.

translation: The process of peptide formation from the instructions carried by an mRNA strand.

transmembrane potential: The potential difference, measured across a plasma membrane and expressed in millivolts, that results from the uneven distribution of positive and negative ions across the plasma membrane.

transudate: A fluid that diffuses across a serous membrane and lubricates opposing surfaces.

transverse tubules: The transverse, tubular extensions of the sarcolemma that extend deep into the sarcoplasm, contacting cisternae of the sarcoplasmic reticulum; also called *T tubules*.

tricuspid valve: The right atrioventricular valve, which prevents the backflow of blood into the right atrium during ventricular systole.

trigeminal nerve: Cranial nerve V, which provides sensory information from the lower portions of the face (including the upper and lower jaws) and delivers motor commands to the muscles of mastication.

triglyceride: A lipid that is composed of a molecule of glycerol attached to three fatty acids.

trochanter: Large process near the head of the femur.

trochlea: A pulley; the spool-shaped medial portion of the condyle of the humerus.

trochlear nerve: Cranial nerve IV, controlling the superior oblique muscle of the eye.

trunk: The thoracic and abdominopelvic regions; a major arterial branch.

T tubules: *See* **transverse tubules**.

tuberculum: A small, localized elevation on a bony surface.

tuberosity: A large, roughened elevation on a bony surface.

tumor: A tissue mass formed by the abnormal growth and replication of cells.

tunica: A layer or covering.

twitch: A single stimulus–contraction–relaxation cycle in a skeletal muscle.

tympanic duct: The perilymph-filled chamber of the internal ear, adjacent to the basilar membrane; pressure changes there distort the round window.

tympanic membrane: The membrane that separates the external acoustic meatus from the middle ear; the membrane whose vibrations are transferred to the auditory ossicles and ultimately to the oval window; also called *eardrum* or *tympanum*.

U

umbilical cord: The connecting stalk between the fetus and the placenta; contains the allantois, the umbilical arteries, and the umbilical vein.

umbilicus: The navel.

unipolar neuron: A sensory neuron whose cell body is in a dorsal root ganglion or a sensory ganglion of a cranial nerve.

unmyelinated axon: An axon whose neurilemma does not contain myelin and across which continuous propagation occurs.

uracil: A pyrimidine; one of the nitrogenous bases in the nucleic acid RNA.

ureters: Muscular tubes, lined by transitional epithelium, that carry urine from the renal pelvis to the urinary bladder.

urethra: A muscular tube that carries urine from the urinary bladder to the exterior.

urinary bladder: The muscular, distensible sac that stores urine prior to micturition.

urination: The voiding of urine; micturition.

uterus: The muscular organ of the female reproductive tract in which implantation, placenta formation, and fetal development occur.

utricle: The largest chamber of the vestibular apparatus of the internal ear; contains a macula important for static equilibrium.

V

vagina: A muscular tube extending between the uterus and the vestibule.

vascular: Pertaining to blood vessels.

vasoconstriction: A reduction in the diameter of arterioles due to the contraction of smooth muscles in the tunica media; elevates peripheral resistance; may occur in response to local factors, through the action of hormones, or from the stimulation of the vasomotor center.

vasodilation: An increase in the diameter of arterioles due to the relaxation of smooth muscles in the tunica media; reduces peripheral resistance; may occur in response to local factors, through the action of hormones, or after decreased stimulation of the vasomotor center.

vasomotion: Changes in the pattern of blood flow through a capillary bed in response to changes in the local environment.

vasomotor center: The center in the medulla oblongata whose stimulation produces vasoconstriction and an elevation of peripheral resistance.

vein: A blood vessel carrying blood from a capillary bed toward the heart.

vena cava: One of the major veins delivering systemic blood to the right atrium; superior and inferior venae cavae.

ventilation: Air movement into and out of the lungs.

ventral: Pertaining to the anterior surface.

ventricle: A fluid-filled chamber; in the heart, one of the large chambers discharging blood into the pulmonary or systemic circuits; in the brain, one of four fluid-filled interior chambers.

venule: Thin-walled veins that receive blood from capillaries.

vertebral canal: The passageway that encloses the spinal cord; a tunnel bounded by the neural arches of adjacent vertebrae.

vertebral column: The cervical, thoracic, and lumbar vertebrae, the sacrum, and the coccyx.

vesicle: A membranous sac in the cytoplasm of a cell.

vestibular duct: A coiled tube filled with perilymph that lies within the bony labyrinth; it is continous with the tympanic duct at the tip of the cochlear spiral.

vestibular nucleus: The processing center for sensations that arrive from the vestibular apparatus of the internal ear, located near the border between the pons and the medulla oblongata.

vestibulospinal tracts: Descending tracts of the medial pathway that carry involuntary motor commands issued by the vestibular nucleus to stabilize the position of the head.

villus/villi: A slender, finger-shaped projection of a mucous membrane.

virus: A noncellular pathogen.

viscera: Organs in one of the subdivisions of the ventral body cavity.

visceral: Pertaining to viscera or their outer coverings.

visceral smooth muscle: A smooth muscle tissue that forms sheets or layers in the walls of visceral organs; the cells may not be innervated, and the layers often show automaticity (rhythmic contractions).

viscosity: The resistance to flow that a fluid exhibits as a result of molecular interactions within the fluid.

viscous: Thick, syrupy.

vitamin: An essential organic nutrient that functions as a coenzyme in vital enzymatic reactions.

vitreous humor: The gelatinous mass in the vitreous chamber of the eye.

voluntary: Controlled by conscious thought processes.

W

white blood cells (WBCs): The granulocytes and agranulocytes of whole blood.

white matter: Regions in the central nervous system that are dominated by myelinated axons.

white ramus: A nerve bundle containing the myelinated preganglionic axons of sympathetic motor neurons en route to the sympathetic chain or to a collateral ganglion.

Wormian bones: *See* **sutural bones**.

X

xiphoid process: The slender, inferior extension of the sternum.

Y

Y chromosome: The sex chromosome whose presence indicates that the individual is a genetic male.

Z

zona fasciculata: The region of the adrenal cortex that secretes glucocorticoids.

zona glomerulosa: The region of the adrenal cortex that secretes mineralocorticoids.

zona reticularis: The region of the adrenal cortex that secretes androgens.

zygote: The fertilized ovum, prior to the start of cleavage.

Chapter 14 Chapter opener top Cecilia Lim/iStockphoto; bottom Olena Druzhynina/iStockphoto; 14.4.3 Annett Vauteck/iStockphoto; 14.4.4 iStockphoto; 14.8 Frederic H. Martini; p.495 top Simone van den Berg/shutterstock; center Jens Stolt/shutterstock; bottom Subbotina Anna/shutterstock.

Chapter 15 Chapter opener Andersen Ross/Getty Images; 15.2.3 Biophoto Associates/Photo Researchers; 15.2.4 iStockphoto; 15.7 Michael J. Timmons; 15.10.1 Ralph T. Hutchings; 15.10.4 iStockphoto; 15.12.2 left Diane Hirsch/Fundamental Photographs, NYC; right Diane Schiumo/Fundamental Photographs, NYC; 15.14 Custom Medical Stock Photo; 15.17 David Kevitch/iStockphoto; 15.18.1 iStockphoto; 15.18.2 Sharon Dominick/iStockphoto; 15.18.3 left Olga Ekaterincheva/iStockphoto; right Micheline Dubé/iStockphoto; 15.18.4 iStockphoto; 15.18.5 Konrad Lange/iStockphoto; p.538 Custom Medical Stock Photo; p.539 top Micheline Dubé/iStockphoto; center right jamstockfoto/fotolia; bottom Lisa F. Young/shutterstock.

Chapter 16 Chapter opener iStockphoto; 16.6. Robert B. Tallitsch; 16.7.2 top Frederic H. Martini; bottom Robert B. Tallitsch; 16.8 Ward's Natural Science Establishment; 16.10 School of Anatomy and Human Biology, UWA, Australia; 16.11 Mikhail Kokhanchikov/iStockphoto; 16.13 left Sharon Dominick/iStockphoto; right Ruhani Kaur/UNICEF India; 16.14.1 Project Masters, Inc./The Bergman Collection; 16.14.2 John Paul Kay/Peter Arnold/PhotoLibrary; 16.14.3 Project Masters, Inc./The Bergman Collection; 16.14.4 Custom Medical Stock Photo; 16.14.5 Biophoto Associates/Science Source/Photo Researchers; p.572 top to bottom Mikhail Kokhanchikov/iStockphoto; Sharon Dominick/iStockphoto; John Paul Kay/Peter Arnold/Photo Library; p.573 Alexkalina/dreamstime.

Chapter 17 Chapter opener Henrik Jonsson/iStockphoto; 17.2.1 left Frederic H. Martini; right Cheryl Power/Photo Researchers; 17.2.3 Ed Reschke/Peter Arnold/PhotoLibrary; 17.4.4 Karen E. Petersen, Department of Biology, Univ. of Washington; 17.6 Robert B. Tallitsch; 17.8 Custom Medical Stock Photo; 17.9.1 Shevelev Vladimir/shutterstock; 17.9.2 p.592 left iStockphoto; right Eye of Science/Photo Researchers; 17.9.2 p.593 top left National Medical Slide Bank/CMSP; bottom left Christopher Badzioch/iStockphoto; top right CC-BY-SA photo: Paulo Henrique Orlandi Mourao; p.595 Kyu Oh/iStockphoto; 17.10 Biophoto Associates/Photo Researchers; 17.12 Timothey Kosachev/iStockphoto; p.621 top Robert B. Tallitsch; center Custom Medical Stock Photo; bottom Shevelev Vladimir/shutterstock; p.623 Susumu Inoue, Michigan State University College of Human Medicine, Wayne State University School of Medicine; and Hurley Medical Center.

Chapter 18 Chapter opener iStockphoto; 18.1 Robert B. Tallitsch; 18.3 Ralph T. Hutchings; 18.6.4 left Science Photo Library/Photo Researchers; right Biophoto Associates/Science Source/Photo Researchers; 18.6.5 top Reprinted with permission of the Department of Pathology, Virginia, Commonwealth University and the VCU Health System (www.pathology.vcu.edu); bottom Medtronic Freestyle Bioprosthesis®, copyright Medtronic, Inc.; 18.7.1 top Ed Reschke/Peter Arnold Images/Photolibrary; bottom B & B Photos/Custom Medical Stock Photo; 18.7.2 Peter Arnold/PhotoLibrary; 18.7.3 both Howard Sochurek/Corbis; 18.7.4 ICVI-CCN/Voisin/Photo Researchers; 18.8 iStockphoto; 18.13 William C. Ober; 18.14 Larry Mulvehill/Photo Researchers; 18.15.2 left Vladislav Mitic/iStockphoto; right top to bottom Greg Nicholas/iStockphoto; Özgür Donmaz/iStockphoto; iStockphoto; 18.19.2 iStockphoto; p.673 Ed Reschke/Peter Arnold Images/Photolibrary; p.674 William C. Ober; p.675 left Yuri Arcurs/fotolia; right Ryan McVay/Thinkstock.

Chapter 19 Chapter opener Henrik Jonsson/iStockphoto; 19.1 Frederic H. Martini; 19.2 Barkley Fahnestock/iStockphoto; 19.4.1 Robert B. Tallitsch; 19.4.2 Biophoto Associates/Photo Researchers; 19.4.3 Ralph T. Hutchings; 19.5 both Robert Tallitsch; 19.6.2 iStockphoto; 19.6.4 Robert Tallitsch; p.691 iStockphoto; 19.7.1 Josh Hodge/iStockphoto; 19.7.4 CC-BY-SA Photo: Etxrge; 19.10 Ronald Bloom/iStockphoto; 19.11 ImageStream image courtesy of Amnis Corporation (www.amnis.com); 19.13 Jackson Egen and James P. Allison; 19.16 Monika Wisniewska/iStockphoto; 19.18 p.714 top left Stan Rohrer/iStockphoto, top right Courtesy of Sentara Healthcare; bottom Ramona Heim/iStockphoto; 19.18 p.715 left C. Goldsmith, P. Feorino, E. L. Palmer, & W. R. McManus/CDC; right Steve Cole/iStockphoto; p.719 top Biophoto Associates/Photo Researchers; center Arie v.d. Wolde/shutterstock; bottom Stan Rohrer/iStockphoto.

Chapter 20 Chapter opener Nikolay Suslov/iStockphoto; 20.1 iStockphoto; 20.2 Ralph T. Hutchings; 20.3 Jason Stitt/iStockphoto; 20.4 Lester V. Bergman/Corbis; 20.5 Ralph T. Hutchings; 20.13.1 Jon Helgason/iStockphoto; 20.13.2 iStockphoto; 20.13.3 top to bottom Dana Spiropoulou/iStockphoto; Jaren Wicklund/iStockphoto; iStockphoto; 20.15 idris esen/iStockphoto; 20.16 iStockphoto; p.758 Dana Spiropoulou/iStockphoto; p.759 both shutterstock.

Chapter 21 Chapter opener Sebastian Kaulitzki/iStockphoto; 21.2 Frederic H. Martini; 21.5 Ralph T. Hutchings; 21.6 left Alfred Pasieka/Peter Arnold/PhotoLibrary; right Astrid and Hanns-Frieder Michler/Science Photo Library/Photo Researchers; 21.8 Ralph T. Hutchings; 21.10 M.I. Walker/Photo Researchers; 21.11 all Ralph T. Hutchings; 21.16 Robert B. Tallitsch; 21.17 Ralph T. Hutchings; 21.20 p.804 top left Richard Nejat (gumsurgery.com); center left Barbara Rice/NIP/CDC; bottom left Gastrolab/Photo Researchers; bottom right PD Image: Stell98; 21.20 p.805 top left David M. Martin, M.D./Photo Researchers; center right Courtesy of Ronald Bleday/BWH; bottom left Joel Mancuso, University of California, Berkeley; p.809 top iStockphoto; bottom CC-BY-SA Image: James P. Gray.

Chapter 22 Chapter opener dreamstime; 22.9 p.830 top Suprijono Suharjoto/iStockphoto; bottom Olga Lyubkina/iStockphoto; 22.9 p.831 top ILYA GENKIN/iStockphoto; 22.10.4 Kelly Cline/iStockphoto; 22.10.5 Joop Hoek/iStockphoto; 22.11 p.834 left dreamstime; top right David Gaylor/shutterstock; bottom B&B Photos/Custom Medical Stock Photo; 22.11 p.835 top left William C. Ober; bottom left Lyle Conrad/CDC; right Myfootshop.com; p.837 APHP-PSL-GARO/PHANIE/Photo Researchers; 22.13 Christopher Bissel/Stone/Getty Images; p.844 APHP-PSL-GARO/PHANIE/Photo Researchers; p.845 top to bottom fotolia; shutterstock; thinkstock.

Chapter 23 Chapter Opener Skip ODonnell/iStockphoto; 23.2 Ralph T. Hutchings; 23.6 Steve Gschmeissner/Photo Researchers; 23.12 Beranger/Photo Researchers; p.875 Photo Researchers; 23.13.2 Ward's Natural Science Establishment; 23.13.3 Frederic H. Martini; 23.13.4 Robert Tallitsch; p.882 top Beranger/Photo Researchers; bottom Photo Researchers; p.883 top Eduardo Jose Bernardino/iStockphoto; bottom Rick's Photography/shutterstock.

Chapter 24 Chapter opener Mediumstock/Veer; p.907 top left shutterstock; top right thinkstock; bottom left thinkstock; bottom right Seelevel.com.

Chapter 25 Chapter opener shutterstock; 25.2 Eye of Science/Photo Researchers; 25.3.1 Frederic H. Martini; 25.3.2 Don W. Fawcett, M.D., Harvard Medical School; 25.3.3 Don W. Fawcett, M.D., Harvard Medical School; 25.4 all Frederic H. Martini; 25.8.2 clockwise Frederic H. Martini; Frederic H. Martini; Frederic H. Martini; Frederic H. Martini; C. Edelmann/La Villete/Photo Researchers; 25.9.2 Frederic H. Martini; 25.9.3 Custom Medical Stock Photo, Inc.; 25.10.3 left, center Frederic H. Martini; right Michael J. Timmons; 25.14 p.936 top left Kim Gunkel/iStockphoto; center left Brent Melton/iStockphoto; bottom left Alamy; top right Getty Images; center right shutterstock; bottom right Alamy; 25.14 p.937 top left Jennifer Trenchard/iStockphoto; top right Maureen Spuhler; 25.15.2 Elizabeth Kreutz, used with permission from the Lance Armstrong Foundation; 25.15.3 iStockphoto; p.942 Brent Melton/iStockphoto; p.943 left shutterstock; right dreamstime.

Chapter 26 Chapter opener Piotr Podermanski/iStockphoto; 26.1 Francis Leroy/Photo Researchers; 26.6 Frederic H. Martini; 26.7.1 Arnold Tamarin; 26.7.2, 26.7.3, 26.7.4, 26.7.5 Photo Lennart Nilsson/Albert Bonniers Forlag; 26.7.6 Photo Researchers; 26.8 Aldo Murillo/iStockphoto; 26.10 left to right Jaroslaw Wojcik/iStockphoto; Jaroslaw Wojcik/iStockphoto; Justin Horrocks/iStockphoto; Jacob Wackerhausen/iStockphoto; p.969 top left Valentin Casarsa/iStockphoto; inset Bob Ainsworth/iStockphoto; bottom right Science Photo Library/Photo Researchers; 26.12.2 Liza McCorkle/iStockphoto; 26.12.4 iStockphoto; 26.13 Science Photo Library/Photo Researchers; 26.14.1 left Science Photo Library/Photo Researchers; right fotolia; 26.14.2 left Science Photo Library/Photo Researchers; right CC-BY-SA Photo: Model A001 for Body Morphology Project, Malcolm Gin photographer; 26.14.3 left Science Photo Library/Photo Researchers; right From: *Robbins Pathologic Basis of Disease*, 6th ed., by Kumar and Hagler, from the Interactive Case Study Companion, Copyright 1999, W. B. Saunders Company, Elsevier; p.976 iStockphoto; p.977 Photo Lennart Nilsson/Albert Bonniers Forlag; p.978 all Science Photo Library/Photo Researchers; p.979 top shutterstock; bottom iStockphoto.

INDEX